Habitat

The Fundamental Unit for Understanding and Conserving Nature

Roger dedicates this book to Margaret, Pamela and Alex
Simone dedicates this book to his past and present cats: Wilson, Tigretta, Grigiotto, Piumina and Rosina

Habitat

The Fundamental Unit for Understanding and Conserving Nature

by

Roger L.H. Dennis and Simone Fattorini

CABI

CABI is a trading name of CAB International

CABI
Nosworthy Way
Wallingford
Oxfordshire OX10 8DE
UK

Tel: +44 (0)1491 832111
E-mail: info@cabi.org
Website: www.cabi.org

CABI
200 Portland Street
Boston
MA 02114
USA

Tel: +1 (617)682-9015
E-mail: cabi-nao@cabi.org

A catalogue record for this book is available from the British Library, London, UK.

ISBN-13: 9781800627482 (hardback)
9781800627499 (ePDF)
9781800627505 (ePub)

DOI: 10.1079/9781800627505.0000

Commissioning Editor: Ward Cooper
Editorial Assistant: Helen Elliott
Production Editor: James Bishop

Typeset by Straive, Pondicherry, India
Printed and bound in the UK by CPI Group (UK) Ltd, Croydon, CR0 4YY

'There is a painting by Pisanello in the National Gallery that catches exactly such a moment: St Hubert in an early Renaissance forest, confronted by birds and beasts. The saint is shocked, almost as if the victim of a practical joke, all his arrogance dowsed by a sudden drench of nature's profoundest secret: the universal parity of existence.'

J.R. Fowles (*The French Lieutenant's Woman*, 1969)

Pisanello, *The Vision of Saint Eustace* (detail) c. 1438–1442. National Gallery, London.

(The Yorck Project (2002) *10.000 Meisterwerke der Malerei* (DVD-ROM), distributed by DIRECTMEDIA Publishing GmbH: https://upload.wikimedia.org/wikipedia/commons/thumb/2/23/Pisanello_018.jpg/2490px-Pisanello_018.jpg)

Contents

Preface xi

Acknowledgements xiii

1 **The Traditional Habitat Model: Utility, Affinities and Failings** 1
Spheres of existence – habitats in a wide perspective 1
Biogeographical regions: adaptation, diversity, convergence and divergence 4
Habitat type – the traditional biome view 9
Habitat without vegetation 17
Microhabitats 18
Habitat edges, ecotones and ecoclines 20
Habitat – a species' view 23

2 **Inside Habitats: Interactions in Space and Time** 28
Units of integration: an ecosystem perspective 28
Population status and changes within habitats 37
Niche and organism interactions 41
Neutral theory, interaction theory and competitive exclusion: extremes in a continuum 44
Guilds and functional groups 52
Traits and life-history strategies 54
Focal species in ecosystems 59
Interconnections in ecosystems: the wood wide web 62

3 **Habitat Changes Over Time** 65
Habitat trajectories – terrestrial vegetation succession 66
Habitat trajectories – heterotrophic succession 76
Microhabitat dynamics: regeneration and decay 80
Geological cycles: succession and regression 88
Seasonal patterns and processes in the habitat space 91
Diurnal cycles in and out of the habitat space 94

4 **Traditional Habitat Principles and a Diversity of Measures** 97
Some generalisations for traditional habitats 100
The pattern of species richness in habitats 106
Limits to assessing species richness in habitats 114
Indices of species diversity and evenness for habitats 116
Coefficients of niche breadth and overlap 119
Measuring relationships across habitat units 121
How distinct are the same species in different habitats? 124
Turnover between habitats with time 126
Diversity and gradients within and beyond habitat units 126
Phylogenetic diversity 131
Functional diversity and redundancy: their significance 131

5 **Basic Principles of the Resource-based Habitat Paradigm** **133**
Resource-based habitat: focusing on a species' habitat 133
Components of a resource-based habitat (utilities, consumables, conditions) 136
The matrix and its significance 137
Elements of habitat suitability and key resources 143
Resource compaction and intersection in habitats 146
Changes in habitat suitability 150
Unravelling relationships between species' traits and species' niche and habitats 150
Beginnings of the resource-based habitat: early case studies 156
The link between lifetime track, home range and the resource-based habitat 157
Distinguishing territorial defence from home range 159
Variation in the resource-based habitat 160
Resource integration and habitat compaction 161

6 **Bounds of the Resource-based Habitat: Dimensions, Movements and Interactions** **163**
A closer look at the individual: the semiosphere 163
Cues and order in habitat selection 165
Computing components of habitat variability 166
Habitat and space-time scales 168
Factors in resource-based habitat dimensions 169
Motility and mobility in the resource-based habitat 172
Interactions in the resource-based habitat 173
Dividing up the resource space: selecting the minutiae of essential needs 179
Resource-based habitat as a working system 181
Generalism and specialism 182
Habitat factors in the evolution and extinction of species 183

7 **The Resource-based Habitat in Unstable Landscapes** **186**
The basic structure and dynamics of populations 187
The metapopulation concept: isolation and area of habitat units 188
Metapopulations and habitat quality 199
Metapopulations on the move: shifting habitats 200
Factors driving local populations to extinction 202
Is the matrix ever entirely empty? 205
Marginality and the status of resource-based habitats 206
Habitat: climate and land use changes 207
Habitats and genetic changes 208
The place of habitat in biodiversity 210

8 **The Resource-based Habitats in the Kingdoms of Life** **212**
Kingdoms of life 212
Bias in biodiversity studies 216
Prokaryote habitats: Bacteria and Archaea 216
Eukaryote habitats: Protista (Protozoa and Chromista) 221
Habitat structures in invertebrates 223
Habitat structures in vertebrates 224
Habitat structures in plants and fungi 226
Resource-based habitat: the basic elements for measurement in case studies 228
Resource-based habitats: some distinct cases to consider 230
Slow-moving and fast-moving organisms 230
Complexity of social organisation: castes and slaves 233

Group size (singletons, pairs to population massing) 235
Occupation of a variety of biomes and biotopes 238
Occupation of different spheres or zones of activity 241
Occupation of a variety of spheres or zones 248
Fidelity to and movement within locations/zones 249
Phagy (consumer) groups 249
Summary points 250

9 Primates and Human Evolution: Expanding Habitat Demands 251
The pongid–hominid evolutionary tree 251
Models of hominid evolution 255
Home ranges and habitats in living primates 259
Prosimians 260
New World monkeys 261
Old World monkeys 261
Apes bar one 263
Primate socioecology 265
Palaeolithic and Mesolithic human home ranges and habitats 265
Neolithic humans and the first settlements 270
The Metal Age, Medieval advances and the beginnings of modern industry 273
The Industrial Revolution and the Anthropocene. Profusion in resource-based habitats 274
Need, greed and war: the resource-based habitat and the human dilemma 276

10 The Impact on Habitats in the Anthropocene 282
Conservation: setting the scene for the Anthropocene 282
The Anthropocene and prime threats: climate change and habitat loss 285
The red list audit: potential extinctions 289
Ecosystem issues 290
Genetic losses 292
Invasives: Williamson and Fitter's 10th of a 10th of a 10th rule 292
Winners and losers in the Anthropocene 296
Should we bother to conserve species? 299
Measuring the human impact on habitats 300

11 Conserving Life: Biodiversity and the Resource-based Habitat 305
Traditional habitat designation and the conservation of biodiversity 306
Conserving what biodiversity? 307
Landscape taming versus landscape wilding 312
Construction opportunities for habitats and biodiversity 313
Making urban sites fit for species 315
A systems approach: rarity and stability 318
The resource-based habitat and species conservation 322
Species' traits and the identification of resource-based habitats 323
Structural and ancillary features of resource-based planning 325
A species' intimate community 328
A species' special landscape 329
Introductions and cascading effects 331
Zoos, botanical gardens and wildlife parks. A foundation for human-mediated translocations 334
Species distributions, limits and variation in their resource-based habitats 335
Climate change and shifts in resource-based habitats 336
Finale 339

Appendices

Appendix 1 Surveying Natural Populations 340
Plants 341
Terrestrial arthropods 343
Freshwater invertebrates 344
Marine invertebrates 345
Fish 345
Amphibians and reptiles 345
Birds 346
Mammals 346
Microscopic organisms 346

Appendix 2 Population Dynamics 348
The exponential growth 348
The logistic model 349
Lotka–Volterra predator–prey equations 351
Volterra equations for competition 353

Appendix 3 Diversity Measures 355
Species richness 355
Non-parametric estimators 356
Asymptotic models for extrapolating species richness 357
Species Abundance Distribution (SAD) models 357
Indices of α-diversity that incorporate species abundance 360
Beta diversity and (dis)similarity measures 363
Functional diversity 365
Phylogenetic diversity 367
Taxonomic diversity and distinctness 367

Appendix 4 Tree of Life 369

Appendix 5 Geological Events in the Succession of Species 372

Glossary 375

References 395

Index 497

Preface

Habitat is the fundamental unit for understanding and conserving nature. However, it is perhaps the most confusing of terms used in ecology. Most scientists and the general public are familiar with its use for uniform parcels of land: a meadow, a wood, a marsh, a pond, etc. Moreover, discussion of conservation of any one species, a species' habitat, is often related to these parcels of land. Since the early 1990s there has been a growing interest in the concept, with the emergence of the resource-based habitat concept (RBH), described for butterflies by the first author in 1996. This concept demonstrates, importantly, how many species extend over numerous traditional habitats, affecting our measure of their ecological preferences, distribution and abundance data obtained from counting methods based on standardised surveys. While originally developed for butterflies, the RBH concept can be extended to any organism, including humans. The concept has not, however, been proposed to replace the traditional habitat concept. It was observed that both have their uses.

The book describes the importance of both concepts. The traditional habitat concept is of great value when dealing with communities in relatively homogeneous landscapes; the latter is of vital importance when hoping to conserve single species, usually key species in communities. The book explores how these concepts relate to species' traits, niches, territories, life-time tracks, home ranges, diversity patterns and other important concepts central to ecology. One thing is clear in the RBH concept: species may extend over a number of, sometimes all, Earth zones from the base of the lithosphere to the top of the troposphere. Wherever a species travels in its lifetime, migratory bird species being typical examples, all zones through which it passes are included in its RBH. Migration paths through the atmosphere are part of their RBH and are often the most dangerous zone through which they pass and sustain themselves.

Humans are no different from all other organisms in having RBHs. We show how RBHs differ between primate species and have evolved through time. RBHs among primates are immensely varied from region to region. Several things have become patently clear. Since their origin, humans have been greatly expanding their RBHs; greed is a critical part of their nature, a powerful feature of their RBHs, such that other members of their species, and as we know very well of other species too, are squeezed into impoverishment or are purposely eliminated. Humans are also changing the nature of the Earth. Of this there is no doubt, the consequences made clear by brilliant specialists in ecology and climatology, and by those who have care for life on the planet. In discussing these topics, we illustrate the changes that occur to both traditional habitats and RBHs over space and through time from less than seconds to millions of years. Organisms of the past had RBHs, a key feature of their existence from the very beginning of life, much as organisms have them now. We illustrate this with examples from the broad range of life forms living on Earth.

This book offers a comprehensive exploration of the traditional and resource-based habitat concepts, emphasising their importance for understanding ecosystem functioning and biodiversity conservation. The book is articulated into 11 chapters. Each chapter examines key aspects of habitat in relation to fundamental ecological arguments.

Chapter 1 introduces traditional concepts of habitats, exploring their spatial and environmental characteristics and the complex interactions shaping their distinctions from microhabitats to biomes. Chapter 2 delves into ecosystems, their structure and function, focusing on the importance of habitat to understand energy flows, trophic interactions and niche spaces that define organisms' ecological roles within communities. Chapter 3 highlights the temporal and spatial changes that drive habitat transformations over varying temporal and spatial scales. Chapter 4 discusses the role of habitat in shaping community composition and diversity, species abundance distribution patterns, and the roles of specialists and generalists in community structure and ecosystem functioning. Chapter 5 discusses in detail the resource-based habitat concept, contrasting it with traditional habitat definitions and exploring its connections to niches, home ranges and territories. Chapter 6 examines the variability of resource-based habitats, emphasising the importance of organisms' characteristics, resource availability and evolutionary processes in shaping their resource-based habitats. Chapter 7 explores the relationships between habitat and population (especially metapopulation) dynamics, highlighting the interplay between local populations, matrix characteristics and resource availability from evolutionary and conservation points of view. Chapter 8 illustrates the resource-based habitat concept for a variety of life forms, addressing how attributes such as mobility, individual lifespan, body size, population size and social organisation require tailored approaches to understanding their resource-based habitats. Chapter 9 applies the resource-based habitat concept to human evolution and culture, analysing how changes in resource use, technology and social structures have shaped the human resource-based habitat and contributed to the current environmental crisis. Chapter 10 examines the role of human activity in driving present extinction rates and ecosystem collapse, and how they relate to habitat concepts, calling for a profound shift in humanity's relationship with nature to safeguard life on Earth. Chapter 11 concludes with the urgent need for integrative approaches to biodiversity conservation, advocating a systems-based view that incorporates the resource-based habitat concept to address the ongoing biodiversity crisis.

The book also includes five Appendices. Appendix 1 briefly illustrates methods used in population surveys for a variety of organisms, with emphasis on those referred to in the various chapters of the book. Appendix 2 presents an overview of basics of population dynamics frequently cited in the book. Appendix 3 offers a description of the variety of measures used to describe how diversity varies across space and time, with emphasis on those cited in the book. Appendices 4 and 5 present a sketch of the tree of life and of the major events in evolution mentioned in the book. Terms that are particularly relevant (key words) or that are used with specific meanings are noted in bold and defined in the Glossary at the end of the book. The short scale was used for large numbers, with a billion equal to one thousand million and a trillion equal to one million million.

Our hope is that this book adds to our understanding of communities on the planet and our role in them. We are intelligent organisms, unfortunately with many unpleasant traits. This book will only have achieved its task if readers begin to treat *all* life forms on the planet with the respect we *should* give to members of our own species — a respect that humanity so evidently fails to demonstrate.

Acknowledgements

Roger L.H. Dennis would like to record his lasting appreciation and gratitude to his close friend Simone Fattorini for his extensive work on the text, especially the appendices, all boxes and many figures, apart from his modifications and additions to the main text. R.L.H.D. produced an initial draft by January 2024. Thereafter, owing to R.L.H.D.'s illness, S.F. took on the task of producing the book. His patience and kindness and readiness to take on so many complex tasks have not been found in any other co-author in R.L.H.D.'s career; his fine attention to detail, especially in meeting the criteria for publishing the text, was extraordinary. The book is, in every way, a joint effort and an amalgam of skills by two close friends. R.L.H.D. also thanks his wife Margaret for all her fine-tuning of the manuscript and acting as a copy editor for the final version. He extends his gratitude to Emeritus Professor Tim R. New for his ever-optimistic support over many years and past colleagues associated with the development of the resource-based habitat concept, including Leonardo Dapporto, Laurence Cook, Hans Van Dyck, Camille Turlure, Tim G. Shreeve, Tim Sparks, John Dover, Nicolas Schtickzelle, Quentin Dubois, Michel Baguette and Peter Hardy.

Simone Fattorini would like to express his heartfelt thanks to R.L.H.D. for inviting him to join this venture. It has been an immense honour to co-operate in producing this book. The experience has not only been a remarkable opportunity for scientific collaboration, but also an extraordinary occasion for personal growth. S.F. deeply appreciates the trust and spirit of friendship that shaped their work together. S.F. is grateful to his colleagues at the University of L'Aquila, especially Maurizio Biondi, Antonio Di Sabatino, Diana M.P. Galassi, Mirco Iotti, Loretta Pace, Marika Pellegrini and Daniele Salvi, for their insightful discussions on topics addressed in this book. He also thanks the many students who, first during his teaching at middle schools in Rome and then at the University of L'Aquila, provided stimulating reflections: Livia De Simoni, Giulia Lin and Alessia Vitozzi are especially acknowledged. Throughout the writing process, S.F. had the pleasure of collaborating with Letizia Di Biase, whose fresh ideas inspired some of the considerations expressed here. Paulo A.V. Borges and Rosalina Gabriel have been an unparalleled wellspring of motivation for S.F. Their immense contributions to the study of the Azorean biotas, alongside their humanity, have provided continual inspiration. Interactions with scientists and students during the three editions of the 'Summer School in Island Biogeography and Macroecology' they organised offered exceptional opportunities for cultural exchange. Special thanks to Isabel R. Amorim, Mário Boieiro, Pedro Cardoso, Rui Bento Elias, Joaquin Hortal, Thomas Matthews, José Maria Fernández-Palacios, Gabor Pozsgai, François Rigal and Ana Santos. Discussions with Alicia T.R. Acosta, Davide Bergamaschi, Maria Laura Carranza, Alessandro Chiarucci, Andrea Di Giulio, Diego Fontaneto, Pierpaolo Franzese, David Johns, Paolo Galli, Ferenc Jordán, Michela Marignani, Wendy Moore, Emiliano Mori, Carlo Rondinini, Luca Salvati, Noelline Tsafack and Werner Ulrich highlighted specific issues that might have otherwise been overlooked. The book greatly benefited from extensive discussions with Giovanni Strona, an inexhaustible source of powerful ideas. A warm and supportive family environment was instrumental in completing this book: for this, S.F. thanks Federica Fedele, Gabriella Petrella, Micaela Fattorini and the little ones, Elena and Arianna Degiorgi – who knows if Silvio Fattorini would have been satisfied with the graphics of some of the figures.

We are very grateful to Margaret Dennis, Peter Hardy, Davide Bergamaschi and Letizia Di Biase for their individual assistance with proof corrections.

We are also deeply grateful to the authors, institutions and publishers for kind permission to include their figures and tables in their original or modified form. We express our gratitude to all authors of images published without copyright or under Creative Commons licences, which made their free use possible in this book. Illustrations are important for enriching the text and such generosity is greatly appreciated.

Appropriate credit is given where reproductions are made. Every effort has been made to contact holders of copyright to obtain permission to reproduce copyright material. However, if any have been inadvertently overlooked, the publishers will be pleased to make the necessary arrangements at the first opportunity. We also extend our sincere gratitude to all individuals who generously contributed pictures to illustrate this book. Copyright is retained by authors contributing them to the book.

We are entirely indebted to the CABI staff for their assistance with awkward matters including permissions. In particular, we would like to thank Kin Yunn Ho (Managing Editor) and James Bishop (Production Editor) for their assistance during the production process; Holly Regan-Jones (Copy Editor) for her invaluable work in identifying errors in the text and for her patience and professionalism throughout the process; and Helen Elliott (Editorial Assistant) for her enormous help with permissions. Grateful thanks also go to Ward Cooper (Commissioning Editor) for his support and encouragement to produce the book. This is the fourth book by R.L.H.D. with which Ward has been associated. Finally, we thank the production and typesetting staff at Straive, whose care and attention to detail were invaluable in preparing the final version of this book.

1 The Traditional Habitat Model: Utility, Affinities and Failings

Abstract

The traditional definition of a habitat is a spatial unit with distinctive environmental characteristics and biological communities, from extremely small spaces to massive regions called biomes. A traditional definition of a species' habitat is a distinctive physical (geolocated) space where it is found because of the presence of the resources it uses and conditions it exploits. Habitats occupy Earth spheres of activity (air, sea, land, water) and biogeographical regions, abiotic and biotic processes within which contribute to their distinctions. The boundaries between habitats are often far from crisp, but can merge imperceptibly into one another as described by ecotones. A species' habitat is rarely limited to, or fully utilises, a traditional (homogeneous) habitat space but is greatly more complex, and described by the resource-based habitat paradigm.

This chapter explores the two traditional uses of the term 'habitat' in biology (Table 1.1) (Hall *et al.*, 1997; Kirk *et al.*, 2018). First, and more typically, **'habitat'** as traditionally applied refers to any identifiable place (e.g. site, locality, situation, residence) or to the environment (e.g. topography, soils, **substrates**, vegetation types, environmental conditions) and its subdivisions (i.e. **microhabitats**) occupied by organisms. This concept of habitat can be used for both land habitats (e.g. meadows, conifer or deciduous woods, fens, mires, moors, heaths, caves, glaciers, urban green spaces) and water habitats (e.g. rivers, lakes, ponds, streams, swamps, estuaries, kelp forests, seagrasses, the **bathypelagic zone** of the oceans, hydrothermal vents). The **atmosphere** represents the aerial habitat for a multitude of organisms (including, for example, microbes, fungal spores, floating seeds, pollen, insects and birds), whereas organisms of many species form the (organismal) habitat of their pathogens and parasites. Second, and closely associated with this interpretation of the word 'habitat' – in effect a projection of it – reference is often made to a particular **species' habitat** or domain, which conjures up the type of environment occupied (e.g. a heath or a meadow) by the species. Occasionally, the array of resources, biotic and physical conditions which the organism requires for development, survival and reproduction is implicit in this use of the term if not explicitly stated (Ryan, 2019).

This chapter considers the context of habitats as traditionally employed, the uses of this widely applied term and associated failings, for instance difficulties in protecting **'red data book' (red list)** species (Mace and Baillie, 2007; Baillie *et al.*, 2008; Juslén *et al.*, 2016). Focus is on variability in habitats, changes that occur over space, particularly on some basic 'island biogeography' factors of habitat dimensions and their disposition that make reference to a particular species' habitat, that has proved an elusive concept. Changes within habitat space over time are discussed in Chapters 2 and 3.

Before we consider these two fundamentally different notions of 'habitat', it is important to place them in some context, and no context is more inclusive and germane than that of our entire planet and its various spheres of activity. These spheres of activity, in different ways, provide the space and conditions for the habitats of all organisms on the planet varying in mass from **micro-organisms** to whales.

Spheres of existence – habitats in a wide perspective

Earth functions as a complex (biogeo-)system, something understood for more than a century (Skinner and Murck, 2011; Martin and Johnson, 2012). Recent advances have focused intently on identifying the links and processes (transfers of mass and energy) that cause changes to the planet.

DOI: 10.1079/9781800627505.0001

Table 1.1. Various definitions of the term 'habitat'.

- The locality in which a plant or animal naturally lives (Darwin, 1872: 434).
- Place, living space, where an organism lives (Odum, 1963).
- The place where an organism lives, characterised by its biotic or physical characteristics (Whittow, 1984).
- Type of environment in which an organism lives (Collin, 1988).
- The locality, site and particular type of local environment occupied by an organism (Lincoln *et al.*, 1982).
- Place where a species normally lives, often described in terms of physical factors such as topography and soil moisture and by associated dominant forms (e.g. intertidal rock pools or mesquite woodland). Definitions in ecological literature vary widely but there is consensus for the following: key environment features related to a species; habitat and vegetation classifications may be concordant, but not always so; subdivisions occur – microhabitats (Calow, 1999).
- Habitat is a suite of resources and environmental conditions that determine the presence, survival and reproduction of a population (Caughley and Sinclair, 1994; Weddell, 2002).
- Habitat (resource-based habitat) is a zone (area) comprising a set of resources (consumables and utilities [including conditions]) for the maintenance of an organism. The resources occur in union and/or intersect and may also be equivalent; links between resource outlets are established by individual searching movements of the organism (Dennis and Shreeve, 1996).
- Habitat [comprises] the resources and conditions present in an area that produce occupancy – including survival and reproduction – by a given organism. Habitat is organism-specific; [...] it is the sum of the specific resources that are needed by organisms (Hall *et al.*, 1997).
- Habitat is an area with a combination of resources (e.g. food, cover, water) and environmental conditions (e.g. temperature, precipitation, presence or absence of predators and competitors) that promotes occupancy by individuals of a given species (or population) and allows those individuals to survive and reproduce (Morrison *et al.*, 2006; Mathewson and Morrison, 2015).
- Habitat is the sum of the abiotic and biotic factors essential to the life and reproduction of the species within its natural geographic range (Haslett, 2007).
- Habitat is a point in environmental space, defined by a set of conditions, resources and risks for the species of interest (Northrup *et al.*, 2022).

These have galvanised the attention of the scientific community, if not the political and economic set, concerned for planetary stability and sustainability and have made us aware of projected changes in Earth global climates and sea levels, water, land and air pollution, mineral and biotic resource conservation and, more recently, pandemics (Skinner and Murck, 2011; Brusseau *et al.*, 2019; Walther, 2014; Qiu *et al.*, 2017; Walker *et al.*, 2020). Mass (element and molecular) cycling and energy transfers involve real units of the globe (spheres of activity) as opposed to temporary and arbitrary ones (political units). Spheres divide up the Earth, almost, as a set of well-known functioning layers including the **biosphere**, **atmosphere**, **hydrosphere** and **lithosphere**, the latter (crust and upper mantle) being part of the **geosphere**. But they are not quite a neat set of layers, as, for instance, gases from the atmosphere penetrate depths in the lithosphere, and water – the distinguishing molecule of the hydrosphere (or cryosphere where frozen) – occurs substantially as a vital component on the surface and within the layers of the lithosphere and in the lowest layer of the atmosphere, the troposphere.

The **biosphere** (**prokaryotes, protists**, plants, animals and fungi), which evolved through an immensity of geological time beginning roughly 700 million years after the Earth's formation 4.6 billion years ago (Oschmann *et al.*, 2002; Dodd *et al.*, 2017), penetrates these layers, and increasingly humans have risen to dominate the whole. Such domination by the human mind is referred to as the **noösphere** (noösystem), a concept first developed by Pierre Teilhard de Chardin, Édouard L.E.J. Le Roy and Vladimir I. Vernadsky (Pitt and Samson, 1998; Wyndham, 2000) (Fig. 1.1). Accompanying the late Cenozoic emergence of the noösphere is the current geological epoch referred to as the **Anthropocene**, distinguished by human impacts on Earth's ecosystems (see Chapter 10). The part of the environment that has been influenced or shaped by human activities may be recognised as a further sphere, the **anthroposphere**. Although the Anthropocene is still not officially recognised as a subdivision of geological time, and a variety of starting dates have been proposed on the basis of different markers (from the development of agriculture, about 12,000 years BP, to the middle of the 20th century, for the radionuclides released into the atmosphere from atomic bomb detonations) (Zalasiewicz *et al.*, 2010; Lewis and Maslin, 2015; Hazen *et al.*, 2017), the term is now widely

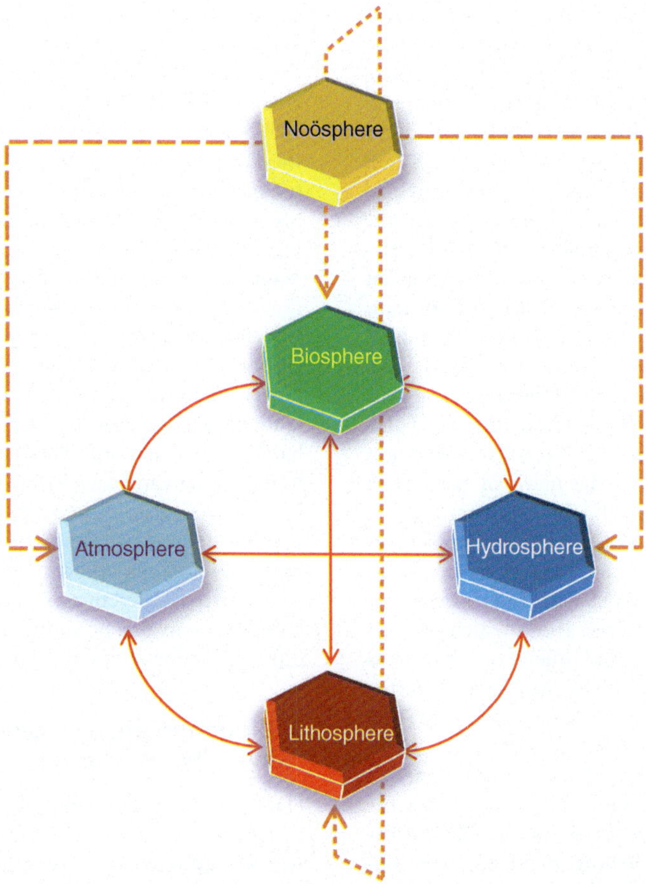

Fig. 1.1. The primary Earth spheres of activity: lithosphere (the Earth's crust), hydrosphere (water – oceans, lakes, rivers, ground water and hydrological cycle), atmosphere (gases – troposphere to exosphere; air below ground), biosphere (prokaryotes, protists, plants, fungi, animals), noösphere (sphere of the mind and human thought) (Bateson, 2000; Wyndham, 2000). Arrows, illustrating effects and feedbacks, indicate transfers of energy and matter.

used and well expresses the centrality of human impacts on the current status of all spheres of activity.

These spheres of activity are all highly relevant to understanding habitat, however the term 'habitat' is defined. Regardless of the fact that some taxa may appear to have a minimal or no impact on one or more Earth spheres (e.g. butterflies have little impact on the lithosphere, hydrosphere and atmosphere), nevertheless they belong to biotic communities that do impact on all spheres which then in turn influence the status of low-impact or near zero-impact species. The interactions of biological, geological and chemical processes through the so-called **biogeochemical cycles** (i.e. the movement and transformation of elements between the organisms and the Earth's abiotic compartments) are a long-recognised example of the impact of life on the atmosphere, lithosphere and hydrosphere (e.g. Vernadsky, 1945; Hutchinson, 1944). However, most ecologists have traditionally assumed that while organisms are strongly influenced by the abiotic conditions they experience, their influence on the abiotic environment would be largely inferior or even minimal.

One of the most popularised views that contrasted this paradigm emerged at the beginning of the 1970s with the formulation of the so-called 'Gaia hypothesis' by James E. Lovelock (Lovelock, 1972, 1987; Lovelock and Margulis, 1974), which postulates that life and environment coevolve, making the Earth a self-regulating complex system. Although this idea (already anticipated at the end of the 19th and the beginning of the 20th centuries by the Russian scientists Peter A. Kropotkin, Rafail V. Rizpolozhensky, Vladimir I. Vernadsky and Vladimir A. Kostitzin; Lapenis, 2002) should be considered more metaphoric than scientific (Kirchner, 1989), it had the merit of helping to emphasise the importance of considering the impact of life on the Earth's physical and chemical environment. An example of the importance of a more integrated view of the relationships between life and the physical environment for the development of new views of Earth's systems is represented by the controversial 'biotic pump' hypothesis, proposed by Anastassia M. Makarieva and Victor G. Gorshkov (2007), that forests are the major drivers of atmospheric circulation on Earth (Pearce, 2020). Independently from the correctness of this hypothesis, it had the merit of promoting new discussions about established paradigms.

In any case, the impact of an organism on spheres of activity depends on the space-time scales of reference, impact increasing with finer scales. For instance, microbial processes, about which perhaps least is known, penetrate and impact on all systems and spheres of activity. This is most evident in the web of microbial interactions occurring in soil, faeces and the phyllosphere (Zhao *et al.*, 2021); these are referred to variously as (micro)biomes or (micro)spheres. For instance, the **phyllosphere**, the total above-ground parts of plants, comprises the **caulosphere** (stems), **phylloplane** (leaves), **anthosphere** (flowers) and **carposphere** (fruits). The **soil** zone around the roots includes the **rhizosphere** (roots) and **laimosphere** (subterranean stems). All these zones or microspheres of activity host diverse **communities** of **micro-organisms** including **prokaryotes, protists** and fungi. These terms are useful for defining broad classes but for empirical study in any locality are better conceived as being part of interconnected **ecosystems** (see Chapter 2).

The typical context of traditional habitats has largely been that of terrestrial environments, which as mentioned above comprise patches of **substrate** and vegetation (e.g. a meadow, a wood, a moss, a heath in temperate latitudes). But, as life on the planet penetrates all the spheres from the depths of the lithosphere (deep cave systems; Romero, 2009) and the hydrosphere (ocean trenches; Todo *et al.*, 2005) to the top of the troposphere (e.g. micro-organisms; Bryan, 2017), then the bounds of habitat must be considered to envelop a much wider domain. Simply, each species has a habitat, thus its habitat must enclose whatever type of space that species occupies including different Earth spheres. Such spaces project a vertical dimension as well as the more familiar horizontal ones. Consequently, we must consider that habitats are (at least) irregularly sized and shaped three-dimensional units of the Earth spheres which contain life. In the traditional map view, habitat units (distinct patches of life and substrate (rock), gas and water), such as a meadow or a wood, are displayed as contiguous, regularly shaped units (with field boundaries) over the landscape. But even here, there is a third dimension for organisms (e.g. shape of the land, herbs, shrubs, trees, depth of water occupied in ponds and lakes) occupying vertical 'space' that is often ignored. The traditional view of any one species' habitat is also referred to such discrete units of human design and ownership. However, later it will become clear that although certain species' habitats may be limited to such discrete units, most comprise mobile, punctuated, overlapping envelopes spread across discrete units described by traditional habitats; many can be shown to be vast and occupy different major zones and Earth spheres.

Biogeographical regions: adaptation, diversity, convergence and divergence

Habitats around the globe are not just distinguished by spheres of existence but also by **biogeographical regions** (or realms) (Cox, 2001; Morrone, 2002). They pack into isolated zones owing to the distribution of continents and the barriers between them, particularly oceans, deserts and mountains. This type of regionalisation is based on taxic distributions, and should not be confused with regionalisations based on plant growth forms and types of vegetation, an approach first introduced by Alexander von Humboldt (von Humboldt and Bonpland, 1814–1829; see Ebach, 2015 for a discussion). Concepts such as those of **vegetational belts**, terrestrial **biomes, life zones** and **biotic provinces** (see below, section *Habitat type – the traditional biome view*) can all be considered as derivatives of von Humboldt's approach, as they are based on the recognition of similarities in the physiognomy of the vegetation structure across space (Fattorini, 2016; Fattorini *et al.*, 2019a; Mucina, 2019), whereas biogeographical regionalisations are based on the recognition of distinct biotas, especially in terms of endemic and/or spatially clearly limited taxa and communities (e.g. Rueda *et al.*, 2010; Morrone, 2015a; Fattorini, 2016, 2017a). Recognition of areas of endemism can be also considered a particular approach to biogeographical regionalisation (Dennis *et al.*, 1991, 1995, 1998; Dennis and Schmitt, 2009; Fattorini, 2016, 2017a). In summary, von Humboldt's approach (and its derivatives) is based on ecomorphological criteria, while biogeographical regionalisation is based on taxic distributions (Fattorini, 2016). Although Udvardy (1969) introduced the word 'faunation' as a zoological counterpart of 'vegetation' (i.e. to indicate the totality of individual animals living together in an area), it remained substantially ignored by zoogeographers, and the possibility of recognising regions based on faunations has not been explored.

The first attempt at a biogeographical regionalisation is represented by de Candolle's (1805) subdivision of France on the basis of plant distribution. At global scales, early examples of **biogeographical regions** for land organisms include de Candolle's (1820) botanical regions, Prichard's (1826) and Schmarda's (1853) zoological regions, and Sclater's (1858) ornithological regions (see Ebach and Goujet, 2006; Parenti and Ebach, 2009; Ebach *et al.*, 2008; Ebach, 2015). A multitude of global regionalisations have been proposed during the past two centuries (see Morrone, 2015a and Morrone and Ebach, 2022 for review), but the best known and most influential is that represented by the **zoogeographical regions** proposed by Wallace (1876) on the basis of mammal distributions (Fig. 1.2a). This regionalisation (with innumerable minor changes introduced by a plethora of subsequent authors) has been a reference for describing animal species distributions at the global scale for more than a century, and even recent revisions (e.g. Cox, 2001; Holt *et al.*, 2013; Falaschi *et al.*, 2023) have not greatly changed the overall scheme (Fig. 1.2b), although for specific groups some substantial differences may emerge (e.g. ants; Wang *et al.*, 2024).

While zoogeographers used Wallace's scheme as a main reference, the most influential scheme in **phytogeography** was that proposed by Engler (1879–1882), which has survived almost to the present day with the introduction of relatively minor modifications, for example by Diels (1908), Walter (1954) and Good (1947, 1974) (see Takhtajan, 1986; Cox, 2001; Pfadenhauer and Klötzli, 2020; Procheş and Ramdhani, 2020; Carta *et al.*, 2022) (Fig. 1.2c). Attempts to surpass the distinction between zoogeographical and phytogeographical regions, in order to produce a unique biogeographical scheme, have been made by numerous authors including Udvardy (1975), Pielou (1979, 1992), Müller (1986), Morrone (2002, 2015a) and Morrone and Ebach (2022). Also, biogeographical regions have been proposed for the **marine environment** (Costello *et al.*, 2017b; Castro-Insua *et al.*, 2018; Kocsis *et al.*, 2018; Victorero *et al.*, 2023; see Cox and Moore, 2010 and Lomolino *et al.*, 2010 for a review of the history of **marine regionalisation**, dating back to the latter half of the 19th century) (Fig. 1.2d), and for the **freshwater environment** (Leroy *et al.*, 2019).

Early attempts to establish a regionalisation of the oceans are illustrative of two important aspects that characterise the **marine environment**: (i) species are distributed both horizontally (as on land) and vertically (i.e. at different depths of the water column) and (ii) there are no obvious barriers that divide the oceans (apart from those represented by the continents), a situation that has long generated the prejudice that the distribution of marine organisms is less constrained than that of terrestrial ones. Edward Forbes, who is recorded as a founder of marine biogeography and a champion of extensionism (i.e. the hypothesis that currently disjunct distributions can be explained by assuming that landmasses had a larger extent in the past; Fattorini, 2017b), first proposed a regionalisation for the marine realm, in which the marine world is divided into nine latitudinal bands of similar faunas subdivided into five zones of depth (Forbes, 1856). He also noticed that in the marine environment, species distributions were mainly influenced by climate, water chemistry and depth, and that depths of distribution were matched by latitudinal distribution (thus echoing the already well-known parallelism between elevational and latitudinal patterns observed on land – see below, section *Habitat type – the traditional biome view*). The influence of temperature as a main driver of species distribution in the oceans was subsequently emphasised by Dana (1853), who divided the marine world into different regions according to lines of equal minimum sea surface temperature (isocrymes). This approach (maintained also in recent regionalisations; see, for example, Briggs and Bowen, 2012) reflects the difficulties of dividing oceans and seas into regions using the uniqueness of species assemblages, especially in the open ocean, as demonstrated by the fact that attempts to identify marine regions on the basis of patterns of endemicity have been largely limited to coastal areas (see Cox and Moore, 2010 and Lomolino *et al.*, 2010 for historical reviews).

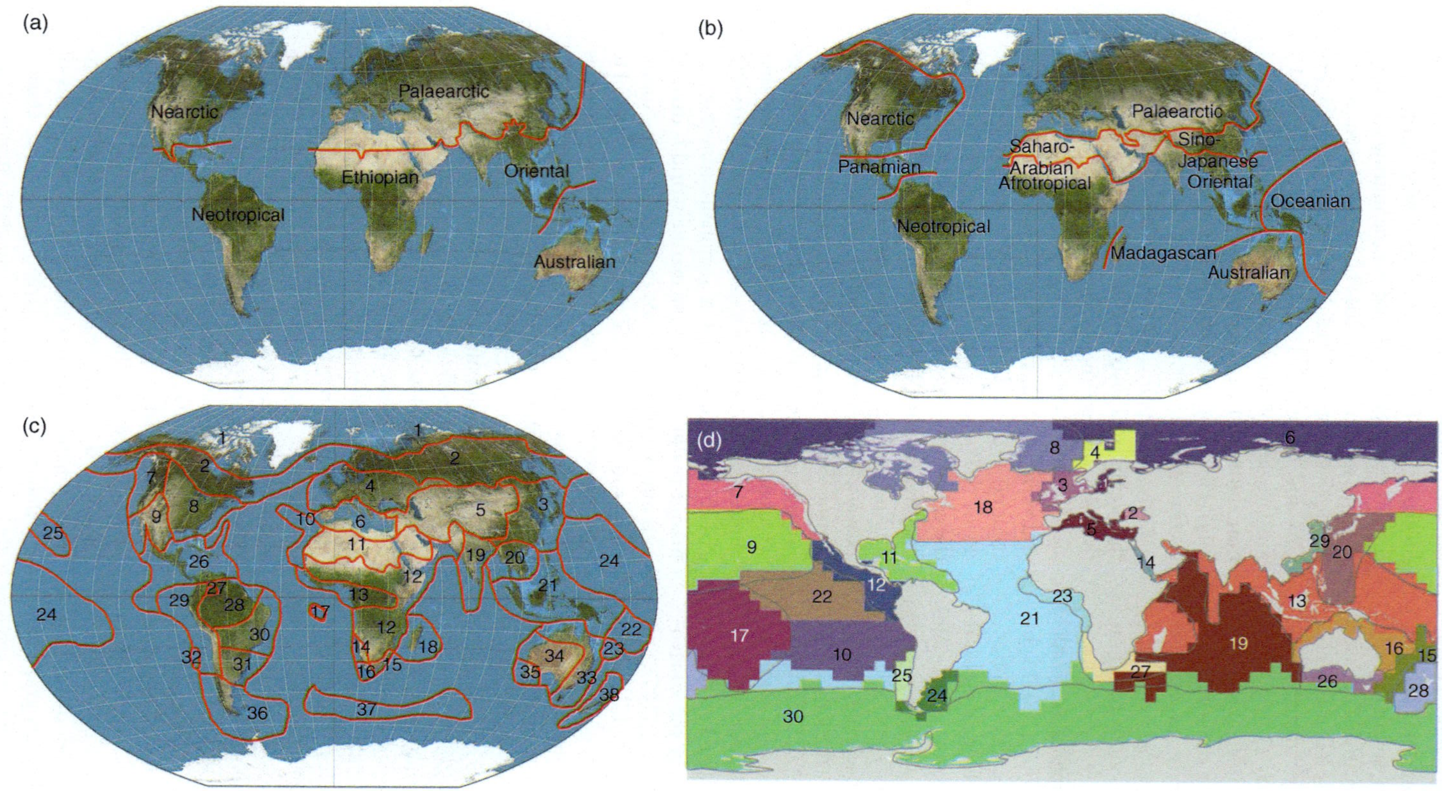

Fig. 1.2. World maps showing the major biogeographical regions. **(a)** The original zoogeographical scheme by Wallace (1867). **(b)** A recent modification of Wallace's zoogeographical regions proposed by Holt *et al*. (2013). **(c)** Phytogeographical regions proposed by Pfadenhauer and Klötzli (2020) on the basis of Mattick (1964), Takhtajan (1986) and Cox (2001): 1 Circum-Arctic, 2 Circumboreal, 3 East Asian, 4 South-Eurosiberian, 5 Irano-Turanian, 6 Mediterranean, 7 Rocky Mountains, 8 North American-Atlantic, 9 Madrean, 10 Macaronesian, 11 Saharo-Arabian, 12 Sudano-Sambesian, 13 Guineo-Congolian, 14 Karoo-Namibian, 15 Zulu, 16 Capensic, 17 Ascension, St. Helena, 18 Madagascan, 19 Indian, 20 Indochinese, 21 Malesian, 22 Fiji, 23 New Caledonian, 24 Polynesian, 25 Hawai'ian, 26 Caribbean, 27 Guianan, 28 Amazonian, 29 Andean Pacific, 30 Brazilian, 31 Argentinean, 32 Fernandezian, 33 Northeast, 34 Central Australian, 35 West Australian, 36 Patagonian, 37 Antarctic, 38 New Zealand. **(d)** Marine biogeographical regions proposed by Costello *et al*. (2017b): 1 Inner Baltic Sea, 2 Black Sea, 3 NE Atlantic, 4 Norwegian Sea (*partim*), 5 Mediterranean, 6 Arctic Seas, 7 N Pacific, 8 N American Boreal, 9 Mid-tropical North Pacific Ocean, 10 South-East Pacific, 11 Caribbean and Gulf of Mexico, 12 Gulf of California, 13 Indo-Pacific seas and Indian Ocean, 14 Gulfs of Aqaba, Aden, Suez, Red Sea, 15 Tasman Sea, 16 Coral Sea, 17 Mid South Tropical Pacific, 18 Offshore and NW North Atlantic, 19 Offshore Indian Ocean, 20 Offshore W Pacific, 21 Offshore S Atlantic, 22 Offshore mid-E Pacific, 23 Tropical E Atlantic, 24 Argentina, 25 Chile, 26 S Australia, 27 S Africa, 28 New Zealand, 29 North West Pacific, 30 Southern Ocean. (Panel **a** based on Wallace, 1867. Panel **b** based on Holt *et al*., 2013. Panel **c** based on Pfadenhauer and Klötzli, 2020. Panel **d** from Costello *et al*., 2017b. Used under licence CC BY 4.0, http://creativecommons.org/licences/by/4.0/. World map used for panels **a–c** by Strebe: https://en.wikipedia.org/wiki/File:Winkel_triple_projection_SW.jpg.)

The global distinctions in taxa have a long geological history related to plate tectonics, movements with the constant shift in continental land masses and seas, though occasionally disrupted by stochastic shocks triggered by mass disturbances through periodic global extinctions (e.g. the Cretaceous-Palaeogene **mass extinction** event, K-PgE: Fastovsky and Sheehan, 2005; Brusatte *et al.*, 2015; Fattorini, 2020) (see Chapter 10). Thus, it is not surprising that a major boundary recovered by global regionalisations distinguishes Gondwanaland and Laurasia which separated about 200 million years ago in the Jurassic period (e.g. Fattorini, 2022). However, detecting boundaries between different biogeographical regions is not straightforward (Kreft and Jetz, 2010; Holt *et al.*, 2013; Ficetola *et al.*, 2017; Morrone, 2018) and **transition zones** can be recognised between adjacent biogeographical units (Breckle, 2002; Kreft and Jetz, 2013; Ferro and Morrone, 2014). It is important to note that these transition zones should not be confounded with **ecotones** (see below, section *Habitat boundaries, edges, ecotones and ecoclines*), since transition zones refer to gradual changes in faunas and floras (e.g. from placentals to marsupials in the Wallacea, the transition zone between the Oriental and the Australian regions), whereas ecotones refer to changes in **communities** (e.g. from trees to herbs at the junction between a forest and a grassland). A recent global regionalisation that explicitly takes into account transition zones has been proposed by Morrone (2015a). Different regionalisation schemes arise not only from the application of different methodologies (Rueda *et al.*, 2010; Kreft and Jetz, 2013; Morrone, 2015a), but also from the considered organisms. For instance, those distinguishing phytogeographical from zoogeographical regionalisations (Morrone, 2002, 2015a) give an indication of varying opportunities for, and differential **mobility** of, organisms. Exemplary is the importance applied by phytogeographers to the unique biota of the south-western portion of South Africa, which is regarded as separate from the rest of sub-Saharan Africa (see, for example, the classical phytogeographical regionalisation of Takhtajan, 1986; see Morrone, 2015a for a review).

Beside global regionalisation schemes, a number of finer biogeographical subdivisions have been proposed at smaller scales. Illustrative are, for Europe, the phytogeographical map proposed by Walter (1954) and adapted to the fauna by Freitag (1962), the subdivision proposed by Illies (1967, 1978) for the **limnofauna**, or the biogeographical regions of Europe based on vegetation types (see Cervellini *et al.*, 2020). Regionalisation schemes proposed in Europe for specific taxa include those suggested for butterflies (Dennis *et al.*, 1991, 1995, 1998; Dennis and Schmitt, 2009), dragonflies (Heiser and Schmitt, 2010) and fish (Reyjol *et al.*, 2007). Outside Europe, examples of regional subdivisions are those proposed for Africa (Linder *et al.*, 2012 based on vertebrates and plants; Droissart *et al.*, 2018 based on plants), North America (Hagmeier *et al.*, 1964; Hagmeier, 1966 based on mammals; Brown *et al.*, 1998 based on plants and animals; Burr and Mayden, 1992 for freshwater fish), South America (Morrone, 2015b based on plant and animal taxa, with a synthesis of previous regionalisations) or Australia (Ebach *et al.*, 2015 based on vegetation).

However, there is much contention regarding how the application of regionalisation procedures should lead to hierarchical classifications, which can be expressed by formal taxonomic arrangements subject to nomenclatorial rules (i.e. International Code of Area Nomenclature) (Ebach *et al.*, 2008; Zaragueta-Bagils *et al.*, 2009). Thus, there is confusion in the use of terms to indicate the different levels of subdivisions (e.g. realms, regions, dominions, provinces and districts) (Morrone, 2009; Morrone and Ebach, 2022). The Interim Biogeographic Regionalisation of Australia (www.dcceew.gov.au/environment/land/nrs/science/ibra) and the European Biogeographical Regions (EBRs; an official reference for the EU Habitats Directive and Birds Directive, and hence a fundamental tool for biodiversity conservation planning in Europe) (Cervellini *et al.*, 2020) are mostly based on vegetation forms, not on species distributions, and hence should not be considered true biogeographical regionalisations. Similarly, the classical schemes of provinces introduced by Udvardy (1975) for the entire world are based on a mix of faunal and floral similarities and types of ecosystems.

The presence of discontinuities in faunas and floras that are at the base of biogeographical regionalisations results in the evolution of distinctions in regional taxa at different levels (classes) from processes related to **vicariance** (geographical separation) and stochastic long-distance dispersal. Distinctions among regional taxa, once the preserve of numerical taxonomy based on morphology, are now being thoroughly explored by molecular phylogeographical techniques (Yang, 2018). The techniques provide the fundamental data on taxonomic relationships among organisms distributed over the globe but, as with mammal and reptile evolution, the explosion of discoveries of complete fossils is adding enormously to our knowledge and exposes failings in reliance purely on molecular techniques (Pickrell, 2019). For instance, molecular techniques alone cannot resurrect failed lines of descent lost from the geological record. The sheer complexity of evolutionary histories in different taxa should warn us that the contents of contemporary habitats are not simply the product of isolation leading to linear distinctions in communities occurring in them today; increasingly, extinct communities are being discovered in the geological record that lack contemporary examples (e.g. Williams and Jackson, 2007; O'Brien *et al.*, 2025; Shupinski *et al.*, 2025).

Even so, many habitat types in different biogeographical regions across the planet share similar conditions of existence – physical and environmental constraints – related to the medium occupied (air, land, fresh water, salt water) and key climatic inputs (e.g. radiation, temperatures, moisture) linked to their global positions (e.g. latitude, altitude, continentality, depth below the surface). Trends in global environments have led to repeated patterns in the evolution of morphological characteristics that are described as **rules** (Table 1.2), affecting taxa at intraspecific and interspecific levels, and species assemblages (**species richness**, abundance, biomass, etc.) (Gaston *et al.*, 2008).

Possibly the best known (and most widely discussed) of these patterns (**ecogeographical rules**) is **Bergmann's rule** (of which Allen's rule can be considered a specific case applied to single parts of the body; see Lomolino *et al.*, 2010), which states that endothermic animals tend to be bigger in colder climates (Fattorini, 2024a). This pattern should arise from the following mechanism: with increasing size, the ratio surface area:volume decreases, and hence there is proportionally less surface from which the heat is lost; as a result, larger animals need to produce less heat in relation to their size to raise their temperature above the ambient temperature. Bergmann (1847) originally illustrated this rule both by comparing the size of different but allied species living in places with different climates, and hence at interspecific levels, and by comparing different races of the same species of domestic animals. More than 170 years since Bergmann's work, it is still debated whether the rule should be applied at the inter- or intraspecific level (or both), and whether it can be valid only for endothermic animals or can be extended to ectothermic ones, although for different reasons (see Fattorini, 2024a for a discussion). For example, for marine crustaceans, increasing body size at higher latitudes can be explained by the fact that a lower

Table 1.2. Some common ecogeographical 'rules' in biological traits within and among species.

Rule	Description
Allen's rule	In closely related endothermic vertebrate forms, those in warmer environments have longer appendages than those in colder environments.
Bergmann's rule	An increase in size of a species occurs towards higher latitudes and lower temperatures.
Foster's (island) rule	Smaller species become larger and larger species become smaller on islands compared with mainland regions.
Gloger's rule	Within a species of endotherms more heavily pigmented forms tend to occur in more humid environments.
Jordan's rule	Fish species develop more vertebrae in colder than in warmer environments.
Rapoport's rule	Among species, there is an increase in range size with latitude.
Thorson's rule	Benthic marine invertebrates at low latitudes tend to produce large numbers of eggs developing to pelagic (often plankton-feeding) and widely dispersing larvae, whereas at high latitudes they tend to produce fewer and larger yolk-feeding eggs and larger offspring, often by viviparity or ovoviviparity (often brooded).

Traits in and across species are found that co-vary with geography (i.e. latitude, altitude, longitude (continentality), depth of lake or sea water) and environmental variables (e.g. precipitation and humidity, plant productivity, radiation, salinity, temperature). The table includes a few commonly referred to rules (studies of this kind are legion and a wide range of relationships can be found at different spatial scales; see Gaston *et al.*, 2008 for references), but it is important to note that there are exceptions to each rule (e.g. Bergmann's and Foster's rules do not apply to treeshrews (Mammalia, Scandentia; Sargis *et al.*, 2018). Some trends with environmental gradients can be very complex as in the case of disparate wing patterns in butterflies across the British Isles, the variation accounted for by differences in butterfly wing-scale structures, thermoregulation, mate location and predator evasion mechanisms (Dennis and Shreeve, 1989; Kang *et al.*, 2021).

temperature leads to an increased cell size and longer life span, and hence a larger body size (Timofeev, 2001). But alternative explanations have been proposed also for endotherms, including the idea that larger animals have a greater ability to tolerate cold and periods of food shortages, for example because of more abundant fat stores (see Ashton *et al.*, 2013). The rule has been evoked to explain clinal variations in body size along both latitudinal and elevational gradients, and the inverse relationship frequently observed in ectotherms like many insects (larger bodies in warmer climates, sometimes referred to as the 'converse Bergmann rule'; Fattorini, 2024a) has been explained in various ways, including the effect of temperature on metabolic and development rates (Shelomi, 2012). Despite the many criticisms that Bergmann's and other ecogeographical rules have received, such trends continue to inspire generations of ecologists and have also long been noticeable among humans, particularly responses to climate (Schlingmann *et al.*, 2021; Pisor and Jones, 2021).

More complex relationships occur in a global patchwork. Similar conditions over the globe in the form of **biomes** (see next section) counter divergences in taxa. Thus, although regional isolation, as in the case of islands, provides opportunity for taxonomic divergence (e.g., Lomolino *et al.*, 2010; Whittaker *et al.*, 2023), this is countered by similarity in conditions leading to convergence of **traits** (e.g. among plants, the similar forms and functions of Euphorbiaceae and Cactaceae; among mammals, the skull shape of the marsupial *Thylacinus cynocephalus* and the placental *Canis lupus*). As it is, **ecosystems** in different biogeographical regions with similar abiotic conditions have a basic convergence in organism interactions and community structures, dynamic features of habitats reinstated after each global extinction (Lowery *et al.*, 2018). This presents complications for taxonomy since convergences can mimic true relationships; thus, a combination of fossil material and molecular phylogeny becomes essential for unravelling true from 'false' similarities, **homologies** from analogies (**homoplasy**) (Wiley, 1981), as the complex diversity of Triassic and Jurassic mammal clades should make clear (Pickrell, 2019).

Habitat type – the traditional biome view

If habitats are made up of units of the globe, the first question is: how do we recognise them? Practical guidance as to designation of such habitat spaces is all too often noticeably lacking. The division must be finer than that of the spheres (air, land and water) but occasionally must also cut across them. Habitats are most typically described for the Earth surface; they have been identified primarily by vegetation and lack of it (e.g. vegetated and 'desert'), and then by different vegetation structures (e.g. forest and grassland, coniferous and deciduous forest) and their seasonal development (evergreen and deciduous) including the timing of seasons. In this way, broad patterns have emerged, referred to as terrestrial global **biomes** (or major **life zones**), a term first coined by Clements (1907, 1916) for a biotic **community** (Egerton, 2017) (Fig. 1.3). They are characterised not only by distinctive abiotic conditions (geology, soils, photoperiods, climate, etc.) but also by distinctive communities of plants and animals, many species of which remain to be identified, even tree species in the fast-diminishing tropical rainforests (Lapola *et al.*, 2008; Slik *et al.*, 2015). Most obviously, the terrestrial biomes correspond with gradients in heat (ambient temperatures) and water availability (precipitation, evapotranspiration) driven by atmospheric processes linked to latitude, altitude and continentality as illustrated in a ternary graph (Fig. 1.4) (e.g. Holdridge, 1947, 1967; Whittaker, 1975; Dennis, 1993a; Sisneros *et al.*, 2011). As would be expected, these biomes largely coincide with climatic regions (e.g. Thornthwaite climates, Köppen-Geiger climates) (Feddema, 2005; Beck *et al.*, 2018).

As climate changes in a roughly similar way along latitudinal and elevational gradients, vegetation on mountains, at least in some cases, tends to show patterns of elevational variation that resemble the latitudinal ones (Fattorini *et al.*, 2019a). This parallelism between vegetation structure along latitudinal and elevational gradients was clearly illustrated by von Humboldt (von Humboldt, 1816, 1817; von Humboldt and Bonpland, 1805), who compared the elevational organisation of vegetation types to the latitudinal zonation (**life zones**) already recognised by Willdenow (1792). Although Tournefort is frequently credited as the first botanist who noticed this pattern, this was due to an error by Linnaeus, who cited Tournefort's work, but there is no trace of this parallelism in Tournefort's work, and the first author who recognised it was actually Linnaeus himself (Fattorini, 2021). This assumed parallelism has profoundly fascinated ecologists and is also reflected in the first ecological regionalisations based

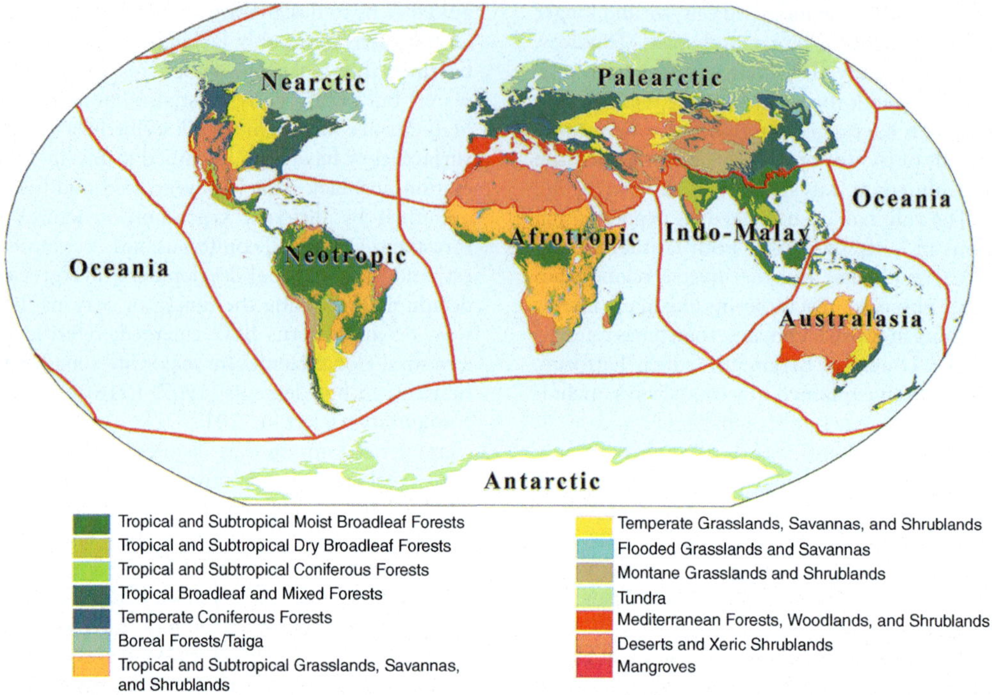

Fig. 1.3. Major biomes (n = 14, colour coded). (From Olson *et al.*, 2001, used with permission of Oxford University Press.)

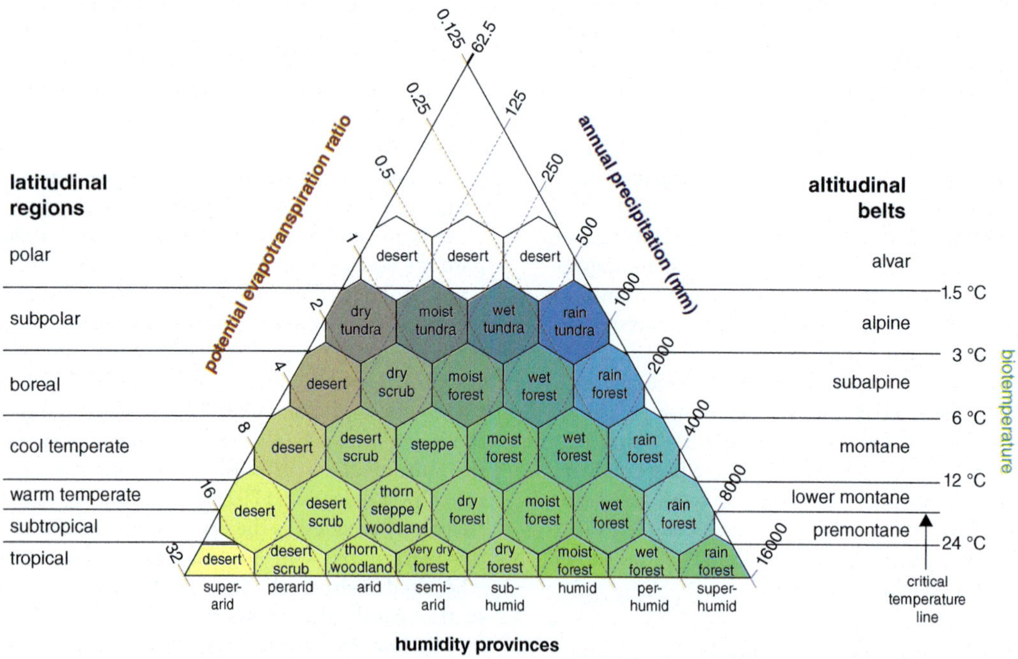

Fig. 1.4. Ternary graph showing Holdridge's life zones. In this representation, life zones are identified at the intersection of lines representing climatic variables. Note the parallelism between latitudinal regions and altitudinal belts. (Image by Peter Halasz: http://en.wikipedia.org/wiki/File:Lifezones Pengo.svg, used under licence CC BY SA.)

mostly on spatial variations in temperature and precipitation (e.g. the life zones proposed by Allen, 1871 and Merriam and Stejneger, 1890). This pattern is commonly evoked in mountain ecology (Fattorini *et al.*, 2019b) and is particularly evident in some systems where, because of steep variations in climatic parameters, different **vegetational belts** (i.e. elevational subdivisions based on climatic or vegetation characteristics; Fattorini *et al.*, 2019a) show physiognomies analogous to those of global biomes (**biome stacking**; see, for example, the Santa Catalina Mountains, where desert, oak, pine and fir vegetation types are recognised from the base to the tip; see Chapter 7, Box 7.1, Fig. B7.1.5).

However, this parallelism reflects only general resemblances in the vegetation structure. For example, zones of high elevation vegetation are often simply regarded as elevational counterparts of the Equator to Arctic latitudinal zones, although such analogy does not consider site-specific conditions of mountain areas, such as their mesoclimate, relief, substrate, slope, exposure and snow cover (Jedrzejek *et al.*, 2012). Because of the peculiar climatic conditions that characterise mountain ranges within a given zone, they represent distinct biomes, for which Walter and Box (1976) introduced the concept of **orobiome**, and used the words **zonobiomes**, subzonobiomes and (sub)zonal biomes to indicate wide climate zones, and their subdivision in transcontinental biomes and (sub)continental biomes. It should also be noted that there are situations in which the major driver of biotic assemblages on land is represented not by climatic characteristics but by other elements, for example rock and soil properties. In these circumstances, it is possible to identify forms of vegetation which deviate from what could be expected on the basis of climate, and that, at the same time, appear similar in different areas with corresponding soils (**azonal biomes**). For example, soils formed on serpentinites, because of their general deficiency of aluminium, have slow rates of clay formation, being therefore more prone to erosion, shallow and poor in nutrients. As a result, they tend to host vegetation settings very different from the surrounding ones (islands of brush and bare rocks in a sea of forest and grasslands). To denote such conditions, Walter and Box (1976) introduced the word **pedobiome** (including lithobiomes on stony soils, psammobiomes on sandy substrates, halobiomes on salty soils, helobiomes in marshes, hydrobiomes on waterlogged soil, peinobiomes on nutrient-poor soils, and amphibiomes on soils subject to intermittent submersion, such as riverbanks and mangroves; Walter and Breckle, 1985).

Independently from the definition and classification schemes that are used for biomes (see, for example, Mucina, 2019 for a review), habitats are divisions of biomes *sensu lato* – that is, they include water bodies such as ponds, lakes and rivers, and the atmosphere – though most authors and educationalists refer to biomes as major habitats, presumably in an attempt to reduce the number of terms and simplify notions of land divisions. As climate shapes animal distributions both directly and indirectly (through its influence on the distribution of vegetation), terrestrial biomes mainly based on a combination of climatic parameters (e.g. Holdridge, 1967 as a classical scheme) are generally used as useful broad habitat categories in animal ecology.

Confusion created by differences between schemes – regardless of the area or region – in classifying biomes and habitats is easily understood. Humans constantly attempt to divide continua into discrete entities. Thus, even the number of terrestrial global biomes illustrated by different authors varies from as few as six to more than ten. If such difficulties emerge in identifying the broad variation of the planet, what then may we expect of the finer divisions? Inevitably, reaching consensus within the scientific community regarding the subdivisions of continents and oceans into finer units, like **ecoregions** (defined by Bailey, 2014 as systems of landscapes, in turn defined as mosaics of ecosystems, being therefore smaller than biomes but larger than ecosystems), is even more challenging (see, for example, Dinerstein *et al.*, 1995; Olson *et al.*, 2001; Spalding *et al.*, 2007; Bailey, 2014 as examples of the variety of definitions, global and regional applications on continents and oceans) (Fig. 1.5). Similar concepts are the **biotic provinces**, as introduced by Dice (1943; large areas characterised by the occurrence of unique biotic communities) for North America and then applied to other parts of the world (see Udvardy, 1969), the **biogeographical provinces** of Udvardy (1975; which are conceptually derived from Dice's biotic provinces and are subdivisions of biogeographical regions on the basis of their biomes) and the provinces of Demarchi (1996; smaller than ecoregions). The word '**ecozone**' is sometimes used to indicate broad, global areas that are defined by specific environmental and climatic conditions, which influence the variety

Fig. 1.5. Map of terrestrial ecoregions of the world according to Olson *et al.* (2001). This map recognises 867 distinct units, roughly a fourfold increase in biogeographical discrimination over that of the 193 units of Udvardy (1975). (From Olson *et al.*, 2001, used with permission of Oxford University Press.)

of ecosystems (biomes) that can exist within them. For example, Schultz (2005) uses the same term ecozone to indicate the nine major ecological zones, larger than the biomes, into which he divides the geosphere (i.e. polar and subpolar, boreal, temperate mid-latitudes, dry mid-latitudes, subtropics with winter rain, subtropics with year-round rain, dry tropics and subtropics, tropics with summer rain and tropics with year-round rain).

Therefore, students should expect to find different habitat classifications in the literature for a specific territory, as illustrated for the UK in Box 1.1, a valuable example as solutions have now been sought to integrate the various schemes. Clearly, there is not room in this book to consider the bewildering variety of subdivisions comprising habitats across the world. However, the basic features of habitats that emerge in every biome are of huge importance in understanding the natural world for the purposes of conserving animal and plant communities. The basic features of land habitat schemes for any region or area are as follows.

(i) *Level of naturalness*. Increasingly throughout the Holocene, habitats have become impacted and demarcated by human influences (e.g. conurbations and built-up areas, industrial zones, cultivated fields; at the finest scales, road surfaces, pavements, street furniture). Therefore, habitats can be natural, semi-natural or artificial (human-made) units of the global spheres. Given the profound impact of humankind on the whole **biosphere**, it is of paramount importance that non-natural habitats are defined, identified and mapped. While this is common practice at small scales, only in the past decades has it been proposed to evaluate human transformation of the Earth's surface at large scales (Lapola *et al.*, 2008). Terrestrial biomes, ecoregions and other similar large-scale units rely heavily on **potential natural vegetation** (see Chapter 3), which can be very different from the true ecological situation. Mapping anthropogenic biomes, which reflect the actual ecological conditions, may be of vital importance to acknowledge the human influence on global ecosystems (Ellis and Ramankutty, 2008).

(ii) *Type of vegetation*. Many habitat classifications use vegetation forms. For example, the European Nature Information System (EUNIS) is a pan-European classification system that defines habitats largely on the basis of vegetation types. Also, the habitat classifications used for the identification of the CORINE (Coordination of Information on the Environment) biotopes (note that in the context of the CORINE project, the word 'biotope' is used to indicate geographical entities of conservation interest) as well as the **CORINE land-cover** classes

Box 1.1. Habitat types of the UK

Traditional habitat types in the UK have been reviewed thoroughly by Lake *et al.* (2015). They illustrate superbly more than 70 habitats (not 64 as stated by them) in ten broad classes (Table B1.1) largely based on the UK Biodiversity Action Plan following the Convention on Biological Diversity 5 June 1992 at the Earth Summit in Rio de Janeiro and which came into force on 29 December 1993 (see Secretariat of the Convention on Biological Diversity, 2011); nine are natural units and one (labelled as 'Other') accommodates under three broad labels all the variants of intensively produced units by humans (i.e. arable, orchards, brownfield). The habitats described include all Annex I habitats (with the exception of marine habitats) of the European Habitats Directive (HD) occurring in Britain or Ireland (Evans, 2010). To illustrate the sheer complexity of habitats in the case of the UK, the authors outline five other classifications. Further than that, they tabulate their own classification in an appendix of 'Habitat Correspondence Tables' and illustrate using different colours four different conditions (complete correspondence, overlap, rough equivalence and lack of equivalence) that link it with five other schemes. Each habitat type has also been the focus of singular attention, as in the case of the delicately balanced and vital peat bogs of northern Europe (Fenton, 2021).

This level of comparison is hugely important as it shows not only how different schemes are produced by different means (i.e. aerial colour photography; multispectral satellite imagery; LiDAR; field survey) and contrast from one another, but also the degree to which they have attributes in common. Changes are inevitable with increasing spatial resolution of remote imagery (e.g. from 4 m to 2.5 m ; Jensen and Cowen, 1999; Mathieu *et al.*, 2007). However, the level of detail can be bewildering and the loss of essential details for particular purposes can be frustrating. Therefore, integrated approaches are necessary. Between 2015 and 2018 there has been an effort by the Centre for Ecology and Hydrology working in conjunction with other stakeholders to produce a unified and comprehensive classification coding system (UKHab) for all terrestrial, freshwater and marine habitats in the UK; this has now been published and the data are open access (for habitat definitions and the user manual, see https://www.ukhab.org/; accessed 30 May 2025). The full classification comprises 213 primary habitats and 296 secondary codes; all habitat components in the 'Habitat User Classification Manual' are letter/number and colour pattern coded for survey and mapping purposes. It is divided into five levels: (i) major **ecosystems** (terrestrial, freshwater or marine); (ii) ecosystems;

Table B1.1. Traditional habitat types in the UK.

Basic habitats	Divisions
Woodlands	Lowland mixed oak and ash wood, lowland dry oak and birch wood, beech wood, yew wood, wet woodland, wood pasture, Atlantic oak wood, upland mixed ash wood, Caledonian forest, Atlantic hazel wood, upland birch wood, coniferous plantation
Scrub	Mixed scrub, hedgerow, bracken
Heathlands	Lowland dry heath, maritime heath, dune heath, Breckland heath, shingle heath, limestone heath, lowland wet heath, upland dry heath, upland wet heath
Grasslands	Lowland calcareous grassland, lowland dry acid grassland, lowland meadow and pasture, coastal and floodplain grazing marsh, upland calcareous grassland, northern hay meadow, upland rush pasture, upland acid grassland, Calaminarian grassland
Mountains	Montane dwarf-scrub heath, high montane heath and snow-bed, mountain ledge, montane scrub
Rocky habitats	Rocky slopes, scree, limestone pavement
Wetlands	Blanket bog, raised bog, upland spring and flush, lowland fen, valley mire, reedbed
Fresh waters	Nutrient-rich (eutrophic) lake, upland lake, loch and tarn, peat-stained (dystrophic) waters, turlough and fluctuating mere, mesotrophic lake, marl lake, lowland nutrient-poor lake, brackish lake, pond, ditch, canal, fast-flowing river and stream, sluggish river and stream, chalk river and stream
Coastal habitats	Mudflat and sandflat, salt marsh, sand dune, machair, coastal vegetated shingle, saline lagoon, rocky shore, soft cliff, hard cliff and cliff slope
Other habitats	Arable, brownfield, traditional orchard, etc.

(After Lake *et al.*, 2015.)

Continued

Box 1.1. Continued.

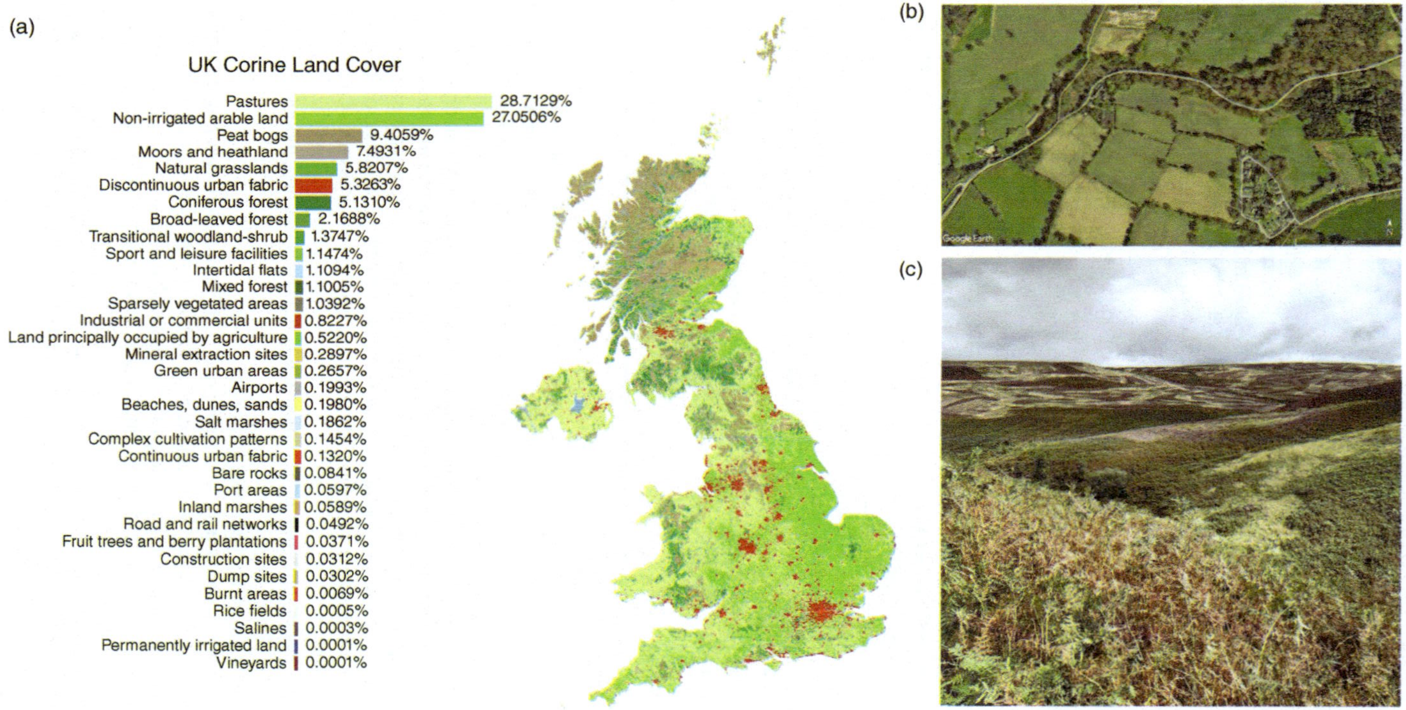

Fig. B1.1.1. Habitat types of the UK. **(a)** CORINE land cover map of the UK. **(b)** Photograph of habitat units in the Cheshire/Derbyshire (UK) landscape. Landscape of the south side of the Alderley Edge escarpment. It shows land artificially divided by hedges and fences as well as areas under semi-natural conditions including bare substrate, the result of past copper mining, with some areas remaining unvegetated. **(c)** The landscape of the Goyt Valley, Buxton. Note changes in vegetation without any apparent boundaries other than a fence high up on the moor. (Panel **a** from Rae, 2017. Produced by the University of Leicester, The Centre for Landscape and Climate Research and Specto Natura and supported by Defra and the European Environment Agency under Grant Agreement 3541/B2012/R0-GIO/EEA.55055 with funding by the European Union, used under licence CC BY 4.0. Panel **b** image from Google Earth. Panel **c** photo by R.L.H. Dennis.)

Continued

Box 1.1. Continued.

(iii) broad habitats; (iv) UK Biodiversity Action Plan Priority Habitats; (v) ED Annex I habitats – a superb achievement. The scheme is intricate, all-inclusive and provides a detailed field key; the latter is designed to cover all habitat types in Europe (Evans, 2010) so that any habitat units (polygon/stand/parcel) large enough to exist as a minimum mappable unit within the UK can be identified and classified. It also incorporates field hydrology used to determine habitat types over much of Europe (Bunce *et al.*, 2011).

Thus, maps and/or data exist for the entire UK and of increasingly higher resolution in raster and vector format. The level of detail depends on the scheme selected and thus the resolution of the surveys (Fig. B1.1.1a). A finer scale still of land divisions is provided by field survey of plant **communities** in the UK undertaken with the collection of vegetation relevés (data for c.35k **quadrats**) using the National Vegetation Classification based on the five volumes of British Plant Communities (Rodwell, 2006). This classification ensures assessment of the complex basic elements (plant species, fungi, growth forms, etc.) and boundaries (**ecotones**) between vegetation categories, including boundary units between larger patches which also contain crucial habitat components (resources, conditions) for animal species. Detailed plant community mapping can be overlaid on maps of basic habitat units enabling vital data for the assessment of biodiversity of the habitat units. Biodiversity data, owing mainly to the vast preponderance and small size of arthropods, and the seclusion and **mobility** of higher animals, are necessarily determined by field survey, though remote survey is becoming increasingly common for many creatures (e.g. birds and small mammals) (see Appendix 1). Such surveys have the potential for linking organisms (presence/absence, population counts) to specific plants and plant/substrate structures. This then allows comparisons of different habitats for specific organisms and biodiversity and the obverse comparison of animal and plant species for different habitats. An aerial photograph of Alderley Edge (Fig. B1.1.1b) and a panorama of the Goyt Valley (Fig. B1.1.1c) illustrate how land units may be artificially divided up (e.g. by hedges or fences) or merge into one another almost imperceptibly.

are largely based on naturalness and vegetation. Of course, some land habitats may lack vegetation (e.g. hot deserts, polar nunataks and montane peaks) but nevertheless be distinctive units with characteristic **life forms**. For example, coastal areas such as the Eyre Peninsula in Australia may be divided into at least ten different habitats with limited vegetation cover (dune shrublands, dunefields, open sandy beaches, sand flats, salt marshes, subcoastal lakes, mangroves, rocky coast, cliffs and cliff tops and offshore islands; Brandle, 2010).

(iii) *Hierarchy*. Habitat units tend to be hierarchical and with nested subsets, with initial (basic) units subject to finer divisions (for example, both the EUNIS and CORINE land-cover systems are hierarchical, with different levels of classification for different scales). Units of the finest level (extreme) are often referred to as **microhabitats** (e.g. the hollow at the foot of a tree plate of a fallen tree, cavities in rotting trees, a puddle, a log, a moss-covered boulder, indentations in a mollusc shell).

(iv) *Size*. The size of habitat units is highly variable. Inevitably, the mean sizes of habitat units are directly related to their position in a hierarchy of habitat classes.

(v) *Delimitations*. The shape of habitat units and the boundaries between habitat units tend to relate to abiotic divisions (e.g. geology, soils, topography, altitude, local climate) and human influences. But for analytic purposes units may also be overlapping entities (e.g. different but related vegetation communities including a number of the same plant species).

(vi) *Boundaries*. Boundaries between units can be 'artificial' (e.g. hedges, fences, walls, plantation lines) or 'natural' (changes in vegetation type, cover or specific plants), and distinct (sharp and visible) or indistinct (ecotones of various kinds) (Fig. 1.6).

The basic features of habitats are repeated in classifications from place to place on the globe. Most classifications tend to produce a nested hierarchy of vector (management) units or have capacity for developing a hierarchy of habitat units. A common issue is the discordance in habitat designations for different regions (areas) sharing similar biomes. An important question is: how does discordance in habitat designation occur? A basic reason is that all classifications are ultimately subjective, both in the sense of the variables chosen and the classification techniques selected for identifying the categories. Another reason is that it is always possible to see finer details in any natural unit of vegetation or substrate that may bias selection of variables for a particular objective. Different variables are often

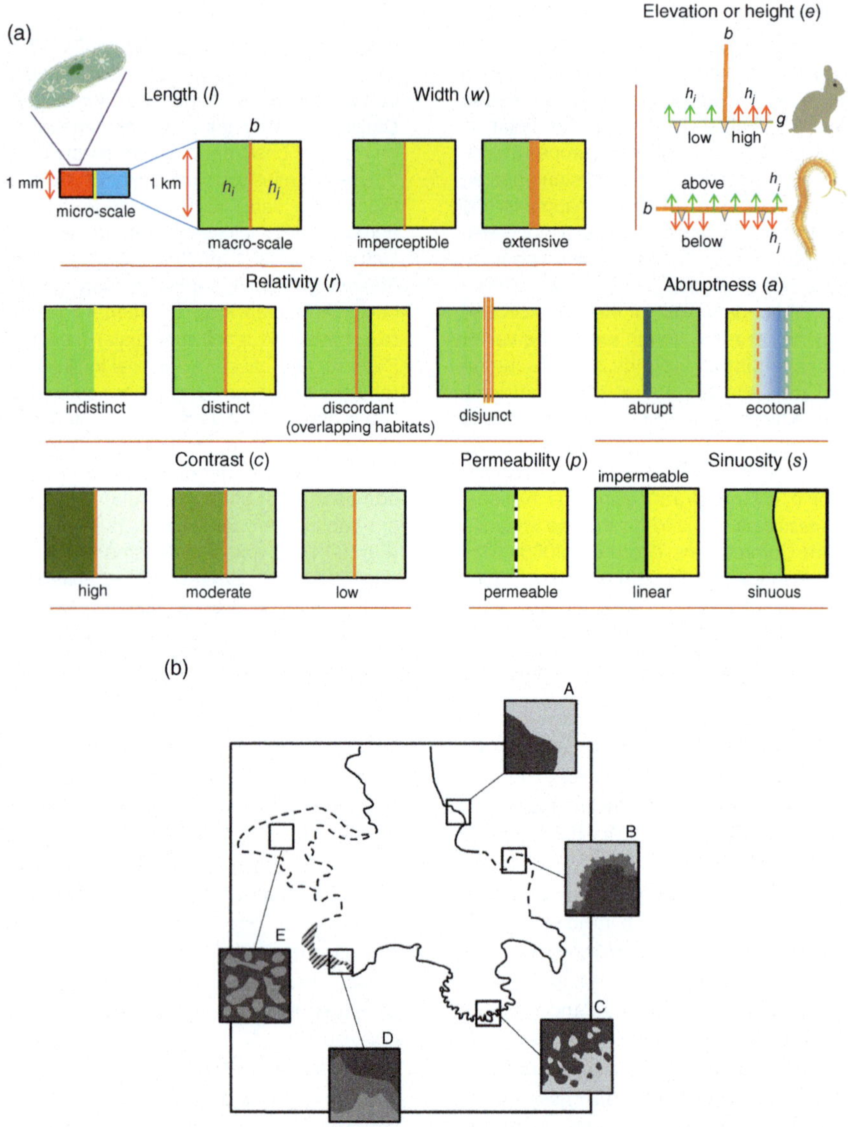

Fig. 1.6. Structural features of habitat boundaries and ecotones. **(a)** Some basic habitat boundary structural contrasts (i.e. for length, width, height or elevation, relativity of boundary to adjacent habitats, abruptness [ecotonal: blue area, the ecotone with dashed lines indicating habitat edges], structural contrasts of adjoining habitats, permeability and sinuosity). Symbols: *h*, habitat *i, j*; *b*, boundary; *g*, ground or substrate surface; *l*, length; *w*, width or breadth; *e*, elevation or height. The scale of boundary features (or barriers) has relevance only to a focal organism (e.g. paramecium, centipede, rabbit). Although all features are capable of direct measurement, the height of a boundary has more significance as a relative measure: $e_r = e/o_d$ or e/o_m, where o_d is organism dimensions and o_m is organism mobility. It is important to appreciate that the home range of any organism may cross a boundary structure or be limited by it, the structure thereby acting as a barrier. Boundaries may be artificial or natural in origin. **(b)** Various ecotones in natural vegetation. The diagram shows a range of possible vegetation boundaries: (A) a sharp boundary; (B) a diffuse boundary; (C) a convoluted boundary with islands; (D) a gradual transition; (E) a complex mosaic. (Panel **a** modified from Strayer *et al.*, 2003, used with permission of Oxford University Press. Sources of drawings: paramecium: https://openclipart.org/detail/215770/paramecium-caudatum; rabbit: https://openclipart.org/detail/173511/rabbit-by-rones; centipede: https://openclipart.org/detail/230212/centipede. Panel **b** redrawn with changes from Rodwell, 2006; used with permission of Elsevier BV.)

used in classifications and, even when they are the same, different weightings (biases) are given to them or different measurements made on them: to land surface shape, geology, soils, climate, vegetation structure and key species in the biota. For the processes of classification, innumerable methodologies exist and continue to emerge also in relation to the development of new technologies for data acquisition (e.g. Díaz Varela *et al.*, 2008; Calvert *et al.*, 2014; Torres and Qiu, 2014; Cooper *et al.*, 2019). Inevitably, different variables and techniques lend themselves to producing different 'habitat' endproducts, if only because those engaged in producing classifications have different views on the functional significance of habitats. Unavoidably, the reason for the different schemes largely reflects their different purposes (i.e. local versus countrywide management; legal protection; monitoring local vegetation, changes in biodiversity and countrywide land-use). This is of huge importance when habitats are considered from a particular species' standpoint (e.g. for red list species), an issue examined later in the section: *Habitat – a species' view*. The complexity involved in habitat definition and mapping is perhaps best displayed for the UK, as habitats in the UK have been a focus of study for decades and are now mapped in fine detail for different schemes (see Box 1.1).

Habitats without vegetation

The concept of habitat can be, and actually is, applied for contexts different from the vegetated land, the classic cases being 'badlands', rock and sand (erg) deserts (Kasanin-Grubin, 2008). Other prominent cases of non-vegetated terrestrial habitats are the **subterranean habitats**. These include both large cavities (caves) and small interstitial cavities. Shallow subterranean habitats include aquatic habitats (seepage springs, **epikarst**, calcrete aquifers and the **hyporheic zone**) and terrestrial habitats (abandoned mines, spaces between rocks, rock cracks and the like, lava tubes and iron ore caves). Within large caves, different zones can be recognised according to the intensity of external influence (Culver and Pipan, 2019) which host different biotic communities. Also, geological complexity may lead to the presence of various habitat units in the cave spaces, such as lakes, dripping pools, stalactites and stalagmites, characterised by different communities (e.g. Galassi *et al.*, 2017). Aquifers host highly specialised animals, but their ecological study is hampered by sampling problems, and recognition of distinct habitats within them is difficult due to the water hydrological continuum (Fattorini *et al.*, 2017d, 2018a); however, there is evidence that different species use different parts of the aquifer (Fattorini *et al.*, 2017d). Although the microbiomes of subterranean environments remain poorly understood due to access difficulties, a substantial amount of Earth's microbial biomass resides beneath the surface – for example, the continental subsurface is estimated to host 2 to 6×10^{29} cells – playing a key role in global biogeochemical cycles (Wegner *et al.*, 2023). Despite the scarce energy supply from rock-water interactions, even multi-million-year-old igneous rocks are habitable (Suzuki *et al.*, 2024).

Soil is also considered a habitat (Lavelle, 2012), as it hosts a multitude of soil organisms (ranging from **micro-organisms** to mammals, such as moles) with different degrees of specialisation for the **edaphic** life. Soil communities not only vary horizontally (because of the presence of soil with different characteristics in different areas; thus, different soils may be different habitats) but also vertically (e.g. Ellers *et al.*, 2018; Mundra *et al.*, 2021). Thus different soil strata may also be considered as different habitats.

In the water and air spheres, the recognition of distinct habitats may be problematic, because these media represent continua that are particularly difficult to subdivide. While it may be relatively easy to recognise entire water bodies (lakes, ponds, streams, etc.) as **aquatic habitats**, it is more difficult to define habitats within them. However, different zones can be identified at different depths of the water column according to the amount of light penetration. On this basis, the water column in the oceans is typically divided into two major zones: the **photic zone** (which receives sunlight) and the **aphotic zone** (perpetual darkness). The photic zone can be divided into a **euphotic zone** (true light zone, in which the rate of photosynthesis is higher than the rate of respiration) and a **dysphotic zone** (where some photosynthesis is possible but the rate of respiration is greater than the rate of photosynthesis). Another commonly used but more articulated subdivision distinguishes, from the surface to the greatest depths, the **epipelagic** (photic and warmer), **mesopelagic** (dysphotic, with a sharp decline in temperature), **bathypelagic** (aphotic, with almost unchanging cool temperatures) and **hadopelagic** (deep trenches) zones. Another vertical subdivision (based on depth and configuration of the ocean bottom) distinguishes an **intertidal** (or **littoral**), a **neritic** (or sublittoral), a **bathyal** and an **abyssal** zone. These subdivisions can be regarded as different habitats which, for their large scale, represent the marine counterparts of the terrestrial biomes and can therefore be considered marine biomes.

It is obvious that while vegetation setting may be very useful to define terrestrial habitats/biomes, it cannot be applied to other contexts like the marine environment. For this reason, defining biomes as major ecosystem types instead of major types of natural vegetation may be more appropriate to embrace the whole variety of ecosystems. The wide use of vertical subdivisions does not mean that there are no structural habitats in the oceans. Coral reefs, sea grass, oyster beds and **rhodoliths** are obvious examples. The existence of habitats on the seabed is often ignored, but seabed habitat structures can effectively be classified on the basis of **substrates** using a combination of techniques assessing bathymetric contours (slope, aspect, rugosity), sediment types and biological communities as in the Gulf of Maine, USA, and north-east Atlantic (Howell, 2010; Valentine, 2019). Based on biotic and abiotic characteristics (such as bathymetry, water chemistry, currents, temperature and geomorphology), a number of classification schemes have been proposed that identify units of different scales in the marine environments (a table presented in Riddle *et al.*, 2011 includes 17 schemes), which can be regarded as marine habitats under different definitions.

In standing water (**lentic**) systems, different vertical zones are also identified on the basis of sunlight penetration: the **littoral zone** (shallow, **photic** waters), the **limnetic zone** (offshore, photic waters) and the **profundal zone** (offshore, **aphotic** waters). For flowing waters (**lotic** systems), several longitudinal zonation schemes have been proposed, based on characteristic fish species, whole communities or structural parameters (see Aarts and Nienhuis, 2003). For example, Illies and Botosaneanu (1963) identified three main zones (in turn divided into smaller zones), on the basis of abiotic and biotic factors: **crenon** (headwaters, divided into eu- and hypocrenon), **rhithron** (mid-reaches, divided into epi-, meta- and hyporhithron) and **potamon** (lower reaches, divided into epi-, meta- and hypopotamon). Although the river continuum concept (introduced by Vannote *et al.*, 1980) does not recognise discretely defined river zones, it explains them as different sections of a gradual change in ecological properties that emerged from longitudinal zonations (Aarts and Nienhuis, 2003). Morphological features also make it possible to identify a variety of habitat units, such as banks, sloughs and pools (see, for example, Baker *et al.*, 1991 for a classification of habitats in the Mississippi, USA).

In general, considering air as a habitat may appear surprising, but the **airspace** perfectly satisfies the **resource-based** definition of habitat (Dennis, 2010). Thus, recognising the **atmosphere** as a habitat is not only legitimate, but may be extremely fruitful in ecological research and nature conservation, making its protection easier (Diehl, 2013). This is of particular importance for the conservation of species that spend a significant portion of their lives in the airspace (e.g. many birds, especially migratory ones). Of course, no organism can spend its entire existence in the air. However, the airspace is home to myriad microscopic organisms, and is occupied by birds, bats and especially insects, which use air for mating, dispersing/migrating and feeding. Being in connection with the **geosphere** and the **hydrosphere**, the atmosphere participates in important biotic exchanges between them. Hundreds of trillions of micro-organisms are exchanged between ocean and atmosphere daily; thus, investigating the atmosphere as a microbial habitat would be of paramount importance for improving our understanding of the complex relationships of the marine domain with the other parts of the Earth system, with important implications for future climate and biogeochemical predictions (Alsante *et al.*, 2021). Although it may appear as a homogeneous space, the atmosphere is a heterogeneous habitat. Aerial organisms may exhibit distinct preferences for specific parts of the airspace (such as different altitudinal layers within the air column; Diehl, 2013). For example, it is well known that although insect density generally decreases with increasing height, patterns vary among groups (e.g. Freeman, 1945; Wellington, 1945; Johnson, 1969). Because of the highly dynamic nature of the airspace, many of its attributes are ephemeral (Diehl, 2013), and this lack of a stable structure obscures the recognition of different habitats within the atmosphere. However, extending our knowledge of how different organisms use different parts/characteristics of the atmosphere would be extremely important for understanding the biology of airborne organisms and the plethora of animals that use the airspace for a variety of activities (Horton *et al.*, 2016).

Microhabitats

It may seem strange to isolate **microhabitats** as a separate topic when, for all purposes, these must be just the smallest units of Earth spheres, usually terrain, that can be considered to hold a community of organisms, but there is much confusion between what are in fact microhabitats and microresources. Many species that occupy what are referred to by many researchers as microhabitats do so for some fraction of their lives, occasionally a short period, but occasionally through much of their life cycle or for their entire lives (e.g. viruses within bacteria (a nanobiome) or small viruses in larger viruses (picobiome); Yong, 2016; Stough *et al.*, 2019). Small to tiny enclaves of any environment that are used by a species for its entire existence can truly be said to be microhabitats. But those species that use the enclave for only a (small) part of their life cycle can only be said to use it as a microresource unit. There is some splitting of hairs in such a distinction as it is a matter of degree, and it is understandable if organisms using a small or tiny enclave in any environment for most (but not all) of their lives are said to inhabit it as a microhabitat (Siitonen, 2012). In fact, they are exploiting it as a microresource. Later in the text, it will become obvious that only creatures that use the same tiny or small enclave for their entire existence can be said to occupy a microhabitat.

Obvious examples of microhabitats are ephemeral rainwater-filled hollows under tree plates, spaces under bark, tree hollows or fallen rotten tree timber occupied by local populations and communities of arthropods, waterborne and **saproxylic** respectively, throughout their life cycles (Fig. 1.7a–d). Winter and Möller (2008) list 20 microhabitats in broken and rotting trees that are relevant for a vast range of arthropods. In a lowland beech forest in Germany, they found on average about 250 microhabitats/ha. An enormous number of species that are not saproxylic also use the spaces under bark and tree cavities to find shelter (Fig. 1.7e). But there are levels below this spatial scale. Classic examples are provided by trapdoor spiders, that create their own burrow habitats, the trap doors and even barriers to prevent incursions from competitors and enemies (Fig. 1.7f–h) (Gupta *et al.*, 2015; Leroy and Leroy, 2005) – an example of how organisms actively modify their environment to create their own (micro)habitats. At the minutest scale, as noted above for viruses and bacteria, one organism can be a habitat for another organism. Not only pathogens but also many parasites (helminths, ticks, lice, etc.) and parasitoids (such as wasps and flies whose larvae develop in the body of other insects) spend much of their lives in another organism (their host), and thus can be argued to have at least a significant part of their habitat in that organism. Some parasites have all their life cycles in their hosts, such as the head louse *Pediculus humanus capitis* or the pubic louse *Pthirus pubis*, which abandon a human body only to disperse to another one. Some organisms have their conspecifics as habitat (e.g. oyster shells for conspecific spats; Nalesso *et al.*, 2008). But the term 'microhabitat' for all attempts at revision in terminology – inclusive of spaces from the expanse of a tree plate to a fly – remains something of a misnomer inasmuch as there are hugely varying scales for such entities from a hedge (macro-scale) down to tiny organisms living entirely within the live or dead remains of other organisms (e.g. bacteria such as the genus *Pseudomonas* and *Escherichia coli* on the fly *Musca domestica* (Diptera); Schlegel and Jannasch, 2006; Nayfach *et al.*, 2021) or the microbes (endolithic micro-organisms) that inhabit the interior part of rocks, minerals, corals, shells, etc. (e.g. Günther, 1990; Golubic *et al.*, 2005; Walker and Pace, 2007; Omelon, 2016; Wegner *et al.*, 2023); these are truly micro-scale.

But there is a more powerful argument for limiting the term to very small spatial entities that hold communities. During the 1970s to 1990s, microhabitat was extended to mean much more than the concept of a place where for a lifetime an organism may reside, and a community at that. It started to be used as small parts, often very small parts, of habitats that are used for certain functions by animals and plants during *part* of their lives. Examples are contrasting nest sites of seven bird species in woodland snowmelt drainages located on the Mogollon Rim in central Arizona (Martin, 1998), and foraging resources in heteromyid rodent species of North American deserts (Price, 1978; Longland and Price, 1991). As indicated above, this particular use of the term 'microhabitat' is a misnomer; it led to the concept of 'microhabitat partitioning' (Jorgensen, 2004). What in fact are often being described are resources (be they abiotic or biotic) and/or conditions used by a species within any traditional habitat space. Morris (1987b) made a crucial distinction but added to the confusion in terminology: he defined macrohabitat as the spatial area in which individuals perform all their biological functions and microhabitat as being composed of environmental variables that affect individual behaviour. Note the use of the term 'variables' as

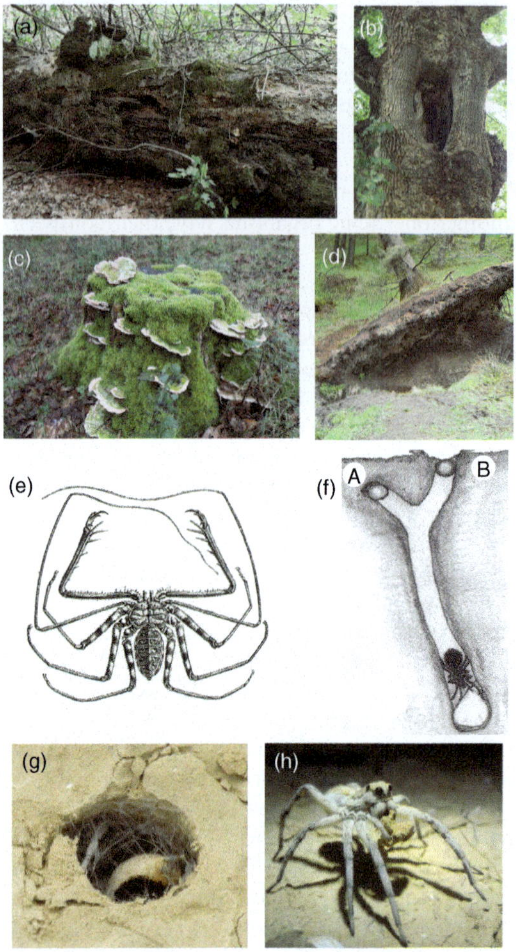

Fig. 1.7. Examples of microhabitats and microhabitat dwellers. Trees offer microhabitats for a wide variety of organisms. Typical tree microhabitats are: **(a)** spaces under bark; **(b)** hollows in trees; **(c)** mosses and fungi on stumps; **(d)** tree plates. Microhabitat dwellers are common among arthropods. Whip spiders (Arachnida: Amblypygi) **(e)** live in tropical and subtropical regions around the world. They are found in caves, between rocks and under pieces of bark and are usually only seen at night since they are nocturnal. These habitats and microhabitats are typical of legions of invertebrates, including arachnids, insects, crustaceans, molluscs, worms, etc. Burrow habitats are typical of trapdoor spiders. Panel **(f)** illustrates a burrow of a trapdoor spider (*Ancylotrypa*) showing positions of the marble (A) stored at the top of the blind arm, and (B) in position to plug and protect the burrow. The presence of a trapdoor in *Lycosa hyraculus* has a negligible effect on the microclimate at the bottom of burrows, its roles being more likely predator avoidance and prevention of flooding and stones and sand from falling in. *L. hyraculus* presents typical responses to trapdoor removal. Panel (**g**) shows a *L. hyraculus* spinning a mesh in its burrow entrance. Panel (**h**) shows a spider carrying a piece of soil in its pedipalps and chelicerae, possibly to replace a trapdoor. These spiders may also carry soil during burrow maintenance when they remove accumulated soil by binding it into balls with silk. (Panel **a**: photo by S. Fattorini. Panels **b, d**: photos by R.L.H. Dennis. Panel **c**: photo by Pamela Macaulay. Panel **e** from Lydekker, 1896: https://commons.wikimedia.org/w/index.php?curid=2018734. Panel **f** from Leroy and Leroy, 2005. Panels **g** and **h** from Steves *et al.*, 2021, used under licence CC BY 4.0.)

that is how different resources and conditions may be perceived at all scales; resources and conditions are indeed variable and variables. This distinction of microhabitats and microresource units, all variables, becomes much clearer when we turn later to the **resource-based habitat** paradigm (see Chapter 5).

Habitat edges, ecotones and ecoclines

Habitats meet at boundaries (**edges**), occasionally at **nodes** (points, small areas along edges at which resources are concentrated) (Dennis, 2010). The form edges take depends much on the dimensions of habitats, their extent and shape, and the grain (units used for study) of observations. For instance, all boundaries may be considered as two- or three-dimensional when the grain of observations is miniscule, but one-dimensional when grain is large and when, in effect, they are observed from a greater height. This becomes evident as one zooms in using Google Earth on features having any lateral extent. Herein, a simple definition of features is applied. Edges between habitats are boundaries between two or more **substrates** and/or vegetation communities. **Ecotones** and **ecoclines** have greater lateral extent (breadth in map view) than edges and may be considered to have one or more edges with adjoining habitats. In general, ecotones are regarded simply as transition zones (areas) between two or more substrate/vegetation communities. By contrast, the word 'ecocline' is used for transition areas between two communities with distinctive gradients for one or more organisms and/or variables. In other words, ecoclines may be considered a subset of ecotones, but note that habitats themselves may have ecological gradients or ecoclines. Edges can be considered to be divided into ecotones and non-ecotones, though a division may not be simply binary. In the former case, the adjoining habitats contribute to the edge species (e.g. flora); in the latter case, the boundary conditions are not transition between the adjoining habitats but largely idiosyncratic as in the case of planted hedges separating two fields.

There are numerous definitions of edges, ecotones and ecoclines (van der Maarel, 1990; Strayer *et al.*, 2003; Kark and van Rensburg, 2006; Hufkens *et al.*, 2009) of varying sophistication and differing in objectives, but all are fundamentally subsets of the simple definitions above. Thus, a more sophisticated definition of an ecotone is: 'a transition zone between adjacent ecological systems, having a set of unique features defined by a spatial-temporal scale and by the strength of the interactions between the systems' (Hansen *et al.*, 1988). Emphasis is often given to their relatively small scale compared to adjoining habitats, and to their space-time dynamism (Martín *et al.*, 2006; Minor, 2007), but consideration also needs to be given to situations where they are effectively **disclimaxes** (transitory conditions that make a site unable to support the expected adjoining **climax** community; see Chapter 3). A number of definitions have explicit causative components to their formulations, but these are considered here to hinder investigation of habitats as they presume (even when these features are regarded as stochastic in origin) *a priori* knowledge of formation of the features intended for investigation. The essential task is to identify features of habitats and their surrounds (bounds) as these present clues of origin, formation and influence on plant and animal populations, providing the basis for understanding the distribution and behaviour of biota using them. Edges, ecotones and ecoclines can vary enormously in a number of ways extending from the dimensions of **biomes** at the upper extreme to the micro-scale at the lower extreme (Bradshaw *et al.*, 2003; Strayer *et al.*, 2003), thus differing widely, for instance, in dimensions (length, breadth and height), in geometry (shape), in contrast between adjoining habitats, in integrity, abruptness and gradient (steepness), and so on (Table 1.3; see Fig. 1.6a).

Distinguishing ecotones and ecoclines can be a complex process requiring multivariate analysis of the variation across and along the features (see Chapter 3). Perhaps the most complex boundaries are associated with natural vegetation units (Rodwell, 2006: 55), which in one heathland area alone may be sharp or diffuse, gradual or convoluted with islets of contrasting habitat, and a complex mosaic perhaps reflecting the influence of a number of factors (see Fig. 1.6b). Ecotones and ecoclines may be contemporary or relict in origin, and maintenance may be endogenous or exogenous and cut across major spheres of activity (i.e. atmosphere/lithosphere, lithosphere/hydrosphere). For example, ecotones have been demonstrated in arthropod communities across the temperature, humidity and light gradients of the hypogean-epigean transition of cave environments (Prous *et al.*, 2004) and the shorelines, edges of lacustrine and coastal environments (Ray and Hayden, 1992). Familiar gradients of ecoclines are those associated with dispersal over distance as

Table 1.3. Some basic attributes of ecological boundaries.

How did the boundary originate and how is it maintained?
Investigative or tangible
Causal or consequential
Contemporary or relict
Endogenous or exogenous in origin
Endogenous or exogenous controls (maintenance or suppression)
What is the spatial structure of the boundary?
Grain size
Extent
Thickness and dimensionality
Geometry of adjacency (neighbourhood)
Interactive or non-interactive
Abruptness, steepness
Patch contrast
Integrity (perforated versus unbroken)
Geometric shape and tortuosity
Number of attributes (single or multiple)
Offsets or congruencies of multiple attributes
What are the functions of the boundary?
Transformation
Transmission
Absorption
Amplification
Reflection
Neutral
How does the boundary change over time?
Changes in any structural or functional properties
Mobility (stationary, directional, oscillating or random)
Age and history

(From Strayer *et al.*, 2003, used with permission of Oxford University Press. See Fig. 1.6.)

with seed dispersal extending forest into prairie or meadow, the shoreline littoral between high and low water, and estuarine assemblages under the counterinfluences of freshwater and salinity tidal ebb and flow (Attrill and Rundle, 2002). Different issues, for instance constraints, emerge at different scales (e.g. plant, patch, population, landscape), as evident in a study at a biome transition in central New Mexico in the Sevilleta National Wildlife Refuge (Gosz, 1993).

Edges, through feedbacks on soil characteristics and hydrology, as well as their structural features and vegetation, have a wide range of influences over the turnover of biota, changes which are fundamental to providing insights into the **resource-based habitat** view (Dennis *et al.*, 2003; Dennis, 2010). Primarily, edges affect movements of organisms across them as they may vary in permeability (i.e. **barriers**; see Box 2.2 in Chapter 2). Poorly permeable edges may function to protect (buffer) adjoining habitats from external influences, but may prevent fluxes occurring between habitats with many associated negative effects (e.g. isolation of individuals and populations). More permeable edges may be useful for improving **connectivity** and hence movements of energy, materials and organisms (thus reducing the effects of isolation), but may also favour penetration of disturbance factors into the patches (e.g. pollutants, light, heat, parasites, predators, competitors, aliens) (e.g. Fagan *et al.*, 1999; López-Barrera *et al.*, 2007; Malt and Lank, 2007; Dennis, 2010: 154; Dennis *et al.*, 2013; Magura and Lövei, 2020).

Edges also vary in length and connectivity, affecting their function as **corridors** (e.g. hedges) (Davies and Pullin, 2007; Dover, 2019) (see Box 2.2). Furthermore, they influence resource use (e.g. nesting sites for birds; courtship and roost sites for butterflies; foraging sites for animals such as Mustelidae; fox dens; rabbit warrens), population density, species occurrences and diversity in adjoining habitat units, gene flow and genetic diversity. Hedges are obvious edge structures. Apparently, there are 390,000 km of hedgerows in England measuring between 1 and 6 metres in height, the greatest concentration in the South West, according to LiDAR scanning by the Centre for Ecology and Hydrology (Briggs, 2024). But the quality and structure of hedgerows matter more than their sheer length. More broadly, an increase in occurrence (density) of individuals and of biodiversity at habitat edges is referred to as the '**edge effect**' (Odum, 1953), but it is important to note that the outcome can be very different. In fact, ecological responses to edges are extremely diversified (Ries *et al.*, 2004, 2017; Ries and Sisk, 2004), and, for example, edge effects on biodiversity can be negative, positive or absent, depending on the organisms and the context (Murcia, 1995; Fattorini, 2014a; Razafindratsima *et al.*, 2018; Tsafack *et al.*, 2023). For example, edges may be richer in generalist species, but may be unfavourable to specialised (e.g. interior) species (e.g. Murcia, 1995; Villard, 1998, Winter *et al.*, 2000). On very wide scales, boundaries between **biomes** or **ecoregions** are considered as wide ecotones, where again biodiversity can be particularly high (Araújo, 2022; Lightfoot *et al.*, 2008; Kark, 2013; Zhou *et al.*, 2023).

Edge effects largely vary according to the edge size and contrast between the adjoining habitats, with responses typically stronger near 'hard' (high-contrast, usually poorly permeable) edges where conditions change abruptly than near 'soft' (low-contrast, usually more permeable) edges where conditions change gradually from one habitat to the adjacent one (Ries *et al.*, 2004, 2017). In fact, soft edges, because of their usually wider size (which is a generally necessary condition for a gradual change), lend themselves to the development of ecotones. In reserve planning and management, the many negative effects of edges lead to recommendations for having reserves with shapes that reduce the perimeter-to-area (P/A) ratio (Primack, 2014). With increasing P/A ratio, there is a decrease in conservation value because the interior portions of the protected area (which are the primary object of conservation) become progressively closer to the edge, and hence progressively exposed to the negative influence of the surrounding space (Helzer and Jelinski, 1999; Wintle *et al.*, 2019). Thus, a circular shape is preferred to an elongated or indented one, as it allows the preservation of a larger amount of interior habitats and species (Sisk *et al.*, 1997; Davies *et al.*, 2001; Yamaura *et al.*, 2008; Kotze *et al.*, 2012). However, this depends on the mass of habitat, as indentations in large habitats provide a variety of local climate contrasts that influence plants and thus animals dependent on them (Box 1.2). Moreover, the effects of edges on protected areas have been mostly investigated on terrestrial reserves, but they are relevant in the marine environment as well (Ohayon *et al.*, 2021, 2023).

A particularly important observation, in view of different concepts of habitat, is that edges may 'house' species not found in adjoining habitat patches, or occurring at much lower densities in these patches (e.g. exotic plants in the Illinois shale barrens; West *et al.*, 2009), and may act as speciation centres or as enclaves for endemic species (Kark, 2013). These latter observations question the integrity of traditional habitats from an individual species' perspective (see the following section, *Habitat – a species' view*).

Habitat – a species' view

Although it might be easy enough to conceive and identify a traditional habitat unit, it is another matter to determine a particular **species' habitat**. Having said that, a traditional view is that particular habitats have their community of organisms; from that it would be supposed that abiotic signals occur (soil, local climate, etc.) for recording of habitat communities and this has been the trend for many decades, as evident in the widely celebrated *Natural History of Selborne* (White, 1981 [1789]). From a scientific vantage, Eugene Odum (1963) described habitat as a place where an organism lives (its address) and importantly compared it to (distinguished it from) a species' ecological **niche**, which he described as the role the organism plays in the community of species (its profession). One thing we can be sure about: a habitat (as traditionally described) will include many species; traditional habitats can provide invaluable audits of species for prescribed units of space at a point in time. It is well appreciated too that the larger the habitat unit and the less isolated it is from similar habitat units, the more species it will contain. Moreover, the greater the heterogeneity in habitats around a focal habitat, the more species will be recorded in the focal habitat. One can go on: the

Box 1.2. Are habitat units ever homogeneous?

Traditional habitats are not homogeneous units. They may look the same to human observers and emerge as uniform units in statistical analyses, but simple structural contrasts can have a substantial impact on the **community** structure within the units, even if the **substrate** (soil) is the same. For example, there is a clear indication in data for woodland rides in Bernwood Forest, Buckinghamshire, UK (Table B1.2, Fig. B1.2.1), that, although the habitat in all rides may be described as belonging to a single category (rides), the contrasts in simple dimensions (ride widths, lengths, height and composition of surrounding trees), therefore in local climate (temperatures, shade, humidity), and the growth and type of plants in the rides, will differ. These differences can have a profound impact on butterfly community structure in each ride. In these rides, the number of butterfly species correlates highly with the increased variety of nectar sources, which in turn corresponds with higher light levels and ride width. The figures should make these contrasts evident to readers. A multivariate analysis of species composition highlights these contrasts and is a warning to researchers to be constantly aware of heterogeneity in habitats. In this book we emphasise the importance of heterogeneity in species composition between what may appear to be homogeneous units. For the preservation of biodiversity, we need to understand the factors responsible for heterogeneity and diversity in life forms.

Table B1.2. Butterfly abundance and diversity in woodland rides in Bernwood Forest, Buckinghamshire, UK.

Ride no.	No. of species	No. of individuals per 100 m	Shannon diversity index
1	30	365.7	2.361
2	21	299.6	2.222
3	21	427.4	2.220
4	20	466.6	2.299
5	25	415.0	2.279
6	27	427.4	2.047
7	3	99.3	0.591
8	14	182.0	1.371
9	13	164.3	1.713
10	12	104.4	1.286
11	13	136.6	1.985
12	24	229.7	2.173

Based on 33 transects over 20 weeks from May 8 to September 24, 1978. Data collected by Caroline A. Steel. Adapted from Peachey (1980). (From Dennis, 1992.)

Continued

Box 1.2. Continued.

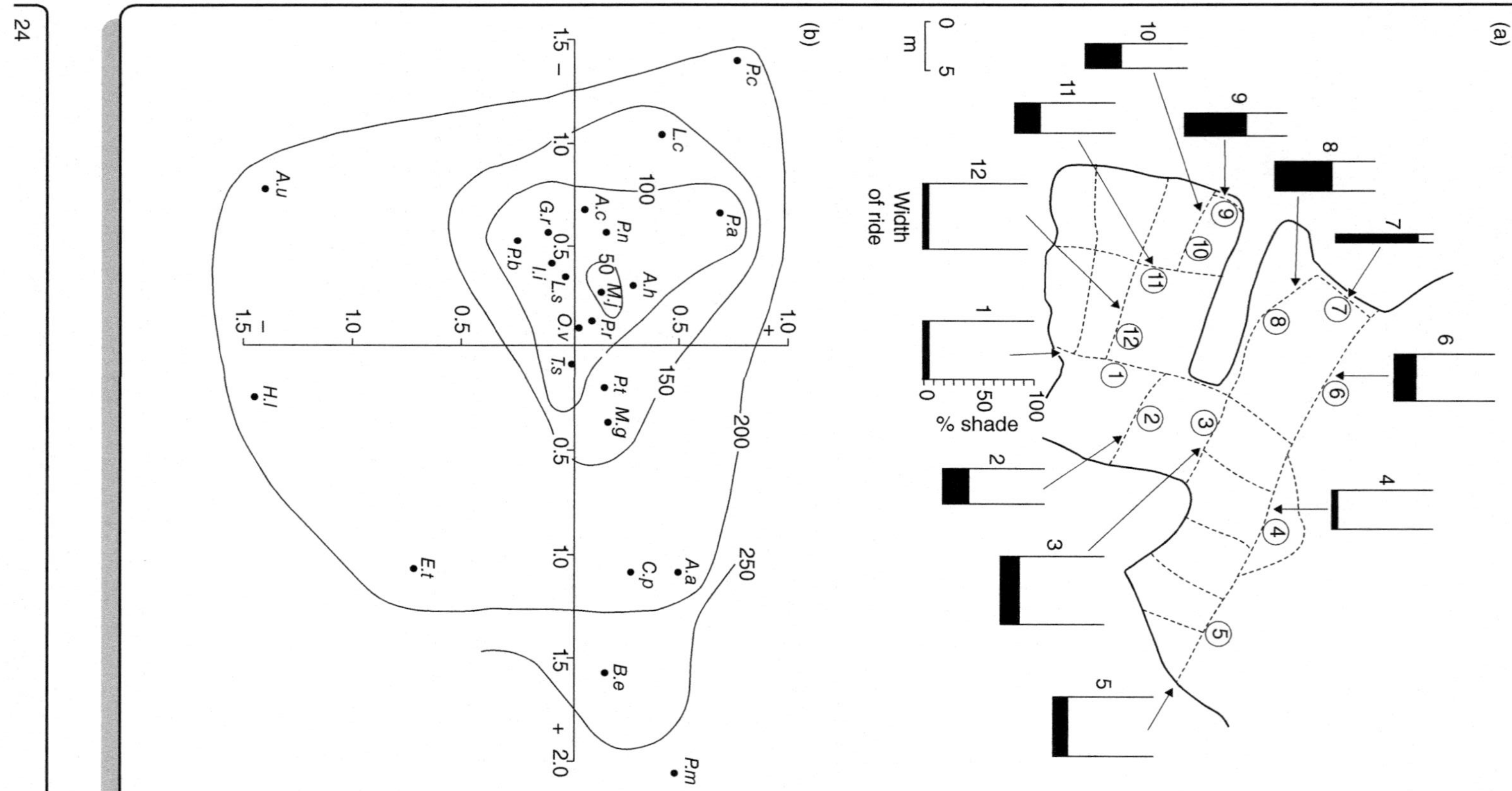

Fig. B1.2.1. Affinity among butterfly species for 12 woodland rides in Bernwood Forest, Buckinghamshire (UK). **(a)** Physical characteristics of the rides. **(b)** Affinities between butterfly species based on their relative frequencies in rides obtained with a multidimensional ordination analysis (non-metric scaling with Euclidean distances; axes have arbitrary values). Isolines give coefficients of variation; low scores indicate species that are spread over numerous rides and high values those that are concentrated in few rides. Species in the upper right part of the plot inhabit wide rides with short vegetation, whilst those in the upper left part of the plot frequent increasingly shaded rides with taller vegetation. Species are noted by generic and species initials (i.e. *T.s., Thymelicus sylvestris; O.v., Ochlodes venata* (now *Ochlodes sylvanus*)*; E.t., Erynnis tages; P.m., Pyrgus malvae; L.s., Leptidea sinapis; G.r., Gonepteryx rhamni; P.b., Pieris brassicae; P.r., Pieris rapae; P.n., Pieris napi; A.c., Anthocharis cardamines; H.l., Hamearis lucina; L.c., Limenitis camilla; A.u., Aglais urticae; I.i., Inachis io* (now *Aglais io*)*; P.c., Polygonia c-album; B.e., Boloria euphrosyne; A.a., Argynnis aglaja* (now *Speyeria aglaja*)*; P.a., Pararge aegeria; M.g., Melanargia galathea; P.t., Pyronia tithonus; M.j., Maniola jurtina; A.h., Aphantopus hyperantus; C.p., Coenonympha pamphilus*). See Table B1.2 for numbers of ride species and diversity statistics. Data collected by Caroline A. Steele. Adapted from Peachey (1980). (From Dennis, 1992.)

richer the habitat unit in terms of organisms at the bottom of the **food chain** (e.g. plants and fungi), the more animal species there will be (Rosenzweig, 1995). These are typical observations long appreciated from island biogeography research (MacArthur and Wilson, 1963; Dennis and Hardy, 2018).

Two distinct terms of importance come into focus in understanding a species' habitat: **habitat use** and **habitat selection** (Jones, 2001). Habitat use determines exactly what members of a species' local population actually use within the habitat. Beyer *et al.* (2010) have an additional view. They define habitat use as the proportion of time that animals spend in a particular habitat. Underlying this are the abiotic and biotic elements making up the **resource-based habitat** that determine the time spent in exploiting any resource or using any set of conditions (Dennis, 2010). Habitat selection may depend on one or more criteria determining the choice of the site as a habitat. This may be an actual resource (a **consumable resource** or a **utility resource**; Dennis *et al.*, 2003), a set of conditions or a surrogate for a resource, resources or conditions, and indicates the need for **mobility** in any organism to exploit resources. Different components of the environment can be used with different intensities, and the same component can be used more or less intensively at different times. The disproportional use of some resources over others represents the **habitat preference** of a species (Beyer *et al.*, 2010). Determination of resource use by any species requires a detailed survey over different habitat types for entire seasons and broods (Dennis, 2010).

However, an interesting feature emerges when the abundance of species in any one habitat unit is investigated: there are many fewer abundant species in it than there are **rare species** (Gaston, 1994); that is, when rare species are defined in terms of occurring in small numbers of individuals (a universal pattern that characterises the most disparate communities and for which there are no known exceptions; Magurran, 2004; McGill *et al.*, 2007; see Chapter 4). It is perhaps obvious that an abundant species in a spatial unit can be said to be occupying its habitat. But can a so-called habitat space be referred to as a particular species' habitat for the rare species found within it? One immediate thought that should occur to readers, raised above, is that as habitat units become smaller, then not only will the number of species be fewer but so will all species be less abundant; rarity (as defined on the basis of low numbers; Gaston, 1994) becomes inevitable for habitat units as they become smaller. As such, typical island biogeography factors (island area, island isolation) are likely to impose rarity as a matter of course (Dennis and Hardy, 2018). As habitats are in effect islands of substrate and vegetation, the same reasoning applies. Thus, a rare species could indeed be in its habitat, but in such small numbers that local extinction is nevertheless highly likely, unless it is matched by increased **mobility**.

We can look more closely at the habitat unit containing the species found there and formulate a few more questions. (i) How abundant is each species in the unit? (ii) How is it distributed in the habitat unit? (iii) If a species is rare, then is it abundant in a different habitat or is it rare everywhere? (iv) A fundamental issue is: how do these species coexist in the habitat unit? Answering these questions should tell us something about the ecological associations and the condition of the species. The last issue (iv) raises the concept of **Gause's principle of exclusion** or **Grinnell's axiom** (complete competitors cannot coexist) and we take a closer look at that in Chapter 2. First of all, it is useful to consider more closely what is involved in referring to a 'species' habitat'. Bearing in mind we are still referring to identifiable units of space (shoreline, seabed, etc.), we may define a **species' habitat** as the occupation of a named habitat unit by a local population (sufficient individuals to breed successfully to produce a following generation) of a particular species. Whether a species occurs uniquely within a habitat unit depends on the criteria (basis) used to classify the habitat unit. Inevitably further issues then arise. As no habitat unit is likely to be entirely homogeneous (unless managed for that very purpose by humans, i.e. a **brotion**; Dennis, 2010), then there is no reason to expect that the species is evenly distributed over the habitat unit. Only part of the unit may actually comprise its habitat. Moreover, the local population may be the consequence of 'mass effects' in the sense of **vicinism** (neighbourship and overflow; Zonneveld, 1995), in which case there will be joint occurrences of plant and animal species in adjoining habitat units despite similarities or differences in vegetation and substrates.

Increasingly, with this reasoning, it becomes evident that a species' habitat may not be singular but plural (multicategorical) involving a range of different habitat types (e.g. the red fox *Vulpes vulpes*, which uses different habitats along an elevation gradient depending on the season; Cagnacci *et al.*, 2004). In any case, this conundrum is accounted for simply by the size or placement of the habitat unit in a hierarchy of habitat units (e.g. based on vegetation classification). With the most highly ranked (thus largest) of the habitats in a classification, a plant or animal species is likely to depend on only part of that habitat space, whereas in one of the lowest ranked (smallest) units, a plant or animal species may occupy all of the habitat space. For example, in the UK, the large heath butterfly *Coenonympha tullia* is found on raised bogs or mosses but, as would be expected, is observed far more frequently in sunny, sheltered, dense tussocky areas of the larval host plant (i.e. the hare's-tail cotton grass *Eriophorum vaginatum*) and adult nectar source (i.e. *Erica tetralix* cross-leaved heath) than over areas, for instance, of ling *Calluna vulgaris* (Dennis and Eales, 1997, 1999). Viewed in this way, it should become clearer that a species' habitat is something different from the traditional habitat; it is this issue – determining the components that constitute a species' **resource-based habitat** – that we focus on in other chapters of this book.

The complexity of what is or is not a species' habitat becomes evident in potential combinations of presence or absence of a species in two adjoining units separated by an edge of some kind (e.g. walls, hedges, ecotones) (Fig. 1.8). The reader will be familiar with vernacular conundrums over habitat: that is, there are species whose names would convey association with specific habitat types. Among British birds, for example, are the wood pigeon *Columba palumbus* and the green woodpecker *Picus viridis*, whose habitats extend well beyond woodland, the former feeding on open cropland and the latter exploiting ants' nests in meadows. Many animal species occupy a variety of habitat types as in the cases of the stone martin *Martes foina*, the fox *Vulpes vulpes* and the hedgehog *Erinaceus europaeus* across the Greater Paris area, France (Capon *et al.*, 2021). Then again, some species seem capable of functioning equally well in different habitats in one region (e.g. the barn owl *Tyto alba*; Meek *et al.*, 2009). But what of creatures that migrate hundreds or thousands of kilometres between contrasting biogeographical regions (e.g. the monarch butterfly *Danaus plexippus*; the Arctic tern *Sterna paradisaea*) and that seasonally occupy different biomes let alone distinct habitats within different biomes? This question has not been tackled, but is a key issue addressed in this book (see especially Chapter 8).

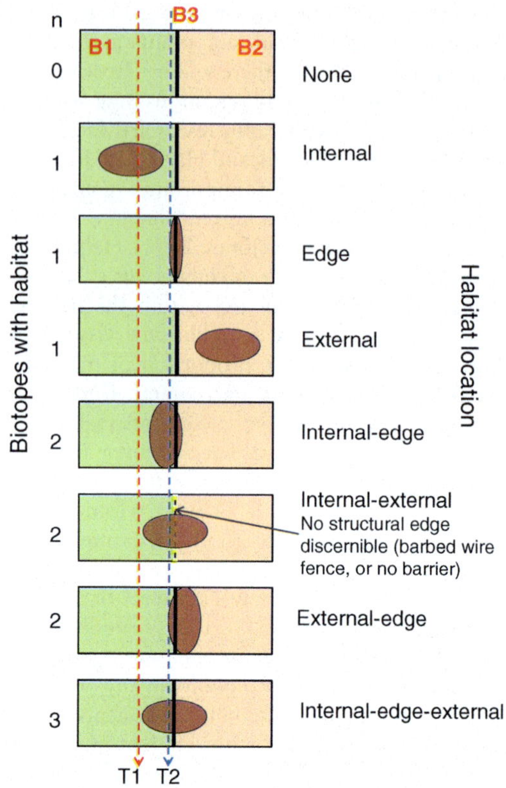

Fig. 1.8. The basic problems in traditional species-specific habitat designation (e.g. for a butterfly species). The habitat (brown) for a single species is shown as occupying none, one, two or three of two main biotopes (B1 light green, B2 dark green) separated by an intervening structural edge biotope or ecotone (B3, continuous black line) absent in one set (6th down, yellow pecked line). Those species whose resource-based habitats incorporate edge biotopes can be referred to as ecotone species. The positions of two transect lines (blue and red) for population assessments (as used by the UK Butterfly Monitoring Scheme or other survey programmes) are shown; these will pick up more or fewer individuals of the species in each case. Transect counts will also be affected by whether the habitat is represented by the full suite of resources for all life-history stages and conditions for the species or, alternatively, just part of the habitat possibly including or excluding the habitat core (containing the larval host plant, thus egg, larval and pupal resources). If researchers designate habitat differently (i.e. only the early stages suite of resources compared to the full suite of resources), this will affect the status of the biotopes if interpreted as habitat. As the resource-based habitat structure differs for species, especially for specialists versus generalists, a transect section will bisect varying fractions and types of habitat components for different species. (From Dennis *et al.* 2014, slightly modified.)

How, then, do we explain such conundrums? Although each piece of ground is in effect unique, it is part of a spectrum of variability in substrate characteristics, structure and biota, and therefore will have characteristics in common with other patches of ground; for this reason, components of patches overlap for biota. This becomes clear in the relationships both among habitats for species and between species for broadly defined habitat types. On close analysis at finer scale, it becomes evident why this is the case: traditional 'habitats' are heterogeneous and species vary widely in tolerances of conditions, as is apparent for any single designated habitat type. A useful example is provided in research by Caroline Steele on rides through Bernwood Forest in Oxfordshire, UK (see Box 1.2). From this example, woodland rides may be classified as a single habitat type, but it is clear that they vary in width and orientation, and thus local climate, and correspondingly in the quality and abundance of plant species and therefore the relative abundance of different butterfly species. The sheer complexity in links between butterfly species and similar habitat types (grasslands) is, for example, well illustrated by butterfly species-habitat networks in north-east Italy (Cappellari and Marini, 2021; Fig. 1.9). The complexity of interactions between and within habitats will be discussed in the next chapter.

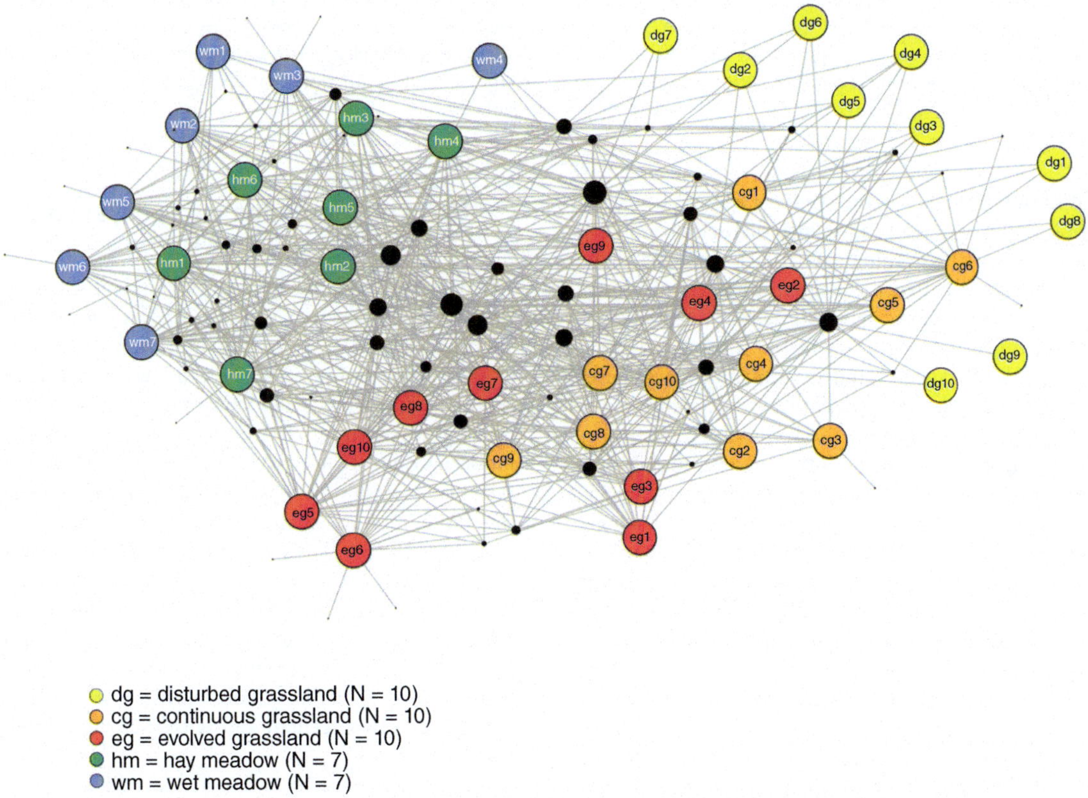

Fig. 1.9. Affinity between butterflies inhabiting multiple grassland types across a heterogeneous protected area in north-east Italy. In this bipartite species–habitat network, coloured nodes represent habitat patches with their codes, black nodes represent butterfly species (with node size reflecting the number of links for each species) and grey links indicate species occurrence. The study system involved 44 patches and 74 butterfly species. (From Cappellari and Marini, 2021, used under licence CC BY 4.0.)

2 Inside Habitats: Interactions in Space and Time

Abstract

Traditional habitats form part of ecosystems characterised by energy flows and trophic cascades through organisms linked by negative (e.g. predation and competition) or positive (e.g. facilitation) interactions, which can be extremely complex and wide (as in the case of the still poorly understood common mycorrhizal networks). Each organism in any traditional habitat has its distinctive but fluctuating niche space, characterised by a volumetric occupancy in a hyperdimensional space of traits which define its ecological role. A combination of stochastic (unpredictable environmental changes) and deterministic (e.g. intraspecific competition) factors influences population dynamics and species assemblages. The assemblage of species as communities is facilitated by contrasting life-history strategies, diverse and distinctive individual traits, and habitat heterogeneity. Based on their niches and traits, species can be grouped in guilds, functional groups or other categories that indicate their particular role in the ecosystem (e.g. keystone species).

With **autecology** (i.e. the study of the relationships that single organisms have with their environment), less so with **synecological** studies (in which the relationships between the environment and species assemblages are investigated), there is a tendency to focus on the conservation of but a few species, or even a single species and its immediate associates (parasites, pathogens, predators, prey, competitors, etc.) in habitats. In fact, the well-being of any species is dependent on the intricate functioning of a vast range of species belonging to different taxa, as well as the abiotic environment which is strongly influenced by the biota. Although much research is still required on which factors make ecosystems stable, biodiversity is generally regarded as fundamental for ecosystem stability (Tilman *et al.*, 2006; Loreau and Claire de Mazancourt, 2013) (Box 2.1). To understand the impact of different species on a target species, directly or indirectly, it is important to appreciate that 'habitat' units are not simply isolated collections of independent biotas and substrate components but part of functioning systems, called **ecosystems**. Inasmuch as all aspects of life occupy habitat space, and are therefore involved in interactions and drive ecosystems, 'habitat' is a concept relevant to all life forms (Ruggiero *et al.*, 2015). In Chapter 5 it is demonstrated that for any species it is a *sine qua non*.

Units of integration: an ecosystem perspective

The concept of ecosystems is ancient, possibly dating back to Theophrastus (Major, 1969; Hughes, 1985; Woodward, 1987). However, it started to assume its own epistemological status only during the 19th century (e.g. Forbes' 1887 view of a lake as a microcosm of interacting parts is clearly in line with the modern concept of ecosystem). The basics of ecosystems, as resulting from the integration between the biota and its environment, were realised by Cowles (1899), Thienemann (1918) and Allee (1934) before the word was introduced in the scientific literature (see Willis, 1997), and naturalists like Alexander von Humboldt, Charles Darwin and Ernest Haeckel were obviously aware of the subtle relationships that the organisms living in a given area have between them and with their non-living environment.

The term 'ecosystem' was introduced by Tansley (1935) but coined by Clapham (see Willis, 1997) and is now far advanced, with deep understanding of the circulation of materials having distinct functions (i.e. specific cycles of water and macronutrients, carbon, sulphur, nitrogen etc.; and micronutrients or trace elements, such as manganese, copper, zinc, boron, molybdenum, cobalt) and of energy flows

DOI: 10.1079/9781800627505.0002

Box 2.1. Networks and stability

Within communities, species interact directly or through their shared resources (such as nutrients, space, light, water) leading to **networks** of interlinked species (Fig. B2.1.1). Interactions can be of various types, including trophic interactions (e.g. predation, herbivory), mutualistic interactions (e.g. pollination, seed dispersal), and competitive interactions (interspecific interference for shared resources) (see Table 2.1).

In network modelling, each species represents a **node** and pairwise interactions between species are expressed by links between nodes (Newman, 2010; Estrada, 2012) (see Fig. B2.1.1). Links can be described by either a binary variable (absence or presence of interaction: unweighted networks) or a number expressing the strength of the interaction (weighted networks). Interactions can be either symmetrical (or undirected: species *i* influences species *j* and species *i* is influenced by species *j* in the same way) or asymmetrical (or directed: species *i* influences species *j* differently from how species *j* influences species *i*) (Newman, 2010; Estrada, 2012; Strona, 2022). Trophic networks are typically asymmetrical, and the direction usually goes from resources to consumers. For example, in host–parasite networks, links go from hosts to parasites. Thus, a trophic network appears as a list of node pairs, where the first node in each couple indicates a resource and the second node indicates the consumer that uses that resource. By contrast, in mutualistic networks, it is not possible to distinguish between resources and consumers. However, things may be more complex, and a link can encompass multiple kinds of interactions (e.g. a pollinator may also forage on a plant). The simultaneous presence of different kinds of links connecting the same nodes may be addressed through the use of multilayer networks (Strona, 2022). To further complicate things, networks are not static entities, as network structures change over time (for example, the structure of plant–pollinator networks in temperate climates shows seasonal variations in both the identity and strength of the interactions) (Strona, 2022).

A common distinction in network analysis is that between unipartite and bipartite networks. In unipartite networks, the same node can be simultaneously a target and a source of interactions to other nodes, whereas in bipartite networks nodes belong to two different categories, and those of one category can only interact with those of the other category (interactions within the same categories are not considered). Food webs are typically unipartite networks (see Fig. B2.1.1), whereas plant–pollinator and host–parasite networks are classical examples of bipartite networks (Fig. B2.1.2).

An ecological network of S species can be mathematically described by the matrix $A=[\alpha_{ij}]$, where each element α_{ij} expresses the link between species *i* and species *j*, i.e. the influence of species *j* on species *i*. The influence expressed by the interaction coefficients may be positive or negative. For

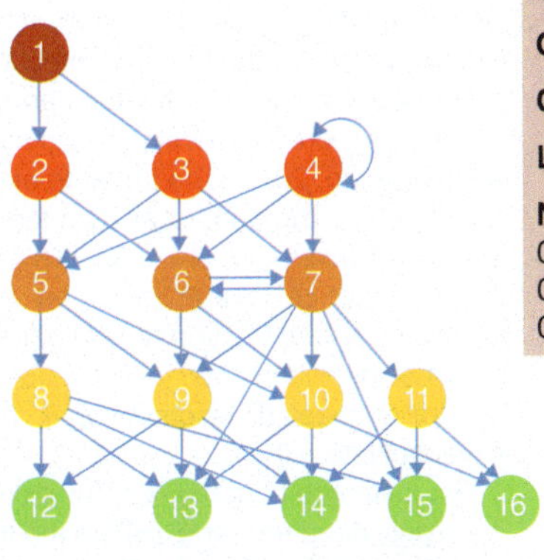

Size $S = 16$

Connectivity $L = 35$

Connectance $C_D = 2.187$

Linkage density $D = 0.457$

Node degree distribution: $P = \{0.125, 0.188, 0.250, 0.188, 0.375, 0.375, 0.500, 0.313, 0.375, 0.375, 0.250, 0.125, 0.250, 0.188, 0.188, 0.125\}$

Fig. B2.1.1. Network representation of a hypothetical food web. Each species (1–16) is a node. Species 1 is a top predator; species 2–11 are intermediate species; species 12–16 are basal species (producers). Species 8–11 are herbivores. Species 2–6 are carnivores. Species 7 is an omnivore. Note that species 6 is both a predator and a prey of 7. A loop in species 4 indicates cannibalism. Values of some characteristics of this network (size, connectivity, connectance, linkage density and node degree distribution) are shown. The network is unweighted.

Continued

Box 2.1. Continued.

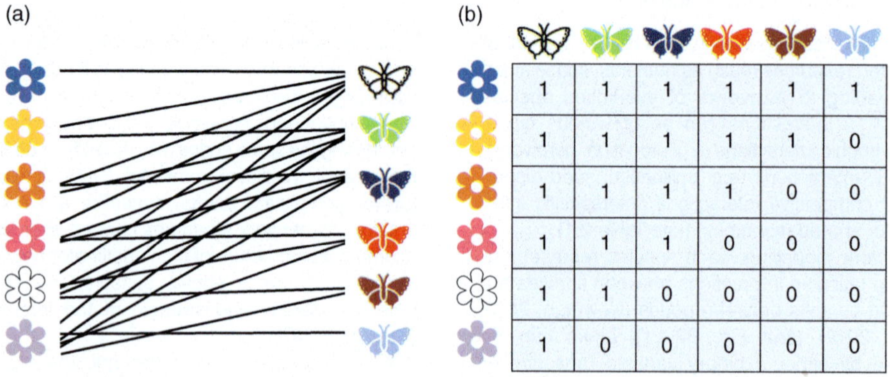

Fig. B2.1.2. A hypothetical bipartite network (plants and their pollinators). Panel **(a)** illustrates links (interactions) between pollinators and plants. Note that in this mutualistic network links are undirected. Panel **(b)** illustrates the same network as a presence/absence matrix (1 indicates presence of interaction, 0 indicates lack of interaction). The matrix is perfectly nested.

example, if a species *i* feeds on a species *j*, the interaction coefficient α_{ij} describing the effect of species *j* on species *i* will be positive, while the coefficient α_{ji} describing the effect of species *i* on species *j* will be negative. In competitive interactions, both coefficients will be negative, while in mutualistic interactions both coefficients will be positive (see Table 2.1).

Network complexity can be characterised with reference to various properties (May, 1972, 1973b; Newman, 2010; Estrada, 2012) (see Fig. B2.1.1), including the following.

- *Size* (*S*): the total number of interacting species (i.e. species richness, that is the number of nodes). For example, in bipartite networks, i.e. those composed of two groups of respectively *m* and *n* species (such as plant–pollinator or host–parasite networks), $S = m + n$.
- *Connectivity* (*L*): the total number of realised interactions.
- *Connectance* (*C*): the number of realised interactions (*L*) divided by the number of all possible interactions, calculated as $C_D = L/S^2$ for directed networks and as $C_U = L/[(S^2–S)/2]$ for undirected networks.
- *Linkage density* (*D*): the average number of interactions per species ($D = L/S$; higher values indicate that the network is dominated by **generalists**, i.e. species with many interactions; low values indicate that the network is dominated by **specialists**, i.e. species with few interactions).
- *Node degree distribution* (*P*): the probability distribution of the number of interactions per species, thus expressing the proportion of nodes with each possible degree; the degree of a node refers to the number of links of a species with its interacting partners.

In some instances, instead of referring to individual taxonomic species, 'trophic species' (i.e. functional groups of species sharing the same set of predators and preys) are used (Schoener, 1989b; Pimm *et al.*, 1991; Goldwasser and Roughgarden, 1993; Martinez, 1994; Williams and Martinez, 2000; Dunne *et al.*, 2004).

To incorporate the strength of interactions, weighted connectance and weighted linkage density can be applied, in which the proportion of biomass flow is used to weight the links (Bersier *et al.*, 2002; Tylianakis *et al.*, 2007; Dormann *et al.*, 2009). As a generalisation of the node degree distribution, the interaction–strength distribution can be calculated to take into account the weights associated with each link (Newman, 2010). The strength of each species is expressed by the sum of all its interaction strengths (Feng and Takemoto, 2014; Suweis *et al.*, 2015). The dependence of a species *i* on a species *j* is expressed as the fraction of interactions between the two species relative to the total number of interactions of species *i* (Bascompte *et al.*, 2006; Vieira and Almeida-Neto, 2015).

The influence of network complexity on its stability is a highly debated issue. Before the 1970s, the common thought was that stability is increased by complexity: richer communities should be more stable. This intuitive idea was challenged by May (1972, 1973b) who, using random community matrices, concluded that complexity tends to destabilise community dynamics. Specifically, he argued that the

Continued

Box 2.1. Continued.

system will tend to sharply change from stable to unstable as the number of species (S), connectance (C) or standard deviation of the strengths of interactions (σ) increase beyond a critical value. The system would be stable only if:

$$\sigma\sqrt{SC} < 1. \tag{B2.1}$$

May's conclusions have been criticised as based on random matrices, and subsequent studies on the relationships between network characteristics and stability provided contradictory results (Landi *et al.*, 2018). Results not only vary greatly according to the type of networks (predator–prey, mutualistic or competitive), but are also influenced by the definition of stability itself (**resilience**, persistence, extinction, cascades) (Landi *et al.*, 2018). Many elements were found to contribute to this variety of results.

- Species richness is considered an important determinant of the stability of mutualistic and competitive networks but, in food webs, it may increase or decrease stability, depending on many factors such as the strength of predator–prey links (which increases the stability of small webs but destabilises larger webs) or the role of redundancy (stability increases with the number of species per **functional group**).
- The relationship between connectance and stability is negative in mutualistic networks, positive in competitive network, and either positive or negative in food webs (in this latter case also depending on whether connectance is weighted or not).
- The influence of the strengths of interactions in food webs may vary according to many factors, including the type of the relationship, the size of the network or the trophic level; in competitive networks, weak interactions tend to favour community stability; in mutualistic networks, the heterogeneity of species strength distribution and the asymmetry of species dependencies tend to increase stability; in competitive networks, intraspecific competition tends to increase stability, while the role of interspecific competition is less clear (Landi *et al.*, 2018).

Stability is also influenced by network architecture. Network architecture is described by **compartmentalisation** and **nestedness**. Compartments are subsets of the ecological network where interactions occur mostly within the compartment and less frequently with other compartments (Fig. B2.1.3). Compartmentalisation tends to stabilise food web networks, but has a negative effect on the persistence and resilience of mutualistic networks (Landi *et al.*, 2018). Compartmentalisation is generally expressed as **modularity**, in which the nodes are partitioned to maximise the number of links between nodes belonging to the same module with respect to a random network (Newman and Girvan, 2004; but see Rosvall and Bergstrom, 2007 and Landi and Piccardi, 2014 for critique and comparisons among different approaches

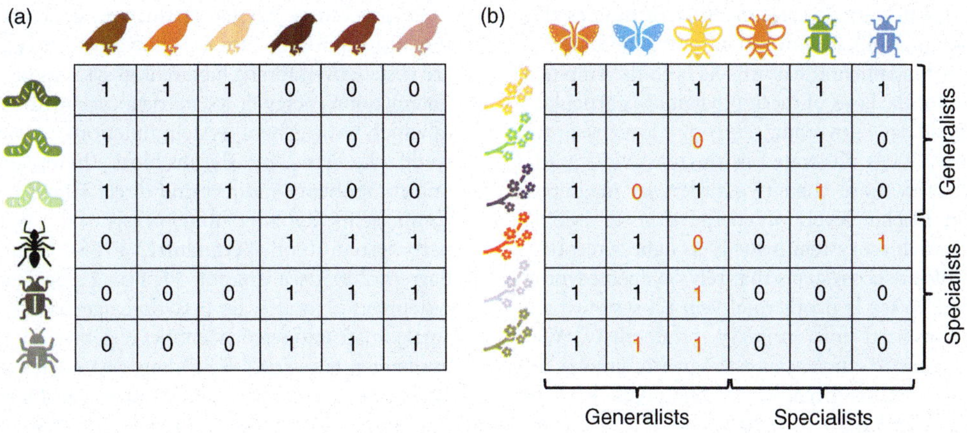

Fig. B2.1.3. Network compartmentalisation and nestedness. Panel **(a)** illustrates a hypothetical food web (birds and their invertebrate prey) in which two compartments are clearly defined. Panel **(b)** illustrates a hypothetical mutualistic network (plants and their pollinators) with distinction between specialists and generalists. Note unexpected absences of interactions (red 0s) and unexpected realised interactions (red 1s). There is only one (unexpected) specialist–specialist interaction. Specialist–specialist interactions are actually rare as they reduce network stability.

Continued

Box 2.1. Continued.

to identify modules). Nestedness is defined as the ordered loss of links in a matrix where rows and columns are sorted according to species richness (see Fig. B2.1.2b). This means that specialists can only interact with a subset of species with which more generalists interact. As a result, generalists tend to interact with both other generalists and specialists, whereas specialist–specialist interactions are rare (Bascompte *et al.*, 2003). For example, in a bipartite mutualistic web of plants and their pollinators, plants that have many pollinators may interact both with pollinators that pollinate many plants and with pollinators that pollinate few plants (and pollinators that use many plants can interact with both plants that use many pollinators and plants that use few pollinators), but interactions between plants that use few pollinators and pollinators that use few plants are rare (see Fig. B2.1.3b). Nestedness is expected to stabilise networks, because it decreases coextinction risks (e.g. the extinction risk of a plant that is pollinated by few pollinators is reduced if they are generalist pollinators, as it is less probable that they will go extinct). However, the importance of nestedness for the stability of ecological networks is debated (Strona, 2022); for example, there is evidence that extreme nestedness may facilitate sequential species extinctions (**cascading coextinctions**) (Campbell *et al.*, 2012). But in host–parasite networks it has been observed that specialised parasites have a lower extinction risk than expected because they are associated with hosts showing a high parasite species richness, which is a reflection of the hosts' **generalism** (e.g. they have larger ranges) and hence of their lower extinction risk (Strona *et al.*, 2013b, 2016). Several measures of nestedness and algorithms to test their statistical significance have been proposed (see Box 4.5 in Chapter 4), and results may vary considerably on the basis of which combination of metric and algorithm is used (Strona *et al.*, 2013b, 2014a,b; Strona and Fattorini, 2014a), making it difficult to interpret contrasting patterns found in nestedness analyses based on different procedures.

Finally, the role of specialisation in determining network stability is debated. There is some evidence of a positive relationship between the degree of ecological specialisation, habitat stability and consumer specialisation. As specialised consumers can be more efficient in using their (few) resources when these are available but have, in principle, a higher coextinction risk than generalist species, specialisation may increase an ecological network's robustness to species loss under stable environmental conditions, but also makes them more vulnerable to potential cascading effects of primary extinctions if environmental conditions become unstable (Strona *et al.*, 2021).

through trophic relationships (**food webs**), operating at a range of spatial scales in different **biomes** (Fig. 2.1) (Odum, 1963; Odum and Barrett, 2005; Putman, 1994). A fundamental feature has been the empirical support for the laws of thermodynamics, particularly the second law governing entropy (Jørgensen and Svirezhev, 2004) illustrated in the massive degradation of even more than two orders of magnitude between **trophic levels** of energy transfer initially entering each ecosystem (usually as light; exceptions to the rule are ecosystems that rely on chemosynthesis, e.g. deep-sea hydrothermal vent ecosystems) and being converted into primary productivity. With trophic transfer efficiencies vary greatly among different ecosystems because of differences in light, nutrients and length of **food chains** (Odum, 1963; Hairston and Hairston, 1993; Dickman *et al.*, 2008).

The concept of ecosystems elaborated by Clapham in the 1930s (*vide supra*) first focused on the functioning of biota in **ecotopes** (used here as a synonym for small-scale habitats; Sørensen, 1936), subsequently expanding to a global view. Ecosystems are assemblages of animals, plants, micro-organisms, non-living things and their shared environment, each ecosystem having distinctive structures and functions (Odum, 1963; Schowalter, 2016). They are nested, overlapping hierarchical systems of activity (interactions, energy flows, nutrient cycling, food webs) in which basic general system functions are evident: feedbacks between different biotic (living) and abiotic components; indirect and direct effects between components; dynamic homeostasis of the system or part system (von Bertalanffy, 1968; Odum and Barrett, 2005; see Chapter 11, Box 11.2). Thus, ecosystems at a local scale (e.g. drainage basins, lakes) amalgamate into regional entities (e.g. broadly marine, freshwater, terrestrial; see Odum and Barrett, 2005) and finally into a single global system (the **biosphere**). The interactions between global ecosystem components account for different aspects of planetary functioning, such as regional climates, distinctive biomes, differential rates of geomorphic activity and drainage basin hydrology. Natural and human impacts on ecosystems trigger changes in states of activity, as in the case of global climate change and habitat creation and degradation (Pullin, 2002).

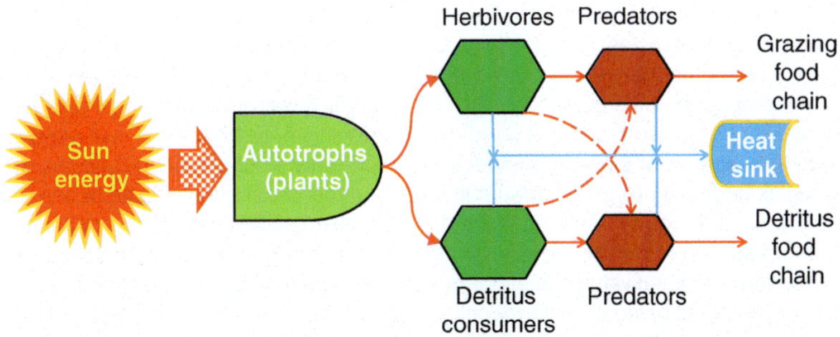

Fig. 2.1. Basic features of the trophic cascade (energy flow) in ecosystems. In every ecosystem, two interconnected types of food chains can be recognised: the grazing chain (transfer of energy starts from autotrophic organisms) and the detritus chain (transfer of energy starts from dead decaying organic matter). Solid red arrows indicate food consumption (energy transfer) within each chain, dotted red arrows food consumption between the two chains. Blue arrows indicate energy loss. In terrestrial ecosystems, the main primary producers are represented by plants, but in other ecosystems other autotrophs are the main primary producers (e.g. phytoplankton in marine ecosystems). (Modified from Odum and Barrett, 2005.)

It would be incorrect to assume that a habitat class (e.g. a meadow, a wood, a pond) coincides neatly with the more open bounds of an ecosystem, if only because of the incessant flows of energy and nutrients in and out of the habitat unit conveyed by organisms and abiotic processes. But, as with oceanic islands, the greater the extent to which a specific habitat type is surrounded by unoccupied (or at least adverse) **matrix**, the more it can be conceived as a distinct ecosystem. Moreover, it is feasible to isolate habitats artificially (theoretically) as ecosystems for study, but they can never be conceived as completely isolated ecosystems (see Wilkinson, 2021: Chapter 5 on lakes, tarns and cwms). Thus, a pond, a lake, an island, a wood can be arbitrarily segregated, treated as a black box, and be the focus of inputs into and outputs from the unit: of energy, nutrients, animals and plants, fungal spores, microbes, pollutants, inorganic materials, etc. On the other hand, the focus can be entirely that of interactions between elements (individuals) and components (taxa or **trophic levels**, i.e. the position of an organism in the sequence, called **food chain**, through which nutrients and energy pass as one organism eats another) within the unit, the increasing resolution of investigations being referred to as grey and white box approaches. For a complete understanding of 'habitat', the primary ecosystem focus is the scale-interdependence, for trophic and other interactions (Table 2.1), among species and higher taxa in the assemblage making up the habitat, as well as the flows and links (i.e. nutrients, energy, genetic material) between biotic and abiotic components within the habitat.

Ecosystems have functions and structures which change over time. Functionally, there are energy and matter fluxes, temporal changes (including ones in net primary productivity, i.e. the conversion of carbon dioxide into biomass driven by photosynthesis), and in basic activities such as **herbivory**, **predation**, decomposition and pedogenesis, local colonisation and extinction. At an individual level there are changes in behaviour. From a structural vantage, ecosystems have both vertical and horizontal extent, distinguished as **ecotopes**, as well as distinctive patterns of biomass (Fig. 2.2) and biodiversity (taxa; see Chapter 4). Biotas within ecosystems fall into well-designated classes: **autotrophs** (organisms that synthesise nutritive organic molecules from inorganic materials, viz. photosynthetic organisms, which use sunlight as a source of energy – plants, algae and cyanobacteria – and chemosynthetic organisms, which use energy released by chemical reactions – many bacteria and archaea) and **heterotrophs** (organisms that obtain nutrients from other organisms), the latter subdivided into finer trophic categories: consumers (primary consumers, which feed on autotrophs; secondary consumers, carnivores that feed on primary consumers; tertiary consumers, carnivores that feed on carnivores), omnivores and reducers (detritivores, typically arthropods, and decomposers, such as fungi and bacteria) (Whittaker, 1975).

Food webs (i.e. networks of interconnected food chains; see Box 2.1) have long been at the heart of

Table 2.1. Terms and relationships associated with direct pairwise interactions between individuals of populations belonging to the same or different species. Competition, each population inhibits the other; predation/lethal defence, each potentially the prey of the other (e.g. humans and wolves; lethal defence is a strategy employed by organisms to deter predators, competitors or herbivores through the use of highly toxic or harmful mechanisms that can kill or severely harm the attacking organism); herbivory, predation, parasitism, individuals of one population benefit (predator or parasite) at the expense of those in another population (prey, host, model in Batesian mimicry); mutualism, individuals of both populations benefit (e.g. Müllerian mimicry); commensalism, individuals of one population benefit (the commensal), individuals of the other being unaffected; amensalism, individuals of one population are inhibited whereas those of another population are unaffected; neutralism (as conceived by neutral theory; Hubbell, 2001), the interaction has no consequences for individuals of either population. + beneficial, 0 neutral, and – negative effect. Relationships are often described as being obligate (or obligatory) or facultative. Organisms have a number of ways of communicating with one another (sight, sound, smell, touch, etc.) which influence their relationships. One area in great need of detailed study is bioluminescence; researchers are becoming increasingly aware of the importance of bioluminescence as revealed by UV torches (e.g. David Atthowe's studies in Welsh woodlands and rock pools, UK; www.bbc.co.uk/news/uk-wales-68301119; see Syed and Anderson, 2021; Gray and Karlsson, 2022). (After Dennis, 2020.)

Effect on organism 2 \ Effect on organism 1	+	0	–
+	**mutualism** (Müllerian mimicry; symbiosis)	commensalism	**herbivory predation parasitism** **(Batesian mimicry)**
0	commensalism	neutralism	amensalism
–	**herbivory predation parasitism** **(Batesian mimicry)**	amensalism	**competition predation/lethal defence**

ecosystem studies inasmuch as trophic relationships between organisms have been regarded as accounting for stability in populations and biodiversity. There has been a tendency, then, for mathematical theory in population dynamics to drive concepts (models) of food web stability (Cohen and Newman, 1985; Lawton and Warren, 1988; Putman, 1994; Polis and Winemiller, 1996), much as energy flow and organism size have encouraged expectations of ecosystem trophic structures (see Figs 2.1 and 2.2) (Putman, 1994; Woodson *et al.*, 2018). This has led to mistaken notions directing the process of constructing food web topologies (e.g. an assumption of the rarity of omnivory; discarding reducers such as fungi; ignoring reciprocal interactions between consumers) and the tendency of mathematical models to determine concepts of ecosystem integration rather than relying on empirical investigation to drive conceptual developments (Paine, 1996; Tanabe and Namba, 2005). The classic case of underestimated interactions is that of **intraguild predation**, in which an animal kills and eats another species, that is also a potential competitor (Arim and Marquet, 2004), both at arthropod (e.g. spiders, scorpions and solpugids; Polis and McCormick, 1986) and at vertebrate level (e.g. Serengeti carnivores: Pallardy, 2021a; raptors: Sergio and Hiraldo, 2008).

If we are to understand habitats fully, then it is important that forms and functions of ecosystems are derived first from empirical investigation and meta-analyses of numerous independent studies carried out in different biomes by different research groups (e.g. as illustrated by the 8092 unique marine predator–prey relationships existing within the North Sea: Thompson *et al.*, 2020; or the Serengeti plant–mammal food web: Baskerville *et al.*, 2011; see below, section *Guilds and functional groups*). These studies need to cater for changes; web

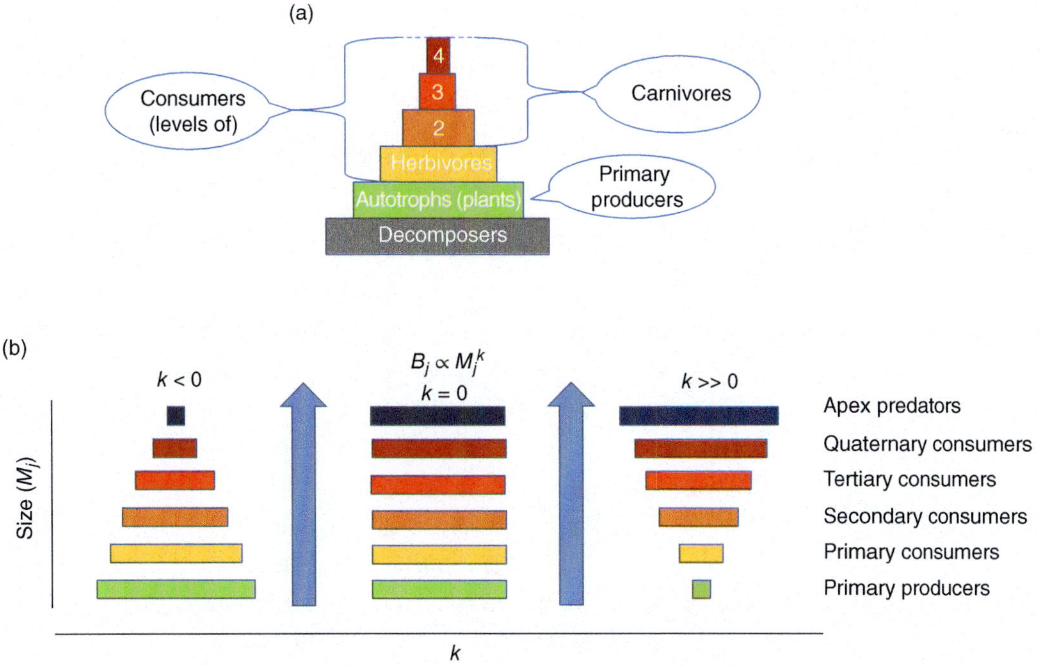

Fig. 2.2. Trophic pyramids. **(a)** Trophic pyramid illustrating biomass of a typical terrestrial ecosystem. Herbivores are consumers of first level. Carnivores that feed on herbivores are consumers of the second level, and so on. **(b)** Variation in trophic pyramids. Relationship between trophic structure and size-spectrum scaling for marine ecosystems when predation is predominantly size based (big fish eat smaller fish). *k* shows the decrease in biomass (*B*) with individual size (*M*); approximately, $k_t = 0.25 + \log(TE)/\log(PPMR)$, where *TE* is trophic transfer efficiency and *PPMR* is the predator–prey mass ratio; the intercept (0.25) arises from allometric scaling laws that suggest biological function scales as individual mass to the ¼ power. Blue arrows indicate flow of energy. Note, there are three types of ecological pyramids: for population numbers, biomass (dry weight) and energy. The pyramid of numbers and biomass can be upright or inverted (or assume the form of a column) depending on the nature of the food chain in the particular ecosystem. When $k < 0$, abundance and biomass are concentrated in basal trophic levels (primary producers and consumers). When $k = 0$, biomass is equally distributed across all body sizes, and top-heavy ecosystems occur when $k > 0$. Pyramids of energy are always upright, as energy is inevitably lost through trophic levels. Trophic transfers invoke the 10% rule: roughly 10% of the previous trophic level's energy is available to the level immediately higher up; 90% is lost as heat (respiration) and incompletely digested food (egesta or waste). (Panel **b** modified from Woodson *et al.*, 2018, used under licence CC BY 4.0.)

structure, function, pattern, and process can change over time (food web dynamics) and these can occur from a variety of processes and patterns of biotic and abiotic factors in a habitat (Bengtsson and Martinez, 1996).

Recently, there has been a shift in emphasis from trophic relationships to the investigation of all kinds of interaction in ecosystems; ecosystem properties such as biomass and production have been shown to depend not only on species non-trophic interactions but also on the diversity of relationships between individuals of species (Goudard and Loreau, 2008). This has led in turn to new methods in investigating and modelling interaction webs (Sander *et al.*, 2015; Strona, 2022). It is increasingly evident that even what might have been regarded as trivial relationships among organisms can have important consequences for the structure and function of ecosystems. As illustrated in Table 2.1, a great variety of reciprocal and non-reciprocal linkages occur among organisms, all of which affect individual fitness, reproductive success and mortality without the agents responsible for various effects being involved in consumption of prey or other resources. Regarding trophism, a vast death rate among animals is caused not by killing and consumption by predators but by pathogens and their vectors, the dead processed by innumerable small arthropods, fungi, bacteria and some protists. Thus, for a focus on habitat, it becomes vital to consider all interactions among individual organisms that affect individual fitness. They may be conceived as emanating from a matrix of interactions among species (Fig. 2.3), the specific links coded as to type of interaction (usually either as antagonistic or mutualistic), the web and web subsets modelled using a recently fast-growing panoply of advanced techniques (Fath *et al.*, 2007; Rosvall and Bergstrom, 2010; Strona and Veech, 2015; Fenu *et al.*, 2017; Miele and Mathias, 2017; Liao *et al.*, 2020; Strona,

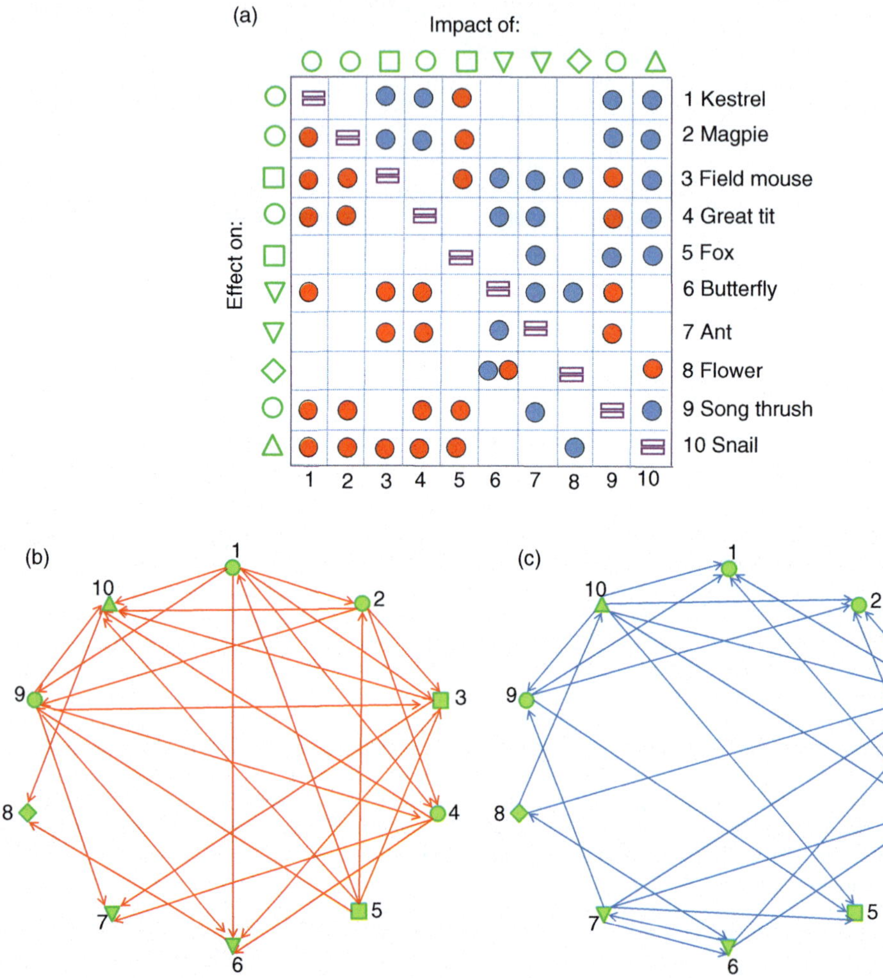

Fig. 2.3. Interaction matrices. A scheme illustrating interactions (both trophic and non-trophic; see Table 2.1) among components (a selection of a few species) in a hypothetical habitat community. **(a)** Impact of any species (column) on any other species (row). Red, negative impact (predation, effective competition, etc.); blue, positive impact (trophic advantage as food supply, protection, ecological engineer, etc.). A double negative impact is indicative of direct competition, a double positive is indicative of symbiosis (these cases are not illustrated in the example). Note, one species may have a negative and a positive impact on another species (butterfly using a flower as both a host plant and nectar source, acting as a pollinator in the latter case). The matrix of interactions can be displayed in a number of other ways, as a network, a hierarchy (i.e. based on size or overall impact on the intrinsic rate of increase on species within the community) or in the form of a two-dimensional plot based on multivariate analysis. Panels **(b)** and **(c)** illustrate network representations of negative and positive links, respectively. Green symbols distinguish type of species (higher taxon) shown in **(a)**. See Lawton and Warren (1988) for a different food web design ordered by body size of animals, and a worked example of an interaction matrix in Sander *et al.* (2015).

2022) (see also Box 2.1). The complexity of the relationships is well illustrated by the links to a single arctic plant (*Dryas octopetala* × *integrifolia*) at Zackenberg, Greenland (Schmidt *et al.*, 2017) (Fig. 2.4).

The impact a species has within an ecosystem depends on two fundamental factors: (i) the size of its population(s) and (ii) the **niche** occupied by the species, thus the interconnections or rather the specific relationships with other species within the ecosystem. These two factors determine the nature and magnitude of impacts that a species has on others in the ecosystem. The following two sections briefly investigate these two aspects of a species' biological significance.

Population status and changes within habitats

Although the relative abundance of species in habitats may have stability (i.e. the **species abundance distribution** pattern of a species assemblage, Chapter 4, does not change over time), as with butterflies on British islands (e.g. Isle of Wight; Dennis and Hardy, 2018), the population of each species is not constant but varies over time, in the short term in relationship to the life cycle and in the long term relating to changing conditions within habitats. Some of the changes to numbers are the consequence of population changes *per se* inasmuch as

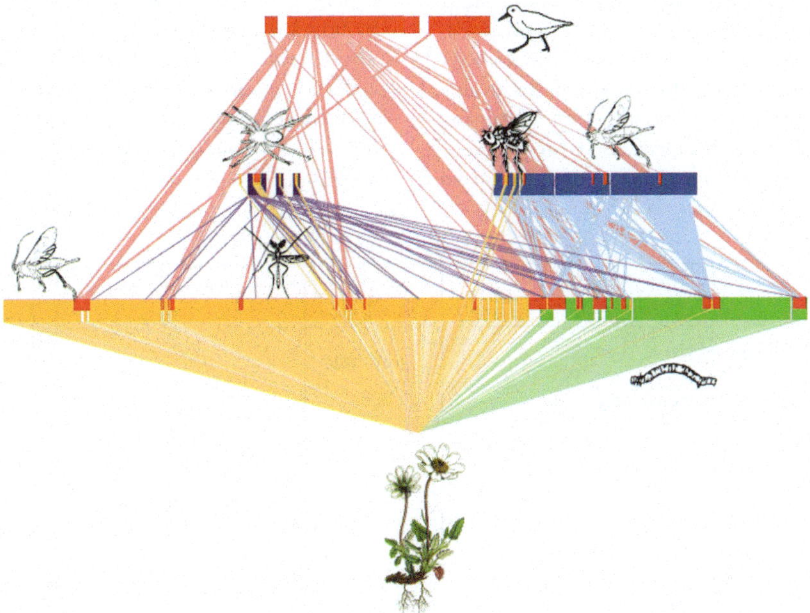

Fig. 2.4. Interaction web illustrating the quantitative biotic links involving a single plant taxon (*Dryas octopetala* × *integrifolia*) in the high arctic food web of Zackenberg, Greenland. Antagonistic interactions: green, blue, purple and red connectors; mutualistic interactions: yellow. Each block represents one species at a trophic level. Green blocks represent lepidopteran larvae (herbivores) found actively feeding; the widths of the blocks represent the numbers of individuals detected. The widths of the light green connectors show the proportion of each herbivore taxa found feeding on *Dryas*, i.e. the relative dependence of this herbivore taxon on *Dryas*. Blue blocks represent parasitoid species attacking the lepidopteran herbivores feeding on *Dryas*; the widths of the blocks represent the total number of interactions in which the species was involved. Widths of the light blue connectors represent the numbers of feeding events involving each herbivore. Purple blocks represent three spider species; widths of these blocks represent the total numbers of feeding events involving each species, with connector width proportional to the specific predator-by-prey interaction. Red blocks represent feeding interactions involving three bird species. Blocks on the upper level show the total numbers of feeding events detected for each bird species, and blocks on the two lower levels represent the total number of interactions involving each *Dryas*-affiliated prey taxon. Widths of the light red connectors represent the numbers of feeding events for each predator-by-prey combination. Yellow blocks represent taxa visiting *Dryas* flowers (i.e. potential pollinators) as trapped by sticky flower mimics. Again, the widths of the blocks represent the numbers of individuals found, with widths scaled to 1/6 of those of the other colours, to accommodate all 185 taxa detected. (From Schmidt *et al.*, 2017, used under licence CC BY 4.0.)

an increase or decrease in population size affects the initial resource base and conditions within habitats. The shift in population from one period to the next (stage to stage) is given by the familiar equation:

$$N_{t+1} = N_t + B + I - D - E \qquad (2.1)$$

where population size at time t is N_t, and at time $t + 1$, N_{t+1}; B, D, I and E are respectively the number of births, deaths, immigrants and emigrants to and from the population. With increasing numbers of births and immigrants the population increases and with deaths and emigration the population decreases. Life-history studies of species within habitats focus on population changes and on the contribution of births, deaths, immigrants and emigrants to changing population size, as well as the factors involved.

Central to these studies are methods for estimating changing population numbers. Methods for attaining population estimates for different organisms differ considerably; techniques, used for different purposes, may vary as to being absolute or relative estimates of population size, involve direct or indirect assessments, and achieve total counts (complete enumeration) or sample estimates from which estimated total numbers are then obtained (Appendix 1). For understanding processes in habitats, of far greater importance than basic changes in numbers of individuals are the factors causing increase and decrease in numbers throughout life at each stage of development or phase of maturity. Linking changing numbers to factors affecting mortality identifies vulnerable stages and phases in life for the population of that species. Techniques for exploring such changes include broadcast indirect (inferential) methods such as age/sex-specific pyramids as for the **demographic transition model** in humans (Lundquist *et al.*, 2015) (Fig. 2.5; see also Chapter 10, Fig. 10.2) and the study of ageing criteria (e.g. growth lines, tooth eruption and wear patterns, cranial suture closure) in skeletal features, for instance in marine gastropods (e.g. *Buccinastrum deforme*; Bökenhans *et al.*, 2016) and vertebrates

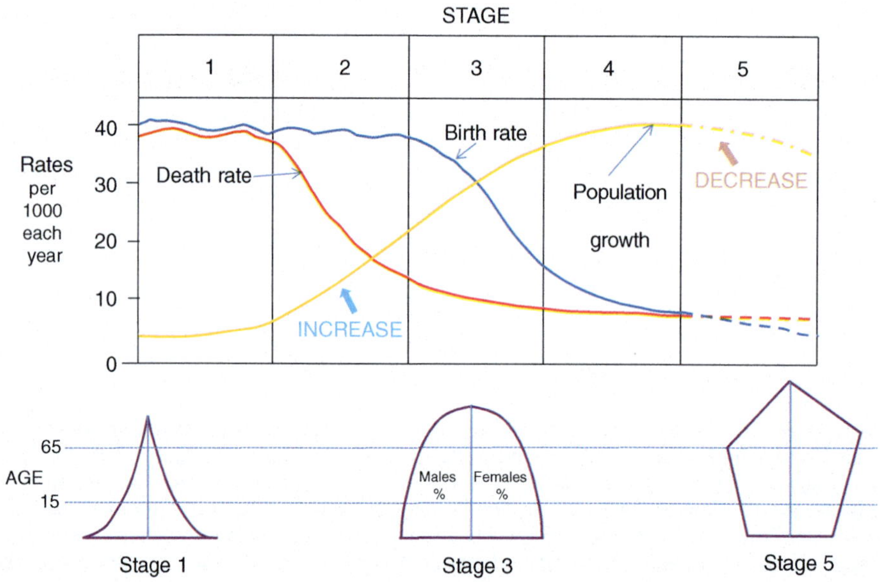

Fig. 2.5. The demographic transition model. This model describes changes in demographic characteristics from high birth and death rates to low birth and death rates as a country develops economically from a preindustrial to an industrialised system. These changes produce different age structure diagrams, also known as age-sex pyramids. The two halves of these pyramids correspond to the numbers (or proportions) of males and females. The first pyramid indicates an increasing population, the second pyramid a stable population, and the third pyramid a declining population. Note, in the last pyramid, the increase in the proportion of females with increasing age. Number codes indicate stages of population change: 1 high birth and high death rate; 2 high birth rate, death rate falling rapidly; 3 birth rate falling and death rate falling more slowly; 4 low birth rate and low death rate; 5 birth rate falling or rising again, death rate stable. Compare with Fig. 10.2. (Based on a number of sources, including Henderson, 2003; Lundqvist *et al.*, 2015; Max Roser: https://commons.wikimedia.org/wiki/File:Demographic-TransitionOWID.png., used under licence CC BY 4.0.)

(e.g. Dall sheep, *Ovis dalli*: Deevey, 1947; humans: Manjunatha and Soni, 2014). More detailed studies involve individual and cohort life tables. These provide the basis for exploring key mortality agents during the life cycle, as for many arthropods (e.g. the field grasshopper *Chorthippus brunneus*: Richards and Waloff, 1954; butterflies: Warren, 1992) (Fig. 2.6).

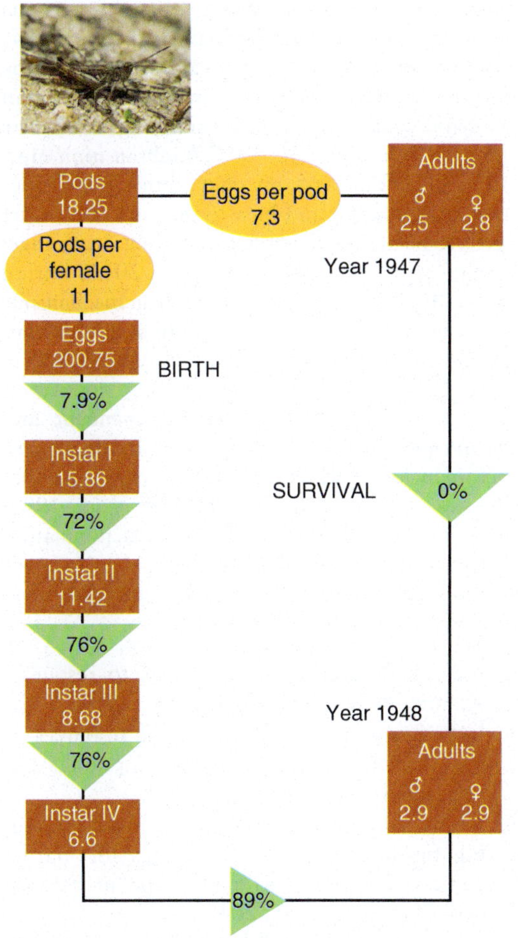

Fig. 2.6. Life table of the field grasshopper *Chorthippus brunneus*. Numbers in triangles are probabilities of survival to the following stage. Numbers in boxes are individuals per 10 m^2 surviving at that stage. The collection of stage data is an essential part of key factor analysis (see Chapter 6) in survival. (Modified from Chapman and Reiss, 1999, used with permission of Cambridge University Press; data from Richards and Waloff, 1954. Photo by Jörg Hempel: http://commons.wikimedia.org/wiki/File:Chorthippus_brunneus_LC0177.jpg, used under licence CC BY SA 3.0 DE.)

These techniques tend to be used in short-term studies over one or few life cycles. The dynamism of populations over the long term requires their constant monitoring. These long-term approaches have been most effective for following changes in human populations but also have become a feature of studies of arthropods (e.g. UK butterflies; Sánchez-Fernández *et al.*, 2021), birds (e.g. North America and Europe; Reif, 2013) and mammals (e.g. Ontario, Canada; Fryxell *et al.*, 1998). Arising from these studies is knowledge that populations rarely remain constant; they grow and decline, some even undergoing distinct cycles (e.g. Begon *et al.*, 1996; May and McLean, 2007; Rockwood, 2015).

The study of population dynamics has fascinated ecologists since Darwin, as his reading of the implications of an **exponential population growth** for humans presented by Thomas R. Malthus (1798) played an essential role in his formulation of the theory of natural selection (Darwin, 1887): the surplus of offspring resulting from exponential growth determines **intraspecific competition** (i.e. individuals of the same population compete for the resources), which is considered a driver of selection (e.g. Vorzimmer, 1969; Bowler, 1976). The exponential growth model (Fig. 2.7; see Appendix 2) assumes that a population indefinitely grows with a fixed, **intrinsic** (natural maximum) **rate of increase**, r. This means that resources (such as space and food) are unlimited, which is an obviously false assumption. Yet, there are circumstances in which populations may follow (at least initially) an uncontrolled exponential growth. For example, introduced species (see Chapters 10 and 11) frequently show exponential growth (Shigesada and Kawasaki,

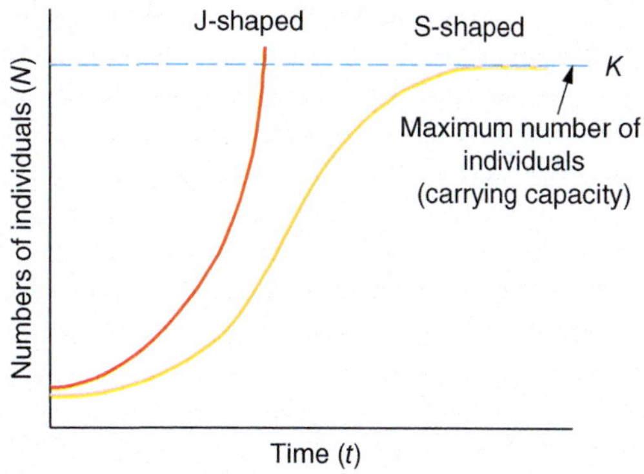

Fig. 2.7. Exponential (J-shaped, in red) and logistic (S-shaped, in yellow) curves of population growth. Dotted line indicates the carrying capacity, K.

1997; Kelly *et al.*, 2021), as in the new area they might find abundant resources and no predators, competitors or pathogens that can limit their growth. However, in general, birth and death rates are not stable but density dependent, and population growth tends to follow a sigmoidal shape, known as the **logistic growth** (Appendix 2). In this case, an ideal population, after a phase that approximatively follows an exponential growth, progressively reduces its growth to an equilibrium value, which is commonly (but incorrectly) identified with the **carrying capacity** of the environment (see Fig. 2.7 and Appendix 2). While the exponential model describes an intrinsically unregulated growth, the logistic model describes populations that are intrinsically regulated by interspecific competition and/or other density-dependent factors. In fact, a wide variety of factors regulate population growth and population size *sensu lato*. Growth rate is affected by any limiting factor operating on the intrinsic rate of growth for individuals in a population.

Factors controlling population size may be described as predictable (e.g. space, light, food and territorial resources) and unpredictable (e.g. catastrophes such as volcanicity, storms, fire and severe frosts). In fact, even predictable and persistent (relatively constant) factors such as light, space and resources are subject to environmental changes and human interference; distinctions are simply a matter of the frequency and magnitude of changes and the length of time over which a population is monitored in a habitat. The magnitude of individual events is related to time, with higher magnitude events having a distinctive distribution over time. Assuming stability (stationarity) in conditions, that is the lack of a trend such as gradual climate change, the recurrence interval or return period (RI) for an event over n_t years (or other spans) of magnitude m (i.e. the upper or lower limits of population size of a species being equalled or exceeded) is crudely given by:

$$RI = (n_t + 1) / m \qquad (2.2)$$

The probability of the event being equalled or exceeded is then (Dennis and Hardy, 2018):

$$P[m] = m / (n_t + 1) \qquad (2.3)$$

Even during the span of a few decades, when environmental conditions might be regarded as relatively static, oscillations in population growth can occur on a regular basis owing to the impact by species on each other in the habitat in their role as herbivores, predators, parasites or parasitoids (see Appendix 2). Examples are the short cyclic population changes in voles (*Microtus* and *Clethrionomys*, which are possibly, but not necessarily, driven by the interaction with predators) and lemmings (*Lemmus lemmus*, which are driven by their interaction with food plants) (Turchin *et al.*, 2000; Selås, 2016) and the longer cycles in snowshoe hares (*Lepus americanus*) and lynxes (*Lynx canadensis*) (which, contrary to what was assumed for a long time, cannot be explained by a simple predator–prey relationship since, for example, the hare is influenced by many predators other than the lynx, while the lynx is primarily influenced by the hare; Stenseth *et al.*, 1997; Krebs *et al.*, 2018). Although the conditions may differ for species undergoing boom-and-bust cycles, and much yet needs to be discovered about them (e.g. lemmings; Ims *et al.*, 2011; Fauteux *et al.*, 2015; Ehrich *et al.*, 2019), they share similarities in mechanisms of population dynamics (i.e. high reproductive rates and density-related mortality; Myers, 2018; Oli, 2019).

An important distinction affecting population changes is that between **density-dependent** and **density-independent** factors. Density-dependent factors include negative effects due to competition, diseases and predation (e.g. Hansen *et al.*, 1999; Hixon *et al.*, 2002). In particular, as population density increases, several detrimental processes intensify in their influence, leading to reduced survival (Tanner, 1966).

- Individuals may have less access to resources (Connor *et al.*, 2000; Wilde *et al.*, 2014).
- Individuals may be forced to use resources of lower quality, with consequences for adults and offspring survival (Hecht, 2023).
- Individuals may have to spend more time searching for food. with consequences for mating success in addition to a reduction in parental care (Prevedello *et al.*, 2013).
- Contacts between individuals increase with increasing density, which can lead to pathogens and parasites spreading more easily (Michel *et al.*, 2016).
- High-density populations under constant conditions (e.g. the pond specialist copepod *Lovenula raynerae* preying on a cladoceran prey, *Daphnia pulex*) also attract predators that are able to build up more efficient search images on prey, and learn how to catch their prey with greater ease (Cuthbert *et al.*, 2021).

- At higher densities, catabolites accumulate in the environment which becomes unhealthy (Ramsey *et al.*, 2007; DeLong *et al.*, 2014; Pettersen *et al.*, 2020).
- High density also increases conflict (for example, between males) with potential for early mortality (Knell, 2009).
- In some situations, higher densities can lead to cannibalism (Fisher *et al.*, 2021), even among herbivores, as in the case of the orange tip butterfly *Anthocharis cardamines* (Courtney, 1980).

Density-dependent factors are associated with predictable agents and thus influence **population regulation** (e.g. classically, in plants the **self-thinning rule** applying to conifer regeneration; Westoby, 1984). Density-independent factors are typically associated with unpredictable changes that affect population growth rate and size (e.g. changes in climatic parameters, storms, fires, volcanic eruptions), with a tendency to cause extinction. For example, while population size is typically related to food supply, and hence to habitat quality, frequent **disturbance** may have unpredictable effects on population abundance (Kunegel-Lion *et al.*, 2022).

It has long been debated (especially from about 1950 until circa 1980) whether populations are mainly regulated by internal (density-dependent) or external (density-independent) factors. The importance of density-independent factors was especially championed by Andrewartha and Birch (1954, 1984), who stressed the importance of climate in regulating populations. This point of view was in contrast with the idea, strongly advocated by Nicholson (Nicholson, 1933, 1954, 1957; Nicholson and Bailey, 1935; see Begon *et al.*, 1996; Hixon *et al.*, 2002), that intraspecific competition (i.e. density dependence) always tends to cause populations to reach, and maintain, stable density values. There is now a tendency to consider density dependence virtually universal, being found also in groups (like insects) where density-independent factors were traditionally considered to be important (e.g. Turchin, 1995; Sale and Tolimieri, 2000; Brook and Bradshaw, 2006; Miller, 2007; Nowicki *et al.*, 2009).

The former emphasis placed on density-independent factors for certain organisms was partially due to the fact that the role of density-dependent factors can be difficult to detect, but they are still operating, in natural populations, and even irregular (chaotic) patterns can be the result of density-dependent processes instead of external factors (see Appendix 2). However, it should be noted that the debate was largely due to some misunderstanding of Andrewartha and Birch's (1954) work, due to use of a formulation for the logistic equation (eq. A2.9 in Appendix 2) which obscures the compound nature of '**carrying capacity**' (K) as a population equilibrium (Mallet, 2012). In fact, Andrewartha and Birch (1954) did not deny the existence and importance of density-dependent factors, but criticised the use of terms like 'regulation', 'balance' and 'control' when applied to density, arguing that equilibrium densities (K) may not be stable but very variable, due largely to climate-induced mortality affecting the parameter r. Actually, in contrast to the common assumption that K is a fixed value, it expresses an equilibrium density as the ratio between r and crowding (see Appendix 2). Thus, the argument that densities are largely determined by density-independent factors is true, in the sense that environmental conditions (e.g. unfavourable climate) may reduce equilibrium densities. In fact, density-dependent factors interact in non-trivial ways with density-independent factors, leading to temporal fluctuations in population abundance/density (Mutshinda *et al.*, 2023). Changes in external conditions are not always unpredictable but may be cyclical, and populations can respond to these changes in various ways. The most predictable pressures on a population occur during extremes of seasonal conditions (too cold or hot, too dry or wet) when all individuals in a population attempt to adopt escape mechanisms such as migration, dormancy or hibernation (diapause) (Milner-Gulland *et al.*, 2011; Hodek, 2012). Cases in which all members of a population permanently or temporarily leave what many describe as their traditional habitats call into question the integrity and cohesion of a population, that is a 'population' as envisaged as a geographical 'individual' occupying a single habitat type (Harvey, 1969).

Niche and organism interactions

The magnitude (size) of a population in an ecosystem or habitat space is one part of its potential impact on other species and the environment. The key concept is the role an organism plays in an ecosystem, its **niche**. The term 'niche' has a long history, being already understood *ante litteram* by pioneers of ecological thinking, such as von Humboldt, Darwin and Haeckel (e.g. Pocheville, 2015; Wulf, 2015). Its development has been accompanied by what has become a confusion of definitions, such as those of Grinnell (1917, 1924, 1928, who conceived the niche as the complex of biotic and abiotic factors that allows the presence of a species), Elton (1927, 1966, who concentrated more on the role, especially the trophic position, of a species in a **community** (i.e. an assemblage of coexisting species; see Chapter 3, Box 3.3), although – contrary to common thought – his view of the niche was not so different from that of Grinnell; see Schoener, 1989a), Kostitzin (1935, who envisaged the niche as the hypervolume occupied by a species in the multidimensional space – i.e. the hyperspace – defined by the various factors that influence its existence, a concept analogous to that formalised and popularised by Hutchinson, 1957) (Fig. 2.8), or MacArthur and Levins (1964, 1967, who focused on the frequency of utilisation of a resource) (see Schoener, 1989a; Pocheville, 2015; Sales *et al.*, 2021, for reviews of niche concepts and their history). The controversies over which definition should be applied have been such that one researcher thought it best that the term not be defined at all (Bell, 1982).

In fact, the confusion has been caused more by what we expect of the term rather than what is basically required: a simple description of a species' relationships and interactions within an ecosystem. Niche is closely tied to ecosystem interactions already discussed above and to the basic components that

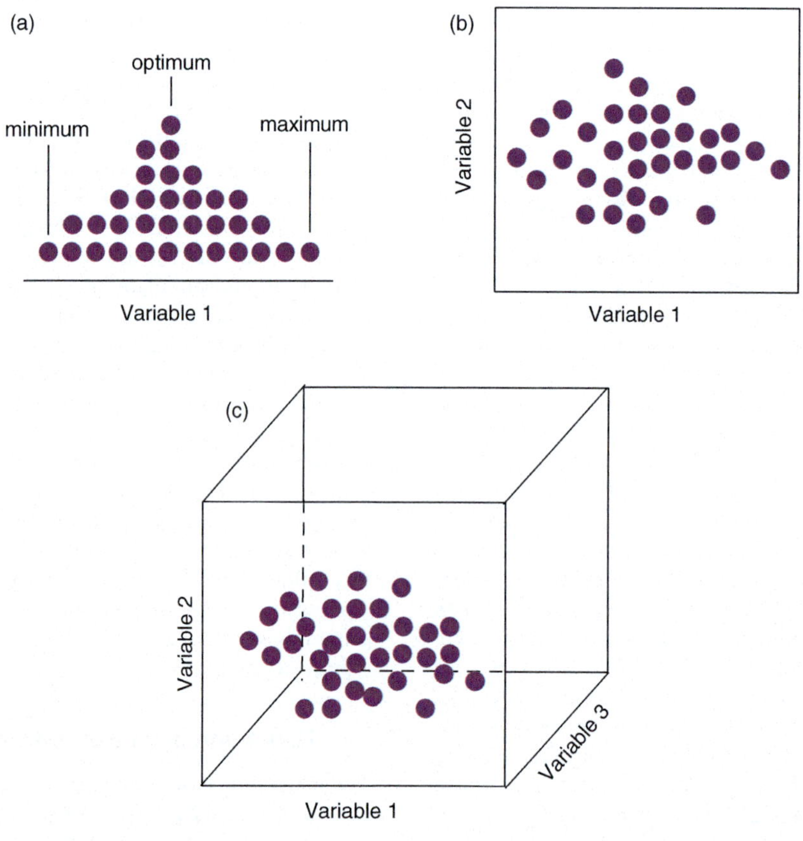

Fig. 2.8. The ecological niche as a hypervolume. For a certain number of individuals of a given species (n = 33 in the figure), preference for a given environmental variable is recorded. Panel **(a)** illustrates how these individuals are distributed along the gradient represented by this variable, from a minimum to a maximum value, with most of them around the optimum. If the same individuals are measured also for a second variable, each individual will occupy a point in the bidimensional space defined by the values recorded for these two variables **(b)**. If a third variable is added, each individual will occupy a specified position in a three-dimensional space **(c)**. If fourth, fifth, sixtieth, etc. variables are introduced, a n-dimensional space is constructed, and the position of the individuals will define the niche hypervolume.

make up an organism's habitat in the sense of the requirements that make space habitable for the species (cf. the **resource-based habitat**; see Chapter 5). Although there have been successors to Hutchinson's formulation of niche (MacArthur and Levins, 1967; Chase and Leibold, 2003; Soberón and Peterson, 2005), here niche is defined in the basic sense of Hutchinson (1957, 1978) but with important provisos: without any implications for causes or effects underlying either interactions within an ecosystem or for the statistical distribution of biodiversity observed within ecosystems and habitat spaces.

A species' niche needs to be viewed from two perspectives: first, in a web of interactions as a node (species) linked to others in a web; and second, in axes for different abiotic and biotic variables, with individuals of different species displayed as points (see Box 2.1 and Fig. 2.8). In this way, Hutchinson (1957) envisaged a geometric definition of species occupying a space (zone) within axes for ≥2 niche variables. When two variables are independent (not correlated), the axes are at right angles to one another and the areas occupied by the species will be rectangular. When correlated, the angle between the axes is <90° (Pearson $r = \cos\theta$, the angle subtended by the axes). With an increase in the measured variables on individuals – and any organism is dependent on a great number of conditions and resources – an *n*-dimensional hypervolume is conceived, in which every point corresponds to an 'individual' (depending on the resolution of focus, an actual individual organism or a local population of a species, the mean for individuals) in the so-called hyperspace of *n* variables or their reduction (combination) on ordination (e.g. by principal component analysis or factor analysis; see Legendre and Legendre, 2012 for description of these methods). The hyperspace is therefore a geometrical concept that simultaneously describes the range of conditions used by a species within a habitat and the distribution of individuals according to variation in those conditions.

Hutchinson's original model conceived competition as a driving force of relationships. His model preceded concepts of stochastic movements and extinctions to local populations embedded in **metapopulation** models (where the overall population, that is the metapopulation, is composed of interacting local populations subject to immigration, colonisation and extinction; Hanski and Gilpin, 1997; see Chapter 7) and **neutral theory** (which postulates that species assembly is driven by random processes; Hubbell and Foster, 1986; Hubbell, 2001; see next section, *Neutral theory, interaction theory and competitive exclusion: extremes in a continuum*). As it is possible to plot individuals from *S* species within the hyperspace, the geometry (distances between individuals) enables direct comparisons to be tested between local populations belonging to different species, for instance using discriminant analyses (see Legendre and Legendre, 2012 for description of this type of multivariate analysis). Hutchinson also distinguished a **fundamental niche** space (a species' ecological properties on the *n* variables) from a **realised niche** (the niche space actually occupied by a species owing to competition, predation, emigration, random effects, etc.). In general, the realised niche is smaller than the fundamental niche, for example because of the effects of **interspecific competition** (see Appendix 2): a species cannot use a given resource that is part of its fundamental niche when it is exploited by a superior competitor. However, the realised niche may be larger than the fundamental niche when a species benefits from the presence of another species (Bruno *et al.*, 2003; Bulleri *et al.*, 2016).

Most authors distinguish two kinds of positive (i.e. beneficial) interaction: **facilitation** (when the presence of one species (facilitator) modifies the environment in a way that enhances growth, survival or reproduction of another species) and **mutualism** (reciprocally positive interactions between pairs of species), but some use the word 'facilitation' in a broader way and consider mutualism as a specific case of facilitation (Stachowicz, 2001; Bronstein, 2009; McIntire and Fajardo, 2014). Since ecological and evolutionary theory has been long dominated by the study of negative interactions (especially competition), the importance of positive interactions has received widespread attention only in the past few decades (Bruno *et al.*, 2003; McIntire and Fajardo, 2014; Bulleri *et al.*, 2016). This is possibly a consequence of cultural biases due to the role of competition in economy thinking (although the idea that there was an overemphasis on competition in Darwin and/or among contemporary Occidental naturalists is probably unfounded; Raerinne, 2023), and it is not a coincidence that one of the first authors who stressed the importance of positive interactions in evolution was the anarchist communist Pyotr A. Kropotkin (1842–1921); however, it should be noted that he did not focus on a critique of interspecific competition but on the importance of positive interactions between conspecifics in opposition to the role assigned to intraspecific competition (Gould, 1988, 2002; Holt, 2009a; Hale, 2014; Pantel *et al.*, 2023).

It should be pointed out that, herein, Hutchinson's original concept of hyperspace is simply used as a descriptive device for the placement of individuals and species. There are two basic components of the hyperspace to consider: (i) the abiotic environment (conditions used or avoided by individuals) and (ii) the biotic environment (i.e. resources exploited by individuals). Often, individual **traits** (i.e. an individual's phenotype, genotype, physiology, life history and behaviour) are included as components of the biotic environment. It should be pointed out, contrary to the expectations of Hutchinson faced with fewer competitive interactions than predicted by his niche theory, that there is no fixed boundary to the hyperspace for the variables describing placement of individuals belonging to different species. Dimensionality is fluid inasmuch as new abiotic and biotic resources are continually being discovered (or compartmentalised) for species as well as novel traits describing the anatomy, genetics, physiology, life history and behaviour of species in the habitat and/or additional species included in an analysis. In fact, by simply adding more species to a community, more trait space is created, particularly if new resources are added for a novel species unshared by other members of the community (e.g. for a parasitic species specialising on a recent colonist). As it is, any technique of ordination alone will distort the original hyperspace. It is important to appreciate that niche is an 'abstract space' but one that has been transferred from measurements on individuals from locations where they are actually found (habitat) compared to locations (physical space) where they are not found (**matrix** space) (Fig. 2.9 and Box 2.2; see also Chapter 5). The separation of habitat and niche in this book accords with observations by Whittaker *et al.* (1973). No causal inferences exist; for those, specific hypotheses and their testing are required.

Neutral theory, interaction theory and competitive exclusion: extremes in a continuum

Over the past few decades there has been heated division over whether the patterns emerging in habitat spaces are the result of competition (so-called **niche theory**) or sheer chance (**neutral theory**) (Sales *et al.*, 2021), the products referred to as niche assembly communities and dispersal assembly communities respectively by Hubbell (2001).

In niche theory, species can coexist only if they have a low **niche overlap** (i.e. if they share small parts of their niche space with each other; see also Chapter 4). A high niche overlap for resources is expected to lead to conflicts (with the final outcome being the exclusion of the less efficient competitor; e.g. Giménez Gómez *et al.*, 2018; Pascual-Rico *et al.*, 2020), whereas a low niche overlap should imply lower levels of interaction, thereby allowing sustainable coexistence of species (Koutsidi *et al.*, 2020; Yang and Hui, 2020). In particular, overlap in habitat use has been invoked as a key factor in shaping ecological communities from the small scale (e.g. close-range interspecific interactions; Lombardo and Mjelde, 2014) to the large scale (e.g. species distribution areas, including evolutionary aspects; Holt, 2009b), with results being affected by environmental heterogeneity; for example, for ground beetles of central Asian grasslands, it has been hypothesised that competition may play some role in community structure and biodiversity maintenance in homogeneous grasslands, whereas species assemblages are more driven by the spatial distribution of resources in grasslands which are more fragmented (Tsafack *et al.*, 2021).

There are undoubtedly fine examples of **competitive exclusion** to match the laboratory experiments of Gause (1934) in his two species of *Paramecium* (e.g. the replacement of red squirrels by American grey squirrels in the UK: Gurnell *et al.*, 2004; Sheehy *et al.*, 2018). But even ecologically closely related species with the same habitat preferences can coexist, if interspecific interferences are not more important than possible intraspecific interferences (Den Boer, 1980; Price *et al.*, 2011; Fattorini *et al.*, 2020c), and a vast number of potentially close relationships do not involve competition at all. For example, the butterflies *Anthocharis cardamines* and *Pieris napi* often use the same crucifer larval host plants but different parts (Dennis, 2010). In general, most insect communities seem not to be regulated by interspecific competition and, even when competition has been reported, severe effects, such as competitive exclusion, have not been demonstrated (Wise, 1981; Shorrocks *et al.*, 1984; Niemelä, 1993; Price *et al.*, 2011; Fattorini *et al.*, 2020c).

Neutral theory as conceived by Hubbell (2001) emerged from research into the importance of neutral (non-selective) processes in population

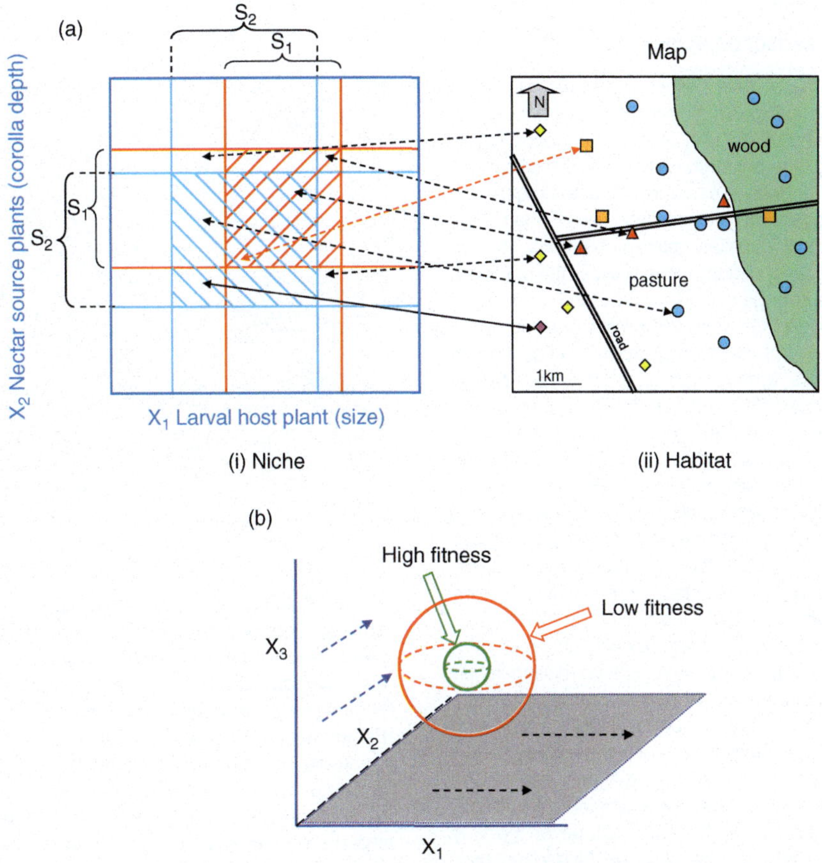

Fig. 2.9. Niche versus habitat. **(a)** Schematic diagrams to illustrate the basic difference between niche and habitat. Fundamental niches for two butterfly species (S_1, S_2) are defined here by just two variables (X_1 and X_2), thus defining a two-dimensional space: (i) niches, and their link to the physical space they occupy in the landscape (ii) habitats. On theoretical grounds, owing to competition, only one species is expected to persist where their potential for survival overlaps. Diagonal shading indicates distinct zones for the two species and cross-hatching where they would overlap, determined from samples taken in the surrounding region. They may actually be found together for a number of reasons (competition not activated). Symbols indicate populations of the species, a triangle for S_1, a circle for S_2 and a square for both species together; a diamond indicates that neither species occurs (pink when this may be owing to chance; yellow where neither the host plant nor the nectar sources are in the suitable range of both species). In the example, two sites are indicated from the niche overlap, one where both species occur and one where only one species occurs. It should be understood that, for simplicity, just two variables are illustrated (here labelled as single measures of host plants and nectar sources to help understanding), but that niche separation envisages many such variables. Such niche separation, and potential competition, typically involves species using the same consumable resources (larval host plants, nectar flowers). However, two butterfly species may use the same host plant individuals and yet not come into any contact (e.g. *Anthocharis cardamines* consume flower pods and *Pieris napi* consume leaves of crucifers). **(b)** Niche shown as an *n*-dimensional hypervolume for just three variables. The placement of species within a niche axis is unlikely to be absolute (unlikely to have the same fitness through the range of the niche variable occupied) but varies and has a frequency distribution with a mean and variance. The larger sphere indicates the zone of the fundamental (potential) niche whereas the smaller sphere indicates a zone of higher fitness (realised niche). (Adapted from Dennis, 2010.)

Box 2.2. Landscape ecology

Landscapes have been defined in various ways. The most popular definition is that of a heterogeneous land area composed of a cluster of interacting ecosystems (Forman and Godron, 1986). Alternative definitions are those of the landscape as a mosaic of heterogeneous landforms, vegetation types and land uses (Urban *et al.*, 1987) or simply as a spatially heterogeneous area (Turner, 1989). Landscape ecology is the study of the structure, function, and change of landscapes (see Farina, 2007, 2022; Turner and Gardner, 2015; With, 2019 for comprehensive treatments of the discipline). Some authors emphasise the broad scale of landscape ecology analyses (Turner, 1989; Forman, 1995), as they involve heterogeneous areas including multiple interacting ecosystems. A fundamental aspect of landscape ecology is its focus on heterogeneity. While other classical ecological disciplines, such as population ecology, community ecology and ecosystem ecology, investigate patterns and processes within homogeneous contexts, landscape ecology concentrates on intrinsically heterogeneous systems (Shreeve and Dennis, 2011).

A landscape includes elements with similar characteristics, called **patches**, that can be distinguished from the **matrix** in which they are interspersed (Fig. B2.2.1). Patches vary in quality in both space and time, as well as in their degree of **connectivity**, which has important consequences on patch interactions and landscape dynamics. For example, the ability of species to move through the landscape or the way disturbances propagate depend on landscape connectivity. Connectivity is influenced not only by the geographical distance between the patches, but also by the qualitative differences between adjacent landscape patches (Acosta *et al.*, 2003), the degree of discontinuity between patches and matrix (that is, by the characteristics of their margins, such as their permeability; see Chapter 1) and the permeability/suitability of the matrix (some matrices can be crossed more easily than others). **Corridors**, i.e. strips of lands that interconnect patches and which are characterised by habitats suitable to the species that inhabit these patches, increase connectivity.

From the topological point of view, landscape features can be divided into two classes: linear features (**edges** or arcs, such as hedges, road verges, river banks, wood margins, walls and roads) and areas (surfaces or faces, such as fields, woods, lakes, moors and urban sprawl). A distinction between a point (**node** or vertex), that is an ideally unidimensional feature, and an area is a matter of spatial scale (nodes are actually small areas). Both edges and surfaces may act as **corridors**, along which movements occur, and **barriers**, where they are prevented (Fig. B2.2.2). Edges and nodes may influence organism movements in various ways, acting as conveyors, barriers, magnets or gateways. Elements acting as conveyors and gateways permit the passage of organisms in transit, whereas barriers cause reflection, deflection and repulsion, and magnets cause attraction and increased length of stay.

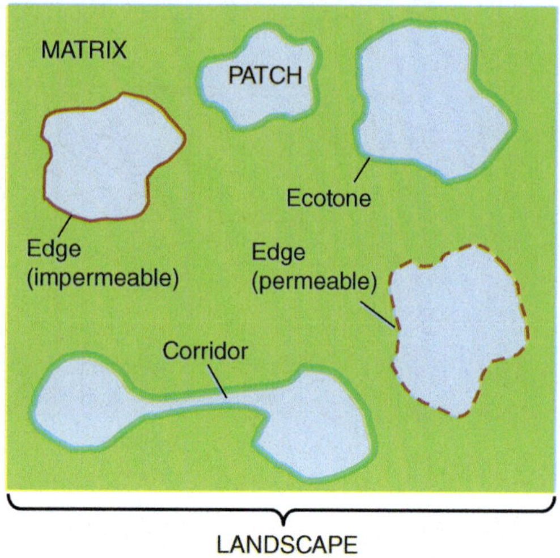

Fig. B2.2.1. Terminology in landscape ecology. The matrix (more or less unsuitable habitat) is in green; patches (suitable habitats) are in blue. Edges are boundaries between different habitat types; ecotones are transitional areas between two habitats. Ecotones can be viewed as wide edges in which communities of the two habitats overlap. The word 'ecotone' is, however, frequently used as a synonym of 'edge' and so ecotones can be both narrow (with sharp environmental changes) or wide (with gradual transition, as defined above). In general, edges may be more or less permeable, also depending on the contrast with the matrix (see Chapter 1).

Landscape ecology is intimately connected with **metapopulation** ecology (see Chapter 7; Dennis 2010) and offers an essential framework for conservation studies, as anthropogenic changes in the structure and function of landscapes affect biological diversity and ecological relationships. For example, as habitat fragmentation, one of the most important causes of biodiversity loss (Fattorini and Borges, 2012b; P.A.V. Borges *et al.*, 2019), transforms previously large areas of continuous habitats into landscapes of suitable patches surrounded by a more or less unsuitable matrix, understanding, maintaining and improving connectivity is a fundamental point of the conservation agenda (Worboys *et al.*, 2010).

Continued

Box 2.2. Continued.

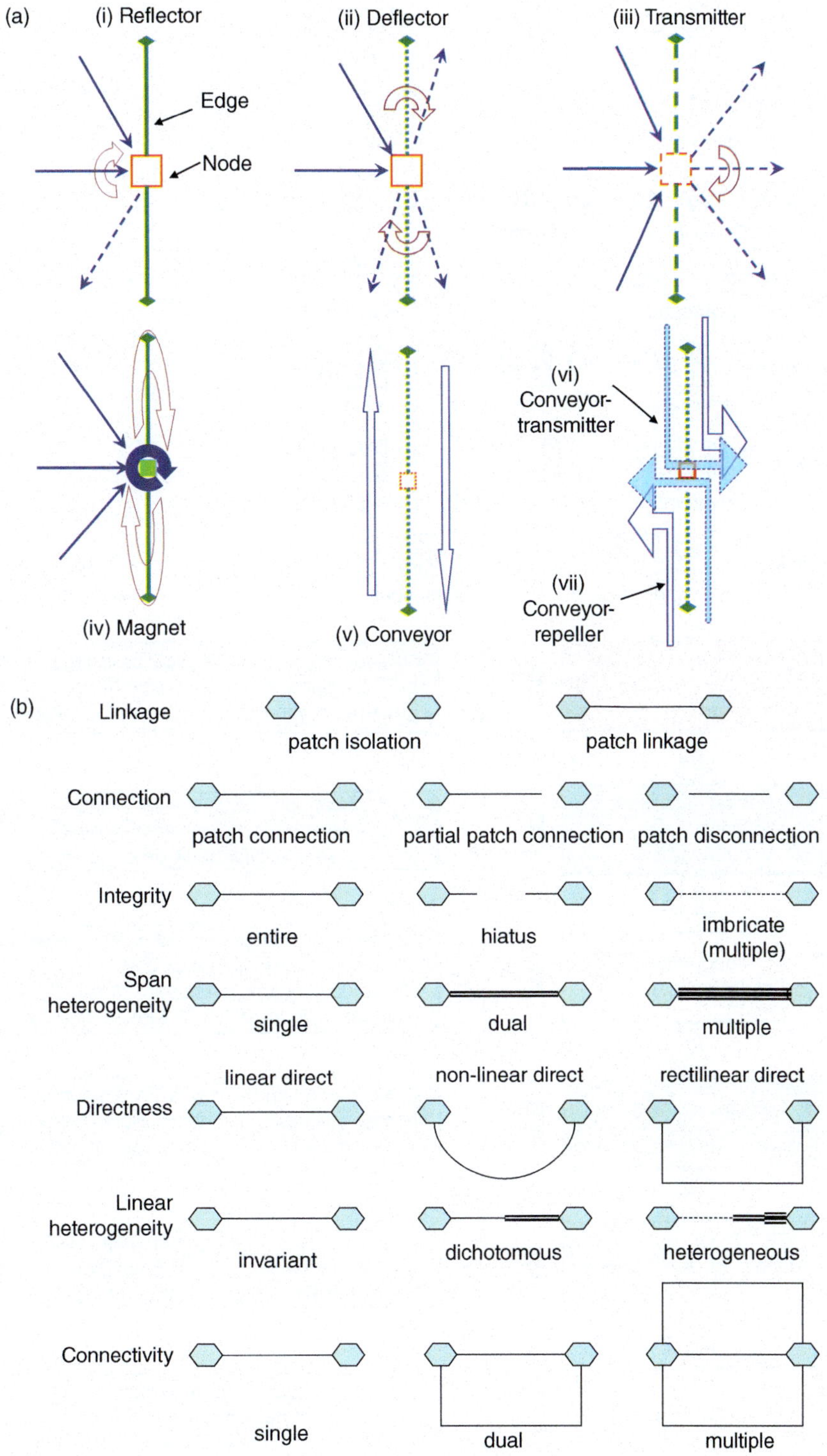

Continued

Box 2.2. Continued.

(c)

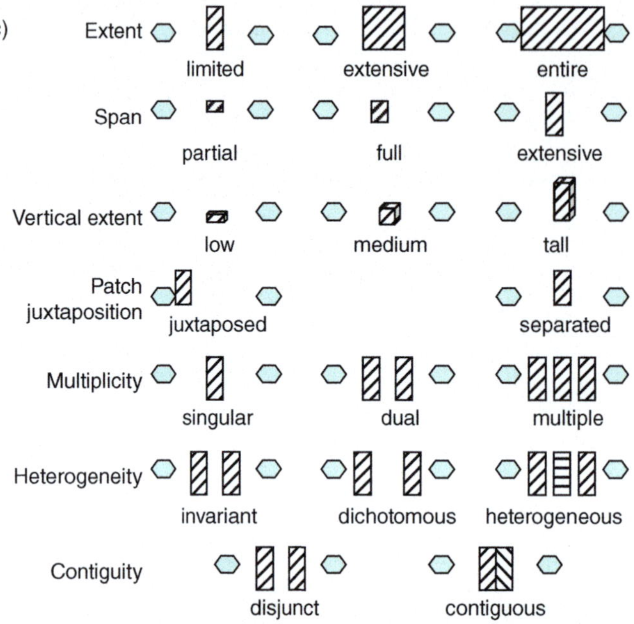

(d)

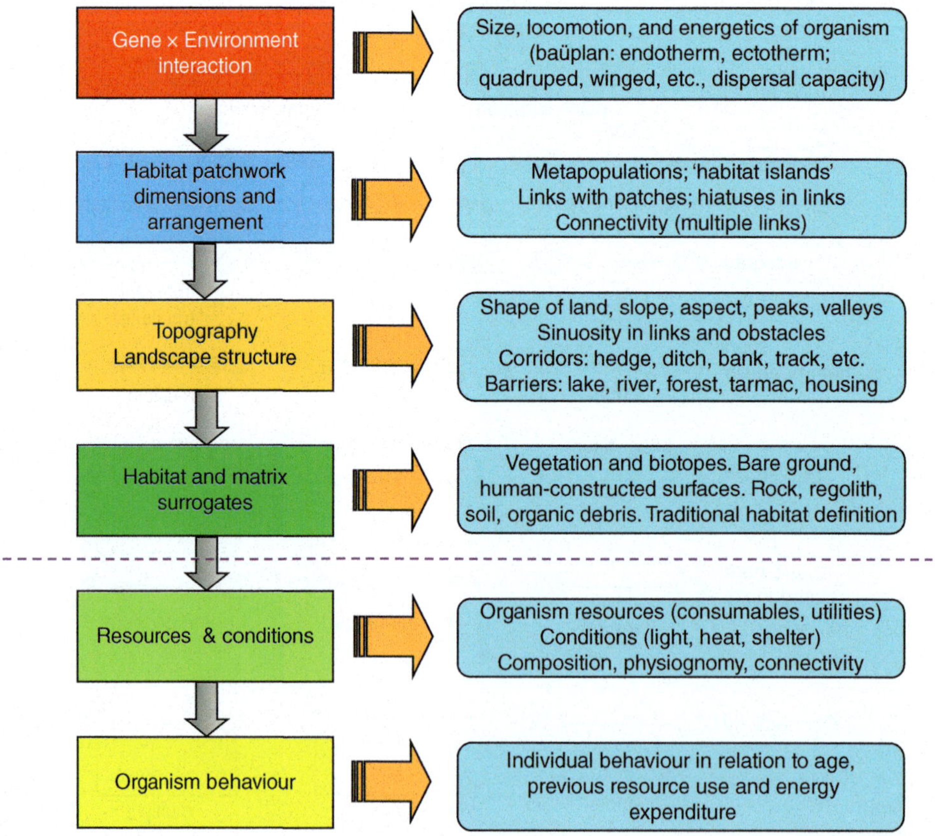

Continued

Box 2.2. Continued.

(e)

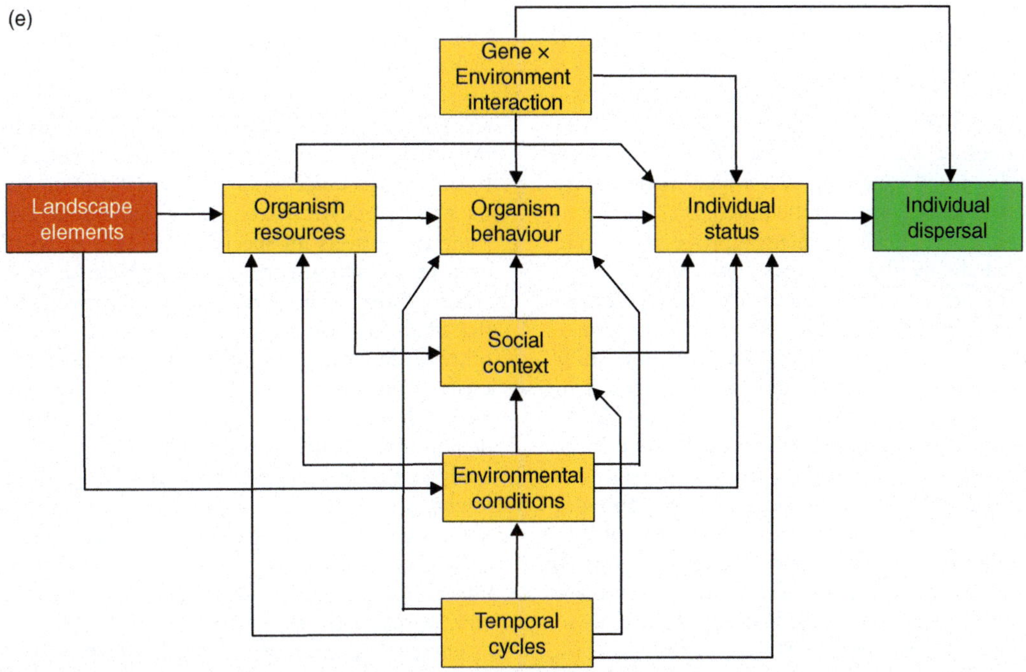

Fig. B2.2.2. Barriers and corridors in the landscape. **(a)** Response of butterflies in transit to nodes and edges. (i) Reflector, entry and return at edge or node greater than critical angle; (ii) deflector, entry greater than critical angle but exit less than critical angle (switch to conveyor) at either node or edge; (iii) transmitter, entry and transmission at node or edge greater than critical angle; (iv) magnet, transit discontinued, remaining locally within node or edge; (v) conveyor, transit along edge (no returns) with movement away from edge less than critical angle; (vi) conveyor-transmitter, disruptor (physical features causing response) causes movement away from edge at greater than critical angle after transmission through edge; (vii) conveyor-repeller, disruptor causes movement away from edge at greater than critical angle. Angles taken from margin of feature at point of interception; critical angle (>22.5°, >2 m of movement away from margin for every 5 m of edge length) taken to be that which indicates an obvious movement away from feature. Barriers are reflectors and conveyor transmitters, and deflectors, conveyor repellers and magnets when movement remains on the same side as that of contact; corridors are conveyors and deflectors; gateways are transmitters and conveyor transmitters where the rate of movement post contact ≥ prior to contact. In v, vi and vii, the node is a disruptor (solid) and non-disruptor (pecked). Some topological and topographic components of corridors **(b)** and barriers **(c)** for a pair of habitat patches. Corridors can take on a variety of basic topologies, on which more complex structures develop. Hexagons, habitat patches and organisms occupying resource unit(s); lines of various forms between adjacent hexagons **(b)** indicate corridor links; black shaded blocks **(c)**, barriers of various kinds. **(d)** 'Layers' of factors and approaches considered in organism transmission between habitat units across matrix. The top four layers are typical factors of varying scales influencing movements through environment and matrix. The lowest two layers add more dimensions to complexity in individual organism transmission across the matrix.
(e) Some of the basic factors taken from **(d)** linked by their logical associations and affecting individual dispersal in the landscape. (Panel **a** from Dennis, 2010, used with permission of John Wiley & Sons. Panels **b** and **c** from Dennis *et al*., 2013, used with permission of Springer Nature BV.)

genetics. The importance of neutral processes in evolution was first illustrated by the role of genetic drift (i.e. changes in allele frequencies due to random chance) shown by Wright (1931), and culminated in the neutral theory of molecular evolution proposed by Kimura (1983; see Leigh, 2007 for a historical review). This envisaged the changes in allele frequencies by way of mutation, demographic stochasticity and migration. Hubbell expanded on this view with a conception of populations in which individuals had identical birth rates, death rates and dispersal, differentiation of communities being dependent on random effects of population size and thus consequently on genetic changes and dispersal rates (Rosindell *et al.*, 2011). Neutral theory, in effect, provides a null model against which niche theory involving interactions can be tested.

In fact, both niche and neutral processes operate within **habitats** and **landscapes** (see Box 2.2), the 'ground' between them described as being a gradient between two extremes of selection pressures associated with competitive exclusion and sheer chance effects leading to differences in species richness and in turn to dispersal capacity (Gravel *et al.*, 2006; cf. **metapopulation** dynamics in Chapter 7). However, although there is a gradient of processes within habitats, there is no single linear continuum between two end points. At one extreme there is certainly Hubbell's neutral theory (random effects). But the other end is not occupied purely by competitive exclusion (see Table 2.1; the –/– cell), as so often portrayed, but by all kinds of interactions, specifically the eight interaction types illustrated in Table 2.1 encircling the 0/0 neutral zone; this is the other extreme to neutral theory, and we may refer to it as **'interaction theory'**, a zone in the hyperspace, its extremities driven primarily by natural selection. It is well to bear in mind that a species is involved in more than one association type with other species under the umbrella of niche theory (natural selection), all inclining towards other extremes from neutral theory (random effects). Importantly, it includes another doubly interactive extreme end point: **mutualistic symbiosis** (e.g. the lycaenid butterfly *Phengaris arion* with the ant *Myrmica sabuleti*: Thomas *et al.*, 1989; the host shrimp *Alpheus bellulus* and its partner goby *Amblyeleotris japonica*: Kohda *et al.*, 2017; corals with microscopic algae and hydrozoans: Muscatine and Porter, 1977; Montano *et al.*, 2017; the + /+ cell in Table 2.1).

In any ecosystem, every species theoretically interacts with every other, through direct or indirect links. Although many of these indirect effects may, in practice, be negligible or difficult to detect, mapping all relevant interactions remains a challenge – particularly in species-rich ecosystems. In such cases, the ecological hyperspace may appear densely filled, with most species exhibiting non-zero interactions. Even examining specific subsets of species and their roles within interaction webs can prove demanding (Fig. 2.10; see also Fig. 2.4). In principle, it is possible to establish the nature of interactions between selected species within such webs. However, the problem remains that there is no single unique solution portraying the relationships in the hyperspace of variables as all ordination methods in multivariate analyses are vulnerable to subjective choice of variables, scales of measurement and analytical techniques. To include everything that affects all species is virtually impossible even in a collapsed space, simply because specific relationships occur (predator–prey, parasite–host) that share no common scale of measurement with other species; the hyperspace at its extreme development is imaginary and Bell (1982) was probably right in not allocating too much responsibility to any definition of niche. But this does not detract from analyses of focal relationships (**guilds**, see the following section *Guilds and functional groups*) that are of interest to researchers, for instance to test the existence of competitive exclusion, as long as it is understood that any analysis is bound to be a partial view of reality (Root, 1973; Colwell, 1979).

There is an additional issue that points to inherent instability in describing the hyperspace. Interaction webs indicate that not only are habitats not stable entities but an organism's niche also changes with space-time, since shifts in abiotic components can cause a cascade of modifications in the ecosystem, consequently altering interactions between organisms (Schmidt *et al.*, 2017). These changes over time are considered in Chapter 3. This has to be expected if only as the logical outcome of modelling competitive exclusion and other interactions. Changes are not just imposed on an organism but are created by its own activities, which then feed back onto the organism itself. Classic examples are described as part of **niche construction theory**, the organisms referred to as **ecological engineers**. This expression is used to indicate the species that directly or indirectly modulate the availability of resources to other species by causing physical state changes in biotic or abiotic materials, and in so doing they modify, maintain and/or create habitats (Jones *et al.*, 1994, 1997; Odling-Smee *et al.*, 2003, 2013). Examples of ecological engineer species include kelps (that build the structure for the kelp forests, thus changing local abiotic processes such as light, sedimentation and water flow and creating complex habitats that support diverse and productive communities; Layton *et al.*, 2019), corals (as their skeleton represents the structure on

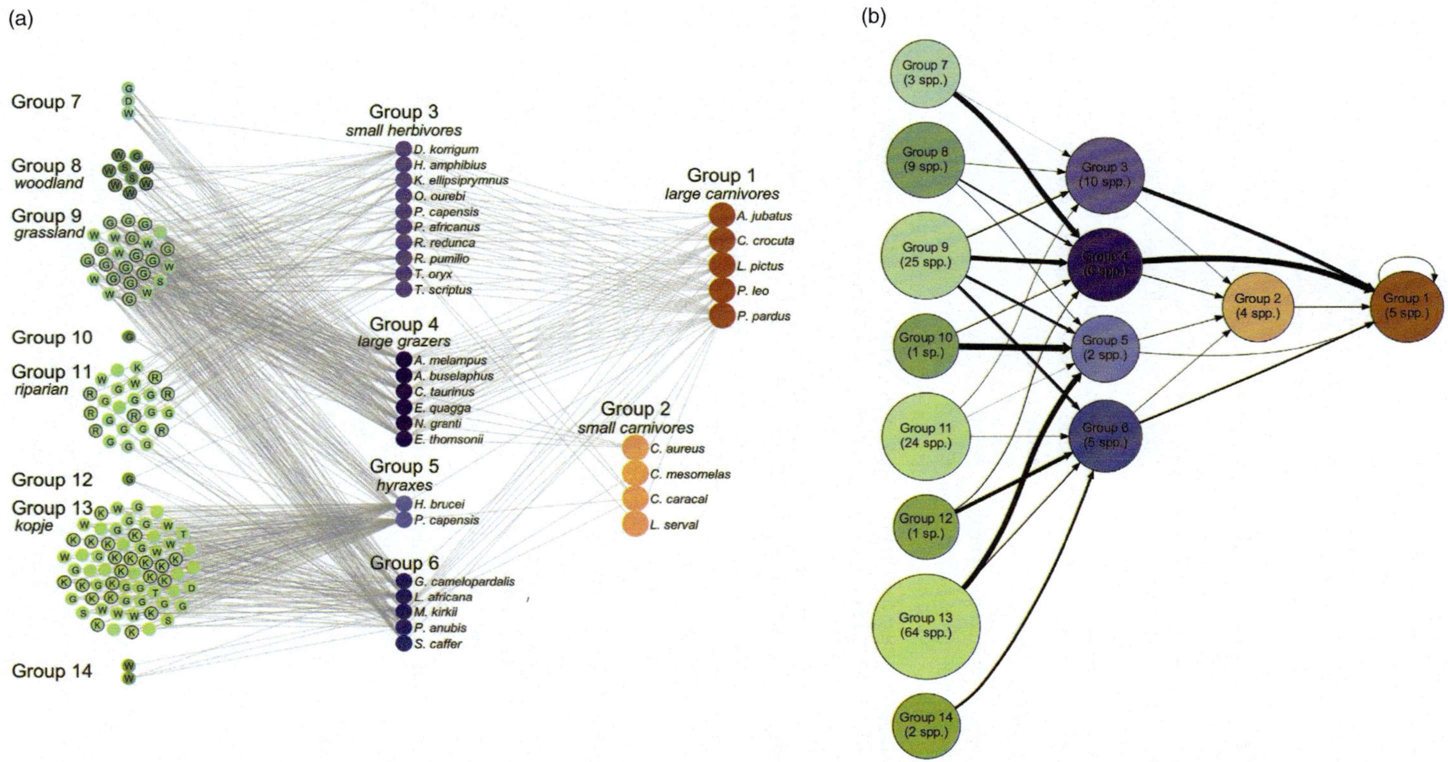

Fig. 2.10. The Serengeti plant–mammal food web. **(a)** The network is shown organised and coloured by group according to a consensus from different partitions (see list in Table 2.2), and arranged by trophic level from left (plants) to right (carnivores). Plants are identified by the first letter of habitat type, if available: (G) grassland, (W) woodland, (R) riparian, (K) kopje, (S) shrubland, (T) thicket, and (D) disturbed. Plant groups are labelled by significantly overrepresented habitat types, and species assigned to the overrepresented type are labelled with black borders. **(b)** Layout of aggregated groups. Nodes in the network are aggregated and coloured by group according to the consensus partition listed in Table 2.2. Line thickness indicates the link density between groups. Node area increases with the number of species in a group. See Table 2.2 for names of species. (From Baskerville *et al.*, 2011, used under licence CC BY 4.0.)

which the entire ecosystem depends; Cheung *et al.*, 2021), beavers (which transform terrestrial ecosystems to aquatic; Brazier *et al.*, 2021), termites (in that they move enormous amounts of matter, rework the soil, construct complex nests, etc.; Jouquet *et al.*, 2019; see also Chapter 8), large herbivores (as they profoundly shape the structure and dynamics of vegetation, even at a landscape scale; Smit and Putman, 2011), moles (as molehills modify the landscape, being for example crucial resources for mate location and thermoregulation for the butterfly *Aglais io*; Dennis and Sparks, 2005). Basically, with their activity, ecological engineers create new habitats for other species.

Guilds and functional groups

When, for a community, species are plotted in niche hyperspace, an important outcome is a non-random collection of points (clusters of means for individuals of species) within *n*-dimensional ellipsoids (variances). The clusters – where demonstrated not to be of stochastic origin – likely describe collections of species that share some important characteristics. These clusters we may call **guilds** and perhaps **functional groups** (see below), the former usually associated more with animals, the latter more with plants (Grime, 1974, 1979; Putman, 1994; Wilson, 1999; Blondel, 2003). A guild is effectively a category in ecosystems if only identified subjectively. Guilds may be hierarchical structures in *n*-hyperspace; a broadly defined guild of species may comprise constituent guilds. To simplify, a guild ranks above a species ecologically, as a genus ranks above a species taxonomically, though they are not **taxocenes** (i.e. taxonomically related sets of coexisting species). The term 'guild' was introduced into animal ecology by Root (1967), during research on the blue-gray gnatcatcher *Polioptila caerulea*, as a group of species that exploit the same class of environmental resources in a similar way (in that specific case, foliage-gleaning birds), and this definition is probably the most popular. Concrete animal examples are collections of insectivorous bird species feeding on oak trees in a habitat space (Simberloff and Dayan, 1991), mammals and birds that use tree cavities (Warakai *et al.*, 2013), carnivore and carcass-feeding species (Van Valkenburgh, 1989; Moleón *et al.*, 2015) and organisms feeding on earthworms (Macdonald, 1983a). Bear in mind that issue has been taken with the phrase 'in a similar way', as this could refer to the mechanisms of exploitation (e.g. the way an insect is found and caught by different insectivorous birds) or to the forms of exploitation (e.g. feeding, nesting, roosting; see references in Korňana and Kropil, 2014).

A great many studies have been carried out in which the term 'guild' is used and they give some idea of the immense variation applied to the term (Simberloff and Dayan, 1991). There is also some tendency in using the word 'guild' in a very broad sense (e.g. as groups of species that are similar in some way that is, or might be, ecologically relevant; Wilson, 1999). It is such looser uses of the word 'guild' that may lead to equating the guild concept with that of functional groups (Duckworth *et al.*, 2000). However, while organisms belonging to the same guild (in the original sense described above) typically share some functional traits (and hence belong to the same functional group), it is important to note that functional classifications can be based on traits unrelated to resource use. For example, the widely used Raunkiær's (1905) scheme classifies plants into functional groups (**life forms**) according to the protection strategies during the unfavourable season and the height of the renewable buds in relation to the soil surface (Di Biase *et al.*, 2021b), characteristics that can hardly be related to habitat exploitation. Although originally introduced in plant ecology (Wilson, 1999), some authors doubt that the concept of guild, if strictly intended as based on resource use, can be properly applied to plants, since plants only require a few resources (viz. water, light and nutrients) (Duckworth *et al.*, 2000). However, the way these resources are used can vary greatly among different plant species, which allows the possibility of recognising plant guilds consistent with a resource-based definition (de Kroon and Olff, 1995). For example, plants of the same height use light in a similar way which is different from that of plants belonging to other strata; plants with similar root systems use water differently from plants with different roots; nutrients can be used differently according to soil depth; and so on. Owing to the complexity of resource exploitation, the temptation has been to subdivide the category of guild into several types and subtypes, for instance exploitation of abiotic factors such as shade or temperatures (β guilds) and biotic factors such as resources (α guilds) (Wilson, 1999). However, it is more useful to define 'guild' in each case and to identify the actual interactions and functions, as this leads on to the scrutiny of traits and life-history strategies by different members of the same and different guilds. Consequently, researchers need to make their definitions of 'guild' clear.

Even under Root's definition, there can be important taxonomic distinctions in the selected or identified clusters. A guild may include anything from a closely related group of species (e.g. violet-feeding fritillary butterflies of the American tallgrass prairies; Swengel, 2017), through more distantly related species, as in the case of insectivorous birds feeding on the same stand of oaks (Almeida and Granadeiro, 2000), to animals from distinct clades such as carnivores operating in the Serengeti (Craft *et al.*, 2015). Consequently, they have also been determined at different spatial scales (i.e. small-scale habitats versus biomes); the larger the scale, the more likely it is that components of guilds display some spatial structure, that is, establishing relationships between guilds and habitats (e.g. Serengeti; Baskerville *et al.*, 2011) (Table 2.2; see also Fig. 2.10). Often multiple guilds are exposed

Table 2.2. Groups identified in the Serengeti food web based on a consensus from different partitions.

Group 1	*Acinonyx jubatus*, *Crocuta crocuta*, *Lycaon pictus*, *Panthera leo*, *Panthera pardus*
Group 2	*Canis aureus*, *Canis mesomelas*, *Caracal caracal*, *Leptailurus serval*
Group 3	*Damaliscus korrigum*, *Hippopotamus amphibius*, *Kobus ellipsiprymnus*, *Ourebia ourebi*, *Pedetes capensis*, *Phacochoerus africanus*, *Redunca redunca*, *Rhabdomys pumilio*, *Taurotragus oryx*, *Tragelaphus scriptus*
Group 4	*Aepyceros melampus*, *Alcelaphus buselaphus*, *Connochaetes taurinus*, *Equus quagga*, *Nanger granti*, *Eudorcas thomsonii*
Group 5	*Heterohyrax brucei*, *Procavia capensis*
Group 6	*Giraffa camelopardalis*, *Loxodonta africana*, *Madoqua kirkii*, *Papio anubis*, *Syncerus caffer*
Group 7	*Digitaria scalarum*, *Dinebra retroflexa*, *Hyparrhenia rufa*
Group 8	*Chloris gayana*, *Combretum molle*, *Digitaria diagonalis*, *Duosperma kilimandscharica*, *Eragrostis cilianensis*, *Microchloa kunthii*, *Sporobolus festivus*, *Sporobolus fimbriatus*, *Sporobolus spicatus*
Group 9	*Vachellia tortilis*, *Andropogon greenwayi*, *Aristida* spp., *Balanites aegyptiaca*, *Bothriochloa insculpta*, *Brachiaria semiundulata*, *Croton macrostachyus*, *Cynodon dactylon*, *Digitaria macroblephara*, *Eragrostis tenuifolia*, *Eustachys paspaloides*, *Grewia bicolor*, *Harpachne schimperi*, *Heteropogon contortus*, *Hibiscus* spp., *Hyparrhenia filipendula*, *Indigofera hochstetteri*, *Panicum coloratum*, *Panicum maximum*, *Pennisetum mezianum*, *Sida* spp., *Solanum incanum*, *Sporobolus ioclados*, *Sporobolus pyramidalis*, *Themeda triandra*
Group 10	*Pennisetum stramineum*
Group 11	*Vachellia seyal*, *Vachellia xanthophloea*, *Andropogon schirensis*, *Chloris pycnothrix*, *Chloris roxburghiana*, *Crotalaria spinosa*, *Cymbopogon excavatus*, *Digitaria milanjiana*, *Digitaria ternata*, *Echinochloa haploclada*, *Eragrostis exasperata*, *Euphorbia candelabrum*, *Hyperthelia dissoluta*, *Kigelia africana*, *Lonchocarpus eriocalyx*, *Olea* spp., *Panicum deustum*, *Panicum repens*, *Phragmites mauritianus*, *Psilolemma jaegeri*, *Sarga versicolor*, *Setaria pallidefusca*, *Setaria sphacelata*, *Typha capensis*
Group 12	*Senegalia senegal*
Group 13	*Abutilon* spp., *Acalypha fruticosa*, *Vachellia robusta*, *Achyranthes aspera*, *Albizia harveyi*, *Albuca* spp., *Allophylus rubifolius*, *Aloe macrosiphon*, *Aloe secundiflora*, *Blepharis acanthodioides*, *Capparis tomentosa*, *Pennisetum ciliare*, *Cissus quadrangularis*, *Cissus rotundifolia*, *Commelina africana*, *Commiphora merkeri*, *Commiphora schimperi*, *Cordia ovalis*, *Croton dichogamus*, *Cyperus* spp., *Cyphostemma* spp., *Digitaria velutina*, *Diheteropogon amplectens*, *Emilia coccinea*, *Eragrostis aspera*, *Eriochloa nubica*, *Ficus glumosa*, *Ficus ingens*, *Ficus thonningii*, *Grewia fallax*, *Grewia trichocarpa*, *Heliotropium steudneri*, *Hibiscus lunariifolius*, *Hoslundia opposita*, *Hypoestes forskaolii*, *Iboza* spp., *Indigofera basiflora*, *Ipomoea obscura*, *Jasminum* spp., *Kalanchoe* spp., *Kedrostis foetidissima*, *Kyllinga nervosa*, *Lippia ukambensis*, *Maerua cafra*, *Ocimum* spp., *Pappea capensis*, *Pavetta assimilis*, *Pavonia patens*, *Pellaea calomelanos*, *Phyllanthus sepialis*, *Pupalia lappacea*, *Rhoicissus revoilii*, *Sclerocarya birrea*, *Senna didymobotrya*, *Sansevieria ehrenbergii*, *Sansevieria suffruticosa*, *Solanum dennekense*, *Solanum nigrum*, *Sporobolus pellucidus*, *Sporobolus stapfianus*, *Tricholaena teneriffae*, *Turraea fischeri*, *Ximenia caffra*, *Ziziphus* spp.
Group 14	*Boscia augustifolia*, *Commiphora trothae*

See Fig. 2.10. (From Baskerville *et al.*, 2011; used under licence CC BY 4.0.)

as in the case of seven predator–prey marine guilds identified from a study of stomach contents in North Sea fish (Thompson *et al.*, 2020). Then, some guilds are the consequence of extensive multifunctional analyses, as in the case of the Australian rainforest frogs that have been classified into nine ecological guilds based on reproduction, habitat use, temporal activity, microhabitat and body size (Williams and Hero, 1998). Thus, the dimension and hierarchical organisation of the guilds will depend on the breadth of the resources and taxa considered.

The expectation of analyses of guilds is **competitive exclusion** and/or **resource partitioning**, that is, the subdivision of limited resources to avoid competition. But there are numerous ways in which a resource base, abiotic or biotic regardless, may not be limiting, as, for example, in the case of pierid butterflies whose larvae feed on different parts of crucifers, as well as on different crucifer species in the same or different parts of a habitat space (Dennis, 2010). Bonsall *et al.* (2004) show how continuously varying life histories and trade-offs in characters (traits) and evolutionary processes can allow multiple competitors to coexist; these attributes and adaptations facilitate flexible dimensions of niche on any hyperdimensional axis and thus shape ecological communities. This observation warns against assuming that resource partitioning is a signal of competitive exclusion operating within a habitat, when in fact it may be the product of cumulative colonisation on sites of species whose specific adaptations to a resource type have deep geological heritage. For instance, Gilbert (1985) revealed extensive resource partitioning for feeding, and traits associated with feeding, in adult British hoverfly species (Diptera: Syrphidae); however, although this may be associated with some competition between different species, their basic mouthpart traits and resource exploitation have largely unknown space-time dimensions. The same is the case with adult and larval traits and resource exploitation in British butterflies, which the authors consider to be of long standing (Dennis and Shreeve, 1988; Shreeve *et al.*, 2001; Dennis *et al.*, 2000, 2004). Although a particular habitat space may be the current arena for a community of related or unrelated species, the origins of adaptations may belong to unknown habitat spaces in geological space and time.

The most favourable locations where current factors and historical factors may be juxtaposed, and where evidence exists for competition leading to resource partitioning and morphological evolution, are islands, from which examples abound (Whittaker *et al.*, 2023). Yet, whatever the origin of competition and resource partitioning in a habitat space, it requires unequivocal demonstration. The most notable examples of competitive exclusion emerge with the invasion of habitats by introduced species (e.g. British bird tree hole nester nuthatches *Sitta europaea* usurped by parakeets *Psittacula krameri*: Peck, 2013; invasive European starlings outcompeting tree cavity nesters in Florida, USA: Diamond and Ross, 2019; the northern Pacific seastar *Asterias amurensis* outcompeting for the plant food of the endangered spotted handfish *Brachionichthys hirsutus* in Tasmania, Australia: Ross, 2001).

Traits and life-history strategies

The assemblage of species as communities within habitats is facilitated by three key features: (i) contrasting life-history strategies, (ii) diverse and distinctive individual traits, and (iii) habitat heterogeneity, part of which is abiotic and part biotic (see Chapter 7). A **life-history strategy** is defined as a grouping of genetic characteristics which recur among populations of species and result in similarities in species' ecology (Grime, 1979). Herein, **traits** describe a wide variety of attributes that enable an organism to respond to environmental pressures; they are typically divided into biological and ecological traits. The former include physical and genetic characters, a species' **bauplan** (body plan), physiological aspects and behaviour; the latter include associations with resources (e.g. specific food types) and conditions (e.g. heat and moisture tolerances). Variations in life history and traits provide the mechanisms for organisms surviving within habitats whereas habitat heterogeneity creates the means for survival, the niche spaces for those organisms.

Fundamental life-history strategies are based on terms in the **logistic equation**, ***r*-strategists** (or *r*-selected species, reminiscent of conditions for maximum growth) being typified by high intrinsic rate of increase but also high juvenile mortality, and ***K*-strategists** (or *K*-selected species, equating with conditions at the carrying capacity of the habitat) by lower productivity and relatively low juvenile mortality. These attributes are closely linked to others that at the extremes contrast species in life history and their traits within habitats (Table 2.3).

Table 2.3. Some basic traits of populations at the extremes of *r*- and *K*-strategies.

Trait	***r*-strategy** ← *r*-*K* continuum	***K*-strategy** →
Environment (successional stage)	Variable, Unpredictable (early stages)	More constant, Predictable (advanced stages)
Mortality	Typically high, density independent	Lower, regular, density dependent
Lifespan	Short	Long
Egg size	Small	Large
Egg number	Numerous	Few
Parental care	None or little	Substantial
Competition	Poor competitor	Good competitor
Population size	Variable	Constant
Survivorship curve	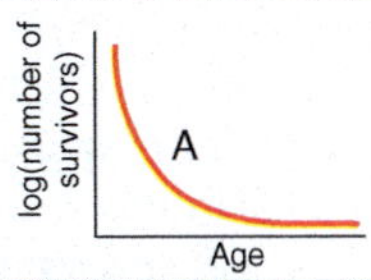	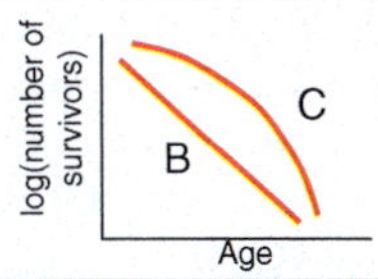

The three survivorship curves are a semi-log plot (log population size (*y*-axis) against time (*x*-axis)): A, High mortality in youth, much less in adults; B, relatively constant mortality rate throughout life; C, substantial survival in youth with increasing mortality with age. (Partially based on Chapman and Reiss, 1999.)

For example, *K*-strategists tend to have fewer offspring but a greater parental investment (e.g. larger eggs, parental care), which reduce juvenile mortality; having a longer lifespan, they tend to have also lower levels of metabolic activity and hence larger body sizes, as all these aspects are interrelated (Southwood, 1976). By contrast, *r*-strategists produce more offspring but have a lower parental investment, and hence a higher juvenile mortality rate; because of their shorter lifespan, they tend to have smaller body sizes (Southwood, 1976).

The formulation of the *r*-*K* selection theory was originally introduced by MacArthur and Wilson (1967) in the context of island colonisation. Early colonists are expected to be *r*-strategists, as in unstable environments highly mobile, opportunistic organisms able to rapidly exploit ephemeral resources are positively selected. By contrast, in more advanced phases of island ecosystem evolution, organisms with more limited dispersal capabilities, but with higher capability to maintain their populations in a resource-limited system approaching the carrying capacity (*K*-strategists), would be favoured. MacArthur and Wilson (1967) elaborated this theory under the assumption that population density may serve as an important selective force on **life-history traits**, i.e. any aspects that influence energy allocation to reproduction and/or fitness. This idea, however, suffers from many problems; not only was it not derived from population models with age structure, and population density is not the only force driving life-history traits (Gotelli, 2008), but it is also based on the misleading idea that *K* may be regarded as a fixed parameter (see Mallet, 2012 and Appendix 2). Based on MacArthur and Wilson's *r-K* selection, Pianka (1970) argued that species could be classified as to whether they fell into *r-* or *K*-selected ecological syndromes, but (as recognised by Pianka himself) it is clear that species occupy a graded scale between these extremes.

Although the original theory of *r-K* selection has been discarded, this terminology is still largely used, as it may synthetically express fundamental differences in species' life history in relative terms, since a species may be qualified as either *r-* or *K*-strategist only in comparison with other, allied species. For example, among dung beetles, aphodines can be considered as *r*-strategists compared to scarabaeines (which are therefore considered as *K*-strategists), because scarabaeines typically have more or less complex parental cares (they supply offspring with food by forming a ball of dung that is rolled away from the pat and buried, or by digging chambers more or less directly underneath the pat), while most aphodiines lay eggs within or under the dung mass where they undergo larval development and do not construct nests (Hanski, 1991; Mantoni *et al.*, 2021b; see Fig. 3.10 in Chapter 3). However, if compared with a dung-breeding fly that lays 30–90 eggs (Blanckenhorn *et al.*, 2010), an aphodiine that lays 4–6 eggs (Borghesio and Palestrini, 2002) would be considered a *K*-strategist. Variation along the *r–K* continuum occurs between and within closely related species and higher taxa (e.g. aquatic planktonic crustacean *Daphnia*: Hartnett, 2019; European freshwater fish: Blanck and Lamouroux, 2007; Chinese and Puerto Rican tree seedlings: Umaña *et al.*, 2015; life history in African dragonflies: Deacon *et al.*, 2021; wing span and body mass in arctic fritillaries: Gentile *et al.*, 2021). It is this variation that is worked on by natural selection (e.g. vertebrates: Zamudio *et al.*, 2016; alpine plants: Henn *et al.*, 2018). Habitat characteristics play an essential role in favouring *K*-strategists or *r*-strategists. In general, *K*-strategists are favoured in stable habitats, whereas *r*-strategists are favoured in ephemeral, dynamic habitats (Southwood, 1976). Long lifespans and high parental investments combine well with stable habitats, i.e. when resources are persistent, but under mutable environmental conditions, *K*-strategists are disadvantaged since the habitat might become less favourable during their life for the development of their offspring. In such circumstances, *r*-strategists, which concentrate on a rapid use of resources, are favoured. For these reasons, *r*-strategists tend to be dominant among pioneer species (i.e. during early phases of the **ecological successions**), whereas *K*-strategists are mainly found in more advanced phases (an argument that will be discussed in Chapter 3).

Life-history strategies can be complex, especially for arthropods, many of which pass through distinct metamorphic stages and appear during different times of the year, some in several broods (e.g. Lepidoptera; Scoble, 1992). It would not be surprising if the neophyte to the ways of the natural world mistook adult and larval stages of butterflies as being different organisms. However, the different stages of British butterflies have long been 'mapped' to different seasons and different conditions, including multi-brooded species such as the Adonis blue *Lysandra bellargus*, all initially fully illustrated in colour by Frohawk (1924). Although adults are characteristic of summer time in mid and high latitudes, though not invariably limited to it (e.g. nymphalids such as overwintering adult small tortoiseshells *Aglais urticae* and peacocks *A. io*), the frequency of stage appearance changes with latitude and altitude, and the same stage for different species may occur at different times of the year (Dennis, 1993a). In plant ecology different strategies have been expanded into the **C-S-R model**: competitive, stress tolerant and ruderal strategies (Grime, 1974, 1979; Grime *et al.*, 1988) (Fig. 2.11). These are associated respectively with plants exploiting conditions that are low in stress and disturbance (C), high in stress (S, i.e. shortages of light, water, nutrients and temperature extremes), and high in habitat disturbance (R, i.e. from soil erosion, fire, wind damage, etc.) and have implications for butterfly-plant associations. As conditions vary from place to place even within habitats, each habitat and biome hosts species of different strategies. Moreover, these strategies do not form exclusive clusters; instead species vary continuously between strategies in axes of ternary graphs, as in

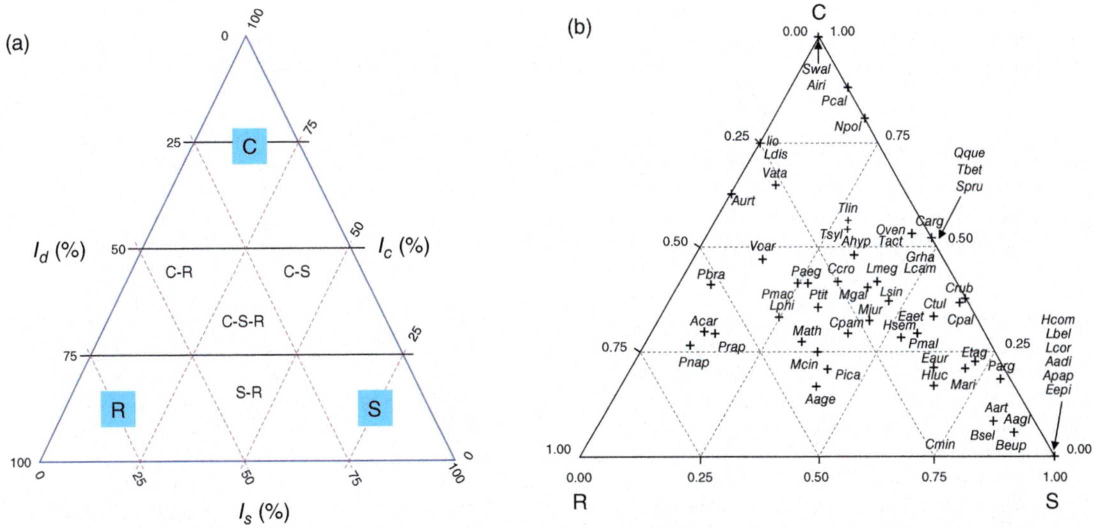

Fig. 2.11. Life-history strategies. **(a)** A model describing the various equilibria between competition, stress and disturbance in vegetation and the location of primary and secondary strategies. Relative importance of competition, stress and disturbance are indicated with *Ic*, *Is* and *Id*, respectively. C competitor, S stress-tolerant, R ruderal, C-R competitive ruderal, S-R stress-tolerant ruderal, C-S stress-tolerant competitor, C-S-R 'C-S-R strategist'. **(b)** Distribution of mean butterfly species scores in ternary axes for C-S-R strategies based on all their host plants; see Table 2.4 for description of traits associated with the three strategies. Abbreviations for species comprise the first letter of the generic (subgeneric) name and the first three letters of the species name: *Carterocephalus palaemon*, *Thymelicus sylvestris*, *Thymelicus lineola*, *Thymelicus acteon*, *Hesperia comma*, *Ochlodes venata* (now *Ochlodes sylvanus*), *Erynnis tages*, *Pyrgus malvae*, *Papilio machaon*, *Leptidea sinapis*, *Colias crocea*, *Gonepteryx rhamni*, *Pieris brassicae*, *Pieris rapae*, *Pieris napi*, *Anthocharis cardamines*, *Callophrys rubi*, *Thecla betulae*, *Neozephyrus* (*Quercusia*) *quercus* (now *Favonius quercus*), *Satyrium w-album*, *Satyrium pruni*, *Lycaena phlaeas*, *Lycaena dispar*, *Cupido minimus*, *Plebejus argus*, *Aricia agestis*, *Aricia artaxerxes*, *Polyommatus icarus*, *Polyommatus* (*Meleageria*) *coridon* (now *Lysandra coridon*), *Polyommatus* (*Meleageria*) *bellargus* (now *Lysandra bellargus*), *Celastrina argiolus*, *Maculinea arion* (now *Phengaris (Maculinea) arion*), *Hamearis lucina*, *Limenitis camilla*, *Apatura iris*, *Vanessa atalanta*, *Vanessa cardui*, *Aglais urticae*, *Nymphalis polychloros*, *Inachis io* (now *Aglais io*), *Polygonia c-album*, *Boloria selene*, *Boloria euphrosyne*, *Argynnis adippe* (now *Fabriciana adippe*), *Argynnis aglaja* (now *Speyeria aglaja*), *Argynnis paphia*, *Euphydryas aurinia*, *Melitaea cinxia*, *Melitaea athalia*, *Pararge aegeria*, *Lasiommata megera*, *Erebia epiphron*, *Erebia aethiops*, *Melanargia galathea*, *Hipparchia semele*, *Pyronia tithonus*, *Maniola jurtina*, *Aphantopus hyperantus*, *Coenonympha pamphilus*, *Coenonympha tullia*. An increased understanding of plant and herbivore strategies is advancing with the development of quantitative measures of both plant and animal attributes. Work to date on butterflies has largely been reliant on Ellenberg's indicator values for plants. These are not entirely suitable. First, because they have not been measured where each butterfly species necessarily occurs. Second, because plants are given single values whereas we know that each is associated with a range of values. Third, butterfly species are often restricted to a narrower band of host plant conditions as in the case of the Scottish chequered skipper *Carterocephalus palaemon* which is biased to more nutrient-rich areas of the purple moor grass *Molinia caerulea*. (Panel **a** from Dennis, 2010. Panel **b** from Dennis *et al.*, 2004, used with permission of John Wiley & Sons.)

the case of UK butterflies classified on their larval host plant associations (Dennis *et al.*, 2004) (Table 2.4; see also Fig. 2.11).

Equally, species' **traits** – physical, physiological, biochemical and behavioural – are extremely varied, with current research focusing ever more intently on their identification with the objective of building data banks on them (Parr *et al.*, 2016). Compilation of trait databases started with plants, but lists are now available for many other organisms from local to global scale (e.g. protists: Giachello *et al.*, 2023; seaweeds: Vranken *et al.*, 2023; lichens: Ellis *et al.*, 2021; plants: Kattge *et al.*, 2011, 2020; Tavşanoğlu and Pausas, 2018; Falster *et al.*, 2021; corals: Madin *et al.*, 2016; ground beetles: Homburg *et al.*, 2014; butterflies: Middleton-Welling *et al.*, 2020;

Table 2.4. Tendencies and comparisons in British butterfly species' biological and geographical traits associated with increasing C-S-R strategy scores.

Biological trait	C	S	R
Wing expanse	Larger	**Smaller**	*Larger*
Proboscis length	Longer	**Shorter**	*Longer*
Oviposition behaviour	*Relatively precise*	**Relatively careless**	*Relatively precise*
Pre-oviposition period	*Relatively longer*	*Various*	**Shorter**
Egg load (total production)	*Various, generally higher*	**Lower**	**Higher**
Egg size	*Relatively larger*	*Relatively smaller*	*Relatively smaller*
Egg batch size	*Relatively larger*	*Relatively smaller*	*Relatively larger*
Larval growth rate	*Relatively faster*	**Slower**	*Relatively faster*
Larval duration	**Shorter**	**Longer**	Relatively shorter
Duration of early stages	*Relatively shorter*	**Longer**	Relatively shorter
Adult life span	**Longer**	Relatively shorter	Relatively shorter
Hibernation stage (egg)	*Several*	*Relatively more*	**Fewer**
Hibernation stage (larva)	Generally less	**Prominent**	*Various*
Hibernation stage (pupa)	*Various*	*Various*	**Increasing**
Hibernation stage (adult)	**Increasing**	**Decreasing**	*Generally decreasing*
Adult hardiness, tolerance of ambient conditions	Decreasing (except overwintering adults)	Increasing	Increasing
Symbiotic (relationship with ants)	Decreasing	**Prominent (Lycaenidae)**	**Decreasing**
Voltinism	*Various*	**Increasingly univoltine**	**Increasingly multivoltine**
Number of host plants	*Various*	**Fewer**	**More**
Mean phagy score (low for monophagy, high for polyphagy)	*Various*	Relatively monophagous	**Relatively polyphagous**
Number of biotopes	*Various*	**Fewer**	**More**
Annual host plants	*Lacking*	**Fewer**	**Increasing**
Biennial host plants	*Lacking*	**Fewer**	**Increasing**
Short-lived perennial host plants	**Generally fewer (exceptions Nymphalidae)**	Generally increasing	**Increasing**
Long-lived perennial host plants	*Usual but various*	*Various*	**Fewer**
Host plant phenology: range	*Generally narrowing*	Narrower	**Increasing**
Nectar sources	*Increasing*	**Fewer**	**Increasing**
Utility resources for pupae and adults	Generally increasing	Fewer	*Various*
Host plant phenology	*Longer-lived plants*	*Longer-lived plants*	**Short-lived plants**
Host plant growth form	Relatively taller and more prominent plants	*Short herbs and shrubs*	**Short plants**
Plant life form association (immature stages)	**Typically tall herbs, shrubs and trees**	**Typically shorter turf to ground substrates**	*Typically tall herbs or below*
Plant life form association (adult stage)	**Taller vegetation, typically trees**	**Shorter vegetation to ground substrates**	**Relatively shorter vegetation**
Minimum habitat space occupied	Relatively larger minimum habitat size	Tendency for minimum habitat size to be smaller	*Various sizes*
Minimum population density	Tendency for typical minimum density to be lower	*Tendency for typical minimum density to be higher*	*Typical minimum densities vary*
Population structure	**Increasingly open structure**	**Increasingly closed structure**	*Open structure expected*
Mobility	**Higher**	**Lower**	**Higher**

Continued

Table 2.4. Continued.

Biological trait	C	S	R
Geographical range	*Various, including large*	**Increasingly restricted**	**Increasingly larger**
Metapopulation type	**Decreasing (none to patchy populations)**	**Increasing (distinctive and typically Levins**; see Chapter 7**)**	*Decreasing (none to patchy populations)*
Distribution cover	Relatively dense cover	**Sparse cover**	**Dense cover**
Incidence on offshore islands	Variable incidence	**Lower incidence**	**Higher incidence**
Change in population status over past 200 years	**Slower losses, persisting and expanding**	**Declining towards extinction**	Slower losses, persisting and expanding
Recent losses to distribution cover	**Relatively fewer**	**Increasing losses**	Relatively few
Conservation status	**Low priority**	**High priority**	**Low priority**

Bold, consistently significant correlations for all and main host plants; normal print, significant for one or other of all and main host plants; *italics*, no significant correlation. See Fig. 2.11.
(From Dennis *et al.*, 2004, used with permission of John Wiley & Sons.)

spiders: Pekár *et al.*, 2021; Mammola *et al.*, 2022; fish: Froese and Pauly, 2023; terrestrial vertebrates: García-Llamas *et al.*, 2019; amphibians: Oliveira *et al.*, 2017; amniotes: Myhrvold *et al.*, 2015; reptiles: Grimm *et al.*, 2014; mammals: Faurby *et al.*, 2018; bats: Cosentino *et al.*, 2023; birds: Tobias, 2022; the Azorean marine biota: Campanyà-Llovet *et al.*, 2023; freshwater macroinvertebrates: EPA, 2012; metacommunity ecology: Jeliazkov *et al.*, 2020; animal energetics: Marques *et al.*, 2018; thermal tolerances for aquatic and terrestrial organisms: Bennett *et al.*, 2018; body mass, metabolic rate and brain size: Herberstein *et al.*, 2022). The objective is increasingly to relate basic life-history strategies and species traits to community functioning and interactions and evolution (e.g. Keddy and Laughlin, 2022). But the key is to build database standards that allow ecological comparisons (Schneider *et al.*, 2019). Many key traits cannot be measured on the dead organisms, those particularly relating to species behaviour and ecology. These may involve strategies for survival, as for instance in the case of eavesdropping among species (listening in to warning and resource-seeking signals), raising awareness of dangers or availability of food and roosts (Stevens, 2013; Goodale *et al.*, 2017; Pallardy, 2021b).

The importance of the study of guilds, traits and life-history strategies to understand ecosystem ecology cannot be overstated. Analyses of ecosystems have advanced in two directions: (i) in complexity (e.g. the recognition of the importance of fungi, bryophytes and bacteria in ecosystem functioning) and (ii) space-time extent. Ecosystems are now not just linked to guilds, traits and life-history strategies but also to evolutionary paths, with convergences and divergences among taxa for adaptations. For instance, using data on traits, phylogeny and range maps of nearly all mammals of the world, Mazel *et al.* (2017) discovered that convergences occur between species among assemblages in different biogeographical regions occupying similar biomes. This approach facilitates predictions, for instance that there should exist marsupial carnivores equating with Old World felids on the Australian continent. Indeed, they once existed into the late Pleistocene (i.e. *Thylacoleo carnifex*; Curry *et al.*, 2014).

Focal species in ecosystems

It is one thing knowing how many species there are in a habitat space (though obtaining reliable estimates can be shown to be no simple matter, especially for inconspicuous organisms in species-rich habitats; see Chapter 4 and Appendix 3). It is another to account for and describe the status of any species in a habitat, that is, its significance to the workings of an ecosystem and its management (say, for conservation purposes) at any one time; extend this issue to changes over time and it becomes even more complex. A number of terms exist for identifying a particular status of one (or few) species within local communities of species. There are **focal species**, cornerstone or **foundation stone species, keystone species, ecological engineers, sentinel species, indicator species, flagship species, umbrella species,** and no doubt many variants of these terms (Box 2.3). They are selected on specific traits (e.g. abundance and/or frequency, trophic significance, physical size, attractiveness, life history, dispersal capabilities). All these concepts are important for conservation and management of sites.

Box 2.3. All species are equal, but some species are more equal than others

A varied terminology is commonly used in conservation biology to refer to those species that are a focus of conservation interest and hence provide the impetus for the conservation of their habitats. These species can be generally indicated as **focal species** as they are the focus of conservation actions. The expression '**indicator species**' is usually used for focal species that are strictly associated with particular communities or ecosystem processes. Preserving a site for indicator species will help to protect the full set of species and ecosystem processes that have the same distribution (Halme *et al.*, 2009). Indicator species may be particularly useful because of differences in detectability among species: monitoring a more detectable (indicator) species may provide indirect information on the conservation status of the associated habitat and hence of other (non-target) species that are more difficult to detect. However, as it is unlikely that a single species may really serve the purpose, the designation of a group of indicator species may better help the protection of a community (Lawton and Gaston, 2001). Identification of indicator species is far from being obvious and usually requires specialised techniques (Indicator Value Analysis (IndVal): Legendre and Legendre, 2012; Market Basket Analysis (MBA): Leote *et al.*, 2020).

Fig. B2.3.1. Charismatic species are frequently used as flagship species. A well-known example is the giant panda (*Ailuropoda melanoleuca*), adopted by the World Wildlife Fund as a symbol for all endangered species and featured in its logo **(a)**. While flagship species have traditionally been large terrestrial mammals, recent years have seen a shift toward considering the potential role of non-mammals – particularly insects – as flagship species. Monarch butterflies (*Danaus plexippus*) **(b)** are a prominent example in insect conservation. Panel **a**: photo by J. Patrick Fischer: https://commons.wikimedia.org/wiki/File:Grosser_Panda.JPG, used under licence CC BY-SA 3.0. Panel **b**: photo by Kenneth Dwain Harrelson: https://commons.wikimedia.org/w/index.php?curid=14917505, used under licence CC BY-SA 3.0.

Flagship species are charismatic organisms that can be particularly useful to attract people's attention and hence to be used as conservation symbols (Fig. B2.3.1). Typical examples are large mammals under severe extinction risk, such as large predators, the African elephant or the giant panda, but increasing attention is also being given to using invertebrates like butterflies as flagships, because of the positive attitude of the general public towards these insects (Fattorini *et al.*, 2023). Protection of indicator and flagship species is expected to have positive effects on entire communities, since their protection automatically extends protection to their habitats and ecosystems, and hence to many other species. For this reason, they are expected to act as **umbrella species** (Ozaki *et al.*, 2006). Umbrella species are selected especially from those which require such extensive resources for their conservation that their protection will extend to many other species by default (Caro and O'Doherty, 1999).

Sentinel species are organisms that reveal the quality of health in their ecosystems by reacting to specific (usually negative) conditions (e.g. by showing population drops, increased deaths, decreased health status or site abandonment) (Aguirre and Tabor, 2004; Tabor and Aguirre, 2004). As for the indicator species, use of suites of sentinel species (including organisms that belong to different trophic levels, have different ecological roles and respond to environmental changes at different spatial and temporal scales) may allow for better environmental monitoring. The concept of sentinel species is therefore strictly associated with that of **bioindicator**. Bioindicators (or biological indicators) are organisms (microorganisms, fungi, animals or plants) used as a tool to monitor changes in environmental health, by recording their presence/absence, changes in abundance or other aspects of their population, changes in behaviour, physiology and morphology, or for the content of certain substances in their bodies. For example, lichens are widely used for monitoring air quality (e.g. Di Biase *et al.*, 2022), soil microarthropods can be indicative of soil quality (e.g. Gardini *et al.*, 2025), birds can be used to evaluate the environmental status of agricultural landscapes (Padoa-Schioppa *et al.*, 2006), and aquatic macroinvertebrates are frequently used to assess water quality (e.g. Sumudumali and Jayawardana, 2021).

Two terms have been claimed as of especial relevance for ecosystems: foundation stone species and keystone species. **Foundation stone species** (in the sense of Ellison *et al.*, 2005) imply effects from the base up, first described by Darwin for kelp in the seas off Tierra del Fuego (Wulf, 2015). The foundation stone is in effect the surface layer of **autotrophs**. Erasing the basal trophic layer would obviously result in the crash of an entire ecosystem; for instance, cut down ancient trees in a monoculture, say, of cork oak and all the hole-nesting species dependent on them are lost (Segura, 2017). There is nothing surprising in this effect. **Keystone species** (as originally defined by Paine, 1969) relate to the analogy of keystones in arches and conversely implicate a top-down effect, where the loss of a higher trophic level species will cause significant imbalances in the ecosystem and the loss of other species. The term **downgrading** (or trophic downgrading) is often applied and is becoming a feature of biomes the world over because of the consequence of the loss of keystone species for entire ecosystems. Keystone species have sequential impacts, a cascade of effects working down the trophic levels (Estes *et al.*, 2011), initially operating on the abundance of one or more other species, and consequently on species diversity. Removal of keystones has potential for massive ecological changes and even ecological collapse (Estes and Palmisano, 1974; Terborgh, 1988; Estes *et al.*, 1998; Macdonald *et al.*, 2013).

The discovery of keystone species, first described by von Humboldt for *Mauritia flexuosa* palms in the Venezuelan Llanos (Wulf, 2015), emerged in several seminal studies on carnivores: (i) initially by Robert T. Paine describing the impact of the starfish *Pisaster ochraceus* on the California mussel *Mytilus californianus*, reducing its abundance and thereby allowing other macroinvertebrates to persist (Paine, 1966; Fig. 2.12); (ii) then by James A. Estes and David O. Duggins (1995) in a three-trophic-level cascade off the Aleutian islands and south-east Alaska, showing the impact of sea otters *Enhydra lutris* on invertebrate herbivores (echinoderms) and their food resource macroalgae (kelp), the abundance of the latter depending on sea otter predation of herbivores; (iii) by Mary E. Power and coworkers (1985), who described a similar freshwater three-trophic relationship, where bass (*Micropterus* spp.) prey on herbivorous minnows *Campostoma anomalum* feeding on river bed algae (Power *et al.* 1985); and most remarkably (iv) by John Terborgh on the impact of the absence of top predators on tropical forest islands in the artificially created Lago Guri in Venezuela, resulting in the chaotic and bizarre dynamics of herbivores and autotrophs (Terborgh, 1988; Terborgh *et al.*, 2001). The keystone paradigm has been extended to herbivores by Tony Sinclair in a study of wildebeest numbers in the Serengeti, where the increase in numbers led to shrub and tree growth and, surprisingly, the increased diversity and abundance of top predators (e.g. Sinclair and Arcese, 1995).

There have been numerous modifications of the original definition (a species whose impact on its

Fig. 2.12. Keystone species. Two of the species studied by Robert T. Paine to determine the effect of keystone species on ecosystem diversity and stability: **(a)** the starfish *Pisaster ochraceus*, and **(b)** the mussel *Mytilus californianus* controlled by it. Paine removed the starfish and documented the impact of the mussel on the ecosystem. The mussels crowd out seven subordinate primary species, which suggests that predation increases diversity by interfering with competition. However, because these mussels are a foundation species, they provide a three-dimensional habitat for over 300 associated species inhabiting the mussel beds; thus, removing sea stars significantly increases community-wide diversity (see Lafferty and Suchanek, 2016). (Panel **a**: photo by D. Gordon E. Robertson: https://commons.wikimedia.org/w/index. php?curid=6434470, used under licence CC BY-SA 3.0. Panel **b**: photo by Sharon Mollerus: https://it.wikipedia.org/wiki/File:California_Mussels_002.jpg, used under licence CC BY 2.0.)

community or ecosystem is disproportionately large relative to its abundance), and a vast range of species has emerged as keystones (see Cottee-Jones and Whittaker, 2012 for examples). The effect on components in ecosystems extends beyond predation (e.g. mutualism: seed dispersal of trees by birds as in the case of acorns by jays; Howe, 1977; Kurek *et al.*, 2019; influence on overall ecosystem functioning (**ecological engineers**): beavers; Naiman *et al.*, 1986; whales; Roman *et al.*, 2014). Numerous studies have been accompanied by modified definitions to the extent that the term has developed multiple applications and, in the view of Cottee-Jones and Whittaker (2012), has become poorly defined, regardless of the efforts of the conference at Hilo, Hawaii, to come to some accommodation in definition (Power and Mills, 1995); in effect it had become a 'panchrestone' (Hardin, 1957; Hurlbert, 1971; Mills *et al.*, 1993). The concept is not simplified by being extended to **keystone structures** and **keystone habitats** (e.g. the impact of trees on biodiversity in savanna landscapes: Tews *et al.*, 2004; *Limaria hians* (Mollusca: Limida: Limidae), a reef-forming keystone species: Hall-Spencer and Moore, 2000) or **keystone communities** (and **keystone ecosystems**) within metacommunities (and ecosystems) (Mouquet *et al.*, 2013). The greatest difficulty has been the measurement of 'disproportionate effects' (see, for example, Jordán, 2009 for a quantitative approach based on species' position in the network of interspecific interactions), a feature of the original definition by Paine, as this can be highly variable (influencing biomass, abundance, diversity, etc.) – and how much larger should an effect be to be classified as disproportionate? Instead, Cottee-Jones and Whittaker (2012) suggest the use of a general definition: 'a keystone species is a species that is of demonstrable importance for ecosystem function'.

An outstanding issue is whether the impact(s) for a particular keystone species is (are) observed in similar communities elsewhere (Lawton, 1999). What is the variation in impact in removing a keystone species on the community occupying a habitat type? More broadly, which, and how many, species can be removed before an ecosystem collapses or enrichment occurs within a habitat space? The chain of effects may be complex; a study of codling moth populations in northern Spanish cider apple orchards raises awareness of both top-down and bottom-up regulation in ecosystems (Martínez-Sastre *et al.*, 2021). From the viewpoint of understanding habitat, the crucial issue is whether such effects are demonstrable at all, that is biologically significant, in habitat spaces; clearly, they can be. In the sense of traditional habitat, theoretically, the loss or gain of a species may be expected to have greater consequences, to lead to the production of a new habitat type during ecosystem evolution (see Chapter 3). Although it is clear that patterns in a habitat space are dependent on regional and landscape processes (the so-called **macroecological viewpoint**; Brown and Maurer, 1989; Lawton, 1999; Lomolino *et al.*, 2010), basic repeatable patterns in habitats (e.g. **species abundance distributions**; see Chapter 4) nevertheless demand that we fully appreciate the processes occurring with the habitat space itself; as later chapters unfold, it is argued that the **resource-based habitat** viewpoint is central to within-habitat patterns and processes (Dennis *et al.*, 2003; Dennis, 2010).

Interconnections in ecosystems: the wood wide web

The above has familiarised the reader with various forms of interspecific interactions that define ecological networks, such as **competition, predation, parasitism, facilitation** and **mutualism**. A classic example is the interaction between the myrmecophyte *Hirtella physophora* (a plant sheltering ants in hollow structures) and other organisms (DeJean *et al.*, 2024). Two *Allomerus* ants build gallery-shaped traps on *Hirtella*. They use hairs from its stems but also the **mycelium** of an ascomycete fungus to capture their prey. They also use their refuse to provide the host plant and the fungus with nutrients (**myrmecotrophy**). Social wasps also build their nests under the *Hirtella* leaves to benefit from the protection provided by *Allomerus* ants against army ants. Fungi are particularly known for forming highly complex relationships with plants. But increasingly topical and of immense importance is a quite different form of interaction represented by the interconnection among plants, messages via 'common mycorrhizal (mycelial) networks' (Fig. 2.13), which Rhodes (2017), Sheldrake (2020) and Hunter (2023) refer to as the 'wood wide web'. In fact, the term 'fungal web' would be more appropriate, as fungi link up more than trees with their mycorrhiza.

Common mycorrhizal (mycelial) networks are continuous networks of mycelia of symbiotic fungi that link up two or more plants of the same or different species under/over ground within the soil and litter surface, and much as any system involve

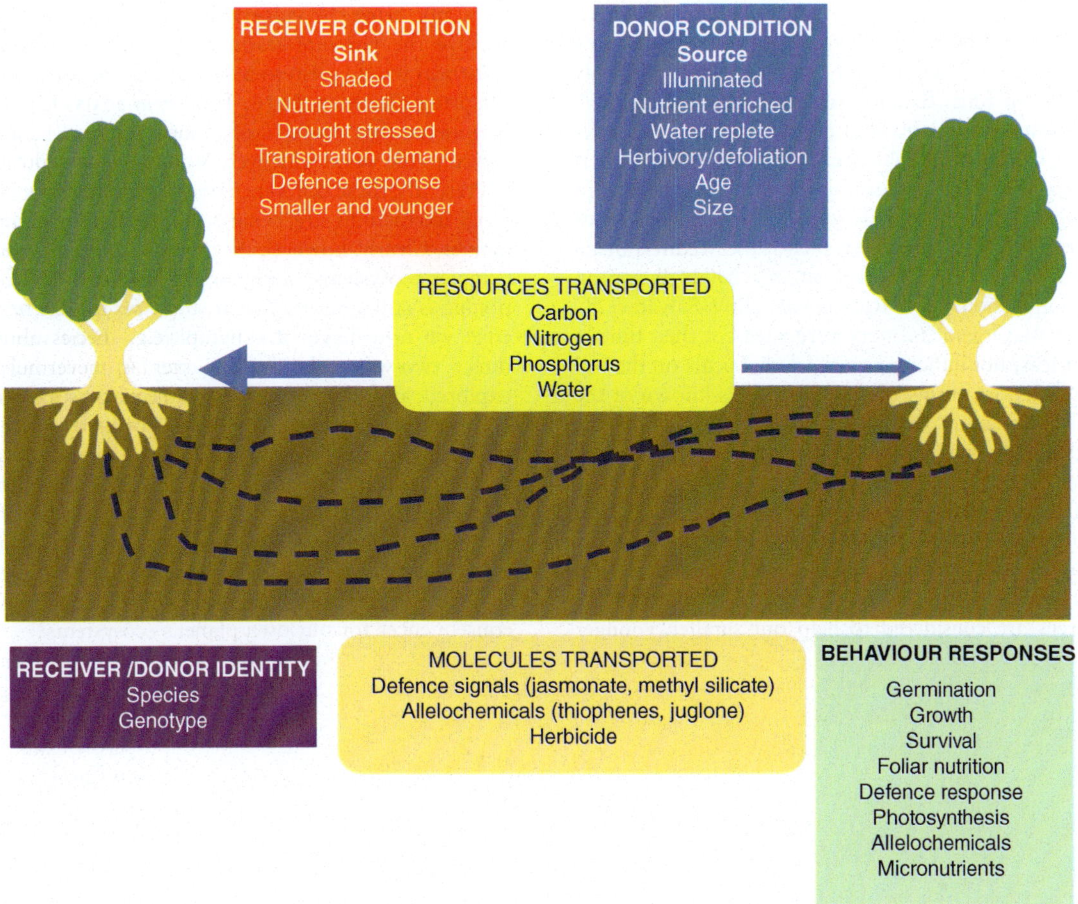

Fig. 2.13. Wood wide web. Basic model of resources and signals noted to travel through a common mycorrhizal (mycelial) network, as well as basic stimuli that assist transfer of molecules in donor and receiver plants. (Redrawn from Gorzelak *et al.*, 2015, used under licence CC BY 4.0.)

complex feedbacks (Gorzelak *et al.*, 2015). **Mycorrhiza** achieve a symbiotic association between the filaments of a fungus and the roots of plants, the mode of transfer undergoing increasing investigation (e.g. electrical signals through mycelium networks; Mayne *et al.*, 2023). The fungus can form its links extracellularly (ectomycorrhizal fungi) or intracellularly (cf. arbuscular mycorrhizal fungi) (Rhodes, 2017). This interconnectedness involves the two-way transfer of nutrients (carbon, phosphorus, nitrogen, water, etc.), and also, in different situations, it involves transfers of defence signals and alert systems, as well as **allelochemicals**. A recent study demonstrated that mycorrhizal fungi build networks as self-regulating travelling waves – pulses of growing tips pull an expanding wave of nutrient-absorbing mycelium, the density of which is self-regulated by fusion (Oyarte Galvez *et al.*, 2025).

Fungi are frequently regarded as the 'Third Kingdom', embracing organisms that were outside the classical dichotomy of animals versus plants. Although likely to be subject to revision, nine phylum-level clades of fungi have recently been recognised: Opisthosporidia, Chytridiomycota, Neocallimastigomycota, Blastocladiomycota, Zoopagomycota, Mucoromycota, Glomeromycota, Basidiomycota and Ascomycota (Naranjo-Ortiz and Gabaldón, 2019). The interconnectedness between plants and fungi is known from deep geological time (e.g. *Aglaophyton major*, Lower Devonian, c. 410 Ma; Remy *et al.*, 1994). Far from being a rare phenomenon, such connections, both positive and negative in effect upon connected plant and fungi, are increasingly recognised to be immense in number and complexity. Yet probably 90% of fungi remain unknown, not documented, despite possibly 90% of plants depending on mycorrhizal fungi (Sheldrake, 2020). Increasingly, it is recognised that disturbing the soil is hazardous, of which anyone familiar with the desertification of soils in long-farmed areas such as the Mediterranean must be aware (Geeson *et al.*, 2002; European Commission, 2003; Rubio *et al.*, 2009; Salvati *et al.*, 2016). Organic farmers were ahead of their time in at least not inflicting artificial chemicals on the soil and its organisms. Currently, little is known of the distribution of various mycorrhizal and soil fungi, especially endophytes compared to ectomycorrhizal and arbuscular fungi (Janowski and Leski, 2022). However, the interdependency of plants and the integrity of soils are in no doubt, nor is the association of particular plants and fungi (*Betula* and fly agaric *Amanita muscaria* mycelia). More mature trees are considered to help out smaller, younger trees, nurturing the vulnerable next generation. To remove the parental standard (e.g. beech, *Fagus sylvatica*) would seem to have consequences for the offspring (Toomey, 2016; Trap *et al.*, 2017). It is considered that 90–95% of all land plants are connected to the wood wide web (Wang and Qiu, 2006).

Presently, these different 'zones' of animal–plant–fungi interactions are being researched as different entities. Clearly, however, they join up in relation to complex ecosystems; our ignorance of their full functioning is still so great that it surpasses the knowledge we now have of other planets. Even more direct two-way relationships are far from fully explored, as the plethora of papers emerging makes clear (e.g. the enhancement of nitrogen by bryophytes in decaying wood via biological interactions makes this clear; Oishi, 2024). The issue is: what will we distribute on other planets from our lack of awareness of what miniscule organisms on Earth may be capable of? What will we inadvertently bring to Earth from other planets that may have serious consequences for our own planet's ecosystems?

3 Habitat Changes Over Time

Abstract

The status of each species and the entire community of species in a habitat space changes over time. Changes in a habitat space vary over a range of time and space scales from seconds to millions of years and from microns to the full scale of the globe. Succession is a key process of all habitat units. The most characteristic and distinctive changes in a terrestrial habitat unit typically occur owing to vegetation succession. Following death of a habitat component (plant or animal), a sequence of bacteria, animals and fungi are involved in its decomposition. Changes to habitats over geological time involve evolutionary transformations as well as shifts in the physical location of the habitat space. Instability in habitat dynamics is most evident to casual human observers with seasonal changes in climate and weather patterns, especially when variations are prominent (e.g. at higher latitudes and altitudes).

Descriptions of habitats may give the impression that any habitat unit is inviolable, unchanging. But this is an artifact of human perception: the gradual changes grow on us and we become used to the 'new' present. This effect is often described in relation to the loss of nature, when it is referred to as the **shifting baseline syndrome** (that is, due to the lack of knowledge of previous conditions, people gradually change their perceptions of what is natural by assuming as normal progressively more deteriorating conditions – a process that can have tremendous negative consequences for environmental conservation) (Soga and Gaston, 2018; Jones *et al.*, 2021; Alleway *et al.*, 2023; see Chapter 10). As an ecological fallacy it equates with the misguided notion that species have all their resources in a traditional habitat unit. In fact, many environments commonly perceived as natural or semi-natural have been deeply shaped by long-term human activity. For example, in most of the European territories, many current **landscapes** are largely the result of millennial human activities, to the point that it may be extremely difficult to reconstruct the natural conditions before human intervention. This is the case for most European grasslands and forests. Most of the former are of anthropogenic origin (Poschlod *et al.*, 2009; Hejcman *et al.*, 2013; Dengler *et al.*, 2014; Fattorini, 2024b) and the distribution of many tree species has been profoundly transformed by humans (for example, although the chestnut *Castanea sativa* was originally present in Europe, the current distribution of this species has been largely favoured by humans; e.g. Conedera *et al.*, 2004, 2016; Conedera and Krebs, 2008); thus, most European forests are far from being in a natural state (Strona *et al.*, 2016).

This altered perception of what is natural does not apply only to environments which were transformed so long ago, but also involves transformations that have occurred even very recently and there is often great social resistance to management (reinstatement) of a habitat that has become transformed from what it was only 25, 50 or 100 years ago. A classic local example for the first author has been the National Trust property at Bickerton Hill in Cheshire, UK, owned by his family and handed over to the National Trust by the author. The hill was in danger of becoming overwhelmed with a dense *Betula* wood, leading to the loss of ground flora, but it has now been restored – despite much local opposition – to its original (preindustrial), scarce lowland heath dominated by open communities of *Calluna* and *Vaccinium* thanks to the enlightened regional National Trust officer Chris Widger in the 1990s and the present millennium (see Chapter 10).

A fundamental principle then is that habitat units are subject to continual transformation; as part of this process, they are subject to 'disturbance' and many of their components are unstable (Sousa, 1984). There are two parts to the changes

DOI: 10.1079/9781800627505.0003

in **biomes** and **habitats**: (i) the integration of changes in space-time; and (ii) the contrasting perspectives of changes by each organism occupying part of the habitat. In the first case (space-time integration), the timing (speed, duration) of events correlates with spatial scale. The pattern may be easily visualised in a double-log linear plot of events in time and space: with every shift in the order of magnitude (or rather of minuteness) of an impulse (input into natural systems), there tends to be an equivalent spatial impact. As an example, the graph of Figure 3.1 illustrates the enormous variation in space-time scales of interactions affecting habitats and biomes, from the activity of microbes and the lightning strike of a Dracula ant (5000 times faster than the blink of an eye) at a scale of a millimetre or so (Larabee *et al.*, 2018) to the creation and break-up of Pangea (a single mega-biogeographical region that existed during the late Palaeozoic and early Mesozoic eras with varied biomes) over scores of millions of years (Stampfli *et al.*, 2013). A habitat type is just a human concept and the different organisms living within a habitat unit have altogether different perspectives of the habitat components and events; a change is noticeable to an organism when the speed of habitat change is slower than the lowest limit of an organism's perception. Perspectives should differ in relation to size (body mass, metabolic rate) and longevity, and therefore **niche** availability.

Habitat trajectories – terrestrial vegetation succession

Regardless of human perception, over time terrestrial habitat units change in a number of obvious respects, such as: (i) **substrate (soils)** and associated underground communities (including **prokaryotes**, **protists**, fungi and animals); (ii) vegetation structure (including total biomass); (iii) the species composition of the epigean communities (including prokaryotic, protist, fungal, animal, lichen and plant species and **holobionts**); (iv) the environmental conditions (e.g. temperature, light and moisture); (v) resources for plants and animals, both **consumable** and **utility resources**; and (vi) the nature and complexity of interactions among organisms, including **functional diversity** (see Chapter 4) and ecosystem processes (e.g. nutrient cycling, energy flow and food web dynamics) (Bautista-Cruz and del Castillo, 2005; Ruprecht, 2005; Lohbeck *et al.*, 2012; Boy *et al.*, 2016; Chang and Turner, 2019).

Habitats that undergo change often follow a complex and gradual transformation in their biotic and abiotic aspects, a process called **ecological succession**. During an ecological succession, different plant communities replace each other, leading to a characteristic temporal sequence of communities known as a **sere** (whereas the different transitional communities within the sere are indicated as **seral stages**) (Fig. 3.2) (Clements, 1904; Gleason, 1926; Miles, 1987; Van Andel *et al.*, 1993). Although most research on ecological succession has been performed in terrestrial ecosystems, similar changes also occur in aquatic ecosystems (e.g. Yu *et al.*, 2019). Successional changes vary as to origin, causes, mechanisms (i.e. biotic interactions), types (pathways), speed and outcome. At one extreme, the habitat unit may change to such a degree that after a time it is no longer recognisable to a casual human observer. At the other extreme, nothing may *seem* to have changed at all. Classic examples are fields retained in an approximately constant state by various forms of management (such as mowing, grazing and burning) that produce **plagioseres** (i.e. deflected successions in which changes in the communities deviate from the expected natural course because of human activity), differing from **orthoseres** (successions lacking human intervention) (Numata, 1969). Of course, the nature of succession varies according to the context (e.g. climate, substrate and topography; Fig. 3.3). The crucial issue here is, however, that if **autotrophic** layers are transformed, so too will be the status of populations of **heterotrophs**, and ultimately the presence or absence of animal species (local extinctions and colonisations) and the pattern of species diversity (Tilman, 1993; Walker and del Moral, 2003).

Successional changes are mediated by many factors (e.g. regional species pool size, distance from seed source and dispersal capabilities) that cause species turnover, leading to changes in many community aspects (e.g. species composition, functional traits and interactions) during a succession (Chang and Turner, 2019). Early views of the influence of biotic interactions on succession were mainly focused on negative interactions (e.g. competition: species that are present in early stages are replaced by superior competitors which then characterise more advanced stages), but in fact facilitation and **life-history traits** may play important roles. For example, Connell and Slatyer (1977) proposed a very successful general framework for ecological succession that integrates three models: facilitation, tolerance and inhibition. With **facilitation**,

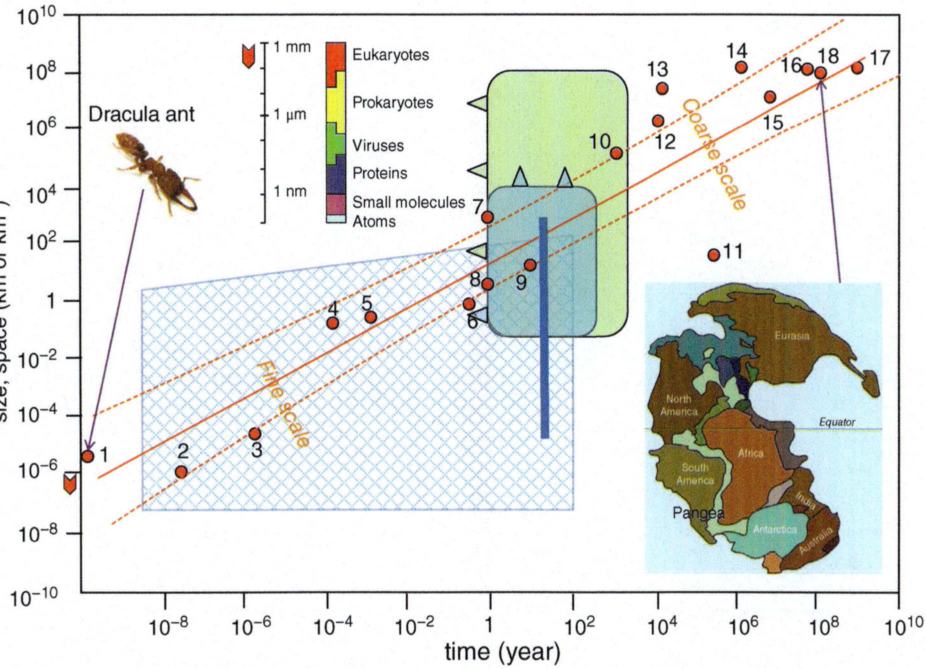

Fig. 3.1. Space-time scaling of selected Earth natural events, including human perception. Generally, the larger a spatial system, the greater the duration of its effect on the planet, even though the initial impact may be virtually instantaneous on the geological time scale (e.g. devastating effects of the Cretaceous-Palaeogene asteroid impact on insect–plant associations; Labandeira *et al.*, 2002). Green area, zone of primary vegetation succession (e.g. after glaciation, volcanic eruption or sea-level fall); blue area, zone of secondary vegetation succession (e.g. abandoned fields or cut tropical forest); cross-hatched area, zone of human perception without instrumentation; thick blue line, grant funding period of typical PhD studentship. Triangles indicate that both primary and secondary succession areas extend well beyond the limits indicated when summed over the globe, critically well beyond a single human's perception without the aid of world databases (e.g. Brown and Lugo, 1990). In such situations chronosequence techniques are employed to determine changes. A chronosequence is a set of sites formed from the same parent material or substrate that differs in the time since they were formed, enabling analysis of changes (i.e. in vegetation) to be determined beyond the period of a single human's perception (Walker and del Moral, 2003; Walker *et al.*, 2010; see Johnson and Miyanishi, 2008 for limitations of this approach that does not affect studies of succession in microbial communities which can be observed in real time). Red dots, a selection of comparative events: 1 strike by the Dracula ant *Mystrium camillae* (23 µsec, with peak at 90 m s^{-1}) (Larabee *et al.*, 2018) and escape response time of <5 milliseconds in a long-legged fly of the genus *Condylostylus* (Dolichopodidae) (Sourkakov, 2011); 2 brief interactions among male butterflies; 3 egg-laying event in a batch-laying butterfly; 4 shifts in arthropod behaviour in relationship to temperature changes; 5 diurnal shifts (night versus day) in behaviour of bats; 6 seasonal shifts in habitat between breeding in summer and overwintering sites; 7 annual migrations by elk, bison, arctic terns, swallows and cuckoos; 8 annual shifts in population sizes of species undergoing boom–crash cycles (e.g. lemmings); 9 significant changes in alpha diversity of species within a habitat or biome; 10 UK forest destruction since the Anglo-Saxon invasion; 11 duration of Yellowstone supervolcano since its appearance; 12 period of cultivation in the Fertile Crescent; 13 global ice cover during the last glacial stage; 14 Pleistocene cycles of glacials and interglacials with vegetation succession and regression; 15 hominid evolution in Africa; 16 the impact on global flora and fauna by the Cretaceous-Palaeogene asteroid impact at around 65 Ma; 17 plant evolution over the Earth's land surface; 18 the period of existence of the supercontinent Pangea. Understanding of events is affected by space-time incompatibility in studies: (i) evolutionary time events have low predictability from high-resolution temporal studies; (ii) short-term studies (in human experience) of extensive spatial phenomena can lead to ecological explanations of evolutionary events as ecology may contain a 'ghost' of those evolutionary events (pseudopredictability). (Inset of Dracula ant from Larabee *et al.*, 2018, used under licence CC BY 4.0. Inset of Pangea by Astroskiandhike: https://commons.wikimedia.org/wiki/File:Pangaea_colors.svg, used under licence CC BY SA 4.0. Inset of relative scale for micro-organisms by Tim Vickers: http://en.wikipedia.org/w/index.php?title=File:Relative_scale.svg&page=1.)

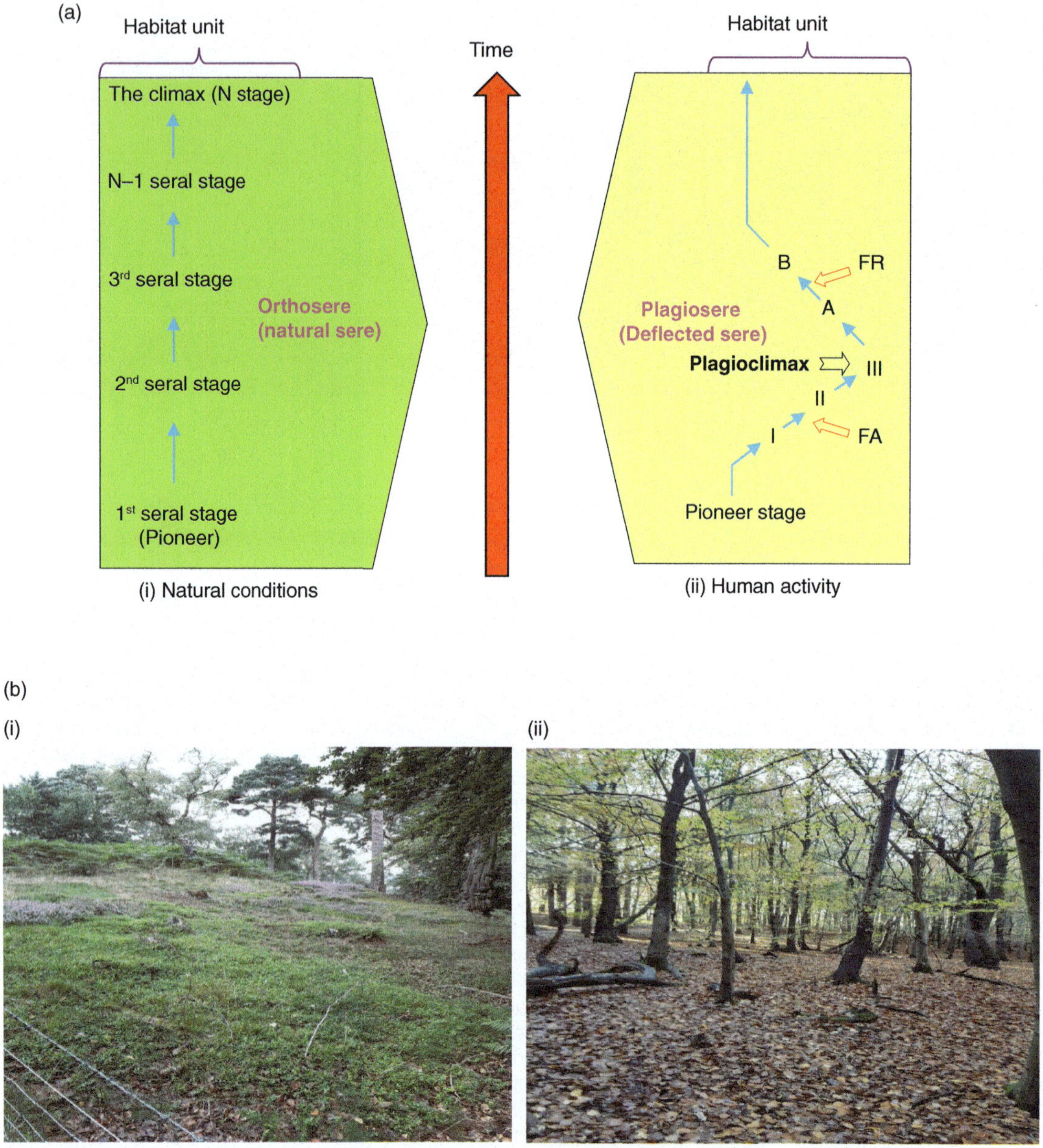

Fig. 3.2. Features of vegetation succession. **(a)** Development of plagiosere compared with that of a normal primary sere via orthoseres (FA, factor creating plagiosere (e.g. grazing; mowing) applied; FR, factor removed; I–III, plagiosere stages; A and B, subsere stages). The mechanisms of succession describe the impact of plants themselves on the substrate and on each other, generally alluded to as distinct phases of activity: nudation (production of bare ground), migration (movement of organisms, seeds, etc.), ecesis (successful establishment of colonising plants), competition (interactions among colonists), reaction (modification of conditions allowing other plants to enter) and stabilisation (homeostasis with regeneration cycles) (Clements, 1916). **(b)** Vegetation succession on Alderley Edge, Cheshire, UK: (i) remnants of the heath that once covered the entire scarp slope and summit during copper mining operations; (ii) current vegetation pattern with beech and birch trees and a ground covered with leaf litter and bracken. (Photos by R.L.H. Dennis.)

(a)

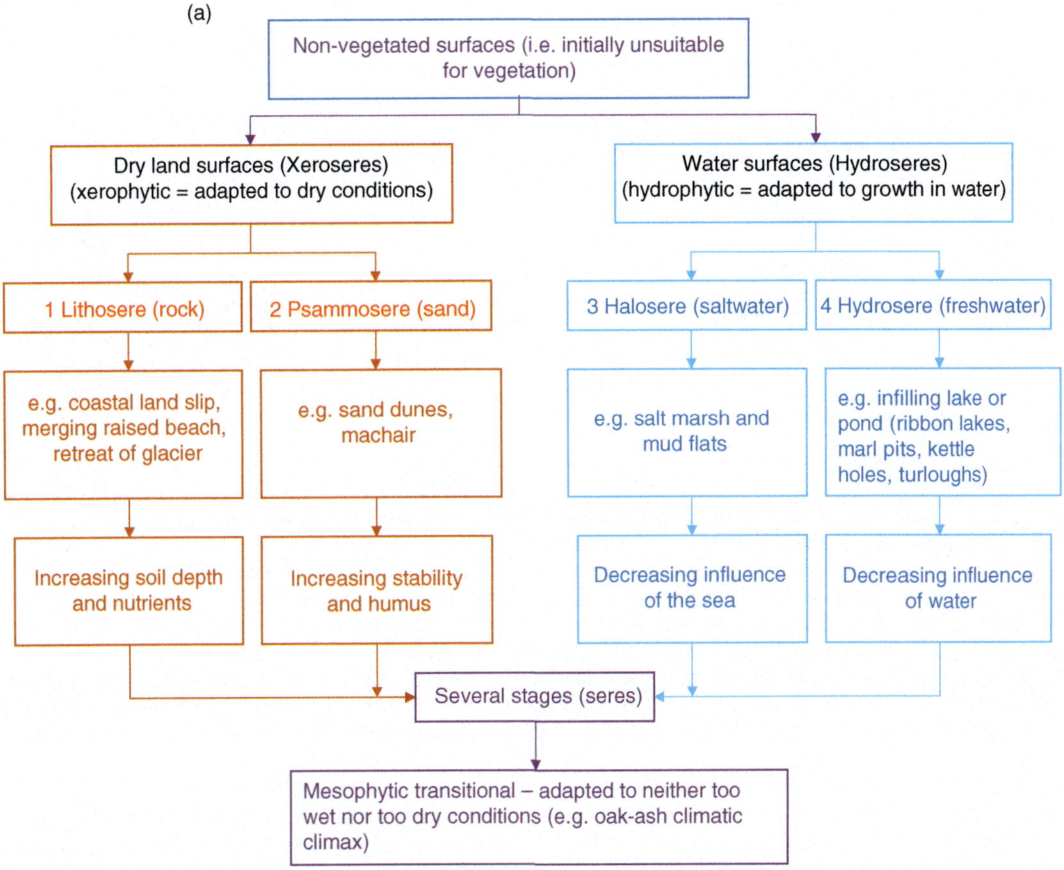

(b)

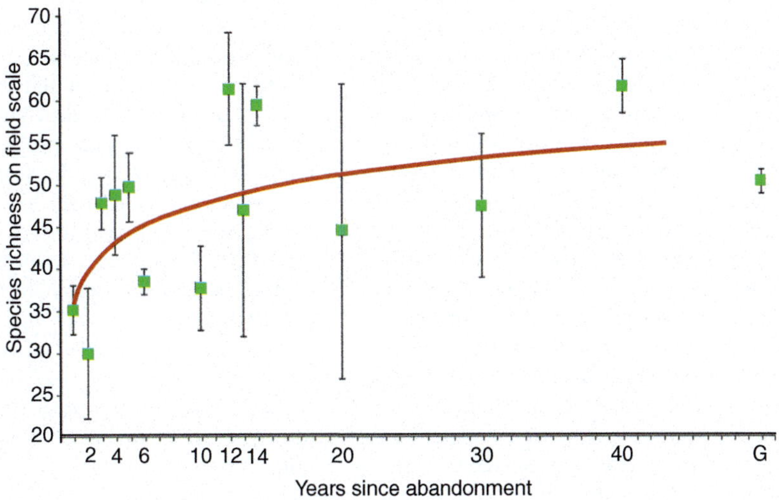

Fig. 3.3. Continued.

(c)

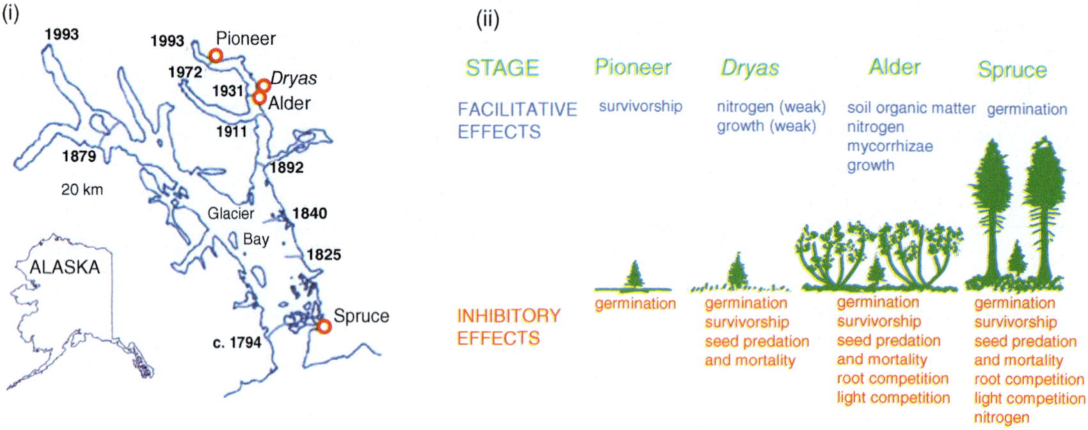

Fig. 3.3. Transformations in different types of vegetation. **(a)** Development of a climatic climax vegetation. Distinct sequences or stages have been described for different starting conditions, termed hydrosere for fresh water, halosere for salt marshes, psammosere for sand dunes and lithosere for bare rock. A reversal of this process is termed retrograde succession, as occurs with developing glaciation over Britain. **(b)** Relationship between species richness and succession in old fields in lowland Romania. Values are means and standard errors. G = reference grassland. Regression line: $Y = 35.17 + 14.66\log_{10}(X)$. **(c)** Succession after a glacier retreat: (i) map of glacier retreat from 1794 to 1993 in Glacier Bay, Alaska; (ii) summary of facilitative and inhibitory effects of each successional stage of vegetation on spruce seedling growth. (Panel **a** from Dennis, 2010, based on Waugh, 2000; courtesy of Nelson Thornes. Panel **b** after Ruprecht, 2005, redrawn and used with permission of Petyr Pysek. Panel **c** after Chapin *et al.*, 1994, redrawn, used with permission of John Wiley & Sons.)

early-successional (pioneer) species change the environment, making it favourable for the subsequent colonisation of later-successional species and, at the same time, less hospitable for their own ecological demands. As a result, pioneer species that can no longer thrive are replaced by new species typical of more advanced phases. In the **tolerance** model, pioneer species neither inhibit nor facilitate other species, and later species may successfully establish themselves if their life-history traits make them more tolerant to lower levels of resources (e.g. ***r*-selected** species, which are favoured in early stages, are progressively replaced by ***K*-selected** species, that are favoured by the environmental conditions of more advanced seral stages, which then outcompete preceding species). Thus, 'late succession' species do not need to be facilitated by early-stage species, but early colonisers make conditions less suitable for themselves. Finally, in the **inhibition** model, early occupants modify the environment so that it is less suitable for later arrivals. Earlier colonists inhibit the growth of later-successional species either directly (e.g. through predation) or indirectly (e.g. through competition for resources) until they die or are damaged.

This three-model framework has been widely used to investigate successional patterns in terrestrial and aquatic systems (Maggi *et al.*, 2011). However, reality is, as ever, more complex and Pulsford *et al.* (2016), for example, proposed a general framework with five main mechanisms of postdisturbance transition: (i) pulse events with rapid state shifts; (ii) stochastic community drift; (iii) facilitation; (iv) competition; and (v) influence of the initial composition of a postdisturbance community. In this framework, stabilising processes (such as biological legacies, inhibition or continuing disturbance) may prevent a transition between community types.

The study of ecological successions of course involves many complex problems and this vitally important concept is an intricate part of ecosystem theory. Actually, dispersal limitation, species pool effects, priority effects, abiotic environmental filtering, stochastic processes, biotic interactions (e.g. competition, facilitation, herbivory) and feedbacks (e.g. between plant and microbial communities) (Chang and Turner, 2019) all influence successional trajectories, with, however, the general result that at the species level resources will have been affected (disrupted, erased or enhanced), as will have conditions for existence (abiotic and biotic, the latter including symbionts as well as enemies). This makes the habitat unit, when defined on the basis of vegetation, an intrinsically unstable system; in turn, so are the dependent animal populations.

Contentious issues still surround the numerous models of potential trajectories and final stages during ecological succession (see Whittaker, 1967, 1975; Walker and del Moral, 2003; Walker *et al.*, 2010). For terrestrial ecosystems, classical views assume that vegetation successions undergo a number of distinctive phases from bare substrate to what is often regarded as the **climax** or **potential natural vegetation** (**PNV**) for the climate (i.e. as related to latitude, altitude and aspect), substrate (i.e. solid and drift geology; marine and freshwater basins) and topology (see Fig. 3.3). Various concepts and theories of climax have been proposed (Box 3.1), but all rely on the basic assumption that seres

Box 3.1. Concepts of climax

Various concepts of **climax** (and associated theories) have been proposed (Meeker and Merkel, 1984; Gibson, 1996). According to the climatic climax (or monoclimax), the climate is the essential driver of vegetation settings and hence, given a certain climate, all **ecological successions** of a region will eventually lead to the same final vegetation, regardless of earlier site differences in topography, parent material of the soil and other factors. Thus, every region has only one climax plant community towards which all communities are developing (Fig. B3.1.1). Although the monoclimax concept is traditionally attributed to Frederic Clements (1916), he was not so apodictic about the process, and did not affirm that a succession must *necessarily* culminate in a particular type of vegetation determined by local climate, but recognised that the climax predicted for an ideal **sere** can be prevented and frequently is prevented from occurring (Kirchhoff, 2020).

A variation of the monoclimax theory is the polyclimatic climax, in which several climaxes can occur in a given region on different **soils** that are produced by different climatic conditions (Tuxen, 1933). According to the polyclimax theory (Tansley, 1939; but see Meeker and Merkel, 1984 for previous authors who contributed to the development of this idea), the climax vegetation of a region may consist of several climax communities controlled by a variety of factors (e.g. soil moisture, soil nutrients, topography, animal activity) other than climate, although one climax (the climatic climax) may be dominant.

All these theories relate the concept of climax to geomorphological equilibrium and soil maturity. By contrast, the climax pattern theory (Whittaker, 1951, 1953) assumes that it is not possible to distinguish plant communities as discontinuous units, but each plant population responds individually to biotic and abiotic conditions (see Box 3.3 for distinction between open and closed **communities**), leading to a final vegetation setting (the climax) whose composition and structure are the result of species responses to the full array of abiotic conditions (e.g. moisture, temperature, nutrients, winds, light) and biotic relationships (e.g. pools of animal and plant species from which the area can be colonised), without reference to geomorphological equilibrium and soil maturity. Similarly, the concept of climax adopted by Dyksterhuis (1949, 1958, referred to as site climax by Meeker and Merkel, 1984) does not refer to geomorphological equilibrium and soil maturity but, differently from the climax pattern theory, considers communities as discontinuous.

Fig. B3.1.1. Analysis of potential natural vegetation indicates laurel forests dominated by *Laurus azorica* as one of the most widely distributed vegetation forms in the Azorean Islands (Elias et al., 2016; the picture illustrates this type of forest in a fragment on Pico Island). Pollen analysis of sediment samples suggests that laurel forests were the vegetation occurring in many areas of the Azorean Islands before extensive alteration occurring after the Portuguese colonisation (Rull *et al.*, 2017). This type of vegetation can be therefore considered the climax vegetation in these areas. (Photo by S. Fattorini.)

evolve towards an 'optimum' vegetation in equilibrium with its environment. This teleological assumption is particularly evident in Clements' (1916) concept of climax (see, for example, Barbara, 2024), but afflicts (at different degrees) all concepts of climax.

The concept of climax vegetation has been severely criticised and it is becoming increasingly evident that there is no one climax vegetation type for a locality (Pyšek *et al.*, 2001; Chang and Turner, 2019); moreover, regression (**retrogressive succession**) can occur from the climax state owing to various causes, such as changes in nutrient status (i.e. depletion of soil phosphorus; Peltzer *et al.*, 2010, Wardle *et al.*, 2004), just as **progression succession** leads towards it. Although the concept of climax as a stable and deterministically predictable vegetation appears philosophically fallacious and empirically unsupported, this does not mean that efforts to hypothesise PNVs (defined as the plant **communities** that would be expected in given areas if all successional sequences were completed under the present local climatic and edaphic conditions without interference by humans) are intrinsically flawed. On the contrary, they can be extremely productive, if viewed as null hypotheses useful to understand successional mechanisms responsible for the observed patterns. As PNVs are distinctive for different **biogeographical regions**, for **biomes** (tropical rainforest, savanna, Mediterranean, deciduous forest, boreal forest, arctic-alpine, etc.) and variation in local climate, geology and geomorphology (e.g. raised bogs and blanket bogs) (Mazziotta *et al.*, 2019), maps of PNVs may represent important tools to orient conservation strategies (Ricotta *et al.*, 2002; Hengl *et al.*, 2018) and valuable sources of baseline information to evaluate qualitative and quantitative discrepancies between the observed and the expected structure of plant communities (e.g. Carranza *et al.*, 2003), and hence to shed light on their level of naturalness (Strona *et al.*, 2016). Moreover, although variants of climax vegetation are unlikely to be observed at large scale because of continual **disturbance** that interferes with succession, at local scale climax communities can actually be recognised as stable communities that remain unaltered for long periods. For example, a tropical forest as a whole can be perturbed by a number of factors (e.g. storms, fires and flooding) that alter the sere, but not all parts of the forest are affected and some can reach and maintain the status of climax communities. In fact, communities are in general repeatedly subject to different kinds of disturbance and the effects of disturbance depend on its severity (Chang and Turner, 2019) (Box 3.2).

Based on the effect of disturbance, two types of ecological succession can be distinguished: **primary succession** and **secondary succession**. Primary successions arise from the formation of new land or from severe disturbances affecting already existing ecosystems to the point of generating virtually completely new ecosystems, often on a regional scale (deglaciation: Crocker and Major, 1955; Chapin *et al.*, 1994, Binney *et al.*, 2017; riparian deposits: Salo *et al.*, 1986; volcanicity: Thornton *et al.*, 1988; Whittaker *et al.*, 2023). When a barren, lifeless habitat is created by geological processes (such as newly created volcanic islands and rocks exposed by retreating glaciers) or soil and organic material are removed from pre-existing ecosystems, ecological successions start *ex novo*. For example, a newly formed volcanic island or an area sterilised by a volcanic eruption is initially completely devoid of life (e.g. Krakatoa; Thornton, 1997), and succession starts from bare rock that needs to be progressively colonised by micro-organisms, lichens, fungi, pioneer plants and so on. Other notable primary successions develop on dune systems and a fall in lake or sea base-level (Olson, 1958; Rydin and Borgegård, 1991). Secondary successions derive from less severe disturbances and are characterised by the retention of a certain amount of previous ecosystem components, e.g. a soil layer with its seed bank. For example, if a forest is affected by a fire, it is possible that some trees (that represent the vegetation of an advanced seral stage) may survive; thus, the new colonisers representative of early stages (such as some grasses) will be mixed in with species typical of advanced phases that survived the disturbance.

Although scale can be an issue (e.g. larger islands in the Thousand Island Lake, China, have more late-successional species than smaller ones; Liu *et al.*, 2019), it is not necessarily so as seen in primary successions in glaciated valleys (e.g. Glacier Bay, Alaska, USA; Buma *et al.*, 2017) and secondary successions in abandoned fields (Cramer and Hobbs, 2007). What has become apparent is that the speed of succession varies according to environmental conditions, being higher in colder (arctic) than in warmer (tropical) latitudes, in humid than in arid environments, and in primary seres (recovery to PNV; lower likelihood of alien species) than

Box 3.2. Levels of disturbance and organisms' responses

In the absence of disturbance events that affect succession trajectories, communities should evolve to the **climax** stage. During the process, there is a species turnover, with many of the species present in the first stages (e.g. ***r*-strategists**) being progressively replaced by other species typical of the more advanced stages (e.g. ***K*-strategists**), until ideally only late-successional species are present. But what happens if the community is affected by disturbance? If the disturbance is severe, the overall diversity will be obviously reduced, since many species will be eliminated. However, a disturbance of intermediate intensity will eliminate some species, but others will be able to survive or rapidly recolonise the habitat. This will create, for example, a mixture of *K*-strategists that survived the disturbance and *r*-strategists which (after being replaced during the succession) were able to recolonise the habitat after the disturbance. As a result, under an intermediate disturbance, diversity can be not only higher than that of severely disturbed communities (where many species are eliminated by the disturbance) but also higher than that observed in undisturbed habitats (where only late-successional species are present). According to this model, known as the **intermediate disturbance hypothesis** (Fig. B3.2.1), there would be an optimum degree of disturbance, which allows the maintenance of a maximum number of species (Dial and Roughgarden, 1988; Blair and Launer, 1997; Mackey and Currie, 2001; England *et al.*, 2008; Bongers *et al.*, 2009; Catford *et al.*, 2012). Note that although Connell (1978) is usually cited as the founder of this model, the idea can be traced back to the 1940s (Eggeling, 1947; Watt, 1947; Tansley, 1949) and it can be also found in Grime (1973); see Wilkinson (1999) for a historical review.

However, the model has been severely criticised and empirical evidence appears equivocal (Fox, 2013); the definition itself of 'intermediate disturbance' is ambiguous, since it may refer to different forms of disturbance (frequency, extent, intensity or duration) as well as to the previous history of disturbances, and hence to what is considered the benchmark. The intermediate disturbance hypothesis proposed for **ecological successions** can be translated into a spatial perspective: along a gradient of disturbance, species richness peaks at an intermediate level of disturbance (**disturbance heterogeneity model**; Porter *et al.*, 2001). Patterns consistent with a disturbance heterogeneity hypothesis have been observed, for example, in urban–rural gradients, where species richness and abundance frequently have their maximum in the middle of the gradient (Germaine and Wakeling, 2001; Fattorini, 2019; Fattorini *et al.*, 2020b). Here, disturbances due to urbanisation (e.g. chemical, noise and light pollution, fragmentation of green spaces and heat island effect – that is, warmer temperatures in highly urbanised areas due to traffic, presence of impervious materials such as asphalt and concrete that absorb and retain heat, buildings that obstruct the natural flow of air, reflections from glass, etc.) are at intermediate levels, which increases habitat heterogeneity and hence resource diversity and abundance.

Different organisms, however, respond differently to the same disturbance. For example, different animal groups show different responses to fire, which is an important form of disturbance in Mediterranean ecosystems, where it plays a pivotal role in vegetation dynamics (e.g. Brotons *et al.*, 2005; Ladd *et al.*, 2005). Effects of fire on ground-dwelling invertebrate

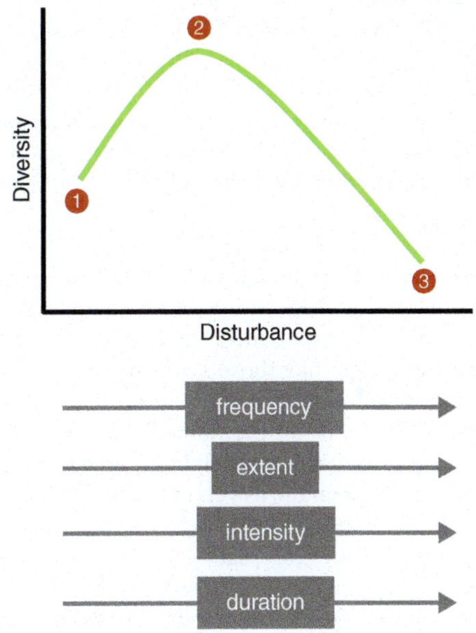

Fig. B3.2.1. According to the intermediate disturbance hypothesis, diversity peaks at intermediate levels of disturbance: 1 at low disturbance levels, diversity is reduced by competitive exclusion; 2 at intermediate disturbance levels, diversity is maximum as competition effects are reduced, allowing more species to coexist; 3 at high disturbance levels, diversity is low as most species are eliminated. Variation in disturbance can be of different forms, including frequency (number of disturbance events: rare to common), extent (the area involved by disturbance: small to large), intensity (slight to severe) and duration (short to prolonged).

Continued

Box 3.2. Continued.

communities in Mediterranean ecosystems are typically severe, especially on those species that have low dispersal abilities (Pitzalis *et al.*, 2005; Fattorini, 2010a; Mantoni *et al.*, 2020), but are less pronounced on insects with good dispersal capabilities and hence high recolonisation capabilities, such as butterflies (Serrat *et al.*, 2015) and dung beetles (Palusci *et al.*, 2021). Similarly, dung beetle communities of savanna habitats in Brazil are resilient to fire and seem not to be structured by this disturbance (Nunes *et al.*, 2019). Because of differences in colonisation capabilities among the species that form the pool of potential colonists, anthropogenic changes in these species pools may influence ecosystem **resilience** negatively. In a study on the recolonisation of an area subject to severe disturbance due to a catastrophic island volcanic eruption, Fattorini and Borges (2012a) found that, after 43 years from the eruption, the resulting arthropod community was dominated by introduced species, whereas native non-endemic species had more difficulty in recolonising the erupted area, and no endemic was able to recolonise it – thus, the alteration in the species pool made the recolonisation process very different from what it would have been under natural conditions, lowering ecosystem resilience.

in secondary seres (Prach and Walker, 2019). The slower recovery of tropical forests may relate to the enormous diversity and complexity of such ecosystems (Chazdon, 2008), which nevertheless are greatly affected by prior and subsequent land use (e.g. pastures versus abandoned clearcuts in the Amazon; Mesquita *et al.*, 2015b). Even so, in mid-latitudinal forests, let alone towards the equator, it is difficult to determine what the appropriate PNV should be.

Ecological successions result from a mixture of deterministic and stochastic factors, allogenic (extrinsic) and autogenic (intrinsic) agents, which makes predictions extremely uncertain. Stochastic models of change (like those expressed by Markov chains, in which the probability of each event of a series depends only on the immediately preceding state) simply indicate probabilities of events the causes of which remain, as yet, undisclosed. Science advances by exposing the causes underlying the probabilities. In any case, each biome is likely to be subject to different patterns of stochastic and deterministic influences (Norden *et al.*, 2015; Li *et al.*, 2016). The truth as to cause-and-effect lies somewhere in between (Putman, 1994; Fridley and Wright, 2012; HilleRisLambers *et al.*, 2012), as events within a habitat unit are produced both by deterministic interactions (i.e. biotic; trophic), such as top-down effects of deer suppressing ground vegetation (Putman *et al.*, 2011) and bottom-up effects of protists on plant performance (Gao *et al.*, 2019), and unpredictable events (i.e. abiotic events), such as broadscale tree felling by storms (Harmer *et al.*, 2004) and woodland fires caused by lightning strikes (Jurskis, 2005). The outcome of succession is usually illustrated as a picture model of changes for local-scale situations (e.g. in a pond, or a mere (small lake) forming in a kettle hole; Klinger, 1996) (Fig. 3.4) or more extensively for regional or biome environments undergoing systematic change, as in the case of the Holocene climax vegetation following deglaciation (e.g. Glacier Bay, Alaska, USA: Goldthwait *et al.*, 1966; Chapin *et al.*, 1994; post-Devensian Britain: Bennett, 1989). The reality of vegetation succession gleaned from detailed surveys is that the patterns of change are far more complex and uncertain. A detailed comparison of contingency and stochastic influences (viz. soil fertility, fire, climate, competition and demography) in old field successional stages at Cedar Creek Ecosystem Science Reserve (Minnesota, USA) exposed consistent influences from contingent factors (Clark *et al.*, 2019).

Despite the 'confounding' effects of stochastic events, vegetation successions are characterised by a number of predictable features and processes (which involve energy flow, community structure, biogeochemical cycles, natural selection and regulation; Odum and Barrett, 2005; Pidwirny, 2006) (Table 3.1), but there are exceptions to the trends outlined. Regarding plants, there are important changes to plant structure and composition (i.e. species diversity, size of individuals or basal stand area, and life cycles, forms and histories), particularly to the complex of plant **traits** (Kleyer *et al.*, 2019). Succession provides potential for the creation of innumerable structural **niches**, e.g. crevices in the bark and chambers (called **domatia**) produced by plants which provide refuges for arthropods, those in a mutualistic relationship with the plants (e.g. ants and mites which combat plant predators and fungi) as well as parasites (e.g. thrips) (Wilson, 1971; O'Dowd and Willson, 1991).

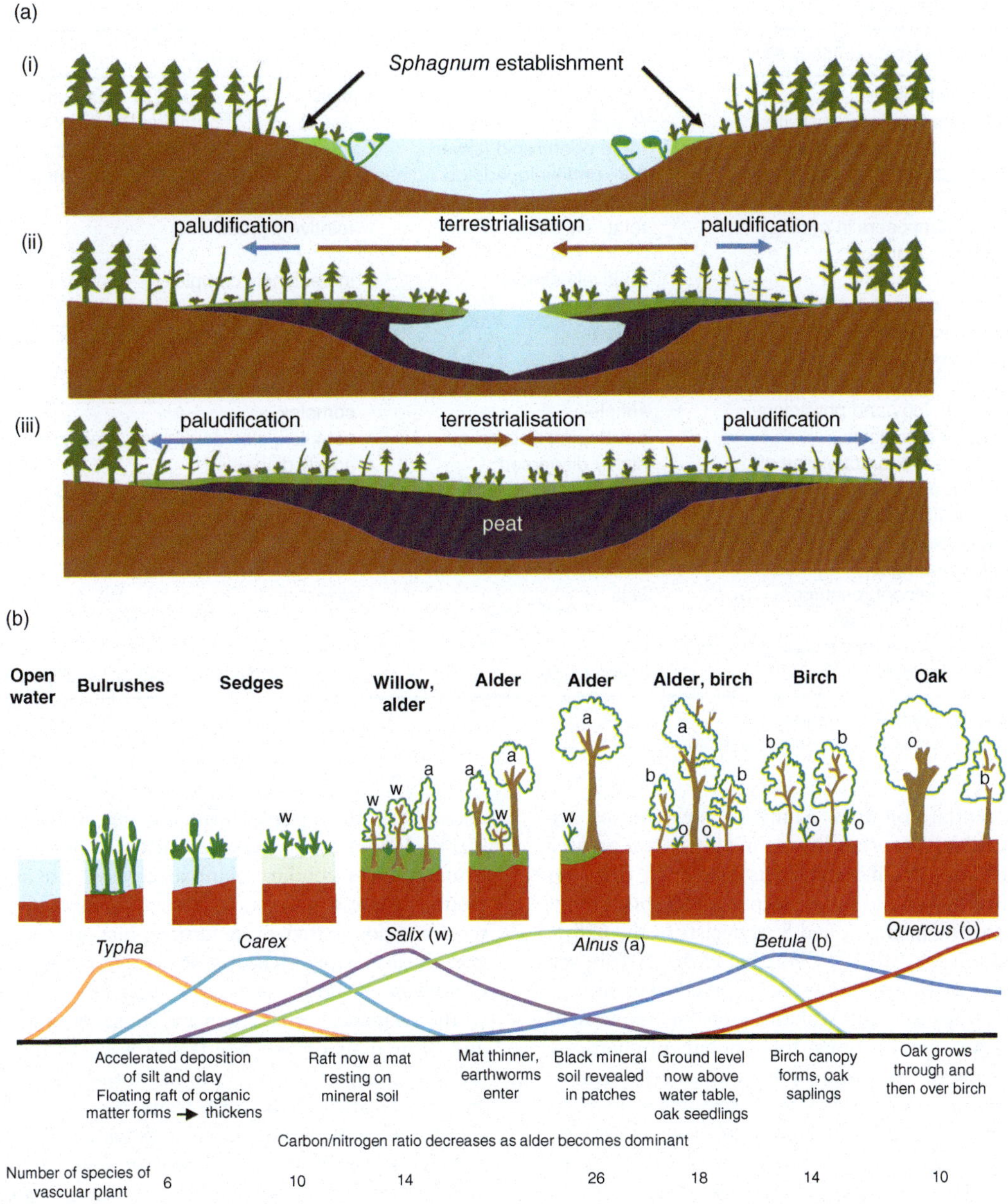

Fig. 3.4. Model of hydroseres. **(a)** Bog vegetation succession as a consequence of both terrestrialisation and paludification: (i) establishment of peat-forming mosses (e.g. *Sphagnum* sp.) along the water body edge; (ii) peat accumulation occurring as a result of both terrestrialisation and paludification causing a rise in water level; (iii) the final development of ombrotrophic bog with surface lying above the surrounding water table. **(b)** Changes during a hydrosere at Sweet Mere kettle hole near Ellesmere, Shropshire, UK. (Panel **a** after Klinger, 1996, redrawn and used under the creative commons licence. Panel **b** after Gilbertson *et al.*, 1985, redrawn.)

Consequently, vegetation succession is a history of loss and gain of traits as well as of species. As for invading species, I.L. Borges *et al.* (2019) and van Breugel *et al.* (2019) show that abiotic conditions interact with both invader traits and resident species' functional diversity to determine the survival

Table 3.1. A selection of plant community and ecosystem characteristics during vegetation succession.

Basic autotrophic community attributes	Early seral stages	Later seral stages
Ecosystem stability	low	high
Rate of net primary productivity	P > R	P ~ R
Biogeochemical cycling (nutrient turnover)	more open/rapid (lower)	more closed/slow (increased)
Soil depth and nutrient status	thin, undeveloped, low nutrient status	deep, well-developed horizons, higher nutrient status
Role of decomposers in cycling nutrients to plants	relatively small	relatively great
Food chain	relatively simple	increasingly complex
Soil humus and non-living organic matter	low	high
Plant species diversity	increasing	decreasing
Plant biomass and size	small	large
Plant longevity	short	long
Plant morphology and physiology	simple	complex
Plant trait associations	developing networks	complex, size dominated, stabilising
Seed dispersal characteristics of dominant plants	highly dispersed	poorly dispersed
Seed longevity (lifespan)	relatively long	relatively short
Photosynthetic efficiency of dominant plants at low light	low	high
Rate of soil nutrient resource consumption by plants	fast	slow
Plant recovery rate from resource limitation	fast	slow
Life-history strategy	*r*-type (ruderals)	*K*-type (stress tolerators and competitors)

P, production (yield); R, respiration.
(Based on Odum and Barrett, 2005, and Pidwirny, 2006.)

of potential colonists. Many of the features are predictable; importantly, one may expect a greater contribution from ***r*-strategists** (ruderals) early on in succession and a greater contribution from ***K*-strategists** (plant C- and S-strategists in the **C-S-R model**; Grime, 1974, 1979; Chai *et al.*, 2016b; see Chapter 2) during later phases. Linked to plant species composition and structure, this has immensely important implications for animal populations occupying or 'vacating' the habitat unit. An important feature is that succession is strongly influenced by regional and local environmental conditions, such as climate, a key component of species turnover in habitat units; for instance, dominance by woody species takes less than a decade in the southern USA compared to 60 years in New England (Fridley and Wright, 2018).

Habitat trajectories – heterotrophic succession

Terrestrial vegetation successions (primary or secondary) are accompanied by enormous changes in soils, micro-organisms, invertebrate and vertebrate species and their populations (Chang and Turner, 2019). As they are based on autotrophic organisms, primary and secondary plant successions are **autotrophic successions**. By contrast, aquatic ecosystems are dominated by **heterotrophic successions**, that is, successions that do not involve autotrophic organisms. In both cases, two issues are involved: (i) the reciprocal effects of seral stages on animals and of animals on the sere trajectory; and (ii) the massive changes (turnover) in life forms (biodiversity, taxa) and **life-history traits** accompanying succession. Some indication of successional changes in progress is evident in the systematic shifts in plant and animal species' microdistributions (i.e. **home ranges**, **territories**, locations of resource exploitation) within habitat units. An obvious example is tree fall and the associated growth of new plants (the so-called **regeneration complex**) filling forest gaps, gaps that allow a range of organisms to colonise the habitat space.

Despite the influences exerted by plants, succession is not simply a plant-determined system (for example, coral ecosystems: Hawaii, Grigg, 1983; Reunion Island, Jouval *et al.*, 2020). **Soil** conditions, with their development (pedogenesis, horizonation and accompanying changes in texture and chemistry; changes in arbuscular mycorrhizal fungi (see Chapter 2) and **micro-organisms**; Sinclair and Sampson, 1931; Trudgill, 1977; Kohls *et al.*, 2003; Gunina *et al.*, 2015; Koziol and Bever, 2019; Wang and Zeng, 2021; Ware *et al.*, 2021), and their influence on controlling colonists (dispersal limitations), are crucial components in plant colonisation and succession (Li *et al.*, 2013; Makoto and Wilson, 2019). As for animals, above-ground invertebrates have a similar effect on plant fitness and success as vertebrates, as both groups can reduce the rate of vegetation succession and shift its trajectory (Brown, 1985). Even though many invertebrates are incapable of colonising denuded surfaces, arthropods, as well as micro-organisms, have a substantial part to play in the fitness of early-successional species and the successional process, as Brown's (1985) study of sea radish *Raphanus raphanistrum* revealed. Further work by Brown and Gange (1992) highlighted how insects with different feeding habits impact on the dynamics of an early-successional plant community. In particular, they found that the reduction of both foliar-feeding insects (Hemiptera: Auchenorrhyncha) and root-feeding insects (Coleoptera: Scarabaeidae; Diptera: Tipulidae) had different, but dramatic, consequences: foliar-feeding insects delay succession by slowing grass colonisation; conversely, root-feeding insects accelerate succession by reducing forb persistence and colonisation.

A fundamental feature of vegetation succession is the turnover of animal species along with the turnover in plant species (e.g. arthropods: Shelford, 1907; Weygoldt, 1969; Evans, 1988; birds: Adams, 1908; Johnston and Odum, 1956; Shugart and James, 1973; Southwood *et al.*, 1986; Raman *et al.*, 1998; Winkler, 2005; Schieck and Song, 2006; Serong and Lill, 2012) (Fig. 3.5); this process is initially evident from shifts in the ranking of the status (population size or density) of animal species within isolated habitats (e.g. islands: Dennis and Hardy, 2018). More than anything, this process – its taxonomic and functional components – should bring home to any reader that habitat is not stationary, a static entity that one has seen, perhaps mapped, and will somehow remain much the same. Even after centuries of a climax state, there will be regression, regeneration and change in taxa; species enter and colonise; others become locally rare and extinct. Much as there are supposedly vegetation stages, so there is a succession of animals. Brown and Southwood (1983), Schowalter (1995) and Schowalter and Ganio (2003) found that sap-sucking insects (Hemiptera) dominated early-successional phases characterised by succulent tissues in temperate and tropical forests, but folivores, predators and detritivores dominated later phases. Many other examples exist of animal successions coinciding with plant succession (e.g. birds: Serong and Lill, 2012; small mammals: Kirkland, 1977; Larkin *et al.*, 2008; Rose *et al.*, 2018; insects: Martinko *et al.*, 2006; spiders: Hurd and Fagan, 1992).

With recent developments in molecular phylogenetic methods, successional processes are also rapidly being assessed for the vast domain of micro-organisms (Fierer *et al.*, 2010; Frouz *et al.*, 2013; Cutler *et al.*, 2014; Nayfach *et al.*, 2021). For example, changes in microbial diversity and biomass during primary succession (on initial bare rock surfaces) have been documented in volcanic and deglaciated landscapes, calcareous lithologies and sand dune systems (Hoppert *et al.*, 2004; Gomez-Alvarez *et al.*, 2007; Nemergut *et al.*, 2007; Schütte *et al.*, 2009; Turner *et al.*, 2019; Garrido-Benavent *et al.*, 2020) (see Fig. 3.6). As for plants, there is much work to do yet on experimental designs to determine the effects and feedbacks of soil fauna on soil and hence plant development (Castle *et al.*, 2016; Gao *et al.*, 2019).

Researchers have sought distinct phases for heterotrophic succession as with vegetation seral stages. A four-stage process emerges in Simberloff and Wilson's (1969) research ((i) non-interactive; (ii) interactive; (iii) assortative (niche partitioning); and (iv) evolutionary), but it is more likely that overlapping phases emerge for different faunas, as in the case of bird species in developing stages from open ground to high woodland (see Fig. 3.5). Punttila *et al.* (1994) found that ant diversity declined during forest succession in Finland. But in the Neotropics ants may dominate later seres (Pérez-Espona, 2021), especially when disturbance has affected ecosystems (as in the case of the Lago Guri islands in Venezuela; Terborg *et al.*, 2001), and ants fascinatingly occupy special **domatia** (myrmecodomatia), as in the case of *Myrmecodia beccarii* which grows on *Melaleuca* trees in Malesia (Huxley, 1978). Such habitat type distinctions pervade all arthropods. What is clear is that resources and conditions within a habitat have to attain certain minimum criteria before a particular species is able not just to enter but to colonise a habitat unit. Appreciating this is of huge importance in being able to understand the concept of the **resource-based habitat** (Dennis *et al.*, 2003).

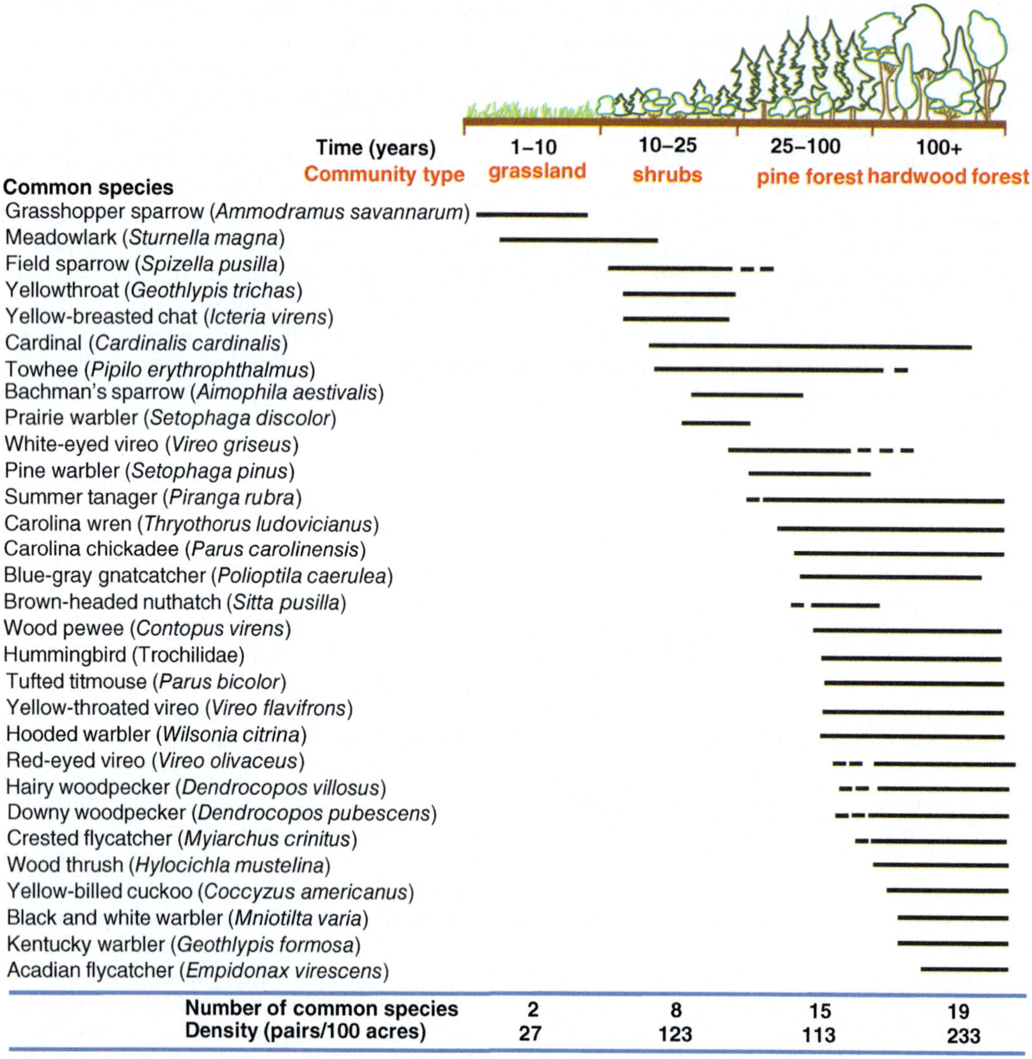

Fig. 3.5. Changes in bird communities during secondary succession on abandoned farmland in the south-eastern USA passing from grassland, through shrubs and pines to hardwoods. Common species: species with at least 5 pairs/100 acres. (Redrawn from Johnston and Odum, 1956, and Odum, 1963.)

Different biomes and successional conditions may have very different patterns ranging from coincidence in development (waxing and waning) of species (the **super-organism concept** of community; Clements, 1916, 1936) to lack of coincidence in colonisation and population development (the **individualistic concept**; Gleason, 1926) (Box 3.3), the latter anticipating Hubbell's (2001) unified neutral theory (see Chapter 2).

A classic study of factors underlying the successional process has been an experiment on small mammals (rodents) following fire on wet heath in eastern Australia (Fox *et al.*, 2003) (Fig. 3.7). From their findings, the authors proposed the **accommodation model** for animal succession; this proposes that an animal species enters a habitat unit when changes in the vegetation succession reach a threshold condition suitable to that species. As succession progresses, the habitat then becomes less suitable for that particular species. Using removal experiments (a retrogression by clipping the vegetation), they revealed a distinct order in response to habitat

Box 3.3. What, if anything, is a community?

There is no universally accepted definition of a **community**. Most authors use the word 'community' to indicate a group of organisms representing multiple species that simultaneously occupy the same place, without any other implications. Other authors reserve the word 'community' to indicate groups of interacting species, and use the word **'assemblage'** when interactions are not necessarily involved. However, some authors use the term 'assemblage' with a more restrictive meaning, referring to a group of phylogenetically related species within a community. This distinction has its roots in the opposition between the **holistic concept** developed by Frederic Clements (1916, 1936: a community is an integrated unit – in the most extreme interpretation, introduced later by Clements himself (Clements and Shelfor, 1939), it is a **super-organism**) and the **individualistic concept** developed by Henry Allan Gleason (1926: a community is nothing more than a group of species that happen to co-occur at one place). Clements' concept may seem similar to the mutualistic organismic community earlier introduced by Johannes Eugenius Bülow Warming (Warming, 1895; Warming and Vahl, 1909), but a fundamental difference exists: Warming assumed mutual dependence as a principle of the organisation of plant communities, whereas Clements considered communities as top-down control-hierarchical entities, in which the subordinate species depend on dominant species (Kirchhoff, 2020).

The holistic and individualistic concepts are usually considered synonymous with those of closed and open communities, respectively (Fig. B3.3.1).

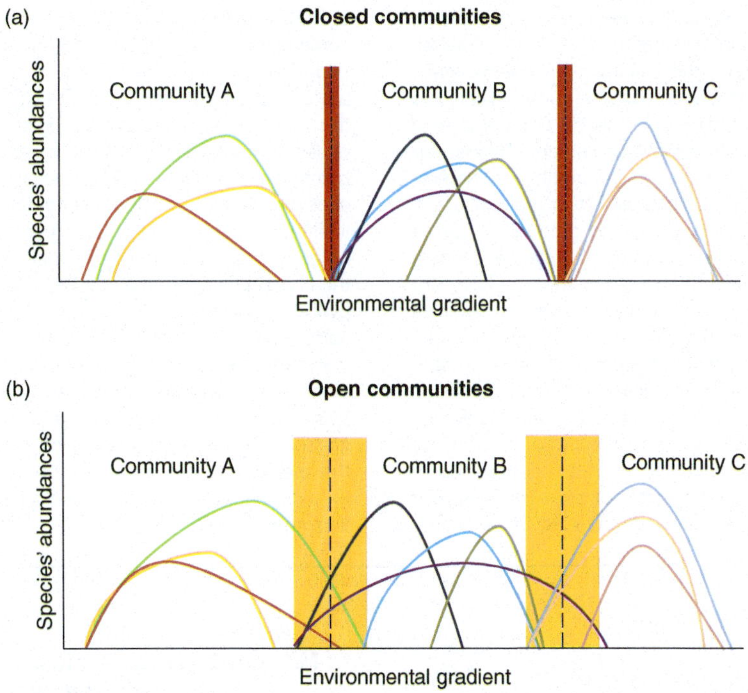

Fig. B3.3.1. Closed **(a)** and open **(b)** communities. In closed communities, distinct clusters of species can be recognised along the environmental gradient. In open communities, changes in species composition and abundance are gradual and groupings are not obvious (although potentially recognisable on statistical basis). The holistic (super-organismal) concept of community predicts that species should be organised into closed communities, whereas the individualist concept should translate into open communities. However, how environmental characteristics change along gradients can play a pivotal role in producing gradual or abrupt changes in species composition/abundance. Presence of impermeable edges (red bars in panel **a**) between different landscape patches will produce well distinct (closed) communities. By contrast, areas of gradual transition (ecotones, yellow bars in panel **b**) will produce gradual changes in species composition/abundance (open communities).

Continued

Box 3.3. Continued.

However, although associated, these concepts are not necessarily equivalent. According to the concept of **open community**, species are distributed along ecological gradients without discontinuities (and this is a consequence of their individual responses, hence the assumed equivalence with the individualistic concept). Although different portions of the gradients can differ for species composition and abundance, it is not possible to identify obvious separations. Communities can therefore be identified only on a statistical basis, with different communities blending into each other. By contrast, the concept of **closed community** assumes that species composition and abundances change abruptly in some parts of the gradients, leading to the possibility of recognising separate segments of the gradient, each corresponding to a distinct community as a result of the unique interactions that make each community an integrated unit (holistic concept), with very limited overlaps in species composition between adjacent segments. Although it seems logical that the individualist concept should translate into open communities, and the holistic concept into closed communities, this assumption is strongly biased by the point of view of organism reactions to the environment. This is certainly important but the environment, and namely how its characteristics change along gradients, can play a pivotal role in producing gradual or abrupt changes in species composition/abundance. Gradual changes in environmental characteristics are expected to produce gradual changes in species composition/abundance, whereas abrupt changes are expected to produce a clear subdivision in species groups. At the same time, different organisms may show different responses to the same gradient: while some organisms may respond to sharp changes in environmental variables with strong changes in composition and abundance, other organisms may be less sensitive to the same changes. For example, a grassland and a pond are expected to define distinct (closed) communities of ground-dwelling and aquatic beetles given the strong contrast between these two biotopes (see discussion about edges in Chapter 1), whereas a grassland and a forest might host open communities of beetles, with many shared species because transition between the two **biotopes** is more gradual; however, changes in animal composition between a grassland and a pond may be less sharp if we consider insects in general (with many species that actually use both biotopes), and those between a grassland and a forest may appear stronger if we consider butterflies (with forests generally hosting fewer species, which are, however, more specialised). Thus, the concepts of open and closed communities cannot be simply equated to those of holistic and individualistic communities, but are more heavily dependent on the environmental structure.

Another issue is whether communities should be considered taxonomically bounded or not. In general, communities should not be considered taxonomically bounded, and this is especially true if communities are recognised as sets of interacting organisms. On the other hand, it is virtually impossible, in most cases, to characterise groups of interacting species belonging to multiple taxa, especially in species-rich ecosystems. As a result, most community ecologists limit their study to the interactions between few taxonomic groups (e.g. plants and their pollinators) or to the characterisation (e.g. diversity, biomass or niche overlap) of co-occurring species belonging to the same taxon (for which the words **taxocenose** or **taxocene** are sometimes used). Another problem is about how species that do not represent a stable part of the community should be considered. Most authors consider as true components of a community only the species that are regularly found in a given space (breeding populations), thus excluding migrants and **vagrants**.

manipulation (viz. *Mus domesticus* → *Pseudomys novaehollandiae* → *P. gracilicaudatus* → *Rattus lutreolus*). None of this is surprising to butterfly and bird researchers in Europe. But, on the wider platform of world habitats, the model forms a basis for examining how novel species are accommodated by specific conditions within habitat units, initially permitting entry, but also how specific conditions place limits on their persistence. The importance of this concept will be even clearer when considering the species abundance distribution within habitats (see Chapter 4).

Microhabitat dynamics: regeneration and decay

Succession is by no means a one-way process, a linear upward trajectory with an eventual outcome; that is obvious in the fall-off in a vegetation community during development, following the peak of a predicted climax state, or over a longer span of time with a downturn in environmental conditions such as the onset of a glacial stage. In fact, downturn includes a complex set of processes embedded in the wider domain of ecosystem theory, involving

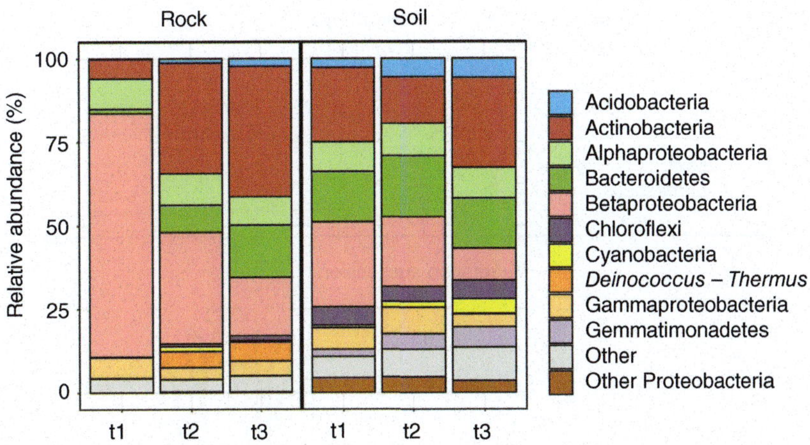

Fig. 3.6. Relative abundance of bacterial phyla in rocks and soil at the Hurd Glacier forefield (Livingston Island, Antarctica) at three successional stages: initial stage (bare soils close to the glacier terminus, t1), c.13 years since glacial retreat (t2) and c.23 years since glacial retreat (t3). Within relatively short time intervals, primary succession processes at polar latitudes involve significant and distinct changes in microbial communities associated with soil development and lichen and moss colonisation. (From Garrido-Benavent *et al.*, 2020, used under licence CC BY.)

community assembly and species interactions and coexistence (see Allee *et al.*, 1949; Chang and Turner, 2019, for references). The impression given by the earliest studies of vegetation succession is a simplistic one of development to a climax. In fact, many of the processes in a habitat unit are regression not progression, decay not construction. This is particularly evident at finer scales within habitats. Classic examples are: (i) the regression complex, for instance the gradual decay of a single tree (plant) from a snag (i.e. a dead or dying standing tree) to a stump, or from a tree felled by windthrow or windsnap to woody debris; (ii) the decomposition of annual leaf litter; (iii) the decay of animal carcasses; and (iv) the breakdown of animal faeces. All these processes of breakdown, creation of necromass (dead plant matter, especially timber), are essential for the further development of soil and plants and the entry of novel species into a habitat space (Beaver, 1984). Mohr (1943) termed these processes as the microseral stages of a microsere and the locations (objects) where the changes were taking place as **microhabitats**, but in fact understood clearly that they were habitats or **biotopes** with communities in their own right.

In all four types of breakdown mentioned above, there are common features: (i) initial contributions of associated fauna (e.g. using a senescing tree as a habitat component, particularly use of the object by animals not linked to breakdown); (ii) the novel immigration of initial colonists of the object; (iii) a turnover (object-specific immigration, colonisation and extinction) in the association and succession of animals and fungi; (iv) the characteristic breakdown by species within different biomes; (v) the particular breakdown taxa relating to the specific attributes of the object (form, size/dimensions, access, type); (vi) the gradual change in state (bulk, chemistry, etc.) of the object itself until, in effect, it disappears. These successions are **heterotrophic successions**.

From being a crucial and distinctive part of a sere, the most readily recognised breakdown of a habitat component involves the regeneration complex, the production of deadwood either through snag or tree fall and tree plate creation (Fig. 3.8; see also Fig. 1.7). **Deadwood** is a key component of ecosystem biodiversity and an important part of the carbon pool, also playing significant roles in maintaining nutrient recycling, regulating slope runoff, and providing sites for seedlings (Harmon *et al.*, 1986; Russell *et al.*, 2015). Many earlier vertebrate colonisers of senescing trees (e.g. squirrels, bats, owls, woodpeckers, jackdaws, nuthatches, titmice; Farris *et al.*, 2004; Smith *et al.*, 2008) use trees as a habitat utility resource though that may have little to do with further breakdown of the tree other than acting as initial excavators and inadvertent hosts of organisms directly involved in decomposition. In many circles of landscape

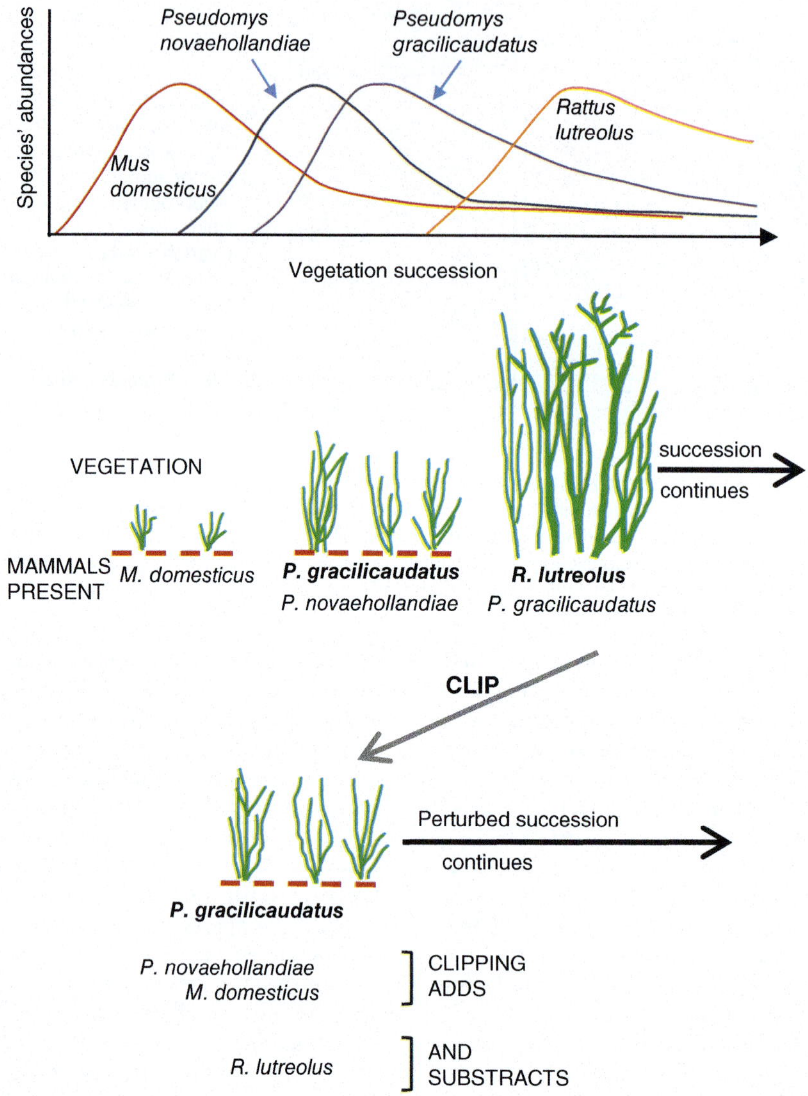

Fig. 3.7. A simplified summary of small mammal succession on wet heath habitat in Australia, altered by clipping, indicating the change of species with different conditions. (After Fox *et al.*, 2003, redrawn, used with permission of John Wiley & Sons.)

planning, it still seems to come as a surprise that deadwood is a microhabitat for a huge variety of fungi and a staggeringly abundant **saproxylic** fauna (i.e. animals that are dependent on or involved in the processes of wood decay) (Rayner and Boddy, 1988; Carpaneto *et al.*, 2015), the former of which primarily link up different fractions of soil, faeces and **phyllosphere** (that is, the aerial part of the plant that can be colonised by micro-organisms) (Zhao *et al.*, 2021). But, for instance, Kenis *et al.* (2004) identified 115 insects (beetles, dipterans) as potential predators of bark beetles (Curculionidae: Scolytinae) alone, a subfamily of saproxylic beetles totalling some 6000 species, and the Agricultural University of Georgia has recorded at least 21 genera of fungi on oaks (Bitsadze *et al.*, 2018).

Deadwood-related biodiversity constitutes about 30% of the global forest biodiversity (Vallauri *et al.*, 2005), and may represent even half of the local

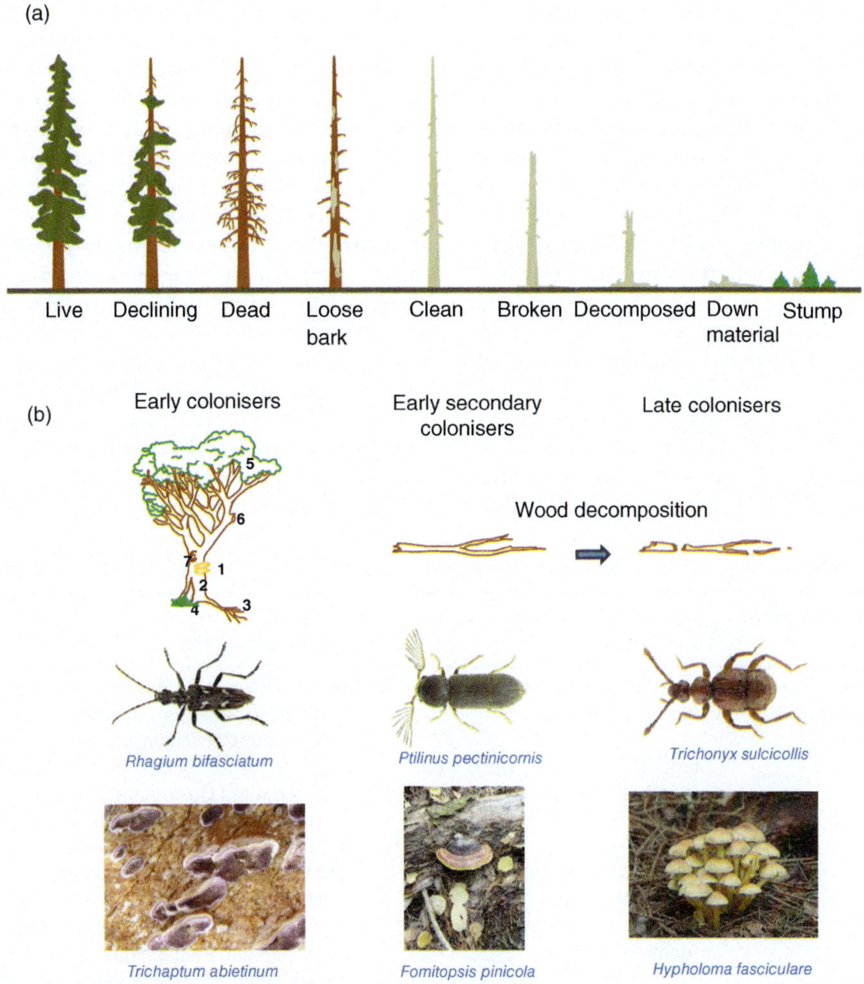

Fig. 3.8. Decomposition during the regeneration cycle in woodlands. **(a)** Development of snags over time and their degradation during the regeneration. Snags are common in situations where self-thinning occurs in self-regenerating damp conifer plantations; decomposition is primarily accomplished by microbes. Note: with decay of deadwood, there is an important distinction between coarse woody debris, fine woody debris and very fine woody debris (see Parisi *et al.*, 2018). **(b)** Saproxylic succession and saproxylic species. Upper section: saproxylic succession. Microhabitats for early colonist saproxylic species that can be provided by a senescing tree: 1 polypore fruiting bodies; 2 open wound; 3 dead root; 4 basal cavity; 5 dead sun-exposed branch; 6 branch cavity; 7 trunk cavity. Deadwood is then consumed by secondary and late colonists. Lower section: examples of saproxylic species of insects and fungi succeeding during wood decomposition. (Panel **a** after Maser *et al.*, 1979, redrawn. Panel **b** based on Parisi *et al.*, 2018, after Siitonen, 2012. Sources of photos: *Rhagium bifasciatum* by Francisco Welter-Schultes: https://commons.wikimedia.org/wiki/File:Rhagium-bifasciatum-06-fws.jpg, used under licence CC0 1.0 Universal. *Ptilinus pectinicornis* by Siga: https://commons.wikimedia.org/wiki/File:Ptilinus_pectinicornis_up.jpg, used under licence CC BY SA 3.0. *Trichonyx sulcicollis* by URSchmidt: https://commons.wikimedia.org/wiki/File:Trichonyx_sulcicollis_(Reichenbach,_1816).png., used under licence CC BY SA 4.0. *Trichaptum abietinum* by Jymm: https://commons.wikimedia.org/wiki/File:Trichaptum_abietinum.jpg. *Fomitopsis pinicola* by Tatiana: https://commons.wikimedia.org/wiki/File:Fomitopsis_pinicola_94751914.jpg, used under licence CC BY 4.0. *Hypholoma fasciculare* by Jörg Hempel: https://en.wikipedia.org/wiki/File:Hypholoma_fasciculare_LC0091.jpg, used under licence CC BY SA 2.0 DE.)

biodiversity of forest beetles (Köhler, 2000). As saproxylic organisms are menaced by the many threats that afflict forest ecosystems (e.g. forest reduction and fragmentation, pollution due to the use of chemicals against forest pests and simplification due to economic forest management), increasing attention is being paid to their conservation, also in order to preserve the essential, but frequently overlooked, roles they play in forest ecosystem functioning through complex processes of deadwood recycling (Carpaneto *et al.*, 2015; Fattorini and Galassi, 2016; Moose *et al.*, 2019). In short, deadwood forms a vital habitat, decomposing to total wastage, but nevertheless remaining readily available in a healthy forest environment due to the constant loss of healthy trees. Deadwood recycling involves a number of stages in a heterotrophic succession, the number depending on the size and position of wood debris. Speight (1989) identified three phases (colonisation phase, decomposition phase and humification phase), but more stages can be recognised depending on specific contexts (Table 3.2). Associated with these phases are distinctive saproxylic organisms (insects and fungi); specialists tend to be associated with one phase whereas generalists range over different phases.

Carrion recycling is operated by vertebrate scavengers (carnivores feeding on carcasses, such as hyaenas, jackals and vultures), invertebrates (especially dipterans and beetles), bacteria and fungi, through a complex process during which the carcass is completely transformed (Scott, 1998; Olea *et al.*, 2019) (Fig. 3.9). The breakdown and succession of invertebrate taxa in carcasses is of long-standing interest as it provides pathologists with the essential information to date the time of death of individuals for

Table 3.2. The different stages of wood decomposition and associated saproxylic taxa in European beech and silver fir.

Heterotrophic succession environment	Decay stage	Saproxylic organisms
European beech (*Fagus*) in France (Dajoz, 1967)	Stage 1: 1 to 4 years since the death of the tree	Coleoptera: Cerambycidae and Buprestidae. Predators: Coleoptera: Cleridae, Cucujidae, Zopheridae (Colydiinae).
	Stage 2: 3 to 7 years since the death of the tree	Coleoptera: Cerambycidae and Anobiidae. Less common: Coleoptera: Buprestidae and Curculionidae (Scolytinae). Predators: Coleoptera: Cucujidae, Trogossitidae, Cleridae.
	Stage 3: 6 to 10 years since the death of the tree	Coleoptera: Cerambycidae, Lucanidae, Tenebrionidae.
	Stage 4: >12 years since the death of the tree	Larvae of Diptera, larvae of Coleoptera (Staphylinidae: Pselaphinae and Ptiliidae) and various other soil invertebrates (Lumbricidae and Enchytraeidae, Oniscidea, Gastropoda).
Silver fir (*Abies alba*) in Austria (Schimitschek, 1952–1953)	Stage 1: The wood is little affected, with no change in colour	Coleoptera: Cerambycidae, Curculionidae Scolytinae, Lucanidae, Elateridae, Lymexylidae. Predators: Coleoptera: Cleridae, Staphylinidae. Diptera.
	Stage 2: The wood is in a more advanced decay and now is brown coloured	Coleoptera: Curculionidae (Scolytinae), Cerambycidae, Elateridae, Lucanidae. Predators: Coleoptera: Carabidae, Staphylinidae, Cleridae. Diptera: Syrphidae. Fungi.
	Stage 3: The bark is largely detached. The wood is reddish and very moist	Coleoptera: Elateridae, Cerambycidae, Staphylinidae. Diptera and ants. Fungi (Polyporales).
	Stage 4: Bark falls	Coleoptera: Pyrochroidae, Lucanidae, Elateridae, Staphylinidae. Soil invertebrates: Collembola, Psocoptera, Diptera, Annelida and Diplopoda.

(Based on Dajoz, 2007.)

Fig. 3.9. Example of decomposition of a pig carcass (*Sus scrofa*): **(a)** bloat stage, **(b)** active decay stage, **(c)** advanced stage, **(d)** dry/remains stage. Notice the extensive maggot feeding and the formation of the 'cadaver decomposition island' around the carcass (dark staining) in the active stage of decomposition. (Photos by Hbreton19: https://commons.wikimedia.org/w/index.php?curid=11817718, used under licence CC BY SA 3.0.)

criminal investigation (see references in Matuszewski *et al.*, 2020). This is an important topic for the stability of society and the research is naturally extensive (see references in Mann *et al.*, 1990 and Javan *et al.*, 2016). Crucial in this research is the identification of indicator species and their number (e.g. Watson and Carlton, 2005; Newsome *et al.*, 2021; Tembe and Mukaratirwa, 2021). Also important is the identification of insect stages, as different stages of development can provide useful insights to estimate the postmortem interval (Anderson and VanLaerhoven, 1996; Bajerlein *et al.*, 2018). Shi *et al.* (2009), in noting four stages in rabbit decomposition (fresh, bloated, decay, dry) between seasons in different years, recorded 49 insect taxa belonging to three orders and 15 families. They found no seasonal or annual differences, unlike other studies (e.g. carrion insects, such as calliphorid and sarcophagid dipterans, exhibit seasonal preferences: Introna *et al.*, 1991; Bourel *et al.*, 1999). However, as in human postmortem studies, they emphasise the importance of identifying stages of decomposers. A key feature, once more, is the sequence of insect species and succession, with an initial peak in dipterans and a later peak in beetles.

Breakdown in carrion depends much on the circumstances (biomes, burial, population dynamics of decomposers, etc.); for instance, the rate of decomposition is significantly affected by changes in temperature and humidity (rainfall) (Lopes de Carvalho and Linhares, 2001; Kŏćarek, 2003; Matuszewski and Szafałowicz, 2013). This is clearly shown by a growing number of decomposition studies in temperate environments (e.g. Archer and Elgar,

2003; Grassberger and Frank, 2004; Tabor *et al.*, 2005; von Hoermann *et al.*, 2023), in the Mediterranean area (e.g. Tantawi *et al.*, 1996; Arnaldos *et al.*, 2001) and in the tropics (e.g. Payne, 1965; Early and Goff, 1986; Oliva, 2001; Wolff *et al.*, 2001; Martinez *et al.*, 2006).

Experiments with cadavers in which the processes are driven only by insects would be regarded as simplistic in a natural context, where predators (e.g. grey wolves, *Canis lupus*) and scavengers often have their fill first before decomposition sets in. From this broader habitat vantage, a fascinating study on the initial wastage of carcasses (mainly ungulates) in the Białowieża Forest (Poland) (Selva *et al.*, 2005) has monitored the initial carrion use by scavengers such as ravens (*Corvus corax*), foxes (*Vulpes vulpes*) and pine martens (*Martes martes*), white-tailed eagles (*Haliaeetus albicilla*), jays (*Garrulus glandarius*), wild boars (*Sus scrofa*), domestic dogs and even great tits (*Parus major*), with clear indications of non-random scavenging in relation to mortality agents (i.e. by predator, natural death or harvest), habitat type (woodland versus open ground) and conditions. When effective, such initial scavenging leaves far less bounty for arthropods.

Studies of the breakdown of animal waste have been mostly focused on domestic animals in relationship to the impact of **dung** on pasture quality and the rate of dung decomposition (Castle and MacDaid, 1972; Dickinson *et al.*, 1981; Aarons *et al.*, 2004; Lee and Wall, 2006; Vadas *et al.*, 2011; Yoshitake *et al.*, 2014). Faeces attract a variety of insects involved in the breakdown of dung and the release of minerals to the soil, especially dung beetles and flies (Fig. 3.10). Dung beetles are universally known for their essential role in recycling dung, with members of the Scarabaeinae (namely the scarab beetle *Scarabaeus sacer* and allied taxa) being prominent for their complex parental care and the importance that they had in the culture of Ancient Egypt (Fattorini *et al.*, 2024). Dung beetles play a pivotal role in dung recycling by accelerating losses in dung moisture and dry matter and increasing nutrients within the top 10 cm of soil (Evans *et al.*, 2019). But dung beetles provide many other essential ecosystem functions, such as seed dispersal and protection of livestock from dipterans that develop in dung and that can be harmful to animals (see references in Palusci *et al.*, 2021).

The importance of dung recycling processes operated by these insects is dramatically illustrated by what happened in Australia after the introduction of cattle. As the native dung beetles of Australia have evolved to feed on the coarse-textured, pellet-like faeces of marsupials, only a few species were able to process cow dung. The persistence of enormous amounts of unburied cow dung created important problems in terms of pasture health, spread of flies, pests and diseases. To address these problems, at the end of the 1960s, the Commonwealth Scientific and Industrial Research Organisation (CSIRO) started a programme of dung beetle introduction lead by George Bornemissza (a Hungarian entomologist who arrived in Australia in 1950 and immediately conceived the idea of importing dung beetles). Bornemissza's original programme was really successful and, after its completion in 1986, it was followed by other similar initiatives, the last one being the Dung Beetle Ecosystem Engineers project (www.dungbeetles.com.au/). Overall, more than 1,700,000 dung beetles belonging to 43 species were released from 1968 to 1996 (23 species established; further introductions are under study) (Bornemissza, 1976; Edwards, 2007; Edwards and Pavri, 2007; Doube, 2018a,b). However, the positive effects of the introduction of dung beetles have been countered by those produced by cane toads (*Rhinella marina*), which were introduced to control scarab beetles eating sugarcane but which consume also the introduced dung beetles, thus reducing the rate of cowpat breakdown (González-Bernal *et al.*, 2013).

Dung is of considerable importance to the maintenance of arthropod and fungal diversity, including the immediate surrounds, which for bovine dung pats are called 'islets' (turf avoided by feeding cattle) and provide microhabitats for spiders and bugs (Hemiptera) (Dittrich and Helden, 2012). Dung and its associated microhabitats are also especially important to organic farming, where maintaining soil health and biodiversity is a central goal. Organic management improves soil biology quality (Mantoni *et al.*, 2021a) and organic systems retain 50% more insects than do conventional farming systems (Geiger *et al.*, 2010). Thus, understanding dung ecology is crucial for advancing more sustainable agricultural practices. For the UK an excellent key is available for organisms colonising cow pats which also illustrates the cow pat ecosystem (Skidmore, 1991). Accounts have been typically made of arthropods and other organisms during decay (Laurence, 1954; Pecenka, 2017; e.g. Coleoptera such as *Aphodius* sp. (Scarabaeidae), Diptera, Hymenoptera: Lee and Wall, 2006; Lumbricidae:

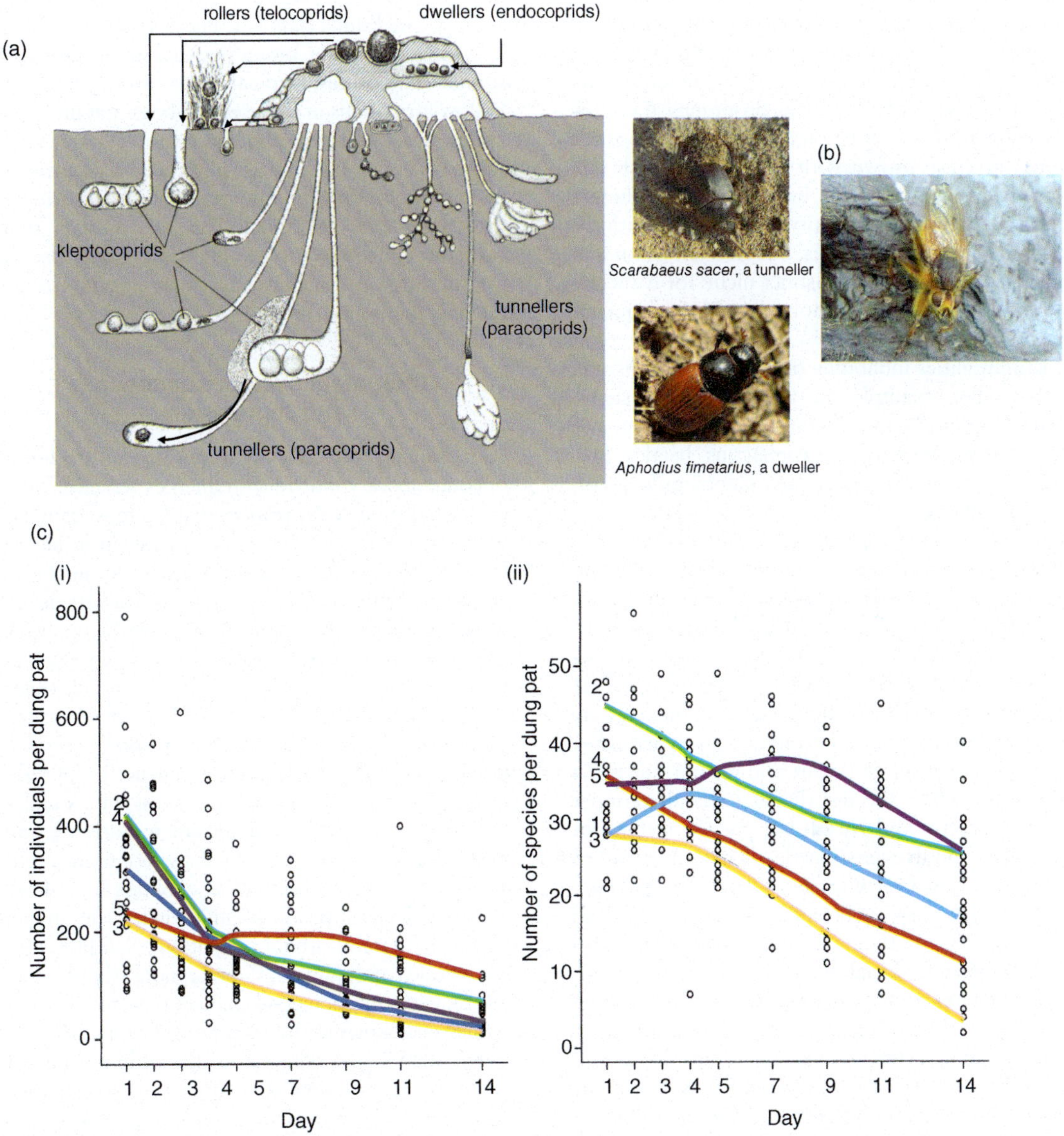

Fig. 3.10. Heterotrophic succession and development in the microsere of cattle dung are largely driven by dung beetles and dipterans. **(a)** Species of dung beetles (Scarabaeidae) are traditionally grouped in three ways: i.e. 'dwellers' (endocoprids), 'tunnellers' (paracoprids) or 'rollers' (telocoprids) based on differences in their nesting behaviour. Kleptocoprids steal the balls from the telecoprids. **(b)** Male yellow dung-fly (*Scathophaga stercoraria*) on cow dung. **(c)** Number of coprophilous beetle individuals (i) and species (ii) per dung pat during succession. Lines are local regression curves for each sampling cycle: 1 early spring; 2 late spring; 3 early summer; 4 late summer; 5 autumn. (Panel **a** redrawn and modified from Doube, 1990. Photo of *Aphodius fimetarius* by James Lindsey at Ecology of Commanster Site: https://it.wikipedia.org/wiki/File:Aphodius.fimetarius.jpg. used under licence CC BY SA 2.5. Photo of *Scarabaeus sacer* courtesy of Letizia Di Biase. Panel **b** photo by Peter O'Connor, anemoneprojectors: https://flickr.com/photos/58414938@N00/34468849956, used under licence CC BY SA 2.0. Panel **c** redrawn from Sladecek *et al.*, 2013, used with permission of Elsevier France.)

Bacher *et al.*, 2018). As with other successions, concentrations and sequences of arthropod and fungal species have been investigated (e.g. Dickinson and Underhay, 1977). Mohr (1943) described six stages in the decay of cattle droppings (i.e. cow pats), from a pioneer stage of freshly deposited dung (greenish-brown with a distinct tarnish and film), through to a light brown, dry chip before they finally lose their identity and merge in with the background substrate. This, like tree fall or an animal carcass, provides a distinct niche for a succession of biotas, in which Mohr plotted the sequence of insects.

Communities inhabiting dung pats show distinct phases. For example, in temperate ecosystems, dung is typically first colonised by dipterans (see Fig. 3.10b), then by coprophagous beetles (e.g. Hydrophilidae, Scarabaeidae, Geotrupidae) followed by carnivorous beetles (Histeridae, Staphylinidae and Carabidae), and finally by earthworms, mites, nematodes and fungi (Menéndez and Gutiérrez, 1999). Different successional patterns of arthropods have been found in different parts of the world, for instance in the dominance and succession of dung beetles, divided up into rollers (beetles that feed in the dung pat as adults and lay eggs within or under the dung mass where they undergo larval development and do not construct nests), tunnellers (beetles that dig chambers more or less directly underneath the pat for feeding or breeding) and dwellers (beetles that form a ball of dung that can be rolled away from the pat and buried for feeding or breeding) (Sabu *et al.* 2006) (see Fig. 3.10a). Dung beetle communities also show important variations in the diversity of rollers, tunnellers and dwellers in relation to climatic conditions that influence pat transformation: in cold and humid climates, excrement remains humid for long periods and dung removal is absent or slow, thus allowing dwellers to dominate, but in tropical and arid areas, where excrement tends to dry out and is removed quickly, rollers and tunnellers are favoured for their ability to protect the offspring through subterranean nests (Mantoni *et al.*, 2021b; Fattorini *et al.*, 2024). Once again, it is a habitat in microcosm with organisms immigrating, colonising, flourishing, then dying or emigrating to seek out new patches. An extremely detailed and intensive study of the assemblage of coprophilous beetles (guilds of coprophagous and associated predatory species) was carried out by Sladecek *et al.* (2013) in which a total of 39,627 beetles belonging to 85 species were recorded using artificially created dung pats (see Fig. 3.10c).

Within the larger biomes or habitats, there are similar microseres; typical ones are hummock-hollow formations in pastures and bogs that undergo cycles of development and dissipation (Watt, 1947; Kershaw, 1973). These hugely influence the occurrence of arthropods; one apposite example is the existence of *Eriophorum* hummocks on raised bogs that form sites for oviposition and larval development for the large heath butterfly *Coenonympha tullia* (Dennis and Eales, 1997, 1999).

Geological cycles: succession and regression

Successional processes extend well beyond the limits of contemporary individual human experience. On a geological time scale, readers will be most familiar with vegetation succession and regression in higher latitudes during the sequence of glacial-interglacial cycles, particularly changes following the last deglaciation (Roberts, 2014). It is now possible to track shifts in entire communities by examination of ancient environmental DNA (eDNA) in substrates (soil, deposits) as in north Greenland (Kjær *et al.*, 2022). Iversen (1958) was the first to produce a distinctive recurring glacial-interglacial model of **soils** and vegetation: development to a forest **climax** under mature brown-earth soils with climate warming, and regressive succession to permafrost conditions and skeletal arctic soils with climatic cooling (Fig. 3.11a). We are still enveloped in Milankovitch cycles driven by Earth's natural orbital oscillations (viz. planetary eccentricity, obliquity and precession; Campisano, 2012), with environmental changes that mirror those witnessed following the recession of world glaciers after a nadir in conditions in the Little Ice Age of recent centuries (Oliva, 2018) but on a longer time scale. The vegetation changes have long been traced using palynological and allied techniques (Green, 2001), the trends clearly detected in the relative abundance of herb-shrub-tree pollen species in fine layers of sediments, typically peat deposits but also soils (Godwin, 1975; Okuniewska-Nowaczyk *et al.*, 2018; Buso *et al.*, 2019). A picture of changes over regions was collated from innumerable dated pollen diagrams (Godwin, 1975; Huntley, 1990) into maps illustrating waves of colonisation and regression over Europe (Huntley and Birks, 1983; Giesecke *et al.*, 2017) and North America (Delcourt and Delcourt, 1987; Ordonez and Williams, 2013).

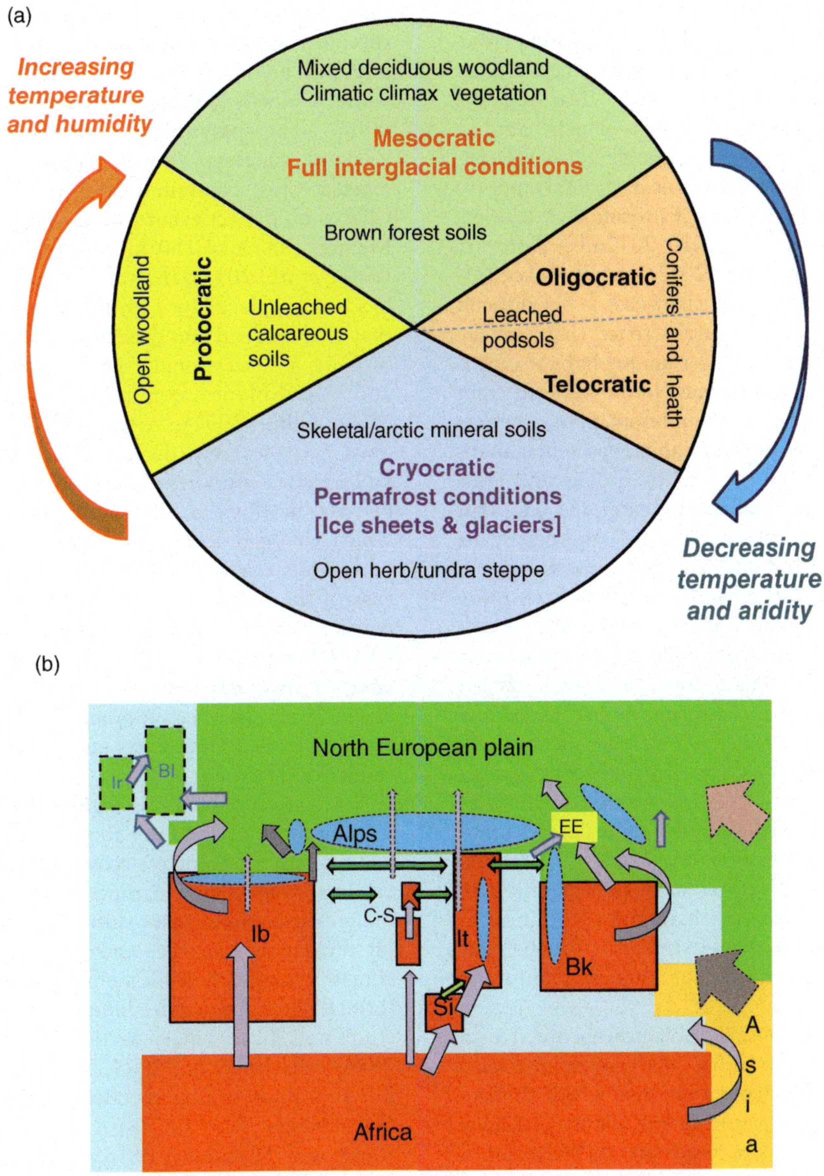

Fig. 3.11. Successional processes and glacial–interglacial cycles. **(a)** Iversen's (1958) model of changing north European environments during a glacial–interglacial cycle. **(b)** Glacial refuges. Topological map of European continental refuges for thermophilous species and other organisms during glacial stages and their tracks of recolonisation immediately with deglaciation. The main refuge centres are in Iberia (Ib), Corsica-Sardinia (C-S), Italy-Sicily (It, Si) and the Balkans (Bk) but with other centres in North Africa and in the Near and Middle East. For more cold-tolerant faunas there were dispersed, limited refuges north of the southern limit of the Alpine chain in Eastern Europe (EE). BI, Britain and Ir, Ireland. Purple tracks, latitudinal movements; green tracks, east–west movements. Blue bands are glaciated major mountain barriers to dispersal; orange, Asian land mass holding other refuges especially for wide-ranging species. It is important to appreciate that refuges were far more finely divided than illustrated in this model, especially within the Alps. (Panel **a** from Dennis and Hardy, 2018.)

A crucial aspect of these cycles is mass migration and isolation, leading to both evolution and extinction of plants and animals. Biogeographical studies have allowed the identification of clear regional distinctions in plants and animals which relate to their isolation in refuges during glacials and interglacials (e.g. European refuges: Dennis *et al.*, 1991; Dapporto *et al.*, 2011a, 2019, 2024; Fattorini and Baselga, 2012; Fattorini and Ulrich, 2012a,b; Fattorini, 2014b; Schmitt and Varga, 2012; Schmitt *et al.*, 2021) (see Fig. 3.11b). With such changes it is not difficult to imagine that any arbitrary space that may be considered momentarily as a unique habitat unit is being continually transformed into a very different habitat space. This is the important point: there is continual turnover in every habitat space that spans the time scale from seconds to epochs and eras. A whole set of novel organisms enter and ones that have been able to colonise the space and persist for a specified period of time have become extinct.

In no circumstance has this process been more impressed on human observers than recovery of the planet after **mass extinctions** (Fastovsky and Sheehan, 2005; Vajda and Bercovici, 2014; Brusatte *et al.*, 2015; Fattorini, 2020). Data now exist which clearly reveal a long-term successional process following the Cretaceous-Palaeogene mass extinction event (K-PgE; Lyson *et al.*, 2019). The major difference with successional processes monitored over terrestrial habitats during the past century, and even the recovery of vegetation in Europe following the last maximum glaciation, is that recovery after mass extinctions involves far more than recolonisation of familiar plants and animals.

Mass extinctions in the geological record (Elewa and Joseph, 2009; Fattorini, 2020) are characterised by an evolutionary chain of events in which old taxa permanently disappear from the planet and new taxa evolve (Alroy, 1999; Fastovsky and Weishampel, 2005; Nicols and Johnson, 2008; Vajda and Bercovici, 2014; Donovan *et al.*, 2016), a slow process to human perception but one much faster than the gradual evolution of life on the planet since the Hadean (Knoll and Nowak, 2017). Solé *et al.* (2002) describe recovery after a mass extinction as being very similar to ecological succession, but with the important distinction that the fossil record has innovative patterns of taxa and **life-history traits** emerging with novel niche availability following the extinction of entire clades (see also Jablonski, 2001, 2005; see Fig. 11.1). These changes also witnessed shifts in the loss and maintenance respectively of **specialist** and **generalist** species, as in the case of insect herbivores in the Williston Basin of southwestern North Dakota, USA (Labandeira *et al.*, 2002). The **palynofloral** signature for New Zealand suggests instantaneous destruction of diverse forest communities as postulated by the Chicxulub impact event (the Yucatán Peninsula in Mexico, diameter 180 km, 66.043 ± 0.043 Ma; Renne *et al.*, 2013), the vascular plant communities being replaced by a proliferation of organisms that feed on dead and decaying matter followed in turn by a successional recovery of pteridophyte and seed-plant communities (Vajda and McLoughlin, 2007). Arthropod taxa will also have been severely disrupted; for example, the DNA record supports the notion of extinctions at 66 Ma in flower-feeding insects such as bees, which emerged along with eudicot plants in the mid-to-late Cretaceous (Rehan *et al.*, 2013). In the case of the K-PgE in Colorado, USA, large mammals only began to appear in the geological record by c.700 ka post K-PgE, with the initial appearance of the nitrogen-rich Leguminosae (Lyson *et al.*, 2019). In effect, a quantum shift in ecosystem processes accompanies mass extinction events (Hull and Darroch, 2013).

But the transformations of these ancient habitats went far beyond all of the above, and new changes will similarly affect today's ecosystems. With geological time, coupled with disruptions in plate tectonics, individual habitat spaces shift in physical location on the Earth's surface, and with accompanying climate changes their biome affiliations are transformed (e.g. tropical habitats in the geological strata now located in Antarctica) (Herngreen *et al.*, 1996; Reguero *et al.*, 2013; Vajda and Bercovici, 2014). It is interesting to reflect on the impact that such changes have on dispersal paths, migration routes and the stability of trophic diversity in past and present communities (Van Valkenburgh, 1988). This may indeed take events beyond the relevance of the contemporary context of habitats to readers, but then pending greenhouse climate changes may well simulate some of these vast changes over the next century (Dennis, 1993a; Loarie *et al.*, 2009; Benito-Garzón *et al.*, 2014; Malhi *et al.*, 2020; Butt *et al.*, 2021; IPCC, 2022). More than anything, this should bring home to readers the instability of habitats, and associated floras and faunas, with time and perspectives beyond their own existence.

Seasonal patterns and processes in the habitat space

Heterotrophic successions of animal species in carcasses or dung involve much faster processes than vegetation (autotrophic) successions. This is true, too, of seasonal patterns of habitat occupancy by animals, changes well within the bounds of individual human perception. Seasonal movements affect land-based organisms as well as sea creatures (see Chapter 8) with a variety of patterns. A basic, though admittedly incomplete, division of seasonal events in habitats separates species that occupy a habitat unit permanently throughout the year from those that change fidelity between different habitat units with the seasons. Since seasonal changes expose organisms to variations in environmental conditions, the perpetual occupancy of habitats through an annual cycle indicates that both conditions and resources, though changing, are nevertheless suitable for the same organisms, if not the same breeding individuals, all year. This is common in equatorial habitats, where conditions remain similar through the years. But a multitude of habitat types are placed within seasonal climate zones, especially when climatic changes are accentuated (e.g. warm versus cold regimes as with arctic to temperate climes; wet versus dry environments as in Mediterranean and tropical seasonal climates; wide variations in high altitude environments); the cold and dry seasons place limits on breeding and survival for most organisms of such regimes. Coping with harsh climatic regimes requires distinctive adaptations.

The adaptations to a downturn in conditions fall into three basic groups. Organisms may (i) stay and adapt (change traits) to a downturn in conditions and remain active (for example, wolves adjust their hunting strategies and behaviour according to the availability of food and the environmental conditions); (ii) stay and adapt to conditions but adopt dormancy (hibernation in winter and aestivation in summer; e.g. reptiles and mammals that hibernate; insects that aestivate) during the downturn in conditions; (iii) move away (migrate) to locate better conditions with suitable resources elsewhere (e.g. migratory birds). An additional group (iv) occurs owing to the effect of animals themselves on the resource base in addition to changes in seasonal environmental conditions; they may engage in perpetual movement over a wide area to ensure a supply of suitable resources and conditions (e.g. zebras, wildebeest). The first two groups, and perhaps the fourth, may be regarded as permanent residents of a habitat unit, the last differing greatly in dimensions and in heterogeneity of conditions and resources. Species that occupy distinct breeding and non-breeding units (group iii) are said to engage in seasonal migration; a feature often forgotten is that they pass through distinct habitats between the extreme locations adopted. The last group (iv) may be argued to occupy extensive and varied habitat types or alternatively be permanent migrants; in effect, their habitat might be an entire biome. This simple division, focusing on seasonal movements, brings into focus what is, or is not, an appropriate habitat unit for an organism.

For migrating species, there is the further issue of fidelity to breeding units as well as non-breeding destinations and the distances involved. If individuals survive the journey, migrating species may (i) return to the same habitat unit the following season (e.g. the barn swallow *Hirundo rustica*) or (ii) return to the same (or similar) region – but not to the same habitat unit – to use a similar habitat unit nearby or at some distance from that used in the previous year (e.g. the Arctic tern *Sterna paradisaea*). Individuals of species which permanently occupy the habitat unit may be regarded as being located in a suitable habitat space but care is needed here, as they may alternatively have difficulty in moving from unsuitable habitats not evident to a human observer. The reader should also be aware that a *species* may occupy the same habitat unit throughout a year but be replaced by different *individuals* (e.g. resident breeding and non-breeding migrating blackbirds *Turdus merula* in UK suburban winter gardens; Ashmole, 1962). Those species that engage in migration and move out of a breeding habitat unit may use an entirely different habitat type abroad during the environmental downturn (e.g. cuckoos, ospreys, chiffchaffs; Berthold, 2001). The question then becomes: what is their true habitat type, the one where they breed, the one ensuring survival during environmental downturn, or both, including the space and conditions in between? In Chapter 5, we argue that all these domains form their habitat, despite the immensity of global space covered by the animal in the course of a year.

Staying in and leaving a habitat unit depends on species' **traits**, the evolution of special attributes that ensure completion of their life history. Those staying require special traits to withstand harsh conditions (e.g. drought or cold) and the lack of resources accompanying those seasonal constraints (e.g. polar bears in Hudson Bay, Canada; Peacock *et al.*, 2010). It is worth noting that species engaged in seasonal movements will tend to be large animals requiring more resources than available in the downturn in conditions. For example, in savanna ecosystems, large gregarious herbivores such as zebras and wildebeest are forced into perpetual movement to seek out adequate resources (especially water) as, of course, are their accompanying predators (lions, hyaenas, etc.) (Strauch, 2013). Species that are more mobile (e.g. those that have a specific mobile stage) are also at an advantage in moving away from difficult conditions. This is particularly the case for species that are alate and may include small animals, as massive long-distance dispersal is known for butterflies (e.g. *Vanessa cardui, Danaus plexippus*; Agrawal, 2017; Stefanescu *et al.*, 2016) and other arthropods (e.g. locusts, spiders; Peng *et al.*, 2020; Woolley *et al.*, 2016), as well as for more powerful birds. Even so, dispersal behaviour from the natal habitat may not involve long-distance movements, for instance individuals moving through similar habitats searching for food (e.g. titmice in woodlands; Morse, 2008). It may also involve solitary as well as mass movements of a species or collections of species.

A simple way of determining the possible combinations of habitat occupancy and abandonment is to construct contingency tables of species' presences and absences within habitats. The value of such an approach may be demonstrated using the classic work of Robert Fuller (1982) on bird habitats in Britain based on data from the Register of Ornithological Sites of the British Trust for Ornithology. The data have been collected for seven basic habitat types: woodland/scrub, 'field' vegetation, wetlands, water bodies, open (non-coastal), coastal and miscellaneous, this latter class including mainly human land uses (such as farmland, hedges, rail tracks, sewage farms, brownfield sites and waste land), each divided into 4–10 subtypes. For every major habitat type and its subdivisions, clear patterns emerge. Records of species for habitat types and specific locations (units) differ dramatically between summer (breeding season) and winter (feeding and roosting). The data reveal changes in the presence of species and changes in bird diversity. Four basic conditions for any species emerge: presence in both seasons, in one or the other only, and in neither. An insight is provided into habitat use with percent occupancy of woodland units in summer and winter (Fig. 3.12). The picture becomes more complicated if an axis is constructed for migration. Just how complicated depends on what is meant by migration, but it is complex enough even if migration is restricted to movement from/to Britain. This gives an eight-cell model (Fig. 3.13). The most typical combination is for species which occupy the breeding habitat in summer to breed, abandon it in winter and migrate, often thousands of miles to warmer climes (cell b in Fig. 3.13; e.g. birds such as swallows, swifts, house and sand martins; Winkler, 2006). Another obvious category includes species that do not breed in British habitats but do occupy them in winter, and have migrated here from northern latitudes (cell c in Fig. 3.13; e.g. Turdidae, such as fieldfares and redwings; Milright, 2003; Skórka *et al.*, 2006). However, two combinations (cells g and h of Fig. 3.13) throw up interesting difficulties from the vantage of understanding traditional habitats. They both record the lack of fidelity (i.e. lack of substantial occupancy of the habitat in either season): one includes rare winter visitors or irregulars (cell g of Fig. 3.13), and the other (cell h of Fig. 3.13) includes birds that appear only occasionally and unpredictably and that we might distinguish as **vagrants**, that is, individuals briefly using a resource (or resources) or just passing through the unit (see Box 3.3). The inference is that the space in which they are recorded is of value in some way, even if only *en passant*.

Most ornithologists tend to restrict their analyses of bird **communities** to the breeding species, thus excluding migrants and vagrants. However, migrants and, to a certain degree, vagrants may play important roles in the community, being (albeit for a short time) prey, predators or competitors for the resident populations. The problem of which species should be considered as part of a community does not affect birds only, but virtually all organisms, given the dynamic nature of communities. Most communities are characterised by the presence of a certain number of species which are represented by few sampled individuals, for which it might be extremely difficult to establish whether they are vagrants or belong to breeding populations of very rare species (see Chapter 4). Similarly, the guidelines developed by the International Union for Conservation of Nature (IUCN, 2012) for assessing species' conservation status at regional level consider vagrants (defined as taxa that are currently found only occasionally within the boundaries of a

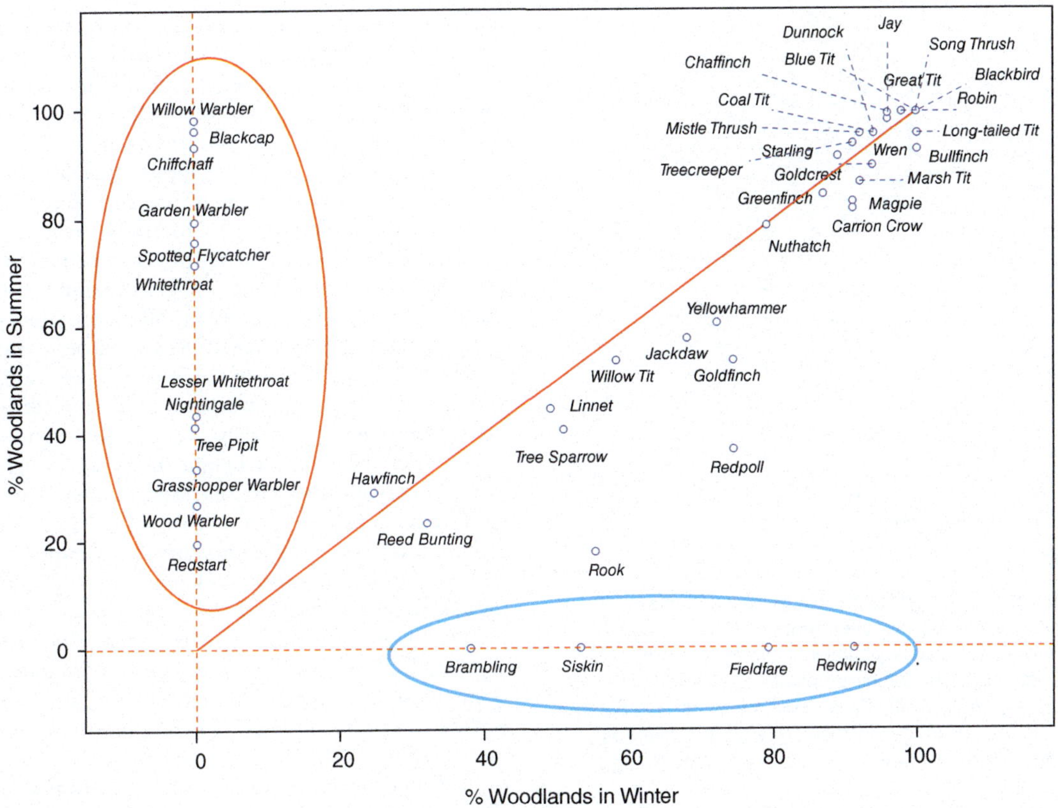

Fig. 3.12. Seasonality in woodland birds. Three broad groups of species are observed: winter visitors (blue ellipse), summer breeding birds (red ellipse) and birds resident all year (species close to the bisector). The latter group, however, consists of individuals of species that are not necessarily located in any single habitat unit, birds replaced by vagrants and/or migrants. (Data from Fuller, 1982.)

region) as ineligible for assessment, whereas **visitors** (defined as taxa that, although they do not reproduce within a region, either regularly occur within its boundaries now or have so occurred during some period of the last century) are eligible, but the dividing line between the two categories cannot be easily established.

Vagrants and visitors may not be considered in the reckoning of habitat components by observers (see Chapter 4), but they have relevance nonetheless for the individual and should make us think about what constitutes habitat for a species and what does not. It is this issue that we explore in some detail when turning to the **resource-based habitat** (see Chapter 5).

Determining combinations of habitat occupancy and abandonment can be particularly satisfying, but nature is not that simple. For instance, there are notable exceptions that contest the strict alternatives of a contingency table on habitat occupancy. We have dealt with one immediately above, that of habitat fidelity (permanency, vagrancy), but there are others. For example, there are species (as noted above) that occupy particular habitats all year but are nevertheless joined by conspecifics that migrate during winter months (e.g. blackbirds arriving from central Europe at much the same time as redwings and competing with UK garden blackbirds). For many species, population **density** in a habitat can fluctuate dramatically during the year. For example, the numbers of individuals change markedly between summer and winter for birds occupying particular habitat units all year. This arises as a result of territoriality in the breeding season, each pair of birds defending spaces (see Chapter 8), whereas in winter they may feed gregariously without contest (e.g. waders, plovers on coastal marshes; Fuller, 1982). For the same reason, diversity (expressed as the number of species or with other indices;

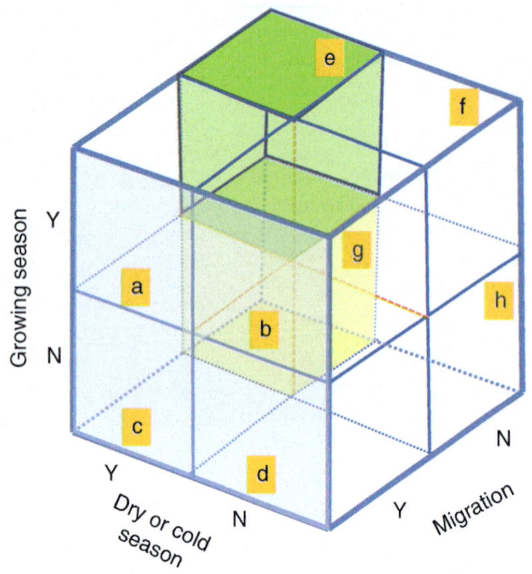

Fig. 3.13. The potential combinations for species of habitat occupancy (summer, winter) and migration behaviour. Eight binary combinations are evident for species, but more combinations are possible when the focus is on individuals. Cells e and g are coloured purely to help readers to identify them. Note: a, b, c and d are all at the front whereas e, f, g and h are all at the back.

see Chapter 4) may actually be much higher during the non-breeding season for the same reason. Additionally, fidelity to a specific habitat unit may differ enormously between seasons. For example, fidelity to a location may be high in the breeding season and fixed on territorial space around nest sites; in winter, however, the same birds (e.g. titmice in woods) may occupy the same habitat type but move between different units to search for food supplies, which tend to be exhausted more rapidly especially as many species and individuals all seek resources, and aggregate to escape predators which lock onto predictable prey. The same individuals (and species) may occupy very different habitat types in summer and winter (e.g. the curlew *Numenius arquata* nests on heaths, bogs, grasslands and high moors in summer but is forced by frozen high ground and inaccessible food supply to occupy low-lying farmland, mudflats and estuaries in winter; Cotter, 1990). Finally, the same species may be found breeding in very different habitat types (e.g. the wren *Troglodytes troglodytes* from gardens and woods to moors and mountain areas; Moss, 2018). The inference is that suitable resources and conditions, though perhaps not of equal standing, are available in different habitat types.

These changes in seasonal fidelity are by no means restricted to birds. Insects such as butterflies are known to have similar patterns of habitat occupancy and vacancy, including species that have little fidelity to a specific habitat type or indeed to a specific habitat unit (precise location) from one generation to the next and shift over large regions, crossing various habitat types (e.g. the painted lady *Vanessa cardui*; Dennis, 1993a, 2020; Stefanescu *et al.*, 2016). One difference between insects and vertebrates, imposed by their size and their behaviour, is that their population structures differ considerably within habitat units they occupy (for example, insects often form localised populations due to tight associations with specific resources, while vertebrates tend to be more dispersed but organised in social groups), an issue we return to later (see Chapters 6 and 8). Moreover, seasonal changes in a habitat, though occasionally distinctive, do not necessarily involve any movement in and out of specific ecosystems. A classic example is the sequence of blooms occurring in lakes owing to nutrient flushes, as at Rostherne in Cheshire, UK, where a bloom of diatoms early on is followed by one of dinoflagellates or colonial cyanobacteria such as *Anabaena* (Reynolds, 2019).

Diurnal cycles in and out of the habitat space

To casual observers, the most obvious changes in temporal patterns of animals are shifts of individuals – associated with changing behaviour – within habitats during a diurnal cycle. Organism behaviour and location within the habitat space occur in relation to abiotic stresses (light, heat, moisture, windspeed, etc.) and biotic resources (food, searching for mates, copulation, rests, roosts, community interactions, etc.). These movements are integral to the concept of a **resource-based habitat** and are investigated closely in Chapters 5 and 6. Most movements may well be within the space identified by researchers as a habitat unit but many movements by nearly all animals extend beyond recognisably homogeneous habitat types, depending, of course, on what a researcher deems to establish as a homogeneous habitat unit (see Chapter 1; Fig. 1.8). Size and mode of movement are critical factors in degree of habitat unit fidelity; the smaller and slower the animal, the more likely it is to remain within a habitat unit during a day. Yet even very slow-moving animals are not only capable theoretically of covering many hundreds of metres in a day (e.g. the mollusc *Helix aspersa* can move at >1 cm/s; Pembury Smith and Ruxton, 2021), but are found in a range of different habitat types. For example, the mollusc *Cepaea nemoralis*, a well-known model organism for adaptation of its colour and banding phenotypes to different habitats including grassland, woodland and **ecotones** such as hedgerows (Ford, 1964; Silvertown *et al.*, 2011), is clearly capable of passing between distinct habitat units (Peltanová *et al.*, 2012).

Within any season, a classic case of changing behaviour relates to differences in activity and location with the nocturnal–diurnal cycle. For instance, bats typically rest up in one habitat by day and at night can range over very different habitat types to obtain food. For example, the Thai nectivore dawn bat *Eonycteris spelaea* roosts in caves by day and extends over agricultural land and forest at night (Acharya *et al.*, 2015). The insectivorous greater horseshoe bat *Rhinolophus ferrumequinum* uses both conifer plantations (27%) and natural forests (73%) as pathways (commuting lines) and feeding habitats in Japan (subspecies *nippon*, Fujioka *et al.*, 2021). In earlier studies conducted in the UK, *R. ferrumequinum* (typical form) preferred to hunt over pastures and forests (Duvergé and Jones, 2003; Flanders and Jones, 2009), whereas the bats in continental Western Europe preferred residential areas and meadow orchards (Dietz *et al.*, 2013). Different environments present different opportunities for resource exploitation.

Shifts within and between different habitat units are regular features of arthropods, birds, mammals and reptiles. For example, foxes (Goldyn *et al.*, 2003), badgers (Piza-Roca *et al.*, 2018), wood pigeons (Fuller *et al.*, 2004; Fey *et al.*, 2015) and carrion crows (Preininger *et al.*, 2019) have their base (roosts) within woodland or at the edge of it, but forage extensively over open ground, particularly farmland, and into the suburbs (Simms, 1990; Parsons *et al.*, 2020). Rooks (*Corvus frugilegus*) may occupy the suburban fringe for nesting, as they did in the first author's undergraduate days in Durham University, UK, though they feed on farmland beyond the urban surrounds (Griffin, 1998; Kasprzykowski, 2003, 2007). The city of Rome, Italy, hosts immense roosts of common starlings (*Sturnus vulgaris*) that may number millions of individuals: these birds spend most of the day in rural areas for feeding (especially on insects), but come back to the city at the sunset, and stay the night in these roosts to take advantage of many favourable conditions offered by the urban environment, such as higher temperature, lower humidity and wind, and fewer predators (Fattorini, 2019). It is much harder to demonstrate diel shifts amongst invertebrates such as butterflies, beetles, flies and wasps – as observations then need support from detailed **mark-recapture** survey, constant tracking or simultaneous sampling in different habitats through time (e.g. at different hours of the day) to infer movements from temporal changes in abundances across habitats. Nevertheless, for many insects, activities are known to extend over different habitat types during a single day, often shifting with available adult food resources (e.g. the meadow brown *Maniola jurtina* and the gatekeeper *Pyronia tithonus*; Dennis, 2004a; west North American grasshoppers; Kemp *et al.*, 1990) or to track changes in environmental conditions, even at very fine scale (e.g. tenebrionid beetles of Mediterranean dunes that move across different sectors of the sea–inland gradient in response to daily changes in temperature and moisture; Fattorini, 2023).

Many sea organisms also occupy different habitat types on a daily basis or over longer periods, showing a variety of movements, such as: (i) the intertidal movements of animals (including amphipod crustaceans and other arthropods, molluscs, fishes and birds; Chelazzi and Vannini, 1988; Gibson, 2003; Holsman *et al.*, 2006; Kimball and Able, 2012; Silva *et al.*, 2014) and micro-organisms (e.g. diatoms; Barnett *et al.*, 2020); (ii) the immense diel vertical migrations of organisms (including crustaceans, molluscs and fishes) that move up to the **epipelagic zone** at night and then back to the **mesopelagic zone** during the day (Robinson *et al.*, 2010), a phenomenon that involves the synchronous movement of such an enormous biomass that, when sonars were firstly widely used during the Second World War, operators were confounded by an apparent solid plane at a depth of 300–500 metres during the day but nearer the surface at night, an unexpected acoustic signature they called the 'phantom bottom' (Carson, 1951); (iii) migrations of sea turtles that use tropical, subtropical and temperate beaches for digging their egg chambers (Eckert *et al.*, 1989; Lutz and Musick, 1997).

All these changes noted for animals in space-time are indicative of exceptions to strict habitat unit-fidelity or habitat unit-community stability viewpoints.

From the above examples alone, it should be obvious that habitats are subject to continual change. For the traditional habitat unit, conditions within the entire unit depend on the immigration and colonisation of plants, animals, fungi, lichens and their accompanying micro-organisms. Most importantly, there is a perpetual turnover of species, novel colonists and species that emigrate or become locally 'extinct'. Depending on the unit taken to be the habitat (from a single organism to a forest), these changes can occur in seconds or repeatedly over millennia. When considering the various measures of diversity in Chapter 4, such changes should constantly be borne in mind. For instance, an important question arises regarding the preponderance of **rare species** in habitats: in these temporal cycles of change in abiotic and biotic conditions, what is the status of rare species? Are they simply the ones that have been left behind, diminishing during the environmental changes and precariously 'hanging on', or, on the contrary, others that infrequently enter the domain temporarily or as new colonists? If the former, how is it that they are indeed maintained at such low frequencies? It would be an interesting exercise to partition rare species within adjoining habitat units into these categories over time, to see if such patterns for individual species can be distinguished and for how long they are maintained. This issue is addressed further in Chapter 11.

4 Traditional Habitat Principles and a Diversity of Measures

Abstract

Key measures of habitat composition are species richness, evenness in species abundances and disparity (differences) among species. Habitats are governed by a set of laws and rules. Principal among them is the occurrence of few common and many rare species in each habitat unit. Species are unevenly distributed in habitat units. Their variable status over the habitat space likely relates to their traits, resources and environmental tolerances. Habitats are open systems exposed to immigrations and emigrations of species. Abundances in habitat units relate to spatial cover inside and outside the habitat space, though this depends on spatial scale, population dynamics and mobility of the organisms. Species can be divided into specialists and generalists in niche breadth, resource use and tolerance of conditions, affecting their status in and out of habitats.

Understanding ecological processes depends on the careful survey of species and the measurement of abiotic factors within **communities**; typically, measurements are taken within the bounds of traditional habitat units. But, whether measurements on biotas are representative of a habitat depends entirely on the degree to which individuals of species forming communities are confined to the habitat space, that is, the extent to which communities in habitat units overlap with one another. There is a tendency to regard measures taken on populations in any habitat unit (e.g. counts of individuals) as applying entirely to populations confined to the habitat unit but – despite notable exceptions (e.g. tree species within isolated woodlands) – this is all too often an unrealistic assumption. As will have been gleaned from the previous three chapters, patterns and processes in habitat units worldwide, across different Earth spheres, have a complex ecosystem foundation resulting in extensive networks stretching across such habitat units at different space-time scales via the interaction of organisms with the abiotic environment and with each other. In this chapter, attention is given to issues affecting community integrity within habitats and some basic measures made on individuals, species and higher taxa in habitats: these measures are applied to determine the degree to which ecological principles are valid.

Initially, it is essential to appreciate scales and fundamental components of **diversity** in habitat ecology. Patterns and processes scale from within-habitats to between-habitats and larger entities (**metapopulations** – i.e. sets of interacting local populations subject to immigration, colonisation, emigration and extinction; see Chapter 7 – to **biomes** and **biogeographical regions**). Biodiversity in landscapes has been attributed to three appropriate scales by Whittaker (1972), occurring: (i) within habitat units (**α-diversity**), (ii) between habitat units (**β-diversity**), and (iii) within landscapes and regions (**γ-diversity**) (Fig. 4.1a,b), though the interpretation of these scales may be extremely variable. Examples, in which all three scales of diversity are considered, are offered by studies on woody plants of oceanic islands (Borges *et al.*, 2018a), spiders in Southern Judea Lowland (Israel) (Gavish *et al.*, 2019), moths in Mongolia (Enkhtur *et al.*, 2021) and hummingbirds in Tlaxcala, Mexico (Martínez-Roldán *et al.*, 2024), among others. The simplest notion is that the α- and γ-scales are simply distinguished by the number of different species found at these scales S_α and S_γ, whereas the β-scale is distinguished by species unique to two or more habitat units being compared. Processes at the three scale levels are not incompatible if only for the reason that it is possible to apply each definition at every spatial scale. For instance, distinctions between habitat units (β) may be envisaged as gradients across given scales, much as gradients for species may be assessed within a habitat space (α) and across much larger regions (γ); spatial scales

DOI: 10.1079/9781800627505.0004

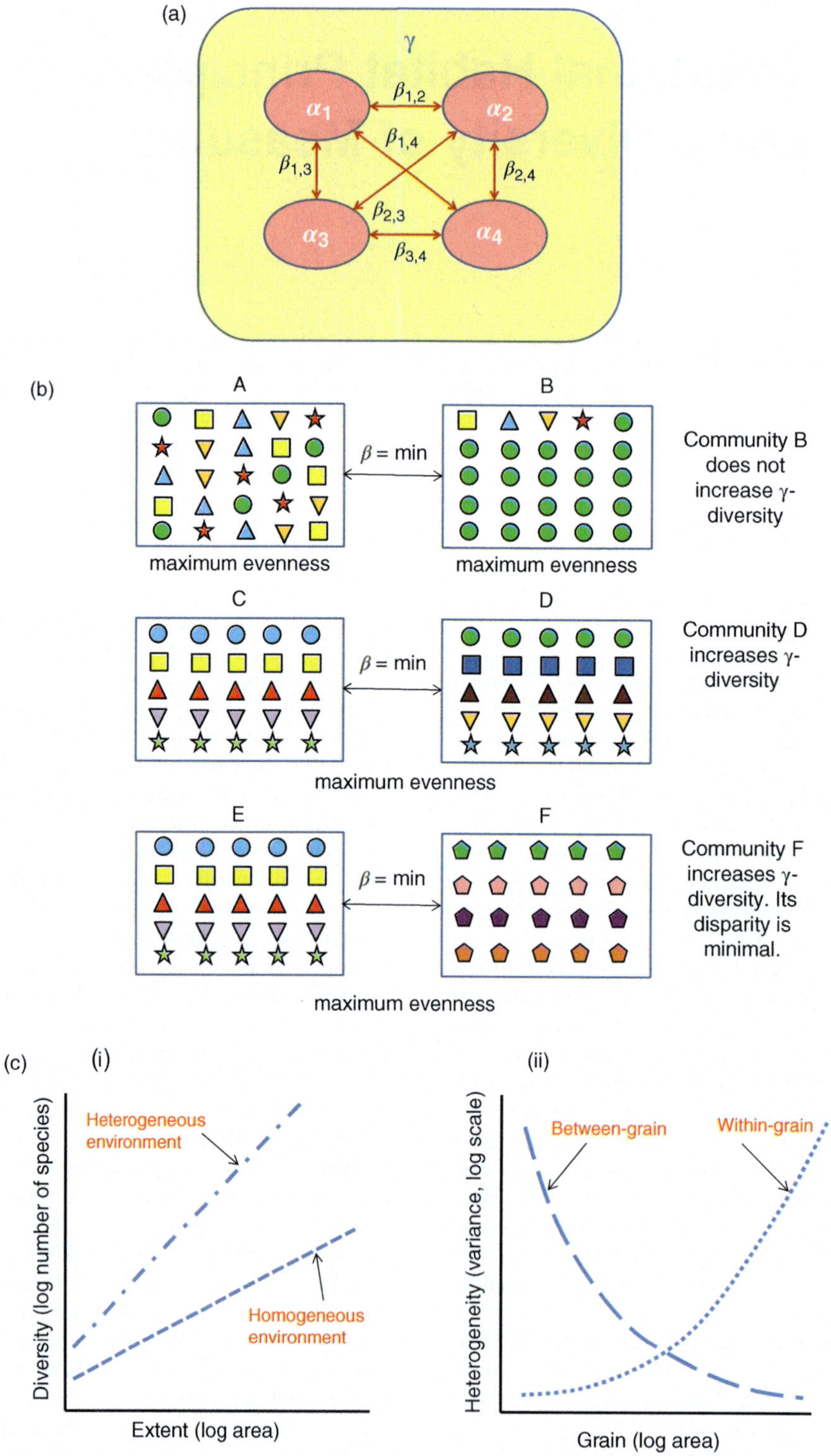

Fig. 4.1. Continued.

are not fixed and absolute for species but involve grades of movement determined by species' **mobility** (Dennis and Hardy, 2018). Mistakes are often made in assuming that resources and conditions within a habitat (α) are homogeneous; this is far from the case (see Chapter 5). Inevitably, if either the scale (extent) of a study (i.e. its spatial bounds) or the mesh (grain) of the sampling units (e.g. cells in raster lattice data, polygons in vector GIS data) of a survey zone is increased, then so too will the heterogeneity of components increase (Fig. 4.1c).

At any spatial scale, diversity has three basic components: **richness** (i.e. number of species in an assemblage), **evenness** (i.e. differences in the number of individuals among species in an assemblage; it is maximal when all species are represented by the same number of individuals) and **disparity** (i.e. differences among the species within the assemblage; the higher the difference among species (morphologically, genetically, phylogenetically, functionally, etc.), the greater the disparity) (Fig. 4.1b). Richness is only influenced by species identity whereas evenness reflects species abundances; disparity is influenced by species characteristics. A variety of diversity measures have been proposed that combine richness and evenness (and, albeit more rarely, disparity) (see Appendix 3). Any assemblage has its own richness, evenness and disparity, independently from those of other assemblages; by contrast, β-diversity characterises sets of assemblages, as it expresses how assemblages differ between them. Thus, two assemblages, say A and B, have their respective richness, evenness and disparity, but not β-diversity: β-diversity is between A and B.

Species and individuals are basic ingredients of habitats, but species can also be allocated to higher taxa: it is the integration of patterns and processes in all these aspects of the natural world that is central to comprehending habitats. Both species and individuals will have unique (changing) spatial distributions and temporal occurrences within a habitat. As the goal of science is to find general explanations, ecologists try to describe diversity patterns and processes through general relationships, based on elements that can be measured and analysed at different spatial scales (from single site to global), temporal scales (from instantaneous to geological time perspectives), aggregation scales (e.g. α-diversity versus γ-diversity) and levels of abstractions (species composition, variation in species composition and 'variation in variation' in species composition; Legendre *et al.*, 2005; Tuomisto and Ruokolainen, 2006; Jurasinski and Koch, 2011; see Appendix 3). Within- and between-habitat studies vary in emphasis. The focus of within-habitat studies tends to be, *de rigueur*, on numbers (abundance) and locations of individuals of different species (typically through measures of species richness and evenness), whereas the emphasis at increasingly coarser scales of landscape is of their differing content of species (diversity and disparity at higher taxonomic level) (Daly *et al.*, 2018). But, at any level, variation in species content is contingent on the abundance of individuals belonging to different species. This is so because the probability of extinction is $P(e) = f(1/N)$ where N is population size (or more strictly $f(1/N_e)$ where N_e is effective population size used to quantify **inbreeding** and **genetic drift**; Frankham *et al.*, 2010) unless countered

Fig. 4.1. Basic relationships in biodiversity at different scales. **(a)** Interrelationships of α-diversity (diversity at the assemblage scale), β-diversity (diversity between assemblages) and γ-diversity (diversity at the landscape level). **(b)** The distinction between species evenness, species diversity and disparity. Each square is a habitat unit containing five species and 25 individuals. Different symbols indicate different genera and different colours illustrate different species. Species belonging to different genera are assumed to be phylogenetically, genetically, morphologically and functionally less similar than species of the same genus. Communities A and B differ in their evenness but are identical in terms of species composition (i.e. there is no β-diversity in terms of species identity); as a result, community B does not contribute to γ-diversity. Communities C and D, and communities E and F, are identical in terms of evenness (maximum in all cases). Communities C (or E), D and F all contribute to γ-diversity. Community F shows minimal disparity compared to the others, as all its species are more similar to each other than to those in the other communities. **(c)** Heterogeneity of phenomena (e.g. taxonomic units, landforms, environmental variables) with changing area: (i) changes in biodiversity with increases in the study area (extent); (ii) changes in variance with changes of sampling units within the study area (grain). With increasing extent, diversity increases more rapidly in heterogeneous than in homogeneous environments. As grain scale increases, less of the variance is attributable to differences between samples, and more of the overall variation is captured within samples. See Wiens (1989) for more extensive treatment.

by immigration from nearby populations (Matthies *et al.*, 2004).

This chapter first considers some basic generalisations that underlie communities in traditional habitats followed by some of the basic relationships within and between them and their measurement. An initial thought relating to data obtained from habitats: study of variation in habitats is far from simple for many taxa if only because of difficulties in determining the number and location of individuals (see Appendix 1), and consequently the numbers of species at any one time. As we have seen in Chapter 3, conditions in habitats are continually changing, and therefore so are numbers of individuals of species and their actual or apparent incidence (occurrence) within habitats, including migrants and vagrants. More than that, bias inevitably enters into our assessment of numbers of both individuals and species depending on sampling regimes and recording methods (moth traps, pit-fall traps, dredge nets, **quadrat** or **transect** observations, etc; King and Porter, 2005; Hancock and Legg, 2012; Kral-O'Brien *et al.*, 2020; Lobo *et al.*, 2021; see Appendix 1 for an overview) and different higher taxa receive very different attention (some taxa are investigated more than others, as taxonomically better known, easier to study, more charismatic, etc.; Cardoso *et al.*, 2011; Wheater *et al.*, 2011; Strona, 2015; Troudet *et al.*, 2017). With increasing growth of crowd-sourced records, new techniques are required for processing the data (Rapacciuolo *et al.*, 2021).

As a final introductory note, we would remark that though we focus on species, the concepts discussed here can be applied to any systematic level. For example, in palaeontological studies it is common practice to use genera (e.g. Sepkoski, 2002; Wang and Dodson, 2006; Alroy *et al.*, 2008; Alroy, 2010; Knope *et al.*, 2020; Flannery-Sutherland *et al.*, 2022) and sometimes even families (Sepkoski, 1984; Benton, 1995) or orders (Sepkoski, 1978) to depict diversity patterns because the fossil record may allow reliable estimates of numbers of genera and higher taxa, but is generally not sufficiently complete to depict variations in species richness (Benson *et al.*, 2021).

Some generalisations for traditional habitats

An interesting feature of ecology is that it has grown without much explicit declaration of its scientific theories or laws, nor what is meant by universal and general laws, rules, theories and principles in the subject (Lawton, 1999; Scheiner and Willig, 2008) (Table 4.1). Put simply, scientific **theory** 'explains' (accounts for) why something happens, whereas scientific **law** describes what happens, hopefully neatly in an equation. Laws should have no exceptions whereas **rules** may be broken; **facts** are proven observations. **Principles** are fundamental guiding concepts or generalisations built into theory. Rules and principles relate to commonly observed patterns and processes. If ecology has no *universal* laws of its own, it is simply because the Earth provides a single sample of our knowledge of a planet with life. However, comparative work over the planet has led to its own 'law-like' relationships and a number of these lend themselves to summary in neat equations. Examples are equations describing population dynamics (see Chapter 2 and Appendix 2), the distribution of the abundance of species in an assemblage (viz. the **species abundance distribution**, an argument that we will discuss in some detail in this chapter and in Appendix 3), or the increase in the number of species with increasing area (the **species–area relationship**; Matthews *et al.*, 2021). Prominent rules and principles certainly underpin habitat ecology (Scheiner and Willig, 2008) (see Table 4.1); many admittedly have hybrid origins in different sciences, but appear repeatedly in ecological assessments of habitat components. Here, attention is focused on basic principles that affect our understanding of communities in habitats in terms of α-, β- and γ-levels, and hence at the respective scales (that is, within habitats, between habitats and inclusive of all habitats, viz. within **landscapes** and regions) (see Table 4.1).

The first observation (see Table 4.1, C1) is the occurrence of few abundant but many **rare species** in a habitat unit. This has long been a focus of ecological research; this important observation is examined in the next section. It has been thought to arise primarily from the interactions of organisms in habitats but there is also the prominent issue of the limitations of suitable resources and conditions within a habitat space exposed by a **resource-based habitat** view (Dennis, 2010). Moreover, heterogeneity in resources extends beyond the habitat space and many species necessarily move from one resource zone to another. Thus, many rare species may be vagrants seeking out resources. This may make it difficult and rather subjective to define which species really occupy a

Table 4.1. Some fundamental principles and rules that influence communities in traditional habitats.

A. Fundamental principles in ecology (modified from Scheiner and Willig, 2008)

1. All organisms are mortal: born, age and die.
2. All organisms move location and/or disperse their offspring.
3. Organisms interact with their abiotic and biotic environments.
4. Ecological traits of species are the result of evolution.
5. Organisms are distributed in space and time in a heterogeneous manner.
6. Environmental conditions are heterogeneous in space and time.
7. Resources are finite and heterogeneous in space and time.
8. The distributions of organisms and their interactions depend on contingencies.

B. Some rules (law-like observations) influencing habitat composition

1. Law of unique individual trait variation (Darwin, 1859; see Jonsson *et al.*, 2021 for recent genetic basis of individual uniqueness). Organisms are unique in genotype and phenotype.
2. Law of the minimum or maximum (Von Liebig, 1840; see Hiddink and Kaiser, 2005 for ecological interpretation). A limiting factor is the one that is the least available or in excess at a particular time (e.g. prey or nutrients).
3. First law of geography (Tobler, 1970). Everything is related to everything else, but near things are more related than distant things.
4. Second law of geography (Tobler, 1999).* Phenomena external to an area of interest affect what goes on inside the area.
5. Island biogeography laws for isolated habitat units (islands) (MacArthur and Wilson, 1963, 1967; see modern synthesis in Hanski and Gilpin, 1997). The relationships between two areas of interest are directly proportional to their size but inversely related to the distance between them (cf. Newton's (1687) law of universal gravitation: every point mass attracts every other point mass in the universe with a force that is directly proportional to the product of their masses and inversely proportional to the square of the distance between their centres).
6. The power law (Taylor, 1961; Cobain *et al.*, 2019; Kilpatrick and Ives, 2003).
   The variance of population density in a habitat is proportional to a fractional power of the mean: $\sigma^2 = \alpha\mu^b$ where σ is the standard deviation, μ is the mean, α is a sampling factor, b is an index of aggregation.
7. Law of habitat quality (cf. Dennis, 2010). The density of a species' population depends on the quality of resources and the suitability of environmental conditions (cf. the resource-based habitat); the first and second moments of a multifactorial distribution of variables describe optima in the distribution of resource qualities and conditions between extremes of Liebig's law of the minimum and maximum.

C. Principles relating to communities in traditional habitat units.

1. Habitats have few common species but many rare ones.
2. Species within a habitat will have different ecological niches.
3. Species and individuals are not equally distributed in the habitat space at any one time.
4. Communities in habitats are constantly influenced by those in other habitat units.
5. Communities in habitats are more similar to those nearby than those further away. As a sequitur, it follows that by increasing the area of focus, there will be an increase in community diversity.
6. Communities on the same substrates (geology, soils) are more similar than those on contrasting substrates.
7. Abundant species have more extensive distributions and wider ranges than scarce species.
8. Many species move between habitat spaces seasonally (are not permanent residents in the breeding habitat space).
9. Communities in habitats are never stable for long and undergo systematic changes in all attributes.

* Foresman and Luscombe (2017) necessarily has to be the third law of geography by date priority and is not relevant to ecology.
[C] principles relate to [A] principles and [B] rules:
C1 → *f* {<u>A2</u>, <u>A3</u>, A4, <u>A7</u>} {B1, <u>B2</u>, <u>B4</u>, B6, <u>B7</u>}
C2 → *f* {A3, <u>A4</u>, A5, <u>A6</u>} {<u>B1</u>, B6}
C3 → *f* {A3, <u>A4</u>, <u>A5</u>, <u>A6</u>, A7} {<u>B1</u>, B2, B6, <u>B7</u>}
C4 → *f* {<u>A2</u>, A7, A8} {B3, <u>B4</u>, <u>B5</u>}
C5 → *f* {<u>A2</u>, A4, A7} {B1, <u>B3</u>, <u>B4</u>, <u>B5</u>}
C6 → *f* {<u>A2</u>, A3, <u>A4</u>, A5, A6, A7, A8} {<u>B1</u>}
C7 → *f* {<u>A2</u>, <u>A3</u>, <u>A4</u>, <u>A7</u>, A8} {B1, <u>B3</u>, <u>B5</u>, B6, <u>B7</u>}
C8 → *f* {<u>A2</u>, A3, A4, A5, <u>A6</u>, <u>A7</u>, <u>A8</u>} {B1, B6, <u>B7</u>}
C9 → *f* {<u>A2</u>, <u>A3</u>, A4, A5, <u>A6</u>, <u>A7</u>, <u>A8</u>} {B1, <u>B2</u>, <u>B3</u>, <u>B4</u>, <u>B5</u>, B6, <u>B7</u>}
A1 is common to all habitat community principles. → 'Relates to'; *f*, 'a function of variability in'. Prominent variables underlined.

habitat and hence constitute a community (see Chapter 3, Box 3.3). Another focal point of habitat communities and connected to the species abundance distribution is that each species has a unique ecological **niche** (see Chapter 2, Fig. 2.8; Table 4.1, C2); this is closely linked to its unique resource-based habitat (see Chapter 5). It is considered to be the outcome of their evolution in heterogeneous environments, but as discussed earlier (see Chapter 2), more attention has been given by researchers to competitive interactions among organisms leading to the evolution of novel niche spaces. Unsuccessful species are likely to become extinct and there are many indications of this in the geological and historical record, especially species that have come into direct conflict with humans (see Chapter 10). Unique niches lead on to (and are the consequence of) the uneven distribution of species within the habitat space (see Table 4.1, C3).

Habitat units are not closed systems (see Table 4.1, C4). Organisms disperse from natal habitats. At least three geographical rules are involved: (i) proximity of a habitat to other units (which facilitates dispersion and colonisation), (ii) interactions between units, which is related to habitat size (and hence population size, number of individuals) and (iii) negatively to level of isolation (e.g. permeability of the **matrix**; see Box 2.2 in Chapter 2 and Chapter 5). At the same time, similar factors affect the relationship between communities in adjacent habitats (see Table 4.1, C5). But this is counteracted by the nature of environmental heterogeneity. Geology is a major influence of habitat substrates so it is unsurprising that habitats with similar (if not identical) lithology and substrates have associated with them similar communities of organisms (see Table 4.1, C6; see also Box 3.1 in Chapter 3 about how substrate evolution is related to the concept of **climax** community).

An often-cited relationship, crucial for species' maintenance if correct, is that between species' abundance and distribution cover (see Table 4.1, C7; Box 4.1) (Gaston, 1994; Lawton, 1999; but see Holt *et al.*, 1997). In such situations there is likely to be a positive feedback between the two variables. The more abundant a species is within a habitat, the more propagules are expected to disperse from it and colonise other habitat units, whereas the wider the distribution of a species, the greater the opportunity for it to colonise vacant sites through dispersal of individuals. Absolute dispersal numbers (emigration) are likely to increase with increased population **density** simply because there are then fewer resources available per individual (and greater intraspecific **competition** for them) in the natal habitat. At the same time, with greater distribution extent the distances to vacant sites will be less, promoting the greater probability of successful colonisation. These observations are neatly described by **metapopulation** models (Hanski and Gilpin, 1997; see Chapter 7). However, by no means does this pattern hold invariably true. Population abundance is affected by many factors independent of position within a species range, including habitat type, habitat quality, interspecific competition, dispersal ecology, etc. The result can be complex textured population abundance patterns compared to a simple theoretical core-edge gradient (Shreeve *et al.*, 1996a; Burgess *et al.*, 2020). There are classic examples of high population density but sparse distribution (e.g. the silver-studded blue butterfly *Plebejus argus*; Thomas and Harrison, 1992), the distributional status of which may be natural and/or the result of habitat fragmentation and reduction (Dennis, 2010). The inference is that the organism requires resources that are not widely distributed; if appropriate resources and conditions are widely distributed then clearly dispersal is a weak trait in the organism confined to few habitat spaces. In such situations there may be other variables that become prominent factors, such as unusual species' traits (see Table 4.1, A4, B1), their resource-based habitats (see Table 4.1, B7) and the distribution of environmental resources (see Table 4.1, A6). Interestingly, resource **specialists** tend to have a greater population density than **generalists** standardised for distribution extent (e.g. aquatic macroinvertebrates, Verberk *et al.*, 2010; butterflies, Dapporto and Dennis, 2013; Dennis *et al.*, 2013).

Another habitat principle (seasonal movement; see Table 4.1, C8) also relates to environmental heterogeneity and the evolution of specialised traits for seasonal long-distance movements to access resources and/or conditions at times of the year when they are inadequate in the natal habitat. A classic example is the huge flocks of birds, particularly waders and waterfowls but also passerines, exploiting winter feeding grounds on salt marshes and estuaries (Dierschke and Bairlein, 2004), as well as the mass migrations of birds from summer to winter feeding and roosting grounds (Fuller, 1982, 2012). Movement may be more prominent among organisms than accepted, as suggested by changes in species abundances from year to year

Box 4.1. Abundance, range and habitat specificity

A virtually universal characteristic of biological communities, discussed in Chapter 4, is that they are constituted of few abundant (dominant) species represented by many individuals and many rare species represented by few individuals. However, a species can be rare for many other aspects beyond abundance (population size). For example, an important attribute that can be used to characterise species rarity is the habitat specificity (also indicated as habitat or ecological breadth, tolerance, specialisation, etc.): species occurring in few habitats are rarer than those with broader ecological preferences (Rabinowitz, 1981; Rabinowitz *et al.*, 1986; Hughes *et al.*, 2000; Crain *et al.*, 2015; Maciel, 2021; Miranda Cebrián *et al.*, 2022). A third aspect that can be considered in evaluating species rarity is their geographical distribution. Starting from the 1990s, it became increasingly evident that the majority of the species in the most disparate taxa are geographically rare, as they have a small **extent of occurrence** (**EOO**, i.e. the area within the limits of their distribution, usually measured as the area contained within the shortest continuous imaginary boundary which can be drawn to encompass all sites of occurrence; polygons in Fig. B4.1.1) and/or a small **area of occupancy** (**AOO**, i.e. the area actually occupied by a species during a particular time interval within its extent of occurrence, usually measured as the number of recorded localities or occupied cells in a grid; filled cells in Fig. B4.1.1) (see Lomolino *et al.*, 2010 for examples, and Gaston, 1990 for exceptions). Thus, to properly reflect the multidimensional nature of rarity, Deborah Rabinowitz (1981; Rabinowitz *et al.*, 1986) suggested that these three aspects (dimensions) of species rarity (geographical distribution, habitat specificity and population size) should be examined simultaneously for each species in an assemblage (note that any concept of rarity is intrinsically relative: a species is not rare in absolute terms but only in comparison with other species for the selected dimension(s), thus species rarity makes sense only within specified assemblages). However, these three forms of rarity do not exhaust all the possible ways in which one species can be rarer than another. For example, it has been stressed that species belonging to very isolated evolutionary branches are phylogenetically rarer than those belonging to evolutionary lines subject to more intense diversification (see, for example, Fattorini *et al.*, 2013a,b, for a discussion), and there is also increasing interest in evaluating rarity in terms of **functional traits** (Violle *et al.*, 2017; Lin *et al.*, 2024).

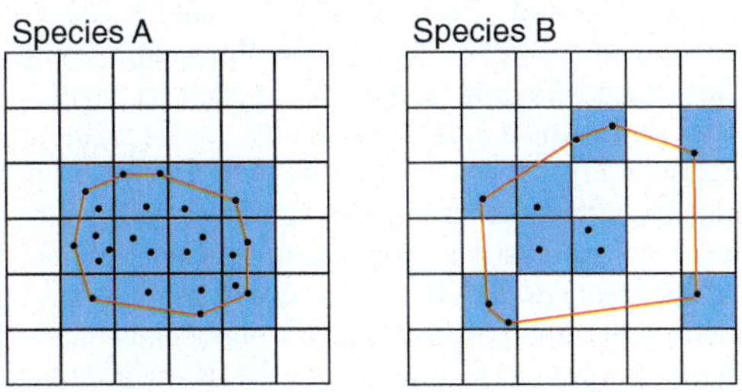

Fig. B4.1.1. Extent of occurrence (area of the polygons) and area of occupancy (number of filled cells in the grid) of two hypothetical species, A and B. Species A has a smaller extent of occurrence but a greater area of occupancy; species B has a larger extent of occurrence but a smaller area of occupancy.

The three forms of rarity considered in Rabinowitz's approach are really continuous variables. But if species are dichotomised for each of these variables (for example, by using median values: species that are below the median for a given rarity measure are rare for that measure; Arita *et al.*, 1990), an eight-celled model is created that reflects different types of rarity and commonness: (1) species that are not rare; (2) scarce species (i.e. species rare in terms of abundance); (3) species with narrow habitat specificity (i.e. species rare by habitat distribution); (4): restricted species (i.e. species rare by range); (5) scarce species with narrow habitat specificity (i.e. species rare for both habitat distribution and abundance); (6) scarce and restricted species (i.e. species rare for both range and abundance); (7) restricted species with narrow habitat specificity (i.e. species rare for both habitat and range); and (8) restricted and scarce species with narrow habitat specificity (i.e. species rare for range, habitat and abundance) (Fig. B4.1.2). Thus, eight forms of rarity are identified into which species can be classified to reflect the different ways in which they are rare, and their relative frequencies in an assemblage can be compared (e.g. Kattan, 1992; Dobson and Yu, 1993; Manne and Pimm, 2001; Fattorini, 2010b,c, 2011). Recently, to construct a continuous measure of rarity that encapsulates the contribution of these three dimensions, an alternative approach has been introduced, in which the inverses of the parameters (geography, habitat and population) are used to construct for each species in a given dataset a synthetic index that ranges from 0 (minimum rarity) to 1 (maximum rarity) (Maciel, 2021).

Continued

Box 4.1. Continued.

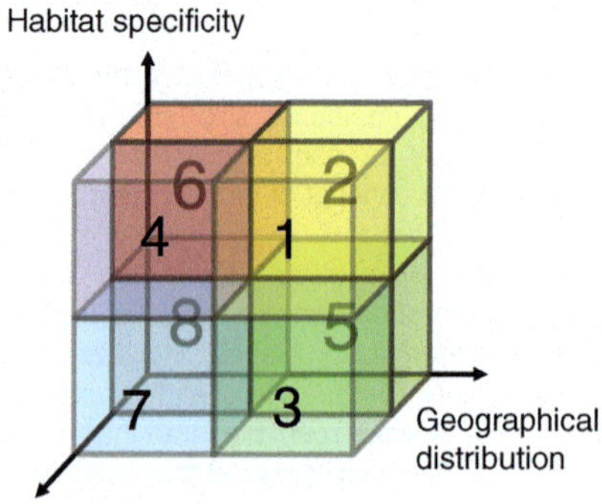

Fig. B4.1.2. Space defined by three rarity dimensions: geographical distribution, ecology (habitat specificity) and abundance (population size). Median values for each axis divide this rarity cube into eight cubes representing various forms of rarity: 1 species that are not rare for any dimension; 2 species rare only for abundance; 3 species rare only for ecology; 4 species rare only for distribution; 5 species rare for both ecology and abundance; 6 species rare for both distribution and abundance; 7 species rare for both ecology and distribution; 8 species rare for distribution, ecology and abundance (Based on Fattorini *et al.*, 2013a.)

The multidimensional characterisation of species rarity based on geography, habitat and population measures, originally proposed for the flora of North America (Rabinowitz, 1981) and the British Isles (Rabinowitz *et al.*, 1986), has been successfully applied not only to vascular plants in the most various geographical context (e.g. Sætersdal and Birks, 1997; Rey Benayas *et al.*, 1999; Pitman *et al.*, 1999; Broennimann *et al.*, 2005; Caiafa and Martins, 2010; Espeland and Emam, 2011), but also to other organisms, including bryophytes (Gabriel *et al.*, 2011), vertebrates (e.g. Kattan, 1992; Dobson and Yu, 1993; Manne and Pimm, 2001; Fattorini, 2010b; Toledo *et al.*, 2014; Birskis-Barros, 2019) and arthropods (Fattorini, 2008, 2010b,c, 2011; Fattorini *et al.*, 2012).

Evaluating species rarity has gained increasing interest in conservation biology, as rare species are considered more vulnerable than common species (that is, they are intrinsically exposed to higher extinction risk; Terborgh and Winter, 1980; Goerck, 1997; Mace and Kershaw, 1997; Manne and Pimm, 2001; Flather and Sieg, 2007; Isaac, 2009; Harnik *et al.*, 2012; Maciel, 2021). For example: (i) species with small populations can decline below the **minimum viable population** more easily than abundant species, because of demographic, genetic and environmental stochasticity; (ii) species with narrow geographical distributions have fewer opportunities to survive catastrophic events (such as fires, pollution peaks and habitat destruction or alteration) than more widespread species, which can survive in unaffected areas of their larger ranges; (iii) habitat specialists have fewer chances of coping with environmental changes than less specialised species, which may remain unaffected by changes in the availability or quality of certain resources or may respond to such changes by using other resources (Pullin, 2002). Table B4.1 illustrates how factors that can enhance extinction risk are associated with forms of rarity. Additionally, rare species may be particularly important in maintaining ecosystem functioning (Lyons *et al.*, 2005) and may be representative of other species of conservation concern (Lawler *et al.*, 2003; Larsen *et al.*, 2007), which further underscores their conservation importance.

In the aforementioned list of the eight types of rarity obtained using Rabinowitz's approach, these were ordered to reflect increasing vulnerability. Species that fall in the cell 'wide range-wide habitat specificity-high abundance' are common in three dimensions (minimum vulnerability), whereas those falling in the cell 'narrow range-narrow habitat specificity-low abundance' are rare in all three dimensions and are supposed to be the most vulnerable to extinction (maximum vulnerability). To rank the remaining six cells, species with narrow geographical distribution are assumed to be more vulnerable at a larger scale, and species with narrow habitat specificity are more vulnerable regardless of their abundance (Kattan, 1992). With this approach, it is possible to prioritise species in terms of conservation concern, as more vulnerable species are potentially more imperilled (Fattorini, 2010b,c, 2011a, 2014c; Fattorini *et al.*, 2012, 2015b). Use of a wide range of dimensions beyond spatial distribution, population size and habitat specificity (such as life span or trophic role), as well as the use of more subtle and multiple measures for the same broad categories (for example, by distinguishing different aspects of habitat specificity or considering various aspects of population rarity, such as both the size of populations and their number) would be undoubtedly useful to obtain rarity scores that better reflect species vulnerability – a hard but not impossible task (Fattorini *et al.*, 2020a).

However, neither rarity nor vulnerability is synonymous with extinction risk, as these concepts, albeit interrelated, express different things. Rarity indicates the position of a species along one or more dimensions (e.g. distribution, population size and habitat specificity) relative to other species. Vulnerability

Continued

Box 4.1. Continued.

Table B4.1. Relationships between rarity and extinction risk.

Factors that can enhance extinction risk	Associated forms of rarity
Restricted distribution	Geography (extent of occurrence)
Fragmented distribution	Geography (area of occurrence)
Reduced niche	Habitat specificity
Small population(s)	Population size
Small number of populations	Population size
Declining population(s)	Population size
Low population density	Population size
Reduced genetic variability	Population size
Low dispersal capabilities	Habitat specificity
Migratory behaviours (need to use different habitats in different moments)	Habitat specificity
Association with stable, rare, small or well-preserved habitats	Habitat specificity
Reduced phenology	Habitat specificity
Broad home range	Population size/area of occupancy
Large body size	Population size
Long life span (*K*-strategy)	Population size

(Modified from Fattorini *et al.*, 2013a.)

expresses how a species may be intrinsically prone to extinction because of its biological characteristics, other things being equal. Extinction risk indicates the probability that a species can be lost as a result of the interrelationships between its biological characteristics and the environmental changes, including anthropogenic factors. A habitat **specialist** (i.e. a species rare for the habitat dimension) is more vulnerable than a **generalist**, because it has an intrinsically higher probability of being extinct if its habitat(s) is (are) transformed, reduced or destroyed in comparison to a habitat generalist; but it is not necessarily imperilled, as long as no change in its habitat(s) occurs (Fattorini *et al.*, 2013b; Fattorini, 2014c). Conversely, a species with large and multiple populations is less vulnerable than one with small or few populations, but might be more imperilled if persecuted or if its habitat is profoundly altered by human activities (Fattorini *et al.*, 2013b; Fattorini, 2014c). The North American bison (*Bison bison*) was originally a very common species (with an estimated population of 60–70 million individuals at its maximum), but was reduced to a few hundred individuals in the 1890s due to human persecution, and counts now some tens of thousands thanks to conservation measures (Isenberg, 2000; Aune *et al.*, 2017). The passenger pigeon (*Ectopistes migratorius*) was largely distributed in North America and was possibly the most numerous bird in the world, with a population that likely reached 3–5 billion individuals, but became extinct at the turn of the 20th century due to massive hunting and habitat destruction (Schorger, 1955; Bucher, 1992; Greenberg, 2014). No one is safe: intense human use of natural and seminatural habitats can have profound impacts even on common species (Conrad *et al.*, 2006; Van Dyck *et al.*, 2009; Fattorini, 2014c).

Many hypotheses have been proposed to explain why most species are rare by geographical distribution, habitat specificity or population size, why different aspects of rarity tend to covary, and why rarity may correlate with certain species traits (for example, body size) (Fattorini *et al.*, 2013b, Fattorini and Di Giulio, 2013). For instance, a positive correlation between distribution and abundance may be a reflection of interspecific differences in ecological specialisation: species able to exploit a wide range of resources become both widespread and locally abundant, whereas habitat specialists have both restricted distributions and low abundances (Brown, 1984; Gaston and Lawton, 1990; Hanski *et al.*, 1993). A positive correlation between body size and range may be a consequence of the fact that larger species have larger **home ranges**, whereas the greater vulnerability of smaller species to density-independent perturbations might reduce the area over which they are able to maintain peak rates of population growth, and thus increases their likelihood of having smaller ranges (Gaston, 1990; Gaston and Blackburn, 1996). A negative correlation between body size and abundance, reported for mammals and birds, may be a consequence of competition, as differences in body size reduce it, and if food is very limited, larger species, which require more food, will have small populations (Fattorini *et al.*, 2013a).

over the UK (e.g. moths and aphids, Taylor, 1978; butterflies, Fox *et al.*, 2006). The last habitat principle (see Table 4.1, C9) relates to habitat instability caused by continuous transformations (see Chapter 3).

Undoubtedly, many factors affect how individuals are distributed in the space (their dispersion), which include both aspects of species biology and habitat characteristics (Box 4.2). Spatial patterns of distribution can be statistically described by an equation relating the variance σ^2 and the mean μ of a species' frequency distribution (e.g. the number of individuals in units of space such as **quadrats** or **transects**; see Appendix 1) through a power law:

$$\sigma^2 = \alpha\mu^b \tag{4.1}$$

which can be linearised by a logarithmic transformation:

$$\log(\sigma^2) = \log(\alpha) + b\log(\mu) \tag{4.2}$$

In this equation, developed initially from the study of different arthropod groups (i.e. moths and aphids; Taylor, 1961), α is considered to be a sampling factor whereas b is a true population statistic. Taylor referred to it as an 'index of aggregation' that describes an intrinsic property of the species in any habitat (Taylor, 1961); it may reflect the 'behavioural interaction between organism and environment' leading to unique spatial dispositions for each organism (including man) *vis á vis* environmental pressures (Taylor, 1965, 1971). The fractional power b relates to extensive areas (e.g. hundreds of km^2) as well as to local habitat areas (e.g. m^2), such that:

$$\sigma_p^{\ 2} = \alpha\mu_p^{\ b}, \tag{4.3}$$

where σ_p^2 is the spatial variance around the mean population density μ_p of a point (habitat) (Taylor and Taylor, 1977).

If the organisms are randomly distributed, $\sigma^2 = \mu$ and $b = 1$. If individuals are clumped (aggregated), b is greater than 1; values of b <1 indicate more uniform distributions. The majority of values of b calculated for natural populations are not only greater than 1 (which suggests that most populations are aggregated), but fall within a relatively small range, typically between 1 and 2 (Hayek and Buzas, 2010). This raised the suspicion that the exponent b does not contain any ecologically relevant information, but might reflect merely statistical properties of the investigated systems (e.g. data series length and sampling error or statistical artefacts of system configuration), a point of view that seems, however, not justified (Cobain *et al.*, 2019). The exponent b is considered an intrinsic biological characteristic of the species: each species occupies a habitat/landscape with its b. However, these b-values might be known only through an extensive knowledge of the species distribution, which typically means many years of sampling over a wide variety of environmental conditions (e.g. Hayek and Buzas, 2010). Thus, the same species may show different, albeit similar, b-values in different geographical regions (e.g. Hayek and Buzas, 2010). Moreover, when different species are compared, it is important that not only b-values but also α-values are examined, since both contribute to distinguish specific lines: while at very high densities the predicted value for the variance depends most heavily on b, the influence of α increases with decreasing densities, and at very low densities the constant α may become more important in the determination of predicted variances than is the constant b (Hayek and Buzas, 2010).

The behavioural interaction purportedly described by b involves individuals in not just escaping competition (mortality) but moving within and beyond the natal habitat; individuals are selected not just for competitive ability but also movement to better locations (for any individual, there is a balance between a centrifugal (leaving) and centripetal (staying) tendency with respect to the natal habitat). Kilpatrick and Ives (2003) demonstrated how fractional powers for b can be produced by negative interactions among species in a community alone. This important concept calls into question the very existence of a traditional habitat space as the isolated domain of a fixed population. Taylor conceives a continual non-random flux of individuals over space-time in which species have characteristic and elaborate behaviours to environmental pressures, both abiotic and biotic (Taylor and Taylor, 1979).

The pattern of species richness in habitats

A basic feature of organisms in habitat units is the occurrence of few common species and many rare species (McGill *et al.*, 2007; Ulrich *et al.*, 2010). Astounding as it may seem, no satisfactory theory has yet emerged to explain this pattern. Many **species abundance distribution (SAD)** models have been proposed to describe the pattern and much contention surrounds their explanatory ability and true meaning (e.g. statistical models versus resource apportionment models; Putman, 1994; Cook

Box 4.2. Patterns of spatial distribution

The distribution of individuals in a space (that is, their dispersion) can be described with reference to the three main patterns: **random** (individuals are distributed at random in the space; some are closer and others more distant only because of change events), **uniform** (also known as regular dispersion, even dispersion or overdispersion, in which individuals are more evenly distributed than expected by chance) and **clumped** (also known as aggregated dispersion, contagious dispersion or underdispersion, in which individuals are closer than expected by chance) (Fig. B4.2.1). These patterns can refer to individuals of the same species (e.g. the distribution of individuals in a population) or different species (e.g. the distribution of individuals of different species in a community). Random dispersion is the least common form of spatial distribution, because biotic interactions and environmental factors tend to produce uniform or clumped distributions (for example, negative interactions will tend to produce more uniform distributions, while positive interactions or patchy distributions of resources will tend to create clumped distributions, as individuals will be necessarily concentrated where they can find the resource). Random distributions often occur in species with seeds that are dispersed by wind and can germinate wherever conditions are favourable. Uniform distributions within populations are typically found in territorial species (see Chapter 5), whereas examples of uniform distributions involving multiple species in communities can be found among plants which release chemicals that inhibit the development of other plants around (a form of negative interaction known as **allelopathy**) or under severe competition for water or light. Clumped distributions are typical of animals that have collective behaviours (e.g. flocking birds, herding ungulates, schooling fish and nest-building ants). However, the same organism may show different dispersion patterns at different scales of observation. For example, territorial birds that nest in a fragmented forest may appear to have a uniform distribution within the fragments, but at the landscape level they will appear clumped, being concentrated in the remaining forest patches. Insect larvae feeding on the leaves of trees may appear randomly distributed within a forest, but they might appear clumped at the level of individual trees.

These three patterns can be statistically distinguished based on variance. In a sample in which individuals have a perfectly uniform distribution, every sampling unit will contain exactly the same number of individuals, thus the variance of the sample will be zero. By contrast, in a very clumped distribution, many sampling units will not contain any individuals, as they will be concentrated in few sampling units, and therefore the variance of the sample will be extremely large. Samples of individuals that are dispersed randomly will have intermediate variance. The random dispersion is statistically described by the Poisson distribution, in which the variance σ^2 equals the mean μ. Thus, in a sample of perfectly randomly distributed individuals, the sample variance (s^2) will equal the mean number of individuals $(\bar{x})$, and hence the ratio (I) of the variance to the mean $s^2/\bar{x}$ will equal 1. This ratio may be used as an index of dispersion. $I \approx 1$ indicates that individuals are randomly dispersed, whereas $I < 1$ in case of uniform distributions and $I > 1$ in case of clumped distributions. While useful, this index provides only a basic characterisation of spatial patterns. These three patterns of dispersions (random, uniform and clumped) are in fact part of a continuum, which is neatly expressed by Taylor's power law described in the main text of Chapter 4. For more detailed modelling – especially when individuals are aggregated – ecologists often use the negative binomial distribution, which can accommodate a wide range of clumped patterns. In this context, the dispersion parameter k may provide a more nuanced measure of aggregation (e.g. Lloyd-Smith, 2007), with small values indicating maximum clumping. Classically, k is estimated as: $k = \bar{x} / (s^2 - \bar{x})$, and the reciprocal ($1/k$) is used as an index of dispersion. However, this index may be affected by sample size and population density

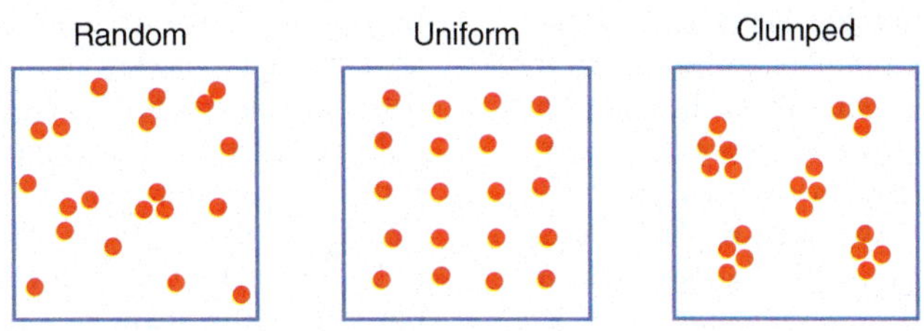

Fig. B4.2.1. Patterns of spatial distribution (diffusion): random, uniform (regular), clumped (aggregated).

Continued

Box 4.2. Continued.

(Krebs, 1999). In addition to these methods, a variety of other approaches have been proposed to express aggregation patterns, including (i) methods based on distances of individuals to the nearest neighbours with associated tests for significance of deviation from randomness (e.g. Clark and Evans' method, Thompson's test, Campbell and Clarke's test), (ii) use of stochastic approaches, and (iii) computation of variance from groups of adjacent quadrats (Hill's method) when the entire population can be mapped; as well as methods based on (iv) measurements from random points to the nearest organism (Hopkins' method, Eberhardt's test, Parker's method), (v) specific coefficients such as those of Green and Morisita, and (vi) methods based on distance to regularity (see Krebs, 1999 for a review).

and Graham, 1996; Hubbell, 1997; Magurran and Henderson, 2003; Sugihara *et al.*, 2003; McGill *et al.*, 2007) (see Appendix 3).

The most familiar portrayals of the data are histograms and rank abundance (Whittaker) plots. In **histogram** representations, the number of species (y-axis) is plotted against the number of individuals (x-axis) (Fig. 4.2a). Since Preston (1948), who arbitrarily converted abundance on the x-axis to $\log_2$, it is common practice to use binning systems based on logarithmic classes (Fig. 4.2b). In the **rank abundance plots**, species are ranked from the most to least abundant, and abundance is plotted against species rank (Magurran, 2004) (Fig. 4.2c). Rank abundance plots are not as intuitive as histograms but have the advantage of distinctly highlighting differences in evenness between datasets, because steeper curves indicate higher dominance (lower evenness). Although in principle any SAD model can be fitted using either the histogram representation or the rank abundance plot, some models are typically fitted using the rank abundance representation, while histograms are preferred for others. For example, histograms are generally preferred to illustrate SADs following the **lognormal distribution** (Fig. 4.3), a type of distribution in which most species have intermediate abundances (see Appendix 3), as using this type of graph immediately makes evident the Gaussian bell shape. By contrast, rank abundance plots are, for example, preferred to illustrate the **geometric series**, a type of distribution in which the abundance of each species in the community corresponds to a sequential, constant proportion (k) of the total number of individuals (see Appendix 3), since it follows a straight line with this representation (Fig. 4.4). This type of representation is also used to illustrate the broken stick model, which should characterise communities with uniform distributions but which has been severely criticised (see Appendix 3).

As a result of the variety of models introduced to describe SADs, differences in their visualisation and variety of potential interpretations, some researchers feel that community patterns can be deciphered best by a top-down (macroecological) view, removing what they consider to be messy detail at fieldwork level (Lawton, 1999). However, it is exactly the details gleaned from careful fieldwork in habitat spaces that indicate vitally important bottom-up views. In fact, despite the frustration emerging from failure to obtain 'universal' models, the study of SAD patterns has been of paramount importance in theoretical ecology, applied ecology and biodiversity management (Matthews and Whittaker, 2014a, 2015). SAD models may be used to predict species extinction risk (Kitzes and Harte, 2015), to evaluate the impact of natural and anthropogenic disturbance (Hill and Hamer 2003; Dornelas *et al.*, 2009; Fattorini, 2010a; Fattorini *et al.*, 2018a; Ulrich *et al.*, 2025) and, hence, to inform management actions (Kim *et al.*, 2013).

Although SADs are expected to vary in response to changes in environmental characteristics (Ulrich *et al.*, 2016), surprisingly few studies have dealt with the ecological drivers that influence SAD patterns (e.g. Silva *et al.*, 2010; Tsafack *et al.*, 2021a; Ulrich *et al.*, 2022). It is generally assumed that ecosystems that may support species with more varied ecological niches have more core species with stable populations with moderate to large numbers of individuals, thus leading to SADs that should be well approximated by lognormal distributions (Hubbell, 2001; Magurran and Henderson, 2003). In contrast, under stressful conditions, communities are expected to deviate more substantially from this pattern, leading to SADs in which species abundances follow a logarithmic distribution (Fisher's **log-series**), with a few very abundant species and many rare species (see Appendix 3). However, if we assume that true SADs are lognormal, the log-series would be nothing but a truncated lognormal due to insufficient sampling (Ulrich *et al.*, 2010). In fact, many ecologists think that skewness commonly found in empirical SADs might be only the result of insufficient sampling

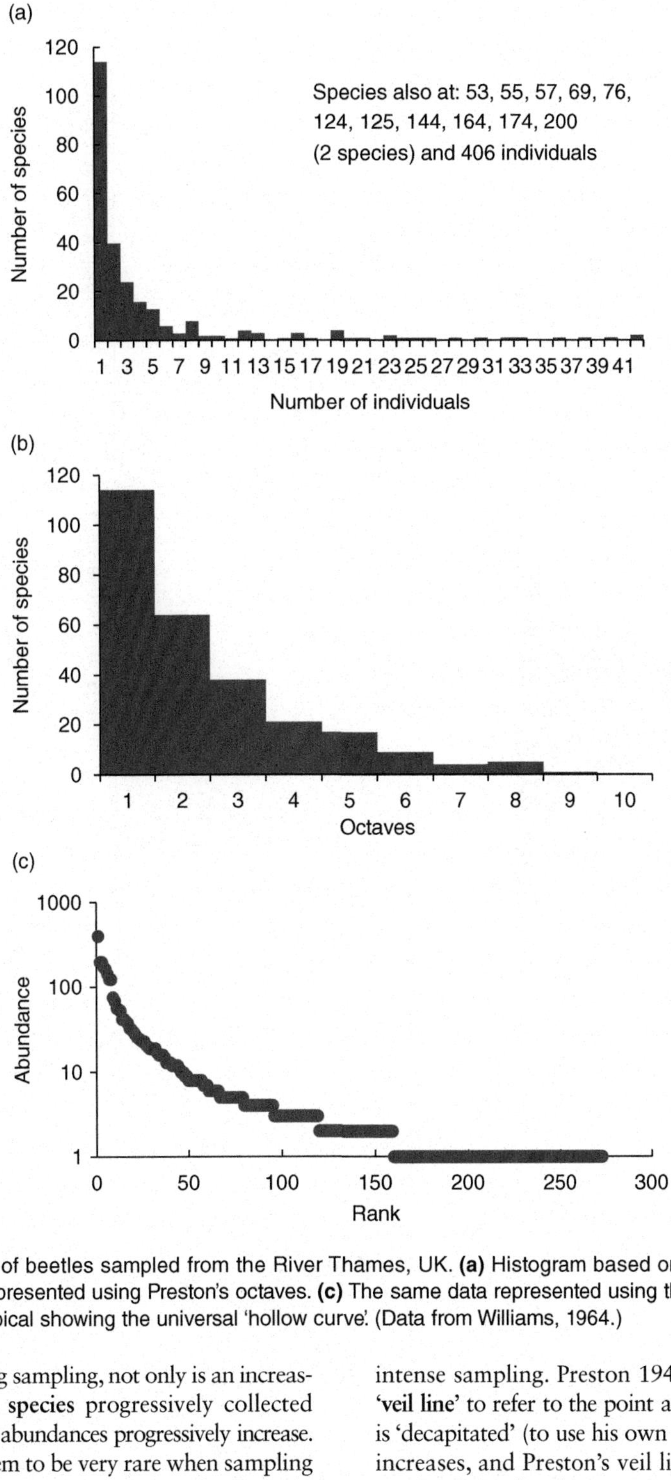

Fig. 4.2. Abundance of beetles sampled from the River Thames, UK. **(a)** Histogram based on an arithmetic scale. **(b)** The same data represented using Preston's octaves. **(c)** The same data represented using the rank abundance plot. The pattern is typical showing the universal 'hollow curve'. (Data from Williams, 1964.)

effort; with increasing sampling, not only is an increasing number of **rare species** progressively collected but also their relative abundances progressively increase. Thus, species that seem to be very rare when sampling is low assume intermediate abundance values at more intense sampling. Preston 1948) used the expression **'veil line'** to refer to the point at which the lognormal is 'decapitated' (to use his own term). As sample size increases, and Preston's veil line is pulled back, the mode of the distribution is revealed (see Fig. 4.3),

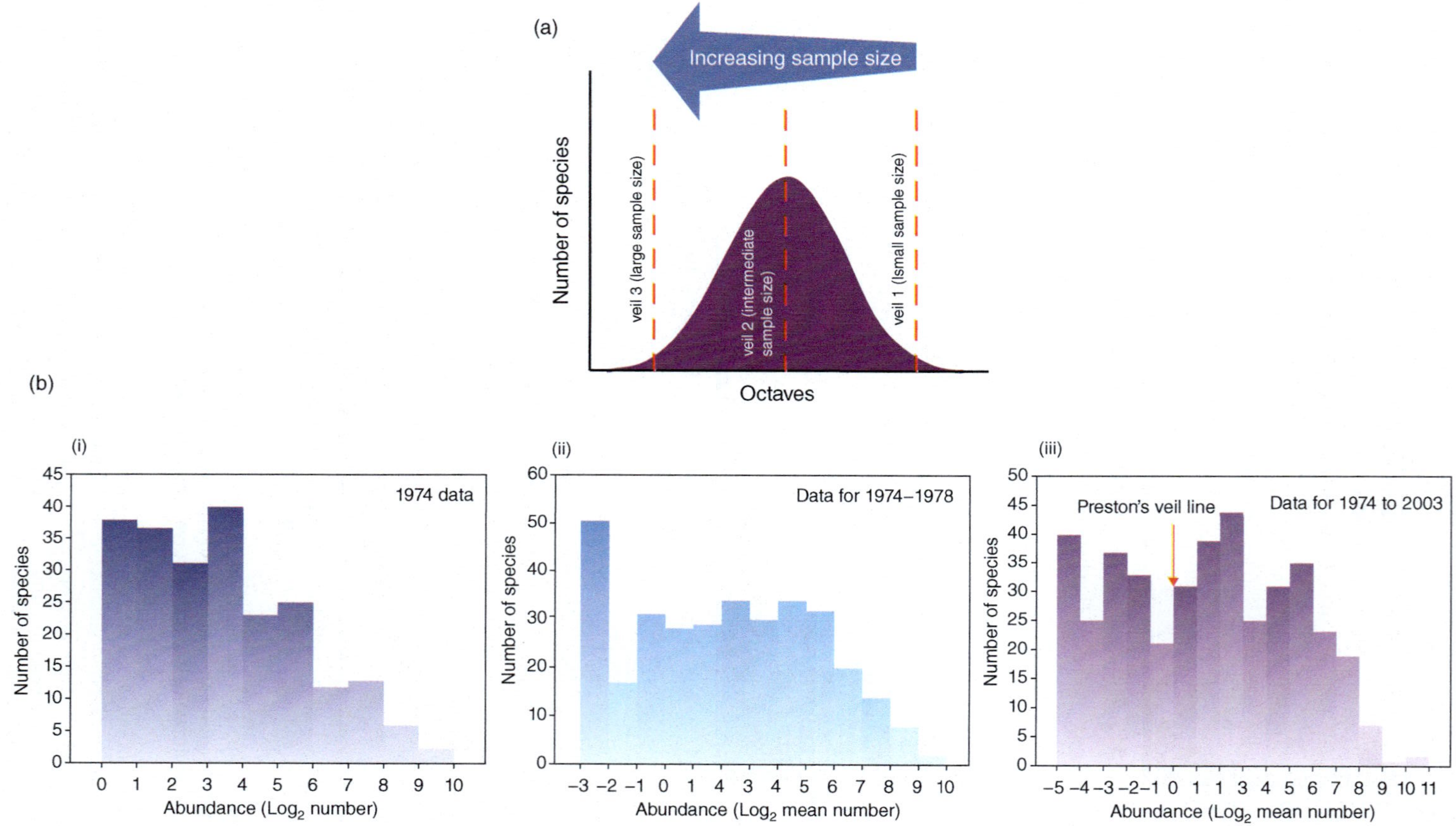

Fig. 4.3. The lognormal model of Preston (1948). **(a)** The shift of Preston's 'veil line' with increased sampling. With increasing sample size, the veil line moves to the left, revealing the mode and eventually the entire lognormal distribution. For example, as data are accumulated for increasingly longer periods, the number of rare species beyond the veil line continues to increase. This occurs if for no other reason than that species observed in a single sampling period become a smaller fraction with increasing sampling periods and the addition of each period leads to a halving of the interval (bin) for the histogram, effectively extending to infinity and never reaching zero. **(b)** A long-term research project based on moth sampling empirically illustrates this phenomenon. Log_2 plots of moth data from Monks Wood Experimental Station (Huntingdon, UK) for three periods: (i) 1974, (ii) 1974–1978, and (iii) 1974–2003. The number of rare species actually increases and the distribution becomes less skewed on the logarithmic scale. This may occur for three reasons: (i) vegetation succession provides new niches for novel species; (ii) chance recording of long-distance migrants entering the trap; (iii) movement of the trap post 1980 from an open to a woodland site. (Panel **b** courtesy of James Bell and Rothamsted Research, Harpenden, Hertfordshire, UK.)

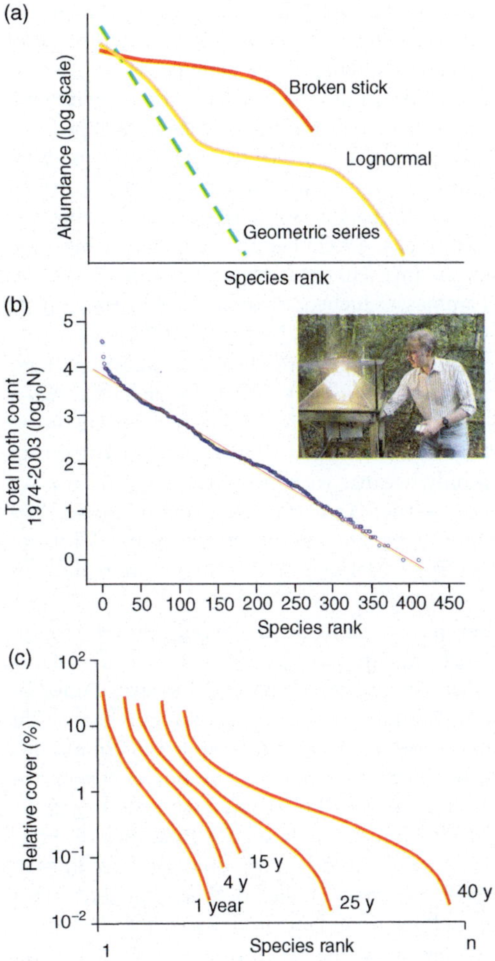

Fig. 4.4. Idealised shape of rank abundance plots. **(a)** MacArthur's broken stick model, lognormal distribution and geometric series. Species are ranked from most to least abundant. The broken stick model implies a relatively uniform exploitation of resources among species in a community; the geometric model, also referred to as the niche pre-emption model, is based on the successive exploitation of resources by species, ranked from the most to the least abundant; the lognormal distribution is typical of the partition of resources in large complex communities and modelled by species having varied niches and subject to different factors and resource use. **(b)** Moth species data recorded cumulatively over 30 years from Monks Wood Experimental Station (see Fig. 4.3b; the inset shows the Monks Wood trap being run by Nick Greatorex-Davies). The distribution does not follow a Preston lognormal distribution, the log abundance-rank distribution being distinctly linear (Pearson $r = -0.997$; $P < 0.0001$) as predicted by the geometric series. In fact, many rare species are probably vagrants visiting the site. With additional years, there is the prospect of recording additional species (see Fig. 4.3a), there being some 900 species of macromoths in the UK (Fox *et al.*, 2021). However, there is also the prospect of losing rare species with time and a moth trap is not a sufficiently sensitive sampling device to determine losses of species though sensitive enough to record newcomers. **(c)** Progression of rank abundance curves approaching a lognormal distribution for plant species literally rooted to their 'natal' habitat for five time intervals over 40 years during vegetation succession following field abandonment in Illinois, USA. (Inset of panel **b** courtesy of J. Nick Greatorex-Davies; data of panel **c** from Bazzaz, 1975.)

and SADs that appeared to be well approximated by a log-series reveal their true lognormal distribution. However, at very large sampling efforts (scales), the log-series can be recovered again due to the addition of many **specialist** species or spatially rare species (Borda-de-Água *et al.*, 2012). Although this veil line may be a true phenomenon (McGill, 2003; Chisholm, 2007), its existence, or its generality, has been questioned (Dewdney, 1997; Green and Plotkin, 2007), and there is abundant evidence that many communities actually follow the log-series (for example, in habitats that host communities that have a high frequency of rare species and that are not particularly species rich; Tsafack *et al.*, 2021a). Even in high diversity systems such as tree communities in tropical forests, local and regional SADs can follow a log-series, if a small share of the species is hyperdominant (ter Steege *et al.*, 2013). The aforementioned geometric series represents the most extreme case of uneven distribution of abundances (it is strictly related to the log-series; May, 1975), and can be found in simple communities characterised by the high dominance of a few species, such as those of very poor ecosystems (e.g. arthropods of arid environments) or those of the initial stages of ecological successions (Fattorini, 2009; Fattorini *et al.*, 2016a, 2017d; Niu *et al.*, 2019, 2020; Tsafack *et al.*, 2022).

However, despite the accumulation of SAD data for the most disparate organisms, some basic problems still afflict the research in this field. SAD models have been conceived for countable organisms, as abundance is expressed as number of individuals, but for many groups it is impossible to obtain suitable data. For example, among vascular plants, SAD data are abundant for tree communities (Hubbell, 1979; Cheng *et al.*, 2012; Baldridge *et al.*, 2016; Ulrich *et al.*, 2016; Arellano *et al.*, 2017; Borges *et al.*, 2018a; Matthews *et al.*, 2019; Antão *et al.*, 2021; Feng *et al.*, 2021), as trees are easy to distinguish and count, although what appears as separate individuals may be ramets of the same genet, but are very rare for grasses where biomass and cover are typically used as measures of abundance (Chiarucci *et al.*, 1999; Ulrich *et al.*, 2010, 2016, 2022; Anderson *et al.*, 2012; Matthews and Whittaker, 2015). This problem applies to other organisms, such as corals, for which number of colonies and cover are used as measures of abundance in SAD analyses (Connolly *et al.*, 2005). The use of continuous probability functions to model species abundance distributions based on cover data is a very recent progress (Moradi *et al.*, 2025). Moreover, important taxonomic biases exist. Whereas extensive research exists for certain groups (such as birds and some arthropods; e.g. Fattorini, 2005; Morlon *et al.*, 2009; Baldridge *et al.*, 2016; Antão *et al.*, 2021, McGlinn *et al.*, 2021), for other groups data are very rare, especially when species are difficult to identify and counting individuals would require an enormous effort. For example, as it is, we have little idea how SADs apply to micro-organisms; work on their diversity is only now getting under way with molecular techniques (see Stoeck *et al.*, 2009; Lara *et al.*, 2020). In discussing conditions near Cow Green Reservoir in the North Pennines, UK, Wilkinson (2021) draws attention to the enormity of the problems of assessing the diversity of small animals: in a square metre, he points out, there can be over 3 million nematodes, 200,000 enchytraeids, 80,000 springtails and 30,000 mites. These figures are as nothing to microbes, the vast number of bacteria and the even greater number of bacteriophages (Harper *et al.*, 2021). This immediately brings into question how many can be expected in each group within a single habitat unit! Once again, an ecosystem vantage seems to be most appropriate bearing in mind feedbacks between the innumerable components at local and regional scales – for instance, the familiarly cited reciprocal feedbacks between local abundances and geographical distributions expected from population dynamics linked to Tobler's laws via Hanski's metapopulation approaches and MacArthur and Wilson's island biogeography generalisations (see Table 4.1, B3-5). Most explanations of the SADs (see Appendix 3) typically fail to recognise common observations made in the field.

- The pattern of SAD curves will depend on the **taxocene** (or taxocenose, i.e. an assemblage of species belonging to the same taxon, e.g. a family) being studied. A collective taxon with limited species and/or food preferences will produce a very different pattern from one with an abundance of species utilising a range of food types (e.g. mammals versus birds versus leaf beetles) (see Krebs, 1999, Figure 12.5 illustrating rank abundance curves for gyrinids from Lake des Fammes, Quebec, Canada, and birds from Quaker Run Valley, New York, USA).
- Species belonging to different taxocenes have contrasting mobility. Thus, for instance, SADs are bound to differ for mobile (e.g. birds), sedentary (e.g. snails) and static (e.g. trees) organisms if only because we cannot be certain what is, or not, restricted to a habitat community (see Box 3.3 in Chapter 3 and Chapters 5 and 6 for difficulties in defining which species are really part of a community).
- Species are unlikely to belong entirely to the focal habitat; only in the case of a small 'island' with a homogeneous habitat and surrounded by uninhabitable **matrix** is this at all likely. The assumption that species included in community assessments have *all* their resources within the habitat space is clearly wrong (see Chapter 5). The habitats of many species cross over different traditional habitat types; some will have most of their resources in the focal habitat, but many 'middling' abundant species will not.
- A number of the 'rare' species recorded in the habitat are simply passing through it (possibly many of the species that are sampled only at high sampling intensity – Preston's veil line). This became obvious when studying woodland and open habitats for butterflies on Alderley Edge in Cheshire (UK); with repeated sampling, gradually every species breeding in the open habitats was discovered to penetrate (pass through) the woodland zones, if sparingly, though in most cases no useful resources existed for the **vagrants** (Dennis, 2000, 2001) (Fig. 4.5). Thus, the randomness and unevenness of species' niches – including intra- and interspecific competition over a habitat's resource base (e.g. plants) – used to explain a SAD (for example, the lognormal model) have also to include the complex patterns of species **resource-based habitats** (their true habitats). A key to SADs is varying fidelity to the focal habitat unit being studied.

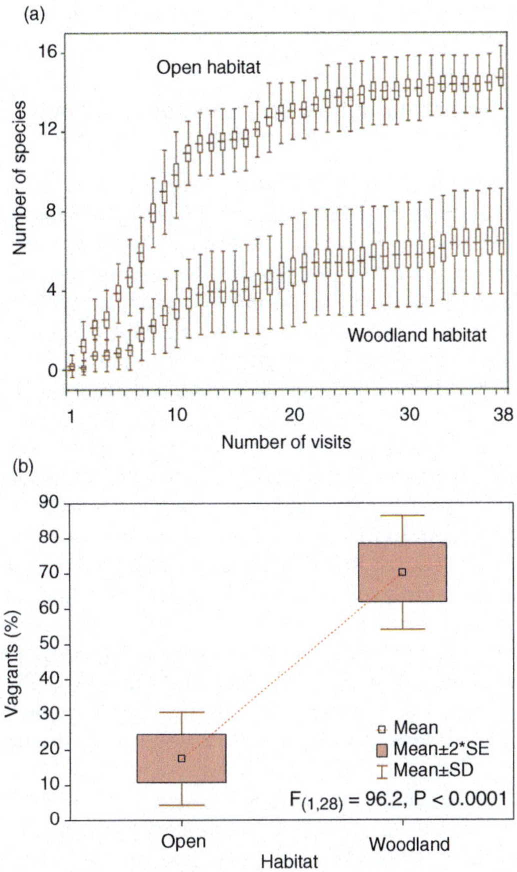

Fig. 4.5. Records of vagrant butterfly species in open (non-woodland) and closed (woodland) biotopes over the Alderley Edge escarpment, Cheshire, UK. Butterflies were surveyed on 38 occasions in 30 units of 1 ha (15 of each broad habitat type). **(a)** Species accumulation curves for the two biotopes (midpoints: means; boxes: standard errors; whiskers: standard deviations). Although both curves approach a plateau, new species are continually recorded in sample units. This pattern suggests the constant arrival of 'rare' species, in this case vagrants from adjoining units and outside the survey zone (see Henderson, 2003 for similar patterns for fish in the Bristol Channel, UK, and Heteroptera on oak trees). Note: losses not recorded. **(b)** Box plots of the relative frequency of vagrants in the two biotopes. Although far fewer species are found in the woodland, the proportion of vagrant species is higher here, owing to a greater bank of species available in the surrounding open biotopes that lack suitable habitats within the woodland. (Based on Dennis, 2000, 2001, used with permission of Springer Nature BV.)

- Communities can change dramatically over time while maintaining the same SAD. As in SAD models species identity is not considered, a community might maintain the same SAD even when all species are changed, if the new species have the same abundance distribution. Also, a certain community might maintain the same species composition and the same SAD, but species might switch their rank (for example, the most abundant species might become the least abundant and vice versa). These changes cannot be detected with SAD analyses.

One of the prominent features recognised for the SAD of a site is that it may well comprise two overlapping distributions; if communities are composed of core and transient species (Magurran and Henderson, 2003), the number of rare species increases with immigration from a large species pool (e.g. Hubbell, 2001; Ter Steege *et al.*, 2013), leading to an 'excess' of rare species found in communities that are well described by the log-series. The occurrence of residents and **vagrants** would certainly produce this pattern.

Residents may vary enormously in abundance. The commonsense inference from field observations is that common resident species are supported by suitable resources (food, hibernation sites, roosts, nest sites, etc.) and conditions (light, heat, moisture, lack of enemies, lack of competitors, etc.) for existence, whereas scarce species lack one or more critical resources and conditions to thrive. This division may relate to the contrasting life-history strategies, particularly **generalists** versus **specialists** (Verbeck *et al.*, 2010). Brown (1984) considered generalists to be at an advantage (more numerous) over specialists. As this depends on resource abundance, compaction and distribution, it could be the other way around. For instance, among butterflies, specialists tend to have highly specific resource requirements and habitat fidelity (thus denser populations) whereas generalists have varied resource requirements and lower habitat fidelity (and hence occur at lower densities) (Dapporto and Dennis, 2013; Dennis *et al.*, 2013). This means that a generalist species, which may be more abundant (greater occupancy) than a specialist one but also more widely distributed, will have a lower **density** at smaller scales. This should be taken into account when species rarity is evaluated (see Box 4.1) and SADs are constructed: SADs are obviously scale dependent. Finally, one thing should be remembered and may be taken as a 'general law': only a fraction of a resource (e.g. food) in a habitat space is actually 'suitable' (Dennis, 2010). This means that a measure of resource availability for any organism is likely to be a fraction of what seems to be available.

Vagrants, too, may be common or rare. A mass migration across the habitat, as for instance occurs in pierid and nymphalid butterflies at higher latitudes, can result in sudden high counts within a habitat unit though these events are sporadic in occurrence (Williams, 1958). More often, vagrants would appear to be infrequent, representing the end of a negative exponential tail of movement out from any habitat unit, especially focal habitats with dense populations (Shreeve *et al.*, 1996a).

Overall, the general pattern is a highly complex picture of factors likely to impinge on abundances. Although an organism can be relatively scarce owing to some obvious predictable factors (for example, large-sized animals tend to have smaller populations because of a rapid depletion of resources; see Box 4.1), species that are abundant are clearly favoured by very appropriate (near-optimal) conditions and high resource abundance. Those species that are relatively rare (e.g. most arthropods) are almost certainly limited by some resource (Liebig's law; see Table 4.1, B2), despite evidence casually obtained to the contrary (e.g. lycaenid butterflies whose larvae parasitise ant nests can be rare even when the host plants where eggs are laid and larvae initially feed on are abundant, because of their association with the ant host; Dennis, 2010). The way towards an explanation of the SADs has its basis in species' traits and life-history strategies of resource use, persistence and reproductive success. The fact that some species are exceptionally common in a community should direct research into determining how this is possible when crowding can have such deleterious consequences for organisms; population crashes after population explosions are common (e.g. *Euphydryas aurinia* (Lepidoptera), Ford and Ford, 1930; *Bibio pomonae* (Diptera), Dennis, 2008a). A further objective would be to determine the pattern of distributions for data collected over long periods (e.g. 30 years) as with the Rothamsted moth data for Monks Wood, Huntingdon, UK, to understand how SADs change over time. For example, vegetation succession should lead to changes in species richness and abundance, from poor communities with some very abundant species (geometric series) to rich communities with many intermediate species (lognormal distribution) (see Fig. 4.4c).

Limits to assessing species richness in habitats

The single most insightful biodiversity measure of a habitat is the number of species (**species richness**, *S*) for any taxon (or a multitaxon community). However, establishing which species should be considered as part of a community or an assemblage and which not is far from trivial. First, it can be extremely difficult to define the community/assemblage, both theoretically and practically. Should we consider only species of the same group (taxocene), all the species belonging to the same **trophic level**, or even all the species that belong to the same **food web**? Or should we consider other forms of relationships? To define any class, we need a property, viz. a characteristic that all its members share. But which is the property that all species must share to be members of a community? Should we count individuals of all stages or only adults? For example, research in insect community ecology typically considers only adults, but omitting larvae (which play fundamental roles in the ecosystem) is actually questionable.

Second, how are communities spatially bounded? We have already seen that it may be extremely difficult to define theoretically and to delimit practically the borders of a community (see the distinction between open and closed communities; Chapter 3, Box 3.3). Third, communities are dynamic: because of continuous processes of immigration and local extinctions, the number and the identity of the species that we find in a certain area, and that share the desired property, are subject to continuous changes. Fourth, it may be difficult to establish whether a species found in an area is a 'relatively' stable component of the community or not. Here, 'relatively' refers to the extent of the temporal phase, the intensity of the relationship and the aspects of the life cycle that we infer a species should have to be considered part of the community. Should we count only species that reproduce successfully in the area? With this logic, a species that does not reproduce in the study area should not be considered a member of the community, even if it may exert an important trophic role (e.g. as a prey or a predator of other species), which may lead to paradoxical results. And, how long should a species occupy an area to be part of the community?

All communities change over time (see Chapter 3) and **ecological successions** are gradual processes: the different communities that form **seres** are not separated by abrupt changes.

In summary, species richness can be constantly shifting because of habitat changes, losses and vagrants (see Henderson, 2003). Individual conspicuousness also plays a role, as the relationships between species and individuals in the habitat space, including their shifting locations, can be more easily determined from direct or remote observation for more conspicuous organisms (see Appendix 1). For some taxa, it is clearly an easy matter to count the number of species and indeed the number of individuals of each species (e.g. breeding birds in a wood, Newson *et al.*, 2005; wildebeest, topis and hartebeests in Serengeti grasslands, Norton-Griffiths, 1973; trees in forests, Borges *et al.*, 2018a). For the vast majority of organisms, however, it is virtually impossible to count all individuals present in a habitat accurately because of their small size and sampling difficulties. In such situations, relative numbers (abundance, cover, biomass or volume) are typically obtained by using equal sampling effort and identical techniques (Hubálek, 2000; see Appendix 1). As conspicuousness and abundance affect detection probability (with more abundant species being easier to find), species richness is a function of sampling effort: the higher the sampling effort, the greater the number of collected species (see Appendix 3). This translates into the questions: how can we establish if we collected all the species of a community (or any other assemblage – for example, the species of fauna or flora of a region) and how can we compare assemblages that were subject to different sampling efforts?

To gain an idea of the completeness of species inventories, one can construct an **accumulation curve**, i.e. a curve that depicts how the number of collected species increases with increasing sampling effort (measured as number of collected individuals or sampling units, such as number of **quadrats**, **transects**, sessions, etc.) (Fig. 4.6; see also Fig. 4.5a and Appendix 3). When a plateau is apparent (i.e. there is no more increase in species even if further individuals or sampling units are added), this means that all species of the sampled assemblage have been collected. However, in most cases, the sampling effort needed to reach the plateau would imply unsustainable sampling effort: when there are many species, collecting all of them (especially the rarest ones) would imply a number of sampled individuals or sampling units that might require enormous efforts in terms of time and resources, not to mention ethical problems if organisms must be killed for their collection and/or identification. To circumvent these problems, it is possible to estimate the true species richness through extrapolation techniques, or to rarefy species richness to a common sampling effort, by means of appropriate **estimators** (e.g. Chao's estimators estimate the true number of species based on the number of the species found in one and two individuals or in one and

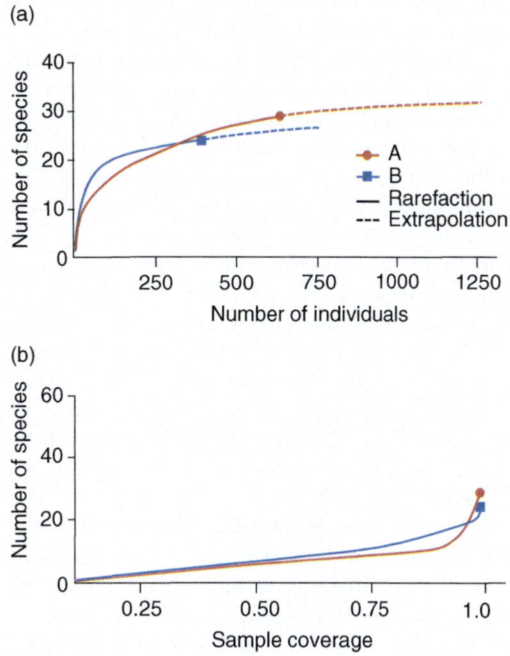

Fig. 4.6. Rarefaction/accumulation curves for two communities (A and B) subject to different sampling effort. The circle and the square indicate the number of species collected in community A and B, respectively. In terms of sampled species, community A appears richer than community B. The two communities, however, were subject to a different sampling effort, which makes comparison difficult. **(a)** If they are rarefied to the same number of 250 individuals, community B is richer than community A. If they are extrapolated to the same number of 750 individuals, community A is richer than community B. These contradictory results indicate the difficulties in comparing communities subject to different sampling efforts. **(b)** The two communities have similar very high coverages (approaching 1 in both cases): at high coverage (~1), it appears clear that community A is richer than community B.

two sample units; see Appendix 3). Rarefaction may appear a safer procedure than extrapolation, yet it may produce misleading results. By reducing the richness of the community with more individuals to that of the community with fewer individuals, we artificially transform a species-rich community into a species-poor one (Tsafack *et al.*, 2021a). Although rarefaction procedures are routinely used to compare communities that differ in the sample size, caution is necessary. To avoid problems arising from comparisons based on sample size, it has been proposed to compare species richness of communities with reference to samples of equal completeness (measured by **sample coverage**) instead of equal size (Chao and Jost, 2012) (see Appendix 3).

Although all these techniques may provide valuable standard measures for comparing species richness within and across habitat units, it is important to understand that all techniques have their limitations (Chiarucci *et al.*, 2018; Khalighifar *et al.*, 2020). From the habitat vantage, long-term repeated sampling will continue to record new species (see Fig. 4.5), particularly for speciose taxa, with consequences for the SADs (see Fig. 4.3). Thus, it is essential to apply (or adjust for) even sampling techniques at the survey stage for comparative research. Indeed, many research programmes do exactly that, attempt an even sampling regime (e.g. Rothamsted moth and aphid traps, see Fox *et al.*, 2013; www.rothamsted.ac.uk/national-capability/the-insect-survey; accessed 30 May 2025).

However, even sampling regimes do not necessarily address the issue of which rare species within habitats are scarce residents and/or frequent vagrants. For instance, the large white butterfly *Pieris brassicae* is typically recorded on Butterfly Monitoring Scheme BMS UK sites each year; in some sites it will breed on taller, bunched garlic mustard *Alliaria petiolata* or cliff cabbage *Brassica oleracea* (Dennis, 2010), but in many of the sites it will simply be a passing vagrant. In the first case, it may be said to be a resident of the habitat space; in the second, it is likely to be excluded as a resident, though recorded each year and occasionally in greater numbers. Assessing number of species inevitably draws attention to the significance of scarce species. Just how important a component are rare species of a traditional habitat unit, community or ecosystem? We will return to this question later (see Chapters 10 and 11). They are more important than may be supposed: the rare species today may become the dominant species tomorrow.

Indices of species diversity and evenness for habitats

It is clear from the above that a crucial part of estimating **species richness** is sampling population abundance. Communities sampled from species pools that have higher evenness and/or more total species will have higher richness (McGlinn, 2021). Thus, species richness is intimately connected with SADs, but cannot be predicted by the different forms of SADs; nor can SADs predict richness. As a measure of species diversity, richness not only assumes species as being equally different (thus ignoring **disparity**) but also ignores SADs, as it regards all species as having the same abundance. While the potential importance of considering disparity has come into focus only recently (see Appendix 3), ecologists have long searched for indices that consider evenness/dominance. A plethora of indices of diversity, dominance and **evenness** have been proposed (some are illustrated in Appendix 3) and usually researchers use a number of them for comparison, as different indices may provide complementary useful information. Their value is in providing different comparative measures of communities, different insights, for species within single habitats and in different habitat spaces. Also these indices have been used to express **'habitat diversity'**, that is, the diversity of habitat types (their number and relative cover) in a given area (Box 4.3). Again, a warning: there are limitations to their effective use, often illustrated in texts describing them (Ludwig and Reynolds, 1988; Hubálek, 2000; Henderson, 2003; Daly *et al.*, 2018).

Researchers are spoilt for (or confounded by) choice in selecting single-value diversity indices (see Table 4.2 and Appendix 3 for some alternatives). A common problem with diversity indices traditionally used in community ecology is that they express diversity in different units. For example, **species richness** S is expressed in units of species, **Shannon diversity** H' (which represents uncertainty about the identity of an unknown individual) is an entropy measure (and therefore has bits of information), and **Simpson index** D (which expresses the probability that two randomly chosen individuals do not belong to the same species) is a probability measure. In such cases, diversity measures are not directly comparable. A second important problem is the non-intuitive behaviour of the indices because of their non-linearity (assuming all species as equally abundant, the values of Shannon and

Box 4.3. Habitat diversity

The concept of '**habitat diversity**' has been widely used in the context of island biology to express the importance of the variety of environmental conditions to explain the number of species occurring on islands. The **species–area relationship** (SAR), i.e. the increase in the number of species with increasing area, is one of the best documented patterns in island ecology (Matthews *et al.*, 2021). Several (mutually non-exclusive) hypotheses have been proposed to explain island SARs. For example, larger islands may host more species because (i) they can intercept a greater number of dispersing individuals (as they are larger targets – passive sampling hypothesis; da Silva *et al.*, 2024) and (ii) have lower extinction rates (as they can support larger populations – equilibrium theory; MacArthur and Wilson, 1967; Ricklefs and Lovette, 1999; Whittaker *et al.*, 2023). However, as we can expect that also the extent and variety of resources the species can use will increase with island area, resource availability, and not area, might be the key to species coexistence (habitat diversity hypothesis; Williams, 1943; Rosenzweig, 1995; Ricklefs and Lovette, 1999; Triantis *et al.*, 2005, 2006; Hortal *et al.*, 2009, 2013). Thus, island area can influence species richness both directly and indirectly, through habitat diversity (Fig. B4.3.1).

Attached to these main hypotheses, several complementary explanations have been proposed (see Whittaker *et al.*, 2023). For example, a greater availability of resources on larger islands may allow individuals to have clumped patterns, which sustain denser populations and reduce extinction rates, thus increasing diversity (resource concentration hypothesis; Connor *et al.*, 2000; Matthews *et al.*, 2015).

One of the most common ways to express 'habitat diversity' is the number of habitat units (= number of **biotopes**; e.g. Kohn and Walsh, 1994; Ricklefs and Lovette, 1999; Triantis *et al.*, 2005, 2006; Fattorini, 2006; Hannus and Von Numers, 2008; Hortal *et al.*, 2009; Proença and Pereira, 2013; Sfenthourakis *et al.*, 2021). Just as **species richness** does not take into consideration the importance of differences in species abundance/biomass as a measure of α-diversity in community ecology, the number of habitats/biotopes as a measure of 'habitat diversity' does not consider the relative extent of different biotopes, a feature that also influences environmental heterogeneity. For this reason, some authors have used 'habitat diversity' indices that take into account the relative area covered by each 'habitat' (biotope) (e.g. Ricklefs and Lovette, 1999; Fox and Fox, 2000). Among these, **Shannon diversity** is probably the

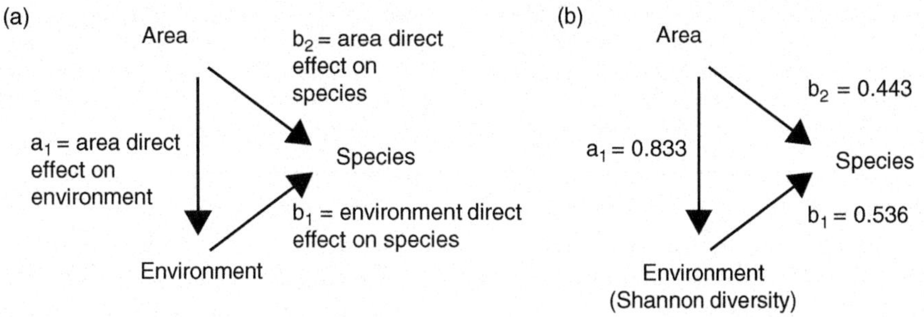

Fig. B4.3.1. Path models of tenebrionid beetle species richness on 25 islands around Sicily as a function of area and environmental heterogeneity measured as Shannon diversity of CORINE land cover types (see Chapter 1). Path analysis allows the relative direct and indirect effects of predictor variables to be assessed according to an *a priori* model. The path model for the effect of area and environmental heterogeneity upon species per island is shown in panel **(a)**. The coefficients a_1, b_1 and b_2 represent direct effects of one variable on another; a_1 is the simple correlation coefficient for the variables environmental heterogeneity and area as obtained from the regression ln(Environmental heterogeneity) = *a*ln(Area) + *c*; b_1 and b_2 are the standardised regression coefficients from the multiple regression model ln(Species richness)= *a*ln(Area) + *b*ln(Environmental heterogeneity) + *c*. Indirect effect of area can be calculated as the product $a_1 \times b_1$. All coefficients reported in panel **(b)** are significant at $P < 0.05$ (see Fattorini *et al.*, 2015a for details). The magnitude of the indirect effect of area (a_1b_1) on species per island was roughly similar to its direct effect (b_2). The direct effect of the environment (b_1) on species richness exceeded the direct area effect on species (b_2), but the total effect of area, when direct and indirect effects are summed ($a_1b_1 + b_2$), was greater than the effect of environment. In this system, environmental heterogeneity is an important predictor of species richness, both directly and through the effect of area, but area alone is also important. (After Fattorini *et al.*, 2015a.)

Continued

Box 4.3. Continued.

most frequently used index, especially in landscape ecology and biogeography (see, for example, Farina, 1998; Luoto *et al.*, 2001; Lobo and Martín-Piera, 2002; Burel and Baudry, 2003; Nogués-Bravo and Martínez-Rica, 2004; Stefanescu *et al.*, 2004; Maes *et al.*, 2005; Dušek and Popelková, 2012; see main text and Appendix 3 for the definition of the Shannon index). However, other indices of diversity may also be appropriate (see Fattorini *et al.*, 2015a).

The expression 'habitat diversity' refers to the traditional idea of habitats as different portions of the **landscape**. However, if we accept that a habitat can only be recognised by referring to the resources and conditions needed by a particular species (i.e. the **resource-based habitat** concept), 'habitat diversity' is not an appropriate definition to qualify general features of areas. More appropriate expressions would be 'environmental heterogeneity' or 'environmental diversity' (see Hortal *et al.*, 2013; Stein and Kreft, 2014; Stein *et al.*, 2014), because they refer to the variety of environmental aspects that may characterise an area and can affect species presence, such as edaphic factors (depth of litter, size of rocks, type of substrates) and climatic factors (Solem *et al.*, 1981; Ricklefs and Lovette, 1999; Kocher and Williams, 2000; Schoener *et al.*, 2001; Stein and Kreft, 2014).

In fact, the expressions 'environmental heterogeneity' or 'environmental diversity' have sometimes been used as synonyms of 'habitat diversity' (e.g. Triantis *et al.*, 2005, 2006), but the use of 'habitat diversity' has been defended on the basis of its relatively wide usage (Panitsa *et al.*, 2006). Fattorini *et al.* (2015a) suggested preferring the expression 'environmental heterogeneity', as it seems more appropriate to express the environmental complexity of an area with reference to any kind of variability in environmental characteristics that can affect species presence (number of biotopes, landscape units, plant associations, soil types, etc.), their proportional contribution, spatial configuration and variation over time, depending on how it is measured. More specific terminologies, such as 'environmental diversity', 'environmental dominance' or 'environmental evenness', may be appropriate according to the index used (see also Stein and Kreft, 2014 for a review of the use of various expressions). In principle, any measure of biodiversity (as those described in Appendix 3) may be applied to quantify 'habitat diversity' in its multiple aspects on the basis of the different types of biotopes (see Fattorini *et al.*, 2015a). Other approaches to quantify 'habitat diversity' are based on the use of topographic and climatic variables. For example, in island biogeography, elevation is frequently considered a proxy of 'habitat diversity' (e.g. Fattorini, 2002a,b, 2006; Traveset *et al.*, 2014; Fattorini *et al.*, 2015a), since elevation is correlated with variation in temperature, precipitation, humidity, wind speed, evaporation and insolation, and hence higher islands have more variability in these parameters. Increasing availability of GIS layers now makes it possible to easily use vast sets of topographic and climatic data to quantify topographic heterogeneity and climatic variability as measures of environmental heterogeneity, thus enhancing our ability to understand their importance in determining variations in species richness (Weigelt *et al.*, 2013; Irl *et al.*, 2015; Keppel *et al.*, 2016).

Simpson indices for a community of 20 species are not the double of those of a community with ten species). For these reasons, there is increasing use of diversity indices known as **Hill numbers** (Jost, 2010; Chao and Jost, 2012; Chao *et al.*, 2014a,b; Chiu *et al.*, 2020; Roswell *et al.*, 2021), which are all expressed as the number of equally abundant species (see Appendix 3). These numbers are produced by a continuous function equation (related to Rényi's, 1961 concept of entropy), which is characterised by a parameter that influences the weight that common and rare species have in giving a Hill number. By plotting Hill numbers as a continuous function of this parameter, a graph is obtained that is known as a **diversity profile**, and which illustrates the contribution of rare and common species in generating diversity values. This approach is especially useful for comparing communities. If the diversity profiles of two communities do not intersect, this means that one community is consistently more diverse than the other. But if the curves cross, this means that one community is more or less diverse than the other according to the importance that we give to rare and common species (Fig. 4.7). Use of diversity profiles, instead of single measures, may therefore appear a very good approach for comparing communities. However, this does not mean that the use of single measures is obsolete. For example, if we want to see how diversity varies in response to variation of some environmental parameter, we need single measures as the response variable. In this case, a well-selected set of traditional indices (e.g. richness, the exponential version of Shannon entropy and Simpson diversity, corresponding to Hill numbers H_0, H_1 and H_3; see Table 4.2 and Appendix 3)

(a)

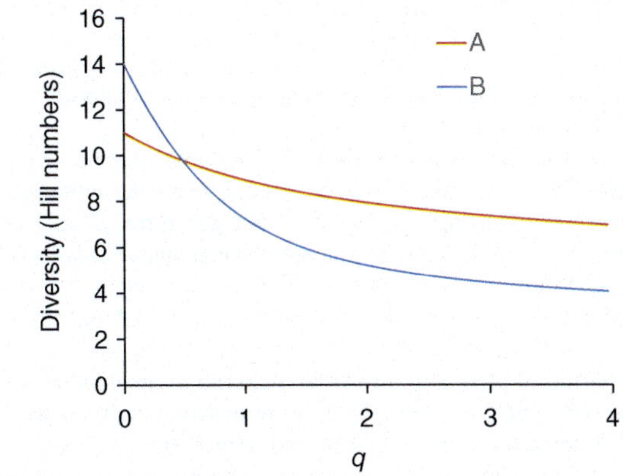

(b)

Hill number	Community A	Community B
Species richness (q= 0)	11	14
Shannon diversity (exponential) (q= 1)	8.95	7.26
Simpson diversity (q= 2)	7.96	5.23

(c) (d)

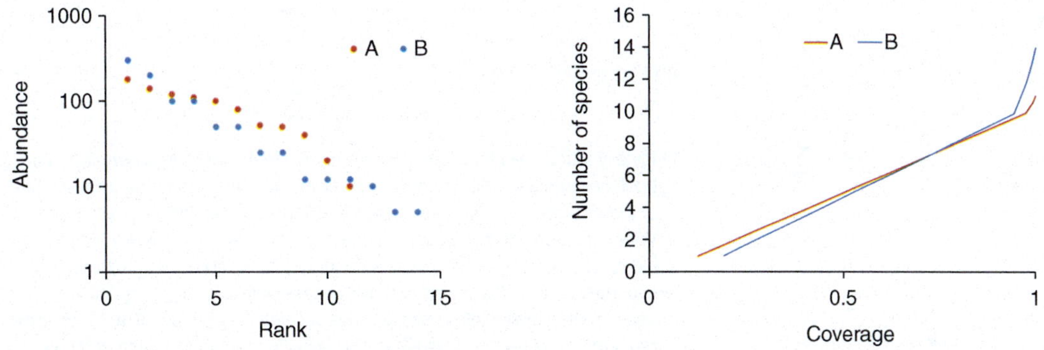

Fig. 4.7. Diversity profiles (Hill numbers) for two hypothetical communities. **(a)** Hill numbers are calculated for the parameter q varying between 0 and 4. **(b)** Values of Hill number for q = 0 (species richness), q = 1 (Shannon diversity) and q = 2 (Simpson diversity). The two curves intersect, which means that one community is more or less diverse than the other according to the importance that we give to rare and common species. Community B is more diverse than community A in terms of species richness (all species are equally weighted, and hence the contribution of rare species is emphasised) but community A is more diverse than community B in terms of Shannon diversity (species are weighted in proportion to their frequencies); the difference between the two communities remains similar when the Simpson index is applied and for even higher values of q (i.e. when the contribution of the most common species is emphasised). This means that community B is more diverse than A for the presence of a greater number of rare species, but community A is more balanced. **(c)** This is confirmed by the rank abundance plots, where the decline in the abundances is steeper for community B. **(d)** The analysis based on coverage indicates that the two communities were adequately sampled.

would allow a decent exploration of how different aspects of diversity are influenced by environmental parameters.

Coefficients of niche breadth and overlap

As envisaged by Taylor's power law, with growing numbers of species and evenness of species (viz. an equal number of individuals) within a habitat, there is the prospect of individuals belonging to those species increasingly coming into contact. Whereas at low densities (individuals/**niche** spaces $\ll 1$) they would be able to exploit resources without competition, at high densities it is considered that they retreat to the appropriate niche spaces best suited to their specific traits (Rosenzweig, 1995), much as nesting sea birds do on different

Table. 4.2. Some indices of diversity. S = number of species, n_i = number of individuals belonging to the ith species, N = total number of collected individuals, x = a parameter of the log-series that can be estimated iteratively; $p_i = n_i/N$.

Name	Formula	Comments
Species richness	S	The number of species that form the study assemblage. Corresponds to the Hill number with $q = 0$.
Margalef richness index	$D_{Mg} = \frac{S-1}{\ln(N)}$	This index was introduced to provide a relative measure of species richness for communities subject to different sampling efforts. Although strongly influenced by the sampling size and effort (Magurran, 2004), it seems to have a good ability to discriminate communities based on richness (Yeom and Kim, 2011).
Menhinick richness index	$D_{Mn} = \frac{S}{\sqrt{N}}$	Similar to Margalef index.
Fisher's α	$\alpha = N(1-x)/x$	Fisher's α has the virtue of providing a reliable measure of diversity when completeness is very low, but it is particularly sensitive to changes in the abundance of dominant species (Beck and Schwanghart, 2010).
McIntosh diversity index	$U = \sqrt{\sum_{i=1}^{S} n_i^2}$	This index describes the sampling unit as a point in an S-dimensional hypervolume and uses the Euclidean distance of this point from the origin as a measure of diversity. This index can be used to obtain a dominance measure: $D = \frac{N-U}{N-\sqrt{N}}$ and an evenness measure: $E = \frac{N-U}{N-\sqrt{S}}$
Shannon diversity	$H' = -\sum_{i=1}^{S} p_i \ln(p_i)$	Originally introduced by Claude Shannon in 1948, and then presented in a book co-authored with Warren Weaver in 1949, leading to the frequent use of the erroneous citation as Shannon–Weaver index. In that book, Norbert Wiener is also cited as source for the development of the index, which led to the expression Shannon–Wiener index. By combining richness and evenness, this index is particularly successful in summarising different aspects of diversity into a single value (Hayek and Buzas, 2010). It assumes random sampling in an infinite population. Its exponential version ($H_1 = \exp^{H'}$) corresponds to the Hill number with $q = 1$.
Pielou evenness	$J = H'/\ln(S)$	This index measures the evenness with which individuals are divided among the species, and it is constrained between 0 (all individuals belong to one species) and 1 (all species are equally abundant) (Magurran, 1988).
Brillouin index	$HB = \frac{\ln(N!) - \sum_{i=1}^{S} \ln(n_i!)}{N}$	This index is similar to the Shannon index, but does not assume random sampling in an infinite population. Thus, it should be preferable to the Shannon index, especially when randomness is not guaranteed (for example, when some species have a greater probability of being captured) or the community is completely censused, with all individuals accounted for (Magurran, 1988). An evenness index is calculated as: $E = HB/HB_{max}$, where HB_{max} is $HB_{max} = \frac{1}{N}\ln\frac{N!}{\{\lfloor N/S \rfloor!\}^{S-r} . \{(\lfloor N/S \rfloor + 1)!\}^{r}}$, with $\lfloor N/S \rfloor$ = integer of N/S, and $r = N - S\lfloor N/S \rfloor$.
Simpson index	$N_2 = 1/\sum_{i=1}^{S} p_i^2$ $D = 1 - \sum_{i=1}^{S} p_i^2$	One of the most commonly used diversity indices. N_2 corresponds to Hill number with $q = 2$. D is an alternative version. Simpson evenness is calculated as: $E = N_2/S$
Berger–Parker index	$D_{BP} = n_{max}/N$	Measures the relative abundance of the most abundant species. It is an index of dominance. The inverse can be used as an index of diversity and corresponds to the Hill number for $q = \infty$.

parts of cliffs or phytophagous insects on distinct plant species or different parts of the same plant. A specific example of appropriate niches is the partition of *Senecio smallii* plants by three hemipteran herbivores (*Neacoryphus bicrucis*, *Harmostes reflexulus* and *Lygus pratensis*) on the basis of bud density and plant patch size (McLain, 1981). When it costs nothing to select habitats and habitats are not monopolised by despots, an **ideal free distribution** is conceived (Fretwell and Lucas, 1969) (see Chapter 6). However, where competition is rampant between two species with overlapping niche requirements, a joint increase in density is perceived to result in the species selecting distinct niche spaces, so that the two species will occupy different habitats and seem not to compete (**isoleg theory**, Rosenzweig, 1979; Brown and Rosenzweig, 1986; Guthrie and Moorhead, 2002; Morris, 2003).

Niche complementarities (i.e. niche differences among species) are important drivers of species coexistence (e.g. Darwin, 1859; MacArthur and Levins, 1967; Silvertown, 2004; Stubbs and Wilson, 2004; Mason and Wilson, 2006). However, local environmental conditions can produce patterns in which coexisting species are ecologically more similar than expected by chance alone, thus weakening the impact of the niche complementarity process (Leibold, 1998; Mayfield and Levine, 2010), for example because of patchiness in resource spatial distribution. Thus, non-random patterns in community assembly are not necessarily driven by competition (Zalewski *et al.*, 2014), but may be influenced by other factors, such as resource partitioning that has evolved over evolutionary periods (Behangana and Luiselli, 2008). Understanding the mechanisms by which species organise themselves into communities is crucial for comprehending how species coexist, thereby maintaining biodiversity.

Researchers have attempted to capture niche occupancy with two broad types of measures: **niche breadth** and **niche overlap** (including preference coefficients). There are, in fact, numerous measures of both niche breadth and overlap (Box 4.4). Despite their use to infer **competition**, overlap indices should only be considered as descriptions of potential interspecific interactions *sensu lato*, unless competitive interactions are known *a priori* (Lombardo and Mjelde, 2014; Fattorini *et al.*, 2017c). High niche overlap may lead to conflicting interactions (such as competition and exclusion) (Giménez Gómez *et al.*, 2018; Pascual-Rico *et al.*, 2020), or only indicate strong interactions in one niche dimension, with coexistence permitted by weak interactions in others (Koutsidi *et al.*, 2020). By contrast, low niche overlap implies lower levels of interaction, thereby allowing sustainable coexistence of species through differential utilisation of resources (Koutsidi *et al.*, 2020; Yang and Hui, 2020). Since past-competitive interactions would result in a differential use of habitat resources, values of niche overlap lower than those expected under models of random utilisation of resources may be suggestive of competition, whereas high values of overlap might indicate that species are variably forced to concentrate in different habitat patches owing to other places being unsuitable (Tsafack *et al.*, 2021b).

To one extent or another, indices of niche breadth and overlap are all affected by the unknown amount of each resource available to a species; unknown, because only careful **autecological** research can determine just what part of a resource is suitable for any species, information often missing from surveys. Additionally, species may use the same resource and not compete for at least three reasons: (i) they use the same resources in different parts of the habitat (e.g. light versus shade; damp versus dry); (ii) they may use completely different parts of the same resource (e.g. small plants versus large plants; leaves versus flowers; young leaves versus old leaves); (iii) they may be using the same resource but at different times of the year. Thus, unless this information is available, only a tentative assessment of niche breadth and overlap can be determined. Indices of niche breadth and overlap may be suggestive of possible relationships, but are really no substitute for a thorough multivariate analysis of resource use by potentially competing organisms based on detailed surveys over different habitat spaces and seasons.

Measuring relationships across habitat units

Any one of the diversity, dominance and evenness and niche breadth or overlap measures applied to a single habitat unit can obviously also be exploited to examine variation across a range of similar or different habitats, and the very same habitat unit as it changes over time with season, succession or other interference factors such as climatic change (see Daly *et al.*, 2018). This is certainly the case with species diversity. However, comparisons between habitat units tend to focus more on the differences

Box 4.4. Measures of niche breadth and niche overlap

Simple measures are discussed below for **niche breadth** and **niche overlap** as they have been largely used in literature on competition and community assembly. More detailed procedures are, however, available (Ludwig and Reynolds, 1988; Krebs, 1999; Swanson *et al.*, 2015) that increasingly involve analyses of density functions (Mouillot *et al.*, 2005) and ecological niche modelling (Warren *et al.*, 2008; Broennimann *et al.*, 2012).

A simple measure of niche breadth is by Levins (1968):

$$B = 1/\sum p_j^2, \text{ or} \tag{B4.4.1}$$

$$B' = Y^2/\sum N_j^2 \tag{B4.4.2}$$

where p_j is the proportion of individuals found in or using resource state j, N_j is the number of individuals found in or using resource state j and $Y = \sum N_j$ is the total number of individuals sampled (Krebs, 1999). Other indices have been introduced to consider resource availability (Hurlbert's niche breadth, which is a variant of Levins' index, and Smith's index; Krebs, 1999). The **Simpson index** of diversity (e.g. Pianka, 1975) and the **Shannon index** (see discussion in Krebs, 1999) have also been suggested as measures of niche breadth.

Many measures of niche overlap have been proposed (see Schatzmann *et al.*, 1986 for a review). A measure of overlap is given by Renkonen's (1938) percentage similarity measure (also known as Schoener's (1974) index):

$$P_{lm} = \left[\sum_{i=1}^{N} \min(p_{il} p_{im})\right] 100 \tag{B4.4.3}$$

where P_{lm} is the percentage overlap between species l and species m, p_{il} and p_{im} are total proportions of resource i used by species l and m, and N is the number of resource states.

A popular niche overlap index is that of Pianka (1974):

$$O_{jk} = \frac{\sum_{i=1}^{N} p_{ij} p_{ik}}{\sqrt{\sum_{i=1}^{N} p_{ij}^2 \sum_{i=1}^{N} p_{ik}^2}} \tag{B4.4.4}$$

where overlaps are symmetrical and p_{ij} and p_{ik} are the proportions of the ith resource used by the jth and kth species (see Krebs, 1999). The index ranges from 0 (absence of overlap) to 1 (complete overlap).

The O_{jk} index is a symmetrical version of Levins' α index of niche overlap (MacArthur and Levins, 1967; Levins, 1968):

$$\alpha_{jk} = \frac{\sum_{i=1}^{N} p_{ij} p_{ik}}{\sum_{i=1}^{N} p_{ij}^2} \tag{B4.4.5}$$

Both the Pianka and Levins formulations express correlations between species distributions over discrete resources (e.g. food types or habitat patches), but the Levins index is asymmetrical (as the overlap of species j over species k does not necessarily equal the reciprocal overlap of k over j) and may be > 1.

The asymmetrical nature of the Levins α index makes it preferable when a high detail of interspecific interactions is sought or known *a priori* (Slobodchikoff and Schulz, 1980; Lombardo and Mjelde, 2014), while the Pianka index is more appropriate when species **autecology** is not adequately known, when a standardised index is desired and/or when analyses are addressed to recover general patterns rather than hyperdetailed interactions (Pianka, 1973; May, 1975).

To assess if mean overlap values in a community are different from those that can be expected by chance alone, observed values can be compared with the distribution of expected values obtained from simulated null assemblages. These simulations are based on matrices that represent resource use. Four randomisation algorithms (commonly referred to as RA1, RA2, RA3 and RA4) have been proposed to construct null matrices that differ in whether utilisations are reshuffled or replaced by a random number, and in whether the zeroes in the matrix are retained or not (Lawlor, 1980; Gotelli and Graves, 1996) (Fig. B4.4.1). RA3 is considered the best existing algorithm. This algorithm retains niche breadth (i.e. the amount of specialisation) for each species but reshuffles zero states, thus randomly varying the particular resources that are used. RA2 retains zeroes (thus preventing species that do not use a certain resource in the true matrix from doing so in the null matrices), but relaxes niche breadth (thus assuming a random equiprobable specialisation). This algorithm may be useful to test for structure in the generalist–specialist nature of the resource utilisation matrix. RA1 replaces every cell in the matrix with a randomly chosen, uniform number between 0 and 1, thus generating null matrices in which all species utilise all resource categories, which is considered too 'disruptive' and hence not recommended. RA4 retains the niche breadth of each species and fixes the zero states to their observed values, thus being too conservative and hence not recommended.

Another approach frequently used to investigate potential competitive interactions is based on the

Continued

Box 4.4. Continued.

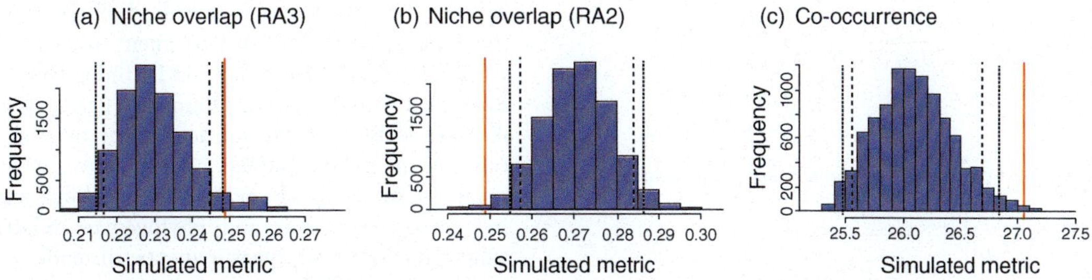

Fig. B4.4.1. Histograms of expected values (blue bars) for spatial niche overlap (Pianka index) and species co-occurrence (*C*-scores) in carabid beetle communities in a Central Asian steppe (Ningxia region, China). Expected values were constructed using the RA3 **(a)** and RA2 **(b)** algorithms for niche overlap, and the fixed-fixed algorithm for the *C*-score **(c)**. In all cases, 10,000 null matrices were constructed. In each graph, the vertical red line indicates the observed value, long-dash lines indicate the one-tailed 95% limits, and the short-dash lines the two-tailed 95% limits. Niche overlap was significantly higher than expected by chance with RA3, but significantly lower with RA2, suggesting that the environment is spatially structured (as removal of zeroes increases the overlap, the unsuitability of some sites for particular species contributes to the pattern of spatial utilisation among species). The significantly high average *C*-score suggests that the presence of a species is directly affected by the presence of other species, and hence that competition may have some role in defining species assemblages in these ecosystems. (After Tsafack *et al.*, 2021b, used under licence CC BY 4.0.)

analysis of **species segregation**. A common procedure is to analyse a matrix of species presence/absences in sites through the calculation, for each species pair, of an index known as the *C*-score (Stone and Roberts, 1990). For any given pair of species *j*,*k*, the *C*-score is calculated as follows:

$$C_{jk} = (R_j - SS)(R_k - SS) \qquad \text{(B4.4.7)}$$

where R_j is the row total for species *j*, R_k is the row total for species *k*, and *SS* is the number of sampling units that contain both *j* and *k*. *C*-scores range from 0 (maximally aggregated) to a maximum of R_jR_k (maximally segregated with no shared sampling units). The average of all the pairwise *C*-scores is the matrix-wide *C*-score. A high average *C*-score indicates a low randomness (i.e. a greater likelihood that the distribution of one species has been directly affected by the presence of other species). Thus, a competitively structured assemblage should have a *C*-score significantly larger than a randomly assembled community (Gotelli, 2000). However, the average *C*-score reflects both positively and negatively associated species pairs, and Sanderson and Pimm (2015) claim to examine single-pairs metrics instead of focusing on ensemble metrics.

Another simple but poorly explored approach to investigate co-occurrence patterns is the use of **generic diversity**. As species belonging to the same genus are expected to be morphologically more similar than species of different genera, if morphological similarity is assumed to be a proxy for similar ecological needs, congeneric species are expected to have more overlapping niches. On this basis, generic diversity (calculated using **Simpson index** of diversity, as the probability that two species taken at random do not belong to the same genus) may be used to draw inferences on species competition and habitat filtering. If competition is stronger between congeneric species, they will tend not to occur, thus increasing generic diversity. However, as congeneric species are also expected to share similar adaptations, highly selective environments should have low generic diversity because only species with similar adaptations can survive there. The balance between the two contrasting selective forces represented by competition and habitat requirements might lead generic diversity to increase with resource limitations, and to decrease with environmental severity (Fattorini *et al.*, 2020c).

in species' content and therefore the similarity (or differences; **β-diversity**) between habitat units in species and higher taxa (Shimatani, 2001). A plethora of indices are available for measuring the similarity (or differences) between habitats for their taxonomic content (Fig. 4.8; see Appendix 3), many of which are currently applied extensively to large geographical areas often based on grid (raster) units

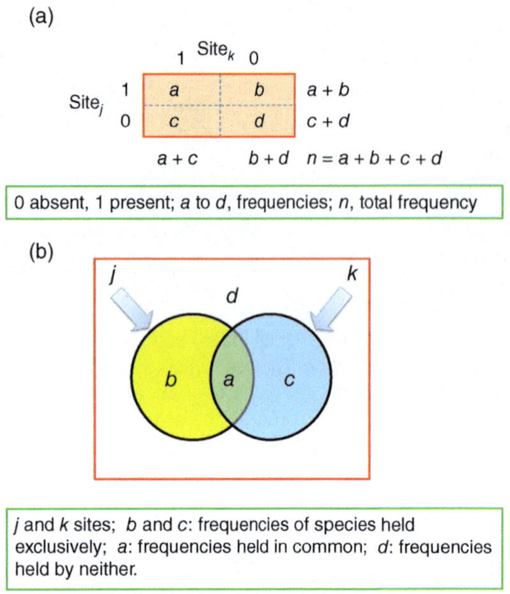

Fig. 4.8. Allocation of binary (presence/absence) observations for calculation of similarity coefficients. **(a)** A 2 × 2 contingency table for the presence or absence of descriptors (e.g. species) for two objects (e.g. sites *j* and *k*), where *a* is the number of descriptors (e.g. species) for which the two objects are coded 1, *d* is the number of descriptors coding the two objects as 0, whereas *b* and *c* are the numbers of descriptors for which the two objects are coded differently (0, 1 or 1, 0), *n* being the total number of descriptors. **(b)** A Venn diagram illustrating the amount of information shared by *j* and *k* (*a*), held by one but not the other (*b* and *c*), and mutually absent (*d*) for the two sites. Many coefficients of similarity have been proposed that use this information in different ways. Jaccard's similarity, for example, is calculated as $S_{J\,(j,k)} = a / (a + b + c)$, which as a dissimilarity is $D_{J\,(j,k)} = (b + c)/(a + b + c)$. The Dice–Sørensen similarity, another widely used coefficient, is calculated as $S_{S\,(j,k)} = 2a/(2a + b + c)$, and the associated dissimilarity as $D_S = (b + c)/(2a + b + c)$ (see Appendix 3 for further details).

(e.g. Kreft and Jetz, 2010; Holt *et al.*, 2013; Whittaker *et al.*, 2013; Dapporto *et al.*, 2015). At regional, inter-island and continental/global scales, differences in endemicity are also a typical focus (e.g. butterflies: Europe, Dennis and Schmitt, 2009; Italian islands, Dapporto *et al.*, 2017; multiple taxa in Brazil: Oliveira *et al.*, 2019; plants in Nepal: Joshi and Joshi, 2022; fungi at global scale: Tedersoo *et al.*, 2022). Many techniques have also been proposed to identify areas of endemism (i.e. areas characterised by the overlapping distributions of two or more taxa), especially in the framework of historical biogeography and conservation biology, such as the Parsimony Analysis of Endemicity, the Cladistic Analysis of Distributions and Endemism, the Endemicity Analysis with Optimality Criterion, the Analysis of Biotic Elements, network analysis methods, and neighbour-joining, as well as to prioritise areas on the basis of species' distributional rarity (weighted endemism) and phylogenetic relationships (phylogenetic measurements of endemism) (see Fattorini, 2017a for a review).

However, one factor remains germane to all analyses: it is important to ensure that all the species included in the analysis 'belong' to the spatial units being compared. The patterns are seductive but a question remains for the researcher: how appropriate is it to treat spatial units as isolated systems of activity? This is a key question researchers should ask themselves.

How distinct are the same species in different habitats?

One expects differences in the species content of contrasting habitat units, though some species may occupy all units surveyed (Fig. 4.9). The study of differences between habitat units has to take on board a pair of important axioms.

- No habitat unit is entirely internally (environmentally) homogeneous, despite their occasional mathematical treatment as such (Daly *et al.*, 2018). Even habitat units that to the human eye appear to be homogeneous (a field sown with couch grass for silage) are heterogeneous at finer scales that may affect small organisms (e.g. arthropods).
- No two habitat units, however apparently similar (e.g. two woods), are identical in every respect, even if they have the same shape and size, have identical neighbouring patches, show the same vegetational setting, etc. Again, there are differences that escape human attention but not the attention of species that exploit components of the habitat (tree trunks, branches, foliage, etc.) as niches.

The two axioms affect all comparisons made between habitat units. We may divide habitats into classes for comparison but there is heterogeneity within the classes, sometimes very considerable heterogeneity. These observations may seem trite to an experienced researcher, but they form the basis of

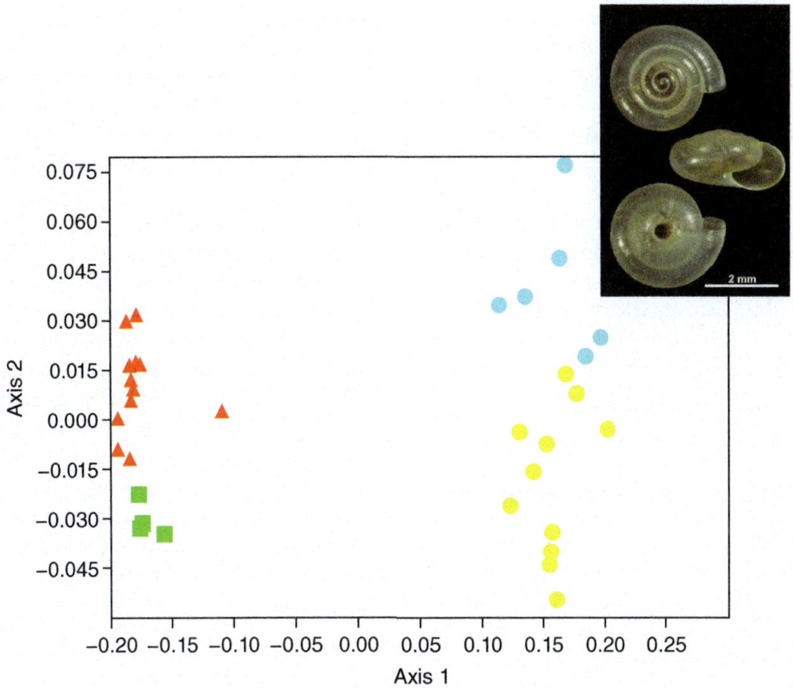

Fig. 4.9. Diversity among habitat communities. Ordination plot (non-metric multidimensional scaling) of similarity (Jaccard coefficient; see Appendix 3) between different habitats from the Malham area of North Yorkshire (UK) based on their mollusc fauna. Circles: wetlands (blue circles, fens; yellow circles, mires). Other symbols: terrestrial habitats (green squares, natural ancient woodlands; red triangles, ungrazed limestone pavements). The plot neatly separates terrestrial from wetland habitats and distinguishes the pairs for each of the basic habitat types. The inset shows *Vitrea crystallina*, a ubiquitous species across different habitats in the Malham area. (Original data from Cameron 1978; photo by O. Gargominy: https://science.mnhn.fr/institution/mnhn/collection/im/item/2010-13267, used under licence CC BY 4.0.)

developing effective (unbiased) sampling regimes to detect similarities and differences between what may appear to be the same or different units. Actually, they are fundamental to our appreciation of what may cause heterogeneity within and between habitat units, as the fundamental part of such research is to identify factors likely to cause differences within and between habitats.

A question then follows: how similar are the individuals that inhabit a certain habitat unit to their conspecifics in adjoining habitat units? They will clearly differ in their numbers (abundance) and – bearing in mind the contrasts that can occur between adjoining habitat units for topography and vegetation cover, plant species, etc. – in many ways that affect their behaviour. The differences may well extend to genetics, as illustrated, for example, by the increase in genetically based dark morphs of moths (e.g. *Semiothisa liturata*, *Alcis repandata*, *Agriopis marginaria*, *Peribatodes rhomboidaria* and *Idaea aversata*) in dark woods (conifer woodlands) compared to light woods (deciduous woodland) or grasslands, with moths light-trapped at sites just 20–50 metres apart at three sites in the UK (Jones *et al.*, 1993). Darker morphs were consistently found to be significantly more abundant in dark woods than in more open conditions, which the authors suggest points to habitat selection (movement behaviour) by distinct morphs in relationship to their needs, specifically **crypsis** during light hours when predators seek them out. Similar findings were discovered for the pygmy grasshopper *Tetrix undulata* in Sweden; in this case the selection by different forms was related to requirements of thermal stability and predator avoidance (Ahnesjö and Forsman, 2006). Likewise, banding in the mollusc *Theba pisana* is strongly associated with habitat, with pale, unbanded morphs being more frequent in open, exposed habitats, while darker, banded snails are more common in shrubby, shady habitats (Johnson,

2012). These pieces of research once again draw attention to the capacity of organisms for selecting habitat spaces and substrates that match their resource requirements and warn against treating habitat units as closed systems and individuals of the same species as identical elements.

An important question emerges: from where does the potential for variation arise? This issue was, at least partially, answered many years ago by a classic study of the marsh fritillary *Euphydryas aurinia* in the Lake District, UK. It was found by Ford and Ford (1930) that variation increased markedly with population increases and settled down to a different variant pattern when the population stabilised once more.

Turnover between habitats with time

By now it should be clear that it is inevitable that habitats change with time (see Chapter 3); the issue is: by how much? The greatest extremes of environmental seasonal change occur in habitats at high latitude and altitude; some groups of animals (e.g. birds) occupying such habitat units are noteworthy for dramatic seasonal changes, occupied in summer by breeding species and in winter by very different species searching for food and roosts (e.g. birds of salt marshes and woodlands; auks on coastal cliffs; Fuller, 1982). With vegetation succession in a habitat unit, one expects a more gradual turnover (i) in the abundance of species resident within the habitat unit, and (ii) in species and higher taxa recorded in the habitat unit. The most compelling evidence for the effect of habitat changes on animal populations emerges with meta-analyses of anthropogenic impacts across the planet, impacts that simultaneously affect changes in populations (extinctions of species) and traits associated with changes in environmental (habitat) conditions (resources and conditions) (Alberti *et al.*, 2017). Such changes in a habitat unit result eventually in a complete change in habitat type (e.g. woodland to shrubland, grassland, bare ground). As noted in Chapter 3, vegetation succession is accompanied by changing numbers of species; changes in the numbers of individuals provide some guarantee that new species are resident.

Systematic changes in the abundance of a species can be studied through use of any sampling regime that provides an absolute count or sound estimates based on consistent criteria (see Appendix 1). However, a direct comparison of species' incidence without population data is less secure. As areal units for comparative research decline in overall dimensions (say, from the dimensions of regions of thousands of km^2 to habitats within them that may occupy but a few m^2), there is a higher probability that any unit will be 'infected' by vagrants from neighbouring units. Thus, it becomes increasingly critical that species included in the comparison are actually resident in any habitat unit (Fig. 4.10). Once certainty has been established for residence (fidelity to the unit) then, as for overall changes in species and higher taxa between habitat units, components of diversity measures can be applied for assessing turnover in any one unit (see, for example, discussion in Appendix 3 about **β-diversity** partitioning into replacement/turnover, richness and nestedness components). One feature should be borne in mind: much as there is spatial autocorrelation between the contents (species and abundances) of adjacent habitat units, there will be temporal autocorrelation for the contents of the same habitat unit through time $t_1, t_2 \ldots t_n$.

Diversity and gradients within and beyond habitat units

γ-diversity at its limit is an extension of α-diversity from a habitat unit to all habitat units within a region. At its simplest, **γ-diversity** is species richness *S* for the entire region. Ecologists have visualised γ-diversity in two different ways: (i) as being multiplicative components of α- and β-diversity ($D\gamma = D\alpha \times D\beta$; Whittaker, 1972) or (ii) as being summed components of the two ($D\gamma = D\alpha + D\beta$; Lande, 1996) where *D* is some diversity measure (Daly *et al.*, 2018; see Fig. 4.1a and Appendix 3 for a discussion of the relationships between γ-, α- and β-diversity measures). There are many ways of conceiving diversity within regions. One way is to divide up a region into a system of neighbouring habitats. If each habitat comprises a different set of species – a highly improbable circumstance – then the network linkage or **connectivity** between them (lines connecting identical species) is zero. The maximum value for connectivity for any region would occur if every species occurred within each habitat. A simple measure of integration (λ) follows from connectivity. Connectivity can be calculated as β_L (standardised connectivity links):

$$\beta_L = a_L / a_T \tag{4.4}$$

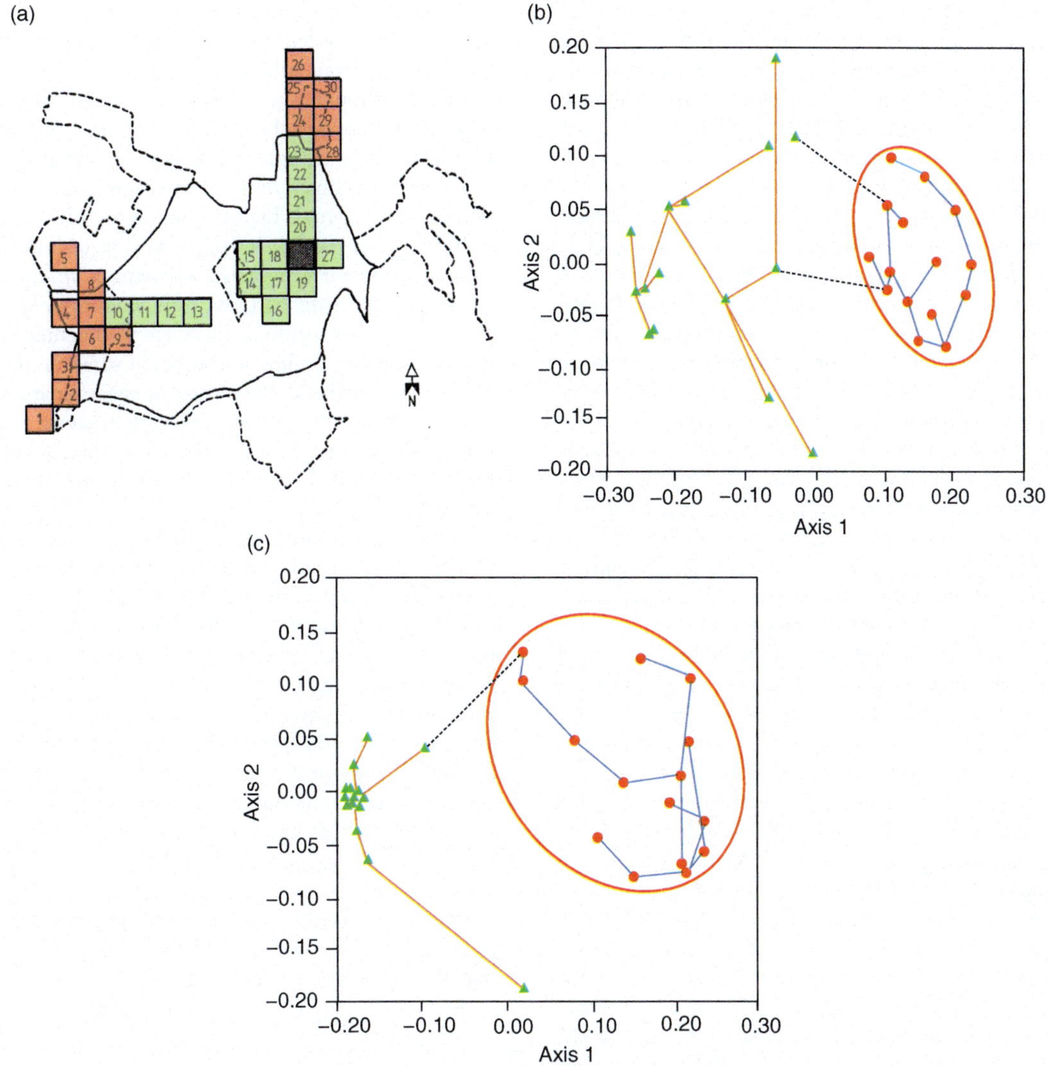

Fig. 4.10. Butterfly species diversity (presence/absence) among habitat units. **(a)** Woodland and open ground habitat units on Alderley Edge, Cheshire, UK (15 individual hectare units each, marked on the ground and sampled 38 times between 1996 and 1998; woodland units 10–23, 27; open units, the remainder). **(b,c)** Ordination plots (non-metric multidimensional scaling of Gower distances D_G with minimum spanning trees) for the aforementioned 30 equally sized and evenly sampled habitat units based on **(b)** all species recorded, and **(c)** resident species with vagrants removed (green triangles, woodland; red dots and red oval, open ground with shrubs and herbs, respectively). Vagrants are defined as lacking suitable larval resources. With the removal of vagrants, the open habitat cluster expands and the woodland cluster becomes more compact with one exception. Note: when vagrants are removed, the two habitat types have one single link between them. The woodland has one widely distributed species (*Pararge aegeria*). (Panel **a** from Dennis, 2000.)

where a_L are arcs or edges (actual links between the vertices, v), $a_T = \frac{1}{2}[v(v-1)]$ is the total links across all habitat units, and v is vertices or species (Dennis, 2010). Thus, for habitat units and species within the units, integration of habitat units within a region (connectivity) is:

$$\lambda = (L \times SU)/(L \times S) = SU/S \qquad (4.5)$$

where: $L = ½[u(u-1)]$ and $SU = \sum^{S}_{i=1} ½[no(no-1)]$, S is species, u is habitat units and no is units occupied by S_i. The index ranges from zero (habitats perfectly disconnected, with species being unique to single habitat units) to unity (habitats fully connected, with every species occurring in each habitat unit).

More important than any algorithm are the implications of uniquely 'disconnected' habitat units in a region at one extreme and of completely 'connected' units at the other. It is not difficult to imagine two distinct habitat units that have no resident species (or higher taxa) in common (e.g. pond and woodland beetles) or two habitats that have all resident species in common (passerine birds in two or more woodlands). In general, it is almost inconceivable that all habitats in a region have either no species in common or all species in common. But then, whatever the value of λ for the region, a single value tells us nothing about the distribution (spatial pattern) of connectivity in the region – whether heterogeneous or even – or how different communities of distinct sets of habitat relate to one another for their species contents. At regional levels, we need to know and explain the variability of species (or higher taxa) and their abundances across different types of habitat.

One answer is to develop more sophisticated metrics that extract further information from the species content of regional habitats and build joint networks of species and individual habitat units. For instance, Cappellari and Marini (2021; see Chapter 1, Fig. 1.9) examine a number of metrics based on species–habitat networks linking butterfly species with habitat units, namely **modularity** (degree of partitioning into distinct modules; Newman, 2006; see Box 2.1 in Chapter 2), **nestedness** (weighted NODF; Almeida-Neto and Ulrich, 2011; Box 4.5), robustness (a measure of network stability against species extinction; Memmott *et al.*, 2004) and **connectance** (measure of network complexity which as in the above description of λ specifies the realised proportion of all possible links in a network; Dunne *et al.*, 2002; Dormann *et al.*, 2008; see Chapter 2, Box 2.1). At interhabitat scale, an approach based on vector units is superior to one built up on raster data, but the latter approach is more usefully applied at countrywide or continental scale (e.g. European butterflies and bumble bees for 50 km × 50 km UTM grid: Settele *et al.*, 2008; Rasmont *et al.*, 2015; global mammals: Schipper *et al.*, 2008; global birds: Jetz *et al.*, 2012; global reptiles: Roll *et al.*, 2017), which adequately captures the variability in major Earth inputs (altitude, latitude, climate). But at a fine scale (100 m to 5 km scale), using a raster approach confounds the effect of variables from adjoining habitat units. In this context, although species **resource-based habitats** may extend beyond the bounds of traditional **habitat** (TH) units (irregular polygons), THs provide much greater consistency for **substrates**, vegetation and environmental factors (see principle C6 in Table 4.1; Burgess *et al.*, 2020). As it is, techniques, such as trend surface analyses (Fortin *et al.*, 2002), exist for exploring patterns over smaller (γ-scale) regions in relationship to basic environmental gradients. The future lies within multiscale (α to γ) approaches to the identification of space-time variation in species and their explanation using environmental and trophic variables and species traits (e.g. Dray and Legendre, 2008; Gavish *et al.*, 2019).

A final key issue deserves to be addressed, one which affects both raster and vector units for **β-diversity** at fine scales: the biological relevance of double zeroes (see Fig. 4.8; d = joint absences) in relationships (e.g. in determining joint incidences of species across different sites). Different measures cater or fail to cater for double zeroes (Legendre and Legendre, 2012) depending on their relevance to overall variation over the study region at the γ-scale. A simple measure is the Sokal and Michener (1958) simple matching coefficient:

$$S_{SM} = (a+d)\ /\ n \tag{4.6}$$

where d = joint absences, a = joint presences and n = total number of occurrences (see Fig. 4.8 and Sneath and Sokal, 1973).

$$S_{SM} \to 0 \text{ as matches / mismatches} \to 0$$
$$\text{and} S_{SM} \to 1 \text{ as mismatches / matches} \to 0$$

Similar is the Yule coefficient:

$$S_Y = (ad - bc) / (ad + bc) \tag{4.7}$$

where b = number of species unique to one unit and c = number of species unique to the other unit, with values ranging from −1 to +1.

Particularly useful is the Baroni Urbani and Buser (1976) coefficient, as it considers mismatches, but limits their impact:

$$S_{BB} = \frac{\sqrt{ad} + a}{\sqrt{ad} + a + b + c}. \tag{4.8}$$

In general, one of the most valuable similarity measures is Gower's (1971) coefficient, as it can handle different types of data (interval-ratio,

Box 4.5. Nestedness

Nestedness is the ordered loss of species from species-rich to species-poor assemblages. In a perfectly nested matrix of species × sites, all species recorded in a given site are a perfect subset of those of the immediately richer site, viz. species occurring at the site of interest always occur in a more species-rich site, whereas species absent from the site of interest are never present in a less species-rich one. The same applies to any type of bipartite network (see Box 2.1 in Chapter 2): for example, in plant–pollinator or host–parasite networks, perfect nestedness means that all pollinators/parasites associated with a plant/host also use plants/hosts that have more pollinators/parasites. Spatially nested patterns are a form of non-random distribution and can arise from selective extinction, colonisation, susceptibility to human disturbance, etc. (e.g. Fattorini, 2007b). Nestedness in bipartite networks such as plant–pollinator and host–parasite networks promotes diversity by making networks more stable, since competition and coextinction risks are reduced (e.g. Bascompte *et al.*, 2003; Bastolla *et al.*, 2009; Sugihara and Ye, 2009; Strona *et al.*, 2013a) (see also Box 2.1).

If rows and columns of a perfectly nested matrix representing a bipartite network are ordered according to their respective totals, it will appear to be divided into two parts: the upper left triangle containing all filled cells and the bottom left triangle containing the empty cells (Fig. B4.5.1a). Of course, we do not expect to find perfectly nested systems in the real world, as species-poor sites may have species not present in the richer ones, plants pollinated by few pollinators may have some pollinators that do not pollinate plants with larger numbers of pollinators, or hosts that are attacked by few parasite species may have parasites not found in hosts that have more parasites. This creates unexpected presences in species-poor assemblages and unexpected absences in species-rich assemblages (Fig. B4.5.1b,c). To measure the degree of nestedness of a matrix several metrics have been proposed.

- *Matrix Temperature (MT)* (Atmar and Patterson, 1993): MT uses the Euclidean distances of unexpected empty or filled cells from the line (isocline) that separates presences (upper left triangle) from absences (bottom right triangle) in a perfectly nested matrix. The sum of these distances is rescaled relative to the maximum possible value for a given matrix size (number of cells) and fill (proportion of filled cells). MT measures the disorder of a matrix: a perfectly nested matrix has a temperature of 0; with increasing disorder (decreasing nestedness) temperature increases. While influential, this metric has been criticised for its sensitivity to matrix size, shape and fill (Ulrich *et al.*, 2008), and it is now used less frequently.
- *Brualdi and Sanderson Discrepancy (BR)* (Brualdi and Sanderson, 1999): BR is a count of the number of discrepancies (unexpected absences or presences) that must be 'corrected' to obtain a perfectly nested matrix. Lower values indicate higher nestedness.
- *Nestedness measure based on overlap and decreasing fill (NODF)* (Almeida-Neto *et al.*, 2008): This is the percentage of occurrences in right columns and species in inferior rows which overlap, respectively, with those found in left columns and upper rows with higher marginal totals for all pairs of columns and of rows. A version of this index (WNODF; Almeida-Neto and Ulrich, 2011) is used for quantitative networks (i.e. with abundance data).
- *Spectral radius* (Staniczenko *et al.*, 2013): This is a mathematical characteristic of matrices based on graph theory. Spectral radius is maximum for a perfectly nested network. This metric can also be used for quantitative networks, and is considered superior to WNODF (Lan *et al.*, 2023).

Since a certain degree of nestedness is expected to occur in most matrices because of chance alone, it is important to assess whether a matrix is significantly more nested than expected by chance. To this end, its nestedness value is compared with the distribution of nestedness values that can be obtained by randomising the original matrix. In other words, the original matrix is rearranged several times, to obtain a large number of random matrices (null matrices), and nestedness is calculated for each null matrix, to obtain a frequency distribution of random nestedness values; then, the real matrix is significantly nested if its nestedness has a value that can be rarely observed among the null matrices (e.g. in less than 5% of the null matrices).

Several algorithms have been developed to construct null matrices, depending on which characteristics of the original we want to preserve in the null matrices. The simplest algorithm (equiprobable-equiprobable, EE) does not have any restriction (apart from that of maintaining the total number of species occurrences in the matrix and the same number of rows and

Continued

Box 4.5. Continued.

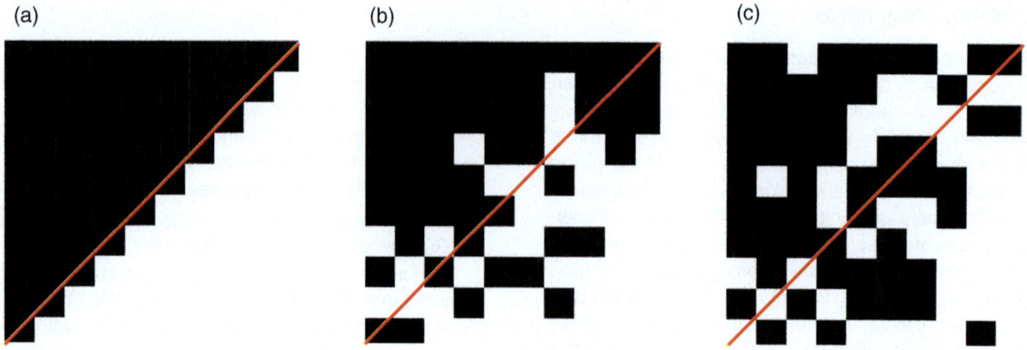

Metric	Value
NODF	100.0
T	0.0
BR	0.0
Spectral Radius	6.691

Metric	Value
NODF	73.791
T	19.515
BR	10.0
Spectral Radius	6.425

Metric	Value
NODF	56.217
T	40.141
BR	16.0
Spectral Radius	5.991

Fig. B4.5.1. Values of some measures of nestedness for three hypothetical matrices. Matrix **(a)** is a perfectly nested matrix. There are neither empty cells above the diagonal (red line) nor filled cells below the diagonal. Matrices **(b)** and **(c)** present unexpected absences (empty cells above the diagonal) and unexpected presences (filled cells below the diagonal). If significance of nestedness is tested using randomisation algorithms, matrix **(a)** was significantly nested for all metrics with both the equiprobable-equiprobable (EE) and the proportional-proportional (CE) algorithms (the fixed-fixed FF algorithm was not applicable because of lack of variance). Matrix **(b)** was significantly nested for all metrics with EE and CE; no metric was significant with FF. Matrix **(c)** was always not nested.

columns); therefore any change is allowed, and both row and column totals vary freely. The result is that the produced null matrices tend to be so different from the original one that there is a high chance of obtaining a significant result (i.e. the algorithm is prone to Type I error: rejecting the null hypothesis when it is true). To avoid this, we can introduce the constraint that the null matrices must have the same row and column totals as the original matrix (e.g. a site with a certain number of species in the original matrix must have the same number of species in all null matrices, although not the same species; and a species occurring in a certain number of sites in the original matrix must continue to occur in the same number of sites, although not the same sites). This algorithm (called fixed-fixed, FF, as it maintains both row and column totals) is, however, so conservative that most of the resulting null matrices tend to have nestedness values equal to or larger than that of the original matrix (being therefore prone to Type II error: accepting the null hypothesis when it is false). A good compromise is the proportional-proportional (CE) algorithm, which assigns to each cell a probability to be occupied proportional to the corresponding row and column totals. Other algorithms are the FE (fixed row totals, equiprobable column totals) and the EF (equiprobable row totals, fixed column totals). FE preserves the observed row totals (e.g. species occurrence frequencies across sites) but allows column totals (e.g. species richness per site) to vary randomly. EF preserves the observed column totals (e.g. species richness per site) but allows row totals (e.g. species occurrence frequencies across sites) to vary randomly. They can be useful to explore the relative importance of species' ecology and environmental constraints in determining nestedness patterns. Nestedness metrics, null models and evaluation of significance are largely discussed in specific literature (e.g. Ulrich and Gotelli, 2007; Ulrich *et al.*, 2009; Johnson *et al.*, 2013; Strona *et al.*, 2013b, 2014a,b; Strona and Fattorini, 2014a).

nominal and binary) within the same dataset, including missing data:

$$G = \frac{\sum_{i=1}^{n} w_{ijk} s_{ijk}}{\sum_{i=1}^{n} w_{ijk}} \quad (4.9)$$

where w_{ijk} are non-negative weights and s_{ijk} is the similarity between the two objects i and j regarding their k-th variable. For binary variables: $w_{ijk} = 1$ and $s_{ijk} = 0$ if $x_{ij} \neq x_{ik}$; $w_{ijk} = s_{ijk} = 1$ if $x_{ij} = x_{ik} = 1$ or if $x_{ij} = x_{ik} = 0$, and double zeroes (mutual absences) are included; $w_{ijk} = s_{ijk} = 0$ if $x_{ij} = x_{ik} = 0$, and the double zeroes are excluded from the comparison. With double zeroes included, the index reduces to the simple matching coefficient; with double zeroes excluded, we obtain the Jaccard index (discussed in Appendix 3) (Sneath and Sokal, 1973; Podani, 1999). The index has been further generalised by Podani (1999) to cater also for ordinal characters, producing the following distance measure, which is now the most commonly applied version of Gower distance:

$$D_{jk} = 1/n\sum\left[\left|x_{ji} - x_{ki}\right| / \left(\max\{x_i\} - \min\{x_i\}\right)\right] \quad (4.10)$$

for objects j,k and variables/attributes i,n.

The finer the spatial scale, the greater the probability that diversity is influenced by unidentified (perhaps unidentifiable) **vagrants**. When dealing with agglomerated habitats in a single γ-space, there is an expectation that all habitat units are accessible to all species. Thus, any one species can infiltrate any habitat unit but vagrants have the capacity of distorting relationships between units; they can invade units but may not stay there, breed there or use any resources. In any study, then, careful thought needs to be given to which species are (or are not) included in any analysis and whether joint absences are relevant or not. In the case of butterflies from Alderley Edge in Cheshire (UK) reported above (see Fig. 4.10), it is clear that entering or omitting vagrants from the database has an impact on habitat unit proximity in the attribute space, as does selecting coefficients that include or exclude double zero associations.

Phylogenetic diversity

From the classical work of Faith (1992), there is increasing interest in understanding how the study of **phylogenetic diversity** (see Appendix 3) can shed light on community assembly and may provide important information for biological conservation (e.g. Webb, 2000; Swenson *et al.*, 2012; Chai *et al.*, 2016a; Pellens and Grandcolas, 2016). Phylogenetic diversity may represent an important attribute of habitats and areas for assessing their conservation importance (e.g. Devictor *et al.*, 2010; Tucker *et al.*, 2017) and β-diversity patterns in phylogenetic diversity may elucidate biogeographical relationships useful for regionalisation (e.g. Holt *et al.*, 2013; Daru *et al.*, 2016, 2017; Pataro *et al.*, 2021). Understanding the phylogenetic structure of communities may be important to decipher the underlying mechanisms. Phylogenetic overdispersion (i.e. species within a community are less closely related than expected by chance) may be indicative of competitive exclusion (as phylogenetically closer species are also ecologically more similar and hence tend to compete), whereas phylogenetic clustering (i.e. co-occurring species are phylogenetically closer than expected) suggests that environmental conditions filter species with similar ecological characteristics (Webb *et al.*, 2002; Cooper *et al.*, 2008). Despite the rapid increase in phylogenetic reconstructions boosted by the development of molecular techniques, phylogenetic relationships (especially at the species level) are available only for a minority of taxa and are frequently characterised by uncertainties. For this reason, as a proxy of phylogenetic relationships, some authors use the Linnean taxonomic arrangement, which allows the construction of indices of taxonomic diversity and distinctness (see Appendix 3), that may be particularly useful to elucidate both community responses to anthropogenic changes and mechanisms underlying macroecological patterns (Heino *et al.*, 2005, 2015).

Functional diversity and redundancy: their significance

Recently, there has been a growing development of measures of **functional diversity** (see Appendix 3) and appreciation of its relevance to communities. The concept is that the greater the diversity of functions played by species, the greater the overall ecosystem functions, including biogeochemical activities, ecological services, transfers of energy and production of biomass. This is considered to be achieved through niche complementarity (Loreau and Hector, 2001). The fundamental question raised is the extent to which losses of individual species impact on ecosystem structure and function. The greater the number of species that fulfil identical functions, the greater is said to be the **functional redundancy** (and hence **resilience**) of a community. For example, through the meta-analysis of 32 functional redundancy–ecological resilience and stability relationships, Biggs *et al.* (2020) found a positive correlation between the two.

The danger of studies of so-called functional diversity and redundancy depends entirely on two factors: (i) the measures of **traits** in species for determining functional diversity and redundancy; a recent paper on British butterflies used only four traits (Lazarina *et al.*, 2023), whereas a vast number have been listed for use by Dennis (2010) and Middleton-Welling *et al.* (2020); (ii) the measures of functional diversity and redundancy applied; a multitude of indices of functional diversity and redundancy exists and the question is: which are most appropriate for a particular study (see Appendix 3)? Experiments that use a single functional trait tend to be biased toward detecting redundancy, as species are more likely to occupy distinct functional niches within a multidimensional functional space (Rosenfeld, 2002). Clearly, the greater the number of traits exploited in any study, the lower the expected measures of functional redundancy. In fact, even within a population of a single species (e.g. humans) there is a vast range of trait values (each trait being a variable to a greater or lesser extent) which influences the extent to which one individual person can replace another for a specific purpose. For example, no two surgeons are of equal ability or adept at a particular specialism (spinal surgery, cancer treatments, joint replacement, digestive tract recovery, etc.) as the first author has experienced 'first hand'. Even so, there is an expectation that functional redundancy will increase with **species richness**, leading to enhanced ecosystem stability (see Chapter 11) and **resilience** (Walker, 1992; Naeem, 1998; Biggs *et al.*, 2020). For example, it may be predicted that the probability that an entire functional group becomes extinct increases with the number of functional groups in a community (functional richness), but decreases with species richness and functional evenness (the distribution of species across functional groups) (Fonseca and Ganade, 2001). However, the relationship between diversity and redundancy is not obvious. There may be ecosystems with few species that are characterised by a high degree of redundancy and others with many species but low redundancy. Ideally, in a pond, we can have some redundancy in the roles of primary consumers represented by planktonic crustaceans (e.g. *Daphnia*) and mosquito larvae, both feeding on algae and detritus, and both prey of small fishes. If one species among these primary consumers were to decline due to environmental pressures or disturbances, another could potentially compensate for its ecological role to some extent. In fact, it has been found that small estuaries, though characterised by lower species numbers compared with large estuaries, show similar functional redundancy (Teichert *et al.*, 2017). On the other hand, even in high-diversity systems such as tropical reefs, functional diversity remains highly vulnerable to species loss if species tend to pack disproportionately into a few particular functions while leaving many functions highly vulnerable (Mouillot *et al.*, 2014).

Regarding measures of functional diversity, Mammola *et al.* (2021) identify two general mathematical frameworks: those based on dissimilarity matrices (e.g. Rao entropy; see Appendix 3) and those subject to multivariate analyses with species displayed in a hypervolume of traits (see Appendix 3). Using this approach, they present a unified scheme to quantify functional diversity under each framework. Regardless of these refinements, one fact is outstanding: different species are unique by definition and thus will differ in the niche space they occupy. Our knowledge of the multitude of taxa is so poor that methods of assessment are ahead of our understanding of species' interactions.

The concept of functional redundancy raises an important issue. Is functional redundancy one reason for the scarcity of species with very low abundance within a compact traditional habitat space (see Chapter 11, Fig. 11.4)? The implication is that competition reduces the status of certain species. Whether right or wrong, it is an issue that deserves investigation. It should not be difficult to establish experimental designs to test this suggestion. However, the researcher should be aware that just 'removing' one species is not necessarily the reason for a growing abundance of another species. Unexpected (unmeasurable) changes in a habitat (TH), such as ceasing to mow a meadow one season or a change in weather such as a wet winter, may be the reason why a different species (the ringlet butterfly *Aphantopus hyperantus*) has suddenly become more abundant following, say, the removal of another (i.e. the meadow brown *Maniola jurtina*). Ecology constantly plays tricks with our expectations.

5 Basic Principles of the Resource-based Habitat Paradigm

Abstract

The concept of the resource-based habitat (RBH), originally proposed for butterflies, is based on the identification of consumable and utility resources used by species and the environmental conditions they experience, with bounds that may fall within or extend beyond traditional habitat units (biotopes). RBHs vary in terms of their composition (types of resources), physiognomy (resource geography) and connectivity (isolation), compaction and intersection, resource breadth (the number of outlets used by species along the specialism–generalism continuum) and matrix (spaces not used by the species). The concept of RBH is related to those of niche (a hyperdimensional appraisal of a species' traits in relation to abiotic and biotic variables), home range (the bounds of the RBH identified by individual lifetime tracks) and territory (a 'defended' space). Resources and conditions not only have independent spatial organisations but also change through time, affecting spatial arrangement of individuals and population dynamics.

Thus far, we have considered what is meant by a traditional **habitat** (a distinctive set of **substrate** units, which corresponds to the concept of **biotope**), and a **species' habitat** in the context of traditional habitats (see Chapter 1, Fig. 1.8). We have seen that traditional habitats are typically parts of wider ecological units, the **ecosystems**, and hence of the **landscapes** and ultimately of the **biosphere**. In fact, each forms part of a hierarchy of integrated systems (see Chapter 11, Box 11.2) and no ecological unit is really isolated, as all parts of the biosphere are interconnected by fluxes of materials, energy and organisms. Even true islands or **habitat islands** (such as lakes, caves, mountain tops, meadows and forest fragments; see Chapter 7, Box 7.1) are subject to influences from the environments beyond their bounds.

Now we can raise the important question: what do we really mean by a species' habitat? Noting a specific traditional habitat unit as characterising a species' domain does not in any way specify what actually constitutes the species' habitat. What does it contain that makes it *that* species' habitat? Refer back to the previous chapter where we learnt that habitat units comprise a few common species but numerous rare ones. The inference is that the common species have most, if not all, of their requirements and conditions for survival and reproduction in the unit. Of course, we have exceptions to this observation as we learnt in Chapter 3, for instance a constant flood or regular recurrence of individuals of certain species (such as migrant birds or butterflies) through spatial units in which they may spend some time, for example for resting or feeding, but that cannot be used for long-term survival and reproduction. The implication is that, although these species may be present in large numbers at certain times, not all conditions and resources necessary for a stable presence are available and suitable in any one area, or they may be available and suitable only temporarily. Similarly, the less common (rare) species tend not to have all their requirements and/or suitable conditions in the unit, at least for crucial periods of their life. But, apart from the geographical limits of such habitat units (size, isolation), what is missing to ensure they are occupied by large populations? Part of the answer lies in the negative interactions among organisms (e.g. interspecific competition; see Chapter 2, Table 2.1). But the reality of the situation can be far simpler and more basic: that resources and conditions are lacking or not suitable, regardless of other occupants of the habitat unit, or indeed that many species space themselves out to avoid detection.

Resource-based habitat: focusing on a species' habitat

The idea for the **resource-based habitat** (**RBH**) of species emerged in a study of butterflies on British

DOI: 10.1079/9781800627505.0005

islands (Dennis and Shreeve, 1996), specifically to account for Baker's (1978) definition of butterfly migration as movement between habitat units. Baker did not define what he meant by habitat units and therefore it became necessary to isolate the factors determining a habitat for a butterfly species. For this task, simple set notation and Venn diagrams were applied. From **autecological** surveys, it was clear that any butterfly species depended on a number of clearly specified consumable and utility resources and conditions for all four stages of development (eggs, larvae, pupae, adults) (Dennis, 2010). Initially, a resource-based habitat was defined for colonial species with closed population structures which have all their resources in proximity (Dennis *et al.*, 2003, 2006a, 2007). The resources could overlap, partially or completely, be contiguous or to a degree separated from one another; in Venn diagram terms, the resources and conditions combine by union and intersection to form a habitat (Fig. 5.1a). Initially, then, the RBH was defined simply as 'that collection of resources and conditions ensuring persistence of a population at a site' (Dennis, 2010).

Even then, however, it was clear that this definition was inadequate for a number of butterfly species that rarely use a specific site more than once, and for only a part of their life cycle, and may migrate over massive distances, with succeeding generations using different locations over a wide region (e.g. *Vanessa cardui*; Stefanescu *et al.*, 2007, 2011, 2013, 2016). Of course, many other organisms not only move much further than a single traditional habitat unit but have considerably longer lifespans with repeated visits to the same region, some even to identical locations (e.g. swifts *Apus apus*). Therefore, a definition of habitats, which takes into consideration what organisms need for their persistence, requires a global view: 'A RBH is that geolocated collection (set) of resources and conditions sufficient to maintain a population or cohort of a species' (Dennis, 2020). This definition refers to all the resources and conditions in *all* abiotic Earth layers of activity (i.e. **lithosphere**, **hydrosphere** and **atmosphere**) used by a species.

To casual human observers, even to scientists engaged in local autecology, this effectively global definition of habitat may seem to be unpalatable. Yet, remove any one of the building blocks of resources or conditions for any stage in development for any species and it is certain to become extinct (Fig. 5.1b). Convenience of a definition for human observers at a local or regional level is superseded by one relevant for organism survival, and we consider what this means for conserving species in Chapter 11. In fact, many small population units of Pliocene and Quaternary hominids, including *Homo sapiens* (see Chapter 9), must have lived in the self-same manner of moving about large areas of the globe; otherwise, we would not have colonised the world in the way we have done during the entire Pleistocene (Beyer *et al.*, 2021; Groucutt *et al.*, 2021). Actually, this global extension to the definition of the RBH should cause few problems of interpretation for human observers and landscape managers, as the RBH for any cohort of individuals can be subdivided (subsets in Venn terms) by seasonal activity (e.g. a summer breeding habitat versus a winter hibernation habitat; Fig. 5.1a (f, g)) and therefore by geography, including the areas in between during migration (i.e. a transfer habitat). Later, we look into this issue of the sky and the open sea as mediums of transfers being part of the habitat for many organisms (see Chapter 8).

The definition of habitat focuses on populations. But, given the impossibility of making observations for all individuals within populations (at least in most cases), habitats are defined on the basis of the activities (behaviour, resource use, changing locations) recorded in statistically representative samples of individuals in local populations, as explained in Chapters 6 and 11. To this end, summary statistics (e.g. mean, variance, skewness and kurtosis) of the variables of interest from samples within local populations may provide appropriate descriptions for local population units within traditional habitats. Through accumulation of data from different population units, this approach ultimately allows us to describe the habitat for an entire species. At this stage, it is important only to appreciate that the term 'resource-based habitat' describes a complex variable: habitat is a multivariable component of an organism's requirements and the level of reference (e.g. individuals, mating pairs, troops of primates, herds of herbivores, flocks of birds, packs of carnivores, shoaling fish, colonies of social insects, social colonies of birds and mammals, such as the nesting colonies of birds and the breeding colonies of bats and sealions, local populations, regional populations and species) needs to be designated in any communication. So, we may refer to an individual's habitat, understanding the limitations implied by an individual, or we may discuss a species'

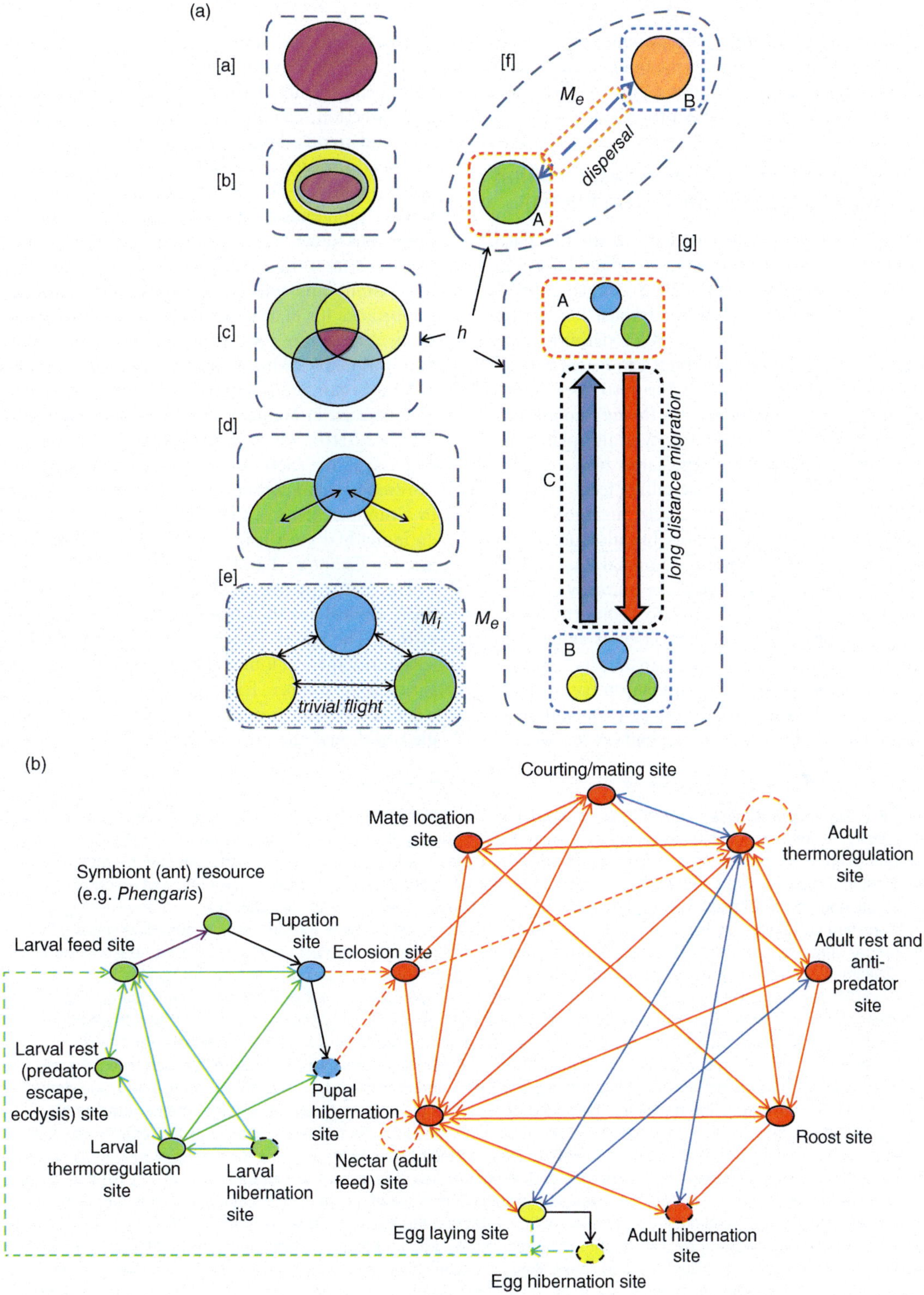

Fig. 5.1. Continued.

habitat, again with full understanding of how this has been determined within the bounds established by traditional taxonomy or at a more sophisticated level from the species' DNA profile.

Components of a resource-based habitat (utilities, consumables, conditions)

All species in the seven kingdoms of life (Archaea, Bacteria, Protozoa, Chromista, Plantae, Fungi, Animalia; Ruggiero *et al.*, 2015; see Chapter 8) depend on resources and conditions. For butterflies, the initial focus of the RBH definition, the essential components of habitat can be seen to fall into three groups: **consumable resources, utility resources** and **environmental (living) conditions.** Consumables are larval food (typically plants but also animal matter; e.g. the large blue butterfly *Phengaris arion*, whose caterpillars feed on larvae of *Myrmica* ants; Dennis, 2020) and adult food (nectar from flowering plants but also other substances including substrates, animal faeces, tree sap, etc., e.g. the purple emperor butterfly *Apatura iris*; Oates, 2020). Utility resources comprise a wide range of substrates used for different stages (e.g. places on which eggs are deposited, larval sites for ecdysis, hibernation sites, pupation sites, adult mate location sites, territorial perches and roost sites) (see Table 2.2 and appendices in Dennis, 2010). Living conditions (environmental conditions) include environmental states, for instance light, temperature, humidity, rainfall, wind speed and direction, edaphic conditions affecting nutrients and toxins, largely abiotic factors which are not only intricately associated with resource types, quality and distributions, but are also key to transfers of individuals between resource outlets. For organisms other than butterflies, different conditions are important; for instance, water currents, pH, salinity and tides are obviously of paramount importance for marine organisms, and factors such as air temperature, pressure and winds are important for many flying animals. Just how complex habitat requirements can be for a butterfly species is illustrated in Figure 5.1b (see also appendices in Dennis, 2010), which outlines the potential stages of passage for each butterfly individual. The state of resources and conditions for any stage may vary in a number of ways that can affect mortality and an individual's intrinsic rate of increase, issues outlined in Chapters 2 and 6.

The status of resources and conditions is obviously critical for all organisms and not only butterflies. The key to understanding their habitats is to identify these resources and necessary conditions for their persistence. Earlier, it was pointed out that a species' habitat is unlikely to coincide with a single traditional habitat unit, especially those drawn

Fig. 5.1. The resource-based habitat as originally defined for butterflies (Dennis, 2010). **(a)** Basic variants of resource disposition making up a habitat are illustrated, each diagram simply illustrating just one spatial relationship between resources in which each resource type is described as a single zone. [a] Equivalence and equality of resources. [b] Nested disposition (equivalence, intersection and union) of resources. [c] Intersection and union of resources. [d] Contiguous union of resources. [e] Disjointed union of resources linked by 'trivial' (daily resource-seeking) flights. [f] Disjointed non-union (likelihood that resource spaces are not used by same individuals in succeeding years) of resources linked by seasonal dispersal (movement across distinct biotopes), each part comprising a seasonal habitat. [g] Disjointed non-union of resources linked by seasonal long-distance migration (e.g. movement between biomes and over large water bodies), each part comprising a seasonal habitat (e.g. for the painted lady *Vanessa cardui*, a migratory butterfly). Field examples of habitats will have combinations of multiple resource dispositions. In each of these diagrams, a maximum of three resources is illustrated: nectar resource (blue), larval resource (green), roost sites (yellow); overwintering sites (brown); habitat core (purple); habitat boundaries are noted by dashed blue lines (*h*). Red summer and blue winter seasonal habitats are marked A and B at the edges respectively of red and blue dotted rounded squares, whereas the transit habitat boundary is dotted black C. Matrices: M_e external matrix (outside *h*); M_i internal matrix blue dotted in diagram [e]. Note: no spatial scale indicated; different species scale resource-based habitats by their 'packing' (group or local population size) and their capacity for movement. Overlap of resources can be measured much as the overlap of home ranges of different species (see indices in Ferrarini *et al.*, 2021). **(b)** Imaginary 'life-time resource track' between resource types for a single individual of a butterfly species. Movements are limited to the larval (green arrows) and adult (red arrows) stages, including the case of a species (such as *Phengaris* (*Maculinea*) *arion*) conveyed (transported) by a symbiont ant (purple arrow). Most adult movements are by flight, though some involve walking as well as/instead of flight (pecked lines). Resource types distinguished for stages (egg, yellow; larva, green; pupa, blue; adult, red). Depending on species, the hibernation stage could be any stage (pecked ovals) with black arrows indicating no movement from previous site. (From Dennis *et al.*, 2014, slightly modified.)

at finer levels in a hierarchy of vegetation units for terrestrial ecosystems (see Chapter 1, Fig. 1.8). In a simple model (see Fig. 5.1a) the habitat area is shown by a pecked outline; this indicates that the resource-based habitat can be regarded as having bounds and can be identified by encapsulating the set of conditions and resources required by a population (in this case). However, the envelope around these resources and conditions may bear little relationship to the boundaries around traditional habitat units (e.g. the small tortoiseshell butterfly *Aglais urticae*, a rather ubiquitous species) (Fig. 5.2; see also Chapter 1, Fig. 1.8). Additionally, if the habitat is broken down into its constitutive resources (consumables and utilities) and environmental conditions, the bounds for these individual resources and conditions also rarely coincide entirely unless occurring in traditional habitat types that contain a random admixture of them (e.g. a meadow under traditional management for the meadow brown butterfly *Maniola jurtina*) (cf. Fig. 5.2a versus b). Different resource requirements of species for different functions (behaviours) often depend on different substrates and vegetation structures (e.g. egglaying on open grassland *Molinia* heath, territorial perches on birch seedlings in open hot spaces, patrolling for mates along the woodland edges, etc., in the large skipper butterfly *Ochlodes sylvanus*; Dennis and Williams, 1987). The crucial extension of this observation is that the bounds of suitable resources and conditions for a local population of a species rarely coincide with a vegetation unit, that is, what is described as a traditional habitat unit (i.e. a **biotope**) (see Fig. 5.2; see also Chapter 1, Fig. 1.8).

The matrix and its significance

For every organism there are resource and non-resource zones. The latter have been referred to as **matrix** and understood to be, effectively, empty space for species, that is, areas and states not used by them. Matrices vary in relevance as to grain and extent for each species (Wiens, 1989). There are perhaps two types of empty space in a butterfly's habitat: (i) areas within the habitat unit avoided or generally ignored but crossed by individuals (internal empty spaces) and occasionally used by them in stressful circumstances (e.g. bare ground used by butterflies to settle on and heat up); and (ii) the empty space beyond the habitat margin across which emigrants and immigrants need to transfer on leaving and entering the habitat respectively (external empty spaces), components of which are also occasionally used (see Fig. 5.1a and Box 5.1). Consider, for example, the ecological needs of the meadow brown butterfly (*Maniola jurtina*) in UK. For this butterfly, patches of bare ground or gorse bushes are examples of the first type of empty space whereas a lake, extensive fields of cereals, root crops or oilseed rape crops are examples of the

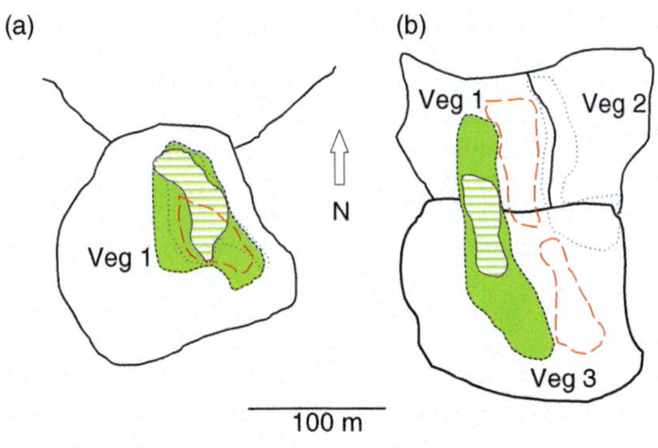

Fig. 5.2. Coincidence and discordance of a population's or a species' resources with a traditional habitat unit as exemplified for butterflies. **(a)** Complementary resources (e.g. mate location and nectar) coincide with the host plant area (but not entirely with the host plant area that is suitable for egg laying) and within a single vegetation unit (Veg 1). **(b)** Complementary resources do not coincide with the host plant area and both host plants and complementary resources cross vegetation units (Veg 1 to Veg 3). Typically, the area over which a host plant is exploited by a butterfly will be a fraction of the total host plant area and resource zones are not expected to coincide entirely with vegetation units. Thick continuous lines, vegetation unit boundaries; green areas, host plant; shaded areas, host plant suitable for oviposition and breeding; red pecked lines, nectar; blue dotted lines, mate location zones. (From Dennis, 2010.)

Box 5.1. Butterfly case studies: elements of the resource-based habitat paradigm

A colleague once admitted to the first author that it was astonishing that no-one had previously promoted the **resource-based habitat (RBH)**, or at least thought of it. That seems highly improbable as he recognised that the concept is so obviously 'right' and simple to understand. Furthermore, the elements that go to make up the RBH have long been studied in the field by numerous researchers. All that was missing was that the vital bits had not been joined up to develop the paradigm; humans, especially in the industrialised world, have difficulty conceiving that other organisms occupy a 'habitat' space larger than fields and woods they can perceive from a hilltop, but it is obvious that, for example, larger animals occupy considerable fractions of entire biomes that evidently form their habitats (e.g. the Bengal tiger *Panthera tigris tigris*; Ahmed and Rahman, 2023). Admittedly, there are practical difficulties in implementing the RBH, but then the main target species should at least be those threatened with extinction. To illustrate the type of research needed to understand a species' RBH, this box considers the features of RBHs in local populations of a few butterfly species.

Functionally, the RBH may be conceived as the ecological equivalent of **-omes** in biology that developed from the 1990s, viz. the sum constituents within a cell, tissue or organism (genome, proteome, transcriptome, epigenome, lipidome, etc.), fields of study in biological sciences that end with -omics (e.g. genomics, proteomics, transcriptomics, etc.) (Beale *et al.*, 2022). The equivalent for habitats in ecology would then be **RBHomics** and one can subdivide the associated sets of resources and conditions with similar suffixes. The ideas first developed – but remained unprinted – in the mind of the first author, when he started to investigate a number of common and conspicuous butterfly species near his home. It was evident that whatever formed their 'habitats' differed considerably from one species to another, even from one individual of a species to another, even if their home ranges were fundamentally similar in outline. At the species level, some of their requirements showed obvious differences, for flight capacity, egg placement, adult food, roosts, territories and other life-history essentials. By focusing on the most crucial aspects of their requirements (resources and conditions), it became evident why they occurred where they were found, and the local limit to their occurrence.

The starting point in the study of restricted, rare or exceptional species is always location: where are individuals found throughout their life cycle? As butterflies typically fly faster than most people can run, recording locations at any time is performed by **mark-recapture** (MR) along a fixed transect that covers the site where they have been seen by casual observers. Examples are offered by studies focused on eight species (*Anthocharis cardamines*, *Pieris napi*, *Aglais urticae*, *A. io*, *Lasiommata megera*, *Maniola jurtina*, *Polyommatus icarus*, *Ochlodes sylvanus*) in Cheshire, UK (the Bollin Valley, Brereton Heath and Alderley Edge) (see Dennis, 2010 for references). The mapped data of captures and recaptures for both sexes provided an outline for the **home range**, minimal daily movements and fidelity to specific zones within the home range (Dennis and Bramley, 1985; Shreeve *et al.*, 1996b). It was also possible to determine behaviour of butterflies, their use of resources and attachment to specific local climates and substrates at the same time.

The study of the orange tip (*Anthocharis cardamines*) in the Bollin Valley (Fig. B5.1.1) revealed that: (i) the butterfly was largely confined to the valley and the sunny north side; (ii) there was some leakage through the woodland on either side of the valley and diminishing dispersal out over the fields above the valley; (iii) barriers existed for the butterfly, particularly a major obstacle in the form of a motorway (the M56) (a specimen marked no. 9 had difficulty even overflying hedges between valley fields and was restricted largely to a single small field); and (iv) there were contrasting recapture rates for males and females. From tracking individual butterflies, it was possible to produce time graphs of activities (Fig. B5.1.2); the tracks revealed complex patterns of movement, occasionally fast and linear, often circular; use of nectar sources such as the bluebell *Hyacinthoides non-scripta* in woodland, inspections and interactions with other butterflies, especially males with females, and egg laying on the two host plants (*Cardamine pratensis* and *Alliaria petiolata*). Subsequent mapping of the main resources (nectar flowers and larval host plants) disclosed that both were widely scattered, occasionally clumped in the valley (host plants) and adjoining woodland (nectar flowers). A more detailed study and sketches of flight paths of some 132 individuals at the M56 bridge/tunnel revealed the difficulties the butterflies had in attempting to continue their flights to the north-west and south-east, effectively breaking the population in two though failing to ensure complete isolation; five (4.8%) managed to lift over the bridge and cross the motorway (Dennis, 1986b).

Similar studies of the wall brown (*Lasiommata megera*) and meadow brown (*Maniola jurtina*) produced much the same findings. The two notable differences were: (i) a pattern of delayed emergences in the former

Continued

Box 5.1. Continued.

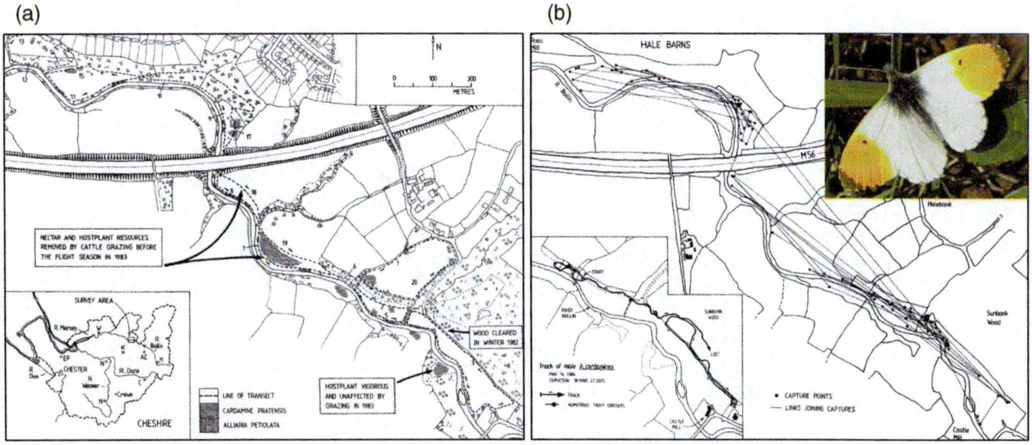

Fig. B5.1.1. The distribution of the orange tip *Anthocharis cardamines* in a section of the Bollin Valley (Cheshire, UK). The population is divided by the motorway M56 bridge that greatly reduces movements north and south. **(a)** The line of transect (all butterflies seen were caught and code marked with different colours north and south of the motorway). The map shows the distribution of the two host plants (*Cardamine pratensis* and *Alliaria petiolata*) scattered in clumps. These and many other plants along the valley were used as nectar sources. 1–20: numbers of sections along the transect. **(b)** Locations of captures and recaptures, mostly of males. Those of any one butterfly are linked with straight lines. A number of butterflies were tracked rather than captured; the path of one is displayed in the inset map. Most butterflies were restricted to the valley but a number 'leaked' through the woodland on the valley sides to enter farmland. (From Dennis, 1986b, used with permission of the Amateur Entomologists' Society. Inset courtesy of Peter B. Hardy.)

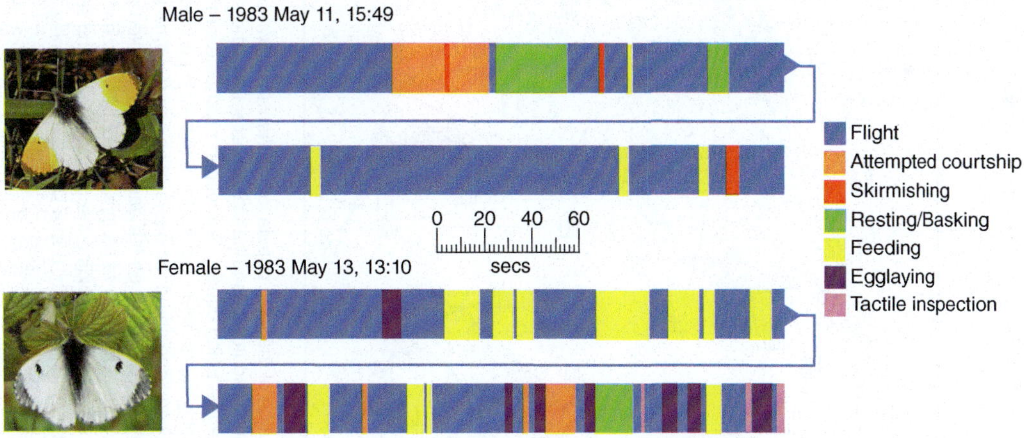

Fig. B5.1.2. Seriation in behaviour in a male and female orange tip butterfly *Anthocharis cardamines* over an 8-minute period. The complexity of behaviour is evident, as was the random switch in behaviour on encountering resources and other individuals. (Redrawn from Dennis, 1986b, used with permission of the Amateur Entomologists' Society. Photos courtesy of Peter B. Hardy.)

butterfly from north to south related to local climate (Dennis and Bramley, 1985), and (ii) a far more complex pattern of resource distributions for the wall brown than the meadow brown. Resources for the wall brown were rather scattered (Fig. B5.1.3), whereas those for the meadow brown tended to be

Continued

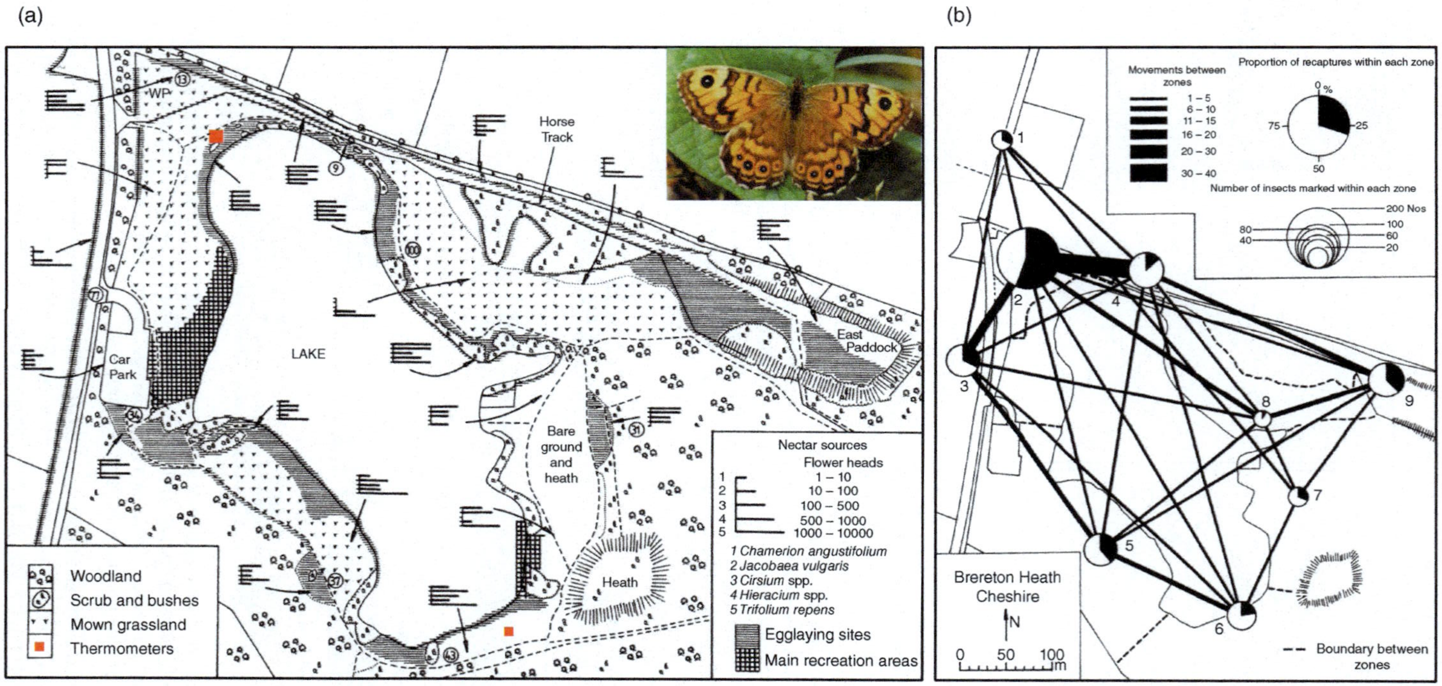

Fig. B5.1.3. The distribution of the local population of the wall brown butterfly *Lasiommata megera* on Brereton Heath country park, Cheshire (UK), in 1983; the home range and resource-based habitat are effectively the surround of the nine zones used as distinct sites for the butterfly. **(a)** Resources are rarely in single packages, being usually clumped in patches over the heath: egg-laying zones and nectar concentrations are shown; territorial perches occur mainly along the northern boundary in the west paddock (WP), along the trees bounding the heath and along the sandy horse run. The butterfly avoids all birch woodland and does not feed on ling (*Calluna vulgaris*). Wind speed is given for eight sites (numbers in circles) as a percentage of that for the maximum value recorded on the east bank (100) taken over 4 days in different wind directions. Overall, *L. megera* prefers warm sunny sheltered sites. **(b)** Simplified illustration of counts, recaptures and movements between the nine zones on the country park; movements occur around or over the central lake, the butterfly avoiding the *Calluna* heath and birch woodland surround but expanding onto the edge of an adjoining rough field and the single-track roadway in the north. (From Dennis and Bramley, 1985, courtesy of the Society for European Lepidopterology. Inset courtesy of Peter B. Hardy.)

Continued

Box 5.1. Continued.

restricted to distinctive zones of meadow grass over the site (Fig. B5.1.4), but a definite linear pattern along hedges, fences, wood and adjoining road and farmland edges existed for egg laying, mate location (territories and patrols) and roosts (a distinct **edge effect**), with females avoiding males following mating. Territories of the wall brown were at strategic locations, whereas the meadow brown engaged in perpetual patrolling, but also opportunistically exploited some nectar sources. In both cases, the butterfly avoided the woods and *Calluna* heath, but meadow browns were observed overtopping woodland while moving from one paddock to another, as well as crossing the lake and open short-grass and sandy areas. Despite the separation and clumping of resources, it was clear that Brereton Heath formed one population in both cases.

Mark-recapture can produce some unexpected findings. An important one was the extremely low value of recaptures for the green-veined white (*Pieris napi*). Most marked one day had vanished from the Bollin Valley, and the field above the wood, by the next day. This proved to be very different from valley bottoms cutting through upland sites in the Goyt Valley, above Macclesfield (Cheshire, UK), and from similar hollows in northern Scotland, UK (Scourie), which have colonies more similar to the orange tip (Dennis, 1993b; Hardy *et al.*, 2010). Another feature has been the rapid change in habitat conditions (see Fig. 3.2b). The wall brown, abundant on Brereton Heath in 1984, was extinct by 1986 and thereafter gradually lost from all sites in the county and over much of the UK. In contrast, with the continual growth of grassland patches, the meadow brown occupied much of the area previously held by the wall brown. Within 3 years, everything had changed with the additional arrival of the gatekeeper *Pyronia tithonus*, the speckled wood *Pararge aegeria* and the small skipper *Thymelicus sylvestris*; finally, the arrival of the ringlet *Aphantopus hyperantus* and (temporarily) the silver-washed fritillary *Argynnis paphia* in the new millennium. The Brereton site was abandoned as a quarry in 1980, the excavation zone filling up as a lake and the site becoming a public country park and a nature reserve. The Alderley Edge site was converted to National Trust and private land following copper mining.

Beyond that, specific resource exploitation can be assessed using dedicated surveys. Such were brought to bear on roosting in the wall brown butterfly (Dennis, 1986a), testing of artificial substrates versus natural features, and diurnal patterns of mate location, in territories of the large skipper (*Ochlodes sylvanus*) (Dennis and Williams, 1987) (Fig. B5.1.5) and wall brown (Dennis 1982–83, unpublished data), nectar use and thermoregulation in the meadow

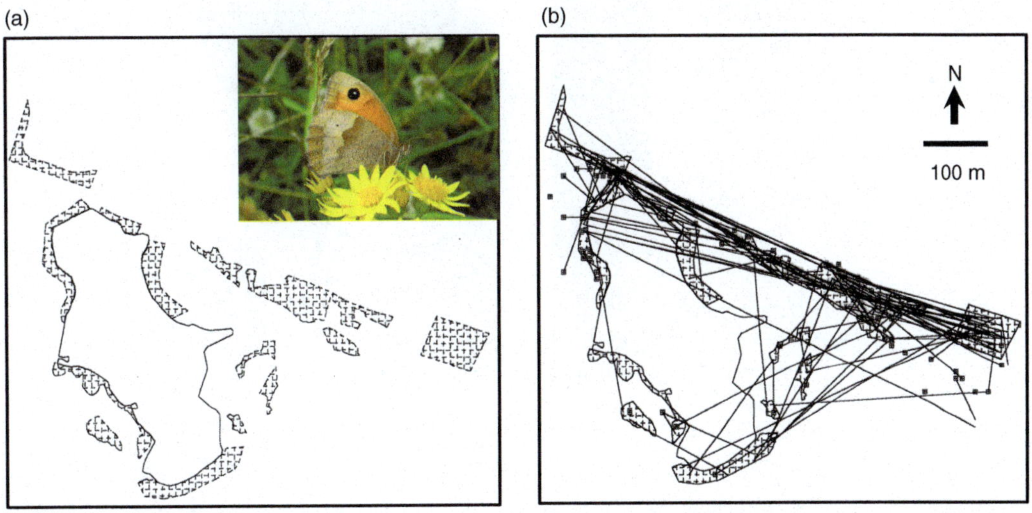

Fig. B5.1.4. The distribution of the meadow brown *Maniola jurtina* on Brereton Heath, Cheshire (UK). **(a)** The fragmentation of resource zones; in this butterfly, most of the resources can be found in single meadows; here many are small units developing as the country park recovers from previous quarrying. **(b)** Movements of the butterfly. A great number of cross-movements occur between 'resource unit' fragments, even across the lake. Small squares record individuals captured but not recaptured. (Redrawn with small changes from Shreeve *et al.*, 1996b, used with permission of the Society for European Lepidopterology. Inset courtesy of Peter B. Hardy.)

Continued

Box 5.1. Continued.

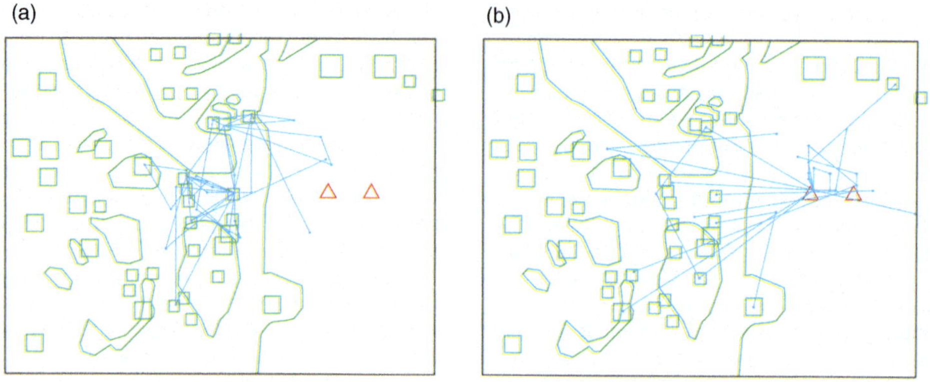

Fig. B5.1.5. Experimental work on the use of features within a territory of a large skipper (*Ochlodes sylvanus*) male. Influence of the placement of a plastic bag on perch sites adopted by large skippers within the centre of a territory on Lindow Common, Wilmslow, Cheshire (UK). Perch sites adopted with and without a white plastic bag, alternated four times over 10-minute periods; **(a)** without presence of plastic bag; **(b)** presence of one plastic bag in one of two positions shown. Area shown: 8 m × 6 m. Triangles, plastic bag; squares, birch seedlings; polygons, other vegetation boundaries. (From Dennis and Williams, 1987; after Dennis, 2010).

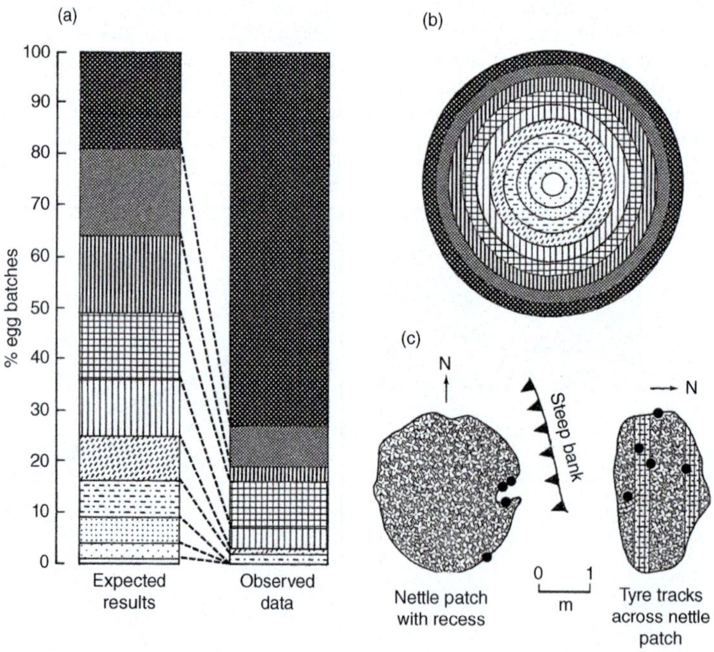

Fig. B5.1.6. The edge effect and recess effect in egg laying in the small tortoiseshell *Aglais urticae* on nettle *Urtica dioica* in the Bollin Valley, Cheshire (UK). **(a)** The relative proportion of egg batches in 10 annuli of a circular nettle patch compared with that expected from the proportion of host plant shoots available in the annuli; significantly, 73% of the egg batches are laid in the outer ring compared with 19% expected from a concentric model. **(b)** The concentric model for egg laying on host plant patches. Outer annuli offer a wider area for egg laying despite being of equal width to inner annuli. **(c)** The use of edges and recesses (two examples: natural enclaves, tyre tracks) for egg laying by *A. urticae* (Dennis, 1984b). (From Dennis, 1992.)

Continued

Box 5.1. Continued.

brown (unpublished data), egg-laying edge effects in the orange tip, the common blue (*Polyommatus icarus*) (Dennis, 1984a), the wall brown (Dennis, 1983b) and two nymphalids, the peacock (*Aglais io*) and the small tortoiseshell (*A. urticae*) (Dennis, 1984b) (Fig. B5.1.6), use of specific hosts among single host plants (e.g. the smallest plants in the green-veined white; Dennis, 1985a), as well as the comparative use of different host plants and excess egg loading on single plants in the orange tip (Dennis and Hardy, 2006). It is in dealing with different host plants for any species that it becomes evident that very different conditions are often required for the plants. For example, in the case of the orange tip butterfly, the cuckoo flower *Cardamine pratensis* prefers wet but exposed conditions such as damp river cut-offs (abandoned meander loops), whereas the hedge garlic *Alliaria petiolata* prefers more shaded but dry soils, as for instance river banks. In the same way, some butterfly species prefer shaded locations (e.g. the speckled wood *Pararge aegeria* uses sun spots in woodlands on Alderley Edge, Cheshire, UK), whereas others like the meadow brown and the gatekeeper prefer exposed sunny environments in the location.

To build RBHomics, such studies need to assess all resources and conditions for a species, not just at one site but at many sites selected judiciously to provide a representative sample of environmental conditions (e.g. for geology, soils, cover, local climate) over a species' range. Part of this work should delve into the 'killing agents' (i.e. predators, parasites, diseases) through the life cycle. Methods for such work are intensive as well as extensive and involve marking individuals through life, as well described by Warren (1992).

second (Shreeve *et al.*, 1996b). However, it is rare indeed that either type of matrix, internal or external, is completely empty of resources (Dennis, 2004a; Dennis and Hardy, 2007). For instance, both settings provide the meadow brown butterfly with places of rest, shelter and thermoregulation during routine flights in the habitat and dispersal beyond it. On the other hand, stretches of water are closer to being true matrix for this butterfly, as settling on them and taking off again is a rare event (Dennis and Shreeve, 1996).

In effect, the so-called matrix is at least two things for an organism: (i) the medium through or around, over or under which it transfers (e.g. rock, soil, water and air for butterflies) between resource concentrations or resource outlets; and (ii) empty or resourceless space in which an organism is unable to extract benefit in terms of its normal resource exploitation (e.g. sea for land animals) but which as an obstacle needs to be passed. It should be borne in mind that some resources are sufficiently commonplace as to be ignored by researchers; for instance, land animals depend on the land surface, but perhaps we should not forget the importance of atmosphere to them, the air they have to breathe, whose conditions are influenced by weather and increasingly by pollution; the impact of car exhaust fumes on the human animal is well known, even if little is done about it. For bird migrants, the matrix is largely the medium(s) of transfer (atmosphere) from which they may or may not land to rest or obtain food, but is also the medium in which many obtain their food (e.g. the peregrine falcon *Falco peregrinus*; Stevens *et al.*, 2009) and which during migratory flights has potential, through weather conditions, for enhancing or reducing their intrinsic rate of increase.

Elements of habitat suitability and key resources

From Figure 5.1, it is evident that core habitats for organisms such as butterfly species may be arranged in many different ways. But this suite of potential alternatives only gives a glimpse of the full complexity of habitat structures, particularly as resource-based habitats are effectively space-time scaled by individual organisms. Simply stated, the composition and structure of each habitat (RBH) for an operational unit (a mating pair, a local population, a troop, a swarm, etc.) of any one species is unique; no two local habitat units for any species are identical – even if the quantity and quality for each resource outlet (consumables and utilities) and abiotic conditions are much the same – simply because there are so many other ways in which they can differ. Just what this habitat unit describes is determined by the space (e.g. **home range** and **lifetime track**) occupied by the individuals of a local population (or other levels of individual aggregations, from pairs to species), and is discussed below. Moreover, as may be gleaned from details of ecological succession discussed in Chapter 3, a point deserving constant repetition, the resources and conditions of a single species' habitat unit are inconstant. This has immense repercussions for understanding the requirements of species across the globe. Thus, to obtain any kind of understanding of a species' habitat, systematic measures need to be made on the hierarchy of elements and subsets constituting that species (e.g. individuals, groups, troops, herds, shoals, local populations and discrete regional units) across its range for the essential life-history traits that ensure persistence. This crucial issue of measurement, assessing a species' requirements, is considered in the penultimate section of this chapter (*Variation in the resource-based habitat*) and in Chapter 6. Here we first explore some simple components that make up the variability of habitat units.

The sheer extent of variation for a single local population of a species' habitat can be illustrated by considering three basic components of habitats: **composition**, **physiognomy** and **connectivity** (Dennis, 2010; Dennis *et al.*, 2006a; Fig. 5.3). The basic features of each of these components are listed in Table 5.1. Regarding **composition**, organisms may be more or less selective for resources,

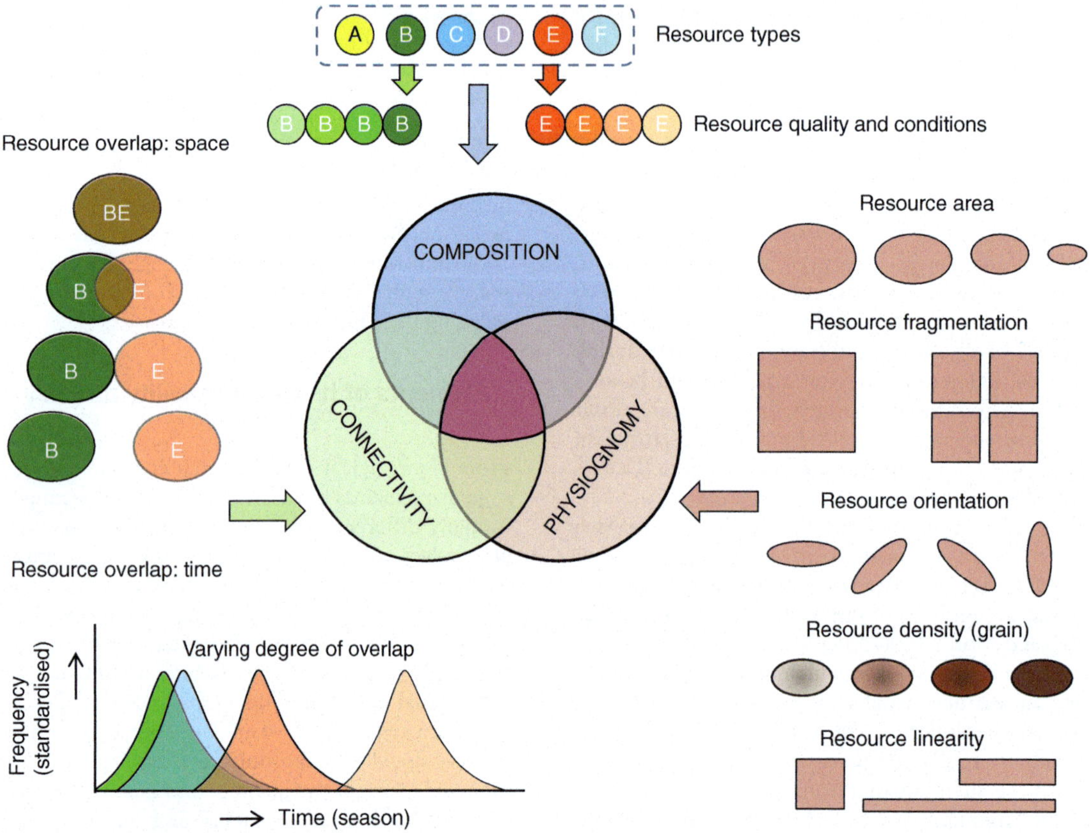

Fig. 5.3. Basic dimensionless model for habitat suitability based on three fundamental habitat components: resource composition, physiognomy and connectivity (Dennis *et al.*, 2006a). Each of these components is described by a number of mutually non-exclusive parameters. The purple zone where the three intersect can be regarded as the collection of conditions for which the functional habitat becomes available or maximised, although in fact the intersection of maximum values for the basic variables could impair function if it also increases competition and exposure to enemies. The functional habitat (resources all accessible to most individuals) depends on individual mobility in each species. Turlure *et al.* (2009, 2019) used a similar set of variables describing suitability of habitats: composition, availability and configuration; availability depends on individual resource type composition (subsets), physiognomy and connectivity, and configuration is described by resource physiognomy and connectivity. (From Dennis *et al.*, 2014, slightly modified.)

Table 5.1. Resource components affecting habitat patch quality for butterflies.

Composition of resources (resource types used)

- Resource types: number of different consumable and utility resources
- Resource complementarity: all resources present or not
- Resource limitation (Liebig's law): a resource in short supply or of suboptimal status
- Resource fraction available: resource elements used by a butterfly (e.g. taller or shorter plants, flowering plants) are always less than resource type
- Resource density: degree to which a resource is packed and therefore suitable or not
- Resource context: suitability of surroundings (e.g. overtopping, shade) and substrates (e.g. dry or wet substrates) for all resources
- Resource substitutability and supplementarity: single versus multiple outlets for resource requirements
- Resource independence: matching of conditions (e.g. moisture, humidity, sunshine, temperatures, wing exposure) for different resource requirements

Physiognomy of resources (placement and configuration in habitat)

- Resource size (area) and variability in composition and physiognomy: do resource outlets meet target organism requirements for size and spatial attributes?
- Resource topography (slope, orientation, altitude) influencing context (e.g. hilltops or edges for mate location)
- Resource substitutability: multiple resource use (e.g. same species used as a host plant and a nectar source) increases habitat compaction
- Resource shapes: effects of linear as opposed to compact resources
- Resource contagion: clumped, random, uniform (e.g. comminution and fragmentation of resource elements)

Resource connectivity (resource proximity)

- Resource proximity and spatial variance (overlap, neighbouring, isolation, ruggedness) and effects on connectivity
- Substitutability and compaction of resources increasing connectivity
- Resource abundance (density) or size and compaction increasing connectivity
- Resource context and topography influencing mobility and connectivity

Temporal aspects (changes in resources over time)

- Resource base changing with time (with effects on composition, physiognomy and connectivity)
- Resource time limitation (resources have limited availability and timing)
- Resource independence (resources respond to different influences over time)
- Resource supplementarity and resource lifespan
- Resource dimensions (area) and relationships to temporal changes
- Contextual influences on temporal changes in resources

Definition of terms: *Composition* refers to the occurrence, density, multiplicity, variation in and context of a resource type. *Physiognomy* describes the geography of a resource type (e.g. location, altitude, orientation, slope, area, height, shape, contagion and fragmentation). *Connectivity* describes contact and isolation of resource types from one another and involves adult and larval mobility. All these variables change over time. No two habitat units are identical in the composition, physiognomy and connectivity of resources for any one species.

some of which can operate as non-obvious limiting factors. For example, it would be thought that the key resource for butterflies is their larval host plants and that is usually the case. But even for a phytophagous insect like a butterfly, larval host plants are not always the limiting factor for a population; adult nectar sources can also be limiting, and no doubt so can hibernation sites, mate location sites, roosts and sites for evading predators and parasites (Dennis, 2020). Roosts and nest sites can certainly be limiting factors for local populations of birds (e.g. starlings *Sturnus vulgaris*; Chapman and Reiss, 1999), as well as caves for bats (Humphrey, 1975), the spaces under bark for many **saproxylic** insects or the stones in a grassland for many ground-dwelling invertebrates (such as many earthworms, centipedes, millipedes, scorpions, spiders, earwigs and carabid beetles). Any specific resource type can be single or multiple (as in the case of the number of host plants used by insect larvae), entire or fractional (larvae may use for food a part or an entire plant), non-transferable or transferable (a plant may be used only by larvae or by both larvae and adults for food), and main or just subsidiary (Wiklund, 1981). Food resources vary enormously in composition for butterfly species (e.g. in size, age, water content and nutrient (nitrogen) status) and the same applies to most animals (consider, for example, how variable in biomass can be the prey of a raptor feeding on small mammals, reptiles, other smaller birds and even invertebrates). Similarly, differences in the proportion of elements make soils extremely variable in terms of resources for plants.

Physiognomy refers to resource geography: location, altitude, slope position, shape, orientation, contagion and fragmentation; these and other attributes influence resource use in butterflies, as well as in the most disparate organisms. **Connectivity** largely depends on the physiognomy of resources, their relative isolation or the barriers between them, but also is dependent on an organism's **mobility** and the availability of the resources over time (see Chapter 3 and below, section *Changes in habitat suitability*) (Dennis, 2010). Different labelling and arrangements of the attributes for these components can be constructed (for example, Turlure *et al.*, 2009, 2019 use a similar classification based on composition, configuration (spatial organisation) and availability in the study of two Lycaenidae (*Lycaena*) and three Nymphalidae (*Boloria*). But the basic message is the same: altering the composition and structural arrangement of the resources or conditions will have an impact on a species' population status; it will become more or less common or rare in the space identified as the resource-based habitat unit and in any traditional habitat unit occupied.

The impact of these and other factors can be illustrated by examining just three resources for eight pairs of resource associations within a compact habitat unit for a butterfly (viz. resource complementarity, sufficiency, supplementarity, density, neighbourhood effects, linear effects, heterogeneity and ruggedness; Fig. 5.4). These conditions (variable situations for resources) can affect any stage of development. One of the key distinctions in resource use is that of **generalism** versus **specialism** (e.g. regarding larval host plant use, **monophagous** versus **oligophagous** versus **polyphagous**; see Chapter 6). Any resource type may be subject to this distinction: the singular or multiple use for any resource type whether it be a consumable (e.g. larval host plant, adult nectar source) or utility resource (e.g. mate location or hibernation sites). The impact of these conditions is most evident in species with so-called closed population structures or **metapopulation** structures of compact, linked local populations, as illustrated in Figure 5.1a (a–e) (e.g. the wall brown *Lasiommata megera*; Dennis and Bramley, 1985; Dennis, 1986a). But, even so, the global habitats of those species which migrate over massive distances and that have very different seasonal resource exploitations can be examined in the same way by compartmentalising their seasonal habitats as indicated beforehand, in Chapter 4 (e.g. summer breeding habitats versus winter feeding habitats) (see Fig. 5.1a (f, g)).

Some basic generalisations emerge from consideration of the composition, physiognomy and connectivity of different butterfly species' resource-based habitats; these are summarised in Table 5.2 and have relevance for other organisms.

Resource compaction and intersection in habitats

The traditional usage of the term 'habitat' envisages resources as being compact, co-occurring within a single **biotope** (traditional habitat) and found at a single site. However, from the foregoing discussion on resource integration, it will be apparent that there is a vast difference between resources being integrated and a compact habitat. **Mobility** ultimately delimits integration whereas resource intersection underpins habitat compaction (Fig. 5.5). Increased resource intersection ensures greater potential for a correspondence between the biotope and the RBH. But resource intersection is not sufficient, in itself, to ensure the existence of a compact habitat: for that, resources should share the same space (i.e. the spatial distributions of different resource types must coincide entirely or form nested subsets, where one is fully contained within another). As a result, degree of compaction increases directly with resource density (i.e. compaction increases as the so-called 'trivial' non-resource space – i.e. internal matrix within compact habitat units – or the internal matrix between resource elements is eliminated). However, resource elements may be spread out widely so that, even though substantial intersection occurs, RBH bounds may be expansive, resource density low and resources permeated with non-resource space (trivial spaces or internal matrix). Moreover, various resource types may be supplemented (same resource) or complemented (novel resource) by additional outlets (Dennis, 2010) which, at one and the same time, may enhance resource integration but also may extend the habitat into a wide array of additional biotopes, rendering it less compact.

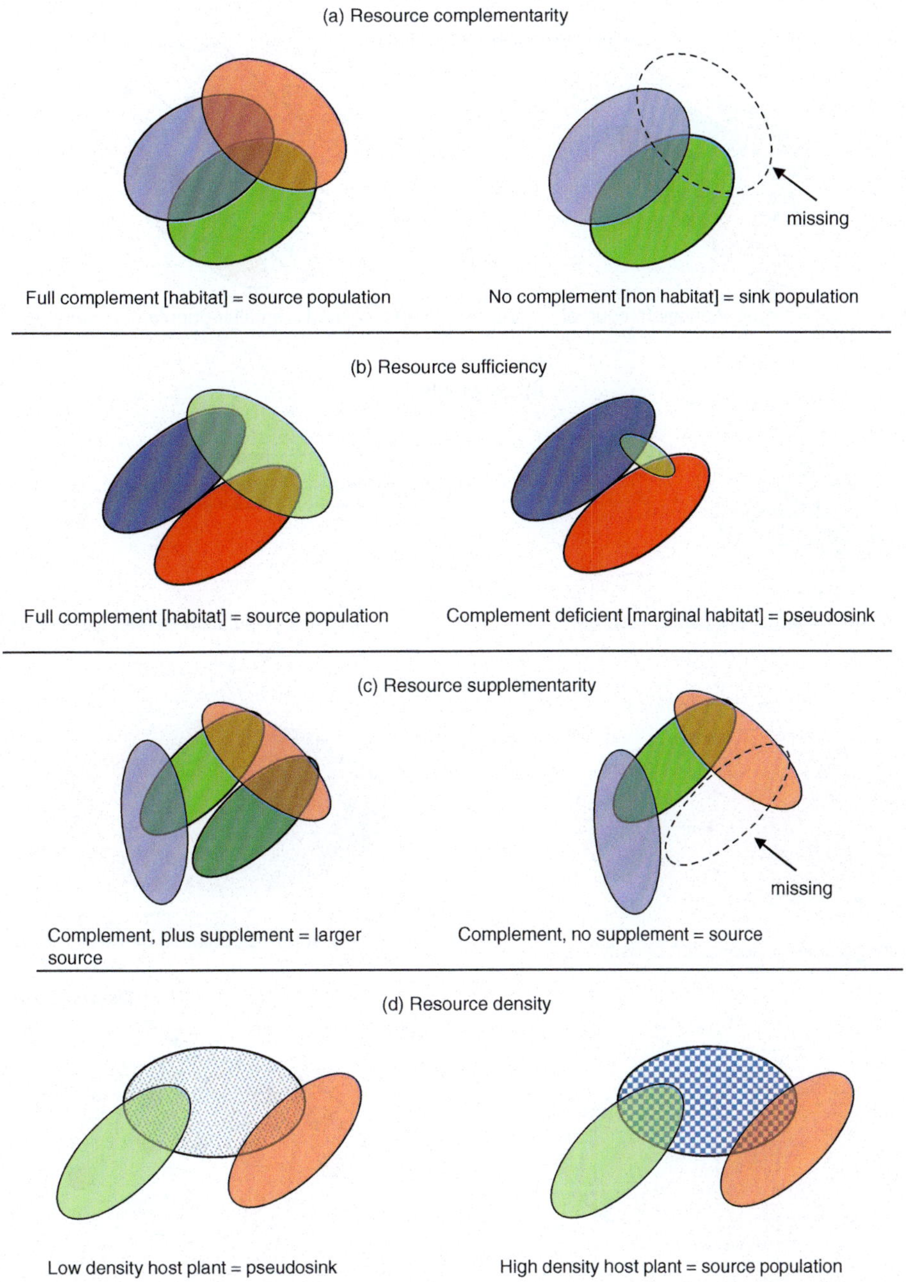

Fig. 5.4. Some basic resource attributes that affect population size, density and status. Schematic diagrams to illustrate contrasting situations for resource attributes and their impact on local populations. **(a)** Resource complementarity: all resources are required for population persistence; absence of resource type may allow temporary existence of a 'population' which will be a sink. **(b)** Resource sufficiency: if a resource type is too small or sparse to support a population, this creates a pseudosink. **(c)** Resource supplementarity: two or more distinct outlets (e.g. host plant species) over a site provide insurance against environmental changes and boost population size. **(d)** Resource density: a key resource (e.g. host plants at higher density or in better condition within a site) supporting a larger population. **(e)** Neighbourhood effects: contiguity and absence of barriers ensure access and exploitation of all resources at a site and larger population growth. **(f)** Resource package shape: linear resources (e.g. host plants) may well be of lower

Continued

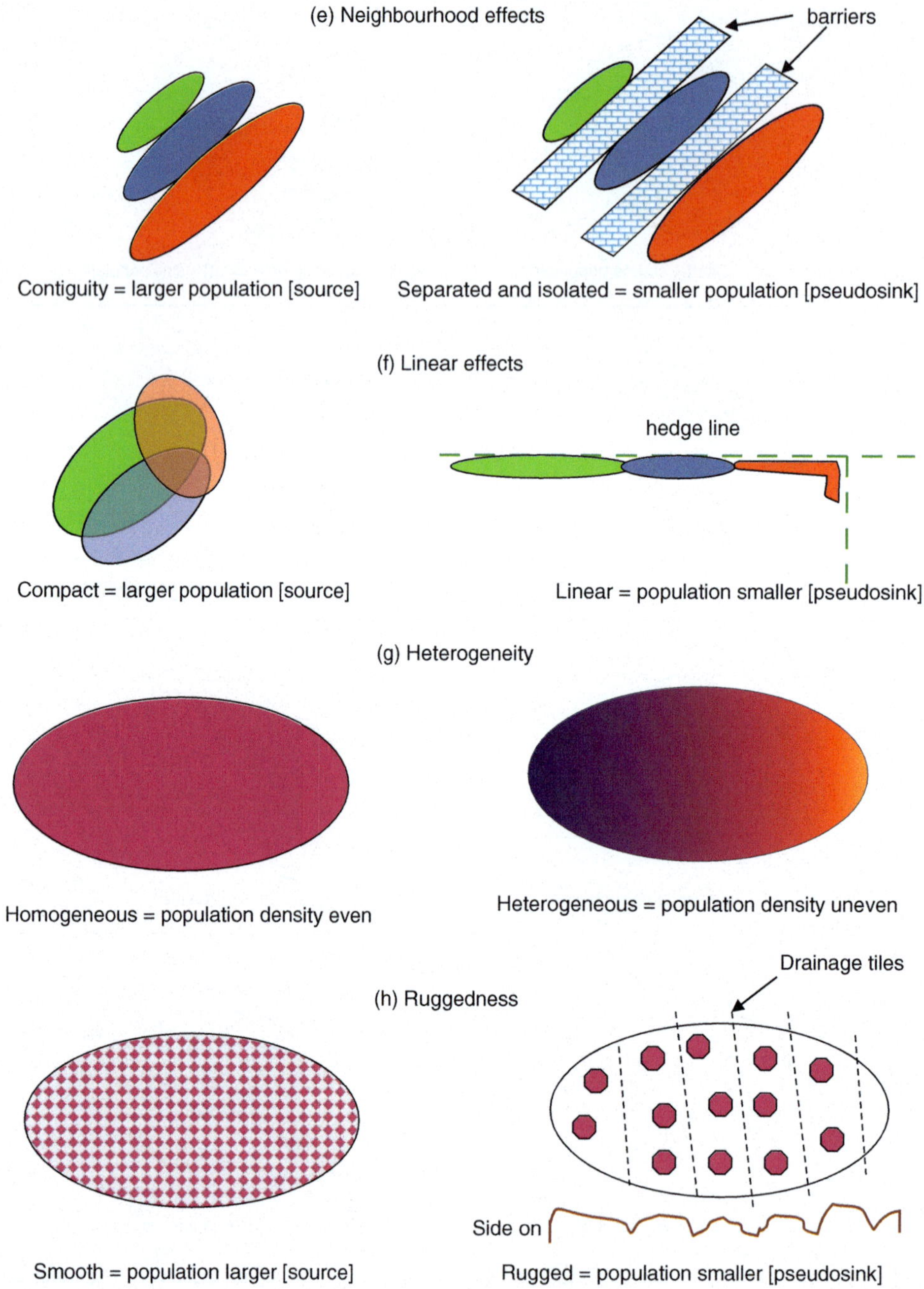

Fig. 5.4. Continued.
quality but also create more intense intraspecific negative interactions, thus depressing populations. **(g)** Resource heterogeneity: heterogeneity results in uneven density or quality of resources over habitats and movement between high-density resource zones and areas of lower density/quality, in effect sources and sinks within a habitat patch. **(h)** Ruggedness: an increase in resource spatial variance adversely affects population size and population persistence. Ellipses are resources (three used with reference to butterfly biology as an example): green, larval host plant; blue, adult nectar source; red, mate location site; purple, multiple resource distributions. (From Dennis, 2010.)

Table 5.2. Some resource-based habitat principles relating to resource composition, physiognomy and connectivity in butterfly biology.

General principles of resource composition
• Only a fraction of potential resources is suitable for a butterfly species within any single biotope. • The suitability of a resource is affected by its context of non-resource items. • The greater the number of larval host plants used by a butterfly species, the greater the number of vegetation units (biotopes) a habitat is expected to cover. • The greater the number of larval host plants a butterfly species uses, the greater the variety of complementary resources a species is expected to exploit. • The more generalist in larval host plant use a species is regionally, the more supplementary resources are expected to occur for it within a single site. • The wider the range in larval host use, the wider the parasitoid (and predatory) community exploiting the butterfly species within a single site.
General principles of resource physiognomy
• The more host plants used by a butterfly within a site, the larger the area of the habitat. • The larger the area of a resource within a site, the more varied will its elements be physiognomically and the more opportunities for exploitation. • The larger a resource within a site, the less critical are factors for exploitation associated with its spatial attributes. • Transferable resources (e.g. flowering plants used for larval host plants and nectar) increase habitat compaction and dispersion of resources within a habitat accompanies non-transferable resources. • Some resources are more likely to be linear (edge resources) than others and influence resource exploitation. • Resource types (and complements of resources) vary in physiognomy, influencing resource status and habitat status for species. • Comminution (fragmentation on a fine scale, compaction and context), generally adversely, affects the quality of a resource.
General principles of resource connectivity
• The more abundant resources are within a site, the less likely are they to be isolated from one another and the more connected the habitat structure. • The greater the transferability of resources (equivalence), the more connected will be the resources making up a habitat. • The more diffuse (less compact, comminuted, fragmented) a resource or a set of resources is, the less connected they are and the more mobile an organism needs to be to exploit them. • The degree of connectivity depends on the structural attributes of the matrix (context) separating resources.

Although these principles were determined for butterfly resource-based habitats, many of them relate to other organisms (Dennis, 2010).

In this way, an atom provides a useful analogy for habitat: just as electrons remain in an uncertain, non-deterministic orbit around the nucleus, so do resources and conditions around zones of primary consumable resources, in the case of many butterflies the main resource being concentrations of larval host plants. Implications hold for habitat limits and the internal matrix among resource items, much like the electron cloud defines the boundaries and structure of an atom. Reference is often made to the habitat boundary but this, like the outer limits of an electron cloud, can be elusive to locate in practice. To further complicate things, however, resources in a habitat, much like electron densities in orbitals, can have different densities, becoming progressively more dense with increasing concentration (convergence and overlap) of favourable environmental parameters affecting resource distributions.

From the above, it becomes evident that the relationships (degree of overlap, equality, nestedness) between a species' resource items in a landscape provide a useful basis for quantifying habitat compaction (Dennis and Shreeve, 1996). For instance, resource items can be identified and mapped to real space compartments (e.g. grid or small vector land units) and their overlap quantified or categorised. Although not all resource outlets for a species may be easily mapped or measured and quantified, a species' biology is massively affected by relative resource dispositions, and that is something that can be assessed without sophisticated measurements or quantitative surveys. For instance, in a simple classification of different species, the habitat may be described (qualified, scored) in the broad terms of the physical relationships among resources typically found at most locations (e.g. a sample of sites from existing autecological surveys). Different categories of resources (consumables and utilities) and conditions may: (i) include the same or different items (in equal or different proportions), and (ii) may intersect, be contiguous or be disjoint, the space for each resource being described by an envelope (e.g. a convex polygon) around each resource type. The degree of feasible intersection is perhaps most appropriately scaled by typical movements: whether by 'trivial' moves (walk, short flight between resource items), dispersal (abandonment of production resource sites) or migration (e.g. over water or through urban sprawl); a simple illustration is presented in Figure 5.5.

Changes in habitat suitability

As pointed out above, do not expect the resources and conditions for a local population to remain stable; they do not and with the ongoing climate change are likely to become increasingly unstable. These positive and negative changes in the resource base are referred to as **resource dynamics**. As would be expected for changes occurring in traditional habitat units (see Chapter 3), resource dynamics occur at different temporal scales (e.g. seconds, hours, days, seasons to longer term **vegetation successions** that occur over years and centuries) with different levels of predictability (Dennis and Sparks, 2006). Variation in resource quality in the daily cycle is linked to predictable agents such as changes in daylight, temperature and humidity, but also to unpredictable weather events. Variation over seasons is closely linked to seasonal climate and in the longer term, over years, resources within a resource-based habitat are affected by vegetation succession. In turn, such resource changes lead to changes in the population size of species, referred to as **population dynamics**, elements of which are outlined in Chapters 2 and 3 and Appendix 2.

Because the RBH concept has its roots in butterfly biology, we will continue to use butterflies as a privileged source of examples, with occasional reference to other organisms to illustrate the generality of the concepts, which are equally valid for the most disparate organisms. For example, a series of principles that have been identified for changes in resource composition, physiognomy and connectivity for butterflies can be applied to a variety of other organisms (Table 5.2). These principles in turn link up to principles relating changes in resources to changes in population status, some basic ones of which are outlined in Table 5.3 and simple constructs of which are illustrated in Figure 5.3.

Changes in the resource base or conditions within the resource-based habitat may result in: (i) changes in the size of a population, and (ii) the timing of its appearance (for example, development stages or broods of arthropods). As the size of the resource base changes, so too will the population size of a dependent organism. Synchronisation of key stages with the period over which a resource is suitable (i.e. butterfly larvae with host plant condition) also affects population size. Moreover, as any fundamental resource for an organism changes in suitability and/or availability, so too will the status of the population, not just in size (number of individuals) over time but also in appearance (locations) over the habitat space. A nice example is presented by Shreeve's (1986) study of the speckled wood butterfly *Pararge aegeria*, which switches in use of open and woodland conditions during spring, summer and autumn broods. Further principles and examples for butterflies are highlighted in Dennis (2010: 76).

How, then, does a resource-based habitat differ from a traditional habitat unit? First, it compartmentalises a habitat on the basis of its resources and suitable conditions for persistence (specified by its bounds) throughout organism life cycles, bounds which may fall within or extend beyond one or any number of traditional habitat units described by their substrates, vegetation or other features. Second, it recognises that any resource or condition has the capacity for independent spatial arrangement and temporal change. Any change within a single resource or environmental condition has the capacity for affecting the spatial arrangement of individuals of a population, a subunit of a population (pairs, groups, troops, swarms, etc.) or higher classes of populations, ultimately influencing the notion of a resource-based habitat for the entire species.

Unravelling relationships between species' traits and species' niches and habitats

In Chapter 2, we have already considered the role of a species' **niche** and how it relates broadly to habitat, that is the traditional **habitat** (see Chapter 2, Fig. 2.8). **Resource-based habitat** studies help us to see more clearly a species' niche and how it differs from notions of a **species' habitat**. First, in a

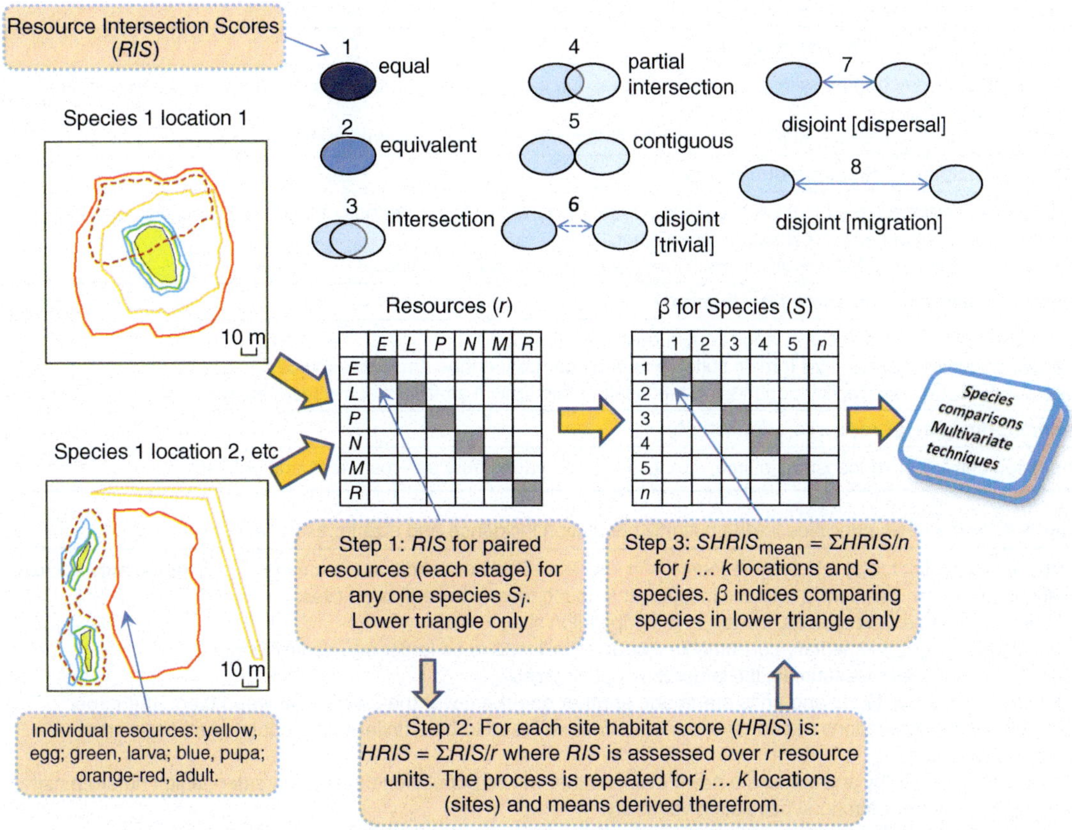

Fig. 5.5. Assessment of resource compaction making up a species' habitat based on the mean sum of resource intersection scores (*RIS* values) over samples of resources and sites. The procedure is illustrated using butterflies that exploit six different resource types during their life cycle. *RIS* values follow an inverse eight-point scale, with *RIS* = 8 indicating little likelihood of an individual (or its progeny) visiting the same site in a following year. *RIS* values obtained for pairs of consecutively used resources can be summed over sites and the average taken across pairs of resources to provide a matrix of mean habitat resource intersection scores for species ($SHRIS_{mean}$). Letter codes for resource use in this simplified example: E, egg site (yellow); L, larval food (green); P, pupal site (blue); N, adult nectar (red, pecked); M, adult mate location (orange); R, adult roosting (dark red, pecked) (see Dennis *et al.*, 2014). Method: (i) Determine 'overlap' of resources *r* (scores) for species *i* at location *j*. (ii) Sum overlap scores over resource units (note: do *not* replicate information by taking E over L and L over E), such that Σ*RIS* for site *j* is obtained. (iii) If replication of resource overlap is done in full, then Σ*RIS* is divided by the number of resources *r* which in a square matrix is *r*/2; this gives *HRIS* (habitat resource intersection score). For example, for six resources (see matrix 'Resources' in the figure) all with score 1, *RIS* is 15/2 = 7.5. *HRIS* = Σ*RIS*/*r*. (iv) Repeat process for *j* … *k* sites. (v) *HRIS* for species (*SHRIS* for *j* … *k* sites) is calculated as *SHRIS* = Σ*HRIS*/(*j* … *k*). For location where many patches occur for each resource type, the average for each pair of resource types is taken. The final matrix for comparison of species can be assessed using indices of β-diversity (see Chapter 4) distinguishing *S* species in any one region. Many measures can be constructed to determine the overlap (intersection) of resources, such as those used for determining the overlap of home ranges between related species (see Ferrarini *et al.*, 2021). (After Dennis *et al.*, 2014.)

resource-based habitat, a range of resources are exploited and conditions encountered in which any individual of a species comes into contact not just with different substrates but also with other occupants of the space, many of which have roles as symbionts, competitors, predators and parasites. Second, each stage or phase of development of an individual (sprout, seedling, vegetative, budding, flowering and ripening phases in vascular plants; eggs, larvae, pupae and adults in holometabolous

Table 5.3. Some basic principles of resource dynamics for butterfly resource-based habitats.

General principles of resource dynamics
• The resource base changes with time, influencing resource composition, physiognomy, connectivity and resource exploitation. • Potential resource types are available for specific time periods and for different lengths of time.
General principles of resource composition
• Generalist species will tend to have access to a resource type for longer periods of time than specialist species. • Some resources have shorter life spans than others. • Consumables will tend to be available for shorter periods of time than utilities.
General principles of resource physiognomy
• Spatial attributes of resources change continuously with time. • Larger resources of a type will remain suitable for a longer period than smaller resources. • Temporal resource availability depends on the location and spatial attributes of the resource type.
General principles of resource connectivity
• As resources are never constant in time, connectivity also varies between resources and habitats.

Table 5.4. Some basic principles linking butterfly population incidence and size to resource status.

- Populations are limited by whatever resource is in shortest supply or condition is exceeded; complementarity applies.
- Habitat quality can be inferred from the location and abundance of butterflies on sites.
- The larger the area of resources, the larger the population size.
- Populations with supplementary consumable resources will tend to be larger and persist longer.
- The denser the area of resources, the larger the population size.
- The closer resource types and units are to one another and the fewer the barriers between them, the larger the population. Populations based on linear resources will tend to be smaller than populations based on more compact resources.
- As resource geography and status are never even, neither will be population density. A greater habitat heterogeneity accompanies a lower temporal population variability.
- Populations are smaller on sites with increased spatial variance (ruggedness) in the resource base.

(From Dennis, 2010: 70.)

insects; eggs, tadpoles, froglets and adults in anurans, etc.) can experience different conditions and encounter other organisms. Their role (niche) changes in their progress through life and the part played by them; the traits employed by different phases and stages of growth are very different and occupy different situations in space and time. Niche, then, like the resource-based habitat, is not a constant through life but a measure of a shape-shifting existence involving the hyperdimensional space of species' traits (see Box 5.2 for functional traits and Appendix 3 for measures of functional diversity, including the concept of functional space).

Therefore, the full niche of a species can only be understood in the context of all stages of all the individuals in a collective unit (i.e. a representative sample of individuals and populations), along with their traits in the different contexts of resource use and the conditions experienced. As different stages or phases of growth (including individual status) may occur across time (e.g. during different seasons), other contexts and niche parameters also change systematically and/or randomly, depending on the interface of organisms with specific environmental changes. Truly, even when not extracted from spatially explicit data over a species' range, niche is a highly complex hyperdimensional appraisal of a species' traits and associations in relation to a constellation of abiotic and biotic variables. By comparison, the resource-based habitat is a relatively simple concept (though across the range of a species it may be immensely time-consuming to determine, unless of course that species has a narrow geographical range, as have many imperilled species): it provides the physical context for each phase or stage of development, the fundamental resources and conditions used or experienced and the bounds in space-time within which these are employed/encountered.

Niche is abstracted from the real Earth space-time in which resources and conditions in a habitat space are structured and function to influence individuals

Box 5.2. Functional traits

Functional traits are characteristics of an organism that influence its fitness. These traits can be of the most varied types, including morphological traits (e.g. plant height, leaf size, root depth, body size and shapes of particular organs or apparatuses, such as bird bills, insect wings, mandibles and legs and mammal teeth), physiological traits (e.g. water use efficiency, nutrient uptake efficiency, metabolic rates), reproductive traits (e.g. seed size, reproductive frequency, timing of reproduction), behavioural traits (e.g. feeding habits, foraging strategies, predator avoidance behaviours), life history traits (e.g. age at maturity, lifespan, number of offspring produced) and trophic traits (e.g. diet, feeding mechanisms, trophic level). Habitat associations have also been proposed as traits (Grimm *et al.*, 2014; Madin *et al.*, 2016; Oliveira *et al.*, 2017), but this is generally discouraged, as traits should be measurable at the individual level, whereas it is assumed that habitat preferences can only be deduced from aggregated data and cannot be measured on a single individual (De Bello *et al.*, 2021). For example, Ellenberg's indicator values (which express plant preferences for temperature, light, continentality, soil moisture, pH and soil nutrients; Di Biase *et al.*, 2023) are not measured on single individuals, but are inferred from species' observed distribution along natural gradients. For this reason, environment–species associations, such as Ellenberg's indicator values for plants and similar indices for animals (e.g., their thermal preferences), are considered 'ecological features', not true traits (De Bello *et al.*, 2021). However, it cannot be excluded in principle that environmental preferences can be measured at the individual level and have a genetic basis (for example, within a species, different individuals may have different thermal preferences, capacity for movement, etc.), and in such cases habitat association might be correctly considered as a trait.

Functional traits play crucial roles in determining how an organism interacts with the biotic and abiotic components of the environment. Therefore, the study of functional traits is crucially important to understand community assembly, intra- and interspecific interactions, ecosystem processes, etc. (Díaz *et al.*, 2004, 2007; Garnier *et al.*, 2004; Violle *et al.*, 2007; Fortunel *et al.*, 2009; Kraft *et al.*, 2015a,b; Carmona *et al.*, 2016).

The study of plant traits has largely contributed to clarify the importance of plant functional characteristics in determining biodiversity patterns and ecosystem functioning (Cornelissen *et al.*, 2003; Flynn *et al.*, 2011; Fry *et al.*, 2014; Jain *et al.*, 2014) (Fig. B5.2.1). For example, functional traits may reflect differences in species roles better than taxonomic diversity, which is why they can be more important than taxonomic diversity to explain several ecological processes, including productivity (Flynn *et al.*, 2011; Roscher *et al.*, 2012; Fu *et al.*, 2014; Jain *et al.*, 2014; Finegan *et al.*, 2015; Tobner *et al.*, 2016; Zhang *et al.*, 2017).

As functional traits capture different aspects of species' resource use and habitat requirements (Suding *et al.*, 2008; Thuiller *et al.*, 2010; Vandewalle *et al.*, 2010), they can provide important insights into the processes that regulate species coexistence (McGill *et al.*, 2006; Ackerly and Cornwell, 2007), community or ecosystem functioning (Tilman *et al.*, 2001; Petchey and Gaston, 2002) and biodiversity maintenance (Garnier *et al.*, 2004; McGill *et al.*, 2006; Shipley, 2010).

A common way to quantify the importance of the traits exhibited by species in a community is the use of trait means weighted by species relative proportions (community-weighted mean, CWM; Garnier *et al.*, 2004; Lepš *et al.*, 2011; Miller *et al.*, 2019; Keddy and Laughlin, 2022; Di Biase and Fattorini, 2025):

$$CWM = \sum_{i=1}^{n} p_i x_i \tag{B5.2.1}$$

where n is the total number of species, p_i is the relative cover/abundance (weight) of species i and x_i is the trait value of species i.

In the CWM the more abundant species have a greater influence than the less abundant ones. Thus, the use of the CWM allows the identification of the prevailing trait values within a community, providing insights into the functional strategies and ecological roles of the different species, and hence selection effects (Roscher *et al.*, 2012; Fu *et al.*, 2014; Li *et al.*, 2015). A number of studies have demonstrated systematic variations of CWM values of plant traits along abiotic gradients (e.g. Cornwell and Ackerly, 2009; Sonnier *et al.*, 2010; Keddy and Laughlin, 2022; Di Biase *et al.*, 2021a). For traits related to physiological mechanisms, changes in CWM values along environmental gradients can reflect how the environment selects functional strategies. For example, the decrease in CWM values for wood density and leaf dry matter content with increasing precipitation reflects physiological trade-offs related to resource availability: when water is abundant, plants may use their resources for faster growth and resource acquisition, thus showing lower wood density and leaf biomass, which are associated with drought resistance and resource conservation (Poorter *et al.*, 2009; Markesteijn *et al.*, 2011). Most applications of the CWM approach come from plant ecology, but studies involving animal communities are also abundant

Continued

Box 5.2. Continued.

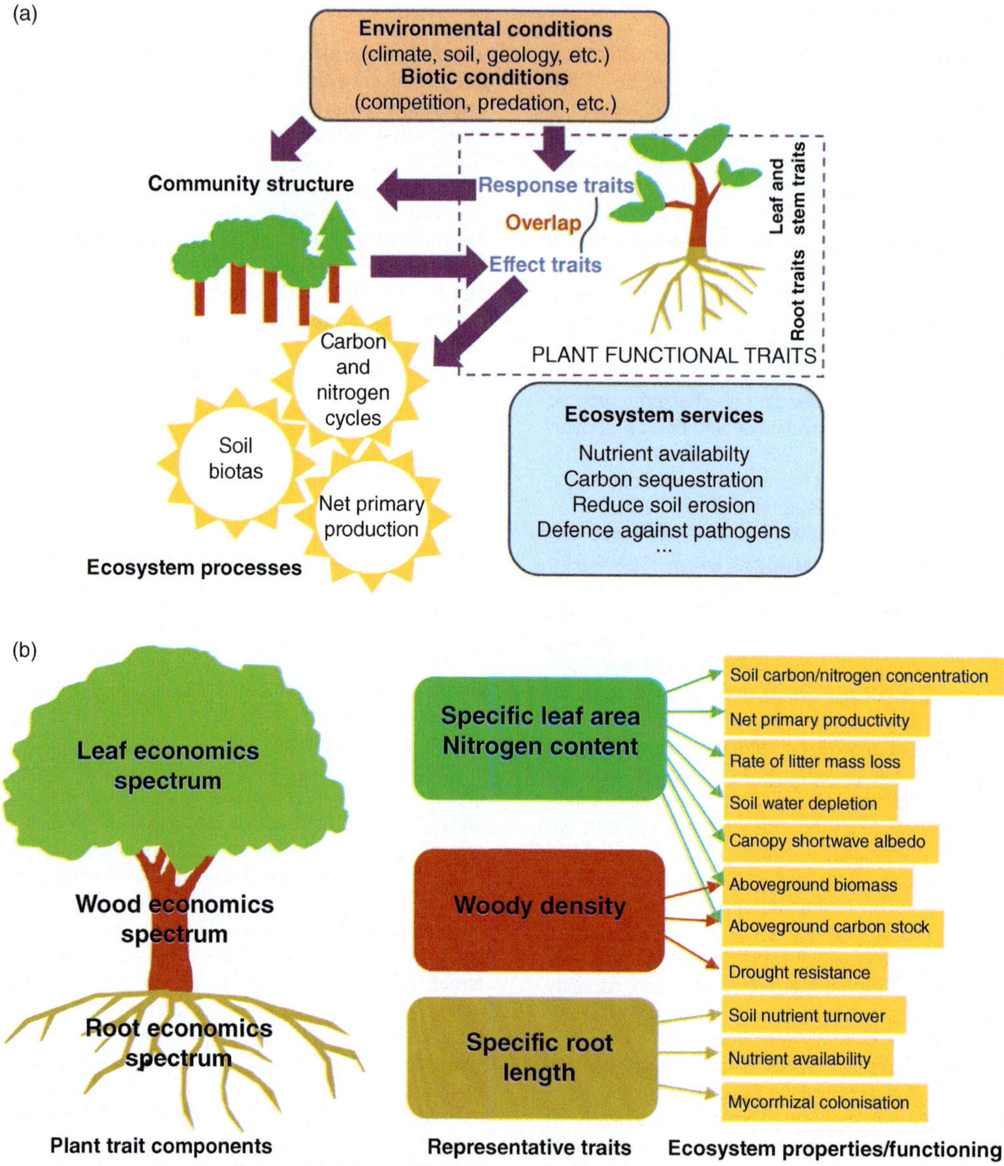

Fig. B5.2.1. Roles of plant functional traits in ecosystem functioning **(a)** and some specific examples **(b)**. (Redrawn with small changes from Liu *et al*., 2021, used under licence CC BY. Panel **a** was adapted from Lavorel and Garnier, 2002 and Faucon *et al*., 2017.)

and rapidly increasing (e.g. Fichaux *et al*., 2019; Carvalho *et al*., 2021; Korányi *et al*., 2021; Palmeirim *et al*., 2021; Rao *et al*., 2021; Filgueiras *et al*., 2022; Téllez-Hernández *et al*., 2023).

Although some traits are fixed within species (e.g. life forms, ability to fix nitrogen, carbon fixation pathways, some behaviours, type of organs), most quantitative species' traits have more or less pronounced intraspecific variation (Albert *et al*., 2010a,b; Thuiller *et al*., 2010; Westerband *et al*., 2021). The assumption that this intraspecific variability is negligible compared to the interspecific one (traits vary much

Continued

Box 5.2. Continued.

more between species than within them) led researchers to overlook its importance. However, there is abundant evidence that intraspecific trait variability is larger than previously assumed (Albert *et al.*, 2010a,b; Jung *et al.*, 2010; Westerband *et al.*, 2021) and can contribute substantially to explain species coexistence, plant community assembly and the overall functional trait response to environmental variation (Cornwell and Ackerly, 2009; Jung *et al.*, 2010; Bolnick *et al.*, 2011; Courbaud *et al.*, 2012; Larson and Funk, 2016; Chelli *et al.*, 2019; Di Biase *et al.*, 2021a; Di Biase and Fattorini, 2025). In some cases, intraspecific variation may be so large as to be as important as, or even more important than, interspecific variation in determining community functional structure (Albert *et al.*, 2010b; Jung *et al.*, 2014; Westerband *et al.*, 2021).

Community-weighted mean values based on trait measures taken on individuals from sampling units (e.g. **plots**) representative of different environmental conditions reflect the contribution of both the interspecific and the intraspecific variation to the functional structure, since not only do the different sampling units vary in species composition (and hence in the associated traits), but the same species can exhibit different trait values in different sampling units. Thus, the CWM calculated using trait values measured at the level of sampling units expresses the total (inter- and intraspecific) variation. By contrast, if trait values are averaged across all sampling units, and hence a given species always receives the same trait value in all sampling units, CWM values will express only variation in species composition/abundance (interspecific variation). By comparing CWM values calculated with these two procedures, it is possible to disentangle the inter- and intraspecific sources of variations in the trait structure of communities (intraspecific variation = total variation - interspecific variation) (Lepš *et al.*, 2011; Zelený *et al.*, 2024; Di Biase and Fattorini, 2025).

If interspecific variation is higher than intraspecific variation, this indicates that, when faced with environmental variations, communities tend to modify the taxonomic identity and abundance of species more than their trait plasticity. By contrast, a higher importance of intraspecific variation indicates that communities are able to cope with environmental variations through trait plasticity, and are hence more resistant to such variations (Albert *et al.*, 2011; Violle *et al.*, 2012; Kichenin *et al.*, 2013).

In general, it is expected that environmental factors that influence species ability to survive in a certain area (such as the presence of necessary resources and idoneous climatic conditions) influence interspecific trait variation (as only species with certain traits can be present in that area), whereas factors that operate at finer scale (local conditions and resource abundance and distribution) select individuals that can thrive under the prevailing environmental conditions (and hence select for functional traits at the intraspecific level) (Albert *et al.*, 2011; Westerband *et al.*, 2021). Similarly, factors operating over long periods are expected to be mostly responsible for interspecific variations, while changes at intraspecific level can be driven by factors operating on a smaller time scale (Auger and Shipley, 2013; Jung *et al.*, 2014). Intraspecific variation may therefore exert a primary role in driving local diversity. For example, variation between members of the same species may allow them to avoid competition (Clark *et al.*, 2007; Clark, 2010). Moreover, if there are conditions that minimise differences in the average fitness between species (for example, a strong pressure exerted by pathogens, predators on prey or herbivores on plants), even small niche differences can be sufficient to allow species coexistence with an increased intraspecific variability (Chesson, 2000; Adler *et al.*, 2007; Mayfield and Levine, 2010; Le Bagousse-Pinguet *et al.*, 2014).

Inter- and intraspecific variation can show positive or negative covariation (Di Biase and Fattorini, 2025). If inter- and intraspecific variations along gradients occur in the same direction (positive covariance), the community-level variation will be highly accentuated; if they occur in opposite directions (negative covariance), and both sources of variation hold similar importance, the overall variation may not exhibit a clear pattern (Lepš *et al.*, 2011). A positive covariation is expected when the same environmental factors that favour species with certain traits also select the same traits at the individual level, thus reinforcing trait selection in the same direction. For example, soil fertility may favour both grassland species that grow tall and tall individuals within the same species (Lepš *et al.*, 2011). Negative covariation arises when individuals with trait values strongly different from the predominant species-mean are locally favoured. This may happen, for example, when individuals with trait values that deviate most from the species-mean trait values might be advantaged under competition, or if the mechanisms that favour some trait values on larger scales can break down at local scales (Zheng *et al.*, 2022).

Despite the importance of disentangling the inter- and intraspecific sources of variation in functional traits, CWM-based studies on the relative importance of inter- and intraspecific trait variation are still relatively rare (Di Biase *et al.*, 2021a; Di Biase and Fattorini, 2025) and mostly restricted to plants. Research in other groups, such as lichens and animals, is still very limited (e.g. Gaudard *et al.*, 2019; Wong *et al.*, 2019; Hurtado *et al.*, 2020; Zheng *et al.*, 2023).

(a group, a population). Autecological studies on species in the field and laboratory expose species' traits, and these can then be used to examine niche space and predict the location of species' habitats. However, many trait studies do not necessarily apply data specific to the resource-based habitat. For instance, they will compare arthropod species for their use of plants in relation to size, phenology, plant part exploited, etc., but too often not the actual plant species used for all behaviours (for example, studies on butterflies tend to concentrate on larval host plants, ignoring nectar sources) and how resources are organised and exploited in the field for population persistence. Thus, most trait measures tend to deal with general attributes and their states, rather than with specific attributes and their states related to movements of individuals within the confines of their resource-based habitats (for example, butterfly activities are often related to the type of vegetation, but the actual plant species being used are ignored). Yet trait studies – for all their failings – are rapidly developing in all organisms and furthering our understanding of the evolutionary and biogeographical trajectories of species in a changing world. For example, a comprehensive review and detailed protocol for 29 broad traits (involving morphology, feeding, life history, physiology and behaviour) has been drawn up to describe the niche space and vulnerability of terrestrial invertebrates (Moretti *et al.*, 2017). In fact, trait studies in butterflies, emerging from autecological surveys, have a long history (e.g. UK butterflies; Dennis *et al.*, 2000, 2004; Shreeve *et al.*, 2001).

Recently, an attempt has been to produce a database for 542 European and Maghreb butterfly taxa (25 traits and 217 attributes; Middleton-Welling *et al.*, 2020), but this work is not as comprehensive as may at first be suggested by the paper title; 76% of traits are ecological variables which when grounded would describe part of the habitat space, but there are very few data on physiology and behaviour of early stages, a factor complex often ignored (for butterfly systems, see Dennis, 2020). As it is, data collected tend to be broad descriptors from the entire range of species and not spatially explicit to local populations. Therefore, the data broadly 'accumulated' from the literature are unrepresentative of the species concerned, with most studies using trait measures at a species level, even though many traits show intraspecific variation (see Box 5.2). Whereas trait studies in vascular plants and some animals are flourishing (see Chapter 2), for most organisms trait research is still in its infancy, and even for abundant and charismatic taxa there is a long way to go. As adaptive traits are genetically determined, with the rapid improvement and wide use of molecular techniques, a future key area of analysis would be to investigate how behavioural and life-history traits are linked to DNA markers.

Beginnings of the resource-based habitat: early case studies

The basic issue underlying both the **home range** and the **resource-based habitat** is the degree of fidelity that an organism has to a particular space. The fundamental questions are why particular species are limited to specific areas during a particular time, and why species occupy very different amounts of global space with the seasons, as does a single species at different times of day or season. Many of the advances in ecology are based on understanding differences between species in their fidelity to space on the planet; together, in any one location they make up biodiversity, again in continual change. The development of the resource-based habitat paradigm emerged from these issues. For almost a century it has been recognised that butterflies provide excellent model organisms for understanding fidelity to specific kinds of space, differences in both abiotic and biotic components, and ultimately movement between different kinds of space.

Initially, it was recognised by Ford (1945) that many amateur entomologists had an uncanny ability to recognise suitable habitats for a particular butterfly species without being able (or unwilling) to articulate their knowledge. Then, towards the end of the last century, the first author realised that it was very much a simple matter of breaking down the 'habitat' into components of the environment used by butterflies during their different stages of development (Dennis and Shreeve, 1996). Earlier, during autecological surveys of butterflies in Cheshire (UK) he had already noticed contrasts in the distributions of organisms within much the same area, related to variable resource distributions, local differences in the environment (local climate, soils, vegetation height, etc.) and changes in the environment over time, even during short time periods. It was obvious that the way to tackle the issues was to study common conspicuous butterflies close to home, rather than rare and exotic species. During the limited time he had available between 1980 and 1986, surveys were then carried out on the orange tip butterfly *Anthocharis cardamines*, the green-veined white *Pieris napi*, the small tortoiseshell *Aglais urticae* and the peacock *A. io* in the Bollin Valley, the wall brown *Lasiommata megera*, the meadow brown *Maniola jurtina* and the common blue *Polyommatus icarus* on Brereton Heath, and the large skipper *Ochlodes sylvanus* on Lindow Common (references cited in Dennis, 2010; see Box 5.1). The species were selected serendipitously to highlight particular issues in life history and behaviour. Only one technique used involved capturing butterfly specimens (**mark-recapture**), in which individual butterflies were given a unique number using a dot code. For the rest, the techniques used avoided disturbing animals. It was increasingly realised that it was important to disturb individual butterflies as little as possible, so as not to disrupt their behaviour. Most importantly, it was seen that an enormous amount can be learnt from just observations and experiments that allow animals to choose a path or resource. These observations provided the first empirical foundations for the **resource-based habitat** concept.

The link between lifetime track, home range and the resource-based habitat

It is one thing to draw a neat surround distinguishing the habitat bounds to a set of attributes in a Venn diagram of a population's resources (see Fig. 5.1a). It is quite another to determine the boundary around the resource-based habitat in the field. What looks to be a simple business of locating areas with suitable resources and conditions and drawing a boundary around them to determine the limits of a local population would at the very least require detailed knowledge of the resources and conditions for that species in that specific location based on autecological survey, or an inventory of resource requirements based on the accumulation of trait-based data. In fact, availability does not mean use. For example, the presence of areas of host plants is insufficient evidence that they are used by a species dependent on that host plant (e.g. nymphalids; Dennis, 2008b). Moreover, the environment is in continual flux. A vegetation map as much as a year adrift of a survey may well not provide an accurate base for a study.

A variety of techniques is available for determining **home ranges**, that is, the bounds of the resource-based habitat of individual animals, in the field. Typically, the minimum convex polygon (MCP) is still widely applied (Van Winkle, 1975; Worton, 1987; Powell, 2000) although more sophisticated methods have been proposed (Laver and Kelly, 2008). Resources, too, rarely occur as neat easily identifiable homogeneous zones (e.g. the wall brown butterfly *Lasiommata megera*; Dennis and Bramley, 1985; see Box 5.1, Fig. B5.1.3a). For most purposes the boundary around known resources and conditions may usefully be considered as adequate for defining limits for management of that population. Even so, the reality of a resource-based habitat is rather different and that would surely be expected from what is known of the speed of environmental changes in a landscape (see Chapter 3): what may be judged to be the habitat today may not be the habitat next year, not even the habitat from day to day or hour to hour (Dennis and Sparks, 2006). The immediate question is: just what does the broad resource-based habitat bound look like during any one cycle or season of development? One possibility is to adopt an area larger than required in any one season, as this will ensure that sufficient needs are available during harsh seasons (e.g. cold, drought, flood, etc.) (Verner, 1977). For example, this idea was tested and favoured by Riechert (1981) in his study of desert spiders.

To determine accurately the habitat boundary for a local butterfly population during any one season, it would be necessary to follow the **lifetime tracks** of a representative random sample of individuals from the moment they are deposited as eggs to the time they expire. Actually, even that might not be entirely adequate, since butterfly species could well be bivoltine or multivoltine (i.e. with two or more than two broods of offspring per year, respectively) and could use very different resources and conditions during each brood, in which case the procedure would have to be carried out throughout the different broods. Focusing for a moment on a sample of lifetime tracks, these provide the potential bounds of the habitat for that season inasmuch as the habitat is contained within a convex polygon – the collective home range for the population (breeding unit) – of the outer points of *all* the lifetime tracks recorded (see Chapter 6): the perimeter for each individual's lifetime track marked out on a map is its home range (see Box 5.1). This process may seem rather straightforward, even if laborious and tedious, to the extent that all that may be required is to have the lifetime tracks of the adults, as eggs are fixed and pupae are immobile, and larvae are also limited in movement – though some larvae can easily move 100 m during their life (Oates, 2020), these distances are much smaller compared to those of the adult butterflies, which can easily fly kilometres (Housharter, 2024). Currently it still remains beyond our ability to follow the movements of a large sample of adult butterflies during every moment of their lives, recording their halts, behaviour and resource use during passage. However, this imaginary zone – one day it might become reality – marked out on a map is an approximation of a population's home range (see Box 5.1). Simply stated: combining data for the home range of each individual (or more realistically the points of capture and recapture in daily surveys; see **mark-recapture** below) provides the home range of the local population, which in effect translates into the habitat bounds for that local population (see Box 5.1; see also Fig. 6.1). Thus, the lifetime tracks of individuals translate into the home range of a breeding unit, effectively marking the habitat boundary of that unit (Fig. 5.6).

A standard surrogate for identifying the habitat boundary for many mobile animals in the field is the application and mapping of **mark-recaptures** (see Appendix 1) linked to recording of their behaviour and resource use at each point of capture, invaluable for determining use of resources and conditions, though perhaps a poor substitute for real-time tracking of individuals.

It should be mentioned here that not all the members of a local population remain entirely within this home range or habitat space. If they did so, the regional population unit would probably become extinct; there is essential leakage, both in and out of the population, that ensures that the species survives within a region (see Chapter 7).

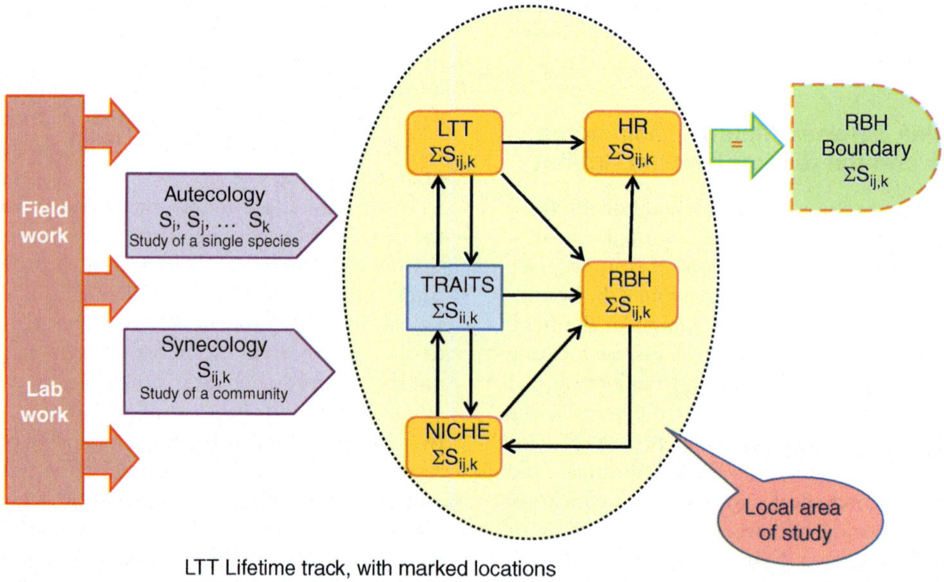

LTT Lifetime track, with marked locations
HR Home range
RBH Resource-based habitat

Fig. 5.6. Relationship between the resource-based habitat (RBH) and home range of a species within a locality. Estimates of the RBH and niche space of a species in a certain locality depend on intensive field and laboratory research, involving the study of both the species in question (autecology) and other species affecting the target species within the area (synecology). Determining the lifetime track (LTT) of individuals (path used, resource use, behaviour) and interactions in the niche space within which a group (individuals, pair group, troop, herd, swarm, etc.) of the species functions provides basic data on a species' ecological traits that in turn provide basic data on determining its RBH. The lifetime track of individuals (locations (loc_1, ... loc_m)) used at time t_1, ...t_m and the resource-based habitat components observed describe the home range (HR) of a species within an area, the HR equating with the boundary to the resource-based habitat, circumvented by locations used.

Even so, mark-recapture and electronic tracking (see Appendix 1) are particularly useful for illustrating how the habitat bound can be determined for practical purposes. For instance, some arbitrary limit to the set of movements can be used, such as the epidermis surrounding 95% of captures from the centre (spatial mean placement of captures) of the distribution. This makes sense if only because a fraction of individuals leaving the habitat are never captured more than once (they die or emigrate) and can be excluded from determining the home range.

All this is very well, and it seems perfectly reasonable to associate the home range with the bounds of a breeding habitat. However, it has already been made clear that many species do not stay in one place during even a single season of the year (see Chapter 3). To take a familiar example from birds, a flight (gulp) of swallows *Hirundo rustica* may have its breeding home range or habitat in the vicinity of a barn or church porch. The focal points of the individual home ranges for the few pairs of birds are the nests in the barn or porch. However, there is a non-breeding habitat during the winter and a transit zone in the autumn and spring. In effect, the full habitat is made up of three zones: (i) a breeding habitat (typically described by the home range of pairs forming the local 'colony' of birds); (ii) a transit zone (transit habitat) during which they fly south to southern Africa (and back) and during which they exploit various resources *en route*; and (iii) an overwintering habitat (a large non-breeding habitat) in which they accumulate resources (feeds) and use resting locations and roosts (Turner, 2010). Occasionally, there will be shifts in behaviour and part of the transit zone (e.g. Portugal) can become the overwintering habitat (van Nus and Neto, 2017). It becomes evident that the true habitat of a species may break up into such distinct units varying enormously in dimensions but, unlike the painted lady *Vanessa cardui*, swallows do not breed in their overwintering quarters. A population's (cohort's) resource-based habitat includes the entire physical zone of resources for individuals throughout their lives, the distinct parts of which can be distinguished by season and activity. To obtain a full understanding of the RBH for an entire species, it would be necessary to obtain data from a representative sample of such 'populations' throughout their global range. The implications for conserving such species, moving over such large distances, are formidable and are examined later (see Chapters 10 and 11).

Distinguishing territorial defence from home range

In Chapter 6 we will look at the factors that influence the size of home ranges. But before doing so, it is useful to distinguish a **territory** (a space [area or volume] 'defended' for a certain time; Maher and Lott, 1995; see Dennis, 2020) from a **home range** (which is the zone occupied by an individual or a group of individuals during their lifetime track). The whole of a home range, a part of it (a single resource or a set of resources with distinct spatial signatures) or no particular part at all may be defended by one, more than one or all of a group of individuals (a single individual, pair, troop, swarm, herd, mob, pod, school, shoal, local population, etc.). Examples of this variability are offered by primates, where group size ranges enormously, from a single individual, as in orangutans *Pongo pygmaeus*, to >800 members in mandrills *Mandrillus sphinx* (Van Schaik, 1999; Abernethy *et al.*, 2002). Among butterflies, the classical sites defended, usually by males, are those used for mate location (Dennis, 2010, 2020). Specific resource territories in butterflies rarely extend to the bounds of an individual's or local population's home range (or RBH) as the sites chosen are typically a limited set of the spaces available in a home range (or RBH), which present particular advantages (viz. oviposition sites, sunny and warm microclimates, prominent landscape features) to individuals defending them at key points in their life cycles.

Defence of a resource space in territoriality may be established against individuals of the same group (intragroup conflict) or against individuals of a separate group (intergroup conflict). It is important to reflect that an organism may be unable to defend its home range against conspecific competitors, but particularly against competing species. Defence may involve a complex suite of behaviours – particularly elaborate are songs in birds (Catchpole and Slater, 2008; Evans and Kleindorfer, 2016), primates such as indris and gibbons (De Gregorio *et al.*, 2021), certain amphibians (Wells, 2007; Rowley *et al.*, 2011), etc. – ranging from mutual avoidance and display to outright physical contact in which individuals may even be killed. The defence of the home range of a group most likely relates to conspecific intergroup competition, which may involve some interactions or the ritual (behavioural) recognition of spatial limits. But when the bounds are vigorously contested, this can escalate into physical interaction (e.g. domestic cats *Felis catus*; Leyhausen, 1979). Even among seemly peaceful animals like butterflies, defence can be very harsh, involving high-speed impacts (e.g. the large skipper butterfly *Ochlodes sylvanus*; Dennis and Williams, 1987). Both the shape and size of territories and home ranges (or RBHs) are known to be altered by competition among different pairs of individuals or neighbouring groups, classically among birds (Simms, 1990; e.g. the robin *Erithacus rubecula*; Lack, 1943). In fact, many factors may influence the determination of territory sizes and shapes. Classical models postulate that residents adjust their behaviour according to rates of intrusion and availability of food (optimality models), but other models assume that territory size and shape result from the balance of pressure exerted at boundaries and local rules of movement and interactions, or from the interactions between territory holders and new individuals attempting to acquire territories (Adams, 2001). Thus the shape, size and habitat contents of the home range may well be distorted by a variety of interactions as well as resources and conditions.

A territory may thus comprise all of the resources (and conditions), including of course mates (e.g. the robin *Erithacus rubecula*; Lack, 1941), within a home range (resource-based habitat) or any smaller number of them. Obviously, the greater the number of resources defended, the more likely it will be that territorial defence coincides with the home range and that defence occupies a longer period of time. An important consideration is that defence behaviours, such as direct combat, threats (e.g. displays) or simple avoidance (through chemical signals), may specifically involve: (i) a particular geographical space, or (ii) a resource not bound to a particular space; in the latter case it is the resource that is important, not the particular location within the home range. But even harem defence, as in the red deer (*Cervus elaphus*), may involve preferred areas within the home range (Carranza *et al.*, 1990). Moreover, defence of a space may be permanent and ongoing or be limited as to time and timing (i.e. mating season). Any resource that is defended is either (i) in short supply for the number of individuals in the group, or (ii) of special value to individuals which are induced to exact more of the resource than actually needed. Again, in the latter case, the example of the red deer is germane, stags attempting to maximise matings. Humans present the ultimate example of acquiring an excess of a resource, often resources unnecessary for survival, a behaviour referred to as greed and the foundation of capitalist societies.

Variation in the resource-based habitat

The key to understanding a species' habitat is to determine what individuals are doing: what resources they are using and conditions they are exploiting/experiencing during their lifespans. In this sense, exploitation of the resource-based habitat is closely bound to the interpretation of a niche. Each individual differs in resource use and tolerance of conditions. In many cases, however, it is difficult even to identify the bounds of an individual. For instance, fungi can range over large areas, a classic example being that of one that ranges over 15 ha, weighs over 100 tonnes and is considered to be over 1000 years old in the forests of Montana, USA (Rayner, 1997). A similar situation occurs in other modular organisms, including plants, sponges and corals, for which the concept of 'individual' is difficult to apply (see Appendix 1). Even so, measures on individuals provide the basis for variation in a group, however large it is, whether a pair of individuals or a large aggregation (e.g. a troop, pod, herd, swarm, culture or colony). In the following chapter, a closer look is taken at differences between individuals. Measurements on these individuals provide the basic data (statistics: means and variances) for local populations, which in turn provide summary statistics for regional populations and the species.

Inevitably, variation for any measure of resource use in a locality will be less than for a regional population and the species as a whole. The crucial point to remember is that although a **resource-based habitat** for an organism may have a limited variability at a local scale, at the species scale there may be considerable variation in resource use and tolerance of conditions. We have only to think of humans. If we measure the resource use and tolerance of climatic conditions for individuals in a limited population, say in the Welsh mountains (UK) where the maternal grandparents of the first author originated, this is but a fraction of the variability in resource use and tolerance of conditions for the human species over the planet. None of the family would find either the cold of the Arctic tundra or the moist heat of a tropical rainforest bearable. As humans are but animals – regardless of their pretensions of being something far superior – this is an entirely suitable example as it illustrates that difficulty of individual members of a species may have in adapting to unusual conditions.

At a broad scale, resource-based habitats vary at the intraspecific and interspecific level. At the intraspecific level, variations in resource use and experience of conditions become evident (i) over the space of traditional habitat units, and (ii) over time, from second to second, minute to minute, hour to hour and so on (see, for example, results obtained for butterflies – Dennis, 2004a; Dennis and Sparks, 2006).

Resource integration and habitat compaction

Resource integration and habitat compaction are likely to have a close relationship with species' resource breadth, the number of outlets used by species. This is evident from what was stated above concerning the impact of supplementary and complementary resources on resource integration. Resource breadth may be measured along the **specialism–generalism continuum** (Dapporto and Dennis, 2013); as different resources are involved, specialism (or generalism) is a complex multi-dimensional concept though it can be simplified. For example, for insects like butterflies, this is frequently done by focusing on the larval host plants and key nectar-producing flowers exploited (Dennis, 2010: 19).

The basic expectation is that **specialists** will tend to have all their resources contained within a single biotope patch and have a recognisable, compact habitat. **Generalists**, on the other hand, will tend to have their resources spread over numerous **biotopes**, and over different parts of the same biotope (e.g. damper versus wetter, lighter versus shadier) and to have indistinct habitat bounds. The resource-structural differences between such specialists and generalists are mirrored in terminology for population structures, traditionally referred to as 'closed' and 'open' population structures respectively (Pollard and Yates, 1993a). These patterns clearly exist in the most disparate organisms (Camus and Lima, 2002; Johnson, 2005; Dennis *et al.*, 2013; McIntosh *et al.*, 2022). For instance, some notable specialist butterflies recognise unfenced biotope margins as intrinsic boundaries to their habitats (e.g. Edith's checkerspot butterfly *Euphydryas editha*: Ehrlich, 1961; Fender's blue *Icaricia icarioides fenderi*: Schultz *et al.*, 2012), thus conforming to closed population models, whereas a number of extreme generalist butterflies opportunistically exploit resources as far apart as different continents (e.g. the painted lady *Vanessa cardui*; Stefanescu *et al.*, 2011, 2013), being therefore extreme examples of open populations (Pollard and Yates, 1993a). The contrast in habitat structure and population spatial dynamics between specialists and generalists is probably driven by the impact of resource distributions and environmental conditions on life histories and movement patterns. Widely scattered resources, especially if varied and abundant, will tend to select for more extensive movements that, in turn, successfully integrate disparate resources. On the other hand, scarce but closely intersecting resources will discourage (select against) dispersal and enhance specialisation and habitat compaction, including the use of visual and other cues for habitat bounds (see Chapter 6, Fig. 6.8).

But there are exceptions, where aspects of specialism may not necessarily lead to habitat compaction and where an open habitat structure may not be associated with host plant generalism. For example, specialism on host plant parts (i.e. young growth), rather than on a host species, may offer access to a wide range of plant species; in the green-veined white butterfly *Pieris napi*, this leads to open habitat structures (Dennis, 1985a), whereas in the silver-studded blue butterfly *Plebejus argus*, owing to ant associations, it does not (Thomas, 1985). In the case where an open habitat structure may not be associated with host plant generalism, abundance, temporal availability and ubiquity of a single host plant can substitute for a wide range of host plant species to ensure population integration (e.g. the small tortoiseshell *Aglais urticae* on *Urtica dioica*; Vandewoestijne *et al.*, 2004).

Thus, the relationship between specialism–generalism, resource integration and habitat compaction depends on resource variation (e.g. host plant and nectar species used), but also on abundance, ubiquity, finer aspects of resource exploitation (e.g. use of different parts of plants, such as buds, flowers, pods, leaves, shoots, roots), space-time availability, as well as other constraints (e.g. the presence of symbionts). The pattern of habitat, dispersed or compacted, depends on the selection for individual movement capacity, movements between different resources and resource items as well as movements between habitat patches, the latter driven by habitat lifespan (i.e. driven by vegetation succession). Although some species stick closely to confined, overlapping resource bounds, most species recognise and respond to resources rather than to the human abstraction of traditional habitat units. There is a distinction in the view of habitat and resource by human observer and target organism. To adapt the Rumsfeldian logic to the two-subject case: a species' resource is a known known – *we* know that they know one when they sense it – compared to which a habitat is a known unknown, that is, *we* know it exists but it is a moot point whether our target species has that insight: some evidently do, but do all?

6 Bounds of the Resource-based Habitat: Dimensions, Movements and Interactions

Abstract

Drawing bounds around individual lifetime tracks gives the bounds of the realised resource-based habitat (which, in the case of animals, represents their home range). Organisms use a variety of senses and cues to find appropriate resources and conditions. As individuals vary in their capacities and preferences (and hence in their fitness), extensive sampling of resource use and conditions experienced over space and time by different individuals is essential to reconstruct the potential resource-based habitat of a population/species. Many factors influence the size of the resource-based habitat dimensions of different organisms, such as their mass, behaviour, group size and life history. Organisms occupy different positions along the specialism–generalism gradient, as a function of their movement capacity and variety of resources used, which leads to different evolutionary processes.

Although there is an inevitable overlap between the traditional habitat and the resource-based habitat concepts, the latter raises questions about how different organisms' habitats are actually measured in the field. Furthermore, it raises issues specific to different organisms, especially related to their mass, group size (whether single, in pairs, extended families, herds, swarms, local populations, etc.), interactions and mobility in a habitat space. The chapter starts by looking more closely at the status of an individual organism and the *modus operandi* of determining the habitat for groups of conspecific individuals. Then, it considers the variety of factors affecting size of the resource-based habitat and finally the interactions of species in overlapping resource-based habitats.

Initially, it should prove helpful to summarise the links between the key terms associated with the **resource-based habitat** (RBH). Following the **lifetime track** (LTT) of individuals allows an enormous amount of data to be built on intra- and interspecific relationships, resource use and exploitation of conditions. This is no easy matter, especially in species-rich ecosystems (e.g. tropical regions). During such tracking of a target species, a wider perspective of the community in which the target species occurs provides invaluable information on the **niche** space of the target species and those species with which it interacts. Different areas or regions differ in community structures, and therefore in the nature of the lifetime track and specifics of the resource-based habitat for any organism. Drawing bounds around the lifetime track(s) of an individual, or a set of individuals, in essence gives the bounds of the realised resource-based habitat (which, in the case of animals, represents their **home range**, HR) (Fig. 6.1); the data on how species react to stimuli, as well as to resources and conditions used or avoided, provide the fundamentals for understanding the resource-based habitat. Gradually such studies, carried out over wider regions, ensure the build-up of a mass of data on species traits, which in turn provides the basis for determining the wider resource-based habitat of the species.

A closer look at the individual: the semiosphere

Although it has long been evident from discoveries in genetic research that each individual is unique in form, biochemistry, physiology and behaviour, there is still a tendency to treat individuals in local populations as replicates of one another in assessing their resource use and tolerance of conditions in the field, as if one individual is as good as another in exploiting resources and in the ability to withstand changing conditions in habitats. Even in the first half of the twentieth century, comparisons

DOI: 10.1079/9781800627505.0006

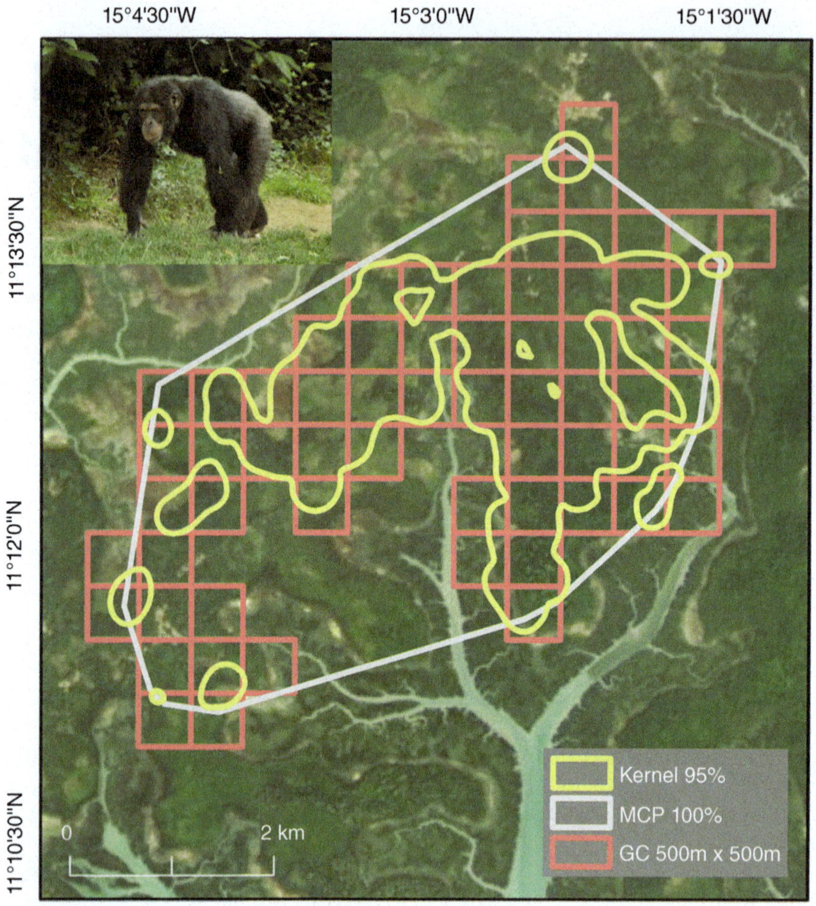

Fig. 6.1. Example of home range measures in a chimpanzee population forest–farm mosaic at Madina in Cantanhez National Park, Guinea-Bissau. Grey line: MCP, minimum convex polygon; red grid: grid cell 500 m × 500 m square occupancy; yellow line: Kernel 95%. Kernel analysis determines the probability of finding a particular individual or group at a particular location by systematically collecting GPS data on observed locations; it uses these data to calculate how intensively the group/individual exploits each area over time and illustrates probability applying isopleths (areas distinguishing spatial units on the map with a given probability of finding individuals), viz. occupation time (Worton, 1995). (From Vieira *et al.*, 2019, used under licence CC BY 4.0. Inset photo by Hans Hillewaert: https://commons.wikimedia.org/wiki/File:Pan_troglodytes_(male).jpg, used under licence CC BY SA 3.0.)

of taxa and populations for physical characteristics clearly recognised the importance of variability among individuals for understanding evolutionary and ecological processes (e.g. Robson and Richards, 1936) and, by the 1970s, the need to use random samples to compare populations from different regions was widely acknowledged (e.g. Constandse-Westermann, 1972 for humans; Dennis, 1977 for butterflies). Yet more recent comparisons of populations tend to use very small samples of individuals on the basis that comparison of a vast number of 'genes' substitutes for number of individuals – though in fact such gene samples fail to represent the entire genome.

Understanding of variability within individuals started first with that of individual humans. A classic case is that of the ancient Egyptians who believed that the human personality had many facets continuing after death, a concept that probably developed early in the Old Kingdom (David, 2003). Thus, a range of terms can be found in ancient Egyptian funerary texts for indicating different parts of the soul (*Khet*, physical body; *Sah*, spiritual body; *Sekhem*, power or form; *Ren*, name or identity). We now have a deep understanding of just how humans vary along many dimensions, including self-appraisal of gender.

However, there is still a tendency to treat any one non-human organism as a substitute for any other, and the description of behaviour and resource use in single instances as representative of a population or species. In this view, individuals are regarded as replicates lacking specific identities in their behaviours. This happens not only for organisms that lack nervous systems (such as plants) but also for those that are considered to have 'simple' nervous systems, even including arthropods, despite some of them being well known for the presence of very sophisticated behaviours (e.g. social insects). This may be surprising, since it is well known at least from 50 years ago that 'simple' nervous systems are actually quite complex (e.g. Usherwood and Newth, 1975), and even animals in evolutionary terms far from us, such as cephalopods, may have extraordinary mental capabilities (Godfrey-Smith, 2016, 2020). Part of the problem emerges with humans substituting their own perceptions for those of the organisms they study, but other organisms 'see' the world from very different perspectives than those of human observers (see, for instance, examples discussed in Scanes, 2018). It has even been proposed to consider plants as intelligent organisms, as they are able to perceive the environment around them and to respond to the inputs they receive in an integrated fashion (a point of view known as 'plant neurobiology' and which is strongly debated, especially when the concept of consciousness in plants is evoked; see, for example, Trewavas *et al.*, 2020 versus Mallat *et al.*, 2021).

However, for the most varied life forms, the concept of the individual as a unique entity in every aspect is now gradually being accepted and it is crucial for the **resource based habitat** (**RBH**). The critical concept here is Umwelt (plural Umwelten; literally environment or surroundings) as related to an individual's lifetime track. The concept of Umwelt was formulated by Uexküll (1957; Sagan, 2010) and refers to an organism's unique perception of its surroundings; this is described as a 'self-centred world' from a self-centred viewpoint of a particular individual. Organisms are said to have different Umwelten even if they share the same environment as a pair or members of a troop. For anyone studying animals such as insects, birds or mammals, this makes perfect sense. No two individuals followed have the same movement paths defining the adult LTT, use identical resources or experience identical conditions. This concept has been further advanced by the more recent distinctly anthropomorphic treatment of individual butterflies as having 'personalities' (Kaiser *et al.*, 2018, 2019). This may seem far-fetched but, for example, among butterflies, individuals clearly have different levels of aggression during mate location and demand for nectar resources, the latter depending on the balance of larval and adult investments (Dennis, 2020). Similar considerations might appear equally obvious for other animals, such as birds or mammals, but, *mutatis mutandis*, might be extended to the most disparate organisms. In effect, an organism establishes and recreates its own Umwelt when it interacts with different components of the world. Therefore, it operates within a dual context: Umgebung (an Umwelt as perceived by another observer) and Innenwelt (the mapping of the self to the world of objects) (Cobley, 2010). Each habitat resource or habitat condition, including other organisms (e.g. conspecifics, enemies), forms a component of Umwelt and interactions between different Umwelten are referred to as the **semiosphere** (Lotman, 1984; see Cobley, 2010). Umwelt is shaped by and shapes an organism's LTT and its notion of resources and thus the delineation of its RBH.

Cues and order in habitat selection

It is one thing to have resources distributed in space-time; it is quite another to be able to locate them. The complex sets of activities that allow organisms to find appropriate conditions represent a process known as **habitat selection**. In animal ecology, habitat selection can therefore be defined as the hierarchical process of innate and learned decisions taken by an animal about where it should be across space and time to survive, but all organisms adopt strategies to find places with suitable conditions and appropriate resources, even if not through active decisions. Habitat selection should not be confounded with **habitat preference** and **habitat use**. Habitat preference, i.e. the disproportional use of some resources over others, is a consequence of the habitat selection process, while habitat use is the way an organism uses or consumes resources (see Mathewson and Morrison, 2015). Of course, different species might have similar habitat preferences but different habitat uses, which allows them to coexist despite the **niche overlap**.

To locate the resources they use, all organisms have a range of senses and cues (these are obvious in most animals, but virtually all organisms have the ability to sense the world around them, through a variety of receptors and sensors). At the same time, they also need to avoid dangers within a habitat space; failure to do so results in higher mortality and lower reproductive success. To avoid dangers, organisms have developed not only a variety of cues and heightened senses but also a range of adaptations (e.g. physical, biochemical, physiological and behavioural) that have evolved through natural selection (Ruxton *et al.*, 2004). Adaptations to potential predators are highly varied, including **crypsis, aposematism** and **mimicry** (as well illustrated, for example, in butterflies and moths; Dennis, 2020). Senses typically involve sight, sound, touch and taste, but also include numerous other stimuli (e.g. hunger, thirst); moreover, some organisms have developed special senses, such as the sensitivity to the magnetic field of migrating birds (Lohmann, 2010; Mouritsen, 2013) and the echolocation (sonar) used by several mammals (including toothed whales, many bat species and, in simpler forms, shrews) and some cave-dwelling birds (cave swiftlets and oilbirds) (Fenton, 1984; Jones, 2005, 2010; Jones and Teeling, 2006; Siemers *et al.*, 2009; Brinkløv *et al.*, 2013; Racicot *et al.*, 2019; He *et al.*, 2021). Senses also vary enormously among organisms for type, magnitude and range. Thus, for instance, some animals are especially sensitive to ultraviolet stimuli (e.g. insects and many vertebrates, including several birds and some amphibians, reptiles and mammals; Honkavaara *et al.*, 2002; Rajchard, 2009; Barghini and Souza de Medeiros, 2012), others to infra-red wavelengths (e.g. the common vampire bat *Desmodus rotundus*, some insects and certain snakes; Campbell *et al.*, 2002; Greppi *et al.*, 2020) and to linearly polarised light (some vertebrates, many insects and certain marine invertebrates; Heinloth *et al.*, 2018).

Cues for seeking resources and averting dangers are fundamentally of two types: (i) abiotic (e.g. light environment, ambient temperatures, wind speed and direction, humidity, precipitation, substrate contrasts, depth in lakes and sea, and altitude up mountain sides and in the atmosphere); and (ii) biotic (e.g. food resources; friends, enemies, mates and competitors). Both these types of cues may be determined singly by any one individual operating alone, or communally, through responses by conspecifics or other species, by a 'domino' reaction often triggered by less dominant individuals peripheral to the group or cluster of individuals (e.g. in the wood pigeon *Columba palumbus*, inexperienced subordinate individuals are forced to the fringe of the flock to act as lookouts for danger; the dominant birds in the centre are then able to feed undisturbed; Murton *et al.*, 1971). The cues trigger responses to enemies or friends typically through behavioural signals and to particular resources or conditions through sensing.

Senses and cues are typical entry points to the use of resources in appropriate conditions during passage along a LTT. For example, in studies of butterflies, the cues are distinctive for adult feeding (nectar flowers), egg laying (host plants), mate location (hot spots, peaks and crossroads), roosting (cover) and other behaviours (Dennis, 2010). However, the path of some species may appear to be random within a traditional habitat space. This is evident in pierid butterflies and the red admiral *Vanessa atalanta* where non-host plants are seen to be inspected by female butterflies (Dennis, 2020) during the course of their LTT, the females landing and scratching the leaf surface to release chemicals. In such situations, it would appear that host plant resources are in effect located randomly, but more often than not, we do not know enough about the sensory cues that a species may use within a habitat space. Even so, it is possible to calculate the probability that an individual will encounter an uncued specific resource (Box 6.1).

Computing components of habitat variability

Resource-based habitat variability has two fundamental kinds of measures: (i) basic summary statistics for exploitation of resources (or avoidance of enemies) and adaptation to conditions, and (ii) spatial statistics and mapping of the lifetime tracks and the home range of individuals that delineate the resource-based habitat. During life, each organism uses various resources and avoids specific conditions; this is particularly evident in mobile animals that can actively avoid unsuitable conditions, but applies to all species, although strategies adopted vary among groups (for example, plants cannot move from unsuitable places, but may avoid germinating during unsuitable ecological conditions

Box 6.1. Locating the target

Search theory has its origins in the work done by Bernard Koopman and his colleagues to aid the US Navy in optimising the chances of locating submarines (concealed enemy 'targets') during the Second World War (Koopman, 1980). This has led to the law of random search, a model which assumes that: (i) the target can occur anywhere in an area searched and need not be static; (ii) the searcher's path is random within an area *A* and can go anywhere; (iii) along any portion of the path the searcher has a lateral capacity for detection of *w*/2 (where *w* is width of perception). Thus the probability of encountering a target (e.g. a resource) is:

$$P = 1 - e^{-(wL)/A} \qquad \text{(B6.1.1)}$$

where *L* (metres) is the path length along which the searcher covers, *w* (metres) the effective width over which the target can be identified and *A* (metres^2) the area searched (Fig. B6.1.1). Clearly, the probability of a successful encounter increases as the path length and path width increase, and decreases as the area to be searched increases (Tilley and Dennis, 2017).

If (as is usually the case with a butterfly's host plants, for instance) a number of targets exist (*n*) within the area, the law of random search can be modified to:

$$P = 1 - e^{-(wL)n/A} \qquad \text{(B6.1.2)}$$

This approach has been successfully applied by Tilley (1995, 2016) to the marsh fritillary butterfly *Euphydryas aurinia*.

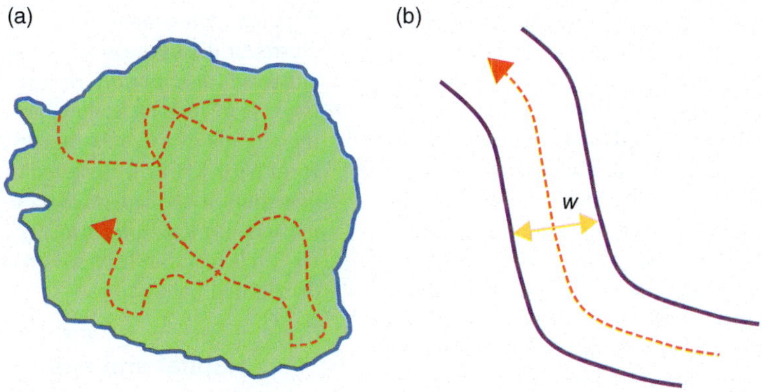

Fig. B6.1.1. Butterfly searching for a resource in a habitat. **(a)** Random search path taken by a butterfly in a habitat patch. **(b)** Path width (*w*) is the width of the path over which a target can be recognised. (From Dennis and Hardy, 2018, redrawn.)

through seed dormancy). Resource exploitation and adaptation to conditions can be summarised as ecological traits each described using simple descriptive statistics (e.g. nominal: frequencies, percentages, mode; ordinal: median, quartiles; interval-ratio: mean, standard deviation, variance, coefficient of variation; Dytham, 2003) but also by recording the seriation of behaviours and resource uses (see seriation in behaviour in a male and female of the orange tip butterfly *Anthocharis cardamines* illustrated in Box 5.1 and Fig. B5.1.2). Spatial distributions and changes in spatial distributions also have their summary statistics (i.e. spatial central tendency, spatial dispersion, spatial homogeneity; Fortin *et al.*, 2002; Fortin and Dale, 2005). Once these measures have been made on individuals, they can be further summarised for collections of them, be they pairs, troops, herds, swarms or local populations. In turn, local population data can be further summarised across regions to give regional resource-based habitat summaries and at the ultimate scale for the species.

The crucial decision is determining what is or is not an individual. This may seem obvious to an entomologist, a mammalogist or an ornithologist, but may be far from obvious for a botanist or a mycologist (see Appendix 1). Botanists think of plants as **modular organisms** (Chapman and Reiss, 1999) and the same concept applies to most fungi and to animals like corals and sponges; but in most cases, the individual is the only unit of which an ecologist can be certain, all further amalgamations of individuals (groups) being both variable and temporary.

For some organisms, it is possible to determine their lifetime track in sufficient numbers to provide a full documentation of ecological traits and their home range. For example, this is commonly done with mammals (especially larger ones) through radiotelemetry, but radio tracking devices also exist for small animals such as insects (Kissling *et al.*, 2014). Locating individuals is facilitated by a number of other possible techniques, such as fin, leg or wing flags and stable isotopes (Jessop *et al.*, 1998; Kenward, 2001; Fuller *et al.*, 2005; Silvy *et al.*, 2005). Even individual butterflies, however delicate they are, can be followed by applying marker pen codes on the wings (Ehrlich and Davidson, 1960; see also Appendix 1 about how animals can be marked). As noted above, the **lifetime track** is the path taken from birth to death of each and every individual, along which their survival and reproductive success are recorded. Understanding a species' **resource based habitat** (**RBH**) and **niche** depends on intricate measures being made on sufficient **traits** and samples of individuals throughout the stages of development over time (minutes, hours, days, seasons, years) and space (sites, locations, regions, even in distinct **biomes**). When individuals exist in social groups (pairs, troops, flocks, local populations) then attention needs to be paid to the space-time autocorrelation in resource use and conditions experienced by individuals, and sampling adjusted accordingly to obtain independent replicates for a thorough determination of the resource-based habitat. The fact that individuals in herds move as one (e.g. the wildebeest *Connochaetes taurinus*) means that fewer of them actually need to be individually radio tracked, but each demonstrates unique behaviour; it is such differences that make them more or less vulnerable to resource uptake and enemies.

Variation in behaviour depends much on the aggregation of individuals in groups. Animals, much as resources and conditions of existence, may occur at any one time in quite different patterns of vertical stratification (e.g. layering in forests or water columns) and horizontal dispersion, ranging from regular distributions through random arrays to clumped distributions (see Box 4.2 in Chapter 4; see also Box 3.1 in Dennis, 2010). These patterns have an effect on behaviour, resource use, vulnerability to conditions and enemies. A wide array of measures is available to determine the dispersion of individuals (see Ludwig and Reynolds, 1988; basic principles are also illustrated in Chapter 4, Box 4.2, of the present book). The crucial point here is that the dispersion of a resource can affect the dispersion pattern of an organism, which in turn may well influence the behaviour and resource use of the individuals so affected (Table 6.1). For example, although all combinations illustrated in Table 6.1 are possible, a clustered distribution of resources will determine a clustered distribution of the individuals that use them (e.g. insects feeding on clustered plants), but individuals of animals that live in groups (e.g. that form herds) have a clustered distribution even if the distribution of their resources is random. Scale also matters; for example, insect eggs may be laid in clusters on leaves, but leaves with eggs may be randomly distributed on a plant.

Habitat and space-time scales

One issue is clear. A single sampling event of even a representative collection of individuals in a local population is insufficient to be certain of having captured the resource-based habitat parameters of a group (e.g. a population). Resources and conditions, as detailed above (Chapter 3), change over time scales, from less than seconds to millennia; thus, individual organisms will be redistributed over space-time as changes occur (Dennis and Sparks, 2006). Crucially, sampling of data on resources and conditions for individuals needs to consider issues of time (time of day, changing conditions such as

Table 6.1. Combination of dispersion patterns for an organism and its resources within a habitat space.

		Individuals		
	Pattern	Uniform (U)	Random (R)	Clumped (C)
Items or patches of a resource	Uniform (U)	UU	UR	UC
	Random (R)	RU	RR	RC
	Clumped (C)	CU	CR	CC

weather, seasons of the year, age and stage of each target environment, etc.; e.g. age of trees for **saproxylic** fungi and beetles, age of carcasses for flies and metamorphic stages of lepidopterans for a parasite) and space (different topologies, topographies and elevations of seemingly identical landscapes, different traditional habitat types, different substrates within seemingly identical landscapes, different landmasses and climatic regions, etc.).

During sampling for space-time variability, a number of issues emerge that can easily be overlooked. The first is an assumption that conditions in an area have remained constant. This may seem unlikely to happen, but – as an exemplary anecdote – we can cite that one of us (R.L.H. Dennis) has been witness to just this kind of thinking, the assumption by a student that a vegetation map of the Great Orme (UK) taken several years beforehand formed an accurate picture of vegetation conditions later on and could be used directly as a template of current patchworks. Conditions in any habitat space are constantly changing, even if to the human eye nothing may appear to have altered. Just changing sward height may alter the environment for competitors, thus the **realised resource-based habitat** for an individual at a particular location. Since the contents of traditional habitat spaces remain largely unknown, it is imperative that sampling extends over **landscapes** that may be classed as identical to one another. Second, there is a danger of assuming that residual environments for a species are optimal environments. For example, the high brown fritillary butterfly *Fabriciana adippe* can be found in bracken-covered areas in northern England (UK), but these are not optimal places for this species (viz. zones of coppiced woodland such as Silverdale, Lancashire, UK, but which only lately have been subject to suitable management; Dennis, personal observation). We know that a variety of insects are able to survive in urban green spaces (Fattorini *et al.*, 2018c; Fattorini, 2019), but these are in general non-optimal habitats and, in some cases, they can even act as ecological traps; for example, they can attract butterflies but might not have adequate resources to support sufficiently large numbers of individuals to contribute to the local abundance and persistence of butterfly populations (Levy and Connor, 2004). In fact, areas of distribution where they no longer occur may actually have been superior in resource-based habitat status but destroyed by human exploitation with the application of modern agricultural methods (Shreeve *et al.*, 1996a). Where a species is spread over a wide range of environments (e.g. substrates, climates), the array of the realised resource-based habitats for a local population provides an insight into what the **potential resource-based habitat** might be like. As such, it may then be feasible to determine the core optimum resource-based habitat.

By extensive sampling of resource use and conditions experienced over space and time, an increasingly sound picture is accumulated of the resource-based habitat of a species, its local populations and individual variability. It becomes evident that species make mistakes in resource use (e.g. a herbivorous insect ovipositing on an unsuitable plant, a parasite attacking an animal that is not among its hosts, a pollinator visiting a non-suitable flower) and conditions exploited. Time is also a crucial variable; an insect may lay its eggs on what seem to be suitable egg-laying sites in a particular moment, but later are not suitable for the larvae, which fail to develop. Drawing on results for survival and reproductive success for an organism in a multidimensional space of resources and conditions is also of paramount practical importance, as it presents the fundamental data for assessing conservation priorities and informing conservation actions (Mace and Collar, 1995; Gärdenfors *et al.*, 2001).

Factors in resource-based habitat dimensions

Many factors influence the size of resource-based habitat dimensions. We have already looked at a number, fundamentally at the arrangement of components within the basic medium of the habitat (air, water, land, etc.), therefore factors influencing the landscape distribution of resources and conditions affecting any organism (Chapter 5). But there are others, specific to the basic biology of each organism: (i) its size and sex (if present); (ii) the number of individuals occupying the habitat unit, that is the group size of the organism; (iii) the behaviour (e.g. territorial defence mechanisms and mating frequency (polyandry, polygyny)) and mobility of the organism (e.g. speed of movement on land or in water, flight ability, particularly the development of wings in animals such as birds; Sheard *et al.*, 2020); (iv) the life history of the organism, especially the degree of fidelity to a specific area (however large the domain of suitable conditions and resources);

and (v) its trophic position (food type) linked to familiarity with landscape features.

Quite obviously, the larger the organism (mass) within a taxon, the more resources are required to maintain it (e.g. compare a lion with a household cat). The relationship is well known; for mammals, McNab (1963) suggested that the area with sufficient resources (resource-based habitat, i.e. **home range**, HR) for an individual is proportional to its mass (M) to the ¾ power:

$$HR \propto M^{0.75} \quad (6.1)$$

This has been found to be the case with some other organisms (e.g. immature ant lions *Myrmeleon* sp.; Swenson *et al.*, 2007) but, in general, mobile animals require a higher power (Schowalter, 2016). Examples of mass effects on home range abound, but perhaps the most obvious are for primates (Milton and May, 1976). Therefore, the larger the organism, the larger the space it is likely to occupy (Fig. 6.2). But the size of home range for different taxonomic groups (e.g. orders or families) of organisms can vary enormously even for animals of similar body mass. For example, butterflies much the same size may occupy varied extents of potential adult home ranges depending on their selection of larval host plants or adult nectar sources (Dennis, 2010). Differences in body mass and home range often relate to the type of food the organism consumes, whether fruit, flowers, nectar, leaves, bark, etc., if only because these are unequally distributed over time and in the landscape. For example, in ungulates, the spatial scale of movements, thus home range size, depends primarily on individual

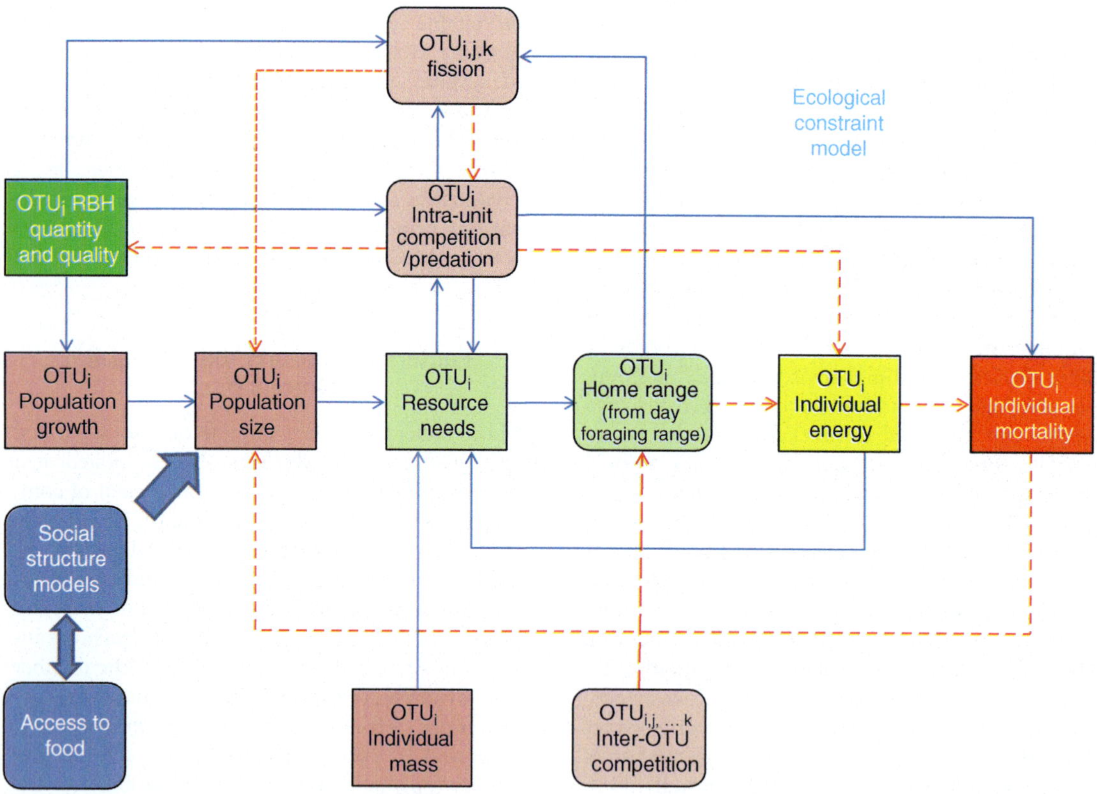

Fig. 6.2. Some basic ecological constraints affecting the size of groups (number) of animals exploiting a resource-based habitat and the home range. Prominent factors affecting group size are: (i) availability (quality and quantity) of resources; (ii) competition from conspecifics and non-conspecifics; (iii) dimensions of the available home range such as an island of resources (thus the size of the resource-based habitat); (iv) individual mass of the species, and therefore food requirements, part of which includes the ability to defend the troop. Blue lines, positive links; red pecked lines, negative links. OTU, operational taxonomic unit.

body mass, group size and the structure of the landscape, with important differences between grazers and browsers, because of difference in food quality, abundance and distribution. In general, grass is a low-quality forage but is abundant over large areas in open landscapes, whereas browsed vegetation is easier to digest but is more patchily distributed, and more abundant in closed habitats (Clauss *et al.*, 2013; Ofstad *et al.*, 2016). In general, for mammals, it is well known that the relationship between body mass and home range differs according to the trophic role: carnivores have larger home ranges than herbivores of the same size (Harestad and Bunnel, 1979; Kelt and Van Vuren, 2001).

Group size, as mentioned earlier, ranges from a single individual to many hundreds, or even billions, as in the case of locusts (see below). A fundamental question affecting these issues is why some organisms occur in groups (i.e. greater in number than a pair and/or their offspring) and others do not. One frequently examined explanation is the **resource dispersion hypothesis (RDH)**: if key resources are heterogeneous in space or time, animals might form groups that share the same space at little cost to each other; although co-operation may be absent, sociality is favoured (Macdonald, 1983b; Johnson *et al.*, 2002; Macdonald and Johnson, 2015). This hypothesis – applicable to humans as well as other animals (including other mammals, birds, fish and crustaceans; Macdonald and Johnson, 2015) – is not independent of others, such as antipredator defence, co-ordinated prey capture and prey seeking (Fig. 6.3). Exploring the factors underlying group size is a valuable starting point to

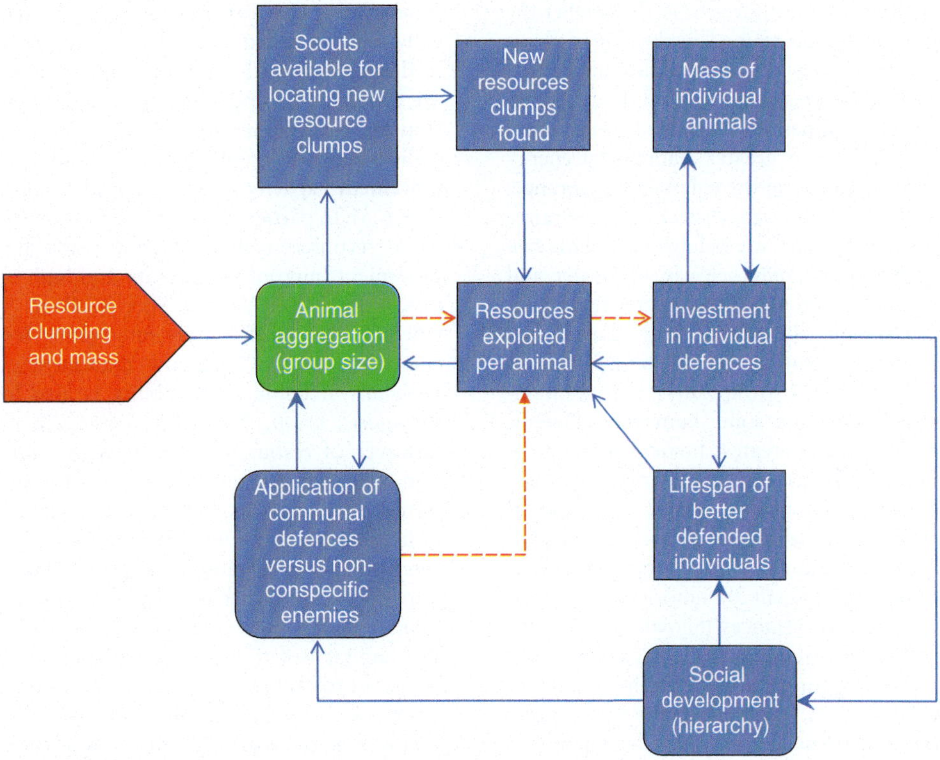

Fig. 6.3. Basic features of the resource dispersion hypothesis. Clumping of resources, of whatever kind, may lead to the aggregation of individuals and the evolution of social behaviours, particularly when resources are contested by conspecifics. Defence of individual items ('bites' of a resource) results in a social hierarchy through dominant males and females. Social advancement occurs when individuals are confronted by other species also gaining access to the same clumped resource(s) through development of defences (see Johnson *et al.*, 2002 for the full development of the resource dispersion hypothesis model). Aggregations also have the advantage of early warning of enemies and of locating additional resources through encouraging 'scout' development in individuals placed lower in the social hierarchy. Blue lines, positive links; red pecked lines, negative links.

an understanding of home range and the resource-based habitat. A crucial feature of the RDH is that clumped resources may be temporary and mobile (e.g. carcasses exploited by hyaenas; Newsome *et al.*, 2013) as well as static but patchy (e.g. favoured fruit trees for frugivorous animals). An obvious relationship is that patch richness sets a maximum ceiling to group size, as in lion (*Panthera leo*) prides (e.g. the number of herbivores as potential prey using waterholes in the Serengeti), which neatly illustrates that landscape ecology is a major driver of ranging behaviour (home range) and limits group sizes through exploitation of resources along lifetime tracks (Valeix *et al.*, 2012).

Group size is an obvious factor affecting lifetime track, home range and the resource-based habitat, as the more mouths to be fed, the larger and/or richer the area required for the group. The inverse is that the weaker the resource base, the smaller the group that can be maintained in the home range. Compensation for low resource quality and quantity by extending the area covered by a group (assuming no competition) requires even more nutrition to survive as greater amounts of energy are expended in covering the enlarged home range.

Among the residual factors affecting home range size, the most prominent are behaviour and life history. **Mobility** is crucial as, clearly, the faster and further an animal can move, the greater the potential for expanding the home range dimensions. Intercontinental migration – as observed, for example, in many birds (Newton, 2007), marine mammals (Boyd, 2004) and some butterflies (Dennis, 2020) – provides the greatest home range dimensions inasmuch as natural boundaries to movements of animals become less significant. Dispersal over extremely long distances is a common phenomenon in many animal species: locust swarms of even billions of individuals are notorious for travelling over enormous distances (Showler, 2008), but even flightless animals have the capacity for long-distance dispersal (for instance, many linyphiid and tetragnathid spiders, which use ballooning to disperse over large areas and may also be capable of using their legs and silken strands to sail across water surfaces; Goodacre, 2020).

Motility and mobility in the resource-based habitat

A study of the biodiversity of any given area reveals an extensive collection of organisms with a wide variety of motilities and mobilities. Two basic elements hold together the resource-based habitat for each organism. The first is the degree to which resources and suitable conditions overlap for each organism. The second is the extent to which stages in the life cycle of any organism can link up those resources and conditions. For any stage of development, the link up for resources and adequate conditions depends on their proximity and the mobility of the organism.

These observations suggest that there will be a significant relationship between a species' position on the **specialism–generalism** resource gradient and its **movement capacity**; also, that species with compact habitats should occur towards one extreme (specialism/weak movement capacity) of the relationship, and those with indistinct (highly disjoint) habitat structures towards the other (generalism/strong movement capacity) (Fig. 6.4). This is basically what is found to be the case for British butterflies (including long-distance migrants): species are distributed along a line describing the relationship between **colonisation ability** (describing specialism/generalism and resource availability) and **movement capacity** (Dapporto and Dennis, 2013) (Fig. 6.5). In particular, species with widely disseminated resources (diffuse habitats) have both high movement (migration) capacity and high colonisation ability (e.g. *Vanessa* species, *Pieris* species), whereas those with closely intersecting (compact) resources tend to have low movement capacity, especially if colonisation ability is also low (e.g. *Phengaris arion*, *Lysandra bellargus*); a faster turnover of resources through vegetation succession requires greater ability for movement, as shown by species able to use relatively early successional stages (e.g. *Thymelicus sylvestris*, *Melanargia galathea*). More generally, there are clear relationships between dispersal capabilities and habitat characteristics; habitat stability, habitat uniformity and habitat isolation reduce the need for dispersing, as neatly illustrated by the evolution of flightlessness in these circumstances (Box 6.2).

However, the issue is more involved than simply considering the landscape components for just one target organism under study. A wider focus on the diversity of organisms in a patch is crucial for determining the suitability of the patch for the target organism. A first issue is the extent to which resources and conditions for different organisms overlap, particularly those that interact with one another. Then, mobility has another profound

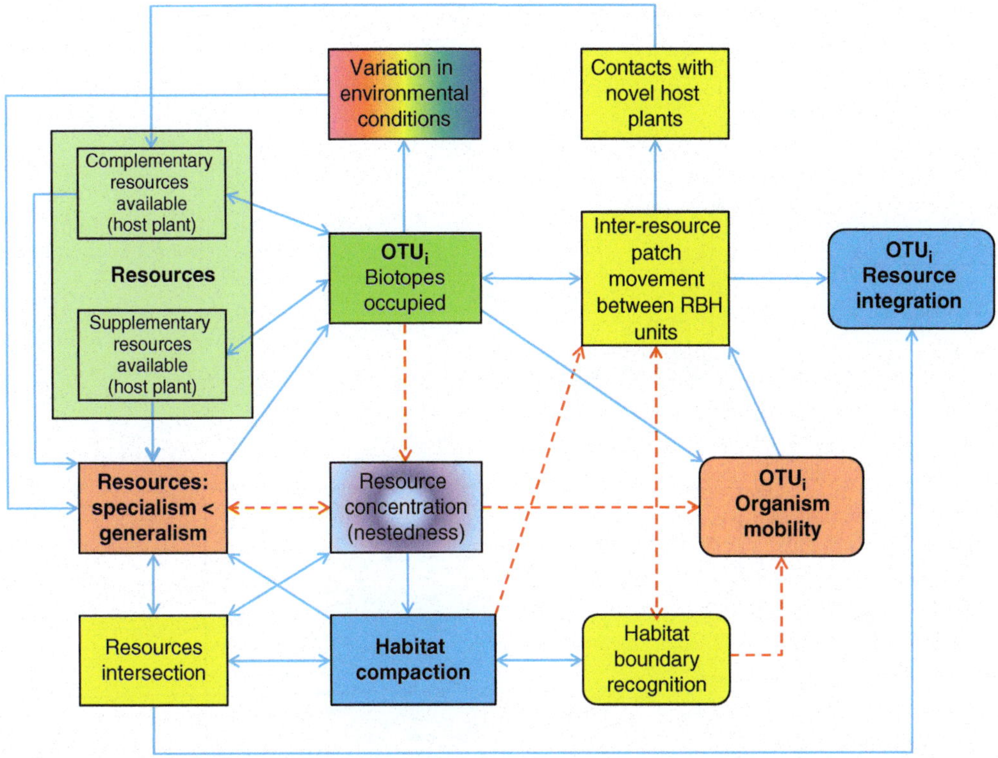

Fig. 6.4. Basic factors leading to resource integration and compaction. Resource integration depends as much on interpatch movements (mobility) as on resource intersection, whereas resource (habitat) compaction is primarily dependent on habitat structuring, thus resource intersection and nestedness. Resource specialism and generalism, via resource supplementarity and complementarity, are key to driving these two outcomes. Blue arrows, positive relationships; red pecked arrows, negative relationships. OTU, operational taxonomic unit; RBH, resource-based habitat. (Modified from Dennis, 2020.)

impact, influencing the ability (or otherwise) that any one organism has to avoid its enemies and to find resources. Two pieces of a landscape, whether terrestrial or aquatic, may look similar in every respect, identical even, but without a full understanding of the target organism amid the community of symbionts, competitors and predators, it is not possible to determine the suitability of the environment that otherwise would make up a favourable habitat. In other words, understanding the resource-based habitat of any one species extends well beyond knowledge of the overlap of resources and conditions for that organism, and of its mobility. Consider for a moment what appears to be a uniform meadow with its diversity of wildlife; then split it in half. Not for one moment is it likely that the diversity of organisms, including a target species of study, will be equally abundant in both halves and have identical distributions. The characteristics of soil, slope, local climate and much else will affect the conditions differently in both halves. This is a reminder that the resource-based habitat of any species needs to be studied over many sampling units (e.g. **plots**) and include the interactions with many other species.

Interactions in the resource-based habitat

Identifying the resources and conditions that support a species within an area is vital for determining its resource-based habitat. However, it is by no means an end point. The focus then must move to a finer scale, determining the shifting distribution and fitness of individuals of the target species over the resources and conditions of the habitat space through their lifetime. In other words, the habitat is not a static (stable) entity and monitoring changes exposes the potential vulnerability of a species to

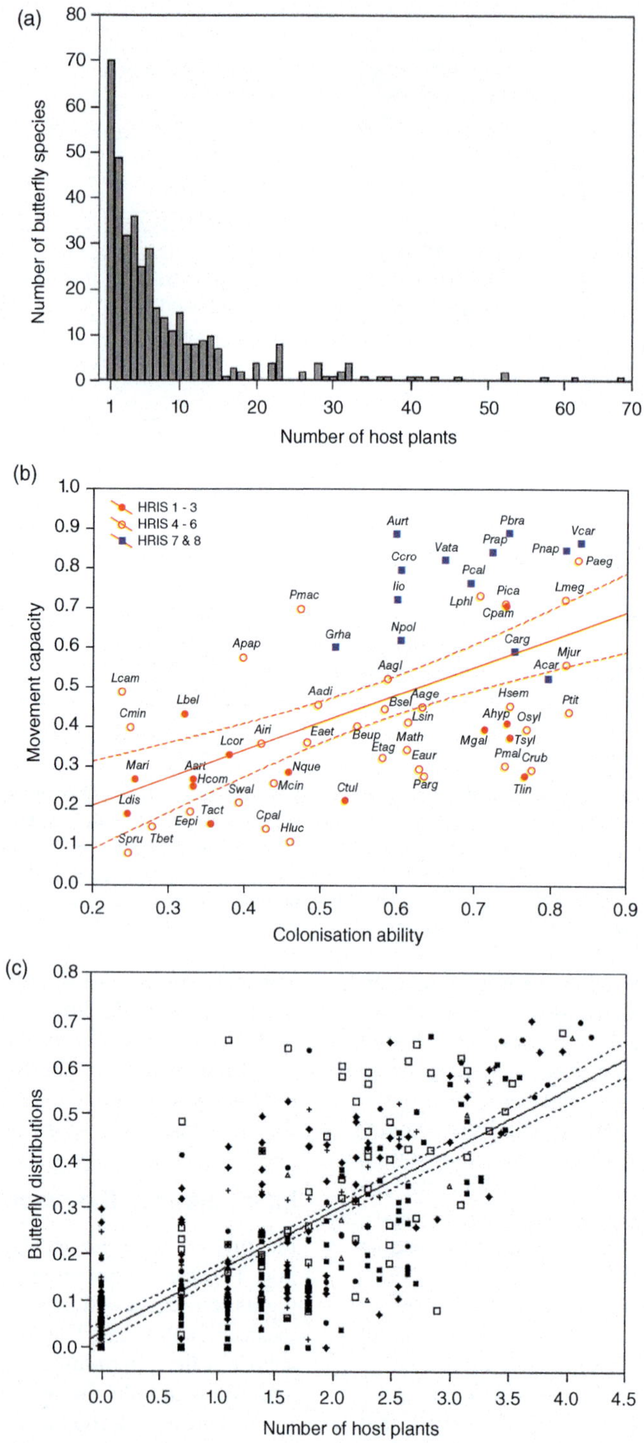

Fig. 6.5. Variation in resource use in butterflies. **(a)** Frequency distribution of the number of larval host plants used by 387 European butterfly species. **(b)** Measures of colonisation ability in British butterflies (which integrates simple

changes in selection of resource items by individuals in relation to the condition of resources and their interactions with conspecifics and with other species. Numerous models are purported to account for the dispersion and success of individuals in a habitat space, some of which unrealistically treat individuals as having equivalent potential fitness. Classic amongst these is the **ideal free distribution (IFD)** model (Fretwell and Lucas, 1969; Fretwell, 1972; Cartar and Abrahams, 1997; Gillis, 2003; Cantrell *et al.*, 2007; Cressman *et al.*, 2004; Cressman and Krivan, 2006; Miller and Coll, 2010; Street *et al.*, 2018) (Fig. 6.6). Under the assumption that fitness increases with patch quality and decreases with conspecific interference, an IFD presents a theoretical way in which a group of individuals (i.e. local populations) distribute themselves among several patches of resources within their environment, in order to minimise resource competition and maximise fitness, so that no individual could improve its fitness by moving to another patch. The idea is that the number of individuals that collect on various patches is directly proportional to the amount of resources available on each patch. Simplistically, it suggests that if resource patch A is twice the size of patch B, there will be twice as many individuals exploiting patch A as patch B. Morris's **isodar** theory (Morris, 1987a,b, 1988; Morris *et al.*, 2000) then follows this thought through by indicating that the fitness of individuals will be the same throughout patches whether the patch is large or small, rich or poor; the theory envisages population densities on patches adjusting to the conditions (resources) such that the fitness of individuals is equivalent (habitat isodars are lines or planes of density where the expected fitness of individuals is the same in all occupied habitats). The same reasoning goes for the impact of habitat quality and quantity.

These models have their seductive features in seeking optimal pattern and process for a species within a 'habitat' space. However, it is important to recall that individuals have unique abilities, capabilities, talents, strength, tolerances, etc. For example, animal personality (in particular, the prevalence of inactive personality types) may reduce the capability of a population to track changes in habitat quality, thus weakening the predictive power of the IFD (DiNuzzo and Griffen, 2020).

The IFD and isodar theories are thoroughly disrupted if 'things' (resource patches and individuals) are not equal, for instance when individuals have different capacities for resource holding. Nevertheless, these notions of individual distribution over resource space are valuable as a starting point as long as they are recognised as such. Individuals cannot be treated as equal units, as would be supposed in these models, even if they adopted a completely neutral environment *vis-à-vis* mortality agents.

The **ideal pre-emptive distribution (IPD**; Pulliam and Danielson, 1991) is one model that accounts

Fig. 6.5. Continued.
measures of the number of resources exploited) relate to their movement (migration) capacity. Each dot is a butterfly species. *HRIS*: habitat resource intersection scores (see Fig. 5.5 for definition). *HRIS* 1 highest, 8 lowest. Species with the highest values of movement (migration) capacity and colonisation ability tend to use widely disseminated resources (diffuse habitats, highest *HRIS* values). **(c)** Relationship between distributional extent in European butterflies and number of host plants used. Distributions, proportion of mapped squares (Kudrna, 2002) occupied, arcsine transformed; number of larval host plants, log transformed. Symbols: cross, Hesperiidae; triangle, Papilionidae; circle, Pieridae; diamond, Lycaenidae and Riodinidae; small square, Nymphalidae (Satyrinae); large square, Nymphalidae (remainder). Species names in panel (**b**): Aadi: *Fabriciana adippe* (formerly *Argynnis adippe*); Aage: *Aricia agestis*; Aagl: *Speyeria aglaja* (formerly *Argynnis aglaja*); Aart: *Aricia artaxerxes*; Acar: *Anthocharis cardamines*; Airi: *Apatura iris*; Apap: *Argynnis paphia*; Aurt: *Aglais urticae*; Beup: *Boloria euphrosyne*; Bsel: *Boloria selene*; Carg: *Celastrina argiolus*; Ccro: *Colias crocea*; Cmin: *Cupido minimus*; Cpal: *Carterocephalus palaemon*; Crub: *Callophrys rubi*; Eaet: *Erebia aethiops*; Eaur: *Euphydryas aurinia*; Eepi: *Erebia epiphron*; Etag: *Erynnis tages*; Grha: *Gonepteryx rhamni*; Hcom: *Hesperia comma*; Hluc: *Hamearis lucina*; Iio: *Aglais io* (formerly *Inachis io*); Lbel: *Lysandra bellargus*; Lcam: *Limenitis camilla*; Lcor: *Lysandra coridon*; Ldis: *Lycaena dispar*; Lmeg: *Lasiommata megera*; Lphl: *Lycaena phlaeas*; Lsin: *Leptidea sinapis*; Mari: *Phengaris (Maculinea) arion* (formerly *Maculinea arion*); Math: *Melitaea athalia*; Mcin: *Melitaea cinxia*; Mgal: *Melanargia galathea*; Npol: *Nymphalis polychloros*; Nque: *Favonius quercus* (formerly *Neozephyrus quercus*); Osyl: *Ochlodes sylvanus*; Paeg: *Pararge aegeria*; Parg: *Plebejus argus*; Pbra: *Pieris brassicae*; Pcal: *Polygonia c-album*; Pica: *Polyommatus icarus*; Pmac: *Papilio machaon*; Pmal: *Pyrgus malvae*; Pnap: *Pieris napi*; Prap: *Pieris rapae*; Spru: *Satyrium pruni*; Swal: *Satyrium w-album*; Tact: *Thymelicus acteon*; Tbet: *Thecla betulae*; Tlin: *Thymelicus lineola*; Tsyl: *Thymelicus sylvestris*; Vata: *Vanessa atalanta*; Vcar: *Vanessa cardui*. (Panels **a** and **c** from Dennis *et al.*, 2011; panel **b** from Dennis *et al.*, 2014.)

Box 6.2. Evolution of flightlessness

As flight is costly, there is strong selective pressure against flight, when the benefits of flight are lower than its cost (e.g. Roff, 1986, 1990). Thus, it is not surprising that flightlessness has repeatedly evolved in the most disparate groups of winged insects. Several environmental conditions may promote flightlessness, including: (i) habitat stability, (ii) habitat uniformity, and (iii) habitat isolation. Habitat stability is generally considered the most important factor for the evolution of flightlessness in insects (Wagner and Liebherr, 1992). In highly unpredictable environments, dispersal may be advantageous, allowing insects to avoid extinction when environmental conditions become inadequate for population persistence, but in stable habitats, non-dispersers may be favoured, as energy can be directly used for reproduction instead of being wasted for dispersing (Roff, 1986). For example, in Hemiptera (Gerridae), the degree of wing reduction is correlated with habitat stability, and flightless orthopterans are distinctly associated with predictable and stable environments (Roff, 1986, 1990).

In heterogeneous environments, maintaining good dispersal capabilities is essential to use resources that are patchily distributed, but if resources are uniformly distributed ground-dwelling insects can exploit them by walking and burrowing, a condition that promotes flightlessness (cf. Roff, 1986). Isolation favours flightlessness as flight might lead insects to move from suitable to unsuitable (and possibly fatal) places (Roff, 1986). For example, for tiger beetles of saline habitats, flight is a waste of valuable energy and even a possible hazard to reproducing, because these habitats are isolated and food is scarce (Kamoun and Hogenhout, 1996). Strand habitats (such as the shorelines of lakes and rivers, or coastal dunes) have often been associated with wing reduction and loss, as they are isolated habitats between highly contrasting environments (e.g. Wagner and Liebherr, 1992; Carpaneto and Fattorini, 2002). The most obvious condition of isolation is that of island populations, and island biotas are notorious for being rich in flightless insects (Whittaker *et al*., 2023). The idea that flightlessness evolves on islands because dispersive forms are carried away on the wind (originally proposed by Darwin, 1859) has been criticised, however, as it cannot be applied to other types of isolated habitats rich in flightless forms (such as caves) where species are not exposed to the wind. Thus, other hypotheses have been proposed for flightlessness on islands, invoking: (i) the reduced benefits of flight (e.g. because of predator and competitor release, as frequently observed on islands), (ii) the increased costs of flight (e.g. because of wind exposure); and (iii) the tendency for less dispersive groups to radiate (cladogenesis) within an island or an archipelago because of reduced gene flow (Whittaker *et al*., 2023).

The increased costs of flight in windy habitats can also explain the evolution of flightlessness in insects living at high elevations, where winds are strong (Mani, 1968). However, low temperature and low atmospheric pressure also contribute to favouring flightlessness at high elevations, and the relationship between elevation and wing reduction is complex, involving a more general trade-off with body size and overall fitness in response to increasingly harsher conditions (Rendoll-Cárcamo *et al*., 2023).

Flightlessness (or at least reduced flight capacity) on islands is also observed in many birds and it was exactly the inability to fly that made many island birds so prone to extinction due to human persecution (Whittaker *et al*., 2023; Fig. B6.2.1). This tendency to reduced flight capacities in island birds may be explained by the absence of predators (mammals and

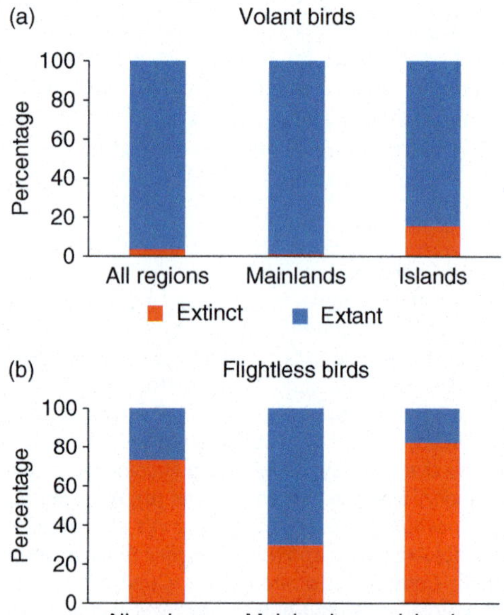

Fig. B6.2.1. Global percentages of species that went extinct since the late Pleistocene in volant **(a)** and flightless **(b)** birds. Most extinctions occurred on islands, and flightless species were the most affected. (Based on Sayol *et al*., 2020.)

Continued

Box 6.2. Continued.

raptors) on islands, but it is also connected to other **traits**, such as gigantism, ground nesting and foraging behaviours, that can co-operate in producing important **bauplans** and ecological shifts (Whittaker *et al.*, 2023). In fact, the evolution of flightlessness cannot be decoupled from the lifestyle of the animals. For example, in insects, flightlessness combines well with fossorial life and feeding on detritus, since the spatial distribution and persistent nature of detritus deposits reduce the need to fly (Holm and Scholtz, 1980). In many cases, the prevalence of flightless species in certain habitats is not due to the evolution of flightlessness there, but to the fact that already flightless lineages were favoured in colonising them (e.g. some insects inhabiting coastal dunes belonging to flightless lineages typical of arid environments; Fattorini, 2023).

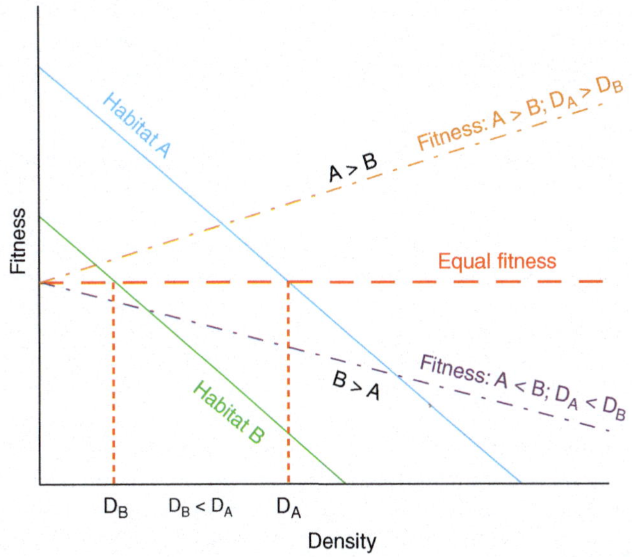

Fig. 6.6. Fitness of individuals in relation to group (population) density in two habitats varying in resource richness (A rich and B poor). Animals displaying an ideal free distribution allocate themselves among resource patches so that each individual receives the same amount of the resource, in which case their fitness (intrinsic rate of increase) remains equal (red pecked lines). Ideal free isodars predict that the density of a species in habitat A will increase linearly with its density in habitat B, such that each individual has the same fitness. In the illustrated example, habitat A will always have more individuals than habitat B, and fitness decreases with increasing densities. The horizontal red pecked line identifies the equal fitness values for certain densities in A and B. If the two habitats present different conditions for the species then both the density of the species and fitness of individuals may also differ (orange pecked line, A > B, e.g. host plant in A superior to that in B; purple pecked line, B > A, e.g. predation in B less than that in A). The word 'isodar' derives from 'iso' (from the Greek *isos*, equal) and 'dar' (that stands for Darwinian fitness).

for differences between individuals. It assumes that individuals are capable of 'defence' and that the first animal to occupy a site can pre-empt it, while later individuals are forced to occupy different sites, possibly of poorer quality (e.g. egg laying in the orange tip butterfly *Anthocharis cardamines*, where succeeding butterflies are 'encouraged' to find novel sites; Wiklund and Ahrberg, 1978; Thomas, 1984). Under ideal pre-emptive selection, average reproductive success declines with increasing density, because each individual is driven to select the best site available from those not yet occupied, leading to a spatial pattern of overdispersion (see Chapter 4, Box 4.2). However, this is not always the case if resources are in short supply, when there is then an excessive loading by individuals on a resource (Dennis and Hardy, 2006; Dennis, 2010). In the case of excessive loading, there will be resource depletion, and – as in the case of orange tip butterflies – an excessive death rate.

A very different model from the IFD would assume that larger (more massive) individuals can oust smaller incumbents, a proposition that would be thought to dominate intraspecific **competition**, but which counterintuitively is not ubiquitous (Glaudas *et al.*, 2020). To account for the role of competition, the **ideal despotic distribution (IDD)** model (Fretwell and Lucas, 1969; Fretwell, 1972; Beckman and Berger, 2003; Oro, 2008) predicts that direct competition for **territories** should result in dominant residents occupying high-quality habitats or areas while forcing weaker competitors into less favourable habitats or areas within habitats. In effect, intraspecific **competition** and territoriality challenge IFD and IDD models by introducing density-dependent habitat selection, which affects how individuals distribute themselves across habitats (Pulliam and Danielson, 1991; O'Neil *et al.*, 2020). The question then arises: what is to be expected in a landscape of uneven fitness and unequal resources and abiotic conditions? At any one moment in time over the 'lifespan' of a habitat unit, there will be a shift in fitness for individuals and for the group of individuals (mean and variance for the local population) as a whole occupying it. A contrast in fitness among individuals in the field is effectively exposed in **key factor analysis** (a technique that predicts fluctuations in population size based on the relative contribution of the key factors of births, deaths, immigration and emigration), in which it becomes clear that abiotic as well as biotic factors influence differences in individual fitness (Dennis, 2010, 2020). When summarised for a whole local population, it becomes evident which factor(s) is/are adversely affecting mean fitness and which combination of factors maximises fitness for the populations of a species. Furthermore, competition occurs not just between individuals of the same species (intraspecific competition) but also between individuals of different species (interspecific competition); undoubtedly, interspecies competition influences the outcome of intraspecies competition and vice versa (e.g. nectar source competition in butterflies; Tudor *et al.*, 2004; Tiple *et al.*, 2009). In the case of interspecific competition, differences in fitness are likely to be more obvious. To reverse the state of interactions, a symbiont or facilitator *raises* the fitness of an individual occupying a particular resource patch inhabited by the symbiont/facilitator (see Chapter 2).

So, what can we expect of the distribution of conspecific organisms over a single resource type in a habitat unit? Here, we take examples from butterflies in which the **resource-based habitat** model was first explored (Dennis, 2010). (i) The resource substrate may vary; it may be singular through different regions or may vary (comprise complementary elements) to the extent that multiple opportunities occur within a single 'habitat' resource type (e.g. different egg-laying and roost sites). (ii) The process of occupying the substrate(s) for a particular activity (behaviour) may be singular or highly varied (e.g. butterfly mate location may involve perching, patrolling, physical defence, lekking without physical defence, etc.; Dennis, 2020). (iii) The behaviour may be complex and have interspecific functions as well as intraspecific objectives. It is useful to consider a few examples. First, mate location by males for females in the large skipper butterfly *Ochlodes sylvanus* on Lindow Common (heath), Cheshire, UK, was found to be based on at least three different substrates: on bramble bushes in flower in which males formed leks, intermittently feeding while awaiting females; along tree-lined edges and zones of *Molinia* grass-dominated wet heath where males actively patrolled for females emerging from pupae; in sunlit clearings on prominent perches where males set up vigorous territories through which females were likely to pass. The different mechanisms reflected female availability during different times of the day and flight season (Dennis and Williams, 1987). Second, the bunching of egg batches by the small tortoiseshell butterflies *Aglais urticae* and peacock butterflies *A. io* under leaves on the edges and along clefts within nettle patches in the Bollin Valley (Cheshire, UK) was found to be a strategy for defence against parasitoids (Dennis, 1992). Often these species would bunch their egg batches together, building up numbers of larvae. When the larvae emerged, they provided multiple defences against parasitoid attacks by bunching together, waving their spine-covered bodies, concealing themselves within denser webs and releasing allomones (molecules released by an individual of one species that affect the behaviour of a member of another species). Third, the wall butterfly *Lasiommata megera* was observed to use the underside of ranch fencing in the last rays of the sun to roost at Brereton Heath (Cheshire, UK) (Dennis, 1986a, 2010). Here it was clear that they avoided the lowest spar of the ranch fencing in reach by predators but clearly made no effort to space themselves on the higher spars where they would occasionally aggregate. Evidently, there was no disadvantage in roosting together, but there was a disadvantage in using the lowest spars in the tall grass cover, particularly as these were more rapidly shaded in the evening.

Dividing up the resource space: selecting the minutiae of essential needs

Butterflies have proved to be very useful model organisms for the development of the **resource-based habitat** concept, since they illustrate not only quite different requirements for the four basic stages of development – eggs, larvae, pupae, adults; more if the different larval phases are considered – but also very precise requirements for each resource, viz.: (i) choice of a resource patch among a number within the habitat; (ii) choice of a suitable place on a selected resource patch (e.g. plant patch edges, as in the case of the small tortoiseshell *Aglais urticae* for egg laying) or even sites off the larval host plant so as not to attract enemies; (iii) choice of a unit (stem, leaf, flower, etc.) of the resource patch; and (iv) modification (engineering) of the unit chosen so as to optimise resource exploitation (see Dennis, 2010, Table 4.1 for 15 factors influencing oviposition bias on host plants in British butterflies). To consider the fundamental resource for larvae at the outset – larval host plant(s) – each species makes quite different demands for: (i) size of plant patch; (ii) size of the plant itself; (iii) thus the age of the plant; (iv) the part of the plant for food (root, stem, leaf, flower, etc.); (v) the chemistry of the plant (as disclosed by scent); (vi) the growth form of the plant; (vii) the abiotic associations of the plant (e.g. soil conditions, light/shade, humidity and moisture availability); and (viii) the biotic context of the plant (growth form of surrounding plants, density of the plant or surrounding plants, other plant and animal species, disturbance, etc.) (Dennis, 2010). A further factor is the capacity of the plant part to be engineered by the butterfly into a suitable site from which eating forays can be conducted, and the larva can escape predation (e.g. larvae of the white admiral butterfly *Limenitis camilla* on *Lonicera* leaves), including a latrine (Fig. 6.7). These and other factors associated with key resources (Fox, 1996; Dennis, 2010) comprise the **microhabitat** or the resource-based microclimate for the species.

These micro-scale factors form a crucial part of a species' resource-based habitat. As such, it is worth illustrating them by way of examples. All too often, only a small fraction of a resource patch is

Fig. 6.7. Micro-resources and features of the white admiral *Limenitis camilla* **(a)** butterfly habitat. **(b)** The complex structure of the larval feeding platform and abandoned latrine. **(c)** Honeysuckle leaf surface showing sticky secretion from glandular trichomes which seriously compromise larval feeding. **(d)** Early stages in the larval hibernaculum (i.e. a shelter for hibernation). **(e)** Later stages (post winter) of the larval hibernaculum, with a close-up of the silk threads **(f)**. (From Fox, 1996; photos courtesy of Barry W. Fox.)

suitable for the development of an organism. For example, a mere 8% of nettle patches on Alderley Edge (Cheshire, UK) were suitable for the development of the peacock *Aglais io* and the small tortoiseshell *A. urticae* butterflies in 1986 (Dennis, 2008b). Nettle host plant sites chosen comprised fresh, young plants in large patches in sunlight. Selection of conditions for obtaining mates can be equally as demanding as food site selection. For the former butterfly (*A. io*), a hierarchy of site selection was observed on the Carrs, Wilmslow (Cheshire, UK) in 2003 and 2004: macro-scale features on southern sun slopes, meso-scale landforms on molehills and micro-scale land features (indentations and slopes on molehills) (Dennis and Sparks, 2005; Dennis, 2010) all provide key perching sites with warmer microclimates. In the case of the large skipper butterfly *Ochlodes sylvanus*, artificial meso-scale landforms (plastic bags) (see Box 5.1, Fig. B5.1.5 in Chapter 5) have been shown to be more effective as perches than natural features, even though they draw the incumbent away from the centre of the territory (10 cm to 0.5 m birch saplings) (Dennis and Williams, 1987; Dennis, 2010). Often edges of host and non-host plants significantly influence the preferences of organisms, as in the case of the orange tip butterfly *Anthocharis cardamines* and the silver-studded blue butterfly *Plebejus argus* respectively, the eggs having an edge bias (Dennis, 1983a,c; Thomas, 1985).

To complete this brief section, a number of principles can be recalled that apply to variation within resource patches for organisms, especially for phytophagous arthropods (Table 6.2) (see Chapter 4 in Dennis, 2010). For example, a number of micro-scale factors are potentially responsible for oviposition bias on host plant patches within any resource-based habitat space (Table 6.3). Of course, similar considerations could be made for the most disparate organisms. For instance, parasites may only use some particular parts/organs of their hosts, soil-dwelling animals may require very specific edaphic conditions to dig, plants may be dependent on very specific environmental conditions (e.g. nitrogen

Table 6.2. Some basic principles affecting resource quality and use in a habitat.

Principles relating to spatial variation in a resource type
• The status of a single resource type (rewards/punishments) varies not only over the habitat but also over any single resource patch. • Geographical location and relative situation of a resource type within a patch of substrate as within a site are key to its composition and constitution. • Spatial variation at fine scales presents opportunities for individuals to compensate for deficiencies in resources at coarser scales. • Aggregations in non-gregarious organisms on resource elements are typically indicative of the limited availability or spatial and temporal concentration of resources of appropriate quality. • The boundaries between two zones (edges) often form sharp gradients in the suitability of a resource type.
Principles relating to temporal variation in single resource types
• The spatial variation in every resource attribute changes over time. • The status of a resource type (rewards/punishments) for a consumer varies over time for any single locality. • Distinctive temporal patterns occur in the use of some resources.
Principles relating to individual preferences and behaviour
• Different (supplementary) outlets of a resource type differ in their suitability for individuals in a population, in terms of reproductive success, development and mortality, and can be arranged in a suitability hierarchy. • Individual preferences differ in threshold and variability, thus range of use, quantitatively, ordinally and categorically. • That only a fraction of any resource is useful and used by an organism is indicative of selectivity for particular resource attributes. • Bias in the placement of any stage of an organism on single resource items is an indication of fine-tuning in exploiting the resource. • A hierarchy of spatial scales exists for selectivity of any resource attribute. • Resource attributes used are not constant but vary with the growth of individuals and changing conditions. • First instar larvae and early phases of phytophagous arthropods nearly always depend on tender young food to feed on regardless of the food requirements of later instars. • Organisms engaged in gregarious activities require larger resource items (individuals, packages, patches) than those carrying out activities singly.

Table 6.3. Factors influencing butterfly oviposition bias on host plant patches within habitat space. For details see Dennis (2020).

1. Plant taxa/type (forms).
2. Plant geography (density, abundance, spacing).
3. Plant size/growth form (synchronisation and timing) – note not just size of a plant but size of individual patches.
4. Plant architecture including colour.
5. Plant composition/chemistry (nitrogen, tannins, etc.; soil water condition, nutrient supply; colour).
6. Plant nutrients and defensive chemicals (N, Ca, Mg – nitrogen being major limiting factor for growth).
7. Plant accessibility/context (overtopping), growth stage synchronisation (bud burst).
8. Plant environment (light, temperature, moisture, exposure, drowning in water).
9. Plant juxtaposition (neighbouring resources: nectar, predators).
10. Inhibitors/degree of exploitation (competitors, predators, parasitoids).
11. Symbionts (ants for Lycaenidae); proximity of egg laying related to feeding strategy ('cuckoo' versus predatory).
12. Plant edges or bare earth edges; plant bare ground.
13. Factors that affect larval growth rate (e.g. plant toughness).
14. Egg avoidance; visual and chemical (e.g. oviposition deterring pheromones).
15. Enemy avoidance (e.g. ants in non-myrmecophilous butterflies).

Note that individual preferences are not considered.

content) to develop, etc., leading to extremely finely tuned diffusion patterns as a function of within-habitat characteristics.

Resource-based habitat as a working system

Now that we have an appreciation of what is meant by a **resource-based habitat (RBH)**, we can move on to understand where a focal organism (for whom we wish to resolve its RBH) stands within an **ecosystem** context. It is inevitable that if we are studying the spatial context (different areas; distinct **biotopes**) in which a focal species resides, a number of independent ecosystems might be involved which comprise the target organism in varying densities and reflect how it is influenced by resources and conditions in different ways. A single organism-focused ecosystem study which concentrates on the RBH is no different from any ecosystem study, except that attention is specifically centred on organisms (i.e. plants, animals, fungi, protists, prokaryotes) that are intimately linked to the target organism. In resolving the system linkages for our focal organism, again it should be borne in mind that no physical boundaries should restrain the study unless the RBH is fortuitously limited within such convenient boundaries.

All organisms form part of a species' ecosystem within a zone that includes the majority of members of the target species and affects the intrinsic rate of increase of its population (see Chapter 2 and Appendix 2). Moreover, the boundary of our ecosystem may be considered to be the limit of the focal organism (e.g. the home range). In situations where many such arbitrary systems' units are studied (in space and/or time; a field, a wood), the boundaries may overlap, and vary in shape, size and other attributes. As for the number and complexity of connections between our target species and the other species with which it is associated, that depends entirely on the sophistication of the target species' system and the community of organisms that affect (either positively or negatively) the focal species. Studying the connections involved in an organism's ecosystem (see Box 2.1 in Chapter 2) clarifies what the component parts of its resource-based habitat are. Species and conditions are included in the system directly or indirectly affecting, or affected by, our target species. As the number of connections scales with species richness, species-rich ecosystems – such as tropical rainforests, where there are so many animal and tree species involved (Pimm *et al.*, 2014; Slik *et al.*, 2015), or coral reefs, which concentrate nearly one-quarter of the total marine biodiversity (Wagner *et al.*, 2020) – may involve highly complex networks of associations (e.g. Schmid *et al.*, 2020; Pozas-Schacre *et al.*, 2021). Greater emphasis is obviously given to those species either most affected or which have the greatest impact on our target species. It is important to appreciate that there may be highly complex shifts in seasonal conditions where the organism might undergo dramatic changes (e.g. insects with complete metamorphosis), especially as a complex of different organisms come into play in interactions with the shift from summer to winter (or dry to wet season, night to day) and back, and with the changes in stage or phase of the target organism. In any such study, it is also crucial to appreciate that a large number of vulnerable species may exist within the home range (habitat boundary) of the focal species being studied.

The impact of the organism (and on that organism) can be quantified using a variety of techniques such as the aforementioned **key factor analysis**. Despite being a long-established approach for the study of arthropods (Varley *et al.*, 1973; Dempster, 1975; Southwood and Henderson, 2000), in recent years key factor analysis has virtually vanished from our ecological repertoire and scientific publications in arthropod ecology, especially work on butterflies. Yet this and allied techniques, locked into the marking and tracking of individuals throughout their lives (see Appendix 1), has provided vital information for understanding life histories and behaviour, population dynamics and ultimately the biogeography of an organism (see, for example, Warren, 1992, for butterflies).

Key factor analysis not only provides vital information on what organisms (therefore species) use in habitats, but also the killing agents that cut short their lives and reduce their fecundity. What kills an organism is of equal importance for understanding life history and behaviour as what an organism uses (Halsey, 2018). To understand life, it generally has to be followed intimately in its habitat. To obtain an adequate measure of the **home range** of individuals that exist in groups (i.e. pair, troop, herd, flock or local population), an appropriate sample of units of individuals needs to be surveyed (e.g. numerous pairs, or local populations, and individuals in the populations) (see Appendix 1). With marking of individuals and drone monitoring, it will soon become a simpler business to determine the home ranges and resource use of individual animals belonging to local populations, including small animals such as butterflies. If sufficient individuals are mapped for their lifetime tracks, the collective home range becomes apparent through a number of measures; typically, it is a convex polygon linking outlying points for individuals on a map (Burt, 1943; Giuggioli *et al.*, 2006). The home range is intimately associated with the RBH, as it incorporates the resources and conditions used by individuals. **Territory** is also a mapped space and identifies those resources that are defended (see Chapter 5 for the distinction between territory and home range); as such, it is a crucial part of the RBH. In butterflies, for example, these are usually mating sites but can also be adult and progeny resources (Dennis, 2010, 2020). Even museum specimens can provide information about an organism's habitat components, from recorded ancillary knowledge (e.g. vegetation) from where the specimens were collected at a particular time, though the populations there are now extinct and the habitat components have now changed (e.g. Hart, 2018; Meineke and Davies, 2019; Schmitt *et al.*, 2018).

Generalism and specialism

One of the key ecological concepts in the resource-based habitat debate is the extent and nature of variability among individuals and all higher units (such as herds, swarms and local populations). We have seen that each individual varies not just in its physical and behavioural attributes (e.g. mass, length, physiology, speed of travel and reactions to stimuli) but also in its breadth and tolerance of specific resources, interactions within the community of organisms, and abiotic conditions broadly common to the species. Different individuals typically vary in their exploitation of different resources (e.g. interindividual differences in the diet) and tolerance of conditions (warmer, cooler, drier, damper microclimates).

From this individual variability (and its assessment across different populations throughout the range of the species), it is but a short step to an understanding of **generalism** and **specialism** in species. The latter points to singularity in tolerance and resource use (with, at its extreme, great specificities of dependence); the former points to broad use of resources and tolerance of different abiotic and biotic conditions of the environment in which the organism is placed. The tendency among researchers, however, has been to consider that the **specialism–generalism continuum** applies to a single key resource (Cui *et al.*, 2018): a specialist would, for instance, use a single plant for sustenance, whereas a generalist would use a variety of plants, perhaps different species in one family or plants in different families (e.g. in the UK the small blue butterfly *Cupido minimus* uses only *Anthyllis* *vulneraria* whereas the holly blue *Celastrina argiolus* uses holly, ivy and many other plants) (Dennis, 2010). In fact, the concept of specialism and generalism is much wider as it refers to the use and tolerance of *all* resources and conditions that individuals of a species could possibly exploit within ecosystems. The critical point is that we need to be clear about the way (for what resources and conditions) in which individuals (or groups of individuals) are specialists or generalists. Only with this information can we determine the extent to which a population (and, in a wider frame, a species) is made up of **generalists** and **specialists**.

A basic issue is that specialism/generalism is not a binary attribute for any trait, but meristic or even quantitative in the sense that averages (means) and variances have meaning. To the extent that these attributes of resource use and tolerance of conditions are capable of measurement, then it is possible to establish a combined or single measure of specialism/generalism; the measure would incorporate all the key variables describing resources and conditions evident for that species. In fact, consideration of generalism/specialism led to the development of models in butterflies, the purpose being to assess their potential for initial colonisation and persistence at sites (i.e. **colonisation ability**) and movement potential or migration ability between habitat units (i.e. **movement capacity**) (Dennis *et al.*, 2012; Dapporto and Dennis, 2013) (see Fig. 6.5). As our knowledge increases, it becomes easier to envisage the combination of key trait models that express the bounds of specialism and generalism for different organisms (Dennis, 2010; Jeliazkov *et al.*, 2020; Middleton-Welling *et al.*, 2020). As we have seen above, the basis for measurement lies with the individual of a given species in any group or population. From measures on individuals can be determined the scale and variability of resource exploitation, and vulnerability to conditions, be they abiotic (light, heat, moisture, etc.) or biotic (community interactions). It is clear that just as every individual varies in its selection and use of resources and conditions, so too will individuals vary in measures of specialism/generalism (thus populations and species will have specialism/generalism means and variances), compound traits that determine the vulnerability of species, their persistence, evolution or extinction – an issue considered briefly below in the last section of the chapter.

The concept of specialism/generalism is not a simple one to digest so it may help to consider an example. British and European butterflies present a well-studied case. For butterflies, relationships have been sought between the exploitation of different resources and associations with different conditions (Dennis *et al.*, 2000, 2004; Shreeve *et al.*, 2001; Middleton-Welling *et al.*, 2020), as well as the importance of non-consumable resources (Dennis, 2004a) and the influence of presumed matrix (Dennis and Hardy, 2007; Jambhekar and Isvaran, 2022). Variation in resource use and experienced conditions was discovered to be important in the use of all resources and conditions, including consumable resources (i.e. larval host plants and adult nectar-feeding plants), utility resources (e.g. mate location, roost, rest or pupation sites) and abiotic (e.g. climatic) conditions. The prime example is that butterfly species differ for the number of the larval host plants used (see Fig. 6.5a), but also for all resource types. In aggregate, species are found to link up closely with the competitive, stress-tolerant, ruderal (**C-S-R**) strategies (see Chapter 2) of their host plants (Dennis *et al.*, 2004; see Fig. 2.11 in Chapter 2) – a demonstration, if one were needed, that organisms are tightly associated in ecosystems (Hodgson *et al.*, 2014). Simple measures of the number of resources exploited in butterflies correlate not only with their capacity for migration (see Fig. 6.5b) but inevitably also with their geographical extent (see Fig. 6.5c) (Dennis *et al.*, 2014; Dennis, 2020 – the list of basic variables comprising colonisation ability and movement capacity in the early work on British butterflies is given in Table 6.4).

Habitat factors in the evolution and extinction of species

Global biodiversity, then, seems to comprise a mix of generalists and specialists in resource use, thus in the breadth of their resource-based habitats; species fall along a gradient between the two extremes of specialism and generalism for every resource and condition, and thus for combined measures of resource variability exploitation (e.g. the gradient in colonisation ability). Emerging clearly is the link of generalism and/or specialism with potential for movement, speciation and extinction (Fig. 6.8). In British and European butterflies, movement capacity seems to be associated more with generalism (Dennis *et al.*, 2013; see also Fig. 6.5b). In North American and British birds, the reverse is the case: specialism in resource use is more closely tuned to

Table 6.4. Variables used to assess aspects of generalism and specialism in British butterflies that were used to determine colonisation ability and movement (migration) capacity.

A. Resource variables (*colonisation ability*, component based on 10 variables)
• Number of biotopes exploited
• Number of larval host plants used
• Host plant abundance in natural vegetation (classification categories)
• Niche breadth (mean of Ellenberg ranges for all larval host plants based on all criteria but salinity; see Chapter 8)
• Larval host plant abundance taken over eight biotopes
• Larval host plant density (records/distribution cover)
• Nectar plant specialism (residuals of weighted regression of nectar sources on nectaring records)
• Number of nectar sources exploited
• Symbiosis with ants (inverted scale: 1–4 with 1 highly dependent on ants)
• Number of utility resources (outlets) used by pupae and adults (excluding nectar plants)
B. Migration variables (*movement capacity*, component based on four variables)
• Mobility index
• Flight period
• Wing span (mm)
• Voltinism (maximum number of broods per year in region)

(From Dapporto and Dennis, 2013, used with permission of Elsevier Science & Technology Journals.)

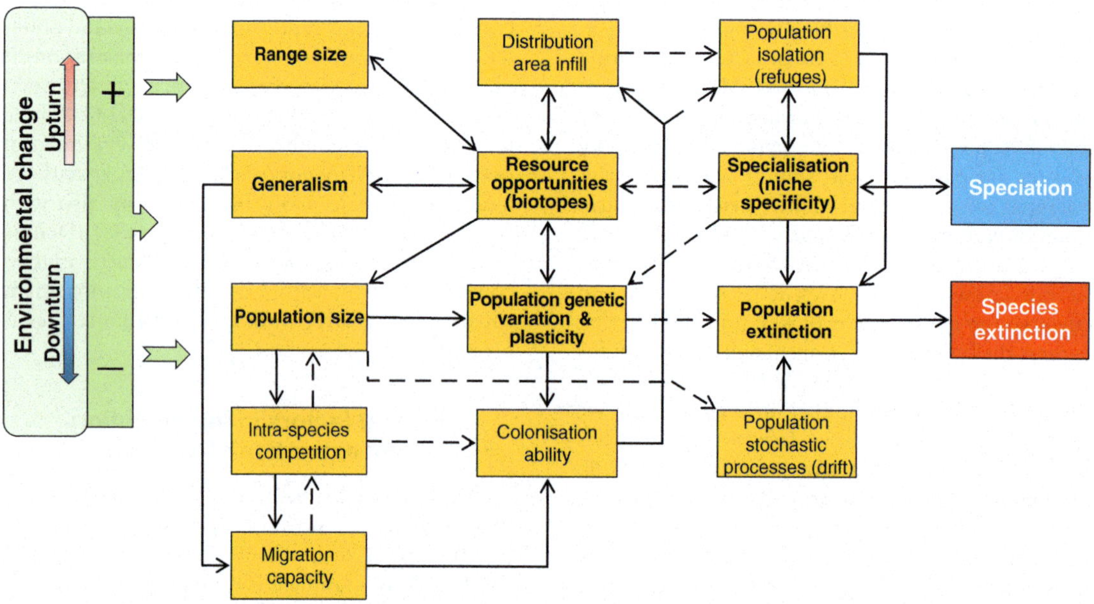

Fig. 6.8. Basic factors in the generalism–specialism cycle linking host plant generalism (specialism) to speciation and extinction in phytophagous insects (butterflies). The full model envisages a cycle of generalisms and specialisms triggered by oscillations in environmental conditions (typically abiotic fluctuations/cycles, particularly climate: e.g. warmer/drier periods causing population increases and expansion in resource use; colder/wetter periods causing population collapses and isolation leading to extinction). Each cycle in abiotic conditions over millennia and geological time uniquely influences range changes and abundances in different taxa, leading to clade furcation (Dennis *et al*., 1998, 2011). Isolated and geographically marginal populations (in which the niche is the furthest from the average environment for a species; Calenge and Basille, 2008) are those most likely to undergo significant changes leading to speciation. The model has its basis in island biogeography in that generalists have greater capacity for colonising new environments and novel plants, persisting and producing specialists that become new species. (Simplified diagram from Dennis *et al*., 2011.)

intercontinental migration (Martin and Fahrig, 2018). The important point is that resource use and migration are correlated to a varying degree (and polarity) in different organisms. Just how generalism/specialism is linked to capacity for movement about the globe has crucial ramifications for the persistence, development or loss of biodiversity. Initially, it is important to distinguish local specialism or generalism (occurring in a single population) from that occurring over the range of a species. In the former case, individuals within a single group (population) may vary in specialism or generalism for a resource. In the latter case, a species may be a generalist in the sense of using a variety of resources over its range, but incline to specialism within single populations.

Generalism and specialism are in fact closely aligned, inasmuch as generalism can lead to speciation both in situations of **sympatry** (occurrence of species in overlapping areas) and **allopatry** (occurrence of species in non-overlapping areas). In the first situation, generalists with broad but suboptimal adaptations to a resource can – through behavioural adjustments and sloppy ecological fitting to different resource elements – result in selection for traits which enhance resource exploitation and result in novel species developing in sympatry (Janzen, 1985; Araujo *et al.*, 2015), much as may be considered to have been the case in Galápagos finches (Grant and Grant, 2008; Lamichhaney *et al.*, 2015). In the second situation, expansion in the range of a species in response to an abiotic trigger (e.g. climate change) with subsequent downturn in conditions can lead to isolation of populations, possibly adapted to different resources and conditions. Again, over long periods of time, allopatric species evolve (see Fig. 6.8) (Dennis *et al.*, 2011). In both cases, although generalism may provide the basis for variation in resource use and changes in behaviour, speciation is the result of increasing specialism. However, specialism carries with it a greater vulnerability, a tendency to extinction, whereas generalists have a greater capacity for adaptation (see Box 4.1 in Chapter 4 about the relationship between vulnerability and ecological specialisation). From the above, it is evident that colonising specialists have the capacity to become generalists as populations expand into novel resource zones, and that generalists can in turn produce new isolated specialists as geographical ranges collapse (or when competition for resources occurs in sympatric areas), a process referred to as the **oscillation hypothesis** (Janz and Nylin, 2008; Dennis *et al.*, 2011).

7 The Resource-based Habitat in Unstable Landscapes

Abstract

Populations differ enormously in respect of the number of individuals, their life cycles, mass, mobility and spatial structure, the occupied area and forms of interactions. Many species are organised as metapopulations, that is, sets of interacting local populations inhabiting patches of suitable habitats interspersed in an inhospitable matrix (e.g. forest fragments). Metapopulation dynamics depend on many factors, including the size, isolation, quality and stability of each local population's patch, as well as the characteristics of the matrix (e.g. how it can be crossed and used by dispersing individuals). Populations are subject to spatial and temporal variations in resource availability and conditions, making the resource-based habitat variable for any species, as reflected by the occurrence of genetic and phenetic distinctions between populations. Habitat, understood as the resource-based habitat, has a crucial place in evolution, in species diversity patterns across space and time, and hence in species conservation.

With so much human pressure on the Earth's resources, the reader should be well versed in the nature of continuously changing landscapes; with the shift in resources and conditions, there will be significant changes to the populations of every organism. 'Population' as a descriptor of the number of individuals is a common term in usage. It is especially widely used in geography, where we might refer to the global population of humans, or of some subset, be it that of a state, county, parish or any other such arbitrary unit of government or land area, where the numbers are assessed with regular censuses. But extend the word to animals and plants, and difficulties become plain. With increasing knowledge, one may happily refer to the global population of lions, elephants, zebras, polar bears or any other large animal that can be counted using drones or other techniques that allow virtually complete censuses (e.g. by **mark-recapture** methods), especially if populations are composed of few individuals concentrated in a small area (see Appendix 1). It is another matter to determine the global population size of small organisms, insects and other arthropods, for instance, unless they are conspicuous and the total area occupied by them is small. But if they are widely distributed, inconspicuous and concealed, then the task is virtually impossible.

Thus, for most organisms we lack an accurate count of the individuals for different species. Moreover, it becomes extremely difficult to determine the bounds of subsets of populations, unless substantial barriers make it plain where their existence may begin and end. Even so, such subsets are of great importance, particularly for organisms whose global populations are clearly fast diminishing. Subsets implicate genetic distinctions that may be found to relate to contrasts in their current or past ecology, and hence to the factors influencing their **resource-based habitats**. As is evident from bias in studies of different life forms (see Chapter 8, Fig. 8.1), the focus has been mainly on larger creatures, not just because larger life forms are more noticeable to observers but also because larger animals require more food and that places limits on population size, thereby increasing the risk of their extinction (see Box 2.3 in Chapter 2 and Box 4.1 in Chapter 4). As estimates of population size are an important parameter for evaluating the conservation status of species and populations, this means that for certain groups we have virtually complete global or regional **red lists**, while for others the conservation status is known only for a minority of the species. For example, 100% of known bird species (11,195 species) are evaluated in the Red List of the International Union for Conservation of Nature (IUCN), while only 12,747 species of insects are evaluated, which means about 1% of the total number of known insects (about 1,054,000; IUCN,

DOI: 10.1079/9781800627505.0007

2024b). This situation is in fact even worse if we consider that virtually all bird species are known, whereas only a minor fraction of the extant number of insect species has been described (the total number of extant insect species might be estimated at around 21 million species (Li and Wiens, 2023), which would mean that less than 0.06% of insects species are known from the conservation point of view; for other estimates of insect diversity, ranging from 5 to 30 million species, see Chapter 8, Box 8.1). The fact remains that population subsets of organisms exist which differ from one another in numerous ways, almost certainly in detail of their resource-based habitats. If we consider that these organisms deserve our protection (for the intelligent among us there can be no doubt) then we need to identify these populations and population subsets and discover the factors that allow them to persist.

The basic structure and dynamics of populations

Populations differ enormously on a number of criteria; some basic ones are: (i) population size (numbers of individuals) and number of populations; (ii) life cycles (age structure, social organisation, seasonal numbers, etc.); (iii) size (mass) of individuals; (iv) area (spread) occupied by the population(s); (v) spatial structure (distribution) of individuals within populations (uniform, random, clumped; see Box 4.2 in Chapter 4); (vi) mobility of individuals, whether static, on the move seasonally or continually; (vii) degree of co-ordination and social integration among individuals (for example, pairs in separate **territories**, as with many birds, versus highly synchronised or aggregative behaviours, such as mass emergence in mayflies and herding in mammals). On the 'ground', these factors determine a variety of structures, some of which form distinctive populations, much as do specific bacteria concentrated in particular body organs (e.g. gut microbiomes). In other cases, as with paired territorial birds, it is difficult to determine the bounds of a population, which in fact may be more extensive than, say, the isolated wood in which a few nests are found. For instance, with blackbirds a population may run from wood to wood, into gardens, scrub along hangers and hedgerows.

At least two things in populations may be mobile or static: the population itself and its habitat. All populations are subject to demographic changes, whether or not bound to a particular substrate. Many processes can contribute to regulate populations, in relation to their **density** or to external biotic and abiotic factors (e.g. inverse density dependence, density independence, **Allee responses**; Fig. 7.1), none of which may be mutually exclusive (see also Appendix 2). Similarly, the habitat (or at least the key components of it, such as the host plant(s) exploited by herbivores or their predators that may be many or few) again is mutable. It is owing to these factors that a population of an organism may adopt a variety of dynamics and take on a number of different forms (spatial structures). The distribution of individuals may be effectively continuous or discontinuous and, where sufficient in number, takes on a distribution mentioned above, that is random, evenly spread (uniform, systematic, regular) or bunched (clumped, aggregated) (see Box 4.2 in Chapter 4). It is in states of apparent continuity, but of low density, that it becomes difficult to establish where a population (viz. a local population) has its bounds. One may then envisage a sequence in pattern, from a continuous, evenly spread distribution to a pattern of isolated local populations. In a variety of contexts (ranging from pre-Neolithic human societies to plants and animals within tropical rainforests), populations of each

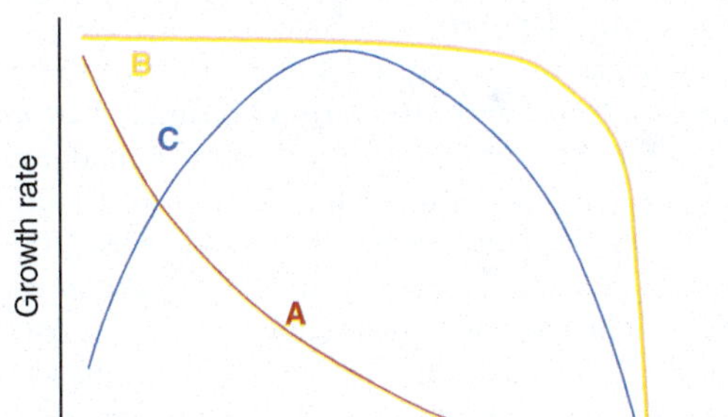

Fig. 7.1. Three patterns of population growth rate (per unit of population) in relation to population density: A, Growth rate decreases as density increases (self-limiting or inverse density dependence). B, Growth rate remains high until density becomes high and factors outside the population become limiting (density independent). C, Growth rate is highest at intermediate densities. In many species, undercrowding, not competition, limits population growth, for example because at very low densities individuals can have difficulty in finding partners for mating (Allee effects; see Kramer *et al.*, 2018).

organism are expected to be widely spread, if of low density. In effect, we have a continuum of interactions that may appear to be multiple interconnected populations; when organisms comprise overlapping, interacting groups with indistinct boundaries, the concept of particulate populations becomes nebulous. Consequently, the capacity of organisms to survive depends considerably on the ability of individuals to track one another down in a complex landscape.

The metapopulation concept: isolation and area of habitat units

Initially, before the preagricultural (Neolithic) revolution, terrestrial **landscapes** were a great deal more complex than most of the landscapes in the current industrial era covering increasing amounts of the planet (e.g. Nentwig, 2007; Hodder *et al.*, 2009; Boivin *et al.*, 2016; Fuller *et al.*, 2017). Amid such natural and contemporary complexity, individuals require traits that allow them to find their resources and one another, to be able to adapt to continual changes of **vegetation succession**, and now the rapidly shifting abiotic and biotic pressures caused by humans. A crucial ecological consequence of this transformation is the increasing fragmentation of natural habitats and, correspondingly, of the populations they support.

In landscapes where organisms are distributed across a broad and irregularly structured environment without sharply defined habitat boundaries, populations may form what is known as a **multipartite population** (Den Boer and Reddingius, 1996) (Fig. 7.2). In this framework, individuals are spread across a heterogeneous and semi-continuous space, where local subgroups may function somewhat independently but remain part of a broader, interconnected whole. However, as habitat patches become increasingly smaller and more delineated from the surrounding ground, and as their mutual isolation becomes obvious and increases, interactions among them concomitantly decline and populations break up into distinct, isolated patchworks (Dempster, 1991). Though disconnected into patches of suitable habitat, they may still be connected by regular, albeit limited, transfers of individuals, depending on organism **mobility**.

As this spatial fragmentation increases, the resulting collection of isolated units, between which occasional transfers occur, is referred to as a **metapopulation** (Levins, 1969, 1970) (Fig. 7.3) Thus, a metapopulation is a set of interacting local populations inhabiting patches of suitable habitats interspersed in a 'dead' ground referred to as a **matrix** (Driscoll, 2005) (see also Box 2.2 in Chapter 2). Metapopulations take on dynamics similar to those of island archipelagos (Dennis, 2010; Dennis and Hardy, 2018). The main difference is that in typical island dynamics, most immigration is from a mainland source (not affected by extinction) to the island and such immigration events tend to be infrequent due to large distances relative to the dispersal abilities of the species. By contrast, in Levins-type metapopulation models there are no major sources but frequent exchanges of individuals between local populations (patches) that are subject to extinction. Extinctions on some patches are matched by colonisations on others, so that the whole system is considered to be in homeostasis. The Boorman–Levitt metapopulation model (Boorman and Levitt, 1973) represents an intermediate case, in which habitat patches (islands) are located within typical dispersal distance from a very large patch (mainland) where the local population never goes extinct.

The concept of a matrix depends on the context. For example, in the case of marine islands, the sea is the matrix, whereas for waterborne organisms, the land is the matrix. The contrast between the habitat that can be used by an organism and the surrounding matrix may be more or less pronounced (see Chapter 5). Habitats surrounded by very contrasting matrices are frequently regarded as **'habitat islands'**, such as forest fragments, lakes and caves (Table 7.1, Box 7.1). It is not surprising that principles of island biogeography have been frequently applied to the study of these habitats, and that they have contributed to the development of metapopulation models. Attention has been mostly focused on habitat islands originated by landscape fragmentation (see, for example, Whittaker *et al.*, 2023). But many habitats are naturally isolated (see Table 7.1), and many organisms have their populations naturally organised as metapopulations. For example, for a freshwater animal, individual ponds are suitable patches in a matrix of unsuitable land. For instance, the caterpillars of the small China mark moth *Cataclysta lemnata,* which feed on duckweed in ponds, form relatively stable metapopulation units (R.L.H. Dennis, personal observation). However, because of increasing fragmentation of natural habitats due to human activities (e.g. forest fragmentation), more and more

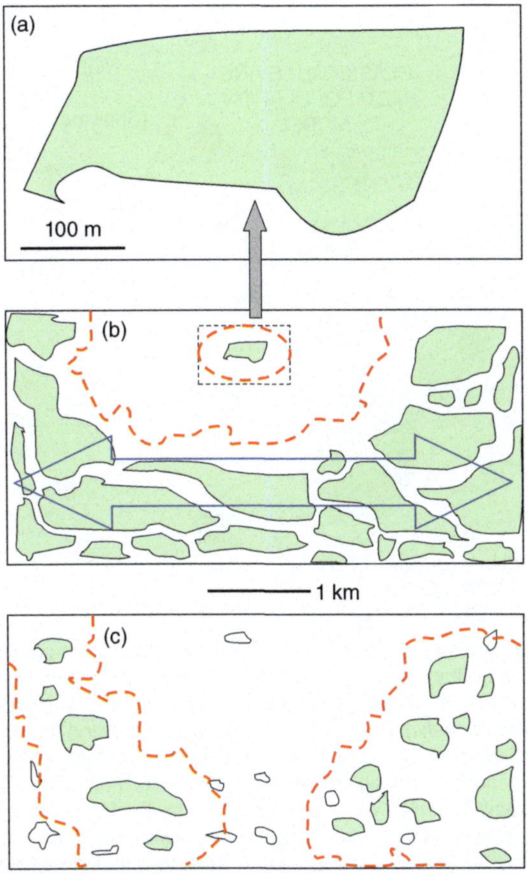

Fig. 7.2. Relationship of population structures to habitat components in multipartite populations. Local population **(a)** on the periphery of a **(b)** multipartite population complex in which there is movement throughout the system, and its loss with the degeneration to **(c)** a metapopulation with habitat fragmentation. Green areas are populated habitat patches and white areas are unpopulated habitat (population persistence below a certain threshold, e.g. <5 years) amid the matrix. Pecked lines indicate bounds of most (e.g. 95%) individuals recaptured. A classic case of an outlying population being lost is that of the Scotch argus *Erebia aethiops* at Grassington in the Yorkshire Pennines, UK. (From Dennis, 2010.)

species that would not naturally occur as metapopulations tend now to be structured as such, making metapopulation theory central to conservation biology (e.g. Opdam, 1991; Akçakaya *et al.*, 2007; Huang *et al.*, 2020).

The basic metapopulation model has the general form:

$$\partial p_m / \partial t = i_m - e_m \qquad (7.1)$$

where p_m is the fraction of homogeneous patches occupied by a single species over time t, and i_m and e_m are the overall immigration (colonisation) and extinction rates, respectively, measured across the entire set of patches (see Fig. 7.3).

By assuming that immigration and extinction rates are constant, Levins (1970) proposed a first simple equilibrium model in which population dynamics within a patch were simplified to the point where only presence and absence are considered (a patch is either occupied or empty):

$$dp/dt = ip(1-p) - ep \qquad (7.2)$$

where i is the immigration (colonisation) rate, e is the extinction rate, and p is the fraction of occupied patches. With this model, at equilibrium, a certain fraction of patches remains always unoccupied.

This equation is mathematically analogous to the logistic model of population growth

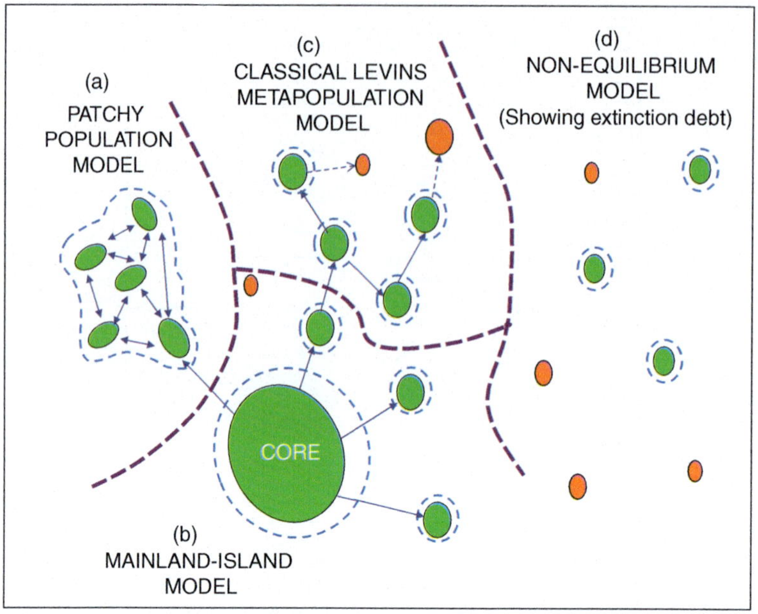

Fig. 7.3. Some basic metapopulation models for habitats on a mainland. **(a)** Patchy population model (all patches are occupied, with continuous flux of individuals among all patches, e.g. the green-veined white *Pieris napi*). **(b)** Mainland–island or Boorman–Levitt model (fluxes of individuals occur from a larger source (the mainland) to smaller patches (islands), e.g. the brown argus *Aricia agestis*). **(c)** Classical Levins metapopulation (patches are subject to continuous immigration and emigration, being temporarily empty or occupied, e.g. the Glanville fritillary *Melitaea cinxia* on the Isle of Wight, UK). **(d)** Non-equilibrium model (showing distinct signs of extinction debt with populations becoming extinct and not replaced, e.g. the large heath butterfly *Coenonympha tullia* in lowland UK). Green patches, colonised; red patches, vacant; dashed blue lines, boundary of a local population; arrows, transfers. No scale is provided; the scale of a metapopulation is established by the mobility of the organism. Thus, the same systems of patches of suitable habitats within a matrix may act as a patchy population to one species and as a non-equilibrium population with gradual extinction and the loss of the species to another organism. (From Dennis and Hardy, 2018.)

Table 7.1. Examples of habitat islands.

Habitat islands	Isolates	Matrix
Woodlands	Forest fragments	Open habitats
Urban green spaces	Parks, gardens, green roofs, playgrounds, urban farms, golf courses, cemeteries, etc.	Impervious surfaces (built-up areas, roads, car parks, etc.)
Lentic water bodies	Lakes, ponds, ditches, seeps, seasonal pools, swamps, basin marshes, marine lakes, etc.	Land
Lotic water bodies	Rivers, streams, brooks and springs	Land
Subterranean environments	Limestone caves, lava tubes, rock fractures, aquifers, underground lakes, etc.	Surface environments, rocks
Sky islands	Mountain tops	Lowland areas
Hosts for parasites	Host bodies	Any other environment different from the host

(see Chapter 2 and Appendix 2), with a carrying capacity K given by:

$$K = 1 - (e/i) \tag{7.3}$$

and a growth rate r:

$$r = i - e \tag{7.4}$$

This equilibrium model has been subject to modifications (e.g. Hanski, 1982; Gotelli and Kelly,

Box 7.1. Habitat islands

Forest fragments represent probably the best-known example of habitat islands. With deforestation, formerly continuous, large areas of forest are transformed into forest fragments, interspersed in a **matrix** of open vegetation, agricultural fields, urban sprawl, etc. At the beginning, even small patches might host a high number of species, as they have temporarily retained most of the diversity that occurred in the forest before fragmentation independently from their characteristics, such as their size and isolation (**supersaturation**), thus creating an **extinction debt**. However, with time, species richness in each fragment will decline (**relaxation**) to the number of species that the patch can really support (Laurance, 2008; Whittaker *et al.*, 2023) (Fig. B7.1.1).

One of the most prominent cases of habitat fragmentation is represented by the transformation of land into habitat islands due to increasing urbanisation. From the perspective of organisms that live in urban green spaces, the urban environment may appear as a set of habitat islands (the green patches where they live) interspersed in a more or less inhospitable matrix of impervious surfaces (built-up areas) (Fattorini, 2019). Many studies conducted in the most disparate organisms (including insects, especially butterflies and beetles, snails, birds, reptiles and vascular plants) showed that species richness in urban green spaces is influenced positively by their area and negatively by their isolation, as commonly found in true islands, while the role of shape and extent of natural habitats within patches was less supported (Adler and Tanner, 2013; Fattorini *et al.*, 2018c; Fattorini, 2019). Moreover, while some systems may be at equilibrium (species numbers in green spaces are regulated by continuous processes of extinction and recolonisation, thus following the equilibrium theory of island biogeography; see below), others are not (Adler and Tanner, 2013; Fattorini *et al.*, 2018c). Variability largely arises from the characteristics of the organisms, since those that have higher dispersal capabilities (e.g. birds) can follow equilibrial models more easily than those with poor dispersal capabilities (e.g. flightless insects) (Adler and Tanner, 2013; Fattorini *et al.*, 2018c). The commonly observed increase in the species richness of green spaces in relation to their size (e.g. Fattorini *et al.*, 2018c for insects; Leveau *et al.*, 2019 for birds) reflects the generality of the **species-area relationship (SAR)**, one of the most documented patterns in ecology (Matthews *et al.*, 2021).

In most cases, at least for isolates, the SAR is well approximated by the power function:

$$S = cA^z \qquad \text{(B7.1.1)}$$

which can be linearised by logarithmic transformation as:

$$\log(S) = \log(c) + z\log(A) \qquad \text{(B7.1.2)}$$

where S is the number of species, A is area, and c and z are estimated parameters (Whittaker *et al.*, 2023). The c parameter expresses the number of species per unit area, while z can be interpreted as a scaling factor describing the speed of the response of species richness to area changes along the SAR curve (the slope of the relationship in the linearised form) (Fattorini *et al.*, 2017b). High z-values are

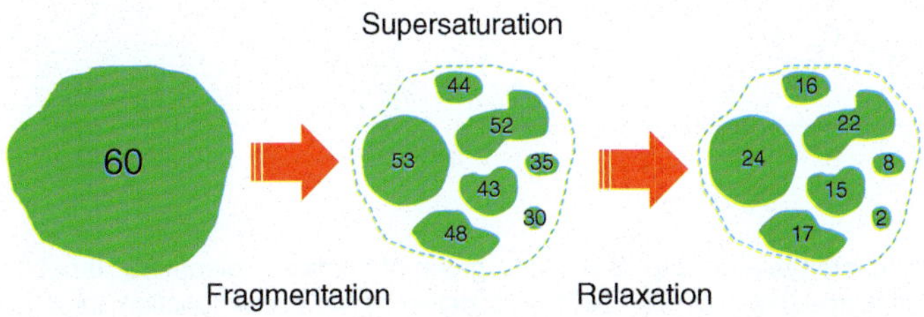

Fig. B7.1.1. Relaxation after fragmentation. Immediately after fragmentation of a previously larger area of a certain habitat (e.g. a forest), the number of species that can be found in the resulting isolated patches may be relatively high and independent from patch characteristics. This means that the new patches will host more species than they can really support (supersaturation), thus creating an extinction debt. With time, species richness in each fragment will decline (relaxation) to the number that it can support according to its characteristics (area, isolation, habitat diversity, shape, etc.). Numbers in the scheme indicate hypothetical numbers of species occurring in the formerly continuous habitat and in the resulting patches immediately after fragmentation (supersaturation) and after relaxation.

Continued

Box 7.1. Continued.

thought to reflect high isolation (Whittaker *et al.*, 2023). As the *z*-values of bird SARs for urban green spaces from different cities around the world ($z \approx 0.2$) are towards the lower end of the range of values theoretically predicted for island archipelagos (Leveau *et al.*, 2019), this suggests low levels of isolation among urban green patches for these animals. A similar value found for very sedentary insects (tenebrionid beetles in Rome, Italy; Fattorini, 2014a) suggests that in general urban green spaces are less isolated than true islands (Fig. B7.1.2).

Although anthropogenic factors play a crucial role in generating habitat islands, many ecosystems are naturally patched and can be considered as habitat islands. Perhaps the most obvious case is that of islands of inland water bodies in a sea of land. Inland water bodies include **lentic** (still) and **lotic** (flowing) waters. Examples of lentic waters are lakes, ponds, marshes, swamps and bogs; lotic waters include rivers, streams, creeks, brooks, springs, etc. As for true islands, diversity in water bodies is positively influenced by area, habitat diversity and age, and negatively by isolation.

In general, research on freshwater lake communities found that species richness of the most diverse organisms increases with lake area (Oikonomou, 2021), although it might also be a consequence of sampling effects for some taxa (Gooriah and Chase, 2020). Moreover, at least in some cases, this relationship might in fact be a consequence of the higher habitat heterogeneity of larger lakes. For example, shoreline length and shoreline development (a measure of basin shape and index of available littoral habitat) are better predictors than lake area for abundance and diversity in the littoral molluscs of nine lakes in the Beaver and Manitou Archipelagos in Lake Michigan, USA (Harris *et al.*, 2011). A study of Chinese lakes (Zhao *et al.*, 2006) showed that lake area is, in general, not correlated significantly with fish species richness, which was mainly driven by variables expressing productivity (temperature, precipitation and evapotranspiration) (Fig. B7.1.3). The slopes (*z*-values) of the power function used to describe the SARs in lake ecosystems tend to be lower than those of true islands (Drakare *et al.*, 2006). As *z*-values tend to increase with isolation, these findings indicate that habitat isolation is not a primary factor driving lake SARs. In general, the SAR patterns for lentic and lotic systems are more similar to expectations for continental habitat patches than for islands (Drakare *et al.*, 2006; Jenkins, 2014). The

(a)

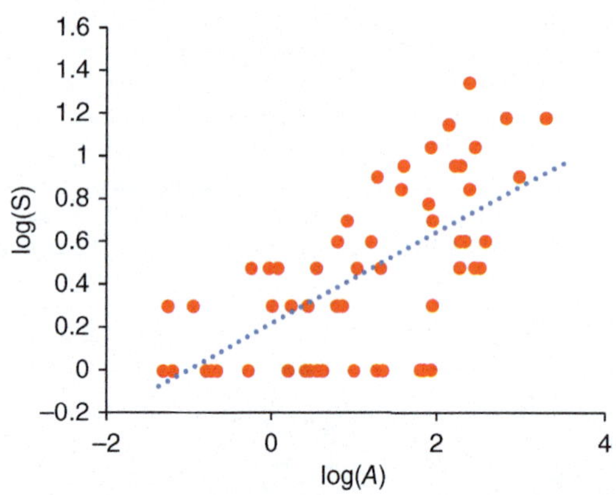

(b)

Fig. B7.1.2. The species-area relationship (SAR) for a group of beetles (Coleoptera: Tenebrionidae) inhabiting green spaces in urban Rome (Italy). **(a)** The SAR is modelled by the power function: $\log(S) = 0.220 + 0.212 \log(A)$, where *S* is the number of species and *A* is the area of the green spaces (in hectares). Decimal logarithms are used. $R^2 = 0.413$, $P < 0.0001$. **(b)** Most tenebrionids inhabiting urban green spaces in Rome have very limited dispersal capabilities (the picture illustrates *Blaps mucronata*, a large-sized (about 22 mm), apterous species mostly found in archaeological sites). The slope ($z = 0.212$) of the SAR is higher than the typical values observed in non-isolated areas (usually, $z = 0.12–0.17$), but close to the lower values found in true island systems (typically, $z = 0.20–0.35$). This suggests that, despite the limited dispersal capabilities of these beetles, their communities in the green spaces are not strongly isolated. (Data for panel **a** were extracted from Fattorini, 2014a. Panel **b** photo by S. Fattorini.)

Continued

Box 7.1. Continued.

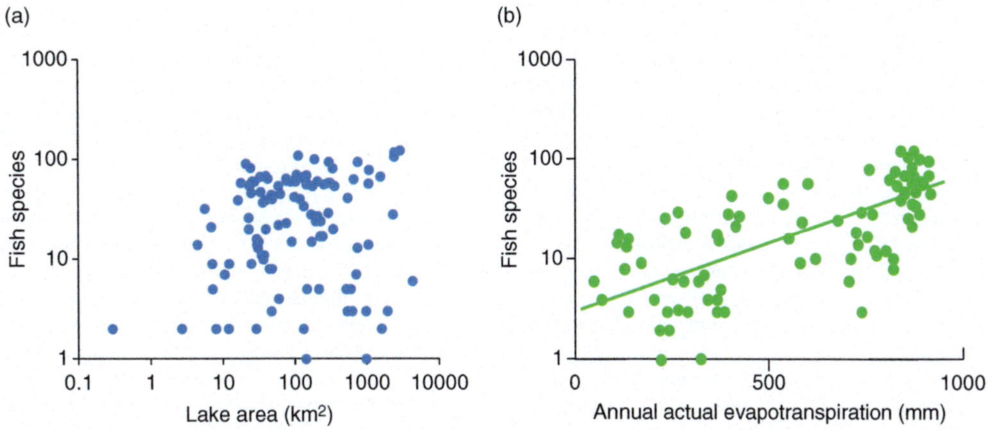

Fig. B7.1.3. Fish diversity in Chinese lakes. Species richness is not correlated with lake area **(a)** but increases with annual actual evapotranspiration **(b)**. Regression line in panel **(b)**: $y = 0.001x + 0.494$, $R^2 = 0.53$, $P < 0.0001$ (decimal logarithms are used). (After Zhao *et al.*, 2006, used with permission of John Wiley & Sons.)

generally low importance of distance as a predictor of richness in inland water bodies may be surprising, as they seem to be very isolated systems. However, even aquatic organisms disperse, sometimes showing very high dispersal capabilities. Large organisms (fish) can disperse between lakes via stream connections, through flooding events that temporarily connect lakes, and tornadoes or waterspouts that take animals from one lake and drop them into others. Micro-organisms, certain small animals and plants can move between lakes through stream connections, flooding events and aerial dispersal by wind, waterfowl or vectors, especially if they have stages resistant to desiccation. Many aquatic insects (odonates, plecopterans, ephemeropterans, many beetles and hemipterans, etc.) have winged adults. Molluscs can disperse through stream connections or by waterfowl, wind or more mobile animals (e.g. waterbirds, reptiles, amphibians or even insects). Finally, lake communities have been profoundly affected by human-assisted **introductions** (voluntary or not, especially fish). In general, lake communities seem to have high temporal turnover rates (changes in species composition), which, however, decrease with increasing species richness, thus indicating that more diverse communities are more resistant to colonisation and extinction (Shurin *et al.*, 2007).

A special, non-freshwater case of lentic water is that of marine lakes. There are some 200 known marine lakes around the world, but most are concentrated in South-East Asia (especially Vietnam, Palau and Indonesia) (Blanchette *et al.*, 2020). Modern marine lakes originated after the Last Glacial Maximum, when increased sea levels forced water and propagules into low-lying inland, and maintain a marine character through narrow submarine connections to the sea. Their ecology is largely shaped by the degree of connection to the adjacent sea, which influences salinity and tidal variations in environmental parameters (Blanchette *et al.*, 2020; Becking *et al.*, 2011). Like true islands, marine lakes often harbour genetically isolated populations, recognised as endemic subspecies or species, frequently characterised by a rapid evolution of novel characters similar to those typical of island syndromes, such as: (i) lowered dispersal capabilities (reduced swimming speed, which parallels the evolution of flightlessness in island animals); (ii) reduced size relative to ocean ancestors (as dwarfism in animals living in true islands); (iii) a reduced colouration (a phenomenon observed, for example, in island birds); and (iv) vestigial morphology (corresponding to the diminution of antipredatory devices found in island animals) (Dawson, 2016; see Whittaker *et al.*, 2023 for discussion of island syndromes). Macroinvertebrates of marine lakes seem to be influenced by area and isolation as predicted for island systems (Dawson and Hamner, 2008). Species diversity of pelagic copepods inhabiting marine lakes varies according to the level of isolation and the local environmental conditions for each marine lake (Saitoh *et al.*, 2011), as typically observed for biotic communities of true islands.

Lotic ecosystems are reciprocally isolated but internally they form a continuous environment, if water flows without interruption. However, river bodies can be naturally subdivided into relatively separate sections

Continued

Box 7.1. Continued.

by ruffles and into very disconnected parts by human interventions (dams). For example, the slope of the SAR for fish species from different parts of the Trent River (Canada) separated by dams is more similar to those of true oceanic islands that tend to have distant colonising source populations than to those of other habitat islands, which are typically less isolated than true islands (Carl *et al.*, 2021). In this system it has also been found that **habitat diversity** calculated by applying the **Shannon index** to vegetation cover (see Box 4.3 and Appendix 3) was an important predictor of fish diversity, while other aspects of habitat diversity (such as depth, substrate and velocity) had less importance.

A very particular case of lotic water is represented by springs. Diatoms of Finnish springs (Teittinen and Soininen, 2015) do not follow biogeographical patterns typical of island communities, as their richness is not influenced by area and isolation, probably because of the different number of microhabitats sampled at different sites and the very high dispersal capabilities of these micro-organisms (passive dispersal via air or by other organisms, e.g. waterbirds). Diatom community similarity among springs was related to their environmental similarity but not to their geographical proximity, which implies that spatial factors are of minor importance for spring diatoms compared with environmental filtering. Generally, species-poor communities were found in springs with harsher growing conditions, while species-rich assemblages occurred in springs with average environmental conditions. Fattorini *et al.* (2016b) investigated whether the **equilibrium theory of island biogeography** (ETIB) could be applied to the copepods of groundwater-fed springs in central Italy. The ETIB (formulated by MacArthur and Wilson, 1963, 1967; see Whittaker *et al.*, 2003 for a review) predicts that species richness on islands is regulated by their size and distance from the species source (e.g. the mainland). Species richness is expected to increase with island area because larger areas host larger, and thus more stable, populations (that is, extinction rates decline with island area). By contrast, immigration rates decrease with distance to the species source, as more distant islands are more difficult to reach. Because of continuous processes of extinction and colonisation, species on islands are subject to temporal turnover, but the average number of species in each island should remain approximately similar and determined by area and isolation. In general, Fattorini *et al.* (2016b) found support for ETIB predictions about copepod species richness in springs, which was positively influenced by area and negatively by elevation (which is inversely related to the size/proximity of the source of immigrants, i.e. the aquifer). Overall high nestedness (calculated as NODF; see Chapter 4, Appendix 3) and significant correlation between inter-spring distances and spatial turnover (expressed by the Simpson coefficient, β_{sim}; see Appendix 3) were consistent with an equilibrium scenario mainly regulated by immigration and extinction. However, separate analyses for stygobites (species that complete their whole life cycle in groundwater) showed no significant nestedness and no correlation between inter-spring geographical distances and β_{sim}. The very different behaviour of stygobites, which include many distributional and evolutionary relicts, can be explained by their low capability to disperse through the aquifers. Differently from non-stygobites, stygobites tend to be mainly confined to the springs where they were transported and remained trapped by spring-bed sediments.

Caves are islands of cavities surrounded by impenetrable rock. However, for some organisms, caves can be interconnected by the 'ocean' of the surface (e.g. bats) and fractures or small solution tubes (e.g. some minute arthropods). Some adaptations in cave animals parallel the evolution of flightlessness (Roff, 1990), gigantism and dwarfism (Howarth and Moldovan, 2018) observed on true islands (Whittaker *et al.*, 2023). Also, like true islands, caves have exceptionally high levels of endemicity, a consequence of their isolation (Culver and Pipan, 2019). Some papers attempted to apply island biogeography principles to cave animals, especially to test the effect of area and isolation on species diversity. A seminal study on the cave fauna of southern Switzerland (Vuilleumier, 1973) found that the cave fauna was strongly affected by resource availability (larger areas, water availability and presence of bats that enrich cave ecosystems with their guano) but not by the climatic conditions of their surroundings (elevation) and isolation (proximity to other caves). Brunet and Medellín (2001) showed that the number of bat species roosting in central Mexican caves was related to cave ceiling area (a surrogate for habitat diversity, especially solution pockets in the ceiling that apparently act to retain body heat of roosting bats), but not to cave isolation (since distance between caves was within the flight range of bats). As regards invertebrates, a study in the Brazilian Atlantic rainforest found that richness correlates

Continued

Box 7.1. Continued.

positively with the cave's linear development possibly because of the increase in microhabitats and food resources (Silva *et al.*, 2011). In the Azorean lava tubes and volcanic pits, cave length is a predictor of troglobite species richness (Borges *et al.*, 2012) (Fig. B7.1.4). Here it was also found that the mean local endemic troglobite species richness per island (**α-diversity**, see Chapter 4) is a linear function of the regional number of species (**γ-diversity**, see Chapter 4) occurring on each island, which indicates that the size of most local cave species assemblages is proportional to the size of the regional pool of potential colonists.

Biological communities on mountaintops are separated from each other horizontally by long distances and vertically from those living at lower elevation by climatic barriers. Because of this isolation, high-elevation habitats are indicated as sky islands. This expression was originally coined by Weldon Heald (1951) in his writings on the mountains of south-eastern Arizona (USA), which harbour isolated montane forest or woodland surrounded by a sea of desert (Fig. B7.1.5), and then expanded to include the most disparate settings in which high-elevation areas are separated by inhospitable lowlands, such as plateaus, mountain ridges, paramos, alpine meadows, flat-topped steep-sided plateaus in northern South America called tepuis, patches of cloud forest at the tops of the eastern arc mountains of Tanzania, volcanic peaks in the Philippines, and even tops of guyots and seamounts which are surrounded by deep ocean (McCormack *et al.*, 2009). Because of their horizontal and vertical isolation, sky islands are generators of diversity characterised by high levels of endemism (McCormack *et al.*, 2009; Moore *et al.*, 2014). Current diversity patterns of mountain sky islands seem to be mainly due to elevational shifts which occurred during glacial cycles, with limited impact of current dispersal (McCormack *et al.*, 2009; Moore *et al.*, 2013). Brown's (1971) seminal work on mountains in Utah and Nevada (USA) revealed that, as predicted for true islands, the number of montane specialist small mammals is correlated with mountaintop habitat island area. However, in contrast with what could be expected under the ETIB, mammal diversity was not influenced by distance from a larger montane forested area. This suggests that modern mountaintop communities are fragments of formerly widespread communities that inhabited lower elevations during the Last Glacial Maximum. Similarly, mammal populations of the Great Basin sky islands (USA) are not in equilibrium but underwent **relaxation** after the Pleistocene, as shown by the lack of distance effect and absence of appropriate species from large areas of apparently suitable habitat (Brown, 1978). By contrast, no significant reduction in numbers of bird species as a function of insular isolation and a strong dependence of species diversity on habitat diversity suggest that the distribution of permanent resident birds in Great Basin sky islands represents an approximate equilibrium between contemporary rates of colonisation and extinction (Brown, 1978). Of course, this is a reflection of the higher dispersal capabilities of birds compared to mammals. Because of their

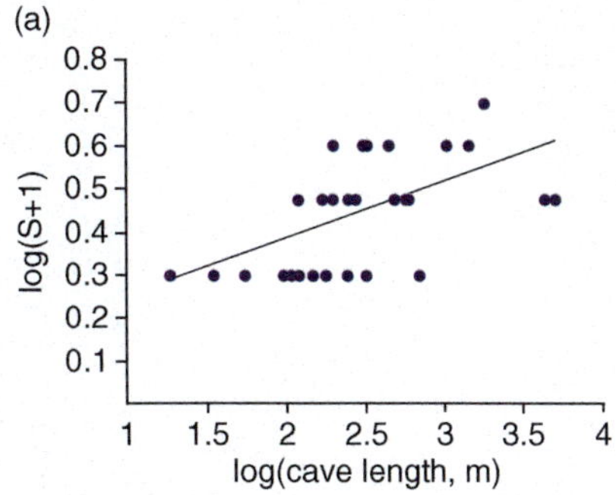

Fig. B7.1.4. Troglobiont species richness (S) in lava tubes and volcanic pits in the Azorean Islands. **(a)** Relationship between richness and cave length: $\log(S+1) = 0.12 + 0.13 \log(Length)$, $R^2 = 0.33$; $P < 0.001$ (decimal logarithms are used). **(b)** Roof formations in the Algar do Carvão cave (Terceira). This very large cave, remarkable for the peculiar mineralogical characteristics of its silicate stalactites, is a prominent tourist attraction. The intensive show-cave activity is, however, markedly impacting the temperature and relative humidity of the cave, and the presence of artificial lights has favoured the spread of green algae, mosses, etc. (inset). (Panel **a** from Borges *et al.*, 2012, redrawn with permission of P.A.V. Borges. Panel **b** photos by S. Fattorini.)

Continued

Box 7.1. Continued.

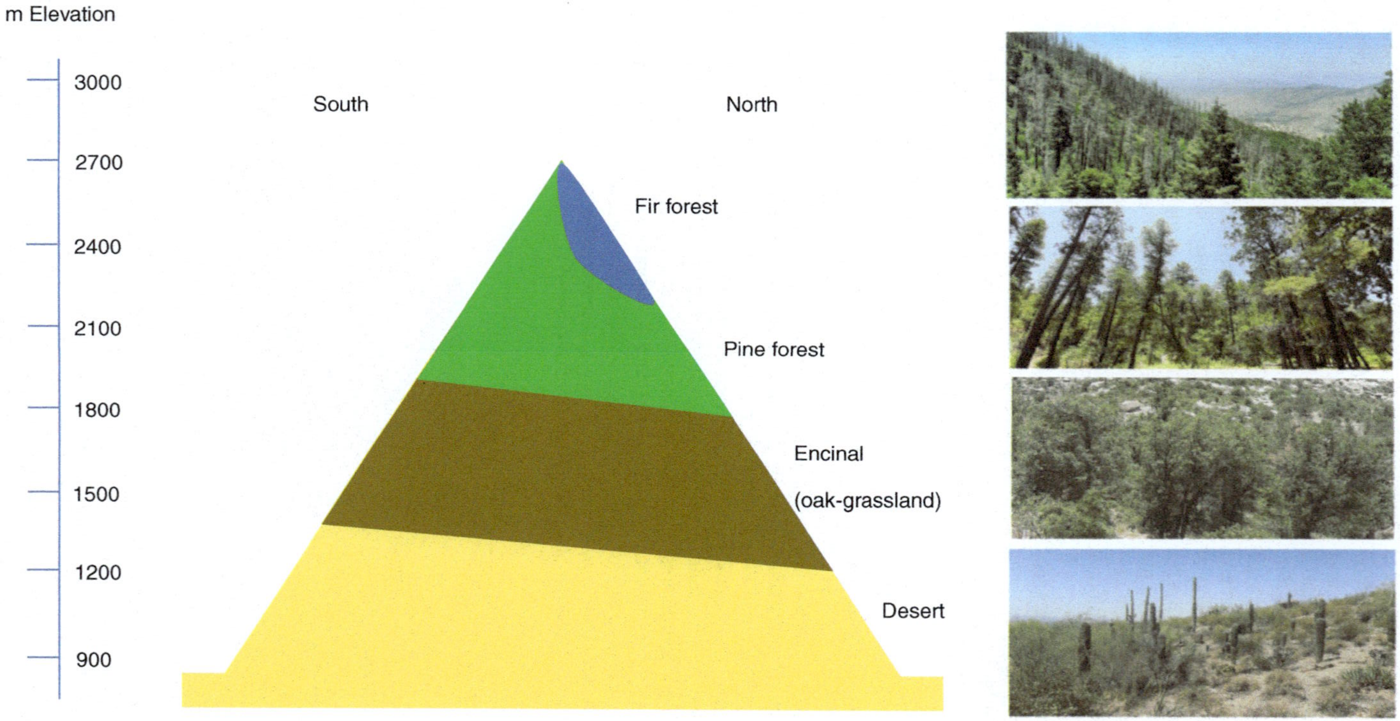

Fig. B7.1.5. Distribution of major vegetation types in the Santa Catalina Mountains (Arizona, USA; see Brusca and Moore, 2013 for a more detailed subdivision). The Santa Catalina Mountains belong to the Madrean Sky Island Archipelago. Forests on mountaintops are isolated horizontally from those of other mountains by geographical distances and from the lower habitats of the same system by strong environmental differences. (Scheme after McGinnies, 1981; photos by S. Fattorini.)

Continued

Box 7.1. Continued.

ability to inhabit and disperse across woodlands and other low-elevation habitats, forest mammals in the American Southwest are influenced by post-Pleistocene immigrations as well as extinctions (Lomolino *et al.*, 1989). In central Africa, the Last Glacial Maximum confined many insect species to single refugial montane areas because of aridification of lowland areas, allowing rapid allopatric speciation in flightless beetles (Fattorini, 2007a). The impact of aridification in Africa due to glaciations at higher latitudes on montane biota is neatly illustrated by the evolutionary history of the African montane bird *Pogonocichla stellata* (Muscicapoidea), with mitochondrial DNA sequences revealing vicariant events related to peaks of Plio-Pleistocene aridification (Bowie *et al.*, 2006). Bird species richness on mountaintops in north-western Argentina (Nores, 1995) increases with the island area and decreases with distance from the Andes, but there is no evidence of turnover over a century. However, a partially nested pattern found in this system seems to result from selective immigration rather than a state of faunal 'relaxation': 70% of species occur on just one or two islands, and the number of species is inversely correlated with island occurrences, with most species occurring in one or two islands.

The conceptual framework of SAR studies has been frequently used to investigate host–parasite relationships, by considering hosts as isolated 'areas' and parasites as species 'inhabiting' these areas. The issue has been reviewed by Strona and Fattorini (2014b). The number of parasite species per host is expected to increase with the size of the host's range and body. The first expectation is based on the assumption that hosts with broader distributions have greater chances to be reached by a larger number of parasite species (parasite–host range relationship). The second expectation derives from the idea that body size drives the number of parasite species per host, as more niches are available on larger hosts (parasite–host size relationship). Most research, however, failed to support these relationships. Reasons for this lack of support can be mainly due to (i) the non-neutrality of host–parasite relationships, and (ii) the evolutionary tendency towards host specificity. Host–parasite networks are not neutral, as they are the result of processes that minimise coextinction: specialist parasites minimise their risk of coextinction by using hosts with low vulnerability. Since large-sized species are typically less abundant and more endangered than small-sized ones (see Box 4.1), this may lead larger hosts to harbour fewer parasite species than smaller-sized ones, in contrast with expectations based on the parasite–host size relationship. Most parasites tend to be highly specialised towards particular hosts, leading to a pronounced skewness also in the distribution of the number of parasites per host. As a result, the number of parasites per host tends to be low, independently from host body size and range (Fig. B7.1.6).

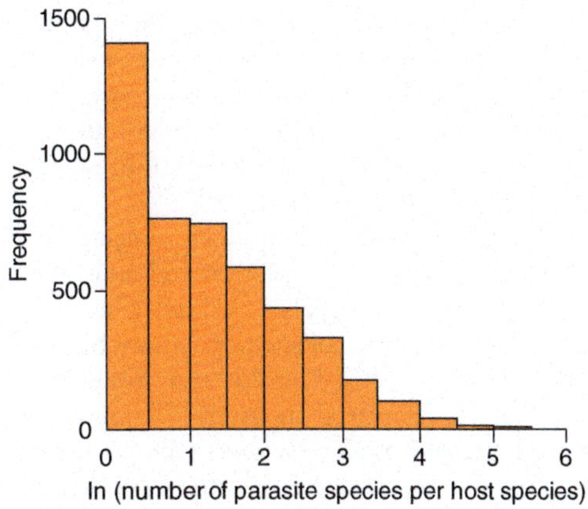

Fig. B7.1.6. Histogram showing the distribution of the number of parasite species per host species in fish. The most recurrent situation is that of hosts used by only one parasite species. (Natural logarithms are used). (From Strona and Fattorini, 2014b, redrawn, used under licence CC BY 3.0.)

1993; Schnell *et al.*, 2013) and local population turnover (extinction/recolonisation) is considered to have a characteristic half-life (Gilpin, 1991). May (1994) described an analogy for the equilibrium model (where the rates of colonisation and extinction of local populations remain constant) in which, as within a city night, lights are being switched on and off (different ones *in perpetuum*) but the overall illumination remains the same. Various models have also been drawn up to describe the nature of fall-off of movements of individuals belonging to each species, with distance. A negative exponential function describing the probability M of an individual moving a certain distance D is:

$$M_{ij} = e^{-kD_{ij}} \tag{7.5}$$

where e is Euler's constant. Assuming an inverse power function, fall-off of movement is described by:

$$M_{ij} = D_{ij}^{-k} \tag{7.6}$$

in which k is a migration constant (describing the rate at which movement declines with distance – the larger the k, the more rapid the decrease) (Hill *et al.*, 1996; Baguette *et al.*, 2000). These differ from a linear expectation in the decay of movements given as:

$$M_{ij} = -kD_{ij} \qquad (7.7)$$

An important difference between the negative exponential and inverse power models is that the latter has a longer tail, viz. a few individuals travel further.

Within a metapopulation, the individual populations may play different roles, acting as **sinks** (local populations that would go extinct in the absence of immigration), **sources** (local populations where growth rates at low density are positive, providing immigrants for sink populations) and **pseudosinks** (patches that can support local populations without immigration, but where density would decrease relative to carrying capacity without immigration from sources) (Dennis, 2010). Differences between sources, pseudosinks and sinks do not represent a clear division but are part of a gradient, along a compensation axis defined by immigration, emigration, birth and death rates (Fig. 7.4).

Modelling and empirical findings indicate that critical limits in patch size, isolation and number of 'habitat' patches are reached below which the metapopulation starts to fail and populations become extinct (Hanski, 1997; Hanski and Simberloff, 1997). 'Habitat' is placed in inverted commas as the models created to date have been based almost entirely on a single resource (e.g. for Lepidoptera, the larval host plant(s)). Expected time to metapopulation extinction has been found to be closely related to the product of $P\sqrt{H_S}$ where P is the fraction of occupied patches (local populations, assuming homeostasis) and H_S is the number of suitable patches (Hanski, 1997). The number of well-connected patches required for long-term persistence is termed the **minimum viable metapopulation (MVM)** (Thomas and Hanski, 1997), but as few organisms have been explored, great caution is required in relying on particular numbers of patches in specific cases. An additional concept for metapopulation persistence, the **minimum amount of suitable habitat (MASH)**, clearly depends on the quality of habitat (Dennis and Eales, 1997; Hanski, 1997). Like any model, metapopulation models are based on certain principles and assumptions, especially about patch area, isolation and habitat quality (see, for example, some made for butterflies listed in Table 7.2).

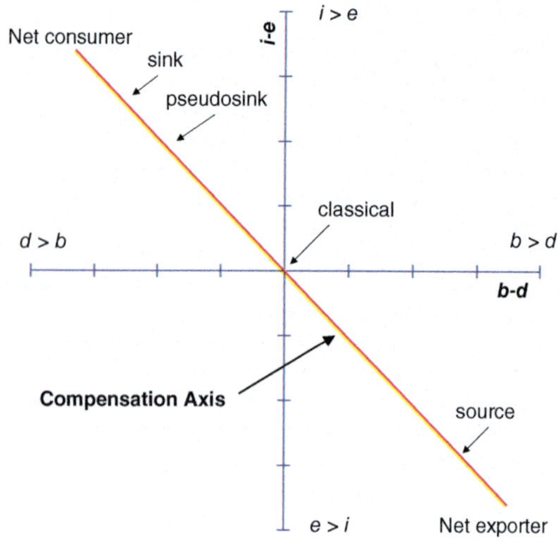

Fig. 7.4. The Compensation Axis for metapopulation units (local populations). *b*, birth; *d*, death; *i*, immigration; *e*, emigration (per capita rates). Classical (Levins) situations have $i = e$ and $b = d$. Thomas and Kunin (1999) have shown how each (meta)population system can be described by two axes based on four population parameters: per capita birth, death, emigration and immigration rates. This also defines the status of individual population units as sources, in equilibrium, pseudosinks or sinks. There is controversy in the literature as to how frequently classical (Levins type) metapopulations (i.e. metapopulations at equilibrium) are expected to be found. Hanski and Kuussaari (1995) consider that as many as 57 of 94 resident Finnish butterfly species may belong to the Levins-type metapopulation, and Hanski and Thomas (1994) regard 50% of the 60 British butterfly species as conforming to this model. (From Thomas and Kunin, 1999, redrawn, used with permission of John Wiley & Sons.)

An empirical approach to metapopulations involves application of logistic regression in which a binary response (dependent) variable (patch incidence: 0, absent; 1 present) is regressed against patch area and isolation. It has been carried out on a number of butterflies species (e.g. the silver-studded blue *Plebejus argus*, the silver-spotted skipper *Hesperia comma*, the Lulworth skipper *Thymelicus acteon* and the heath fritillary *Melitaea athalia*; Thomas and Harrison, 1992; Thomas *et al.*, 1992; Thomas and Jones, 1993; Hanski and Gilpin, 1997). In these studies, patch area is a zone of host plant suitable for egg laying and development; isolation of patches has been measured in two ways: to the nearest occupied patch and to the nearest patch occupied or not. The model, with isolation

Table 7.2. Basic principles of metapopulation models for day-flying Lepidoptera.

Generalisations based on patch area and isolation
• The larger the patch size (habitat), the larger the population and the longer it is expected to persist.
• The larger the patch size, the larger the number of emigrants expected to emanate from it as potential colonists for other patches. Smaller patches may nevertheless emit more emigrants per capita of their populations.
• Larger patches, compared with small patches, are expected to receive more immigrants from the same source if equally isolated from it.
• The greater the distances separating patches, the fewer individuals expected to migrate successfully between them.
• The persistence of populations within a metapopulation, and the metapopulation itself, depends on the number of patches and their spatial arrangement as well as on patch size and isolation.
Principles of habitat quality for metapopulations
• Patches are not equal in resource composition, structure and connectivity (between resource outlets) and thus not equal in habitat quality.
• Some resources other than consumables (larval host plant and nectar) can be limiting for any life-history stage. Utilities (e.g. roost sites, mate location sites) should be included in assessments of habitat quality.
• Differences between local populations in basic inputs (topography, substrates, climate, land management) will inevitably cause differences in habitat quality.
• Even patches with identical starting points will be unlikely to retain the same habitat quality trajectory owing to stochastic influences affecting inputs on sites.
• Connectivity (between patches) can affect habitat quality as enemies and competitors are influenced (possibly differently to the target prey organism) by connectivity between habitat patches.
• Although large patches are more likely to have high-quality habitats, this is certainly not an invariable law.

(From Dennis, 2010.)

measured to nearest occupied patches, proved to be highly significant for all the above mentioned four species, though patch area was significant only for *P. argus* and *H. comma*. More sophisticated measures of isolation have been introduced over the years (Moilanen and Nieminen, 2002). However, Hanski (1994a, b) recommends use of a measure for isolation (I) that takes into account distances to all patches weighted by their areas (their potential for emigrants and colonists):

$$I = -\sum \exp(-\alpha D_{ij})A_j \tag{7.8}$$

where $A_{j\neq i}$ is patch area, D_{ij} is distance between patches, and α is a constant describing the survival rate of migrants over distance D_{ij}.

A fundamental assumption of metapopulation models is that larger patches support larger populations. However, there are a number of reasons why the assumed relationship between local population size and patch size in a metapopulation can break down. Some obvious ones are: (i) a disproportionate increase in the extent of suboptimal habitats within patches with increasing patch size; (ii) inequality in economic potential for land units of different size leading to increased intensity in the exploitation of larger units; (iii) increased exposure – lack of shelter – of larger land units relative to boundary vegetation (Dover *et al.*, 2000; Matter and Roland, 2002; Matter *et al.*, 2003). Even so, since patches for a studied species can range in size (area) over three or more orders of magnitude (i.e. 0.1 ha to 100 ha), a relationship between population size (or species incidence) and patch area is generally expected, if only because increments in population size will tend to be associated with increases in patch size despite any lowering in the quality of larger patches. The reason for this is that with big increases in patch size, resource quantity can partly offset reductions in resource quality. However, if differences in patch areas are in orders of magnitude but differences in their populations are not, and vice versa, then quality and composition of the resource base very likely lie at the root of the difference.

Metapopulations and habitat quality

However sophisticated such models appear to be, they are nothing like reality, where the *quality* of each local population's 'patch' (its complex of resource-based habitat components, including conditions of existence) plays a fundamental role in determining (meta)population dynamics (Dennis and Eales, 1997, 1999; Dennis, 2010); also through its influence on the traits of each species which are subject to selection, with respect to both predictable (e.g. vegetation succession) and unpredictable (e.g. weather) factors. The first example to demonstrate the significance of habitat quality in an effective patchwork was the large heath butterfly *Coenonympha tullia* in raised bogs; it was found that habitat quality (expressed by an index including several attributes) ranked above patch area and isolation in accounting for occupied sites, despite various measures being used to describe both patch area and isolation (Dennis and Eales, 1997, 1999; Dennis, 2010) (Fig. 7.5). Various other studies of metapopulations have also demonstrated the importance of habitat quality (Gutiérrez *et al.*, 1999; J.A. Thomas *et al.*, 2001; Matter *et al.*, 2003; Välimäki and Itamies, 2003; Auckland *et al.*, 2004).

Where key resources (e.g. host plants for butterflies) are fragile, limited and substantially isolated (e.g. the monocarpic kidney vetch *Anthyllis vulneraria* for larvae of the small blue butterfly *Cupido minimus*; Morton, 1985), an extreme form of metapopulation may be found in which substantial occupancy and virtually total abandonment may occur in pulses over years dependent on oscillating suitable and unsuitable conditions. The pattern emerging is of distant migration, advancement over the countryside, discovery of small resources and colonisation during stable hot, sunny weather but then a breakdown in mating success and reproduction, extinction of these small, usually tiny local populations, and collapse of the distribution/geographical range during summers with obscured skies, cool and wet weather, conditions inimical for activity and flight. A prime example is the 'coming and going' of the silver-washed fritillary *Argynnis paphia* in Cheshire, probably colonised from more permanent strongholds in neighbouring Derbyshire, UK (R.L.H. Dennis, personal observation) (Fig. 7.6). As a useful term, we might refer to such situations as **ephemeral metapopulations**.

A further aspect of resource availability (consumables and utilities) is the number of potential resources. In an earlier text (Dennis, 2010: 145), the first author demonstrated that with an increase in the number of resources (i.e. variety of larval host plants and adult nectar sources) used by a butterfly species, there is increased size of habitat and a decrease in the proportion of terrain that is completely vacant **matrix**. That is, the matrix is likely to incorporate some resources that can be used. By applying set theory one can establish, by way of example, that for five distinct resource types (e.g. the larval host plant, which is the key resource, and four further resources: nectar source, pupation sites, mate location sites, roosts), there are 16 habitat combinations that include the key resource, and a further 16 combinations that lack the key resource but which include some other resource(s) (thus, they are a matrix, but a matrix with resources).

Metapopulations on the move: shifting habitats

Each one of these local population models has a stable **landscape** (vegetation, **biotope**, traditional habitat) patchwork, within which each individual, and the entire local population, has its **resource-based habitat** (**RBH**). The bounds of each patch form limits to the home range, the boundary of its resource-based habitat and that of the local population. Again, in accepting these assumptions, we engage in highly simplistic notions of stability for the patch. However, patches are rarely stable. Even if they are bounded by barbed wire and fences, the conditions within the enclosed areas undergo constant change. To those walking their dogs past them each day, changes may not be that noticeable because they occur through continuous, subtle shifts. But to the organisms that inhabit the patches, the conditions may undergo profound changes, not just annually but also seasonally and daily. Casual observation may not pick up the loss of bare substrate essential for creating a microclimate of high ground temperatures near a host plant, a condition that might be fundamental for a butterfly such as the grizzled skipper *Pyrgus malvae* (Brereton, 1997). Such situations can only be accessed scientifically by expert skills and methods, repeated systematic sampling with care taken to avoid as much human interference as possible. For instance, for ground invertebrates (e.g. carabid beetles), a fixed pattern of pitfall traps (see Appendix 1) stationed in exactly the same spots is required. For other organisms, alternative techniques need to be chosen depending on their ecology, with the shared aim of maintaining the same precision potentially *ad infinitum* (e.g. Wheater *et al.*, 2011). We never cease to be surprised by remarks that relate to the apparent constancy of a heath, a wood or some other biotope and how the narrator would be pleased to see it left

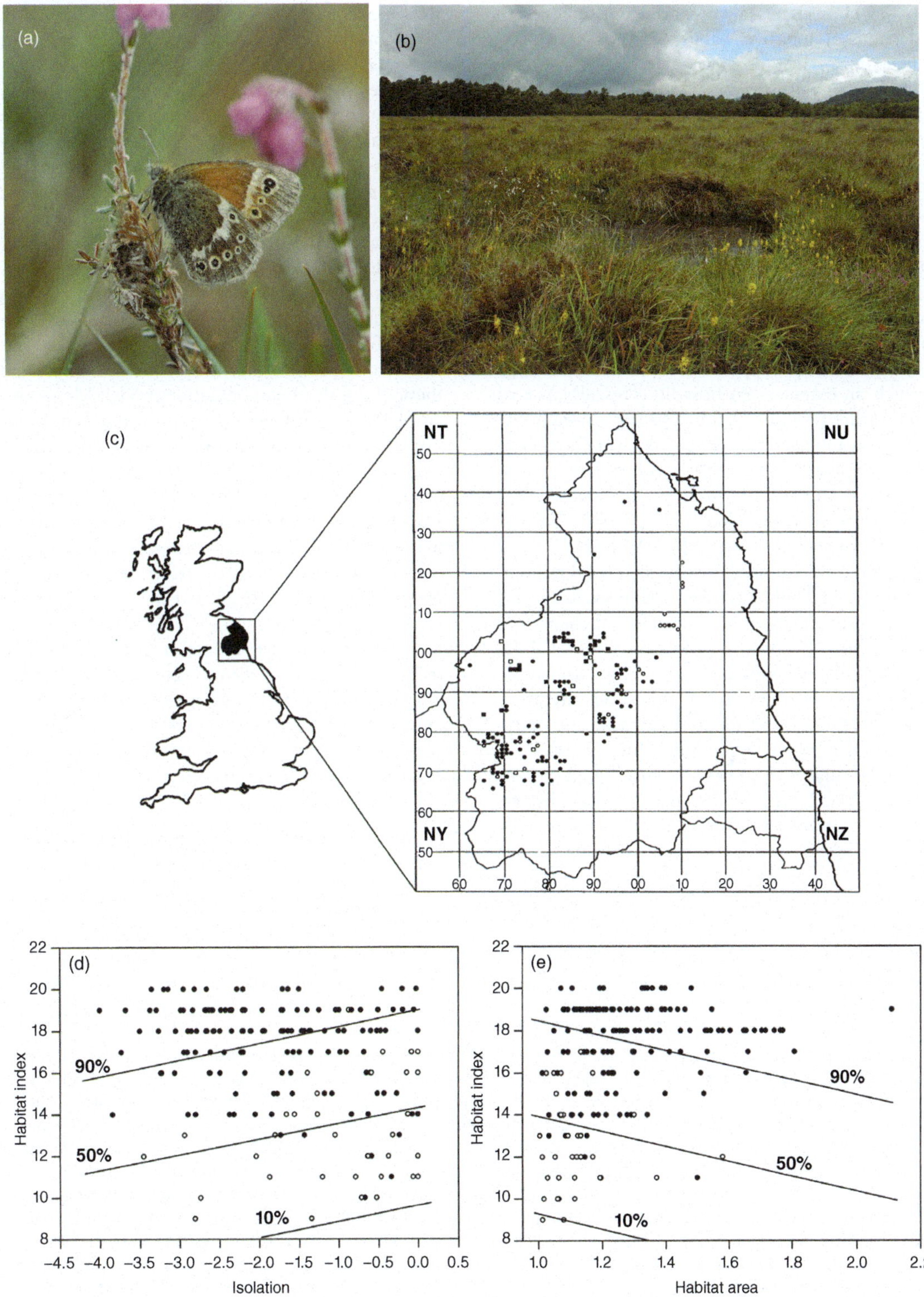
(a)
(b)
(c)
NT
NU
NY
NZ
50
40
30
20
10
00
90
80
70
60
50
60
70
80
90
00
10
20
30
40
(d)
Habitat index
Isolation
90%
50%
10%
22
20
18
16
14
12
10
8
−4.5
−4.0
−3.5
−3.0
−2.5
−2.0
−1.5
−1.0
−0.5
0.0
0.5
(e)
Habitat index
Habitat area
90%
50%
10%
22
20
18
16
14
12
10
8
1.0
1.2
1.4
1.6
1.8
2.0
2.2

Continued

as it is! How it has always been the same! Such reflections of apparent stability are nonsense.

In a Levins metapopulation (see above, section *The metapopulation concept: isolation and area of habitat units*), extinction and colonisation are assumed to be part of a stochastic process. In fact, extinctions may well be the consequence of deterministic agents such as **ecological succession** (because originally favourable conditions change to unfavourable ones) and colonisation by other, superior competitors, not to mention concomitant anthropogenic causes, such as intentional changes to the substrate by the landowner without thought for the organisms occupying the patch. In time, the whole patchwork supporting a species may change to such an extent as to bear little resemblance to some earlier period of habitat survey although the species of interest may retain its stronghold in a new patchwork covering the same area or region. A prime example is the shifting patchwork occupied by the silver-studded blue butterfly *Plebejus argus* on the Great Orme (North Wales, UK) from the 1960s to 2015, and its associated local population status affected by rock fall defences, fires, goat and sheep grazing, public pressures and changes in local climate (Dennis and Sparks, 2006; R.L.H. Dennis, personal observation). With broadscale climate shifts we may expect substantial alteration to local climates and to the vegetation on slopes (Dennis, 2020), including the plants selected as food by primary consumers (herbivores). A case in point is the changes in the patches occupied by local populations of the silver-spotted skipper butterfly *Hesperia comma*, which now use taller patches of sheep's fescue *Festuca ovina* during hotter summers than in the 1980s (Davies *et al.*, 2006; Lawson *et al.*, 2013). In fact, local climate transformations can be far faster not only than those arising from the ongoing climate change, but also than those perceived from season to season or day to day. Work on the silver-studded blue butterfly *Plebejus argus* has demonstrated massive modifications in its resource-based habitat within minutes. With the sun fully out in calm conditions, *Plebejus argus* occupies open bare slopes well above the base of the Orme (Dennis and Sparks, 2006). However, with passing cloud and the loss of sun, increased windspeed and rain, the butterflies will rapidly take cover in gorse, cotoneaster and bramble along the path lower down the slopes, and bracken areas on upper slopes (Dennis and Sparks, 2006). What may appear to constitute the butterfly's RBH one moment may no longer be part of its RBH the next.

It is obvious that not all species fall under the umbrella of the metapopulation model. Many species have more or less continuous (unfragmented) distributions and an extreme situation is the case of those that for much of their lives are in constant motion, such as the monarch butterflies (*Danaus plexippus*). Where monarchs congregate for diapause in Mexican forests, they form enormous mixed gatherings of roosting individuals from different parts of North America (Agrawal, 2017). Similarly, many birds occupy different sites in northern latitudes following their winter feeding in Africa, and where they form pairs are not in the characteristic format of metapopulations (Newton, 2007). At the other extreme, there are species represented by a single, very small population, such as those endemic to individual caves or mountaintops.

Factors driving local populations to extinction

Local population extinction is the result of death (D) and emigration (E) rates exceeding birth (B) and immigration (I) rates ($D + E > B + I$) (see Chapter 2). When a set of local populations, a metapopulation,

Fig. 7.5. Metapopulation of the large heath butterfly *Coenonympha tullia*; prominence of habitat quality over habitat patch size and isolation. **(a)** *Coenonympha tullia*; **(b)** Meathop Moss (Cumbria, UK), a valley raised bog habitat patch for the butterfly. **(c)** Distribution of *Coenonympha tullia* in Northumberland (UK) during 1995–1997 (1 km squares): 1995–1996 (circles); 1997 (squares); vacant sites and unknown status (open symbols); occupied sites (closed symbols). Inset map illustrates position of Northumberland in the UK. **(d)** Distribution of sites in axes for habitat quality and isolation: colonised (black circles) and vacant (open circles) sites; probabilities from logit regression of status on habitat quality index, isolation and habitat area are as follows: $P < 0.0001$, $P < 0.01$, $P > 0.05$. **(e)** Distribution of sites in axes for habitat quality and habitat area: colonised (black circles) and vacant (open circles) sites; probabilities from logit regression of status on habitat quality index, isolation and habitat area are as follows: $P < 0.0001$, $P < 0.01$, $P < 0.01$. $Isolation = -\Sigma\exp(-D_{ij})$, where D_{ij} is the edge-to-edge distance (km) to the five nearest neighbouring occupied sites. Lines give fitted 90%, 50% and 10% predictions for occupancy. (Panels **a** and **b** courtesy of Jim Asher. Panel **c** from Dennis and Eales, 1999, used with permission of Elsevier Science & Technology Journals. Panels **d** and **e** from Dennis, 2010.)

Fig. 7.6. Sporadic metapopulations. **(a)** Silver-washed fritillary butterfly *Argynnia paphia,* with a female form *valezina.* **(b)** *Argynnis paphia* violet (*Viola* sp.) host plant along a road side in Macclesfield Forest, Cheshire, UK. **(c)** Locations where *Argynnis paphia* has been observed between 2006 and 2021 in the Macclesfield Forest. Each year is shown by a different coloured dot. The yellow line links the marginal dots indicating the possible home range limit. The Ranger station is the control centre for the National Peak Park. The butterfly is colonising the forest from Derbyshire, which is hotter and drier, and has better resources. **(d)** A process-response model illustrating the interaction of factors leading to the colonisation, range extension, extinction and range collapse in sporadic (ephemeral) metapopulations (e.g. *A. paphia*). Continuous arrows indicate a positive link, dashed arrows a negative link. Butterfly metapopulations in temperate regions may be extremely vulnerable to weather conditions from year to year. During cool wet summers there tends to be a collapse of metapopulation units. During hot dry summers there tends to be an increase in the number of metapopulation units and their ability to maintain populations. Butterflies such as *A. paphia* with strong flight can rapidly gain ground and build metapopulations during hot dry summers, but collapse with cold and wet seasons. (Panel **a** courtesy of Jean McMahon. Panel **b** photo by R.L.H. Dennis.)

Continued

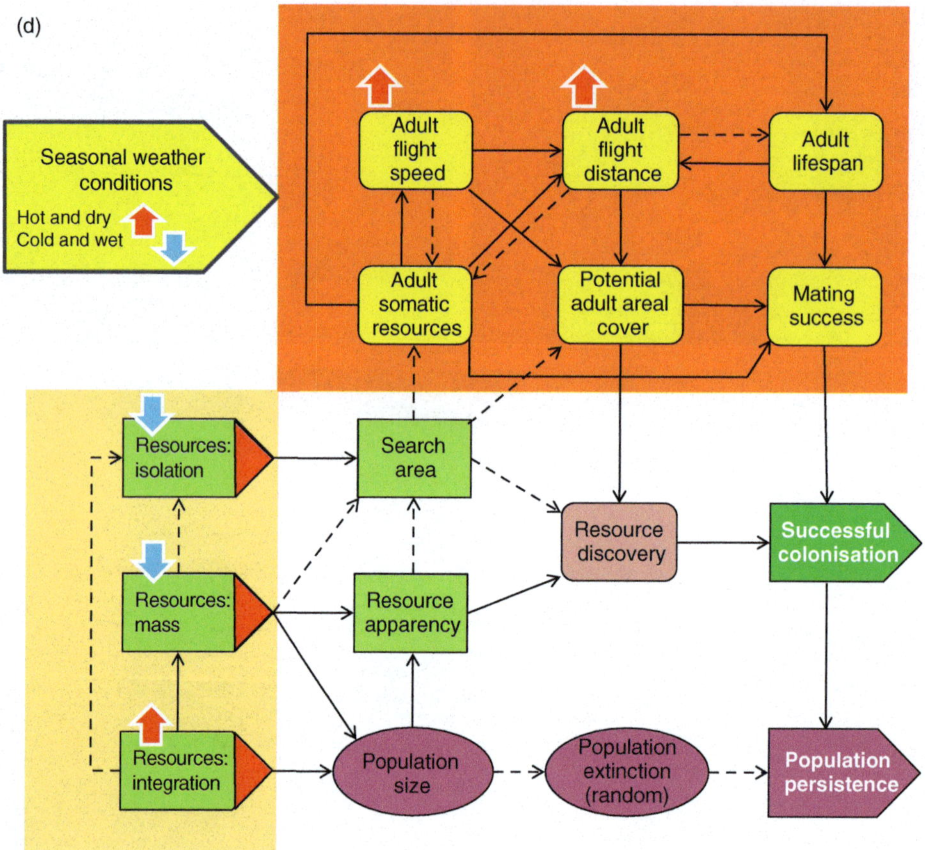

Fig. 7.6. Continued.

becomes extinct, it is the consequence of the systematic loss of local populations exceeding the colonisation of 'new' ones. The system is out of balance and is degrading faster than recovery: in every patch, deaths exceed births and losses exceed immigration into them. Only in the mainland-island metapopulation model (Boorman–Levitt) is there hope of full recovery, with colonisations from the large 'mainland' patch compensating for the extinction in the remaining patches (Dennis, 2010). A number of models have been built to illustrate that metapopulation persistence often depends on the area of patches (islands) and their proximity to one another in terms of mobility of the organism (Legendre *et al.*, 2008). Fundamentally, the larger the patches and the shorter the distances, the more likely it is that the metapopulation will persist. Populations vary in size from year to year and it takes a chance 'poor year' or chance 'sequence of years' of over-wet, excessively cloudy, cold weather – or some other unfavourable condition – to bring the entire system crashing to extinction.

On the other hand, factors impinging on the metapopulation may be systematic; a change of climate in the region over a number of years or decades, a gradual change in vegetation (i.e. **ecological succession**) or systematic changes in land use due to anthropogenic factors are classic examples. Just what effect changes in the environment have on any organism depends on the traits of the organisms occupying the patches, as well as on the parameters describing net production and mobility. An analysis of changes affecting metapopulations of different species occupying the same patches can produce very different results over different periods of years. In a north German study (10 years; 359 populations of eight short-lived plants: *Lepidium campestre*, *Thlaspi perfoliatum*, *Rhinanthus minor*, *R. serotinus*, *Melampyrum arvense*, *M. nemorosum*, *Gentianella* *ciliata* and *G. germanica*), the probability of survival of local populations increased significantly with patch size in all but one species (*R. minor*) (Matthies *et al.*, 2004). Even so, many populations were subject to powerful stochastic forces operating over the years. For instance, nearly 30% of populations of the investigated plants became extinct over the study period. Estimated population sizes required for 90% probability of survival over 10 years varied widely among species, without any apparent influence on **life-history traits**.

Globally, when all organisms are taken into consideration, one might expect that a metapopulation of a large animal (e.g. the blue wildebeest *Connochaetes taurinus* that feeds on short grasses) has a more precarious future than a metapopulation of a small animal (e.g. a grass-feeding butterfly). This is basically true (see, for example, the concept of vulnerability discussed in Chapter 4, Box 4.1). But bear in mind always that the outcome of change depends on the type of changes affecting metapopulations and the **traits** of organisms, such as their size, motility, breadth of food supply and number of enemies (the fewer the better). Thus, synchronised extinctions and colonisations can occur across population units (Sutcliffe *et al.*, 1997) and warn against too great a reliance on basic geography alone (number of population units, patch area, number of patches, isolation of patches). Rohde (2006) would impress on us that systems in non-equilibrium are more widespread than those in equilibrium. One crucial control in attaining at least a homeostatic, if not equilibrium, state is habitat heterogeneity (see Box 4.3 in Chapter 4); increased heterogeneity results in increased persistence of local populations, because a greater heterogeneity counteracts the impacts of stochastic changes, such as habitat quality changes in relation to weather (as observed, for example, in the bush cricket *Metrioptera bicolor* in Sweden; Kindvall, 1996).

Is the matrix ever entirely empty?

This question was once again first investigated in relationship to Lepidoptera (Dennis, 2004a; Dennis and Hardy, 2007; Hardy and Dennis, 2010; Shreeve and Dennis, 2011; Spalding *et al.*, 2012). Two types of area may be regarded as **matrix**: (i) zones vacant of resources *within* the **resource-based habitat** bounds; and (ii) regions beyond the RBH patch (e.g. outside the home range). However, the matrix may be considered to be completely inimical to survival (as typical of island systems, e.g. butterflies or birds flying over extensive oceans), or actually contain substrates usually not considered a resource but occasionally of use, more particularly of vital importance for survival (e.g. substrate used for settling and resting between two patches that make up habitat and have all the resources required by the species) (Table 7.3).

Regarding 'empty' space within a resource-based habitat, there is a danger of assuming that zones lacking obvious resources in the RBH (e.g. bare soil without larval host plants and flowers lacking nectar) are of no value to a species such as a butterfly. In fact, such spaces may be of great unknown value. Bare ground dries and heats up rapidly, providing potential substrate for resting and rapidly attaining heat for flight, as well as sustaining puddles with valuable salts for mud puddling. Trees around meadows generally not used by many butterflies nevertheless present areas where butterflies rest and roost (e.g. the wall brown butterfly *Lasiommata megera*; Dennis and Bramley, 1985), and where mate seeking occurs. This has been witnessed time and again for pierid butterflies not thought to have any use for trees at all (Dennis and Hardy, 2007). Moreover, butterflies rarely find heavily mown grass of use yet the green-veined white butterfly *Pieris napi* has been observed searching for tiny shoots of Cruciferae (e.g. the cuckoo flower *Cardamine pratensis*) and then laying eggs on them (Hardy and Dennis, 2010). The same observations relate to matrix outside a resource-based habitat. The existence of usually ignored substrates can prove to be of great value for individuals dispersing between habitat patches, in effect providing a **corridor** or set of stepping stones for the butterfly (Dennis, 2004a). Most astonishing are observations of the large white butterfly *Pieris brassicae* landing

Table 7.3. Principles of matrix quality for metapopulations.

- The matrix is only distinct from habitat by degree and is rarely absolutely lacking in any resources.
- The matrix is not an isotropic plane (a smooth level surface) with movements equivalent in all directions. Matrix structure affects the direction, distance and speed of movements.
- Matrix resources, even in small quantities, affect survival, maintenance and reproduction within metapopulations.

(From Dennis, 2010: 142.)

on the sea to rest and then taking off again (Dennis and Hardy, 2018). Similarly, in urban ecosystems, in which green spaces are suitable patches interspersed in a matrix mainly occupied by built-up surfaces, resources in the matrix may help to sustain individuals that are dispersing towards more suitable areas (Fattorini *et al.*, 2018c); even a pot of flowers on a balcony can provide a resting place and food source for a pollinator insect in an **urban area**. A pot can even provide a breeding site, as with the geranium bronze butterfly *Cacyreus marshalli* (John *et al.*, 2018).

These few examples should raise concern for us when we intend to conserve and manage landscapes for species in **red lists** (e.g. the IUCN Red List: www.iucnredlist.org, accessed 13 May 2025): that we have all the information about the spaces they occupy (land, water and air) necessary to ensure their persistence and growth, without placing other species in danger.

Marginality and the status of resource-based habitats

Marginality has a number of meanings depending on the context. Regarding the broad range of a species, we might refer to the peripheral (marginal) populations located at the boundary of its range (typical usage). In situations when the range is changing, the leading edge of populations that are growing, that is, colonising new land (expanding), may be said to become less marginal, whereas the rear edge of populations in decline and becoming extinct (the range contracting) are said to become more marginal. At the population level, marginal populations are those in decline; the number of individuals is decreasing and/or the bounds of the population are contracting. At the local population level, too, marginality can be understood more clearly in the sense of the resource-based habitat; in this case, marginality would relate to a factor (one or more resources or conditions) causing decline in the population (see Liebig's law of the minimum and/or maximum; see Table 4.1 in Chapter 4).

Marginality, then, is clearly an issue of concern for conservation and management. Among butterflies, most often the decline in quantity or quality of the two fundamental consumable resources, larval host plants and adult nectar sources, is typically responsible for populations declining and becoming more marginal. However, there have been very few, if any, tests on the impact of other resources (utilities such as rest sites, sites to avoid predation, roosts, hibernation sites and mating sites; Dennis, 2010, 2020) on marginality in butterflies. Although utilities may appear less important than resources, they can have a significant impact on reproduction and survival of individuals in local populations in their quantity, quality and organisation, and therefore their degree of marginality (Dennis, 2010) (e.g. *Lycaena* sp. and *Boloria* sp.; Turlure *et al.*, 2019).

Marginality is fundamentally related to variations in conditions and resources along spatial gradients over the globe, such as those represented by latitude, longitude (continentality), altitude and depth in the sea. Over a species range, the magnitude of temperatures, radiation, precipitation, salinity, wind speed, etc. and their temporal patterns (seasonality) affect the resource base and conditions in which species exist (Shreeve *et al.*, 1996a). Even when resource use appears to be much the same from margin to margin of a species' range, the quality and persistence of resources, and the associated biodiversity (thus the interactions with other organisms), will change and impact differently on local populations. The evidence for this becomes apparent with the study of phenology of a species (i.e. the timings of cyclical events, such as migrations, mating, egg laying, flowering and hibernation; e.g. flowering patterns that change along elevational gradients; Bucher and Römermann, 2020), its physiology (number of broods, young born, size or condition of organs, tolerances, etc.; e.g. increase of species' thermal tolerance breadths with latitude; Sunday *et al.*, 2011), colouration (e.g. increasing melanism with altitude and latitude: Berry, 1977; Guppy, 1986; Alho *et al.*, 2010; and along rural to urban gradients, as in the peppered moth *Biston betularia*: Cook, 2003; Cook and Saccheri, 2013) as well as other species-specific attributes.

Marginality is evident even at a local scale, say within the span of a single local population. Where a single local population crosses a heterogeneous landscape (with hills and valleys, variation in geological strata, soil characteristics and human impacts, etc.), there will be changes in abiotic and biotic conditions, including consumable and utility resources. Many animals change their elevational distribution in response to seasonal changes in climate and food availability. For example, this is a common phenomenon among ungulates, such as mountain goats (*Oreamnos americanus*), which show seasonal movements, moving to lower elevations and on southerly aspects during the winter (Poole and Heard, 2003; Rice, 2008). But it occurs in many other groups, including birds (e.g. Inouye *et al.*, 2000) and insects (Kimura, 2021). The north-facing slopes may be highly unsuitable during a cold summer season, but of immense value as a refuge during a hot season affected by drought (Dennis, 2010).

Moreover, as the global climate becomes hotter, so too will differences in the landscape over which a species is found. The extent to which local populations of a species within a small region differ depends on the structure of the landscape occupied; even if all are similar in the occupied area, each will differ in its degree of marginality, that is, in its susceptibility to changing conditions. Any study of a species' patchwork should map the vulnerability of resources and the changing conditions in each patch. It would be a serious mistake by management to treat patches as of identical potential only because of similar dimensions; substrate and topography alone will make for differences. The basic ploy is to determine the weak resource zones in each patch and improve them without creating a deleterious impact on other resources and conditions for the target species and others on the various sites.

Habitat: climate and land use changes

Changes in animal and plant distributions are inevitable. The inference is that there are shifts in resource-based habitats of species undergoing movements in species' distributions, and changes in species' traits. There are two basic views taken of the process: first, changes dating to the past; second, changes that are predicted to occur in the future. The study of past changes usefully contributes to our understanding of future possible changes. Most importantly, the constellation of traits held by species determines the trajectory of species subject to changes in abiotic and biotic factors (e.g. marine fauna: Jones and Cheung, 2015; terrestrial fauna: Gaston, 2007; Zografou *et al.*, 2021). Even so, there are fundamental spatial patterns in global diversity (Lomolino *et al.*, 2010), and with climate warming there is likely to be a shift in maximum diversity of plants and animals from lower latitudes to higher latitudes (Dunne *et al.*, 2018).

Assessing past changes in ranges depends on two basic forms of evidence: (i) movements of species may be determined from fossils; (ii) the past distribution of contemporary species may have left no remains or little useful fossil evidence; determination of past distributions of such species depends on ancillary data. Regarding range changes determined from fossils, we can be reasonably certain that the species occupied, at least approximatively, the location or surrounding area indicated by fossils. However, fossils may occur in a wide variety of different contexts (exposed on substrates; within sub-substrates of earth and sediments; in caves or other concealed locations, etc.), each context affecting the usefulness of the finds. Past distributions of many plants are usefully determined from pollen (**palynology**) or macroscopic remains from peat bogs and even soils (Godwin, 1975; Huntley and Birks, 1983; Macphail *et al.*, 2004; Dawson *et al.*, 2019). However, some habitats are particularly problematic. For example, caves are not evenly distributed and thus the animals associated with them (birds, mammals; Cvetković and Dimitrijević, 2014) may have had a much wider distribution than the spread of caves. Moreover, many of the organisms found in caves, including ancient hominids, have been brought into the caves by predators (Pickering *et al.*, 2008); in which case, the cave is not necessarily part of their resource-based habitats. Similarly, long-distance pollen dispersal can significantly bias palynological interpretations by allowing pollen to appear in sedimentary records far from where the parent plants actually grew (Hjelmroos and Franzén, 1994). Finally, it is important to remember that it would be wrong to assume that the climatic and habitat conditions of the fossil's location matched that of any of the areas where the species is currently found; environmental conditions change over time and species evolve, possibly changing their preferences.

Butterflies are a classic example of organisms that leave very few fossils, though a number have been discovered dating back to the Cretaceous–Palaeogene boundary, supporting notions of their lengthy evolution (De Jong, 2017; Dennis, 2020). For the vast majority of butterfly and other animal species there are no fossil remains. In this case, their past distributions can only be determined from the conditions they currently occupy and the resources they use today. With this information, past distributions during slices of time can be assessed based on the current presence of the conditions and resources that these species use. It is not a matter of proof but of likelihood, as the distribution and resource use of a focal species may now be different in some respects from what they were once. This method was used by the first author to determine the origin and occurrence of butterfly species in Britain as a whole during 500-year periods for the past 20 ka, based on climate, biotopes and host plants (Dennis, 1977, 1993a). A finer analysis is now possible for raster units over much larger regions such as Europe. Turning to the future, as our technical ability improves, so we will be able to use finer details of a species' resource-based habitat across its current range to assess its likelihood of past occurrence over former landscapes. Of course, the process depends on equivalent advances in climatology, palaeontology, archaeology and geology used to reconstruct conditions and resources over the ancient landscapes.

The whole process of determining changes in future distributions of species with climate and habitat changes is simply a reverse of the historical process, but with the ability to use finer details determined for species from **autecological** surveys and sophisticated modelling of future landscapes (Poloni *et al.*, 2022). Undoubtedly, advancements in statistical and machine learning techniques of **species distribution modelling (SDM)**, which produce empirical descriptions of species ranges by combining species occurrence data with environmental predictors, will improve the accuracy of prediction of current, past and future ranges of species enormously (Lee-Yaw *et al.*, 2022; Franklin, 2023; Benavides Rios *et al.*, 2024). However, determining past distributions is likely to be a crude process compared to that of predicting future distribution changes as proof is absent and the reconstructions will remain speculative. The obvious difference is that predicting future changes is supported by proof as changing conditions for future dates are crossed; those living in the future will know if we are correct or wrong. With greenhouse climate changes occurring so rapidly, we will not have to wait long for records which support or undermine our models of species distribution shifts in response to projected climate changes. However, one has to be prepared for genetic changes with changes in climate as species adapt to new conditions (Dennis, 1977, 1993a; Hill *et al.*, 2008; Kellermann and Heerwaarden, 2019); these changes will affect our predictions.

Testing predictions may also serve to evaluate modelled retrodictions. For instance, traits considered to drive increased dispersal in butterflies at the end of the last glaciation (Dennis, 1977: 128; Dennis, 1993a) are supported by predicted changes with current climate warming (Hill *et al.*, 2008). This does not imply that all 'observations' on species' dispersal are trustworthy. Aspects of 'interference' occasionally occur, such as **introductions** (see Chapter 10). Moreover, predictions can be affected by taxonomic confusion (e.g. failure to distinguish and mixing cryptic species), unexpected changes in biodiversity relationships, and evolutionary shifts in resource use, typically host plants (e.g. the brown argus butterfly *Aricia agestis*; Kemp *et al.*, 2008). The stage has been reached where predicted distributions are now being tested. A classic example is research into the speckled wood butterfly *Pararge aegeria*, applying a model incorporating climatic data and potential habitat (biotopes, e.g. woodland cover). Hill and colleagues found that observed and simulated expansion rates were in very close agreement, being however 42–45% slower in an area that had 24% less woodland (Hill *et al.*, 2001). Our predictions may have statistical support, but as statistical significance is easily attained, the crucial thing is the degree to which our predictions are supported.

Habitats and genetic changes

Perhaps the most distinctive evidence that the **resource-based habitat** is variable for any species is the occurrence of genetic (and phenetic) distinctions between populations. More broadly, the RBH of a species using different resources (e.g. occupying different substrates) in different regions can have a substantial impact on its life history and on wider aspects of its biology (Asher *et al.*, 2001; Dennis, 1993a, 2020). For example, regional differences in wing phenotype are indicative of different selection pressures – past and present – on butterflies. The earliest observations were made on British butterflies in the first half of the 20th century by James W. Tutt, Edward B. Poulton, Edmund B. Ford, John W. Heslop-Harrison and Bryan P. Beirne (Dennis, 1977). For many years, the highly variable patterns among species in the British Isles were explained as evolutionary changes which occurred in isolation during the last (Devensian) glaciation (Ford, 1945; Beirne, 1947). One of us (Dennis, 1977) showed that this was not the case as most species could not have persisted during the severe cold conditions, dense ice and sea cover of supposed refugia. Instead, it became obvious that much of the variation had either evolved in refugia in southern Europe during the last glacial period (a process repeated with every glaciation; see Fig. 3.11b in Chapter 3) or during forest cover and climate changes in the Holocene (Dennis, 1977, 1992, 1993a; Dennis and Hardy, 2018). A model was established to account for wing phenotypic changes, complex contributory and conflicting pressures on butterflies in their Holocene habitats (Dennis and Shreeve, 1989; Dennis, 1993a; Dennis and Hardy, 2018) (Fig. 7.7). In this model, pattern changes depend not just on the pressures from thermoregulation, mate location/courtship and escaping predation, but also on the postures adopted at rest and when warming up after being inert.

The elements of habitat create different pressures on the wing patterns, and the changes to wing patterns are under complex genetic control (Dennis, 2020). There is nothing simple in assessing the impact of habitat resources and conditions on butterfly species or other organisms. If just one element of butterfly wing pattern is selected (wing spots), the placement, structure, patterning and size of wing spots are indicative of different pressures applied to each species in each habitat. The most distinctive changes are observed in the spot clines observed in the large heath butterfly *Coenonympha tullia*. These have been given a relatively simple explanation in relationship to climate and predation (populations in warmer biotopes, where individuals are more active, have numerous larger spots, and unique wing spots too, facilitating predator deflection and confusion, whereas populations in colder biotopes, which are less active, have but tiny remnants of

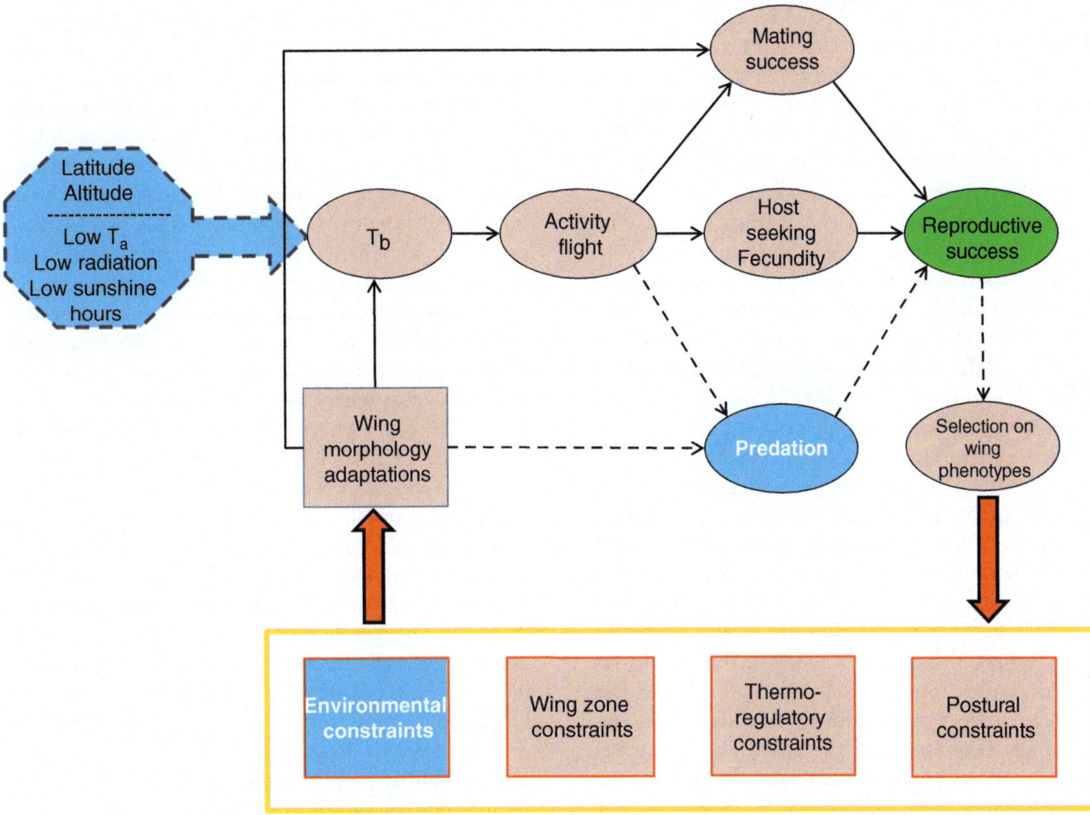

Fig. 7.7. Generalised model for butterfly wing pattern transformations in cooler climates. The type of transformation depends on a number of constraints, the most important of which are body and wing postures and thermoregulatory mechanisms (reflectance versus absorbance baskers) which have a phylogenetic basis. Octagon, zonal factors (T_a, ambient temperature); rectangle, butterfly morphology; ellipses, biological processes (T_b, body temperature); squares, constraints on morphological adaptations. Continuous lines, positive relationships; pecked lines, negative relationships; large red arrows, various relationships. (After Dennis, 1993a.)

marginal spots, and the spot pupils shine as silver in sunlight, thus improving **crypsis** in addition to deflection; Dennis *et al.*, 1984, 1986). The patterns of spots in the meadow brown butterfly *Maniola jurtina* are more difficult to account for, but nevertheless have been modelled (integrating startling, deflective or cryptic functions, depending on activity level and sex, as males are more active) (Ford, 1964; Brakefield, 1979; Brakefield *et al.*, 1992; Dennis, 2020), allowing potential testing in the field. However, conundrums still persist, such as the extraordinarily reduced spotting in the ringlet butterfly *Aphantopus hyperantus* in its variant form *arete* (Ford, 1945); this is relatively frequent only in northern Cumbria and south-west Scotland (UK), the reasons for which are unknown, a topic ripe for research.

The place of habitat in biodiversity

At the end of the previous chapter, we considered the role of habitat in evolution and speciation. It should be evident now that habitat, understood as the **resource-based habitat**, has a crucial place in evolution and hence in species diversity. The basis of diversity is that individuals in a population of each species display a variety of traits and varying potential for resource exploitation and ability to adjust to environmental conditions. Over a species' range (from the intrapopulation to the interpopulation scales), the distinctions between populations subject to different environments have potential for being substantial. The variation is expected to involve gradients of **specialism** and **generalism** in resource exploitation, as well as differences in robustness to changing conditions. The consequences are that individuals in populations are varyingly adapted to the resources they exploit and the conditions they can withstand. The distinguishing feature of *different* species in a single community (i.e. in a single place) is that although they share some resources and a great many of the conditions of 'place', they must, virtually by definition, differ in some respect for resources (utilities and consumables) exploited and the specificities of traits used to access them. Furthermore, although the species that live in the same place will basically share the same environmental conditions, this does not imply that even closely related species share exactly the same range of conditions – for example, differences in micro-scale or local conditions (climatic, pedological, chemical, etc.) experienced during different stages of development at that place.

The big question is: how do these differences between species (speciation) come about? Species may emerge in two broadly distinct ways: either (i) they diverge (evolve) into distinctly new species at a single place where these resources exist, the so-called **sympatric speciation**; or (ii) they develop in different places, notably by **allopatric speciation**, subsequently coming into contact at a number of places when conditions change and permit contact and where their resources coexist (Dennis, 2020). The crucial thing is that at some point in time, populations of a species take different paths (diverge). There is a division in resource use: one group (or more) switches its use of resources or some part of the resources available (flowers, fruits, leaves, soft wood, etc.), possibly in distinct abiotic conditions (temperature, humidity, salinity, etc.) that vary over a site or between two or more sites. Intermediate conditions may also apply in **parapatry** (a geographical distribution in which ranges of allied taxa do not significantly overlap but are immediately adjacent to each other, with the possible presence of a hybrid zone). This process leads to a multiplicity of species, those most closely related we might suggest to be sister species that have most recently differentiated. This process has potential for increasing biodiversity; small and contracting populations will have the reverse effect. The extent to which a site has more similar species would suggest a more recent evolution of the biodiversity of the site and the surrounding region. By contrast, areas where more divergent taxa are concentrated (i.e. where **phylogenetic diversity** is higher; see Chapter 4 and Appendix 3) are expected to have a longer evolutionary history.

The basic question is: just what is it in resource-based habitat terms that allows for diverse biotas? This is by no means a simple question but one that is crucial to ask for the conservation of global biodiversity. The simple fact is that the distribution of any species, and therefore its RBH, almost certainly includes other organisms, some related, others unrelated. These may interact with the target species or have little to do with it. But, nevertheless, particular **landscapes** lend themselves to a varying multiplicity of organisms. It is inevitable that in determining the link between habitat and biodiversity, the researcher will be concerned with a specific portion of the landscape. This may be the RBH of a target species or just a portion of the landscape (a traditional habitat space; **biotope**) that for some reason is rich in our target species. It is then not difficult for the geographer to imagine the factors that could lead to an increase of species at that location. Obvious factors include, among others: (i) geology; (ii) topographic heterogeneity (slope and aspect); (iii) elevation and elevational range; (iv) area of the **home range**; (v) degree of human (or other animal) disturbance; (vi) degree of warmth, moisture, humidity; (vii) light and shadow; (viii) plant diversity, a main driver of animal diversity; (ix) the range of both utility and consumable resources of the target organism and of others that become the focus of attention; (x) the intricate spatial variation and space-time arrangement of resources and conditions. However, as sites are never exactly equivalent for resources and conditions, though biotopes may appear to be so or are treated as such, populations of different species occupying them have the potential to respond differently in relationship to changes in the resource availability and environmental conditions, that is, the ability of individuals in those local populations to contend with them. A target for research is to determine why species respond in different ways to what seem in all respects to be identical biotopes. A classic example is the meadow brown *Maniola jurtina*, a long-time subject of evolutionary genetics (Ford, 1964; Brakefield, 1984, 1992) which varies enormously in numbers from meadow to meadow. It is extraordinary that professional lepidopterists claim to know all that is needed on butterfly species when they cannot answer this simple question about a very common one. Since any research programme on species ecology is draining on funds that most people would claim crucial for human well-being, it is sensible that effort should be first expended on species that are imperilled or of critical importance to ecosystems in which humans have a major stake (e.g. pollinators of crops) (Ollerton *et al.*, 2014; Breeze *et al.*, 2020). However, we should remember that every species may play important roles in maintaining ecosystem functioning (see Chapter 11).

8 The Resource-based Habitats in the Kingdoms of Life

Abstract

The resource-based habitat (RBH) concept can be applied to all life forms (prokaryotes, protists, plants, fungi, animals) and viruses. Because of the enormous diversity of life, RBHs contrast extensively between different organisms in relation to attributes such as individual lifespan, body size, population size and mobility/movement capacity. Differences among organisms in reproduction and development, behaviours, resource use, responses to environmental conditions, geographical distribution, dispersal capabilities and zones of activity are all attributes that require very different approaches and techniques for the study of their RBHs. Issues of particular relevance in the application of the RBH concept across different organisms include movement speed (slow-moving versus fast-moving organisms), complexity of social organisation (castes and slaves), territoriality, occupation of a variety of biomes and biotopes (migratory species), occupation of different spheres (land surface, marine water, freshwater, soil, etc.; sometimes, even the same individual in different life phases) and trophic category. The application of the RBH differs among species, i.e. for contrasting life phases and trophic categories, and across different landscapes and spheres (life zones) of activity.

Although butterflies have provided the model organisms for the development of the **resource-based habitat (RBH)** concept (Dennis *et al.*, 2003; Dennis, 2010), it is evident that the multitude of organisms on the planet display a bewildering and contrasting variety of traits beyond the butterfly realm, attributes that require very different approaches and techniques for the study of their respective resource-based habitats. Is this an obstacle for researchers? Of course not. The term 'habitat' is already broadly applied to *all* organisms and extending the resource-based habitat concept to all life forms on the planet – from prokaryotes (and possibly viruses) to Hominidae, including protists, fungi, animals and plants (Table 8.1) – is an obvious step forward. However, for only a small fraction of species have even basic habitat associations been fully investigated, that is, their occurrence in traditional habitats (**biotopes**). Far less attention has been paid to their resource-based habitats, yet knowledge of their RBH components (i.e. the resources they use and the conditions they can tolerate) is crucial for their existence, and hence for their conservation. The RBH concept underscores and expands on crucial knowledge (e.g. species characteristics, such as population size, generation lengths, extent of occurrence and area of occupancy; see Box 4.1 in Chapter 4) that Georgina Mace demonstrated as critical for an objective evaluation of their extinction risk when first establishing the criteria for the IUCN Red List of Threatened Species that are still in use (e.g. Mace *et al.*, 2008).

Kingdoms of life

First, a view of life forms. A basic problem has been, and still is, how life on the planet is classified. Even at the highest levels of division (kingdoms), this is by no means clear. From the Linnean biological classification scheme to the current day, there have been various approaches to describing the variety of life, starting with a binary life kingdom solution (Plantae and Animalia; Linnaeus, 1735) to five (Monera, Protista, Plantae, Fungi, Animalia; Whittaker, 1969), six (Eubacteria, Archaebacteria, Protista, Plantae, Fungi, Animalia; Woese and Fox, 1977; Woese *et al.*, 1977; Bacteria, Protozoa, Chromista, Plantae, Fungi, Animalia; Cavalier-Smith, 1998, 2004, 2010), seven (Bacteria, Archaea, Protozoa, Chromista, Plantae, Fungi, Animalia; Ruggiero *et al.*, 2015), and eight kingdom classifications (Eubacteria, Archaebacteria, Archezoa, Protozoa, Chromista, Plantae, Fungi, Animalia; Cavalier-Smith, 1993). In all modern classifications, however, a basic division is maintained between **eukaryotes** and **prokaryotes**, organisms having and lacking

DOI: 10.1079/9781800627505.0008

Table 8.1. Domains, kingdoms (according to Ruggiero *et al.*, 2015) and major groups of life. Similar to bacteria, archaea (Archaebacteria) do not have interior membranes but both have a cell wall and may have flagella to swim. Archaea differ in the fact that their cell wall does not contain peptidoglycan and cell membrane uses ether linked lipids as opposed to ester linked lipids in bacteria. Chromista is a biological kingdom of unicellular or multicellular eukaryotic species including chromophyte algae (those with chlorophyll c, not a) and many groups traditionally included among protozoans. The boundary between Chromista and Protozoa is unclear. Protozoa are typically heterotrophs (the only photosynthetic protozoans are Euglenophyceae).

Domains	Kingdoms	Major traditional groups
Prokaryota	Bacteria	Monera
	Archaea	
Eukaryota	Protozoa	Protista
	Chromista	
	Plantae	Plantae
	Fungi	Fungi
	Animalia	Animalia

internal cellular structures such as a nucleus and organelles, respectively. Prokaryotes include Bacteria and Archaea (collectively indicated as Monera), whereas eukaryotes include all the other groups (Plantae, Fungi, Animalia, and the organisms variously classified as Protista, Protozoa, Chromista and Archezoa, depending on the different schemes reported above).

With the advancement of molecular techniques and phylogenetic methods, even the most recent classifications have been found wanting, owing to the lack of **monophyly** in traditional kingdoms (Hinchliff *et al.*, 2015; Hug *et al.*, 2016; Burki *et al.*, 2020; Lara *et al.*, 2020). For instance, the eukaryotes do not cluster any more as a separate domain opposed to prokaryotes. Instead, they fall within the domain of Archaea (Hug *et al.*, 2016). As it is, there is still the question of whether viruses should be included within a 'kingdom' arrangement (Raoult *et al.*, 2004; Hegde *et al.*, 2009; Luketa, 2012). The result is effectively a certain tendency to abandon the kingdom concept. However, a classification of life into seven kingdoms (in turn grouped into five major traditional groups and two domains) still has practical use and the advantage of conveying life diversity in a simple manner (see Table 8.1). The simple seven kingdom structure of Ruggiero *et al.* (2015) provides an adequate guide to the variation in life forms and their habitat structures. Perhaps the most important point to emerge from all recent research is the enormous diversity of life on Earth among prokaryotes (Bacteria and Archaea) and protists (Protozoa and Chromista), making diversity in arthropods and vertebrates seem very small indeed (Box 8.1).

The tree of life would suggest that resource-based habitats have been developing for as long as life has existed and been evolving on the planet. Although different life forms have emerged during different geological epochs and eras, nevertheless the basic demands for appropriate resources and conditions indicate that the resource-based habitat differs only in detail – what a species needs, can tolerate and avoids – and not in any fundamental way that denies the universality of the RBH concept. It is clear that prokaryotes and viruses have their own resource-based habitat demands, not only existing wherever other life can exist but also in extreme abiotic conditions (**extremophiles**; Rainey and Oren, 2006; Schlegel and Jannasch, 2006; Hartemink *et al.*, 2015). The basic pattern in the evolution of life forms is provided in Appendix 4. The massive gap waiting to be filled – even in current (contemporary) traditional habitats – is the diversity and evolution of Prokaryota.

Box 8.1. How many species are there?

Although the oldest known fossils – represented by morphological remains of micro-organisms, biofilms and stromatolites (structures created by bacteria) – are about 3.5 billion years old (Westall *et al.*, 2023), there is some (controversial) evidence of life in rocks that are at least 3.77 billion (and possibly 4.28 billion) years old (Dodd *et al.*, 2017). Thus, it can be supposed that life originated in the Hadean (about 4.5 billion years ago), with replication and synthesis of functional RNAs already occurring during this eon (Salditt *et al.*, 2023; Westall *et al.*, 2023). This would mean that life evolved relatively soon after the formation of the Earth (about 4.5–4.6 billion years ago; Dalrymple, 2001; Kleine *et al.*, 2009; Connelly *et al.*, 2012).

Prokaryotes are not only the first organisms that appeared on Earth, but continue to be one of the most diverse (if not the most diverse) living forms (Table B8.1). The number of known species (intended as organisms distinct genomically) among prokaryotes is rapidly increasing. Although only about 20,000 species have received a validly published name (Parte *et al.*, 2020), more and more species are being discovered even if indicated by placeholder names only (Pallen *et al.*, 2021). Yet, the known diversity of prokaryotes is just a very small fraction of the extant diversity, which is estimated to be 2.2–4.3 million species (Hugenholtz *et al.*, 2021), if not 1.7 billion species (Larsen *et al.*, 2017) or even 1 trillion (10^{12}) (Locey and Lennon, 2016). A similar situation occurs in protists (Table B8.1). We know only a few thousands of **protist** species (Pawlowski *et al.*, 2012), but the number of discovered species is rapidly increasing (even if possibly inflated by intraspecific variability mistaken for species diversity; Caron and Hu, 2019) and the number of existing species is probably in the order of millions (Adl *et al.*, 2007; estimates of Larsen *et al.*, 2017 range between 15.3 and 163.2 million species). Algae include about 50,600 species (Guiry, 2024). They are considered protists in many classifications, but in others they are included among the Plantae. Fungi (Table B8.1) are a hyperdiverse **clade** of **heterotrophic eukaryotes**, including about 155,000 described species, but the number of existing species is estimated between 1.5 and 3.8 million (possibly even more than 165 million); thus, only less than 10% of species have been documented (Mueller and Schmit, 2007; Larsen *et al.*, 2017; Sheldrake, 2020; Phukhamsakda *et al.*, 2022; Antonelli *et al.*, 2023).

Plants (kingdom Plantae) are photosynthetic eukaryotes (only a minority of species do not photosynthesise) that include about 380,000 known species (some 350,000 are flowering plants) (Guiry, 2024; WFO, 2025). Joppa *et al.* (2011a) estimated that there are probably about 450,000 species of flowering plants (Table B8.1). This means that, at least for flowering plants, the current taxonomic knowledge is relatively good (more than 77% of species diversity documented). Overall, plants are the best known life forms taxonomically.

The number of described animal species is around 1,560,000, most of which (>70%) are insects (IUCN, 2024b). The estimated number of living animals varies considerably from one study to another (Table B8.1). For example, Mora *et al.* (2011) estimated ~7.77 million species of animals, whereas May (1988) calculated 10–50 million species. Using different sets of assumptions, Larsen *et al.* (2017) estimated 15.3–163 million animal species (with the most realistic assumptions giving 102–136 million). Most of them are insects, with an estimated 1.84–2.57 million species (Hodkinson and Casson, 1991), 5 million species (Gaston, 1991), 21 million species (Li and Wiens, 2023) or even 30 million species (Erwin, 1982). Stork *et al.* (2015) estimated ~6.8 million arthropod species (consisting mostly of insects; ~1.5

Table B8.1. Estimates of species richness. Data compiled from references cited in the text.

Major group	Estimated number of known species	Estimated number of living species
Prokaryotes	20,000	2.2 million – 1 trillion
Protists*	26,000–74,400	15.3–163.2 million
Fungi	155,000	1.5–165 million
Plants	380,000	c.500,000
Animals	1,560,000	7.8–163 million
Viruses	14,700	>1.7 million

*Estimated number of known protist species varies mostly because of different definitions of the group.

Continued

Box 8.1. Continued.

million beetle species). Larsen *et al.* (2017) estimated 52–81.6 million arthropod species. Although estimates of the total number of existing insect species – and arthropods more generally – are highly uncertain, it is clear that we currently know only a small fraction of their diversity.

As regards viruses, there are currently about 14,700 known species (International Committee on Taxonomy of Viruses, 2024), but the number of existing species is certainly much higher (Table B8.1). Mammals and birds alone probably host more than 1.7 million undiscovered types of viruses (Carroll *et al.*, 2018). Especially impressive is the diversity of bacteriophages (Ackermann, 2006; Turner *et al.*, 2021), viruses that infect and devour Bacteria and Archaea for their replication (Padilla-Sanchez, 2021). The number of existing phage species is probably around 100 million (Rohwer, 2003; Cobián Güemes *et al.*, 2016), although it is possible that there are 'only' 3.9 million (Ignacio-Espinoza *et al.*, 2013). However, only some thousand bacteriophages have been examined in the electron microscope and molecularly characterised (Ackerman, 2011; R.H. Wang *et al.*, 2023). Bacteriophages are composed of proteins that encapsulate a DNA or RNA genome, and may have structures that are either simple or elaborate. Their genomes vary from few a kilobases (phage MS2) to >700 kilobases (Al-Shayeb *et al.*, 2020). Bacteriophages are the most abundant form of 'life', as the number of phage particles on Earth is estimated at 4.8×10^{31} (the number of prokaryotic cells is estimated at about 4.15×10^{30}) (Cobián Güemes *et al.*, 2016). There are many more viruses and bacteria on the Earth than stars in the observable universe (estimated at 10^{24}; Marov, 2015). Borin *et al.* (2023) demonstrated how a bacteria–phage multiscale network structure can evolve rapidly even under simple ecological conditions lacking initial spatial structure. Within 21 days, laboratory coevolution linked *Escherichia coli* and bacteriophage φ21 in a coevolutionary nexus bound in a black box (closed) experiment; diversification forms elaborate cross-infection networks. At first, nestedness emerges as bacteria and phages establish increasing resistance and infectivity via multiple cycles, effectively an arms race, whereby bacteria evolve mutations that functionally remove receptors and phages innovate to use new receptors. It is the speed with which such systems can emerge that is surprising. It is an unexpected finding and one that warns against assuming that evolutionary processes are slow, and should temper enthusiasm for bringing 'materials' back from outlying planets before we fully understand our own Earth systems (Marks, 2023).

In this chapter a brief outline is first given of different life forms. Then, using a small selection of species whose traits differ markedly from those of butterflies, issues are explored of applying the RBH to organisms across different kingdoms. Needless to say, resource-based habitats contrast extensively between different organisms, even widely for the same organism over the Earth. However, some crucial factors ensure that the dimensions of the RBH follow some basic principles. The dimensions of the RBH are driven by a number of factors, the most obvious being organism size, social grouping/population size and mobility capacity (movement, migratory, dispersal and/or colonisation capabilities). Thus, if A is the area of a RBH (or the home range) of any species, then:

$$S \rightarrow A\ [+]$$
$$G \rightarrow A\ [+]$$
$$M \rightarrow A\ [+]$$

Therefore $S + G + M \rightarrow A$ of RBH

where S is size or mass of the organism, G is the size of social group or population size, and M is mobility or movement capacity.

Other factors also influence the area of the RBH, such as the organs of the species under study, the stages and phases of development, the ability to run, fly, dig or swim, the use of plants or not (herbivores, carnivores, omnivores), the defences against enemies, and a parasite or prey life history. Moreover, the abiotic and biotic environment provisioning the resources also varies as to spacing over time. Thus:

$$SRD \rightarrow A\ [+]$$

where SRD is seasonal resource dispersion (that is, the inverse of resource integrity), with resources used and conditions experienced depending on an organism's (variable) occupation of land, sea or air.

Basically, then, for any i-species, we can imagine:

$$(aS_i + bG_i + cM_i) + (dSRD_i) \rightarrow A_i \text{ of } RBH_i$$

where a, b, c and d are hypothetical coefficients expressing the relative weight (importance) of these factors in determining the RBH.

Other crucial factors affect the complexity of the resource-based habitat; none, perhaps, more mysteriously than **sociality**. Human populations, with their highly variable social structures and degrees of complexity (Levinson and Enfield, 2006), represent a particularly distinctive case, an issue considered in Chapter 9.

First, then, after some caution about bias in studies, a brief outline of the different kingdoms and their phyla. Inevitably, little can be said of the 'simplest', or earliest, forms (viruses, bacteria, etc.) as only recent developments have made it possible to begin to grasp the complexity of their diversity. Much more is known of larger, multicellular organisms (since they come within the orbit of normal human vision), though it would be a mistake to say we understand them; they are more complex, if only because they are inhabited by a multitude of prokaryotes that largely determine their existences. For example, a recent study recovered about 5000 species in the human microbiome, most of which (77%) were never described before (Pasolli *et al.*, 2019).

Bias in biodiversity studies

It is perhaps inevitable that some bias affects the outcome of most studies of biodiversity, though efforts are now being made to reduce or eliminate potential biases (Johnston *et al.*, 2021) and/or to manipulate data to reduce them (e.g. Rocchini *et al.*, 2023). Typically, there are taxonomic, spatial and temporal biases in biodiversity studies (e.g. Dennis and Thomas, 2000; Beck *et al.*, 2014; Girardello *et al.*, 2019; Sánchez-Fernández *et al.*, 2021; Bowler *et al.*, 2022). Crucially, in studies of biodiversity, taxonomic bias is the most frustrating feature to emerge in research (Fontaneto *et al.*, 2009; Trimble and van Aarde, 2012; Troudet *et al.*, 2017; Stahl *et al.*, 2020; Yang *et al.*, 2021), leading to representations mismatching the occurrence of the species on patches of the planet. For example, Troudet *et al.* (2017) used 626 million occurrences from the Global Biodiversity Information Facility (GBIF) to assess the taxonomic bias in 24 animal classes in publications (recovered from Web of Science) and web pages (from Bing searches). Unsurprisingly, mammals, birds, amphibians and reptiles are overrepresented, and insects, arachnids and molluscs are underrepresented in biological research (Fig. 8.1); inevitably bias will enter studies of habitats too. Biases in our knowledge of life unavoidably afflict the following pages reporting aspects of the RBH through the diversity of life. However, here, the objective throughout has been to illustrate how the RBH concept can be applied to any organism; it is inevitable that the selected species are familiar ones (model organisms) that are widely studied in the field, and possess specific characteristics that make it easier to illustrate their distinctive resource-based habitats and associated methodological issues.

Crucially, researchers into resource-based habitats will need to adjust surveys for the organisms they study. Massive differences exist across organisms: (i) in scale from microns (nanometres for viruses) to metres, (ii) in lifetime mobility from virtual stasis to constant mobility (degree of residence, distances covered from <1 mm to 1000s km), (iii) in climatic regimes (e.g. from hot to cold, wet to dry landscapes), (iv) within and across spheres of activity (land, sea and air), (v) in social organisation from singular-living to massively concentrated and co-ordinated populations, (vi) in degree and expanse of territoriality, (vii) in food type (herbivore, carnivore, omnivore), and (viii) in a vast array of morphological, biochemical, physiological and behavioural traits. Even so, though many distinctions among life forms exist, all organisms, be they prokaryotes or eukaryotes, have resource-based habitats, but the nature of these RBHs differs vastly between different organisms. The key to conserving what may become a **red data book** species is to understand the type of the RBH to which that species belongs. In the following sections on different taxa, we move from a selection of the smallest to the largest organisms and pick out the features that influence their RBHs. Although the classification of organisms is far from being finalised, this has no effect on the concept of the RBH, simply because each species has its own (shifting) sets of resources and conditions and thus placement in space-time.

Prokaryote habitats: Bacteria and Archaea

Prior to the third millennium the focus of biologists in the field was mainly on organisms as large as, or larger than, the size of insects. Only recently with advances in molecular biology has it dawned on ecologists that most of the creatures on the planet are **prokaryotes** (Bacteria and Archaea; Woese

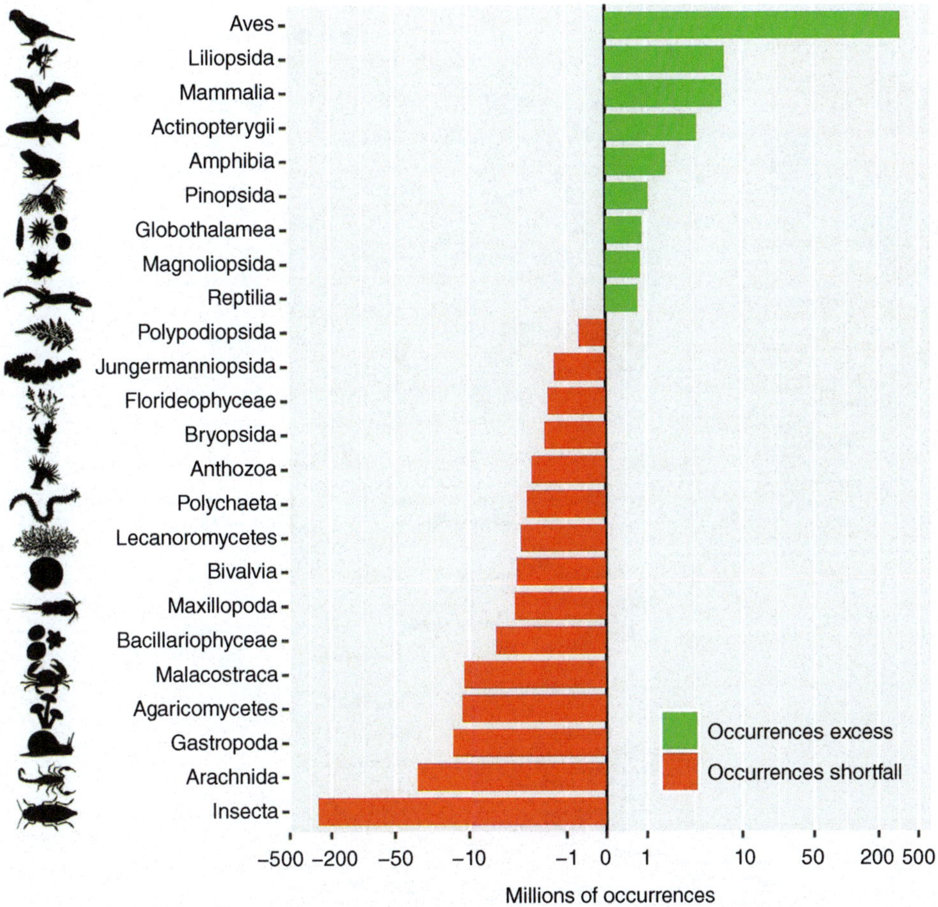

Fig. 8.1. Taxonomic bias in biodiversity occurrence data. The vertical line at $x = 0$ depicts the 'ideal' number of occurrences per class, where each class is sampled proportionally to its number of known species. Green and red bars show the classes that are over- and underrepresented in the Global Biodiversity Information Facility mediated database compared to this 'ideal' sampling, respectively. Insects lack >200 million occurrences and birds have an excess of >200 million occurrences compared to an unbiased taxonomic sampling. (From Troudet *et al.*, 2017, used under licence CC BY 4.0.)

et al., 1990), the vast majority of which are still unidentified (Fig. 8.2a,b; see also Box 8.1). In point of fact, the function of the **biosphere** depends profoundly on the activities of the microbial world (Pace, 1997). It has been calculated that possibly more than 10^{14} bacterial cells occur in the human gastrointestinal tract (Backhed, 2005; Gill *et al.*, 2006), that is, more than 10 times the total number of human cells in an adult body (28 trillion in the female, 36 trillion in the male; Hatton *et al.*, 2023). Although another estimate suggests that the ratio of human:bacterial cells is approximatively 1:1 (Sender *et al.*, 2016), it remains clear that the number of bacteria hosted in our bodies is enormous. More generally, some 2200 prokaryotic species have been isolated from human beings (human skin and mucosal sites, such as the oral cavity, intestine and vagina) (Hugon *et al.*, 2015), notably including 'good' bacteria that promote our health, helping to fight diseases, produce vitamins and digest complex carbohydrates. Frequently quoted is the number of bacterial organisms in small samples of soil: 10^6–10^{10} bacteria, belonging to 4×10^3–5×10^4 species, within a single gram of soil (e.g. Roesch *et al.*, 2007; Barillot *et al.*, 2013). These creatures measure between 0.2 μm (e.g. *Mycoplasma genitalium*) and ~2 cm (*Thiomargarita magnifica*), most within 0.5 μm and 5 μm, and differ fundamentally in external and internal structure from eukaryotes (Staley, 1999; Volland *et al.*, 2022). Internally, they are

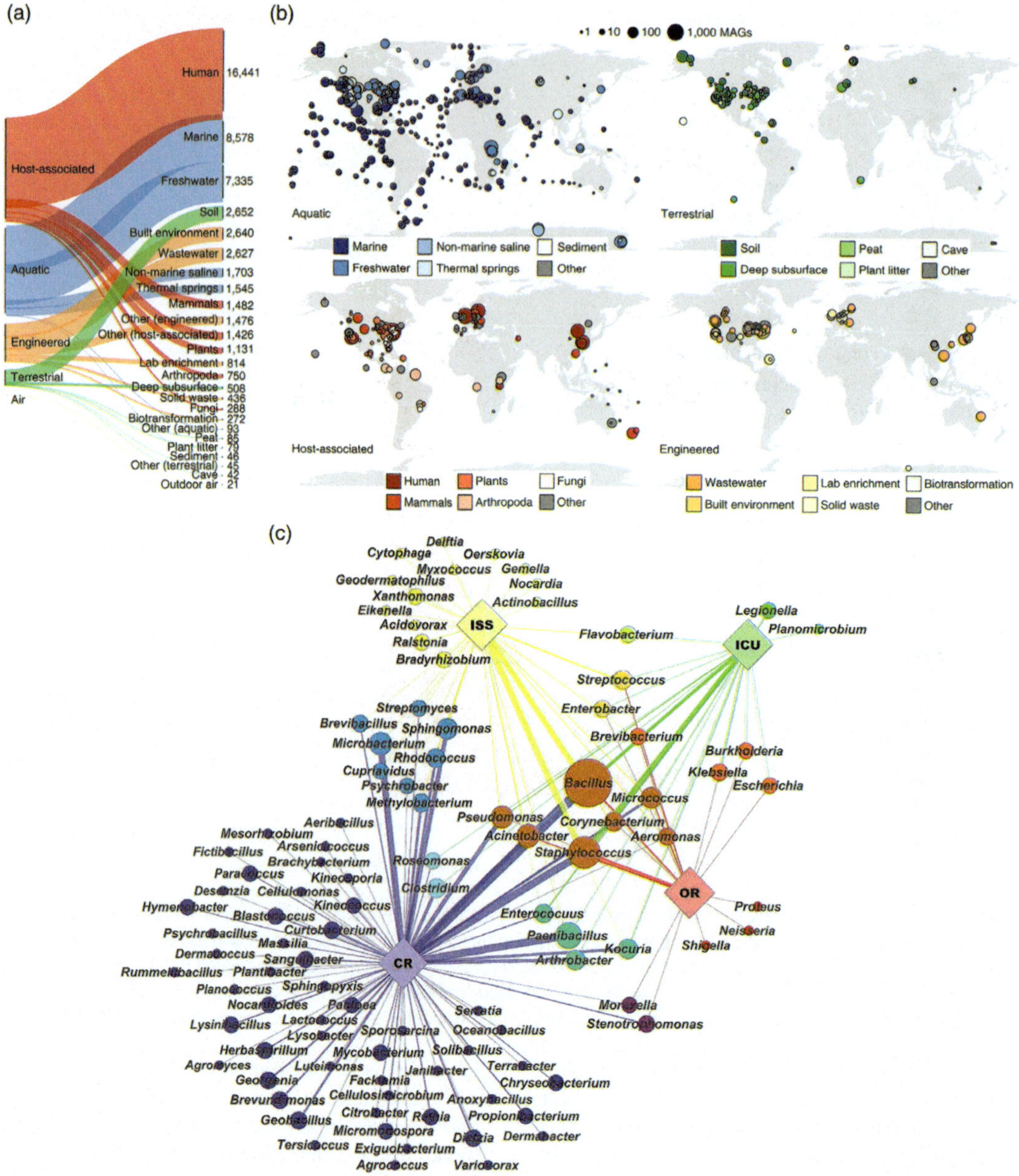

Fig. 8.2. Emerging aspects of prokaryote diversity. **(a,b)** Environmental and geographical distribution of metagenome-assembled genomes (MAGs). MAGs are microbial genomes reconstructed from metagenome data. Metagenomics is the study of the structure and function of entire nucleotide sequences isolated and analysed from all the organisms (typically micro-organisms) in a bulk sample (e.g. of soil or water). **(a)** Distribution of MAGs across biomes and sub-biomes. The number of MAGs associated with each sub-biome is indicated next to the plot. **(b)** Geographical distribution of MAGs within each biome. **(c)** A network showing microbial profiles (diversity at genus level) of selected habitats based on a range of isolates obtained from these environments. Nodes and edges were coloured by colour mixtures of their respective environments. International Space Station (ISS) – yellow; intensive care units (ICU) – green; cleanrooms (CR) – blue; and operating rooms (OR) – red. (Panels **a** and **b** from Nayfach *et al.*, 2021, used under licence CC BY 4.0. Panel **c** from Mora *et al.*, 2016, used under licence CC BY 4.0.)

typically distinguished by lacking a nucleus to the cell, mitochondria or any other membrane-bound organelles. However, heightened resolution is now discovering new elements of their internal structure (bacterial microcompartments; Kerfeld *et al.*, 2005; Murat *et al.*, 2010; Volland *et al.*, 2022). Externally they adopt a variety of shapes, including spherical (cocci), rod-shaped (bacilli), spiral (spirilla), comma-shaped (vibrios) or corkscrew-shaped (spirochaetes). They can exist as single cells, in pairs, chains or clusters.

Ecologically, prokaryotes are distinguished by a number of key attributes. (i) Being so small, they have the immense advantage of inhabiting **micro-habitats** (see Chapter 1) within human-identified traditional habitats (**biotopes**), a resolution of space evident only at a microscopic level. (ii) They may function individually or as part of a colony. (iii) They have **motility**. (iv) They occupy by far the greatest range of **biomes** and biotopes on the planet. Many are **extremophiles**, that is, they are variously adapted to extreme conditions: temperatures, pressures, salinity, acidity (pH levels), oxygen levels, etc., being able to colonise polar ice environments, hot springs, deep lithospheric environments, hypersaline ponds, organs within other living organisms, radioactive waste, etc. (v) They are the most ancient of life forms, having experienced some 4 billion years of evolution (see Appendix 5) and adaptation to abiotic conditions and biotas on the planet. As such, they have had the potential to occupy all life forms evolving subsequent to their own evolutionary origins.

The resources and conditions dictating success in specific prokaryotes have yet to be determined. As the state of microbes is dependent on the high surface area to volume ratio, they should react rapidly to specific nutrients and physico-chemical conditions (Schlegel and Jannasch, 2006). Ecological studies on prokaryotes are rapidly progressing, for example in the genus *Agrobacterium*, whose populations live in different habitats (i.e. bare soil, rhizosphere and host plants), and hence face different environmental constraints (Tzfira and Citovsky, 2008). They have evolved the capacity to exploit diverse resources and to escape plant defences and competition from other microbiotas. By modifying the genome (via horizontal gene transfer) of their host, *Agrobacterium* populations exhibit the remarkable ability to construct and exploit the ecological niche of the plant tumours they induce (Dessaux and Faure, 2018). This niche is distinguished by the accumulation of specific low molecular weight compounds called **opines** which have a critical role in *Agrobacterium*'s functioning. *Agrobacterium* are **Gram-negative bacteria** which are found in virtually all land environments that support plant life. The Gram-negative bacteria include the model organism *Escherichia coli*, as well as many pathogenic bacteria (e.g. *Pseudomonas aeruginosa*). They present a challenge to medicine as their outer membrane protects them from many antibiotics (e.g. penicillin), detergents that would normally damage the inner cell membrane and lysozyme, an antimicrobial enzyme produced by animals and plants which forms part of the innate immune system (Ferraboschi *et al.*, 2021).

Three aspects of microbe habitats are worth considering here. First are the supposedly microbe-free human environments (such as intensive care units, operating rooms, cleanrooms and space stations), which increasingly prove difficult to attain the status of 'clean environments', particularly as specific bacteria are building antibiotic resistance. Bacterial communities of these 'clean' environments mainly derive from human microbiomes and indicate how new habitat structures may generate new bacterial communities composed of survival specialists, particularly as the bacteria adapt to antibiotics (Mora *et al.* 2016) (see Fig. 8.2c). Second, as with any other organisms, microbial communities change with successional stages (see Chapter 3) in all aspects of the natural environment (see Schütte *et al.*, 2009). Bacteria carry out key processes in the development of soil (e.g. Schulz *et al.*, 2013), in biogeochemical cycling (e.g. Hicks *et al.*, 2018; Masson-Boivin and Sachs, 2018; Wang *et al.*, 2019), and in the facilitation of plant colonisation and development (e.g. Lynch, 1990; Hayat *et al.*, 2010; Gómez-Godínez *et al.*, 2023). These findings indicate that bacteria are suited to invading specific conditions, limiting the bounds of their resource-based habitats. Third, many are engaged in enzootic cycles involving transfer between vectors and hosts, as in the case of *Borrelia burgdorferi* spirochaete bacteria, which are associated with ticks (*Ixodes*) in different phases (nymphs, larvae and adults), parasitise mammals such as deer and can pass on to humans (Fig. 8.3) (Iyer *et al.*, 2015; Samuels and Samuels, 2016; Samuels *et al.*, 2021). Thus, their resource-based habitats shift in relation to the phase of the cycle. To adapt to its changing environment, to survive in each phase of its enzootic cycle, *B. burgdorferi* uses three functional systems to regulate the expression of genes (Samuels and Samuels, 2016).

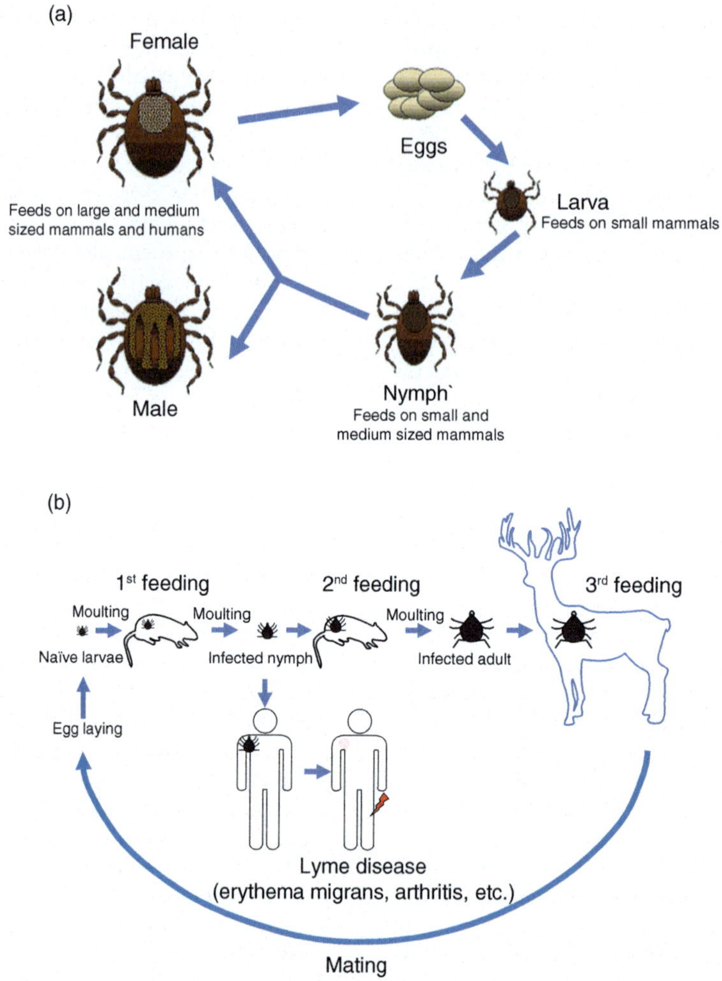

Fig. 8.3. The life cycle of the *Ixodes* tick **(a)** and transmission of *Borrelia burgdorferi* spirochaetes **(b)** passing through animals in different phases and affecting humans with Lyme disease. (Panel **a** partially redrawn from Centers for Disease Control and Prevention, United States Department of Health and Human Services: https://commons.wikimedia.org/wiki/File:Life_cycle_of_ticks_family_ixodidae.PNG; panel **b** partially redrawn from Troxell and Yang, 2013, used under licence CC BY 3.0.)

For a prokaryote, then, what is its resource-based habitat? It is the physico-chemical space, with its dead and alive elements, however minute (invisible to the naked human eye), that embraces and flows in and out of each individual cell, with the flowing out of materials affecting their flowing in. For application of the RBH concept, researchers necessarily sample members of the same species, and resources and conditions are measured to understand variations and express statistical preferences. All this may seem incogitable in the study of prokaryotes. Yet time and again, what seems impossible to do now will become commonplace, and very soon. Thus, gradually, the complexities underlying the resource-based habitat of bacterial disease agents such as *Borrelia* species, associated with Lyme disease and relapsing fever (with aggregation of blood cells known as erythrocyte rosetting) (Guo *et al.*, 2009; Samuels *et al.*, 2021), will become a matter of commonplace laboratory investigation. What may seem extraordinary is that having such limited knowledge of the taxonomy and impact of prokaryotes on our own planet, we should even be considering looking for micro-organisms on other planets such as Mars.

A vitally important recent finding is the discovery of bacteria that 'eat' (degrade) plastics, thus expanding their resource-based habitat (Asiandu *et al.*, 2021). Yoshida *et al.* (2016) have discovered a bacterium that degrades and assimilates poly(ethylene terephthalate) and, more recently, Liu *et al.* (2024) developed stable and reproducible plastic-associated biofilm communities derived from the gut microbiome of larvae of the tenebrionid beetle *Zophobas atratus*. The importance of these findings cannot be overestimated (see Dey *et al.*, 2023). The usual techniques to reduce plastic waste are degradation in landfill, incineration and recycling, all of which will prove less effective than the ultimate capacity of bacteria to break down plastics and other hazardous materials we are dumping on the environment. Undoubtedly, this capability will lead to the evolution of new 'species' of bacteria. This evolutionary step goes way beyond the recent discovery of hermit crabs using plastic and other human refuse as 'shells' for their protection (Jagiello *et al.*, 2024), or the finding that a large number of marine invertebrates – including crabs and anemones – can survive and reproduce on the plastic debris of the Great Pacific Garbage Patch (Haram *et al.*, 2023), a gyre of 45–129 thousand tonnes of ocean plastic floating in an area of 1.6 million km^2 (Lebreton *et al.*, 2018). However, like many other species that use artifacts of any types (e.g. species that use man-made structures for resting, recovering or nesting, such as reptiles and birds in urban environments; Fattorini, 2019), it is also a clear example of species expanding their resource-based habitat through the use of artificial components in their environments.

Eukaryote habitats: Protista (Protozoa and Chromista)

Habitat issues pertaining to the prokaryotes also apply to the Protista, the largely microscopic unit of the eukaryotes (unicellular, colonial or simple multicellular organisms), whose origins date back further than 1 billion years (the first fossils indicating the presence of eukaryotic life are 1600–800 million years old, but molecular data suggest that the last eukaryotic common ancestor (LECA) had possibly already emerged more than 1800 million years ago; Brocks *et al.*, 2023). These organisms do have a nucleus to their cells, but nevertheless are usually minute compared to the eukaryotes that have been so extensively studied: fungi, plants and animals. They are a **polyphyletic** grouping of several highly heterogeneous independent **clades** (Cavalier-Smith, 2003; Adl *et al.*, 2005). The one common factor is that all members have characteristics that exclude them from the remaining traditional groups. Otherwise, they have diverse body forms, types of reproduction, life histories and **motility** (modes of movement including none, flagella, cilia and pseudopods), including both **autotrophic** and **heterotrophic** organisms. As such, they occupy a variety of traditional habitats and microhabitats and exhibit diverse modes of obtaining 'food' (many ingest food by endocytosis, a cellular process in which particles are engulfed into the cell). Inevitably, these organisms feed on smaller food items than themselves; for instance, most ciliates ingest bacteria or other tiny protists, which may nevertheless have larger organisms as habitats. Therefore, their resource-based habitats are wonderfully variable. Even so, two forms are often isolated for comparison: (i) Protozoa (animal-like protists, such as amoebae, apicomplexans, trichonymphas, actinopods, foraminiferans and ciliates) and (ii) Algae (plant-like protists, such as euglenas, dinoflagellates, diatoms, golden algae (Chrysophyceae), brown algae (Phaeophyceae), red algae (Rhodophyta) and green algae (Chlorophyta)). Note that these groups do not necessarily correspond to clades. They range from the giant multicellular kelp (*Macrocystis pyrifera*) to unicellular organisms, which are a primary component of plankton; the former (giant kelp) is a **keystone species** for other organisms and has a distribution tightly tied into geographical conditions (Bushing, 1997). In many classifications, algae are included in the kingdom Plantae.

There are some basic issues or problems when determining the RBHs of micro-organisms, be they eukaryotes or prokaryotes. The first is to isolate the population of study, the bounds of individuals which provide datum points for statistical analyses of requirements and conditions. This is not entirely a micro-organism problem, as many invertebrates, even when abundant, are also so small that it can be difficult to find them in sufficient number to assess variability in resources and conditions. Second, the researcher of these groups has to bear in mind that not only are these organisms affected by the resource-based habitat factors of the microhabitat, but also – in the case of parasitic/symbiotic life styles – the habitat of the organisms in which they are implanted. This is plainly self-evident in the eukaryote *Plasmodium falciparum* passing through mosquitos and humans (Ke *et al.*, 2015) (Fig. 8.4) which has a similar parasitic history as the enzootic cycle of prokaryote spirochaete *Borrelia burgdorferi* (see above).

A fascinating, though somewhat horrifying element of the micro-organism world is how humans have become a 'space station' for their existence. We think we are pre-eminent in our 'cleverness'; we are pathetic compared to the multitudes of creatures that invest in us. Even creatures we hold dear may transmit these micro-organisms which have consequences for humans, such as *Toxoplasma gondii*, an obligate intracellular parasitic protozoan transmitted by cats and responsible for toxoplasmosis. Found worldwide, *T. gondii* is capable of infecting virtually all warm-blooded animals but felids, such as domestic cats, are the only known definitive host in which it may undergo sexual reproduction (Dardé *et al.*, 2011). Through neurological and biochemical actions, these parasites modify antipredatory behaviours in rodents (for example, by reducing their avoidance of cat habitations) and hence increase their chances of being preyed on by felids (Berdoy *et al.*, 2000; Berenreiterová *et al.*, 2011; Worth *et al.*, 2014; Watson and Davis, 2019; Tong *et al.*, 2021). When *T. gondii* infects humans, it can

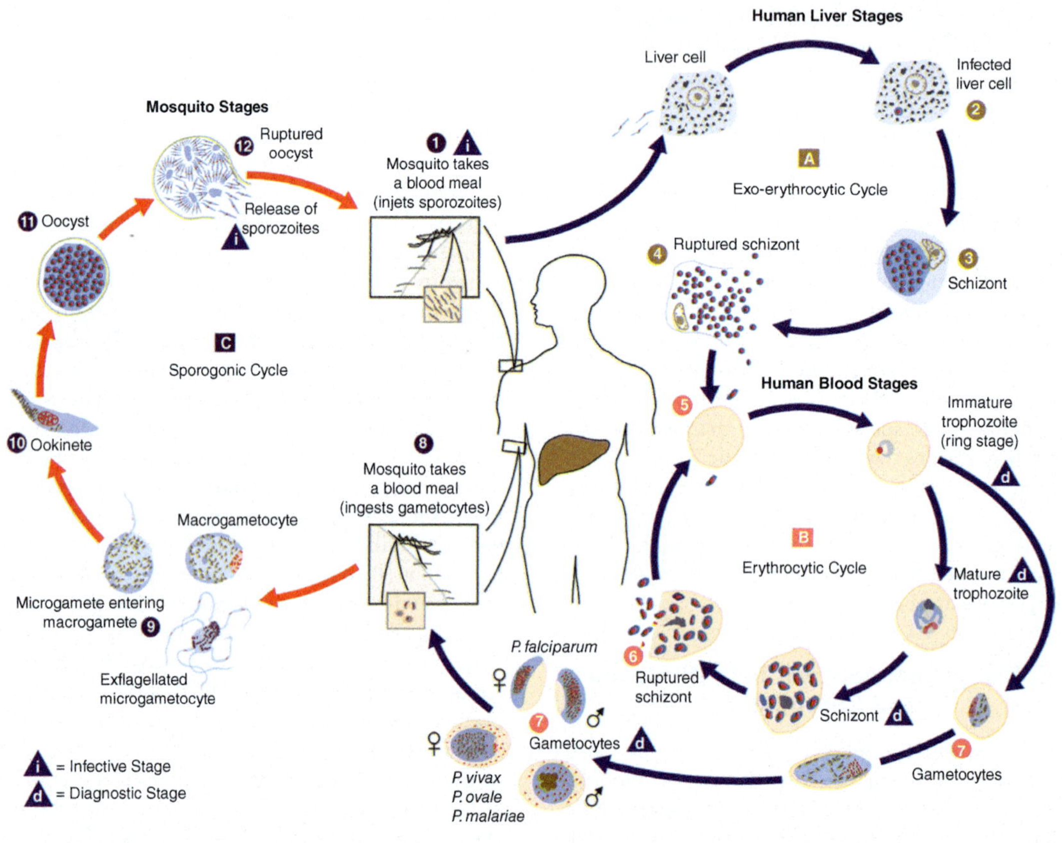

Fig. 8.4. *Plasmodium* life cycle causing malaria in humans. A mosquito causes an infection by a bite (1). The sporozoites (the infective stage of the malaria parasite) enter the bloodstream and migrate to the liver. They infect liver cells (2), where they multiply into a new form (called schizonts, 3), rupture and release merozoites (4). The merozoites infect red blood cells (5), where they develop into ring forms, trophozoites (feeding stage) and schizonts (during which nuclei multiply) that in turn produce further merozoites (6). Sexual forms (gametocytes) are also produced (7), which, if taken up by a mosquito with the blood (8), mature in the mosquito gut to gametes. The male and female gametes fuse generating zygotes (9). The zygotes become motile and elongated (ookinetes) (10). They are able to cross the midgut wall and develop into oocysts (11). Oocysts develop new sporozoites (12) that migrate to the insect's salivary glands, ready to infect a new human host. A, exo-erythrocytic schizogony; B, erythrocytic schizogony; C, sporogonic cycle. (From CDC: Alexander J. da Silva and Melanie Moser: https://phil.cdc.gov/details.aspx?pid=3405.)

induce behavioural changes, including some related to risk-taking (Johnson *et al.*, 2018; Desmettre, 2020), in addition to various health problems including miscarriage and birth defects in pregnant women (Dubey, 2010).

Habitat structures in invertebrates

The majority of invertebrates are spread across eight main phyla (Table 8.2). However, numerous other 'minor' phyla also exist, especially in secluded aquatic or terrestrial microhabitats, and, as parasites, concealed in other animals or plants (Garey and Schmidt-Rhaesa, 1998; Kinchin, 2000). An example is the fascinating Tardigrada (water bears or moss piglets), segmented micro-animals (typically 0.5 mm long) living in the most disparate environments, with some species that can withstand extremely cold and extremely hot temperatures, very high pressures, high doses of ionising radiation and even the vacuum of outer space; they are also one of the few animals that can suspend their metabolism and survive in a dehydrated state for years (Kinchin, 1995; Jönsson *et al.*, 2008). Equally, there are numerous different groups or population structures within each invertebrate phylum with habitat issues much the same as those outlined in examples for vertebrates below and butterflies in Chapters 5 and 6. As such, herein just a brief outline of the major groups of invertebrates and some issues involved in determining their habitats is given below.

Invertebrates are animals lacking a vertebral column (backbone or spine) derived from the notochord. The most familiar invertebrates are arthropods (such as insects, arachnids, crustaceans and myriapods), molluscs (such as chitons, snails, bivalves and cephalopods), annelids (such as earthworms and leeches), cnidarians (such as hydras, jellyfishes, sea anemones and corals) and roundworms (nematodes). Numerous insects (the most diversified animal class, including about 30 orders; Table 8.3) and their arthropod relatives have much the same social and resource-seeking structure as those described for butterflies and moths (i.e. **presocial**: associate for courtship, mating, roosting), at least those closely examined for resource-based habitat issues (Dennis, 2010; Crous *et al.*, 2014; Turlure *et al.*, 2019). But many other insects live in much tighter relationships than butterflies, even tighter than those butterfly species that have all their resources in one vegetation patch (e.g. *Lysandra coridon* on calcareous downland; Krauss *et al.*, 2005). Classic among these are Coleoptera, especially Chrysomelidae. For example, a researcher will find all stages of the alder beetle (*Agelastica alni*) occurring on or below alder trees at the same time, including egg batches under leaves already perforated by a previous larval brood, courtship and mating activity, occasional pupae and feeding larvae (most pupation occurs in an earth cell just below the ground surface, or in leaf litter); trees are often virtually saturated with the beetle, in effect a tight local population including a mass of individuals occupying a single tree or small copse of alder trees (Stenhouse, 2011). No similar example occurs among butterflies. Many insects use the trunks of trees or burrow into the soil and again their

Table 8.2. Diversity of main invertebrate phyla. Numbers of species are based on Zhang (2013). Estimates for some groups are controversial and may be very different from those reported here. For example, Rosenberg (2014) estimates 70,000–76,000 described species of living Mollusca.

Phylum	Examples	Estimated number of species (approximate values)
Porifera	Sponges	8700
Cnidaria	Jellyfish, corals, anemones, hydras	10,200
Platyhelminthes	Flatworms, flukes, tapeworms	29,500
Annelida	Earthworms, leeches, lugworms	17,400
Mollusca	Snails, clams, squids, octopuses, other molluscs	85,000
Nematoda	Ascarides, vinegar eels, hookworms, nematodes, pinworms	25,000
Arthropoda	Crabs, barnacles, scorpions, spiders, ticks, insects, millipedes, centipedes	1,257,000
Echinodermata	Sea stars, brittle stars, sea urchins, sea cucumbers	7600

Many other 'minor' phyla of invertebrates have been named which are not reported here. Only the most easily recognisable are indicated. Because of disagreement on the taxonomic status of certain groups, there is no consensus about the number of extant animal phyla, but most zoologists admit that the animal kingdom contains about 30 phyla.

Table 8.3. List of described extant insect orders. Species numbers based on estimates reported by Zhang (2013).

Order	Common name*	Approximate number of extant species described
Archaeognatha	Jumping bristletails	500
Zygentoma	Silverfish, firebrats	550
Ephemeroptera	Mayflies	3100
Odonata	Dragonflies and damselflies	6000
Orthoptera	Grasshoppers, locusts and crickets	24,000
Phasmida	Stick insects	3000
Embioptera	Webspinners	460
Grylloblattodea	Icebugs	32
Mantophasmatodea	Gladiators	17
Plecoptera	Stoneflies	3700
Dermaptera	Earwigs	1900
Zoraptera	Angel insects	36
Mantodea	Mantises	2400
Blattodea	Cockroaches and termites	7600
Psocoptera	Booklice	5600
Phthiraptera	Lice	5100
Thysanoptera	Thrips	5900
Hemiptera	True bugs (e.g. cicadas, aphids, planthoppers, leafhoppers, assassin bugs)	102,000
Hymenoptera	Sawflies, wasps, bees and ants	153,000
Strepsiptera	Stylops	600
Coleoptera	Beetles	390,000
Neuroptera	Lacewings, mantidflies, antlions	5500
Megaloptera	Alderflies, dobsonflies and fishflies	360
Raphidioptera	Snakeflies	180
Trichoptera	Caddisflies	14,500
Lepidoptera	Butterflies and moths	155,000
Diptera	Flies and mosquitos	157,000
Siphonaptera	Fleas	2100
Mecoptera	Scorpionflies	400

*For orders that do not have common names, examples are given.

resources differ hugely from butterflies. Most bees and most solitary, flower-visiting wasps nest in the ground, whereas comparatively few butterflies use subterranean structures (e.g. *Phengaris arion*, a parasite of *Myrmica* ant species; Hayes, 2015). Many insects (termites, bees, wasps, hornets and ants) have exceedingly tight and role-specific colonial structures, referred to as **eusociality** (Wilson, 1971). A fascinating example is the fire ant (*Solenopsis invicta*) which self-assembles into waterproof rafts to survive floods (Mlot *et al.*, 2011). Exemplary cases of the impact of eusociality on the RBH are presented by termites, some of which are particularly notorious for constructing enormous mounds. The space between mounds may depend on environmental conditions (as in *Syntermes dirus*; Martin *et al.*, 2018) or intra- and interspecific competition (Grohmann *et al.*, 2010). Many species of termites form regular groups of individuals within their mounds occupying the same basic traditional habitat (Fig. 8.5) (Engel *et al.*, 2009), whereas other insects may occur loosely, often singly, within the same area (square km) but in very different traditional habitats (e.g. wasps). Each insect group brings with it different problems of study and sampling (see Appendix 1). If we consider that insects are, in general, the most investigated and easy to collect invertebrates, the study of resource-based habitats demands great ingenuity from researchers who intend to unearth the secrets of less well-known creatures they study (e.g. Porifera, Platyhelminthes, Nematoda, Annelida and Echinodermata).

Habitat structures in vertebrates

These complex patterns of habitat based on resources and conditions also extend throughout the vertebrates, in (i) fish (Helfman *et al.*, 2009; Jin

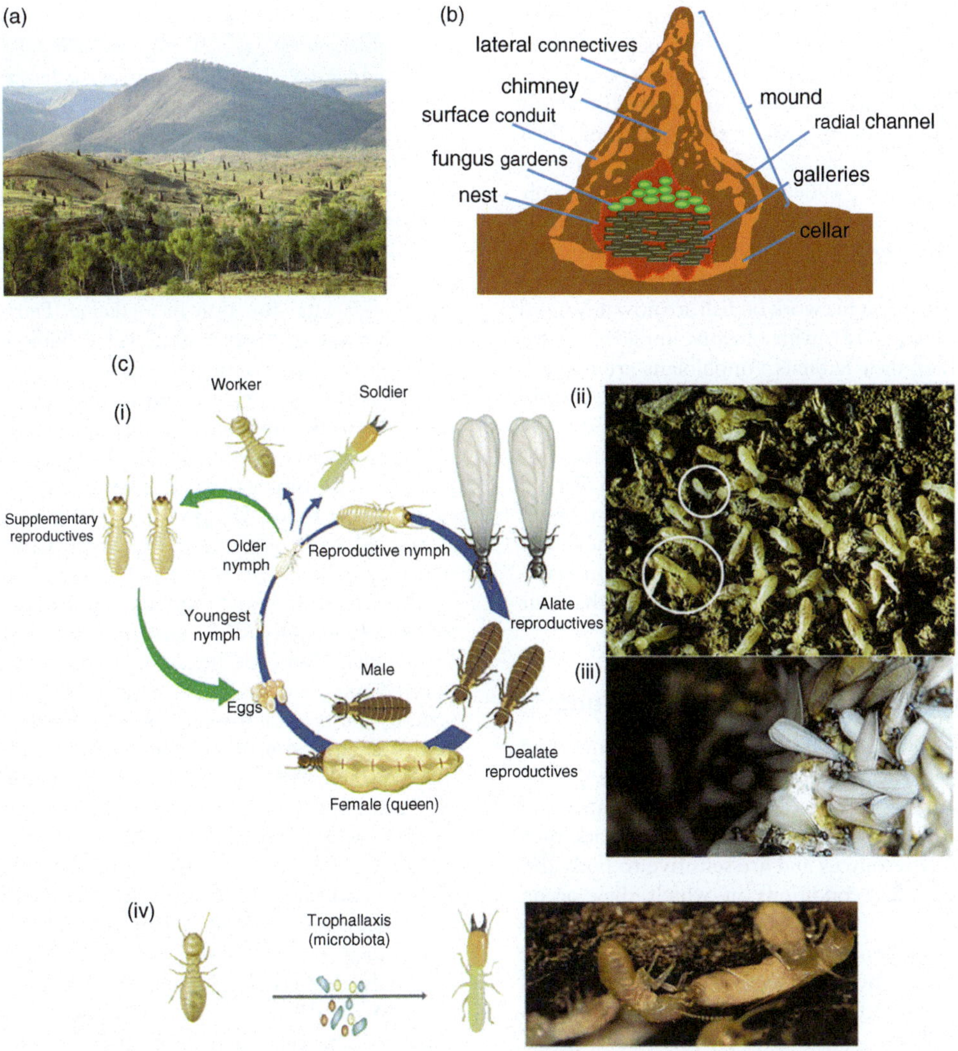

Fig. 8.5. Termites and termite mounds. **(a)** Termite mounds in the Bungle Bungle Range (Western Australia). **(b)** Cross-section through a mound and nest of *Macrotermes michaelseni*, showing locations of nest and fungus garden, and basic layout of the network of ventilatory tunnels. **(c)** Eusocial termites: (i) the termite life cycle; (ii) photo showing a colony of the 'lower' termite *Reticulitermes grassei* (not a mound-building termite): a nymph and soldier are inside circles, the rest of the termites are part of the worker caste; (iii) alates of *R. grassei*; (iv) trophallaxis and coprophagy allow the direct or indirect transmission of symbiont micro-organisms among workers and other castes. (Panel **a** photo by Ouderkraal: https://commons.wikimedia.org/wiki/File:Termite_Mounds_in_the_Bungle_Bungle_Range_in_Western_Australia.gif, used under licence CC BY SA 3.0. Panel **b** after Turner, 2001, redrawn. Panel **c** from Berlanga and Guerrero, 2016, used with permission of Springer Nature; photos by Rubén R. Duro.)

et al., 2021; Mackay-Hope, 2021), (ii) amphibians (Hofrichter, 2000), (iii) reptiles (Cogger and Zweifel, 2000), (iv) mammals (Burgin *et al.*, 2020) and (v) birds (Alderton, 2017). Taxonomically, the five groups above belong to the phylum Vertebrata, with Urochordata (also known as Tunicata, e.g. ascidians) and Cephalochordata (lancelets) being phyla in their own right (Irie *et al.*, 2018) within the broader clade of chordates (Chordata). Mammals are well known for their varied **social structures** from single adults (e.g. solitary old male lions) to single families (e.g. vulnerable lionesses with their cubs), to substantial prides and packs (e.g. wolves), even to massive, continuously moving herds such as African wildebeest and American bisons (before being decimated by humans). Birds have long been known for the shifting habitat space (**territories**) of their resource-based habitats (e.g. the song thrush *Turdus philomelos*; Lack, 1966b). They are also the only vertebrates, apart from bats, to have extensively adapted to life in the air. Fishes are the most diversified and less known vertebrates. However, suggestions for molecular taxonomic work on fish are now advanced (Weigt *et al.*, 2012), with obvious implications for the study of their habitats. Amphibians are remarkable for having RBHs that extend over freshwater and land environments. Reptiles exhibit an array of habitat specialisations from deserts to aquatic environments, with many species strongly tied to thermal conditions and basking behaviour. Some vertebrates even have eusocial structures (e.g. humans, mole rats; Foster and Ratnieks, 2005; Wilson, 2013). In mammals it is a characteristic ploy to track behaviour throughout life cycles (see Appendix 1).

Habitat structures in plants and fungi

Although little may be said herein of plants and fungi, these two kingdoms run on the same 'principles' of the resource-based habitat. All plants and fungi have their own resource-based habitats: they require key resources and are sensitive to both abiotic and biotic conditions in which they occur. Lacking chlorophyll, fungi depend on organic matter as saprobes or decomposers to acquire their nutrients by absorption; they release enzymes and nutrients are absorbed subsequently through **hyphae**. Their taxonomy is fast developing. According to Naranjo-Ortiz and Gabaldón (2019), on current knowledge, nine phylum-level clades can be defined: Opisthosporidia, Chytridiomycota, Neocallimastigomycota, Blastocladiomycota, Zoopagomycota, Mucoromycota, Glomeromycota, Basidiomycota (Fig. 8.6a) and Ascomycota. Many fungi are closely associated with particular plants (e.g. 90% of *Auricularia* spp. grow upon the wood of deciduous trees and shrubs, favouring elder, though they also occur on deadwood in wet evergreen forests of the Western Ghats, India; Priya *et al.*, 2016) and most plants depend on them for their existence (Sheldrake, 2020). Maples in New York City, USA, have distinct signals of fungal invasion (Luley *et al.*, 2009). Some are of colossal dimensions (e.g. an *Armillaria* in Michigan, USA, extends over 70 hectares and is about 2500 years old; Andersen *et al.*, 2018) and they have an ancient ancestry (e.g. *Potteromyces asteroxylicola* dating to the Devonian, about 407 Ma; Strullu-Derrien *et al.*, 2023), with indications that fungal **mycelia** were responsible for the invasion of land by plants and that **mycorrhizal** mycelium acts as a global carbon pool (Day, 2023; Hawkins *et al.*, 2023). Increasingly, fungi are considered as the route to sustainable agriculture (Kalamulla *et al.*, 2022) and have a vital function in medicine (Petri *et al.*, 2014; Schwartz-Marin *et al.*, 2021; Whinkin *et al.*, 2023) and pollutant removal (Vaksmaa *et al.*, 2023).

Fungal diversity is only now beginning to be realised and the number of fungal species that can be found associated with plant species at regional level may be impressive. For example, Bitsadze *et al.* (2018) found 59 species of 34 fungal genera associated with oak trees in Georgia. Close relationships also occur between fungi and animals (e.g. the zombie-ant fungus *Ophiocordyceps unilateralis*/*Hirsutella formicarum* on formicine ants, which alters the behavioural patterns of the infected ants, forcing them to leave their habitat (canopy nests and foraging trails) for the forest floor – an environment with temperature and humidity favourable to the fungal growth; Mongkolsamrit *et al.*, 2012). Trees affected by fungi are regularly seen throughout woodland (see Fig. 8.6b). Ambrosia beetles (Curculionidae: Scolytinae and Platypodinae) and tree-killing bark beetles (Curculionidae: Scolytinae) have complex symbiotic relationships with fungi (see Fig. 8.6c). Ambrosia beetles are obligately dependent on their ambrosia fungi as they provide them with a broad range of essential nutrients, ensuring their survival in an extremely nutrient-poor environment. Overall, at least 12 lineages within Scolytinae and Platypodinae, and one lineage of ship-timber beetles (Lymexylidae), have evolved this mutualistic behaviour independently (Lehenberger *et al.*, 2021). In tree-killing bark beetles, fungi may play an important role in the death of bark beetle-attacked trees, although phytopathogenicity is not required to aid bark beetle hosts in overcoming tree defences (Six and Wingfield, 2011).

The kingdom Plantae includes several conventional major groups, such as Bryophyta (mosses), Pteridophyta (ferns), Coniferophyta (conifers) and Magnoliophyta (flowering plants) – algae are also frequently included among Plantae. Plants (primarily terrestrial species) account for 80% of the total biomass on the Earth; bacteria across all ecosystems

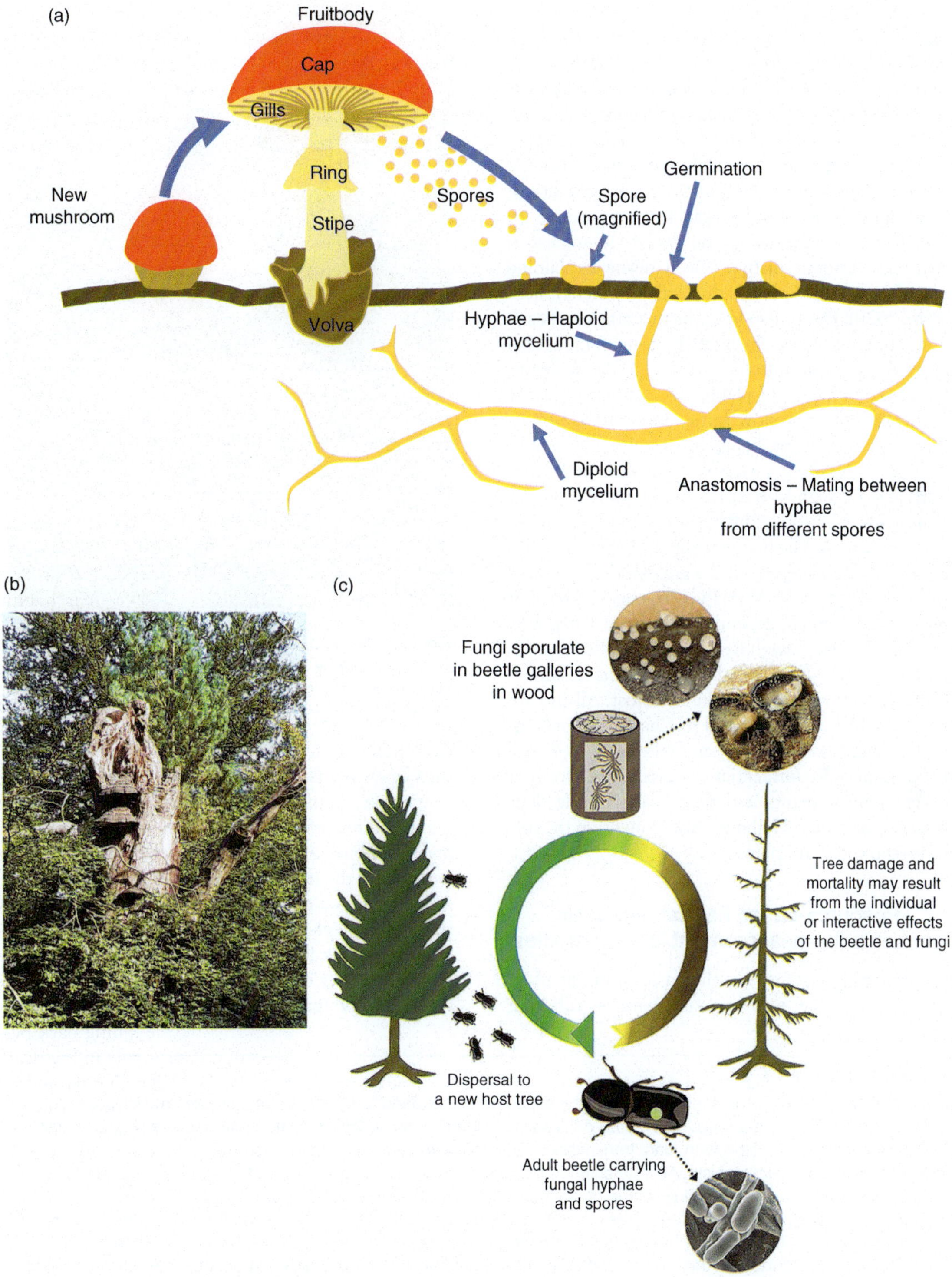

Fig. 8.6. Sphere of fungi. **(a)** Morphology and life cycle of a mushroom (Basidiomycota). Of the many spores produced by a mushroom, only a few will land where they can produce hyphae. Hyphae grow in soil and later form

rank second at 15% (Bar-On *et al.*, 2018). Plants require key inputs in a range of specific temperature, moisture/humidity and light regimes, soil/substrate structures (their soil requirements placed within the space of rock, and determined by the ternary combination of clay, sand and silt), soil nutrients and other (symbiotic) organisms, especially fungi, their distributions reflecting climate, topography and stability of conditions. There are several features unique to plants (e.g. polyploidy, mating system, sessile habit) that may lead to distinct mechanisms explaining variation in range size and thus habitat extent (Sheth *et al.*, 2020). Plants also have distinct effects on each other and not just on the light environment, but also on root space and nutrient capture. A classic example of the influence of plants on one another is the occurrence of the yellow rattle (*Rhinanthus minor*) (Fig. 8.7), a selective hemiparasite plant of grasses that functions by drawing out nitrogen and other nutrients through a transfer organ known as the **haustorium** (Gibson and Watkinson, 1989; Cameron and Seel, 2007). The hemiparasite modifies the competitive relationships between grassland component species, thereby affecting the diversity pattern of plants and animals dependent on them (Chaudron *et al.*, 2021). Here we have one of the subtlest influences in ecology, a crucial factor in the resource-based habitat of both parasite and host. All this we have touched on beforehand in Chapter 2 concerning ecosystems. For every animal's resource-based habitat, there are influential resources in the form of other animals, plants and fungi, each having their own resource-based habitats, not to mention the role of prokaryotes and protists.

Fig. 8.7. The hemiparasitic yellow rattle (*Rhinanthus minor*). (Photo by Sannse: https://it.m.wikipedia.org/wiki/File:Yellow-rattle_close_700.jpg, used under licence CC BY SA 3.0.)

Resource-based habitat: the basic elements for measurement in case studies

Before taking a glimpse at the basic components of the RBH of a few well-known organisms differing from butterflies, it is as well to have a simple list of the key elements likely to be involved in each study. The emphasis in each study should be to build an understanding of the interface between the organism and the abiotic/biotic environment. Mapping is a key tool as the mapped space used by an organism, at different scales, indicates the known bounds of survival. The elemental traits of an organism's life history and its RBH provide the key for assessing potential wider domains. For example, only in this way can the ground away from detailed study

Fig. 8.6. Continued.

a tiny mushroom button. This then grows, and the stipe and cap expand to become a mushroom. **(b)** Polypore fungi in ancient beech tree (*Fagus sylvatica*) on Alderley Edge (Cheshire, UK). **(c)** Life cycle of tree-killing bark and ambrosia beetles and their role in fungi transmission. Tree damage caused by fungi associated with these insects is inherently linked to their life cycle. After hatching from eggs laid in maternal galleries under the tree bark, beetle larvae remain within these tunnels, feeding on the phloem and creating additional burrows. During this phase, fungi grow and produce spores, sometimes serving as a nutritional resource for the larvae. As the larvae mature, they pupate and develop into adults under the bark before emerging to seek new host trees, carrying fungal spores with them. These spores, transported on the beetles' exoskeletons, sometimes within specialised structures called mycangia, or via other associated organisms, are then introduced into new trees. (Panel **b** photo by R.L.H. Dennis. Panel **c** from Linnakoski and Forbes, 2019, used under licence CC BY 4.0.)

zones be assessed as possible sites for an organism in cases of crisis (e.g. loss of land from human demands). However, an important proviso: many species are capable of surviving in non-natal areas and this may suggest overlooked traits in any study. Among birds, prominent recent examples are the rose-ringed parakeet (*Psittacula krameri*) in the UK (Butler, 2021) and the Alexandrine parakeet (*Psittacula eupatria*) in Italy (Viviano and Mori, 2021). These allochthonous (= alien, exotic or introduced) birds are exemplary cases of tropical species that have successfully colonised temperate zones. The rapidly growing number of introduced species everywhere in the world is a matter of concern because they represent serious threats to native biodiversity (see Chapter 10).

The basic elements of a study on any organism are as follows.

- Individuals (i) making up a 'colony' (pair, small troop, local population, etc.) for species S^1; that is, a set ($S^1_{i...j...n}$) of conspecific individuals, interacting (or not) with individuals of other species. S^1_i and S^1_j are conspecifics; S^2 … S^n are any interspecifics in contact (or not) with $S^1_{i...j...n}$ over regions, biomes, biogeographical regions, etc.
- Type (t) of sex and gender (modern human categorisation, *LGBTQIA+*); also caste (in **eusocial** animals; e.g. workers, queens and males in ants and termites). Stage of development (*St*; e.g. eggs, larvae, pupae, adults in insects with complete metamorphosis; larval instars; multiple larval stages in parasitic flatworms), age (*Age*; e.g. years from birth), phase (*Ph*; e.g. solitarious and gregarious individuals in locusts).
- Lifetime track of the individual (LTT_i).
- Behaviours of the individual(s) ($B_{i...j...n}$). Varied notations are needed for feeding, mate searching, copulation, resting, roosting, preying, parasitising, speed of movement, aggregation, etc., depending on the organism.
- Resource use of individuals. Note that all data on resource use are gathered from studies on behavioural attributes $B_{i...j...n}$.
- Conditions ($C_{i...j...n}$) to which an organism is subjected, abiotic and biotic, with their respective effects (negative, neutral, positive). Abiotic conditions include climate and weather (temperature, wind direction, wind speed, rainfall, cloud cover, radiation, etc.), soil type, slope, aspect, chemical composition of substrates, physicochemical parameters of water (salinity, pH, electrical conductivity, dissolved oxygen, etc.), light, etc. Biotic conditions include any intra- and interspecific interactions.
- Geographical location ($GL_{i...j...n}$) in three dimensions, x, y, z. Key here is the location (mapping) of resources and specific conditions within the space used/occupied (e.g. the **home range**) by all individuals ($\Sigma GL_{i...j...n}$).
- Organic location of individuals ($OL_{i...j...n}$), especially parasites.
- Movement and migration ($M_{i...j...n}$) of individuals during transfers between resources and during passage between subsets of the resource-based habitat (e.g. summer breeding habitat and winter feeding/resting habitat in birds).
- Sphere (zone) of activity. The essential feature here is *zone* (geographical sphere: terrestrial epigean (land), subterranean, aquatic (saline water, freshwater, depth), aerial (atmosphere), etc.; see Chapter 1).
- Traditional habitat (*TH*) (**biotope, biome, biogeographical region**, similar divisions for air and water, etc.). This encompasses the chemistry and conditions of the medium (e.g. soil, vegetation, salinity) discussed above.

Useful distinctions in habitats can be measured by using Ellenberg values (e.g. Ellenberg, 1988; Hill *et al.*, 1999), a score system that allows multidimensional characterisation of the environment on the basis of plant preferences for temperature, light, continentality, soil moisture, pH and soil fertility (nutrients) (see Di Biase *et al.*, 2023 for a discussion). We are still a long way from designing classifications and algorithms for key elements and compounds that affect species, but the path has been established for British plants by Grime *et al.* (1988), and the rapidly growing literature on **functional traits** (see Chapters 2 and 4 and Box 5.2 in Chapter 5) gives us hope that similar approaches will be extended in the future to animal groups. Improving our knowledge of the basic elements of species RBHs is essential, if only for their conservation, as information on habitat loss, population decline and range (distribution) contraction (which have their roots in RBH measurements) is crucial for determining a species' status (Red List Index, RLI; Butchart *et al.*, 2007).

Below, some basic features are outlined of the RBH for organisms that differ from Lepidoptera, for which the concept was first conceived. Groups that differ markedly from Lepidoptera: (i) are slower/faster (on land, in water and in the air); (ii) differ in complexity of social organisation (e.g. castes and slaves); (iii) differ in group (population) size (from pairs to masses); (iv) vary more in degree of cross-biome/biotope use (viz. traditional habitat); (v) differ in the occupation of spheres (or life zones) of activity (water, air, soil); (vi) differ in occupation of cross-spheres or zones; (vii) contrast in fidelity to locations/zones (non-migrants, constant migrants); and/or (viii) exploit different trophic levels (food) or combinations thereof (herbivores, omnivores, carnivores, decomposers).

Resource-based habitats: some distinct cases to consider

Slow-moving and fast-moving organisms

To illustrate some issues involved in the study of the RBH in slow-moving animals, we consider here snails as an exemplary case. Although land and water snails may have some habitat characteristics in common with Lepidoptera – for example, many species in both groups exhibit 'closed' populations (i.e. a traditional habitat occupancy and a compact population structure) – they also have very different attributes that make them a useful contrasting subject. Obvious distinctions are: (i) relative immobility (very slow movement, nil to 0.004 m/s, i.e. 14.4 m/h in *Cornu aspersum*; Cameron, 2016); however, it is now well known that snails' eggs may be transferred alive through the guts of birds, ensuring rapid long-distance dispersal (Rutherford, 2011; Cameron, 2016); (ii) small **home ranges** and a homing instinct (birth location fidelity) (e.g. *Cornu aspersum*; Dunstan and Hodgson, 2014); (iii) polyphagy and omnivory; (iv) hermaphrodite reproduction; (v) frequent egg production (even monthly), three stages of development (egg, larva/juvenile, adult) and a growing adult stage; (vi) a potentially relative long life span (0–10 years; Heller, 1990); (vii) an almost invariable dependence on calcium (a European exception is *Zonitoides excavatus*; e.g. Boycott, 1934). These features require very different approaches to acquiring information on key aspects of home range habitats, and the need for one particular human attribute, patience, in the study of snail behaviour (Cameron, 2016). Much as Lepidoptera, populations of snail species may occupy a single, few or many traditional habitats. An example of a UK traditional habitat generalist is *Clausilia bidentata* (from fens to mountaintops; Boycott, 1934; Cameron, 2016); by contrast, *Theba pisana* (an Atlanto-Mediterranean species introduced in regions around the world) is rather strictly associated with coastal environments, especially dunes (Cowie, 2009).

For the researcher into land mollusc resource-based habitats, the Oregon forestsnail (*Allogona townsendiana*) provides a useful example, as early research highlights issues that would benefit from a RBH approach (Fig. 8.8a). *A. townsendiana* is a gastropod that occurs extensively in Washington and Oregon (USA) but is classified as an endangered species in Canada (Edworthy *et al.*, 2012; COSEWIC, 2013). This species is mainly associated with forests with high growth productivity, including riparian areas, ravines, gullies and depressions containing both permanent and ephemeral watercourses; the wooded edges of streams, marshes, seasonally flooded and wet lowland areas; forest interfaces, and edge habitats where moisture is retained (COSEWIC, 2013). In a detailed study of the snail (Edworthy *et al.*, 2012), 21 adults were tagged and tracked, and observations logged at regular intervals for 3 years in Langley (British Columbia, Canada). For a subset of 15 individuals, their home ranges (Figure 3 in Edworthy *et al.*, 2012) have been mapped and their basic traditional habitat associations recorded. All home ranges overlapped and from these it was possible to draw a home range for the specimens sampled. However, to obtain a population home range or local population RBH, a larger area needs to be explored over different substrates. Loosely, *A. townsendiana* is noted to occupy humid deciduous and mixed woodlands and woodland/meadow edges (traditional habitats) less than 360 m above sea level (COSEWIC, 2013). Food was seen to include stinging nettle (*Urtica dioica*), but otherwise it is thought that coarse woody debris supplies potential food items such as fungi, bacteria and algae. Although nettles are key attributes of the habitat of these snails, many other plants occur in their home ranges and coarse woody debris also plays critical roles (e.g. as a mating site; Steensma *et al.*, 2009). Stinging nettle is known to have high levels of calcium necessary for healthy shell growth in snails, and indicates moist, soft, rich mesic deep soils (pH 6.4–6.9; temperature 9.9–13°C) with high amounts of nitrogen and phosphorus (Steensma *et al.*, 2009; COSEWIC, 2013). Again, these soil attributes very probably have critical importance for the snail and egg laying (eggs are laid singly or in clusters). Movement was rarely more than 15 m (maximum 28.9 m) from their original site over 3 years of study (home ranges extended from 18.4 to 404.4 m^2) (Edworthy *et al.*, 2012).

Fig. 8.8. Land and sea gastropods. **(a)** Oregon forestsnail (juvenile) (*Allogona townsendiana*). **(b)** Common periwinkles (*Littorina littorea*) on seaweed. (Panel **a** photo by Walter Siegmund, identification by Robert Forsyth: https://commons.wikimedia.org/wiki/File:Allogona_townsendiana_1205.JPG, used under licence CC BY SA 3.0. Panel **b** photo by Ryan Hodnett: https://commons.wikimedia.org/wiki/File:Common_Periwinkles_%28Littorina_littorea%29_on_Seaweed_-_Nesodden,_Norway_2020-09-20_%2801%29.jpg, used under licence CC BY SA 4.0.)

A problem with the research is that there has been little firm data logging of behavioural and life-history attributes linked with the environment through tracking of home ranges. Although the spatial bounds of the 21 individuals were followed closely, there is no detailed record of actual behaviour from minute to minute, hour to hour and day to day – as done for butterflies (Dennis, 2010) – when the snail is active (surface activity of the snails appears to peak in spring and early summer, coinciding with mating and oviposition), nor what and why hibernation sites were selected. Predators probably include small mammals, birds and other molluscs (eggs and juveniles have been seen being eaten by the robust lancetooth snail *Haplotrema vancouverense*) (Steensma *et al.*, 2009). Although it is clear that human-driven development (commercial and residential) forms the greatest enemy of the species in Canada's Frazer Valley (COSEWIC, 2013), the reader is left wondering why this apparently rather undemanding species is so limited geographically in Canada, though it may well be close to a climatic limit. According to the COSEWIC assessment, the Canadian population for the Oregon forestsnail is estimated to be in six figures. However, 52 of the 66 known sites (78.8%) assessed are fragmented and isolated; more than 50 of the 75 sites (2 km × 2 km squares) examined are less than 5 ha in size and/or are subject to future urban development.

The research on *A. townsendiana* illustrates an example of a mollusc which requires intensive behavioural surveys, linking life-history and behavioural traits to use or avoidance of specific substrates and conditions in home range assessments. A resource-based approach would require a thorough multivariate analysis of traits, resources and conditions within a systematic grid of such sample sites over the snail's range, and a comparison within each 'site' of locations having and lacking the organism. Concurrently with home range tracking, independently and subsequently, behavioural studies are an imperative.

Many other land gastropod species could be cited for comparison. Perhaps the most remarkable is *Cepaea nemoralis*, in which colour and banding have been assessed in relationship to geography and a range of traditional habitats, both present and historical (archaeological sites). In this case, frequencies in colour and banding patterns give insights into selection pressures on this gastropod in its exploitation of traditional habitat and resources (see references in Cook, 2008b; Silvertown *et al.*, 2011; Davison *et al.*, 2019). Still an issue is the rate at which this and other land snails can spread (Peltanová *et al.*, 2012). Until this issue is fully understood, the potential bounds of snail resource-based habitats remain unknown.

Water snails and shore gastropods pose very different problems compared to land snails (Underwood, 1979). One issue is how to track individuals and their behaviour in a medium (i.e. sea water with incoming tide and waves) that prevents access much of the time to the locations chosen by individuals and their capacity for relocating chosen stations on rocks (Chapperon and Seuront, 2012). One thing is clear: the density of gastropods such as the common periwinkle *Littorina littorea* (see Fig. 8.8b) varies enormously both up and down, as well as across beaches (e.g. North Atlantic intertidal variability in density: 0 to >600 m^{-2}; subtidal density: 0–19 m^{-2}) (Carlson *et al.*, 2006). This variability relates to surfaces (i.e. rugosity, salt marsh, sand sediments, bare rock exposure), prey and predators (Eschweiler *et al.*, 2009). Moreover, the form (shape, shell thickness and size) also relates to wave action (Trussell *et al.*, 1993). A complexity of sampling is that *Littorina* have a capacity for finding their way back to initial stations following dislodgement through movements that probably combine responses to the force of gravity, chemical stimuli and water current (Seuront *et al.*, 2018).

The common periwinkle is one of nine species in the genus *Littorina* mainly found on rocky shores in the higher and middle intertidal zone in the northern hemisphere (Jackson, 2008). *Littorina littorea* is widely distributed on rocky coasts, in all except the most exposed areas, from the upper shore into the sublittoral (maximum depth 60 m), food availability being dependent on habitat affiliation much as in the case of another littorinid, *Echinolittorina lineolata* (Apolinário *et al.*, 1999), which is mainly distributed in the tropical zone of the western Atlantic. In sheltered conditions, *L. littorea* can also be found in sandy or muddy habitats such as estuaries, salt marshes and mudflats. The species is fairly tolerant of brackish water. Classically, it is seen in small tidal pools on wave-cut platforms, these different 'habitats' indicating the complexity of the species' resource-based habitat. Local populations can exceed hundreds to thousands of animals/m^2 (Carlson *et al.*, 2006). *L. littorea* can be found at considerable depths (Fretter and Graham, 1980), but predation (especially by the crab *Cancer borealis*) limits its abundance subtidally (Perez *et al.*, 2009).

Any study of the RBH of *Littorina* necessarily has to consider the need for stratified sampling of the very different traditional habitats within which the species is found. Physical distinctions (i.e. in size and shape) between local populations are associated with differences in substrates and exposure to wave action (Trussell *et al.*, 1993; Eschweiler *et al.*, 2009). Although adults may be easy to locate and measure, early stages are a different matter. The snail has a planktonic development; females release planktonic egg capsules that are pelagic; the eggs hatch into veliger larvae which remain in a planktonic stage for 4–7 weeks and subsequently settle (Fish, 1979; Cummins *et al.*, 2002; Hohenlohe, 2002). As with many animals, finer issues of resource-based habitats must necessarily be gleaned using experimental conditions, in this case tanks (Newell, 1958; Castelo Branco *et al.*, 2014). An important finding is that the duration of the pelagic phase is closely temperature dependent (Castelo Branco *et al.*, 2014).

These land and marine gastropods indicate just how complex the study is of different species in contrasting conditions. Although much the same stages have to be passed through, the difficulties any one stage presents differ enormously from species to species, land to water. Identification of population bounds may also be very much more difficult than for Lepidoptera. As the periwinkle is a commercial species and one in danger of population decline, a considerable amount of data has accumulated on its biology, its geography and, particularly, on its sensitivity (to substrate loss, desiccation, smothering, temperature, water flow, turbidity, wave exposure, and a range of chemical pressures); its ability to accumulate chemicals and consequential behavioural changes makes this species also a valuable **bioindicator** (Jackson, 2008).

On the other hand, many animals move faster than butterflies. From the RBH perspective, this means that they can have broader home ranges, although this does not necessarily translate into a wider array of resources used and conditions experienced. It is obvious that larger animals can move faster than smaller, but they might perceive the landscape at a coarser grain. For example, gazelles are known to run very fast and have large home ranges, which however are composed of relatively uniform savanna and grassland landscapes. The type of resource and sensitivity to conditions matter. Because of their minute size, micro-organisms may appear as slow-moving in absolute terms, but they can be very fast in relative terms. For example, the protist ciliate *Mesodinium rubrum* (size 38 μm) may reach 7 mm/s (Lisicki *et al.*, 2019); this means that in 1 second it can cover a distance of almost 200 times its size. For these microscopic animals, important environmental differences occur at extremely fine scales; for example, different species of ciliates occupy different depths in the water column of a pond (Finlay, 2004). Some birds, bats and insects are able to reach very high speeds, in some cases even for relatively long periods, which allows them to move over long distances and hence to exploit geographically distant resources.

Finally, there are organisms that lack any means of self-locomotion (sessile organisms). These include many plants, fungi and several animals (e.g. sponges, corals, bivalves, some annelids, barnacles, scale insects, tunicates). Most sessile animals, however, have a motile phase in their development. Sessile organisms may define the (traditional) habitat for other species. For example, dominant plants are frequently used for (traditional) habitat categorisation for animals, and coral reefs are indicated as the habitat of many fish species. These organisms act as **ecological engineers** that construct the habitat for a multitude of other organisms (see Chapter 2).

Complexity of social organisation: castes and slaves

Social organisms like hominins (e.g. *Homo neanderthalensis*, *Homo sapiens*) or social insects present a new range of problems for the determination of their resource-based habitats. There are different forms of sociality along a continuum (Keller and Perrin, 1995; Wilson, 2000). In **eusociality** (where some individuals renounce direct reproduction to rear their siblings), beyond stages or phases of development, or growth and ageing, are castes of varying status differing in numbers, behaviour, resource needs, conditions for functioning; complex traits and interactions in mappable space-time. Eusociality evolved several times in insects (termites, ants, bees and wasps; Wilson, 1971, 2000), and it is also found in crustaceans (the sponge-dwelling shrimp *Synalpheus*; Duffy *et al.*, 2000) and mammals (the bathyergid genera *Heterocephalus* and *Fukomys*), but eusocial behavioural tendencies are also shown by meerkats (*Suricata suricatta*) and dwarf mongooses (*Helogale parvula*; Williams and Shattuck, 2015) – humans too can be considered eusocial mammals (Foster and Ratnieks, 2005).

We illustrate some aspects of the RBH in eusocial animals with reference to termites (Order: Blattodea, Suborder: Isoptera), the first social insects to evolve a caste system more than 100 million years ago (Engel *et al.*, 2016). Termites are in general small insects, most measuring between 4 and 15 mm (but queens of *Macrotermes bellicosus* may exceed 100 mm when physogastric, i.e. when the abdomen swells in order to hold enlarged ovaries, thus increasing fecundity); the usual colonial structure contains nymphs (semi-mature young), workers, soldiers and reproductive individuals of both sexes (numerous kings and one or more queens). A simple life cycle illustration of relationships between these castes is presented in Figure 8.5c. Termites are not Hymenoptera, as might be expected by the casual observer (i.e. they are not allied with ants, bees and wasps that also have social structures), but are closely related to cockroaches (sharing a common ancestor with the cockroach genus *Cryptocercus*, from which they branched off probably in the Jurassic; Evangelista *et al.*, 2019) and capable of immense damage to crops (plantation forests) and buildings (Krishna *et al.*, 2013). More than 3000 species are known and 4000 suspected, with equally varied life histories (Bignell *et al.*, 2010; Krishna *et al.*, 2013). Inasmuch as they exist in 'tribes' or colonies, each unit is in effect a local population (equivalent to a tribe in humans), often separated from others by a few metres and between which there may be substantial conflict (Tarnita *et al.*, 2017).

The structure of termite nests is legion. Simply, termite nests can be described (mapped) as being of two parts: (i) inanimate (nest structure) and (ii) animate (termites) (Bignell *et al.*, 2010). The nests are divided into three categories: (i) subterranean (completely below ground), (ii) epigeal (protruding above the soil surface as mounds made out of earth and mud), and (iii) arboreal (built above ground but linked to the surrounding ground via shelter tubes). A nest has many functions such as providing a protected living space and shelter against predators. Most termites construct underground colonies rather than multifunctional nests and mounds (Eggleton *et al.*, 1996). Perhaps the most extraordinary aspect of termite existence is this remarkable variety of epigeal (nest) ground structures (see Fig. 8.5a,b) with much the same internal organisation of caste functions. The question facing any study of a resource-based habitat of termites, then, is why the nests vary in the way they do. Issues that readily come to mind are specific predators and parasites, but also abiotic pressures involving temperature, humidity, ventilation, hydrothermal circulation, gas exchange, suitable conditions for raising fungi (food) and bacterial defences (e.g. *Pseudomonas* in *Odontotermes obesus*) (King *et al.*, 2015; Ocko *et al.*, 2019; Agarwal *et al.*, 2021). In conveying and nurturing their own food in fungi gardens, termites such as *Macrotermes natalensis* are effectively engaged in farming (Kim *et al.*, 2014; Roberts *et al.*, 2016). *Termitomyces* on which *M. natalensis* feed is a genus of basidiomycete fungi belonging to the family Lyophyllaceae. There are 30–40 species in the fungal genus, all of which are completely dependent on termites to survive.

Termites are renowned for their abundance, their existence as a **superorganism** (that is, a group of synergistically interacting organisms of the same species) and their construction of micro-landforms. Some 1.1 million live *Macrotermes* termite mounds have been estimated to occur in the northern Kruger National Park (South Africa) alone (Meyer *et al.*, 1999). Much research has been carried out on termite mounds. Some identical coned mini-hills and intervening hollows or vales would suggest the existence of similar internal structure, with a basal nest and territorial surround, but this has been shown to be entirely wrong; others gravitate to underground 'vales' between the hillocks, the latter acting as waste ground for detritus (sediment) excavated from the tunnels underground. The diversity of mound structures is driven by the integration of environmental physics with organismal behaviour; a mathematical model that couples environmental physics to insect behaviour illustrates how the advection and diffusion of heat and pheromones through a porous medium are modified by the mound geometry and, in turn, modify that geometry through a minimal characterisation of termite behaviour (Ocko *et al.*, 2019). Termite mounds have substantial ecological importance, and not simply because they are extreme examples of ecological engineering by non-humans (see Chapter 2 on **ecological engineers**), but also because of the impact on other organisms (some on a massive scale; e.g. 4000-year-old soil mounds constructed by *Syntermes dirus* covering an estimated 230,000 km^2 of seasonally dry tropical forest of north-east Brazil; Martin *et al.*, 2018), controlling the space-time organisation and dynamics of complete ecosystems (Tarnita *et al.*, 2017; Muvengwi and Witkowski, 2020). Determining their RBH requires sampling within individual nests and intervening ground, and then repeating the process across and between numerous nests through full cycles (life and seasonal), in different regions where the genome exists and the biome impact differs. Although the architecture of many termite nest structures has been accessible due to road cuts and other human activity, it is inevitable that understanding of termite nest structure and function has involved excavating mounds and surrounding ground. However, employing non-destructive techniques should be a priority.

Some ants (Hymenoptera, Formicidae) present an additional complexity to caste structures through the enforced use of slave ants (**dulosis**; D'Ettorre and Heinze, 2001). Apparently, there are six origins of slavery in formicoxenine ants (Beibl *et al.*, 2005). Some species are obligate, others are facultative slave makers (e.g. Savolainen and Deslippe, 2001). The interesting twist to the resource-based habitat involves signs of a coevolutionary arms race between a slave maker and its host (e.g. *Temnothorax americanus* and its host *Leptothorax longispinosus*; Foitzik *et al.*, 2001). However, enslaved ants are not as helpless as they are often thought to be, as they may act also in favour of their own maternal population through various forms of rebellions (Czechowski and Godzińska, 2015). Study of RBH across the range will undoubtedly result in different findings (balance in numbers of slaves to hosts) in different regions (e.g. obligate versus facultative slave makers; Savolainen and Deslippe, 2001). The crunch is that an RBH for any of these species cannot be determined from a single colony or colonies in one region and, furthermore, determining traits for survival and reproduction requires establishing colonies in 'laboratory' conditions.

Dulosis is a particular form of **social parasitism**, that is, the coexistence in the same 'nest' of two species of insects, one of which is parasitically dependent on the other (D'Ettorre and Heinze, 2001). Slaver ants are a case of social insects that use other social insects. But insect societies are infiltrated by a number of other non-social animals, such as butterfly caterpillars, beetles, spiders and even snails, that exploit hosts' altruistic behaviour or their resources (Kistner, 1982; Fiedler, 2001; Witte *et al.*, 2002; Als *et al.*, 2004; Geiselhardt *et al.*, 2007; von Beeren *et al.*, 2011; Cushing, 2012; Parker, 2016; Grüter *et al.*, 2018; Mizumoto *et al.*, 2022). To infiltrate insect societies, these social parasites evolved adaptations in order to avert aggressive behaviours from the host, including sophisticated mimicry (for example, the carabid beetle *Paussus favieri* is able to mimic caste-specific stridulations to dupe the workers of its ant host *Pheidole pallidula* and to be treated as royalty; Di Giulio *et al.*, 2015; Fattorini *et al.*, 2021). For these parasites, the RBH encompasses the complex conditions and resources determined by their hosts.

Group size (singletons, pairs to population massing)

Shifting up scale to vertebrates, especially territorial species, brings a whole range of new problems for determining the RBH. The simplest division of the landscape is the possession of a space by the smallest fraction of individuals of a species: a single individual or a male/female pair. The work on British birds has been classical in this respect (Fig. 8.9), such as the seminal research by David L. Lack on European robins *Erithacus rubecula* (Lack, 1943) and song thrushes *Turdus philomelos* (Lack, 1966b). The basic pattern is the carving up of a traditional habitat (e.g. a woodland) by pair groups in the breeding season followed by its division by singletons during winter (e.g. European robins). An alternative to a breeding season's territorial pair of birds is the widespread use of the landscape in winter (i.e. different substrates and biotopes, different locations) typically by large groups of individuals, such as families or family groups of thrushes (Turdidae: fieldfares *Turdus pilaris* and redwings *Turdus iliacus*; Paridae: blue tits *Cyanistes caeruleus*, great tits *Parus major*, etc.) for feeding, resting, roosting and survival from predators, the individuals moving about together and alerting each other to dangers (Fuller, 1982). Even then, some resources (locations) are in regular use day after day (i.e. long-tailed tit *Aegithalos caudatus* roosts; McGowan *et al.*, 2006). The gathering of many different species in feeding grounds has the advantage of more eyes watching for predators; different species have the ability to pick up warning signals from each other in such situations (Stevens, 2013; Magrath *et al.*, 2015; Goodale *et al.*, 2017).

Some birds form colonies and take over a large fraction of a traditional habitat unit (e.g. a wood); examples among these may be a clamour of rooks *Corvus frugilegus* in deciduous woodland (Fig. 8.10a; colony size varies considerably; within 15 km of Durham City (UK) in 1996 the range was 5–136 nests; Griffin, 1999) or a colony of cormorants (*Phalacrocorax carbo*, Fig. 8.10b) on cliffsides, although they can also occupy trees inland (e.g. Trentabank larches, Macclesfield Forest, UK; R.L.H. Dennis, personal observation). The extreme is herds of ungulates (see *Fidelity to and movement within locations/zones* below) noted for mass migration across suitable biotopes. This contrasting use of biotope is important for determining the factors influencing the nature of the RBH of a species (see *Occupation of a variety of biomes and biotopes* below).

A variety of hypotheses have been proposed for colonial behaviour compared with exploiting the space in pairs (Griffin, 1998).

- Nest site limitation hypothesis (there is a limitation to the number of safe sites for nesting; e.g. American magpies *Pica hudsonia* tend to form diffuse colonies in areas with a limited number of trees or with abundant food resources; Birkhead, 1991).
- Predation hypothesis (coloniality may help to combat predation; e.g. colonial breeding in the great crested grebe *Podiceps cristatus* is an effective defence against the American mink *Neogale vison*; Brzeziński *et al.*, 2018).
- Central-place-foraging hypothesis (individuals tend to aggregate in places that are, on average, at the minimum distances from the resources, when these are dispersed and territories cannot be adequately defended; a model proposed, for example, for the red-winged blackbird *Agelaius phoeniceus*; Orians and Pearson, 1979).
- Information-centre hypothesis (living in colonies offers the advantage of obtaining information from other individuals about the location of unevenly distributed resources; a theory originally developed by Ward and Zahavi, 1973 and applied to several bird species; e.g. in the Eurasian griffon vultures (*Gyps fulvus*) individuals that have no knowledge of a currently available food source follow 'informed' individuals to food sources; Harel *et al.*, 2017).
- Recruitment-centre hypothesis (group foraging presents many benefits like a reduction in the risk of predation by the simple dilution effect or by earlier detection of a predator, which can explain why successful foragers return to a communal place (Richner and Heeb, 1996); for example, this theory seems to be suitable to explain raven roosts; Danchin and Richner, 2001).

Variation in bird pairs per colony probably relates to the number of opportunities for both nesting

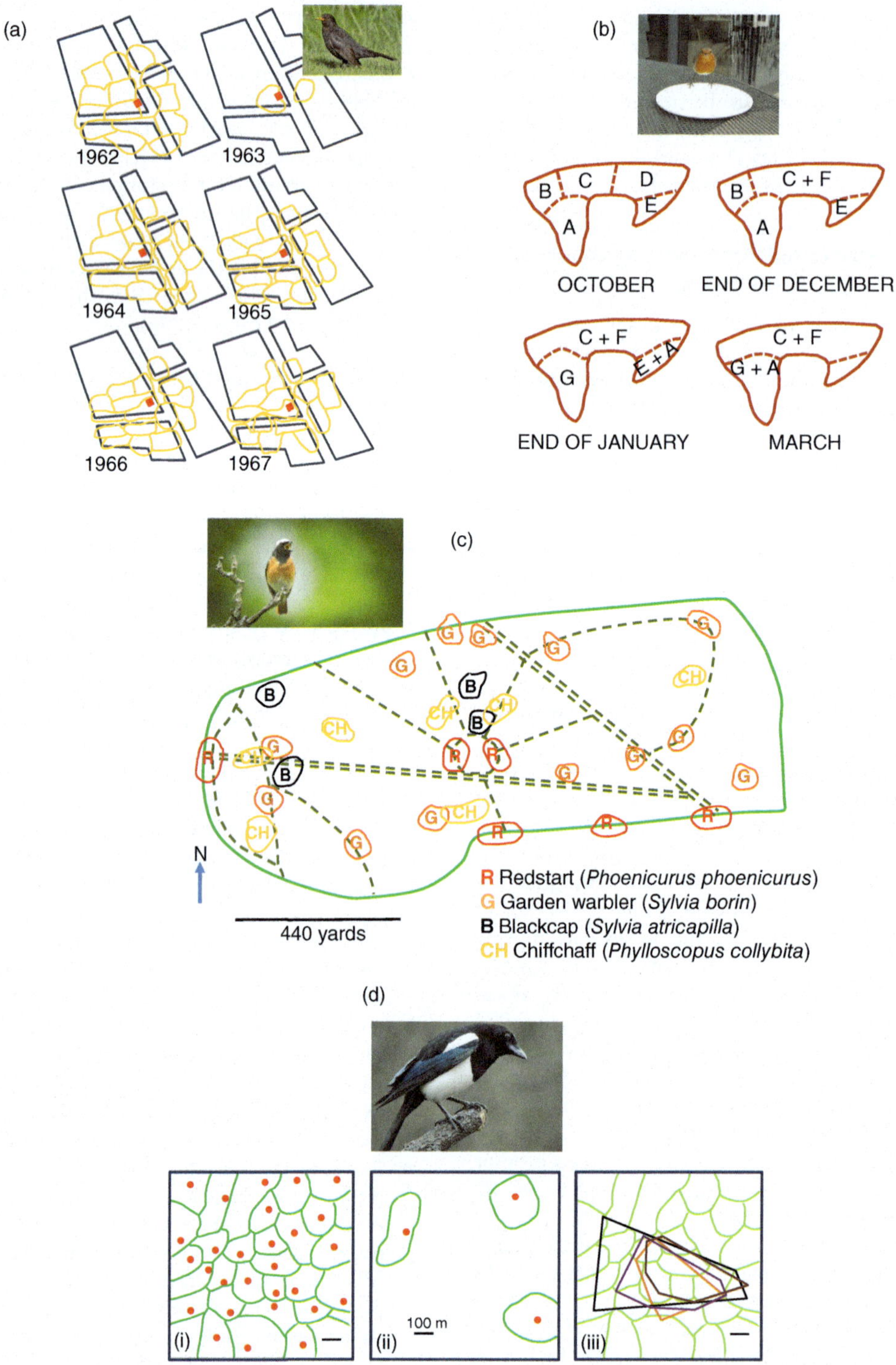

Fig. 8.9. Territories and home ranges of selected British birds. **(a)** Blackbird (*Turdus merula*) territories (yellow polygons) around Eric Simms' house (red square) at Dollis Hill, London, between 1962 and 1967; the crucial features here are

Fig. 8.10. Communal nesting. Rooks (*Corvus frugilegus*, **a**) and cormorants (*Phalacrocorax carbo*, **b**) that nest communally with individuals sharing a home range and resource-based habitat. (Panel **a** photo by Derek Harper: https://commons.wikimedia.org/wiki/File:Rookery,_Chard_churchyard_-_geograph.org.uk_-_755937.jpg, used under licence CC BY SA 2.0. Panel **b** photo by MirMar: https://commons.wikimedia.org/wiki/File:NSG_Insel_Imchen,_Kladow,_Berlin,_Germany_06.JPG, used under licence CC BY SA 3.0.)

and feeding in the surrounding location, bearing in mind that nests in poor positions 'fail' and that long flights out to feeding grounds increase vulnerability of eggs and young from competitors and predators, as predicted by the **ideal free distribution** (see Chapter 6), which focuses on the conditions (i.e. distribution of individuals among patches of resources) in which fitness returns for all individuals are equal (Brown *et al.*, 1990). For example, Kasprzykowski (2003, 2007) found that, in colonies of rooks (*Corvus frugilegus*) in Poland, flocks of less than 10 individuals were dominant (maximum 132) and stayed within 0.5–1 km of the rookery (maximum foraging distance 2.8 km), the distance dependent on the size of a rookery. Breeding success in these rookeries increased with the area of preferred crops: spring cereals and meadows/pastures. Interestingly, in Britain, carrion crows (*Corvus corone*) generally nest and function in isolated pairs but this bird is capable of developing sociality, as in Spain (co-operative breeding), and may function in groups of many individuals (maximum of nine individuals; mode 3; Baglione *et al.*, 2002; Canestrari *et al.*, 2011).

To take a British woodland as an example of traditional habitat space, a large number of passerine bird species pairs (songbirds and other perching birds) occupy and defend different portions of it during the summer breeding season (see Fig. 8.9). Their nest sites are rarely at the centre of the space which is defended (a **territory**) by song, display and/or fighting. The territories tend not to be of the same size and shape, nor do individual portions necessarily approximate the same shape and size in different seasons and years (Lack, 1943) (see Fig. 8.9a,b). Furthermore, there are often contested overlapping spaces as well as 'dead' spaces (**matrix**; see Chapter 2,

Fig. 8.9. Continued

changes in number and shape of territories from year to year, to be expected with shifts in individual resource holding power and changes in the quality of habitat spaces. **(b)** Seasonal changes in robin (*Erithacus rubecula*) territories in a copse in South Devon. **(c)** Territories of four species breeding in Bagby Wood, Northamptonshire, in 1949 (dashed lines are tracks and paths); note the contrast in the number of territories, occasional overlaps and the sparse nature of territories in the east area with conifers. **(d)** Comparison of magpie (*Pica pica*) territories in the Peak District (dots are magpie nests): (i) dense small territories in an area of high-quality resources; (ii) sparse territories in an area with poor resources (high-altitude zone near Sheffield); note dead ground unused by any birds and lacking territorial interactions; (iii) home ranges of four non-breeding juvenile magpies in the space of birds with nesting territories. (Panel **a** from Simms, 1978, redrawn. Panel **b** from Lack, 1943, redrawn. Panel **c** from Simms, 1990, redrawn. Panel **d** from Birkhead, 1991, redrawn with permission of rightsholder. Inset photographs: blackbird by Andreas Trepte: www.avi-fauna.info; https://en.wikipedia.org/wiki/File:Common_Blackbird.jpg, used under licence CC BY SA 2.5; dominant robin at Dunham Massey tea room territory (Manchester, UK) by R.L.H Dennis; common redstart by hedera.baltica: https://commons.wikimedia.org/wiki/File:Common_redstart_%2851215029141%29.jpg, used under licence CC BY SA 2.0; magpie by Tim Watts: https://commons.wikimedia.org/wiki/File:Magpie_%2815563869424%29.jpg, used under licence CC BY 2.0.)

Box 2.2) unused by any of the pairs of birds (Lack, 1966b) (see Fig. 8.9c,d). These observations indicate an uneven spread of resources and unique changes in resource quality over space-time for each species. Individuals of the same species often trespass in other birds' territories, as Lack (1943) discovered for robins, and quickly leave when they notice the presence of the incumbent; if they stay, then their presence triggers defence behaviour from the incumbent. Life spans vary from less than a season to several years; with death comes a shift in allegiances of the surviving member and changes in territory space (size and shape) in relationship to their **resource-holding potential** (also referred to as resource-holding power – that is, the capacity of an animal to maintain or gain control of a resource; Parker, 1974; see Allen and Krofel, 2017 for a review).

As noted above, different species (such as various warblers and titmice) use contrasting (including overlapping) spaces in the same traditional habitat unit (meadow, wood, hedgerow, etc.) in any year, communicating with members of their own species through species-specific songs (see Fig. 8.9c). This feature nicely illustrates that different species require different variants of resources, classically demonstrated by exploitation of different food types (seeds, fruit, earthworms, bark and leaf arthropods, etc.). The availability of nest sites is typically the crucial resource for the existence of a suitable territory in a traditional habitat; as a prominent example of the problems this can create, the first author observed the repeated use of an old telegraph post, adjacent to the spruce wood at Chapel (Macclesfield Forest, Cheshire, UK), by a great spotted woodpecker (*Dendrocopos major*), which evidently lacked suitable nesting trees. A range of disturbances can put some species off setting up a territory more than others. For example, the first author noticed how uncontrolled dogs discourage birds, such as wood warblers (*Phylloscopus sibilatrix*), in shrubby woodlands in Alderley Edge (Cheshire, UK).

At the individual level, varying territory size and trespassing indicate that territories occupy smaller spaces than **home ranges**. However, the larger the territory, the greater the energy required to defend it and to exploit resources. The home range equates better with the bounds of an individual's resource-based habitat; a resource-based habitat that produces successful broods may represent a minimum viable territory. A crucial feature of organisms in a landscape is that areas may or may not be defended. However, when an animal is numerous, empty spaces would indicate that some key resource or favourable condition (i.e. absence of predator, presence of prey) is missing from certain places. The very fact that robin territories have sole occupants from late summer into the winter indicates that many subdominant individuals, for instance young, need to seek out resources in appropriate traditional habitat spaces during the winter months while continuously on the move (Lack, 1943).

There is, of course, no escape from complexity. To determine the RBH, the student needs to sample many pairs, including mobile winter flocks over the countryside, from one spatial unit (traditional habitat) and region to another. Complexity is continually increased by anthropogenic transformations, which change the spatio-temporal distribution of resources and conditions. For example, the gradual incursion of urban and suburban landscapes has profound impacts on the population structure of birds; for instance, population size in carrion crows increases with the development of human food waste sites as in Austria (Preininger *et al.*, 2019); kestrels (*Falco tinnunculus*) show high breeding density in Rome, Italy, because facades with large areas and scaffolding holes allow many pairs to occupy all sides of buildings, thus obtaining a better spacing from neighbouring pairs (Salvati *et al.*, 1999); similarly, jackdaws (*Corvus monedula*) show increased breeding density and reduced colony spacing in Rome as scaffolding holes are particularly safe nests (Salvati *et al.*, 2002).

Occupation of a variety of biomes and biotopes

Moving on from pairing birds that divide up a relatively homogeneous **biotope** (traditional habitat) into **territories** throughout the breeding season, there are many species that, having set up a home (i.e. a nest site) in one traditional habitat, have a **home range** extending over other traditional habitats, often **landscapes** of contrasting types of vegetation cover. Classic examples are the wood pigeon *Columba palumbus* and the carrion crow *Corvus corone* (Murton, 1965; Gregory and Marchant, 1996). Both set up their nests in trees within woodlands or in isolated trees in hedgerows, the wood pigeon typically at the wood edge and preferring hawthorn (Inglis *et al.*, 1994; Terraube *et al.*, 2016). However, for food, they exploit the open countryside and suburban settings as well as woodland.

The wood pigeon is often seen stripping young growth from beech trees in spring and beech nuts from the same trees in autumn, but also raids cereal and pasture fields (clover) throughout the year, often in colossal numbers (Murton, 1965).

Among British birds, some species occupy a variety of contrasting biotopes. Coots *Fulica atra* (Cavé *et al.*, 1989; Brinkof *et al.*, 1993) and moorhens *Gallinula chloropus* (Relton, 1972; Huxley and Wood, 1976) have their headquarters (nest sites) on the edge of ponds and lakes, or on islets of floating or fixed debris. However, although they feed on and under the water, they also forage along field hedges and in fields, and roost in short trees and shrubs such as hawthorn. It is clear that coots prefer larger ponds and lakes than moorhens (Cheshire, UK; R.L.H. Dennis, personal observation, 2019) and this represents an interesting case study of the resource-based habitat in these species. Among coots, old pairs have larger territories, measured as water area, than young ones (Cavé *et al.*, 1989). However, territory size is not the only important feature for territoriality in this species: vegetation elements along the shore are crucial constituents of the territory (nest sites). It appears that the composition of the territory, measured as length of unprotected shore, depends on the age of the male, not on that of the female.

In Chiayi County (Taiwan), eight crucial factors were found by Lai *et al.* (2018) to affect the habitat suitability of ponds for moorhens. In descending order, they are: (i) the proportion of farmland area within 200 m around the pond; (ii) pond area; (iii) pond perimeter; (iv) aquatic plant coverage of the pond surface; (v) duration of drought period; (vi) coverage of high and low shrubs around the pond bank; (vii) bank type; and (viii) water surface-to-bank distance. Different regions for these species reveal different predators (e.g. the mink *Neogale vison* on the Upper Thames, UK; Ferreras and MacDonald, 1999; probably the viperine snake *Natrix maura*, rats *Rattus* spp., the swamphen *Porphyrio porphyrio* and the marsh harrier *Circus aeruginosus*, which may raid bird nests, in Algeria; Samraoui *et al.*, 2013). These species also demonstrate other complex aspects of life history affecting habitat use. In the Wildfowl and Wetlands Trust Centre Llanelli (South Wales), Forman (2001) revealed that although monogamy was the dominant breeding system, many individuals bred annually in polyandrous (one female with multiple males), polygynous (one male with multiple females) and polygynandrous (multiple males with multiple females) groups. Also, some females laid eggs parasitically in the nests of neighbouring conspecifics. As many as three broods can be attained in one season, the young birds staying until the late winter; and from thence they are persuaded to leave the pond (R.L.H. Dennis, personal observation, Stockton Farm pond, Wilmslow, Cheshire, UK; see Leonard *et al.*, 1988 for parent–offspring interactions). Nest sites can vary substantially as discovered in a study in Algeria (Samraoui *et al.*, 2013); most nests (92%) at Boussedra were located in dense stands of the lesser bulrush *Typha angustifolia*, whereas a few were found on lower branches of the tamarisk *Tamarix gallica* trees; in contrast, at Lake Tonga nests were predominantly found on lakeshore bulrush *Schoenoplectus lacustris* (41%) or associated with alder trees *Alnus glutinosa* (32%).

Many species switch biotopes during different seasons; for example, species that, as well as setting up relatively limited home ranges and territories in the breeding season in one biotope, migrate (move over the landscape landing at propitious sites) often huge distances during the rest of the year – feeding, resting, roosting and escaping predators as they move about the landscape between suitable biotopes. For instance, curlews (*Numenius arquata*) nest inland and on uplands such as the Pennine moors (England), feeding in the soft summer peaty substrate under heather (*Calluna vulgaris*) and blanket bog plants (*Erica* sp., *Eriophorum* sp.), but as their nesting ground becomes frozen and therefore food is unavailable on the moors in the winter, they move to the coast, to the salt marshes and mudflats of estuaries, where food is plentiful and accessible from autumn to spring (e.g. Morecambe Bay in Lancashire and Cumbria) (Wilson, 1973; Bainbridge and Minton, 1978). A study conducted between 2013 and 2019 revealed that along the Atlantic seaboard the spring migration occurred between April 4 to 14 (average ± SD: 10.2 ± 8.4 days) while the autumn migration occurred between 18 to 29 June (10.9 ± 9.9 days) using, on average, about five stopovers lasting approximately 30 hours (Pederson *et al.*, 2022; Fig. 8.11). In a study of a male long-billed curlew *Numenius americanus* over three migratory cycles using satellite telemetry in North America, the bird showed strong fidelity to stopover, breeding and wintering sites, not only on a geographical scale but also on a local scale across the years (Olalla-Kerstupp *et al.*, 2015).

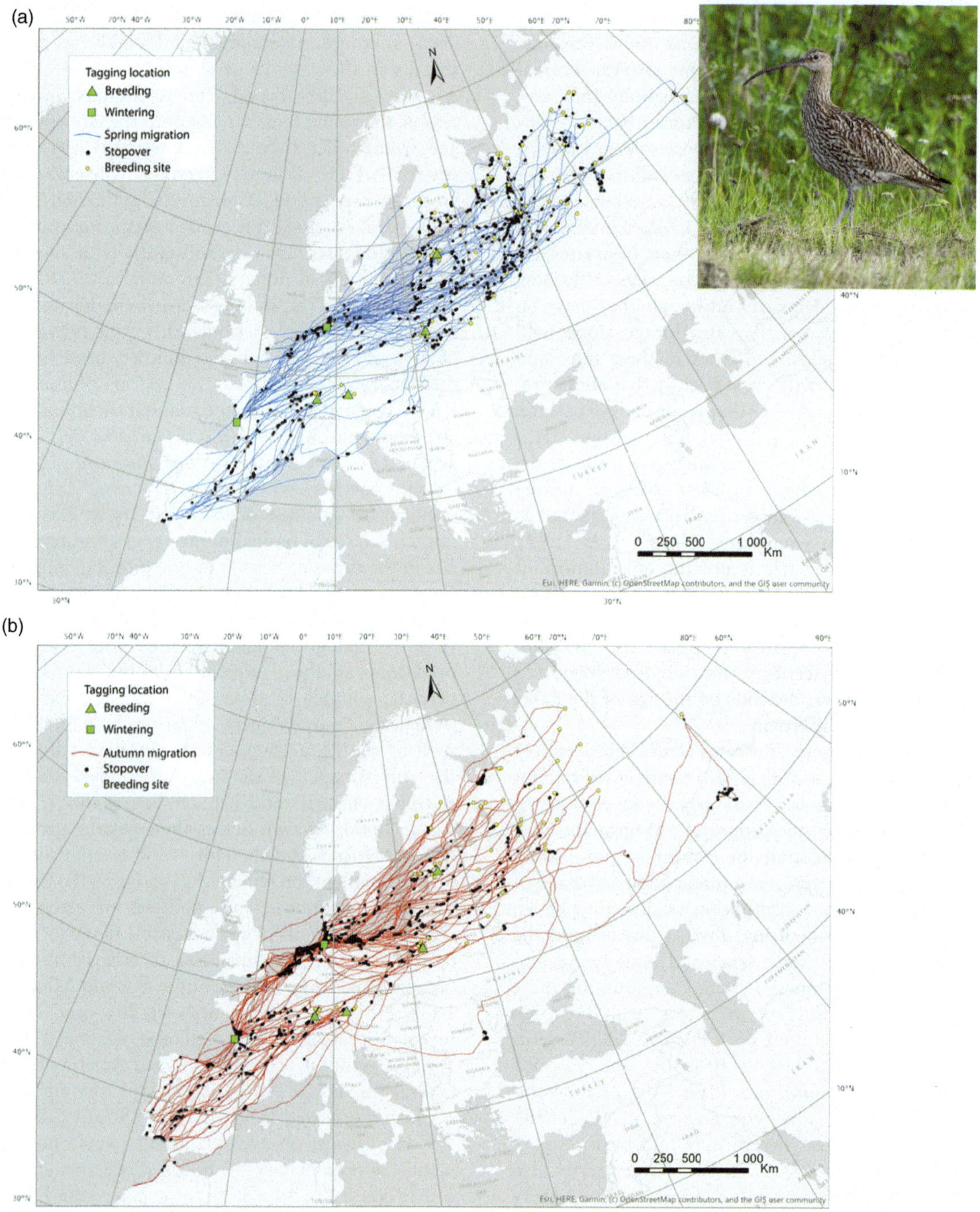

Fig. 8.11. Migration tracks of curlews (*Numenius arquata*) with stopovers (black) and breeding sites (yellow) based on **(a)** spring and **(b)** autumn periods. Green triangles: curlew caught at breeding sites; green rectangles: curlew caught at wintering sites. (From Pederson *et al.*, 2022, used under licence CC BY 3.0. Inset photo by Andreas Trepte: www.avi-fauna.info, https://en.wikipedia.org/wiki/Eurasian_curlew#/media/File:Eurasian_Curlew.jpg, used under licence CC BY SA 2.5.)

Others move hundreds or thousands of miles away from the European summer nesting sites across oceans and continents during the non-breeding season. The barn swallow *Hirundo rustica* uses breeding sites in northern Europe; its wintering sites in Africa differ for vegetation patterns but are closely correlated with departure sites for thermal conditions, the birds probably using some information on the temperatures at destination while they are still on their wintering grounds to modulate their timing of departure (Pancerasa *et al.*, 2018). Species that have a similar pattern of seasonal habitat change from summer to winter include the parasitic cuckoo (*Cuculus canorus*; Jacobsen *et al.*, 2017) and a number of predatory birds, some of which exploit resources in the **hydrosphere** (lakes, rivers, sea, etc.) as well as the land (e.g. the osprey *Pandion haliaetus*; the white-tailed eagle *Haliaeetus albicilla*; Anderwald *et al.*, 2021).

In the aquatic realm, particularly remarkable are migratory fish. Many fish species show migratory behaviours (Secor, 2015), the most impressive being those of species that migrate from the sea up into freshwater to spawn (**anadromous** species, Box 8.2) or that live in freshwater and go to the sea to spawn (**catadromous** species, Box 8.3). These species may have extremely wide home ranges characterised by highly contrasting ecological conditions. As terrestrial animals, we humans tend to perceive aquatic environments as relatively similar, underestimating the profound differences between freshwater and marine water, but they are completely different environments for fish, for example because of enormous contrasts in salt concentration, which profoundly influence fish physiology. Diel vertical migration occurs in a wide variety of marine animals (crustaceans, molluscs, fishes, etc.) and it is the largest synchronous migration in the world (see Chapter 3). Of course, the different depths in the water columns exploited by these animals are very different habitats.

Occupation of different spheres or zones of activity

One can go much further than countryside biotopes and find species that concentrate their activity in Earth spheres other than on the land surface above sea level (**hydrosphere, atmosphere, lithosphere**). We have already discussed, throughout the book, examples of organisms that occupy these other spheres. We concentrate here on some cases particularly instructive of the peculiarities of the species that inhabit these spheres from the perspective of the RBH concept.

A common feature of many organisms that occupy the hydrosphere and the atmosphere is their extremely large **home ranges**. The longest and heaviest organism recorded on the planet is the blue whale *Balaenoptera musculus* (Mizroch *et al.*, 1984). Together with dolphins and porpoises, whales form some of the most remarkable animals of the oceans and seas (Reeves *et al.*, 2003), species under distinct threat from human activity (Cooke, 2018). The rorquals, such as blue whales, fin whales (*Balaenoptera physalus*) and humpback whales (*Megaptera novaeangliae*), are lunge feeders (that is, they feed on plankton by a kind of inverted suction feeding, during which they engulf an enormous volume of water containing prey that is subsequently filtered through the baleen; Goldbogen *et al.*, 2011; Goldbogen and Madsen, 2021); their entire life is spent in the oceans, migrating with the seasons. One of their main prey is krill (small crustaceans); the Antarctic krill *Euphausia superba* comprises an estimated biomass of c.380 million tonnes (Atkinson *et al.*, 2009), one of the dominant species on the planet and half of which is consumed by whales, seals, penguins, squids and fish annually. Increased drag in larger humpback whales requires a lot of energy, but they have adopted an extraordinary communal mode of attack that ensures enormous krill consumption, the bubble net feeding (that is, the directed use of air bubbles to capture herring), an astonishing example of highly intelligent co-ordinated activity (Friedlaender *et al.* 2011) (Fig. 8.12) (www.bbc.co.uk/programmes/p005lnvw, accessed 15 May 2025). The massive consumption of krill by baleen whales is a critical part of the ocean ecosystem functioning. Since whales eat more than previously thought (Savoca *et al.*, 2021), they also discharge more excrement. By scooping up food and pumping out faeces, whales help to keep key nutrients suspended close to the ocean surface, which in turn generates blooms of the carbon-absorbing phytoplankton that form the basis of ocean food webs, thus playing a fundamental ecological role in the capacity of ocean ecosystems to absorb planet-warming carbon dioxide (Pearson *et al.*, 2023). The global ecosystem aspects involved are sufficient reason for understanding the RBH of rorquals. The basic problem of understanding their home range is self-evident: the RBH for these creatures includes the entire ocean realm and depth

Box 8.2. Anadromous species

Anadromous species are known among lampreys (e.g. *Geotria australis*, *Petromyzon marinus*, *Entosphenus tridentatus*), sturgeons (*Acipenser* spp.), salmonids, smelts (Osmeridae), clupeids (*Alosa* spp. and *Tenualosa* spp.) and other fish groups.

The best known, and possibly most remarkable, case is that of the Atlantic salmon *Salmo salar* (see, for example, Mills, 1989; Aas *et al*., 2010; COSEWIC, 2010; Darwall, 2023, for a synthesis of its biology) (Fig. B8.2.1). In this species reproduction takes place in the upper reaches of rivers but, after 1–6 years, the juveniles move to the sea where they grow to sexual maturity and then come back to their spawning sites. After hatching, the larvae (lucifugous) remain hidden in the gravel until complete consumption of the yolk sac (alevin stage). They then emerge from the sediment and begin to feed on microscopic organisms. When alevins reach a certain size (6–7 cm in length), they begin to feed on larger prey (this stage is called fry): after a year, with acquisition of distinct skin markings, they are called parr. Male parr, which exhibit highly territorial behaviour, may reach sexual maturity before descending into the sea, in which case they may mate with ascending adult females. In freshwater, juveniles mainly inhabit the fast-flowing stretches of rivers with cold water (at least 3 months with water temperatures above 10 °C). Salmon can also live in lakes, as either juveniles or adults, ascending the tributary rivers of the lakes. Some populations do not make reproductive migrations. When they reach a certain stage of development (10–15 cm in length) parrs undergo a physiological change called smoltification. This prepares them for the transition from freshwater to the saltwater environment of the ocean. Once they are smolts, salmon migrate downstream by taking advantage of floods. Migration takes place at night, in dense shoals. Once at sea, they usually settle near estuaries, where they remain for some months feeding on crustaceans that give their flesh its reddish colour. Finally, they move away from the coast, become ichthyophagous (i.e. fish-eating) and adopt a pelagic lifestyle, feeding mainly in the surface layers of the water column. In the ocean, salmon spend a significant portion of their lives feeding and growing. This phase can last from one to several years, depending on the species. Adult salmon return to their natal rivers to spawn, driven by chemical cues. Reproduction takes place in the same upper reaches of rivers where they were born. As a result, each river has a genetically different salmon population. Each individual mates with several other individuals. After breeding, the majority

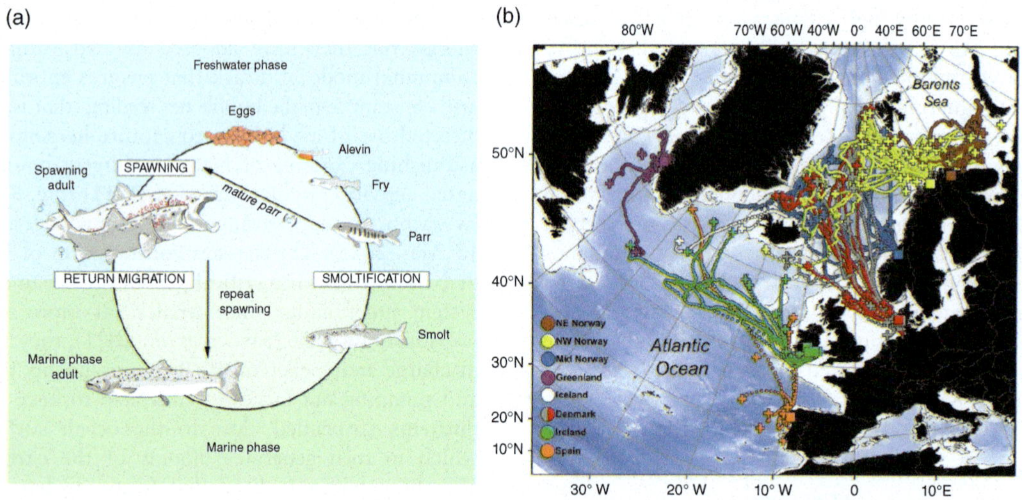

Fig. B8.2.1. Atlantic salmon phases and migration routes. **(a)** Salmon life cycle. **(b)** Migrations of Atlantic salmon tagged in eight different geographical areas. Release locations post tagging are shown by squares (from 11 northeast Atlantic river catchments and at-sea at western Greenland). Crosses show the pop-up location of the tags of these salmon individuals. The dashed line shows the 500 m depth contour, while darker blue shading indicates increasing depth. (Panel **a** from Mobley *et al*., 2021, used under licence CC BY 4.0, http://creativecommons.org/licenses/by/4.0/. Panel **b** from Rikardsen *et al*., 2021, used under licence CC BY 4.0, http://creativecommons.org/licenses/by/4.0/.)

Continued

Box 8.2. Continued.

of males die, while many females survive, but only a small minority undertake a new reproductive migration. The journey back to freshwater can be incredibly challenging and covers long distances, sometimes spanning even thousands of kilometres. Salmon often travel 50 kilometres per day on their spawning journeys. They face numerous obstacles during migration, including predators, dams, pollution and changes in water temperature and flow.

Salmon migration is crucial for maintaining healthy ecosystems as they transport nutrients from the ocean to freshwater habitats. The salmon population is declining, and the species is considered near threatened by the IUCN (Darwall, 2023). Although causes of decline are not entirely clear, they include overfishing, habitat destruction/alteration (for example, in its freshwater phase the species is strongly affected by hydropower development, water pollution and sedimentation, primarily caused by logging and agricultural activities), invasive species and climate change. Conservation efforts focus on restoring habitats, managing fisheries sustainably, and improving migration routes.

Box 8.3. Catadromous species

Catadromy occurs in fish species from a wide range of families, including Anguillidae (eels), Mugilidae (mullets), Galaxiidae (galaxiids and southern whitebaits), Cottidae (sculpins), Gobiidae (gobies) and Pleuronectidae (flounders). The European eel (*Anguilla anguilla*) is a classic example. The extraordinarily complex life cycle of this species (Fig. B8.3.1) puzzled naturalists for centuries and remains imperfectly understood. Another 18 eel species are recognised, but their biology is in general even less known (Deelder, 1984; Tesch, 2003; Arai, 2020; Cresci, 2020; Pike *et al.*, 2020).

Adult European eels use a wide variety of aquatic biotopes (rivers, canals, estuaries, lakes, ponds and lagoons) throughout Europe and North Africa. They are generalist, very voracious predators feeding on the most disparate aquatic small prey, especially on the bottom. All European eels are born in the Sargasso Sea (a region in the North Atlantic Ocean), the place where the reproduction of this species takes place and adults die. Thus, all individuals of this species form a single enormous, virtually panmictic population (Dannewitz *et al.*, 2005; Pujolar *et al.*, 2009). Eels undertake a long and arduous migration from freshwater or coastal habitats in Europe to the Sargasso Sea to spawn. This migration starts in autumn and covers thousands of kilometres. The reproductive instinct is so strong that eels living in

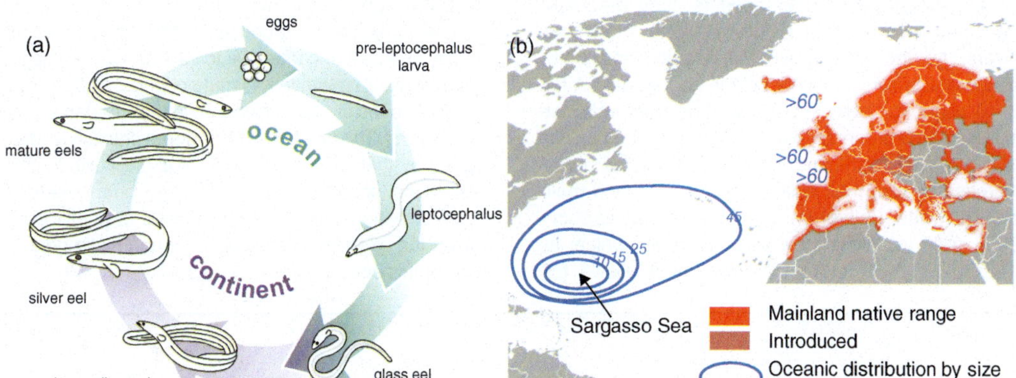

Fig. B8.3.1. European eel (*Anguilla anguilla*) phases and migration. **(a)** Eel life cycle. **(b)** Mainland and oceanic distribution of the European eel. (Panel **a** from Henkel *et al.*, 2012, used under licence CC BY. Panel **b**: source of the map of the world with mainland distribution: Alexander-no1: https://commons.wikimedia.org/wiki/File:Rangemap_Anguilla_anguilla.png, used under licence CC BY SA-3.0. Mainland distribution was derived from IUCN data (Freyhof and Kottelat, 2008). Oceanic distribution was taken from Schmidt, 1922.)

Continued

Box 8.3. Continued.

closed lakes or ponds do not hesitate to get out of the water, crawling on the ground like snakes; this happens at night, especially in rainy conditions or when the grass is wet with night dew (which allows migrating fish to avoid dehydration) and when there is no moon (given the lucifugal character of the species). Eels are able to exist for relatively long times out of water thanks to cutaneous respiration. Eel migration may take several months.

Before departing, adult eels undergo profound morphological changes, such as an increase in eye size, changes in pigmentation (their bellies become white and their flanks silver, earning them the name 'silver eels'; it is assumed that migration occurs at great depths, where there is little, blue light) and the degeneration of the digestive system (migrating eels stop feeding). Through still little known itineraries (probably with a combination of chemical, geomagnetic and geographical cues), eels reach the Sargasso Sea, where spawning takes place, after which they die. The fertilised eggs develop into larvae called leptocephali, which are carried by ocean currents across the Atlantic Ocean towards the coasts of Europe (it is generally assumed that this passive migration lasts probably 2 years; Zenimoto *et al.*, 2011, but it might occur in less than 1 year; Arai *et al.*, 2000). Leptocephali appear to feed on marine snow (tiny particles floating freely in the ocean). As the leptocephali approach the European coastlines, they transform into glass eels, which are transparent and resemble miniature eels. These glass eels enter the estuaries and continue their upstream migration, strenuously overcoming any obstacle they encounter on the way. In freshwater habitats, they gradually transform into pigmented juvenile eels known as elvers. As the eels grow, their flanks and belly assume a brownish-yellow colour (yellow eel). After 6–10 years (or even more) spent in freshwater or brackish environments, the eels reach maturity. This is marked by further morphological changes: eyes increase in size, their sides become silver and their bellies white ('silver eels'). Silvering is associated with increased levels of cortisol, which is needed for their return migration as it allows for mobilisation of energy during migration (van Ginneken *et al.*, 2007a,b). Unlike salmons, young eels do not return to the same places their mothers came from. Some individuals never enter freshwater but remain in their marine environment for their entire life, or move between fresh, brackish and saltwater habitats throughout the growth phase (Durif *et al.*, 2023).

Despite eels having fascinated naturalists since Aristotle, their life cycle was discovered only in relatively recent times. As fishermen had never found any fish whose juveniles resembled eels, their biological cycle remained obscure until modern times. For example, Sigmund Freud, at the beginning of his career, searched desperately for the male sexual organs and found them only after examination of an enormous number of eels (Lee, 2020; Perkins-McVey, 2022) – the ovaries had instead been identified at the end of 1700 by the Italian Carlo Mondini, although the discovery remained long controversial (Eigenmann, 1902). These difficulties in finding eel gonads can now be clearly understood as maturation is completed in the sea. Leptocephali were originally described as a different species (*Leptocephalus brevirostris*), and only at the end of the 19th century were they recognised as the larvae of the eel by the French zoologist Yves Delage and the Italian zoologist Giovanni Battista Grassi and his pupil Salvatore Calandruccio (Grassi is famous for having clarified the life cycle of the malarial parasite; the same subject was simultaneously investigated by Ronald Ross, who – because of a defamatory campaign against Grassi – was the only one of the two to receive the Nobel Prize in 1902 for this discovery; Capanna, 2006).

The European eel population has declined significantly in recent decades due to various factors, including barriers to migration, climate change, habitat loss, invasive species, parasitism, pollution, predation and overfishing. The species is considered critically endangered by the IUCN (Pike *et al.*, 2020). Efforts to conserve the species have traditionally been made at local or national level, which reduced their efficiency, given the migratory behaviour and panmictic population of this species (Pike *et al.*, 2020).

they cover, zones exploited by most national states, a global expanse including all the associated resources and conditions essential for their survival.

An animal that can be considered the invertebrate counterpart of whales in many respects is the giant squid (*Architeuthis dux*) (Fig. 8.13). This creature is the largest extant invertebrate, with a documented maximum body length (from posterior fins to the tip of two long tentacles) at around 12 m (McClain *et al.*, 2015) (previous estimates exceeding 20 m (see Ellis, 1998) seem unrealistic). However, it is not the heaviest invertebrate: the colossal squid (*Mesonychoteuthis hamiltoni*, total length 6 m; Rosa *et al.*, 2017) has a greater body mass (about 500 kg; estimates for *A. dux* are 200–280 kg; McClain *et al.*, 2015). Like rorquals, the giant

Fig. 8.12. Bubble net feeding in humpback whales. A group of 15 humpback whales bubble net fishing in North Pass between Lincoln Island and Shelter Island in the Lynn Canal north of Juneau, Alaska, on 18 August 2007. (Photo by Evadb; edit by jjron: https://en.wikipedia.org/wiki/Bubble-net_feeding#/media/File:Whales_Bubble_Net_Feeding-edit1.jpg.)

Fig. 8.13. The giant squid. **(a)** A giant squid found at Ranheim in Trondheim (Norway) 2 October 1954. The specimen was measured to a total length of 9.2 metres. **(b)** This marvellous creature might have inspired myths and stories about marine monsters. Illustration for J. Verne, *Vingt Mille Lieues sous les mers* (Edition Hetzel, 1871). (Panel **a** from NTNU Museum of Natural History and Archaeology: www.wikidata.org/wiki/Q204646#/media/File:Giant_squid_Ranheim.jpg, used under licence CC BY 2.0. Panel **b** from https://en.m.wikipedia.org/wiki/File:20000_squid_holding_sailor.jpg.)

squid is widespread, occurring in all oceans, although records from tropical and polar latitudes are rare; its vertical distribution might be 300–1000 metres (Roper, 2016). This enormous predator is in turn preyed on by toothed whales, especially the sperm whale *Physeter macrocephalus* (as shown by remains of giant squid found in their stomachs); giant squid sucker scars on sperm whales are a proof of epic fights between the prey and its predator (Ellis, 1998). Because of its vulnerability to ecosystem stresses resulting from anthropogenic effects, including ocean acidification and warming, it has been proposed to consider this charismatic creature as an emblematic species to represent concern for the conservation of marine invertebrate biodiversity (Guerra *et al*., 2011). Despite the prominent role of the giant squid in popular culture, probably being the inspiration for the legendary kraken and many tales of marine monsters (Ellis, 1998), the scarcity of information available on its biology (e.g. Roper and Shea, 2013) illustrates how little we know about life in the deep oceans.

Now to the air above. The ultimate creature of the air (atmosphere) is perhaps the common swift (*Apus apus*). During their 10-month wintering period, swifts fly for all but 1% of the time. They fly at an average speed of 570 km/day (max. 832 km/day) (Hedenström *et al*., 2016). Thus, their RBHs are expansive, massive spaces (shifting colossal sinuous spirals) in the air, stretching from northern Sweden and Finland (north of the Gulf of Bothnia) to the Gulf of Guinea and the Congo Basin in Africa (Figure 1 in Åkesson and Bianco, 2021). Their placement in the air space varies from day to day; even so, nearly all their resources, but for nesting, occur above the ground in that air space with 'stopovers' above insect-rich areas for feeding, especially in the autumn flight south. Trees and buildings can be used for roosting, especially by juveniles, perhaps during periods of food shortage (Lack, 1966a; Holmgren, 2004; Hedenström *et al*., 2016). Without highly developed techniques, it would not be possible to trace their daily and annual home ranges, techniques that are bound to improve with time (Hedenström *et al*., 2016; Åkesson and Bianco, 2021). The fact stands: as well as nesting sites in house eaves, cliffs and caves, their RBH includes all the resources and conditions in the massive air space mentioned above, from ground level to at least 4000 metres. Even water is taken on the wing by skimming over river and lake surfaces (it has also been reported that swifts are able to drink by swallowing rain drops; Bersot, 1931). This RBH for the swift, extending far beyond the bounds of a village, town or river valley, is extremely difficult for the researcher to identify; it exists nonetheless. Failing to protect the Earth's atmosphere, its resources and conditions, will have serious consequences for this bird and much else (see Chapter 1).

A vast number of organisms seemingly exist more in the earth (**soil**) rather than on it, including an enormous amount of bacteria, protists, fungi and invertebrates, but even some vertebrates, which are classified as 'megafauna' by soil biologists (Coleman and Crossley, 2004). Classic examples of soil mammals are the giant armadillo *Priodontes maximus* (Desbiez and Kluyber, 2013), the pink fairy armadillo *Chlamyphorus truncatus* (Borghi *et al*., 2011), and the mole *Talpa europaea* (Mellanby, 1967; Funmilayo, 1977; Fellowes *et al*., 2020). These are elusive animals and ecologists are increasingly turning their attention to biodiversity below the soil surface, for instance in the study of microorganisms discussed earlier in this chapter (section *Prokaryote habitats: Bacteria and Archaea*). It should be mentioned at this point that surficial and subsurface edaphic issues affect many butterflies that spend their early stages near or below the soil and whose RBHs have been studied (Dennis, 2010). In fact, soil is part of the RBHs of a number of flying insects that live there as immatures for most of their life. A prominent example is the North American periodical cicadas (*Magicicada* spp.) (Alexander and Moore, 1962; Marshall and Cooley, 2000; Campbell *et al*., 2015). These cicadas have a long larval period underground (13 years in some species, 17 years in others, such as *Magicicada septendecim*), followed by synchronised mass emergence and quick transformation to adults. Mating, egg laying and the death of all adult cicadas occur within weeks. These animals are also an example of organisms that occupy different spheres during their life cycle (see *Occupation of a variety of spheres or zones* below).

Here, attention is given to earthworms, a group of invertebrates which primarily live in or below the soil, of which about 6000 species are known (Mısırlıoğlu *et al*., 2023). These creatures have an unusual biology, not least of which is that they are hermaphrodites. Ecologically, they are placed into two basic groups: megadrilli (soil living) and microdrilli (these are the aquatic species), the former further separated into three functional divisions (epigeic: species that live in the topsoil and deposit wormcasts on the surface; endogeic: species that build complex lateral burrow systems in all layers of the soil; and anecic: species that build deep vertical burrows and surface to feed) (Kamboj *et al.*, 2021) (Fig. 8.14). It is these organisms that present some of the most difficult issues in interpreting resource-based habitats. The bounds of home ranges of individuals and of local populations of earthworm species require intensive sampling of all habitats and microhabitats (deadwood, dung, spaces under stones, leaf litter, compost, wood piles, gravel and shingle, stream beds, tree holes, molehills, etc.) within the vicinity of a study zone, before expanding the study out to other regions (for techniques, see Brown, 2017).

Here brief focus is on the lob worm *Lumbricus terrestris* (a member of the Lumbricidae, a family which includes some 690 species/subspecies in 47 genera; Mısırlıoğlu *et al.*, 2023); it is the largest earthworm in Britain, and probably the primary focus of Darwin's attention (Darwin, 1881). Ecologically, *L. terrestris* is an anecic species, which lives in deep self-made burrows and forages by night on rotten leaves. Mating occurs between individuals that are simultaneously anchored to their burrows (Nuutinen and Butt, 1997; Kutschera and Elliott, 2010). However, the worm can also be epigeic in forests with large litter layers and persist without the construction of a burrow (Lamparski, 1985). We know a great deal about its foraging and food, invasiveness, predators, conditions of existence, basic habitats and mating, and the crucial part it and other earthworms and microorganisms play in soil ecosystems on which human populations depend (Curry and Schmidt, 2007; James, 2010). But we still have insufficient data on its home ranges, number and pattern of underground structures, the churning of sediment and organic matter (worm casts), and the distribution of nutrients (minerals) throughout soil horizons. It is far from clear what type of ground fails to support this and other worms. Regional studies and analysis of such data are important for a full understanding of the RBH of this and other worms across different **biogeographical regions** (Phillips *et al.*, 2019; Tóth *et al.*, 2020; Simon *et al.*, 2022). Such complications draw attention to the need for careful study.

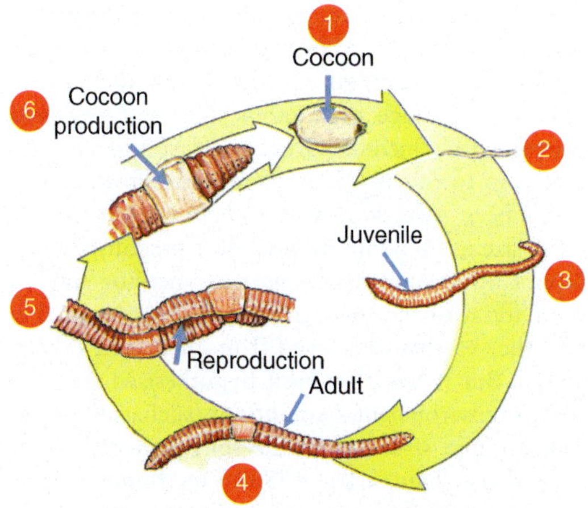

Fig. 8.14. Life cycle of earthworms. 1. Cocoons (egg casings) may contain a few to dozens of eggs. Cocoons tend to be 'lemon' shaped, but the specific shape varies between species. The amount of time that they take to hatch is also very variable and can vary by species, but also by environmental conditions. For example, for some species the cocoons may hatch quicker in warmer conditions than in cooler conditions, while other species may 'wait out' undesirable drought conditions within the soil as the cocoon stage. 2. Hatchlings look just like mini-earthworms, they are only smaller and paler. They are often produced in large numbers in spring. It can be easy to confuse these with potworms (Enchytraeidae), which are small-segmented worms closely related to earthworms. As the hatchling feeds and grows, it will gain the same colour as an adult earthworm. 3. Juvenile earthworms look very much like the adults but are missing the saddle (clitellum). 4. Adult (sexually mature) earthworms can be easily recognised through the presence of the saddle. 5. Worms can reproduce asexually (parthenogenesis) or sexually. Sexual reproduction (shown here) involves two earthworms. The two earthworms produce a slime tube and grip onto each other using paired glandular swellings, known as *tubercula pubertatis*, which are located on the saddle. The slime tube provides the right environment for the two earthworms to exchange sperm, with each earthworm storing the sperm of its partner for later use. Because both earthworms are performing the function of both a male and female during sexual reproduction, they are known as simultaneous hermaphrodites. Following this sperm exchange the earthworms separate. 6. A mucous sheath is formed around the clitellum and is moved along the earthworm until it comes off the head end. Along this journey it picks up the egg(s) and the sperm of the other earthworm with which it mated. This mucous sheath forms the cocoon and fertilisation of the egg occurs within the cocoon. (Modified from Keiron Derek Brown; illustration by Rick Kollath: www.earthwormsoc.org.uk/lifecycle, used under licence CC BY 4.0.)

In the case of *Lumbricus badensis* (a giant species endemic to the Black Forest, south-west Germany), there is the problem of why this worm, right at the edge of the range of another species, *Lumbricus friendi*, with which it shares close relationship, is so huge (up to 60 cm) (Kutschera and Elliott, 2010). It has been speculated that *L. badensis* evolved its large size as an adaptation against predators when (about 8000–6000 years ago) its ancestor entered the habitat of the Black Forest, where presumably no other earthworms lived (Kutschera and Elliott, 2010). But it has also been hypothesised that the specific environmental conditions (such as the composition and structure of the soil) favoured larger individuals (Lamparksi, 1985). One thing is clear: longer worms can access more ground for food and mates while still being anchored to their holes. It is possible that size in *L. badensis* relates to the limited quantity of sufficient quality food in the pine-spruce-beech woodland of the Black Forest. As is evident from the work of Lamparksi (1985), Kobel-Lamparski and Lamparski (2004), Kutschera and Elliott (2010), and Butt *et al.* (2021), this is an ideal organism to study.

Occupation of a variety of spheres or zones

There are, of course, obvious examples of animals that occupy different spheres to fulfil different functions. Among vertebrates, exemplary are turtles (Fig. 8.15) and crocodiles. Turtle, tortoise and terrapin are all names for the hard-shelled, egg-laying reptiles in the taxonomic order Testudines (348 species in 14 families; Thomson *et al.*, 2021). These include land turtles (tortoises), freshwater terrapins and sea turtles (Eckert and Eckert, 1987; Eckert *et al.*, 1989; Spotila *et al.*, 1996; Turtle Taxonomy Working Group, 2021). Sea turtles are classic examples of organisms that move between the **hydrosphere** and the **lithosphere**; these animals range widely over the sea, but return to specific beaches to lay and bury their eggs (e.g. the leatherback turtle *Dermochelys coriacea*; Hendrickson, 1980). Some occupy land areas and freshwater wetlands and adjoining land, an example being the spotted turtle *Clemmys guttata*. Chandler *et al.* (2019) found that home ranges of individual spotted turtles were generally small, varying from 0.38 to 8.21 ha at two sites (95% minimum convex

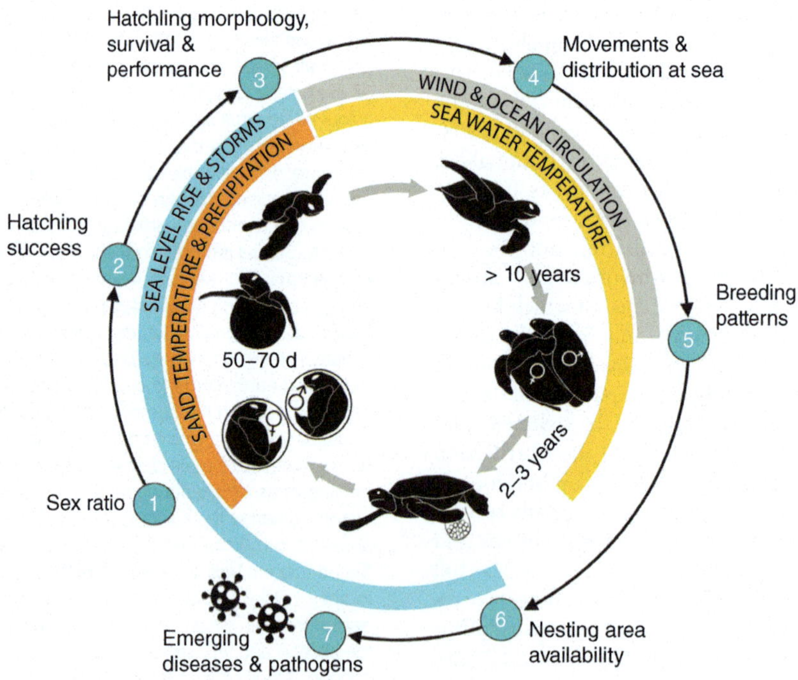

Fig. 8.15. Life cycle of sea turtles with parameters expected to be affected by climate change. (From Patrício *et al.*, 2021, used under licence CC BY 4.0.)

polygons), and that estimates for the population as a whole varied from 11.1 to 49.4 ha.

Crocodiles, alligators and caimans (Order Crocodilia) are small to massive semi-aquatic carnivorous egg-laying predators of vertebrates and invertebrates, occupying freshwater (i.e. rivers, lakes, wetlands), brackish water and saltwater habitats (Gatesy and Amato, 2008). These ancient Cretaceous creatures have home ranges that extend across distinct biotopes and Earth spheres, land and water, and thus highly variable resource-based habitats. Some even climb trees (Brochu, 2003; Dinets *et al.*, 2013).

Of course, home ranges and RBHs have to include the space, conditions and resources of both spheres of activity for the examples given above in the reptilian class. Similar considerations can also be applied to amphibians (e.g. frogs, toads, salamanders) and to the multitude of other animals that use different spheres in different phases of their life cycles (e.g. insects that are aquatic as immatures and terrestrial as adults, such as dragonflies, stoneflies, mayflies, caddisflies and mosquitos).

A peculiar situation among mammals is represented by bats which roost in caves, where some species can form enormous colonies (e.g. the Mexican free-tailed bat *Tadarida brasiliensis mexicana* forms colonies that often exceed several million individuals; the large number of bats in these colonies makes it difficult for females to selectively nurse their own pups, leading to a certain percentage of 'non-parental' (indiscriminate) nursing; McCracken, 1984). Bats can use caves throughout the whole year, or for hibernation. Many bats spend the daytime in caves, but at night they fly for long distances. Thus, their RBH may be extremely large. The Mexican free-tailed bats (which are the fastest flying animals (horizontally) at 160 km/h) perform nightly movements of 50 km from the cave roosts, which translates into an idealised circular area of influence around caves of about 8000 km^2 (Medellin *et al.*, 2017). They are also a migratory species, able to cover very long distances (a study recorded a maximum distance from banding site to point of recovery of 1840 km; Glass, 1982). Since many cave-roosting bats feed on flying insects, their RBH extends across the lithosphere and the atmosphere; some species feed on ground-dwelling insects (making the lithosphere important for foraging) and others on aquatic ones (hydrosphere) (Lanza, 2012).

Fidelity to and movement within locations/zones

Movement has already emerged as a prime feature of some organisms. Some are in perpetual motion, driven by the need for constant replenishment of food supplies, their movements typically in huge herds, considered an adaptation to protect themselves against predators. Classic examples are the South African buffalo's (*Syncerus caffer*) recursive patterns that must have been mirrored by American bison (*Bison bison*) when they were at their zenith (Bar-David *et al.*, 2009), as well as those of African wildebeest (*Connochaetes gnou* and *C. taurinus*; Carmichael, 2018), zebras (e.g. *Equus quagga*; Schmitt *et al.*, 2014) (Fig. 8.16) and African elephants (*Loxodonta africana*; Thomas *et al.*, 2008). For the latter species, Thomas *et al.* (2008) determined the home range of a matriarchal group based on the continuous tracking of one animal. These studies are essential for understanding the precise needs of these wandering animals, which is by no means a random indeterminate path. Interestingly, the elephants studied by Thomas *et al.* (2008) did not use traditional habitat randomly, the pattern of activity being faithfully repeated within their summer and winter 5-year home ranges using well-defined corridors. An indication of resource needs is that the summer home range was in an area more than four times that of the winter home range. Moreover, the impact of abiotic conditions was clear in the distance walked in relationship to minimum daily temperatures and water availability. A study based on **transects** through the home range was used to determine the fine-scale exploitation of the landscape within the home ranges. Precise documentation of behaviour during these studies builds the databank needed to define the creatures' resource-based habitats.

Phagy (consumer) groups

Butterflies – the model animals on which the resource-based habitat was formulated – are largely herbivores, both as larvae and adults. But there are significant exceptions that have been accounted, as in the large blue *Phengaris* (*Maculinea*) *arion*, a social parasite of ants (Fiedler *et al.*, 1996; Dennis, 2010, 2020). However, a glance at ecosystems (see Chapter 2) reveals a wide range of organisms that are purely carnivores or herbivores, including parasites of animals or plants (e.g. gall makers;

Fig. 8.16. Herbivore herds. Zebras (*Equus quagga*) in Masai Mara, Kenya. (Photo by Kandukuru Nagarjun: https://flickr.com/photos/64924693@N00/31966157887, used under licence CC BY 2.0.)

Felt, 1940; Roskam, 2019), but also omnivores (creatures which forage on an array of plant and animal matter, e.g. pigs, badgers, bears, coatis, civets, hedgehogs, opossums and rats; Pollan, 2006), saprophages (organisms that directly absorb nutrients through external processes, e.g. fungi) and detritivores (organisms that digest dead matter internally, e.g. woodlice, earthworms and insects such as dung beetles and carrion beetles; but see Getz, 2011 for a different, more refined classification of consumer categories). Comprehension of their RBHs involves full documentation of these webs of interactions and the conditions *en passant*.

Summary points

By now, it should have become obvious that a vast number of organisms have RBHs that differ substantially in structure and complexity from Lepidoptera, none more so than those occupying more than a single sphere (zone) of activity and/or are continually on the move. The very fact that all organisms depend on elements of all the Earth's spheres presses home the undoubted importance of their part and dependence in natural ecosystems. There is still much to be learnt, particularly regarding the deep sea where, for example, holothurian densities may be in a constant state of flux with no uniform response in relation to changing food supply (Lemon *et al.*, 2022). Even so, nothing can subsist without sunlight, heat, water, gases and nutrients provisioned by the Earth's spheres (i.e. the lithosphere, the hydrosphere and atmosphere) and sun energy. All these points and the foregoing examples may appear trite to those trained in biology, as is the poisoning of the planet by human activity and greenhouse-induced climate change (Dennis, 1993a). Nothing is perhaps more perplexing than why some animals suddenly expand their ranges (e.g. the collared dove *Streptopelia decaocto*; Fujisaki *et al.*, 2010) whilst others retract or remain approximately stable in their distributions. In all this, what never ceases to astonish the authors is the crass ignorance of those to whom we give responsibility for decision making on actions that affect each portion of the planet; how in many cases it is evident their care is more for personal aggrandisement than stability in resources that affect not just humans but all organisms that in turn influence our own persistence.

9 Primates and Human Evolution: Expanding Habitat Demands

Abstract

The application of the notions of the resource-based habitat (RBH) and home range (HR) to non-human primates illuminates many aspects of human biological and cultural evolution. A study of human history and geography illustrates how human RBHs and HRs changed in relation to cultural evolution (e.g. changes in social organisation, technology, diet, communication, settlements and beliefs). Human evolution was marked by major transitions in resource exploitation that culminated in the current post-industrial era characterised by unprecedented environmental crises. Humans are subject to all the fundamental activities carried out by other animals but, at the same time, have developed unique behaviours and needs that affect their RBHs and HRs. Particularly remarkable is the role of human greed as a source of conflicts: humans – probably as a consequence of sheer numbers, group mass and demands for so many more resources – frequently resort to war. However, other animals (e.g. ants) also engage in war-like behaviours.

The pongid–hominid evolutionary tree

One group of organisms holds a special place in science as it is the extensive **clade** that led along one of its branches to humanity (Perelman *et al.*, 2011; Tuttle, 2014) (Fig. 9.1). It should now (but is not) be common knowledge that apes are our closest relatives and that humans have evolved from an ape branch (i.e. the human–chimpanzee Last Common Ancestor (LCA): possibly *Graecopithecus*; Moorjani *et al.*, 2016; Fuss *et al.*, 2017; but see Benoit and Thackeray, 2017 for a critique and Fuss *et al.*, 2018 for a defence). Gestures of communication appear in both lines of descent (Graham and Hobaiter, 2023). The genomes of humans and chimpanzees differ by just a few percent, but this corresponds to over 30 million single nucleotide changes (Chimpanzee Sequencing and Analysis Consortium, 2005). In addition to single nucleotide substitutions, deletions and duplications of DNA fragments of different size, insertion of transposable elements and chromosomal rearrangements have all contributed significantly to the differences between the two species (Suntsova and Buzdin, 2020). Proximity of humans to other apes, then, is highly complex, differing in allele states, amino acids, insertions or deletions, gene duplication, etc. (Jones *et al.*, 2018). However, even in the most 'advanced' countries, such as the USA, a large fraction of resident people still do not *believe* in evolution (USA, 2005 respondents: 40% accepted evolution, 39% rejected it and 21% were uncertain; Tuttle, 2014).

Science has determined that humans and apes are linked to Old and New World monkeys, and prosimian groups such as lemurs (restricted to Madagascar), lorises and galagoes, some also including tarsiers among them (see Fig. 9.1 and Table 9.1). Extant primates (currently 533 recognised species according to the IUCN/SSC Primate Specialist Group: www.primate-sg.org/who_ares_the_primates/; accessed 19 May 2025) are largely confined to tropical forests, savanna and grasslands of the Old and New Worlds (the only free-living non-human primate in Europe is the Barbary macaque *Macaca sylvanus* in Gibraltar, where it was introduced), those occupying the coldest environment being the Japanese macaques *Macaca fuscata* (Enari and Sakamaki-Enari, 2013); many fossils have also been found in Africa, Asia and Europe. It is common scientific knowledge, too, that humans are more closely related to chimpanzees (*Pan* spp.) than to gorillas (*Gorilla* spp.) (Almécija *et al.*, 2021). The Asian apes are more distant still (orangutans, *Pongo* spp.; hylobatids including four genera: *Hylobates*, *Hoolock*, *Nomascus*, *Symphalangus*). When the first author was an anthropology student in Durham (UK), in the 1960s, the great apes (Pongidae) were placed in a separate family to humans (Hominidae) based on contrasts in ecology (thus, on the basis of home

DOI: 10.1079/9781800627505.0009

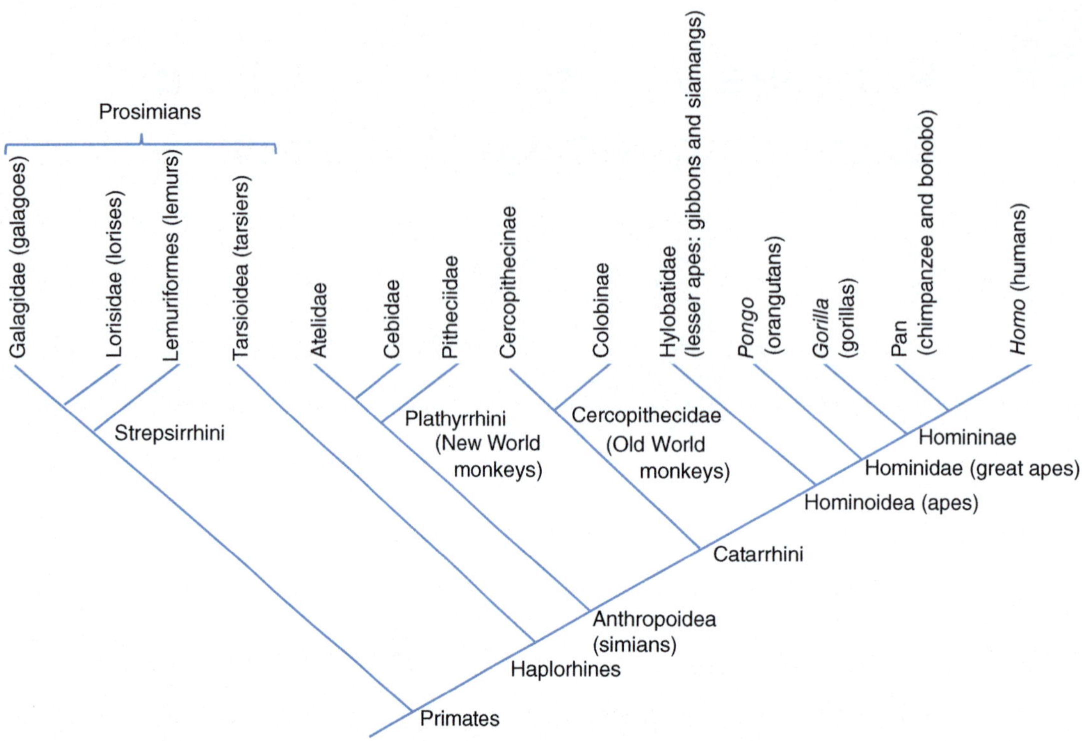

Fig. 9.1. The primate tree showing vernacular terms for, and basic relationships between, contemporary primates. Cladogram based on Herlyn (2016). Cladistically, humans are apes.

ranges and niches) and physical traits, as well as the fundamental genetic differences that have led to the quantum shift that has produced dramatically divergent paths of evolution. A very different picture is presented currently by Almécija *et al.* (2021) where all apes (gibbons, siamangs, orangs, gorillas, and chimpanzees) are regarded as hominoids, all but hylobatids (gibbons and siamangs) are hominids, orangs are pongines, humans, gorillas and chimpanzees are hominines and *Homo* is the only extant hominin (Table 9.1) (although some authors consider *Pan* as a hominin (Mann and Weiss, 1996), this point of view seems to be in the minority). The term 'human' is generally applied to members of the genus *Homo*.

This chapter presents a foundation for understanding the notions of the **resource-based habitat (RBH)** and **home range (HR)** as relating to humans and their close relatives; it should not shock us that these concepts are as relevant to humans as to all other animals. A study of human history and geography illustrates just how varied human RBHs have become with the development of the modern industrial era. The forms which the evolutionary development of humans has taken are not an issue here as, however humans have developed, they share in the concept of the RBH. The authors are aware that it may be an unpleasant reality for many humans to accept that they are simply part of the animal kingdom. Many would like to think that they (humans) are 'superior', but their organs perform much the same beautiful – crude – complex functions as those of any beast. A few moments reflection should make the reader aware that they are subject to all the fundamental activities carried out by other animals, functions that many humans would readily conceal in communication with one another and in print (i.e. fighting, moving between home bases (places of rest and meeting), searching for resources, resting, sleeping, feeding, copulating, defaecating, urinating, vomiting, becoming parasitised and ill, dying; a sweaty, stinking mass of protoplasm), thus almost invariably excluded in any diary, biography or novel with but few brave exceptions (e.g. Samuel Pepys' 17th-century diary; Knighton, 2004) and sanitised by biologists. True enough, humans have

Table 9.1. Classification of extant primates.

Rank	Order Primates
Suborder	Strepsirrhini
Infraorder	Lemuriformes
Superfamily	Lemuroidea
Family	Cheirogaleidae: dwarf and mouse lemurs (genera *Cheirogaleus*, *Microcebus*, *Mirza*, *Allocebus*, *Phaner*)
Family	Daubentoniidae: aye-aye (genus *Daubentonia*)
Family	Lepilemuridae: sportive lemurs (genus *Lepilemur*)
Family	Lemuridae: true lemurs (genera *Lemur*, *Eulemur*, *Hapalemur*, *Prolemur*, *Varecia*)
Family	Indriidae: woolly lemurs, sifakas, and indri (genera *Avahi*, *Propithecus*, *Indri*)
Infraorder	Lorisiformes
Family	Lorisidae
Subfamily	Perodicticinae: angwantibos and pottos (genera *Arctocebus*, *Perodicticus*)
Subfamily	Lorisinae: lorises (genera *Loris*, *Nycticebus*, *Xanthonycticebus*)
Family	Galagidae: galagoes (bush babies) (genera *Euoticus*, *Galago*, *Galagoides*, *Otolemur*, *Paragalago*, *Sciurocheirus*)
Suborder	Haplorrhini
Infraorder	Tarsiiformes
Family	Tarsiidae: tarsiers (*Carlito*, *Cephalopachus*, *Tarsius*)
Infraorder	Simiiformes (Anthropoidea)
Parvorder	Platyrrhini (New World monkeys)
Superfamily	Ceboidea
Family	Callitrichidae: marmosets and tamarins (genera *Saguinus*, *Leontocebus*, *Leontopithecus*, *Callimico*, *Mico*, *Cebuella*, *Callithrix*)
Family	Cebidae
Subfamily	Cebinae: capuchins (genera *Cebus*, *Sapajus*)
Subfamily	Saimiriinae: squirrel monkeys (genus *Saimiri*)
Family	Aotidae: night monkeys (owl monkeys) (genus *Aotus*)
Family	Pitheciidae
Subfamily	Callicebinae: titis (genera *Plecturocebus*, *Cheracebus*, *Callicebus*)
Subfamily	Pitheciinae: sakis and uakaris (genera *Pithecia, Chiropotes*, *Cacajao*)
Family	Atelidae
Subfamily	Alouattinae: howler monkeys (genus *Alouatta*)
Subfamily	Atelinae: spider and woolly monkeys (genera *Ateles*, *Brachyteles*, *Lagothrix*)
Parvorder	Catarrhini (Old World monkeys and apes)
Superfamily	Cercopithecoidea
Family	Cercopithecidae
Subfamily	Cercopithecinae
Tribe	Cercopithecini: Allen's swamp monkey, talapoins, patas monkeys, vervet monkeys, guenons (genera *Allenopithecus*, *Miopithecus*, *Erythrocebus*, *Chlorocebus*, *Allochrocebus*, *Cercopithecus*)
Tribe	Papionini: macaques, mangabeys, baboons, gelada, drill and mandrill (genera *Macaca*, *Lophocebus*, *Rungwecebus*, *Papio*, *Theropithecus*, *Cercocebus*, *Mandrillus*)

Continued

Table 9.1. Continued.

Order Primates				
		Subfamily		Colobinae
			Tribe	Colobini: colobus monkeys (genera *Colobus*, *Piliocolobus*, *Procolobus*)
			Tribe	Presbytini: lutungs, surilis, gray langurs, doucs, snub-nosed monkeys, proboscis monkey, pig-tailed langur (genera *Trachypithecus*, *Presbytis*, *Semnopithecus*, *Pygathrix*, *Rhinopithecus*, *Nasalis*, *Simias*)
Superfamily	Hominoidea			
	Family	Hylobatidae: lesser apes (gibbons and siamangs; genera *Hylobates*, *Nomascus*, *Hoolock*, *Symphalangus*)		
	Family	Hominidae		
		Subfamily		Ponginae: Orangutan lineage (genus *Pongo*)
		Subfamily		Homininae
			Tribe	Gorillini: Gorilla lineage (genus Gorilla)
			Tribe	Panini: Chimpanzee lineage (genus *Pan*)
			Tribe	Hominini: Human lineage (genus *Homo*)

developed distinctions from all other animals, even the most closely related primates (creation of sophisticated technologies, writing, music making, drawing and sculpting, communicating complex concepts, perceptions of the past and future and belief systems using a variety of languages, and much more). This has achieved even more complex resource-based habitats for this remarkably talented, variably 'humane', at times generous, more often selfish and greedy, occasionally lovable but fundamentally despicable, vile and unnecessarily cruel Machiavellian animal. The last comments emanate from the tendency to kill and torture animals, including fellow humans, for no other reason than the pure pleasure of witnessing tortured flesh. The simple truth, exposed especially by Darwin, is that humans are but a part of the animal kingdom, distinguished indeed by their greater intelligence but little else, as history records a regular tendency to mass destruction as much as to self-preservation. Even the complex concept of 'God' (or plural) can be simply regarded as a mechanism generated by humans to facilitate and control daily functions, explain the inexplicable and retain order.

This chapter focuses primarily on humans in ecology, as an element in ecosystems, admittedly having a complex set of functions beyond all other creatures on this planet, but also having all those attributes of survival and reproduction shown by all other animals.

Models of hominid evolution

However humans evolved from the LCA (e.g. *Graecopithecus* at c.7.2 Ma; *Sahelanthropus tchadensis* at 7 Ma; *Orrorin tugenensis* at 6 Ma, predating *Ardipithecus ramidus* at 4.4 Ma: Moorjani *et al.*, 2016; Fuss *et al.*, 2017; White, 2018; Barras, 2018), each fossil found presents an organism with its own set of **traits, habitats** and **niches** in the varying contexts recorded in the lithology and substrates. As each taxon evolved, so too did its traits, context and RBH differ. As RBHs undoubtedly change with contrasts in physical traits, it is worth noting briefly some of the basic attributes distinguished by fossils that have been identified along the anastomosing evolutionary network of hominids. But first, a warning, before attributing names to the many hominid fossils. A vast array of hominid and hominin species (more than 10 *Homo* species named) have been discovered (supposedly), but each of these is based on very few specimens and hominin taxonomy remains strongly debated (Wood and Boyle, 2016; Bobe and Wood, 2022; van Holstein and Foley, 2022). The distinctions drawn, especially the naming of a multitude of species, would not pass muster with population scientists and ecologists who would require a thorough assessment of variation based on adequate samples of individuals from sites covering the range of each animal, their ecological and behavioural traits, fixed precisely in space and time.

As few fossils had been found by the 1970s, it was long assumed that the evolutionary path for both apes and humans was a simple linear one. However, during the 1990s and the new millennium, a vast array of Hominidae (including Ponginae) fossils were discovered, which created uncertainty about the so-called 'missing links' in each epoch for both extant non-human and human **clades** within the Hominoidea. A large number of ape fossils date to the Miocene, when Earth might well have been called the 'Planet of the Apes'. It is now clear that we are dealing with an extremely complex evolutionary history, one that is difficult to untangle due to limited fossil evidence and the confounding effects of **homoplasy** in morphological traits. Briefly, such taxonomic complexity, as in fossil and extant primates, arrived via evolutionary steps through the Cenozoic, following the mass extinctions of organisms with the Yucatán asteroid impact at 66.5 Ma (K–Pg boundary). Distinct primate fossils date from the Eocene but molecular data would suggest earlier origins predating the K–Pg event at 85 Ma (Ocampo *et al.*, 2006; Martin, 2012). The main pressures for evolutionary shifts were later Miocene changes in climate (drier) and ecology (open grassland landscapes). Trait distinctions in hominins (*Australopithecus*, allied genera and *Homo*) from other apes fundamental for human development are: (i) bipedalism from an orthograde posture associated with arboreal habitats (Fig. 9.2); (ii) freedom of hands for carrying objects, wielding weapons and precision manipulation of tools (dexterity); (iii) big brain development (control of fire, cooking of food, concept creation, representative and decorative imagery, as well as music); (iv) reduction of the sagittal crest and associated musculature which permitted cranial expansion and reduction in jaw dimensions and dentition; (v) language, communication enhancement and concept development; (vi) loss of fur, increased activity and varied sexual signalling (Jones *et al.*, 2018).

These changes relate to a movement out of dense woodland habitats prior to effective bipedalism

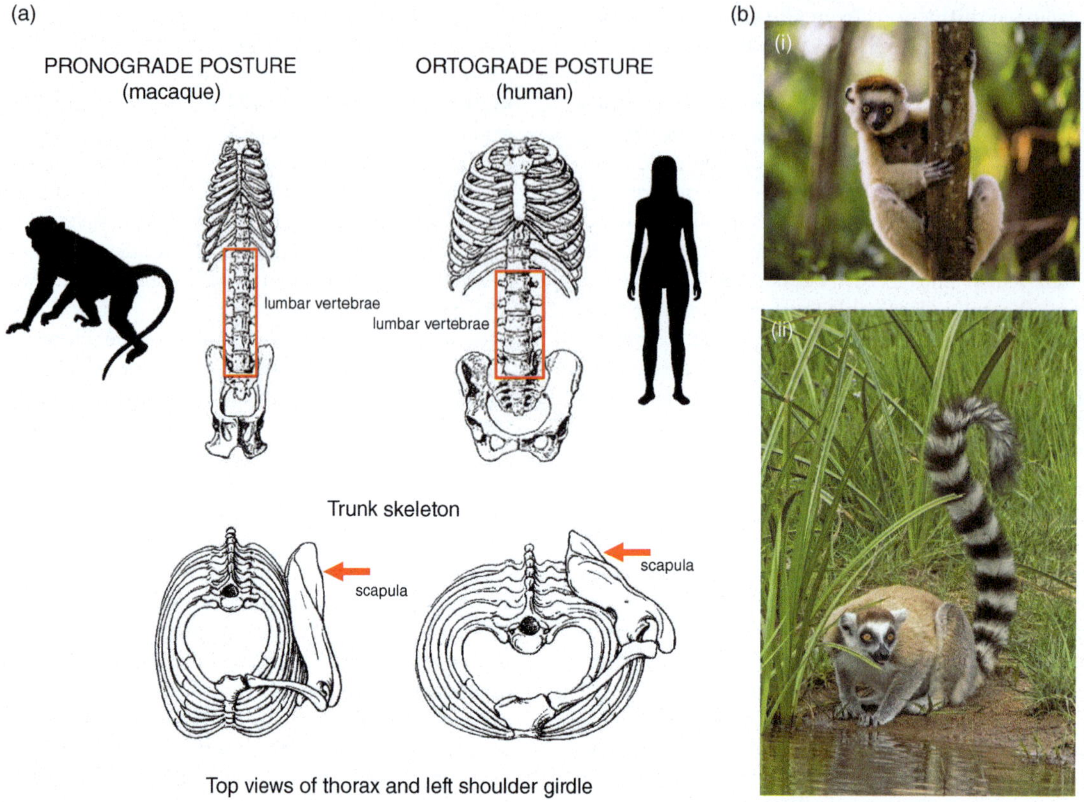

Fig. 9.2. Primate postures. **(a)** Pronograde and orthograde primate postures for mobility. Pronograde posture: vertebral column held horizontally; primarily quadrupedal movement; thorax narrow and elongated; scapulae positioned laterally on the thorax; generally more lumbar vertebrae (often more than five), contributing to greater spinal flexibility. Orthograde posture: vertebral column more vertical; quadrupedal or bipedal movement; thorax broad and short; scapulae positioned more dorsally; fewer lumbar vertebrae, leading to a shorter and less flexible lower back. **(b)** Current prosimians falling into the two postural groups: (i) arboreal orthograde Verreaux's sifaka *Propithecus verreauxi*; (ii) terrestrial pronograde ring-tailed lemur *Lemur catta*. (Panel **a** skeletons from Burwell *et al.*, 2009, used under licence CC BY 2.0; silhouettes by T. Michael Keesey, used under licence CC PDM 1.0, and Carlo De Rito, used under licence CC0 1.0 Universal, from www.phylopic.org/. Panel **b** photos: Verreaux's sifaka by Ron Knight: https://commons.wikimedia.org/wiki/File:Verreaux%27s_Sifaka_(Propithecus_verreauxi)_(8603104141).jpg, used under licence CC BY 2.0. Ring-tailed lemur by Charles J. Sharp: https://en.wikipedia.org/wiki/File:Ring-tailed_lemur_(Lemur_catta).jpg, used under licence CC BY SA 4.0)

within more open landscapes (savanna) in which bipedalism was enhanced and subsequent traits emerged and developed. The contrast in resources and conditions is profound. These animals (*Australopithecus,* allied genera and *Homo*) were highly varied and apparently speciose (Humphrey and Stringer, 2018), with both robust and gracile forms, suggesting differences in diet and therefore environment, but the precise evolutionary track between *Homo* and other Hominini (especially *Australopithecus*) remains unknown (de Ruiter *et al.*, 2017). Although, of course, it is not possible to determine the detailed RBHs of these fossil hominins, it is possible to distinguish differences in their habitats and range of movements. Inevitably, they would have occupied very different RBHs and home ranges, leading to the first dispersions over Africa (e.g. *Australopithecus afarensis*, 3.2 Ma; White, 2018) and ultimately beyond the continent (*Homo erectus*: Antón *et al.*, 2016; *Homo sapiens*: Tuttle, 2014) and the emergence of distinct species in Asia and Europe (Upper Pleistocene: Neanderthal, *Homo neanderthalensis*: Endicott *et al.*, 2010; Nielsen *et al.*, 2020; Denisovans: Reich *et al.*, 2010; Karlsson, 2022; *Homo floresiensis*: Sutikna *et al.*, 2016; Rogers *et al.* 2017) (Fig. 9.3). The essential features

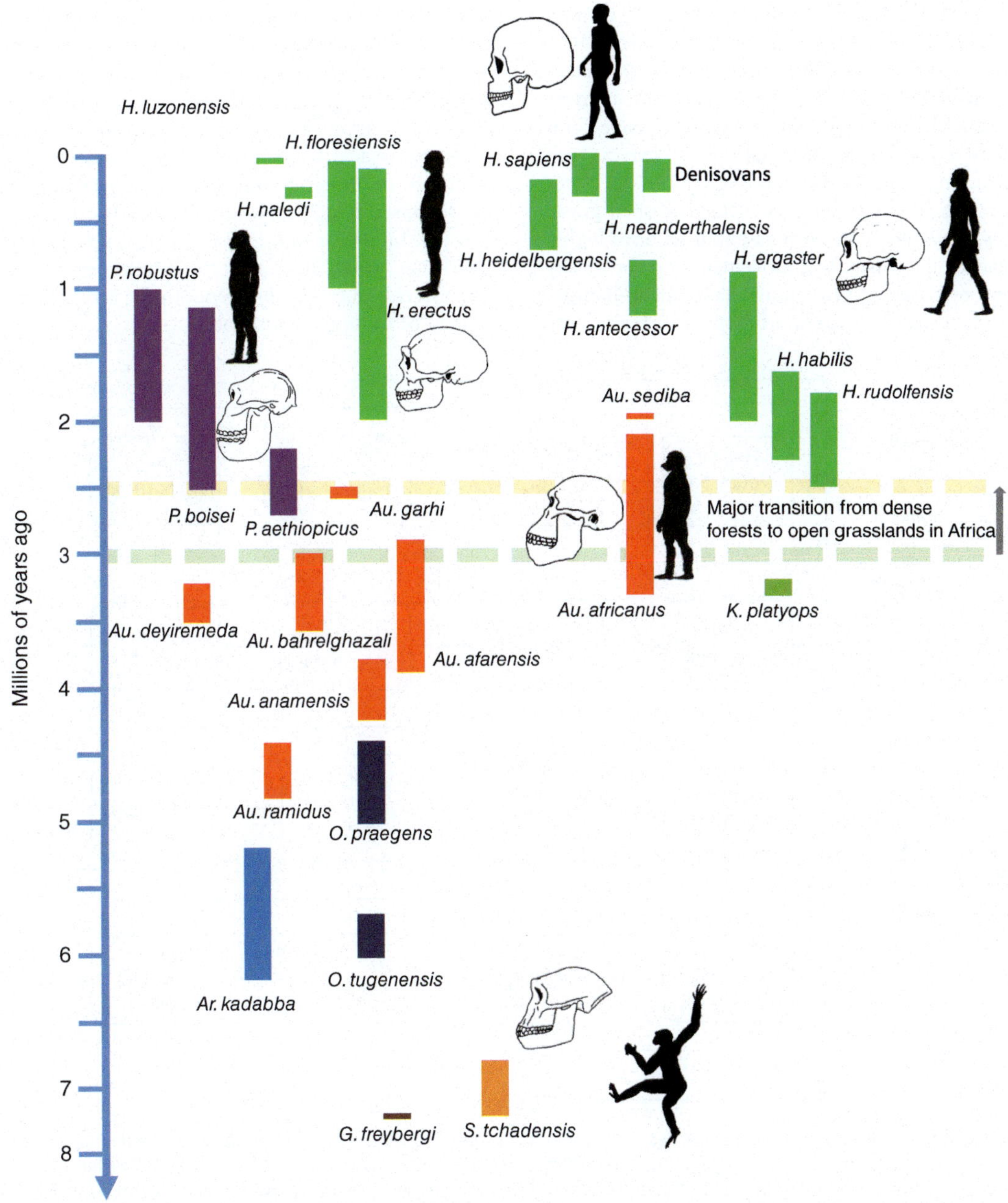

Fig. 9.3. Fossil markers in human evolution. Evolution from an orthograde forest-dwelling animal (*Ardipithecus*) to a distinctly bipedal grassland/scrubland-dwelling animal (*Australopithecus*; *Homo*). Abbreviations: *G.*, *Graecopithecus*; *S.*, *Sahelanthropus*; *O.*, *Orrorin*; *Ar.*, *Ardipithecus*; *Au.*, *Australopithecus*; *P.*, *Paranthropus*; *K.*, *Kenyanthropus*, *H.*, *Homo*. (Sources of skull images: *S. tchadensis* and *P. boisei*: original. *Au. africanus*: https://en.m.wikipedia.org/wiki/File:Australopithecus_Skull.png, used under licence CC0 1.0 Universal Public Domain Dedication; *H. ergaster*: https://en.m.wikipedia.org/wiki/File:Ergaster_Skull.png, used under licence CC0 1.0 Universal Public Domain Dedication; *H. sapiens*: https://en.m.wikipedia.org/wiki/File:Sapiens_Skull.png, used under licence CC BY SA 3.0. Silhouettes by T. Michael Keesey: www.phylopic.org/, used under licence CC0 1.0.)

are the contexts of the fossils; from the context (associations of plants, animals, tools, etc.) it is possible to determine their biotopes. *Ardipithecus*, as a woodland animal, would have had a more restricted home range limited by the biotope complexity and resource distribution (i.e. leaves, fruit and animals) in a 3D forest environment. Once *Australopithecus* and *Homo* occupied open grasslands, it would have been essential to have highly efficient bipedalism, fast movement and the ability to protect themselves against large pack-hunting predators on the grassland floor. Dexterity (use and production of weapons, as the canine teeth were useless for defence), use of fire and excellent communications would have been critical for survival. The need for food and water would have required searches by larger troops (also needed for defence) over a wider ground area. Some of these features are possible to determine; others can only be surmised (Fig. 9.4). Undoubtedly, the crucial factor in human evolution was the expansion of Miocene grasslands and the ability to extract resources and establish defence in the very different world of an open landscape.

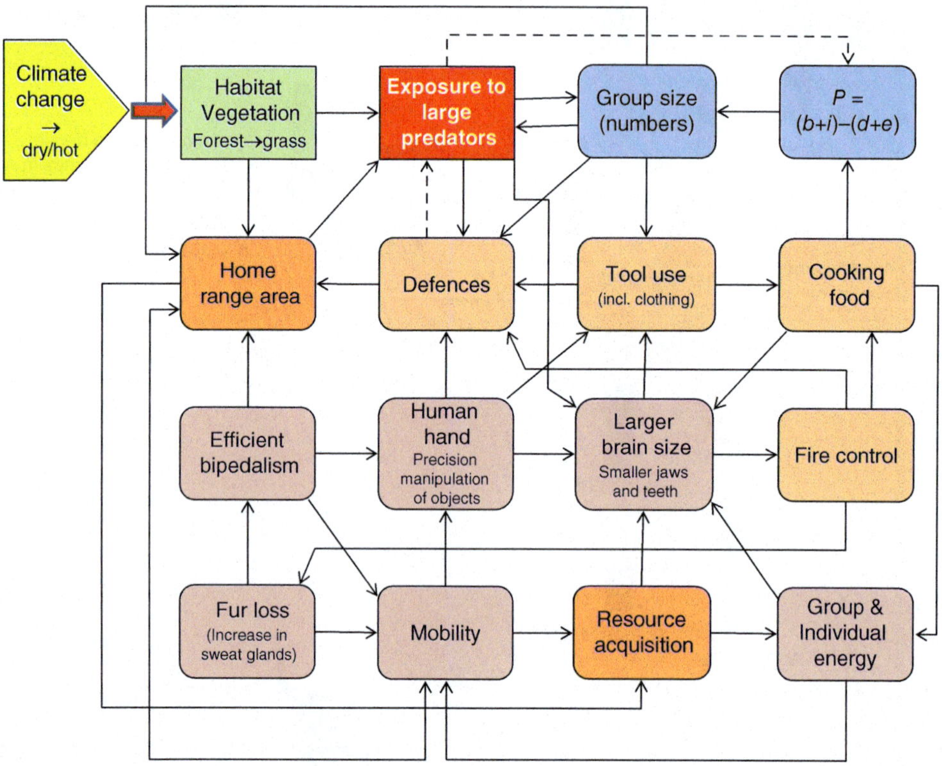

Fig. 9.4. Some of the basic evolutionary steps envisaged in human evolution leading to *Ardipithecus* and *Australopithecus*. Changes occurring with the shift in the Miocene climate (yellow box, becoming drier) leading to a movement out of the orthograde Last Common Ancestor (LCA) into grassland and shrublands (light green box). Such a move would require a search for food by groups over a wider area (home range), and larger groups to provide defences against top predators (red box). Enhanced bipedalism was essential for greater mobility, for freeing up the hands for making tools, manipulating fire (for defence and cooking) and leading to a larger brain size and adaptation to what would have been a dangerous environment. Loss of fur and increase in sweat glands made it possible for hominins to hunt during hot daylight hours in grasslands; use of fire and development of cover at night (hides) was a possible trigger for fur loss. Dark brown boxes, home range and resource-based habitat; blue boxes, population variables; pale pink boxes, traits; light brown boxes, evolutionary changes in use of the environment. Arrows designate effects of one variable on another: unbroken line, positive effect, pecked line, negative effect. *P* = population size; *b* = number of births, *i* = number of immigrants, *d* = number of deaths, *e* = number of emigrants. (Figure created by R.L.H. Dennis; information from White, 2018; Humphrey and Singer, 2022.)

Home ranges and habitats in living primates

To turn to living primates (Rowe, 1996), before feeling our way to ancient and modern humans, we are faced with a plethora of contrasts – variation born of size (mass), group number, **social structure**, lifestyle and food (on trees or the soil surface) – affecting the form of a multitude of traits: arms, legs, tails, eyes, teeth and many more biological attributes. As such, there is insufficient space to inspect all their influences on resource exploitation and needs. But with consideration of a few species from each evolutionary **clade**, we can begin to understand the sheer complexity of **resource-based habitats**, **home ranges** and **niches** in these animals.

A first valuable step to an understanding of each animal's resource-based habitat is to consider the relationships between key variables that influence the use of its 'space', particularly home range: population density, group size, individual mass, social structure (hierarchies, bonding, territoriality), inter- and intraspecific interactions and **landscape** used (open versus closed **biotopes**), all of which influence diet. Two aspects of home range are intimately linked to resource use and therefore the resource-based habitat of any species: (i) the size of home ranges, and (ii) the overlap of home ranges between groups of the same and different species.

Among primates, **home range** might be expected to increase in area respectively with: (i) increased mass of the animal, metabolism and dietary habits (for frugivores and omnivores more than for folivores for their body weight; Milton and May, 1976; Harvey and Clutton-Brock, 1981); (ii) the size of groups (Clutton-Brock and Harvey, 1979); and (iii) open versus closed landscapes (Milton and May, 1976; Wrangham, 1980), though volume of space used is perhaps a better measure. On the other hand, home range may well be smaller when population **density** is higher, if this leads to greater **competition** for resources. Among primates, group or troop size has been found to have a regular scaling pattern (with preferred sizes of about 2.5, 5, 15, 30, 50 individuals), a scaling ratio similar to that found in human social networks (Dunbar, 2017; Dunbar *et al.*, 2018). Grouping has advantages and disadvantages. Advantages include: (i) enhanced protection against enemies, especially predators such as leopards (McLester *et al.*, 2018) – an example of bulk defence is the surmised revenge killing of dogs by monkeys (McCullough *et al.*, 2012); (ii) enhanced protection against infanticide and of subdominants (bonding; Hamilton, 1971; Wrangham, 1979); (iii) greater ability to find food and other resources (Swedell, 2012); (iv) greater ability to exploit resources than members of smaller conspecific groups (Cody, 1971). However, large groups also have negative implications: (i) increased interactions (competition) among members of a larger group, which may lead to greater expenditure of energy and injury in accessing food items and bonding; (ii) contests for supremacy again leading to injury or death; (iii) difficulty in finding sufficient food (and other) resources for the group, with inevitably greater expenditure of energy; (iv) greater conspicuousness to enemies and higher costs of defence (the latter being expected to increase with body mass and home range size; Jetz *et al.*, 2004).

These and other issues are all part of the **ecological constraints model** (which establishes that ecological factors constrain the size of groups by influencing movement patterns and foraging efficiency; Chapman and Chapman, 2000; Teichroeb and Sicotte, 2009). Competition is likely to increase when food supplies decline, are clumped (or other resources exist in discrete patches) and vary in quality, *ergo* quantity of the best food; thus, it is not difficult to imagine that group size will vary in relationship to the varying quality and quantity of food supplies (cf. fruit > foliage > insects > grasses) and other resources resulting in a **fission–fusion social organisation** with shifts in conditions and social structure. Thus, groups may vary enormously in number (e.g. spider monkeys *Ateles geoffroyi* from 1 to 35 in Costa Rica's Santa Rosa National Park; red colobus *Piliocolobus rufomitratus* from 12 to 150; chimpanzees *Pan troglodytes* from 20 to 150) (Lehmann and Boesh, 2004; Chapman and Teichroeb, 2012). Social factors (such as female reproductive success or infanticide) should not be underestimated in affecting group size. For example, female *Colobus guereza* reproductive success is maximised in small to medium groups (Chapman and Teichroeb, 2012).

Regarding the degree of overlap of home ranges among groups of primates, this depends on both conspecific and heterospecific interactions. Pearce *et al.* (2013) in a meta-analysis of 100 primate species found that, unlike territoriality, home range overlap is highly plastic between and within species. Overlap increases: (i) with home range size; (ii) with lower rainfall (and thus lower habitat productivity, and more so in terrestrial than arboreal species); and

(iii) larger bodied species. Overlap is predicted to decrease with high population density (Jetz *et al.*, 2004), but only if home ranges are actively defended – equating with **territories** – against intruders (Pearce *et al.*, 2013). However, many primates are not territorial versus conspecifics and heterospecifics (e.g. the proboscis monkey *Nasalis larvatus*; Boonratana, 2000). Thus, other factors (e.g. social, such as loud calls of gibbons: Cowlishaw, 1992; scent marking of lemurs: Mertl-Millhollen, 2006) need to be considered for the occurrence or absence of territoriality and home range overlap. Even so, areas of overlap may be zones of avoidance, if the costs of intergroup conflict create a landscape of fear (**risk hypothesis**; Wrangham *et al.*, 2007). Although Wrangham *et al.* (2007) did not find evidence of this mechanism in redtail monkeys (*Cercopithecus ascanius*), white-faced capuchins (*Cebus capucinus*) or chimpanzees (*Pan troglodytes*), a later study (Tórrez-Herrera *et al.*, 2020) suggested it could operate in white-faced capuchins, though behavioural evidence was mixed. The use of peripheral areas (overlap zones) may occur because they contain resources whose value may outweigh the cost of intergroup encounters.

Below are outlined some issues of resource-based habitats noted in the major primate groups: prosimians, New World monkeys, Old World monkeys and hominids. Prosimians (minus tarsiers) are strepsirrhines (moist noses); tarsiers are haplorhines, having simple, dry noses.

Prosimians

A number of these mammals (e.g. *Propithecus*) most closely simulate the orthograde root of hominids, including humans. The group comprises tarsiers, lorises, galagoes (bush babies) and lemurs, the latter entirely restricted (as an evolutionary product) to Madagascar. Lemurs are fascinating, having produced an extraordinary variety of species (>100 extant species grouped into 15 genera, plus a number of extinct taxa) on a single island isolated for 88 million years (Samonds *et al.*, 2013). In mass they vary from 25 g (Madame Berthe's mouse lemur *Microcebus berthae*) to 9 kg (the indri *Indri indri*). Among the most remarkable are the aye-ayes (*Daubentonia madagascariensis*, lemurs with rodent-like teeth that perpetually grow and a special long, thin middle finger that is used for extracting insect grubs from holes in trees); in some ways they simulate various species of tarsier (Asian haplorhine prosimians, predatory carnivores, with long hindlimbs for rapid vertical transfer between trees) in their large eyes and nocturnal habit. Food (the key consumable resource) amongst lemurs differs across the expansive **clade**, including fruit, foliage, insects, even poisonous bamboo; movement varies from pronograde (the ring-tailed lemur *Lemur catta*) to orthograde (sifakas *Propithecus* spp.; indri *Indri indri*); substrates used are terrestrial and arboreal; some are diurnal, others nocturnal, and strategies for defending territories can be physical, vocal or based on dispersing characteristic scent. Key features are the orthograde leaping and 'dancing' posture of sifakas, and their communal defence against predators, almost human in its ingeniousness, determination and bonding, including aftercare (such as defence against snakes; Gardner *et al.*, 2015).

The crucial feature of lemurs is then the astonishing mix of **generalism** and **specialism** in **traits**, physical, biological and social, all having emerged on a single large island. With destruction of natural **biotopes** by humans, it is difficult to recreate the circumstances of their development, but there is clear support for the cyclical evolutionary paradigm postulated by the **oscillation hypothesis** (i.e. specialisations and speciation are promoted by repeated upturns and downturns in environmental conditions; Janz and Nylin, 2008; Dennis *et al.*, 2011); for instance, closely related species tend to occupy different climatic niches (Kamilar and Muldoon, 2010) suggesting a basis of generalism leading to expanding distributions followed by specialisation and speciation. These animals (mostly matriarchal) provide an invaluable model taxon for the study of the expansion and contraction of resource-based habitats and the flexibility of home ranges, in the occupation and use of different biotopes (Jolly, 1966; Sussman *et al.*, 2003; Jolly *et al.*, 2006; Sussman and Ratsirarson, 2006; Sauther *et al.*, 2013, 2015; Gould and Gabriel, 2014). It becomes evident in the study of *Lemur catta* that the size of home ranges depends not just on body mass and group size, but also on the flexibility of resource exploitation. *Lemur catta* (mass: 2.3–3.5 kg; lifespan: 16–19 years; group number: 5–24 individuals; recorded home range: 6–23 ha; vocal and scent communication) having access to ground and trees – spending more time on the ground (one-third of a day) than other lemurs – and being distinctly omnivorous, has great potential for expanding its home range and resource-based habitat (Koyama *et al.*,

2006; Sauther *et al.*, 2015) depending, of course, on human impact on the Madagascan environment.

New World monkeys

Existing species of New World monkeys (Platyrrhini; broad flat nose) fall into five families, 20 genera and 156 species (204 taxa including subspecies) (Mittermeier *et al.*, 2013; Lynch Alfaro *et al.*, 2015; Dunn and Cristóbal-Azkarate, 2016). More than two dozen extinct genera extend back in the 30 million years of the Cenozoic of South America (Rosenberger and Hartwig, 2013). Much as with the lemurs of Madagascar, the New World monkeys have radiated into the Neotropical forest ecosystem in a wide variety of habitats (upper canopy to lower trunks), exploiting a range of food resources: leaves (howler monkeys, *Alouatta* spp.), ripe fruit (spider monkeys, *Ateles* spp.), hard-shelled fruits (saki and uakari monkeys; *Pithecia* spp. and *Cacajao* spp., respectively), exudates (marmosets: *Callithrix*, *Cebuella*, *Callibella* and *Mico*), insects (callitrichids), fungi (Goeldi's monkey *Callimico goeldii*), etc. Although all species are arboreal, they vary enormously in form, body size and locomotion (i.e. from the pygmy marmoset *Cebuella* at 100 g to the woolly spider monkey *Brachyteles* at 10,000 g; arboreal quadrupedalism to vertical clinging and leaping (marmosets and tamarins) and prehensile tails as with the atelines). All but one (the nocturnal genus *Aotus*, owl monkeys) are diurnal. New World monkeys have a great capacity for learned behaviour; they occur in relatively large troops but have small litters (1–2 offspring) and long lactations. Mating strategies vary (Dunbar, 1995), ranging from pair-bonded and monogamous (owl (*Aotus*), titi (Callicebinae) and some callitrichid monkeys) to mutual promiscuity (woolly spider monkeys), and contrasting degrees of sexual dimorphism.

New World monkeys present admirable models for the study of home range extent and overlap between conspecific and interspecific troops (e.g. spider monkeys; Fedigan *et al.*, 1988; Cawthon-Lang, 2007; Takahashi, 2008; Ramos-Fernandez *et al.*, 2013; Schneider and Sampaio, 2015). One study of the **risk hypothesis** in four troops of white-faced capuchins (*Cebus capucinus*) revealed that although members of troops socialised (viz. activity, vocalisation, foraging) less in peripheral (overlapping) areas compared to the centre of home ranges, the differences were not statistically significant (Tórres-Herrera *et al.*, 2020). An interesting issue emerging from the study is that, despite fruit trees containing more ripe fruit near home range borders – and troops depleting these trees more than in home range centres – there was no mention of an apparent shift in home ranges to centre on changes in the location of rich trees (Fig. 9.5). The exploitation of food resources was not outweighed by fear of intergroup encounters. This study ran over 100 h of 'focal follows' but it is clear that any study of home range dynamics requires data of related species extending over days, months and many years, to determine **niche overlap** as well as spatial overlap (e.g. frugivorous gray woolly monkeys *Lagothrix cana* and black-faced black spider monkeys *Ateles chamek*; Cavalcante *et al.*, 2020). The obvious questions are: How much, and what spread of, tracking time is required to determine factors that result in biologically substantial shifts in home ranges? What determines the centre of home ranges? Clearly the size of home ranges depends largely on resource requirements as, if insufficient for the troop, then pressures are applied to troop membership and existence. However, knowledge is fast growing on extensive co-operation and coalitionary behaviour, a reliance on social learning and tool use in New World monkeys (e.g. capuchin monkeys *Cebus* spp.: Perry, 2011; common marmosets *Callithrix jacchus*: Zhao *et al.*, 2019).

Old World monkeys

The Old World monkeys (Family Cercopithecidae; subfamilies Cercopithecinae – which are mainly African but include the diverse genus of macaques, which are Asian and North African – and Colobinae; more than 150 extant species included in more than 20 genera) are in many ways more diverse still than their New World relatives and are distinct in physical traits (e.g. eight to 12 premolars; lack of prehensile tails). Some exhibit exotic features (e.g. the highly colourful mandrill *Mandrillus sphinx*). Development (gestation) takes 5–7 months. As with humans, births are typically single but twins also occur. In the majority of species, female offspring stay with their mothers for life (viz. the basic social group among Old World monkeys is a matrilineal troop). Juvenile males leave the group and join a different troop. In many species, each group has a single male which drives off rivals. In other species, males are more tolerant, establishing hierarchical relationships between dominant and

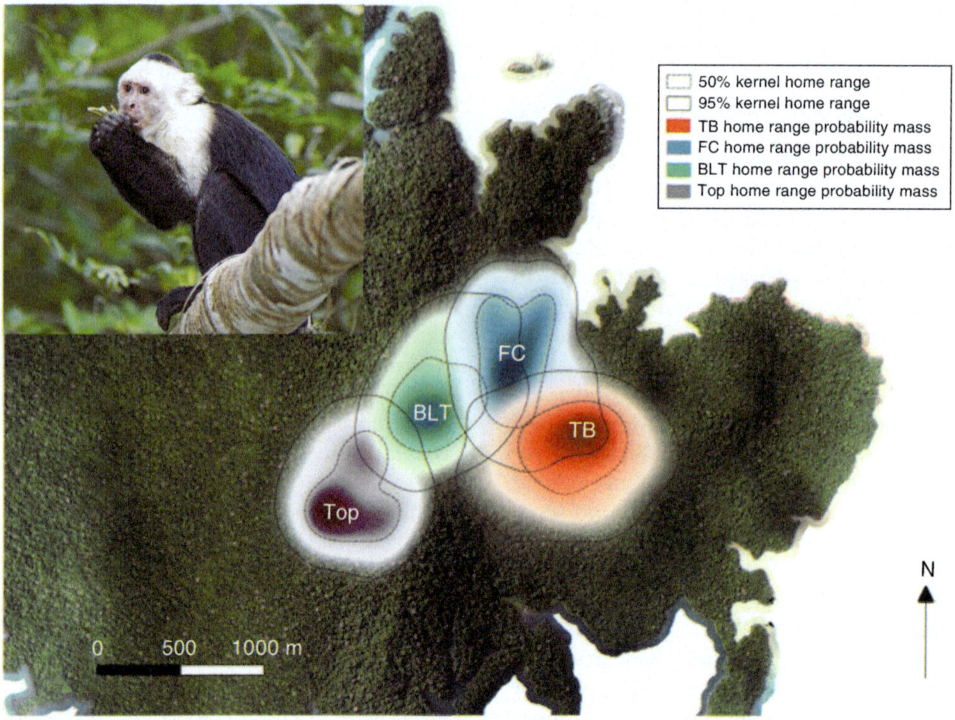

Fig. 9.5. Overlapping home ranges for four white-faced capuchin (*Cebus capucinus*) study groups (BLT, FC, TB and Top) on Barro Colorado Island (Panama) from February 2009 to January 2010. The probability mass (kernel) for each group's home range is represented by colour gradients (darker colours show higher probability mass). The 50% kernel home ranges are marked by dashed black lines, and 95% kernel home ranges are represented by solid black lines. (From Tórrez-Herrera *et al*., 2020, used under licence CC BY 4.0. Inset photo by Joseph C. Boone: https://commons.wikimedia.org/wiki/File:White_faced_Capuchin.jpg, used under licence CC BY SA 3.0.)

subordinate males. They occupy a wide range of biomes from dense tropical rainforest, shrublands and mountainous terrain to savannas, bare but for a cover of grass. Most are distinctly opportunistic, eating fruit but consuming almost any available food items, such as flowers, leaves, bulbs and rhizomes, insects, snails and small mammals; also, garbage and handouts from humans. Many (e.g. baboons *Papio* spp.; gelada *Theropithecus gelada*, Fig. 9.6) feed mainly on grass. Group sizes are highly variable, even within species, depending on the availability of food and other resources.

The geladas (mainly grass feeders) are of particular interest as they are exposed and vulnerable to predators (e.g. leopards, hyaenas, dogs, humans) on open grassland (Cowlishaw, 1994). Sheer numbers of individuals and a complicated social hierarchy provide mechanisms for survival in this situation; most significant is the shifting structure of bands of

Fig. 9.6. A male gelada (*Theropithecus gelada*) in the Simien Mountains National Park, Ethiopia. (Photo by Bernard Gagnon: https://commons.wikimedia.org/wiki/File:Gelada_04.jpg, used under licence CC BY SA 3.0.)

geladas and the variation in home range size between the wet and dry season (much larger) in relation to food and water supplies (Kifle *et al.*, 2013; Kifle and Bekele, 2021). The **resource-based habitat** for geladas is likely to be very complex owing to a multilevel social structure (i.e. reproductive units containing a single adult male and his followers, bands of two or more adult males with their followers, all-male units containing only bachelor sub-adult males, and a large community (troop) made up of bands; Kifle and Bekele, 2021). Each of these units (as with individuals) has its home range and individual track in which it exploits a range of food types and other resources. There is a question of how these units adapt to climate change and human impact as farming extends into their home ranges (Kifle and Bekele, 2021). A key to survival is roost sites on cliffs (gorge sides), to which they flee and on which their agility remains unmatched (Lin *et al.*, 2020). They are also capable of active counterattacks and mobbing of predators (e.g. dogs and leopards) as at Arsi, Ethiopia (Iwamoto *et al.*, 1996). Geladas have herd structure in movements and raid human farmland which therefore puts their existences under threat (Crook, 1966; Yihune *et al.*, 2009).

A glance at the literature on geladas indicates the sheer complexity of establishing the resource-based habitat of the species, essential for conserving them (Iwamoto and Dunbar, 1983); it is well to appreciate that a species such as this of least concern (IUCN: www.iucnredlist.org/es/species/21744/217754712; accessed 19 May 2025) may quickly enter higher ranks of concern. Moreover, one thing has become clear from studies of Old World monkeys and that is resource-based habitats are never static. Species have discovered how to use unusual 'tools' (e.g. Japanese macaques *Macaca fuscata* using hot springs; Zhang and Watanabe, 2007), tools for extracting food (e.g. use of stone tools for opening molluscs in *Macaca fascicularis*; Malaivijitnond *et al.*, 2007); the transmission of tool use as in Tonkean macaques *Macaca tonkeana* (Ducoing and Thierry, 2005); robbing and bartering to obtain food (e.g. long-tailed macaques *Macaca fascicularis* at Uluwatu Temple, Bali, Indonesia; Brotcorne *et al.*, 2019). Researchers should be aware of growing complexity in studies of their animals (their capacity for learning tool use), and hence the difficulties of linking up social structure with behaviour and resource use while mapping their fine movements over a range of seasons and years.

Apes bar one

Apes, as noted above, are our closest living relatives. All earlier species of *Homo* remain as remnant DNA in *Homo sapiens*. The task of ecology working through the resource-based habitat of great apes and humans can be no less than to thoroughly understand the needs of these species (*Pan* (2 species), *Gorilla* (2), *Pongo* (3), *Hylobates* (9), *Hoolock* (3), *Symphalangus* (1), *Nomascus* (7); Fig. 9.7), so that we can best conserve their habitats, their living conditions in natural areas, essentially to provide space for them on the planet. This task involves an immense amount of complex eco-analysis. Bit by bit, during the past 60–70 years, the most dedicated research has been undertaken and collated by primate ecologists, sociobehavioural scientists, biogeographers and their many subfields of endeavour. Foremost among these were the 'explorers' of their **ecosystems** (e.g. chimpanzees: Jane Goodall, 1969, 1971, 1986; Goodall and van Lawick, 1971; gorillas: George Schaller, 1963; Dian Fossey, 1972, 1983; orangutans: Birutė Mary Galdikas, 1975).

To understand humans, their evolution, their capabilities and built-in failings (greed, exploitation, law breaking, violence, homicide, destruction, war), there is no more important work, before we ever attempt to launch a **metapopulation** of our species across the solar system, than the study of African apes as our closest relatives. In broad outline, what do we need to know? Basic requirements are knowledge of: (i) groups, troops, populations as social systems over time units; (ii) group and population size limits and the capacity and ingredients for compromise between units; (iii) basic consumable resources (food and the issue of carnivory); (iv) environmental limitations on activity and life (i.e. suitable biomes/biotopes; places to rest or roost); and (v) spatial requirements and crucial components of that space.

Here, the focus is on our closest relatives – resources and conditions for chimpanzees and gorillas. Knowledge of these components of survival has emerged piecemeal from independent studies. However, the resource-based habitat is substantially underestimated without studies being extended over the entire geographical range and each troop's activities and behaviour locked into an explicit space-time **lifetime track** and **home range** (Pusey *et al.*, 2007). Both chimpanzees and gorillas occupy a range of contrasting habitats (in the traditional

Fig. 9.7. Great apes of Africa and Asia. **(a)** Chimpanzee *Pan troglodytes*. **(b)** Western lowland gorilla *Gorilla gorilla gorilla*. **(c)** Sumatran orangutan *Pongo abelii*. **(d)** White-handed gibbon *Hylobates lar* (Sources of photos: chimpanzee, Dakshin90: https://commons.wikimedia.org/wiki/File:Attitude_of_Chimp.jpg, used under licence CC BY SA 4.0. Western lowland gorilla, Ltshears: https://it.wikipedia.org/wiki/File:Gorilla_496.jpg, used under licence CC BY SA 3.0. Sumatran orangutan, Kabir Bakie: https://en.m.wikipedia.org/wiki/File:Orangutan_01.jpg, used under licence CC BY SA 2.5. White-handed gibbon, Thomas Tolkien: https://commons.wikimedia.org/wiki/File:Hylobates_lar_gibbon.jpg, used under licence CC BY 2.0.)

sense), even at the biome scale (i.e. tropical rainforest ⟷ secondary forest ⟷ savanna ⟷ anthropogenic cultivated landscapes, with ⟷ indicating 'merging into', so a mix of, intermediate or pure conditions; van Leeuwen *et al.*, 2020) and this is the cue to their ability to exploit a range of resources and withstand different conditions (Bryson-Morrison *et al.*, 2017). Inevitably they come into contact and conflict (Robbins, 2008; Caillaud *et al.*, 2020); in some areas, gorillas and man have become very closely associated, which may cause other forms of conflict (Oishi, 2013). The different biomes place different pressures on chimpanzees and gorillas, if only because of contrasting availability of resources, as well as pressures from predators and parasites (Scull *et al.*, 2012). Understandably, the home ranges (with biome-dependent variations on diel, seasonal and geographical bases) of spatially overlapping and adjacent troops also differ, though it is not clear why this occurs (e.g. in *Gorilla gorilla diehli*; Sunderland-Groves *et al.*, 2009). Food was initially thought to be plants (leaves, fruit, etc.), but it is now known that both species consume meat, adopting carnivory, especially chimpanzees (Newton-Fisher, 2014). Research has engaged a number of different techniques: (i) 'follow' or trace techniques (for determining social interactions such as dominance hierarchies, mating relationships, affiliations, nest building and selection of food items including hunting monkeys, e.g. chimpanzees: Newton-Fisher, 2014; *Gorilla beringei*: Robbins and McNeilage, 2003; Robbins, 2008), (ii) transect recording behaviour (e.g. *Pan troglodytes schweinfurthii*: Hakizimana *et al.*, 2015); (iii) specific surveys of constructions (i.e. study of night and day nests, that act as surrogates for individual and troop movements, including analysis of dung, e.g. Laing *et al.*, 2003; Sunderland-Groves *et al.*, 2009; Hakizimana *et al.*, 2015). Other studies have undertaken investigation of population density, dispersion, group size, home range, intergroup and interspecies relationships (e.g. chimpanzees: Robbins and McNeilage, 2003; Vieira *et al.*, 2019; gorillas: Robbins *et al.*, 2004; Sanz *et al.*, 2022).

Some of the studies may surprise readers, as for the extensive analyses of dung at nest sites; this enables determination of population (troop) size and composition, food exploited, the age of nest sites from systematic changes to dung, such as fungal growth on the surface of dung boli (Laing *et al.*, 2003); moreover, to establish whether nests are day or night nests, dung acting as a marker of night nests (Sunderland-Groves *et al.*, 2009; van Dijk *et al.*, 2021). The analysis of excrements also provided the first evidence that non-human primates consume therapeutic plants, thus paving the way for the study of animal self-medication (Laumer *et al.*, 2024). *Pan*, *Gorilla* and *Pongo* all make nests, and there are clear preferences for nest sites (*Gorilla*: Haurez *et al.*, 2014).

Perhaps the most surprising finding is that chimpanzees are active predators of other mammals. Films have been taken of the highly co-ordinated hunting techniques used to kill monkeys, their dismemberment and 'sharing out of the remains' (www.youtube.com/watch?v=ffRJaDSzNeU; accessed 19 May 2025). They use collective planning to corner and ambush the monkeys, then tear them to pieces before consuming them (Newton-Fisher, 2014). This collaborative interplay followed by consumption of monkey meat gives some idea of how such behaviour arose in ancestors of *Australopithecus* and *Homo* species. Such observations draw chimpanzees behaviourally close to humans, Palaeolithic *Homo* as well as modern humans. In this violence, chimpanzees are quite capable of killing gorillas (Southern *et al.*, 2021).

Primate socioecology

These last few commentaries on primate species have but touched on isolated complexities in ecology and habitats but make no attempt to draw them together in a synthetic model of the diversity in primate social relationships. Over the years, attempts have been made to establish such a social model, unsuccessfully (Thierry, 2008). Wrangham's (1979, 1980) attempt focused on competition (viz. benefits of between-group competition outweigh the costs of within-group competition). Van Schaik (1999) pointed to predation as the overriding reason for group living. Since the 1980s, a complex of other factors has been added to the model by Van Vuren and Van Vuren (1996) and Sterck *et al.* (1997). Thierry (2008) drew these factors up in an 11 × 4 contingency table. However, he exposed a number of weaknesses that illuminate the variables of group existence (e.g. impact of males and other organisms such as parasites as well as variation in cognitive abilities). The simple fact is that group existence and hierarchies have a multivariate basis and different species are influenced differently by variables affecting social cohesion at a range of different space-time scales. It is a problem that lends itself to interspecific and species-specific multivariate analyses rather than simplistic notions of ecological models. Even so, it is not difficult to envisage a balance between the need for individual defence provided by groups and conflict for individual resources within the group as conditions change and young grow into a position of seeking power. Thus, we need to follow animals over long periods and measure space-time changes in any species.

Palaeolithic and Mesolithic human home ranges and habitats

The most astonishing feature of evolution is the technological rise of humans and related hominin species from 3.3 million years ago, dating from the appearance of the oldest known stone tools (Lomekwi site, Kenya; Harmand *et al.*, 2015) (Fig. 9.8). Though the greatest developments have occurred in the Holocene from c.12 ka BP with the beginning of consolidated settlements, pastoralism and cultivation, specialisation in trades, sophistication in social structure – and what we now refer to as the **Anthropocene** (the technological epoch; see Chapter 10) – vitally important changes preceded these periods with shaping of stone tools and control of fire in the long interval from the Miocene to the end of the last glaciation. Advances include, among others, campsites growing in size and duration of occupation, larger dwellings made of hides, wood and mammoth bones, clay working, rope development and weaving. Complexity of fire use increased with stone hearths, ovens and stone oil

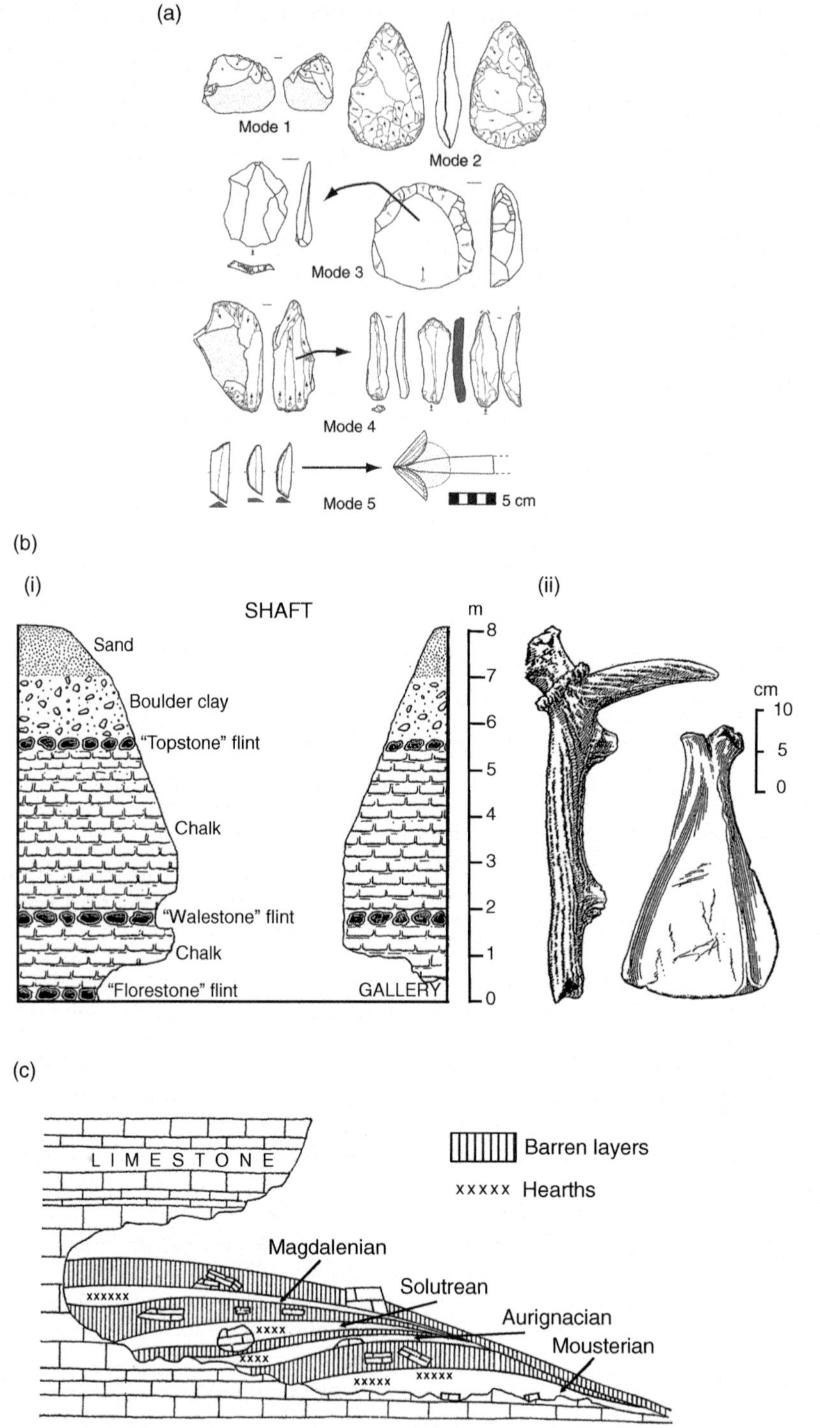

Fig. 9.8. Palaeolithic stone tools. **(a)** Graham Clark's model of Palaeolithic stone technology development. Because of the lack of a universal chronology of Palaeolithic technological change, Clark (1969) proposed a series of five

lamps (Stutz, 2018). Interestingly, evidence of stone technology has been recorded among other living primates, such as long-tailed macaques (*Macaca fascicularis*) in Khao Sam Roi Yot National Park, Thailand (Luncz *et al.*, 2017). The Palaeolithic extends over a long period and it is inevitably associated with changes in many aspects of hominin habitats, such as: (i) the nature of camps and the extent of cover (affecting mobility over the ground, thus home range); (ii) food (including prey) and its preparation; (iii) combating enemies; (iv) variability and complexity of tools; (v) conceptual development, rites, burial practices and communication; (vi) social structure and hierarchies of larger social groups; thus, changes in complexity, variability and interaction across all aspects of *Homo* evolution.

Each species over this long period would have had its **home range** and **resource-based habitat**. These would differ enormously in nature for different species, and for 'populations' ranging in size from a nuclear family to familial groups and larger interrelated entities, depending also on the level of development (evolutionary transformation), location, **biome** and **biotope**, and the wealth of resources and conditions at each location. Humphrey and Stringer (2018) give an excellent summary of the changes accruing with shifts in evolution and biotope occupancy occurring with climate change. Importantly, HRs and RBHs would also depend on the degree of **competition** between groups of hominins, as well as the level and nature of predation on them, key pressures often ignored by archaeologists and anthropologists dealing with Early and Middle Pleistocene sites (but see Adler and Bar-Oz, 2009).

With no apology for repetition, it is crucial to appreciate that *Homo* is an animal; during this long phase of hominin evolution, humans were subject to all the pressures experienced by other animals, including predation, much as they are today still in parts of tropical India (by leopards, tigers) and arctic Canada (polar bears), as well as acting as predators themselves on their own kind (cannibalism). Therefore, their capacity to exploit space depended much on the number of individuals in a group, their desperation for food, levels of competition, defences and weapons adopted (tool use), and familiarity with the ground. Not only that, but changes in social and cultural development (e.g. development of 'coming of age' rituals, initiation into manhood) led on to dangerous practices in tribal societies that effectively placed predators (e.g. humans versus felids) against one another – social developments intended to provide experience for facing natural dangers (Tindale, 1974; Tonkinson, 1991).

To understand the RBH and HR of Palaeolithic hunters and gatherers, there are two spatial levels to examine. The first and most obvious level is the places (sites) where groups had some kind of base, even if temporary (i.e. overnight), and many groups throughout prehistory would indeed have used temporary 'camps' while they tracked prey. The fundamental problem is that it is not possible to determine site use on a day-to-day basis for samples of decades. Archaeological sites provide selective details on coarser slices in time, sometimes capturing multiple periods of use (e.g. Neanderthal (Mousterian) layers in the Kebara Cave on Mount Carmel; Bar Yosef *et al.*, 1992). A further problem is the interpretation of the multitude of artefacts, such as lithic tools, and their spatial organisation at sites, alongside other features including site structures (stone blocks), remains of food (vegetable and animal matter), evidence of fire use (burnt ground, hearths), bedding (considered to include ash cover from fires), hominin burials, art/sculpture forms, site bounds (usually lithic) and the bones of hominins and other animals found there. The current issue is

Fig. 9.8. Continued.

fundamental modes of stone tool production, numbered in order of first appearance: (1) pebble core and flake technology (exemplified by Oldowan industries); (2) large bifacial cutting tool technology (including hand-axes and cleavers); (3) prepared-core technologies such as Levallois flake, point and blade production; (4) prismatic blade and bladelet production (which emphasises blade production using indirect punch techniques); and (5) production of retouched microlithic points and cutting edges on bladelets and flakes (which yields pieces inserted into compound projectile or cutting tools). **(b)** Archaeology of the Stone Age: (i) Grime's Graves in Norfolk (UK) illustrating beds of flint used for lithic manufactures in the Palaeolithic and Neolithic; (ii) tools used by Neolithic and early Bronze Age miners: shoulder-blade of ox used as shovel (Harrow Hill, Sussex, UK) and red deer antler pick (Grime's Graves). **(c)** Section of ideal rock shelter (abri) (e.g. SW France) showing a stratification of Palaeolithic cultures from Mousterian (Middle Palaeolithic, c.160,000–40,000 BP) to Magdalenian (Upper Palaeolithic–Mesolithic (c.17,000–12,000 BP). (Panel **a** drawing by John J. Shea, from Stutz, 2018, used with permission of John Wiley & Sons. Panels **b** and **c** from Oakley, 1972.)

the nature of these sites at different times in the Palaeolithic. Are they merely temporary meeting places or more permanent sites (homes)? Caves and abris form particularly useful sites as protection from the elements as well as enemies, but may well have had other uses (burials and defaecation, as in the Holocene native Americans; Poinar *et al.*, 2001). Many open sites are thought by some to be little more than 'nests' (overnight roosts) emanating from earlier stages of evolution, like those used by related living primates such as chimpanzees and gorillas (Kolen, 1999). The sites certainly contain 'hearths' *sensu lato*; these would have been essential to provide warmth and cooking, but also to offer light on dark nights while sleeping, to ensure warning of predators and approaching unrelated groups of humans, and to facilitate tool making, especially the production of slivers of flint or of igneous rocks, sharp as scalpels, as produced in the Levalloisian industry (e.g. Adler *et al.*, 2014). It is these locations that give insights into the basics of the RBH as well as the HR, since the tools can be tracked to geological sites.

The sites for *Homo sapiens* and *Homo neanderthalensis* can be distinguished by tools, but there is also clear evidence of 'exchange' in technologies (Zaidner *et al.*, 2025). There must almost certainly have been division of labour, inasmuch as individuals (different sexes and age groups) vary in their ability to engage different tasks, especially the crucial business of tool making and hunting, which increased in complexity throughout the Pleistocene (Stiner and Kuhn, 2009). A vast number of such sites have now been discovered but interpretation of their functions, including conceptual activity, and social structure remains speculative (Maher and Conkey, 2019). Towards the Mesolithic in Europe, increasingly complex implements appeared including blades, knives, chisels, scrapers, arrowheads in bone and flint (Zhilin, 2016; Waddington, 2004) and many weapons, such as a wooden spear in an elephant skeleton at Lehringen, Germany (Movius, 1950). Towards the end of the Palaeolithic, composite tools (harpoons, spear throwers or atlatls) appeared; of course, many more objects made out of wood, ivory and bone have been lost, the loss more evident from the Upper Palaeolithic (Oakley, 1972).

The second level is the 'outer' space (beyond camps) where resources existed within a home range, including food (animals and plants), water, lithic sources (especially flint or basalt for cutting through and scraping hides), covering and bedding (especially hides). The higher the latitude occupied by a group, the more resources were required to keep warm at night and especially in winter. Exploited sites, or sites where activities have been recorded, are numerous. However, many sites still reflect biased excavations inasmuch as the greater the distance (expressed as a radius, r) surveyed out from the centre (core), the greater the area to be covered (area of the resulting circle = πr^2), and the likelihood of finding resource exploitation decreases away from 'hearths'. It is notably difficult to establish the full extent of the home range, incorporating the entirety of group activity. Even then, we may be underestimating the movement of earlier Palaeolithic groups as they follow shifting resources (e.g. herds) with the seasons. Attempts have been made to determine the home range of groups of hominins in three ways: (i) based on energy expenditure and movements of adult human foragers in modern hunter-gatherer societies (some 30–60 kcal per km; 50–70 km of movement per week while carrying loads; Binford, 2001; Van Andel and Davies, 2003; Watson *et al.*, 2008; Churchill, 2014); (ii) from home ranges of large-bodied carnivores such as grey wolves *Canis lupus* (Churchill *et al.*, 2016); (iii) using estimation of lithic transport for tool making (c.10,000 km^2; Féblot-Augustins, 1993; but for exceptional circumstances – lithic movements due to geomorphological agents – see Slimak and Giraud, 2007). Thus, estimates of home range areas for Neanderthal groups range widely from 1400 to >27,000 km^2 (Burke, 2006; Churchill *et al.*, 2016). However, all these methods have significant limitations. In the first case, the circumstances (context) around such contemporary hunter-gatherer groups have changed dramatically in modern times (because of contact with the industrial and agricultural world and unknown social and intellectual changes). In the second, no other animal hunts as a biped with freed-up hands. In the third, we have no idea to what extent exchange of goods occurred between different groups (e.g. stone for finished weapons or food).

However, very clear evidence exists for changes in technology (tools – from the crude Oldowan culture at c.2 million years ago to the finely sculptured Mesolithic arrowheads and harpoons) (Oakley, 1972). Along with increases in cranial capacity, skeletal remains of increasingly large mammals indicate the potential for changes where activity has become concentrated at sites (e.g. hunting of

West Caucasian tur *Capra caucasica* at Ortvale Klde in the southern Caucasus by sympatric *H. sapiens* and *H. neanderthalensis*: Adler and Bar-Oz, 2009; mammals of southeast Iberian sites: Scott and Gibert, 2009). However, there is a definite indication of a shift from a 'nest' mode to a 'hearth' mode, to impermanent 'house' (hut) structures, and to more permanent settlement (houses) by the Neolithic. The shift in stone tools has been little short of astonishing (i.e. Acheulian → Levalloisian → Aurignacian → Gravettian → Solutrean → Magdalenian in *Homo sapiens*; Châtelperronian and Mousterian being distinctly Neanderthal creations; Coolidge and Wynn, 2004; although it is not always possible to determine what species are involved because of their overlap in some areas; Shea, 2014). The primitive Oldowan culture lasted a long time but, subsequently, tool preparation was amazingly sophisticated. In Africa, the earliest known Acheulean hand-axes and bifacial chopping tools date from c.1.5 Ma ago (Chazan *et al.*, 2008); in Europe, 0.9 Ma ago (Scott and Gibert, 2009). The most important shift in flint tools was the truly sophisticated Levalloisian industry (Neruda and Nerudová, 2000). Technological development is illustrated by Graham Clark's (1969) five modes (see Fig. 9.8a; see Oakley, 1972; Stutz, 2018, table 1 for Palaeolithic terminology; and Shea 2013).

One of the fundamental problems of interpreting the past is that crucial data to adopt a thorough ecological approach to the lives of Palaeolithic hunter–gatherer groups are scarce. Instead, physical anthropology and palaeoanthropology have moved ever closer to social anthropology, sociology and psychology; recent accounts increasingly devolve to a complex of imaginative suppositions and we are fed reams of egregious speculations (e.g. on the nature of home, house, settlement, etc.), obfuscating hard reasoning (instead of using logic, mathematics and ecological approaches) about what is clearly evident (or not) at sites. Although cultural dimensions cannot be ruled out when speaking about homes ('Homes themselves may even be better understood as containers for myths and memories, and concepts of home for hunter-gatherers themselves were likely as varied as their imaginations and experiences took them'; Maher and Conkey, 2019), a set of fabricated mystic conclusions is not helpful and readers are directed to studies of related primates to get an impression of what is crucial to existence. A more direct and thorough approach to Palaeolithic sites was presented by Kolen (1999) some 25 years previously.

In fact, humans are just animals, behaviourally volatile, vulnerable to fluctuating resources, a range of predators and a mass of parasites. Crucial insights are needed about: (i) likely predators (lions, tigers, hyaenas, grizzly bears, crocodiles, etc.), enemies and competitors (rhinos, elephants), evidence for which may be possible to extract from site substrates; (ii) the nature of sites selected for stationary work (viz. topophilia (positive affective bonds between human beings and environments) versus topophobia (the dislike or fear of places); tools, hide preparation for clothes and roosts; meat and plant preparation for food making, etc. (Kolen, 1999); (iii) lines of movement over 'fields' for resource procurement. All these would have played crucial roles and adequate strategies were probably developed and applied with great care, giving hominins a clear view of approaching trouble (predators), shelter from adverse weather and from competitors too. Extraordinary as it may seem, there is little discussion about the fundamental involuntary functions such as waste disposal, urination and defaecation (see Yang *et al.*, 2014) – where these actions are carried out when the group is 'in camp' and when the group or a select 'team' is actively resource seeking; how a family group or an element of a tribal unit avoids polluting itself, its food and water supplies, deals with illness, and avoids alerting predators and competitors to their presence are all crucial issues. The deaths from disease much later in Georgian and Victorian UK cities bring home this repugnant feature of human existence.

This disgusting aspect is with us in many modern human settlements. Some subjects are difficult to discuss simply because they can offend readers but the scientific viewpoint would be to regard hominins as suggested here: as animals that can be studied like any other animal group. So, to raw issues: from a more recent period, over 1000 human faecal remains have been found in dry caves dating to 2000 years BP (e.g. Hinds Cave, Chihuahuan Desert, SW Texas; Shafer and Bryant, 1977; Poinar *et al.*, 2001). To what extent may this reflect Palaeolithic behaviour? The oldest identified human faecal remains date to 50 ka BP (Neanderthals, El Salt, Spain; Sistiaga *et al.*, 2014). Their value is in giving the location of evacuations and the contents of meals; hominins (viz. *H. sapiens* and *H. neanderthalensis*) were by no means solely meat eaters, eating also fruits and vegetables (Bocherens, 2009;

Richards and Trinkaus, 2009; Ecker *et al.*, 2013; Humphrey and Stringer, 2018). What we do not know is how bodily functions affected perception of sites by their occupants, and how this influenced their selection of sites and 'stay time' at sites. As noted above, what we do have is a sense of technological advancement in the sophistication of stone tools. **Lithic reduction**, refitting and density analyses have proved useful for determining the organisation of space at sites, as in the case of seven Middle Palaeolithic open-air sites (<2000 m^2 in area) in France (Clark, 2016). Such approaches are valuable for investigating the use of sites but determining the home range of groups when occupying 'camp' sites presents profound difficulties for future research, as noted above.

Neolithic humans and the first settlements

It may seem strange to separate the Palaeolithic (Upper, Middle and Lower) and Mesolithic from the Neolithic, based, as they all are, fundamentally on lithic industries. But nevertheless, there was a great change in land use and holding, if deeper in origins and more gradual than once assumed: the Palaeolithic regarded as linked to a string of temporary camps and continual movement (migrations), the Neolithic to permanent (i.e. stone and mudbrick) settlements and a more sedentary mode of life. Even so, clear evidence of long-distance migrations in the Neolithic exists (Price *et al.*, 2011). Change was indeed gradual, the earliest Neolithic type cultures being the **Natufian culture** (15,000–11,500 BP; type site: Shuqba cave at Wadi en-Natuf, Palestine; Bar-Yosef, 1998a). The beginning of stone-built settlements is well illustrated by Çatalhöyük settlement in the Konya Plain in southern Turkey (Bar-Yosef, 1998b; Fig. 9.9) dated to 9050–7900 BP (Chyleński *et al.*, 2019), an earlier date than previously cited (7790 ± 50 BP; Marciniak *et al.*, 2015). The production of pots occurred when humans were still hunter-gatherers (Balter, 2005), c.14 ka BP in modern Israel and Jordan, *before* they turned to domestication of animals (e.g. sheep, goat, pig, cattle, horse) and cultivation of plants (e.g. cereals, peas, olives). At least this would appear to be the case in the Near East. However, permanent settlement brought with it a host of other changes linked to a sedentary existence and to the development of a domestic economy. Even so, the spark of inventiveness that distinguishes *Homo sapiens* was already evident in the Upper Palaeolithic, in the extraordinary cave paintings of Lascaux and related cave sites (15,000–17,000 BP; Bahn, 2007; Mauriac, 2011), especially an ecological calendar for prey (Bacon *et al.*, 2022), illustrative of brain power sufficient to generate all the concepts that have emerged in modern times and that represent fundamental steps in the scientific view of the world (e.g. gravity, Isaac Newton, 1687; origin of species, Charles Darwin, 1859; quantum theory, Max Planck, 1900; plate tectonics, Alfred Wegener, 1912; general relativity, Albert Einstein, 1915; information theory, Claude Shannon, 1948). The crucial issue explored here is the degree to which, and the nature thereof, the adoption of a Neolithic economy and society affected the resource-based habitat and home range of the only surviving hominin species, *Homo sapiens*, and its interactions (niche space) with other organisms.

The underlying question is: how did the Neolithic Revolution, particularly growth of settlements and domestication, come about? Many ideas have been presented. The **resource-based habitat** view would suggest that a key factor would have been the diminution of resources and the need to control the quantity of both plants and animals (Stiner, 2001). There is evidence that the success of the new economy would have led to human population growth; in turn, this would have stimulated further development of ideas and production and development of settlements. Even so, the basis for the Neolithic economy and society was human ingenuity (psychology), cognition and religion (worship of the Mother Goddess) already clearly evident in the Upper Palaeolithic (Hodder, 2007). Changes in technologies (e.g. the wheel and pottery) stimulated new demands for new resources, thereby increasing competition.

Regarding the resource-based habitat, two issues are uppermost: (i) siting and growth of settlements; and (ii) construction and organisation of settlements and what they can tell us about society and economy. The first issue is the extent to which siting and growth of settlements are related to changes in the exploitation of food resources. During the amelioration of climate in the early Holocene, populations of humans were increasing, no doubt at the expense of animals hunted and plants gathered for consumption, and for the production of other materials (e.g. hides for clothing and for camp sites). A key feature of domestication and cultivation is likely to have been the proximity

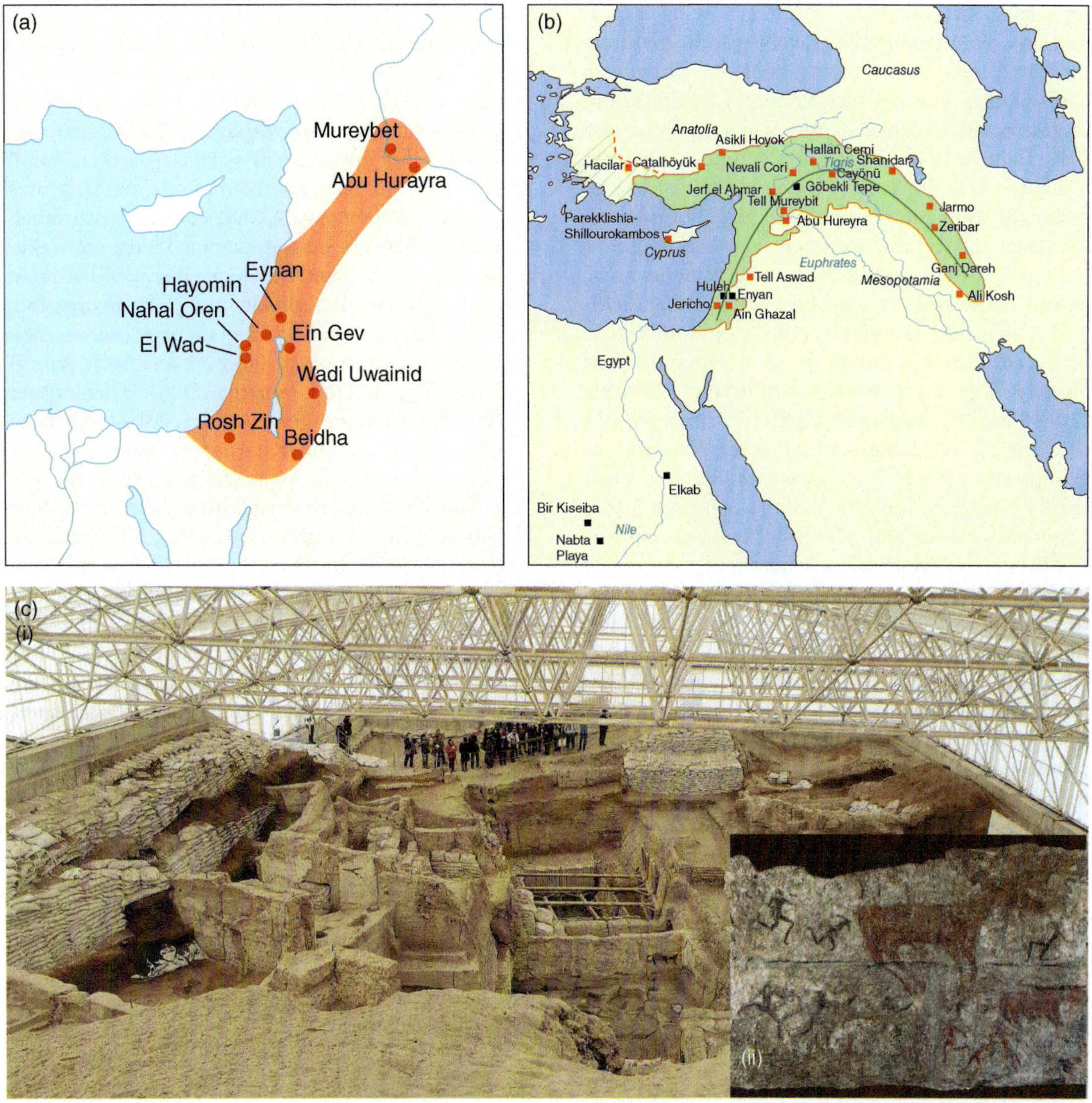

Fig. 9.9. Neolithic culture and settlement. **(a)** Natufian culture: a late Epipalaeolithic archaeological culture of the Levant, dating to c.15,000–11,500 BP. The Natufian communities are regarded by some as the originators of the first sedentary Neolithic settlements of the region, considered to be the earliest in the world. There is evidence of deliberate cereal cultivation, specifically rye, by the Natufian culture, at Tell Abu Hureyra (Syria) and Shubayqa 1 (Jordan), the site of earliest evidence of agriculture in the world, 4000 years before the emergence of agriculture in south-west Asia. (**b**) Map of the Fertile Crescent showing the main archaeological sites of the Pre-Pottery Neolithic period, 9500 BP. Black squares indicate pre-agricultural sites. (**c**) Çatalhöyük settlement (Turkey), 9500–8400 BP: (i) excavation site; (ii) deer hunt, detail of a wall painting from level III of Çatalhöyük (c.7700 BP). (Panel **a** by Crates: https://en.wikipedia.org/wiki/Natufian_culture#/media/File:NatufianSpread.svg, used under licence CC BY 3.0. Panel **b** by Bjoertvedt: https://commons.wikimedia.org/wiki/File:Fertile_crescent_Neolithic_B_circa_7500_BC.jpg, used under licence CC BY SA 3.0. Panel **c**(i) photo by Murat Özsoy 1958: https://commons.wikimedia.org/wiki/File:%C3%87atalh%C3%B6y%C3%BCk,_7400_BC,_Konya,_Turkey_-_UNESCO_World_Heritage_Site,_08.jpg, used under licence CC BY SA 4.0. Panel **c**(ii) photo from Digital Collections – University at Buffalo Libraries: https://digital.lib.buffalo.edu/items/show/34870.)

of settlements to these resources owing to: (i) mobility restrictions, with travel depending entirely on foot or use of animal draught (i.e. drawing a cart load, e.g. equines such as horses, donkeys, and mules; bovines such as cattle, water buffalos and yaks; Outram *et al.*, 2009); (ii) the time available in a day to carry out tasks; (iii) the danger of loss of resources from competition with neighbouring settlements; and (iv) the need to protect crops against wild animals grazing on them. If settlements arose before cultivation, it may be expected that fields and settlements were not as close geographically as observed. But one thing is clear: humans had already domesticated dogs before the Neolithic (20–40 ka BP; Botigué *et al.*, 2017); consequently, the process of domestication and protection of groups and their belongings was under way prior to the Neolithic and it is a short conceptual step from pastoralism and plant gathering to cultivation. As it is, evidence is accumulating of varied plant use in food, much of which may have been drawn from the surrounding countryside of a settlement and not necessarily from sown fields (Evershed *et al.*, 1991; Carretero *et al.*, 2017; Bishop, 2021; Bishop *et al.*, 2022; Romero, 2022). Furthermore, exploitation of food resources by hunter-gatherers may have been improved by pottery used for cooking, in addition to its use as cremation pots (Courel *et al.*, 2021; Vandiver, 2022).

The second matter is the construction and organisation of settlements, the development of lithic entities that may have been associated with social rites (e.g. Skara Brae and Stonehenge; Field *et al.*, 2015; Darvill, 2016, 2022; Bayliss *et al.*, 2017), defence (the headless people in wall art give clues to conflict; Iversen, 2016; Ahlström, 2012) and the development of a social hierarchy (Bogucki, 1999). 'Houses' do not seem to be solely places of dwelling (for resting, sleeping, eating, copulation and child bearing, home industry such as ceramics, wood and lithic implements, etc.) but also places for ancestors. Burying the dead under houses was common in early agricultural **villages** in the Near East (e.g. at Çatalhöyük, one dwelling alone in the honeycomb-like maze of dwellings had 64 skeletons: www.catalhoyuk.com/archive_reports/1997/ar97_12.html?utm_source=chatgpt.com/; accessed 19 May 2025). An interesting question is: how varied were Neolithic stonebuilt and woodenbuilt structures across the world, not just in the Near and Middle East (the Fertile Crescent), North Africa and Europe, but also diachronously across China, Mesoamerica (Mexico and Central America), South America (the Andes and the adjacent Amazon Basin), and North America (eastern United States) (Sadowski, 2017)? It is well known just how varied 'house' structures were within regions of Europe at this time. For instance, in northern Greece, small settlements on 'tells' (raised mounds marking sites of ancient settlements, <2 ha) contrast with much larger settlements on flat-extended sites and lakeside sites (e.g. Makriyalos in Pieria; Kleitos in Kozani; 10–50 ha; Bickle and Kalogiropoulou, 2017), whereas wooden rectangular houses, isolated or in clusters of various numbers, were typical of Neolithic Ireland (Smyth, 2006). Then, other structures (standing stones, circles, cists and cairns) and caves are considered to have been used for other aspects of daily living (religion: veneration of the Mother Goddess; waste disposal: cisterns; burial of community members; Hodder, 2007, but see Meskell *et al.*, 2008 for a reassessment of the figurines at Çatalhöyük). Settlements are often demarcated in some way, in northern Greece usually by ditches. In the Paris Basin (France), the early Neolithic longhouses, much as in modern New Guinea, are outlined by causewayed and palisaded enclosures (Marshall, 1979; Bickle, 2013); they tend to be constructed on gravel terraces above the floodplain on fertile soils, close to the main rivers. As for ecological surroundings, regions of both northern Greece and the Paris Basin were dominated by woodlands (Ntinou and Badal, 2000; Chartier, 2010).

There is a common factor running through all Neolithic settlements: their similarity to **metapopulation** models (i.e. each settlement may not necessarily have been permanently occupied, and effectively a patch for reproduction and survival, but used for a certain time, abandoned and possibly reused), a concept originally developed for other animals, especially butterflies (Hanski, 1997; see Chapter 7). Most activity probably occurred within the bounds of the settlements or just beyond them in fields and pastures (Furholt, 2018). However, constant movement also occurred between adjoining settlements linked to trade, 'marriages', meeting relatives, co-ordinated rituals, disputes among land holders or co-ordination of defences against 'outsiders' (e.g. Greece; Runnels *et al.*, 2009). Longer distance 'migrations' almost certainly occurred in trading between communities further afield (e.g. the hand axes in microdiorite from the Graig Lwyd (North Wales) factory and those in epidotised greenstone from the Langdale Pike (Lake District) factory are found as far as the Pennines and Midlands in the UK; Pitts, 1996). In the case of other animals, especially butterflies, food supply typically determines the outline of patches for local populations. In the case of humans, the animal itself makes a choice of a patch for settlement construction. Even so, the patch itself provides critical resources such as a suitable place to rest from daily activities, mating, preparation of food and its consumption, preparing tools and planning activities. But the choice of site for the settlement is almost bound to have been made on the basis of 'riches' (agricultural resources, growing food supplies) within easy access. The fascinating affiliation with metapopulation-like structures in invertebrates (see Chapter 7) is that the Neolithic may well form the beginning of a hierarchy of settlement sites. The opposite of such a metapopulation model would be the 'wandering model' (i.e. shifting camps), which takes into account temporal variation in patch arrangements (an aspect not considered in metapopulation models, which assume patches as spatially stable units; Dennis, 2020). This describes a hierarchy of home ranges for a **hamlet**, a township, specialist centres such as mines, a regional centre, a capital city, and thus resource-based habitats for each population, tribe, family and, most importantly, for the individuals occupying a region. It is perhaps the beginning of the development of walls, of which now there are numerous varieties for housing, boundaries, terraces, etc. (e.g. Algarve drystone walls in Portugal; Feroldi *et al.*, 2021), all of which provision a constellation of microhabitats as well as providing an epidermis for units, whether stable in shape and size or not.

The Metal Age, Medieval advances and the beginnings of modern industry

The Neolithic triggered the start of the rapid rise in population size, invention and discovery. It is not difficult to follow the broad pattern of settlements, industrial and administrative constructions in Europe emanating from Asia, then Greece and Rome; however, technological developments display a diachronous pattern over the Eurasian continent (Renfrew *et al.*, 1974). For instance, in Britain, the Stone Age gradually merged into the Bronze Age well after the Near East, the transition phase referred to as the Chalcolithic culture (Oakley, 1972). The situation becomes even more complex and varied if we enlarge our perspective to include cultural evolution in human populations throughout the world. As an example of how the **resource-based habitats** and **home ranges** changed through time in connection with socioeconomic evolution, we will briefly discuss only some major steps in British history after the Neolithic because this illustrates many key points well, but – *mutatis mutandis* – histories of other peoples and regions could serve the purpose equally well.

In Britain, the basic 'phases' are referred to as the Copper/Bronze age (c.5300–2700 BP), the (First) Iron Age (c.2700–2000 BP), the Romano-British Period (43–410 CE), the Anglo-Saxon-Viking Age (410–1150 CE), which extended into the Middle Ages, and then the Renaissance (14th century to the 16th/early 17th centuries), with which the Modern Age begins (with great distinctive social changes in most fields and disciplines, including art, architecture, politics, literature, exploration and science; Livet and Mousnier, 1980; Monfasani, 2015). Throughout this period, there were marked developments in technology and resource-based habitat demands, with the appearance of powerful hierarchies and 'tribal' defences (e.g. hill forts, such as Maiden Castle in Dorset and Old Oswestry in Shropshire; Lock and Ralston, 2022) becoming a distinct feature from the Bronze Age. The peak of development was attained in the Romano-British phase, when sophistication attained its zenith in Britain for the Roman commanders using slave and indentured labour of the native British (e.g. underfloor heating, saunas, indoor plumbing, beautifully tessellated floors and walls in stone-built villas, universal Roman coinage for exchange of goods and a money economy replacing natural economies and bartering; Birley, 2005; Cunliffe, 2005; Moorhead, 2013; Black, 2017). The RBHs and HRs were entirely different for rulers and ruled. Religion (beliefs), the form of constructions (tracks, roads) and burial rites all underwent changes in each phase, the most distinctive Roman settlements having an orthogonal layout and links via well-constructed straight roads. Organisation of 'forces' and weapon technology also reached a peak with the 'art' of warfare. Movement via trade and conquest covered the entirety of Europe (e.g. Bell Beaker pottery in the Bronze Age: Hamilton, 2002; Orton and Hughes, 2013; Romano-Britain Samian ware in 1st–3rd centuries: Willis, 2011), though the greatest mobility was restricted largely to the elite with their forces. New skills abounded and native populations were keen to benefit from discovery, but received but small measure of the Roman success, other than security.

With the decline of Rome, Britain entered a long devolutionary phase, dominated by mass incursions and violence from the continent and across Europe, a distinctive period of wide variances in RBHs and HRs and hence in living standards. Even so, restrictions on development and habitats were determined by limits to transport (i.e. by boat, ship, foot and horse) and territorial zonation (territories of clans of different peoples, many of whom prevented movement). Europe, a single unit under Rome in terms of resources and the population's demand for them, became fragmented into warring factions, a feature of post-Roman life. Most significant in Britain is the number of skills that seem to have been forgotten following the collapse of the Roman empire, with the loss of its highly developed culture (Ward-Perkins, 2006). Clearly, there would have been a contraction of RBHs to absolute essentials for most people in many places (i.e. restriction of walled buildings only for the superior elements of the social hierarchy, e.g. Norman castles; Renn, 1968), with the majority generally living in squalor (Gies and Gies, 2016). Territoriality, which focused on relatively small regions commanded by hillforts in the Bronze and Iron Age, had evolved into much larger mobile zones of antagonism in post-Roman times. Interestingly, recent findings (Cassidy *et al.*, 2025) indicate that matrilocal societies dominated in Iron Age Britain, suggesting a very distinctive pattern of RBH, in which female lines established the RBH bounds with scattered men fighting to maintain them. But the greatest RBH requirement was peace, without which it became extremely difficult for populations to establish settled living and equable trade.

In Britain, a peaceful period followed the defeat of the Vikings by Alfred the Great (after the Battle of Edington on 6–12 May 878); then again later in 1066 with the Norman conquest and the strict hierarchy that followed under the Norman hegemony (e.g. Grant and Fink, 1958). In this latter historic stage, there is a detailed record of **villages** and land ownership (Domesday Record; Baxter and Lewis, 2017). The majority of individuals were effectively indentured labourers (serfs) under the demesne of their lord (Fig. 9.10a), living in squalid conditions with a layer of nobles and church enjoying a much higher standard of living. Otherwise, a variety of occupations were concentrated in townships, leading gradually to a hierarchy of places, which seems to conform to the **central place theory** of geography (that is, settlements (**hamlet** to conurbation) are spatially organised in a hierarchical way, with 'central places' providing economic services to surrounding areas; Christaller, 1933; Lösch *et al.*, 1954) (Fig. 9.10b).

The Industrial Revolution and the Anthropocene. Profusion in resource-based habitats

The Industrial Revolution (including a 'Second Iron Age' or 'Steel Age'), from the 18th to the 21st centuries in Europe, propelled discovery and development, driving a massive surge of changes in human resource-based habitats and home ranges, and culminating in the **Anthropocene**. There is no room to expound the changes in detail, but to take the UK, the following features are prominent: (i) the dramatic changes in machine-based industry located in factories; (ii) the rapid nature of discovery; (iii) the sharp increase in novel 'goods' and luxuries; (iv) the huge improvement in constructions compared to Tudor buildings, with distinct architectural phases such as Georgian, Victorian, Edwardian and modern buildings, increasingly accompanied by indoor bathrooms and toilets, kitchens with gas/electric ovens and hot plates; (v) the astonishing advances in communications (metalled roads, then macadam and asphalt, canals, rail and locomotives, flights, especially commercial flights post Second World War, private cars, etc.), including communication technologies (telegraph, telephone, mobiles, radio, television, etc.); (vi) huge changes in classes, with access to public education (with literacy leading to increases in publications), votes for lower-class men and then women, and the recognition of LGBTQIA+ gender identities; (vii) extraordinary changes in health especially with the National Health Service; (viii) fundamental shifts in belief systems (e.g. from fixism to evolutionism; from almost universally shared religious beliefs to irreligion); (ix) significant changes in the treatment of miscreants, the end of capital punishment; (x) the huge demand for entertainment, especially sport; (xi) access to wealth through credit, freeing up talent in the population and generating greed (diffusion of **Veblen goods,** such as a Rolls Royce Phantom or carbuncle-sized diamond rings; Trigg, 2001) (Fig. 9.11); (xii) the spread of fashion in clothing, houses, vehicles and much else to the entire population. These changes are not exclusive

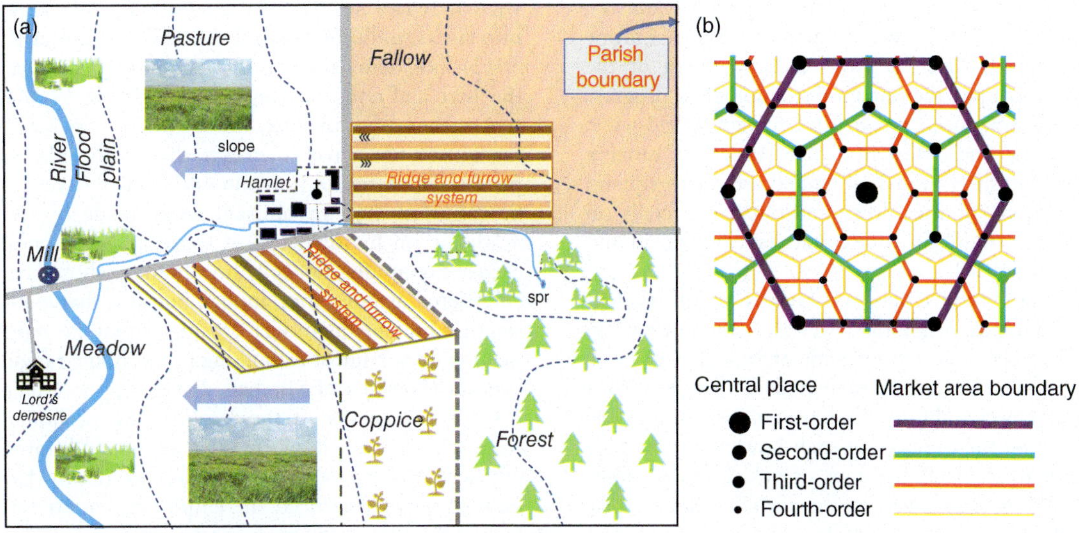

Fig. 9.10. Settlement patterns. **(a)** Basic model of Medieval British agricultural village system with strips of land shared out amongst villagers, and the land worked with communal effort under the overall ownership of the demesne lord. **(b)** Schematic model of settlement hierarchy termed Central Place Theory. Christaller (1933) pictured places occurring in such a hierarchy, divisions between 'places', occurring as hexagons, a shape most closely approximating a circle of land but avoiding 'empty spaces'. Settlements of higher order (1) are envisaged to cater for the needs of lower order settlements (i.e. 2, 3, 4, etc.) which lack goods at higher ordered locations (Goodall, 1987). Christaller's model is very simple, making a number of assumptions about places, resources, purchasing power and transport, etc. Although some landscapes approximate to the model, of course, diversity for these components and topography influence the pattern of settlements.

Fig. 9.11. Cityscape of the centre of first-order places: midtown Manhattan, New York, a city of Veblen goods and conspicuous consumption, where new resources are likely to be available for 'customers'. (Photo by S. Fattorini.)

to British modern history, but – in more or less different forms – involved many European countries, with the practices and culture of western Europe and North America now reaching much of the world. Despite the immense technological progress of our era, social inequalities remain world-wide, with the variability in living standards ranging from the sheer luxury of landed estates to squalid conditions of the 'down-and-outs', especially in low- and middle-income countries (e.g. Padgett and Priyam, 2017).

Variability in RBHs and HRs is plainly evident (e.g. amongst neighbours even in a single suburban middle-class *cul-de-sac* south of Manchester, UK). In most **urban areas**, even houses that date to a similar period differ in extensions, size of gardens (marking the territory of each household), boundary features, facilities (type of vehicles and equipment), etc., according to the social status of their inhabitants. Activities within and between urban areas follow complex patterns. 'Shopping' (purchase) usually falls within the nearest product-specific town and city resource outlets, but there are varying links between specific resource outlets and entertainment categories (e.g. sporting outlets: football teams). Occupations (e.g. education from basic to higher levels of academia, business, political outlets, religious places of worship) present various degrees of stress as they differ in salaries, places of work, amount of business and holiday home/foreign travel, if any, and the degree of interaction amongst neighbours. It is at this level that human families can be seen to have different **territories, home ranges** and **resource-based habitats**. Across the world, these components vary enormously (one-room flats to massive estates). Darwin was astonished by how rapidly we had developed as an intelligent organism, but equally horrified at the conditions adopted by Fuegian Indians on his circum-global expedition, with individuals curled up naked while sleeping on a beach in cold rain and sleet (Browne, 2002; Carey, 2005). This surely is no different from those pavement dwellers in London or Manchester or Delhi now. Even homes scattered in **villages** may lack obvious resources usually associated with them (latrines), as Lawrence shockingly described for a Sardinian village (Italy) on his tour there in the 1920s (Lawrence, 1997 [1921]); in fact, at a global level, most people use extremely simple, and often unhygienic, shared toilets, or have no access to a toilet at all (Fattorini, 2019). In this increasingly complex world, the unspeakable, human faeces, are abused resources, even in the UK where raw sewage is often discharged directly into rivers, polluting coastal areas (Waltner-Toews, 2012). For many people, it is enough to consider the 'here and now'. Among humans, thanks to current means of transport, home ranges may vary from just a few square metres to thousands of kilometres. The track of individual humans has yet to be followed, but an unusual study made by an academic on his own use of space and resources clearly exemplifies not only how humans should be considered as any other animal, but also that considering humans as model organisms may be productive for a fruitful rethinking of the home range concept (Powell and Mitchell, 2012).

Of course, many of the resources in the home come from outside it. During this long period starting with copper and tin goods, inventions by humans have produced an enormous variety of 'goods' and artefacts, many of which include technologies beyond the understanding of the majority of the population; even those used by extremely common devices like smartphones are so sophisticated that most of us do not have any idea of how they really work. A striking feature is that humans now regard as critical resources components that were unheard of in the 19th century and earlier (e.g. rare earth elements used in high-tech devices, such as laptops, mobiles and televisions).

One thing is clear to the ecologist: the alarming impact of population growth. In the current fast-developing age where population is growing to a point beyond the exponentials (Taagepera and Nemčok, 2023), human settlement has certainly taken on a hierarchy (Christaller, 1933; Lösch *et al.*, 1954) (see Fig. 9.10b). The home range for many humans now includes movement across the entire planet and the mixing of populations that developed in relative isolation for millennia.

Need, greed and war: the resource-based habitat and the human dilemma

The main human dilemma is the steeper than exponential growth in human population. All else is dependent on it: the needs of families, the greed beyond need that accompanies human behaviour, the fall-out when resources fail. Human 'needs' have been the focus for a number of researchers. Most familiar is probably Maslow's **theory of hierarchy of needs** (Maslow, 1943, 1987), set out as a pyramid (Fig. 9.12). The pyramid progresses from

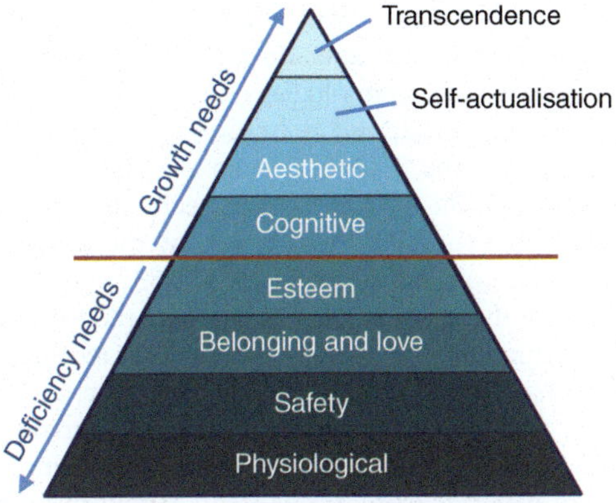

Fig. 9.12. Maslow's pyramid, a theory of the hierarchy of human needs (Maslow, 1943, 1987). Maslow's hierarchy of needs is a simple model to explain motivations for different aspects of human behaviour. Each rising level represents different human needs (from physiological needs to transcendence). Achieving transcendence implicates achieving conditions beyond ordinary limitations (i.e. a spiritual, religious state or drugged state beyond normal reality). Deficiency needs concern basic survival, arise due to deprivation and motivate people when they are unmet; growth needs are more psychological, do not stem from a lack of something and are associated with the desire to grow as a person. (Based on McLeod, 2024: www.simplypsychology.org/maslow.html.)

basic physiological needs at the base towards social and psychological needs at higher levels. In resource-based habitat terms, these needs may be reduced to essentials: (i) consumable resources (such as air, water, food); and (ii) utilities (the most typical are places to rest, find shelter, copulate or work – to produce cash or immediate products on which to survive or to be used for education, entertainment, healthcare, etc.). Of course, Maslow's pyramid includes necessities that were not relevant when butterflies were used as the model organism by one of the authors (Dennis, 2010) (e.g. technologies, holidays, friendships, social and self-esteem, confidence, respect, inventiveness), but these are now intimate elements in human health. Thus, the RBH tied into home range becomes a far more complex notion for humans than when dealing with arthropods, the initial target of interest for the development of the RBH concept.

Greed extends beyond need. It may be visualised as an intense intraspecies competitive strategy for survival. It has a *raison d'être* in storing up resources for 'drought' periods (viz. the so-called maximisation resource-utilisation principle). Earlier it was noted that butterflies engage the principle by building up resources for winter hibernation and summer aestivation (Dennis, 2010). Humans never 'stop' grasping, nor do they have enough of any resource; wealth is required, beyond all necessity for survival, as an indication of power and success, but then to lack it can be to lose all. It is not difficult to conceive how communal 'need', extending to communal 'greed', in a growing population is causing problems for human planetary stability. It is worth mentioning here that we do not 'destroy' the Earth, but we do destroy what is essential for ourselves and a vast number of other creatures.

Yet, not only is excess population growth in humans leading to the unravelling of the resource base and the environmental conditions on which humans depend – the stable state of the **hydrosphere, lithosphere** and **atmosphere**; it is also regarded as the underlying cause of the loss of global biodiversity and potentially human extinction, the latter the result of resource exhaustion and wars. Wars have accompanying disasters: famine, epidemic diseases, the poisoning of water supplies, etc. Greed leads to conflict, which on a grand scale is war. At its nuclear extreme, we will drown in our own blood, urine, excrement and carbonised remnants; the planet, like a stopwatch, after our disappearance will be reset ready for a stage of saprophages and detritivores. Evolution will probably 'rekick' on fungi, **carrion** eaters and greater radiation-tolerant animals feeding on decaying matter, as after the last major extinction event at about 66 Ma (Solé *et al.*, 2002; Fattorini, 2020). The current failures in diplomacy and widespread famine would lead us to predict an increase in wars, in number and scale. In this section, attention is given to this unique human property, our hatred but also love of war, and we focus on how it may relate to the human resource-based habitat. The **Anthropocene** will be recognisable to any residual intelligent beings in several hundred million years as a thin layer of distinctive molecules (especially plastic materials) and a bone bed from mass war graves.

War has antiquity in human populations; it dates back probably to the Palaeolithic and certainly the Neolithic (Price *et al.*, 2006; Bowles, 2009; Pinker, 2011; Ferguson, 2013). The Bible is full of accounts of war. The prime elements of war are need, greed and aggression leading to violence (Kissel and Kim, 2019). Regardless of the triggers of war, the significant endproducts are, for the victors – presuming there to be victors – an expansion of their resource-based habitat, the treasures of success, *Schadenfreude* (pleasure in another's misfortune); for the losers, shame, slavery and, and in extreme cases, cultural or even demographic extinction (e.g. the near-eradication of the Tasmanian aboriginals, the profound cultural disruption and depopulation of the Taino people of the Caribbean, and the large-scale destruction of indigenous groups during the colonial expansion into the Americas, Africa and Australia). Success may result in land gain or some other component of wealth (humans as slaves; Fig. 9.13), as retribution for the other side's failure. The sick reality of it all is that wars often rage between closely related peoples (e.g. families fighting on opposite sides of the 17th-century British Civil War and the 19th-century American Civil War). A weak population rapidly becomes an extinct population; the evidence is in the virtual extinction of American natives and Australian aborigines, and the subjugation and removal of many of the Earth's peoples, a term suitably covered by 'genocide'. An understandable reaction is revenge, which becomes a palpable and understandable 'resource' in the human species. Currently, considerable effort is being given to understanding the beginning of wars, why they arise and how they may be ended with joint satisfaction to the conflicting sides; there is much in what John Fowles has to say about the root of violence in '*nemo*', the sense of nothingness that humans feel in the face of finiteness (Fowles, 1998).

The extent to which researchers agree on the labelling of conflict depends substantially on the definition of the key terms 'aggression', 'violence' and 'war'. In this work, war is defined simply as intercommunity violence and includes intertribal 'displays' as in New Guinea tribesmen (Sillitoe, 1972). We have purposely left out 'humans' in the definition of war – although the focus of the chapter is 'humanity' – as wars are not exclusive to humans, being for example common in ants (Box 9.1). Battles in ants (between colonies of the same species or of different species) are legendary (Hölldobler and Wilson, 1990; Whitehouse and Jaffe, 1996; Lymbery *et al.*, 2023) and their similarities with human wars are fascinating. For example, Henry David Thoreau (1817-1862) was profoundly impressed by what amounted to a state of ruthless war between two ant colonies, small red and large black ants, in his wood pile – a war he compared with those fought by humans (Thoreau, 1854). War can be subdivided into various types, with classification schemes depending on which specific aspect is considered, such as the aims, technologies used, extent and participants involved, and the cultural context (see, for example, the transition from the 'Wars of the Knights' to the 'Wars of the Technologists' in Europe; Howard, 2009). Ultimately, the absolute war (Carl von Clausewitz, 1943 [1832]) has the intention of complete extermination of the foe. The victor takes all the spoils (remaining resources, assuming any are left).

The focus here is actually something narrower than a review of all the complex factors leading to human war (Jackson and Morelli, 2009): in brief, the resource-based habitat factors involved in aggression, violence and warfare. To understand this viewpoint, we need to appreciate the significance of evolutionary steps leading to violence: interindividual aggression (e.g. neighbours arguing over a few centimetres of their garden boundary, exploding into abuse and physical tit-for-tat action, whose consequences would be only ridiculous if they were not a manifestation of the most depreciable aspects of our behaviours, like the case that hit the headlines of a boundary tree on a front drive sliced in half over a dispute about bird song, www.independent.co.uk/news/uk/home-news/sheffield-tree-bird-neighbour-dispute-b1871812.html; accessed 19 May 2025), intrafamily unit membership aggression, intergroup aggression, interspecies aggression, the escalation of inter-group aggression to inter-group violence, and the place of 'sharing' resources and conditions in triggering conflict (e.g. the sense of being 'cheated' of chattels following family deaths). Greed begins with the human child: 'I want what he's got'. To which a sage relative would reply 'Much wants more'.

From the above issues, some simple points emerge: (i) aggression, however defined, is a basic feature of existence (i.e. predators violently kill prey; prey attempt to injure predators; closely related herbivore individuals compete for limited, or superior, resources (e.g. the largest fruit, the best nest site, the most attractive mate); (ii) aggression often leads to violence, especially in well-matched individuals and tribes. The violence is often recorded in personal damage (e.g. bruises and bone breakages in a pub brawl) and occasionally death. Even butterflies suffer damage from territorial bouts (e.g. impacts of male large skipper *Ochlodes sylvanus*; Dennis and Williams, 1987). These points are plainly obvious from observations in primate

(a)

(b)

Fig. 9.13. Particular human means of action and communication for resource acquisition. **(a)** War, typically when disputes occur over home ranges and resources. *The Phantom Horseman* (1870–1893) by Sir John Gilbert (1817–1897) gives an unforgettable impression of hell on the battlefield. **(b)** One of the numerous justifications for American slavery was the belief that the supposed 'benevolent paternalism' of planters was beneficial or essential. The illustration is taken from the *American Anti-Slavery Almanac* for 1840. (Panel **a** photo from Birmingham Museums Trust, used under licence CC by 0PD, https://dams.birminghammuseums.org.uk/asset-bank/action/viewAsset?id=17338&index=0&total=1&view=viewSearchItem. Panel **b** from https://commons.wikimedia.org/wiki/File:Anti-slavery_almanac_1840_detail.jpg.)

research (i.e. from play to lethal aggression between chimpanzees and gorillas; Sanz *et al.*, 2022). War is simply an extension of such aggression and violence, action conveyed over evolutionary time to a larger scale of groups, troops, clans, tribes, ultimately in the modern era to spatially unbounded (or bounded) religious sects and political states. Konrad Lorenz and André Leroi-Gourhan argued that the development

Box 9.1. War in non-human animals

War is typically defined as an organised, armed conflict between groups, often driven by political, territorial or ideological motivations. Key characteristics of war include co-ordination among members of a group, intentional aggression and a shared objective. This definition presupposes a degree of planning, strategy and often symbolic meanings that vary among human cultures, all aspects that are absent in other animals. However, many species of non-human animals exhibit behaviours that resemble some aspects of human warfare, particularly in terms of group conflict. For example, chimpanzees (*Pan troglodytes*) are known to engage in lethal intergroup raids, where a group of males from one community will invade another's territory, targeting individuals they outnumber (e.g. Sandel and Watts, 2021). These raids often result in death and are believed to be motivated by the desire to expand territorial control or access resources such as food and mates. Other animals, such as ants, also exhibit large-scale conflicts that appear strikingly war-like (Whitehouse and Jaffe, 1996; Moffett, 2011; Champer and Schlenoff, 2024). These conflicts involve organised attacks, the use of specialised roles within the colony (e.g. soldiers) and significant casualties (Fig. B9.1.1). However, these behaviours lack the cultural and symbolic elements usually implied in human war.

The cognitive abilities of non-human animals are a key consideration in determining whether their behaviours may really qualify as 'war'. Humans plan wars with foresight, devise strategies based on shared goals, and often imbue warfare with cultural and ideological meaning. While chimpanzees and other animals exhibit planning and co-operation, their actions are typically driven by immediate environmental pressures rather than abstract concepts or long-term strategies. Moreover, war in human societies often involves alliances, diplomacy and the manipulation of social constructs, which are absent in non-human animal interactions. For example, while ant colonies may fight over resources, they do not negotiate treaties or form alliances with other colonies. For these reasons, many zoologists prefer not to use the word 'war' when describing animal conflicts. Although labelling animal behaviours as 'war' can lead to anthropomorphism, recognising parallels between human and non-human conflicts can deepen our understanding of the evolutionary roots of aggression and co-operation.

Fig. B9.1.1. Ant colonies engage in battles in which they use various tactics. In battles over food, the closer colony can have an advantage because of faster recruitment and shorter travel time. As a result, a smaller colony (black) may secure a close food source against a larger colony (red). (Based on Champer and Schlenoff, 2024.)

of hunting techniques and technology in primitive societies played an essential role in reducing inhibitions to intraspecific killing, to develop 'human hunting', ultimately to war (cf. Berruti and Ruzza, 2016). Often more than two countries are involved in wars as both seek allies in their quarrel. Once started, they are ferociously difficult to terminate as attempts to do so may well result in 'a loss of face' (reputation, honour, resources), a humiliating prospect for an aggressor. In war, a key feature is complete lack of concern for the basic resource of action – the troops; bear in mind that the larger the state, the weaker the familial link between those contending an issue and those used as cannon fodder expended to force it (Fig. 9.13). The only thought given to the troops by those engineering war is to minimise their losses and maximise their fitness to fight. Real concern for the key resource (troops) by the leadership is as irrelevant as it was for Thoreau's ants, namely to ensure sufficient 'force' to follow through on future battles.

War is not something that humans have invented in recent times, it is a disease embedded in the primate psyche (i.e. our omnivorous ape ancestors); Victorians faced with Darwin's 'Origins' had every right to see hopelessness in our primate ancestry.

It is the product of the way *all* organisms have behaved, one to another, through geological time. However, differently from other creatures, humans have undergone a cultural evolution which made them able to reflect on their own actions. As a species, we have virtually complete control over our behaviours and so we are responsible for actions, although circumstances may force individuals towards particular conducts. Thus, our biological origins cannot be used as an excuse for our deplorable behaviours (violence, egoism, greed, etc.). Furthermore, it would be completely wrong to think of a dichotomy between nature and culture; culture, after all, is nothing more than the product of our neurons' interactions with our surroundings and therefore is itself part of our biology. It is our responsibility to decide whether to follow virtuous paths suggested by millennia of ethical reflection or to continue to justify our most abject behaviours as 'natural'.

The difference displayed by humans and other primates is one of extremes. Whilst apes are capable of substantial intergroup violence, humans – probably as a consequence of sheer numbers, group mass and demands for so many more resources – frequently resort to war. The success or failure of different populations in contests is put down by Diamond (1997) to geographical advantages/disadvantages, but it is more complex than environmental determinism, a set of physical factors (resources or conditions), and relates to personal opportunity and degree of greed, aggression and violence, which may emerge anywhere stochastically; no population of humans is a delightful community. While we may reasonably argue that significant numbers of individual humans are compassionate 'givers' or 'doves', many are latent psychopaths or 'hawks' (e.g. the homicidal maniacs of the French Revolution); possibly as many as 30% of the global population have psychopathic tendencies with substantially reduced empathy (according to the association PsychopathyIs; DeAngelis, 2022). But under certain circumstances, virtually entire peoples may lose any empathy and turn into highly organised societies of torturers and exterminators, with their enthusiastic or silent supports. Wars, with their accompanying obscenities (genocides, Holocaust camps), earn for humans the title of the most selfish, thoughtless, vicious, vile and revolting species on the planet. These characteristics of human behaviour feature highly in popular music and films; so much of our entertainment is based on violence and war. We should perhaps reflect on Milton's warning to the Parliamentarian General Fairfax: 'For what can war but endless war still breed?' (Carey, 2020). The resources gained may soon be lost. A sign that we grow in civilisation will be a rejection of violence and war. But the evidence would suggest that there is no hope of that occurring, despite some optimism outlined by Sigmund Freud in a reply to a letter from Albert Einstein in the summer of 1932: 'Perhaps our hope that these two factors – man's cultural disposition and a well-founded dread of the form that future wars will take – may serve to put an end to war in the near future, is not chimerical. [...] Meanwhile we may rest on the assurance that whatever makes for cultural development is working also against war' (Einstein and Freud, 1933).

In the end, the question is one of inequality, as powerfully expressed by Gandhi.

> *I suggest that we are thieves in a way. If I take anything that I do not need for my own immediate use and keep it I thieve it from somebody else. [...]. If only everybody took enough for himself and nothing more, there would be no pauperism in this world, there would be no more dying of starvation in this world. But so long as we have got this inequality, so long we are thieving.*
>
> (Mahatma Gandhi, 1960)

10 The Impact on Habitats in the Anthropocene

Abstract

Current extinction rates match those of mass extinctions in the geological past. Human imposition of a colonial relationship on the non-human world is the cause of this ongoing mass extinction and alteration of ecosystem functioning. Anthropogenic transformations (habitat loss, overexploitation of natural resources, pollution, climate changes, species introductions, etc.) negatively affect resources and conditions for species survival. Loss of species may profoundly alter ecosystem functioning, eventually causing ecosystem collapse with catastrophic consequences for our own species. The most important obstacles to biological conservation are political, not difficult-to-solve scientific problems. The conservation of the planetary diversity, and therefore of our own species, requires a profound revolution of thought; the human attitude towards nature (one of exploitation and domination) is entirely incompatible with the preservation of life on Earth.

Conservation: setting the scene for the Anthropocene

It should be obvious to the reader by now that there is a clear distinction between what is traditionally referred to as 'habitat' (traditional habitat, TB) and the concept of a **'resource-based habitat'** (RBH). Even so, despite the distinction between them, every species within a community has its own resource-based habitat, varying over time and space (viz. having a temporal and spatial variance); each species also features in various traditional habitats, and in them may be more or less abundant. The crux is that both traditional and resource-based habitats have fundamental parts to play in conserving any species and the biodiversity of sites; both may be subject to losses (or gains) induced by human activity more than by other organisms (Table 10.1).

Prior to an appraisal of the impact of humans on global wildlife, it is useful to raise some basic points concerning conservation that emerge from the above observations. First, the resource-based habitat of a species can only be extracted by very detailed and expansive survey, over the different spaces (localities, regions) where it occurs. Therefore, it is a 'method' best directed at 'rare' species that cover small areas or regions, the equivalent of 'A & E' units in hospitals for individual human beings.

Second, in the long term, the key directive is to preserve as many species as possible in a period in which so many species are being lost so quickly. This is no longer disputed, but accepted as fact; with 7.5–13% (150,000–260,000) of c.2 million known species having already become extinct since 1500 CE (Cowie *et al.*, 2022), we are facing a global, anthropogenic biodiversity crisis, representing a sixth **mass extinction** event, after the Ordovician–Silurian, Late Devonian, Permian–Triassic, Triassic–Jurassic and Cretaceous–Palaeogene extinction events (Barnosky *et al.*, 2011; Kolbert, 2014; Fattorini, 2020; see Appendix 5). This has not just affected birds and mammals (the most documented cases) but has undermined invertebrates (for which, in general, information is less adequate to estimate extinction rates; Fig. 10.1) and plants, though they may have suffered less (Cowie *et al.*, 2022). As it is impossible to preserve species without knowing their habitats, it is imperative to improve our understanding of species rapidly. Perhaps the most degrading example of human behaviour to other animals is trophy hunting. The demand for animal body parts is a component of a huge international 'trade' (Lindsey *et al.*, 2007): as many as 1.7 million hunting trophies were traded between nations between 2004 and 2014, and at least 200,000 trophies from threatened taxa, or an average of 20,000 trophies per year, were traded between nations in the same period (IFAW, 2016). Up to 25,000 animal parts have been brought into the UK as hunting trophies since the 1980s, according to the database of the Convention on International Trade in Endangered Species (CITES) (McKiernan, 2024).

  Habitat (R.L.H. Dennis & S. Fattorini)
DOI: 10.1079/9781800627505.0010

Table 10.1. Key areas of human activity regarded as causing biodiversity loss.

- **Climate change.** Human-induced climatic warming and changing rainfall patterns; these alter species ranges and the underlying water and chemical cycles which define current ecosystems.
- **Habitat loss.** Through pervasive incremental encroachment such as that caused by urban sprawl.
- **Hunting and the overexploitation of species.** For instance, the wildlife trade and wild capture of fishes.
- **Deforestation.** Tropical rainforests are particularly rich in biodiversity and are being destroyed.
- **Intensive farming.** Many species are threatened by agriculture, removal of woodland and hedges, etc.
- **Water use.** For instance, in some of the largest water catchments in the world, dams and irrigation reduce water flows.
- **Invasives.** The spread of invasive species and diseases through trade and travel.
- **Chemical pollution.** For instance, that associated with widespread pesticide and herbicide use, overuse of fertiliser and radioactive substances.
- **Materials pollution.** The virtually indestructible plastic waste (microplastics now pervade all systems); the long-term effects on biodiversity are unclear.

(Adapted from The Royal Society: https://royalsociety.org/news-resources/projects/biodiversity/human-impact-on-biodiversity/; accessed 20 May 2025.)

Third, it would be trite to point to the necessity of survey and study before managing a **landscape** for a flora/fauna, unless due to ignorance and/or laziness, lack of funding or acceptance of sheer complexity, randomly wilding a landscape is chosen as a technique over a planned deterministic taming of it (cf. **agricultural wilding**, that is, preserving wild crops and plants for agricultural purposes, versus **rewilding**, that is, restoring the natural uncultivated state of an area; Macdonald and Feber, 2015; Vogt, 2021; Tree and Burrell, 2023). Carver *et al.* (2021) identified ten guiding principles for rewilding that recognise that this ploy exists on a continuum of scale, connectivity and level of human influence, the objective being to restore **ecosystem** structure and functions in order to achieve a self-sustaining autonomous nature. These principles are: (i) utilise wildlife to restore trophic interactions; (ii) employ landscape-scale planning that considers core areas, connectivity and coexistence; (iii) focus on the recovery of ecological processes, interactions and conditions based on reference ecosystems; (iv) recognise that ecosystems are dynamic and constantly changing; (v) anticipate the effects of climate change and where possible use this information as a tool to mitigate impacts; (vi) promote local engagement and support; (vii) integrate science, traditional ecological knowledge and other local knowledge; (viii) adopt adaptive strategies based on monitoring and feedback; (ix) recognise the intrinsic value of all species and ecosystems; (x) change paradigms for the coexistence of humans and nature.

Although there is a powerful feeling that there is no hope of fully restoring wildlife and ecosystems on a very large scale, studies are being increasingly focused on the ecosystems contained within the envelope of habitat spaces, and rewilding projects are gaining much attention (Carver *et al.*, 2021); the crux of this is determining what makes for a particular diversity. Without solid information on ecosystem characteristics and functioning, how can we try to restore them? Examples exist of detailed sampling and analytical strategies (e.g. Bässler *et al.*, 2009) that should be applied before and after any management programme. We remind the reader that changes will always occur for any single species or a diverse **community** in any space even if 'man' is not present (responsible for the changes). The very nature of **vegetation succession** ensures this as a basic law of ecology (see Chapter 3). The potential of these natural changes must be carefully considered in management programmes (e.g. using/altering ecological succession in restoration; Walker *et al.* 2007). Scientific findings, scientific judgement and the precautionary principle are important elements in restoration programmes, but the behaviour of people and the timing of decision making will drive the process (Johns, 2019).

Fourth, and there is no doubt about it, the human species in its various cultures is responsible for the rapid alterations in water and air and on land, including changes in the distribution of species, with species **introduction** being one of the major threats to biodiversity conservation (Primack, 2014; Strona, 2022). Therein lies the paradox of losses and gains in our chaotic world: the poignant human equivalent is a raging war while medics desperately try to save innocent victims and combatants. We now need to medicate what we have injured, before the organism dies. By that, we mean the loss of biodiversity; the planet will continue whatever we do to it. Thus, another crucial question

Fig. 10.1. Current sixth mass extinction casualties. Much attention is paid to vertebrate extinction but invertebrates are menaced as well; simply, they are less known. The legendary giant earwig (*Labidura herculeana*) is an extinct species endemic to St Helena, an island rich in imperilled insects (Gerlach *et al.*, 2024). On St Helena, there are about 390 native terrestrial insect species, 310 of which are endemic; 70 endemics have been red list assessed and 87% are at risk of extinction. Although only two species are listed as extinct (the giant earwig and the St Helena darter *Sympetrum dilatatum*), six species are potentially extinct. Principal threats are habitat loss, invasive predators, competitors and pathogens, and climate change. (**a**) Male and female of *Labidura herculeana*. Scale bar: 1 cm. Specimens from the Manchester Museum, UK. (**b**) Prosperous Bay (a place once inhabited by this insect) during the construction of St Helena Airport. Habitat loss, by the removal of nearly all surface stones for construction, is a potential cause of the extinction of this species. (Panel **a** from Miles, 2015, used under licence CC BY 4.0. Panel **b** photo by S. Stanely: https://en.wikipedia.org/wiki/Prosperous_Bay_Plain#/media/File:Prosperous_Bay_Plain_Airport_(16512903591).jpg, used under licence CC BY 2.0.)

is: do we have the resources, even the concerns (is it even necessary?), for ensuring global ecosystem homeostasis (holding a steady state in biodiversity), or should we just do what we do – mostly destructive – and let all wildlife adapt or more likely die at our hand? Chris D. Thomas (2017) bravely if brutally addresses this issue, arguing that many human-driven changes may actually increase biodiversity by creating opportunities for species redistribution and accelerating evolutionary processes, particularly in fragmented and human-altered landscapes (see sections below: *Invasives: Williamson and Fitter's 10th of a 10th of a 10th rule* and *Winners and losers in the Anthropocene*). If we do intend (have the will and resources) to conserve biodiversity and single species, what planned methods (techniques) are needed to establish suitable habitats for **diversity** and whole **ecosystems**, and suitable **resource-based habitats** for any one particular species? Techniques and approaches are legion (see Soanes *et al.*, 2023 and their audit of management of Australian cities, 318 actions across nine contexts!). Of course, variable approaches are necessary in different settings, but do the involved parties (research centres, stakeholders, local administrations, governments, environmental organisations, etc.) communicate with one another?

Fifth, both approaches (the TB and RBH) (should) involve a cross-national effort. Imagine, if you will, the state of neighbour relationships on the Earth in our alleged 'civilised times' (at the time of writing there are 56 active conflicts; Institute for Economics & Peace, 2024) and answer the question: are different states prepared to join in any conservation endeavour when it is plainly obvious that habitat destruction and air pollution continue unabated? Personally, we think this unlikely, even in the long term; existential selfishness dominates collective concerns, though as global crises threaten stability we are prepared to be – we hope to be – proved wrong.

As a final comment, a very important issue should strike the reader: the matter of legality (Suprunova, 2023). It is absolutely clear that virtually every substantial parcel of land (and associated inland waters and air space) falls legally or illegally within the domain of a 'state', whether directly administered or indirectly through private, public or communal ownership (exceptions are Antarctica, international waters and international airspace). Practically every meadow, every wood, every lake, thus every traditional habitat, soon the ocean floor, etc., is owned/exploited by an individual, a state or an organisation (a parish, a county council, a charity, a business company, etc.). Thus, it is possible to protect the organisms on these sites only if the 'owners' can be encouraged or compelled to have concern for the wildlife of their domain. However, no such neat boundaries, be they physically obvious (fences, walls) or drawn up officially (e.g. national sea zones officially depicted on NOAA nautical charts; www.noaa.gov/maritime-zones-and-boundaries, accessed 20 May 2025), can encircle the space-time shifting resource-based habitat for a vast number of species (consider, for example, the obvious case of migratory species; see Chapter 8). The human element of co-operation is paramount but sadly not much in evidence, and yet advice has been plentiful (Millhauser and Earle, 2022).

Now, a view of the Anthropocene and human impact on the planet.

The Anthropocene and prime threats: climate change and habitat loss

The **Anthropocene** is a geological unit, much like the Holocene, though there are those who would throw doubt on this distinction (Walker *et al.*, 2024) and it is not formally recognised by the International Commission on Stratigraphy or the International Union of Geological Sciences (see Box 10.1 for definitions based on various start dates). The 'bottom line' is that the Anthropocene is an episode – the temporal scale ahead of us must remain doubtful – distinguished by fundamental changes to the Earth system (**lithosphere-hydrosphere-atmosphere**) (Zalasiewicz *et al.*, 2010), primarily in response to the massive growth in numbers of the dominant species (which might more suitably be named *Homo insapiens*) into billions (with an estimated 9.2 billion by 2050), with their megacities and agricultural-industrial infrastructure demolishing natural ecosystems (Bongaarts, 2009) (Fig. 10.2a,b). It is true that with increasing economic development, there is a tendency in human populations towards a demographic stabilisation, from high birth and death rates to low birth and death rates, a phenomenon described as the **demographic transition model** (DTM; see Chapter 2, Fig. 2.5). Causes for decreasing fertility rates in many countries are multiple, including broader societal shifts regarding parenting and gender roles – see, for example, Hegarty (2024). But although the global population is predicted to follow this model (Fig. 10.2c), we cannot rely on the demographic transition to address the ongoing ecological crisis, if only because the global human population has already pushed many global systems beyond recovery (Samway, 2019). Humans, billions of them, are the ultimate cause of the 'sixth mass extinction'.

The Anthropocene is witness to massive terraforming and mining (erosion and sedimentation) of the Earth's surface. Satellite imagery has established a global dataset of more than 6000 surface excavations between 2000 and 2017 for a wide range of

Box 10.1. Defining the Anthropocene

It seems that the word 'Anthropocene', or something equivalent (anthropogenic era), appeared for the first time in the Soviet Union around 1920, possibly introduced by the geologist Alexei Pavlov (1854–1929) (see Häusler, 2018; Brookes and Fratto, 2020). It may well be a synonym for the Anthropozoic (Stoppani, 1873), but it is not a synonym for the **noönsphere** (Vladimir Vernadsky in Le Roy, 1927), which has a philosophical basis (sphere of cognition and reason) and which in many ways describes the human ecosystem. The Anthropocene has a physical basis and can therefore be expected to emerge in geological strata, the start date for which is considered by many to be 1950, approximately the date at which so many unique human markers emerge in deposits (e.g. lead, plutonium, plastics, etc.; Zalasiewicz *et al*., 2010; Steffen *et al*., 2011; Gałuszka and Migaszewski, 2017). There is, however, good reason to apply the date to the first atomic test in 1945, with the world-wide dissemination of the 'golden spike' of plutonium-239 best encountered in all undisturbed sediments between 30 and 60 degrees north of the equator (Waters *et al*., 2015) (Fig. B10.1.1). Crutzen and Stoermer (2000) triggered the modern use of the term and concept, regarding the beginning of the Anthropocene as the start of the Industrial Revolution, with massive burning of coal, changes to atmospheric gas composition (i.e. increases in sulphur dioxide, carbon dioxide and methane, and synthetically fixed nitrogen) (Zottola and de Majo, 2022). The initial domestication of plants and animals, and the development of agricultural economies and landscapes (roughly 11,000 to 9000 BP), have also been proposed as marking the beginning of the Anthropocene epoch, as these transformations provide evidence of significant human capacity for ecosystem engineering or niche construction behaviours (Smith and Zeder, 2013). It is possible that the period of degradation in biodiversity began even earlier, as it now seems clear that megafaunal species richness and community biomass have dropped by about 71% and 95% respectively since the end of the Last Interglacial (c.127,000 BP), that is, before *Homo sapiens* fully occupied Europe (Davoli *et al*., 2023).

Fig. B10.1.1. The radiological signature in the geological record left by detonations of nuclear weapons has been proposed as a marker for the Anthropocene, with the first A-bomb test in 1945 as a starting point (Monastersky, 2015). The picture is a screen capture from 1959 colour footage of atomic bomb tests at Camp Desert Rock, Nevada Test Site, Nevada. (From US Army. Prelinger Archives: https://commons.wikimedia.org/wiki/File:U.S._Army_-_Atom_Bomb_Effects.webm, used under licence CC0 1.0 PD.)

minerals covering some 57,277 km^2 (Maus *et al.*, 2020). However, as humans have been responsible for scraping, covering and disturbing (directly or indirectly) virtually every square metre of the Earth's surface, agriculture being a prime example, many 'excavations' are too indefinite in resolution to be included in this audit. At the lowest end of the scale are the numerous pits, hollows and humps created by humans; examples range from burials, archaeological digs and geological excavations to the creation of ditches, ponds and gardens; levelling ground for building; long barrows; hummocks; an infinity of micro-landforms and so on. Extraction of marl from terrains of lacustrine origin (the location for charophytes, macroscopic green algae that played an important role in the precipitation of lacustrine calcium carbonate; Pełechaty *et al.*, 2013; Schneider *et al.*, 2015) is a classic example of small excavations in Cheshire (UK); these excavations subsequently formed important ponds for great-crested newts *Triturus cristatus* and dragonflies (Grayson, 1992-93; Pentecost, 2010). At the other end of the scale are quarries of gigantic proportions created in the process of extracting minerals (e.g. the massive opencast Rhenish lignite extraction at Ville, Germany; Henselowsky *et al.*, 2021; vast opencast coal mines in China; Zhang *et al.*, 2020). Then, it is easy to forget the innumerable shafts and adits for the same purpose (e.g. coal, lead, copper, tin extraction; global lithium

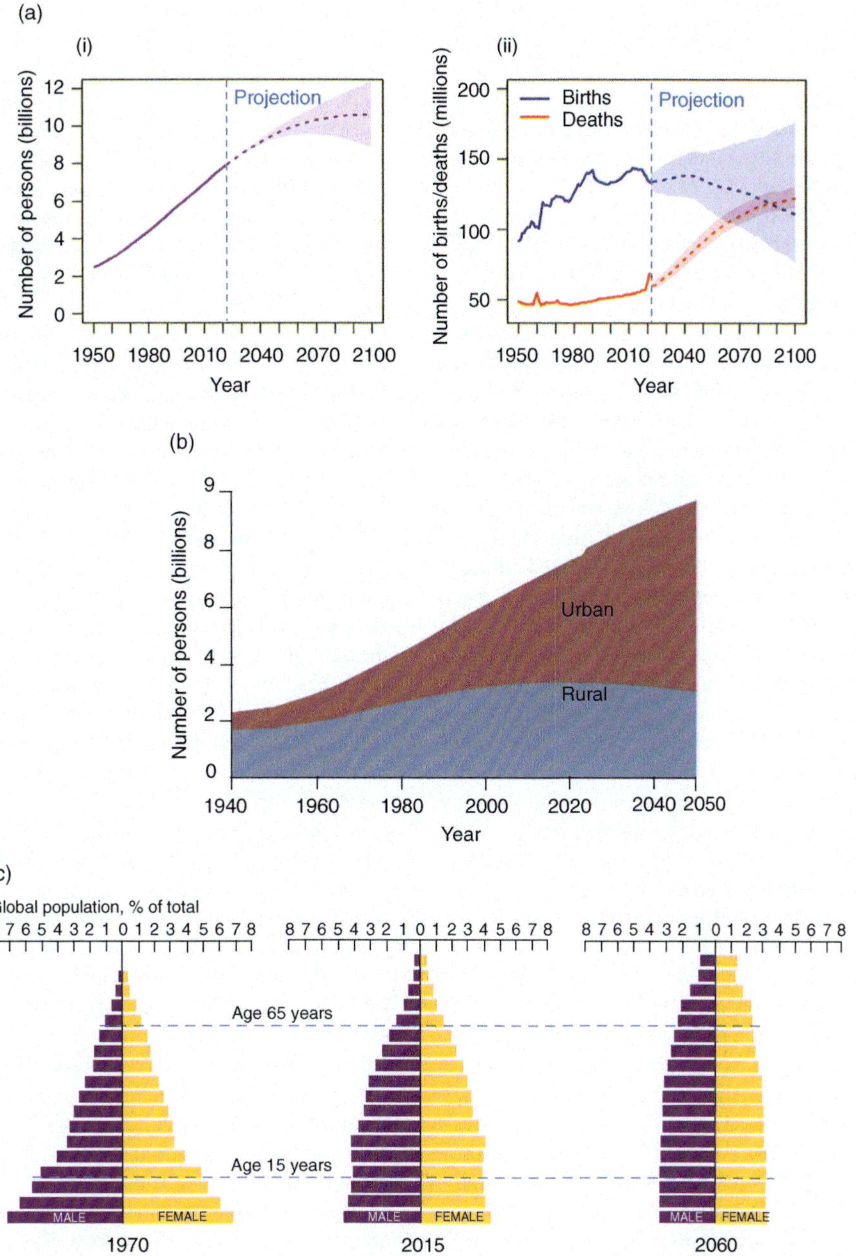

Fig. 10.2. Human population status. **(a)** Global human population numbers (i) and the total number of human births and deaths (ii), estimates 1950–2022 and projection with 95% prediction intervals 2022–2100. **(b)** Global urbanisation growth. Total urban and rural population, given as estimates to 2023, and UN projections to 2050. Projections are based on the UN world urbanisation prospects and its median fertility scenario. **(c)** Changes in the form of population pyramids 1970–2050. From pyramid (1970) to bell (2015) to barrel (2060). These changes in age structure indicate a stabilisation, as predicted by the demographic transition model. Compare with Fig. 2.5. (Panel **a** from United Nations Department of Economic and Social Affairs, Population Division, used under licence CC BY 3.0 IGO. Panel **b** redrawn from https://ourworldindata.org/urbanization, used under licence CC BY 4.0. Panel **c** redrawn from Galiana and Haseltine, 2019, used under licence CC BY 4.0.)

mines; Shaw, 2021). Yet, such 'cavities' or 'toppings' provide ample opportunity for ecosystem creation (Dennis, 2010; see Chapter 11). Indeed, a key task of conservation is the conversion of mined and quarried zones to recompense populations forced from such sites and the creation of growth areas for biodiversity (Lei *et al.*, 2016).

The chemical pollution of land, sea and air (gaseous emissions, herbicides, pesticides, microplastics, etc.) is possibly the most iniquitous human contribution, because it is less evident than extensive excavations of the bedrock. The impact is nevertheless obvious, as in the case of the insecticide DDT (dichloro-diphenyl-trichloroethane), now banned in most countries (with an exception for malaria control; www.epa.gov/ingredients-used-pesticide-products/ddt-brief-history-and-status; accessed 20 May 2025). But the spiral of loss continues unabated; visitors to Stormy Point on Alderley Edge (Cheshire, UK) have a view of the green fields to the north. It gives the impression of life, but it is dead ground, killed by chemical inputs; much of Britain is the same. All chemicals, as well as apparently inert materials used or released on the land, in agricultural fields, in cities and in the oceans (fertilisers, herbicides, pesticides, microplastics; e.g. Auta *et al.*, 2017; Sheela *et al.*, 2022), carry with them death, even if the intended purpose for their use in crops is to create a profusion of food (Sud, 2020; see Choudri *et al.*, 2018, for a review of active ingredients in chemicals and references).

Increasingly, the solution is considered to lie in not disturbing the soil covering at all, but seeding within the turf. Ploughing since 'ancient times' has led to irreversible **soil** erosion (Baumhardt *et al.*, 2015); for instance, the entire Mediterranean region has lost much of its original soil cover with the removal of extensive forests of holm oak; a reversal of this process would take centuries (Ferreira *et al.*, 2018). Another impressive example is the badlands of America – highly erosive landforms whose origin and development may have been determined by anthropogenic factors (Moreno de las Heras and Gallart, 2018). Contrasting soil erosion and restoring eroded soils is extremely difficult, especially on a large (landscape) scale. For example, part of management must include remodelling the land surface, a stage in conservation as yet virtually untried, beyond modifying the 'softer' (biotic) elements (e.g. vegetation and fungi) of the **landscape** to counter effects of climate change on soil conservation (Prober *et al.*, 2019).

Huge, significant changes to the **atmosphere** due to anthropogenic causes (e.g. increased concentration of greenhouse gases such as carbon dioxide and methane) characterise the ongoing climate change; consequences of changing temperatures and weather patterns include droughts (potentially leading to starvation), ice cap ablation, rising sea levels, island and coastal drowning, and much else (Wong and Woo, 2010; Cazenave and Le Cozannet, 2014). For example, climate change causes shifts, extremes and variation in precipitation levels and heat loads, resulting in flooding and fires (Seneviratne *et al.*, 2021; Gorta *et al.*, 2023), with severe consequences for biodiversity. Attention is also increasingly focused on the impact of climate change on sea chemistry, particularly rising acidity (e.g. significant drops in oceanic pH which stress calcifying organisms such as corals; Kleypas *et al.*, 2006). It is the sheer speed of change, and the possibility of irreversibility before disastrous consequences are experienced, that is most concerning. The impact on human populations alone should have already galvanised counter-action yet the global response remains slow and fragmented, and these consequences are increasingly beyond human control (e.g. megafires in Australia; Gorta *et al.*, 2023). But beyond the obvious impacts on human life, climate change interferes with the 'unseen', intricate aspects of species' biology, their traits and capacity to respond to changes, all of which affect their resource-based habitats and geography (see section *The Anthropocene and prime threats: climate change and habitat loss* below).

The irreversible changes to the **biosphere** (i.e. TH and RBH loss and poisoning; invasive species), the focus of this chapter, are leading to a sharp rise in extinction rates that are considered to match the seminal **mass extinctions** in the geological past (sixth mass extinction, see above). For example, according to Pimm *et al.* (1995) and Chapin *et al.* (2000), current estimates place extinction rates at 100–1000 times greater than the background level. With increasingly challenging changes to climate and **biotopes**, species in any one traditional habitat space have three 'choices' (Dennis, 1993a; Poloni *et al.*, 2022; Fattorini, 2024c):

- to adapt to new conditions *in situ* (viz. evolution)
- to move in line with shifting conditions and resources (e.g. species that shift their range poleward to track increasing temperatures)
- to become extinct (much as high-elevation species that, being already on the top of mountains, cannot move further up to track increasing temperatures).

The fundamental problem is that critical changes to one organism in an **ecosystem** can put a gamut of other species under pressure for survival.

Basic issues are the impacts on all species, on their resources and conditions for survival: (i) population decline despite habitats appearing intact due to alterations that do not change habitat physiognomy (e.g. introduction of specific parasites or pathogens); (ii) reduction of populations due to reduction of habitat space; (iii) erosion of genetic diversity as a consequence of population decline; (iv) impacts on population characteristics, life history, resource use and vulnerability to conditions, with the fragmentation of their habitats; (v) spatial and temporal shifts in habitats, understood in the individualistic sense of habitats for each species (Dennis, 2010, 2018; Primack, 2014; Exposito-Alonso *et al.*, 2022; Pinto *et al.*, 2024). Ultimately, the goal is the prevention of the loss of species. As stated above, there are too many species to consider them individually, unless their requirements coincide or they are of 'special interest' (a loaded expression if ever there was one), in which case efforts for survival must focus on maintaining and offsetting losses or 'growing' biodiversity. As stated throughout this book, the attention to 'special' species (**flagship, foundation stone, keystone**; see Chapter 2) requires an intensely focused resource-based habitat (some would argue 'niche') approach. Moreover, it is necessary to impress on conservation bodies that species responses to environmental changes may be different in different parts of their ranges. For example, climate change may have very different effects on species at the 'polar' (opposite) margins of their ranges (Li *et al.*, 2023). This should be carefully considered when ecological niche modelling (Sillero *et al.*, 2022) is used to forecast future species distributions, as results might be strongly biased by the geographical distribution and resolution of the data used to derive species preferences (e.g. Passos *et al.*, 2024).

The red list audit: potential extinctions

The next issue must be: what has been lost and what are we about to lose (Hooper *et al.*, 2012; WWF, 2024)? It may appear to many casual observers that little has been lost over the years, decades, centuries or millennia. First, then, a reminder of an often-quoted truism: the **shifting baseline syndrome** (Soga and Gaston, 2018): 'In the absence of past information or experience with historical conditions, members of each new generation accept the situation in which they were raised as being normal' (see also Chapter 3). The many visitors to the National Trust's (NT) dense wood above Alderley Edge (north Cheshire, UK) or to the once densely wooded Bickerton Hill (south Cheshire) have no perception that the entire sandstone scarps were heathland little more than a century ago (see Chapter 3, Fig. 3.2b). The proof is contained in photographs and records of the butterfly species (e.g. the silver-studded blue *Plebejus argus*; the green hairstreak *Callophrys rubi*) characteristic of lowland heath. Little more than two decades ago, there were still patches of heather (*Calluna vulgaris*) and bilberry (*Vaccinium myrtillus*) on Alderley Edge; you will find but few remnants now in the publicly accessible NT property; a patch of bilberry in pine and birch wood remains in the cordoned-off area at Stormy Point and on the south side of Alderley Edge (Logunov and Dennis, 2016). Over the entire country (UK), the long-term abundance trends of larger moths are disturbing; 41% (175 species) of 427 species have decreased, whereas only 10% (42 species) have increased (Fox *et al.*, 2021) over 50 years: this indicates that most species experienced long-term population declines. But in the absence of this type of information, the rarity of many species might seem a normal situation.

Consequently, many people are unaware that we have lost any 'biodiversity' at all and probably do not care anyway. However, the few who do care dedicate themselves to research with an unfailing intensity. Thanks to their efforts, we have a clear idea of the larger elements of our fauna and flora in crisis. The global **red list** for species is directed by the International Union for Conservation of Nature (www.iucnredlist.org; accessed 20 May 2025). In the IUCN's own words: 'Far more than a list of species and their status, it is a powerful tool to inform and catalyze action for biodiversity conservation and policy change, critical to protecting the natural resources we need to survive. It provides information about range, population size, habitat and ecology, use and/or trade, threats, and conservation actions that will help inform necessary conservation decisions'. In other words, it forms the very basis for understanding the resource-based habitat of endangered species, identifying those heading towards extinction. The proponents are well aware of its shortcomings: 'Currently, there are more than 169,000 species on the IUCN Red List, with more than 47,000 species threatened with extinction, including 44% of reef building corals, 41% of amphibians, 38% of trees, 37% of sharks and rays, 34% of conifers, 26% of mammals, 26% of freshwater fishes and 12% of birds' (www.iucnredlist.org/about/background-history; accessed 20 May 2025). They appreciate that in establishing a 'Barometer of Life', they need assessments on at least 260,000 species (www.iucnredlist.org/about/barometer-of-life; accessed 20 May 2025). Moreover, since species assessments become outdated after 10 years under IUCN rules, there is a serious risk that much of the red listing effort may soon be wasted without adequate plans to keep the list up to date (Rondinini *et al.*, 2014).

The basic problem is one of failure to maintain diversity; using 31 indicators of change (covering species population trends, extinction risk, habitat extent and condition, community composition), Butchart *et al.* (2010) have demonstrated failure in the prevention of biodiversity loss for all 31 markers since 2002, when world leaders committed – through the Convention on Biological Diversity (Secretariat of the Convention on Biological Diversity, 2003) – to achieve a significant reduction in the rate of biodiversity loss by 2010.

Numerous negative biological markers now exist that parallel equally the failure of halting the human-driven forces of change. The earliest losses that can be blamed on human activity predate the Holocene, with excessive hunting using Stone Age technology occupying a prime position (Surovell, 2008); the path to mass extinction has been a long and inexorable one. Wild mammal biomass has declined by some 85% since the rise of humans 100,000 years ago (Ritchie, 2021), and during the Pleistocene and Holocene, numerous large and small-bodied animals, known only from the fossil record or other remains, have suffered extinction (Surovell, 2008) (Fig. 10.3). Such alarming extinction rates have led to global assessments of danger to faunas, which have shown a steady deterioration in their conservation status worldwide (e.g. amphibians and reptiles: Alroy, 2015; birds: Lees *et al.*, 2022; mammals: Hoffmann *et al.*, 2011).

One of the most distinctive markers of change is insects, especially Lepidoptera (Leather, 2017; Reynolds, 2023). An analysis conducted at the beginning of the 2000s revealed that the worst case scenario for butterfly diversity loss in Europe was Flanders (Belgium), with habitat destruction, fragmentation and nitrogen deposition being the major causes (Maes and Van Dyck, 2001). Twenty years later, a more detailed analysis found that, despite ongoing policy focus and conservation efforts, sedentary species of nutrient-poor biotopes such as heathlands and semi-natural grasslands continued in a decreasing trend, whereas woodland specialists and **polyphagous** species (i.e. species whose larvae are dietary **generalists**) showed the strongest expansion (Maes *et al.*, 2022). The most severe examples of moth declines are from northern hemisphere regions of high human population density and intensive agriculture; the principal continental-scale factors probably include reductions in habitat quality and quantity resulting from land use change and climate change (Wagner *et al.*, 2021). In general, insects, which are the largest and most widely distributed terrestrial animal taxon, are experiencing a fast and unprecedented decline worldwide (C.D. Thomas *et al.*, 2019; Eggleton, 2020; Fattorini *et al.*, 2023), with potentially global catastrophic consequences, for which the word *Insectageddon* has been introduced (van der Sluijs, 2020). However, as current information on insect decline is strongly biased by the general lack of knowledge on the abundance of species and their trends in space and time (Cardoso and Leather, 2019), a rigorous quantification of global insect decline and extinction remains extremely challenging (Montgomery *et al.*, 2020).

Ecosystem issues

In **ecosystems**, species have two causes of death: (i) their own selves, viz. a genetic extinction from the mortality of individuals, and (ii) the death of life forms on which they depend, and thus loss of a crucial component in an ecosystem. The loss of a species or multiple species has consequences way beyond their own status as unique genetic entities.

Within an ecosystem, a sudden loss or gain (e.g. through human introduction) can throw the system into chaos. The crunch with species loss is: what effects does their dismissal bring to the community in which they were found and from which they are dismissed? Inevitably, that depends on (i) their abundance, and (ii) the nature of their interconnectedness with other species (see Chapter 2). The greater their number and intensity of linkages with other organisms and associations (negative or positive), the greater the impact on the community (see Box. 2.1 in Chapter 2). Storch *et al.* (2022) show that: (i) biodiversity dynamics depend on nonlinear relationships between number of species, community abundance and population size-dependent

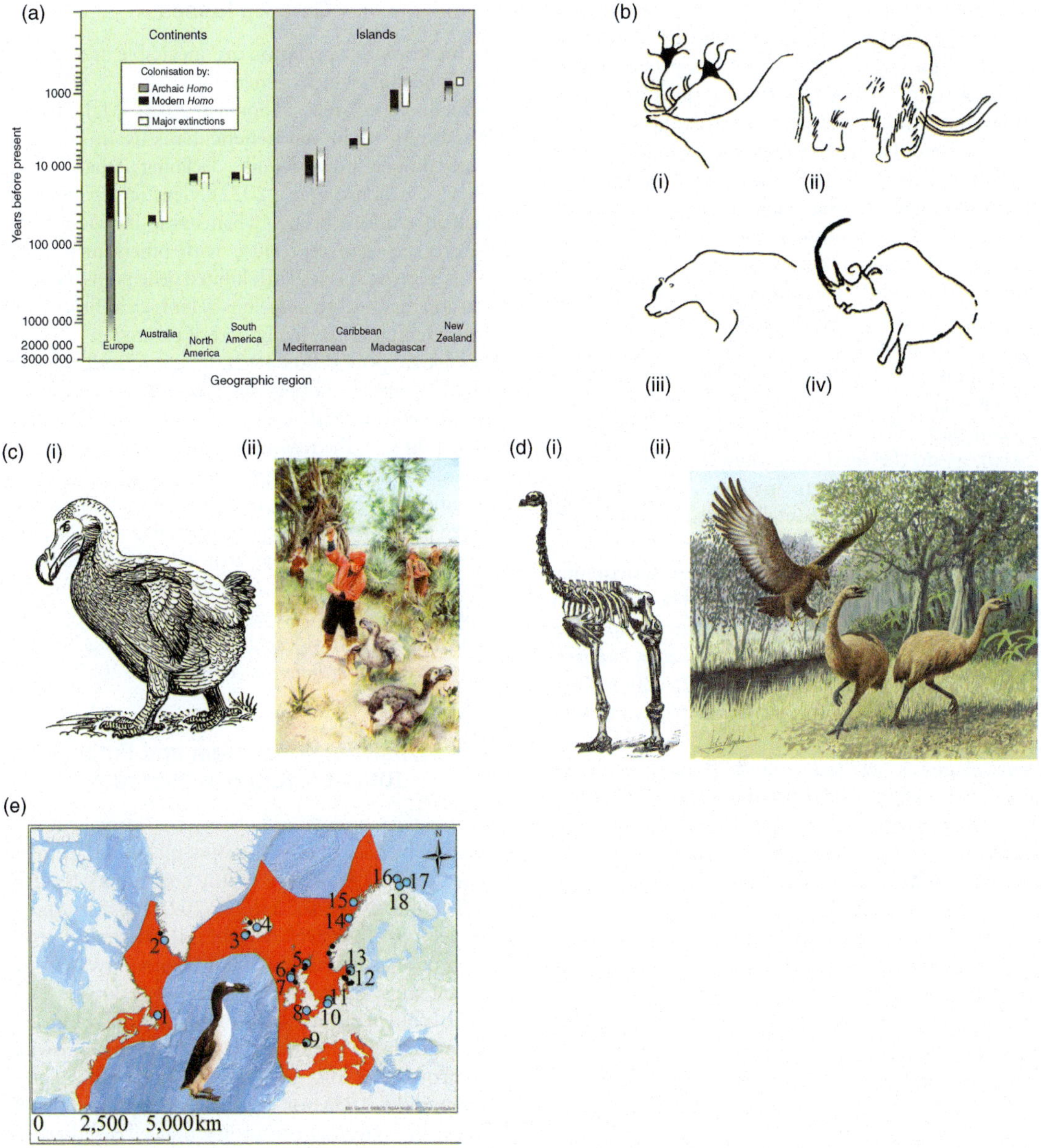

Fig. 10.3. Past extinctions. **(a)** The relative timing of human colonisation and the major wave of animal extinctions for various continents (left) and islands (right). Solid colours and lines indicate certainty about age. Dashed lines and colours indicate uncertainty about age. Age axis is log-scaled. **(b)** Line drawing reproductions of Upper Palaeolithic cave paintings of extinct Pleistocene fauna in France: (i) Irish elk (*Megaloceros giganteus*), Lascaux cave; (ii) woolly mammoth (*Mammuthus primigenius*), Rouffignac cave; (iii) cave bear (*Ursus spelaeus*); and (iv) two-horned rhinoceros (*Coelodonta antiquitatis*), Chauvet cave. **(c)** Dodo (*Raphus cucullatus*) extinct in Mauritius in the 17th century: (i) drawing of presumed form; (ii) illustration of presumed sailors killing dodos on visits to the island, regarded by some as not the main reason for their extinction. **(d)** Extinct moas of New Zealand: (i) heavy-footed moa (*Pachyornis elephantopus*); (ii) moas attacked by Haast's eagles (*Hieraaetus moorei*), also extinct. **(e)** The great auk (*Pinguinus impennis*) and its former distribution in the North Atlantic. Red shading indicates the geographical distribution of the great auk. Sites marked with blue dots represent samples used in a genetic analysis; black dots

processes of species extinction and origination (speciation or colonisation); and (ii) non-linear effects of biodiversity on ecosystem processes create shifts in the balance of relationships within the community. In this way they illustrate how interactive dynamics have particular attractors (i.e. stable equilibria determined by environmental conditions). It is well to consider that a change may lead to multiple biodiversity equilibria; the smallest change in one attribute can redirect the system into a different homeostasis, before the next change takes control. These attractors set boundaries on biodiversity, for instance by limiting carrying capacity for community biodiversity, that could well differ from local to global scales. In this context, stochasticity plays a crucial part in changes (Ovaskainen and Meerson, 2010), as do the increasing number of new immigrants (reference here is to any species, not necessarily involving a human context), species from other regions that occasionally dominate ecosystems and can profoundly impact on their functioning (Williamson, 2011; Roy *et al.*, 2012; Thomas, 2017; e.g. the Japanese knotweed *Reynoutria japonica*: Fennell *et al.*, 2018, Fig. 10.4a; the grey squirrel *Sciurus carolinensis*: Gurnell *et al.*, 2016, Fig. 10.4b). Thus, the presence of alien species can render predictions of ecosystem stability based on current elements of fauna and flora completely unreliable. Because of this complexity, unpredictable sudden changes in interspecific relationships and hence in biodiversity dynamics can result from changes in variables that control the system in subtle ways that are difficult to understand, including variables intrinsic (internal) to an organism's functioning (Dennis, 2020). Each system has its own threshold levels for key variables; if these thresholds are crossed, the system shifts into a new state with no or very limited capabilities of recovery. This also applies to alien species; identification of thresholds of potential concern is an important strategy to decide when to implement appropriate management options (e.g. Foxcroft, 2009).

Genetic losses

Extinction and speciation are inevitable and natural processes in which genetic variation is lost and gained respectively (Thomas, 2017; Maheshwari, 2020). As such, the genetic bank is in constant flux, though now steadily declining (Singh, 2012; Exposito-Alonso *et al.*, 2022). **Genetic depletion** (or genetic erosion) occurs when fertile individuals die before they can reproduce with others in populations, leaving species with limited gene pools. Genetic erosion can lead to a decline in biological fitness and hence increases the likelihood of extinction of populations or species (Box 10.2). It is most serious when unique genes are lost in small populations, typically located in restricted or fragmented habitats (Primack, 2014). In fact, the term is sometimes used in a number of contexts, for instance describing the loss of a specific allele or gene at one extreme to the loss of an entire phenotype or species at the other. Even so, loss of a species does not mean that its genetics disappear altogether. This is the case with *Homo sapiens*: we still carry about us, in the bulk of humanity if not individually, a considerable amount of the gene bank of *Homo neanderthalensis* (Reilly *et al.*, 2022).

Invasives: Williamson and Fitter's 10th of a 10th of a 10th rule

A complete understanding of ecosystems, thus of the biodiversity of communities, includes the effects that **invasives** may have on the community (Ludsin and Wolfe, 2001; see Strona, 2022 for a thorough discussion). Invasives may be: (i) internal, derived from other communities some distance away in the same region (continent or island) but outside their local ecological limits, but managing to colonise a **biome** or local traditional habitat (**vicinism**) owing to propagules dispersing from favourable zones (Zonneveld, 1995) or conveyed with human assistance; (ii) external, species introduced accidentally or on purpose from other communities or biomes,

Fig. 10.3. Continued.

denote other sites from which material was obtained, but for which samples were not used for genetic analysis. Genetic data, GPS-based ocean current data and analyses of population viability do not provide any evidence that great auks were at risk of extinction prior to the onset of intensive human hunting in the early 16th century. (Panels **a** and **b** from Surovell, 2008, used with permission of Elsevier Science and Technology Journals. Panel **c**(i) from Archives of Pearson Scott Foresman: https://commons.wikimedia.org/wiki/File:Dodo_2_%28PSF%29.png. Panel **c**(ii) from Johnston, 1914; drawing by Walter Paget: https://archive.org/details/pioneersinsoutha00johnuoft/page/n140/mode/1up?view=theater. Panel **d**(i) from https://commons.wikimedia.org/wiki/File:Pachyornis_elephantopus.jpg. Panel **d**(ii) artwork by John Megahan, from Bunce *et al.*, 2005, used under creative commons licence from Public Library of Science. Panel **e** from J.E. Thomas *et al.*, 2019, used under licence CC BY 4.0.)

Fig. 10.4. Invasives affecting native flora and fauna in UK. **(a)** Japanese knotweed *Reynoutria japonica*. **(b)** Grey squirrel *Sciurus carolinensis*. **(c)** Himalayan balsam *Impatiens glandulifera*. **(d)** Ash die-back (*Hymenoscyphus fraxineus*) effects. **(e)** Harlequin ladybird beetle *Harmonia axyridis*. **(f)** Signal crayfish *Pacifastacus leniusculus*. (Panel **a** photo by David Short: https://flickr.com/photos/14583963@N00/31881337434, used under licence CC BY SA 2.0. Panel **b** photo by BirdPhotos.com: https://en.wikipedia.org/wiki/File:Eastern_Grey_Squirrel.jpg, used under licence CC BY 3.0. Panels **c** and **d** photos by R.L.H. Dennis. Panel **e** photo by Bruce Marlin: https://commons.wikimedia.org/wiki/File:Harmonia_axyridis.jpg, used under licence CC BY SA 2.5. Panel **f** photo by David Pérez (DPC): https://en.wikipedia.org/wiki/Signal_crayfish#/media/File:Pacifastacus_leniusculus_01_by-dpc.jpg, used under licence CC BY 4.0.)

Box 10.2. Genetics and conservation

Large gene pools exhibit extensive genetic diversity, which is associated with robust populations that can survive intense selection; fluctuations in population size coincide with fluctuations in genetic variability, but the trough in numbers may often be associated with a new stabilisation, before a subsequent upturn, as demonstrated since the first investigations in ecological genetics (see, for example, a classical study by Ford and Ford, 1930 showing phenotypic variation in a fluctuating colony of the marsh fritillary butterfly *Euphydryas aurinia*). A common effect of population **bottlenecks** (i.e. reductions in population size) is the loss of genetic variation (e.g. Nei *et al.*, 1975; Amos and Harwood, 1998). In small populations resulting from bottlenecks, the alleles inherited by offspring are only a random subsample of those present in the larger parental generation (Fig. B10.2.1); thus, after some time, some alleles will be lost and others will be fixed (**genetic drift**), independently of selection pressures, a stochastic evolutionary process long ignored or misunderstood by evolutionists focused on the role of selection as the force driving changes in allele frequencies (e.g. Gould, 2002; see Kliman *et al.*, 2008 for an elementary treatment of genetic drift). Thus, bottlenecks generally reduce variability through intense genetic drift; even if the population regains a large size after the bottleneck, only the alleles that were randomly present in the small population during the bottleneck can be represented in the new generations.

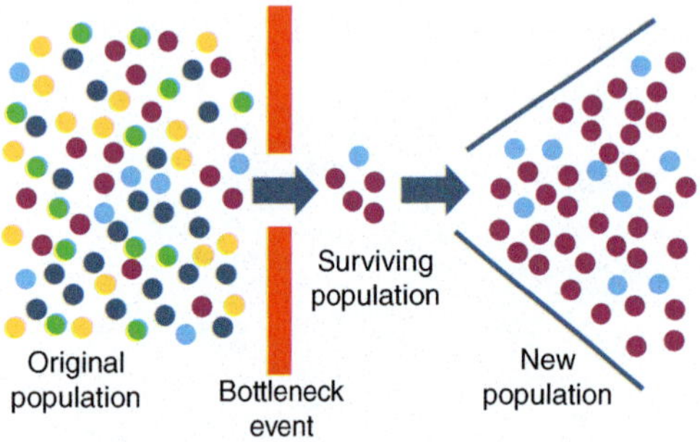

Fig. B10.2.1. Schematic representation of genetic drift because of population bottleneck. Dots of different colours represent individuals with different alleles. Genetic diversity is strongly reduced by the bottleneck, since the new population will contain only the alleles of the few survivors. During this process, an allele rare in the original population can become common in the new population by chance alone, even if it is disadvantageous.

Genetic drift is a common phenomenon in small and isolated populations, and represents an important element of concern for conservation as it (i) allows harmful alleles to rise in frequency by chance alone, and (ii) reduces the capability of the population to evolve adaptively to new environmental challenges (Fenster *et al.*, 2018). Another source of loss of genetic variability is **inbreeding** (non-random mating), which leads to an increase in alleles identical by descent, thus reducing survival and reproduction (**inbreeding depression**) (Frankham *et al.*, 2019). Both genetic drift and inbreeding cause a decline in heterozygosity, but while genetic drift occurs even with completely random mating (albeit at a low rate), and leads to changes in both genotype and allele frequencies, inbreeding leads to changes in genotype frequencies but not in allele frequencies. As habitat fragmentation and isolation reduce population size and increase population isolation, these environmental transformations decrease local genetic variability and increase genetic differentiation between populations through genetic drift, reduced gene flow and inbreeding, which can all increase extinction risk (e.g. Saccheri *et al.*, 1998, 1999).

More generally, decreased variability reduces the capabilities of populations to adapt to environmental and demographic change (Young *et al.*, 1996; Frankham *et al.*, 2010), thus reducing their ability to use certain components of **resource-based habitats**. Conservation measures can be used to counter the negative effects of drift and inbreeding by assisted breeding in which 'unrelated' individuals are mated, for example through **augmentation**, i.e. the intentional movement and release of an organism into an existing population. Augmentation (also known as reinforcement, restoration or supplementation) is frequently used as a conservation measure to enhance population viability by increasing population size and genetic diversity (Frankham *et al.*, 2011; Liddell *et al.*, 2021), although its efficacy has been questioned (Manlick *et al.*, 2017). An important issue with augmentation is that crosses between individuals from genetically distant groups or populations may result in a reduction of fitness, a phenomenon known as **outbreeding depression** (Frankham *et al.*, 2011; Byrne and Silla, 2020). Although it seems that the risks of outbreeding depression have been overstated (Frankham *et al.*, 2011; Weeks *et al.*, 2011), conservation programmes should try to minimise both inbreeding and outbreeding risks, for example by mixing slightly differentiated populations instead of mixing extremely differentiated, locally adapted populations (Marsden *et al.*, 2013).

including those from overseas, by humans or other animals (e.g. birds; Viana *et al.*, 2016). The latter have important implications for native flora and fauna. An unexpected observation is that of Williamson and Fitter (1996a): for a variety of British invasive animals and plants, their statistical rule holds that:

- 1 in 10 of those imported escape to the wild (introduced or casual)
- 1 in 10 of those that make it to the wild become naturalised (established); and
- 1 in 10 of those established become invasive (and possibly a pest).

Of course, this reasoning is totally unrealistic, as firmly underpinned by metadata from 53 publications collated by Jeschke and Pyšek (2018), which showed greater success of aliens. The basic 10th 'ruling' of Williamson and Fitter (1996a) also differs substantially from the carefully evaluated observations by the Centre of Ecology and Hydrology for the UK (Roy *et al.*, 2012): it states that 'there has been a dramatic increase in the number of species arriving [in the UK] and those becoming established over the last 400 years and there is no indication of this trend slowing'. Indeed, over the last 200 years there have been, on average, 6.8 additional non-native species arriving and establishing per year in contrast to 0.8 additional non-native species arriving and establishing per year from 1600 to 1799 (Roy *et al.*, 2012). But there is a great difference between those species becoming established and causing little or no damage, those that produce important damage, and those that bring with them some benefits. For instance, in the UK, among plants, the Himalayan balsam (*Impatiens glandulifera*; see Fig. 10.4c), which can smother river banks, provides bountiful nectar for bees as well as other benefits, such as supporting herbivory and offering structures for resting places and concealment from predators for small mammals and birds (Coakley and Petti, 2021), whereas the Japanese knotweed *Reynoutria japonica* (see Fig. 10.4a) has undoubted negative impacts on the British environment, even breaking through the foundations of houses (references in Serniak *et al.*, 2017 and Renčo *et al.*, 2021; but see Fennell *et al.*, 2018 for a study showing that there is no evidence supporting the claim that *R. japonica* causes damage in excess of the norm for many other plants). Among fungi, the ascomycete *Hymenoscyphus fraxineus* causes ash die-back (see Fig. 10.4d), a lethal disease of ash trees (*Fraxinus* spp.) (Enderle *et al.*, 2019). In the case of insects, the harlequin ladybird beetle (*Harmonia axyridis*; see Fig. 10.4e) has had a serious impact on native ladybird beetles due to its spread over the country (Brown and Roy, 2018; Brown *et al.*, 2018) and the history of the loss of great elm trees (mainly caused by the fungus *Ophiostoma novo-ulmi* which is spread by elm bark beetles, especially beetles in the genus *Scolytus*) is well known (www.forestresearch.gov.uk/tools-and-resources/fthr/pest-and-disease-resources/dutch-elm-disease-ophiostoma-novo-ulmi/; accessed 20 May 2025). The American signal crayfish (*Pacifastacus leniusculus*; see Fig. 10.4f) has negative impacts on the native white-clawed crayfish (*Austropotamobius pallipes*): it carries the deadly crayfish plague which kills the native crayfish, preys on them, out-competes them for food and habitat and can wipe out native fish stocks (https://cdn.buglife.org.uk/2019/08/White-clawed-crayfish_0.pdf; https://assets.publishing.service.gov.uk/media/5a7bfbe0e5274a7318b9051a/genw0606blai-e-e.pdf; accessed 20 May 2025).

Using the hierarchy of hypotheses (HoH) approach (in which hypotheses are conceptually structured in a hierarchically nested way, where the lower branches can be directly connected to empirical results), Jeschke and Pyšek (2018) attempted to refine the tens rule scheme by dividing it into invasion tens rule and impact tens rule as follows.

- Invasion tens rule: about 10% of species successfully take consecutive steps of the invasion process.
  - (i) Transport → introduction: about 10% of transported non-native species are released or escape.
  - (ii) Introduction → establishment: about 10% of the introduced species establish themselves.
  - (iii) Establishment → spread: about 10% of the established species substantially spread from their point(s) of introduction.
- Impact tens rule.
  - (iv) About 10% of established non-native species cause a significant detrimental impact (i.e. they have harmful ecological, socio-economic or human health effects); this subhypothesis thus relates to the transition establishment → impact/pest species.
  - (v) About 1% of all introduced non-native species cause a significant detrimental impact; this subhypothesis thus relates to the transition introduction → impact/pest species.

The authors found that the tens rule lacks empirical support and neither of its subhypotheses is supported by currently available evidence.

The fact is that the impact of invasives is a great deal more complex than assumed in any simple model developed by Williamson and Fitter (1996a) or Jeschke and Pyšek (2018), and used uncritically by Thomas (2017), when all aspects of biodiversity are accounted (see Jarić and Cvijanovic, 2012 and Strona, 2022). For instance, how much of the country do the 1875 foreigners mentioned by Thomas cover? How do they fit into British ecosystems? How many have been subjected to intensive study? If not causing extinctions, there is no doubt that invasives can have hugely devastating effects (e.g. Dutch elm disease described above: Dennis, 2010; Potter *et al.*, 2011; ash die-back: Enderle *et al.*, 2019). Unquestionably, this 10th ruling suggests a much smaller impact than invasive species may have. We would add to the concern of numerous researchers who have argued that the 10% rule might be more of an indicator of our lack of understanding of the impacts that established introduced species produce than their real negative impact (actual ratio). Jarić and Cvijanović (2012) advise that in such a state of affairs, it is crucial to adopt the precautionary principle. A pressing issue is a ubiquitous pattern in the species abundance distribution: the small number of common species yet the large number of **rare species** (see Chapter 4); that there are so many apparent vacant **niches** for invasives should concentrate attention on the niches of species in ecosystems not used by the uncommon component. Clearly, the vacant 'space' does not have suitable resources or conditions for the native species, but might be suitable for aliens (see Strona, 2022).

The ability of species to invade foreign ecosystems is probably owing to the uniqueness of each species' resource-based habitat and niche association (or to their escape from the predators and parasites present in their native ranges; enemy release hypothesis; Keane and Crawley, 2002; Strona, 2022), enabling invasive species to locate 'empty ecological spaces', or their capacity to undermine the niches of species already present. Invasives tend to exhibit unusual **traits** (sometimes distinctly unrelated to native species) (Callaway and Aschehoug, 2000; Cappuccino and Arnason, 2006; Strauss *et al.*, 2006) and with 'armoury' we fail, as yet, to appreciate. Gauzere *et al.* (2020) suggest that empty volumes in trait space are common; they can arise from both abiotic and biotic processes and are more likely in cold climates. They demonstrate that most vacant niches are also invasible. Studies of plant invasives indicate that they have distinctive traits (Williamson and Fitter, 1996b; Mathakutha *et al.*, 2019) that support the **vacant niche hypothesis** (i.e. that not all 'niche spaces' are being used by the species that are found in a particular site at a given moment; Herbold and Moyle, 1986; Gibson-Reinemer, 2015). These observations should boost reflections on the impact invasives may have on the statistical distribution of species within traditional habitat communities (few common, many rare species; see Chapter 4). But they still do not explain why native species have not evolved to adopt the empty 'spaces'. Clearly, a careful examination of each species' resource-based habitat in relation to its community associates and its many traits is fundamental to explain why some aliens become invasive and others do not.

Winners and losers in the Anthropocene

Extinction is inevitable in an increasingly globalised world with the prime objective of capital growth (Maheshwari, 2020). Invasives hold the key to winners and losers. The fundamental question is: what makes some species or populations more vulnerable than others? A number of traits and conditions influence vulnerability.

- *Larger species versus smaller species.* Despite opportunities concomitant with dominance of size (for example, large body size enhances fecundity and survival; see Baker *et al.*, 2015), larger species require more space and their populations tend to be smaller with lower genetic diversity (e.g. De Kort *et al.*, 2021; see below).
- *Small populations as opposed to large populations.* Larger populations have greater resistance and resilience to stochastic effects (Primack, 2014). The inference is that, for example, strict carnivores feeding on limited prey may be more vulnerable than herbivores owing to small population effects (e.g. low heterozygosity level and high **inbreeding** coefficient; Lodé, 2006).
- *Spatially limited compared to widespread populations.* Widely distributed species, species with larger ranges, occupying a greater variety of traditional habitats have a greater chance of surviving local and regional catastrophes (see Chapter 4, Box 4.1). Widespread populations are more

likely to have greater variability in relationship to environmental variability that increases with occupancy of wider areas (Tobler's first law of geography; Waters, 2017; see Table 4.1 in Chapter 4). It follows that species vulnerable to habitat loss (particularly those susceptible to human demands) will also be more likely to become extinct. For example, **saproxylic** species are one of the most threatened organism groups, since they are suffering from the dwindling of forests (Siitonen *et al.*, 2012).

- *Habitat specialists versus habitat generalists.* **Specialists** are vulnerable to the loss of a crucial resource element whereas **generalists** have more ecological 'escape routes' (Dennis *et al.*, 2011). The exceptions are species whose specialisms coincide with 'human habitats' (e.g. synanthropic species, urbanophilous species; Fattorini, 2019) or that use widespread habitats (in the traditional sense). In the latter case, although the species is associated with a particular habitat, it is not vulnerable because its habitat is abundant (Crisfield *et al.*, 2024).
- *Species with poor dispersal capabilities versus species with high dispersal capabilities.* Poor dispersal capabilities may increase extinction risk. Flightless birds are more vulnerable than flying birds (the same for insects), especially if carnivores enter their habitat (e.g. on islands), some literally clubbed to death by *Homo sapiens* for human consumption or purely for 'fun' (Turvey and Cheke, 2008; J.E. Thomas *et al.*, 2019; e.g. the dodo *Raphus cucullatus* on Mauritius; the moa *Euryapteryx geranoides* in New Zealand; the great auk *Pinguinus impennis* in the Atlantic fringe; see Fig. 10.3c–e) (see also Chapter 6, Box 6.2).
- *Species dependent on symbionts.* Species that live in strict association with others are particularly vulnerable for a variety of reasons to do with life-history functional efficiency (e.g. complex life cycles and specialisation in parasites; Strona, 2022).
- *Species occupying compact or contiguous biotopes.* Finally, and most important of all, those species with all their resources and suitable conditions found in a compact traditional habitat (**biotope**) space, or which can use traditional habitats uniformly widespread across the **landscape**, have a better chance of survival than species whose resources (consumable and utilities) are isolated from one another and challenge their dispersal capacity.

Those that pass these tests are probable winners but those that fail these tests are likely to be losers. Much depends on what surrounds the target space of interest for any one key species of 'interest' (a key habitat zone). Human impact, by changing species habitats, also changes species rarity and vulnerability (Table 10.2).

Table 10.2. Six anthropogenic reasons why species become rare. All these factors involve reasoning from the resource-based habitat concept (Dennis, 2010).

Habitat loss. Species originally associated with widespread habitats and abundant become geographically rare and scarce when their habitat is reduced or degraded. Under habitat transformation, species may be forced to occupy suboptimal habitats.

Invasive species. Alien species may reduce abundance, geographical distribution and habitat distribution of native species because they can act as predators and competitors. Alien species can also act negatively on populations of native species because they can introduce new diseases or alter the habitats.

Pollution. Species can be affected directly by the toxicity of contaminants, or indirectly by the influence that pollutants have on their food resources.

Persecution and exploitation. Recreational activities, commercial uses and indiscriminate killing due to some perceived threat or undesirable quality can reduce species ranges, population size and habitat use.

Disease. Introduced diseases or increase in native diseases because of environmental alterations (for example, environmental changes that favour parasites and vectors) can contribute to the decline of native species both directly (as individuals die because of the disease) or indirectly (sick/parasitised individuals are more vulnerable to other factors, e.g. more exposed to predation).

Global climate change. Climate change can impact species distribution, population size and habitat use as they can render some areas or habitats unsuitable, alter species behaviours (e.g. circadian rhythms) and disrupt species interactions (e.g. phenological mismatches).

(Partially based on: www.dnr.state.mn.us/rsg/whyrare.html; accessed 20 May 2025.)

As noted above, Thomas (2017) is a singular adherent and champion of increased diversity in the **Anthropocene** through subdivision of land into a variety of terrains, owing to facilitation of species movement across the bounds dividing finer landscape mosaics. Thus, Flanders would be regarded as such a place, surrounded as it is by so-called 'richer' traditional habitat zones, though it is clear from Maes and Van Dyck (2001) that this region does not benefit in any material way from **vagrants** coming from surrounding areas comprising full sets of resources. Here it is sufficient to refute this notion if only because the only way such diversity changes can be assessed is through time, sampling without bias the very same areas involved, an issue completely ignored by Thomas. There is something distinctly repugnant in the loss of native (frequently endemic) overlooked species in isolated areas and their swamping by Anthropocene invasives. In the cited case of New Zealand (Thomas, 2017), do we really want (and it would be arrogant not to interpret 'we' foremost as New Zealanders) New Zealand to be a phenocopy of the British Isles? New Zealand has already lost several of its remarkable creatures, such as the moas (Dinornithiformes) and Haast's eagle (*Hieraaetus moorei*) (Bunce *et al.*, 2005) (see Fig. 10.3d); do we really want its original components one day only to be neatly arrayed under glass in zoo-like conditions? The conversion of New Zealand grasslands using grass species from Europe for sheep rearing would suggest that this process is well under way. How can Thomas' 2 ha own home property be considered informative of the biodiversity of a much larger area, when no mention is made of the habitat components of the 22 butterflies observed within his groundspace? Moreover, there is always the matter of how many are just vagrants or have temporary occupancy. Biodiversity depends on having all the resources and all the suitable conditions available in the target space and neighbouring areas; Thomas at least should be aware that butterfly species can travel substantial distances, thus being frequently recorded from places where they do not find really suitable resources and conditions (see Dennis, 1977, 1992, 2010; Dennis and Hardy, 2018). The cautionary note has long been available for commonness versus rarity, but even common species may become vulnerable; the extinction in North America of the passenger pigeon *Ectopistes migratorius* from a population that numbered 3–5 billion (see Box 4.1 in Chapter 4) should make us aware of the **resource-based habitat** demands of each species.

Proponents of biodiversity conservation of any group at any scale must take these points into consideration. In fact, Thomas made a call for more robust data and rigorous analyses very soon after his book appeared in the public domain (C.D. Thomas *et al.*, 2019), though it was a call that had been made years before by the first author (Dennis *et al.*, 2008) and by colleagues subsequently (Lobo *et al.*, 2021; Sánchez-Fernández *et al.*, 2021). Environmental changes may create opportunities for many species by producing a variable resource base that affects different groups in different ways. For example, landscape changes due to artificialisation of **substrates** (e.g. buildings, concrete, tarmac) may create a **matrix** with new opportunities for many species (e.g. Dennis and Hardy, 2007; Forman, 2014; Cariveau *et al.*, 2019). Anthropogenic transformations, however, change through time, producing continuous variations in their effects; for example, agrarian landscapes must have been initially favourable to herb-dependent lepidopteran species in Medieval times, species now extinguished by modern intensive techniques (Dennis, 1977). Thus, assuming a universally beneficial role for human intervention is, at best, misleading. Nor has Thomas addressed to the full extent the mixing of flora and fauna via 'natural processes' (e.g. glacial-interglacial cycles (G-IG); Dennis, 1977; Huntley, 1991; Dennis *et al.*, 1991, 1995, 1998; Dapporto *et al.*, 2009, 2011a,b, 2012, 2017). For temperate and tropical species, each IG induces spread and hybridisation, whereas each G phase causes isolation, extinction and speciation under totally different conditions (Dennis, 1977, 1992, 1993a, 2010; Dennis and Hardy, 2018; see also Box 7.1 in Chapter 7). The movement of species in the Anthropocene is regarded simply as a complex and extreme version of this process, as climate warming brings with it highly variable conditions, drought and floods, as well as increasing temperatures, including bouts of critical heat for many organisms (Fattorini, 2023).

If there is a lack of interest in the effect climate change has on our faunas and floras, then of course humans, being what they are, will find that the situation becomes critical when it impacts more directly on human survival. For too many in human society, other organisms are regarded with complete indifference; such indifference carries the warning that to allow the extinction of overlooked **keystone** species to go ahead may also mean the loss of entire ecosystems. Who knows the full value of such systems to our future?

Should we bother to conserve species?

With so many species to conserve, many as yet unnamed (see Chapter 8, Box 8.1), and the huge difficulty in conserving them despite our growing knowledge of ecosystems and many individual species, the immediate question is: are we capable of conserving them amid the maelstrom gathering with the growing human population? Worse still, in a world driven by personal and social needs and greed, those dedicated to maintaining individual species and biodiversity are forced to face a fundamental question: why should we bother to conserve other species (singular or plural) at all? To the dedicated minority, and currently there is a minority of the 'caring', this question is as close to heresy as one can get. This is a question implicit (if not explicit) in Thomas (2017), born of the pressures placed by humans on land, sea and air regardless of the multitude of other organisms. Despite what would seem to be an obvious answer to many decent people ('Yes, we should conserve all species!'), based on a sense and duty of care that humans might have developed for other creatures (biophilia: Wilson, 1984; Fattorini *et al.*, 2017c), too many humans in powerful positions have little or no regard, no awareness even, for wildlife of any kind, other than those suitable for shooting. Even the most benign of conservationists have developed their own pecking order of conservation – a bird before a butterfly before a snake! (see, for example, Fattorini *et al.*, 2023; see also Chapter 2, Box 2.3). Few hold to a parity of other creatures with 'man', a creature who has immense transformative power but rarely makes good use of it. This is hardly surprising when the often dysfunctional interrelationships of humans are explored in any detail.

On the 'yes' side for conservation, the points that arise are these.

- *The emotional argument.* We like them, they are creatures like us and their existence improves our sense of well-being.
- *The resource argument.* Many are invaluable for human continuance (for example, as resources for human use; especially the 'medical argument', e.g. organisms as sources of medicines like alkylating agents and antibiotics) and should in effect be 'farmed'.
- *The cautionary argument.* We have no idea of the value of many of the species for the benefit of planetary stability and thus for our own benefit.

On the 'no' side are the following.

- Emotional attachments are simply twaddle, an indication that we have not yet 'grown up'.
- They are simply resources that should be exploited, but without regard for maintenance, as something else always emerges to replace what is lost.
- Conservation, anyway, is almost impossible, as our knowledge of species resource-based habitats is largely unknown. Even for one species, it involves unnecessary expense. In many cases, long-term conservation is virtually impossible owing to the sheer costs, disagreements between states whose boundaries are crossed by the species, etc.
- Conservation will be time-limited anyway owing to the increasing costs of maintenance (management). With human populations growing at such a high yearly rate (0.85%; www.worldometers.info/world-population/; accessed 20 May 2025), the resources available for conservation will decline and preserving species will become more expensive. Already, there is a desperate need for more land and resources for the Earth's enlarging impoverished peoples, triggering inevitable conflict and destruction. An important negative to the plea for conserving plants and animals is the vast numbers of humans living in desperate conditions, barely surviving; from this fact the (false) argument can be made that we need all the resources available for provisioning our own species and it is just unfortunate if there is little 'room' left in our world for other life forms.

A number of issues follow these conflicting thoughts. Foremost: (i) should we not control our own productivity, and limit the population size on the planet and therefore the pressure placed on planetary resources and biodiversity (Bongaarts, 2009; Shaykheeva *et al.*, 2016; Johns, 2019; United Nations Department of Economic and Social Affairs, Population Division, 2022; see Fig. 10.2a,b)? (ii) Does erasing other species (particular ones or multiples in diverse communities) actually benefit our own species? Certainly not (Chami *et al.*, 2022). (iii) How seriously are global processes and Earth ecosystem maintenance affected by the loss of species, particular organisms, small (**prokaryotes**, **protists**, fungi) to large (animals, plants), and species in bulk (Hooper *et al.*, 2005, 2012; Millennium Ecosystem Assessment, 2005)?

The answers to these questions take the issue of conservation beyond the merely emotional one of decency and treating other organisms with due respect. A simple but dreadful realisation arises if greed for the land, sea and air (the other creatures themselves on and in it) outweighs any sense of concern we have for wildlife. A crude anti-conservation argument presented is that resources (hard cash, manufactured materials and muscle) have to be found to aid the survival of wildlife, the expense extending not just for a year or two, a decade, a half century or more, but *in perpetuum*. Protecting wildlife in perpetuity, including dealing with invasives, requires vast resources and individual states are unprepared to justify the costs. A suggested alternative is to gain money through recognition of specific creatures as **focal species** (i.e. species that can provide the impetus for the conservation of their habitats; see Chapter 2, Box 2.3) or as tourist attractions (Curtin, 2005; Rizzolo, 2021).

Coarsely, conservation has its strongest opponents in both human need and greed (we need such-and-such a plant to survive for any reasons/type of argument). Anthropocentrically, the species have to hold crucial positions in the human niche space to attract our attention: losing it (them) will lead to fundamental crises for humans. This is not a new perspective (though a vile one; is it true?); the crux of the situation is that we understand so little of the vast number of organisms; as stated above, many remain undiscovered. Thus, we lack a basis for constructing rational arguments for retaining individual species. Thomas (2017) seems unconcerned about the loss of species; he would argue that 'diversity' is increasing by an effective new 'Pangea' driven by invasions (recalling that Pangea was a supercontinent that incorporated nearly all of Earth's landmasses between the Late Carboniferous, c.335 Ma, and the Early Jurassic, c.175 Ma; see Fig. 3.1). One of his arguments runs that even if many species (invasives) are finding their way to New Zealand from Europe, the native flora and fauna of New Zealand are effectively untouched by these invasives. But the loss of native species, which will continue, is not simply a matter of invasives pushing them out, but other (human impelled) factors eating up that special landspace holding them. Thomas' contention that humans are 'helping nature grow and change', though not necessarily wrong, is distinctly anomalistically aberrant because it fails to investigate the complex systems so affected; the inevitable disruption and impoverishment of ecosystems have catastrophic impacts on all biotas and (ultimately) human populations. To consider for one moment its potential nadir (human internecine, global-wide nuclear warfare), the planet, of course, will survive with a motley of residual species and ultimately develop new forms from novel evolutionary pressures, but few would argue that this is desirable.

Measuring the human impact on habitats

The fundamental measure of human impacts on the landscape (land and ocean floor) is habitat loss, a subject that is growing steadily in the academic press (Dennis, 2015; Mugume *et al.*, 2015). The impact greatly affects migratory species (Hardesty-Moore *et al.*, 2018; Dennis, 2020), if only since the loss of a spatial subset may well result in the loss of essential resources/conditions and hence in the loss of the species. The evidence of loss is clear in reductions of animal biomass and species (Hussain *et al.*, 2012; Hallmann *et al.*, 2017; P.A.V. Borges *et al.*, 2019). Habitat loss is a complex subject, and involves at least three major components: (i) habitat reduction; (ii) habitat fragmentation; (iii) habitat degradation (Fattorini and Borges, 2012b; Primak, 2014; P.A.V. Borges *et al.*, 2019) (Fig. 10.5a); not one of these attributes excludes the other two. Typically, these different components of habitat loss have an integrated impact on an individual species' habitat (RBH) and on biodiversity (Fig. 10.5b). Thus, any one of these three types of loss to traditional habitat space compounds the effect on the other two. Losses for one species may also result in gains for other species with these changes to the landscape but even so there is bound to be loss of species (Dennis, 2015).

A simple formula has been devised to describe the impact of human activities on the environment (Chertow, 2000):

$$I = P \times A \times T \tag{10.1}$$

where I is impact, P is population numbers (density), A is affluence (defined as a per capita measure of wealth, consumption, or production), and T is technology. Growth of rich, densely populated areas is responsible for the greatest damage. The expectation is that **specialist** species are hit harder than **generalist** species, but this depends entirely on what has been introduced into a landscape subjected to habitat reduction, fragmentation and degradation. For example, conversion of a landscape into nitrogen- and

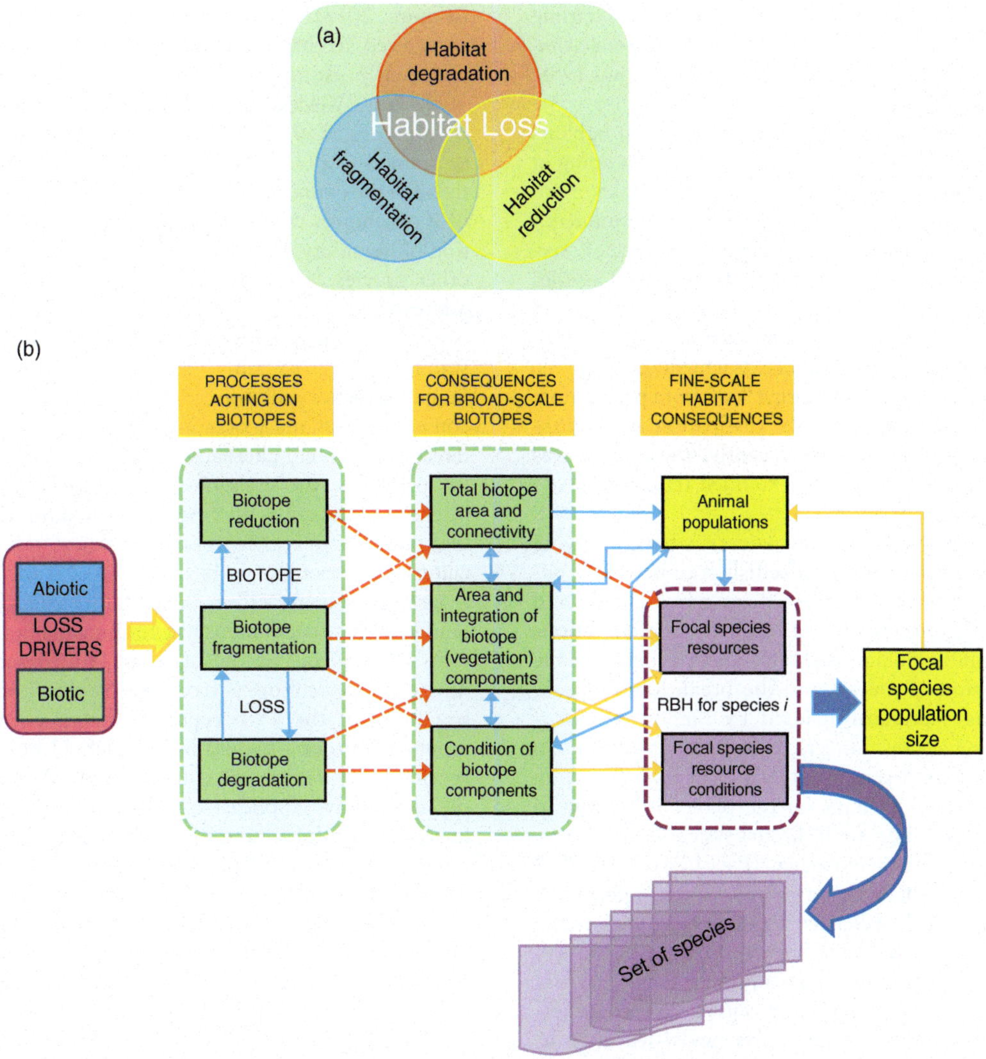

Fig. 10.5. Habitat loss. **(a)** Habitat loss expressed as combinations of reduction (destruction), fragmentation and degradation of the habitat space. For the traditional definition of habitat, the habitat space is biotope; for the resource-based habitat (RBH) concept, the space can involve any resource used by the organism. **(b)** Concatenation of biotope loss for habitat status of focal species. Abiotic (climate) and biotic (human) factors driving three processes of biotope loss have consequences for biotopes (i.e. area, connectivity and subunit integrity) and in turn for the resources and conditions (the RBH) of focal species and their populations; the consequence for each species is considered individually, the multidocument at the base indicating that a multiplicity of species will be affected in different ways. Boxes: biotope (green), resource-based habitat (purple), populations (yellow). Arrows: positive links (blue, continuous), negative links (red, pecked), positive or negative links (orange, continuous).

phosphorus-rich grassland is entirely deleterious (e.g. van Dobben *et al.*, 2017); on the other hand, changes to a **city** landscape are often associated with the presence of plants which provide food for many species, such as buddleia (*Buddleja davidii*), an excellent nectar source for butterflies, especially nymphalids (Lehner *et al.*, 2022), and ivy species (*Hedera helix* and *H. hibernica*), a source of food for many pollinators which use their nectar in autumn (Garbuzov and Ratnieks, 2014), and for various birds (including the blackbird *Turdus merula*, the blackcap *Sylvia atricapilla*, the European robin *Erithacus rubecula*, the European starling *Sturnus vulgaris* and the wood pigeon *Columba palumbus*) which use its berries (Snow and Snow, 1988; Bottema, 2001).

However, the overall impact of 'loss' on habitats (TB and RBH) is obvious: the rapid downgrading of biodiversity on the planet. A series of complex models have been constructed to measure the influence of spatial changes to **landscape** on a species' survival. For example, Dennis (2010) described a series of fragmentation measures developed by Jaeger (2000) and others; each term includes a number of potential effects; for instance, fragmentation may be visualised as a set of increasing 'loss' from perforation, through incision, dissection, dissipation and shrinkage, to attrition. More recently, these measures have been reviewed and extended (Uuemaa *et al.*, 2009; Cushman *et al.*, 2010).

We cannot rely on human decency for conserving ecosystems but there is a selfish argument for conservation action, one based on human need and greed, the preservation of resources and conditions for suitable human habitats. The 'payback' in the extinction of species and the breakdown of ecosystems – primarily caused by human actions – ultimately poses dangers to the survival of human populations. It is now patently clear that the action of billions of humans on the planet is leading to enormous changes to the Earth's systems that, in turn, threaten human existence (Fig. 10.6). Table 10.1 lists some components of global changes inflicted by humans which will have serious consequences for human life. It has taken many years for 'authority', even farmers operating on many thousands of acres of land and making handsome profits in crop production, to accept the reality of these changes and to realise the full consequences of human exploitation, especially the consequences of climate change (Dennis, 1993a; Meehl *et al.*, 2007; Collins, 2013; Wanders and Wada, 2015) and destruction of natural ecosystems (e.g. mass deforestation; Runyan and D'Odorico, 2016).

The global climate is not just warming but is becoming increasingly unpredictable, with expanding variance (extremes) in conditions (e.g. in October 2024, Valencia, Spain, received its annual total of rainfall in just one month), leading to critical effects on the **biosphere**, **atmosphere** and **hydrosphere** (drought, waterlogging and flooding, and the melting of polar ice, glaciers and tundra permafrost, which in turn causes rising sea levels and consequently creates serious threats to coastal cities and whole islands) (Millennium Ecosystem Assessment, 2005; Mimura, 2013). Gradually, human impact has been recognised as having deep 'roots': a progressive chemical poisoning of land, sea and air (Mishra *et al.*, 2016; Weldeslassie *et al.*, 2018; Landrigan *et al.*, 2020; Wang *et al.*, 2022), ultimately the poisoning of human populations (McMichael *et al.*, 2003), through the entire history of humanity. We face a set of stark choices that amount to either the continuation of humanity or its loss along with that of many other species. These issues are continually being addressed but, currently, there is little evidence for sincerity among those who have the power to ensure stability of ecosystems. Biodiversity-compatible human societies will not come about automatically, conservationists must organise, apply pressure, take direct action, build alternative institutions and mobilise people on the basis of common concerns and interests (Johns, 2019). Note that the most important obstacles to conservation are political, not difficult-to-solve scientific problems (Johns, 2019).

Strange as it may seem, some earlier changes to land brought benefits, not just for humans but also for wildlife. Many years ago, Dennis (1977) described how early agriculture – from the Neolithic to the beginning of the 19th century's era of chemical inputs in Europe – would have benefited butterfly species in the UK as well as humans. But then came the competitive application of chemicals, the ecologically harmful timing of agrarian practices, the intensive turnover of 'crops' and the replacement of forest with plantations, all of which have been accompanied by the withdrawal of nature. We are now witnessing the beginning of our own downfall, concomitant with the loss of soil from land, significant shifts in atmospheric composition (e.g. increasing concentrations of carbon dioxide and methane) and fouling of our water supplies with our own sewage (both freshwater and the oceans; Moe and Rheingans, 2006). We have become the terminators of our own existence simply because competition for making a profit dominates western cultures, driven by greed. Human well-being depends on material welfare, health, good social relations, security and freedom, which are affected by changes in ecosystem services (Millennium Ecosystem Assessment, 2005; see Fig. 10.6). Materials, of immense value to human well-being, are being grossly overexploited to the extent that ecosystem failure is becoming a reality. But once an ecosystem has undergone an abrupt change, recovery to the original state is slow, costly and sometimes impossible. The economic value of **ecosystem services** is immense. For example, in their seminal paper, Costanza *et al.* (1997) estimated that the combined value of 17 ecosystem services in the USA amounted to US$16–54 trillion per year.

There is now a rapidly growing literature on the notion of value in relation to natural capital and ecosystem services (Häyhä and Franzese, 2014). Yet the importance of preserving natural capital is far from being at the heart of current development policies. It is inevitable that the globe faces massive land use changes from expanding urbanisation, industrialisation, intensification of production and (loosely) land use fragmentation. The future is not one of sustainability in natural ecosystems but of continued competition for exploitation, and the endproduct is devastation, politicians not taking on board advice from pundits or caring what the impact will be on each successive generation as long as they acquire their slice of 'flesh and blood'. Most correctly, Johns (2019) defined conservationism as 'the last anti-colonial battle'. The current ecological crisis is best understood as human imposition of a colonial relationship on the natural world. Human economic activity, and hence consumption, doubled between 1900 and 1950, then quadrupled between 1950 and 1992, and continues to grow (Johns, 2019).

Where will human interference stop? Nowhere if we intend to persist for millions of years without changing our attitudes. What does that imply? Theoretically, the colonisation of other planets (possibly even outside our solar system), each colonised planet a patch in an extraterrestrial human **metapopulation**. Some see the process of terraforming distant planets as a feasible process (Sandu *et al.*, 2017; Bortoni-Anzures and Herrera-Rivas, 2019; Bortoni-Anzures and Ortiz-Moctezuma, 2019), but it is a highly complex business and currently beyond our technology (Todd, 2006; Pazar, 2018). Our planet has a number of unique attributes, for example a liquid iron outer core generating a magnetic field (magnetosphere) that is an

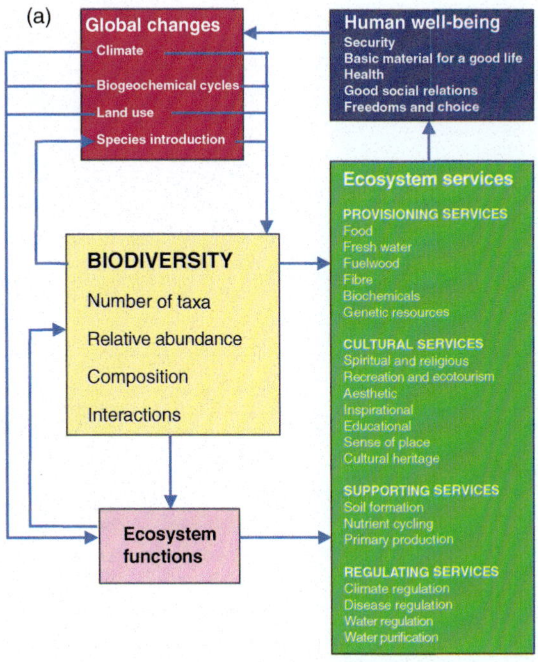

Fig. 10.6. Humans impacting biodiverse ecosystems. **(a)** Relationships between biodiversity, ecosystem functioning and ecosystem services based on Millennium Ecosystem Assessment (2005) typologies. Biodiversity is both a response variable affected by global change drivers and a factor modifying ecosystem processes and services and human well-being. **(b)** Recovering ecosystem services by means of restoration and reclamation. **(c)** Model of different stadiums of a cultural landscape disturbed by lignite mining activities and subsequent restoration or reclamation with potential land use scenarios for the postmining landscape; examples of ecosystem services in the foreground of the respective stadium of development are shown in the tables below. (Panel **a** based on Mandal *et al.*, 2011. Panels **b** and **c** from Gerwin *et al.*, 2023, used under licence CC BY 4.0.)

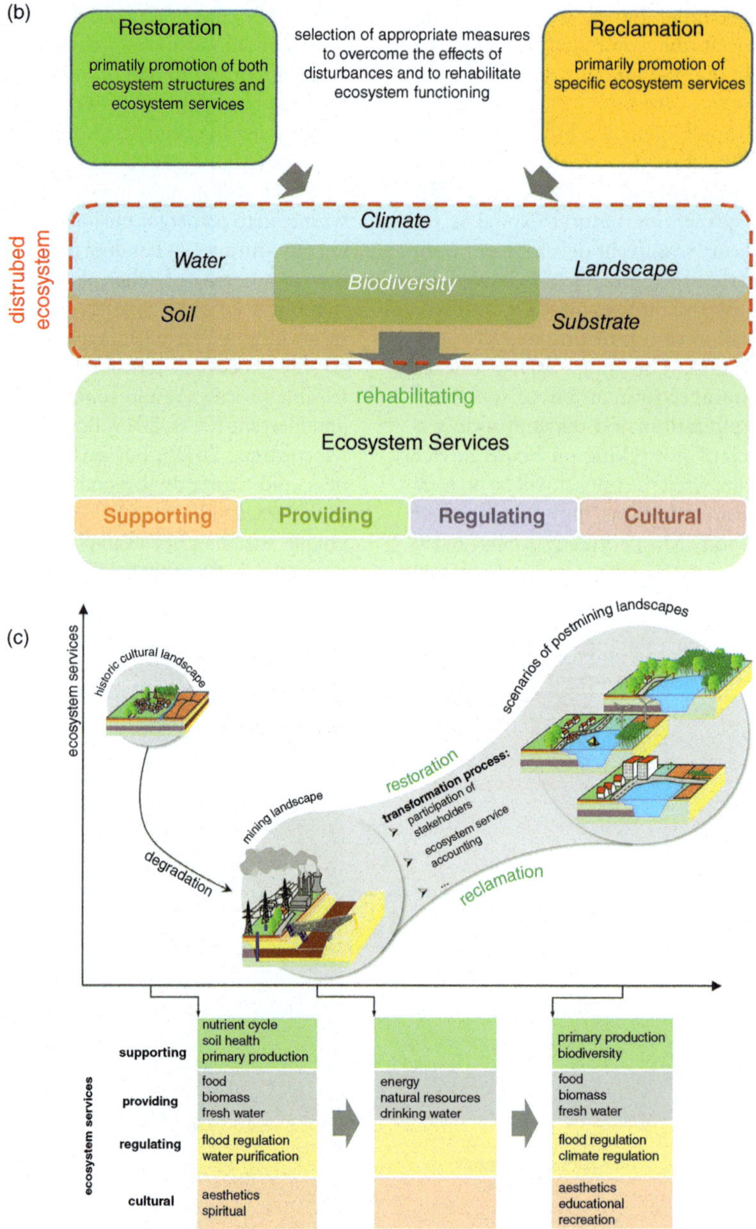

Fig. 10.6. Continued.

essential component for protecting the planet from harmful solar particles (Zhang, 2011; Dormy, 2014; Borovsky and Valdivia, 2018), and water in its distinct states of solid, liquid and gas, crucial for life. Even so, as regards extraterrestrial exploration, what may appear impossible now might in the future become practicable (for example, with the assistance of artificial intelligence; Russo and Lax, 2022); we are still a young species. However, our survival depends on our capacity to co-operate – but can we offset the mutual hatred and jealousy that are a fundamental part of human existence and replace war with intelligent compromise? A view of current warfare, of neighbourhood hostility, of criminal parasitism and predation, would perhaps suggest to the reader that this is an unlikely outcome.

11 Conserving Life: Biodiversity and the Resource-based Habitat

Abstract

Major scientific challenges to biodiversity conservation arise from the enormous number of species involved in ecosystem functioning and our generally poor understanding of the underlying mechanisms. Counteracting the current massive extinction rate due to the impacts of humans on the biosphere is imperative. Conservation requires integrative approaches that should be founded on a solid knowledge of how species interact through their habitats; thus strategies for biodiversity conservation based on traditional habitats need to be expanded through the development of approaches centred on the resource-based habitat concept. The adoption of a broader, systems view of biodiversity should incorporate information on species responses to environmental changes through their peculiar traits across their ranges. An ecosystem approach, including all micro-organisms, fungi, plants and animals, is the way forward to an understanding of how species interact within communities and hence to the preservation of Earth's biodiversity.

Conservation has two basic targets: (i) the management of biodiversity at all scales, from small areas (e.g. traditional **habitats** such as a meadow, a wood, a marsh, a wave-cut platform, an urban space) to large zones (an island, a county or state, a biome, a continent) – viz. a biotope or a vast area into which biotopes fit; (ii) the preservation (conservation) of a particular species through its **resource-based habitat**. Although these two elements (components) of the natural world are finely integrated, it is useful to deal with them separately below, if only because the techniques differ, especially the costs of maintaining a whole ecosystem (the community in a traditional habitat) and the space required with all the essential ingredients for a single species.

First, a *résumé* of biodiversity issues. Gaston (2010) distinguishes between three basic aspects of biological diversity (**genetic, organismal** and **ecological**). Genetic diversity encompasses the various aspects of genetic differences (nucleotides, genes and chromosomes) within and between populations. The organismal diversity encompasses the diversity through the taxonomic hierarchy and its components (from individuals upwards to species and beyond, e.g. genera, families, orders, classes and phyla). Ecological diversity refers to the variety of ways in which species interact with one another and their environment (e.g. diversity of **niches, communities, biotopes, ecosystems, landscapes** and **biomes**).

The most typical measure of biological diversity at any scale rests on the number of species, a form of organismal diversity. Globally, estimates of species diversity vary enormously for different taxa (see Box 8.1 in Chapter 8). Overall, there may be between about 5 million and more than 160 million species on Earth, but only about 2 million species have been formally named (Singh, 2012; see Box 8.1). Over evolutionary time scales, biodiversity (B) is a balance between rates of speciation (S) and extinction (E) ($B = \delta S - \delta E$), with a normal background rate of species extinction of about 0.1 extinction per million species per year, and speciation rates of 0.05–0.2 new species per million species per year, depending on the taxonomic group (De Vos *et al.*, 2015). In general, speciation rates tend to be higher than extinction rates, leading to a progressive increment of life forms through time; for example, the diversity of marine and continental life (measured by the number of families) has increased exponentially since the end of the Cambrian (Benton, 1995, 2001).

Fossil records show that extinction rates tend to remain fairly constant over time, but there are occasions during which a large proportion of species is lost in a relatively short period. These events of biological crisis are known as **mass extinctions** and are triggered by large-scale and rapid changes in environmental conditions (Fattorini, 2020). Estimates of the number of mass extinctions that occurred in

DOI: 10.1079/9781800627505.0011

the Phanerozoic (541 Ma to the present) vary greatly, depending on the criteria used to measure past diversity and the proportion of species lost to qualify an extinction event as major. However, most authors recognise five such mass extinctions (Fattorini, 2020; see Chapter 10). Notably, current extinction rates are estimated to be 1000 times higher than natural background rates, and future rates are likely to be 10,000 times higher (De Vos *et al.*, 2015). This means that we are experiencing a new (sixth) mass extinction, an anthropogenic one. As extinction rates are higher than discovery rates, we are losing a diversity of which we have no knowledge. Even assuming that taxonomists will survive their object of study, with a current rate of species description of c.8000 species per year, an extinction rate of c.0.72% per year and a global diversity of c.8 million species, global diversity will have been reduced to 33% or even 15% before we will have described all extant species (Strona, 2022). Though the number of taxonomists is increasing, the number of species per taxonomist is decreasing, which stresses the need for a rapid increase in the number of taxonomists (highly specialised scientists) who can seek out species that have evaded discovery before it is too late (Joppa *et al.*, 2011b). Who knows what the consequences might be of the extinction of species whose existence is not even known about?

Traditional habitat designation and the conservation of biodiversity

It is a mistaken notion to regard a list of species as sufficient to describe the biodiversity of any traditional habitat space (area), although species listing functions as an essential starting point. Certainly, in any ecological study, lists of species accompanied by measures of abundance are necessary to describe community structure (see Chapter 4). It is obviously crucial that any assessment for the maintenance of a site includes a list of species and, if comprising a mix of units of different traditional habitat types, an 'association' between the different **substrate** units based on data for the taxa. For example, availability of accurate lists of species in their habitat units is essential for revisitation (resurveying) studies, in which historically surveyed communities are resurveyed some years later to investigate changes in species composition, especially to detect species loss (Kéry, 2004; Sperandii *et al.*, 2019; Holderegger and Spillmann, 2022).

The link between biodiversity, resources and conditions forms the next stage. The study of such interconnectivity is central to understanding biodiversity as each habitat unit often forms a part **ecosystem** (see Chapter 2); we know that species in any recognisable spatial unit live on top of, under, beside, within, on or in those of other species. Some connections are strong, others weak or non-existent; some are negative, others positive. A fundamental truth is that no two communities are identical (in species composition, **species abundance distribution**, interspecific relationships, etc.; see Chapters 2 and 4), not even if we artificially compartmentalise each traditional habitat unit into finer divisions of apparently uniform space for survey and comparison; the space may look homogeneous but that is so rarely the case. Such 'internal' (within-habitat) differences, along with species' traits, can create asymmetry in behaviour and life history, and thus in resource use and vulnerability to abiotic conditions. Sampling within habitats gives us a measure of internal habitat diversity that may not be apparent to the 'eye'; sampling between habitat units measures habitat diversity at a higher (e.g. **landscape**) scale (see also Box 4.3 in Chapter 4 for the concept of habitat diversity). The results may be surprising, if only because each artificial unit (e.g. grid squares) established in a traditional habitat will present different resources and conditions for the 'x' species.

Much uncertainty remains, if only because of the following factors.

- The limited time of observations. Human lifespan is short compared to the length of experiments that need to be carried out (e.g. the 24-year study of Mongolian grassland under natural turnover by Bai *et al.* (2004) is an indication of just how much of one's life can be gobbled up by these studies). However, long-term ecological research (LTER) is essential to understanding ecosystem functioning and distinguishing natural changes from anthropogenic modifications (Haberl *et al.*, 2006; Müller *et al.*, 2010; Mirtl *et al.*, 2018; Willig and Walker, 2016; Waide and Kingsland, 2021). The rapid growth of LTER projects between the end of the 20th century and the beginning of the 21st century, indicated by the development of international and national networks (e.g. https://lternet.edu/; https://elter-ri.eu/; https://www.nceas.ucsb.edu/lter-network-office; accessed 21 May 2025), gives rise to hope that in the future we will have a greater understanding of ecological phenomena that take place over large time scales. At present, however, our understanding of long-term ecological dynamics remains limited.

- The differences between **resilience** and **resistance** of ecosystems (Box 11.1) in confined (closed) and open traditional habitat units. Confined and open habitats differ in the degree of input and output, and hence in the way they are exposed to and can react to perturbations. Compared to open habitats, closed habitats have, in general, lower resistance and resilience for various reasons. First, because of their small size, confined habitats have a limited ability to 'absorb' disturbances (lower resistance) and a reduced capacity for species migration and recolonisation after a disturbance (lower resilience). Second, confined habitats often host fewer species, many of which are adapted to specific, stable conditions. Thus, their **functional redundancy** is low (which weakens resistance) and recovery depends on specialised species that may not return or reproduce quickly (undermining resilience). Third, closed habitats host isolated populations with lower genetic diversity and hence decreased adaptive capacity; consequently, recolonisation possibilities are reduced (low resilience) and populations may lack phenotypes adapted to the new conditions (low resistance).
- The lack of knowledge about the availability of niches within any traditional habitat space and the multitude of traits held by species within and outside a habitat unit (the so-called Hutchinsonian shortfall; Cardoso *et al.*, 2011; Hortal *et al.*, 2015). There is growing interest in the study of functional traits, but for many organisms this knowledge remains very limited (see Chapter 2, Box 5.2 in Chapter 5 and Appendix 3).

These forms of uncertainty underscore just how little we truly know about how ecosystems function and respond to change.

Conserving what biodiversity?

Turning to the question of what the consequences might be of losing species we do not even know exist, the preceding section adds further cause for concern. Thus, it seems more than sensible to preserve what we do not yet understand. However, when conserving biodiversity the first thought is typically: what is it in a particular traditional habitat's biodiversity that needs the most attention from management? It is certainly more than a limited selection of taxa (e.g. exclusively orchids, leaf beetles, small mammals or butterflies), one of the reasons why single-taxon conservation societies may play a limited role in conserving biodiversity: they may get 'things' right for one class of taxa (e.g. butterflies), but not for others. To answer such a question fully, land managers must not only think like fully qualified ecologists, statisticians and politicians, but also be aware of stochastic events. They will find conundrums entering their assessments, including statistical pitfalls (such as **regression to the mean**, a stochastic phenomenon typically disregarded in vegetation research, making analyses of the drivers of change potentially misleading; Mazalla and Diekmann, 2022) and effects of recording bias linked to species behaviour or seasonality (Dennis, 2010). But they must be especially aware of what should be incorporated in the study of a particular space; which abiotic and biotic components maintain or undermine ecosystem stability and how their interactions impose variability in biodiversity. Just how complex the process can become is shown, for example, by the many issues associated with species translocations (**reintroduction**, **reinforcement** and **conservation introduction**; see Chapter 10 and section *Introductions and cascading effects* in this chapter for examples of problems caused by introduced species), which makes an audit of the processes involved in 'release events' of 'key' species essential (Daniels *et al.*, 2018) (Table 11.1).

Minimally, diversity in any habitat space or part habitat space (e.g. butterfly species; Dennis, 2010) has a critical base of resources and conditions. This rapidly becomes a hyperdimensional display of species and their **traits** against measures of conditions and potential resources (e.g. food, roosts and mate location sites); further axes can identify predators, competitors, parasites, facilitators, etc. of the target species. The 'points' closest together in the multidimensional space are those units containing taxa most similar in their relationships to the array of conditions and resources present in their environment (Dray and Legendre, 2008; Dapporto *et al.*, 2015). They can therefore be treated in much the same way in the process of management. One fundamental problem is some inborn reaction against particular taxa – yet ones that may have a crucial part to play in ecosystems – in those carrying out a study, which results in their being ignored or even eliminated, for example when insects are overlooked

Box 11.1. Resistance and resilience

While **resilience** describes the speed with which a system returns to its original status after a perturbation (that is, the system's capacity of recovery), **resistance** is the ability of the system to persist or withstand a disturbance (Nimmo *et al.*, 2015; Van Meerbeek *et al.*, 2021) (Fig. B11.1.1). Both resistance and resilience contribute to define ecosystem stability. The degree of ecosystem resilience and resistance varies considerably according to the involved system and the characteristics (type, intensity, duration) of disturbance. For example, fire is a prominent form of disturbance in Mediterranean ecosystems, but its effects on communities depend on the intensity, frequency and season of burning, the characteristics of the burned habitat and the organisms involved (Fattorini, 2010a; Mantoni *et al.*, 2020; Palusci *et al.*, 2021).

Stability is a property of all biological levels, from genes to communities. But our understanding of the processes that enable components to resist or revert to baseline is still very limited: new, important efforts combining eco-evolutionary knowledge with ecosystem-level concepts of resilience are required (Thorogood *et al.*, 2023). According to Van Meerbeek *et al.* (2021), stability does not result from resistance and resilience alone but has three other properties: recovery (i.e. the ability to return fully to the reference state), latitude (i.e. the maximum amount a system variable can be changed before losing its ability to recover) and tolerance (i.e. the ability of a system to tolerate perturbations). All five components of stability — resistance, resilience, recovery, latitude and tolerance — are separately necessary but not sufficient measures for assessing stability, and their measurement has been proven to be challenging, also because of conceptual and methodological disagreements (Van Meerbeek *et al.*, 2021). Rapid changes due to human activities are now profoundly affecting ecosystem stability by crossing threshold levels for key variables, causing ecosystems to shift into new, undesirable states, with limited (if any) possibility of recovery. It has been observed that, at global level, three interlinked planetary boundaries (rate of biodiversity loss, climate change and human interference with the nitrogen cycle) have already been overstepped, generating environmental changes that pose unacceptable risks to humanity (Rockström *et al.*, 2009).

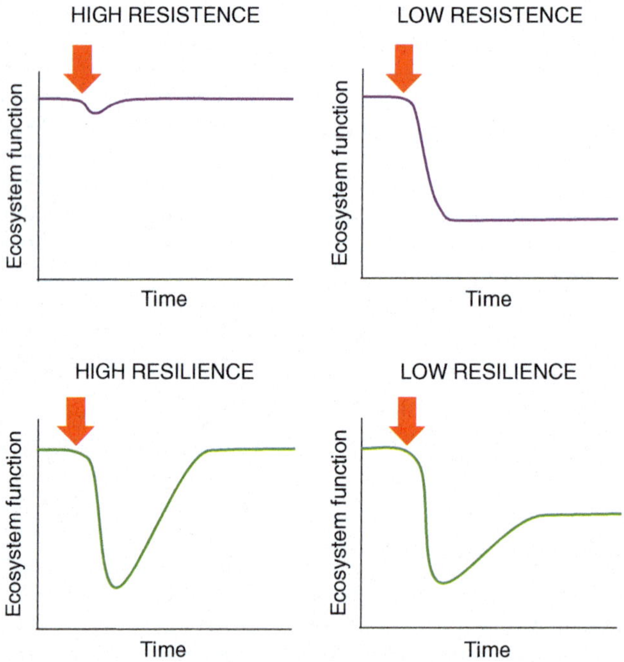

Fig. B11.1.1. Schematic representation of the effects of disturbance (red arrows) on ecosystem functioning. The graphs illustrate the difference between ecosystems with high and low resistance and with high and low resilience.

Table 11.1. List of recommended actions and types of information to be recorded prior to and after releases of butterfly species.

1. To be documented prior to all release events for planning purposes.

Information should include:

Taxon

Proposed location (conservation land name) and climate

Proposed site(s) within larger land area (i.e. management unit or other designation)

Proposed schedule of release (i.e. month, other indicators of appropriate timing)

Recipient site(s) restoration and management history

Recipient site landscape structure and characteristics. Includes patch area and spatial structure, barriers to dispersal, surrounding matrix

Release plan and objectives

a. Type of release (introduction, reintroduction, reinforcement/augmentation)
b. Type of livestock (wild, captive-bred, head-started)
c. Numbers to be released and reason
d. Life stage to be released and reason
e. Genetic and health screening for source population and recipient population
f. Condition of livestock (freshly eclosed, mated, etc.); treatment of livestock prior to release (time period held, conditions, resources provided, etc.)
g. Proposed frequency of releases and reason
h. Proposed duration of release efforts
i. Type of release (i.e. soft, hard)
j. Proposed measures of success

Postrelease monitoring plan

Affiliated organisation/agency and personnel conducting release

Overseeing body/permitting agency

2. Establish a plan to publish/report results.

3. Document *a priori* questions in release and monitoring objectives.

Consider reintroduction as an experiment to assess different methods.

4. Postrelease monitoring (short-term). This would include active monitoring of released organisms and any later life stages.

Behaviour upon release.

Behaviour in subsequent days following release. This would include signs of reproduction (i.e. mating, oviposition), residence in site, dispersal to neighbouring patches or sites (if organisms are marked), and other ecologically relevant observations such as ant–larval interactions (for myrmecophilous taxa) and predation/mortality.

5. Postrelease monitoring (long-term). This would include active monitoring of subsequent generations resulting from organism releases.

Employ standardised methods to estimate relative abundance or population density in order to track population trends over time.

Develop appropriate monitoring schedule based on the target taxon's phenology and voltinism.

Survey broadly across release site(s) and any neighbouring habitat areas to document occupancy and potential dispersal/colonisation.

Conduct regular monitoring on an annual basis.

Document signs of reproduction (i.e. mating, oviposition, presence of immature stages).

Document opportunistic ecologically relevant observations, if possible, including ant–larval interactions (for myrmecophilous taxa), use of specific floral resources, and any predation/mortality.

Continued

Table 11.1. Continued

6. Habitat monitoring.
Conduct baseline vegetation surveys of all recipient sites prior to any release event. Surveys should focus on mapping, quantifying and characterising identified key landscape resources for the target organism (i.e. larval host densities, patch sizes, patch configurations, plant quality, plant phenology; nectar source diversity, density, phenology and condition).
Include regular basic vegetation surveys concurrent with organism monitoring where possible.
Document impact of and photograph any observed habitat disturbances (invasive species removal, fire, herbicide treatment, storm damage, etc.).
7. Project evaluation and documentation.
Meet with recovery team annually to discuss efforts and findings/results. Revise efforts and protocols as needed using an adaptive management cycle.
Include legacy monitoring data if available and of appropriate quality in analyses for projects involving reinforcement/augmentation.
Evaluate project after fifth year (and at regularly established intervals if organism releases and monitoring are extended) and report all outcomes in a scientific journal or at least in a report. Individual recovery plans generally have specific recommended metrics of success based on population size, area of occupancy and/or population trends/stability over time.

(**entoalexia**) or anthropocentrically regarded as either useful or harmful (**entomyopia**) (Leather, 2021).

Time is the one axis not mentioned in the above paragraph. One prominent feature of any space is that resources and conditions, and thus the associated taxa (species), change over time. This is most obvious over geological time, particularly after **mass extinctions** (e.g. end Ordovician, late Devonian, end Permian, end Triassic and end Cretaceous (Cretaceous–Palaeogene event); Barnosky *et al.*, 2011; McCallum, 2015; Fattorini, 2020; Ritchie, 2022; see Chapter 10 and Appendix 5). Classic examples of changing faunas and floras occur as far back as the Palaeozoic, for example the explosion in trilobite species from the Cambrian and their extinction in the Permian (Webster, 2007; Fig. 11.1a); the development of *Prototaxites* fungus communities dating from the Silurian to the Devonian (Fig. 11.1b), communities in which a fungus towered >8 m over land plants (Honegger *et al.*, 2017); then the mid-Devonian to Carboniferous first trees, with the progymnosperms *Wattieza* and *Archaeopteris* reaching 10 m tall (Stein *et al.*, 2007; Austen and Austen, 2016; Fig. 11.1c), and the development of progressively more complex forests (Fig. 11.1d, e) hosting extraordinary arthropods such as giant dragonflies (e.g. *Meganeurites gracilipes*; Nel *et al.*, 2018; Fig. 11.1f). The Triassic was characterised by crocodylomorphs, such as the fearsome bipedal *Carnufex carolinensis*, the 'Carolina Butcher' (Zanno *et al.*, 2015; Fig. 11.1g), which held a prominent position among terrestrial reptiles, including early dinosaurs, the latter of which became dominant in the Jurassic and mostly went extinct with the Cretaceous–Palaeogene extinction event (avian dinosaurs (birds) survived and continue today) (Fig. 11.1h, i). Notably, these different periods were characterised not only by changes in the climatic conditions but also by variations in the gaseous content of the atmosphere, as neatly reflected in patterns of insect size variation in the fossil record (Chapman and Karr, 2012). Oxygen concentration is a key physiological control of body size in flying insects as they have (i) high metabolic oxygen demands and (ii) a respiratory system in which oxygen diffuses through a network of tiny tubes (tracheae) – a system that works well for small bodies but becomes inefficient as body size increases. Thus, insect size in the pre-Cretaceous fossil record appears to be driven by atmospheric oxygen content (hyperoxic conditions): the greater the percent of atmospheric oxygen, the larger the insects. But this relationship held only for the first 150 Myr of insect evolution. Subsequent drop in size has an ecological explanation in the rise of birds and the importance of manoeuvrability of flight, with smaller size enabling more agile flight (Chapman and Karr, 2012) – as ever, abiotic and biotic factors interact in driving evolutionary trajectories.

At a much shorter temporal scale, we have already seen (Chapter 3) how communities can change very quickly in our own limited ecological time through natural **vegetation succession** and in response to anthropogenic **disturbance** (e.g. for butterflies, see

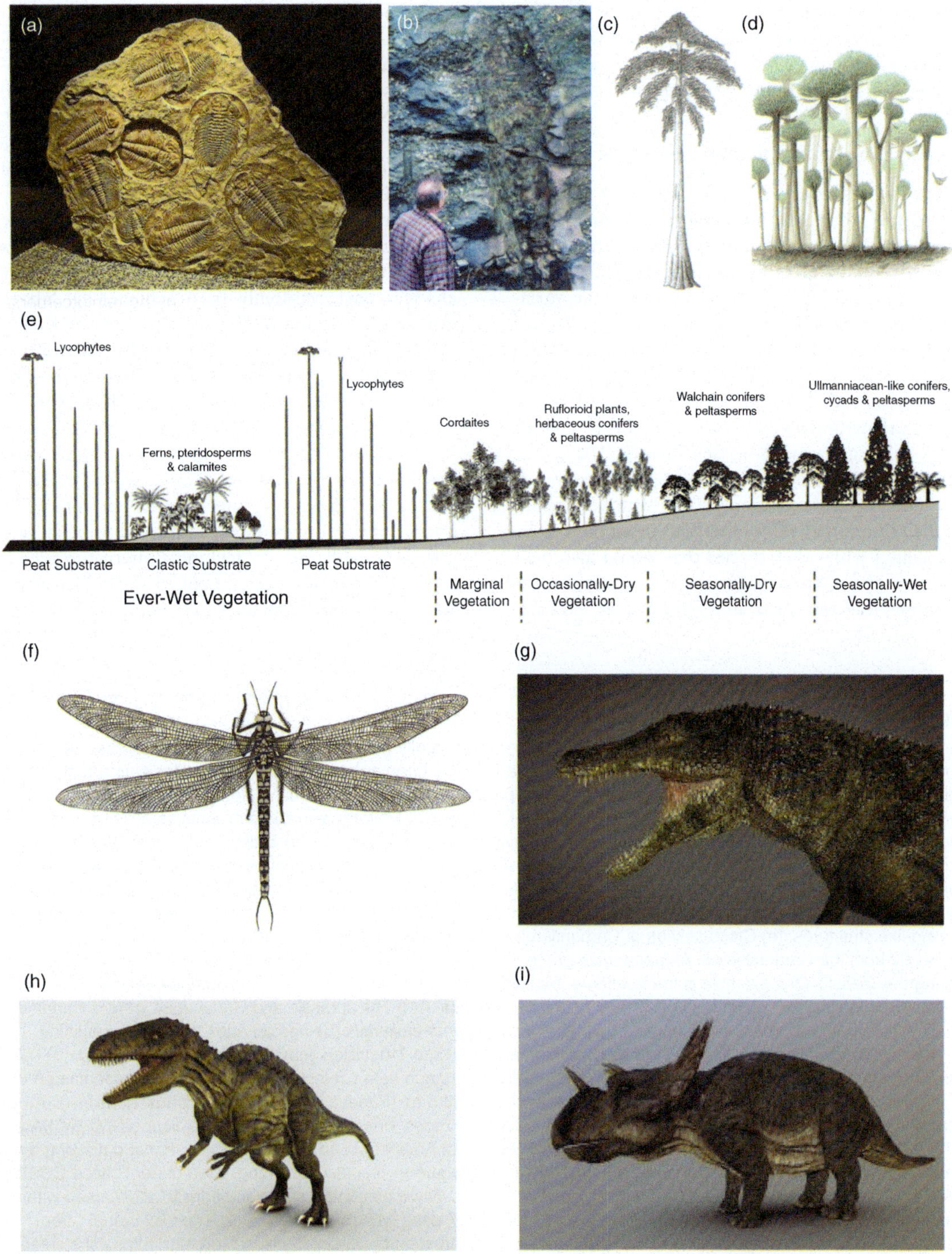

Fig. 11.1. Life forms between mass extinctions. **(a)** Cambrian trilobites (*Acadoparadoxides briareus* from Morocco). **(b)** A mysterious Devonian fungus, *Prototaxites milwaukeensis*. According to some reconstructions, the *Prototaxites* fungi towered more than 8 m over most of the land plants of the Silurian and Devonian periods and were as much

Dennis, 2010; for birds, see Fuller, 2012). This dynamic nature of communities is a critical factor and one that management has constantly to bear in mind, as discussed immediately below.

Landscape taming versus landscape wilding

Currently, there are two fundamentally different approaches to raising biodiversity: (i) management based on results of careful and thorough surveys, which we may call 'taming' landscape; and (ii) land abandonment but in such a way that some variation is maintained between different components of the landscape (e.g. habitat types from bare earth to complete dense tree and shrub cover) using large herbivores and their predators (deer, wild boars, wild cattle with wolves, lynxes, etc.); this is referred to as wilding or **rewilding** (see Chapter 10). In the first approach, an (alleged) well-determined endproduct is the focus of attention, usually with key species as the main targets, but ones that are known to be associated with other species that would also benefit from the same management technique (see, for example, the concept of **umbrella species**, Box 2.3 in Chapter 2). A classic case is the work carried out in the UK, the Italian Alps and the Czech Carpathians to maintain the resident and reintroduced large blue butterfly *Phengaris* (*Maculinea*) *arion* (Simcox *et al.*, 2005; Spitzer *et al.*, 2009; Casacci *et al.*, 2011) (see section *A species' intimate community* below). Really the objective is a resource-based habitat one, but it is increasingly being tailored to improve biodiversity based on the community of creatures associated with the butterfly. In the second approach, massive biodiversity changes have randomly been achieved by 'abandonment' of 'traditional agriculture', as at Knepp Castle, UK (3500 acres) recounted by Tree (2018) and Tree and Burrell (2023). Owing to The Dangerous Wild Animals Act 1976 (UK) it was not possible to release wild boars (*Sus scrofa*), wolves (*Canis lupus*) or lynxes (*Lynx lynx*) but they could and did release fallow deer (*Dama dama*), roe deer (*Capreolus capreolus*) and Tamworth pigs. The impact of this programme has proved to be largely self-sustainable (despite lacking a major top predator such as wolves; deer are culled and sold as venison) and more remunerative compared to the

Fig. 11.1. Continued.

as 1 m wide. However, some consider that *Prototaxites* were actually giant lichen-like structures (Selosse, 2002), comprising a union of fungi and photosynthetic algae, and it has been suggested that *Prototaxites* was a fungal rhizomorph that grew on its side and probably at least partially underground (Vajda *et al.*, 2022). According to a recent analysis of the internal structure and molecular fingerprint of *Prototaxites*, these organisms are not really fungi but should be best considered a member of a previously undescribed, entirely extinct group of eukaryotes (Loron *et al.*, 2025). **(c)** A Devonian-Carboniferous tree, *Archaeopteris*. **(d)** The oldest fossilised forest known on Earth (390 Ma) on sandstone cliffs along the Devon and Somerset coasts in south-west England, UK (Davies *et al.*, 2024); the fossilised trees (*Calamophyton* sp.) had hollow thin trunks and lacked leaves, with branches covered in hundreds of twig-like structures. **(e)** Cross-section of Carboniferous forest, showing different forms of vegetation. **(f)** A giant dragonfly from the Carboniferous, *Meganeurites gracilipes*. **(g)** Crocodylomorphs were dominant terrestrial predators during the Triassic. One example is the fearsome, bipedal, 3 m long crocodile ancestor *Carnufex carolinensis* (known as the 'Carolina butcher') from the Late Triassic of North America. **(h,i)** The Jurassic and Cretaceous were dominated by dinosaurs both carnivorous and herbivorous. *Acrocanthosaurus atokensis* **(h)** was an apex predator surpassing 11 m in length that existed in what is now North America from 113 to 110 million years ago (early Cretaceous). *Regaliceratops peterhewsi* **(i)**, which lived about 68 million years ago (late Cretaceous) in an area corresponding to Alberta (Canada), was a herbivore with an estimated length of 5 m. (Panel **a** photo by Hectonichus: https://commons.wikimedia.org/wiki/File:Paradoxididae_-_Acadoparadoxides_briareus.JPG, used under licence CC BY SA 3.0. Panel **b** from Boyce *et al.*, 2007; ©The Geological Society of America, used with permission. Panel **c** drawing by Falconaumanni: https://commons.wikimedia.org/wiki/File:Archaeopteris_reconstruccion.jpg, used under licence CC BY SA 3.0. Panel **d** drawing by Peter Giesen and Chris Berry: https://www.eurekalert.org/multimedia/1018227, used with rightsholder's permission. Panel **e** from Thomas and Cleal, 2017, used with permission of Springer BV. Panel **f** from Nel *et al.*, 2018, used under licence CC BY 4.0. Panel **g** reconstruction by Julian Johnson-Mortimer: https://sketchfab.com/3d-models/carolina-butcher-head-8aa35d1fab1f411e843d876e41b17043, used under licence CC BY 4.0. Panel **h** reconstruction by Julian Johnson-Mortimer: https://sketchfab.com/3d-models/acrocanthosaurus-06b1114bc5d747aca2664c69444831c8, used under licence CC BY 4.0. Panel **i** reconstruction by Julian Johnson-Mortimer: https://sketchfab.com/3d-models/regaliceratops-fb292480dc88483f919c3b11637da3da, used under licence CC BY 4.0.)

earlier farming of this area, which is composed of a boggy heavy clay. It has delivered extraordinary results. A surprising number of unexpected species have arrived at the site (e.g. the nightingale *Luscinia megarhynchos*, the turtle dove *Streptopelia turtur* and the purple emperor butterfly *Apatura iris*; Oates, 2020 – to mention but a few).

Rewilding has now become a standard strategy for conserving biodiversity (Corlett, 2016). Following the initial adoption of somewhat 'anarchic' approaches, attention has recently been given to more formal frameworks (variants of the process). Massenberg *et al.* (2022) particularly emphasise the temporal dynamics of changes in landscapes, which are often long term and therefore difficult to monitor, the openness of rewilding processes and the related uncertainty about end states, as well as the complexity of human–nature relations and the associated value of the methods one might refer to as pluralism. In other words, they feel that rewilding should approach a more formal management system with distinct endproducts in mind. Note how different this approach is from the original objective established at Knepp, where the approach was entirely random.

Applying **Q methodology** (a research method used in psychology and social sciences to study differences in people's perspectives about a certain issue), Holmes *et al.* (2020) have analysed the views of rewilding practitioners and advocates across Europe. They have identified two distinct approaches, one focusing on extensive (random) radical transformation of rural landscapes towards wilder states and the other based on pragmatism (determinism), and hence on different, evidence-based management forms of rewilding in different locations. They also found that, in general, people tend to consider human presence as a legitimate part of wild nature, but identified disagreements among people over how to decide what level of human manipulation and presence is acceptable. These approaches are not as far apart as may seem to be the case, and it will become increasingly evident just what endproducts are produced when applying any one rewilding approach. Practitioners will select the approach which they feel best suits the **landscape** and **substrates** of that landscape, and the surrounding human population.

Unless a more or less 'random' approach is selected (with rewilding), the future is bound to induce the development of large companies involved in biodiversity engineering. The maintenance of biodiversity requires extensive space and complex modelling and thus businesses involved in engineering will need large capital investment and numerous well-trained staff. The two essentials are size (area of the space engineered) and variability (variation in substrates and plant communities, ranging from bare ground to mature forest) (Shreeve and Dennis, 2011). But Shreeve and Dennis also argue that the behavioural component of how species respond to resources and conditions needs to be taken fully into account in modelling species responses to landscape; this, in turn, determines how limited resources for conservation should be deployed. In general, the majority of species may be best served if conservation effort is additionally focused on increasing landscape heterogeneity. For example, Habel *et al.* (2023) find that landscape homogenisation and simplified butterfly community structures go hand in hand in northern Austria. **Disturbance** is also a critical part of this agenda. Complexity is the keyword for engineering, not only for the fundamental reason of acquiring 'space', investment and returns to stakeholders, but also ensuring that targets are met in determining the security of species richness and density. Just how complex this can be is well illustrated in the influence different landscape elements have on a single group – wild bee pollinators in relation to floral resource availability and/or diversity (Ammann *et al.*, 2024), or even on only two species – the blue tit *Cyanistes caeruleus* and the great tit *Parus major* (two insectivorous birds) in relation to urban tree species composition (Jensen *et al.*, 2023). But the more frequently (re)wilding methods are used, the easier it will become to predict their outcome, 'data' that can be built into engineering projects. Wilding will become less of a 'wild' (stochastic) approach'!

Construction opportunities for habitats and biodiversity

Following on from the above, an important question is: just how does it become increasingly evident that a (re)wilding approach attains a level of predictability? It is not only a matter of substrate, geographical location and the organisms released on any site. Most crucial is increased knowledge of the **traits** of species that are released (thus their trophic position or **niche** associations in any ecosystem), and the traits of species that arrive and establish with a designated approach. Among plants, abiotic conditions have been noted to interact with both invader traits and resident **functional diversity** to determine invader survival (I.L. Borges *et al.*, 2019). Plant **functional traits** have also been suggested to underlie changes in novel ecological strategies for dominant species in stages of forest succession (Chai *et al.*, 2016b). Thus, it is to be expected that traits and the position of species within an ecosystem (their niches and trophic status) determine the outcome of a project, but this will depend on how many 'controlling' species are released and in turn maintained. Despite the increasing interest in trait studies (see Chapter 2), in some groups (e.g. insects; Gentile *et al.*, 2021) they are in their infancy. Moreover, while research on intraspecific trait variation is advanced in plants (Di Biase and Fattorini, 2025), studies on intraspecific variation in animal traits remain more limited (see Box 5.2 in Chapter 5), although intraspecific variability is obviously also important in these organisms, with some traits varying more between populations than between species (e.g. growth rate, mortality rate and length of breeding season in European freshwater fish; Blanck and Lamouroux, 2007). Essential for future management are global databases (Jeliazkov *et al.*, 2020), including traits related to species vulnerability to abiotic agents (e.g. FLAMITS, a global database of plant flammability traits; Ocampo-Zuleta *et al.*, 2024). Fortunately, trait databases are rapidly expanding (see Chapter 2).

The key lesson to be learnt in management is that almost all aspects of a landscape are capable of being changed; moreover, they will change whether or not the land is managed. The limit to management is imposed by cost-effectiveness and the returns (benefits) to society. The landscape, substrates, vegetation and fauna are all capable of being individually changed. The conditions too (cover, light environment, local climate, hydrology, etc.) can be changed but any change in one component will change others through direct effects and feedbacks. These changes will affect both biodiversity and **ecosystem services** (Dennis, 2020). However, the objectives of managing biodiversity and supplying ecosystem services may not always be in concert, resulting in trade-offs (Xiao *et al.*, 2018). At a global level, this translates into the need for an integrated conservation planning framework to allocate resources optimally to achieve the long-term preservation of the multiple dimensions of biodiversity and ecosystems services (Girardello *et al.*, 2019).

The entire landscape is exposed to engineering, above and below sea level; hills, valleys, trenches, slopes varying in angles and different aspects, river beds and banks, the abyssal sea plain, etc. are all capable of modification (Dennis, 2010; Szabó *et al.*, 2010), as is impressed on any observer of the gigantic quarries and mines that humans have formed to extract mineral wealth (Fig. 11.2a; see Chapter 10). The past impact of mining is not always obvious as the bulldozed slag heaps of coal

Fig. 11.2. Suitable structures for colonisation in industrial and urban environments. **(a)** Abandoned quarries may represent important habitats for many species. Paul da Pedreira do Cabo da Praia (Terceira, Azores, Portugal) is a wetland that originated with the abandonment of the Cabo da Praia quarry, whose aggregates were mainly destined for the construction of the Praia da Vitória Oceanic Port. The wetland is maintained by a dynamic of periodic floods associated with the tidal cycle, and represents an important habitat for various water birds, particularly waders. **(b)** Green wall at the Li Ka Shing Library, Singapore Management University, Singapore. Green walls (also known as living walls or vertical gardens) are a form of vertical built structure intentionally covered by vegetation to deliver ecosystem services in urban areas. (Panel **a** photo by S. Fattorini. Panel **b** by Project Manhattan: https://commons.wikimedia.org/wiki/File:Green_wall,_Li_Ka_Shing_Library,_Singapore_Management_University_-_20140205.jpg, used under licence CC BY SA 3.0.)

mines south of Durham city (England) are witness, now being forested and built over. But mining activities profoundly alter the landscape, creating new, highly dynamic habitats both during their exploitation and when they are abandoned and possibly subject to restoration projects (Williamson *et al.*, 2003; Salgueiro *et al.*, 2019, 2020; Telea *et al.*, 2019; Zhang *et al.*, 2020; see also Chapter 10). For example, in the case of the enormous mine of Hambach (Rhineland, Germany: www.rwe.com/en/the-group/countries-and-locations/hambach-mine-site; accessed 21 May 2025), the potential ecosystem services this mine could provide are enormous. One possibility for the restoration of this mine, once mining is complete (probably around 2040), is that of creating an artificial lake. If the site were entirely flooded using the waters of the Rhine, the future Lake Hambach would be about 40 km^2 in size and up to 360 m deep (with a volume of about 4 billion m^3, thus exceeding the volumes of all natural lakes in Germany, with the sole exception of Lake Constance: www.georesources.net/cms.php/en/news/1815/Hydro-power-in-the-Hambach-opencast-mine-after-lignite-phase-out-in-Germany; accessed 21 May 2025). From a broader perspective, Gerwin *et al.* (2023) provide a timeline of changes in the Rhineland from slash-and-burn agriculture in the Neolithic to the colossal 20th-century lignite mines, also envisaging a set of interacting ecosystem services post mining that might have benefits to human populations (Fig. 11.2a). But even then, it is difficult to forecast the quality of the new habitats that can be obtained after mining (for example, water quality: Blanchette and Lund, 2016) and there are limits to the changes that can be achieved, such as recreating mountains destroyed by mining, though perhaps the future will prove that wrong.

On a relatively small scale there is a vast range of opportunities. Just one example is the lakes and marl pits in the county of Cheshire, UK, many of them zones of marl, sand and gravel extraction. The county is studded with old marl pits which are now ponds, excellent habitats for great-crested newts (*Triturus cristatus*; Lewis *et al.*, 2007) and moorhens (*Gallinula chloropus*; Forman, 2001). A classic example of a 'managed' site used for studying the behaviour and populations of the wall brown *Lasiommata megera* and the meadow brown *Maniola jurtina* is Brereton Heath near Holmes Chapel in Cheshire (see Chapter 5). This was an abandoned sand and gravel quarry, taken over and then bulldozed to form a park by Cheshire County Council. The depths of the pit are now a lake full of fish and birds (Dennis and Bramley, 1985; Shreeve *et al.*, 1996b). Substrates can also be changed; throughout Macclesfield Forest in the Pennines (UK), paths have been created out of limestone chippings (basic conditions) covering the Carboniferous coarse-grained sandstones (the so-called 'Millstone Grit') and crumbling shales (acid conditions). It is clear that it is feasible to cover expanses of ground with alien substrates to determine a base for a specific biodiverse flora and fauna (Bach *et al.*, 2020). However, changing the vegetation requires frequent rotational management (removal, thinning, replanting, planting key host plants of invertebrates, control of invasives; Simões *et al.*, 2014; Winter *et al.*, 2018). Changes occur quickly and protection of particular elements of the fauna/flora requires careful study (Breshears *et al.*, 2019). Following such changes, species can be reintroduced from zoos, livestock parks and original native sites directly by adopting a variety of tactics (see, for example, Batson *et al.*, 2015 for a review of **translocation** tactics used for birds and mammals), or can be left to rely on natural migration.

Making urban sites fit for species

With the rise of humans, the most prominent feature on the planet has been urban development, ranging from **hamlets** to massive conurbations and the megalopolises (Gottmann, 1961, 1976; Fattorini, 2019; Ross, 2019). For the majority of human history, humans have lived in low-density groups, which then developed into denser rural settlements, such as hamlets and **villages**. Intensive urban development began with **towns**, which had a 'core' of activities and were involved in trade (Fattorini, 2019). Despite many ancient civilisations being characterised by the presence of important urban centres (towns and **cities**), massive urbanisation is a modern phenomenon. Up to the 19th century, population increased very slowly and most people lived in rural areas; the rapid increase in human population after the Industrial Revolution began the exponential curve of urban growth, with urban dwellings surpassing rural dwellings in 2007 (now more than 4 billion people live in **urban areas**; by 2050 it is estimated that 66% (7 billion) of the world population will occupy them) (see Fig. 10.2b).

Settlements of highly varied structure exist and a large variety of management techniques and tools have developed to enhance nature in these areas (Fattorini, 2019), for example some 318 distinct actions alone in Australian cities (Soanes *et al.*, 2023). Urban areas vary enormously in their horizontal and vertical expansion (Forman, 2014, 2019; Fattorini, 2019). The largest settlements can be flat in profile but with advancement typically grow to include increasingly high central skyscrapers (see Fig. 9.11). Those with flat profiles, characteristic of many tropical states in Africa, have limited trade hinterlands and are characterised by 'urbanisation without industrialisation' and manufacture barely touching global trade (Gollin *et al.*, 2016), whereas those with massively high-rise structures are typical of international trade and accumulating wealth (Venables, 2018). Cities and conurbations have distinctive structures and are highly complex socio-economic/political systems (Fattorini, 2019). Many theoretical models have been proposed to describe city structures. The simplest model is the monocentric city model, with a central business district (CBD) surrounded by residential areas on land with lower costs and rent values (Alonso–Mills–Muth model; Duranton and Puga, 2015). Although appealing for its simplicity, this model has obvious limitations (for example, individuals are assumed to be identical in terms of income and preferences). As a result, an array of more realistic models has been developed to depict the evolution, devolution and destruction of city structures, elements in systems driven by highly dynamic socio-economic/political forces, and competition leading to war; see, for example, the concentric zone model (which depicts urban land usage as a series of concentric rings; Burgess 1967), the sector model (a modification of the concentric zone model in which different sectors – wedges – instead of rings are recognised; Hoyt, 1939), the multi-nuclei model (where the urban area includes more than one CBD or a single CBD and many secondary outlying business districts; Harris and Ullman 1945), the residential and retail location model (in which retail trade and services are located in relation to residential demands, and residences located in relation to combined retail and basic employment; Lowry 1964; Harris and Wilson, 1978). Antipova (2018) provides an exacting review of these various models. No single model is universally applicable, as different cities may conform more closely to different structural patterns.

Towns and cities are distinguished by a wide range of characteristic structures (walls, roofs, eaves, gardens, yards, drains, urban streams, alleys, etc.) that can support ecological functions, with many opportunities for enhancing biodiversity, as discussed by Francis (2010), Forman (2014, 2019), Dover (2015), Threlfall *et al.* (2017) and Fattorini (2019), amongst others (see Fig. 11.2b; see also Box 7.1 in Chapter 7). All these structures affect the way humans live and the massive range of other organisms occupying them. A city or conurbation includes a complex range of utilities (e.g. tall structures for nesting, as observed for peregrine falcons *Falco peregrinus* nesting on Salisbury Cathedral: Pagel *et al.*, 2018; and kestrels *Falco tinnunculus* using scaffolding holes in buildings in Rome: Salvati *et al.*, 1999; floorboards and timbers for cockroaches, termites and beetles: Ebeling, 1978), resources (e.g. foods in bins and waterways for raccoons *Procyon lotor*, cats *Felis catus*, foxes *Vulpes vulpes*, jackdaws *Corvus monedula* and an enormous range of arthropods; Fattorini, 2019) and conditions (e.g. differences in radiation, heat and moisture; Forman, 2014; e.g. warm roosting sites for pied wagtails *Motacilla alba* in parks: Chandler, 1979). Human activity in urban areas can have a variety of effects on other organisms: negative, neutral or positive (Fattorini, 2019; see Table 2.1 for the forms of interactions). These effects apply not only to wildlife but also to the humans occupying urban areas (e.g. cities may provide economic opportunities but can also generate poverty and increase violence). The effects may be intentional (planned) or unintentional. In particular, there is one 'effect' that has long been known to cause damage to all life: the release of elements and molecules into the air, soil and water (e.g. Forman, 2014; Fattorini, 2019; Briffa *et al.*, 2020; Di Biase *et al.*, 2022). Though there is hope for pollution reduction (for example, bacteria may have significant potential to remove pollutants from human-impacted environments; Malik *et al.*, 2023), the high pollution levels that typically characterise urban areas remain a major element of concern for human health (Fattorini, 2019).

Biodiversity in urban areas challenges the usual view that humans impact all other life negatively. A whole range of microbes, plants, fungi and animals can benefit from the urban scene (Forman, 2014; Fattorini, 2019). Of course, many of these are 'pests' for humans (e.g. Norwegian rats *Rattus norvegicus*; Colvin *et al.*, 1996), and their control can be difficult, but there are top predators in cities much as in rural ecosystems (e.g. domestic cats). However, it is felicitous when a completely unintentional 'event' can aid an organism that poses no threat to human existence, and may even be beneficial to it (e.g. many plants). Different organisms respond differently to urban settings: plant richness tends to increase in urban areas (especially because of the presence of introduced species), whereas animal species richness declines (Faeth *et al.*, 2011). However, responses vary greatly among taxa (Fattorini, 2019). For some groups, urban areas may even be as rich in species as natural areas of the same size (e.g. birds; Fattorini *et al.*, 2018b), but urban faunas are generally of lower conservation interest because most of their species are ubiquitous (e.g. McKinney, 2008).

Butterflies are a classic example of plant-dependent organisms that include both positive and negative responder species, well demonstrated by Hardy (1998) for Manchester (UK) and by Deschamps-Cottin *et al.* (2023) for Marseille (France), where a plant engineering project led to a doubling of species. Many cities in the tropics, especially the Indian subcontinent, are now testing these findings in quite different regional climates (e.g. Tiple *et al.*, 2007; Chowdhury *et al.*, 2021). Butterfly ability to occupy urban areas obviously depends on their **resource-based habitats** (RBHs), especially nectar and host plant sources, but vegetation is also essential for many other animals in urban ecosystems. The key to most urban biodiversity is plant life, albeit much of it exotic/planted, acting as a 'template upon which natural ecological and evolutionary processes act' (Faeth *et al.*, 2011) (Fig. 11.3).

Although urbanisation is in general a threat for species survival, there is an increasing appreciation of the potential role of urban green spaces in biodiversity conservation if appropriately conceived and managed (Fattorini, 2019). For example, urbanisation has, in general, a negative impact on insects but many species (including some of conservation concern) can benefit from urban green spaces (e.g. Fattorini, 2011, 2014a, 2019). Understanding which characteristics of urban green spaces may favour insect diversity is of course an essential step to informing conservation actions in urban areas (Fattorini, 2014a; Fattorini and Galassi, 2016; Fattorini *et al.*, 2018c). Fundamental also are the characteristics of the **matrix** surrounding the green spaces (e.g. proportion of impervious surfaces), as well as their size and level of isolation (e.g. distance from rural areas bordering the city), with variable

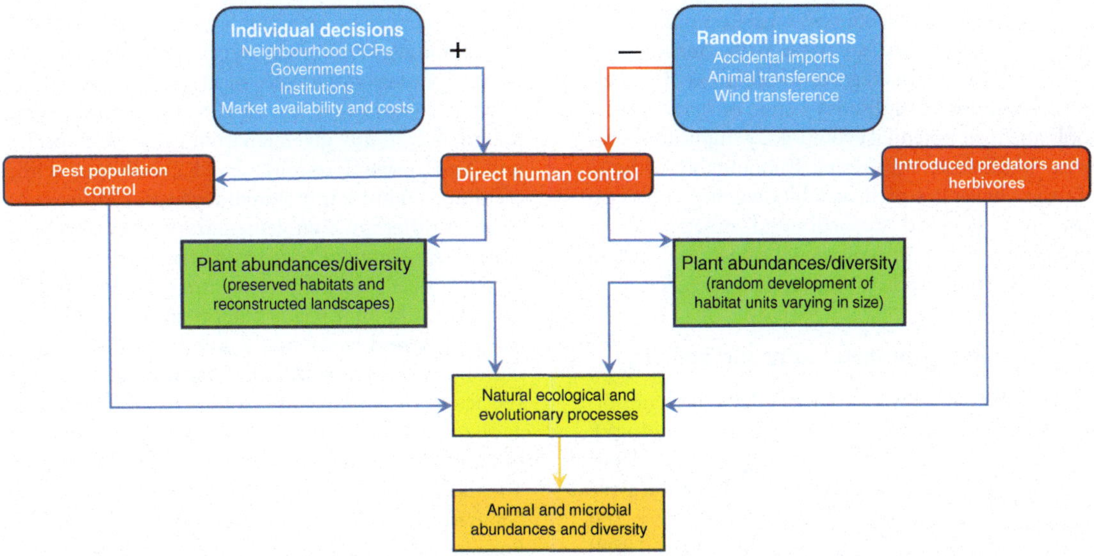

Fig. 11.3. Vegetation is an obvious feature of the urban environment. This figure illustrates a conceptual model of human-determined influences and random effect dominance of plant abundance and diversity control over ecosystems and thus biodiversity in urban environments. CCRs are covenants, conditions and restrictions. Signs: blue positive, an increase or decrease in one factor influences another in the same way; red negative, an increase in one factor has the opposite effect in the other; orange both signs, both negative and positive effects on different animal components possible. (Model modified and extended from Faeth *et al.*, 2011, used with permission of John Wiley & Sons.)

effects depending on the RBH of the different organisms (Fattorini, 2019). For example, while the diversity and abundance of some groups that can benefit from urbanisation (e.g. pests) increase from the periphery to the city centre, those of the species sensitive to urbanisation show an opposite pattern; in some cases, peaks are observed in the middle of the rural–urban gradient, where the presence of an intermediate level of urbanisation allows the coexistence of species belonging to both categories (Forman, 2014; Fattorini, 2019; Fattorini *et al.*, 2020b; Di Pietro *et al.*, 2021). As for the entire globe, understanding the complex systems of human-dominated and 'natural' processes affecting biodiversity is crucial for conserving urban biodiversity, a template steadily overtaking the semi-natural and natural arena; crucial components are scale effects (city and 'substrate unit' patch size), as envisaged in **metapopulation** models (Uchida *et al.*, 2021). In a meta-analysis of intra-urban biodiversity variation across an extensive variety of taxonomic groups in 75 cities worldwide, Beninde *et al.* (2015) noted that large sites (>50 ha) are necessary to prevent a rapid loss of area-sensitive species. Preserving and enhancing urban biodiversity clearly requires a combination of ecosystem and RBH approaches. Emanating from this issue is what and how to conserve or eradicate in a context where humans work, live, play, assist and compete with one another (Nilon, 2011). In this regard, for example, Lambert and Donihue (2020) discuss a suite of management tools and their potential evolutionary implications for urban biodiversity engineering. Undoubtably, addressing the challenges of urban biodiversity conservation will demand co-ordinated efforts across disciplines, blending ecological science with urban planning, policy and community engagement.

A systems approach: rarity and stability

For decades now, different sciences have been adopting a systems approach (ecosystems in ecology), a paradigm replacing the 'normal' positivist, deterministic, mechanistic, piecemeal, comparative approach based on the direct fact-gathering and experimentation model predating it (Lister, 1998) (Box 11.2). A systems approach may provide important insights into understanding why, in any unit area, a few species are common (abundant) and most species are uncommon (scarce), a ubiquitous pattern of **species abundance distribution** (see Chapter 4 and Appendix 3). Certainly, this could well be a sampling issue, as implicated in a meta-analysis of data from the Global Biodiversity Information Facility of 39 taxonomic classes of eukaryotic organisms from 1900 to 2019 (Callaghan *et al.*, 2023): with increasing sampling effort, the predominance of the most common species (number of occurrence records) is progressively reduced (i.e. Preston's veil is progressively lifted; see Chapter 4 and Appendix 3). This is consistent with previous research (e.g. Fattorini *et al.*, 2013b) showing that the frequency distribution of range sizes is approximated by a **log-normal distribution** (see Chapter 4 and Appendix 3). It is also an issue of scale. For example, Crisfield *et al.* (2024) distinguish three dimensions of rarity measurement: range size, occupancy and local abundance (but see Box 4.1 in Chapter 4 for a different classification of rarity into range size, habitat and abundance dimensions). They identify four processes affecting rarity (environmental filtering, movement, demography and interactions) and hypothesise that range size and occupancy are primarily driven by environmental filtering and movement, whereas local abundance is more strongly influenced by demography and interactions. It is at the scale of traditional habitat units that RBH factors affect local abundances. At this scale, there are good reasons from a RBH view why a few species are abundant whereas many are scarce. Limits in resources (types and abundance) as well as predation, parasitisation and disease are all obvious factors that act negatively on population abundance: in any given environment, a few species will find optimal resources and conditions, while conditions will be suboptimal for most species.

But a different, complementary view is also possible. Little thought is given to the idea of 'rarity' actually being an adaptation for survival. Species sparse in a habitat space will be far less evident to any enemy and hence better able to evade extermination. However, it may be disadvantageous for a species to be too rare in traditional habitat spaces, as then males and females will experience difficulties in finding each other (see **Allee effect**, Chapter 7). To address this issue, species can adopt peculiar behaviours that enhance encounter probabilities. For example, butterfly biologists have long shown how, in such situations, individuals use landscape features as meeting points and non-visual attributes (scents) as cues (Dennis, 2020). Rarity may be advantageous in many respects. If a species is uncommon or rare, not only will predators (enemies) come across it less frequently, but also the

Box 11.2. Systems theory

Although some basic notions of systems theory arguably go back to the pre-Socratic Greek philosopher Heraclitus (c.500 BCE) and the reflexive logic of Immanuel Kant (1724–1804), the most important contributions to the definition of a systems theory originated in the latter half of the 19th century with the work of German philosopher Ludwig Noiré (1829–1889), and then in the early 20th century with the work of the Russian and later Soviet physician, philosopher and science fiction writer Alexander Alexandrovich Bogdanov (born Alexander Malinovsky, 1873–1928) (Vane-Wright, 2021).

Bogdanov, who was a Bolshevik revolutionary whose ideas were strongly opposed by Lenin (Geymonat, 1976), attempted to unify social, biological and physical sciences by viewing them as systems of relationships and aiming to identify organisational principles common to all (Vane-Wright, 2021). While striving to establish the groundwork for a new culture and science beyond capitalism, Bogdanov envisaged a transdisciplinary science bridging the gap between natural and human sciences, empowering workers to co-ordinate their labour and production and, more generally, to liberate individuals from the distorted mindset fostered by class and gender divisions, allowing them to understand the limitations and the significance of their environments and other forms of life (Gare, 2000).

Bogdanov's ideas were, however, largely overlooked, especially in the West, and a General Systems Theory was subsequently (and possibly independently) developed by the Austrian biologist Ludwig von Bertalanffy (1950, 1968; see Dennis, 2020; Vane-Wright, 2021). The principles of a systems theory advocated by von Bertalanffy were then adopted by the most diverse disciplines, especially with the rising of cybernetics.

In the General Systems Theory, systems models are symbolic models, usually *a priori* models that help in the understanding of reality, and incorporate complex interactions and feedbacks, random as well as deterministic influences. Systems theory is based on the notion that a system is 'more than the sum of its parts', viz. a system has properties that are not evident in its individual components but show up when combining those components, so they are called **emergent properties**. Any systems approach focuses on the relationships between the elements that interact in, and hence form, a system. Key principles of systems theory are therefore as follows.

- Components are interdependent. Systems are composed of interconnected parts or components, and the behaviour of the system depends on the relationships and interactions among these parts. Changes in one part of the system can influence other parts and the system as a whole.
- Systems are integrated units. A system cannot be understood by analysing its parts in isolation.
- Systems are hierarchical. Systems consist of subsystems that themselves can be seen as systems. These nested structures reflect how systems operate on multiple levels simultaneously.
- Systems are typically open. They exchange energy, matter and information with their environment.
- Systems are flexible and adaptable. In open systems a given end state can be reached by many potential means. The same end state can be reached from different initial conditions and through different pathways.
- Systems are regulated by feedback mechanisms. Open systems are dynamic and maintain themselves through interactions with their surroundings. Feedback loops (both positive and negative) are essential for the regulation and stability of systems. Negative feedback stabilises the system, while positive feedback can amplify changes and drive system evolution.
- Systems may maintain an equilibrium. Systems strive for a state of balance or equilibrium while adapting to changes in their environment. This balance is dynamic, allowing for both stability and flexibility.
- Behaviours are frequently non-linear. Systems often exhibit non-linear behaviours, where small changes in one part can lead to significant impacts elsewhere.
- Systems have the capacity to organise themselves without external control, adapting to changing conditions through internal processes.
- Emergence is a hallmark of systemic complexity. Complex systems generate properties and behaviours that cannot be predicted or explained solely by analysing their individual parts.

General Systems Theory emphasises the need to study systems across disciplines, as similar principles and patterns can be observed in biological, ecological, social and technological systems.

A champion of the application of systems theory to ecological research was the American ecologist Eugene P. Odum (1913–2002). His book *Basic Ecology* (1983) offers a clear discussion of the concepts of systems and emergent properties in ecology. Each

Continued

Box 11.2. Continued.

level of organisation of life (from molecules to the biosphere) has peculiar emergent properties which define different scientific domains, but levels are interconnected and understating the functioning of lower levels is essential (albeit not sufficient) to understand the higher levels (Fig. B11.2.1).

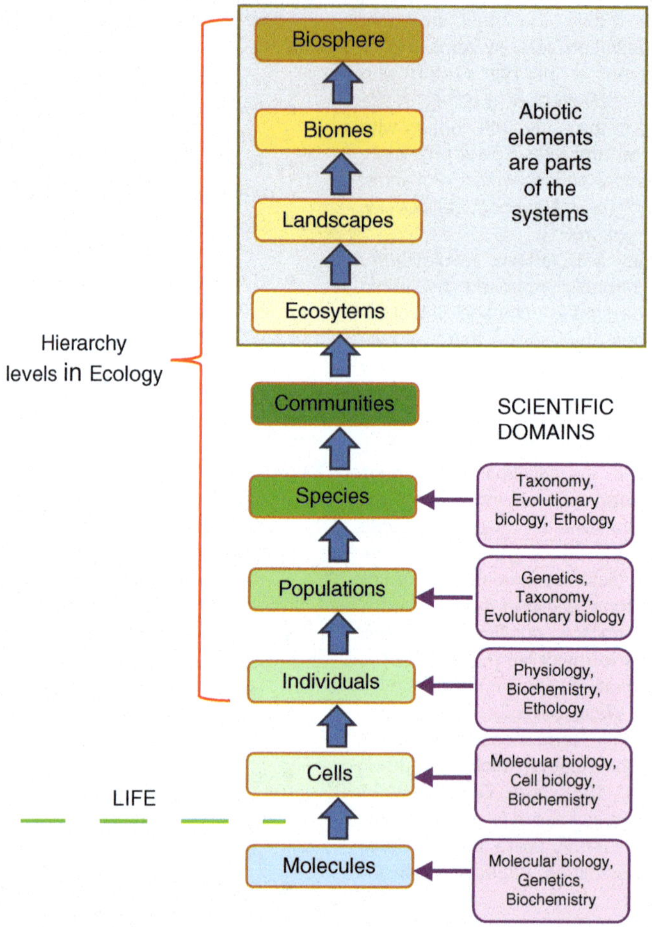

Fig. B11.2.1. Schematic representation of the hierarchical organisation of life and the associated major scientific domains.

locations of such species become less predictable. In these situations, the species is likely to be more mobile and the RBH will be more dispersed and perhaps less specialised than for species which develop large populations on single substrates (traditional habitats, TBs) (Fig. 11.4). Species that are abundant obviously have all their required resources and conditions close at hand so their RBHs tend to be compact. In addition, either enemies are lacking or they have evolved sophisticated mechanisms for avoiding them. The interplay between all these elements may well contribute to the observed patterns of species abundance distributions.

As species with small ranges are exposed to fewer predators, predation may also favour spatial rarity, although other factors may explain the correlation between range and abundance (Gaston, 1990). Brown (1984) suggested that a positive

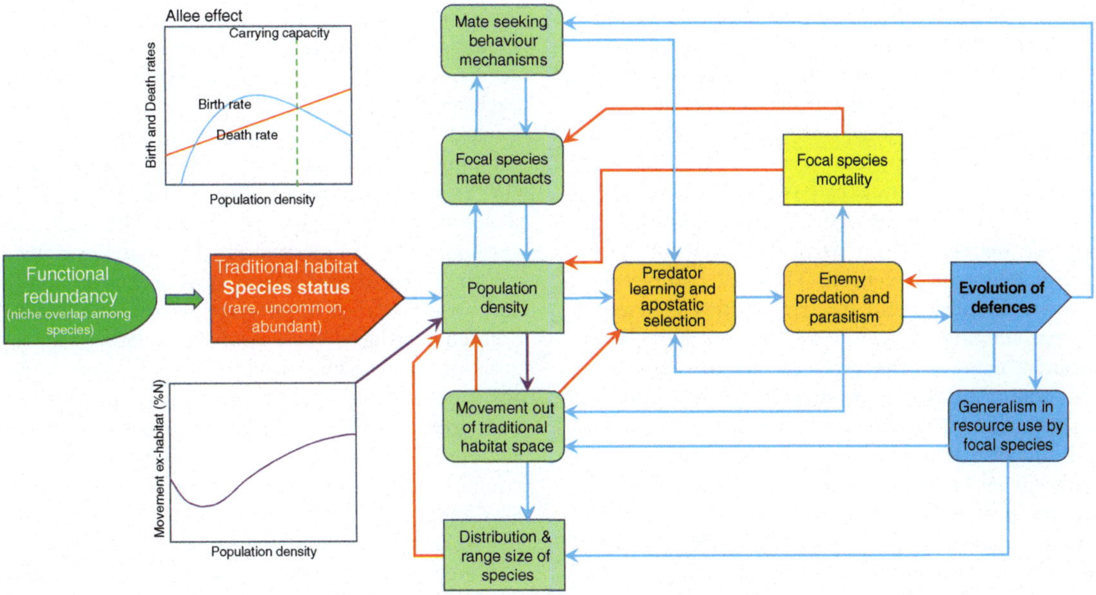

Fig. 11.4. Factors influencing population size of a species within a traditional habitat. Species may be 'using' uncommonness or rarity as an adaptation versus predation (apostatic selection, i.e. frequency-dependent selection by predators; see Weir, 2021). One issue worth considering is that uncommon species may actually 'use rarity' as an enemy escape mechanism, whereas abundant species are probably subject to apostatic selection and to remain abundant must not only have all the appropriate conditions on site but continually evolve defence mechanisms against enemies. The interesting issue is that both abundance and 'rarity' (uncommonness) may lead to extensive distributions, the former from mass effects and emigration, the latter as a mechanism with generalism to escape enemies. Allee effect: death rate increases monotonically with increasing population density; before decreasing with population density, birth rate shows an initial increase, for example because at very low densities individuals can have difficulties in finding partners for mating (see also Fig. 7.1 in Chapter 7).

correlation between range and abundance may be a reflection of interspecific differences in ecological specialisation. According to Hanski *et al.* (1993), the underlying mechanism is that species able to exploit a wide range of resources become both widespread and locally abundant, whereas ecological **specialists** have both a restricted distribution and a low abundance (see also Gaston and Lawton, 1990). This equates to an admission that widely distributed species have a wider RBH and a lower ecological specialisation, while species with smaller ranges may, being associated with single substrates (TBs), have compact RBHs and hence higher ecological specialisation. This is also consistent with the observation that only few species in a given community are ecologically **generalists,** most species being ecologically specialised (e.g. Dapporto and Dennis, 2013). But both uncommon species (that spread widely to avoid extinction) and abundant species (that are encouraged to disperse out of site; Dennis and Hardy, 2018) may well have widespread distributions. This would certainly explain why, although a *significant* correlation may (or may not) be demonstrated to exist between local abundance (density) and range size at different scales, in fact the effect size and explained variance (correlations, R^2 values) may not be that strong with increasing spatial scale (Cowley *et al.*, 2001; Fattorini, 2008).

Understanding the causes of rarity has important implications for a systems approach to ecosystem stability, and hence conservation. Actually, it is debated whether ecosystems are stabilised by species richness *per se* (e.g. Baskin, 1994; Cleland, 2011) or species identity — and particular species **traits** through **functional redundancy (insurance hypothesis**; e.g. Baskin, 1994; Yachi and Loreau, 1999; Dennis, 2004a; Ives and Carpenter, 2007). It is obvious that the most abundant species have a prominent role in ecosystem stability, but Ehrlich and Ehrlich (1981) thought that even **rare species** may act as important components (elements) of ecosystems; as such, losing a few has no demonstrable effect on the functioning of the ecosystem but there will be a crisis point when a certain number are lost – the ecosystem will then start to fail or function differently (a concept known as the 'rivet popper' model: removal of rivets by passengers on an aeroplane does not have any effect initially but eventually the removal of one particular rivet will have catastrophic consequences). Thus, all biodiversity components, even rare species, contribute to ecosystem stability. Baskin (1994) did not agree and thought many of the rarer species are in effect redundant, much like an aircraft that has built-in safety redundancy. Our view is that every species will function as an element in different subsystems, and will be important as a replacement for other species in times of crises (local extinction of a species) depending on the random shifts in ecosystem functioning: a species which is rare today may become common tomorrow. Ecosystems are dynamic units and differences in species abundance across time may contribute to their stability. For example, Powell *et al.* (2024) found that insect species with different traits tend to have asynchronous population dynamics and such asynchrony in population abundance can buffer the effects of environmental change, leading to greater community and ecosystem stability. Adopting a systems approach, linking species rarity to species interrelations and their variation through time, is therefore essential for an understanding of ecosystem stability (see also Box 2.1 in Chapter 2). Moreover, if all species do matter, including all sparse ones at sites, then it follows that in conserving biodiversity all sites matter too, for they may contain the glue binding together a community as a whole.

Although species are the basic units in biology, to understand ecosystem functions – and hence to understand and preserve their stability – a systems approach must consider the different scales involved, from genes to landscapes, through populations and communities; each level of the hierarchy is characterised by its own **emergent properties**, to which the cultural (informational) perspective must be added (Lister, 1998). A working problem is that of the human observer whose presence can change behaviours in the organisms being monitored, as well as create biases in data collection (Johnston *et al.*, 2021; Hoy and Brereton, 2022). In fact, biases pervade data collection for all organisms and are increasingly problematic with public involvement in data collection (Sánchez-Fernández *et al.*, 2021). But they affect taxa to a different degree depending on their traits (Lobo *et al.*, 2021; Backstrom *et al.*, 2024), leading to favouritism towards limited branches of the Tree of Life (Mammola *et al.*, 2023) (see also Chapter 8). Lister (1998) had problems in dealing with the different scales of the ecological hierarchy. Although the development of modern multivariate analyses (e.g. fourth corner analysis; Dray and Legendre, 2008) and an acceptance of overlapping units in space-time have reduced these problems, she is right in pointing to the immense difficulty of determining any kind of diversity from the natural world. This, with gross sampling techniques subjected to DNA analysis (e.g. via metabarcoding and genome skimming), is now becoming less of a difficulty and we may look forward to a time when whole genomes can be sampled in bulk and be determined quickly and inexpensively (Theissinger *et al.*, 2023). This development is absolutely essential if we are to understand and audit the multitude of the smallest forms of life (archaea, bacteria, protists, microscopic animals, etc.). However, enumerating and characterising the units of diversity in a landscape/ecosystem (be they genes, populations, species or communities) cannot reveal the mechanisms that maintain diversity if relationships among the parts are not understood.

The resource-based habitat and species conservation

When conserving single **keystone**, **flagship** or **umbrella species** (see Chapter 2, Box 2.3), it is one thing to appreciate the centrality of the RBH; it is quite another to assess how the RBH is determined. It can be an extremely costly business, both in time and effort, to discover nature's elegant tricks. But this is not always the case; it may well be worthwhile if, for example, an umbrella species (or species plural) is easy to locate and if it acts as an effective indicator of biodiversity (as in the case of five palms in the region of the Purus and Madeira rivers, in Brazilian Amazonia, which may serve as a biodiversity umbrella for the avifauna; Menger *et al.*, 2024). Even so, because there are so many keystone or flagship species that we may wish to conserve for a variety of reasons, methods need to be sought for reducing time, effort and cost. The starting point is to ascertain as many of the species traits as possible from the literature: biological, physiological, behavioural, ecological, etc. (see Chapter 2 and Box 5.2 in Chapter 5). Once these data are available from an adequate sample of sites over the entire species range, then a better judgement can be made of how to study the RBH of particular sites through appropriate fieldwork, and of the likely pressures from shifting conditions such as climate change or habitat destruction.

One point deserves emphasis here: there are creatures which tend to be conveniently forgotten, much to our disgrace – animals that we have domesticated over the past 40,000 years, used and abused, occasionally treated with immense kindness (e.g. dogs, cats, pigs, sheep, cattle – including Chillingham wild cattle, gayal, zebu and a variety of breeds such as Holsteins, Friesians and Herefords – horses, donkeys, camels, reindeer *Rangifer tarandus*, yaks *Bos grunniens*; many small creatures such as ferrets *Mustela furo* and hamsters *Mesocricetus auratus*; large animals such as Indian elephants *Elephas maximus indicus*, e.g. Sukumar, 2006; Teletchea, 2019). These creatures are ones we should know and understand intimately and for their benefit not ours; increasingly their RBH should be raised to a standard where they are provided with all that they need to have a decent life.

Species' traits and the identification of resource-based habitats

Building **trait** databases for species is a labour-intensive task but is a crucial foundation for determining the RBH of species (e.g. Dennis *et al.*, 2004; Dennis, 2020). Note that it is not enough to acquire traits from a single site or from scattered casual observations over the species' distribution or range, though there are many useful analyses of ecological associations and community affiliations built on such data. Classic species databases exist for butterflies (UK: Dennis, 2010; Dennis *et al.* 2000, 2004; Shreeve *et al.*, 2001; Europe and Maghreb: Middleton-Welling *et al.*, 2020; world: 75,103 traits for 12,448 of about 19,000 global species: Shirey *et al.*, 2022), other arthropods (e.g. ground beetles: Homburg *et al.*, 2014; spiders: Pekár *et al.* 2021; Mammola *et al.*, 2022), freshwater macroinvertebrates (EPA, 2012), birds (e.g. Sheard *et al.*, 2024), mammals and reptiles (e.g. Myhrvold *et al.*, 2015; Faurby *et al.*, 2018), amniotes (Myhrvold *et al.*, 2015), amphibians (e.g. Gallagher *et al.*, 2021), fish (e.g. Blanck and Lamouroux, 2007; Froese and Pauly, 2023), seaweeds (Vranken *et al.*, 2023) and plants (e.g. Kattge *et al.*, 2011, 2020; Gallagher and Leishman, 2012; Tavşanoğlu and Pausas, 2018; Falster *et al.*, 2021; Alfaro *et al.*, 2023). The process of database building from accessible sources (i.e. the processing workflow used to compile, scan, digitise, extract, atomise and compile species trait records from literature resources) is well illustrated by Shirey *et al.* (2022). But the approach is very different when building the data from systematic field research, which needs to be intensive and extensive to capture the variability of species traits across their whole ranges and entire life cycles.

Clear rules, standardised terminology (vocabulary) and open science principles are starting points for database building and application (Schneider *et al.*, 2019; Gallagher *et al.*, 2020). However, data must already be available – and therefore collected. Early in the process stage, researchers find they have to 'get their hands dirty' and dig deeper into a species' traits in order to build up an understanding of the species' life cycle and organisms 'touching' that life cycle (Dennis, 2020). This may involve understating multiple and subtle relationships. Classic features filling the literature are, for example, those defined as elements of **satyric mimicry** (or imperfect mimicry) and partial mimicry. In satyric mimicry, the colour patterns of an animal may make its identity ambiguous, and this ambiguity interferes with the process of perception in predators for a sufficient time to allow the potential prey to take evasive action (Howse and Allen, 1994; Howse, 2013). In partial mimicry, the component being mimicked has high fidelity but only to a specific salient body component, e.g. flies and moths that mimic the frontal appearance of jumping spiders (Muñoz-Amezcua *et al.*, 2023). In such cases, the evolution of these traits implicates a very close space-time relationship of the mimics with both the creatures they are mimicking and those they are attempting to escape. The examples among butterflies are numerous (Dennis, 2020).

Keller *et al.* (2023) outline ten relatively simple rules that encompass the entire life cycle of trait data to guide researchers in making critical decisions, from data selection to information preservation and sharing (Fig. 11.5). Further extensions of trait databases are their relationships with information about species abundance and distribution. For example, Jeliazkov *et al.* (2020) have collated 80 datasets from trait-based studies into a global database for MetaCommunity Ecology (referred to as

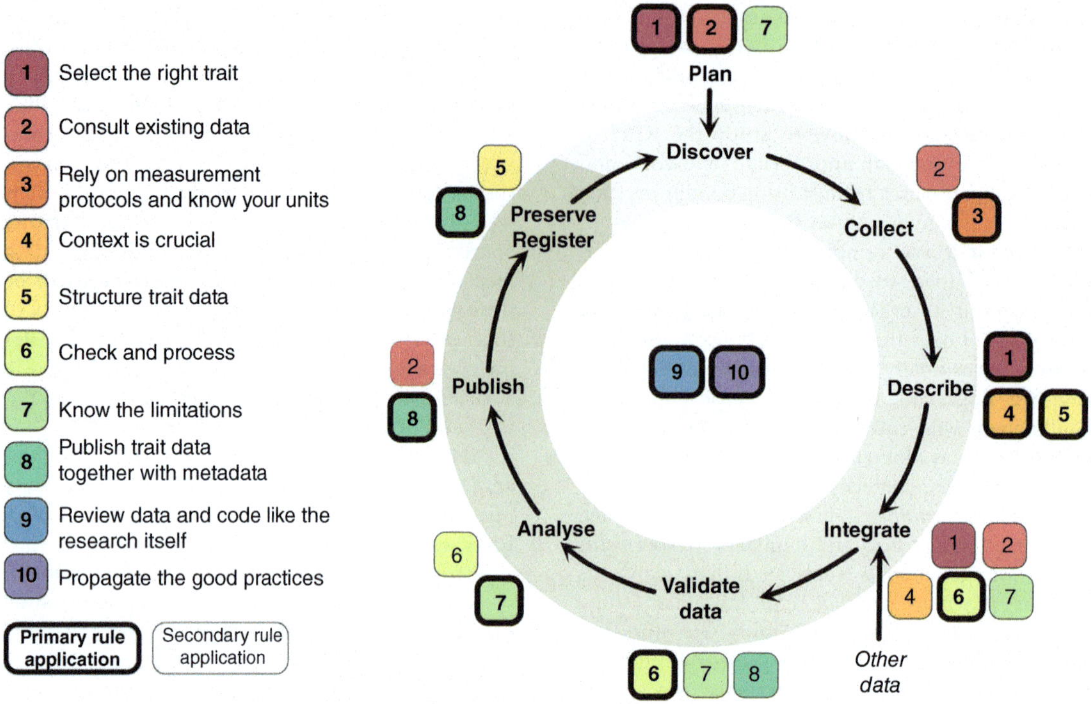

Fig. 11.5. Ten basic rules and their application in the overall trait data life cycle. Each rule is primarily applied to a specific element of the cycle (in bold) but can also be necessary to other elements (secondary application). Rules 9 and 10 apply to the whole cycle. (From Keller *et al.*, 2023, used under licence CC BY 4.0.)

CESTES) encompassing species information on community abundances or presences/absences across multiple sites, species traits, environmental variables and spatial co-ordinates of the sampling sites. Crucially, for determination of a species' RBH traits, data must be spatially explicit, obtained from sampling widely in traditional habitats over a species' (DNA unit's) range (Gentile *et al.*, 2021). Once these data are available, then traits for species can be assessed more confidently against abiotic and biotic drivers of diversity (e.g. moths in the tropical rainforest of Costa Rica; Rabl *et al.*, 2020). This knowledge is essential for appropriate site management and **translocation** projects.

But this is just a start to an understanding of the RBH of a focal species. The RBH ideally demands that we understand *all* the resource requirements (consumables and utilities) and conditions that are suitable for the species across its range. In many cases, however, the term 'range' is biased towards the regions to which conservationists have access – largely because much of the globe is 'cut off', e.g. for political reasons. But as made clear earlier in this text (see Chapter 8), conserving the breeding site for a species does not ensure its persistence if states over which the species may travel on migratory routes are not persuaded to let it pass unhindered (Sokos *et al.*, 2013; Balmori, 2019), and if overwintering sites are not also protected (Sun *et al.*, 2020). It is also important to recall here that the term 'conditions' includes not just spatially explicit substrate, climate and community associations, but also the way resources (many affected by other organisms) interact to facilitate the survival and success of the species. These change with time (see Chapter 3) and we are certainly aware of that fact from the effects of the currently fast-moving climate change (Dennis, 1993a; see also below, section *Climate change and shifts in resource-based habitats*).

The methods used to obtain life-history and ecological data on species are contained in publications on standard **autecological** survey techniques (see Appendix 1); these developed from the early 20th century (Allee *et al.*, 1949; Jørgensen *et al.*, 1996; Krebs, 1999; Pastorok *et al.*, 2001; Walter and Hengeveld, 2014), and have been greatly advanced recently by the inclusion of electronic surveillance, remote recording (drones, fixed cameras; Gillespie *et al.*, 2015; Ivošević *et al.*, 2015; Steenweg *et al.*, 2017) and a thorough understanding of the importance of behaviour from experimental research both in the field and in an outdoor laboratory setting (Danchin *et al.*, 2008; Marcantonio *et al.*, 2023) (Fig. 11.6). Butterfly and primate biologists, for instance, have an intimate knowledge of the key part played by behaviour in an animal's life history (e.g. Wich and Marshall, 2016; Dennis, 2020). Understanding the behaviour of animals is central to the conservation of all species (e.g. anole lizards *Anolis brasiliensis*: Mesquita *et al.*, 2015a; Gulf Coast box turtles *Terrapene carolina major*: Meck *et al.*, 2020; great bustards *Otis tarda tarda*: Janó and Végvári, 2016; Goliath frogs *Conraua goliath*: www.bbc.com/news/science-environment-64745722, accessed 21 May 2025; Fig. 11.7), as an animal's behaviour may lie at the centre of evolutionary changes (see Davies *et al.*, 2012; Vane-Wright, 2014).

Structural and ancillary features of resource-based planning

Planning measures for a species depend greatly on accessibility to suitable space. Crucial issues include the following.

- *Scale and funding*. Are areas for a species' habitat (consequently biodiversity) engineering sufficiently large and potentially 'linked up' to other sites to ensure movement and long-term persistence (Hanski and Gilpin, 1997)? Land (including water spaces) is expensive and there are many competing demands for it. Therefore, engineering for a single species must have more in mind (biodiversity) than just that species alone. Typically, large land areas can be allocated to a single non-human species only if it has substantial value and the land is marginal for human use; however, even when land is gifted to a single species, many others will occupy it, most of them invisible to the naked eye or casual scrutiny.
- *Identification of suitable substrates as a foundation for a specific RBH*. Not all land areas are suitable for conversion for the RBH of a particular species. Classic examples include butterfly species that require alkaline substrates (such as well-drained calcareous bedrock and rendzina soils for their host plants) when the available land is based on acid sands or impermeable clays (e.g. many of the UK lycaenid butterflies; Dennis, 2010); or organisms associated with mature forest with rotting trees (see Chapter 3) where the actual substrate is in an early successional stage

(a) (b)

Fig. 11.6. Outdoor experimental cages for determining behaviour in insects (butterflies). **(a)** An outdoor flight set-up consisting of four cross-shaped greenhouse aluminium tunnels covered with insect mesh, placed in Bois de Lauzelle forest (Ottignies-Louvain-la-Neuve, Belgium), an experimental site belonging to the Université Catholique de Louvain, and an inset representing the study species *Limenitis camilla*. **(b)** Close-up of the inside of the cages. (Photos courtesy of Raluca Vodă. Panel **a** from Marcantonio *et al.*, 2023 used under licence CC BY 4.0.)

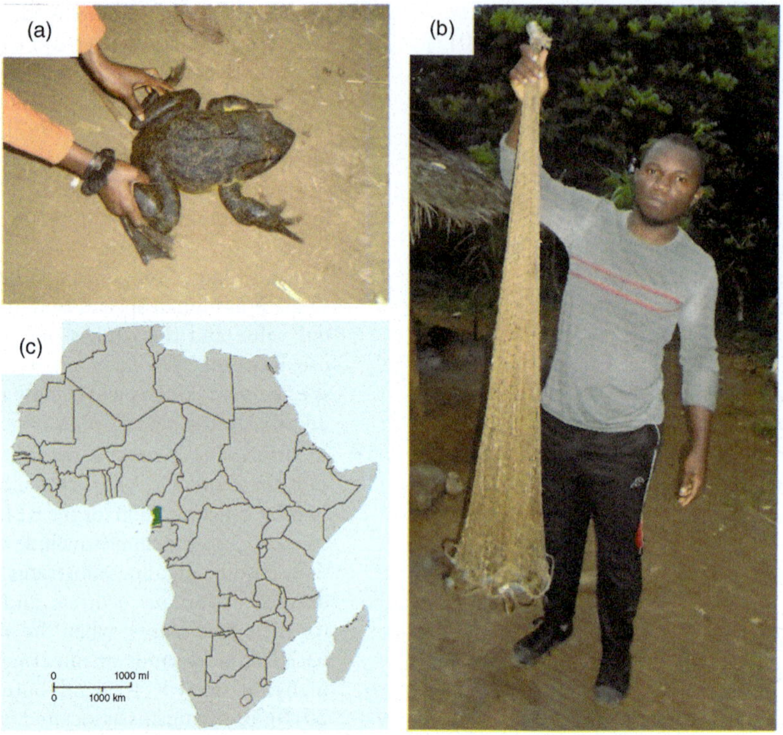

Fig. 11.7. A classic example for a resource-based habitat approach and remedial action. Conservation of the Goliath frog *Conraua goliath* (**a**), the largest extant anuran. The habitat of the Goliath frog is lacustrine (rivers and streams which are rapidly being destroyed) and the species is now classified as endangered on the IUCN Red List. For decades the Goliath frog has been overhunted for food (**b**; the photo shows a net used by hunters to collect these animals). This species is distributed in tropical western Africa within the states of Cameroon and Equatorial Guinea (**c**). Some people in the area regard themselves as blessed to have this remarkable creature; to them it has a cultural value. (Panels **a** and **b** from Tasse Taboue *et al.*, 2024, used with permission of Cambridge University Press. Panel **c** map by Christian Fischer: https://commons.wikimedia.org/wiki/File:RangeMapConrauaGoliath.png, used under licence CC BY SA 3.0.)

(such as grassland or scrubland). Habitat fragmentation can strongly reduce suitable habitats, as is the case for species associated with mature trees, which are typically present only in well-preserved forests. For example, pied woodpeckers *Dendrocopos major* and nuthatches *Sitta europaea* need trees with cavities or in which cavities can be 'drilled' or crafted, although they show some plasticity in the use of alternative materials (Cantarero *et al.*, 2015). In a young spruce forest in Cheshire, UK (Macclesfield Forest), it was revealing that a pied woodpecker had repeatedly used a telegraph pole in an open field near the village of Forest Chapel as its nesting site above the forest (R.L.H. Dennis, personal observation). It had access to all the resources in the forest and a ready-made nesting site thanks to humans.

- *Landscape variation.* Aspect and altitude are important components of successful colonisation (see Chapter 7), not only because of variation in wet/dry and hot/cold years (Dennis, 1993a, 2010), but also because upland areas provide refuges for many species erased by increasingly intensive agriculture on more accessible land at lower elevations (Kleijn *et al.*, 2009; Fattorini *et al.*, 2019b). Organisms 'shift' their distributions in relationship to both short-term and long-term local climate changes and contrasts in seasonal weather (Taheri *et al.*, 2021; Fattorini, 2024c; although many species are not demonstrating range shifts in the expected directions: Rubenstein *et al.*, 2023). Species occupying adjoining low-level and upland areas have an important advantage. A classic example (Hardy *et al.*, 2010) is the orange tip butterfly

Anthocharis cardamines that, along with its main host plant, the cuckoo flower *Cardamine pratensis*, occupies ground from sea level to >500 m from east Cheshire into the Pennines, UK. During years when the butterfly emerges much earlier than the appropriate plant stage is available, the butterfly tends to load its eggs on a few plants on Alderley Edge in east Cheshire. As the butterfly larvae are cannibalistic (Dennis, 2020), the loss of individuals can be substantial. However, on higher ground in the Goyt Valley, near the southern edge of the Pennines, both butterfly and plant tend to emerge coincidentally. In effect, the high ground provides a refuge for population losses at lower altitudes. A similar problem (Dennis, 1985b) arises when a nymphalid such as the small tortoiseshell *Aglais urticae* lays its eggs in early March and then substantial losses of larvae occur in subsequent cold Aprils and Mays. Early egg laying is rare at higher altitudes and the larvae tend to develop more rapidly as the sun angle and temperatures are higher in May and June. Altitude once again comes to the rescue of butterflies that respond too readily to early sunny springs followed by cold or wet Aprils.

- *Building a team.* Ensuring the persistence of a species and the community to which it belongs is usually not a 'one-person' job; it may require a team of researchers (even from different countries) to cover the whole range. Classic cases emerge with birds such as the cuckoo *Cuculus canorus*, one of the many species that have intercontinental ranges (Davies, 2016). A key interest is expansions (e.g. Cetti's warbler *Cettia cetti*; Robinson *et al.*, 2007) and retractions (e.g. the icterine warbler *Hippolais icterina*; Engler *et al.*, 2016) in ranges. Researchers should be aware that species ranges are not static and co-ordinated efforts may be needed to track their changes.
- *Understanding their part in communities.* This point follows on from the first above. An ecosystem view of biodiversity would require that we envisage the modification of **landscapes** to benefit not just a single species but an entire **community**. The trick is to ensure that the focal species benefits from a wider perspective of management. For instance, an intriguing part of arthropod conservation is the way that management of suitable landscapes for the large blue butterfly *Phengaris arion* has potential for benefiting other arthropod and plant species, and not just in the UK following its extinction there (Spitzer *et al.*, 2009; Hayes, 2015; see below, section *A species' intimate community*). Conservation policy requires a substantially wider target than simply preserving **keystone** and **flagship** species (see Chapter 2 and Box 2.3), and should not focus solely on a limited group of organisms. This demands that researchers have a wider perspective than simply focusing on an imperilled species. At the same time, they should also be thinking how they can build a future for those species; that is, the best policy is for a target species also to be an **umbrella species** (see Box 2.3 in Chapter 2). Much has been written about umbrella species (Breckheimer *et al.*, 2014; see examples in Roberge and Angelstam, 2004) and umbrella taxa (e.g. macrophytes and odonates used as umbrella potential for wetlands; Bried *et al.*, 2007). A classic case is the sage grouse (*Centrocercus* spp.): this bird (object of the largest public–private conservation effort for a single species in the USA) was found to be a suitable surrogate for the majority of species identified as sage associates, though not all species benefited equally from its conservation (Runge *et al.*, 2019).
- *Ability and access for modifying the substrate.* For example, a conservation approach for species spending early stages in ground vegetation is to modify the height of sward to make the habitat suitable, but this assumes a substantial degree of predictability. **Substrate** conditions shift with climate changes and management needs to keep a close eye on such variations with repeated surveys. For instance, in the case of the silver-spotted skipper butterfly *Hesperia comma*, between 1982 and 2001–2, the optimum percentage of bare ground within a traditional habitat space used for egg laying shifted from 41% to 21%; this substantial decrease in the requirement for bare ground has resulted in an improvement in the suitability of most habitat patches (Davies *et al.*, 2006). More in general, climate warming increased the availability of thermally suitable habitat for *H. comma* at the cool, northern edge of the species' distribution. Consequently, this increased: (i) egg production and population growth; (ii) the potential habitat area for the butterfly; (iii) the microhabitat availability, as a wider range of substrates can be used; (iv) the number of traditional habitat patches available in the landscape open to colonisation, including those with northern aspects

(C.D. Thomas *et al.*, 2001; Davies *et al.*, 2006). Of course, climate warming can have opposite effects in southern regions of Europe.

A species' intimate community

It can take many years for single researchers to determine all the elements making up a species' life history and it is clear that it is not a welcome strategy. On the other hand, an immense amount of knowledge can be gathered on a single species by one dedicated researcher, as famously demonstrated by the work of Jane Goodall on chimpanzees, Dian Fossey on gorillas and Biruté Mary Galdikas on orangutans, each supported by Dr Louis Leakey (see Chapter 9).

A classic case of a thoroughly complex organism that is familiar to people involved in insect conservation is the large blue butterfly *Phengaris arion* in Britain, a species with a complex biology involving a symbiotic relationship with ants. Although previously studied by entomologists in the early 20th century, the initial intricate picture of this species' **resource-based habitat** was unearthed by Jeremy A. Thomas, following up observations by Thomas Algernon Chapman, Edward Bagwell Purefoy and Frederick William Frohawk (Thomas and Lewington, 1991). He discovered why it had become extinct in Britain (Thomas, 1980) and reintroduced it from Sweden in 1983–1992 (Thomas *et al.*, 2009). It had little to do with the abundance of the host plant – the wild thyme *Thymus serpyllum* – that was used by ovipositing females and larvae in their initial stages; it certainly had a great deal to do with the loss of rabbits from myxomatosis, allowing an increase in the height of the sward and hence the creation of unsuitable conditions for an ant on which the larvae prey. Thomas further discovered that this ant was also responsible for conveying the larva to the ant nest. Together with his colleagues, he determined the critical ant in Britain (*Myrmica sabuleti*) that functions as both its symbiont and prey (Thomas *et al.*, 1989; Thomas and Elmes, 1993); the ant benefits from exudates from the larva while the larva eats the prey in early stages: food for food. Rare features of the **clade** of large blue butterflies go further inasmuch as different species (or population units) have different ant prey and, in turn, parasitoids living on them (e.g. *Neotypus pusillus* (Ichneumonidae), endoparasitoid of the dusky large blue *Phengaris* (*Maculinea*) *nausithous*: Stankiewicz *et al.*, 2004; *Ichneumon eumerus* (Ichneumonidae), a specialist parasite of the larva of the mountain Alcon blue butterfly *Phengaris rebeli*: Dziekańska *et al.*, 2020). These endoparasitoids are inevitably rarer than the butterflies, the latter determining a precarious balance of numbers that an ant nest can maintain without losing both symbionts, butterfly parasites and ants, there being a limit to the number of large blue larvae that can exploit an ant nest (Thomas and Wardlaw, 1990, 1992). The nature of the required substrate (a complex landscape of high rugosity) is also an important feature for population maintenance, providing varied environmental conditions that counter unpredictable conditions (Hochberg *et al.*, 1992; Thomas, 1995; Thomas and Elmes, 1998; Thomas *et al.*, 1998a, b; Johst *et al.*, 2006). Thomas also found that larval niche selection and evening exposure enhance adoption of the large blue larvae (Thomas, 2002) and that the butterfly makes sounds that match the queen ant (acoustic mimicry; Barbero *et al.*, 2009). The matching of chemical exudates is also critical in these butterfly–ant relationships (e.g. cuticular chemistry in *Phengaris* (*Maculinea*) *alcon*, Nash *et al.*, 2008).

Once the butterfly was studied elsewhere in Europe, it was found that other species of ant were also used (Sielezniew *et al.*, 2010), indicating just how complex the resource-based habitat of a species can be. A study involving 529 ant nests in Poland revealed that nine ant species were parasitised by *P. arion* (Sielezniew *et al.*, 2010). This demonstrates clearly that the study of a species' RBH must extend widely over its range (Schönrogge *et al.*, 2004; Tartally *et al.*, 2019). Detailed studies (using DNA) have now been carried out on many relatives of the large blue and their relationships intimately determined, showing the evolution of various parasitic life histories in this butterfly clade (*Phengaris*-*Maculinea*) (Als *et al.*, 2004; Sielezniew *et al.*, 2015).

There are at least three things to be learnt from the study of *Phengaris* (*Maculinea*). First, if you think you are studying a **rare species**, it is very likely that there is an even rarer species dependent on it. Second, as the work has progressed, the number of closely related species under observation has increased, each with its unique manner of persistence; some are directly predatory, others effectively 'milk' their prey (**trophallaxis**); the latter method ensures survival of the food base. Complexity lies at the basis of organism persistence. Third, the work was started by individuals working alone, often in one small area or state, but then extended to cover larger areas within the ranges of these butterflies. Currently, there is a small army of individuals working on this extraordinary clade (e.g. Als *et al.*, 2004; Di Salvo *et al.*, 2019; Tartally *et al.*, 2019); this is not work that can be completed by one individual – biologists need to co-operate in their tasks. Here, complications have been discussed for a group of butterflies. Just because some (other) animals are very large, it would be a mistake to imagine their life histories are much simpler!

A species' special landscape

If a scientist, amateur or professional, starts their study of an organism that captures their interest, it is not long before their investigations are adopted by other students, spreading out to other 'landscapes'; comparative data soon accumulate. Every species occupies a **landscape**, more accurately in the plural, *landscapes*; these may be envisaged like Chinese boxes, smaller units encapsulated in larger and ever larger units of space. As these units get larger, they are associated with distinct modes of movement (trivial movements, dispersion between metapopulation units and long-distance migration), causing increasing difficulty for the organisms, and their associations with different resources and conditions will in all likelihood increase in variability and in the demands they make on adaptability. For any species to persist in any landscape, two things are crucial: (i) landscape 'units' or 'elements', including artificial ones, are heterogeneous (Estrada-Carmona *et al.*, 2022; see Box 2.2 in Chapter 2), the scale of occupancy and degree of heterogeneity experienced being potentially unique to each species; (ii) each species must be able to adapt to variability in resources (e.g. food supply, roosts, predator escape features) and conditions (e.g. local climate, soil moisture, light), within these heterogeneous units. Species incapable of movement and incapable of adaptability to variable resources and conditions are candidates for extinction. These points are reminders of **metapopulation** models (see Chapter 7); one of the key elements of variability is the range of variation in distances between habitat units; suitable distances depend on the mobility of different species.

At a landscape scale, which is much cruder than the intimate associations between species discussed in the preceding section, what do we mean by components of heterogeneity? Fundamentally, we mean 'rocks' (lithology), topography, soil, vegetation structure and the like. A glance at a map of Britain will disclose many species that relate to geology (calcareous (limestones), siliceous (gritstones, sandstones, silt, etc.), mudrocks (clays and shales), various igneous and metamorphic rocks) or to different spans of altitude (e.g. plants, Vincent, 1990; butterflies, Dennis, 1992). At a more local scale, heterogeneity is most evident in topography, slopes and aspects. This variability in the land is crucial as it affects the degree of surface moisture (water), local climate, shelter, radiation, etc., which influence organism distribution. At even finer scales, differences in the position – on a cut, rotting tree trunk, spalling tree bark, stones and boulders in rock pools or cow pats on open ground – may be of paramount importance, especially for small organisms (e.g. fungi, mosses, invertebrates such as molluscs and arthropods). The significance of this variability at multifarious scales will become all too plain with climate change. A species that can readily move from one slope to another, from one aspect to another, from one side of a boulder to another has a greater chance of survival during periods of increased radiation, heat, drought, moisture or flood, and hence will be less vulnerable to the increasingly unpredictable climatic conditions. Landscapes are forever changing, and the changes occurring are driving species at all scales to extinction (Estrada-Carmona *et al.*, 2022).

A classic example of the impact of scales of landscape variability is illustrated by our knowledge of the imperilled silver-studded blue butterfly *Plebejus argus* (Fig. 11.8). This species was initially studied on the headland of the Great Orme (North Wales, UK) (Dennis, 1972, 1977), work subsequently expanded to cover the country (Ravenscroft, 1990; Dennis, 1992; Thomas *et al.*, 1999). At the scale of the entire headland (some 3 km long and 2 km wide, rising from sea level to 207 m), the butterfly is mainly found on the drier southerly and westerly slopes, where its several host plants occur. The contrast with the north and east slopes is profound; here, the denser vegetation (dense grasses, bracken) and damp, cold conditions are inimical to suitable growth of the host plants and to the occurrence of the ants with which the butterfly is associated (Dennis and Hardy, 2018). The relationship the butterfly has with the vegetation structures on the south and west slopes is fascinating. The shrubs (bramble, clumps of cotoneaster, gorse) shelter ants and butterflies, providing roosts,

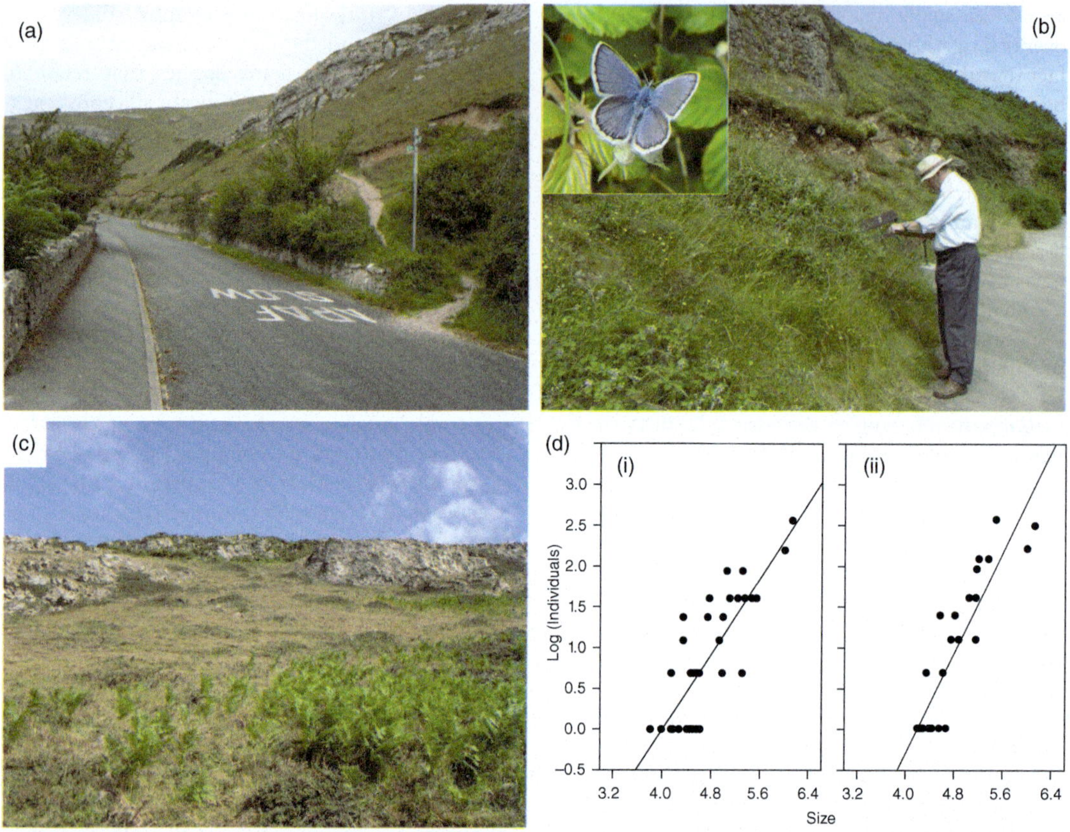

Fig. 11.8. The Great Orme, North Wales (UK), habitat of the silver-studded blue butterfly *Plebejus argus* showing one of the key larval host plants (*Helianthemum* sp.) and shrubs and bushes on the lower slopes used for roosting, day shelter, mate location and egg-laying sites in their proximity. **(a)** South-west side of the Orme and open habitat. **(b)** South side of the Orme, with bushes low down used for roosting, mate location and egg laying underneath near ant nests (R.L.H. Dennis depicted in photo). **(c)** Open slopes above Invalid's Walk on south side but with substantial cover of *Cotoneaster*, bramble and gorse shrubs used especially for nectar feeding, shelter, mate location and roosting. **(d)** Relationship of numbers of *P. argus* resting on *Cotoneaster* shrubs with size of shrubs on the Great Orme: (i) 24 June 2004, >11.00 a.m., ambient temperature (Ta) 12.5°C, wind speed mean 8.5 m per sec. $Y = -4.608 + 1.152X$, $R^2 = 67\%$; (ii) 2 July 2005, >11.30 a.m., Ta 15.9°C, wind speed mean 7.6 m per sec. $Y = -6.469 + 1.538X$, $R^2 = 79\%$; $Y = \log_{10}$ number of *P. argus*; X = shrub size ($\log_{10}(\pi \times \text{length} \times \text{breadth})$). Research by the first author in 2003–2005 demonstrated the importance of shrubs as habitat components (the resource-based habitat of the species) that escaped the attention of a university research group from Leeds and York between 1983 and 2015. As a consequence, advice was given for rangers on the Orme Country Park to clear (spray) all shrubs off slopes, instead of selective clearance to prevent complete cover by shrubs. It is a typical example of broadcast management (total clearance) which should only ever be carried out in special circumstances (e.g. Japanese knotweed *Reynoutria japonica*) after careful thought or experiment, taking into account the impact on other species (including insects and birds) on the headland. One of the shrubs may well be a unique species (for *Cotoneaster cambricus*, see www.ukwildflowers.com/Web_pages/cotoneaster_cambricus_wild_cotoneaster.htm; accessed 22 May 2025). (Panels **a–c** photos by R.L.H. Dennis. Panel **d** from Dennis and Sparks, 2006.)

mate location sites and oviposition sites. The shrubs therefore are crucial resource-based habitat elements. The butterfly flies over higher slopes in sunshine and when conditions are relatively calm, but retreats to the cover of shrubs on lower slopes in cold, overcast, wet weather and at night (Dennis and Sparks, 2006). Thus, the vegetation structure is crucial for maintaining the population. But more recent trips (R.L.H. Dennis, personal observation, 2015) revealed that nearly all the shrubs and the bramble had been erased from the southern slope overlooking the town of Llandudno, despite the fact that this imperilled species has been known to be present there in abundance since the 2000s (Dennis, 2004b; Dennis and Sparks, 2006). More sensible conservation practice would have left areas (patches) with low shrub cover. Inflicting a single management regime over the entire space for any slope or aspect is a grossly unintelligent measure.

Plebejus argus also provides a classic example of a species that occupies completely different biotopes in different parts of Britain: dry limestone grasslands (including f. *cretaceae* in Kent), dry *Erica* and *Calluna* heaths and wet *Calluna* mosslands (f. *masseyi* in Cumbria and Llyn Hafod-y-llyn, Wales), and dry sand dunes (Dennis, 1992; Thomas *et al.*, 1999; Adrian Spalding, personal communication). These populations differ in phenotypes, genotypes, host plant use (*Helianthemum nummularium*, *Lotus corniculatus*, *Erica cinerea*, *Calluna vulgaris*, *Ulex europaeus*, *Sarothamnus scoparius*, etc., plants belonging to different families) and ant dependency (*Lasius alienus* and *L. niger*; Thomas *et al.*, 1999). The butterfly is a host plant quality **specialist** (exploiting specific parts of plants in early stages of growth) but also a host plant **generalist** (using different species), with a biology strictly related to that of its ant symbionts. As Thomas *et al.* (1999) point out, the implications for conservation are that **introduction** or **reintroduction** programmes should ensure that the same *Lasius* species which are associated with *P. argus* in donor populations are also abundant in receptor habitats.

The significance of heterogeneity should never be overlooked in any management plan for the diversity of donor species in any taxon. General adoption of less intensive (but variable) management of marginal, unproductive land, such as verges, field margins and linear features, and incorporation of disturbance to encourage natural resource changes should be considered in management plans to promote heterogeneity (Shreeve and Dennis, 2011) and hence diversity (e.g. butterflies in the Drawieński National Park area of north-west Poland, comprising various habitats of cultivated fields, fallow land, forests and marshes; Szyszko-Podgórska *et al.*, 2023) as well as providing refuges for a focal species which prefers a more limited set of biotopes.

Mapping the precise location and traits of a **keystone** or **umbrella** species should be the starting point for any RBH research (Fig. 11.9). Once its location, numbers and/or trends have been identified, a decision can be made concerning its status (see, for example, criteria defined by the International Union for Conservation of Nature; IUCN, 2024a). From a mapping programme the fundamental features of the landscape(s) occupied should become obvious, but this is a task which is a great deal more complex in water than on land. At this point, for threatened species, the work should include detailed analyses from trained researchers. Crucially, important traits, resources used, conditions experienced, climate, symbionts, parasites, predators, life cycle features and population dynamics fall into this next stage, determining the vulnerability of the species to different threats, 'natural' and human. To understand how a population responds to environmental factors, however, data from repeated surveys over a period of years are needed, as shown by shifts in the microscale distribution of *P. argus* in north Wales (Hodgson *et al.*, 2015). In situations where vulnerability is evident, monitoring programmes, management and the need for maintenance follow. Bearing in mind the pressure on resources for research, careful attention needs to be applied to the selection of species in any RBH approach. They may be selected simply for the reason that they can act as focal species, e.g. highly apparent, charismatic species, species useful as a model or umbrella species for others (see Box 2.3 in Chapter 2).

Introductions and cascading effects

Initially, the authors have to declare that the issue of legality associated with species **translocation** is not discussed in this book. For example, every state will have its own rulings about what may and what may not be introduced into (or moved within) its bounds, and the procedures which have to be addressed for surveys and land management. This topic is not discussed here as it is too extensive, but it is a crucial barrier/facilitator of 'engineering' habitats.

There are three types of **translocation** of a species: (i) translocation of individuals into an existing population to enhance population viability by increasing population size and genetic diversity (**augmentation**, reinforcement, restoration or supplementation); (ii) replacement of an extinct species in a place or places where it once occurred (**reintroduction**); (iii) placement of a species, novel to the 'state', or in novel locations for it ((**conservation**) **introduction**). In all cases, translocation efforts

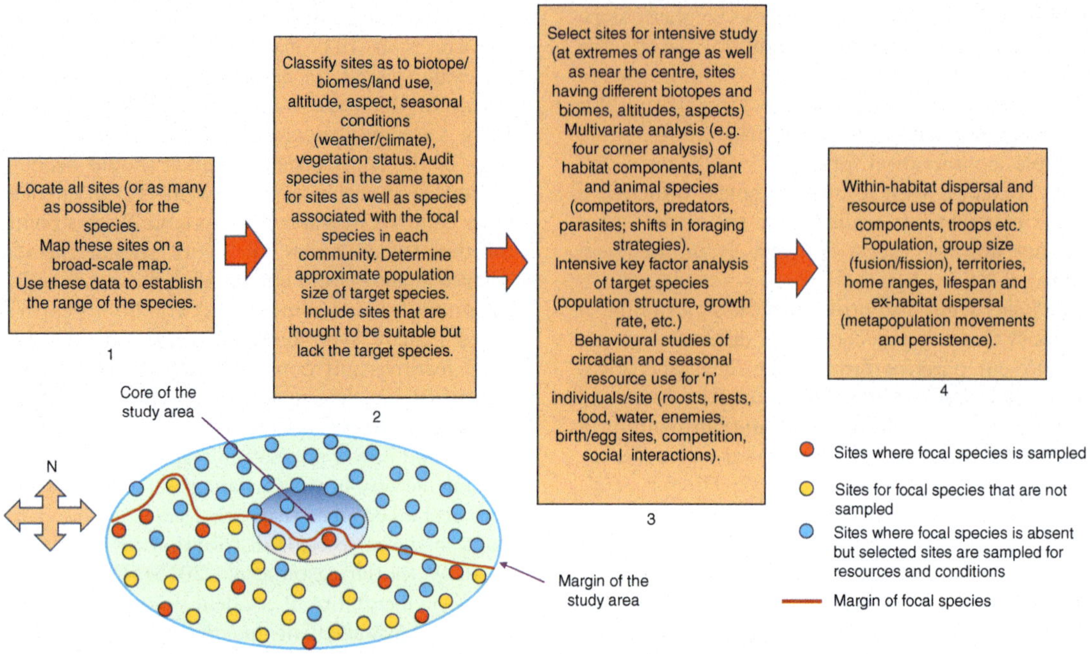

Fig. 11.9. A simple guide to recording data on a focal species' resource-based habitat. Several features are to be expected or should at least be tested. For example, it is likely that a species has contrasting responses, especially during different developmental stages, as well as different responses to substrates under different conditions (local temperature and wind; contrasting attributes at opposing edges of their thermal range, different altitudes and aspects, etc.) (Evans *et al.*, 2023). Also, animals may have different strategies when obtaining prey, especially when these strategies change between sites (S. Wang *et al.*, 2023). Sampling needs to be representative of changing conditions (altitude, local climate) over species' local and regional range.

must be supported by clear scientific justification and comply with relevant legal frameworks, including cases of passive or accidental introduction, where a lack of scrutiny can still carry significant ecological risks. These programmes are inherently delicate and complex, as they involve ecological, genetic and often socio-political considerations. Consequently, all translocation operations must adhere strictly to established, evidence-based protocols to minimise risks and maximise the likelihood of conservation success (see, for example, Table 11.1 for butterflies).

In the first case (augmentation), the main element of concern is that crosses between individuals from genetically distant groups or populations may result in an **outbreeding depression**. In the second case (reintroduction), there is the all-important question: why did the former population become extinct in the first place? (This also partially applies to the first case: why did the population decline?) Here, we can return to the work of J.A. Thomas (Thomas, 1995) on the large blue butterfly *Phengaris* (*Maculinea*) *arion* and his discovery that the host plant – the initial target for the butterfly – was not a critical factor in itself but that the conditions in which the host plant occurred were central to the loss of the butterfly: crucially, a short sward with suitable conditions for the ant symbiont. Then, there is the fundamental issue of space. Is the area planned for the reintroduction (as well as in the case of augmentation and introduction) large enough for the organism to persist on it? Here, it is a matter of defining what is meant by persistence (a decade, a century, etc.). A key part of this is: are there multiple locations also large enough and in close enough proximity, for the organism to form a **metapopulation** by natural dispersal (a single site may be insufficient; see Chapter 7)? If not, are there locations linking up potential sites that can be engineered for 'stepping stones' or corridors (see Dennis, 2010: 279; Dennis *et al.*, 2013)? The third case is even more problematic, as will be discussed shortly.

One factor stands out in reintroductions/augmentations. The location that once housed the species is *bound* to have changed to the extent that it is no longer suitable for a reintroduction (or augmentation), but in what way has it changed? Thus, landscape engineering based on knowledge of the RBH of the species elsewhere, *sensu lato*, becomes an essential part of any reintroduction or augmentation programmes. Engineering may seem a strange term to use for managing the environment (semi-natural and human-transformed), but engineering is what we are good at as a species. However, in this case, engineering should be for the benefit of other creatures and not solely for humans. Once life-history, biochemical, behavioural and physical traits have been determined for a species, its resource-based habitat bounds and abiotic limits determined, (re-)introductions can be placed where best suited for the species, or the sites can be modified to improve conditions for it.

The placement of a species in locations where it has not occurred before (introductions *sensu lato*) may arise as an accidental or intentional release of individuals; this has often been the case and the legality of the latter event has to be considered; there are examples where legality has been ignored much to the frustration of work being carried out by conservation bodies. These species are typically abundant in their natural setting and may become a 'pest' in their new locations (see Chapter 10). These alien species are not the foci of RBH approaches, but they do indicate the fragility of the natural world and the degree to which 'wanted' species, 'associated' species and keystone species require close scrutiny before active management.

Intentional translocations require an ecosystem approach before release, as with passing time their presence may have an unexpected impact. Rodriguez-Cabal *et al.* (2013) examined the impact of invasions of exotic species (ungulates and wasps) in Patagonia, and found they caused a **disassembly** of an interaction web (see O'Neill, 2016) including the southernmost hummingbird species (*Sephanoides sephaniodes*), a mistletoe (*Tristerix corymbosus*), the dominant understorey shrub (*Aristotelia chilensis*), which is the most common host of the mistletoe, and an endemic marsupial (*Dromiciops gliroides*). This process they refer to as one of **cascading coextinctions** (Rodriguez-Cabal *et al.*, 2013) (Fig. 11.10).

Cascade effects of introduced species can be very difficult to anticipate, as in the case of the cane toads (*Rhinella marina*) which were introduced to Australia to consume scarab beetles eating sugarcane, but which also consume the dung beetles introduced for cowpat breakdown (see Chapter 3). We have already discussed (see Chapter 10) the impact of alien species introduced unintentionally (e.g. many pests introduced with the plants they feed on; rats) or intentionally (for the most diverse purposes, such as economic gain, aesthetics, biocontrol, etc.) and which subsequently spread beyond the place of introduction, with detrimental effects on the recipient ecosystems (e.g. the mosquitofish *Gambusia* spp. introduced from North America to many parts of the world as a biocontrol to lower mosquito populations: Jordan *et al.*, 2021; the coypu *Myocastor coypus* introduced from South America to North America, Europe and Asia primarily by fur farmers: Schertler *et al.*, 2020).

The significant negative impacts often associated with introduced species greatly constrain their potential use in conservation strategies. Introductions used in conservation are called **conservation** (or benign) **introductions**, and can be defined as the intentional translocation of a species outside its native range to improve its overall conservation status. A classic example is the translocation of the critically endangered Nihoa millerbird (*Acrocephalus familiaris kingi*) (Freifeld *et al.*, 2016). This subspecies, endemic to Nihoa Island in the remote northwestern Hawaiian Islands, was introduced to Laysan Island, an island formerly occupied by the closely related, ecologically similar Laysan millerbird (*Acrocephalus familiaris familiaris*), which went extinct in the early 20th century when the island was denuded by introduced rabbits. As in the meantime Laysan recovered substantially, now offering ample habitat and a rich prey base for millerbirds, it was considered a good place to introduce the Nihoa millerbirds. The introduction was successful; most of the 50 translocated birds were known to have survived 1 year after the translocation, and the population was estimated at 164 birds after 2 years from the introduction. The use of conservation introductions should be considered, however, as an *extrema ratio* and the unexpected, frequently negative consequences of introductions represent a salutary warning about introducing species, even if these are labelled as reintroductions and thought entirely beneficial to the web of species at sites.

We are at the stage where we are finding ecosystems increasingly complex to understand fully; thus, the notions of assembly and disassembly of

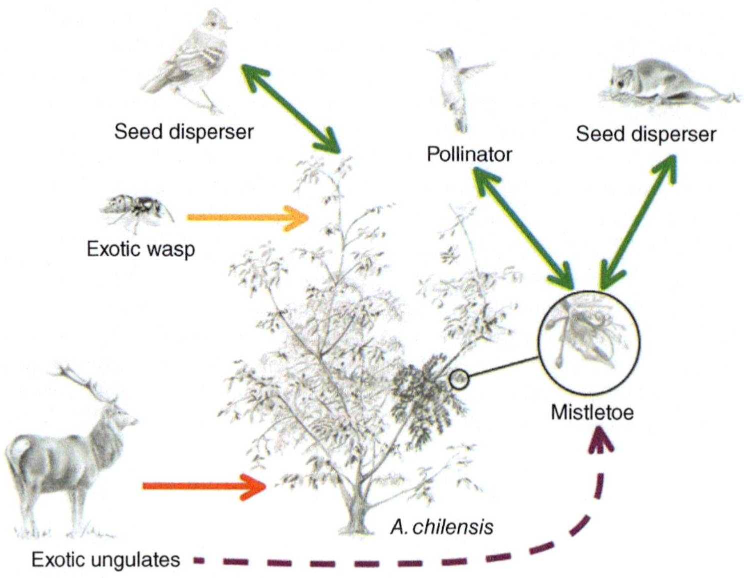

Fig. 11.10. Community disassembly. An interaction web showing the direct effects (solid arrows: orange, fruit predation; red, herbivory) of exotic species on *Aristotelia chilensis* and the indirect effect (dashed arrow, violet) on the keystone interactions (solid arrows, green: pollination and seed dispersal) between a hummingbird (*Sephanoides sephaniodes*), a mistletoe (*Tristerix corymbosus*), a marsupial (*Dromiciops gliroides*) and a bird (*Elaenia albiceps*) in Patagonia, Argentina. Exotic ungulates affect the entire plant community, causing changes in habitat structure and food availability that affect the marsupial. Because mistletoe seeds must pass through the gut of the marsupial to germinate and be deposited on branches of *A. chilensis* to establish, the local extinction of marsupials affects mistletoe recruitment. Because the mistletoe is the primary source of nectar for the hummingbird during the austral winter, ungulates affect hummingbirds by affecting *A. chilensis*, and in turn the mistletoe. As an exotic wasp (*Vespula germanica*) feeds on the fruits of *A. chilensis*, it has negative effects on the bird and the marsupial, which also feed on these fruits. The impact of exotic ungulates and the exotic wasp cascaded through the interaction web to influence the number of seeds removed by both the marsupial and *E. albiceps*. (From Rodriguez-Cabal *et al.*, 2013; drawings by Ezequiel Rodriguez-Cabal, used with permission of rightsholders.)

ecological networks require much in-depth investigation (Bascompte and Stouffer, 2009). Currently, it is as well to move cautiously. Knowing how an umbrella or 'important' species fits within **food webs** at different locations increases our understanding of how it actually functions in different communities, even ones that casually appear to have much the same basic components (see Bascompte and Stouffer, 2009). However, we must be prepared for surprises; our key species may fit into the food web in different ways and have a different status in different communities.

Zoos, botanical gardens and wildlife parks. A foundation for human-mediated translocations

The idea of having a zoo with the purpose of breeding rare and imperilled animals so as to return them to the wild belongs to Gerald Durrell (MacNiven, 1998). It was a ground-breaking use of zoos, whose original purpose was to titillate tourists and make money – visitors to World Association of Zoos and Aquariums (WAZA) number over 700 million every year, an annual investment of US$350 million (Gusset and Dick, 2011). By October 2010, the Convention on Biological Diversity (Secretariat of the Convention on Biological Diversity, 2011) in Nagoya, Japan, still faced with the task of halving the loss of natural habitats and preventing the extinction of threatened species, realised the importance of Durrell's idea. The contracting parties became aware that *in situ* conservation of habitats would need to be supported by *ex situ* conservation actions. Such conservation actions require co-operation programmes with zoos, wildlife research facilities, botanical gardens and institutions that preserve germplasm collections (e.g. seed banks) (see also IUCN, 2002). In the following year, Conde *et al.* (2011) noted that captive breeding played a major role in species' recovery (e.g. for 17 of 68 vertebrate species whose threat level was reduced). Examples were Przewalski's wild horse (*Equus ferus przewalskii*), the blackfooted ferret (*Mustela nigripes*) and the California condor (*Gymnogyps californianus*). Captive breeding can be especially important to save populations on small islands, such as the case of recent projects on endangered snails (www.chesterzoo.org/news/rescuing-the-desertas-island-land-snail-from-the-edge-of-extinction; accessed 22 May 2025) and lizards (https://associazionetriton.it/progetti/life-eolizard/; accessed 22 May 2025). Even so, the number of threatened species saved by zoos from extinction is still far too low (Fa *et al.*, 2011, 2014).

Zoos, wildlife parks and aquariums have a critical part to play in animal conservation, not least by changing societal perceptions of the natural world, encouraging funding, training zoologists in conservation ecology, and executing and publishing research, as well as acting as a temporary station for animals while habitats are being 'engineered' for them and other species (Loh *et al.*, 2018). Association of Zoos and Aquariums (AZA) members published some 5175 papers on zoology and veterinary science between 1993 and 2013 (Loh *et al.*, 2018). This indicates that it is time for research productivity and conservation outcomes to be explored so as, at least, to attract further funding for successful transfers to re-established natural habitats.

Botanical gardens may play similar roles in assisting biological conservation, by enabling *ex situ* conservation of imperilled species (Cano *et al.*, 2025), with the potential for translocation, by developing research on plant conservation (including species responses to climate change) and promoting educational activities and societal awareness of the importance of plant conservation (e.g. Donaldson, 2009; Neves, 2024). Botanical gardens can also contribute to animal conservation, both directly (by favouring the conservation of animals, especially insects, that use plants cultivated in the gardens) and indirectly (e.g. by promoting awareness of insect conservation) (Fattorini *et al.*, 2023). A further form of *ex situ* conservation is germplasm collections, which allow the long-term preservation of large amounts of genetic material in small spaces with low risk of degradation (e.g. the Svalbard seed bank). However, although *ex situ* conservation offers many opportunities, it also has important limits and problems, ranging from the small number of individuals that can be maintained to adaptation to artificial conditions, behavioural changes, ethical issues, etc., as discussed by Primak (2014).

Species distributions, limits and variation in their resource-based habitats

Case issues for single species in the study of RBHs vary enormously in complexity across any species' range, and are further obfuscated in the face of habitat loss and climate change. There is the need for many 'heads' and 'hands' in their study. Though programmes are now available for representing and manipulating species range data (e.g. Cobos *et al.*, 2022), and many facilities (e.g. platforms such as iNaturalist and GBIF) make it possible to collect a lot of information on species distributions, a profound knowledge of the species biology is essential to evaluate data quality and interpret the results, as a number of taxonomic and ecological biases (not to mention errors) affect available data (e.g. Beck *et al.*, 2014; Troudet *et al.*, 2017; Hughes *et al.*, 2021; Mammola *et al.*, 2023; Rubio-López *et al.*, 2023; van Tongeren *et al.*, 2023; see also Chapter 8). The scale of investigation extends from minutiae to broad geography. For instance, at the large-scale level, not all species are so abundant or accessible that they can easily be tracked to specific resource uses, but a spatial survey linked to physical attributes of the surrounding environments can reveal a great deal about their limitations. For instance, the stock dove (*Columba oenas*), the scarcer of the doves in urban Manchester (UK), does not take as easily to urban settings as do the wood pigeon (*Columba palumbus*) and the feral rock dove (*Columba livia domestica*). Richardson *et al.* (2023) found that stock doves tended to be associated with traditional habitats with more surface water during the non-breeding season than during the breeding season. They demonstrate the importance of large trees within urban green spaces for this cavity-nesting species, and 'softer' boundaries (composite edge habitats) around urban sites for its presence in proximate areas.

Once species have been tracked to precise locations and their ranges determined, this provides the framework for sampling **traits** in the species (see Fig. 11.9). Sampling criteria here are crucial. Too often, life-history details are still built on individuals which have been obtained from one of few locations. To understand trait variation in a species, it is essential that such intimate studies and experiments are adopted from adequate numbers of individuals from sites throughout its entire range. Inevitably, this means co-operation between scientists and interested 'parties' from these different locations. This is not the place to expound on the critical issues of standard sampling techniques (some examples are given in Appendix 1) but it is essential that any student is fully aware of all sampling issues, biases that can emerge in any study, and the effects of changing conditions from one year to the next. For instance, a butterfly such as the comma (*Polygonia c-album*) will produce one or two generations depending entirely on when and where breeding experiments are carried out; moreover, individuals producing a second brood in any one year (f. *hutchinsoni*) are distinctive in wing pattern and other traits from those that produce only a single brood (Nylin, 1988, 1992; Nylin *et al.*, 1996).

A major factor, already broadly addressed above, is the 'takeover' of land by urban growth. Many species that could occupy growing urban areas around the globe would benefit from engineered parks with a range of vegetation settings (Fattorini, 2019). Apart from a full understanding of species' traits, and the interaction with other species in the community, it is instructive to explore the broad links between plants and animals inasmuch as the impact of engineering habitats may not benefit all species. For example, a study on the plant–butterfly networks in urban agricultural and forest sites in central Mexico revealed that a native forest site had a higher diversity of butterfly species than agricultural and urban sites; however, the urban site had a higher plant richness due to the higher incidence of alien plants, as well as a higher number of links and modules (see Box 2.1 in Chapter 2 for this terminology), but a smaller number of interactions (Muñoz-Galicia *et al.*, 2023) (Fig. 11.11 and Table 11.2). Thus, land transformation reshapes communities in non-obvious ways.

Climate change and shifts in resource-based habitats

Climate change is now raising the bar and challenging the limits of our knowledge and ability to respond (e.g. IPCC, 2022). An example obvious to most naturalists for many years is the poleward and upward shifts in the distributions of many organisms (see Fattorini, 2024c). At the finer scale, increasingly it is being found that climate change has direct influences on species, so as to alter their capacity for exploiting the components of their **resource-based habitats** (Hodkinson *et al.*, 2011; Fattorini, 2023; Harvey *et al.*, 2023). These changes critically affect species biological traits (life history, behaviour, food exploitation). Evidence from insects clearly demonstrate how many aspects of their biology are controlled by climatic conditions. For example, Sales *et al.* (2018) found that heatwaves reduce male fertility and sperm competitiveness in the red flour beetle (*Tribolium castaneum*), with successive heatwaves almost sterilising males – an advance on work by Johnson *et al.* (2004). Thermal experiments on sub-Antarctic and Arctic insects in line with climate change revealed a taxonomic influence, with physiological sensitivity to heat differing between higher level taxa (beetles versus flies), but not between native and non-native species within the same higher taxon (Renault *et al.*, 2022). These findings would suggest the restructuring of biodiversity with climate change, and impress upon those with responsibility for keystone species the unexpected impact on them by the biotic components affected by climatic agents.

Changes in abiotic and biotic circumstances for a focal species can alter the traits of that species either directly or indirectly, and in turn affect its resource-based habitat. Just the process of biological invasions (see Chapter 10) and pathogen spreading can have impacts of similar importance. For instance, Scheele *et al.* (2022) discovered that an invasive pathogen drives directional niche contractions in amphibians. They quantified the geographical range and realised-niche contractions for Australian frog species affected by a globally expanding chytrid fungus (*Batrachochytrium dendrobatidis*); the amphibians are contracting toward lower elevations, warmer temperatures, smaller diurnal temperature ranges and zones of higher annual precipitation. Epidemic diseases have a special place in changing the ecological niche structure and ecological dynamics of communities, including humans (e.g. the recent COVID-19 pandemic, https://en.wikipedia.org/wiki/COVID-19; accessed 22 May 2025). Examples are numerous for domesticated animals (e.g. **fasciolosis** in cattle, a re-emergent parasitic disease; Alba *et al.*, 2023).

Classic impacts of climate changes on animal traits are on hibernation and phenology. In a meta-analysis

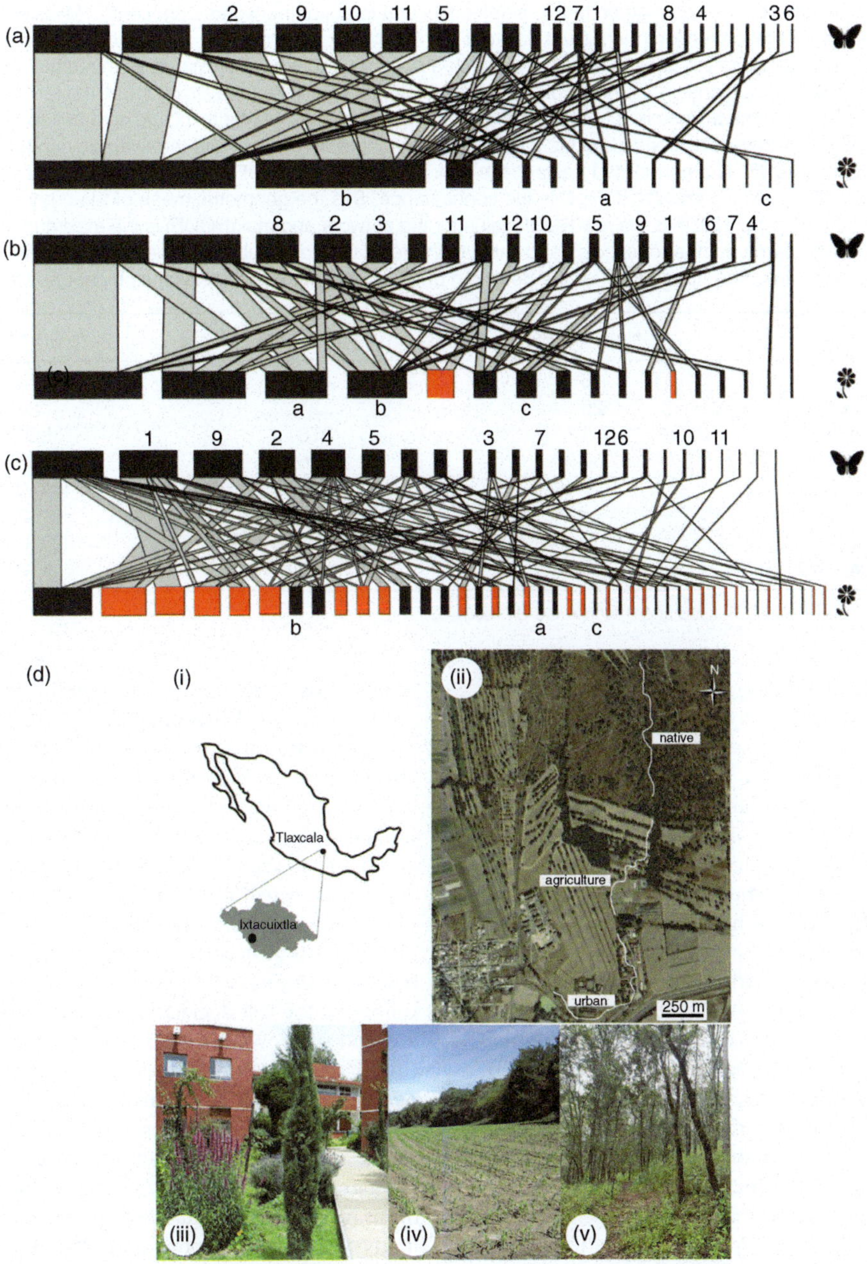

Fig. 11.11. Bipartite plant–butterfly interaction networks in three sites with different land use in central Mexico. In each network the nodes on the top represent butterfly species and those on the lower side different plant species. Red nodes represent alien plant species. The thickness of each link (grey lines) indicates the frequency of each pairwise interaction (butterfly flower visits). **(a)** Network for the native forest site. **(b)** Network for the agricultural site. **(c)** Network for the urban site. Same numbers and letters on the nodes denote shared species among the three study sites. **(d)** The location for the study and the three vegetation types: (i) geographic location of the study site in the state of Tlaxcala; (ii) bold lines indicate the survey transects established to record plant-butterfly interactions in (iii) urban, (iv) agricultural and (v) native forest site. (From Muñoz-Galicia *et al.*, 2023, used under licence CC-BY 4.0.)

Table 11.2. Impacts of land use change on native plant–butterfly interaction networks from central Mexico. Values of network metrics for the interaction matrices of different plant–butterfly assemblages in three different habitats. Connectance is a measure of the density of connections in a network and was determined as the actual proportion of links of both butterflies and plants in the network compared to all possible links (see Box 2.1 in Chapter 2). Modularity is a measure of how the interactions in a network are structured, with high modularity indicating that interactions are concentrated within specific subgroups of nodes (see Box 2.1). Vulnerability expresses the average number of links of butterflies per plant. $NODF_c$ is a normalised nestedness metric based on the NODF measure: $NODF_c = NODF_n/(C \times \log(S))$, where $NODF_n = NODF/\max(NODF)$, C is the connectance, S is the geometric mean of the number of species in each level of the network, NODF is the raw NODF value for the network and max(NODF) is the maximum nestedness of a network with the same number of species and links as the focal network, subject to the constraint that every species has at least one link. See Appendix 3 for the definition of NODF. Higher $NODF_c$ indicates higher nestedness.

Metric	Native forest	Agricultural	Urban
Number of plant species	16	18	44
Number of butterfly species	25	21	24
Number of interactions	531	378	524
Number of links	50	46	83
Connectance	0.12	0.12	0.07
Vulnerability	7.24	3.95	2.28
Modularity	0.54	0.56	0.61
Nestedness ($NODF_c$)	2.63	1.91	2.29

(From Muñoz-Galicia *et al.*, 2023; used under licence CC-BY 4.0.)

of papers on the subject of trait changes, Findlay-Robinson *et al.* (2023) found that in most studies and species, dates of emergence from hibernation became earlier, the size of litters increased and the frequency of **interbout arousals** increased with increasing ambient air temperature. The overall impact must be to place greater pressure on resource exploitation and overall population size or the size of progeny. Change in phenology is one of the best documented aspects of insect response to climate change, with shifts toward earlier seasonal activity being widely recorded (Forrest, 2016). However, increased temperatures may also change overwintering strategies (Fattorini, 2023). A delay in the winter diapause induction might lead to a complete or partial additional generation in the autumn that cannot survive or enter diapause (Harvey *et al.*, 2023). For many insects, an advanced phenology translates into an increase in the length of the activity period (Menéndez, 2007), but some species do not show the expected advance in springtime (Forrest, 2016) and this might lead to a contraction in the activity period if summers are too hot. In the case of insects living in arid and hot environments, overly high temperatures in summer might lead many species, especially the summer specialists, to become inactive earlier, and if this is not compensated for by a phenological advancement, their phenology will be shortened. This contraction in phenology might reduce the possibility of finding food and water (Fattorini, 2023).

A key research area is experimental work on the effects of solar radiation, temperatures, humidity and other climate-related factors on species traits. It is one thing to determine the RBH of a species in a short time-slice, quite another to predict how it will respond to future climate and other changes. One needs only think of the massive shifts in flora and fauna that occurred after each of the previous **mass extinctions** (see Appendix 5), or more recently following the last deglaciation, to see this effect on a grand scale. We have yet to understand fully the causes of the loss of Palaeozoic trilobites, the loss of those wonderfully gigantic Carboniferous dragonflies and the clubmosses and horsetails, as well as the earliest gymnosperms (seed-bearing plants) through which they flew; the eventual takeover by dinosaurs as dominant land predators, replacing many crocodylomorphs after the Triassic; and finally, the emergence of primates after the last mass extinction, leading to the two specimens belonging to the alleged dominant species *Homo sapiens* writing this book. One should always bear in mind the power of **micro-organisms** with climate change. The importance of the microbiotas in the chain of events triggered by climatic change, or related to human activity, has recently lead to the development of research into the resource-based habitats of micro-organisms (e.g. niche shifts in bacteria in soils; Yates *et al.*, 2022).

Finale

Since its original development in the field of butterfly biology (e.g. Dennis and Shreeve, 1996; Dennis *et al.*, 2003), the resource-based habitat paradigm has been expanded to include organisms other than Lepidoptera (e.g. tick-borne diseases; Vanwambeke *et al.*, 2016) even though the concept is not named (e.g. bittern *Botaurus stellaris* conservation in the UK; Brown *et al.*, 2012). With this book, our aim is to illustrate how the adoption of the RBH concept can be useful for understanding and conserving nature. We are aware that extending the use of the RBH concept from focal species for which a lot of information is available (or can be realistically gained) to entire communities may be discouraging. However, an understanding of ecosystems should make the reader aware that there is no fundamental difference between maintaining biodiversity and conserving a single focal species. Maintaining communities demands our attention to the interconnections established by 'key' species, which may well include those to which we attribute focal attention for any number of reasons, even merely aesthetic functions. The novelist John Fowles (1969) would argue from the vision of St Hubert (page v of this book) that this includes all species on the planet. Then again, a focal species may act as a valuable umbrella species for many others in a community, or at least as a guide to the adoption of adequate measures to preserve ecosystem stability.

The strong inference is that an ecosystem approach, including all micro-organisms, fungi, plants and animals, is the way forward to an understanding of how species interact within communities in traditional habitat spaces or over wider zones, biotopes, biomes and biogeographical regions, and hence to the preservation of Earth's biodiversity. Each holds a position of importance in a community. The way forward is rapidly being explored. For example, Zeuss *et al.* (2024) developed a networked sensor system for integrated biodiversity monitoring (the Nature 4.0 project) to address these challenges. This approach considers organisms not only as targets of investigation but also as parts of the modular sensor network by carrying sensors. Their networked sensor system consists of three main closely interlinked components having a modular structure: (i) sensors, (ii) data transmission, and (iii) data storage, which are integrated into pipelines for automated biodiversity monitoring and management. Such new approaches might represent important progress towards building a complete model of different communities in distinct biomes, and determine the place of humans in each system.

The increasing availability of sophisticated and economical technologies will obviously play a pivotal role in boosting ecological research. But we can benefit from the invaluable aids that the new technologies can offer only if we can use them within appropriate conceptual frameworks. The few examples in this book should illustrate to future researchers the sheer complexity involved in determining the traditional habitat types occupied and the resource-based habitats in the ranges of species faced with the rapid changes humans are now imposing on the globe. What we protect and enhance is entirely up to humanity, but *Homo sapiens*, in its various forms, must first accept, learn and grow into what humanity means in all its aspects. This must primarily encompass withdrawing from conflict and protecting and nurturing the offspring of all creatures, including our own, through a caring, selfless, generous husbandry of the planet.

Appendix 1 Surveying Natural Populations

In most cases, counting all individuals of a population is practically impossible. For this reason, population size is usually estimated from sampling (e.g. Krebs, 1999; Henderson, 2003; Sutherland, 2006). Moreover, if in some circumstances knowing the total number of individuals in a population is needed (e.g. in planning harvesting strategies or modelling energy flows or nutrient cycles), in many cases measures of **density**, instead of total numbers, may be appropriate. There are two measures of density: (i) absolute density (the number of individuals per unit area or volume) and (ii) relative density (the abundance of a specific species compared to the total number of individuals of all species in a **community**). For example, for monitoring population trends of a certain species (for pest control as for conservation purposes), estimating variations in its absolute density may be appropriate. On the other hand, in community ecology, we are usually more interested in the relative abundance (density) of the species (for example, to calculate diversity measures, see Chapter 4) than in their population size.

The crucial issue in surveying natural populations is that different organisms require very different sampling techniques. Population estimates are just that: estimates of a population. Papers abound dealing with bias in sampling population size or density and will continue to emerge in the literature, especially as analyses depend increasingly on amateur contributions to the recording of species and accumulation of data (Dennis and Thomas, 2000; Lobo *et al.*, 2021; Sánchez-Fernández *et al.*, 2021).

To avoid biases in population estimates, sampling must follow appropriate designs (for example, it is necessary to establish if sampling should be conducted with or without replacement; if it should be random, stratified or systematic; the requested sample size and number of replicates, etc.). Choice of appropriate designs should be guided not only by general statistical principles (illustrated, for example, by Elzinga *et al.*, 1998; Krebs, 1999; Sutherland, 2006; Hayek and Buzas, 2010) but also by considerations that vary according to the specific organisms and habitats. It is obvious that surveying orchids in a grassland would require sampling designs and techniques very different from those that would be appropriate for herbivorous insects living in the canopy of a tropical forest.

A common procedure to estimate the population size of mobile organisms (such as most animals) is the **mark-recapture** technique (Powell and Gale, 2015; Royle and Converse, 2020). The basic procedure of mark-recapture methods is to mark a number of captured individuals over a short time (n_1), release them, recapture a certain number of individuals (n_2) and count how many of them have the mark (m_2). Under the assumption that the marked individuals dispersed randomly in the population, and that sampling was equally random, by a proportionality argument we have:

$$\frac{\hat{N}}{n_1} = \frac{n_2}{m_2} \qquad \text{(A1.1)}$$

and hence

$$\hat{N} = n_1 \frac{n_2}{m_2}, \qquad \text{(A1.2)}$$

where $\hat{N}$ is the estimate of the population size.

This index (known as the Lincoln or Petersen index), however, tends to overestimate population size; to reduce this bias, an improved version (Chapman index) has been recommended by Seber (1982):

$$\hat{N} = (n_1 + 1)\frac{(n_2 + 1)}{(m_2 + 1)} - 1 \qquad \text{(A1.3)}$$

The 95% confidence limits can also be calculated for this estimate.

A number of variants exist (e.g. Schnabel, Burnham–Overton, Jolly–Seber; Krebs, 1999; Henderson, 2021) with various assumptions about the conditions

DOI: 10.1079/9781800627505.app1

of the populations, the numbers sampled and the capture probability. While the Petersen method is based on a single episode of marking animals and a single episode of recapture, the Schnabel method uses multiple samples; both assume closed populations. The Burnham–Overton method is also conceived for closed populations and uses multiple samplings but allows capture probability to vary among animals. The Jolly–Seber method is designed for open populations.

The way animals are marked may be extremely varied, depending on the characteristics of the animals themselves; marks may be represented by rings (e.g. birds, bats), plastic tags (e.g. mammals, reptiles), staining (alphanumeric codes, symbols, colour dots, e.g. insects, reptiles), morphological modifications (e.g. toe clipping in reptiles; branding in reptiles and amphibians), radiotelemetry (from large mammals to insects), etc. Some animals are naturally marked when they can be recognised by individual phenotypic features (e.g. colour and scale patterns in reptiles; spots on beetle elytra). Non-invasive methods for conducting capture–recapture surveys without physically capturing animals include the use of digital photography, digital sound recording, DNA sampling and encounter histories (encounters are classified as 'recaptures' or 'new marks') (Manning and Goldberg, 2010).

Density measures are usually obtained by counting (collecting or recording) individuals using methods that can be classified into two main categories: line **transects** (a route, along which individuals are counted) and **plots** (usually **quadrats** but they can also be rectangles or circles; counting points, in which the observer counts individuals within a certain radius, can be considered a form of plot). These two approaches have different underlying statistical aspects that must be carefully considered in relation to the organisms and the habitats (see, for example, McClintock and Thomas, 2020). Both approaches can be applied at the most disparate scales (e.g. populations within a management/treatment area, a habitat, an administrative boundary or for the entire range of the species) (Fig. A1.1a–d). Transects are paths (not necessarily straight) that can be of any length and width, depending on the organisms and the habitats, ranging from a few metres for small organisms (e.g. butterflies in an urban park) to kilometres (e.g. large mammals in the savanna). While transects are used for both sedentary and more mobile animals, as well as for plants, plots are commonly used for plants or animals that remain (at least relatively) immobile during the counting period (e.g. slow-moving insects, molluscs, echinoderms, etc.). Plots can be of any size, from very small (e.g. grids used for counting bryophytes and lichens on stones and trees) to very large (e.g. plots for counting trees). In some cases, instead of counting individuals, the surface covered by any given species is estimated. Counting individuals may be impractical when they are extremely numerous, and even theoretically questionable in the case of modular organisms (e.g. sponges, corals, many plants and fungi, etc.; in this case, ramets – physiologically independent units – might be regarded as individuals from many functional aspects but technically the individual is the genet, which is the entire genetic individual formed from a single zygote). In some cases, density measures are obtained by sampling standard volumes of soil (e.g. microarthropods living in the soil), sediments (e.g. **meiofauna**), water (aquatic small organisms, e.g. plankton) or air (airborne organisms, e.g. insects). Aerial photographs can be used in some cases (e.g. large mammals). Automatic techniques used for population surveys include, for example, recording acoustic signals (e.g. amphibians) and camera trapping (cameras that take videos or still images when a light beam is broken or when a switch is mechanically triggered, e.g. mammals, birds, insects).

Since different organisms and habitats require different sampling designs and techniques, there is an immense variety of sampling procedures. Moreover, although standardisations are obviously desirable, details of sampling procedures tend to vary case by case, because of differences in environmental conditions, economic restrictions, temporal limits, available materials, etc. In the following sections, we schematically illustrate a selection of procedures that portray the enormous diversity of methods used to sample organisms in different habitats.

Plants

Estimates of plant density (number of individuals or stems or any other counting unit per unit area), cover (the amount of ground covered by the vertical projection of plant matter) or biomass are usually obtained using quadrats (plots; Fig. A1.1a–d), which represent the sampling units. Another approach typically used in plant population surveys is the point-intercept method: pins are lowered at regular intervals along a line, hits are recorded, and percent cover is calculated as the number of hits (for each

Fig. A1.1. Sampling methods. **(a)** Use of a quadrat to collect plant data. **(b)** Microhabitat sampling: Rosalina Grabiel (University of the Azores) collecting mosses on a stone within a plot. **(c)** Materials used to collect moss data at a very small scale: a small sample frame, envelopes to contain samples, forms to record environmental data, etc. **(d)** Data collection on trees: Rui Bento Elias (University of the Azores) measuring the size of a tree within a plot. **(e)** Berlese funnels used to extract arthropods from soil or litter samples. **(f)** A pitfall trap with a roof to sample ground-dwelling arthropods (the scheme illustrates how it works). **(g)** Use of sweep nets to collect arthropods from herbs and medium-height vegetation: the second author sampling insects along a transect. **(h)** Flying insects can be actively sampled with nets: the second author collecting pollinators in an urban area. **(i)** A SLAM (Sea, Land and Air Malaise) trap used for sampling flying insects across diverse habitats. This modified Malaise trap design captures insects from multiple directions and is suitable for terrestrial, aquatic and transitional environments. **(j)** Pan traps to collect flower-visiting insects. **(k)** Beating used to sample arthropods living on shrubs and small trees: Paulo A.V. Borges (University of the Azores) showing the white cloth used to collect arthropods that fall from the foliage. **(l)** Lightsheeting, a technique used to sample moths and other nocturnal insects. **(m)** Hand sampling of arthropods in a microhabitat: the second author collecting insects living under bark with an aspirator. **(n)** A sampler used to collect invertebrates from riffles in shallow streams: flowing water carries the invertebrates into the collecting net as the stream bottom is disturbed. **(o)** A grab sampler used to collect sediments from which organisms can be extracted. **(p)** Mist nets used for bird ringing. **(q)** Bird hides used to collect data on bird biology. **(r)** A water sampler used to collect water (and hence plankton) at various depths. (Photos **a–f**, **i–l** and **n–r**, S. Fattorini; photos **g**, **h** and **m**, Federica Fedele.)

plant species) divided by the total number of pins. Although cover measures obtained by points are considered the least biased and most objective, species with low cover values are often undersampled because points rarely intersect them. Both plots and points can be disposed along transects but, if they are not sufficiently distant to be the sampling units (i.e. they are too close to be independent of one another), the entire transect is the sampling unit.

When estimating numbers of individuals or proportional cover, classes (instead of exact values) are frequently used, e.g. 1–3, 4–10, 11–30, 31–60, 61–100, 101–200, 201–500, 501–1000, 1001–5000, and so on for individuals; <1%, 1–5%, 6–25%, 26–50%, 51–75%, >75% for cover, usually assigned to species by visual estimates. A multitude of scales have been proposed, including, for example, Braun-Blanquet, Londo, van der Maarel (Londo, 1976; van der Maarel, 1979; Pätsch *et al.*, 2019). A disadvantage of cover measures is that cover can change dramatically over the course of a growing season. However, as plants are immobile, use of permanent plots (i.e. plots that can be repeatedly visited) makes it possible to record annual species that only appear above ground once every few years, or even once every few decades, and to track temporal changes in species composition, density, cover or biomass (much more easily than for animals). Yet the presence of long-lived species may make it difficult to detect changes even across long time spans. Taking photos of plots (photoplots) may be a practical tool to track changes in vegetation (and hence in habitat characteristics) through time (invasion by aliens, successional changes, grazing, trampling, etc.). At larger scales, aerial photographs and remote sensing data can be used to estimate many vegetation characteristics and their temporal changes.

Comprehensive treatments of methods used in plant population censuses and surveys are provided by Elzinga *et al.* (1998) and Sutherland (2006), while Hayek and Buzas (2010) present useful general insights.

Terrestrial arthropods

Given the immense variety of terrestrial arthropods and their habitats, the number of methods used for sampling these animals is virtually uncountable. Comprehensive presentations of methods used for surveying terrestrial arthropods are provided by Leather (2005), Samways *et al.* (2009), Gibb and Oseto (2019), Montgomery *et al.* (2021) and Santos and Fernandes (2021). Borges *et al.* (2018b) provide a detailed description of various protocols adopted for long-term monitoring of arthropod populations.

Taxonomic difficulties and inconspicuousness make the adoption of visual surveys for most arthropods generally impossible (notable exceptions are butterflies and dragonflies, which in many areas can be easily recognised and counted via visual surveys, but different species contrast for visual conspicuousness; Dennis *et al.*, 2006b).

Soil or litter arthropods (especially microarthropods) can be sampled by taking predetermined amounts of soil or litter, from which animals may be extracted using Berlese funnels (devices in which a heat source, for example a light bulb, is used to dry the soil sample, forcing the animals to move deeper into the funnel, until they slip down the funnel into a reservoir containing a preserving liquid; Schinner *et al.*, 2012) (Fig. A1.1e). Litter arthropods may also be collected using a device known as a Winkler extractor: litter and rotten wood samples are first vigorously agitated above a coarse mesh screen; arthropods and small materials are thus concentrated in the material that passes through the screen; this material is finally placed in thin mesh sacks suspended and enclosed within a bag that ends with a jar containing a fixative; arthropods that fall from the litter are collected in the jar (Ivanov and Keiper, 2009; Sabu *et al.*, 2011; Owens and Carlton, 2015). Sand-dwelling arthropods can be sampled by sieving a predetermined amount of sand.

Ground-dwelling arthropods (e.g. ground beetles, rove beetles, centipedes, wandering spiders, etc.) are frequently collected by pitfall traps (Brown and Matthews, 2016; Hohbein and Conway, 2018). These traps (Fig. A1.1f) are cans or buckets buried flush with the surface of the ground; animals simply fall into them and cannot escape. There is, however, an immense variety of models that differ for a number of details (material used, shape, size, presence and type of roofs, internal funnels and grids, etc.). Pitfall traps can be used with or without killing fluids/preservatives (e.g. water, saltwater solution, alcohol, acetic acid, propylene glycol, formalin, etc.) and baits. Liquid baits (e.g. beer, vinegar, honey solution) are placed directly in the vessel, while solid baits (e.g. carrion, fruit, dung) are placed on a platform in the middle of the trap or suspended immediately above it (for example, dung on a grid; Mantoni *et al.*, 2021b). Killing fluids/preservatives and baits are sources of potential biases, attracting or repelling certain species more than others, which makes it difficult to compare population densities of different species. Traps can be placed randomly, in a grid layout or along transects, at distances that may or may not maintain statistical independence. Duration of sampling is obviously important, since abundance of collected animals will increase with increasing time of trap activity. Moreover, although it is assumed that arthropods are sampled according to their relative abundance, more mobile species have a higher probability of being captured; for this reason, instead of abundances, researchers using pitfall traps sometimes calculate the 'activity density' (number of collected individuals standardised by the number of traps and days of functioning) (Fattorini *et al.*, 2020b).

Arthropods of low understorey vegetation (including herbs and grasses) can be collected using vacuum samplers (e.g. converted garden leaf vacuums). Arthropods of medium vegetation (including shrubs) are frequently sampled using sweep nets that are vigorously moved through the vegetation (Fig. A1.1g), along transects, within quadrats or randomly walking in a certain area; number of sweeps or duration of sampling are used as a measure of sampling efforts.

Flying insects (e.g. butterflies, odonates, dipterans, hymenopterans) can be actively sampled with nets (Fig. A1.1h) along transects, within quadrats or by staying or randomly walking in a certain area for a prefixed time. Malaise traps (tent-like traps made of fine mesh material and with a collecting bottle at the highest point) are widely used to sample flying insects, especially dipterans and hymenopterans. There is a vast variety of models; that known as the SLAM trap (an acronym for Sea, Land and Air Malaise) (Fig. A1.1i) is particularly versatile, as it is completely freestanding and self-supported. Colour traps (sticky surfaces or water filled traps: pan traps) (Fig. A1.1j) are largely used to collect flying insects that are attracted by different colours (Vrdoljak and Samways, 2012).

Arthropods living on shrubs and small trees are frequently collected by beating (Fig. A1.1k): a white cloth stretched out using a frame is held under a tree or shrub and branches are shaken or beaten; arthropods fall from the foliage onto the cloth and can be manually collected (standardisation of sampling effort is achieved by using a pre-established time or number of hits). Plastic bottles baited with attractive/preserving liquids and suspended on trees are a type of aerial trap frequently used to collect forest insects, especially saproxylic species. Chemical knockdown (by fogging or misting) is used for sampling canopy insects. A multitude of types of light traps are used to sample nocturnal insects; despite the variety of models, they all use the same principle: insects attracted by light enter a receptacle from which they cannot escape (catches are influenced by the type of light and sampling should be time-constrained). With lightsheeting, a light is placed in front of a white sheet, where moths and other flying insects land (Fig. A1.1l).

Suction traps are extensively used for surveying small flying insects (e.g. dipterans and aphids). Use of suction and light traps since 1964 at Rothamsted Research (UK) has provided unique data for detecting seasonal and annual variations in aphid and moth population density respectively (www.rothamsted.ac.uk/national-capability/the-insect-survey, accessed 23 May 2025).

Because many arthropods use their environments at a fine spatial scale, sampling is frequently conducted by hand at microhabitat level (e.g. under bark, under stones, and in tree holes and nests of vertebrates or other insects) (Fig. A1.1m). Soil emergence traps that cover a portion of the ground are used to capture ground-nesting bees and wasps in order to evaluate their nesting preferences (Cope *et al.*, 2019). Colonial insects pose specific problems. Methods for ants are presented by Agosti *et al.* (2000).

Freshwater invertebrates

Freshwater macroinvertebrates are sampled using a number of methods (Hauer and Lamberti, 2007; Cummings *et al.*, 2016; Callisto *et al.*, 2021). Active sampling uses sweep nets and dippers (abundances are quantified as the number of animals per sweep or unit volume), area samplers (a variety of devices to collect a known area of a habitat, especially used for benthos fauna in lotic waters) (Fig. A1.1n), core samplers (which collect known volumes of substrate, used especially for hyporheic, i.e. substrate dwelling, species) and column samplers (which collect known volumes of water). Pumps are also used to collect samples of interstitial waters for extracting hyporheic species. Gee minnow traps (made of metal mesh with inverted funnels on both ends, with or without bait) are also frequently used. Emergence traps are used to sample insects when they are emerging from water to make their last moult. Light traps can attract freshwater insects, especially caddisflies, some stoneflies and beetles. For microscopic invertebrates, see the section *Microscopic organisms* in this Appendix).

Marine invertebrates

Quadrat and transect sampling methods are used to survey benthic (sessile or at least slow-moving) invertebrates (e.g. many molluscs, some crustaceans, worms, cnidarians, echinoderms). Transect methods can be combined with contact sampling, by counting the number of individuals found directly underneath or in the near vicinity of points along transects. For species that are easily visible and recognisable, researchers can directly count individuals during underwater surveys (or surface sampling, for animals inhabiting shoreline habitats); in other cases, samples are taken for laboratory examination. Benthic invertebrates (especially inconspicuous ones, including the meiofauna) can be sampled by collecting sediments (Fig. A1.1o), from which organisms are then extracted in the laboratory (population density is estimated based on the number of individuals per volume or weight of sediment). Baited traps and cages are also used to sample some marine invertebrates. Analysis of environmental DNA (eDNA) from water or sediment samples may provide insights into the presence and relative abundance of species (Taberlet *et al.*, 2018). Underwater video techniques can be used to capture images of some species. Nets and plankton samplers are used for pelagic species. For more information on various techniques used for sampling and monitoring marine invertebrates, see Murray *et al.* (2006), Eleftherious (2013), Unsworth *et al.* (2014), Costello *et al.* (2017a), Ardura (2019) and He *et al.* (2023).

Fish

Fish sampling in marine and freshwater environments can use a wide variety of techniques based on nets (e.g. gill nets and trammel nets; trawls, dredges or purse seines, especially for deep-water fish; round haul seines, especially for coral fish; beach seines, for coastal habitats) and traps (e.g. hoops, fykes, trap nets, pots, weirs; baited or not). Freshwater fish populations are frequently sampled using electrofishing, a technique that uses direct current electricity flowing between a submerged cathode and anode: fishes tend to swim toward the anode, where they can be caught or stunned. Due to the conductivity of saltwater, in general, electrofishing is not performed in saltwater because electricity spreads through the water, rather than through the fish.

A common problem with methods used to sample fish is that they tend to be selective (e.g. the mesh size and sensitivity to electric current influence which species are collected). Methods also vary in terms of postcapture mortality (e.g. high for gill nets and low for trap nets and seines; electrofishing mortality varies according to the species, the size of the animals, the temperature and conductivity of water). Also, sampling can be biased by environmental conditions (e.g. debris may make seines less effective).

Fish sampling data are generally reported as sampled individuals per unit effort (e.g. number of counted individuals per electroshocking interval, seine haul or trawl distance). Underwater video techniques and eDNA analyses are recent promising technologies to assist fish population surveys. Detailed accounts of techniques for sampling and monitoring fish populations can be found in Hilborn and Walters (1992), Royce (1996), Powell and Gale (2015), Aoki *et al.* (2018), Vea Salvanes *et al.* (2018), Yao *et al.* (2022) and Nalmpanti *et al.* (2023).

Amphibians and reptiles

Ground-dwelling amphibians and reptiles are usually surveyed by searching for individuals that are active and inspecting their potential refugia. Surveys are conducted along transects or within quadrats. The number of individuals observed/captured depends not only on the length of the transect or the area of the quadrat, but also the time spent in the field. In some cases, animals are searched by randomly walking in a certain area for a predetermined amount of time. A variety of traps is also used, including, for example, coverboards (sheets of plywood or other material that are laid on the ground to create artificial refugia where reptiles can be subsequently searched for), tubes placed horizontally on the ground, glue traps, funnel traps (cylinders with a funnel-shaped throat at one or both ends, used both for terrestrial and aquatic habitats), pitfall traps, etc. Sweep samples, stovepipe sampling, kick sampling, egg mass and nest counts, electroshocking and snorkelling surveys are used in aquatic habitats. In most cases, there is no need to sacrifice animals for identification, since most species can be identified in the field, or a tissue sample can be taken for genetic identification in the laboratory. Comprehensive overviews of methods used to sample and monitor amphibian and reptile populations are provided by Heyer *et al.* (1994), Dodd (2009), McDiarmid (2012) and Graeter *et al.* (2013).

Birds

Point counts and transects are the two main methods used to estimate bird abundance. In point counts, the researcher, located at a fixed position, counts the birds seen and/or heard in a predetermined amount of time within a certain radius. Transects can be of any length and shape, although linear transects are generally preferred. In strip-transects, birds are visually and/or acoustically recorded while walking along the transects within a certain distance from the path. Transects should be walked for a standard time to avoid bias in the number of birds estimated related to time sampled. In line-transect distance sampling, bird distances from the transect area are also recorded and used to estimate detection probability (as closer individuals are easier to detect) and hence to calculate bird density. Other techniques consist in locating and counting nests or singing males (territory mapping), which are especially useful for gathering other data on bird populations, such as nest spacing and **territory** size (e.g. Salvati *et al.*, 1999). In some cases, playback response techniques are used, which may be especially useful for nocturnal or secretive species (Salvati *et al.*, 2000). Mist nets are used especially in combination with mark-recapture activities (Fig. A1.1p). Bird hides are sometime used to collect data on species biology (Fig. A1.1q).

As many bird species migrate, there is the question whether counts should include all species, including migrants and vagrants (migrants found outside their normal range), or only the breeding populations. This is a special case of a fundamental problem concerning how species that are unrepresentative of a permanent part of the community should be considered. Although there is a certain tendency only to focus on breeding/resident species, migrants may also be important as they can represent (albeit temporarily) prey, predators or competitors for the resident populations, and hence may occupy an important role in **food webs**. Details on methods used in bird censusing can be found in Franzreb (1981), Bibby *et al.* (2000), Sutherland *et al.* (2004), Hyrenbach *et al.* (2007), Thomas *et al.* (2010) and Taulman (2013).

Mammals

Large mammals can be surveyed using transects, standardised visual search and even aerial photographs and remote sensing data. For mammals that live in groups (e.g. cave-dwelling bats, large herbivores), visual surveys can be relatively easy. For certain species, abundance data can be deduced from foot tracks, scat counts and hair traps. In some cases, playback is used. A number of traps are frequently used. Passive traps for small mammals include, for example, pitfall traps, snap traps, and box traps (a commonly used device is the Sherman trap, a folding, easily transportable, metallic device: when an animal enters the trap and steps on the treadle, its weight causes a treadle latch to release and the door shuts). Radiotelemetry and camera trapping are commonly used to track mammals' movements, and hence habitat use, through their life. Methods used in mammal censusing and monitoring are reviewed in Wilson *et al.* (1996), Long *et al.* (2008), Boyd *et al.* (2010), Boitani and Powell (2012).

Microscopic organisms

The small size and high numbers of microscopic organisms make estimating their populations particularly challenging. Sample collection may require the use of plankton nets, particular devices to collect water (Fig. A1.1r), sediment or soil samples, pumps and extractors, like the Baermann funnel for nematodes; in this device, a muslin/paper filter bag containing the sample, e.g. a certain amount of soil, is immersed in water in a funnel ending with a rubber tube closed with a clamp; nematodes penetrate the muslin and sink to the bottom; the funnel outlet is opened and a drop of water containing the nematodes is collected (Tintori *et al.*, 2022).

After samples are taken, methods and techniques for species identification and abundance estimations vary greatly depending on the characteristics of the specific organisms. For some microscopic organisms, such as unicellular algae, protists, microscopic animals (e.g. plankton and meiofauna: rotifers, nematodes, small crustaceans, etc.), fungi and some **prokaryotes**, direct microscopic count of individuals in a known volume of sample is typically used. For **micro-organisms** that can be cultivated *in vitro* (like some prokaryotes and fungi), serial dilution and plating may be applied: a sample is serially diluted and each dilution is placed on a culture plate; after incubation, colonies are counted and the original population size is estimated based on the dilution factor. For certain aquatic micro-organisms (mainly bacteria), membrane filtration is used: the sample is passed through a filter with known pore size, retaining micro-organisms on the filter; the filter is spread onto a culture plate and colonies are counted. The Most Probable Number (MPN) method is a statistical approach used especially for samples with low microbial concentrations: samples are serially diluted and inoculated into multiple culture tubes or wells. The presence or absence of microbial growth is recorded for the different dilutions; the number of positive and negative samples is used to estimate the original concentration within the appropriate order of magnitude.

All culture methods have important limits; they can be applied only to culturable organisms (while most microbes are not culturable or require extremely specific conditions) and may provide biased estimates, since they stimulate the growth of only a limited part of the microbial diversity present in a sample.

Flow cytometry (a technique that uses lasers and fluorescent markers to count and distinguish individual cells in a liquid sample) is used for a wide range of micro-organisms (including, for example, bacteria, unicellular algae and planktonic organisms). Direct counts with DAPI (4',6-diamidino-2-phenylindole) staining are used for many microscopic organisms, like bacteria and eukaryotic microbes; micro-organisms are stained with DAPI and counted using epifluorescence microscopy. Quantitative polymerase chain reactions quantify micro-organism abundance by measuring the abundance of specific DNA (qPCR) or RNA sequences (qRT-PCR). CARD-FISH (Catalysed Reporter Deposition Fluorescence *In Situ* Hybridisation) is a technique widely used in environmental microbiology and microbial ecology to identify and quantify specific microbial populations in water, soil or sediment samples (including those from challenging ecosystems such as deep-sea hydrothermal vents). All these molecular methods, however, use specific DNA probes and hence require prior knowledge of the target organism's genetic material. Metagenomic sequencing can provide information about the diversity and abundance of microscopic organisms, even in complex microbial communities. For some micro-organisms (prokaryotes, algae, fungi), indirect biomass-based methods can be used. Micro-organism biomass is estimated by measuring the carbon content in a sample and then, using established biomass-to-carbon ratios, estimates of population density are obtained. For fungi, it is possible to estimate biomass by measuring ergosterol (a sterol found in fungal cell membranes) and then, using conversion factors, estimates of population density are obtained. Respiration and enzyme activity-based methods may provide information on the metabolic activity of micro-organisms and hence on their abundance. Technical details on the aforementioned and other methods used for studying microscopic organisms can be found in Atlas and Bartha (1998), Pansu and Gautheyrou (2006), Hurst (2007), Filion (2012), Somerfield and Warwick (2013), Wilkinson (2015), Gérikas Ribeiro *et al.* (2016), Suthers *et al.* (2019), Pascoal *et al.* (2021) and Green and Goldman (2023).

Appendix 2 Population Dynamics

Modelling population dynamics, i.e. how population size and density vary in time and space, is central to ecological theory and its applications, from pest management to conservation biology. Finding mathematical models that describe population growth (or decline) is at the core of population studies and has attracted the interest of mathematicians, statisticians and biologists since the inception of ecological thinking. One example is the role played by Malthus' theory of human population growth in triggering in Darwin's mind the idea for a causal mechanism of natural selection (although the influence of Malthus on Darwin may have been overstated by Darwin himself; see, for example, Vorzimmer, 1969).

I happened to read for amusement Malthus on Population, and being well prepared to appreciate the struggle for existence which everywhere goes on from long-continued observation of the habits of animals and plants, it at once struck me that under these circumstances favourable variations would tend to be preserved and unfavourable ones to be destroyed (Darwin, 1859).

The basics of population dynamics models were largely founded within the first half of the 20th century. These early attempts were characterised by excessive simplifications, but by virtue of their mathematical simplicity they served as a foundation for subsequent reasoning and development of more realistic (but also more complex) models. The subject is addressed in many textbooks to which the reader is referred for a comprehensive view (e.g. Rhodes *et al.*, 1996; Vandermeer and Goldberg, 2013; Newman, 2014; Rockwood, 2015; Murray and Sandercock, 2020; Seno, 2022). The aim of this Appendix is to familiarise the reader with basics that are particularly relevant to the scope of the present book.

The exponential growth

The cornerstone of population dynamics is the **exponential growth** which can be easily derived from the general equation of population dynamics (eq. 2.1 in Chapter 2) under some restrictive assumptions. In general, the change in population size (ΔN, that is, the difference in population size between the two periods, N_{t+1} and N_t) can be written as:

$$\Delta N = N_{t+1} - N_t = N_t + B + I - D - E - N_t = B + I - D - E, \tag{A2.1}$$

where B, D, I and E are respectively the number of births, deaths, immigrants and emigrants.

Under the assumption that the population is closed (hence, $I = 0$ and $E = 0$):

$$\Delta N = B - D \tag{A2.2}$$

If, for mathematical convenience, we also assume that the growth is continuous (i.e. time intervals are infinitely small), the population growth rate (i.e. the change in population size dN that occurs in an infinitely small interval of time dt) can be written as:

$$dN / dt = B - D \tag{A2.3}$$

As birth and death rates are proportional to N, B and D can be expressed as $B = b \times N$ and $D = d \times N$, where b and d are, respectively, the instantaneous birth and death rates, i.e. the number of births per individual per unit time, and the number of deaths per individual per unit time. Hence:

$$dN / dt = b \times N - d \times N = (b - d)N \tag{A2.4}$$

The difference $b - d$ represents the instantaneous rate of increase, r, and eq. A2.4 can be written as:

$$dN / dt = rN \tag{A2.5}$$

which, after integration, gives:

$$N_t = N_0 e^{rt} \tag{A2.6}$$

where N_t is the population size at time t, N_0 is the initial population size, and e is the Euler number (the base of natural logarithms, $e \approx 2.718$). When $r > 0$, the population increases exponentially, while for $r < 0$ the population declines exponentially to extinction. This equation, which expresses an **exponential**

DOI: 10.1079/9781800627505.app2

growth, becomes a straight line, whose slope corresponds to r, if population size is ln-transformed.

The exponential model assumes not only that the population is closed but also that there are no differences in b and d among individuals due to differences in their genetic characteristics, size or age; that there are no time lags (i.e. that individuals born and die continuously, without discrete generations); and that b and d are constant through time. Asexual populations composed of genetically identical individuals which are able to reproduce immediately after they are born (e.g. experimental cultures of bacteria or protists) may closely meet these assumptions, but they are obviously violated in organisms that reproduce sexually, and in which individuals have different birth and death rates according to their genetic characteristics, age or size. Additionally, for many organisms, time does not behave as a continuous variable (for example, in seasonal environments, many insects reproduce only once, a situation that can, however, easily be accommodated by using a discrete difference equation, instead a continuous differential equation, to produce a discrete growth model). Finally, if b and d were constant through time, the population would increase indefinitely with a constant rate r, which implies that resources (such as space and food) are unlimited. Although no resource is infinite in the real world, some resources may be considered practically infinite when the population is very small, and thus populations can show an initial growth that may be approximated by the exponential model.

Even though the exponential model may serve as a realistic representation of population growth only in some circumstances, it represents an essential conceptual reference and the mathematical foundation for modelling population dynamics.

The logistic model

The exponential model assumes that the rate of increase is fixed. However, because of the multiple negative effects of crowding, we expect that birth and death rates are not stable but will depend on population density. Specifically, with increasing density, three different scenarios are possible: (i) birth rate decreases but death rate remains stable; (ii) death rate increases but birth rate remains stable; (iii) birth rate decreases and death rate increases. Under the assumption that b declines and d increases linearly as the population grows, all three scenarios lead to the same model, in which the instantaneous rate of increase diminishes with time, leading to the equation:

$$\frac{dN}{dt} = (r - \alpha N)N = rN - \alpha N^2 \qquad \text{(A2.7)}$$

where r is the intrinsic rate of increase and α expresses the crowding effect.

By integrating eq. A2.7, it is possible to obtain an equation that describes how population size changes with time, as done for the exponential model (eq. A2.6):

$$N_t = \frac{rN_0 e^{rt}}{r + N_0\alpha(e^{rt} - 1)} \qquad \text{(A2.8)}$$

These two equations represent a growth model known as the **logistic model**. This model was introduced by Pierre François Verhulst (1804–1849, who also coined the name *logistique*, although the etymology remains unclear) in a series of papers (1838, 1845, 1847) to describe the growth of human populations. Though Du Pasquier (1918) called attention to Verhulst's work, it was ignored until 1920, when Pearl and Reed (1920), without knowing of Verhulst's contribution, rediscovered the logistic curve as an empirical curve to describe the growth of a population in an area of fixed limits (in fact, the curve was already rediscovered by McKendrick and Kesava Pai, 1911). In 1925, a rational, as distinguished from an empirical, derivation of the curve was given by Lotka (see below).

Readers familiar with ecology textbooks will have noticed that the formulation used in equation A2.7 is different from that commonly used to introduce the logistic function. Because of a series of historical accidents (see below), the logistic model is usually expressed with an alternative formulation:

$$\frac{dN}{dt} = rN\left(1 - \frac{N}{K}\right) = rN\left(\frac{K - N}{K}\right) \qquad \text{(A2.9)}$$

where K is the asymptotic value (maximum N), corresponding to the equilibrium value ($\hat{N} = r/\alpha$) of eq. A2.7.

After integration, eq. A2.9 gives:

$$N_t = \frac{K}{1 + \left[(K - N_0)/N_0\right]e^{-rt}} \qquad \text{(A2.10)}$$

This formulation may appear appealing, as it clearly illustrates how (eq. A2.9) when the population density is very low ($N << K$), the population increases almost exponentially, since the fraction N/K is very small, and hence the term in parentheses is approximately 1; but increasing N further, the fraction N/K increases and when $N = K/2$, the term in parentheses equals 0.5, thus the increase becomes half that predicted by the exponential model. Increasing N further, the term in parentheses becomes << 1 and the increase is strongly reduced. When $N = K$, the growth is 0 (see Fig. 2.7 in Chapter 2). Mathematically, $K/2$ is the inflection point of the function (when $N < K/2$ the function is convex, whereas when $N > K/2$ the curve is concave) and K is its asymptote. If the population exceeds K, the term in parentheses becomes negative and the population declines towards K.

The formulation of the logistic equation presented in eq. A2.9 was promoted by Raymond Pearl (1879–1940) in the second edition of his influential book on medical biometry (1930) and adopted in the famous works of the Russian biologist Georgy Frantsevich Gause (1910–1986). Gause was a student of Vladimir V. Alpatov, who in turn worked with Pearl in the USA as a Rockefeller fellow before returning to the USSR (Gall and Konashev, 2001), and this clearly influenced his thinking about population logistic growth. The superb results obtained by Gause with his experiments on population growth and competition (Gause, 1932, 1934; Gause and Witt, 1935) and the popularity of Pearl's book contributed to the virtual fixation of eq. A2.9 instead of eq. A2.7 in the ecological literature.

Although equations A2.7 and A2.9 represent the same underlying model, their different formulation is not without consequences (Mallet, 2012). The parameter r in eq. A2.9 is usually indicated as 'the intrinsic rate of increase', as in the exponential growth model, which seems to imply that it is density independent. However, while r is really a density-independent parameter in eq. A2.7, in eq. A2.9 it is density dependent. This is clearly illustrated by using partial derivatives. If, for eq. A2.9, we pose:

$$\frac{1}{n}\frac{dN}{dt} = R = r\left(1 - \frac{N}{K}\right) \qquad \text{(A2.11)}$$

where R is the per capita growth rate, by taking partial derivatives it appears immediately clear that r is a density-dependent parameter, because $\partial R/\partial r = 1 - N/K$, and $\partial R/\partial K = rN/K^2$. By contrast, if, for eq. A2.7, we pose:

$$\frac{1}{n}\frac{dN}{dt} = R = r - \alpha N \qquad \text{(A2.12)}$$

the equivalent partial derivatives are: $\partial R/\partial r = 1$ and $\partial R/\partial \alpha = -N$. Thus in eq. A2.7, there is a clear separation between the density-independent r and the density-dependent α parameters of population growth (the density-independent effects of birth and death remain separate from the effects of crowding), while they are convoluted with them in eq. A2.9. This separation not only makes eq. A2.7 more intuitive and natural than eq. A2.9, but avoids paradoxes that afflict eq. A2.9 (Mallet, 2012). When $r = 0$, eq. A2.9 predicts that the population neither grows nor declines, no matter how large N is, even if larger than K, which is paradoxical, as for values larger than K the population is expected to decline. If $r < 0$, with $N > K$, the population is also expected to decline to extinction rather rapidly, but, as the product of two negatives in eq. A2.9 is a positive, this equation would predict an infinite growth, which is an absurd outcome. By contrast, eq. A2.7 does not suffer from these paradoxes: when $r = 0$, eq. A2.7 correctly predicts that the population will decline, at a rate $dN/dt = -\alpha N^2$, as expected, and when $r < 0$, the decline will take place even faster, with speed that increases with increasing N, as it should. To complicate things further, the erroneous identification of the equilibrium value K with a fixed resource limit leads to a misunderstanding of eq. A2.9. Eq. A2.7 has equilibrium $\hat{N} = r/\alpha$, which seems to be intuitively reasonable as a dynamic balance between density-independent growth (r) and its density-dependent regulation (α). K, which is the equilibrium value for eq. A2.9, if seen as a **carrying capacity** parameter, confuses a fixed resource limit with equilibrium density. This confusion was introduced by Odum (1953), who linked the concept of carrying capacity, already present in the ecological literature, with the logistic equilibrium K. The enormous success of Odum's book led to the identification of the carrying capacity with K, so that K is used as a synonym of carrying capacity in virtually all textbooks of ecology. A correct understanding of the compound nature of K as a population equilibrium instead of a fixed parameter explains why eq. A2.9 produces misleading results.

Being derived from the exponential model, the logistic model shares the same assumptions of no time lags, migrations, genetic variation, size or age structure. As with the exponential model, the logistic model can be modified to introduce time lags and formulated in a discrete version. Logistic curves with time lags may show dampened oscillations and convergence on equilibrium, or stable limited cycles, depending on the combined effect of r and the lag. In the discrete model, the lag is constant (it is always 1, i.e. one step), and the dynamics depend only on the discrete growth factor r_d. Interestingly, when r_d is small, the population approaches K with damped oscillations, but with larger r_d the population shows cyclic oscillations; when r_d further increases, the population grows in a chaotic way (see, for example, May, 1976a; Coulson and Godfray, 2007). This, however, does not mean that variations are stochastic. The model is still deterministic, since the erratic fluctuations are entirely determined by the initial conditions (viz. once K, r_d and N_0 are specified, the same erratic pattern will be obtained invariably each time the model is run). By contrast, in stochastic populations, K and/or r_d change with each time step (see Gotelli, 2008 for a basic introduction; Lande *et al.*, 2003 for model discussion).

Lotka–Volterra predator–prey equations

The exponential and logistic models attempt to describe how an isolated population of a single species grows. However, species do not live in isolation but interact in various ways (see Table 2.1 in Chapter 2). The first steps in explicitly accounting for species interactions in population dynamics involved attempts to model mathematically **predation** and **competition** between two species. Central for the development of these first models was the work of the mathematician Vito Volterra (1860–1940). In the 1920s and 1930s, Volterra, stimulated by Umberto D'Ancona (1896–1960), developed a simple mathematical model to describe the population dynamics of a predator–prey system by using differential equations (e.g. Volterra, 1926a,b, 1931; D'Ancona, 1927; Volterra and D'Ancona, 1935). Specifically, D'Ancona, who at that time was studying fluctuations in fish populations in the Adriatic, was surprised by the increase in the percentage of predatory fish in the total catch in the years of the First World War, a phenomenon that he hypothesised to be the consequence of the disruption of fishing caused by naval warfare. Volterra threw himself at the problem and in a few months developed an elegant theory that studied the general case of the coexistence of animal species that interact negatively by competition or predation.

In Volterra's model for the predator–prey relationship, the growth of the prey population (P) is expressed by the equation:

$$\frac{dP}{dt} = rP - \alpha HP \qquad \text{(A2.13)}$$

where r is the intrinsic rate of increase of the prey population, H is the predator population (hunters), and α measures the capture efficiency (thus, the number of killed prey is proportional to the encounter rate, expressed by the product of predator and prey number, and the effect of the predator on the per capita growth rate, expressed by α).

The growth of the predator population (H) is modelled by the equation:

$$\frac{dH}{dt} = \beta PH - qH \qquad \text{(A2.14)}$$

where β measures the conversion efficiency (the ability of predators to convert each new prey into new predators) and q is the per capita death rate (in other words, the increase of predators is seen as the number of prey that are converted into new predators, which is proportional to the encounter rate expressed by the product of predator and prey number, minus the predators that die). The equations show that both populations cycle periodically, because they reciprocally control one another's growth (Fig. A2.1a).

The same predator–prey equations were, however, already determined by the US mathematician, physical chemist and statistician Alfred J. Lotka (1880–1949) a few years before Volterra, who was initially not aware of Lotka's works and therefore formulated these equations independently. This independent formulation created a question of priority (see Israel, 1988 for a historical reconstruction). Lotka started to focus on population biology stimulated by Pearl (a champion of the logistic equation, see above) and initially formulated these equations in 1920 using a hypothetical herbivore–plant system as an example (but the paper is very wide in scope, considering oscillating systems of any nature, and actually has its roots in some considerations already developed in 1910 about autocatalytic chemical reactions). Then, Lotka developed these equations in the famous book *Elements of Physical Biology* (1925, where he referred to a parasite–host system), written while working in Pearl's laboratory. Volterra's first works appeared in 1926 (Volterra, 1926a,b). Thus, it is not surprising that Lotka claimed (and deserved) priority in formulating the equations. However, it is

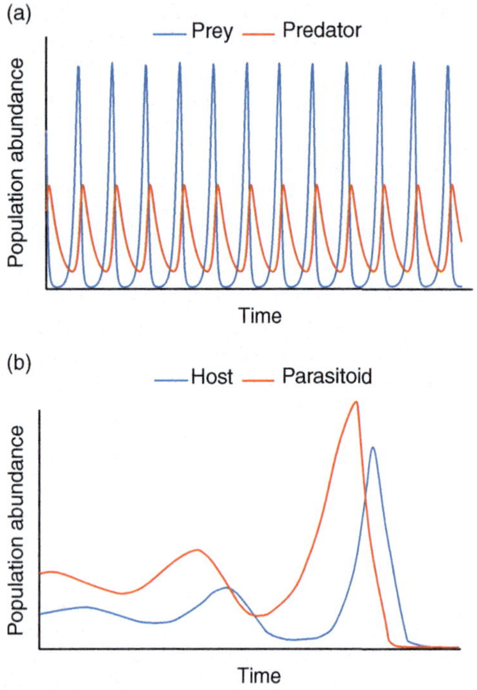

Fig. A2.1. Lotka–Volterra model **(a)** and Nicholson–Bailey model **(b)** of species interactions. In the Lotka–Volterra model, the dynamics are neutral cycles: the populations oscillate indefinitely, with constant amplitude over time. Predators lag prey by one-quarter of a cycle. In the Nicholson–Bailey model, the dynamics show divergent oscillations with overexploitation by the parasitoid leading to the extinction of both hosts and parasitoids.

important to note that Volterra placed these equations in a much broader ecological context. For example, he formulated these equations in conjunction with those for modelling interspecific competition (see section *Volterra equations for competition*), a subject that was mentioned but not formally addressed by Lotka in his 1925 work. Moreover, while Volterra derived the equations from a strictly rational basis rooted in a theoretical approach to population dynamics (and only later considered D'Ancona's empirical data), Lotka derived them by analogy, i.e. he applied an ecological system to the equations he had already introduced in chemistry. These two approaches reflect two basically different research programmes: while Volterra aimed at formulating biological questions using classical physico-mathematical approaches, Lotka aimed at founding a new discipline (which he termed 'physical biology') by eclectically combining biology and physical sciences (Israel, 1988, 1995; Kingsland, 2015).

Although conceived for predator–prey or parasite–host relationships, the Lotka–Volterra model can be equally used for herbivore–plant relationships (in this case, the herbivore is the predator and the plant is the prey); that is, it is theoretically applicable to any system in which a species feeds on another.

An alternative, equally simple model, which was originally developed especially for entomophagous parasites (parasitoid–host interactions), is the Nicholson–Bailey model (Bailey, 1931; Nicholson and Bailey, 1935). While the Lotka–Volterra equations describe a continuous model, the Nicholson–Bailey model is a system of difference equations:

$$x_{t+1} = \lambda x_t e^{-ay_t} \tag{A2.15}$$

$$y_{t+1} = cx_t\left(1-e^{-ay_t}\right) \tag{A2.16}$$

where x_t and y_t are host and parasitoid populations at time t, λ is the number of new hosts produced by one unparasitised host (equivalently $\lambda = e^r$ where r is the intrinsic growth rate), a is the searching efficiency of the parasitoid, and c is the number of new parasitoids that hatch from an infected host. In these equations, e^{-ay_t} is the probability that a host escapes parasitism (under random encounters), whereas $(1 - e^{-ay_t})$ is the probability that a host becomes infected. An equilibrium combination of the two populations is possible, but it is unstable, and the dynamics are characterised by divergent oscillations of both populations, with overexploitation by the parasitoid leading to the rapid extinction of both hosts and parasitoids (Fig. A2.1b). This model has been subsequently modified to incorporate density dependence in the host population growth (Beddington *et al.*, 1975) and to switch from a random distribution of the host–parasitoid encounters to an aggregated one (May, 1978). The Nicholson–Bailey model has, however, gained less success than the Lotka–Volterra model, probably because of the population instability predicted by the former. This instability results from the inclusion of a time lag, that is, from the transition from continuous to discrete time (May, 1973a; Bonsall and Hassell, 2007), but appears in general biologically problematic because it does not allow for species persistence.

Because of their mathematical simplicity, the Lotka–Volterra predator–prey equations are obviously appealing, but they have a number of characteristics that make them somewhat unrealistic. For example, these equations refer to a system in which the predator feeds on only one prey, and the prey is eaten by a single predator. In general, this appears unlikely, since usually a prey is consumed by more than one predator and a predator exploits more than one prey. Even when a predator has a certain preferred prey, if it becomes too rare/difficult to capture, the predator can shift towards more accessible/remunerative prey (**optimal foraging theory**: natural selection will favour foraging behaviours that provide the most benefit (energy) for the lowest cost, taking into account the dependence of energy intake rate on the predator's ability to find, capture, and handle each prey; Stephens and Krebs, 1987; Pyke, 2019; Pyke and Star, 2021; Ritvo, 2022). Other unrealistic assumptions of these equations are that: (i) there are no density effect processes for the prey; (ii) predation rate increases in direct proportion to the increase in prey density (a functional response known as Holling type I); (iii) there is no effect of predator density on predators themselves; (iv) there is no handling time, which means that predators instantly consume prey when they encounter them. Grounded on the Lotka–Volterra model, more realistic models (e.g. Rosenzweig–MacArthur models) were proposed that surpass these limits by incorporating: (i) a carrying capacity (a resource limitation) for prey, which can affect the prey's population dynamics; (ii) a non-linear functional response (e.g. a Holling type II functional response: as prey density increases, the predation rate also increases but at a decreasing rate; or a Holling type III response, in which predators become increasingly efficient as a prey increases, until saturation eventually sets in – a sigmoidal pattern); (iii) the effect of predator density on the predator's functional response (predators may become more satiated and less efficient at capturing prey when their population is high); (iv) the handling time, representing the time it takes for a predator to capture and consume prey (Rosenzweig and MacArthur, 1963; May, 1976b; Grunert, 2021).

These models are clearly less unrealistic but may still be very far from true natural conditions. For example, in certain environments, at very high prey density, predators might be negatively affected by the waste produced by the prey, which makes the environment toxic; in the long run, the predator may ameliorate its efficiency in capturing the prey but the prey can develop increasingly better techniques to avoid the predator, which leads to continuous changes in the parameters of the equations. In general, more elaborate models that take into account the many factors that make the original Lotka–Volterra formulation unrealistic are inevitably more complex and require the knowledge of more parameters, with the consequence that they are more realistic but also more difficult to apply in practice.

Despite its many limits, thanks to its simplicity and generality, the Lotka–Volterra predator–prey model remains an extremely vital framework in ecology, and a century after its introduction, it continues to serve ecologists as a very valuable tool for gaining insights into the dynamics of the most diversified interaction systems (e.g. Haerter *et al.*, 2016; Joseph *et al.*, 2020; Remien *et al.*, 2021; Davis *et al.*, 2022; Singh, 2022; Xu *et al.*, 2023).

Volterra equations for competition

As he had done for the predator–prey interactions, Volterra developed differential equations for **interspecific competition**. Based on the logistic model, Volterra derived the following equations that describe the population growth of two species, N_1 and N_2, that compete for the same resource:

$$\frac{dN_1}{dt} = \left(r_1 - \gamma_1\left[h_1N_1 + h_2N_2\right]\right)N_1 \qquad \text{(A2.17)}$$

$$\frac{dN_2}{dt} = \left(r_2 - \gamma_2\left[h_1N_1 + h_2N_2\right]\right)N_2 \qquad \text{(A2.18)}$$

where r_1 and r_2 express the intrinsic rates of increase of the two populations, the coefficients h_1 and h_2 express the impact of the two species on resource availability (namely, the presence of a certain number of individuals of the first species N_1 decreases the quantity of resource by h_1N_1, and the presence of N_2 individuals of the second species decreases the quantity of resource by h_2N_2, so that both species together decrease the quantity of resource by $h_1N_1 + h_2N_2$), while γ_1 and γ_2 express the ability of the species to use the resource.

By combining γ and h into a single coefficient, the original formulation can be rewritten as:

$$\frac{dN_1}{dt} = \left(r_1 - \alpha_{11}N_1 + \alpha_{12}N_2\right)N_1 \qquad \text{(A2.19)}$$

$$\frac{dN_2}{dt} = \left(r_2 - \alpha_{22}N_1 + \alpha_{21}N_2\right)N_1 \qquad \text{(A2.20)}$$

From these equations it is possible to derive the conditions that allow the two species to coexist or that lead one species to exclude the other (Mallet, 2012).

(1) If $r_1/\alpha_{12} > r_2/\alpha_{22}$ and $r_1/\alpha_{11} > r_2/\alpha_{21}$, species 1 wins and excludes species 2

(2) If $r_2/\alpha_{22} > r_1/\alpha_{12}$ and $r_2/\alpha_{21} > r_1/\alpha_{11}$, species 2 wins and excludes species 1

(3) If $r_2/\alpha_{22} > r_1/\alpha_{12}$ and $r_1/\alpha_{11} > r_2/\alpha_{21}$, there is an unstable equilibrium; which species wins depends on their initial densities

(4) If $r_1/\alpha_{12} > r_2/\alpha_{22}$ and $r_2/\alpha_{21} > r_1/\alpha_{11}$, the two species will coexist (stable equilibrium)

By posing $\alpha_{12}/\alpha_{11} = \alpha$ and $\alpha_{21}/\alpha_{22} = \beta$, equations A2.19 and A2.20 can be rewritten as follows:

$$\frac{dN_1}{dt} = r_1 N_1 \left(\frac{K_1 - N_1 - \alpha N_2}{K_1} \right) \quad \text{(A2.21)}$$

$$\frac{dN_2}{dt} = r_2 N_2 \left(\frac{K_2 - N_2 - \beta N_1}{K_2} \right) \quad \text{(A2.22)}$$

where K_1 and K_2 are the carrying capacities of the two populations, and α and β are the respective competition coefficients. This formulation was introduced by Gause (1932, 1934) and parallels that used for the logistic model. As Gause (1934) became the primary reference for ecologists working on competition, equations A2.21 and A2.22 rather than equations A2.19 and A2.20 are the form now virtually universal in textbooks, although equations A2.19 and A2.20 are more intuitive and as suitable as equations A2.21 and A2.22 (Mallet, 2012)

In Gause's formulation, α is the per capita effect of species 2 on the population of species 1, and β is the per capita effect of species 1 on species 2. If $\alpha > 1$, the per capita effect of interspecific competition is greater than that of intraspecific competition, whereas if $\alpha < 1$, the per capita effect of intraspecific competition is greater than that of interspecific competition; if $\alpha = 0$ there is no competition and the model reduces to the single-species logistic model (the same reasoning applies to β). From these equations it is possible to derive the conditions that allow the two species to coexist or that lead one species to exclude the other, as already seen for equations A2.19 and A2.20.

(1) If $K_1/\alpha > K_2$ and $K_1 > K_2/\beta$, species 1 wins and excludes species 2

(2) If $K_1/\alpha < K_2$ and $K_2/\beta > K_1$, species 2 wins and excludes species 1

(3) If $K_2 > K_1/\alpha$ and $K_1 > K_2/\beta$, there is an unstable equilibrium; which species wins depends on their initial densities and relative growth rates

(4) If $K_1/\alpha > K_2$ and $K_2/\beta > K_1$, the two species will coexist (stable equilibrium)

These equations for competition are commonly referred to as Lotka–Volterra equations, but Lotka started working on competition (Lotka, 1932) only after Volterra's first papers (and, as a matter of fact, Lotka claimed priority for the predator–prey equations, not for the competition equations). Thus, it would be more appropriate to call them Volterra equations (or Volterra–Gause equations, as done by Hutchinson, 1978), and to use the expression Lotka–Volterra equations only for the predator–prey model.

Both the equations used to model predator–prey interactions and those that describe competition can be seen as special cases of potentially more general formulations that can describe any type of interaction (including co-operation or **neutralism**) for any number of interacting species. This has led to the development of the generalised Lotka–Volterra (gLV) equations, which are an extension of the logistic growth model, in which the interaction of one species with another is encoded in the corresponding coefficient in the equation. Although gLV-based models have some important limits (they do not account for stochastic fluctuations and spatial structures), they provide an extended fruitful framework for more advanced community models, being especially useful in network analyses to express the strength of the interactions (see Box 2.1 in Chapter 2; Berlow *et al.*, 2004; Säterberg *et al.*, 2013; Fort, 2018; Kuntal *et al.*, 2019; Morrison, 2021; Xu *et al.*, 2023).

Appendix 3 Diversity Measures

The study of how **diversity** varies across habitats is one of the central themes of community ecology (see Chapter 4). This Appendix provides an overview of how diversity is measured.

Species richness

Species **richness** is the number of species present in an assemblage (e.g. a community) and it is the simplest and most intuitive measure of diversity. However, when samples are used to compare richness for different assemblages, it is important to control for the effect of differences in sampling effort. Species richness increases with increasing sampling effort until all species have been collected and no further species can be included with further sampling (see Figs 4.5 and 4.6 in Chapter 4). This process is illustrated by graphs (known as accumulation curves, species discovery curves or collector's curves) in which the cumulated number of collected species is plotted against the number of individuals (individual-based curves) or sampling units (such as quadrats, areas, volumes, transects, traps, sampling sessions, etc.; sample-based curves). Individual-based and sample-based curves have analogous statistical treatments and the choice of whether to use one or the other depends mainly on practicality. In many circumstances, communities are sampled with a single sampling unit, which is assumed to be a collection of individuals taken at random from the habitat. In this case, only individual-based curves can be constructed. Additionally, if communities differ in the average density of individuals across sampling units, only individual-based curves are appropriate (Gotelli and Colwell, 2001). The concept of accumulation curve is strictly related to that of rarefaction curve, and in the current usage they are practically synonymous. Rarefaction is obtained through algorithms that calculate the expected number of species for any given sample size smaller than the total sample (Krebs, 1999; Colwell *et al.*, 2004, 2012; Chao *et al.*, 2014b). Thus, rarefaction makes it possible to estimate how many species would be found in a smaller subset of a given sample, helping to compare species richness across different samples with varying sizes. The opposite process – extrapolation – involves estimating the number of species that would be observed in a larger sample than the one collected. Extrapolations are made possible by methods that calculate the expected number of species for any number of individual/sampling units, even beyond the observed maxima (see below *Non-parametric estimators*).

Comparisons among communities can be made in three ways.

- Rarefaction to the smallest sample. The community with fewest individuals/sampling units is used as reference, and the expected number of species in the communities with more individuals/sampling units is rarefied to the number of individuals/sampling units of the least sampled community. This approach can introduce bias if well-sampled communities are rarefied to match a poorly sampled one.
- Extrapolation to the largest sample. The community with most individuals/sampling units is used as reference, and the expected number of species in the communities with fewer individuals/sampling units is extrapolated to the number of individuals/sampling units of the most sampled community. However, this may produce unreliable estimates when the extrapolation exceeds the observed data by a large margin.
- Rarefaction/extrapolation to twice the minimum sample size. A common reference is defined as twice the number of individuals or sampling units in the least-sampled community. Communities with fewer than this threshold are extrapolated up to it, while those with more are rarefied down to it. Extrapolating up to twice the observed sample size is generally considered statistically safe (Chao *et al.*, 2014b), making this the most balanced and robust approach.

DOI: 10.1079/9781800627505.app3

The curves, which show the extrapolated part up to the desired number of individuals/sampling units, are known as rarefaction/extrapolation curves (see Fig. 4.6 in Chapter 4). Extrapolation/rarefaction to a common reference is considered the most appropriate method for comparing species richness of communities subject to different sampling efforts (Colwell *et al.*, 2012; Chao *et al.*, 2014a; Hsieh *et al.*, 2016). However, using a given sample size as a reference to compare communities subject to different sampling efforts may produce misleading results, because a given size may be sufficient to fully characterise the lower diversity community, but insufficient to characterise the richer community. For this reason, it has been proposed that it is better to compare species richness of sets of communities with reference to samples of equal completeness (measured by sample coverage) instead of equal size (Chao and Jost, 2012). Coverage expresses the proportion of the total number of individuals in a community that belong to the species represented in the sample (or the number of sampling units in which they are found). These approaches can be used not only for species richness, but also for other Hill numbers (see below *Hill numbers*) (Hsieh *et al.*, 2016).

Non-parametric estimators

Non-parametric estimators allow the species richness of a community to be estimated from samples without assuming any particular statistical distribution.

Chao1. This abundance-based estimator calculates the true number of species on the basis of the number of species in the sample (S_{obs}), the number of singletons (i.e. the number of species represented by only one individual in the sample, F_1) and doubletons (i.e. the number of species with two individuals, F_2):

$$S_{Chao1} = S_{obs} + \frac{F_1^2}{2F_2} \quad \text{(A3.1)}$$

Chao2. This estimator uses incidences instead of abundances (i.e. the number of sampling units in which a species is present). The original estimators Chao1 and Chao2 have been variously modified by Chiu *et al.* (2014), Chiu and Chao (2016) and Böhning *et al.* (2019).

Jackknife 1. This estimator, known as the first-order jackknife, is based, like Chao 2, only on species incidences. It is calculated as:

$$S_{jack1} = S_{obs} + Q_1\left(\frac{m-1}{m}\right) \quad \text{(A3.2)}$$

where Q_1 is the number of species sampled in only one sampling unit, and m is the number of sampling units. As the maximum possible value for a first-order jackknife is twice the observed number of species (which occurs when all sampled species have been sampled in a single sampling unit), this method cannot be applied to communities with exceptionally high numbers of rare species, or to communities that have been sampled inadequately (so that the number of observed species is less than half of the species present) (Krebs, 1999).

Jackknife 2. This estimator, known as second-order jackknife, is also based only on species incidences and is calculated as:

$$S_{jack2} = S_{obs} + \left[\frac{Q_1(2m-3)}{m} - \frac{Q_2(m-2)^2}{m(m-1)}\right] \quad \text{(A3.3)}$$

where Q_2 is the number of species sampled in two sampling units. The second-order jackknife estimator was found to be the least biased estimator in some comparative studies (Meier and Dikow, 2004; Guralnick and Van Cleve, 2005).

Bootstrap. This method is based on p_k, the proportion of sampling units that contain a given species k:

$$S_{boot} = S_{obs} + \sum_{k=1}^{S_{obs}}(1-p_k)^m \quad \text{(A3.4)}$$

Both jackknife and bootstrapping methods are affected by the rarest species, so they are not recommended for samples in which there are many rare species. The bootstrap estimator usually performs poorly and is not recommended (Chazdon *et al.*, 1998).

ACE and ICE estimators. These estimators are based on the encounter probabilities of the observed species taking into account the unobserved but present species. When some species are very common and others are very rare, all the information about species not yet sampled resides among the rarer species. The Abundance-based Coverage Estimator (ACE) estimates the total number of species by estimating the number of unobserved species using the species sampled with ten or fewer individuals (using ten individuals is an arbitrary choice). The Incidence-based Coverage Estimator (ICE), which is the corresponding estimator for incidence data, is based on species found in ten or fewer samples.

The ACE estimator of species richness is therefore:

$$S_{ACE} = S_{abund} + \frac{S_{rare}}{C_{ACE}} + \frac{F_1}{C_{ACE}}\gamma_{ACE}^2 \quad \text{(A3.5)}$$

where S_{abund} is the number of abundant species when all sampled individuals are pooled, S_{rare} is the number of rare species when all sampled individuals are pooled, F_1 is the number of species sampled with a single individual and C_{ACE} (the estimated value of sample coverage) and γ^2_{ACE} (the coefficient of variation for rare species) are calculated as follows:

$$C_{ACE} = 1 - \frac{F_1}{N_{rare}} \quad (A3.6)$$

$$\gamma^2_{ACE} = max\left[\frac{S_{rare}}{C_{ACE}}\frac{\sum_{i=1}^{10} i(i-1)F_i}{(N_{rare})(N_{rare}-1)} - 1, 0\right] \quad (A3.7)$$

where N_{rare} is the total number of individuals belonging to rare species (i.e. those with a maximum of ten collected individuals), and F_i is the number of species that have exactly i individuals when all sampled individuals are pooled.

ICE is calculated as:

$$S_{ICE} = S_{freq} + \frac{S_{infr}}{C_{ICE}} + \frac{Q_1}{C_{ICE}}\gamma^2_{ICE} \quad (A3.8)$$

where S_{freq} is the number of frequent species (each found in more than ten sampling units), S_{infr} is the number of infrequent species (each found in ten or fewer sampling units), and C_{ICE} and γ^2_{ICE} are calculated as follows:

$$C_{ICE} = 1 - \frac{Q_1}{N_{infr}} \quad (A3.9)$$

$$\gamma^2_{ICE} = max\left[\frac{S_{infr}}{C_{ICE}}\frac{m_{infr}}{(m_{infr-1})}\frac{\sum_{j=1}^{10} j(j-1)Q_j}{(N_{infr})^2} - 1, 0\right] \quad (A3.10)$$

where Q_1 is the number of species sampled in only one of the sampling units, N_{infr} is the total number of incidences for the infrequent species (i.e. species found in ten or fewer sampling units), m_{infr} is the number of sampling units that have at least one infrequent species, and Q_j is the number of species that occur in exactly j sampling units. A modified version of γ^2_{ACE} has been proposed for highly heterogeneous communities (species richness >1000) (Chao and Lee, 1992; Chao and Chiu, 2016).

Asymptotic models for extrapolating species richness

Accumulation curves can be used to calculate asymptotic values of richness through appropriate fitting equations. Although these methods are now less popular than non-parametric estimators, they have been used in the past to predict the total number of species present in a given habitat/area. These approaches are based on the observation that accumulation curves decelerate until a plateau is reached. For mathematical convenience, this plateau can be seen as the asymptote of a function that best fits the empirical curve. Many models have been proposed to fit these curves, and various authors have evaluated their performance in comparative works (Soberón and Llorente, 1993; Colwell and Coddington, 1994; Melo *et al.*, 2003; Thompson *et al.*, 2003; Hortal *et al.*, 2006; Jiménez-Valverde *et al.*, 2006; Beck and Kitching, 2007; Fattorini, 2013). These approaches are, however, considered problematic and inferior to the use of non-parametric estimators (Chao and Chiu, 2016).

Species Abundance Distribution (SAD) models

Species Abundance Distribution (SAD) models describe the relative abundance of the species within an assemblage (Fattorini *et al.*, 2016c; Antao *et al.*, 2021; Tsafack *et al.*, 2021a; Callaghan *et al.*, 2023; Moradi *et al.*, 2025).

SADs are universally characterised by the same basic pattern, in which most species have low or intermediate abundances, and only a few are common (see Fig. 4.2 in Chapter 4). However, communities can largely differ in the shape of their SADs, because of differences in species richness and the proportion of rare, intermediate and abundant species (Magurran, 2004; Gray *et al.*, 2006; McGill *et al.*, 2007, 2019; Anderson *et al.*, 2012; Callaghan *et al.*, 2023). This led to the formulation of a plethora of models that attempt to fit empirical SADs (McGill *et al.*, 2007; Ugland *et al.*, 2007; Ulrich *et al.*, 2019). Among the most frequently used models, we can cite the **log-series**, the **lognormal distribution**, the zero-sum multinomial distribution, the gambin model, the **geometric series**, the broken stick model, the Zipf distribution, the Weibull distribution and the gamma distribution (see Fig. 4.4 in Chapter 4 for a graphical representation of some of them).

Log-series. The log-series was derived by Fisher *et al.* (1943) as a purely statistical distribution, but was subsequently interpreted in relation to ecological processes (Hubbell, 2001; Volkov *et al.*, 2003; Pueyo *et al.*, 2007; Harte *et al.*, 2008). The α parameter of

Fisher's log-series is a measure of diversity, and reflects the number of singleton species in the community (Kempton and Taylor, 1974; Magurran and Henderson, 2003). The log-series distribution is known to fit a wide range of communities, particularly those that have a high frequency of rare species and that are not particularly species-rich (Tsafack *et al.*, 2021a).

Lognormal distribution. The lognormal distribution assumes that most species have intermediate abundances (Preston, 1948). If these SADs are represented using **histograms** in which abundance is expressed using logarithmic classes (called octaves by Preston), they will show a Gaussian shape (hence the name 'lognormal'). Although Preston used doubling classes of abundances (log_2), any logarithm can be used (for example, classes based on log_{10}). There is also no consensus about binning. For example, with log_2, it is possible to use classes as follows: bin1: number of species with one individual, bin 2: number of species with 2 individuals, bin 3: number of species with 3-4 individuals, etc. However, another binning system might be: bin1: number of species with one individual, bin2: number of species with 2-3 individuals, bin 3: number of species with 4-7 individuals, etc. The result is that the same data can appear to have different shapes when different logarithms and binning systems are used (Gray et *al.*, 2006). The lognormal is one of the most commonly used models to describe SADs (McGill, 2003). As this model can be seen as a null form of the distribution resulting from the central limit theorem of statistics (May, 1975), it is classified among the purely statistical models (McGill *et al.*, 2007), but can have ecological interpretations with reference to population dynamics (Engen and Lande, 1996) or niche partitioning (Bulmer, 1974; Sugihara, 1980). Lognormal shapes have mainly been associated with 'fully censused' communities regulated by a large number of biotic and abiotic factors that together produce a lognormal abundance distribution (Magurran, 2004; Connolly *et al.*, 2005; Šizling *et al.*, 2009). The lognormal SAD is commonly preferred to represent species-rich communities that are well balanced in terms of rare and common species (Magurran, 2004). The lognormal distribution is generally considered to best describe SADs of undisturbed communities, whereas SADs of disturbed communities tend to follow the Fisher log-series (Matthews and Whittaker, 2015; Ulrich *et al.*, 2016; Haddad *et al.*, 2019). A common assumption in community ecology is that ecosystems that may support species with more varied ecological **niches** have more core species with stable populations with moderate to large numbers of individuals, thus leading to SADs that should be well approximated by lognormal distributions (Hubbell, 2001; Magurran and Henderson, 2003). In contrast, communities under stressful conditions are characterised by higher species dominance and are expected to deviate more substantially from this pattern, approaching a log-series. However, even in high diversity systems such as tree communities in tropical forests, local and regional SADs can follow a log-series, if a small share of the species is hyperdominant (ter Steege *et al.*, 2013). As the lognormal is a continuous distribution, it has been proposed to modify it using Poisson sampling of individuals, since count data are used (Matthews and Whittaker, 2014a).

Zero-sum multinomial (ZSM) distribution. This model has its roots in the **neutral theory** of biodiversity, which assumes that the differences between members of the same trophic level in a community do not influence their success (Hubbell, 2001; Etienne *et al.*, 2007). The ZSM distribution describes the SAD of a sample taken from a metacommunity (i.e. a set of interacting local communities) subject to merely stochastic ecological processes (Alonso and McKane, 2004; Matthews and Whittaker, 2014b). The ZSM assumes that the number of individuals is constant: when an individual dies, the portion of the resource that becomes available will be immediately used by an immigrant from other local communities of the same metacommunity (zero-sum assumption). Thus, resources remain fully saturated at all times. The ZSM distribution assumes the presence of more rare species than predicted by the lognormal. The ZSM has three parameters: the number of individuals in the local community (J), the probability that a dead individual is replaced by a randomly selected immigrant from the metacommunity (m), and the 'fundamental biodiversity number' (θ), a dimensionless parameter that tends to Fisher's α (see below, *Indices of α-diversity that incorporate species abundance*) as the number of individuals in the sample increases. The log-series distribution is a limiting case of the ZSM, and the two models typically provide similar fits to SADs (Prado *et al.*, 2018). In addition, He and Hu (2005) found an analytical relationship between the parameter θ and **Simpson diversity** index (see below for an explanation of this index).

Gambin model. The gambin model has been proposed as a combination of the gamma distribution with a binomial sampling method and represents a very flexible model able to fit a wide range of distributions (Ugland *et al.*, 2007; Matthews *et al.*, 2014; Matthews and Whittaker, 2014a; Ibanez *et al.*, 2020). The shape of the curve in the gambin model is described by a parameter called α (which should not be confounded with Fisher's α), with low α-values indicating log-series distributions and higher values indicating more lognormal distributions (Matthews and Whittaker, 2014a; Matthews *et al.*, 2019). It has been suggested that gambin α could be considered as a diversity index based on species equitability within the community (Ugland *et al.*, 2007; Fattorini *et al.*, 2016c). The gambin model may be uni- or multimodal. Multimodal gambin models have multiple α-values (i.e. one α per mode) (Matthews *et al.*, 2019). Multimodal patterns may indicate that communities are composed of different **guilds** (Marquet *et al.*, 2004) or groups of species with different dispersal abilities (Borda-de-Água *et al.*, 2017).

Geometric series. Among all proposed SAD models, the geometric series represents the least equitable distribution. In the geometric series the number of the individuals belonging to each species is a constant fraction (k) of the remaining individuals. If abundances are assumed to be proportional to resource utilisation, this model expresses the niche pre-emption hypothesis, in which the sizes of the **niche** hypervolumes (measured by species relative abundances) are sequentially pre-empted by the most abundant to the least abundant species. The parameter k, which gives the decay rate of abundance, is the pre-emption coefficient. If in the **rank-abundance plot** a log scale is used for abundance, the species exactly fall along a straight line (Fattorini, 2005). The geometric series is known to provide a good fit to simple communities characterised by the high dominance of a few species (Giller, 1984; Magurran, 1988, 2004).

Broken stick. The broken stick model (MacArthur, 1957) assumes that species abundances in a community vary at random producing a SAD characterised by a high equitability. If S is the number of species, the resources can be considered to be distributed along the length of a stick, on which S-1 points are picked at random and the stick broken at these points. The length of the resulting segments represents the abundance of each species, which is in turn considered to be proportional to their niche size. The broken stick model has been theoretically questioned and it provides good fits only rarely (Wilson, 1993; Hayek and Buzas, 2010). In the broken stick model there are no fitted parameters.

Zipf distribution. The Zipf distribution is a type of power law probability distribution (McGill *et al.*, 2007) with one fitting parameter, the decay coefficient. Ulrich *et al.* (2010) found that a power model produced good fits for incompletely sampled communities, but a more recent study (Baldridge *et al.*, 2016) suggests that in general the Zipf model is a poor characterisation of SADs.

Weibull distribution. The Weibull distribution is a very flexible model that can accommodate a variety of empirical distributions, being able to describe adequately very different SADs (Ulrich *et al.*, 2018b, 2022). The two parameters of the Weibull distribution are ecologically meaningful, as the shape parameter φ expresses an excess of either highly abundant species (low φ) or rare species (high φ), while the scale parameter λ expresses variability in species abundances (Ulrich *et al.*, 2019).

Gamma distribution. The gamma distribution is another flexible function. It has the capacity to accommodate shapes close to the lognormal SADs, as well as unimodal and right skewed distributions (Schmidt and Garbutt, 1985; Callaghan *et al.*, 2023). In fact, the log-series distribution is a special discrete case of the gamma distribution (Kendall, 1948; Caswell, 1976).

As SAD models have been conceived to use counts as a measure of abundance, most work on SADs has been focused on certain animals, such as vertebrates (especially birds) and arthropods (e.g. Morlon *et al.*, 2009; Baldridge *et al.*, 2016; Antão *et al.*, 2021), and trees (Hubbell, 1979; Cheng *et al.*, 2012; Baldridge *et al.*, 2016; Ulrich *et al.*, 2016; Arellano *et al.*, 2017; Matthews *et al.*, 2019; Antão *et al.*, 2021; Feng *et al.*, 2021), while shrubby and herbaceous plants, individuals of which may be difficult to distinguish or count, have largely been overlooked (Anderson *et al.*, 2012; Ulrich *et al.*, 2016, 2022). For these organisms, continuous variables, such as biomass or cover, can be used as a measure of abundance, but these approaches have rarely been adopted (Chiarucci *et al.*, 1999; Ulrich *et al.*, 2010, 2022; Anderson *et al.*, 2012; Matthews and Whittaker, 2015). A recent advancement in this direction is the application of continuous probability functions to model species abundance distributions based on cover data (Moradi *et al.*, 2025).

Indices of α-diversity that incorporate species abundance

There is an immense literature on indices of diversity that incorporate species abundance, and many reviews illustrate their theoretical bases, properties, usefulness, limits, etc. (Magurran, 1988, 2004; Krebs, 1999; Hayek and Buzas, 2010; Tuomisto, 2012). We mention here some of the most commonly used (see also Table 4.2 in Chapter 4).

Fisher's α. If the SAD of a community follows the log-series, the relationship between the total number of species (*S*) and the total number of individuals (*N*) in the sample is:

$$S = \alpha \log_e(1 + N/\alpha) \qquad \text{(A3.11)}$$

where α is a constant describing diversity (see Williams, 1964; Rosenzweig, 1995; Krebs, 1999).

This index can be estimated as:

$$\alpha = N(1 - x)/x \qquad \text{(A3.12)}$$

where x is a parameter of the log-series that can be estimated iteratively.

This index may be a satisfactory measure of diversity even for communities that do not follow the log-series; however, being only based on *S* and *N*, it cannot distinguish situations where *S* and *N* are constant, but evenness changes (perhaps more a theoretical situation than a true problem in real communities). Fisher's α is particularly sensitive to changes in the abundance of dominant species.

In a simulation study, Beck and Schwanghart (2010) found that this index has the virtue of providing a reliable measure of diversity when completeness is very low, but Fisher's α distribution values lose their stability when approaching full completeness sampling; so they prefer the **Shannon index** (see below) or biased-controlled indices to assess biodiversity in full sampled communities. However, Fisher's α can be as reliable as Shannon index to measure diversity in almost completely sampled communities when distributions are correctly rarefied (Tsafack *et al.*, 2021a).

Shannon index. This is possibly the most popular diversity index. The index has its foundations in information theory and represents the uncertainty about the identity of an unknown individual (Morris *et al.*, 2014). The Shannon index is calculated as:

$$H' = -\sum_{i=1}^{S} p_i \ln(p_i) \qquad \text{(A3.13)}$$

where p_i is the proportional abundance of the *i*th species. Any logarithm can be used in the calculation of Shannon index. When p_i is unknown in the statistical population, an estimate is used as n_i/N. This produces a bias, for which a version of the index based on samples has been proposed (Hutcheson, 1970; Bowman *et al.*, 1971). However, the bias in the use of the original version is generally negligible (Krebs, 1999). Shannon index ranges from 0, when the community includes only one species, to high values for communities with many species, each with a similar number of individuals. By combining **richness** and **evenness**, this index is particularly successful in summarising different aspects of diversity into a single value (Hayek and Buzas, 2010). However, the index is sensitive to sample size.

Recently, there has been increasing use of the exponential version of the Shannon index, which is calculated as $e^{H'}$. It is particularly useful as it expresses the number of equally common species required to produce the observed value of *H'* (Magurran, 1988, 2004), thus corresponding to one of the Hill numbers, if natural logs are used (see below).

An evenness index (Pielou evenness) can be derived as the ratio between the Shannon index and the logarithm of species richness:

$$J = H'/\ln(S) \qquad \text{(A3.14)}$$

This is an index of relative evenness, as ln(*S*) corresponds to the maximum value that the Shannon index can assume for a given community. This value (H_{max}) corresponds to a situation in which all species have the same abundance, as in this case *H'* = ln(*S*). The Pielou index measures the evenness with which individuals are divided among the species, and it is constrained between 0 (all individuals belong to one species) and 1 (all species are equally abundant) (Magurran, 1988).

Simpson index. Various versions of this index exist, which may create confusion. The index calculated with the formula:

$$D = \sum_{i=1}^{S} p_i^2 \qquad \text{(A3.15)}$$

represents the probability that two randomly chosen individuals belong to the same species. It is therefore a probability index, which varies from 0 (all species are equally abundant) to 1 (all individuals belong to the same species). Thus, the index increases with decreasing diversity. For this reason, in this formulation, the index is frequently referred to as a dominance index. An alternative formulation, which expresses diversity in a more intuitive way, is the multiplicative inverse of D:

$$N_2 = 1/D \quad \text{(A3.16)}$$

which has the advantage that it increases with increasing diversity. This version of the index corresponds to one of the Hill numbers (see below). A further alternative is taking the complement of D with respect to 1:

$$D_1 = 1 - D \quad \text{(A3.17)}$$

This formulation (known as the Gini–Simpson index) has the advantage that the index increases from 0 to 1 and is defined even when there is only one species. For a finite community, Simpson index is expressed by the formula:

$$D = \sum_{i=1}^{S}\left(\frac{n_i(n_i-1)}{N(N-1)}\right) \quad \text{(A3.18)}$$

Simpson evenness is calculated as:

$$E = N_2/S \quad \text{(A3.19)}$$

Whereas J is mathematically correlated with H', E is mathematically independent of N_2, which makes it a more useful measure of evenness in many contexts (Morris *et al.*, 2014).

Occasionally, survey in the field proves extremely difficult for one reason or another and even rapid assessments prove useful for gaining insights into community diversity. Based on the probabilistic diversity of the Simpson index, Cook (2008a) and Myler and Cook (2010) described a simple technique for determining S and evenness based on sequential sightings for the purpose of a rapid non-invasive survey:

$$D = a/n \quad \text{(A3.20)}$$

Here $n = a + b$ and in a survey a is the total number of like pairs and b the total number of unlike pairs in a sample series (D declines as diversity increases). An approximation of evenness is given by:

$$Evenness = (1 - D)/(1 - 1/S_r) \quad \text{(A3.21)}$$

where S_r is the number of species recorded.

Berger–Parker index. The Berger–Parker index is simply the relative abundance of the most abundant species:

$$D_{BP} = n_{max}/N \quad \text{(A3.22)}$$

Like the Simpson index D, the Berger–Parker index increases when diversity in the community decreases. This index is strongly influenced by sample size and richness, and does not make use of all the information available from samples.

Hill numbers. Diversity indices traditionally used in community ecology are affected by two important problems: (i) they express diversity in different units (for example, richness is a count, and it is expressed in units of species; Shannon diversity is an entropy measure, and therefore has bits of information; and Simpson index is a probability measure) and (ii) they have non-intuitive behaviours. For example, we would expect that a community A with 20 equally common species would be twice as diverse as a community B with ten equally common species. But community A has a Shannon entropy (natural logarithms) H' of 2.996, while community B has a Shannon entropy H' of 2.303; similarly, A has a Simpson diversity (calculated as 1-D) of 0.950, while community B has a Simpson diversity of 0.900. To overcome these difficulties, an increasing number of studies adopt conversions to effective numbers of species (number of equally abundant species). This approach involves calculating how many species a certain community should have to obtain the same value observed with a given index if all species were equally abundant. For example, consider a community of 5 species with the following abundances: 49, 26, 12, 9, 11. This community has a Shannon entropy of 1.389. Converting this index to the respective number of equivalent species means calculating how many species this community should have to produce the observed Shannon entropy value if all the species were equally abundant. The answer is about four species (precisely, 4.01 species). The relationship that makes it possible to convert Shannon entropy to the effective number of species is simple: Shannon entropy can be converted to the effective number of species by taking its exponential ($e^{H'}$). It is also possible to convert Simpson diversity D_1 to the effective number of species by the formula 1/(1–D_1). These relationships arise from the fact that Shannon diversity (in its exponential form) and Simpson diversity indices are, in fact, special cases of a family of indices, known as Hill numbers (Jost, 2010; Chao and Jost, 2012; Chao *et al.*, 2014a,b; Chiu and Chao, 2020; Roswell *et al.*, 2021).

Hill numbers are produced by the equation:

$${}^qD = \left(\sum_{i=1}^{S} p_i^q\right)^{\frac{1}{1-q}} \quad \text{(A3.23)}$$

where the order q defines the resulting diversity measure. When $q = 0$, the corresponding Hill number is the species richness ($N_0 = S$); when $q = 1$ the measure is undefined, but its limit as q tends to unity is the exponential of the Shannon entropy ($N_1 = e^{H'}$); when $q = 2$, the resulting Hill number is the inverse of Simpson dominance ($N_2 = 1/(1 - D_1) = 1/D$); and when $q = \infty$, the resulting Hill number is the inverse of the Berger–Parker index. As Hill numbers are all expressed in the same units (i.e. as number of equally abundant species) they are always perfectly comparable. The trio N_0, N_1, and N_2 provides a very comprehensive description of a community, and it has been proposed to combine them into a single measure called absolute effective diversity (AED; Cazzolla Gatti *et al.*, 2020):

$$AED = N_0 + \frac{N_1^2}{2N_2} \qquad \text{(A3.24)}$$

Contrary to Hill's numbers (which account for only the observed diversity of the studied community), this Chao-like index (see above for Chao estimators of species richness) can easily account for the 'missing effective species' that contribute to the diversity of the community.

As Hill numbers are produced by a continuous function, they may allow a very comprehensive description of the diversity of a community through the construction of a diversity profile (see Fig. 4.7 in Chapter 4) by plotting Hill numbers as a continuous function of the parameter q (Chao *et al.*, 2014a; Chiu and Chao, 2020). As special cases, when $q = 0$ species abundances do not count at all, and hence the contribution of rare species is emphasised; when $q = 1$ species are weighed in proportion to their frequencies and the measure can be interpreted as the effective number of common or typical species in the assemblage; when $q = 2$, the importance of most abundant species is emphasised and the measure can be interpreted as the effective number of dominant or very abundant species in the assemblage (Chao and Jost, 2012; Chao *et al.*, 2014a,b). Diversity profiles are usually constructed for q that varies from $q = 0$ to $q = 4$, as beyond $q = 3$ or 4 there is little variation, whereas values of diversity with $q < 0$ are not used, as they are dominated by the abundances of rare species and have poor statistical sampling properties (Chiu and Chao, 2020). The Hill numbers are the only series of diversity indices that satisfy the six mathematical axioms identified by Daly *et al.* (2018).

Ratios between Hill numbers have been proposed as measures of evenness (Hill, 1973; Alatalo, 1981; Ludwig and Reynolds, 1988; Smith and Wilson, 1996; Borcard *et al.*, 2018). Hill's (1973) N_2/N_1 index and Alatalo's (1981) modified version $(N_2 - 1)/(N_1 - 1)$ are generally unaffected by richness, which make them among the most suitable (Ludwig and Reynolds, 1988), but do not decrease when the abundance of the least abundant taxon in an assemblage is slightly reduced, a property that should be guaranteed by an evenness index (Smith and Wilson, 1996). The N_2/N_0 ratio performs well in Smith and Wilson's (1996) tests. Other suggested evenness measures based on Hill numbers are $\ln(N_1)/\ln(N_0)$ (= Pielou's J), N_1/N_0 (Shannon regularity) and N_2/N_0 (Simpson's regularity). These ratios are affected by the number of taxa in the assemblage, unless N_0, N_1 and N_2 are estimated through rarefaction (see above for rarefaction techniques).

Similarity-sensitive indices. All the aforementioned indices do not consider **disparity**, that is, differences among species. Similarity-sensitive indices incorporate a distance matrix as an additional input of the index along with species abundances. This matrix may express pairwise differences between species of any type (e.g. genetic, phylogenetic, morphological, functional).

The most commonly used formulation is Rao's quadratic diversity index:

$$H_R = \sum_{i,j=1}^{S} d_{ij} p_i p_j \qquad \text{(A3.25)}$$

where d_{ij} is the dissimilarity between species i and species j (if $d_{ij} = 1$ for all $i \neq j$ and $d_{ii} = 0$ for all i, H_R reduces to the Gini–Simpson diversity). This metric, however, contradicts the assumption that diversity is maximal for a perfectly even community (Daly *et al.*, 2018). A similarity-sensitive extension of the Hill numbers is the Leinster and Cobbold (2012) family of diversity indices:

$$^qD^z = \left(\sum_{i=1}^{S} \left((Zp)_i \right)^{q-1} \right)^{\frac{1}{1-q}} \qquad \text{(A3.26)}$$

where

$$(Zp)_i = \sum_{i=1}^{S} Z_{ij} p_j \qquad \text{(A3.27)}$$

with **Z** being a similarity matrix.

Leinster–Cobbold estimators for undersampled data and their decomposition in richness, evenness and disparity are the object of recent explorations (Marcon *et al.*, 2014; van Dam, 2019; Chen and Grinfeld, 2024).

Beta diversity and (dis)similarity measures

A plethora of indices is available for measuring the similarity/dissimilarity between habitats for their taxonomic content (see, for example, Magurran, 2004; Baselga, 2010; Anderson *et al.*, 2011 for reviews). One of the simplest and most intuitive measures of dissimilarity is the one-complement of Jaccard similarity coefficient S_J (see Fig. 4.8 in Chapter 4). Thus, the Jaccard coefficient (see Legendre and Legendre, 2012):

$$S_J = a/(a + b + c) \qquad \text{(A3.28)}$$

becomes a dissimilarity metric (D_J) that can be used to quantify **β-diversity** (Colwell and Coddington, 1994):

$$D_J = (b + c)/(a + b + c) \qquad \text{(A3.29)}$$

and *a*, *b* and *c* are as in Fig. 4.8.

It has been shown (Podani and Schmera, 2011; Carvalho *et al.*, 2012, 2013; Podani *et al.*, 2018) that D_J is formed of two components, referred to as species richness component ($Rich_J$) and species replacement component (or turnover) ($Repl_J$), such that:

$$D_J = Repl_J + Rich_J \qquad \text{(A3.30)}$$

where $Repl_J$ is the replacement (= species turnover) between two sites; that is, *n* species for one site are substituted by *n* species at another site. The number of possible substitutions is determined by the minimum number of species exclusive to *b* and *c*: min(*b*,*c*), which in effect is 2 × min(*b*,*c*) as one substitution always involves two species. Thus:

$$Repl_J = 2 \times \min(b,c)/(a + b + c) \qquad \text{(A3.31)}$$

Species richness of each site is contained in *b* + *a* and *c* + *a*. Accordingly, the absolute difference in richness between two sites can be determined as:

$$Rich_J = |(b + c) - (c + a)|/(a + b + c) = |b - c|/(a + b + c) \qquad \text{(A3.32)}$$

An analogous approach was used to decompose the total β-diversity expressed by Dice–Sørensen's coefficient of dissimilarity (D_S) into a replacement component and a richness difference index:

$$D_S = (b + c)/(2a + b + c) \qquad \text{(A3.33)}$$

$$Repl_S = 2 \times \min(b,c)/(2a + b + c) \qquad \text{(A3.34)}$$

$$Rich_S = |b - c|/(2a + b + c) \qquad \text{(A3.35)}$$

These formulations are known as Podani–Schmera–Carvalho decomposition (Podani and Schmera, 2011; Carvalho *et al.*, 2012, 2013; Podani *et al.*, 2018).

There is, however, another available form of β-diversity partitioning, based on decomposition into a pure replacement (turnover) and a **nestedness** component (Baselga, 2010) (nestedness is a type of richness difference pattern in which the species present in a site are a strict subset of the species occurring in a richer site; see Box 2.1 in Chapter 2 and Box 4.5 in Chapter 4). With this approach, Baselga (2010) partitioned the Dice–Sørensen coefficient of dissimilarity (D_S) into another form of species replacement ($Repl_{BS}$, corresponding to the Simpson coefficient of dissimilarity; Lennon *et al.*, 2001) and species nestedness (Nes_{BS}), such that:

$$D_S = Repl_{BS} + Nes_{BS} \qquad \text{(A3.36)}$$

where

$$Repl_{BS} = \min(b,c)/(a + \min(b,c)) \qquad \text{(A3.37)}$$

A similar approach (Baselga, 2012) was also used for partitioning Jaccard index into replacement and nestedness components:

$$D_J = Repl_{BJ} + Nes_{BJ} \qquad \text{(A3.38)}$$

where

$$Repl_{BJ} = 2 \times \min(b,c)/(a + 2 \times \min(b,c)) \qquad \text{(A3.39)}$$

In literature, $Repl_{BS}$ and Nes_{BS} are frequently indicated as β_{sim} and β_{nest} respectively, and $Repl_{BJ}$ and Nes_{BJ} as β_{jtu} and β_{jne}; D_S is commonly indicated as β_{sor}. It has been found that $Repl_{BS}$ tends to be positively correlated with **γ-diversity** (Ulrich *et al.*, 2018a).

Note that the nestedness components in Baselga's formulations are not measures of nestedness themselves, but measures of how dissimilar two communities are because of nestedness, and hence quantify dissimilarity due to richness difference provided the assemblages are nested (Baselga, 2012; Murray and Baselga, 2015). By contrast, the richness components of the Podani–Schmera–Carvalho approach quantify how dissimilar two assemblages are because of richness differences, these being nested or not (Murray and Baselga, 2015). These approaches, developed for presence-absence data, have also been extended to dissimilarity indices that consider species abundances (abundance-sensitive Sørensen- and Jaccard-type coefficients) (Baselga, 2013; Podani *et al.*, 2013; Legendre, 2014).

We concentrated here on β-diversity coefficients for pairwise comparisons. However, it is possible to have multiple-site measures that express the total compositional dissimilarity across a set of sites. Within the aforementioned Baselga's framework, it is possible to calculate Jaccard-based and Sørensen-based multiple-site dissimilarities, and the respective multiple-site analogues for their turnover and nestedness components (Baselga, 2010, 2012).

Many other multiple-site indices of β-diversity have been proposed, from which pairwise measures can in turn be derived (reviewed by Koleff *et al.*, 2003). The first to be introduced (and possibly most popular; Tuomisto, 2010a) was the Whittaker index:

$$\beta_M = \gamma/\alpha \tag{A3.40}$$

where γ is the regional richness and α is the mean of the richness at individual sites. If the Whittaker index is reformulated in terms of matching components, it becomes a pairwise index that, if multiplied by 2 (Cardoso *et al.*, 2009), corresponds to $Repl_J$ (Legendre, 2014). This measure of β-diversity is, in fact, related to both Jaccard dissimilarity ($\beta_M = 2/(2 - D_J)$) and Sørensen dissimilarity ($\beta_M = 1 + D_S$) (see Tuomisto, 2010a).

The Whittaker index of β-diversity corresponds to a multiplicative partitioning of total species diversity $\gamma = \beta_M \times \alpha$ (hence the subscript M indicating 'multiplicative'), and quantifies how many times as rich in species an entire (e.g. regional) dataset is than its constituent sampling units (e.g. local communities) are on average (Tuomisto, 2010a). However, if γ-diversity is partitioned into additive rather than multiplicative components, β-diversity is calculated as:

$$\beta_{At} = \gamma - \alpha \tag{A3.41}$$

and $\gamma = \beta_{At} + \alpha$ (the subscript A indicates 'additive'). β_{At} quantifies the amount by which the species richness of the entire dataset exceeds that of a single sampling unit of mean species richness, or the absolute amount of species turnover among the sampling units of the dataset (hence the subscript *t*) (Tuomisto, 2010a). The relative species turnover is expressed as:

$$\beta_{Mt\text{-}1} = (\gamma - \alpha_t)/\alpha_t \tag{A3.42}$$

which corresponds to Whittaker's species turnover ($\beta_{Mt\text{-}1} = \beta_W = \gamma/\alpha - 1$) and quantifies how many times the species composition changes completely (i.e. the number of complete effective species turnovers) among the sampling units; for a pair of sampling units, it corresponds to Sørensen dissimilarity (Tuomisto, 2010a).

If β_{At} is divided by γ, a new kind of turnover measure is obtained, called proportional species turnover:

$$\beta_{Pt} = (\gamma - \alpha_t)/\gamma \tag{A3.43}$$

which quantifies the proportion of the species of the entire dataset that is not contained in a single average sampling unit, and, for a pair of sampling units, it corresponds to Jaccard dissimilarity divided by 2 (see Tuomisto, 2010a). It is also related to Sørensen dissimilarity, as $\beta_{Pt} = 1 - (1 / (1 + D_s))$ (see Tuomisto, 2010a). Tuomisto (2010a, 2010b) remarked that β_{At}, $\beta_{Mt\text{-}1}$ and β_{Pt} do not quantify true β-diversity, which is only expressed by β_M.

Although we have referred here simply to species richness (thus discarding differences in abundances), Tuomisto (2010a, 2010b) discussed the above reported formulations in the framework of equally abundant virtual species (effective species; Hill's numbers – see above). Thus, the same equations can be applied using the exponential version of the Shannon index or Simpson index as measures of diversity instead of species richness (Tuomisto, 2010a).

An issue underlying β-diversity measures is their statistical significance. Significance of relationships in contingency tables between two units is typically carried out using χ^2 (chi-square) statistics. Pearson's point correlation coefficient (φ) is the usual correlation coefficient computed for binary (presence/absence) data (Legendre and Legendre, 2012). A suitable coefficient which transforms the χ^2 statistic so as to range between 0 and 1 is the contingency coefficient (Legendre and Legendre, 2012). However, contingency coefficients can only be used to compare tables of the same size. The analysis of multi-way contingency tables can be undertaken using Wilks's χ^2 statistics (Legendre and Legendre, 2012).

The 'orthodox' way to examine relationships between spatial units, including habitat units, is to use multivariate analyses. Both classification and ordination techniques (such as cluster analysis and non-metric multidimensional scaling) can be used to describe relationships between sites based on β-diversity coefficients, or their components, in the search for distinctive patterns (see Fig. 4.9 in Chapter 4). With this approach, the raw matrix of sites × species is used to construct a matrix of intersite distances (dissimilarities) using, as measure of distance, one of the aforementioned β-diversity measures, or its components. However, to explain and analyse β-diversity, a second matrix of sites × environmental variables must be included in the study, which requires other statistical approaches. As discussed by Tuomisto and Ruokolainen (2006), 'explaining β-diversity' means using community composition as a response variable and what gets explained is variation in community composition (i.e. β-diversity), while 'analysing β-diversity' means using β-diversity as a response variable (i.e. variation in community composition), and what gets explained is variation in β-diversity. To explain β-diversity, multivariate techniques that operate on the matrices of sites × species and sites × environmental variables, such as redundancy analysis and canonical correspondence analysis, are appropriate (Tuomisto and Ruokolainen, 2006). To analyse β-diversity, multivariate techniques that use matrices of biotic distances (β-diversity or its components) between sites and intersite environmental distances (such as Mantel test and multiple regressions) are appropriate (Tuomisto and Ruokolainen, 2006).

Functional diversity

The rapidly increasing interest in functional ecology in the last two decades (Magneville *et al.*, 2022) has led to a proliferation of methods addressed to measure how functionally diverse (functional richness), how regularly diverse (functional evenness) and how different (functional difference) are biological communities (Pavoine and Bonsall, 2011; Legras *et al.*, 2018; Schmera *et al.*, 2023). A further level of functional diversity investigation is represented by the spatial heterogeneity in species functional traits, that is the functional β-diversity (Villéger *et al.*, 2013). The vast number of available approaches is indicated by the increasing numbers of reviews and attempts to organise the multitude of concepts and statistical approaches into coherent frameworks (e.g. Schleuther *et al.*, 2010; Pavoine and Bonsall, 2011; de Bello *et al.*, 2021; Mammola *et al.*, 2021; Magneville *et al.*, 2022; Schmera *et al.*, 2023). No doubt, functional diversity measures will continue to increase in number and in sophistication. We report here a brief description of some of the most commonly used.

Functional dendrograms. With this approach, the trait space is represented as a dendrogram of the functional distances among species; thus, the total functional richness can be measured as the total branch length of the tree (Petchey and Gaston, 2002). The choice of the metrics used to calculate distances (e.g. Euclidean, Mahalanobis, Gower distances) and the algorithm applied for agglomeration (e.g. single linkage, complete linkage, UPGMA, Ward's method) influence the results. However, subjectivity in the use of distance metrics is not exclusive of this approach, but is a common problem in functional diversity analyses based on distance matrices. Dendrogram-based functional diversity indices may lead to inconsistent ecological conclusions and should be used after cautionary checking of dendrogram quality (Magneville *et al.*, 2022).

Rao's quadratic entropy (Q; Rao, 1982). This index (already presented as an index that takes into account disparity, see above) represents the expected dissimilarity between two different sample units (individuals, populations or species) taken at random, thus assessing the multidimensional divergence in trait space. It considers both species distance and species weight (i.e. relative biomass or abundance). The Rao entropy approach has been extended in the framework of Hill numbers (Chao *et al.*, 2019) to quantify the effective number of functionally equally distinct species (or 'virtual functional groups'). It also allows a multiplicative partitioning of diversity into functional α-diversity (effective numbers of functionally equally distinct species in an assemblage), γ-diversity (effective numbers of functionally equally distinct species in the pooled assemblages), and β-diversity (effective number of equally large assemblages with no shared species and all species in the assemblages being equally distinct). This β-diversity metric can be transformed to obtain functional (dis)similarity profiles based on abundance-sensitive Sørensen- and Jaccard-type indices. De Bello *et al.* (2007) proposed the difference between the Simpson index (D_1) and Rao's quadratic diversity (Q), as a measure of functional redundancy ($R = D_1 - Q$). In a similar fashion, Ricotta *et al.* (2016) proposed an index of redundancy measured as 1- (Q/D_1) or $(D_1 - Q)/D_1$. Mouillot *et al.* (2014) express functional redundancy as the ratio between species richness and the total number of functional entities (i.e. unique combinations of functional traits), and Strona *et al.* (2021) as one minus the ratio between the number of unique functional entities and total species richness. In general, functional redundancy remains loosely defined in the literature (de Bello *et al.*, 2021).

Functional richness (*FRic*; Villéger *et al.*, 2008). This represents the total amount of functional space occupied by the species within a community. In analogy with the Hutchinsonian niche (see Chapter 2), trait spaces can be seen as hypervolumes, in which each trait is an axis. The simplest approach is the construction of a convex hull, that is, the smallest convex polyhedron encompassing the total set of observations (Cornwell *et al.*, 2006; Villéger *et al.*, 2008). This convex-hull shaping the considered species is the functional richness *FRic*. If a single trait is used, *FRic* is simply the range of trait values present in the dataset community. It is important to remark that convex-hulls assume that the space is uniformly occupied. If one wants to consider that some parts of the space are more filled than others, techniques can be adopted to construct probabilistic hypervolumes (see Lu *et al.*, 2021 and references therein).

Functional identity (*FIde*; Garnier *et al.*, 2004, Mouillot *et al.*, 2013). This is the average of the position of species along each functional axis weighted by abundance/biomass (i.e. it uses a CWM approach; see Box 5.2 in Chapter 5).

Functional dispersion (*FDis*; Laliberté and Legendre, 2010). This index expresses functional divergence as the weighted mean distance (deviation) of individual species to the centroid (centre of gravity, defined by the *FIde* values) of all species. As *Q* and *FDis* are both based on variance, they are substantially equivalent (de Bello *et al.*, 2021) and both tend to increase with richness (Carmona *et al.*, 2017). Any distance or dissimilarity measure can be used. Although *FDis* tends to be correlated with *Q*, it may allow statistical tests for presence/absence data that are not possible with *Q* (Anderson, 2006).

Mean pairwise dissimilarity (*MPD*; Weiher and Keddy, 1995). *MPD* expresses the expected dissimilarity between two randomly selected species without replacement. Differently from *Q*, *MPD* cannot consider individuals, and hence intraspecific variability, but can take into account species abundance/biomass.

Functional divergence (*FDiv*; Villéger *et al.*, 2008). *FDiv* represents how abundance or biomass values are distributed toward the extremities of the occupied functional trait space. It is calculated as the deviation of abundance or biomass to the centre of gravity of the convex-hull of the species present. Functional divergence is high when a species with extreme trait values also has the highest abundance/biomass.

Functional evenness (*FEve*; Villéger *et al.*, 2008). This index describes the regularity of the abundance/biomass distribution of the species in the trait space represented by a minimum spanning tree (i.e. the undirected tree-like graph connecting all species of the assemblage with the lowest cumulative branch length) (Villéger *et al.*, 2008). *FEve* approaches 0 when most species and abundances/biomasses are concentrated in a portion of the trait space, and approaches 1 when species are regularly distributed (with even abundances/biomasses) in the trait space.

Functional originality (*FOri*; Mouillot *et al.*, 2013a). *FOri* expresses how isolated a species is in the functional space occupied by a given community and is calculated as the weighted mean distance to the nearest species from the total species assemblage.

Functional specialisation (*FSpe*; Bellwood *et al.*, 2006; Mouillot *et al.*, 2013). *FSpe* is the mean distance (weighted by abundance/biomass) of a species from the rest of the species in the total pool (i.e. from the centre of the functional space).

Functional mean pairwise distance (*FMPD*; Weiher *et al.*, 1998). This is the mean abundance/biomass weighted distance between all pairs of species. It gives an overall measure of diversity and is considered particularly useful when investigating habitat filtering and **competitive exclusion** questions (Swenson *et al.*, 2020).

Functional mean nearest neighbour distance (*FNND*; Weiher *et al.*, 1998). *FNND* is the abundance-weighted distance to the nearest neighbour within the assemblage and it is considered an indicator of **niche** differences and limits to similarity (Swenson *et al.*, 2020).

Functional β-diversity (Villéger *et al.*, 2013). Functional β-diversity expresses the variability in the functional traits of species among sampling units (communities) in a given area (e.g. a **landscape**). Following Baselga's partitioning approach (see above), it can be decomposed into functional turnover and nestedness components (Villéger *et al.*,

2013). Functional β-diversity may increase either because of high levels of functional turnover (when communities are composed of species with different trait values) or because of a high level of functional nestedness (when a community includes a subset of strategies nested in functional strategies of a larger scale) (Villéger *et al.*, 2013). Within the Podani–Schmera–Carvalho approach (see above) it can be partitioned into replacement and richness difference components (Cardoso *et al.*, 2014).

Phylogenetic diversity

Measures of phylogenetic diversity attempt to reflect the importance of phylogenetic uniqueness; species without close relatives contribute more diversity than do species with many close relatives.

Faith's Phylogenetic Diversity (*PD*; Faith, 1992, 2016). This is the sum of the branch lengths of the phylogenetic tree that span all species of a given assemblage. *PD* is the most used approach to quantify phylogenetic diversity and is particularly useful in biogeography and conservation (e.g. Pellens and Grandcolas, 2016). Assemblages (communities, sites, etc.) with high *PD* have species that are evolutionarily more divergent than those with lower *PD* values. *PD* can be regarded as a phylogenetic generalisation of species richness (Chao *et al.*, 2010). To compare communities with different sample size/coverage, rarefaction/extrapolation techniques (similar to those described for **species richness**, see above) have been developed also for *PD* (Chao *et al.*, 2015).

Net relatedness index and Nearest taxon index (*NRI* and *NTI*; Webb, 2000). The net relatedness index (*NRI*) is a standardised index (varying between -1 and 1) that expresses the phylogenetic structure of an assemblage based on the mean phylogenetic distance, obtained from pairwise phylogenetic distances between the co-occurring species calculated with a null-model algorithm. The nearest taxon index (*NTI*) is similar to *NRI*, but uses the mean nearest neighbour phylogenetic distance (*MNTD*), i.e. the mean phylogenetic distance of each taxon of an assemblage to its nearest neighbour in a tree. In both indices, 0 indicates that the observed value is not different from randomised results. Negative values indicate phylogenetic overdispersion (i.e. taxa within an assemblage are less closely related than expected by the null model). Positive values indicate phylogenetic clustering (i.e. co-occurring taxa are more closely related than expected by the null model).

Phylogenetic β-diversity. Phylogenetic β-diversity measures the dissimilarity between assemblages in terms of their evolutionary lineages. Phylogenetic β-diversity can be quantified using a phylogenetic analogue of Sørensen index, called the PhyloSor index (Bryant *et al.*, 2008). PhyloSor is based on the total length of the branches shared and unshared between assemblage pairs. Similarly to Baselga's partitioning of β-diversity into pure turnover and nestedness components (see above), PhyloSor can be partitioned into 'true' phylogenetic turnover and phylogenetic diversity gradients (Leprieur *et al.*, 2012). An alternative measure of phylogenetic β-diversity is the unique fraction (UniFrac) metric, which measures the fraction of the branch length of the tree that is unshared between a pair of assemblages (Lozupone and Knight, 2005; Lozupone *et al.*, 2011). PhyloSor and UniFrac are nearly identical (Swenson, 2011). Metrics for weighted (abundance) data have been introduced by Ricotta and Burrascano (2009) and Swenson (2011).

Phylogenetic diversity based on Hill numbers. *PD* does not incorporate information on species abundance/biomass. For this reason, phylogenetic generalisations of Shannon and Simpson indices have been proposed; these indices, however, have the same interpretational problems as their parent measures (Chao *et al.*, 2010). To surpass them, Hill numbers were recently extended to incorporate phylogeny and it has also been shown how they can be decomposed into independent alpha and beta components with both additive and multiplicative decompositions (Chao *et al.*, 2010, 2016; Faith and Richards, 2012; Faith, 2013). The resulting classes of diversity measures include generalisations of the Jaccard and Sørensen indices discussed above. For two assemblages, these classes also include UniFrac and PhyloSor as special cases.

Taxonomic diversity and distinctness

Warwick and Clarke (1995) and Clarke and Warwick (1998) proposed two different formulations to express taxonomic relatedness, referred to as taxonomic diversity and taxonomic distinctness, respectively.

Taxonomic diversity (Δ) uses Linnean taxonomic ranks (species, genus, etc.) and is calculated as:

$$\Delta = \frac{\sum\sum_{i<j} w_{ij} x_i x_j}{\sum_{i<j} x_i x_j + \sum_i x_i (x_i - 1)/2} \quad \text{(A3.44)}$$

where x_i and x_j are the abundances of species i and j respectively, and w_{ij} is the weight given to the path length linking species i and j, such that $w_{ij} = 0$ if i and j are the same species, $w_{ij} = 1$ if they are the same genus, etc. It expresses the average taxonomic 'distance' between any two organisms, chosen at random from the sample, as the length of the path connecting them, traced through the Linnean classification of the full set of species involved. When the taxonomic information is ignored, this index reduces to a particular form of Simpson index of diversity,

$$\Delta = \frac{1-\Sigma p_i^2}{1-N^{-1}} \tag{A3.45}$$

Taxonomic distinctness (Δ^*) is calculated as:

$$\Delta^* = \frac{\Sigma\sum_{i<j} w_{ij}x_ix_j}{\Sigma\sum_{i<j} x_ix_j} \tag{A3.46}$$

and expresses the average path length between any two randomly chosen individuals belonging to different species. This is obtained by dividing taxonomic diversity, Δ, by the value it would take were there to be no taxonomic hierarchy (i.e. if all species belonged to the same genus). Unlike Δ, which mixes taxonomic relatedness with the evenness properties of the abundance distribution, Δ^* is a measure of pure taxonomic relatedness and is invariant to scale change, being applicable not only to counts of individuals, but also to other ways of expressing abundances (e.g. biomass).

Both indices can be used with both species abundance and presence/absence data. When applied to species presence/absence data (ω), Δ and Δ^* reduce to a simpler form (Δ^+):

$$\Delta^+ = \frac{\Sigma\sum_{i<j} \omega_{ij}}{S(S-1)/2} \tag{A3.47}$$

where S is the number of species.

Taxonomic diversity is considered to be a better indicator of anthropogenic perturbations on communities than species richness, which is more prone to be affected by variation in natural environmental factors (e.g. habitat heterogeneity) (Heino *et al.*, 2005).

Clarke and Warwick (2001) proposed a further biodiversity index, termed 'variation in taxonomic distinctness' (Λ^+), which is the variance of the path lengths through the taxonomic tree connecting every pair of species in the list and reflects the unevenness of the taxonomic tree. Whereas higher Δ^+ values reflect higher taxonomic diversity, higher Λ^+ values reflect lower taxonomic diversity, thus the two indices tend to be negatively correlated (García-Martínez *et al.*, 2015).

These indices gained some popularity among community ecologists working with aquatic organisms (see, for example, Mistri *et al.*, 2000; Bevilacqua *et al.*, 2009, 2021; Tolimieri and Anderson, 2010; Sutton and Beckley, 2017, and references therein), whereas they are rarely used with terrestrial animals (Krasnov *et al.*, 2005; Guerrero *et al.*, 2011; Fattorini and Vigna Taglianti, 2015; García-Martínez *et al.*, 2015, Heino *et al.*, 2015; Leyte-Manrique *et al.*, 2019; Munian *et al.*, 2020; De *et al.*, 2023) and plants (e.g. da Silva and Batalha, 2006; Moreno *et al.*, 2009; Domínguez *et al.*, 2010; Gwali *et al.*, 2010; Martínez *et al.*, 2019; Jog and Bried, 2021).

Appendix 4 Tree of Life

The term 'habitat' applies to all life on Earth as from its very beginnings; some of the major dates are illustrated in Table A4.1. Further details on the geological time scale are provided in Appendix 5.

Although the term 'habitat' is typically applied to animals, the habitat concept is relevant to all life forms. However, for only a small fraction of species have their habitats been in any way investigated. Yet, their **resource-based habitat** components (resources and conditions) are crucial for their existence. A basic problem has been, and still is, how life on the planet is classified. Even at the highest levels of division (kingdoms) this is by no means clear. From Linnaeus' binomial designation of species to the current day, there have been various solutions to describing the variety of life, initially with a binary life kingdom solution (plants versus animals: Linnaeus, 1735) to five (Whittaker, 1969), six (Woese and Fox, 1977; Woese *et al.*, 1977; Cavalier-Smith, 1998), seven (Ruggiero *et al.*, 2015), and even eight kingdom classifications (Cavalier-Smith, 1993). With the advancement of molecular techniques and phylogenetic methods, even the most recent arrangements have been found wanting, owing to the lack of monophyly in traditional groups (Hug *et al.*, 2016; Burki *et al.*, 2020; Lara *et al.*, 2020). For instance, the **eukaryotes** do not cluster any more as a separate domain; instead, they fall within the domain of Archaea (Hug *et al.*, 2016) (Fig. A4.1a). As it is, there is still the question of whether viruses should be included within a 'kingdom' arrangement (Raoult *et al.*, 2004; Hegde *et al.*, 2009; Luketa, 2012) (Fig. A4.1b). However, the kingdoms and the three domain classification still have practical use. Throughout this development the basic division has also been maintained between **eukaryotes** and **prokaryotes**, organisms having and lacking internal cellular structures such as a nucleus and organelles. Here, we adhere to a simple seven-kingdom structure of Ruggiero *et al.* (2015; see Table 8.1), as this provides an adequate guide to the variation in life forms and their habitat structures.

Perhaps the most important point to emerge from all recent research is the enormous diversity of life on Earth among prokaryotes (Bacteria and Archaea) and **protists** (Protozoa and Chromista),

Table A4.1. Significant dates in the history of the universe and humans. Estimates can be very controversial (see, for example, estimates of land colonisation by arthropods presented in Brookfield *et al.*, 2022; Tihelka *et al.*, 2022; Braddy, 2024). YBP, years before present; CE, Common Era.

Event	Date
Origin of the universe	13.7 billion YBP
Origin of the Earth	c.4.6 billion YBP
Earliest fossil evidence of life on Earth (stromatolites)	3.5 billion YBP
First multicellular organisation (cyanobacteria-like organisms)	c.2.5 billion YBP
Oldest multicellular organisms	c.600 million YBP
First animals	700–600 million YBP
Origin of most animal phyla	539–514 million YBP
First land animals (arthropods)	c.500 million YBP
First land plants	470 million YBP
Earliest hominins	c.7 million YBP
Early modern *Homo sapiens*	c.300,000 YBP
Development of agriculture	12,000 YBP
First human-made self-sustaining nuclear chain reaction	1942 CE

DOI: 10.1079/9781800627505.app4

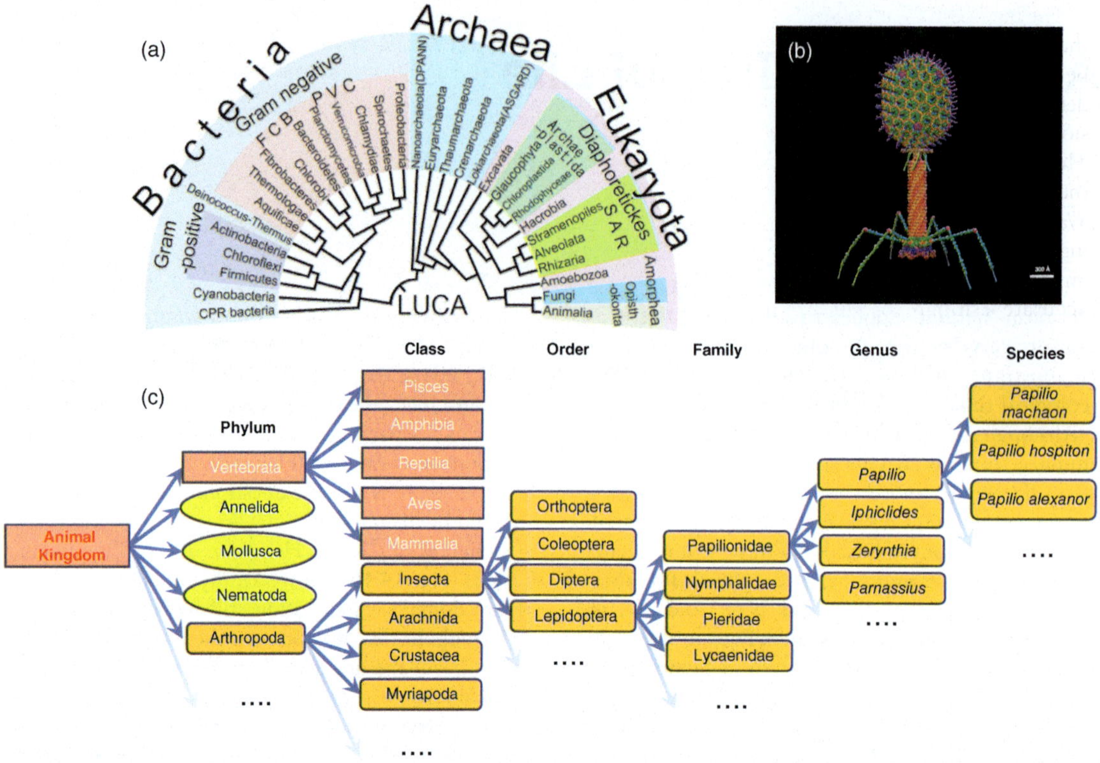

Fig. A4.1. The tree of life. **(a)** Evolutionary tree showing the common ancestry of bacteria, archaea and eukaryotes (LUCA: last universal common ancestor; ~4.2 Ga; Moody *et al*., 2024). Relative positions of major groups are shown around the tree. Archaeplastida correspond to Plantae *sensu lato*, including red algae (Rhodophyta), green algae, land plants and other minor groups. Protists do not form a monophyletic group but are represented by various clades (Excavata, SAR: stramenopiles, alveolates and rhizarians, Hacrobia, Amoebozoa). **(b)** Structural model at atomic resolution of bacteriophage T4. Whether viruses are alive is a topic of scientific debate, and the answer depends on the definition of life. New technologies are changing profoundly the way we see the complexity of life and its evolution. Impressive are, for example, the possibility of obtaining 3D whole cell models (*Mycoplasma genitalium*; Maritan *et al*., 2022) or mapping all cells in an organ (neurones in an insect brain; Dorkenwald *et al*., 2024). **(c)** A diagram illustrating the main nomenclatorial ranks used in taxonomy. Only a few groups of animals for each rank are shown as examples. The phylum Arthropoda includes invertebrate animals (such as insects, arachnids and crustaceans) with a segmented body and jointed appendages, a chitinous exoskeleton which is moulted at intervals, and a dorsal anterior brain connected to a ventral chain of ganglia. Increasingly, it is being realised that animal individuals comprise a multitude of symbiotic organisms. Each, including humans, is in effect a micro-ecosystem (a holobiont), dependent on the micro-organisms for survival within their bodies. This notion, emerging from the study of lichens (Grube *et al*., 2015; Aschenbrenner *et al*., 2016; Cernava *et al*., 2019; Sheldrake, 2020), is hardly surprising when it is considered how long micro-organisms have occupied the planet. We have no idea what we may bring back from future trips to outlying planets; obviously, it would be sensible to determine the nature of life on Earth before inadvertently bringing back to Earth 'materials' from other planets. (Panel **a** drawing by Crion: https://commons.wikimedia.org/wiki/File:Phylogenetic_Tree_of_Life.png, used under licence CC BY 3.0. Panel **b** image by Victor Padilla-Sanchez: https://commons.wikimedia.org/wiki/File:T4_Bacteriophage.gif, used under licence CC BY-SA 4.0.)

making diversity in arthropods and vertebrates seem very small indeed. A recent estimate is of a trillion (10^{12}) microbial species on Earth (Locey and Lennon, 2016).

Species are arranged into hierarchical groups corresponding to different nomenclatorial ranks. A simplified diagram (Fig. A4.1c) illustrates the major ranks as applied to the animal kingdom.

The big question remains unanswered: 'What were the origins of life on Earth?'. A first problem must be the numerous definitions of life; Trifonov (2011) determined some 123 such definitions! His conclusion is that: 'Life is self-reproduction with variations'. The first modern steps taken were in the 1950s with the discovery of the structure of DNA by Franklin, Watson and Crick (in 1953), the simple but astonishing experiment of Miller showing (in 1953) how amino acids (glycine) could be generated, and an accurate estimate of the age of the Earth from meteorites (Patterson, 1956). The Earth is now considered to be c.4566 million years old, the universe three times older at 13,700 million years (see Marshall, 2023). The wide distribution of small-molecule autocatalytic networks in prokaryotes (Xavier and Kauffman, 2022) indicates that molecular reproduction started over 4 billion years ago, much earlier than the last universal common ancestor, with small-molecule autocatalytic networks in constrained yet favourable geochemical settings on Earth, and perhaps much earlier all over the Universe.

Appendix 5 Geological Events in the Succession of Species

The large spans of time (eons, eras, periods) from the beginning of the Earth to the present are depicted by the geological time scale (Fig. A5.1a). Earth formed around 4.6 billion years ago, approximately one-third the age of the universe (roughly 14 billion years ago). The earliest undisputed evidence of life on Earth dates at least from 3.5 billion years ago, but might have originated more than 4 billion years ago, in the Hadean. Photosynthetic organisms appeared probably around 3 billion years ago, in the Archean. During the Proterozoic, major steps in evolution included the origin of **eukaryotes**, the origin of multicellularity, the diversification of eukaryotes and the origin of animals. Figure A5.1b illustrates one of the earliest known Precambrian (Ediacaran) fossils, found in the Charnwood Forest strata in Leicestershire, England. From then on, increasingly complex life forms occupied the planet.

The Phanerozoic is divided into three eras: the Palaeozoic, Mesozoic and Cenozoic. The Palaeozoic (the longest era) witnessed the most rapid and widespread diversification of life in Earth's history, including the 'explosion' of animal life in the Cambrian (when all major animal phyla appeared), the evolution of the first terrestrial animals (arthropods, also in the Cambrian), the origin of land plants (between the end of the Cambrian and the beginning of the Ordovician), the origin of vascular plants (in the Silurian), the origin of seed plants and the colonisation of land by fungi and tetrapods (in the Devonian), the origin of amniotes (in the Carboniferous) and the origin of the therapsids (ancestors of mammals, in the Permian). During the Mesozoic, the first mammaliaforms (which include the ancestors of current mammals) appeared (in the Triassic), non-avian dinosaurs dominated the land (especially during the Jurassic) and flowering plants appeared (in the Cretaceous). In the Cenozoic, mammals radiated: primates originated in the Palaeogene, hominins in the Neogene and *Homo sapiens* in the Quaternary.

The evolution of life has been characterised by repeated events of **mass extinction** (see Chapters 10 and 11). A model has been proposed (Fig. A5.1c) illustrating the basic features of extinction and recovery. In large part this may be simple in terms of loss and increase of species, but ecosystem changes can be highly complex. One thing appears clear: no single factor is responsible for a major extinction event (Fig. A5.1d and Table A5.1). It seems clear to many ecologists that we are entering a sixth major extinction and the cause is a single species, *Homo sapiens*. Figure A5.1d illustrates the 'known' losses of species and ecosystems at each of the major extinction events.

DOI: 10.1079/9781800627505.app5

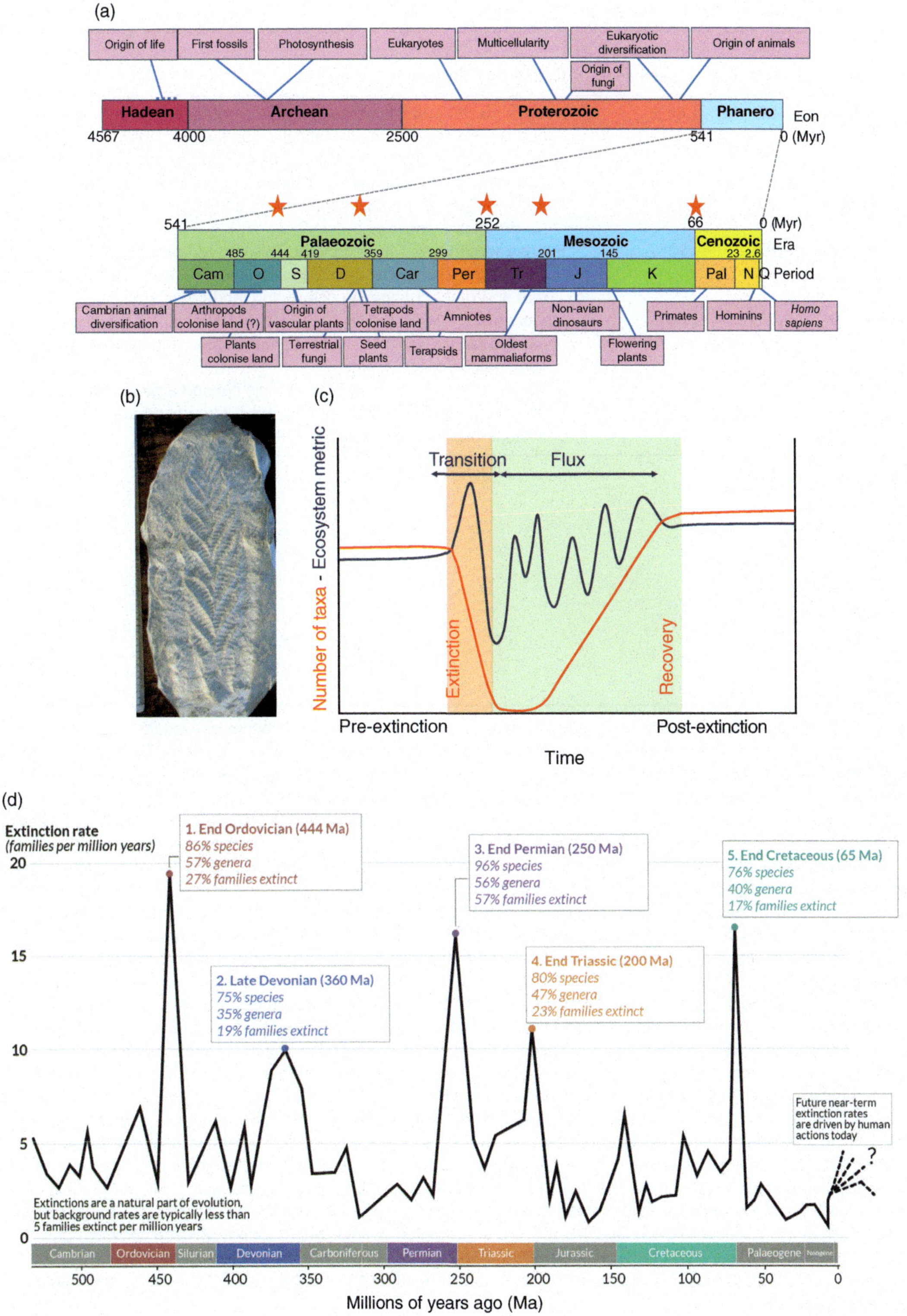
(a)
Origin of life
First fossils
Photosynthesis
Eukaryotes
Multicellularity
Eukaryotic diversification
Origin of animals
Origin of fungi
Hadean
Archean
Proterozoic
Phanero
Eon
4567
4000
2500
541
0 (Myr)
541
252
66
0 (Myr)
Palaeozoic
Mesozoic
Cenozoic
Era
485
444
419
359
299
201
145
23
2.6
Cam
O
S
D
Car
Per
Tr
J
K
Pal
N
Q
Period
Cambrian animal diversification
Arthropods colonise land (?)
Origin of vascular plants
Tetrapods colonise land
Amniotes
Non-avian dinosaurs
Primates
Hominins
Homo sapiens
Plants colonise land
Terrestrial fungi
Seed plants
Terapsids
Oldest mammaliaforms
Flowering plants
(b)
(c)
Number of taxa - Ecosystem metric
Transition
Flux
Extinction
Recovery
Pre-extinction
Post-extinction
Time
(d)
Extinction rate
(families per million years)
1. End Ordovician (444 Ma)
86% species
57% genera
27% families extinct
2. Late Devonian (360 Ma)
75% species
35% genera
19% families extinct
3. End Permian (250 Ma)
96% species
56% genera
57% families extinct
4. End Triassic (200 Ma)
80% species
47% genera
23% families extinct
5. End Cretaceous (65 Ma)
76% species
40% genera
17% families extinct
Future near-term extinction rates are driven by human actions today
?
Extinctions are a natural part of evolution, but background rates are typically less than 5 families extinct per million years
20
15
10
5
0
Cambrian
Ordovician
Silurian
Devonian
Carboniferous
Permian
Triassic
Jurassic
Cretaceous
Palaeogene
500
450
400
350
300
250
200
150
100
50
0
Millions of years ago (Ma)

Continued

Table A5.1. Extinction events on the geological time scale for Earth.

Extinction event	Age (Ma)	Percentage of species lost	Cause of extinction
End Ordovician	444	86%	Intense glacial and interglacial periods created large sea-level swings and moved shorelines dramatically. The tectonic uplift of the Appalachian mountains created lots of weathering, sequestration of CO_2, and with it, changes in climate and ocean chemistry.
Late Devonian	360	75%	Rapid growth and diversification of land plants generated rapid and severe global cooling.
End Permian	250	96%	Intense volcanic activity in Siberia. This caused global warming. Elevated CO_2 and H_2S levels from volcanoes caused ocean acidification, acid rain and other changes in ocean and land chemistry.
End Triassic	200	80%	Underwater volcanic activity in the Central Atlantic Magmatic Province caused global warming and a dramatic change in the chemical composition of the oceans.
End Cretaceous	65	76%	Asteroid impact in Yucatán, Mexico. This caused a global cataclysm and rapid cooling. Some changes may have already predated this asteroid, with intense volcanic activity and tectonic uplift.

(From Ritchie, 2022; CC BY 4.0.)

Fig. A5.1. Evolution of life on Earth. **(a)** The evolutionary time table, showing the course of evolution over the past 5 billion years as inferred from fossils, environmental proxies and high-resolution geochronology. The diagram emphasises the tiny amount of time humans have occupied the Earth compared to many other creatures, time nevertheless enough to have carried out a considerable amount of damage to the planet. Phanero: Phanerozoic; Cam: Cambrian; O: Ordovician; S: Silurian; D: Devonian; Car: Carboniferous; Per: Permian; Tr: Triassic; J: Jurassic; K: Cretaceous; Pal: Palaeogene; Neo: Neogene. Stars indicate times of major mass extinctions. **(b)** *Charnia*, genus of frond-like lifeforms belonging to the Ediacaran biota with segmented, leaf-like ridges branching alternately to the right and left from a zig-zag medial suture (thus exhibiting glide reflection or opposite isometry). The genus *Charnia* was named after Charnwood Forest in Leicestershire, England, where the first specimen was found. *Charnia* is significant because it was the first Precambrian fossil to be recognised as such. *Charnia* was among the earliest complex multicellular life forms. **(c)** Proposed model with nomenclature for separate phases of purely taxonomic richness dynamics, and all other ecological changes (called ecosystem metric) in response to an extinction-causing perturbation. The ecological transition interval coincides with the extinction-causing perturbation, a period that may (but not necessarily) coincide with the extinction interval, depending on the significance of ecosystem feedbacks from taxonomic loss. This is followed by an ecological flux interval, distinguished by the presence of short-lived, often uniquely structured ecosystems. Flux ecosystems are subsequently replaced by stable, postextinction communities typical of the interval that follows. One advantage of separating the terminology between the loss and gain of taxa and all other changes in ecosystem structure and function is that it allows for possible differences in the timing of phases. Note that the ecosystem metric can hypothetically correspond to any aspect of ecosystem structure or function.
(d) Mass extinctions not just of species, but of ecosystems and habitat structures, on the geological time scale. Each mass extinction has led to the development of new ecosystems, habitats and organisms. We are in the process of entering the sixth major extinction event owing to human disruption of planetary ecosystems. Mass extinction is defined as the loss of at least 75% of species within a short period of time (about 2 million years). (Panel **a** partially based on Knoll and Nowak, 2017. Panel **b** photo by Smith609: https://commons.wikimedia.org/wiki/File:Charnia.png, used under licence CC BY 2.5. Panel **c** based on Hull and Darroch, 2013. Panel **d** from Ritchie, 2022, used under licence CC BY 4.0.)

Glossary

Abyssal zone: In deep ocean regions, the **aphotic** zone in which temperature remains at around 2–3 °C year-round. The abyssal zone is between 4 and 6 km beneath the surface.

Accommodation model: A process in which animal species enter an ecological succession when changes in vegetation reach a threshold that provides a suitable habitat for them. As vegetation succession progresses, the habitat becomes less favourable for those particular species, and they are gradually outcompeted by others better adapted to the new conditions.

Accumulation curve: A curve showing the cumulative number of species recorded in a particular environment as a function of the cumulative effort (e.g. number of collected individuals or samples) expended searching for them.

Agricultural wilding: The introduction and conservation of wild crops and plants for agricultural purposes, as wild productive systems, useful for policy or other approaches to guide human–natural environment interactions (e.g. coffee plantations; see Vogt, 2021).

Airspace: The space lying above the Earth, or above a certain area of land or water. Also, the airspace in plants and animals involved in plant and animal functions.

Allee effect (or response): A small population phenomenon in which population growth rate is reduced by undercrowding. There are two main types of Allee effects: strong and weak. Strong Allee effect occurs when below a critical population size or density the population cannot sustain itself and will decline to extinction. In weak Allee effect, there is no critical population size, but growth rates still increase with population density, making smaller populations less resilient. Common causes include mate limitation (difficulty in finding mates at low population densities), co-operative behaviours (social species may rely on group activities like hunting, defending and offspring rearing, which are less effective in small groups), **genetic drift** (smaller populations are more prone to **inbreeding** and loss of **genetic diversity**, reducing overall fitness and hence population growth), predation and safety (larger groups may provide protection against predators through mechanisms like dilution effects or co-ordinated defence).

Allelochemical: A chemical produced by a living organism that exerts a detrimental physiological effect on individuals of another species when released into the environment (effecting **allelopathy**).

Allelopathy: A biological phenomenon by which an organism produces one or more biochemicals (**allelochemicals**) that negatively influence the growth, survival and reproduction of other organisms.

Allopatric, allopatry: Occurring in separate, non-overlapping geographical areas.

Allopatric speciation: The formation of new species that occurs when populations become geographically isolated from each other to an extent that prevents or interferes with gene flow. Compare with **sympatric speciation**.

Alpha diversity (α-diversity): The diversity (e.g. species richness) within a single local sample (assemblage, community).

Amensalism: An interaction in which one organism adversely affects a second organism, but the second has no effect (positive or negative) on the first.

Anadromous: Ascending rivers from the sea for breeding.

Anthosphere: A **phyllosphere** of flowers.

Anthropocene: A proposed (unofficial) geological epoch that denotes the period in Earth's history during which human activities have become the dominant influence on the planet's climate and ecosystems. It has no universally accepted start date, but it is often considered to start with the Industrial Revolution.

Anthroposphere: The anthroposphere, also known as the human sphere, refers to the part of the environment that has been influenced or shaped by human activities. It encompasses all human-made or modified systems, structures and processes, as well as their interaction with natural ecosystems.

Aphotic zone: The lower zone in a water column that receives little to no sunlight, making it impossible for photosynthesis to occur. It contrasts with the **photic zone**, where light is sufficient to support photosynthetic organisms.

Aposematism: An anti-predator adaptation in which an animal or plant advertises its unpalatability, toxicity or other defence mechanisms through conspicuous colours, sounds or other warning signals, thereby warning potential predators to avoid it.

Aquatic habitats: Environments in which organisms live in or depend on water for survival. These include freshwater habitats (e.g. rivers, lakes, streams, ponds, wetlands), marine habitats and brackish water habitats (e.g. estuaries, mangrove forests, brackish marshes).

Area of occupancy: The area within the range of a species that is actually occupied by its population(s).

Assemblage: A group of taxonomically related species that co-occur in the same area. An assemblage may be part of a **community** (e.g. butterflies within a vegetation unit). In general, the species in an assemblage may not necessarily interact strongly with one another; instead, they are grouped based on shared characteristics, such as habitat use, environmental conditions or taxonomic similarity.

Atmosphere: The gas and aerosol envelope that extends from the ocean, land and ice-covered surface of the Earth outward into space.

Augmentation: See **reinforcement**.

Autecology, autecological: The study of an individual organism, population or species in relation to its environment. Compare with **synecology**.

Autotroph, autotrophic: Organisms that require only carbon dioxide or carbonates as a source of carbon and a simple inorganic nitrogen compound for metabolic synthesis of organic molecules.

Autotrophic succession: An ecological succession characterised by early and continued dominance of autotrophic organisms (green plants and algae). Compare with **heterotrophic succession**. See also **ecological succession**.

Azonal biomes: Ecosystems (forms of vegetation) that are not primarily defined by climatic factors (such as temperature or precipitation) but instead by unique local conditions, such as soil type, hydrology or geological features. These **biomes** can occur across multiple climatic zones, hence the term 'azonal' (not tied to a specific zone).

Barrier: In **landscape** ecology, any environmental characteristic that prevents movements of individuals.

Bathyal zone: A benthic (seafloor) region of the ocean, located on the continental slope between the shallower continental shelf and the deeper abyssal plain. Typically extends from 200 to 2000 metres.

Bathypelagic zone: A deep layer of the ocean that lies beneath the **mesopelagic zone**. It extends from approximately 1000 metres to 4000 metres below the ocean surface. **Aphotic**, with almost unchanging cool temperatures.

Bauplan: The structural plan, or body plan, of an organism.

Bergmann's rule: An ecogeographical pattern in which animals living in colder climates tend to have larger bodies than their relatives living in warmer areas.

Beta diversity (β-diversity): The dissimilarity in species composition among samples (assemblages, communities).

Biogeochemical cycles: The movement and transformation of elements between the organisms and the Earth's abiotic compartments.

Biogeographical provinces: Subdivisions of biogeographical regions on the basis of their **biomes**.

Biogeographical regions (or realms): Large, distinct areas of the Earth that are characterised by specific patterns of animal and plant species distribution.

Bioindicator: An organism or a group of organisms used to monitor the health of an environment or ecosystem, providing information about the presence of pollutants or changes in ecological conditions.

Biome: One of the major (large-scale) categories of the world's distinctive plant and animal communities that have formed in response to the physical environment (soil, climate, physical landscape, etc.) in which they are found (e.g. polar desert, tundra, taiga or northern coniferous forest, Mediterranean vegetation, savanna, sandy desert, rainforest). The same biome can be distributed in different parts of the world with different evolutionary histories, thus containing different species, but sharing similar dominant plant life forms in response to similar climatic characteristics.

Biome stacking: The layering or overlapping of different biomes in a specific geographical area due to variations in altitude, latitude or other environmental factors. This phenomenon results in the presence of multiple biomes in close proximity to one another, creating a complex mosaic of **ecosystems** (e.g. a mountain might have a tropical forest at its base, followed by a temperate forest at a higher elevation, and then a boreal forest or alpine tundra at the summit).

Biosphere: The global sum of all **ecosystems**, encompassing all living organisms (plants, animals, fungi, protists, prokaryotes).

Biotic provinces: Large areas characterised by the occurrence of unique biotic **communities**.

Biotope (synonym for traditional habitat): An area of approximately uniform environmental conditions providing a living space for a specific community (of plants, animals, etc.), in which the biological environment is regarded by observers as uniform, and in which the flora and fauna distribution is also uniform.

Bottleneck: In evolutionary biology, a sharp reduction in the size of a population due to environmental events, such as earthquakes, volcanic eruptions, floods, fires, diseases, droughts, habitat loss, etc. This reduction can lead to a loss of **genetic diversity**.

Brotion: An **ecological succession** resulting from human activity.

Carposphere: A **phyllosphere** of fruits.

Carrion: The decaying flesh of dead animals.

Carrying capacity: The maximum population size (number of individuals) that can be supported indefinitely by a given environment.

Cascading coextinctions: The phenomenon of the loss or decline of a species resulting in the loss or endangerment of another species depending on it, potentially leading to cascading effects across trophic levels.

Catadromous: Living in fresh water and going to the sea to spawn.

Caulosphere: A **phyllosphere** of stems.

Central place theory: An urban geographical theory that seeks to explain the number, size and range of market services in a commercial system or human settlements in a residential system based on the role of central places in providing goods and services to surrounding areas.

City: Typically, a highly organised centre of population, of greater size or importance than a **town** or **village**.

Clade: A group of organisms that includes a single common ancestor and all of its descendants (a clade is a **monophyletic** group of organisms).

Climax: The final, stable, and mature stage of **ecological succession** in which a community has reached equilibrium with its environment.

Closed community: A **community** characterised by strict, clearly defined boundaries that separate it from other communities. Compare with **open community**.

Clumped distribution: A spatial distribution pattern in which individuals are closer together than they would be if they were randomly or evenly distributed (see **random distribution** and **uniform distribution**).

Colonisation ability: The capacity of a species to establish itself in a new area or patch of suitable habitat. In butterfly studies, measures of colonisation ability have been proposed based on resource breadth, with higher values indicating a greater capacity to exploit a hypothetical vacant patch. In this context, colonisation ability is regarded as a measure of **generalism** (Dapporto and Dennis, 2013).

Commensalism: An interaction in which one organism beneficially affects a second organism, but the second has no effect (positive or negative) on the first.

Community: A set of coexisting species.

Compartmentalisation: A tendency in **community** to be organised into structural subunits (compartments) within which interactions are strong but between which interactions are weak.

Competition: A negative interaction between organisms that vie for the same limited resources (such as food, water, shelter, mates or space) within an environment. Competition may involve individuals of the same species (intraspecific competition) or individuals of different species (interspecific competition).

Competitive exclusion: See **Gause's principle of exclusion**.

Composition (of habitats): The variety of resources in a habitat.

Connectance: In network analysis, the number of realised interactions divided by the number of all possible interactions.

Connectivity: In landscape ecology, connectivity refers to the degree to which different habitat patches, **ecosystems** or populations are linked or interact with one another within a landscape. It involves the flow of species, energy and nutrients between these components, facilitating ecological processes such as migration, pollination and the movement of resources. Similarly, in habitat studies, connectivity of resources refers to the proximity of resources in a habitat. In network analysis, connectivity refers to the total number of realised interactions among the species in the network.

Conservation introduction: The intentional movement of a species from its current location to an area outside the species' native range to improve its conservation status. Also known as benign introduction, assisted colonisation or managed relocation. Compare with **reinforcement** and **reintroduction**.

Consumable (resource): Any material or substance in any component of the environment that organisms can use and deplete through consumption (e.g. host plant parts, oxygen, water, minerals, salts, nectar, pollen, sap extraction, organic soups from rotting corpses).

CORINE Land Cover: The CORINE Land Cover (CLC) is a European programme providing consistent and thematically detailed information on land cover and land cover changes across Europe. CLC products are based on the classification of satellite images.

Corridors: Parts of the **landscape** that allow organisms to move between **patches** of suitable habitat.

Crenon: The rocky, fast-flowing part of a stream or river, typically found in the upper reaches where the water is oxygen rich and the current is strong. See also **rhithron** and **potamon**.

Crypsis: An anti-predator adaptation in which an animal or plant avoids detection by blending into its surroundings through camouflage, coloration, shape or behaviour.

C-S-R model (or strategies): A conceptual framework that categorises plant strategies based on three key ecological factors: competitiveness (C), stress tolerance (S) and ruderal (R) characteristics. The model highlights trade-offs between different life strategies in response to environmental conditions. Competitive species (C) are adapted to stable environments with abundant resources (such as light, water and nutrients). Stress-tolerant species (S) thrive in harsh, resource-poor environments where they face environmental stress (e.g. drought, extreme temperatures or nutrient deficiencies). Ruderal species (R) are adapted to disturbed environments (e.g. by fire, floods or human activity).

Deadwood: All the non-living woody biomass, whether standing, on the ground or in the soil, but not yet incorporated into the litter. The concept of deadwood includes both tree trunks no longer alive and the decaying parts of still alive trees, such as dead branches and woody debris accumulated in hollow trees or scattered in the litter.

Demographic transition model: A model describing the shift from high birth rates and high death rates to low birth rates and low death rates as societies experience advances in technology, education and economic development.

Density: The number of individuals per unit area (or, in some aquatic populations, per unit volume). Absolute density refers to the total number of individuals of a species per unit area or volume in a specific location, independently from the density of other species in the same location. Relative density refers to the abundance of a species relative to the abundance of all species within a given area.

Density-dependent factor: Any force that affects the size of a population in response to its density (e.g. disease, competition, predation).

Density-independent factor: Any force that affects the size of a population regardless of its density (e.g. climatic factors).

Disassembly: The breakdown in components of an ecosystem (Bascompte and Stouffer, 2009). Community disassembly is the non-random process of progressive species declines and losses (O'Neill, 2016).

Disclimax: A relatively stable ecological **community** often including kinds of organisms foreign to the region and displacing the **climax** because of **disturbance**.

Disparity: The extent to which species in a community are *not* similar (e.g. because of morphological, genetic, phylogenetic or functional differences).

Disturbance: The relatively abrupt loss of biomass or structure from an **ecosystem** that creates opportunities for establishment through alteration of resources or the physical environment (Walker *et al.*, 2010).

Disturbance heterogeneity model: A spatial model of species diversity variation along **disturbance** gradients, which predicts that species richness is highest at an intermediate level of disturbance.

Diversity (of community): A measure of the variety and abundance of species within a community. It can be broken down into **species richness, evenness** and **disparity**.

Diversity profile: A representation of species diversity in an ecological **community** that takes into account both **species richness** and **evenness** through the use of **Hill numbers**.

Domatia: Small structures produced by plants to house arthropods, namely ants (myrmecodomatia) and predacious and fungivorous mites (acarodomatia).

Downgrading (trophic downgrading): The disruption of ecosystems through the elimination of species at the top of food webs, frequently predators. A cascade of losses may occur through **trophic levels** as now evident over the globe in **biome** ecosystems (Estes *et al.*, 2011).

Dulosis: The practice by slave-making ants of capturing the pupae and larger larvae of other ant species and rearing them as workers of their own colony.

Dung: Faeces of animals, particularly herbivores and omnivores, which serve as a source of organic material and nutrients for a variety of organisms.

Dysphotic zone: The dysphotic zone, also known as the twilight zone, is a part of the **photic zone** located between the sunlit **euphotic zone** above and the completely dark **aphotic zone** below. In this zone, light penetration is minimal; there is enough light for visual predators to see, but not enough for the rate of photosynthesis to be greater than the rate of respiration.

Ecocline: See **ecotone.**

Ecogeographical rules: General patterns that describe how ecological and geographical factors influence the distribution, adaptation and morphology of species across different regions. See, for example, **Bergmann's rule.**

Ecological constraints model: A model that establishes that ecological factors constrain the size of groups of primates by influencing movement patterns and foraging efficiency. The ecological constraints model suggests that food resources are found in depletable **patches** and that larger group sizes result in a greater day range due to faster patch depletion. Smaller groups become more advantageous when the increase in travel costs is not compensated by a corresponding increase in energy intake or other fitness benefits.

Ecological diversity: Specifies ecological differences within and between populations (**habitats, ecosystems, ecoregions, biotic provinces, biomes** and **biogeographical realms**) (Gaston, 2010).

Ecological (or ecosystem) engineers: Organisms that directly or indirectly modulate (regulate) the availability of resources to other species, causing physical state changes in abiotic or biotic materials (Jones *et al.*, 1994).

Ecological succession: A change in species (**autotrophs** and **heterotrophs**) over time, either *ex novo* or by **disturbance** (complete removal of organic material, leading to **primary succession**; incomplete removal of organic material, leading to **secondary succession**). Biomass, nutrient availability and vegetation height and structure can either increase (**progressive succession**) or decrease (**retrogressive succession**) (Walker *et al.*, 2010).

Ecoregions: Relatively large units of land containing a distinct set of natural communities and species, with boundaries that approximate the original extent of natural communities prior to major land use changes (Olson *et al.*, 2001).

Ecosystem: A biological community of interacting organisms and their physical environment.

Ecosystem services: The various benefits that humans derive from healthy **ecosystems**. Ecosystem services are grouped into four broad categories: provisioning services (e.g. the production of food and water), regulating services (e.g. the control of climate and disease), supporting services (e.g. nutrient cycles and oxygen production) and cultural services (e.g. recreational benefits).

Ecotone: A transitional zone (area) between two or more substrate/vegetation habitat types, thus **communities** or **ecosystems. Ecocline** describes a gradient of change between the two habitat types.

Ecotopes: The smallest ecologically distinct landscape features in a **landscape** mapping and classification system, representing relatively homogeneous, spatially explicit landscape functional units that are useful for stratifying landscapes into ecologically distinct features (**ecosystems**) for the measurement and mapping of landscape structure, function and change (Sørensen, 1936; Tansley, 1939). In effect, they are the formal units of the traditionally coined 'habitat'.

Ecozones: Major ecological zones, larger than the **biomes.**

Edaphic: Relating to soil, especially as it affects the growth, distribution and development of plants and other organisms. Edaphic factors include the physical, chemical and biological properties of the soil, such as texture, pH, mineral content, moisture levels and microbial activity.

Edges: Boundaries between different habitat types.

Edge effect: An increase in occurrence (density) of individuals and of biodiversity at habitat or resource patch **edges.**

Emergent properties: System properties that arise from the interactions and organisation of simpler components within a system, which cannot be predicted or understood by looking at the individual parts in isolation. Those collective properties are critical attributes of biological systems, as understanding the individual parts alone is insufficient to understand or predict system behaviour.

Entoalexia: A condition in which a person or organisation is totally oblivious to the importance of entomology and insects (entomological blindness; Leather, 2021).

Entomyopia: A condition in which insects are viewed either as pollinators or as nuisances; a lack of foresight or discernment as to the importance of entomology: a narrow view of entomology (Leather, 2021).

Environmental (living) conditions: Environmental states that make the life of an organism possible (e.g. adequate values of light, temperature, humidity, rainfall, wind speed and direction, edaphic conditions).

Ephemeral metapopulation: An extreme form of metapopulation in which substantial occupancy and virtually total abandonment may occur in pulses over time dependent on oscillating environmental conditions. See also **metapopulation**.

Epikarst: The epikarst, also referred to as the subcutaneous zone, consists of extensively weathered carbonate bedrock located directly below the surface or exposed at the surface. Epikarst is highly fractured and porous, resulting from chemical weathering (mostly dissolution by slightly acidic water) of soluble rocks such as limestone, dolomite and gypsum.

Epipelagic zone: The uppermost (**photic**) layer of the ocean, extending from the ocean surface to a depth of about 200 metres.

Equilibrium theory of island biogeography: A dynamic model that explains the number of species found on an island or in an isolated habitat as a balance between immigration (the arrival of new species) and extinction (the loss of species) (MacArthur and Wilson, 1967).

Estimators (of species richness): Statistical methods (formulas) to estimate the total number of species in a community or ecosystem based on samples.

Eukaryotes: Unicellular or multicellular organisms that possess membrane-bound organelles. Compare with **prokaryotes**.

Euphotic zone: The layer of water in which there is enough light for net photosynthesis (i.e. the rate of photosynthesis is higher than the rate of respiration).

Eusocial, eusociality: Any colonial animal species that lives in multigenerational family groups in which the vast majority of individuals co-operate to aid relatively few (or even a single) reproductive group members.

Evenness: The equitability of the species abundance distribution of individuals within a sample (assemblage, community).

Exponential population growth: A continuous increase in a population in which the per capita growth rate stays the same, regardless of population size. Compare with **logistic growth**.

Extent of occurrence: The area contained within the shortest continuous imaginary boundary which can be drawn to encompass all the sites of occurrence of a taxon, excluding cases of **vagrancy**.

Extinction debt: The predicted eventual species loss from an area due to habitat loss (reduction, fragmentation or degradation).

Extremophile: An organism capable of surviving in extreme environments, where conditions approach or exceed the limits of what known life can adapt to, such as extreme temperatures, pressure, radiation, salinity or pH levels.

Facilitation: In species interactions, a process in which one species has a positive effect on the growth, survival or reproduction of another species, typically by altering the environment in a way that benefits the second species. In **ecological succession**, the facilitation model describes a process in which early successional (pioneer) species change the environment, making it favourable for the subsequent colonisation of later successional species, and, at the same time, less hospitable for their own ecological demands. As a result, pioneer species that can no longer thrive are replaced by new species typical of more advanced phases. Compare with **inhibition** and **tolerance**.

Fact: An objective piece of information. Facts are seen as propositions that are true regardless of individual perspectives. Compare with **law, rule** and **theory**.

Fasciolosis: A parasitic worm infection caused by the trematodes *Fasciola hepatica* and *Fasciola gigantica*.

Fissionfusion social organisation: Refers to a fluid pattern of spatiotemporal cohesion in animals where the larger group splits up into subgroups of varying composition that range separately and may join up together or break apart depending on resource availability (Chapman and Teichroeb, 2012).

Flagship species: A charismatic species acting as a symbol for a defined habitat, context, research design or environmental cause.

Focal species: A species that is the primary subject of a study or conservation effort. This species is chosen because it plays a key role in the ecosystem, serves as an indicator of environmental health, or is of particular interest due to its rarity, vulnerability or importance in a specific habitat.

Food chain: A linear sequence of organisms through which energy and nutrients are transferred from one **trophic level** to the next.

Food web: A network of interconnected **food chains**, illustrating how different species are linked through their feeding relationships.

Foundation stone species: Species that are disproportionately important to the continued maintenance of an existing community structure dependent on them.

Freshwater environments: Ecosystems that are primarily composed of water with low concentrations of dissolved salts (e.g. rivers, lakes, streams, ponds, wetlands and freshwater marshes).

Functional diversity: The diversity of species **functional traits** in a community.

Functional group: A group of species that share similar ecological roles or functions, regardless of their taxonomic classification.

Functional redundancy: The presence in the same **ecosystem** of multiple species filling similar roles, which results in a sort of insurance in the ecosystem, with one species able to replace a similar species in the same functional **niche** (Biggs *et al.*, 2020).

Functional trait: Any morphological, physiological, phenological or behavioural characteristic that is measurable on a single individual organism and can affect its fitness.

Fundamental niche: The entire set of environmental conditions and resources that allow a species to persist. See also **niche**.

Gamma diversity (γ-diversity): The total species diversity in a landscape/region.

Gause's principle of exclusion (Grinnell's axiom): Two species which compete for the same limited resource in a stable environment cannot coexist at constant population values: one competing species will eliminate or exclude the other. See also **competition**.

Generalism: The ability of an organism or species to thrive in a variety of environmental conditions and to utilise a wide range of resources. Compare with **specialism**. See also **specialism–generalism continuum**.

Generalist: An organism or species that can make use of a variety of different resources and/or has a broad range of ecological tolerances, thus being able to thrive in a variety of environmental conditions. Compare with **specialist**. See also **specialism–generalism continuum**.

Generic diversity: The 'diversification' of species into genera, calculated as the inverse of the probability that two species selected at random from a community belong to the same genus. As congeneric species are expected to be morphologically more similar than non-congeners, they should also have more overlapping niches than species belonging to different genera. Since **niche overlap** is assumed to be higher in congeneric species, competition should increase generic diversity. On the other hand, generic diversity might be lower in highly selective environments, where only species with similar adaptations can survive.

Genetic depletion: A process in which the already limited gene pool of a threatened species further decreases when reproductive individuals die before having the chance to mate with others in their small, threatened population.

Genetic diversity: Includes the components of the genetic coding that structures organisms (nucleotides, genes, chromosomes) and variation in the genetic make-up between individuals within a population and between populations (Gaston, 2010).

Genetic drift: The change in the frequency of an existing gene variant (allele) in a population due to random chance.

Geometric series: A **species abundance distribution** model in which the number of individuals in each species decreases geometrically from the most abundant species to the least abundant. Compare with **lognormal distribution** and **log-series**.

Geosphere: The geosphere encompasses all the Earth's solid layers, from the outer crust to the innermost core. It is one of the Earth's major systems, interacting with the **atmosphere, hydrosphere** and **biosphere**.

Gram-negative bacteria: Bacteria that do not retain the crystal violet stain used in the Gram staining method of bacterial differentiation. Gram-negative bacteria have a thin peptidoglycan layer and an outer lipid membrane. Bacteria that retain the crystal violet stain are indicated as Gram positive. Gram-positive bacteria have a thick peptidoglycan layer and no outer lipid membrane.

Grinnell's axiom: See **Gause's principle of exclusion**.

Guild: A group of species that exploit the same class of environmental resources in a similar way. This term groups together species without regard to taxonomic position, that overlap significantly in their niche requirements. The guild has a level comparable in the classification of exploitation patterns to the genus in taxonomic schemes. For example, to be considered a member of the foliage-gleaning guild in the oak woodland, the major portion of a bird species' diet has to consist of arthropods obtained from the foliage zone of oaks; as a result, birds that occasionally use the foliage zone would be excluded even though they may exert some influence on the guild's food supply (Root, 1967).

Habitat (or Traditional habitat type): A patch (place, area) of distinctive biotope and substrate, occasionally marked off by human land use. This is the traditional use of the term.

Habitat diversity: The diversity of habitat types (their number and relative cover), and hence the variety of environmental conditions, that can be observed in a given area.

Habitat islands: Isolated patches of suitable environment surrounded by unsuitable areas (**matrix**).
Habitat preference: The relative use of different resources by an individual or species.
Habitat selection: The hierarchical process of behavioural responses to the environment resulting in the choice of space and time of activity.
Habitat use: The way(s) in which the individuals of a species use/consume the resources within the habitat space.
Hadopelagic zone: The deepest part of the ocean, extending from around 6000 metres to the deepest points of the seabed. This zone is named after Hades, the Greek god of the underworld, reflecting the extreme conditions and remoteness of this deep ocean layer.
Hamlet: A small, rural settlement that is typically smaller than a **village**. It usually consists of just a few houses or dwellings.
Haustorium: A root-like structure that grows into or around another structure to absorb water or nutrients.
Herbivory: The consumption of living plant material.
Heterotroph, heterotrophic: An organism that derives its intake of nutrition from plant or animal sources of organic carbon; heterotrophs are secondary and tertiary consumers. Compare with **autotroph**.
Heterotrophic succession: The ecological process by which a community of **heterotrophic** organisms changes over time in response to environmental conditions. Unlike **autotrophic successions**, which primarily involve changes in plant communities and the availability of light, water and nutrients, heterotrophic successions result from the dynamics of consumer organisms and their roles in the ecosystem as conditions evolve. See also **ecological succession**.
Hill numbers: Hill numbers (^qD) are a mathematically unified family of **diversity** indices which differ among themselves only by the exponent q (which is the diversity order): as q can assume any value, it is possible to calculate, for a given assemblage of species, a diversity profile plotting ^qD versus q. Hill numbers include, for example, **species richness** (when $q = 0$), the exponential version of **Shannon index** ($q = 1$) and the inverse of **Simpson index** ($q = 2$).
Histogram (of species abundances): A graphical representation of **species abundance distributions** in which the number of species is plotted against abundance categories, usually in powers of two (octaves).
Holistic concept (of community): A view that considers a community as a complex, integrated system of interacting species. See also **super-organism**.
Holobiont: An assemblage of a host and (often) many other species living within or around it, which together form a discrete ecological unit through symbiosis.
Home range: The physical area covered during regular movements by an animal in the course of resource exploitation in the environment, including land, sea or air.
Homology, homologous: A character shared by a group of taxa due to inheritance from a common ancestor.
Homoplasy: The development of organs or other anatomical structures within different species, which resemble each other and have the same functions, but do not have a common origin. Such organs develop via convergent evolution and are thus analogous, not **homologous** to each other.
Hydrosphere: The layer of liquid and frozen water at or near Earth's surface. It includes the waters of the ocean; rivers, lakes and other bodies of surface water in liquid form on the continents; snow, ice and glaciers; and liquid water, ice and water vapour in both the unsaturated and saturated zones below the land surface.
Hyphae: See **mycelium**.
Hyporheic zone: The region of sediment and porous space beneath and alongside a stream or riverbed where surface water and groundwater mix.
Ideal despotic distribution: A model in behavioural ecology that describes how individuals distribute themselves among different resource patches when there is unequal access to resources due to competition for territories. In this model, dominant individuals (the 'despots') control the best patches, leaving subordinate individuals (weaker competitors) with access to poorer-quality patches. See also **isodar** and **ideal free distribution**.
Ideal free distribution: A model in behavioural ecology that describes how individuals distribute themselves among different resource patches based on their access to resources and the relative quality of those resources. The model assumes that individuals are free to move between patches, and that they choose patches in a way that maximises their fitness (e.g. access to food, shelter or mates) and reduces competition. See also **isodar** and **ideal despotic distribution**.
Ideal pre-emptive distribution: A model in behavioural ecology that describes how individuals distribute

themselves among different resource patches, where individuals actively defend or pre-emptively claim resources before others can access them.

Inbreeding, inbreeding depression: The production of offspring from the mating of individuals that are closely related genetically through common ancestry, as opposed to **outbreeding**, which is the mating of unrelated individuals. Inbreeding reduces fitness, a phenomenon known as inbreeding depression.

Indicator species: Species that are strictly associated with particular communities or ecosystem processes.

Individualistic concept (of community): A view that considers a community as the sum of individual species that occur together due to similar environmental conditions.

Inhibition (in ecological succession): A process in which early occupants modify the environment so that it is less suitable for late successional species. Compare with **facilitation** and **tolerance**.

Insurance hypothesis: The hypothesis that biodiversity provides a 'buffer' or 'insurance' against environmental variability and unexpected changes, ensuring the stability and **resilience** of ecosystems. This hypothesis posits that a greater diversity of species increases the likelihood that some species will perform critical ecological functions under changing conditions, even if others fail.

Interaction theory: A framework for understanding the multiple ways in which species interact.

Interbout arousals: Periodic, short-term returns to euthermy that are experienced by hibernators between bouts of torpor.

Intermediate disturbance hypothesis: This hypothesis states that biodiversity is highest in communities subject to a moderate degree of **disturbance**. At low disturbance levels, dominant species eliminate less competitive ones, leading to their dominance in the ecosystem. At high disturbance levels, all species face an increased risk of extinction. Thus, biodiversity reaches its peak at moderate disturbance levels, as conditions allow species from both early and late successional stages to coexist.

Interspecific competition: See **competition**.

Intertidal zone: The coastal area between the high tide and low tide marks. It is the region that is alternately submerged and exposed to air as tides rise and fall.

Intraguild predation: A type of ecological interaction in which species within the same **guild** prey on one another in addition to competing for shared resources.

Intraspecific competition: See **competition**.

Intrinsic rate of increase (*r*): The difference between the instantaneous natality rate (rate per time per individual) and the instantaneous mortality rate.

Introduction (of species): The process of intentionally or unintentionally moving a species to a new geographical area outside its native range. Introduced species are also known as non-native, allochthonous, exotic or alien species.

Invasive: An organism that establishes itself in a new environment and spreads rapidly, possibly causing harm to the local ecosystem, economy or human health.

Isodar: An ecological concept used to represent the relationship between population densities in two different habitats when individuals distribute themselves in a way that maximises their fitness. Isodars model density-dependent habitat selection for one or two species in two habitats according to the **ideal free** and **ideal despotic distributions**. The term comes from 'iso' (equal) and 'dar' (short for Darwinian fitness).

Isoleg theory: The isoleg is a line in a two-dimensional state-space of two species densities along which the choice of habitat is equal. On one side of the line, individuals of a species optimise their foraging by being strict habitat selectors, whereas on the other side, they do so by using at least a bit of a poorer patch. The isoleg theory states that the zero isoclines of the species are warped into non-linear forms capable of producing competitive coexistence. At the equilibrium point of this coexistence, there may be no overt competition remaining.

***K*-strategists:** Organisms that follow a reproductive strategy characterised by producing fewer offspring but investing more time and energy in each one's survival and development. Compare with ***r*-strategists**.

Key factor analysis: A technique for predicting variations in population size based on identifying the relative contributions of the key factors of births, deaths, immigration and emigration.

Keystone communities: Ecological **communities** that play a disproportionately large role in maintaining the structure, function and biodiversity of an ecosystem.

Keystone ecosystems: Ecosystems that play a crucial role in maintaining the structure, function and biodiversity of larger ecological systems.

Keystone habitat: Specific types of habitats that play a critical role in maintaining the structure, function and biodiversity of an ecosystem

Keystone species: A species that is of demonstrable importance for ecosystem function (see Cottee-Jones and Whittaker, 2012; they provide a tabulation of eight additional keystone definitions).

Keystone structure: A distinct spatial structure providing resources, shelter or 'goods and services' crucial for other species. A 'keystone structure' is allied to but distinct from the concept of **keystone species** (see Cottee-Jones and Whittaker, 2012). For instance, **deadwood** in mixed beech-spruce forests can be considered a keystone structure, as its removal (e.g. due to forest management) could greatly diminish the diversity of **saproxylic** insects. Keystone structure should be clearly distinguished from the term **keystone habitat**, as (traditional) habitats usually refer to broad vegetation types (see Tews *et al.*, 2004).

Laimosphere: The microbiologically enriched zone of soil that surrounds below-ground portions of plant stems.

Landscape: A spatially heterogeneous area composed of a cluster of interacting **ecosystems/habitats**.

Law: Scientific laws are statements that describe a consistent, universal relationship observed in nature. They are based on empirical evidence and have been repeatedly tested and confirmed under specific conditions. Scientific laws are typically expressed in concise mathematical or qualitative terms and describe how certain phenomena behave in predictable ways. Compare with **fact**, **principle**, **rule** and **theory**.

Lentic: Referring to standing water systems, such as lakes, ponds, swamps and reservoirs. Compare with **lotic**.

Lethal defence: A strategy employed by organisms to deter predators, competitors or herbivores through the use of highly toxic or harmful mechanisms that can kill or severely harm the attacking organism.

Life forms: In general, all living things, regardless of their complexity or environment. In specific contexts, life forms refer to groups of organisms based on shared characteristics such as their structure, mode of life, ecological role and how they interact with the environment.

Life-history strategy: A grouping of similar genetically determined characteristics which recur among populations of species and result in similarities in species' ecology. A life-history strategy encompasses the biological (including behavioural) traits that an organism exhibits throughout its life cycle to maximise its reproductive success in a given environment.

Life-history trait: An aspect of behaviour, physiology or anatomy that influences an organism's energy allocation to reproduction and/or fitness (Endler, 1986).

Lifetime track: The trajectory of an individual organism's life history, from birth to death, with a focus on the key events that influence its development, changes in behaviour, resource use, interactions with the environment, and hence its fitness and survival.

Life zones: The distinct characteristic changes in vegetation composition and form that occur along a latitudinal gradient.

Limnetic zone: The open water area of a lake or pond, away from the shore and above the profundal zones. This zone is characterised by sufficient sunlight for photosynthesis, but typically does not have plants rooted in the sediment.

Limnofauna: The animals that live in freshwater.

Linkage density: In network analysis, the average number of interactions per species.

Lithic reduction: The process by which stones or rocks are systematically chipped, flaked or shaped to create tools or artifacts.

Lithosphere: The rigid, outermost rocky layer of the Earth, consisting of the crust and the uppermost part of the mantle.

Littoral zone: The shallow, nearshore area of aquatic environments, both in marine and inland waters (such as lakes, ponds and rivers). In marine environments, the littoral zone is the coastal area that extends from the high tide mark to the edge of the continental shelf, often influenced by the ebb and flow of tides (**intertidal zone**). In inland waters, the littoral zone refers to the shallow, nearshore areas where light can penetrate to the bottom, promoting plant growth.

Logistic growth (model, equation): A model of population growth in which the growth rate slows as the population approaches the **carrying capacity** of its environment. Compare with **exponential population growth**.

Lognormal distribution: A **species abundance distribution** model in which most species have moderate abundances, a few species are very abundant, and a large number of species are rare. Compare with **geometric series** and **log-series**.

Log-series: A **species abundance distribution** model in which a small number of species are highly abundant and a large number of species are represented by just a few individuals. Unlike the **lognormal distribution**, the log-series distribution is highly skewed, with a greater concentration of species at the low end of abundance and fewer species with high abundance. Compare with **geometric series**.

Lotic: Referring to running water systems, such as rivers, streams and creeks. Compare with **lentic**.

Macroecology (macroecological viewpoint): The broad-scale study of ecological patterns, processes and relationships of organisms across large geographical areas, typically at the level of entire **ecosystems** or **biomes**.

Marine environment: The vast interconnected system of Earth's oceans and seas, encompassing all their physical, chemical, biological and geological features.

Marine regionalisation: The subdivision of the world's oceans and seas into distinct regions based on their biological and/or environmental characteristics.

Mark-recapture: A method to estimate the size of animal populations. It involves capturing individuals from a population, marking them and then releasing them back into their habitat. After allowing time for the marked individuals to mix back into the population, a second sample is captured. The proportion of marked to unmarked individuals in this second sample is used to estimate the total population size.

Mass extinction: A widespread and rapid decrease in biodiversity, characterised by the sudden loss of a large proportion of species across various taxa and ecosystems, as shown by the fossil record.

Matrix: The dominant, most continuous and widespread land cover that surrounds and permeates **patches** of different **habitats** or **ecosystems** within a given **landscape**.

Meiofauna: Minute animals (typically 10^{-3} to 10^{-5} m) living in soil and aquatic sediments.

Mesopelagic zone: The (**dysphotic**) layer of the ocean that lies just below the **epipelagic** zone, extending from 200 metres to about 1000 metres in depth.

Metapopulation: A set of local populations (subpopulations) of a particular species that interact with one another through dispersal. Compare with **multipartite population.**

Microhabitat: A place where an organism lives (spends its entire life cycle), but on a much smaller scale (a small or tiny enclave in any environment). The term is often, perhaps more commonly, applied to enclaves where organisms may spend most but not all of their lives. Examples of microhabitats include fallen tree plates, rotting logs, ephemeral pools, hot springs, hedges, spaces beneath stones, tree bark crevices, leaf litter, moss patches, animal burrows and soil cracks.

Micro-organisms: Organisms too small to be observed with the human eye, including bacteria, fungi, algae and protozoa. Most are unicellular.

Migration capacity: See **movement capacity**.

Mimicry: The close external resemblance of an animal or plant (the mimic) to another animal or plant (the model) for an adaptive advantage. This resemblance can deceive predators, prey or competitors (the receiver), providing benefits such as protection or increased access to resources.

Minimum amount of suitable habitat (MASH): The minimum habitat area/number of suitable habitat **patches** that a (meta)population requires to persist in a given environmental setting for a long time.

Minimum viable population (MVP): The smallest population size of a species that is capable of surviving and maintaining a healthy, viable population over time, without the risk of extinction due to factors like **inbreeding, genetic drift**, environmental changes or catastrophic events.

Mobility: The ability to move; capacity for movement.

Modular organism: An organism in which the zygote develops into a discrete unit which then produces more units like itself (e.g. plants, sponges, corals, many species of fungi).

Modularity: The degree to which a network can be divided into distinct subgroups (modules) that are more internally connected to each other than to the rest of the network.

Monophagy, monophagous: Describing an organism, particularly a herbivore or predator, that feeds on only one type of food (e.g. butterflies using only one species or genus of plants). See also **oligophagous** and **polyphagous**.

Monophily, monophyletic: A group of organisms is monophyletic when it includes an ancestral taxon and all of its descendant taxa. Compare with **polyphyly**.

Motility: The capability of motion (including contraction).

Movement (migration) capacity: All the factors, including opportunities (e.g. flight periods, population density, numbers of broods) as well as **bauplan** (viz. the anatomical and physiological traits that affect dispersal ability), which influence the numbers moving out from habitats as well as the distances a species can travel.

Multipartite population: A single, large population consisting of multiple, overlapping, interacting groups. As the habitat patch on which a population is based becomes larger, it is less likely to be uniform in the type, quality and quantity of resources. Consequently, the population is increasingly likely to break down into areas of higher and lower population density, and thus local panmictic units centred on areas where resources overlap or there is a peak in resource quality, separated by larger areas lacking obvious resources. Compare with **metapopulation**.

Mutualism, mutualistic: An interaction between the individuals of two or more species in which each species has a net benefit.

Mutualistic symbiosis: A symbiotic relationship between individuals of different species in which both individuals benefit from the association.

Mycelium: A network of fungal threads (hyphae). Mycelia often grow underground but can also flourish in other places such as rotting tree trunks. A single spore can develop into a mycelium. The fruiting bodies of fungi, such as mushrooms, can sprout from a mycelium.

Mycorrhiza: A **mutualistic symbiosis** between a green plant and a fungus.

Myrmecotrophy: The ability of plants to obtain nutrients from ants that live in or around the plant. Thus, it involves **mutualistic** relationships between plants and ants.

Natufian culture: A Late Epipalaeolithic archaeological culture of the prehistoric Levant in western Asia, dating to around 15,000 to 11,500 years ago. The Natufian culture is regarded as a precursor to food-producing societies in the southern Levant. Its significance stems from the profound cultural shifts associated with increasingly permanent human settlements and its pivotal role in paving the way for the transition to agriculture-based societies. Notable features of the Natufian culture include significant developments in subsistence strategies, settlement patterns, technological advancements, social organisation and ritual practices, all of which were instrumental in the shift from Epipalaeolithic hunter-gatherer groups to Neolithic farming communities.

Neritic zone: The shallow water extending from the low tide mark to a depth of about 200 metres. This zone is characterised by warm waters with abundant sunlight.

Nestedness: The degree to which smaller sets are subsets of larger ones (e.g. species occurring in sites with fewer species are present in sites with more species).

Network: A system of interconnected entities (**nodes**) that interact or are linked in some way.

Neutral theory: A conceptual framework that explains the diversity and abundance of species in ecological communities based on the assumption that all individuals of all species are ecologically equivalent. This contrasts with traditional **niche theory**, which emphasises differences between species and their interactions with the environment.

Neutralism: The absence of any interaction between two organisms; neither has any effect on the other.

Niche: The geometric space comprising conditions and resources measured in a habitat within which a species can persist for a time *t*. In the **fundamental niche,** it occupies all the space to which it is adapted. In the **realised niche**, the space is typically reduced for a variety of reasons ranging from chance effects to interactions (predation, parasitisation, competition, etc.).

Niche breadth: The range of environmental conditions and resources that a species can use or tolerate.

Niche construction theory: An ecological and evolutionary framework that emphasises how organisms actively modify their environment, shaping both their own ecological niche and the selective pressures acting on them.

Niche overlap: The situation in which co-occurring species share parts of their **niche** space with one another.

Niche theory: A conceptual framework that explains the diversity and abundance of species in ecological communities by emphasising the role of ecological **niches**. Compare with **neutral theory**.

Node: In network analysis, a node (also called a vertex) represents a distinct entity or object within a **network**. Nodes are the fundamental building blocks of a network and are connected to other nodes via links, which represent relationships, interactions or connections between them. In **landscape** ecology, they correspond to points, small areas along **edges** at which resources are concentrated.

Noösphere: The sphere of human thought, consciousness and knowledge, encompassing all intellectual and cultural activities. It is a conceptual layer of Earth associated with human cognition, creativity and shared knowledge, integrating mental and cultural phenomena with the physical and biological aspects of the planet.

Oligophagy, oligophagous: Describing an organism, particularly a herbivore or predator, that feeds on a limited number of species, typically within a specific taxonomic group (e.g. butterflies feeding on a few plant species within a single family).
-ome: A suffix used to describe the complete set of components of a particular type in a biological system (e.g. genome, proteome, metabolome, etc.).
Open community: A community characterised by the lack of strict, clearly defined boundaries that separate it from other communities. Compare with **closed community**.
Opines: Low molecular weight compounds found in plant crown gall tumours, or hairy root tumours, produced by pathogenic bacteria of the genera *Agrobacterium* and *Rhizobium*.
Optimal foraging theory: A behavioural ecology model that explains how animals maximise their energy intake while minimising the costs associated with foraging. Although obtaining food provides the animal with energy, searching for and capturing the food also require both energy and time. To maximise fitness, an animal adopts a foraging strategy that provides the most benefit (energy) for the lowest cost, maximising the net energy gained, thereby increasing its survival and reproductive success.
Organismal diversity: Relates to the full taxonomic hierarchy and its components (e.g. individuals, populations, subspecies, species, genera, families, orders, classes, phyla, kingdoms, domains) (Gaston, 2010).
Orobiomes: Ecological regions found in mountainous areas across **zonobiomes**, characterised by unique biological communities and environmental conditions shaped by altitude, slope, climate and geological factors.
Orthosere: The ordinary subsere in abandoned fields lacking biotic suppression(s). See **sere** and **ecological succession**.
Oscillation hypothesis: The hypothesis that colonising specialists have the capacity to become generalists as populations expand into novel resource zones, and that generalists can in turn produce new isolated specialists as geographical ranges collapse (or in sympatry with competition for resources).
Outbreeding, outbreeding depression: Matings between individuals from different genetic lines, populations, subspecies or species. Outbreeding can result in a decline in reproductive fitness, known as outbreeding depression. Compare with **inbreeding**.
Palynoflora, palynofloral: The pollen and spores of a region or site, considered as a whole.
Parapatry: A distributional pattern in which pairs of taxa have separate but contiguous distributions, bordering along common boundaries.
Parasitism: An interspecific relationship in which one species (the parasite) depends on the other species (the host), which is negatively affected by the interaction.
Patches: The distribution of a resource or **biotope** in a discrete, concentrated area, separated by space (**matrix**) from other areas (e.g. patches may be represented by a single tree that is producing food items or a cluster of trees of the same species in the same area, all producing food items).
Pedobiomes: Biomes determined by soil.
Photic zone: The upper layer of a body of water into which daylight penetrates in sufficient amounts to allow photosynthesis to take place.
Phylloplane: The leaf surface habitat.
Phyllosphere: The total above-ground surface of a plant that can be colonised by micro-organisms.
Phylogenetic diversity: A general term to indicate biodiversity measures that take into account the evolutionary relationships between species.
Physiognomy (of resources in habitats): The placement and configuration of resources in habitat.
Phytogeography, phytogeographical: The study of the distribution of plants.
Plagiosere: The gradient of **seral stages** (**vegetation succession**) held under biotic pressure by activities such as grazing, mowing and burning. See also **ecological succession** and **sere**.
Plot: A specific, defined area of land or water that is selected and marked out for the study of ecological variables (e.g. physical variables, population abundance, species composition and biomass).
Polyphagy, polyphagous: Describing an organism, particularly a herbivore or predator, that feeds on a wide variety of food sources, typically across multiple species or taxonomic groups (e.g. butterflies feeding on plant species from various families). See also **monophagous** and **oligophagous**.
Polyphily, polyphyletic: A group of organisms is polyphyletic when it includes organisms that have multiple origins and thus do not share a common ancestor. Compare with **monophyly**.
Population dynamics: The variation in time and space of population size and density for one or more species.
Population regulation: The ecological processes by which the growth of populations is limited, due to the effects on birth and/or death rates.

Potamon: The lower, slower-flowing reaches of a river or stream, typically located downstream in its course. See also **crenon** and **rhithron.**

Potential natural vegetation: The natural vegetation that would theoretically develop in an area under prevailing environmental (especially climatic) conditions, assuming there were no human interferences, such as agriculture, urban development or deforestation.

Potential resource-based habitat: The set of resources and conditions within a site that can potentially maintain a local population of a species.

Predation: An ecological interaction in which one organism (the predator) kills and consumes another organism (the prey).

Presocial, presociality: Animals that exhibit more than just sexual interactions with members of the same species, but fall short of qualifying as **eusocial.** That is, presocial animals can display communal living, co-operative care of young or primitive reproductive division of labour, but they do not display all of the essential traits of eusocial animals. Adult generations might not overlap, the reproductive dominance might be temporary, or some other requirement is not met (Ross and Matthews, 1991). Some sociobiologists further distinguish two different types of presociality: (i) subsocial: parents interact with young; (ii) parasocial: individuals of the same generation live in a single, co-operative dwelling and interact with one another. Parasocial behaviours can be further classified as: (i) communal: each individual cares exclusively for its own young; (ii) quasisocial: individuals care co-operatively for the entire brood; however, all members of the colony are reproductive; (iii) semisocial: a limited group of individuals reproduces, yet the arrangement is not quite eusocial.

Primary succession: Newly exposed or newly formed rock is colonised by living things for the first time. Compare with **secondary succession**. See also **ecological succession.**

Principle: A fundamental, well-established concept that explains how certain phenomena consistently behave under specific conditions. Compare with **fact, law, rule** and **theory**.

Profundal zone: The deeper, darker (**aphotic**) part of a lake or pond.

Progressive succession: The natural process of gradual change in the structure and composition of a biological **community** over time, resulting in an **ecosystem** that becomes increasingly complex and stable. Compare with **retrogressive succession**. See also **ecological succession.**

Prokaryotes: Unicellular organisms that lack a nucleus and other membrane-bound organelles. Their genetic material is not enclosed in a nuclear envelope but is located in a region called the nucleoid. Prokaryotes are classified into two domains: Bacteria and Archaea. Compare with **eukaryotes**.

Protists: Traditionally, a eukaryotic organism that is not an animal, plant or fungus. The taxonomic kingdom Protista is a collection of mostly single-celled organisms that do not fit into any other category. It is likely that protists share a common ancestor, but the exclusion of other eukaryotes means that protists do not form a natural group, or **clade**. There is no consensus on which groups should be included among protists. As commonly understood, protists are a group made up of protozoans, (unicellular) algae and slime moulds.

Pseudosinks: Patches that can support a local population without immigration, but where density would decrease relative to carrying capacity without immigration from sources.

Q methodology: A methodology that combines qualitative and quantitative methods to investigate the subjective views of those directly involved in a particular topic (Millar *et al.*, 2022).

Quadrat: A frame used to isolate a standard unit of area for the study of the distribution of an organism over a large area. See also **plot**.

***r*-strategists:** Organisms that follow a reproductive strategy characterised by producing a large number of offspring, but investing relatively little time or energy into each one's survival and development. Compare with ***K*-strategists**.

Random distribution: A spatial arrangement of individuals within a population where each individual's location is independent of the others (some individuals may be close together while others may be far apart). Compare with **clumped distribution** and **uniform distribution**.

Rank abundance curves (or rank abundance plots): A representation of **species abundance distribution**, in which species' abundances (*y*-axis) are plotted against their rank order (*x*-axis).

Rare species: Species that occur in small populations, in limited geographical areas or in particular habitats.

RBH: See **resource-based habitat**.

RBHomics: The study of the sum constituents within habitats.

Realised niche: The set of resources actually used and the conditions experienced by a given organism after accounting for interactions with other organisms (such as predation and especially competition) and random factors. See also **niche**.

Realised resource-based habitat: The set of resources and conditions within a site that actually maintain a local population of a species.

Red list (or red data book): A public document mainly created for recording imperilled species of plants, animals and fungi. A global red list is produced and regularly updated by the International Union for Conservation of Nature (IUCN). It contains colour-coded information about species' conservation status: red for species that are critically endangered, orange for endangered, yellow for vulnerable, light green for near threatened and dark green for least concern. Extinct species are coded as black, species extinct in the wild as dark purple, regionally extinct species as medium purple-pink. Light grey is used for species for which there is no sufficient data. White indicates not evaluated species. The same scheme is used in regional red lists, with the addition of the category 'not applicable' (light greyish brown; e.g. introduced species).

Regeneration complex: The turnover of plants within a **sere** (e.g. growth of new plants in the form of a mini-succession with the fall of trees).

Regression to the mean: The statistical phenomenon (also known as reversion toward the mean or reversion to mediocrity) where if one sample of a random variable is extreme, the next sampling of the same random variable is likely to be closer to its mean. When numerous random variables are sampled and the most extreme results are intentionally picked out, it refers to the observation that (usually) a second sampling of these picked-out variables will result in 'less extreme' results, closer to the initial mean of all of the variables.

Reinforcement: The deliberate introduction of additional individuals into an existing population of a species to increase its genetic diversity and population size. Also known as augmentation, supplementation, restocking or enhancement (plants only). Compare with **conservation introductions** and **reintroduction**.

Reintroduction: The deliberate release of individuals of a species into an area where it previously existed but has become extinct. Compare with **conservation introductions** and **reinforcement**.

Relaxation: The gradual reduction in species richness and population sizes towards an eventual (hypothetical) lower equilibrium, after a habitat/area has been isolated, fragmented or otherwise disturbed. This process is often a consequence of the loss of connectivity between **ecosystems** and the inability of species to sustain viable populations due to diminished suitable habitat area, reduced resources or increased negative effects along the edge.

Resilience (or ecosystem resilience): The ability of an ecosystem to return to its original state (*sensu lato*) following perturbation (**disturbance**). Compare with **resistance**.

Resistance (or ecosystem resistance): The ability of an ecosystem to persist or withstand a disturbance. Compare with **resilience**.

Resource-based habitat (RBH): The set of resources and conditions within a site sufficient to maintain the local population of a species (Dennis and Shreeve, 1996). This is similar to the definition provided for 'habitat' alone by Block and Brennan (1993). The concept can be more broadly formulated as follows: a RBH is that geolocated collection (set) of resources and conditions sufficient to maintain a population or cohort of a species. This refers to resources and conditions in *all* Earth layers of activity (**lithosphere, hydrosphere, atmosphere**).

Resource dispersion hypothesis (RDH): The RDH asserts that where resources are heterogeneous in space or time, group living might be less costly than living in smaller social units (i.e. a pair) regardless of whether individuals gain direct benefits from group membership (Johnson *et al.*, 2002).

Resource dynamics: The temporal and spatial variations in the availability, quality and distribution of resources that are essential for the survival and reproduction of local populations. These dynamics reflect both predictable patterns (e.g. daily and seasonal cycles) and unpredictable changes (e.g. extreme weather events or long-term ecological succession).

Resource-holding potential (or resource-holding power): The capacity of an animal to maintain or gain control of a resource, such as territory, food or mates, against rivals. The resource-holding potential is often determined by factors such as size, strength, stamina, fighting ability, social status and previous experience.

Resource partitioning: The process by which different species or individuals within a population use different resources, or the same resources in different ways or at different times to reduce competition. By partitioning resources, species can coexist in the same habitat, despite sharing similar ecological **niches**, because they minimise direct **competition** for the same resources.

Retrogressive succession: A type of ecological succession in which the community becomes less complex and less productive over time, often due to environmental degradation or disturbances. Compare with **progressive succession**. See also **ecological succession**.

Rewilding: The restoration of an area of land to a natural (uncultivated) state; often used with reference to the **reintroduction** of species of wild animals that have been driven out or exterminated from the area.

Rhithron: The upper, fast-flowing and cooler reaches of a river or stream. See also **crenon** and **potamon**.

Rhizosphere: The soil region in the vicinity of plant roots in which the chemistry and microbiology are influenced by their growth, respiration and nutrient exchange.

Rhodoliths: Free-living nodules that are dominantly composed of calcifying red algae (i.e. Corallinaceae, occasionally Peyssonneliaceae) commonly associated with other encrusters (i.e. foraminifera, bryozoans, bivalves, serpulid worms or corals). They vary in expanse from less than 30 mm to about 300 mm (Bosellini and Ginsburg, 1971).

Richness: See **species richness**.

Risk hypothesis: The hypothesis that, in species where territorial confrontations can be lethal, the risk of fatal encounters leads individuals to avoid areas where they might meet neighbouring groups.

Rule: A principle or statement that describes a consistent, observed pattern or relationship in nature, often based on empirical evidence or repeated experimentation. A rule is typically seen as a generalisation or guiding concept that applies under specific conditions, but it may not always explain the underlying mechanisms or processes. A rule often applies to specific situations and may not be universally applicable (rules have exceptions). Compare with **fact, law** and **theory**.

Sample coverage: The proportion of individuals that belong to observed species in a sample. Equal sample coverage is a widely accepted standard for comparing diversity across multiple samples (assemblages, communities). Abundance-based sample coverage, also known as Turing's sample coverage, calculates the proportion of individuals in a community that are part of the observed species within a sample. With a fixed size of the sampling unit (e.g. plot), incidence-based sample coverage evaluates the proportion of overall occurrences attributable to identified species.

Saproxylic: Related to the **deadwood** in some way. Saproxylic organisms are a wide trophic category that includes not only saproxylophagous (deadwood-eater) species, but also fungi that live on deadwood, predators, parasites and parasitoids of all the organisms living together in the same **microhabitat**, as well as several sap-feeding insects associated with yeasts and bacteria on living trees wounded by xylophagous (wood-eater) insects.

Satyric mimicry: A form of mimicry in which an animal combines features very imperfectly resembling a model with harmless traits to confuse predators, causing delays in their decision making and giving the prey more time to escape.

Secondary succession: Vegetation succession restarts from an earlier **sere**, with later seres removed from the site by some form of disturbance. Compare with **primary succession**. See also **ecological succession**.

Self-thinning rule: A process of plant death and takeover, such that biomass increases as **density** of plant individuals decreases, classically observed in conifers. The log-log relationship is a straight line following 'canopy' closure, the slope of the line (negative) being species-specific and dependent on conditions (White, 1980; Westoby, 1984; Brunet-Navarro *et al.*, 2016).

Semiosphere: The sphere of semiosis (meaning making) in which sign processes operate in the set of all interconnected Umwelten (an organism's unique perception of the world).

Sentinel species: Organisms that reveal the quality of health in their ecosystems by reacting to specific (usually negative) conditions (e.g. by showing population drops, increased deaths, decreased health status or site abandonment in response to land transformation, presence of pollutants and anthropogenic disturbance).

Seral stage: See **sere**.

Sere: A complete **succession** of plant **communities**, which ideally results in the **climax** community. These different communities are indicated as seral stages or seral communities.

Shannon diversity (index): A mathematical measure used to quantify the diversity of a set of species (e.g. a community), considering both the number of species (**richness**) and the **evenness** of their abundances (how equally individuals are distributed among species). The Shannon index is also widely applied in other contexts to assess heterogeneity or complexity (for example, in evaluating environmental diversity).

Shifting baseline syndrome: The gradual change in the accepted norms for the condition of the natural environment due to a lack of experience, memory or knowledge of its past conditions (Pauly, 1995; Dayton *et al.*, 1998; Papworth *et al.*, 2008).

Simpson index: A measure of biodiversity that quantifies the probability that two randomly selected individuals from a sample (e.g. a community) do not belong to the same species.

Sink (population): A local population that would go extinct in the absence of immigration. Compare with **source**.

Sociality: The degree to which individuals in an animal population tend to associate in social groups (gregariousness) and form co-operative societies. See also **eusociality**.

Social parasitism: A form of **parasitism** where one insect species exploits the social structure and resources of another species (typically a closely related social insect species) for its own benefit. The parasitic insect relies on the host's labour, resources or care, without contributing to the maintenance of the colony or providing any benefits in return. This type of parasitism is common in ants, social bees and social wasps.

Social structure: Refers to the patterns of social relationships between individuals in the same group, which is influenced by which sex disperses from the natal group. Usually, a species with femalebiased dispersal will show stronger relationships between males compared to females, while a species with malebiased dispersal will show strong bonding between females. If both sexes disperse, there may be strong relationships between the sexes.

Soil: The uppermost layer of the Earth's crust. It is a complex and dynamic medium composed of a mixture of minerals, organic matter, water and air, which interact to create a living system.

Source (population): A local population where growth rates at low density are positive, providing immigrants for sink populations. Compare with **sink**.

Specialism: The adaptation of an organism or species to a narrow set of environmental conditions and/or a specific type of resource. Compare with **generalism**. See also **specialism–generalism continuum**.

Specialism–generalism continuum: The range of ecological strategies that organisms or species can exhibit, from being highly specialised (specialists) to being broadly adaptable (generalists). This continuum helps to categorise organisms and species based on their ability to utilise resources, their habitat requirements and their adaptability to changing environmental conditions.

Specialist: An organism or species that relies on a limited set of resources (or even a single particular resource) and/or has a narrow range of ecological tolerances, thus being able to thrive only under specific environmental conditions. Compare with **generalist**. See also **specialism–generalism continuum**.

Species abundance distribution (SAD): The distribution of abundances of species within a given assemblage or community. A consistent general pattern in SAD is the presence of many **rare species** and few abundant species.

Species–area relationship (SAR): The relationship between area and the number of species found in that area. Typically, the number of species increases with the size of the area. This is one of the most universal ecological patterns.

Species distribution modelling (SDM): An approach used to predict the distribution of a species across geographical areas based on environmental (especially climatic) variables and known occurrences of the species. It involves the use of computational tools to identify the areas where a species is likely to be found under current or future environmental conditions. The approach can also be used to reconstruct past distributions.

Species' habitat: The physical and environmental space occupied by a local population of a species. Thus, the occupation of a named traditional habitat unit(s) by a local population (sufficient individuals to breed successfully to produce a following generation) of a particular species.

Species richness: The absolute number of species in a sample, assemblage, community, habitat, **landscape** or region at any point in time; often shortened to '**richness**'. See also **diversity** and **Hill numbers**.

Species segregation: The spatial separation or non-overlapping distributions of different species within a habitat or ecosystem.

Substrate: A surface (either biotic or abiotic) on which an organism lives and may use as food.

Subterranean habitats: Habitats located beneath the Earth's surface, including caves, underground water systems (e.g. aquifers, subterranean rivers and lakes, springs) and soil beneath the surface.

Super-organism, super-organism concept: A highly integrated, collective entity composed of multiple individual organisms that function together as a cohesive unit. Colonies of ants, bees, termites and wasps are classic examples of superorganisms. In these **eusocial** insects, a superorganism is a species that functions as an organised society. Other examples of superorganisms are coral reefs and siphonophores.

The superorganism concept in community ecology is the idea that an ecological **community** functions as an integrated and highly co-ordinated 'organism', much like a single, living organism. In this view, the interactions among species within a community are so tightly interwoven that the community behaves as a collective entity with emergent properties that cannot be understood simply by examining individual species. It can be considered an extremisation of the **holistic concept** of community.

Supersaturation: A condition where an environment contains more species than the available resources can support.

Sympatric, sympatry: Occurring in the same or overlapping geographical areas.

Sympatric speciation: The formation of a new species from a surviving ancestral species while both continue to inhabit the same geographical area. Mechanisms involved may include resource specialisation, behavioural isolation and hybridisation. Compare with **allopatric speciation.**

Synecology, synecological: The study of the relationships between groups of organisms or species and between them and their environment. Compare with **autecology**.

Taxocene, taxocenose: A group of taxonomically related organisms in a community that do not necessarily share the same or similar niches (e.g. an insect assemblage).

Taxocenose: See **taxocene**.

Territory: A physical area no larger than the **home range** in which some resource(s) (biotic such as food and mates, or abiotic such as nest sites and water) is/are defended by one or more individuals (troop, group, herd, etc.) of a species.

Theory: A well-substantiated explanation of some aspect of the natural world that is based on a body of evidence and has undergone extensive testing and validation through observation, experimentation and logical reasoning. Compare with **fact**, **law** and **rule**.

Theory of hierarchy of needs: A psychological theory that suggests human beings have a set of needs that are arranged in a hierarchical order. According to this theory, individuals must satisfy lower-level needs before they can move on to higher-level needs. This theory is often depicted as a pyramid, with basic, fundamental needs at the bottom and more complex, self-fulfilling needs at the top.

Tolerance (in ecological succession): A model of ecological succession in which early successional (pioneer) species neither inhibit nor facilitate the growth and success of later-successional species. Later species can grow to maturity in the presence of early species, eventually outcompeting them. The sequence of succession is entirely dependent on species life-history characteristics. Compare with **facilitation** and **inhibition**.

Town: A human settlement larger than a **village** but smaller than a **city**.

Trait: A specific characteristic of an organism that can be observed or measured. Traits are determined by genetic factors, environmental factors or a combination of both. See also **functional traits**.

Transect: A straight line or path along which ecological observations or measurements are made.

Transition zones: Areas where two distinct **biogeographical regions** meet and intergrade.

Translocation: Human-mediated movement of living organisms from one area, with release in another.

Trophallaxis: The exchange of food between two individuals. The food exchanged may be salivary secretions or regurgitated gut contents.

Trophic level: The position an organism occupies in a **food chain**, determined by the number of energy transfer steps to reach that level. Five trophic levels are usually recognised. (1) Primary producers: **autotrophic** organisms (such as plants, algae and some bacteria) that produce their own energy through photosynthesis or chemosynthesis. (2) Primary consumers: herbivores that feed directly on primary producers. (3) Secondary consumers: carnivores or omnivores that consume primary consumers. (4) Tertiary and higher-level consumers: predators at the top of the food chain that feed on other carnivores or omnivores. A further category is represented by decomposers: organisms (such as fungi and bacteria) that recycle nutrients by breaking down dead organic matter.

Umbrella species: Species assumed or known to correlate (positively) with a suite of species in a particular area or traditional habitat space. They are of particular value in designation of nature reserves (Naskrecki, 2013).

Uniform distribution: A spatial arrangement of individuals where they are evenly spaced. This pattern arises when individuals maintain consistent distances from one another, often due to competition for limited resources or territorial behaviour. Compare with **clumped distribution** and **random distribution**.

Urban area: A developed area usually with a central business district core, including human structures, such as houses, commercial buildings, industry, roads, bridges and railways.

Utility resource: Any environmental element that organisms can use without depleting it (e.g. physical sites for mating, basking in the sun, resting and taking shelter from predators).

Vacant niche hypothesis: A vacant niche implies that niche spaces exist that are not being used. In other words, in ecosystems or habitats more species could exist than are present at a particular point in time, because all 'niche spaces' are not being used by existing species (Rohde, 2006; Gibson-Reinemer, 2015).

Vagrant, vagrancy: An individual of a species found far outside its normal range.

Veblen goods: A type of luxury goods for which demand increases as their price increases, contrary to the typical law of demand in economics, which states that as the price of a good rises, the demand for it generally falls. Named after economist Thorstein Veblen (1857–1929), who introduced the concept in the early 20th century, Veblen goods are considered to have a status symbol value, where higher prices make them more desirable because they signal wealth, prestige and social status. Examples of Veblen goods include exclusive real estates, luxury cars, yachts, designer clothing and accessories (especially jewellery) and celebrity-endorsed perfumes.

Vegetational belts: Distinct vegetation types distributed along an elevational gradient.

Vegetation succession: A subset of ecological succession that specifically focuses on the changes in plant communities over time. See **ecological succession**.

Veil line: A boundary or 'veil' in a **histogram** representing a **species abundance distribution** that separates observable species from those that are too rare to detect due to sampling limitations.

Vicariance: A biogeographical process in which a species' range is divided into separate, isolated populations due to the formation of a geographical or environmental barrier. This separation often leads to speciation over time, as the isolated populations adapt to their distinct environments and accumulate genetic differences.

Vicinism: The occurrence of species in areas that are outside, but close to, those where they are typically found, due to the dispersal of propagules from the most favourable zones.

Village: A small human settlement, typically located in a rural area. It is larger than a **hamlet** but smaller than a **town**.

Visitor: A taxon that occurs in a particular area without being a permanent resident. These taxa do not reproduce or establish stable populations in the area but are present for specific activities or during certain times.

Zonobiomes: Biomes influenced by macroclimate.

Zoogeographical regions: Large, distinct areas of the Earth that are characterised by specific patterns of animal distribution.

Zoogeography, zoogeographical: The study of the distribution of animals.

References

Aarons, S.R., O'Connor, C.R. and Gourley, C.J.P. (2004) Dung decomposition in temperate dairy pastures. I. Changes in soil chemical properties. *Australian Journal of Soil Research*, 42, 107–114.

Aarts, B.G.W. and Nienhuis P.H. (2003) Fish zonations and guilds as the basis for assessment of ecological integrity of large rivers. *Hydrobiologia*, 500, 157–178.

Aas, Ø., Klemetsen, A., Einum, S. and Skurdal, J. (eds) (2010) *Atlantic Salmon Ecology*. John Wiley and Sons, Chichester, UK.

Abernethy, K.A., White, L.J.T. and Wickings, E.J. (2002) Hordes of mandrills (*Mandrillus sphinx*): Extreme group size and seasonal male presence. *Journal of Zoology*, 258, 131–137.

Acharya, P.R., Racey, P.A., Sotthibandhu, S. and Bumrungsri, S. (2015) Home-range and foraging areas of the dawn bat *Eonycteris spelaea* in agricultural areas of Thailand. *Acta Chiropterologica*, 17, 307–319.

Ackerly, D. and Cornwell, W. (2007) A trait-based approach to community assembly: Partitioning of species trait values into within- and among-community components. *Ecology Letters*, 10, 135–145.

Ackermann, H.W. (2006) Classification of bacteriophages. In: Calendar, R. (ed.) *The Bacteriophages*. Oxford University Press, New York, USA, pp. 8–17.

Ackermann, H.-W. (2011) Bacteriophage taxonomy. *Microbiology Australia*, 32, 90–94.

Acosta, A., Blasi, C., Carranza, M. L., Ricotta, C. and Stanisci, A. (2003) Quantifying ecological mosaic connectivity and hemeroby with a new topoecological index. *Phytocoenologia*, 33, 623–631.

Adams, C.C. (1908) The ecological succession of birds. *The Auk*, 25, 109–153.

Adams, S.E. (2001) Approaches to the study of territory size and shape. *Annual Review of Ecology, Evolution, and Systematics*, 32, 277–303.

Adl, S.M., Leander, B.S., Simpson, A.G., Archibald, J.M., Anderson, O.R. *et al.* (2007) Diversity, nomenclature, and taxonomy of protists. *Systematic Biology*, 56, 684–689.

Adl, S.M., Simpson, A.G.B., Farmer, A.M., Andersen, R.A., Anderson, O.R. *et al.* (2005) The new higher level classification of eukaryotes with emphasis on the taxonomy of protists. *Journal of Eukaryotic Microbiology*, 52, 399–451.

Adler, D.S. and Bar-Oz, G. (2009) Seasonal patterns of prey acquisition and inter-group competition during the Middle and Upper Palaeolithic of the Southern Caucasus. In: Hublin, J.-J. and Richards, M.P. (eds) *The Evolution of Hominin Diets: Integrating Approaches to the Study of Palaeolithic Subsistence*. Springer Science+Business Media, Dordrecht, The Netherlands, pp. 127–140.

Adler, D.S., Wilkinson, K.N., Blockley, S., Mark, D.F., Pinhasi, R. *et al.* (2014) Early Levallois technology and the Lower to Middle Paleolithic transition in the Southern Caucasus. *Science*, 345, 1609–1613.

Adler, F.R. and Tanner, C.J. (2013) *Urban Ecosystems: Ecological Principles for the Built Environment*. Cambridge University Press, Cambridge, UK.

Adler, P.B., Hillerislambers, J. and Levine, J.M. (2007) A niche for neutrality. *Ecology Letters*, 10, 95–104.

Agarwal, R., Gupta, M., Antony, A., Sen, R. and Raychoudhury, R. (2021) In vitro studies reveal that *Pseudomonas*, from *Odontotermes obesus* colonies, can function as a defensive mutualist as it prevents the weedy fungus while keeping the crop fungus unaffected. *Microbial Ecology*, 84, 391–403.

Agosti, D., Majer, J.D., Alonso, L.E., Schultz, T.R. (2000) *Ants: Standard Methods for Measuring and Monitoring Biodiversity*. Smithsonian Institution Press, Washington DC, USA.

Agrawal, A.A. (2017) *Monarchs and Milkweed. A Migrating Butterfly, a Poisonous Plant, and Their Remarkable Story of Coevolution*. Princeton University Press, Princeton, USA.

Aguirre, A.A. and Tabor, G.M. (2004) Marine vertebrates as sentinels of marine ecosystem health. *EcoHealth*, 1, 236–238.

Ahlström, T. (2012) *Sticks, Stones, and Broken Bones: Neolithic Violence in a European Perspective*. Oxford University Press, Oxford, UK.

Ahmed, S.K. and Rahman, M.S. (2023) Protecting the Bengal tiger in the Sundarbans, Bangladesh: Conservation measures reconsidered. *International Journal of Environmental Studies*, 80, 1283–1293.

Ahnesjö, J. and Forsman, A. (2006) Differential habitat selection by pygmy grasshopper color morphs: interactive effects of temperature and predator avoidance. *Evolutionary Ecology*, 20, 235–257.

Akçakaya, H.R., Mills, G. and Doncaster, C.P. (2007) The role of metapopulations in conservation. In: Macdonald, D.W. and Service, K. (eds) *Key Topics in Conservation Biology*. Blackwell Publishing, Oxford, UK, pp. 64–84.

Åkesson, S. and Bianco, G. (2021) Wind-assisted sprint migration in northern swifts. *iScience*, 24, 102474.

Alatalo, R.V. (1981) Problems in the measurement of evenness in ecology. *Oikos*, 37, 199–204.

Alba, A., Grech-Angelini, S., Vazquez, A.A., Alda, P., Blin, Q. *et al.* (2023) Fasciolosis in the Mediterranean island of Corsica (France): Insights from epidemiological and malacological investigations. *Food and Waterborne Parasitology*, 30, e00188.

Albert, C.H., Thuiller, W., Yoccoz, N.G., Douzet, R., Aubert, S. and Lavorel, S. (2010a) A multi-trait approach reveals the structure and the relative importance of intra- vs. interspecific variability in plant traits. *Functional Ecology*, 24, 1192–1201.

Albert, C.H., Thuiller, W., Yoccoz, N.G., Soudant, A., Boucher, F. *et al.* (2010b) Intraspecific functional variability: Extent, structure and sources of variation. *Journal of Ecology*, 98, 604–613.

Albert, C.H., Grassein, F., Schurr, F.M., Vieilledent, G. and Violle, C. (2011) When and how should intraspecific variability be considered in trait based plant ecology? *Perspectives in Plant Ecology, Evolution and Systematics*, 13, 217–225.

Alberti, M., Correa, C., Marzluff, J.M., Hendry, A.P., Palkovacs, E.P. *et al.* (2017) Global urban signatures of phenotypic change in animal and plant populations. *Proceedings of the National Academy of Sciences of the United States of America*, 114, 8951–8956.

Alderton, D. (2017) *The Complete Illustrated Encyclopedia of Birds of the World: A Detailed Visual Reference Guide to 1600 Birds and Their Habitats*. Lorenz Books, London, UK.

Alexander, R. and Moore, T. (1962) *The Evolutionary Relationships of 17-Year and 13-Year Cicadas, and Three New Species (Homoptera, Cicadidae, Magicicada)*. Miscellaneous Publications of the Museum of Zoology, University of Michigan No. 121. University of Michigan Press, Ann Arbor, USA.

Alfaro, E., Pérez-Tello, V., Acevedo, M., Ovalle, J., Segovia, R., and Craven, D. (2023) Rasgos-CL: A functional trait database of Chilean woody plants. *Global Ecology and Biogeography*, 32, 2072–2084.

Alho, J.S., Herczeg, G., Söderman, F., Laurila, A., Jönsson, K.I. and Merilä, J. (2010) Increasing melanism along a latitudinal gradient in a widespread amphibian: Local adaptation, ontogenic or environmental plasticity? *BMC Evolutionary Biology*, 10, 317.

Allee, W.C. (1934) Concerning the organization of marine coastal communities. *Ecological Monographs*, 4, 541–554.

Allee, W.C., Park, O., Emerson, A.E., Park, T. and Schmidt, K.P. (1949) *Principles of Animal Ecology*. W.B. Saunders, London, UK.

Allen, J.A. (1871) On the mammals and winter birds of East Florida. *Bulletin of the Museum of Comparative Zoology*, 2, 161–450.

Allen, M.L. and Krofel, M. (2017) Resource holding potential. In: Vonk, J. and Shackelford, T. (eds) *Encyclopedia of Animal Cognition and Behavior*. Springer, Cham, Switzerland.

Alleway, H.K., Klein, E.S., Cameron, L., Douglass, K., Govia, I. *et al.* (2023) The shifting baseline syndrome as a connective concept for more informed and just responses to global environmental change. *People and Nature*, 5, 885–896.

Almécija, S., Hammond, A.S., Thompson, N.E., Pugh, K D., Moyà-Solà, S. and Alba, D.M. (2021) Fossil apes and human evolution. *Science*, 372, eabb4363.

Almeida, J. and Granadeiro, J.P. (2000) Seasonal variation of foraging niches in a guild of passerine birds in a cork-oak woodland. *Ardea*, 88, 243–252.

Almeida-Neto, M. and Ulrich, W. (2011) A straightforward computational approach for measuring nestedness using quantitative matrices. *Environmental Modelling and Software*, 26, 173–178.

Almeida-Neto, M., Guimaraes, P., Guimaraes, P.R., Loyola, R.D. and Ulrich, W. (2008) A consistent metric for nestedness analysis in ecological systems: Reconciling concept and measurement. *Oikos*, 117, 1227–1239.

Almeida-Neto, M., Frensel, D.M.B. and Ulrich, W. (2012) Rethinking the relationship between nestedness and beta diversity: A comment on Baselga 2010. *Global Ecology and Biogeography*, 21, 772–777.

Alonso, D. and McKane, A.J. (2004) Sampling Hubbell's neutral theory of biodiversity. *Ecology Letters*, 7, 901–910.

Alroy, J. (1999) The fossil record of North American mammals: Evidence for a Paleocene evolutionary radiation. *Systematic Biology*, 48, 107–118.

Alroy, J. (2010) The shifting balance of diversity among major marine animal groups. *Science*, 329, 1191–1194.

Alroy, J. (2015) Current extinction rates of reptiles and amphibians. *Proceedings of the National Academy of Sciences of the United States of America*, 112(42), 13003–13008.

Alroy, J., Aberhan, M., Bottjer, D.J., Foote, M., Fürsich, F.T. *et al.* (2008) Phanerozoic trends in the global diversity of marine invertebrates. *Science*, 321, 97–100.

Als, T.D., Vila, R., Kandul, N.P., Nash, D.R., Yen, S-H. *et al.* (2004) The evolution of alternative parasitic life histories in large blue butterflies. *Nature*, 432, 386–390.

Alsante, A.N., Thornton, D.C.O. and Brooks, S.D. (2021) Ocean aerobiology. *Frontiers in Microbiology*, 12, 764178.

Al-Shayeb, B., Sachdeva, R., Chen, L.X., Ward, F., Munk, P. *et al.* (2020) Clades of huge phages from across Earth's ecosystems. *Nature*, 578, 425–431.

Ammann, L., Bosem-Baillod, A., Herzog, F., Frey, D., Entling, M.H. and Albrecht, M. (2024) Spatio-temporal complementarity of floral resources sustains wild bee pollinators in agricultural landscapes. *Agriculture, Ecosystems and Environment*, 359, 108754.

Amos, W. and Harwood, J. (1998) Factors affecting levels of genetic diversity in natural populations. *Philosophical Transactions of the Royal Society B: Biological Sciences*, 353, 177–186.

Andersen, S.B., Bruhn, J.N., Kasimer, D., Wang, H., Rodrigue, N. and Smith, M.L. (2018) Clonal evolution and genome stability in a 2,500-year-old fungal individual. *Proceedings of the Royal Society B: Biological Sciences*, 285(1893), 20182233.

Anderson, B.J., Chiarucci, A. and Williamson, M. (2012) How differences in plant abundance measures produce different species-abundance distributions. *Methods in Ecology and Evolution*, 3, 783–786.

Anderson, G.S. and VanLaerhoven, S.L. (1996) Initial studies on insect succession on carrion in southwestern British Columbia. *Journal of Forensic Sciences*, 41, 617–625.

Anderson, M.J. (2006) Distance-based tests for homogeneity of multivariate dispersions. *Biometrics*, 62, 245–253.

Anderson, M.J., Crist, T.O., Chase, J.M., Vellend, M., Inouye, B.D. *et al.* (2011) Navigating the multiple meanings of β diversity: A roadmap for the practicing ecologist. *Ecology Letters*, 14, 19–28.

Anderwald, D., Czajka, Ł., Rubacha, S., Zygmunt, M. and Mirski, P. (2021) Autumn migration of ospreys from two distinct populations in Poland reveals partial migratory divide. *Avian Research*, 12, 46.

Andrewartha, H.G. and Birch, L.C. (1954) *The Distribution and Abundance of Animals*. University of Chicago Press, Chicago, USA.

Andrewartha, H.G. and Birch, L.C. (1984) *The Ecological Web*. University of Chicago Press, Chicago, USA.

Antão, H.A., Magurran, E.A., and Dornelas, S.M. (2021) The shape of species abundance distributions across spatial scales. *Frontiers in Ecology and Evolution*, 9, 626730.

Antipova, A. (2018) *Urban Environment, Travel Behavior, Health, and Resident Satisfaction*. Palgrave Macmillan, London, UK.

Antón, S.C., Taboada, H.G., Middleton, E.R., Rainwater, C.W., Taylor, A.B. *et al.* (2016) Morphological variation in *Homo erectus* and the origins of developmental plasticity. *Philosophical Transactions of the Royal Society B: Biological Sciences*, 371, 20150236.

Antonelli, A., Fry, C., Smith, R.J., Eden, J., Govaerts, R.H.A. *et al.* (2023) *State of the World's Plants and Fungi*. Royal Botanic Gardens, Kew, Richmond, UK.

Aoki, I., Yamakawa, T., and Takasuka, A. (2018) *Fish Population Dynamics, Monitoring, and Management*. Springer, Tokyo, Japan.

Apolinário, M., Coutinho, R. and Baeta-Neves, M. (1999) Periwinkle (Gastropoda: Littorinidae) habitat selection and its impact on microalgal populations. *Revista Brasileira de Biologia*, 59, 211–218.

Arai, T. (2020) Ecology and evolution of migration in the freshwater eels of the genus *Anguilla* Schrank, 1798. *Heliyon*, 6, e05176.

Araújo, M.B. (2022) Biodiversity hotspots and zones of ecological transition. *Conservation Biology*, 16, 1662–1663.

Araujo, S.B.L., Braga, M.P., Brooks, D.R., Agosta, S.J., Hoberg, E.P. *et al.* (2015) Understanding host-switching by ecological fitting. *PLOS One*, 10, e0139225.

Archer, M.S. and Elgar, M.A. (2003) Yearly activity patterns in southern Victoria (Australia) of seasonally active carrion insects. *Forensic Science International*, 132, 173–176.

Ardura, A. (2019) Species-specific markers for early detection of marine invertebrate invaders through eDNA methods: Gaps and priorities in GenBank as database example. *Journal for Nature Conservation*, 47, 51–57.

Arellano, G., Umaña, M.N., Macía, M.J., Loza, M.I., Fuentes, A. *et al.* (2017) The role of niche overlap, environmental heterogeneity, landscape roughness and productivity in shaping species abundance distributions along the Amazon-Andes gradient. *Global Ecology and Biogeography*, 26, 191–202.

Arim, M. and Marquet, P.A. (2004) Intra-guild predation: A widespread interaction related to species biology. *Ecological Letters*, 7, 557–564.

Arita, H.T., Robinson, J.G. and Redford, K.H. (1990) Rarity in neotropical forest mammals and its ecological correlates. *Conservation Biology*, 4, 181–192.

Arnaldos, I., Romera, E., García, M.D. and Luna, A. (2001) An initial study on the succession of sarcosaprophagous Diptera (Insecta) on carrion in the southeastern Iberian Peninsula. *International Journal of Legal Medicine*, 114, 156–162.

Aschenbrenner, I., Cernava, T., Berg, G. and Grube, M. (2016) Understanding microbial multi-species symbioses. *Frontiers in Microbiology*, 7, 180.

Asher, J., Warren, M., Fox, R., Harding, P., Jeffcoate, G. and Jeffcoate, S. (2001) *The Millennium Atlas of Butterflies in Britain and Ireland*. Oxford University Press, Oxford, UK.

Ashmole, M.J. (1962) The migration of European thrushes: A comparative study based on ringing recoveries. *Ibis*, 104, 522–559.

Ashton, K.G., Tracy, M., de Queiroz, A. and Garland, T. Jr. (2013) Is Bergmann's rule valid for mammals? *American Naturalist*, 156, 390–415.

Asiandu, A.P., Wahyudi, A. and Sari, S.W. (2021) A review: Plastics waste biodegradation using plastics-degrading bacteria. *Journal of Environmental Treatment Techniques*, 9, 148–157.

Atkinson, A., Siegel, V., Pakhomov, E.A., Jessopp, M.J. and Loeb, V. (2009) A re-appraisal of the total biomass and annual production of Antarctic krill. *Deep Sea Research Part I: Oceanographic Research Papers*, 56, 727–740.

Atlas, R.M. and Bartha, R. (1998) *Microbial Ecology: Fundamentals and Applications*, 4th edn. Benjamin/Cummings, San Francisco, USA.

Atmar, W. and Patterson, B.D. (1993) The measure of order and disorder in the distribution of species in fragmented habitat. *Oecologia*, 96, 373–382.

Attrill, M.J. and Rundle, S.D. (2002) Ecotone or ecocline; ecological boundaries in estuaries. *Estuarine, Coastal and Shelf Boundaries Science*, 55, 926–936.

Auckland, J.N., Debinski, D.M. and Clark, W.R. (2004) Survival, movement, and resource use of the butterfly *Parnassius clodius*. *Ecological Entomology*, 29, 139–149.

Auger, S. and Shipley, B. (2013) Inter-specific and intra-specific trait variation along short environmental gradients in an old-growth temperate forest. *Journal of Vegetation Science*, 24, 419–428.

Aune, K., Jørgensen, D. and Gates, C. (2017) *Bison bison* (errata version published in 2018). *The IUCN Red List of Threatened Species 2017*, e.T2815A123789863. Available at: www.iucnredlist.org/species/2815/123789863 (accessed 1 June 2025).

Austen, P. and Austen, J. (2016) Carboniferous coal forests. *Hastings and District Geological Society Journal*, 22, 19–26.

Auta, H.S., Emenike, C.U. and Fauziah, S.H. (2017) Distribution and importance of microplastics in the marine environment: A review of the sources, fate, effects, and potential solutions. *Environment International*, 102, 165–176.

Bach, E.M., Ramirez, K.S., Fraser, T.D. and Wall, D.H. (2020) Soil biodiversity integrates solutions for a sustainable future. *Sustainability*, 12, 2662.

Bacher, M., Fenton, M.G., Bondi, O., Creamer, G., Karmarkar, R.E. and Schmidt, O. (2018) The impact of cattle dung pats on earthworm distribution in grazed pastures. *BMC Ecology and Evolution*, 18, 59.

Backhed, F. (2005) Host-bacterial mutualism in the human intestine. *Science*, 307, 1915–1920

Backstrom, L.J., Callaghan, C.T., Leseberg, N.P., Sanderson, C., Fuller, R.A. and Watson, J.E.M. (2024) Assessing adequacy of citizen science datasets for biodiversity monitoring. *Ecology and Evolution*, 14, e10857.

Bacon, B., Khatiri, A., Palmer, J., Freeth, T., Pettitt, P. and Kentridge, R. (2022) An Upper Palaeolithic proto-writing system and phenological calendar. *Cambridge Archaeological Journal*, 33, 371–389.

Baglione, V., Marcos, J.M. and Canestrari, D. (2002) Cooperatively breeding groups of carrion crow (*Corvus corone corone*) in Northern Spain. *The Auk*, 119, 790–799.

Baguette, M., Petit, S. and Quéva, F. (2000) Population spatial structure and migration of three butterfly species within the same habitat network: Consequences for conservation. *Journal of Applied Ecology*, 37, 100–108.

Bahn, P.G. (2007) *Cave Art: A Guide to the Decorated Ice Age Caves of Europe*. Frances Lincoln. London, UK.

Bai, Y., Han, X., Wu, J., Chen, Z. and Li, L. (2004) Ecosystem stability and compensatory effects in the Inner Mongolia grassland. *Nature*, 431, 181–184.

Bailey, R.G. (2014) *Ecoregions – The Ecosystem Geography of the Oceans and Continents*, 2nd edn. Springer, New York, USA.

Bailey, V.A. (1931) The interaction between hosts and parasites. *Quarterly Journal of Mathematics*, 2, 68–77.

Baillie, J.E.M., Collen, B., Amin, R., Akcakaya, H.R., Butchart, S.H.M. *et al.* (2008) Toward monitoring global biodiversity. *Conservation Letters*, 1, 18–26.

Bainbridge, I.P. and Minton, C.D.T. (1978) The migration and mortality of the curlew in Britain and Ireland. *Bird Study*, 25, 39–50.

Bajerlein, D., Taberski, D. and Matuszewski, S. (2018) Estimation of postmortem interval (PMI) based on empty puparia of *Phormia regina* (Meigen) (Diptera: Calliphoridae) and third larval stage of *Necrodes littoralis* (L.) (Coleoptera: Silphidae) – Advantages of using different PMI indicators. *Journal of Forensic and Legal Medicine*, 55, 95–98.

Baker, J., Meade, A., Pagel, M., and Venditti, C. (2015) Adaptive evolution toward larger size in mammals. *Proceedings of the National Academy of Sciences of the United States of America*, 112, 5093–5098.

Baker, J.A., Kilgore, J.K. and Kasul, R.L. (1991) Aquatic habitats and fish communities in the lower Mississippi river. *Aquatic Sciences*, 3, 313–356.

Baker, R.R. (1978) *The Evolutionary Ecology of Animal Migration*. Hodder and Stoughton, London, UK.

Baldridge, E., Harris, D.J., Xiao, X., andWhite, E.P. (2016) An extensive comparison of species-abundance distribution models. *PeerJ*, 4, e2823.

Balmori, A. (2019) Endangered bird mortality by gunshots: Still a current problem. *Biodiversity and Conservation*, 28, 2555–2564.

Balter, M. (2005) First settlements: The Seeds of Civilization. *Smithsonian Magazine*, 5. Available at: www.smithsonianmag.com/history/the-seeds-of-civilization-78015429/ (accessed 30 April 2025).

Barbara, L. (2024) Resilience and the shift of paradigm in ecology: A new name for an old concept or a different explanatory tool? *History and Philosophy of the Life Sciences*, 46, 2.

Barbero, F., Thomas, J.A., Bonelli, S., Balletto, E. and Schönrogge, K. (2009) Queen ants make distinctive sounds that are mimicked by a butterfly social parasite. *Science*, 323, 782–785.

Bar-David, S., Bar-David, I., Cross, P.C., Sadie, J.R. Knechtel, C.U. and Wayne, M.G. (2009) Methods for assessing movement path recursion with application to African buffalo in South Africa. *Ecology*, 90, 2467–2479.

Barghini, A. and Souza de Medeiros, B.A. (2012) UV radiation as an attractor for insects. *Leukos – Journal of the Illuminating Engineering Society of North America*, 9, 47–56.

Barillot, C.D., Sarde, C.O., Bert, V., Tarnaud, E. and Cochet, N. (2013) A standardized method for the sampling of rhizosphere and rhizoplan soil bacteria associated to a herbaceous root system. *Annals of Microbiology*, 63, 471–476.

Barnett, A., Méléder, V., Dupuy, C. and Lavaud, J. (2020) The vertical migratory rhythm of intertidal microphytobenthos in sediment depends on the light photoperiod, intensity, and spectrum: Evidence for a positive effect of blue wavelengths. *Frontiers in Marine Science*, 7, 212.

Barnosky, A.D., Matzke, N., Tomiya, S., Wogan, G.O., Swartz, B. *et al.* (2011) Has the Earth's sixth mass extinction already arrived? *Nature*, 471, 51–57.

Bar-On, Y.M., Phillips, R. and Milo, R. (2018) The biomass distribution on Earth. *Proceedings of the National Academy of Sciences of the United States of America*, 115(25), 6506–6511.

Baroni-Urbani, C. and Buser, M.W. (1976) Similarity of binary data. *Systematic Zoology*, 25, 251–259.

Barras, C. (2018) Losing the plot. In: Brown, J. (ed.) *Becoming Human*, 2nd edn. New Scientist: The Collection, London, UK, pp. 10–15.

Bar-Yosef, O. (1998a) The Natufian culture in the Levant, threshold to the origins of agriculture. *Evolutionary Anthropology*, 6, 159–177.

Bar-Yosef, O. (1998b) On the nature of transitions: The Middle to Upper Palaeolithic and the Neolithic revolution. *Cambridge Archaeological Journal*, 8, 141–163.

Bar-Yosef, O., Vandermeersch, A., Arensberg, B., Belfer-Cohen, A., Goldberg, P. *et al.* (1992) The excavations in Kebara Cave, Mt. Carmel. *Current Anthropology*, 33, 497–550.

Bascompte, J. and Stouffer, D.B. (2009) The assembly and disassembly of ecological networks. *Philosophical Transactions of the Royal Society B: Biological Sciences*, 364, 1781–1787.

Bascompte, J., Jordano, P., Melián, C.J. and Olesen, J.M. (2003) The nested assembly of plant-animal mutualistic networks. *Proceedings of the National Academy of Sciences of the United States of America*, 100, 9383–9387.

Bascompte, J., Jordano, P. and Olesen, J.M. (2006) Asymmetric coevolutionary networks facilitate biodiversity maintenance. *Science*, 312, 431–433.

Baselga, A. (2010) Partitioning the turnover and nestedness components of beta diversity. *Global Ecology and Biogeography*, 19, 134–143.

Baselga, A. (2012) The relationship between species replacement, dissimilarity derived from nestedness, and nestedness. *Global Ecology and Biogeography*, 21, 1223–1232.

Baselga, A. (2013) Separating the two components of abundance-based dissimilarity: Balanced changes in abundance vs. abundance gradients. *Methods in Ecology and Evolution*, 4, 552–557.

Baskerville, E.B., Dobson, A.P., Bedford, T., Allesina, S., Anderson, T.M. and Pascual, M. (2011) Spatial guilds in the Serengeti food web revealed by a Bayesian group model. *PLOS Computational Biology*, 7, e1002321.

Baskin, Y. (1994) Ecologists dare to ask: How much does diversity matter? *Science*, 264, 202–203.

Bässler, C., Förster, B., Moning, C. and Müller, J. (2009) The BIOKLIM Project: Biodiversity research between climate change and wilding in a temperate montane forest – The conceptual framework. *Waldökologie, Landschaftsforschung und Naturschutz*, 7, 21–34.

Bastolla, U., Fortuna, M., Pascual-García, A., Ferrera, A., Luque, B. and Bascompte, J. (2009) The architecture of mutualistic networks minimizes competition and increases biodiversity. *Nature*, 458, 1018–21.

Bateson, G. (2000) *Steps to an Ecology of Mind*. Chicago University Press, Chicago, USA.

Batson, W., Gordon, I.J., Fletcher, D.B. and Manning, A.D. (2015) Translocation tactics: A framework to support the IUCN Guidelines for wildlife translocations and improve the quality of applied methods. *Journal of Applied Ecology*, 52, 1598–1607.

Baumhardt, R.L., Stewart, B.A. and Sainju, U.M. (2015) North American soil degradation: Processes, practices, and mitigating strategies. *Sustainability*, 7, 2936–2960.

Bautista-Cruz, A. and del Castillo, R.F. (2005) Soil changes during secondary succession in a tropical montane cloud forest area. *Soil Science Society of America Journal*, 69, 906–914.

Baxter, S. and Lewis, C.P. (2017) Domesday Book and the transformation of English landed society, 1066–86. *Anglo-Saxon England*, 46, 343–403.

Bayliss, A., Marshall, P., Richards, C. and Whittle, A. (2017) Islands of history: The Late Neolithic timescape of Orkney. *Antiquity*, 91, 1171–1188.

Bazzaz, F.A. (1975) Plant species diversity in old-field successional ecosystems in southern Illinois. *Ecology*, 56, 485–488.

Beale, D.J., Jones, O.A.H., Bose, U., Broadbent, J.A., Walsh, T.K. *et al.* (2022) Omics-based ecosurveillance for the assessment of ecosystem function, health, and resilience. *Emerging Topics in Life Sciences*, 6, 185–199.

Beaver, R. (1984) Insect exploitation of ephemeral habitats. *South Pacific Journal of Natural Science*, 8, 3–47.

Beck, H.E., Zimmermann, N.E., McVicar, T.R., Vergopolan, N., Berg, A. and Wood, E.F. (2018) Present and future Köppen-Geiger climate classification maps at 1-km resolution. *Scientific Data*, 5, 180214.

Beck, J. and Kitching, I.J. (2007) Estimating regional species richness of tropical insects from museum data: A comparison of a geography-based and sample-based methods. *Journal of Applied Ecology*, 44, 672–681.

Beck, J. and Schwanghart, W. (2010) Comparing measures of species diversity from incomplete inventories: An update. *Methods in Ecology and Evolution*, 1, 38–44.

Beck, J., Boller, M., Erhardt, A. and Schwanghart, W. (2014) Spatial bias in the GBIF database and its effect on modeling species' geographic distributions. *Ecological Informatics*, 19, 10–15.

Becking, L.E., Renema, W., Santodomingo, N.K., Hoeksema, B.W., Tuti, Y. and de Voogd, N.J. (2011) Recently discovered landlocked basins in Indonesia reveal high habitat diversity in anchialine systems. *Hydrobiologia*, 677, 89–105.

Beckmann, J.P. and Berger, J. (2003) Using black bears to test ideal-free distribution models experimentally. *Journal of Mammalogy*, 84, 594–606.

Beddington, J., Free, C. and Lawton, J. (1975) Dynamic complexity in predator-prey models framed in difference equations. *Nature*, 255, 58–60.

Begon, M., Mortimer, M. and Thompson, D.J. (1996) *Population Ecology: A Unified Study of Animals and Plants.* Blackwell Science Ltd, London, UK.

Behangana, M. and Luiselli, L. (2008) Habitat niche community-level analysis of an amphibian assemblage at Lake Nabugabo, Uganda. *Web Ecology*, 8, 125–134.

Beibl, J., Stuart, R.J., Heinze. J. and Foitzik, S. (2005) Six origins of slavery in formicoxenine ants. *Insectes Sociaux*, 52, 291–297.

Beirne, B.P. (1947) The origin and history of the British Macrolepidoptera. *Transactions of the Royal Entomological Society*, 98, 275–372.

Bell, G. (1982) *The Masterpiece of Nature: The Evolution and Genetics of Sexuality.* Biddles Ltd, London, UK.

Bellwood, D.R., Wainwright, P.C., Fulton, C.J. and Hoey, A.S. (2006) Functional versatility supports coral reef biodiversity. *Proceedings of the Royal Society B: Biological Sciences*, 273, 101–107.

Benavides Rios, E., Sadler, J., Graham, L. and Matthews, T.J. (2024) Species distribution models and island biogeography: Challenges and prospects. *Global Ecology and Conservation*, 51, e02943.

Bengtsson, J. and Martinez, N. (1996) Causes and effects in food webs: Do generalities exist? In: Polis, G.A. and Winemiller, K.O. (eds) *Food Web. Integration of Patterns and Dynamics.* Springer Science+Business Media, Dordrecht, The Netherlands, pp. 179–184.

Beninde, J., Veith, M. and Hochkirch, A. (2015) Biodiversity in cities needs space: A meta-analysis of factors determining intra-urban biodiversity variation. *Ecology Letters*, 18, 581–592.

Benito-Garzón, M., Leadley, P.W. and Fernández-Manjarrés, J.F. (2014) Assessing global biome exposure to climate change through the Holocene–Anthropocene transition. *Global Ecology and Biogeography*, 23, 235–244.

Bennett, J.M., Calosi, P., Clusella-Trullas, S., Martínez, B., Sunday, J. *et al.* (2018) GlobTherm, a global database on thermal tolerances for aquatic and terrestrial organisms. *Scientific Data*, 5, 180022.

Bennett, K.D. (1989) A provisional map of forest types for the British Isles 5000 years ago. *Journal of Quaternary Science*, 4, 141–144.

Benoit, J. and Thackeray, F.J. (2017) A cladistic analysis of *Graecopithecus. South African Journal of Science*, 113, 11–12.

Benson, R.B.J., Butler, R., Close, R.A., Saupe, E. and Rabosky, D.L. (2021) Biodiversity across space and time in the fossil record. *Current Biology*, 31, R1225–R1236.

Benton, M.J. (1995) Diversification and extinction in the history of life. *Science*, 268, 52–58.

Benton, M.J. (2001) Biodiversity on land and in the sea. *Geological Journal*, 36, 211–230.

Berdoy, M., Webster, J.P. and Macdonald, D.W. (2000) Fatal attraction in rats infected with *Toxoplasma gondii. Proceedings of the Royal Society B: Biological Sciences*, 267, 1591–1594.

Berenreiterová, M., Flegr, J., Kuběna, A.A. and Němec, P. (2011) The distribution of *Toxoplasma gondii* cysts in the brain of a mouse with latent toxoplasmosis: Implications for the behavioral manipulation hypothesis. *PLOS One*, 6, e28925.

Bergmann, C. (1847) Ueber die Verhältnisse der Wärmeökonomie der Thiere zu ihrer Grösse. *Göttinger Studien*, 1, 595–708.

Berlanga, M. and Guerrero, R. (2016) The holobiont concept: The case of xylophagous termites and cockroaches. *Symbiosis*, 68, 49–60.

Berlow, E.L., Neutel, A.M., Cohen, J.E., De Ruiter, P.C., Ebenman, B. *et al.* (2004) Interaction strengths in food webs: Issues and opportunities. *Journal of Animal Ecology*, 73, 585–598.

Berruti, G.L.F. and Ruzza, S. (2016) The emergence of war in human societies. In: Coimbra, F., Delfino, D., Sîrbu, V. and Schuster, C. (eds) *Late Prehistory and Protohistory: Bronze Age and Iron Age. Proceedings of the XVII UISPP World Congress (1–7 September 2014, Burgos, Spain) Volume 9 / Sessions A3c and A16a.* Archaeopress Archaeology, Oxford, UK, pp. 27–38.

Berry, R.J. (1977) *Inheritance and Natural History*, Collins, London, UK.

Bersier, L.F., Banašek-Richter. C. and Cattin, M.F. (2002) Quantitative descriptors of food-web matrices. *Ecology*, 83, 2394–2407.

Bersot, E. (1931) Comment boivent les martinets. *Nos Oiseaux*, 11, 106–107.

Berthold, P. (2001) *Bird Migration. A General Survey.* Oxford University Press, Oxford, UK.

Bevilacqua, S., Fraschetti, S., Terlizzi, A. and Boero, F. (2009) The use of taxonomic distinctness indices in assessing patterns of biodiversity in modular organisms. *Marine Ecolology*, 30, 151–163.

Bevilacqua, S., Anderson, M.J., Ugland, K.I., Somerfield, P.J. and Terlizzi, A. (2021) The use of taxonomic relationships among species in applied ecological research: Baseline, steps forward and future challenges. *Austral Ecology*, 46, 950–964.

Beyer, H.L., Haydon, H.T., Morales, J.M., Frair, J.L., Hebblewhite, M. *et al.* (2010) The interpretation of habitat preference metrics under use–availability designs. *Philosophical Transactions of the Royal Society B: Biological Sciences*, 365, 2245–2254.

Beyer, R.M., Krapp, M., Eriksson, A. and Manica, A. (2021) Climatic windows for human migration out of Africa in the past 300,000 years. *Nature Communications*, 12, 4889.

Bibby, C.J., Burgess, N.D., Hill, D.A. and Mustoe, S. (2000) *Bird Census Techniques*. Academic Press, London, UK.

Bickle, P. (2013) Of time and the house: The Early Neolithic communities of the Paris Basin and their domestic architecture. In: Hofmann, D. and Smyth, J. (eds) *Tracking the Neolithic House in Europe: Sedentism, Architecture, and Practice*. Springer, New York, USA, pp. 151–81.

Bickle, P.F. and Kalogiropoulou, E. (2017) Very like the Neolithic: The everyday and settlement in the European Neolithic. In: Bickle, P., Cummings, V., Hofmann, D. and Pollard, J. (eds) *The Neolithic of Europe: Papers in Honour of Alasdair Whittle*. Oxbow Books, Oxford, UK, pp. 7–23.

Biggs, C.R., Yeager, L.A., Bolser, D.G., Bonsell, C., Dichiera, A.M. *et al.* (2020) Does functional redundancy affect ecological stability and resilience? A review and meta-analysis. *Ecosphere*, 11, e03184.

Bignell, D.E., Roisin, Y. and Lo, N. (2010) *Biology of Termites: A Modern Synthesis*. Springer, Dordrecht, The Netherlands.

Binford, L.R. (2001) *Constructing Frames of Reference: An Analytical Method for Archaeological Theory Building using Hunter-gatherer and Environmental Data Sets*. University of California Press, Berkeley, USA.

Binney, H., Edwards, M., Macias-Fauria, M., Lozhkin, A., Anderson, P. *et al.* (2017) Vegetation of Eurasia from the last glacial maximum to present: Key biogeographic patterns. *Quaternary Science Reviews*, 157, 80–97.

Birkhead, T.R. (1991) *The Magpies: The Ecology and Behaviour of Black-billed and Yellow-billed Magpies*. Poyser, London, UK.

Birley, A.R. (2005) *The Roman Government of Britain*. Oxford University Press, Oxford, UK.

Birskis-Barros, I., Alencar, L.R.V., Prado, P.I., Böhm, M. and Martins, M. (2019) Ecological and conservation correlates of rarity in New World pitvipers. *Diversity*, 11, 147.

Bishop, R.R. (2021) Hunter-gatherer carbohydrate consumption: Plant roots and rhizomes as staple foods in Mesolithic Europe. *World Archaeology*, 53,175–199.

Bishop, R.R., Kubiak-Martens, L., Warren, G.M. and Church, M.J. (2022) Getting to the root of the problem: New evidence for the use of plant root foods in Mesolithic hunter-gatherer subsistence in Europe. *Vegetation History and Archaeobotany*, 32, 65–83.

Bitsadze, N., Beruashvili, M., Pavliashvili, K., Khazaradze, R., Jorjadze, A. *et al.* (2018) Main oak species and fungi associated with oak trees described in Georgian mycological herbarium. *Annals of Agrarian Science*, 16, 432–435.

Black, J. (2017) Pre-Roman and Roman Britain. In: Black, J. *A History of the British Isles*. Bloomsbury Academic, London, UK, pp. 1–10.

Blair, R.B. and Launer, A.E. (1997) Butterfly diversity and human land use: Species assemblages along an urban gradient. *Biological Conservation*, 80, 113–125.

Blanchette, F., Montroy, S., Patris, S. and Dawson, M.N. (2020) Marine lakes as biogeographical islands: A physical model for ecological dynamics in an insular marine lake, Palau. *Frontiers of Biogeography*, 12, e47736.

Blanchette, M.L. and Lund, M.A. (2016) Pit lakes are a global legacy of mining: an integrated approach to achieving sustainable ecosystems and value for communities. *Current Opinion in Environmental Sustainability*, 23, 28–34.

Blanck, A. and Lamouroux, N. (2007) Large-scale intraspecific variation in life-history traits of European freshwater fish. *Journal of Biogeography*, 34, 862–875.

Blanckenhorn, W.U., Pemberton, A.J., Bussière, L.F., Roembke, J. and Floate, K.D. (2010) A review of the natural history and laboratory culture methods for the yellow dung fly, *Scathophaga stercoraria*. *Journal of Insect Science*, 10, 11.

Block, W.M. and Brennan, L.A. (1993) The habitat concept in ornithology: Theory and applications. *Current Ornithology*, 11, 35–91.

Blondel, J. (2003) Guilds or functional groups: Does it matter? *Oikos*, 100, 223–231.

Bobe, R. and Wood, B. (2022) Estimating origination times from the early hominin fossil record. *Evolutionary Anthropology*, 31, 92–102.

Bocherens, H. (2009) Neanderthal dietary habits: Review of the isotopic evidence. In: Hublin J.-J. and Richards, M.P. (eds) *The Evolution of Hominin Diets: Integrating Approaches to the Study of Palaeolithic Subsistence*. Springer, New York, USA, pp. 241–250.

Bogucki, P. (1999) *The Origins of Human Society*. Blackwell Publishing, Oxford, UK.

Böhning, D., Kaskasamkul, P. and van der Heijden, P.G.M. (2019) A modification of Chao's lower bound estimator in the case of one-inflation. *Metrika*, 82, 361–384.

Boitani, L. and Powell, R.A. (2012) *Carnivore Ecology and Conservation. A Handbook of Techniques*. Oxford University Press, Oxford, UK.

Boivin, N.L., Zeder, M.A., Fuller, D.Q., Crowther, A., Larson, G. *et al.* (2016) Ecological consequences of human niche construction: Examining long-term anthropogenic shaping of global species distributions. *Proceedings of the National Academy of Sciences of the United States of America*, 113(23), 6388–6396.

Bökenhans, V., Bigatti, G. and Averbuj, A. (2016) Age estimation methods in the marine gastropod *Buccinanops globulosus* comparing shell marks and opercula growth rings. *Marine Biology Research*, 12, 881–887.

Bolnick, D.I., Amarasekare, P., Araújo, M.S., Bürger, R., Levine, J.M. *et al.* (2011) Why intraspecific trait variation matters in community ecology. *Trends in Ecology and Evolution*, 26, 183–192.

Bongaarts, J. (2009) Human population growth and the demographic transition. *Philosophical Transactions of the Royal Society B: Biological Sciences*, 364, 2985–2990.

Bongers, F., Poorter, L., Hawthorne, W.D. and Sheil, D. (2009) The intermediate disturbance hypothesis applies to tropical forests, but disturbance contributes little to tree diversity. *Ecology Letters*, 12, 798–805.

Bonsall, M.B. and Hassell, M.P. (2007) Predator–prey interactions. In: May, R. and McLean, A.R. (eds) *Theoretical Ecology: Principles and Applications*, 3rd edn. Oxford University Press, Oxford, UK, pp. 47–61.

Bonsall, M.P., Jansen, V.A.A. and Hassell, M.P. (2004) Life history trade-offs assemble ecological guilds. *Science*, 306, 111–114.

Boonratana, R. (2000) Ranging behavior of proboscis monkeys (*Nasalis larvatus*) in the Lower Kinabatangan, Northern Borneo. *International Journal of Primatology*, 21, 497–518.

Boorman, S.A. and Levitt, P.R. (1973) Group selection on the boundary of a stable population. *Theoretical Population Biology*, 4, 85–128.

Boorstin, D.J. (1983) *The Discoverers*. Random House, New York, USA.

Borcard, D., Gillet, F. and Legendre, P. (2018) *Numerical Ecology with R*, 2nd edn. Springer, New York, USA.

Borda-de-Água, L., Borges, P.A.V., Hubbell, S.P. and Pereira, H.M. (2012) Spatial scaling of species abundance distributions. *Ecography*, 35, 549–556.

Borda-de-Água, L., Whittaker, R.J., Cardoso, P., Rigal, F., Santos, A.M.C. *et al.* (2017) Dispersal ability determines the scaling properties of species abundance distributions: A case study using arthropods from the Azores. *Scientific Reports*, 7, 3899.

Borges, I.L., Forsyth, L.Z., Start, D. and Gilbert, B. (2019) Abiotic heterogeneity underlies trait-based competition and assembly. *Journal of Ecology*, 107, 747–756.

Borges, P.A.V., Cardoso, P., Amorim, I.R., Pereira, F., Constancia, J.P. *et al.* (2012) Volcanic caves: Priorities for conserving the Azorean endemic troglobiont species. *International Journal of Speleology*, 41, 101–112.

Borges, P.A.V., Cardoso, P., Fattorini, S., Rigal, F., Matthews, T.J. *et al.* (2018a) Community structure of woody plants on islands along a bioclimatic gradient. *Frontiers of Biogeography*, 10(3–4).

Borges, P.A.V., Cardoso, P., Kreft, H., Whittaker, R.J., Fattorini, S. *et al.* (2018b). Global Island Monitoring Scheme (GIMS): a proposal for the long-term coordinated survey and monitoring of native island forest biota. *Biodiversity and Conservation*, 27, 2567–2586.

Borges, P.A.V., Gabriel, R. and Fattorini, S. (2019) Biodiversity erosion: Causes and consequences. In: Leal Filho, W., Azul, A., Brandli, L., Özuyar, P. and Wall, T. (eds) *Encyclopedia of the UN Sustainable Development Goals. Life on Land*. Springer, Cham, Switzerland.

Borghesio, L. and Palestrini, C. (2002) Reproductive behavior and larval development in *Agrilinus rufus* (Moll, 1792) and *Oromus alpinus* (Scopoli, 1763) (Coleoptera: Scarabaeoidea: Aphodiidae). *Elytron*, 16, 75–81.

Borghi, C.E., Campos, C.M., Giannoni, S.M., Campos, V.E. and Sillero-Zubiri, C. (2011) Updated distribution of the pink fairy armadillo *Chlamyphorus truncatus* (Xenarthra, Dasypodidae), the world's smallest armadillo. *Edentata*, 12, 14–19.

Borin, J.M., Lee, J.J., Lucia-Sanz, A., Gerbino, K.R., Weitz, J.S. and Meyer, J.R. (2023) Rapid bacteria-phage coevolution drives the emergence of multiscale networks. *Science*, e382, 674–678.

Bornemissza, G.E. (1976) The Australian dung beetle project 1965–1975. *AMRC Review of the Australian Meat Research Committee*, 30, 1–30.

Borovsky, J.E. and Valdivia, J.A. (2018) The Earth's magnetosphere: A systems science overview and assessment. *Surveys in Geophysics Surveys in Geophysics*, 39, 817–859.

Bortoni-Anzures, L. and Herrera-Rivas, H. (2019) Space migrants. Migrantes espaciales. *Ecorfan Journal – Republic of Paraguay*, 5, 1–6.

Bortoni-Anzures, L. and Ortiz-Moctezuma, M.J. (2019) Earth Terraforming. Terraformar la Tierra. *Ecorfan Journal – Republic of Paraguay*, 5, 7–12.

Bosellini, A. and Ginsburg, R.N. (1971) Form and internal structure of recent algal nodules (Rhodolites) from Bermuda. *Journal of Geology*, 79, 669–682.

Botigué, L.R., Song, S., Scheu, A., Gopalan, S., Pendleton, A.L. *et al.* (2017) Ancient European dog genomes reveal continuity since the Early Neolithic. *Nature Communications*, 8, 16082.

Bottema, S. (2001) A note on the pollen representation of ivy (*Hedera helix* L.). *Review of Palaeobotany and Palynology*, 117, 159–166.

Bourel, B., Martin-Bouyer, L., Hedouin, V. and Cailiez, J.C. (1999) Necrophilous insect succession on rabbit carrion in sand dune habitats in northern France. *Journal of Medical Entomology*, 36, 420–425.

Bowie, R.C.K., Fjeldså, J., Hackett, S.J., Bates, J.M. and Crowe, T.M. (2006) Coalescent models reveal the relative roles of ancestral polymorphism, vicariance and dispersal in shaping phylogeographical structure of an African montane forest robin. *Molecular Phylogenetics and Evolution*, 38, 171–188.

Bowler, D.E., Callaghan, C.T., Bhandari, N., Henle, K., Benjamin Barth, M. *et al.* (2022) Temporal trends in the spatial bias of species occurrence records. *Ecography*, 2022, e06219.

Bowler, P.J. (1976) Malthus, Darwin, and the concept of struggle. *Journal of the History of Ideas*, 37, 631–650.

Bowles, S. (2009) Did warfare among ancestral hunter-gatherers affect the evolution of human social behaviors? *Science*, 324, 1293–1298.

Bowman, K.O., Hutcheson, K., Odum, E.P. and Shenton, L.R. (1971) Comments on the distribution of indices of diversity. In: Patil, E.C., Pielou, E.C. and Waters, W.E. (eds) *Statistical Ecology*. Pennsylvania State University Press, University Park, USA, pp. 315–366.

Boy, J., Godoy, R., Shibistova, O., Boy, D., McCulloch, R. *et al.* (2016) Successional patterns along soil development gradients formed by glacier retreat in the Maritime Antarctic, King George Island. *Revista Chilena de Historia Natural*, 89, 6.

Boyce, C.K., Hotton, C.L., Fogel, M.L., Cody, G.D., Hazen, R.M. *et al.* (2007) Devonian landscape heterogeneity recorded by a giant fungus. *Geology*, 35, 399–402.

Boycott, A.E. (1934) The habitats of land Mollusca in Britain. *Journal of Ecology*, 22, 1–38.

Boyd, I.L. (2004) Migration of marine mammals. In: Werner, D. (ed) *Biological Resources and Migration*. Springer, Berlin, Germany, pp. 203–210.

Boyd, I.L., Bowen, W.D. and Iverson, S. (2010) *Marine Mammal Ecology and Conservation: A Handbook of Techniques*. Oxford University Press, Oxford, UK.

Braddy, S.J. (2024) Euthycarcinoid ecology and evolution. *Neues Jahrbuch für Geologie und Paläontologie – Abhandlungen*, 312(1), 15–30.

Bradshaw, G.A., Marquet, P.A. and Ronnenberg, K.L. (2003) *How Landscapes Change: Human Disturbance and Ecosystem Fragmentation in the Americas*. Springer, Dordrecht, The Netherlands.

Brakefield, P.M. (1979) An experimental study of the maintenance of variation in spot pattern in Maniola jurtina. PhD thesis, University of Liverpool, Liverpool, UK.

Brakefield, P.M. (1984) The ecological genetics of quantitative characters of *Maniola jurtina* and other butterflies. In: Vane-Wright, R.I. and Ackery, R.R. (eds) *The Biology of Butterflies*. Academic Press, London, UK, pp. 167–190.

Brakefield, P.M., Shreeve, T.G. and Thomas, J.A. (1992) Avoidance, concealment, and defence. In: Dennis, R.L.H (ed.) *Ecology of Butterflies in Britain*. Oxford University Press, Oxford, UK, pp. 93–119.

Brandle, R. (ed.) (2010) *A Biological Survey of the Eyre Peninsula, South Australia*. Science Resource Centre Information, Science and Technology Directorate, Department for Environment and Heritage, Adelaide, SA.

Brazier, R.E., Puttock, A., Graham, H.A., Auster, R.E., Davies, K.H. and Brown, C.M.L. (2021) Beaver: Nature's ecosystem engineers. *WIREs Water*, 8, e1494.

Breckheimer, I., Haddad, N.M., Morris, W.F., Trainor, A.M., Fields, W.R. *et al.* (2014) Defining and evaluating the umbrella species concept for conserving and restoring landscape connectivity. *Conservation Biology*, 28, 1584–1593.

Breckle, S.W. (2002) *Walter's Vegetation of the Earth – The Ecological Systems of the Geo-biosphere*. Springer, Heidelberg, Germany.

Breeze, T.D., Bailey, A.P., Balcombe, K.G., Brereton, T., Comont, R. *et al.* (2020) Pollinator monitoring more than pays for itself. *Journal of Applied Ecology*, 58, 44–57.

Brereton, T.M. (1997) Ecology and conservation of the butterfly Pyrgus malvae (grizzled skipper) in South-East England. PhD thesis, University of East London, London, UK.

Breshears, D.D., Field, J.P., Law, D.J., Villegas, J.C., Allen, C.D. *et al.* (2019) Rapid broad-scale ecosystem changes and their consequences for biodiversity. In: Lovejoy, T.E. and Hannah, L. (eds) *Climate Change and Biodiversity*. Yale University Press, New Haven, USA, pp. 80–90.

Bried, J.T., Herman, B.D. and Ervin, G.N. (2007) Umbrella potential of plants and dragonflies for wetland conservation: A quantitative case study using the umbrella index. *Journal of Applied Ecology*, 44, 833–842.

Briffa, J., Sinagra, E. and Blundell, R. (2020) Heavy metal pollution in the environment and their toxicological effects on humans. *Heliyon*, 6, e04691.

Briggs, H. (2024) England's hedges would go around Earth ten times. Available at: www.bbc.co.uk/news/science-environment-68132688 (accessed 30 April 2025).

Briggs, J.C. and Bowen, B.W. (2012) A realignment of marine biogeographic provinces with particular reference to fish distributions. *Journal of Biogeography*, 39, 12–30.

Brinkløv, S., Fenton, M.B. and Ratcliffe, J.M. (2013) Echolocation in oilbirds and swiftlets. *Frontiers in Physiology*, 4, 123.

Brinkof, M.W.G., Cavé, A.J., Hage, F.J. and Verhulst, S. (1993) Timing of reproduction and fledgling success in the coot *Fulica atra*. Evidence for a causal relationship. *Journal of Animal Ecology*, 62, 577–587.

Brochu, C.A. (2003) Phylogenetic approaches toward crocodylian history. *Annual Review of Earth and Planetary Sciences*, 31, 357–397.

Brocks, J.J., Nettersheim, B.J., Adam, P., Schaeffer, P., Jarrett, A.J.M. *et al.* (2023) Lost world of complex life and the late rise of the eukaryotic crown. *Nature*, 618(7966), 767–773.

Broennimann, O., Vittoz, P., Moser, D. and Guisan, A. (2005) Rarity types among plant species with high conservation priority in Switzerland. *Botanica Helvetica*, 115, 95–108.

Broennimann, O., Fitzpatrick, M.C., Pearman, P.B., Petitpierre, B. Pellissier, L. *et al.* (2012) Measuring ecological niche overlap from occurrence and spatial environmental. *Global Ecology and Biogeography*, 21, 481–497.

Bronstein, J.L. (2009) The evolution of facilitation and mutualism. *Journal of Ecology*, 97, 1160–1170.

Brook, B.W. and Bradshaw, J.A. (2006) Strength of evidence for density dependence in abundance time series of 1198 species. *Ecology*, 87, 1445–1451.

Brookes, A. and Fratto, E. (2020) Towards a Russian literature of the Anthropocene. Introduction. *Russian Literature*, 114–115, 1–22.

Brookfield, M.E., Catlos, E.J. and Suarez, S.E. (2022) Vertebrate lies? Arthropods were the first land animals! *Geology Today*, 38, 65–68.

Brotcorne, F., Holzner, A., Jorge-Sales, L., Gunst, N., Hambuckers, A. *et al.* (2019) Social infuence on the expression of robbing and bartering behaviours in Balinese long-tailed macaques. *Animal Cognition*, 23, 311–326.

Brotons, L., Pons, P. and Herrando, S. (2005) Colonization of dynamic Mediterranean landscapes: Where do birds come from after fire? *Journal of Biogeography*, 32, 789–798.

Brown, A., Gilbert, G. and Wotton, S. (2012) Bitterns and bittern conservation in the UK. *British Birds*, 105, 58–87.

Brown, C.R., Stutchbury, B.J. and Walsh, P.D. (1990) Choice of colony size in birds. *Trends in Ecology and Evolution*, 5, 398–403.

Brown, D.E., Reichenbacher, F. and Franson, S.E. (1998) *A Classification of North American Biotic Communities*. University of Utah Press, Salt Lake City, USA.

Brown, G.R., and Matthews, I.M. (2016) A review of extensive variation in the design of pitfall traps and a proposal for a standard pitfall trap design for monitoring ground-active arthropod biodiversity. *Ecology and Evolution*, 6, 3953–3964.

Brown, J.H. (1971) Mammals on mountaintops: Non-equilibrium insular biogeography. *American Naturalist*, 105, 467–478.

Brown, J.H. (1978) The theory of insular biogeography and the distribution of boreal birds and mammals. *Great Basin Naturalist Memoirs*, 2, 209–277.

Brown, J.H. (1984) On the relationship between abundance and distribution of species. *American Naturalist*, 124, 255–279.

Brown, J.H. and Maurer, B.A. (1989) Macroecology. The division of food and space among species on continents. *Science*, 243, 1145–1150.

Brown, J.S. and Rosenzweig, M.L. (1986) Habitat selection in slowly regenerating environments. *Journal of Theoretical Biology*, 123, 151–171.

Brown, K.D. (2017) *Earthworm Recorder's Handbook (UK National Earthworm Recording Scheme)*. Earthworm Society of Britain. Available at: www.earthwormsoc.org.uk/handbook (accessed 30 April 2025).

Brown, P.M.J. and Roy, H.E. (2018) Native ladybird decline caused by the invasive harlequin ladybird *Harmonia axyridis*: Evidence from a long-term field study. *Insect Conservation and Diversity*, 11, 230–239.

Brown, P.M.J., Roy, D.B., Harrower, R.C.A., Dean, H.J., Rorke, S. and Roy, H.E. (2018) Spread of a model invasive alien species, the harlequin ladybird *Harmonia axyridis* in Britain and Ireland. *Scientific Data*, 5, 180239.

Brown, S. and Lugo, A.E. (1990) Tropical secondary forests. *Journal of Tropical Ecology*, 6, 1–32.

Brown, V.K. (1985) Insect herbivores and plant succession. *Oikos*, 44, 17–22.

Brown, V.K. and Gange, A.C. (1992) Secondary plant succession: How is it modified by insect herbivory? *Vegetatio*, 101, 3–13.

Brown, V.K. and Southwood, T.R.E. (1983) Trophic diversity, niche breadth and generation times of exopterygote insects in a secondary succession. *Oecologia*, 56, 220–225.

Browne, J. (2002) *Charles Darwin. I. Voyaging; II. The Power of Place*. Jonathan Cape, London, UK.

Brualdi, R.A. and Sanderson, J.G. (1999) Nested species subsets, gaps, and discrepancy. *Oecologia*, 119, 256264

Brunet, A.K. and Medellín, R.A. (2001) The species-area relationship in bat assemblages in tropical caves. *Journal of Mammalogy*, 82, 1114–1122.

Brunet-Navarro, P., Sterck, F.J., Vayreda, J., Martinez-Vilalta, J. and Mohren, G.M.J. (2016) Self-thinning in four pine species: An evaluation of potential climate impacts. *Annals of Forest Science*, 73, 1025–1034.

Bruno, J.F., Stachowicz, J.J. and Bertness, M.D. (2003) Inclusion of facilitation into ecological theory. *Trends in Ecology and Evolution*, 18, 119–125.

Brusatte, S.L., Butler, R.J., Barrett, P.M., Carrano, M.T., Evans, D.C. *et al.* (2015) The extinction of the dinosaurs. *Biological Reviews*, 90, 628–642.

Brusca, R.C. and Moore, W. (2013) *A Natural History of the Santa Catalina Mountains, Arizona with an Introduction to the Madrean Sky Islands*. Arizona-Sonora Desert Museum Press, Tucson, USA.

Brusseau, M.L., Pepper, I.L. and Gerba, C.P. (eds) (2019) *Environmental and Pollution Science*, 3rd edn. Academic Press, London, UK.

Bryan, N.C. (2017) Microbial distributions and survival in the troposphere and stratosphere. PhD thesis, Louisiana State University, Baton Rouge, USA.

Bryant, J.B., Lamanna, C., Morlon, H., Kerkhoff, A.J., Enquist, B.J. and Green, J.L. (2008) Microbes on mountainsides: Contrasting elevational patterns of bacterial and plant diversity. *Proceedings of the National Academy of Sciences of the United States of America*, 105 (Supplement 1), 11505–11511.

Bryson-Morrison, N., Tzanopoulos, J., Matsuzawa, T. and Humle, T. (2017) Activity and habitat use of Chimpanzees (*Pan troglodytes verus*) in the anthropogenic landscape of Bossou, Guinea, West Africa. *International Journal of Primatology*, 38, 282–302.

Brzeziński, M., Chibowski, P., Gornia, J., Górecki, G. and Zalewski, A. (2018) Spatio-temporal variation in nesting success of colonial waterbirds under the impact of a non-native invasive predator. *Oecologia*, 188(4), 1037–1047.

Bucher, E.H. (1992) The causes of extinction of the Passenger Pigeon. *Current Ornithology*, 9, 1–36.

Bucher, S.F. and Römermann, C. (2020) Flowering patterns change along elevational gradients and relate to life-history strategies in 29 herbaceous species. *Alpine Botany*, 130, 41–58.

Bulleri, F., Bruno, J.F., Silliman, B.R. and Stachowicz, J.J. (2016) Facilitation and the niche: Implications for coexistence, range shifts and ecosystem functioning. *Functional Ecology*, 30, 70–78.

Bulmer, M.G. (1974) On fitting the Poisson lognormal distribution to species-abundance data. *Biometrics* 30, 101–110.

Buma, B., Bisbing, S., Krapek, J. and Wright, G. (2017) A foundation of ecology rediscovered: 100 years of succession on the William S. Cooper plots in Glacier Bay, Alaska. *Ecology*, 98, 1513–1523.

Bunce, M., Szulkin, M., Lerner, H.R.L., Barnes, I., Shapiro, B. *et al.* (2005) Ancient DNA provides new insights into the evolutionary history of New Zealand's extinct giant eagle. *PLOS Biology*, 3(1), e9.

Bunce, R.G.H., Bogers, M.M.B., Roche, P., Walczak, M., Geijzendorffer, I.J. and Jongman, R.H.G. (2011) *Manual for Habitat and Vegetation Surveillance and Monitoring: Temperate, Mediterranean and Desert Biomes*. Alterra Report 2154. Wageningen University and Research, Wageningen, The Netherlands.

Burel, F. and Baudry, J. (2003) *Landscape Ecology: Concepts, Methods, and Applications*. Science Publishers Inc., Enfield, USA.

Burgess, E.W. (1928) Residential segregation in American cities. *Annals of the American Academy of Political and Social Science*, 140, 105–115.

Burgess, E.W. (1967) The growth of the city: an introduction to a research project. In: Park, R.E. and Burgess, E.W. (eds) *The City. Suggestions for Investigation of Human Behavior in the Urban Environment*, 2nd edn. University of Chicago Press, Chicago, USA.

Burgess, M.D., Eaton, M.A. and Gregory, R.D. (2020) A review of spatial patterns across species ranges to aid the targeting of conservation interventions. *Biological Conservation*, 251, 108755.

Burgin, C.J., Wilson, D.E., Mittermeier, R.A., Rylands, A.B., Lacher Jr., T.E. and Sechrest, W. (2020) *Illustrated Checklist of the Mammals of the World*. Lynx Edicions, Barcelona, Spain.

Burke, A. (2006) Neanderthal settlement patterns in Crimea: A landscape approach. *Journal of Anthropological Archaeology*, 25, 510–523.

Burki, F., Roger, A.J., Brown, M.W. and Simpson, A.G.B. (2020) A new tree of eukaryotes. *Trends in Ecology and Evolution*, 3, 43–55.

Burr, B.M. and Mayden, R.L. (1992) Phylogenetics and North American freshwater fishes. In: Mayden, R.L. (ed.) *Systematics, Historical Ecology and North American Freshwater Fishes*. Stanford University Press, Stanford, USA.

Burt, W.H. (1943) Territoriality and home range concepts as applied to mammals. *Journal of Mammalogy*, 24, 346–352.

Burwell, R.G., Aujla, R.K., Grevitt, M.P., Dangerfield, P.H., Moulton, A. *et al.* (2009) Pathogenesis of adolescent idiopathic scoliosis in girls – a double neuro-osseous theory involving disharmony between two nervous systems, somatic and autonomic expressed in the spine and trunk: possible dependency on sympathetic nervous system and hormones with implications for medical therapy. *Scoliosis*, 4, 24.

Bushing, W.W. (1997) Monitoring the persistence of giant kelp around Santa Catalina island using a geographical information system. *International Journal of Marine Geodesy*, 20, 205–234.

Buso Junior, A.A., Pessenda, L.C.R., Mayle, F.E., Lorente, F.L., Volkmer-Ribeiro C. *et al.* (2019) Paleovegetation and paleoclimate dynamics during the last 7000 years in the Atlantic forest of Southeastern Brazil based on palynology of a waterlogged sandy soil. *Review of Palaeobotany and Palynology*, 264, 1–10.

Butchart, S.H.M., Resit Akçakaya, H., Chanson, J., Baillie, J.E.M., Collen, B. *et al.* (2007) Improvements to the Red List Index. *PLOS One*, 2, e140.

Butchart, S.H.M., Walpole, M., Collen, B., van Strien, A., Scharlemann, J.P.W. *et al.* (2010) Global biodiversity: Indicators of recent declines. *Science*, 328, 1164–1168.

Butler, C.J. (2021) Naturalised parrots in the United Kingdom. In: Pruett-Jones, S. (ed.) *Naturalized Parrots of the World: Distribution, Ecology, and Impacts of the World's Most Colorful Colonizers*. Princeton University Press, Princeton, USA.

Butt, K.R., Lang, F., Ehrmann, O., Kobel-Lamparski, A., Lamparski, F. and Nuutinen, V. (2021) Field and laboratory investigations of *Lumbricus badensis* ecology and behaviour. *Pedosphere*, 31, 471–474.

Butt, N., Chauvenet, A.L.M., Adams, V.M., Beger, M., Gallagher, R.V. *et al.* (2021) Importance of species translocations under rapid climate change. *Conservation Biology*, 35, 775–783.

Byrne, P.G. and Silla, A.J. (2020) An experimental test of the genetic consequences of population augmentation in an amphibian. *Conservation Science and Practice*, 2, e194.

Cagnacci, F., Meriggi, A. and Lovari, S. (2004) Habitat selection by the red fox *Vulpes vulpes* (L. 1758) in an Alpine area. *Ethology Ecology and Evolution*, 16, 103–116.

Caiafa, A.N. and Martins, F.R., (2010) Forms of rarity of tree species in the southern Brazilian Atlantic rainforest. *Biodiversity and Conservation* 19, 2597–2618.

Caillaud, D., Eckardt, W., Vecellio, V., Ndagijimana, F., Mucyo, J.-P. *et al.* (2020) Violent encounters between social units hinder the growth of a high-density mountain gorilla population. *Science Advances*, 6, eaba0724.

Calenge, C. and Basille, M. (2008) A general framework for the statistical exploration of the ecological niche. *Journal of Theoretical Biology*, 252(4), 674–685.

Callaghan, C.T., Borda-de-Água, L., van Klink, R., Rozzi, R. and Pereira, H.M. (2023) Unveiling global species abundance distributions. *Nature Ecology and Evolution*, 7, 1600–1609.

Callaway, R.M. and Aschehoug, E.T. (2000) Invasive plants versus their new and old neighbors: A mechanism for exotic invasion. *Science*, 290, 521–523 [Erratum: *Science*, 2000, 290, 2075b].

Callisto, M., Mugnai, R., Castro, D.M.P., Linares, M.S. (2021) Sampling methods for aquatic insects. In: Santos, J.C. and Fernandes G.W. (eds) *Measuring Arthropod Biodiversity*. Springer, Cham, Switzerland, pp. 523–543.

Calow, P. (1999) *Blackwell's Concise Encyclopedia of Ecology*. Blackwell Sciences, Oxford, UK.

Calvert, J., Strong, J.A., Service, M., McGonigle, C. and Quinn, R. (2014) An evaluation of supervised and unsupervised classification techniques for marine benthic habitat mapping using multibeam echosounder data. *ICES Journal of Marine Science: Journal Du Conseil*, 72, 1498–1513.

Cameron, D.D. and Seel, W.E. (2007) Functional anatomy of haustoria formed by *Rhinanthus minor*: Linking evidence from histology and isotope tracing. *New Phytologist*, 174, 412–419.

Cameron, R.A.D. (1978) Terrestrial snail faunas of the Malham area. *Field Studies*, 4, 715–728.

Cameron, R.A.D. (2016) *Slugs and Snails*. Collins, London, UK.

Campanyà-Llovet, N., Bates. A.E., Cuvelier. D., Giacomello, E., Catarino, D. *et al.* (2023) FUN Azores: A FUNctional trait database for the meio-, macro-, and megafauna from the Azores Marine Park (Mid-Atlantic Ridge). *Frontiers in Ecology and Evolution*, 11, 1050268.

Campbell, A.L., Naik, R.R., Sowards, L. and Stone, M.O. (2002) Biological infrared imaging and sensing. *Micron*, 33(2), 211–225.

Campbell, C., Yang, S., Shea, K. and Albert, R. (2012) Topology of plant-pollinator networks that are vulnerable to collapse from species extinction. *Physical Review E*, 86, 02192.

Campbell, M.A., Van Leuven, J.T., Meister, R.C., Carey, K.M., Simon, C. and McCutcheon, J.P. (2015) Genome expansion via lineage splitting and genome reduction in the cicada endosymbiont Hodgkinia. *Proceedings of the National Academy of Sciences of the United States of America*, 112(33), 10192–10199.

Campisano, C.J. (2012) Milankovitch cycles, paleoclimatic change, and hominin evolution. *Nature Education Knowledge*, 4, 5.

Camus, P.A. and Lima, M. (2002) Populations, metapopulations, and the open-closed dilemma: The conflict between operational and natural population concepts. *Oikos*, 97(3), 433–438.

Canestrari, D., Marcos, J.M. and Baglione, V. (2011) Helpers at the nest compensate for reduced maternal investment in egg size in carrion crows. *Journal of Evolutionary Biology*, 24, 1870–1878.

Cano, Á., Powell, J., Aiello, A.S., Andersen, H.L., Arbour, T. *et al.* (2025) Insights from a century of data reveal global trends in ex situ living plant collections. *Nature Ecology & Evolution*, 9(2), 214–224.

Cantarero, A., López-Arrabé, J. and Moreno, J. (2015) Selection of nest site and nesting material in the Eurasian nuthatch *Sitta europea*. *Ardea*, 103, 91–94.

Cantrell, R.S., Cosner, C., DeAngelis, D.L. and Padron, V. (2007) The ideal free distribution as an evolutionarily stable strategy. *Journal of Biological Dynamics*, 1, 249–271.

Capanna, E. (2006) Grassi versus Ross: Who solved the riddle of malaria? *International Microbiology*, 9(1), 69–74.

Capon, M., Lysaniuk, B., Godard, V., Clauzel, C. and Simon, L. (2021) Characterizing the landscape compositions of urban wildlife encounters: The case of the stone marten (*Martes foina*), the red fox (*Vulpes vulpes*) and the hedgehog (*Erinaceus europaeus*) in the Greater Paris area. *Urban Ecosystems*, 24, 885–903.

Cappellari, A. and Marini, L. (2021) Improving insect conservation across heterogeneous landscapes using species–habitat networks. *PeerJ*, 9, e10563.

Cappuccino, N. and Arnason, J.T. (2006) Novel chemistry of invasive exotic plants. *Biology Letters*, 2, 189–193.

Cardoso, P. and Leather, S.R. (2019) Predicting a global insect apocalypse. *Insect Conservation and Diversity*, 12, 263–267.

Cardoso, P., Borges, P.A.V. and Veech, J.A. (2009) Testing the performance of beta diversity measures based on incidence data: the robustness to undersampling. *Diversity and Distributions*, 15, 1081–1090.

Cardoso, P., Erwin, T.L., Borges, P.A.V. and New, T.R. (2011) The seven impediments in invertebrate conservation and how to overcome them. *Biological Conservation*, 144(11), 2647–2655.

Cardoso, P., Rigal, F., Carvalho, J.C., Fortelius, M., Borges, P.A.V. *et al.* (2014) Partitioning taxon, phylogenetic and functional beta diversity into replacement and richness difference components. *Journal of Biogeography*, 41, 749–761.

Carey, J. (2005) *The Faber Book of Science*. Faber and Faber, London, UK [The Devil's Chaplain, pp. 114–128].

Carey, J. (2020) *A Little History of Poetry*. Yale University Press, New Haven, USA.

Cariveau, A.B., Anderson, E., Baum, K., Hopwood, J., Lonsdorf, E. *et al.* (2019) Rapid assessment of roadsides as potential habitat for monarchs and other pollinators. *Frontiers in Ecology and Evolution*, 7, 386.

Carl, L.M., Esselman, P. C., Sparks-Jackson, B. L and Wilson, C. C. (2021) The species–area relationship for a highly fragmented temperate river system. *Ecosphere*, 12, e03411.

Carlson, R.L., Shulman, M.J. and Ellis, J.C. (2006) Factors contributing to spatial heterogeneity in the abundance of the common periwinkle *Littorina littorea* (L.). *Journal of Molluscan Studies*, 72, 149–156.

Carmichael, L.E. (2018) *Wildebeest: A Savanna Journey*. Weigl Publishers, Calgary, Canada.

Carmona, C.P., de Bello, F., Mason, N. and Leps, J. (2016) Traits without borders: Integrating functional diversity across scales. *Trends in Ecology and Evolution*, 31, 382e394.

Carmona, C.P., Guerrero, I., Morales, M.B., Oñate, J.J., and Peco, B. (2017) Assessing vulnerability of functional diversity to species loss: A case study in Mediterranean agricultural systems. *Functional Ecology*, 31, 427–435.

Caro, T.M. and O'Doherty, G. (1999) On the use of surrogate species in conservation biology. *Conservation Biology*, 13, 805–814.

Caron, D.A. and Hu, S.K. (2019) Are we overestimating protistan diversity in nature? *Trends in Microbiology*, 27, 197–205.

Carpaneto, G.M., and Fattorini, S. (2002) Flightlessness in psammophilous beetles inhabiting a Mediterranean coastal area: Ecological and biogeographical implications. *Biogeographia*, 23, 71–80.

Carpaneto, G.M., Baviera, C., Biscaccianti, A.B., Brandmayr, P, Mazzei, A. *et al.* (2015) A Red List of Italian saproxylic beetles: Taxonomic overview, ecological features and conservation issues (Coleoptera). *Fragmenta Entomologica*, 47, 53–126.

Carranza, J., Alvarez, F. and Redondo, T. (1990) Territoriality as a mating strategy in red deer. *Animal Behaviour*, 40, 79–88.

Carranza, M.L., Ricotta, C., Fortini, P. and Blasi, C. (2003) Quantifying landscape change with actual vs. potential natural vegetation maps. *Phytocoenologia*, 33, 591–601.

Carretero, L.G., Wollstonecroft, M. and Fuller, D.Q. (2017) A methodological approach to the study of archaeological cereal meals: A case study at Çatalhöyük East (Turkey). *Vegetation History and Archaeobotany*, 26, 415–432.

Carroll, D., Daszak, P., Wolfe, N.D., Gao, G.F., Morel, C.M. *et al.* (2018) The Global Virome Project. *Science*, 359(6378), 872–874.

Carson, R. (1951) *The Sea Around Us*. Oxford University Press, Oxford, UK.

Carta, A., Peruzzi, L. and Ramírez-Barahona, S. (2022) A global phylogenetic regionalization of vascular plants reveals a deep split between Gondwanan and Laurasian biotas. *New Phytologist*, 233, 1494–1504.

Cartar, R.V. and Abrahams, M.V. (1997) Predicting the distribution of organisms among a few patches: Problems with detecting departures from the ideal free distribution. *Oikos*, 78, 388–393.

Cartmill, M. and Smith, F.H. (2011) *The Human Lineage*. John Wiley and Sons, Hoboken, USA.

Carvalho, J.C., Cardoso, P. and Gomes, P. (2012) Determining the relative roles of species replacement and species richness differences in generating beta-diversity patterns. *Global Ecology and Biogeography*, 21, 760–771.

Carvalho, J.C., Cardoso, P., Borges, P.A.V., Schmera, D. and Podani, J. (2013) Measuring fractions of beta diversity and their relationships to nestedness: a theoretical and empirical comparison of novel approaches. *Oikos*, 122, 825–834.

Carvalho, W.D., Mustin, K., Farneda, F.Z., de Castro, I.J., Hilário, R.R. *et al.* (2021) Taxonomic, functional and phylogenetic bat diversity decrease from more to less complex natural habitats in the Amazon. *Oecologia*, 197, 223–239.

Carver, S., Convery, I., Hawkins, S., Beers, R., Eagle, A. *et al.* (2021) Guiding principles for rewilding. *Conservation Biology*, 35, 1882–1893.

Casacci, L.P., Witek, M., Witek, F., Francesca, B. and Bonelli, S. (2011) Habitat preferences of *Maculinea arion* and its *Myrmica* host ants: Implications for habitat management in Italian Alps. *Journal of Insect Conservation*, 15, 103–110.

Cassidy, L.M., Russell, M., Smith, M., Delbarre, G., Cheetham, P. *et al.* (2025) Continental influx and pervasive matrilocality in Iron Age Britain. *Nature*, 637(8048), 1136–1142.

Castelo Branco, R., Antas, P. and Cunha, I. (2014) Preliminary data on *Littorina littorea* development under rearing conditions. *Frontiers in Marine Science, Conference Abstract: International Meeting on Marine Research*. Available at: www.frontiersin.org/10.3389/conf.fmars.2014.02.00029/event_abstract (accessed 1 May 2025).

Castle, M.E. and MacDaid, E. (1972) The decomposition of cattle dung and its effect on pasture. *Grass and Forage Science*, 27, 133–138.

Castle, S.C., Lekberg, Y., Affleck, D. and Cleveland, C.C. (2016) Soil abiotic and biotic controls on plant performance during primary succession in a glacial landscape. *Journal of Ecology*, 104, 1555–1565.

Castro-Insua, A., Gómez-Rodríguez, C. and Baselga, A. (2018) Dissimilarity measures affected by richness differences yield biased delimitations of biogeographic realms. *Nature Communications*, 9, 5084.

Caswell, H. (1976) Community structure: A neutral model analysis. *Ecological Monographs*, 46, 327–354.

Catchpole, C. and Slater, P. (2008) *Bird Song: Biological Themes and Variations*. Cambridge University Press Cambridge, UK.

Catford, J.A., Daehler, C.C., Murphy, H.T., Sheppard, A.W., Hardesty, B.D. *et al.* (2012) The intermediate disturbance hypothesis and plant invasions: Implications for species richness and management. *Perspectives in Plant Ecology, Evolution and Systematics*, 14, 231–41.

Caughley, G. and Sinclair, A.R.E. (1994) *Wildlife Ecology and Management.* Blackwell Scientific Publications, Oxford, UK.

Cavalcante, T., de Souza Jesus, A., Rabelo, R.M., Messias, M.R., Valsecchi, J. *et al.* (2020) Niche overlap between two sympatric frugivorous Neotropical primates: Improving ecological niche models using closely-related taxa. *Biodiversity and Conservation*, 29, 2749–2763.

Cavalier-Smith, T. (1993) Kingdom protozoa and its 18 phyla. *Microbiological Reviews*, 57, 953–994.

Cavalier-Smith, T. (1998) A revised six-kingdom system of life. *Biological Reviews*, 73, 203–266.

Cavalier-Smith, T. (2003) Protist phylogeny and the high-level classification of Protozoa. *European Journal of Protistology*, 39, 338–348.

Cavalier-Smith, T. (2004) Only six kingdoms of life. *Proceedings of the Royal Society B: Biological Sciences*, 271(1545), 1251–1262.

Cavalier-Smith, T. (2010) Kingdoms Protozoa and Chromista and the eozoan root of the eukaryotic tree. *Biology Letters*, 6, 342–345.

Cavé, A.J., Visser, J. and Perdeck, A.C. (1989) Size and quality of the coot *Fulica atra* territory in relation to age of its tenants and neighbours. *Ardea*, 77, 87–97.

Cawthon-Lang, K. (2007) Primate Factsheet: Black spider monkey (*Ateles paniscus*). In: Youlatos, D. (ed.) *Primate Info Net, Wisconsin National Primate Research Center.* University of Wisconsin, Madison, USA. Available at: https://primate.wisc.edu/primate-info-net/pin-factsheets/black-spider-monkey/ (accessed 1 May 2025).

Cazenave, A. and Le Cozannet, G. (2014) Sea level rise and its coastal impacts. *Earth's Future*, 2, 15–34.

Cazzolla Gatti, R., Amoroso, N. and Monaco, A. (2020) Estimating and comparing biodiversity with a single universal metric. *Ecological Modelling*, 424, 109020.

Cernava, T., Aschenbrenner, I., Soh, J., Sensen, C.W. Grube, M. and Berg, G. (2019) Plasticity of a holobiont: Dessication induces fasting-like metabolism within the lichen microbiota. *ISME Journal*, 13, 547– 556.

Cervellini, M., Zannini, P., Di Musciano., M, Fattorini, S., Jiménez-Alfaro, B. *et al.* (2020) A grid-based map for the Biogeographical Regions of Europe. *Biodiversity Data Journal*, 8, e53720.

Chai, Y., Yue, M., Liu, X., Guo, Y., Wang, M. *et al.* (2016a) Patterns of taxonomic, phylogenetic diversity during a long-term succession of forest on the Loess Plateau, China: Insights into assembly process. *Scientific Reports*, 6, 27087.

Chai, Y., Yue, M., Wang, M., Xu, J., Liu, X. *et al.* (2016b) Plant functional traits suggest a change in novel ecological strategies for dominant species in the stages of forest succession. *Oecologia*, 180, 771–783.

Chami, R., Cosimano, T., Fullenkamp, C., Berzaghi, F., Español-Jiménez, S. *et al.* (2022) The value of nature to our health and economic well-being: A framework with application to elephants and whales. In: Paganetto, L. (ed.) *Economic Challenges for Europe After the Pandemic.* Springer Proceedings in Business and Economics, Dordrecht, The Netherlands, pp. 117–162.

Champer, J. and Schlenoff, D. (2024) Battles between ants (Hymenoptera: Formicidae): A review. *Journal of Insect Science*, 24, 25.

Chandler, H.C., Stegenga, B.S. and Stevenson, D.J. (2019) Movement and space use in southern populations of spotted turtles (*Clemmys guttata*). *Southeastern Naturalist*, 18, 602–618.

Chandler, R.J. (1979) Two urban Pied Wagtail roosts. *British Birds*, 72, 299–313.

Chang, C.C. and Turner, B.L. (2019) Ecological succession in a changing world. *Journal of Ecology*, 107, 503–509.

Chao, A. and Chiu, C.-H. (2016) Species richness: Estimation and comparison. In: Balakrishnan, N., Colton, T., Everitt, B., Piegorsch, W., Ruggeri, F. and Teugels, J.L. (eds) *Wiley StatsRef: Statistics Reference Online.* Available at: https://doi.org/10.1002/9781118445112.stat03432.pub2 (accessed 1 May 2025).

Chao, A. and Jost, L. (2012) Coverage-based rarefaction and extrapolation: Standardizing samples by completeness rather than size. *Ecology*, 93, 2533–2547.

Chao, A. and Jost, L. (2015) Estimating species diversity and entropy profiles via discovery rates of new species. *Methods in Ecology and Evolution*, 6, 873–882.

Chao, A. and Lee, S.-M. (1992) Estimating the number of classes via sample coverage. *Journal of the American Statistical Association*, 87, 210–217.

Chao, A., Chiu, C.-H. and Jost, L. (2010) Phylogenetic diversity measures based on Hill numbers. *Philosophical Transactions of the Royal Society B: Biological Sciences*, 365, 3599–3609.

Chao, A., Chiu, C.-H. and Jost, L. (2014a) Unifying species diversity, phylogenetic diversity, functional diversity, and related similarity and differentiation measures through Hill numbers. *Annual Review of Ecology, Evolution, and Systematics*, 45, 297–324.

Chao, A., Gotelli, N.J., Hsieh, T.C., Sander, E.L., Ma, K.H. *et al.* (2014b) Rarefaction and extrapolation with Hill numbers: A framework for sampling and estimation in species diversity studies. *Ecological Monographs*, 84, 45–67.

Chao, A., Chiu, C.-H., Hsieh, T.C., Davis, T., Nipperess, D.A. and Faith, D.P. (2015) Rarefaction and extrapolation of phylogenetic diversity. *Methods in Ecology and Evolution*, 6, 380–388.

Chao, A., Chiu, C.H. and Jost, L. (2016) Phylogenetic diversity measures and their decomposition: A framework based on Hill numbers. In: Pellens, R. and Grandcolas, P. (eds) *Biodiversity Conservation and Phylogenetic Systematics. Topics in Biodiversity and Conservation*, vol. 14. Springer, Cham, Switzerland.

Chao, A., Chiu, C.-H., Villéger, S., Sun, I.-F., Thorn, S. *et al.* (2019) An attribute-diversity approach to functional diversity, functional beta diversity, and related (dis)similarity measures. *Ecological Monographs*, 89, e01343.

Chapin, F.S., Walker, L.R., Fastie, C.L. and Sharman, L.C. (1994) Mechanisms of primary succession following deglaciation at Glacier Bay, Alaska. *Ecological Monographs*, 64, 149–175.

Chapin, F.S., Zaveleta, E.S., Eviner, V.T., Naylor, R.L., Vitousek, P.M. *et al.* (2000) Consequences of changing biotic diversity. *Nature*, 405, 234–242.

Chapman, C.A. and Chapman, L.J. (2000) Determinants of group size in primates: The importance of travel costs. In: Boinski, S. and Garber, P.A. (eds) *On the Move: How and Why Animals Travel in Groups*. University of Chicago Press, Chicago, USA, pp. 24–41.

Chapman, C.A. and Teichroeb, J.A. (2012) What influences the size of groups in which primates choose to live? *Nature Education Knowledge*, 3, 9.

Chapman, J.L. and Reiss, M.J. (1999) *Ecology. Principles and Applications*. Cambridge University Press, Cambridge, UK.

Chapman, M.E. and Karr, J.A. (2012) Environmental and biotic controls on the evolutionary history of insect body size. *Proceedings of the National Academy of Sciences of the United States of America*, 109, 10927–10930.

Chapperon, C. and Seuront, L. (2012) Temporal shifts in motion behaviour and habitat use in an intertidal gastropod. *Journal of the Marine Biological Association*, 93, 1025–1034.

Chartier, M. (2010) Choix et évolution de l'implantation des sites d'habitat au Néolithique ancien dans la vallée de l'Aisne (France). *Bulletin de la Société Préhistorique Française*, 107, 85–95.

Chase, J.M. and Leibold, M.A. (2003) *Ecological Niches: Linking Classical and Contemporary Approaches*. University of Chicago Press, Chicago, USA.

Chaudron, C., Mazalová, M., Kuras, T., Malenovský, I. and Mládek, J. (2021) Introducing ecosystem engineers for grassland biodiversity conservation: A review of the effects of hemiparasitic Rhinanthus species on plant and animal communities at multiple trophic levels. *Perspectives in Plant Ecology, Evolution and Systematics*, 52, 125633.

Chazan, M., Ron, H., Matmon, A., Porat, N., Goldberg, P. *et al.* (2008) Radiometric dating of the Earlier Stone Age sequence in Excavation I at Wonderwerk Cave, South Africa: Preliminary results. *Journal of Human Evolution*, 55, 1–11.

Chazdon, R.L. (2008) Chance and determinism in tropical forest succession. In: Carson, W.P. and Schnitzer, S.A. (eds) *Tropical Forest Community Ecology*. Wiley-Blackwell, Oxford, UK.

Chazdon, R.L., Colwell, R.K., Denslow, J.S. and Guariguata, M.R. (1998) Statistical methods for estimating species richness of woody regeneration in primary and secondary rain forests of NE Costa Rica. In: Dallmeier, F. and Comiskey, J.A. (eds) *Forest Biodiversity Research, Monitoring and Modeling: Conceptual Background and Old World Case Studies*. Parthenon Publishing, Paris, France, pp. 285–309.

Chelazzi, G. and Vannini, M. (1988) *Behavioral Adaptation to Intertidal Life*. NATO Science Series A: Life Sciences. Springer Nature, Cham, Switzerland.

Chelli, S., Marignani, M., Barni, E., Petraglia, A., Puglielli, G. *et al.* (2019) Plant–environment interactions through a functional traits perspective: A review of Italian studies. *Plant Biosystems*, 153, 853–869.

Chen, B. and Grinfeld, M. (2024) Decomposing diversity into measures of evenness, similarity, and richness. *Ecology and Evolution*, 14, e10952.

Cheng, J., Mi, X., Nadrowski, K., Ren, H., Zhang, J. and Ma, K. (2012) Separating the effect of mechanisms shaping species-abundance distributions at multiple scales in a subtropical forest. *Oikos*, 121, 236–244.

Chertow, M.R. (2000) The IPAT equation and its variants. *Journal of Industrial Ecology*, 4, 13–29.

Chesson, P. (2000) Mechanisms of maintenance of species diversity. *Annual Review of Ecology, Evolution and Systematics*, 31, 343–366.

Cheung, P.-Y., Nozawa, Y. and Miki, T. (2021) Ecosystem engineering structures facilitate ecological resilience: A coral reef model. *Ecological Research*, 36, 673–685.

Chiarucci, A., Di Biase, R.M., Fattorini, L., Marcheselli, M. and Pisani, C. (2018) Joining the incompatible: Exploiting purposive lists for the sample-based estimation of species richness. *Annals of Applied Statistics*, 12, 1679–1699.

Chiarucci, A., Wilson, J.B., Anderson, B.J. and De Dominicis, V. (1999) Cover versus biomass as an estimate of species abundance: Does it make a difference to the conclusions? *Journal of Vegetation Science*, 10, 35–42.

Chimpanzee Sequencing and Analysis Consortium (2005) Initial sequence of the chimpanzee genome and comparison with the human genome. *Nature*, 437, 69–87.

Chisholm, R.A. (2007) Sampling species abundance distributions: Resolving the veil-line debate. *Journal of Theoretical Biology*, 247, 600–607.

Chiu, C.-H. and Chao, A. (2016) Estimating and comparing microbial diversity in the presence of sequencing errors. *PeerJ*, 4,e1634.

Chiu, C.H. and Chao, A. (2020) Distance-based functional diversity measures and their decomposition: A framework based on Hill numbers. *PLOS One*, 9, e100014.

Chiu, C.-H., Wang, Y.-T., Walther, B.-A. and Chao, A. (2014) An improved nonparametric lower bound of species richness via a modified Good–Turing frequency formula. *Biometrics*, 70, 671–682.

Choudhary, N.L., Chishty, N., Parveen, R. and Sharma, P. (2020) Abundance and diversity gradient of butterflies from urban to rural habitats in Udaipur District, Rajasthan, India. *Environment and Ecology*, 38, 903–911.

Choudri, B.S., Charabi, Y. and Ahmed, M. (2018) Pesticides and herbicides. *Water Environment Research*, 90, 1662–1678.

Chowdhury, S., Shahriar, S.A., Böhm, M., Jain, A., Aich, U. *et al.* (2021) Urban green spaces in Dhaka, Bangladesh, harbour nearly half the country's butterfly diversity. *Journal of Urban Ecology*, 7, 1–11.

Christaller, W. (1933) *Die zentralen Orte in Süddeutschland.* Fischer, Jena, Germany.

Churchill, S.E. (2014) *Thin on the Ground: Neandertal Biology, Archeology and Ecology.* Wiley-Blackwell, Hoboken, USA.

Churchill, S.E., Walker, C.S. and Schwartz, A.M. (2016) Home-range size in large-bodied carnivores as a model for predicting Neandertal territory size. *Evolutionary Anthropology*, 25, 117–123.

Chyleński, M., Ehler, E., Somel, M., Yaka, R., Krzewińska, M. *et al.* (2019). Ancient mitochondrial genomes reveal the absence of maternal kinship in the burials of Çatalhöyük people and their genetic affinities. *Genes*, 10, 207.

Clark, A.E. (2016) Time and space in the Middle Paleolithic: Spatial structure and occupation dynamics of seven open-air sites. *Evolutionary Anthropology*, 25, 153–163.

Clark, A.T., Knops, J.M.H. and Tilman, D. (2019) Contingent factors explain average divergence in functional composition over 88 years of old field succession. *Journal of Ecology*, 107, 545–558.

Clark, G. (1969) *World Prehistory: A New Outline*, 2nd edn. Cambridge University Press, Cambridge, UK.

Clark, J.S. (2010) Individuals and the variation needed for high species diversity in forest trees. *Science*, 327, 1129–1132.

Clark, J.S., Dietze, M., Chakraborty, S., Agarwal, P.K., Ibanez, I. *et al.* (2007) Resolving the biodiversity paradox. *Ecology Letters*, 10, 647–659.

Clarke, K.M., Fisher, B.L. and LeBuhn, G. (2008) The influence of urban park characteristics on ant (Hymenoptera, Formicidae) communities. *Urban Ecosystems*, 11, 317–334.

Clarke, K.R. and Warwick, R.M. (1998) A taxonomic distinctness index and its statistical properties. *Journal of Applied Ecology*, 35, 523–531.

Clarke, K.R. and Warwick, R.M. (2001) A further biodiversity index applicable to species lists: Variation in taxonomic distinctness. *Marine Ecology Progress Series*, 216, 265–278.

Clauss, M., Steuer, P., Muller, D.W.H, Codron, D. and Hummel, J. (2013) Herbivory and body size: Allometries of diet quality and gastrointestinal physiology, and implications for herbivore ecology and dinosaur gigantism. *PLOS One*, 8, e68714.

Cleland, E.E. (2011) Biodiversity and ecosystem stability. *Nature Education Knowledge*, 3, 14.

Clements, F.E. (1904) *Development and Structure Of Vegetation.* Report of the Botanical Survey of Nebraska, 7. University of Nebraska, Lincoln, USA.

Clements, F.E. (1907) *Plant Physiology and Ecology.* Holt and Co., New York, USA.

Clements, F.E. (1916) *Plant Succession: An Analysis of the Development of Vegetation.* Carnegie Institution, Washington DC, USA.

Clements, F.E. (1936) Nature and structure of the climax. *Journal of Ecology*, 24, 252–284.

Clements, F.E. and Shelford, V.E. (1939) *Bio-Ecology.* Wiley, New York, USA.

Clutton-Brock, T.H. and Harvey, P.H. (1979) Home range size, population density and phylogeny in primates. In: Bernstein, I.S. and Smith, E.O. (eds) *Primate Ecology and Human Origins*, Garland Press, New York, USA, pp. 201–214.

Coakley, S. and Petti, C. (2021) Impacts of the invasive *Impatiens glandulifera*: Lessons learned from one of Europe's top invasive species. *Biology*, 10, 619.

Cobain, M.R.D., Brede, M. and Trueman, C.N. (2019) Taylor's power law captures the effects of environmental variability on community structure: An example from fishes in the North Sea. *Journal of Animal Ecology*, 88, 290–301.

Cobián Güemes, A.G., Youle, M., Cantú, V.A., Felts, B., Nulton, J. and Rohwer, F. (2016) Viruses as winners in the game of life. *Annual Review of Virology*, 3, 197–214.

Cobley, P. (2010) *The Routledge Companion to Semiotics.* Routledge, London, UK.

Cobos, M.E., Barve, V., Barve, N., Jiménez-Valverde, A. and Nuñez-Penichet, C. (2022) rangemap: An R package to explore species geographic ranges. *Biodiversity Informatics*, 17, 59–66.

Cody, M.L. (1971) Finch flocks in the Mojave desert. *Theoretical Population Biology*, 2, 141–158.

Cogger, H.G and Zweifel, R.G. (2000) *Encyclopedia of Reptiles and Amphibians.* University of Chicago Press, Chicago, USA.

Cohen, J.E. and Newman, C.M. (1985) A stochastic theory of community food webs I. Models and aggregated data. *Proceedings of the Royal Society B: Biological Sciences*, 224, 421–448.

Coleman, D.C. and Crossley, D. (2004) *Fundamentals of Soil Ecology*, 2nd edn. Elsevier Academic Press, London, UK.

Collin, P.H. (1988) *Dictionary of Ecology and the Environment.* Peter Collin Publishing, Teddington, UK.

Collins, M. (2013) Long-term climate change: projections, commitments and irreversibility. In: Stocker, T.F., Qin, D., Plattner, G.-K., Tignor, M., Allen, S.K. *et al.* (eds) *Climate Change 2013: The Physical Science Basis. Contribution of Working Group I to the Fifth Assessment Report of the Intergovernmental Panel on Climate Change.* Cambridge University Press, Cambridge, UK.

Colvin, B.A., Degregorio, R. and Fleetwood, C. (1996) Norway rat infestation of urban landscaping and preventative design criteria. In: Timm, R.M. and Crabb, A.C. (eds) *Proceedings of the 17th Vertebrate Pest Conference.* University of California, Davis, USA, pp. 165–171.

Colwell, R.K. (1979) Toward a unified approach to the study of species diversity. In: Grassle, J.F., Patil, G.P., Smith, W. and Taillie, C. (eds) *Ecological Diversity in Theory and Practice*. International Cooperative Publishing House, Pairland, USA, pp. 75–91.

Colwell, R.K. and Coddington, J.A. (1994) Estimating terrestrial biodiversity through extrapolation. *Philosophical Transactions of the Royal Society B: Biological Sciences*, 345, 101–118.

Colwell, R.K., Mao, C.X. and Chang, J. (2004) Interpolating, extrapolating, and comparing incidence-based species accumulation curves. *Ecology*, 85, 2717–2727.

Colwell, R.K., Chao, A., Gotelli, N.J., Lin, S.-Y., Mao, C.X. *et al.* (2012) Models and estimators linking individual-based and sample-based rarefaction, extrapolation, and comparison of assemblages. *Journal of Plant Ecology*, 5, 3–21.

Conde, D.A., Flesness, N., Colchero, F., Jones, O.R. and Scheuerlein, A. (2011) An emerging role of zoos to conserve biodiversity. *Science*, 331, 1390–1391.

Conedera, M. and Krebs, P. (2008) History, present situation and perspective of chestnut cultivation in Europe. *Acta Horticulturae*, 784, 23–28.

Conedera, M., Krebs, P., Tinner, W., Pradella, M. and Torriani, D. (2004) The cultivation of *Castanea sativa* (Mill.) in Europe, from its origin to its diffusion on a continental scale. *Vegetation History and Archaeobotany*, 13, 161–179.

Conedera, M., Tinner, W., Krebs, P., de Rigo, D. and Caudullo, G. (2016) Castanea sativa in Europe: Distribution, habitat, usage and threats. In: San-Miguel-Ayanz, J., de Rigo, D., Caudullo, G., Houston Durrant, T. and Mauri, A. (eds) *European Atlas of Forest Tree Species*. Publication Office of the EU, Luxembourg.

Connell, J.H. (1978) Diversity in tropical rain forests and coral reefs. *Science*, 199, 1302–1310.

Connell, J.H. and Slatyer, R.O. (1977) Mechanisms of succession in natural communities and their role in community stability and organization. *American Naturalist*, 111, 1119–1144.

Connelly, J.N., Bizzarro, M., Krot, A.N., Nordlund, Å., Wielandt, D. and Ivanova, M.A. (2012) The absolute chronology and thermal processing of solids in the solar protoplanetary disk. *Science*, 338, 651–655.

Connolly, S.R., Hughes, T.P., Bellwood, D.R. and Karlson, R.H. (2005) Community structure of corals and reef fish at multiple scales. *Science*, 309, 1363–1365.

Connor, E.F., Courtney, A.C. and Yoder, J.M. (2000) Individuals-area relationships: The relationship between animal population density and area. *Ecology*, 81, 734–748.

Conrad, O.K.F., Warren, A.M.S., Fox, O.R., Parsons, A.M.S. and Woiwod, O.I.P. (2006) Rapid declines of common, widespread British moths provide evidence of an insect biodiversity crisis. *Biological Conservation*, 132, 279–291.

Constandse-Westermann, T.S. (1972) *Coefficients of Biological Distance*. Anthropological Publications, Oosterhout, The Netherlands.

Cook, L.M. (2003) The rise and fall of the *carbonaria* form of the peppered moth. *Quarterly Review of Biology*, 78, 399–417.

Cook, L.M. (2008a) Diversity and evenness from sequential sightings. *Insect Conservation and Diversity*, 1, 263–265.

Cook, L.M. (2008b) Variation with habitat in *Cepaea nemoralis*: The Cain and Sheppard diagram. *Journal of Molluscan Studies*, 74, 239–243.

Cook, L.M. and Graham, C.S. (1996) Evenness and species number in some moth populations. *Biological Journal of the Linnean Society*, 58, 75–84.

Cook, L.M. and Saccheri, I.J. (2013) The peppered moth and industrial melanism: Evolution of a natural selection case study. *Heredity*, 110, 207–212.

Cooke, J.G. (2018) Blue whale; *Balaenoptera musculus* (errata version published in 2019). *IUCN Red List of Threatened Species 2018*. Available at: https://dx.doi.org/10.2305/IUCN.UK.2018-2.RLTS.T2477A156923585.en (accessed 1 May 2025).

Coolidge, F.L. and Wynn, T. (2004) A cognitive and neuropsychological perspective on the Châtelperronian. *Journal of Anthropological Research*, 60, 55–73.

Cooper, K.M., Bolam, S.G., Downie, A.-L. and Barry, J. (2019) Biological-based habitat classification approaches promote cost-efficient monitoring: An example using seabed assemblages. *Journal of Applied Ecology*, 56, 1085–1098.

Cooper, N., Rodríguez, J. and Purvis, A. (2008) A common tendency for phylogenetic overdispersion in mammalian assemblages. *Proceedings of the Royal Society B: Biological Sciences*, 275(1646), 2031–2037.

Cope, G.C., Campbell, J.W., Grodsky, S.M. and Ellis, J.D. (2019) Evaluation of nest-site selection of ground-nesting bees and wasps (Hymenoptera) using emergence traps. *Canadian Entomologist*, 151, 260–271.

Corlett, R.T. (2016) The role of rewilding in landscape design for conservation. *Current Landscape Ecology Reports*, 1, 127–133.

Cornelissen, J., Lavorel, S., Garnier, E., Diaz, S., Buchmann, N. *et al.* (2003) A handbook of protocols for standardised and easy measurement of plant functional traits worldwide. *Australian Journal of Botany*, 51, 335–380.

Cornwell, W.K. and Ackerly, D.D. (2009) Community assembly and shifts in plant trait distributions across an environmental gradient in coastal California. *Ecological Monographs*, 79, 109–126.

Cornwell, W.K., Schwilk, D.W. and Ackerly, D.D. (2006) A trait-based test for habitat filtering: Convex hull volume. *Ecology*, 87, 1465–1471.

Cosentino, F., Castiello, G. and Maiorano, L.A. (2023) A dataset on African bats' functional traits. *Scientific Data* 10, 623.

COSEWIC (2010) COSEWIC assessment and status report on the Atlantic Salmon *Salmo salar* (Nunavik population, Labrador population, Northeast Newfoundland population, South Newfoundland population, Southwest Newfoundland population, Northwest Newfoundland population, Quebec Eastern North Shore population, Quebec Western North Shore population, Anticosti Island population, Inner St. Lawrence population, Lake Ontario population, Gaspé-Southern Gulf of St. Lawrence population, Eastern Cape Breton population, Nova Scotia Southern Upland population, Inner Bay of Fundy population, Outer Bay of Fundy population) in Canada. Committee on the Status of Endangered Wildlife in Canada. Ottawa, Canada. Available at: www.canada.ca/en/environment-climate-change/services/species-risk-public-registry/cosewic-assessments-status-reports/atlantic-salmon.html (accessed 1 May 2025).

COSEWIC (2013) COSEWIC assessment and status report on the Oregon Forestsnail *Allogona townsendiana* in Canada. Committee on the Status of Endangered Wildlife in Canada. Ottawa, Canada. Available at: www.registrelep-sararegistry.gc.ca/virtual_sara/files/cosewic/sr_escargot_forestier_oregon_forestsnail_1213_e.pdf (accessed 1 May 2025).

Costanza, R., d'Arge, R., de Groot, R., Farber, S., Grasso, M. *et al.* (1997) The value of the world's ecosystem services and natural capital. *Nature*, 387, 253–260.

Costello, M.J., Basher, Z., McLeod, L., Asaad, I., Claus, S. *et al.* (2017a). Methods for the study of marine biodiversity. In: Walters, M. and Scholes, R.J. (eds) *The GEO Handbook on Biodiversity Observation Networks*. Springer, Cham, Switzerland, pp. 129–163.

Costello, M.J., Tsai, P., Wong, P.S., Cheung, A.K.L., Basher, Z. and Chaudhary, C. (2017b) Marine biogeographic realms and species endemicity. *Nature Communications*, 8, 1057.

Cottee-Jones, H.E.W. and Whittaker, R.J. (2012) The keystone species concept: A critical appraisal. *Frontiers of Biogeography*, 4, 117–127.

Cotter, G. (1990) *The Curlew*. Shire Publications, Princes Risborough, UK.

Coulson, T. and Godfray, H.C.J. (2007) Single-species dynamics. In: May, R.M. and McLean, A.R. (eds) *Theoretical Ecology: Principles and Applications*. Oxford University Press. Oxford, UK, pp. 17–34.

Courbaud, B., Vieilledent, G. and Kunstler, G. (2012) Intra-specific variability and the competition–colonisation trade-off: Coexistence, abundance and stability patterns. *Theoretical Ecology*, 5, 61–71.

Courel, B., Meadows, J., González, L.G., Lucquin, A. and McLaughlin, R. (2021) The use of early pottery by hunter-gatherers of the Eastern European forest-steppe. *Quaternary Science Reviews*, 269, 107143.

Courtney, S.P. (1980) Studies on the biology of the butterflies *Anthocharis cardamines* (L) and *Pieris napi* (L), in relation to speciation in Pierinae. Doctoral thesis, Durham University, Durham, UK. Available at: http://etheses.dur.ac.uk/8085/ (accessed 1 May 2025).

Cowie, R. (2009) *Theba pisana* (white garden snail). Available at: https://www.cabidigitallibrary.org/doi/10.1079/cabicompendium.62094 (accessed 1 May 2025).

Cowie, R.H., Bouchet, P. and Fontaine, B. (2022) The Sixth Mass Extinction: Fact, fiction or speculation? *Biological Reviews*, 97, 640–663.

Cowles, H.C. (1899) The ecological relations of the vegetation on the sand dunes of Lake Michigan. *Botanical Gazette*, 27, 95–117.

Cowley, M.J.R., Thomas, C.D., Roy, D.B., Wilson, R.J., León-Cortés, J.L. *et al.* (2001) Density-distribution relationships in British butterflies. I. The effect of mobility and spatial scale. *Journal of Animal Ecology*, 70, 410–425.

Cowlishaw, G. (1992) Song function in gibbons. *Behaviour*, 121, 131–153.

Cowlishaw, G. (1994) Vulnerability to predation in baboon populations. *Behaviour*, 131, 293–304.

Cox, C.B. (2001) The biogeographic regions reconsidered. *Journal of Biogeography*, 28, 511–523.

Cox, C.B. and Moore, P.D. (2010) *Biogeography: An Ecological and Evolutionary Approach*, 8th edn. John Wiley and Sons, Chichester, UK.

Craft, M.E., Hampson, K., Ogutu, J.O. and Durant, S.M. (2015) Carnivore communities in the Greater Serengeti ecosystem. In: Sinclair, A.R.E., Metzger, K.L., Mduma, S.A.R. and Fryxell, J.M. (eds) *Serengeti IV: Sustaining Biodiversity in a Coupled Human-Natural System*. Chicago University Press, Chicago, USA.

Crain, B.J., Sánchez-Cuervo, A.M., White, J.W. and Steinberg, S.J. (2015) Conservation ecology of rare plants within complex local habitat networks. *Oryx*, 49, 696–703.

Cramer, V.A. and Hobbs, R.J. (eds). (2007) *Old Fields Dynamics and Restoration of Abandoned Farmland*. Washington DC, USA.

Cresci, A. (2020) A comprehensive hypothesis on the migration of European glass eels (*Anguilla anguilla*). *Biological Reviews*, 95, 1273–1286.

Cressman, R. and Krivan, V. (2006) Migration dynamics for the ideal free distribution. *American Naturalist*, 168, 384–397.

Cressman, R., Krivan, V. and Garay, J. (2004) Ideal free distributions, evolutionary games, and population dynamics in multiple-species environments. *American Naturalist*, 164, 473–489.

Crisfield, V.E., Blanchet, F.G., Raudsepp-Hearne, C. and Gravel, D. (2024) How and why species are rare:

Towards an understanding of the ecological causes of rarity. *Ecography*, e07037.

Crocker, R.L. and Major, J. (1955) Soil development in relation to vegetation and surface age at Glacier Bay, Alaska. *Journal of Ecology*, 43, 427–448.

Crook, J.H. (1966) Gelada baboon herd structure and movement: A comparative report. *Symposium of the Zoological Society of London*, 18, 237–258.

Crous, C.J., Samways, M.J. and Pryke, J.S. (2014) Rockiness determines meso-scale conservation of butterflies in Afro-montane grassland. *Journal of Insect Conservation*, 18, 77–83.

Crutzen, P. and Stoermer, E. (2000) The Anthropocene. *Global Change Newsletter*, 14, 17–18.

Cui, Z-W., Wang, Z-L., Shao, Q., Raubenheimer, D. and Lu, J-Q. (2018) Macronutrient signature of dietary generalism in an ecologically diverse primate in the wild. *Behavioral Ecology*, 29, 804–813.

Culver, D.C. and Pipan, T. (2019) *The Biology of Caves and Other Subterranean Habitats*, 2nd edn. Oxford University Press, Oxford, UK.

Cummings, K.S., Jones, H.A. and Lopes-Lima, M. (2016) Rapid bioassessment methods for freshwater molluscs. In: Larsen, T.H. (ed.) *Core Standardized Methods for Rapid Biological Field Assessment*. Conservation International, Arlington, VA, pp. 185–207.

Cummins, V., Coughlan, S., McClean, O., Connolly, N., Mercer, J. and Burnell, G. (2002) *An Assessment of the Potential for the Sustainable Development of the Edible Periwinkle*, Littorina littorea, *Industry in Ireland*. Marine Resource Series, Marine Institute. Available at: https://oar.marine.ie/handle/10793/218 (accessed 2 May 2025).

Cunliffe, B. (2005) *Iron Age Communities in Britain: An Account of England, Scotland and Wales from the Seventh Century BC until the Roman Conquest*, 4th edn. Routledge, London, UK.

Curry, M., Reed, E.H. and Bourne, S. (2014) Catching the marsupial 'lion' by the tail: *Thylacoleo carnifex* and the Naracoorte Caves. *Australian Cave and Karst Management Association Journal*, 97, 6–16.

Curry, P. and Schmidt, O. (2007) The feeding ecology of earthworms – A review. *Pedobiologia*, 50, 463–477.

Curtin, S. (2005) Nature, wild animals and tourism: An experiential view. *Journal of Ecotourism*, 4, 1–15.

Cushing, P.E. (2012) Spider-ant associations: An updated review of myrmecomorphy, myrmecophily, and myrmecophagy in spiders. *Psyche: A Journal of Entomology*, 2012, 151989.

Cushman, S., Gutzweiler, K., Evans, J.S. and McGarigal K. (2010) Landscape metrics, scales of resolution. In: Cushman, S.A. and Huettmann, F. (eds) *Spatial Complexity, Informatics, and Wildlife Conservation*. Springer, Tokyo, Japan.

Cuthbert, R.N., Dalu, T., Wasserman. R.J., Sentis, A., Weyl, O.L.F. *et al.* (2021) Prey and predator density-dependent interactions under different water volumes. *Evolutionary Ecology*, 11, 6504–6512.

Cutler, N.A., Chaput, D.L. and van der Gast, C.J. (2014) Long-term changes in soil microbial communities during primary succession. *Soil Biology and Biochemistry*, 69, 359–370.

Cvetković, N.J. and Dimitrijević, V.M. (2014) Cave bears (Carnivora, Ursidae) from the Middle and Late Pleistocene of Serbia: A revision. *Quaternary International*, 339/340, 197–208.

Czechowski, W and Godzińska, E.J. (2015) Enslaved ants: Not as helpless as they were thought to be. *Insectes Sociaux*, 62, 9–22.

Dajoz, R. (1967) Écologie et biologie des Coléoptères xylophages de la hêtraie. *Vie et Milieu*, 17, 523–763.

Dajoz, R. (2007) *Les Insectes et la Forêt. Rôle et Diversité des Insectes dans le Milieu Forestier*, 2nd edn. Lavoisier, Paris, France.

Dalrymple, G.B. (2001) The age of the earth in the twentieth century: A problem (mostly) solved. *Geological Society London Memoirs*, 190, 205–221.

Daly, A.J., Baetens, J.M. and De Baets, B. (2018) Ecological diversity: Measuring the unmeasurable. *Mathematics*, 6, 119.

Dana, J.D. (1853) On an isothermal oceanic chart, illustrating geographic distribution of marine animals. *American Journal of Science and Arts*, 66, 153–167, 314–327.

Danchin, E. and Richner, H. (2001) Viable and unviable hypotheses for the evolution of raven roosts. *Animal Behaviour*, 61, F7–F11.

Danchin, É., Giraldeau, L.-A. and Cézilly, F. (2008) *Behavioural Ecology*. Oxford University Press, Oxford, UK.

D'Ancona, U. (1927) Intorno alle associazioni biologiche e a un saggio di teoria matematica sulle stesse con particolar riguardo all'idrobiologia. *Sonderdruck aus Internationale Revue der gesamten Hydrobiologie und Hydrographie*, 17, 189–225.

Daniels, J.C., Nordmeyer, C. and Runquist, E. (2018) Improving standards for at-risk butterfly translocations. *Diversity*, 10, 67.

Dannewitz, J., Maes, G.E., Johansson, L., Wickström, H., Volchaert, F.A.M. and Järvi, T. (2005) Panmixia in the European eel: A matter of time? *Proceedings of the Royal Society B: Biological Sciences*, 272(1568), 1129–1137.

Dapporto, L. and Dennis, R.L.H. (2013) The generalist-specialist continuum: Testing predictions for distribution and trends in British butterflies. *Biological Conservation*, 157, 229–236.

Dapporto, L., Bruschini, C., Baracchi, D., Cini, A., Gayubo, S. *et al.* (2009) Phylogeography and counter-intuitive inferences in island biogeography: Evidence from morphometric markers in the mobile butterfly *Maniola jurtina* (Linnaeus) (Lepidoptera, Nymphalidae). *Biological Journal of the Linnean Society*, 98, 677–692.

Dapporto, L., Habel, J.C., Dennis, R.L.H. and Schmitt, T. (2011a) The biogeography of the western Mediterranean: Elucidating contradictory distribution patterns in butterflies. *Biological Journal of the Linnean Society*, 103, 571–577.

Dapporto, L., Schmitt, T., Vila, R., Scalercio, S., Biermann, H. *et al.* (2011b) Phylogenetic island disequilibrium: Evidence for ongoing long-term population dynamics in two Mediterranean butterflies. *Journal of Biogeography*, 38, 854–867.

Dapporto, L., Bruschini, C., Dincă, V., Vila, R. and Dennis, R.L.H. (2012) Identifying zones of phylogenetic compression in West Mediterranean butterflies (Satyrinae): Refugia, invasion and hybridization. *Diversity and Distributions*, 18, 1066–1076.

Dapporto, L., Ciolli, G., Dennis, R.L.H., Fox, R. and Shreeve, T.G. (2015) A new procedure for extrapolating turnover regionalization at mid-small spatial scales applied to British butterflies. *Methods in Ecology and Evolution*, 6, 1287–1297.

Dapporto, L., Cini, A., Menchetti, M., Vodă, R., Bonelli, S. *et al.* (2017) Rise and fall of island butterfly diversity: Understanding genetic differentiation and extinction in a highly diverse archipelago. *Diversity and Distributions*, 23, 1169–1181.

Dapporto, L., Cini, A., Vodă, R., Dincă, V., Weimer, M. *et al.* (2019) Integrating three comprehensive data sets shows that mitochondrial DNA variation is linked to species traits and paleogeographic events in European butterflies. *Molecular Ecology Resources*, 19, 1623–1636.

Dapporto, L., Menchetti, M., Dincă, V., Talavera, G., Garcia-Berro, A. *et al.* (2024) The genetic legacy of the Quaternary ice ages for West Palearctic butterflies. *Science Advances*, 10(38), eadm8596.

Dardé, M.L., Ajzenberg, D. and Smith, J. (2011) Population structure and epidemiology of *Toxoplasma gondii*. In: Weiss, L.M. and Kim, K. (eds) Toxoplasma gondii: *The Model Apicomplexan. Perspectives and Methods*. Elsevier, Amsterdam, The Netherlands, pp. 49–80.

Daru, B.H., Bank, M.V., Maurin, O., Yessoufou, K., Schaefer, H. *et al.* (2016) A novel phylogenetic regionalization of phytogeographical zones of southern Africa reveals their hidden evolutionary affinities. *Journal of Biogeography*, 43, 155–166.

Daru, B.H., Holt, B.G., Lessard, J.P., Yessoufou, K. and Davies, T.J. (2017) Phylogenetic regionalization of marine plants reveals close evolutionary affinities among disjunct temperate assemblages. *Biological Conservation*, 213, 351–356.

Darvill, T. (2016) Houses of the holy: Architecture and meaning in the structure of Stonehenge, Wiltshire, UK. *Time and Mind*, 9, 89–121.

Darvill, T. (2022) *Keeping time at Stonehenge*. Antiquity, 96, 319–335.

Darwall, W.R.T. (2023) Atlantic Salmon: *Salmo salar*. *IUCN Red List of Threatened Species* 2023. Available at: https://dx.doi.org/10.2305/IUCN.UK.2023-1.RLTS.T19855A67373433.en (accessed 2 May 2025).

Darwin, C. (1859) *On the Origin of Species by Means of Natural Selection*. John Murray, London, UK.

Darwin, C. (1872) *The Origin of Species by Means of Natural Selection*. John Murray, London, UK.

Darwin, C. (1881) *The Formation of Vegetable Mould, Through the Action of Worms*. John Murray, London, UK.

Darwin, C. (1887) *The Life and Letters of Charles Darwin, Including an Autobiographical Chapter* (Darwin, F. ed.). John Murray, London, UK.

da Silva, I.A. and Batalha, M.A. (2006) Taxonomic distinctness and diversity of a hyper seasonal savanna in central Brazil. *Diversity and Distributions*, 12, 725–730.

da Silva, M.A.F., Mendes, C.B. and Prevedello, J.A. (2024) How important is passive sampling to explain species-area relationships? A global synthesis. *Landscape Ecology*, 39, 50.

David, R. (2003) *Religion and Magic in Ancient Egypt*. Penguin Books, London, UK.

Davies, K.F., Gascon, C. and Margules, C.R. (2001) Habitat fragmentation: Consequences, management, and future research priorities. In: Soulé M.E. and Orians G.H. (eds) *Conservation Biology. Research Priorities for the Next Decade*. Island Press, Washington DC, USA pp. 81–97.

Davies, N.B. (2016) *Cuckoo. Cheating by Nature*. Bloomsbury, Oxford, UK.

Davies, N.B., Krebs, J.R. and West, S.A. (2012) *An Introduction to Behavioural Ecology*. Wiley-Blackwell, Oxford, UK.

Davies, N.S., McMahon, W.J. and Berry, C.M. (2024) Earth's earliest forest: Fossilized trees and vegetation-induced sedimentary structures from the Middle Devonian (Eifelian) Hangman Sandstone Formation, Somerset and Devon, SW England. *Journal of the Geological Society*, 181, 4.

Davies, Z.G. and Pullin, A. (2007) Are hedgerows effective corridors between fragments of woodland habitat? An evidence-based approach. *Landscape Ecology*, 22, 333–351.

Davies, Z.G., Wilson, R.J., Coles, S. and Thomas, C.D. (2006) Changing habitat associations of a thermally constrained species, the silver-spotted skipper butterfly, in response to climate warming. *Journal of Animal Ecology*, 75, 247–256.

Davis, J.D., Olivença, D.V., Brown, S.P. and Voit, E.O. (2022) Methods of quantifying interactions among populations using Lotka-Volterra models. *Frontiers in Systems Biology*, 2, 1021897.

Davison, A., Jackson, H.J., Murphy, E.W. and Reader, T. (2019) Discrete or indiscrete? Redefining the colour polymorphism of the land snail *Cepaea nemoralis*. *Heredity*, 123, 162–175.

Davoli, M., Monsarrat, S., Pedersen, R.Ø., Scussolini, P., Karger, D.N. *et al.* (2023) Megafauna diversity and functional declines in Europe from the Last Interglacial to the present. *Global Ecology and Biogeography*, 33, 34–47.

Dawson, A., Paciorek, C.J., Goring, S.J., Jackson, S.T., McLachlan, J.S. and Williams, J.W. (2019) Quantifying trends and uncertainty in prehistoric forest composition in the upper Midwestern United States. *Ecology*, 100, e02856.

Dawson, M.N. (2016) Island and island-like marine environments. *Global Ecology and Biogeography*, 25, 831–846.

Dawson, M.N. and Hamner, W.M. (2008) A biophysical perspective on dispersal and the geography of evolution in marine and terrestrial systems. *Journal of the Royal Society Interface*, 5(19), 135–150.

Day, M.E. (2023) A fungal renaissance. *The Linnean*, 39, 25–30.

Dayton, P.K., Tegner, M.J., Edwards, P.B. and Riser, K.L. (1998) Sliding baselines, ghosts, and reduced expectations in kelp forest communities. *Ecological Applications*, 8, 309–322.

De, K., Singh, A.P., Sarkar, A., Singh, K., Siliwal, M. *et al.* (2023) Relationship between species richness, taxonomic distinctness, functional diversity, and local contribution to β diversity and effects of habitat disturbance in the riparian spider community of the Ganga River, India. *Ecological Processes*, 12, 13.

Deacon, C., Samways, M.J. and Pryke, J.S. (2021) Relative importance of ecological versus biological traits in driving range sizes of African dragonflies. *Journal of Biogeography*, 48, 1309–1321.

DeAngelis, T. (2022) A broader view of psychopathy. *American Psychological Association*, 53, 2.

de Bello, F., Lepš, J., Lavorel, S. and Moretti, M. (2007) Importance of species abundance for assessment of trait composition: An example based on pollinator communities. *Community Ecology*, 8, 163–170.

de Bello, F., Carmona, C.P., Dias, A., Götzenberger, L., Moretti, M. and Berg, M. (2021) *Handbook of Trait-Based Ecology: From Theory to R Tools*. Cambridge University Press, Cambridge, UK.

de Candolle, A.P. (1805) Explication de la carte Botanique de la France. In: de M. Lamarck, J.B.P.A. and de Candolle, A.P. (eds) *Flore française, ou descriptions succinctes de toutes les plantes qui croissent naturellement en France, disposées selon une nouvelle méthode d'analyse, et précédées par un exposé des principes élémentaires de la botanique*, 3rd edn. Desray, Paris, France, pp. v–xii.

de Candolle, A.P. (1820) *Essai élémentaire de géographie botanique. Dictionnaire des Sciences Naturelles* (Vol. 18). F. Levrault, Paris, France.

Deedler, C.L. (1984) Synopsis of biological data on the eel *Anguilla anguilla* (Linnaeus) 1758. FAO Fisheries Synopsis No. 80. Revision 1. FAO, Rome, Italy.

Deevey, E.S. Jr. (1947) Life tables for natural populations of mammals. *Quarterly Review of Biology*, 22, 283–314.

De Gregorio, C., Carugati, F., Estienne, V., Valente, D., Raimondi, T. *et al.* (2021) Born to sing! Song development in a singing primate. *Current Zoology*, 67, 597–608.

DeJean, A., Leroy, C., Corbara, B., Azemar, F. and Orivel, J. (2024) Biological interactions involving the myrmecophyte *Hirtella physophora* and its associates. *Biological Journal of the Linnean Society*, 141, 1–16.

De Jong, R. (2017) Fossil butterflies, calibration points and the molecular clock (Lepidoptera: Papilionoidea). *Zootaxa*, 4270, 1–63.

De Kort, H., Prunier, J. G., Ducatez, S., Honnay, O., Baguette, M. *et al.* (2021) Life history, climate and biogeography interactively affect worldwide genetic diversity of plant and animal populations. *Nature Communications*, 12, 516.

Delcourt, P.A. and Delcourt, H.R. (1987) *Long-term Forest Dynamics of the Temperate Zone: A Case Study of Late-Quaternary Forests in Eastern North America*. Springer, New York, USA.

DeLong, J.P., Hanley, T.C. and Vasseur, D.A. (2014) Competition and the density dependence of metabolic rates. *Journal of Animal Ecology*, 83, 51–58.

Demarchi, D.A. (1996) *An Introduction to the Ecoregions of British Columbia*. Wildlife Branch, Ministry of Environment, Lands and Parks, Victoria, Canada.

Dempster, J.P. (1975) *Animal Population Ecology*. Academic Press, London, UK.

Dempster, J.P. (1991) Fragmentation, isolation, and mobility of insect populations. In: Collins, N.M. and Thomas, J.A. (eds) *The Conservation of Insects and their Habitats*. Academic Press, London, UK, pp. 143–154.

Den Boer, P.J. (1980) Exclusion or coexistence and the taxonomic or ecological relationship between species. *Netherlands Journal of Zoology*, 30, 278–306.

Den Boer, P.J. and Reddingius, J. (1996) *Regulation and Stabilization Paradigms in Population Ecology*. Chapman and Hall, London, UK.

Dengler, J., Janišová, M., Török, P. and Wellstein, C. (2014) Biodiversity of Palaearctic grasslands: A synthesis. *Agriculture, Ecosystems and Environment*, 182, 1–14.

Dennis, R.L.H. (1972) *Plebejus argus* (L.) *caernensis* Thompson (Lep., Lycaenidae). A stenoecious geotype. *Entomologist's Record and Journal of Variation*, 84, 100–108.

Dennis, R.L.H. (1977) *The British Butterflies. Their Origin and Establishment*. Classey Ltd, Farringdon, UK.

Dennis, R.L.H. (1982–83) Mate-location strategies in the wall brown butterfly, *Lasiommata megera* L. (Lep., Satyridae). Wait or seek? *Entomologist's Record and Journal of Variation*, 94, 209–214; 95, 7–10.

Dennis, R.L.H. (1983a) Hierarchy and pattern in the spatial responses of ovipositing *Anthocharis cardamines* (Lep.). *Vasculum*, 68, 27–43.
Dennis, R.L.H. (1983b) Egg-laying cues in the wall brown butterfly, *Lasiommata megera* (L.) (Lep., Satryidae). *Entomologist's Gazette*, 34, 89–95.
Dennis, R.L.H. (1983c) Spatial patterning in butterfly oviposition. Density influences and the 'edge effect' in the orange tip butterfly (*Anthocharis cardamines*) (L.) (Lep., Pieridae). *Bulletin of the Amateur Entomologists' Society*, 42, 55–60.
Dennis, R.L.H. (1984a) Egg-laying sites in the common blue butterfly, *Polyommatus icarus* Rott., (Lep., Lycaenidae). The 'edge effect' and beyond the edge. *Entomologist's Gazette*, 35, 85–93.
Dennis, R.L.H. (1984b) The 'edge effect' in butterfly oviposition: Batch siting in *Aglais urticae* (L.) (Lep., Nymphalidae). *Entomologist's Gazette*, 35, 157–173.
Dennis, R.L.H. (1985a) Small plants attract attention! Choice of egglaying sites in the green-veined white butterfly (*Artogeia napi*) (L.) (Lep., Pieridae). *Bulletin of the Amateur Entomologists' Society*, 44, 77–82.
Dennis, R.L.H. (1985b) Voltinism in British *Aglais urticae* (L.) (Lep., Nymphalidae). Variation in space and time. *Proceedings and Transactions of the British Entomological and Natural History Society*, 18, 51–61.
Dennis, R.L.H. (1986a) Selection of roost sites by *Lasiommata megera* (L.) (Lep., Satyridae) on fencing at Brereton Heath country park, Cheshire. *Nota Lepidopterologica*, 9, 39–46.
Dennis, R.L.H. (1986b) Motorways and cross-movements. An insect's 'mental map' of the M56 in Cheshire. *Bulletin of the Amateur Entomologists' Society*, 45, 228–243.
Dennis, R.L.H. (ed.) (1992) *The Ecology of Butterflies in Britain*. Oxford University Press, Oxford, UK.
Dennis, R.L.H. (1993a) *Butterflies and Climate Change*. Manchester University Press, Manchester, UK.
Dennis, R.L.H. (1993b) Predation in a northern population of *Pieris napi* (L.) (Lep., Pieridae). Evidence from wing fragments. *Entomologist's Gazette*, 44, 157–159.
Dennis, R.L.H. (2000) Impoverished status of butterfly species among woodland biotopes of a northern English SSSI. *Entomologist's Gazette*, 51, 257–273.
Dennis, R.L.H. (2001) Progressive bias in species status is symptomatic of fine-grained mapping units subject to repeated sampling. *Biodiversity and Conservation*, 10, 483–494.
Dennis, R.L.H. (2004a) Butterfly habitats, broad scale biotope affiliations and structural exploitation of vegetation at finer scales: The matrix revisited. *Ecological Entomology*, 29, 744–752.
Dennis, R.L.H. (2004b) Just how important are structural elements as habitat components? Indications from a declining lycaenid butterfly with conservation status. *Journal of Insect Conservation*, 8, 37–45.
Dennis, R.L.H. (2008a) Population outbreaks and scramble mating in *Bibio pomonae* (Fabricius) (Dipt., Bibionidae) during two years in the Cheshire Peak District. *Entomologist's Monthly Magazine*, 144, 130.
Dennis, R.L.H. (2008b) How abundant are key resources for common butterflies? Insights from nymphalid butterflies in a conservation area. *Entomologist's Gazette*, 59, 79–84.
Dennis, R.L.H. (2010) *A Resource-based Habitat View for Conservation. Butterflies in the British Landscape*. Wiley-Blackwell, Oxford, UK.
Dennis, R.L.H. (2015) A decade of the resource-based habitat paradigm: The semantics of habitat loss. In: Elias, S. (ed.) *Reference Module in Earth Systems and Environmental Sciences*, Elsevier SclTech Connect, Waltham, USA, pp. 1–12.
Dennis, R.L.H. (2018) A decade of the resource-based habitat paradigm: The semantics of habitat loss. In: DellaSala, D.A. and Goldstein, M.I. (eds) *The Encyclopedia of the Anthropocene*, vol. 3. Elsevier, Oxford, UK, pp. 85–96.
Dennis, R.L.H. (2020) *Butterfly Biology Systems: Connections and Interactions in Life History and Behaviour*. CABI, Wallingford, UK.
Dennis, R.L.H. and Bramley, M.J. (1985) The influence of man and climate on dispersion patterns within a population of adult *Lasiommata megera* (L.) (Satyridae) at Brereton Heath, Cheshire. *Nota Lepidopterologica*, 8, 309–324.
Dennis, R.L.H. and Eales, H.T. (1997) Patch occupancy in *Coenonympha tullia* (Lepidoptera: Satyridae): Habitat quality matters as much as patch size and isolation. *Journal of Insect Conservation*, 1, 167–176.
Dennis, R.L.H. and Eales, H.T. (1999) Probability of site occupancy in the large heath butterfly *Coenonympha tullia* determined from geographical and ecological data. *Biological Conservation*, 87, 295–301.
Dennis, R.L.H. and Hardy, P.B. (2006) Excessive *Anthocharis cardamines* (L.) (Pieridae) egg load on cuckoo flower hostplants. *Entomologist's Gazette*, 57, 13–15.
Dennis, R.L.H. and Hardy, P.B. (2007) Support for mending the matrix: Resource seeking by butterflies in apparent non-resource zones. *Journal of Insect Conservation*, 11, 157–168.
Dennis, R.L.H. and Hardy, P.B. (2018) *British and Irish Butterflies: An Island Perspective*. CABI, Wallingford, UK.
Dennis, R.L.H. and Schmitt, T. (2009) Faunal structures, palaeogeography and historical inference. In: Settele, J., Van Dyck, H., Konvicka, M. and Shreeve, T.G. (eds) *Ecology of Butterflies in Europe*. Cambridge University Press, Cambridge, UK, pp. 250–280.
Dennis, R.L.H. and Shreeve, T.G. (1988) Hostplant-habitat structure and the evolution of butterfly mate-location behaviour. *Zoological Journal of the Linnean Society*, 94, 301–318.

Dennis, R.L.H. and Shreeve, T.G. (1989) Butterfly wing morphology variation in the British Isles. The influence of climate, behavioural posture and the hostplant-habitat. *Biological Journal of the Linnean Society*, 38, 323–348.

Dennis, R.L.H. and Shreeve, T.G. (1996) *Butterflies on British and Irish Offshore Islands: Ecology and Biogeography*. Gem Publishing, Wallingford, UK.

Dennis, R.L.H. and Sparks, T.H. (2005) Landscape resources for the territorial nymphalid butterfly *Inachis io*: Micro-site landform selection and behavioural responses to environmental conditions. *Journal of Insect Behaviour*, 18, 725–742.

Dennis, R.L.H. and Sparks, T.H. (2006) When is a habitat not a habitat? Dramatic resource use changes under differing weather conditions for the butterfly *Plebejus argus*. *Biological Conservation*, 129, 291–301.

Dennis, R.L.H. and Thomas, C.D. (2000) Bias in butterfly distribution maps: The influence of hot spots and recorder's home range. *Journal of Insect Conservation*, 4, 73–77.

Dennis, R.L.H. and Williams, W.R. (1987) Mate-location behaviour in the butterfly *Ochlodes venata* (Br. and Grey) (Hesperiidae). Flexible strategies and spatial components. *Journal of the Lepidopterists' Society*, 41, 45–64.

Dennis, R.L.H., Porter, K. and Williams, W.R. (1984) Ocellation in *Coenonympha tullia* (Mueller) (Lep., Satyridae). I. Structures in correlation matrices. *Nota Lepidopterologica*, 7, 199–219.

Dennis, R.L.H., Porter, K. and Williams, W.R. (1986) Ocellation in *Coenonympha tullia* (Mueller) (Lep., Satyridae). II. Population differentiation and clinal variation in the context of climatically-induced anti-predator defence strategies. *Entomologist's Gazette*, 37, 133–172.

Dennis, R.L.H., Williams, W.R. and Shreeve, T.G. (1991) A multivariate approach to the determination of faunal structures among European butterfly species (Lepidoptera: Rhopalocera). *Zoological Journal of the Linnean Society*, 101, 1–49.

Dennis, R.L.H., Shreeve, T.G. and Williams, W.R. (1995) Taxonomic differentiation in species richness gradients among European butterflies (Papilionidea, Hesperiodea): contribution of macroevolutionary dynamics. *Ecography*, 18, 27–40.

Dennis, R.L.H., Williams, W.R. and Shreeve, T.G. (1998) Faunal structures among European butterflies: Evolutionary implications of bias for geography, endemism and taxonomic affiliations. *Ecography*, 21, 181–203.

Dennis, R.L.H., Donato, B., Sparks, T.H. and Pollard, E. (2000) Ecological correlates of island incidence and geographical range among British butterflies. *Biodiversity and Conservation*, 9, 343–359.

Dennis, R.L.H., Shreeve, T.G. and Van Dyck, H. (2003) Towards a functional resource-based concept for habitat: A butterfly biology viewpoint. *Oikos*, 102, 417–426.

Dennis, R.L.H., Hodgson, J.G., Grenyer, R., Shreeve, T.G. and Roy, D.B. (2004) Host plants and butterfly biology. Do host plant strategies drive butterfly status? *Ecological Entomology*, 29, 12–26.

Dennis, R.L.H., Shreeve, T.G. and Van Dyck, H. (2006a) Habitats and resources: The need for a resource-based definition to conserve butterflies. *Biodiversity and Conservation*, 15, 1943–1966.

Dennis, R.L.H., Shreeve, T.G., Isaac, N.J.B., Roy, D.B., Hardy, P.B. *et al.* (2006b) The effects of visual apparency on bias in butterfly recording and monitoring. *Biological Conservation*, 128, 486–492.

Dennis, R.L.H., Shreeve, T.G. and Sheppard, D.A. (2007) Species conservation and landscape management: A habitat perspective. In: Stewart, A., Lewis, O. and New, T. (eds) *Insect Conservation Biology. Proceedings of the Royal Entomological Society's 23rd Symposium*, CABI, Wallingford, UK, pp. 92–126.

Dennis, R.L.H., Hardy, P.B. and Shreeve, T.G. (2008) The importance of resource databanks for conserving insects: A butterfly biology perspective. *Journal of Insect Conservation*, 12, 711–719.

Dennis, R.L.H., Dapporto, L., Fattorini, S. and Cook, L.M. (2011) The generalism-specialism debate: The part played by generalists in the life and death of species. *Biological Journal of the Linnean Society*, 104, 725–737.

Dennis, R.L.H., Dapporto, L. and Hardy, P.B. (2012) Colonisation ability and migration capacity of butterflies on islands offshore Britain and Ireland: Relationships to species richness and island geography. *Journal of Biogeography*, 39, 1412–1426.

Dennis, R.L.H., Dapporto, L., Dover, J.W. and Shreeve, T.G. (2013) Corridors and barriers in biodiversity conservation: New perspectives from the resource-based habitat viewpoint for butterflies. *Biodiversity and Conservation*, 22, 2709–2734.

Dennis, R.L.H., Dapporto, L., Dover, J.W. (2014) Ten years of the resource-based habitat paradigm: The biotope-habitat issue and implications for conserving butterfly diversity. *Journal of Insect Biodiversity*, 2(8), 1–32.

de Ruiter, D.J., Churchill, S.E., Hawks, J. and Berger, L.R. (2017) Late Australopiths and the emergence of *Homo*. *Annual Review of Anthropology*, 46, 99–115.

Desbiez, A.L.J and Kluyber, D. (2013) The role of giant armadillos (*Priodontes maximus*) as physical ecosystem engineers. *Biotropica*, 45, 537–540.

Deschamps-Cottin, M., Jacek, G., Seguinel, L., Le Champion, C., Robles, C. *et al.* (2023) A 12-year experimental design to test the recovery of butterfly biodiversity in an urban ecosystem: Lessons from the Parc Urbain des Papillons. *Insects*, 14, 780.

Desmettre, T. (2020) Toxoplasmosis and behavioural changes. *Journal Francais d'Ophtalmologie*, 43(3), e89–e93.

Dessaux, Y. and Faure, D. (2018) Niche construction and exploitation by *Agrobacterium*: How to survive and face competition in soil and plant habitats. *Current Topics in Microbiology and Immunology*, 418, 55–86.

D'Ettorre, P. and Heinze, J. (2001) Sociobiology of slave-making ants. *Acta Ethologica*, 3, 67–82.

Devictor, V., Mouillot, D., Meynard, C., Jiguet, F., Thuiller, W. *et al.* (2010) Spatial mismatch and congruence between taxonomic, phylogenetic and functional diversity: the need for integrative conservation strategies in a changing world. *Ecology Letters*, 13, 1030–1040.

De Vos, J.M., Joppa, L.N., Gittleman, J.L., Stephens, P.R. and Pimm, S.L. (2015) Estimating the normal background rate of species extinction: Background Rate of Extinction. *Conservation Biology*, 29, 452–462.

Dewdney, A. (1997) A dynamical model of abundances in natural communities. *Coenoses* 12, 67–76.

Dey, S., Veerendra, G.T.N., Babu, P.S.S.A. Manoj, A.V.P. and Nagarjuna, K. (2023) Degradation of plastics waste and its effects on biological ecosystems: A scientific analysis and comprehensive review. *Biomedical Materials and Devices*, 2, 70–112.

Dial, R. and Roughgarden, J. (1988) Theory of marine communities: The intermediate disturbance hypothesis. *Ecology*, 79, 1412–1424.

Diamond, J.M. (1997) *Guns, Germs, and Steel: The Fates of Human Societies*. W.W. Norton, New York, USA.

Diamond, J.M. and Ross, M.S. (2019) Exotic parrots breeding in urban tree cavities: Nesting requirements, geographic distribution, and potential impacts on cavity nesting birds in southeast Florida. *Avian Research*, 10, 39.

Diaz, S., Hodgson, J.G., Thompson, K., Cabido, M., Cornelissen, J.H.C. *et al.* (2004) The plant traits that drive ecosystems: Evidence from three continents. *Journal of Vegetation Science*, 15, 295–304.

Diaz, S., Lavorel, S., de Bello, F., Quetier, F., Grigulis, K. and Robson, M. (2007) Incorporating plant functional diversity effects in ecosystem service assessments. *Proceedings of the National Academy of Sciences of the United States of America*, 104, 20684–20689.

Díaz Varela, R.A., Ramil Rego, P., Calvo Iglesias, S. and Muñoz Sobrino, C. (2008) Automatic habitat classification methods based on satellite images: A practical assessment in the NW Iberia coastal mountains. *Environmental Monitoring and Assessment*, 144, 229–250.

Di Biase, L. and Fattorini, S. (2025) Disentangling the role of inter- and intraspecific trait variations in plant community assembly processes. In: Kumar, M., Bussmann, R.W. and Swenson, N.G. (eds) *Plant Functional Traits – Linking Climate and Ecosystem Functioning*. Elsevier, Amsterdam, The Netherlands, pp. 239–259.

Di Biase, L., Fattorini, S., Cutini, M. and Bricca, A. (2021a) The role of inter- and intraspecific variations in grassland plant functional traits along an elevational gradient in a Mediterranean mountain area. *Plants*, 10, 359.

Di Biase, L., Pace, L., Mantoni, C. and Fattorini, S. (2021b) Variations in plant richness, biogeographical composition, and life forms along an elevational gradient in a Mediterranean mountain. *Plants*, 10, 2090.

Di Biase, L., Di Lisio, P., Pace, L., Arrizza, L. and Fattorini, S. (2022) Use of lichens to evaluate the impact of post-earthquake reconstruction activities on air quality: A case study from the city of L'Aquila. *Biology*, 11, 1199.

Di Biase, L., Tsafack, N., Pace, L. and Fattorini, S. (2023) Ellenberg indicator values disclose complex environmental filtering processes in plant communities along an elevational gradient. *Biology*, 12, 161.

Dice, L.R. (1943) *The Biotic Provinces of North America*. University of Michigan Press, Ann Arbor, USA.

Dickinson, C.H. and Underhay, V.H.S. (1977) Growth of fungi in cattle dung. *Transactions of the British Mycological Society*, 69, 473–477.

Dickinson, C.H., Underhay, V.H.S., and Ross, V. (1981) Effect of season, soil fauna and water content on the decomposition of cattle dung pats. *New Phytologist*, 88, 129–141.

Dickman, E.M., Newell, J.M., González, M.J. and Vanni, M.J. (2008) Light, nutrients, and food-chain length constrain planktonic energy transfer efficiency across multiple trophic levels. *Proceedings of the National Academy of Sciences of the United States of America*, 105, 18408–18412.

Diehl, R.H. (2013) The airspace is habitat. *Trends in Ecology and Evolution*, 28, 377–379.

Diels, L. (1908) *Pflanzengeographie*. Göschen'sche Verlagshandlung, Leipzig, Germany.

Dierschke, J. and Bairlein, F. (2004) Habitat selection of wintering passerines in salt marshes of the German Wadden Sea. *Journal of Ornithology*, 145, 48–58.

Dietz, M., Pir, J.B. and Hillen, J. (2013) Does the survival of greater horseshoe bats and Geoffroy's bats in Western Europe depend on traditional cultural landscapes? *Biodiversity and Conservation*, 22, 3007–3025.

Di Giulio, A., Maurizi, E., Barbero, F., Sala, M., Fattorini, S. *et al.* (2015) The Pied Piper: A parasitic beetle's melodies modulate ant behaviours. *PLOS One*, 10(7), e0130541.

Dinerstein, E., Olson, D.M., Graham, D.J., Webster, A.L., Primm, S.A. *et al.* (1995) *A Conservation Assessment of the Terrestrial Ecoregions of Latin America and the Caribbean*. World Bank, Washington DC, USA.

Dinets, V., Britton, A. and Shirley, M. (2013) Climbing behaviour in extant crocodilians. *Herpetology Notes*, 7, 3–7.

DiNuzzo, E.R. and Griffen, B.D. (2020) The effects of animal personality on the ideal free distribution. *Proceeding of the Royal Society B*, 287(1934), 20201095.

Di Pietro, S., Mantoni, C. and Fattorini, S. (2021) Influence of urbanization on the avian species-area relationship: insights from the breeding birds of Rome. *Urban Ecosystems*, 24, 779–788.

Di Salvo, M., Calcagnile, M., Talà, A., Tredici, S.M., Mafei, M.E. *et al.* (2019) The microbiome of the *Maculinea-Myrmica* host-parasite interaction. *Scientific Reports*, 9, 8048.

Dittrich, A.D.K. and Helden A.J. (2012) Experimental sward islets: The effect of dung and fertilisation on Hemiptera and Araneae. *Insect Conservation and Diversity*, 5, 46–56.

Dobson, F.S. and Yu, J. (1993) Rarity in Neotropical forest mammals revised. *Conservation Biology* 7, 586–591.

Dodd, C.K. (ed.) (2009) *Amphibian Ecology and Conservation. A Handbook of Techniques*. Oxford University Press. Oxford, UK.

Dodd, M.S., Papineau, D., Grenne, T., Slack, J.F., Rittner, M. *et al.* (2017) Evidence for early life in Earth's oldest hydrothermal vent precipitates. *Nature*, 543, 60–64.

Domínguez, L.F., Price, J., Otto, R. and Fernández-Palacios, J.M. (2010) Using taxonomic and phylogenetic evenness to compare diversification in two island floras. *Perspectives in Plant Ecology, Evolution and Systematics*, 12(2), 93–106.

Donaldson, J.S. (2009) Botanic gardens science for conservation and global change. *Trends in Plant Science*, 14, 608–613.

Donovan, M.P., Iglesias, A., Wilf, P., Labandeira, C.C. and Cúneo, N.R. (2016) Rapid recovery of Patagonian plant–insect associations after the end-Cretaceous extinction. *Nature Ecology and Evolution*, 1, 0012.

Dorkenwald, S., Matsliah, A., Sterling, A.R., Schlegel, P., Yu, S.-C. *et al.* (2024) Neuronal wiring diagram of an adult brain. *Nature*, 634, 124–138.

Dormann, C.F., Gruber, B. and Fründ, J. (2008) Introducing the bipartite package: Analysing ecological networks. *R News*, 8, 8–11.

Dormann, C.F., Fründ, J., Blüthgen, N. and Gruber, B. (2009) Indices, graphs and null models: Analysing bipartite ecological networks. *Open Ecology Journal*, 2, 7–24.

Dormy, E. (2014) The origin of the Earth's magnetic field: Fundamental or environmental research? *Europhysics News*, 37, 22–25.

Dornelas, M., Moonen, A.C., Magurran, A.E., and Bàrberi, P. (2009) Species abundance distributions reveal environmental heterogeneity in modified landscapes. *Journal of Applied Ecology*, 46, 666–672.

Doube, B.M. (1990) A functional classification for analysis of the structure of dung beetle assemblages. *Ecological Entomology*, 15, 371–383.

Doube, B.M. (2018a) Ecosystem services provided by dung beetles in Australia. *Basic and Applied Ecology*, 26, 35–49.

Doube, B.M. (2018b) Rearing Onthophagus vacca and Bubas bubalus for release in southern Australia. Meat and Livestock Australia Limited, North Sydney, Australia. Available at: www.mla.com.au/research-and-development/reports/2018/rearing-onthophagus-vacca-and-bubas-bubalus-for-release-in-southern-australia/ (accessed 12 May 2025).

Dover, J.W. (2015) *Green Infrastructure*. Routledge, London, UK.

Dover, J.W. (ed.) (2019) *The Ecology of Hedgerows and Field Margins*. Routledge, London, UK.

Dover, J.W., Sparks, T., Clarke, S., Gobbett, K. and Glossop, S. (2000) Linear features and butterflies: The importance of green lanes. *Agriculture Ecosystems and Environment*, 80, 227–242.

Drakare, S., Lennon, J.J. and Hillebrand, H. (2006) The imprint of the geographical, evolutionary and ecological context on species-area relationships. *Ecology Letters*, 9.2, 215–227.

Dray, S. and Legendre, P. (2008) Testing the species traits-environment relationships: The fourth-corner problem revisited. *Ecology*, 89, 3400–3412.

Driscoll, D.A. (2005) Is the matrix a sea? Habitat specificity in a naturally fragmented landscape. *Ecological Entomology*, 30, 8–16.

Droissart, V., Dauby, G., Hardy, O.J., Deblauwe, V., Harris, J.D. *et al.* (2018) Beyond trees: Biogeographical regionalization of tropical Africa. *Journal of Biogeography*, 45, 1153–1167.

Dubey, J.P. (2010) *Toxoplasmosis of Animals and Humans*, 2nd edn. CRC Press, Boca Raton, USA.

Duckworth, J.C., Kent M. and Ramsay, P.M. (2000) Plant functional types: an alternative to taxonomic plant community description in biogeography? *Progress in Physical Geography*, 24, 515–542.

Ducoing, A.M. and Thierry, B. (2005) Tool-use learning in Tonkean macaques (*Macaca tonkeana*). *Animal Cognition*, 8, 103–113.

Duffy, J.E., Morrison, C.L. and Rios, R. (2000) Multiple origins of eusociality among sponge-dwelling shrimps (*Synalpheus*). *Evolution*, 54, 503–516.

Dunbar, R.I.M. (1995) The mating system of callitrichid primates: I. Conditions for the coevolution of pair bonding and twinning. *Animal Behaviour*, 50, 1057–1070.

Dunbar, R.I.M. (2017) Breaking bread: The functions of social eating. *Adaptive Human Behavior and Physiology*, 3, 198–211.

Dunbar, R.I.M., MacCarron, P. and Shultz, S. (2018) Primate social group sizes exhibit a regular scaling pattern with natural attractors. *Biological Letters*, 14, 20170490.

Dunn, J.C. and Cristóbal-Azkarate, J. (2016) New World monkeys. *Nature Education Knowledge*, 7, 1. Available at: www.nature.com/scitable/knowledge/library/new-world-monkeys-148121150/#:~:text=The%20New%20World%20monkeys%20are,habitats%2C%20from%20Mexico%20to%20Argentina.&text=The%20formal%20term%20for%20the,Monkeys%20(NWMs)%20is%20Platyrrhini (accessed 12 May 2025).

Dunne, E.M., Close, R.A., Button, D.J., Brocklehurst, N., Cashmore, D.D. *et al.* (2018) Diversity change during the rise of tetrapods and the impact of the 'Carboniferous rainforest collapse'. *Proceedings of the Royal Society B: Biological Sciences*, 285, 20172730.

Dunne, J.A., Williams, R.J. and Martinez, N.D. (2002) Food-web structure and network theory: The role of connectance and size. *Proceedings of the National Academy of Sciences of the United States of America*, 99, 12917–12922.

Dunne, J.A., Williams, R.J. and Martinez, N.D. (2004) Network structure and robustness of marine food webs. *Marine Ecology Progress Series*, 273, 291–302.

Dunstan, D.J. and Hodgson, D.J. (2014) Snails home. *Physica Scripta*, 89, 068002.

Du Pasquier, L.G. (1918) Esquisse d'une nouvelle théorie de la population. *Vierteljahrsschrift der Naturforschenden Gesellschaft in Zürich*, 63, 236–249.

Duranton, G. and Puga, D. (2015) Urban land use. In: Duranton, G., Henderson, J.V. and Strange, W. (eds) *Handbook of Regional and Urban Economics*. Elsevier, Amsterdam, The Netherlands, pp. 467–560.

Durif, C.M.F., Arts, M., Bertolini, F., Cresci, A., Daverat, F. *et al.* (2023) The evolving story of catadromy in the European eel (*Anguilla anguilla*). *ICES Journal of Marine Science: Journal Du Conseil*, 80(9), 2253–2265.

Dušek, R. and Popelková, R. (2012) Theoretical view of the Shannon index in the evaluation of landscape diversity. *AUC Geographica*, 47(2), 5–13.

Duvergé, P.L. and Jones, G (2003) Use of farmland habitats by greater horseshoe bats. In: Tattersall, F. and Manley, W. (eds) *Conservation and Conflict*. Westbury Publishing, London, UK, pp. 64–81.

Dyksterhuis, E.J. (1949) Condition and management of range land based on quantitative ecology. *Journal of Range Management*, 2, 104–115.

Dyksterhuis, E.J. (1958) Ecological principles in range evaluation. *Botanical Review*, 24, 253–272.

Dytham, C. (2003) *Choosing and Using Statistics. A Biologist's Guide*. Blackwell Publishing, Oxford, UK.

Dziekańska, I., Nowicki, P., Pirożnikow, E. and Sielezniew, M. (2020) A unique population in a unique area: The Alcon blue butterfly and its specific parasitoid in the Białowieża Forest. *Insects*, 11,687.

Early, M. and Goff, M.L. (1986) Arthropod succession patterns in exposed carrion on the island of O'ahu, Hawaiian Islands, USA. *Journal of Medical Entomology*, 23, 520–531.

Ebach, M.C. (2015) *Origins of Biogeography: The Role of Biological Classification in Early Plant and Animal Geography*. Springer, New York, USA.

Ebach, M.C. and Goujet, D.F. (2006) The first biogeographical map. *Journal of Biogeography*, 33, 761–769.

Ebach, M.C., Morrone, J.J., Parenti, L.R. and Viloria, A.L. (2008) International code of area nomenclature. *Journal of Biogeography*, 35, 1153–1157.

Ebach, M.C., González-Orozco, C.E., Miller, J.T. and Murphy, D.J. (2015) A revised area taxonomy of phytogeographical regions within the Australian Bioregionalisation Atlas. *Phytotaxa*, 208, 261–277.

Ebeling, W. (1978) *Urban Entomology*. University of California Press Berkeley, USA.

Ecker, M., Bocherens, H., Julien, M.A., Rivals, F., Raynal, J.P. and Moncel, M.-H. (2013) Middle Pleistocene ecology and Neanderthal subsistence: Insights from stable isotope analyses in Payre (Ardèche, southeastern France). *Journal of Human Evolution*, 65, 363–373.

Eckert, K.L and Eckert, S.A. (1987) Pre-reproductive movements of leatherback sea turtles (*Dermochelys coriacea*) nesting in the Caribbean, *Copeia*, 2, 400–406.

Eckert, S.A., Eckert, K.A., Ponganis, P. and Kooyman, G.L. (1989) Diving and foraging behaviour of leatherback sea turtles *Dermochelys coriacea*. *Canadian Journal of Zoology*, 67, 2834–2840.

Edwards, P.B. (2007) *Introduced Dung Beetles in Australia 1967–2007: Current Status and Future Directions*. Landcare Australia, Sydney, Australia.

Edwards, P.B. and Pavri, C. (2007) Pastures – dung beetles. In: Bailey, P. (ed.) *Pests of Field Crops and Pastures: Identification and Control*. CSIRO Publishing, Collingwood, Australia.

Edworthy, A.B., Steensma, K.M.M., Zandberg, H.M. and Lilley, P.L. (2012) Dispersal, home-range size, and habitat use of an endangered land snail, the Oregon forest snail (*Allogona townsendiana*). *Canadian Journal of Zoology*, 90, 875–884.

Egerton, F.N. (2017) History of ecological sciences, Part 59: Niches, biomes, ecosystems, and systems. *Bulletin of the Ecological Society of America*, 98, 298–337.

Eggeling, W.J. (1947) Observations on the ecology of the Budongo rain forest, Uganda. *Journal of Ecology*, 34, 20–87.

Eggleton, P. (2020) The state of the world's insects. *Annual Review of Environment and Resources*, 45, 61–82.

Eggleton, P., Bignell, D.E., Sands, W.A., Mawdsley, N.A., Lawton, J.H. *et al.* (1996) The diversity, abundance, and biomass of termites under differing levels of disturbance in the balmayo Forest Reserve, southern Cameroon. *Philosophical Transactions of the Royal Society B: Biological Sciences*, 351, 51–68.

Ehrich, D., Schmidt, N.M., Gauthier, G., Alisauskas, R., Angerbjörn, A. *et al.* (2019) Documenting lemming population change in the Arctic: Can we detect trends? *Ambio*, 49, 786–800.

Ehrlich, P.R. (1961) Intrinsic barriers to dispersal in a checkerspot butterfly. *Science*, 134, 108–109.

Ehrlich, P.R. and Davidson, S.E. (1960) Techniques for capture-recapture studies of Lepidoptera populations. *Journal of the Lepidopterists' Society*, 14, 227–229.

Ehrlich, P.R. and Ehrlich, A.H. (1981) *Extinction: The Causes and Consequences of the Disappearance of Species*. Random House, New York, USA.

Eigenmann, C.H. (1902) The annual address of the president: The solution of the eel question. *Transactions of the American Microscopical Society*, 23, 5–18.

Einstein, A. (1915) Die Grundlage der allgemeinen Relativitätstheorie. *Annalen der Physik*, 354, 769–822.

Einstein, A. and Freud, S. (1933) *Why War?* International Institute of Intellectual Cooperation, League of Nations, Paris, France.

Eleftherious, A. (ed.) (2013) *Methods for the Study of Marine Benthos*, 4th edn. Wiley Blackwell, Chichester, UK.

Elewa, A.M.T. and Joseph, R. (2009) The history, origins, and causes of mass extinctions. *Journal of Cosmology*, 2, 201–220.

Elias, R.B., Artur, G., Luís, S., Fernández-Palacios, J.M., Azevedo, E.B. and Reiz, F. (2016) Natural zonal vegetation of the Azores Islands: Characterization and potential distribution. *Phytocoenologia*, 47, 107–123.

Ellenberg, H. (1988) *Vegetation Ecology of Central Europe*. Cambridge University Press, Cambridge, UK.

Ellers, J., Berg, M.P., Dias, A.T.C., Fontana, S., Ooms, A. and Moretti, M. (2018) Diversity in form and function: Vertical distribution of soil fauna mediates multidimensional trait variation. *Journal of Animal Ecology*, 87, 933–944.

Ellis, C.J., Asplund, J., Benesperi, R., Branquinho, C., Di Nuzzo, L. *et al.* (2021) Functional traits in lichen ecology: A review of challenge and opportunity. *Microorganisms*, 9(4), 766.

Ellis, E.C. and Ramankutty, N. (2008) Putting people in the map: Anthropogenic biomes of the world. *Frontiers in Ecology and the Environment*, 6, 439–447.

Ellis, R. (1998) *The Search for the Giant Squid: The World's Most Mysterious and Elusive Creature—Its Biology and Mythology*. Lyons Press, New York, USA.

Ellison, A.M., Bank, M.S., Clinton, B.D., Colburn, E.A., Elliot, K. *et al.* (2005) Loss of foundation species: Consequences for the structure and dynamics of forested ecosystems. *Frontiers in Ecology and the Environment*, 3, 479–486.

Elton, C. (1927) *The Animal Community*. Macmillan, New York, USA.

Elton, C.S. (1966) *The Pattern of Animal Communities*. Methuen, London, UK.

Elzinga, C.L., Salzer, D.W. and Willoughby, J.W. (1998) *Measuring and Monitoring Plant Populations*. Bureau of Land Management National Business Center, Denver, USA.

Enari, H. and Sakamaki-Enari, H. (2013) Influence of heavy snow on the feeding behavior of Japanese macaques (*Macaca fuscata*) in northern Japan. *American Journal of Primatology*, 75, 534–544.

Enderle, R., Stenlid, J. and Vasaitis, R. (2019) An overview of ash (*Fraxinus* spp.) and the ash dieback disease in Europe. *CABI Reviews*, 14(025), 1–12.

Endicott, P., Ho, S.Y. and Stringer, C. (2010) Using genetic evidence to evaluate four palaeoanthropological hypotheses for the timing of Neanderthal and modern human origins. *Journal of Human Evolution*, 59, 87–89.

Endler, J.A. (1986) *Natural Selection in the Wild*. Princeton University Press, Princeton, USA.

Engel, M.S., Grimaldi, D.A. and Krishna, K. (2009) Termites (Isoptera): Their phylogeny, classification, and rise to ecological dominance. *American Museum Novitates*, 3650, 1–27.

Engel, M.S., Barden, P., Riccio, M.L. and Grimaldi, D.A. (2016) Morphologically specialized termite castes and advanced sociality in the Early Cretaceous. *Current Biology*, 26, 522–530.

Engen, S. and Lande, R. (1996) Population dynamic models generating species abundance distributions of the gamma type. *Journal of Theoretical Biology*, 178, 325–331.

England, P.R., Phillips, J., Waring, J.R., Symonds, G. and Babcock, R. (2008) Modelling wave-induced disturbance in highly biodiverse marine macroalgal communities: Support for the intermediate disturbance hypothesis. *Marine and Freshwater Research*, 59, 515–520.

Engler, A. (1879–1882) *Versuch einer Entwicklungsgeschichte der extratropischen Florengebiete der nördlichen Hemisphäre*. W. Engelmann, Leipzig, Germany.

Engler, J.O., Secondi, J., Dawson, D.A., Elle, O. and Hochkirch, A. (2016) Range expansion and retraction along a moving contact zone has no effect on the genetic diversity of two passerine birds. *Ecography*, 39, 884–893.

Enkhtur, K., Brehm, G., Boldgiv, B. and Pfeifer, M. (2021) Alpha and beta diversity patterns of macro-moths reveal a breakpoint along a latitudinal gradient in Mongolia. *Scientific Reports*, 11, 15018.

EPA US Environmental Protection Agency (2012) *Freshwater Biological Traits Database (Traits)*. US Environmental Protection Agency, Washington, DC, USA. Available at: www.epa.gov/risk/freshwater-biological-traits-database-traits (accessed 13 May 2025).

Erwin, T.L. (1982) Tropical forests: Their richness in Coleoptera and other arthropod species. *Coleopterists' Bulletin*, 36, 74–75.

Eschweiler, N., Molis, M. and Buschbaum, C. (2009) Habitat-species size structure variations in periwinkle populations (*Littorina littorea*) caused by biotic factors. *Helgoland Marine Research*, 63, 119–127.

Espeland, E.K. and Emam, T.M. (2011) The value of structuring rarity: The seven types and links to reproductive ecology. *Biodiversity and Conservation*, 20, 963–985.

Estes, J.A. and Duggins, D.O. (1995) Sea otters and kelp forests in Alaska: Generality and variation in a community ecological paradigm. *Ecological Monographs*, 65, 75–100.

Estes, J.A. and Palmisano, J.F. (1974) Sea otters: Their role in structuring nearshore communities. *Science*, 185, 1058–1060.

Estes, J.A., Tinker, M.T., Williams, T.M. and Doak, D.F. (1998) Killer whale predation on sea otters linking oceanic and nearshore ecosystems. *Science*, 282, 473–476.

Estes, J.A., Terborgh, J., Brashares, J.S., Power, M.E., Berger, J. *et al.* (2011) Trophic downgrading of planet Earth. *Science*, 333, 301–306.

Estrada, E. (2012) *The Structure of Complex Networks: Theory and Applications*. Oxford University Press, Oxford, UK.

Estrada-Carmona, N., Sánchez, A.C., Remans, R. and Jones, S.K. (2022) Complex agricultural landscapes host more biodiversity than simple ones: A global meta-analysis. *Proceedings of the National Academy of Sciences of the United States of America*, 119(38), e2203385119.

Etienne, R.S., Alonso, D. and McKane, A.J. (2007) The zero-sum assumption in neutral biodiversity theory. *Journal of Theoretical Biology*, 248, 522–536.

European Commission (2003) *Mediterranean Desertification. Framing the Policy Context.* Office for Official Publications of the European Communities, Luxembourg.

Evangelista, D.A., Wipfler, B., Béthoux, O., Donath, A., Fujita, M. *et al.* (2019) An integrative phylogenomic approach illuminates the evolutionary history of cockroaches and termites (Blattodea). *Proceedings of the Royal Society B: Biological Sciences*, 286, 20182076.

Evans, C. and Kleindorfer, S. (2016) Superb Fairy-Wren (*Malurus cyaneus*) sons and daughters acquire song elements of mothers and social fathers. *Frontiers in Ecology and Evolution*, 4, 1–10.

Evans, D. (2010) Interpreting the habitats of Annex I: Past, present and future. *Acta Botanica Gallica*, 157, 677–686.

Evans, E.W. (1988) Community dynamics of prairie grasshoppers subjected to periodic fire: Predictable trajectories or random walks in time? *Oikos*, 52, 283–292.

Evans, K.S., Mamo, M., Wingeyer, A., Schachta, W.B., Eskridge, K.M. *et al.* (2019) Soil fauna accelerate dung pat decomposition and nutrient cycling into grassland soil. *Rangeland Ecology and Management*, 72, 667–677.

Evans, L.C., Melero, Y., Schmucki, R., Boersch-Supan, P.H., Brotons, L. *et al.* (2022) Bioclimatic context of species' populations determines community stability. *Global Ecology and Biogeography*, 31, 1542–1555.

Evershed, R.P., Heron, C. and Goad, L.J. (1991) Epicuticular wax components preserved in potsherds as chemical indicators of leafy vegetables in ancient diets. *Antiquity*, 65, 540–544.

Exposito-Alonso, M., Booker, T.R., Czech, L., Gillespie, L., Hateley, S. *et al.* (2022) Genetic diversity loss in the Anthropocene. *Science*, 377, 1431–1435.

Fa, J.E., Funk, S.M. and O'Connell, D.M. (2011) *Zoo Conservation Biology*. Cambridge University Press, Cambridge, UK.

Fa, J.E., Gusset, M., Flesness, N. and Conde, D.A. (2014) Zoos have yet to unveil their full conservation potential. *Animal Conservation*, 17, 97–100.

Faeth, S.H., Bang, C. and Saari, S. (2011) Urban biodiversity: Patterns and mechanisms. *Annals of the New York Academy of Sciences*, 1223, 69–81.

Fagan, W.F., Cantrell, R.S. and Cosner, C. (1999) How habitat edges change species interactions. *American Naturalist*, 153, 165–182.

Faith, D.P. (1992) Conservation evaluation and phylogenetic diversity. *Biological Conservation*, 61, 1–10.

Faith, D.P. (2013) Biodiversity and evolutionary history: Useful extensions of the PD phylogenetic diversity assessment framework. *Annals of the New York Academy of Sciences* 1289, 69–89.

Faith, D.P. (2016) The PD phylogenetic diversity framework: Linking evolutionary history to feature diversity for biodiversity conservation. In: Pellens, R. and Grandcolas, P. (eds) *Biodiversity Conservation and Phylogenetic Systematics. Preserving our Evolutionary Heritage in an Extinction Crisis.* Springer, Cham, Switzerland.

Faith, D.P. and Richards, Z.T. (2012) Climate change impacts on the tree of life: Changes in phylogenetic diversity illustrated for acropora corals. *Biology* 1(3), 906–932.

Falaschi, M., Marta, S., Lo Parrino, E., Roll, U., Meiri, S. *et al.* (2023) Global bioregions of reptiles confirm the consistency of bioregionalization processes across vertebrate clades. *Global Ecology and Biogeography*, 32, 1272–1284.

Falster, D., Gallagher, R., Wenk, E.H., Wright, I.J., Indiarto, D. Falster, D., Gallagher, R., Wenk, E.H., Wright, I.J., Indiarto, D. *et al.* (2021) AusTraits, a curated plant trait database for the Australian flora. *Scientific Data*, 8, 254.

Farina, A. (1998) *Principles and Methods in Landscape Ecology*. Chapman and Hall, London.

Farina, A. (2007) *Principles and Methods in Landscape Ecology. Towards a Science of the Landscape.* Springer-Verlag, New York, USA.

Farina, A. (2022) *Principles and Methods in Landscape Ecology. An Agenda for the Second Millennium.* Springer Nature, Berlin, Germany.

Farris, K.L., Huss, M.J. and Zack, S. (2004) The role of foraging woodpeckers in the decomposition of ponderosa pine snags. *The Condor*, 106, 50–59.

Fastovsky, D.E. and Sheehan, P.M. (2005) The extinction of the dinosaurs in North America. *GSA Today*, 15, 4–10.

Fastovsky, D.E. and Weishampel, D. (2005) *The Evolution and Extinction of the Dinosaurs.* Cambridge University Press, Cambridge, UK.

Fath, B.D., Scharler, U.M., Ulanowicz, R.E. and Hannon, B. (2007) Ecological network analysis: Network construction. *Ecological Modelling*, 208, 49–55.

Fattorini, S. (2002a) A comparison of relict versus dynamic models for tenebrionid beetles of Aegean Islands (Greece) (Coleoptera: Tenebrionidae). *Belgian Journal of Zoology*, 132, 55–64.

Fattorini, S. (2002b) Biogeography of the tenebrionid beetles (Coleoptera, Tenebrionidae) on the Aegean Islands (Greece). *Journal of Biogeography*, 29, 49–67.

Fattorini, S. (2005) A simple method to fit geometric series and broken stick models in community ecology and island biogeography. *Acta Oecologica*, 28, 199–205.

Fattorini, S. (2006) Spatial patterns of diversity in the tenebrionid beetles (Coleoptera, Tenebrionidae) of the Aegean Islands (Greece). *Evolutionary Ecology Research*, 8, 237–263.

Fattorini, S. (2007a) Historical relationships of African mountains based on cladistic analysis of distributions and endemism of flightless insects. *African Entomology*, 15, 340–355.

Fattorini, S. (2007b) Non-randomness in the species-area relationship: Testing the underlying mechanisms. *Oikos*, 116, 678–689.

Fattorini, S. (2008) A multidimensional characterization of rarity applied to the Aegean tenebrionid beetles (Coleoptera Tenebrionidae). *Journal of Insect Conservation*,12, 251–263.

Fattorini, S. (2009) Darkling beetle communities in two geologically contrasting biotopes: Testing biodiversity patterns by microsite comparisons. *Biological Journal of the Linnean Society*, London, 98, 787–793.

Fattorini, S. (2010a) Effects of fire on tenebrionid communities of a *Pinus pinea* plantation: A case study in a Mediterranean site. *Biodiversity and Conservation*, 19, 1237–1250.

Fattorini, S. (2010b) Biotope prioritisation in the Central Apennines (Italy): Species rarity and cross-taxon congruence. *Biodiversity and Conservation*,19, 3413–3429.

Fattorini, S. (2010c) Use of insect rarity for biotope prioritisation: The tenebrionid beetles of the Central Apennines (Italy). *Journal of Insect Conservation*, 14, 367–378.

Fattorini, S. (2010d) The influence of geographical and ecological factors on island beta diversity patterns. *Journal of Biogeography*, 37, 1061–1070.

Fattorini, S. (2011) Insect rarity, extinction and conservation in urban Rome (Italy): A 120-year-long study of tenebrionid beetles. *Insect Conservation and Diversity*, 4, 307–315.

Fattorini, S. (2013) Regional insect inventories require long time, extensive spatial sampling and good will. *PLOS One*, 8, e62118.

Fattorini, S. (2014a) Island biogeography of urban insects: Tenebrionid beetles from Rome tell a different story. *Journal of Insect Conservation*, 18, 729–735.

Fattorini, S. (2014b) Tenebrionid beetle distributional patterns in Italy: Multiple colonisation trajectories in a biogeographical crossroad. *Insect Conservation and Diversity*, 2014, 144–160.

Fattorini, S. (2014c) Relations between species rarity, vulnerability, and range contraction for a beetle group in a densely populated region within the Mediterranean biodiversity hotspot. *Conservation Biology*, 28, 169–176.

Fattorini, S. (2016) A history of chorological categories. *History and Philosophy of the Life Sciences*, 38, 12.

Fattorini, S. (2017a) Endemism in historical biogeography and conservation biology: Concepts and implications. *Biogeographia - The Journal of Integrative Biogeography*, 32, 47–75.

Fattorini, S. (2017b) The Watson-Forbes biogeographical controversy untangled 170 years later. *Journal of the History of Biology*, 50, 473–496.

Fattorini, S. (2019) *Ecologia Urbana*. Ediesse, Rome, Italy.

Fattorini, S. (2020) Climate change and extinction events. In: Elias, S. and Alderton, D. (eds) *Encyclopedia of Geology*, 2nd edn. Elsevier, London, UK, pp. 585–595.

Fattorini, S. (2021) Caveat lector; or, the Linnean origin of the myth of Tournefort as a precursor of von Humboldt. *Biogeographia –The Journal of Integrative Biogeography*, 36, a011.

Fattorini, S. (2022) Historical biogeography of earwigs. *Biology*, 11, 1794.

Fattorini, S. (2023) Adaptations of tenebrionid beetles to Mediterranean sand dune environments and the impact of climate change (Coleoptera: Tenebrionidae). *Fragmenta Entomologica*, 55, 1–20.

Fattorini, S. (2024a) An interspecific assessment of Bergmann's rule in tenebrionid beetles (Coleoptera, Tenebrionidae) along an elevation gradient. *Insects*, 15, 673.

Fattorini, S. (2024b) The role of vegetation in elevational diversity patterns of tenebrionid beetles in Central Italy. *Diversity*, 16, 110.

Fattorini, S. (2024c) Upward and poleward (but not phenological) shifts in a forest tenebrionid beetle in response to global change in a Mediterranean area. *Insects*, 15, 242.

Fattorini, S. and Baselga, A. (2012) Species richness and turnover patterns in European tenebrionid beetles. *Insect Conservation and Diversity*, 5, 331–345.

Fattorini, S. and Borges, P.V.A. (2012a) Biogeographical kinetics on an island volcano (Capelinhos, Azores): Fast colonization rates and dominance of arthropod exotic species. *Insect Conservation and Diversity*, 5, 358–366.

Fattorini, S. and Borges, P.V.A. (2012b) Species-area relationships underestimate extinction rates. *Acta Oecologica*, 40, 27–30.

Fattorini, S. and Di Giulio, A. (2013) Should we correct rarity measures for body size to evaluate arthropod vulnerability? Insights from Mediterranean tenebrionid beetles. *Biodiversity and Conservation*, 22, 2805–2819.

Fattorini, S. and Galassi, D.M.P. (2016) Role of urban green spaces for saproxylic beetle conservation: A case study of tenebrionids in Rome, Italy. *Journal of Insect Conservation*, 20, 737–745.

Fattorini, S. and Ulrich, W. (2012a) Spatial distributions of European Tenebrionidae point to multiple postglacial colonization trajectories. *Biological Journal of the Linnean Society*, 105, 318–329.

Fattorini, S. and Ulrich, W. (2012b) Drivers of species richness in European Tenebrionidae (Coleoptera). *Acta Oecologica*, 43, 22–28.

Fattorini, S. and Vigna Taglianti, A. (2015) Use of taxonomic and chorological diversity to highlight the conservation value of insect communities in a Mediterranean coastal area: The carabid beetles (Coleoptera, Carabidae) of Castelporziano (Central Italy). *Rendiconti Lincei – Scienze Fisiche e Naturali*, 26(Suppl. 3), 625–641.

Fattorini, S., Cardoso, P., Rigal, F. and Borges, P.V.A. (2012) Use of arthropod rarity for area prioritisation: Insights from the Azorean Islands. *PLOS One*, 7(3), e33995.

Fattorini, S., Di Giulio, A. and Dapporto, L. (2013a) Measuring insect rarity: Practical issues, pragmatic approaches. *Journal of Insect Biodiversity*, 1, 1–21.

Fattorini, S., Sciotti, A., Tratzi, P. and Di Giulio, A. (2013b) Species distribution, ecology, abundance, body size and phylogeny originate interrelated rarity patterns at regional scale. *Journal of Zoological Systematics and Evolutionary Research*, 51(4), 279–286.

Fattorini, S., Dapporto, L., Strona, G. and Borges, P.A.V. (2015a) Calling for a new strategy to measure environmental (habitat) diversity in Island Biogeography: A case study of Mediterranean tenebrionids (Coleoptera: Tenebrionidae). *Fragmenta Entomologica*, 47(1), 1–14.

Fattorini, S., Maltzeff, P. and Salvati, L. (2015b) Use of insect distribution across landscape-soil units to assess conservation priorities in a Mediterranean coastal reserve: The tenebrionid beetles of Castelporziano (Central Italy). *Rendiconti Lincei – Scienze Fisiche e Naturali*, 26(S3), 353–366.

Fattorini, S., Bergamaschi, D., Mantoni, C., Acosta, A.T.R. and Di Giulio, A. (2016a) Niche partitioning in tenebrionid species (Coleoptera: Tenebrionidae) inhabiting Mediterranean coastal dunes. *European Journal of Entomology*, 113, 462–468.

Fattorini, S., Borges, P.A.V., Fiasca, B. and Galassi, D.M.P. (2016b) Trapped in the web of water: Froundwater-fed springs are island-like ecosystems for the meiofauna. *Ecology and Evolution*, 6, 8389–8401.

Fattorini, S., Rigal, F., Cardoso, P. and Borges, P.A.V. (2016c) Using species abundance distribution models and diversity indices for biogeographical analyses. *Acta Oecologica*, 70, 21–28.

Fattorini, S., Bergamaschi, D., Galassi, D.M.P., Biondi, M., Acosta, A.T.R. and Di Giulio, A. (2017a) Spatial organization of an insect ensemble in a Mediterranean ecosystem: The tenebrionid beetles (Coleoptera Tenebrionidae) inhabiting an Adriatic coastal sand dune area. *Acta Zoologica Bulgarica*, 69, 201–208.

Fattorini, S., Borges, P.A.V., Dapporto, L. and Strona, G. (2017b) What can the parameters of the species-area relationship (SAR) tell us? Insights from Mediterranean islands. *Journal of Biogeography*, 44, 1018–1028.

Fattorini, S., Gabriel, R., Arroz, A.M., Amorim, I.R., Borges, P.A.V. and Cafaro, P. (2017c) Children's preferences for less diverse greenspaces do not disprove biophilia. *Proceedings of the National Academy of Sciences of the United States of America*, 114(35), E7215.

Fattorini, S., Lombardo, P., Fiasca, B., Di Cioccio, A., Di Lorenzo, T. and Galassi, D.M.P. (2017d) Earthquake-related changes in species spatial niche overlaps in spring communities. *Scientific Reports*, 7, 443.

Fattorini, S., Di Lorenzo, T. and Galassi, D.M.P. (2018a) Earthquake impacts on microcrustacean communities inhabiting groundwater-fed springs alter species-abundance distribution patterns. *Scientific Reports*, 8, 1501.

Fattorini, S., Lin, G. and Mantoni, C. (2018b). Avian species-area relationships indicate that towns are not different from natural areas. *Environmental Conservation*, 45, 419–424.

Fattorini, S., Mantoni, C., De Simoni, L. and Galassi, D.M.P. (2018c) Island biogeography of insect conservation in urban green spaces. *Environmental Conservation*, 45(1), 1–10.

Fattorini, S., Di Biase, L. and Chiarucci, A. (2019a) Recognizing and interpreting vegetational belts: New wine in the old bottles of a von Humboldt's legacy. *Journal of Biogeography*, 46, 1643–1651.

Fattorini, S., Mantoni, C., Di Biase, L. and Pace L. (2019b) Mountain biodiversity and sustainable development. In: Leal Filho, W., Azul, A., Brandli, L., Özuyar, P. and Wall, T. (eds) *Encyclopedia of the UN Sustainable Development Goals. Life on Land.* Springer Nature, Berlin, Germany, pp. 1–31.

Fattorini, S., Fiasca, B., Di Lorenzo, T., Di Cicco, M. and Galassi, D.M.P. (2020a) A new protocol for assessing the conservation priority of groundwater-dependent ecosystems. *Aquatic Conservation: Marine and Freshwater Ecosystem*, 30, 1483–1504.

Fattorini, S., Mantoni, C., Bergamaschi, D., Fortini, L., Sánchez, F. J., Di Biase, L. and Di Giulio, A. (2020b) Activity density of carabid beetles along an urbanisation gradient. *Acta Zoologica Academiae Scientiarum Hungaricae*, 66(Suppl), 21–36.

Fattorini, S., Mantoni, C., Di Biase, L., Strona, G., Pace, L. and Biondi, M. (2020c) Elevational patterns of generic diversity in the tenebrionid beetles (Coleoptera Tenebrionidae) of Latium (Central Italy). *Diversity*, 12, 47.

Fattorini, S., Maurizi, E. and Di Giulio, A. (2021) Interactional behaviors of the parasitic beetle *Paussus favieri* with its ant host *Pheidole pallidula*: The mimetic role of the acoustical signals. *Insect Science*, 28, 548–554.

Fattorini, S., Mantoni, C., Dapporto, L., Davini, G. and Di Biase, L. (2023) Using botanical gardens as butterfly gardens: Insights from a pilot project in the Gran Sasso and Monti della Laga National Park (Italy). *Conservation*, 3, 109–126.

Fattorini, S., Vitozzi, A., Di Biase, L. and Bergamaschi, D. (2024) Macroecology of dung beetles in Italy. *Insects*, 15, 39.

Faucon, M.-P., Houben, D. and Lambers, H. (2017) Plant functional traits: Soil and ecosystem services. *Trends in Plant Science*, 22, 385–394.

Faurby, S., Davis, M., Pedersen, R.Ø., Schowanek, S.D., Antonelli, A. and Svenning, J.-C. (2018) PHYLACINE 1.2: The Phylogenetic Atlas of Mammal Macroecology. *Ecology*, 99, 2626.

Fauteux, D., Gauthier, G. and Berteaux, D. (2015) Seasonal demography of a cyclic lemming population in the Canadian Arctic. *Journal of Animal Ecology*, 84, 1412–1422.

Féblot-Augustins, J. (1993) Mobility strategies in the late Middle Palaeolithic of central Europe and western Europe: Elements of stability and variability. *Journal of Anthropological Archaeology*, 12, 211–265.

Feddema, J.J. (2005) A revised Thornthwaite-type global climate classification. *Physical Geography*, 26, 442–466.

Fedigan, L.M., Fedigan, L., Chapman, C. and Glander, K.E. (1988) Spider monkey home ranges: A comparison of radio telemetry and direct observation. *American Journal of Primatology*, 16, 19–29.

Fellowes, M.D.E., Acquaah-Harrison, K., Angeoletto, F., Santos, J.W.M.C., da Silva Leandro, D. *et al.* (2020) Map-A-Mole: Greenspace area influences the presence and abundance of the European Mole *Talpa europaea* in urban habitats. *Animals*, 10, 1097.

Felt, E.P. (1940) *Plant Galls and Gall Makers*. Comstock Publishing Company, New York, USA.

Feng, G., Huang, J., Yue, X., Li, J. and Zang, R. (2021) Disentangling environmental effects on the tree species abundance distribution and richness in a subtropical forest. *Frontiers in Plant Science* 12, fpls.2021.622043.

Feng, W. and Takemoto, K. (2014) Heterogeneity in ecological mutualistic networks dominantly determines community stability. *Scientific Reports*, 4, 5912.

Fennell, M., Wade, M. and Bacon, K.L. (2018) Japanese knotweed (*Fallopia japonica*): An analysis of capacity to cause structural damage (compared to other plants) and typical rhizome extension. *PeerJournal*, 6, e5246.

Fenster, C.B., Ballou, J.D., Dudash, M.R., Eldridge, M.D.B., Frankham, R. *et al.* (2018) Conservation and genetics. *Yale Journal of Biology and Medicine*, 91, 491–501.

Fenton, J.H.C. (2021) *An Illustrated Book of Peat: The Life and Death of Bogs. A New Synthesis*. Fenton Publishing, Argyll, UK.

Fenton, M.B. (1984) Echolocation: Implications for ecology and evolution of bats. *Quarterly Review of Biology*, 59, 33–53.

Fenu, G., Pau, P.L. and Dessì, D. (2017) Functional models and extending strategies for ecological networks. *Applied Network Science*, 2, 10.

Ferguson, R.B. (2013) The prehistory of war and peace in Europe and the Near East. In: Fry, D.P. (ed.) *War, Peace, and Human Nature: The Convergence of Evolutionary and Cultural Views*. Oxford University Press, New York, USA, pp. 191–240.

Feroldi, A., da Silva, E. and Gonçalves, M.M. (2021) Drystone walls in the Algarve, Portugal. Characterization and interconnection with the geology and lithology. *IOP Conference Series: Materials Science and Engineering*, 1203, 022129.

Ferraboschi, P., Ciceri, S. and Grisenti, P. (2021) Applications of lysozyme, an innate immune defense factor, as an alternative antibiotic. *Antibiotics*, 10, 1534.

Ferrarini, A., Giglio, G., Pellegrino, S.C. and Gustin, M. (2021) A new general index of home range overlap and segregation: The lesser kestrel in Southern Italy as a case study. *Avian Research*, 12, 4.

Ferreira, C.S.S., Seifollahi-Aghmiuni, S., Destouni, G., Ghajarniaa, N. and Kalantari, Z. (2018) Soil degradation in the European Mediterranean region: Processes, status and consequences. *Science of the Total Environment*, 805, 150106.

Ferreras, P. and MacDonald, D.W. (1999) The impact of American mink *Mustela vison* on water birds in the upper Thames. *Journal of Applied Ecology*, 36, 701–708.

Ferro, I. and Morrone, J.J. (2014) Biogeographical transition zones: A search for conceptual synthesis. *Biological Journal of the Linnean Society*, 113, 1–12.

Fey, K., Vuorisalo, T., Lehikoinen, A. and Selonen, V. (2015) Urbanisation of the wood pigeon (*Columba palumbus*) in Finland. *Landscape and Urban Planning*, 134, 188–194.

Ficetola, G.F., Mazel, F. and Thuiller, W. (2017) Global determinants of zoogeographical boundaries. *Nature Ecology and Evolution*, 1, 89.

Fichaux, M., Béchade, B., Donald, J., Weyna, A., Delabie, J.H.C. *et al.* (2019) Habitats shape taxonomic and functional composition of Neotropical ant assemblages. *Oecologia*, 189, 501–513.

Fiedler, K. (2001) Ants that associate with Lycaeninae butterfly larvae: Diversity, ecology and biogeography. *Diversity and Distributions* 7, 45–60.

Fiedler, K., Hölldobler, B. and Seufert, P. (1996) Butterflies and ants: The communicative domain. *Experientia*, 52, 14–24.

Field, D., Anderson-Whymark, H., Linford, N., Barber, M., Bowden, M. *et al.* (2015) Analytical surveys of Stonehenge and its environs, 2009–2013. *Proceedings of the Prehistoric Society*, 81, 125–148.

Fierer, N., Nemergut, D., Knight, R. and Craine, J.M. (2010) Changes through time: Integrating microorganisms into the study of succession, *Research in Microbiology*, 161, 635–642.

Filgueiras, B.K.C., Peres, C.A., Iannuzzi, L., Tabarelli, M. and Leal, I.R. (2022) Functional reorganization of dung beetle assemblages in forest-replacing sugarcane plantations. *Journal of Insect Conservation*, 26, 683–695.

Filion, M. (ed.) (2012) *Quantitative Real-time PCR in Applied Microbiology*. Caister Academic Press, Norfolk, UK.

Findlay-Robinson, R., Deecke, V.B., Weatherall, A. and Hill, D.L. (2023) Effects of climate change on life-history traits in hibernating mammals. *Mammal Review*, 53, 84–98.

Finegan, B., Peña-Claros, M., de Oliveira, A., Ascarrunz, N., Bret-Harte, M.S. *et al.* (2015) Does functional trait diversity predict above-ground biomass and productivity of tropical forests? Testing three alternative hypotheses. *Journal of Ecology*, 103, 191–201.

Finlay, B.J. (2004) Protist taxonomy: An ecological perspective. *Philosophical Transactions of the Royal Society B: Biological Sciences*, 359, 599–610.

Fish, J.D. (1979) The rhythmic spawning behaviour of *Littorina littorea* (L.). *Journal of Molluscan Studies*, 45, 172–177.

Fisher, A.M., Le Page, S., Manser, A., Lewis, D.R., Holwell, G.I. *et al.* (2021) Relatedness modulates density-dependent cannibalism rates in *Drosophila*. *Functional Ecology*, 35, 2707–2716.

Fisher, R.A., Corbet, A.S. and Williams, C.B. (1943) The relation between the number of species and the number of individuals in a random sample of an animal population. *Journal of Animal Ecology*, 12, 42–58.

Flanders, J. and Jones, G. (2009) Roost use, ranging behavior, and diet of greater horseshoe bats (*Rhinolophus ferrumequinum*) using a transitional roost. *Journal of Mammalogy*, 90, 888–896.

Flannery-Sutherland, J.T., Silvestro, D. and Benton, M.J. (2022) Global diversity dynamics in the fossil record are regionally heterogeneous. *Nature Communications*, 13, 2751.

Flather, C.H. and Sieg, C.H. (2007) Species rarity: Definition causes and classification. In: Raphael, M.G. and Molina, R. (eds) *Conservation of Rare or Little-Known Species*. Island Press, Washington, DC, USA, pp. 40–66.

Flynn, D.F.B., Mirotchnick, N., Jain, M., Palmer, M.I. and Naeem, S. (2011) Functional and phylogenetic diversity as predictors of biodiversity-ecosystem-function relationships. *Ecology*, 92, 1573–1581.

Foitzik, S., DeHeer, C.J., Hunjan, D.N. and Herbers, J.M. (2001) Coevolution in host–parasite systems: Behavioural strategies of slave-making ants and their hosts. *Proceedings of the Royal Society B: Biological Sciences*, 268, 1139–1146.

Fonseca, C.R. and Ganade, G. (2001), Species functional redundancy, random extinctions and the stability of ecosystems. *Journal of Ecology*, 89, 118–125.

Fontaneto, D., Kaya, M., Herniou, E.A. and Barraclough, T.G. (2009) Extreme levels of hidden diversity in microscopic animals (Rotifera) revealed by DNA taxonomy. *Molecular Phylogenetics and Evolution*, 53, 182–189.

Forbes, E. (1856) Map of the distribution of marine life. In: Johnston, A.K. (ed.) *The Physical Atlas of Natural Phenomena.* Lea and Blanchard, Philadelphia, USA.

Forbes, S.A. (1887) The lake as a microcosm. *Bulletin of the Peoria Science Association*, 77–87. (Reprinted in *Bulletin of the Illinois State Natural History Survey*, 1925, 15, 537–550.)

Ford, E.B. (1945) *Butterflies* Collins, London, UK.

Ford, E.B. (1964) *Ecological Genetics*. Methuen, London, UK.

Ford, H.D. and Ford, E.B. (1930) Fluctuation in numbers and its influence on variation in *Melitaea aurinia*. *Transactions of the Royal Entomological Society of London*, 78, 345–351.

Foresman, T. and Luscombe, R. (2017) The second law of geography for a spatially enabled economy. *International Journal of Digital Earth*, 10, 979–995.

Forman, D.W. (2001) The breeding ecology of the moorhen, Gallinula chloropus, in an artificially created wetland environment at WWT Llanelli, South Wales. PhD thesis, Swansea University, Swansea, UK. Avaialble at: https://typeset.io/pdf/the-breeding-ecology-of-the-moorhen-gallinula-chloropus-in-y7u3834j77.pdf (accessed 14 May 2025).

Forman, R.T.T. (1995) *Land Mosaics. The Ecology of Landscapes and Regions*. Cambridge University Press, Cambridge, UK.

Forman, R.T.T. and Godron, M. (1986) *Landscape Ecology*. Wiley, New York, USA.

Forman, R.T.T. (2014) *Urban Ecology: Science of Cities*. Cambridge University Press, Cambridge UK.

Forman, R.T.T. (2019) *Towns, Ecology, and the Land*. Cambridge University Press, Cambridge UK.

Forrest, J.R. (2016) Complex responses of insect phenology to climate change. *Current Opinion in Insect Science*, 17, 49–54.

Fort, H. (2018) On predicting species yields in multispecies communities: Quantifying the accuracy of the linear Lotka-Volterra generalized model. *Ecological Modelling*, 387, 154–162.

Fortin, M.J. and Dale, M.R.T. (2005) *Spatial Analysis. A Guide for Ecologists*. Cambridge University Press, Cambridge, UK.

Fortin, M.J., Dale, M.R.T. and ver Hoef, J. (2002) Spatial analysis in ecology. In: El-Shaarawi, A.H. and Piegorsch, W.W. (eds) *Encyclopedia of Environmetrics*, vol. 4. John Wiley and Sons, Chichester, UK, pp. 2051–2058.

Fortunel, C., Garnier, E., Joffre, R., Kazakou, E., Quested, H. *et al.* (2009) Leaf traits capture the effects of land use changes and climate on litter decomposability of grasslands across Europe. *Ecology*, 90, 598–611.

Fossey, D. (1972) Living with mountain gorillas. In: Allen, T. (ed.) *The Marvels of Animal Behavior*. National Geographic Society, Washington, DC, USA, pp. 208–229.

Fossey, D. (1983) *Gorillas in the Mist*. Houghton Mifflin, Boston, USA.

Foster, K.R. and Ratnieks, F.L.W. (2005) A new eusocial vertebrate? *Trends in Ecology and Evolution*, 20, 363–364.

Fowles, J. (1969) *The French Lieutenant's Woman*, ch. 29. Jonathan Cape, London, UK.

Fowles, J. (1998) *Wormholes. Essays and Occasional Writings. An Interview with Dianne Vipond*. Vintage/Random House, London, UK.

Fox, B.J. and Fox, M.D. (2000) Factors determining mammal species richness on habitat islands and isolates: Habitat diversity, disturbance, species interactions and guild assembly rules. *Global Ecology and Biogeography*, 9, 19–37.

Fox, B.J., Taylor, J.E. and Thompson, P.T. (2003) Experimental manipulation of habitat structure: A retrogression of the small mammal succession. *Journal of Animal Ecology*, 72, 927–940.

Fox, B.W. (1996) Alternative feeding strategies of the White Admiral (*Ladoga camilla* L.) and the Broad-bordered Bee-Hawk moth (*Hemaris fuciformis* L.) on Honeysuckle (*Lonicera periclymenum* L.). PhD thesis, University of Southampton, Southampton, UK.

Fox, J.W. (2013) The intermediate disturbance hypothesis should be abandoned. *Trends in Ecology and Evolution*, 28, 86–92.

Fox, R., Asher, J., Brereton, T., Roy, D. and Warren, M. (2006) *The State of Butterflies in Britain and Ireland*. Pisces Publications, Newbury, UK.

Fox, R., Parsons, M.S., Chapman, J.W., Woiwod, I.P., Warren, M.S. and Brooks, D.R. (2013) *The State of Britain's Larger Moths 2013*. Butterfly Conservation and Rothamsted Research, Wareham, UK.

Fox, R., Dennis, E.B., Harrower, C.A., Blumgart, D., Bell, J.R. *et al.* (2021) *The State of Britain's Larger Moths 2021*. Butterfly Conservation, Rothamsted Research and UK Centre for Ecology and Hydrology, Wareham, UK.

Foxcroft, L. (2009) Developing thresholds of potential concern for invasive alien species: Hypotheses and concepts. *Koedoe*, 51(1), 1–6.

Francis, R.A. (2010) Wall ecology: A frontier for urban biodiversity and ecological engineering. *Progress in Physical Geography: Earth and Environment*, 35, 43–63.

Frankham, R., Ballou, J.D. and Briscoe, D.A. (2010) *Introduction to Conservation Genetics*, 2nd edn. Cambridge University Press, Cambridge, UK.

Frankham, R., Ballou, J.D., Eldridge, M.D., Lacy, R.C., Ralls, K. *et al.* (2011). Predicting the probability of outbreeding depression. *Conservation Biology*, 25(3), 465–475.

Frankham, R., Ballou, J.D., Ralls, K., Eldridge, M., Dudash, M.R. *et al.* (2019) *A Practical Guide for Genetic Management of Fragmented Animal and Plant Populations*. Oxford University Press, Oxford, UK.

Franklin, J. (2023) Species distribution modelling supports the study of past, present and future biogeographies. *Journal of Biogeography*, 50, 1533–1545.

Franklin, R.E. and Gosling, R.G. (1953) Molecular configuration in sodium thymonucleate. *Nature*, 171, 740–741.

Franzreb, K.E. (1981) The determination of avian densities using the variable-strip and fixed-width transect surveying methods. *Studies in Avian Biology*, 6, 139–145.

Freeman, J.A. (1945) Studies in the distribution of insects by aerial currents. The insect population of the air from ground level to 300 feet. *Journal of Animal Ecology*, 14, 128–154.

Freifeld, H.B., Plentovich, S., Farmer, C., Kohley, C.R., Luscomb, P. *et al.* (2016). Long-distance translocations to create a second millerbird population and reduce extinction risk. *Biological Conservation*, 199, 146–156.

Freitag, H. (1962) *Einführung in die Biogeographie von Mitteleuropa unter besonderer Berücksichtigung von Deutschland*. Fischer, Stuttgart, Germany.

Fretter, V. and Graham, A. (1980) The prosobranch molluscs of Britain and Denmark. Part 5-Marine Littorinacea. *Journal of Molluscan Studies*, 7(Suppl), 242–284.

Fretwell, S.D. (1972) *Populations in a Seasonal Environment*. Princeton University Press, Princeton, USA.

Fretwell, S.D. and Lucas, H.L. Jr. (1969) On territorial behavior and other factors infuencing habitat distribution in birds. *Acta Biotheoretica*, 14, 16–36.

Freyhof, J. and Kottelat, M. (2008) *Anguilla anguilla. The IUCN Red List of Threatened Species 2009.2*. International Union for Conservation of Nature and Natural Resources, Gland, Switzerland.

Fridley, J.D. and Grime, J.P. (2010) Community and ecosystem effects of intraspecific genetic diversity in grassland microcosms of varying species diversity. *Ecology*, 91, 2272–2283.

Fridley, J.D. and Wright, J. (2012) Drivers of secondary succession rates across temperate latitudes of the Eastern USA: Climate, soils, and species pools. *Oecologia*, 168, 1069–1077.

Fridley, J.D. and Wright, J.P. (2018) Temperature accelerates the rate fields become forests. *Proceedings of the National Academy of Sciences of the United States of America*, 115, 4702–4706.

Friedlaender, A., Bocconcelli, A., Wiley, D., Cholewiak, D. and Ware, C. (2011) Underwater components of humpback whale bubble-net feeding behaviour. *Behaviour*, 148, 575–602.

Froese, R. and Pauly, D. (eds) (2023) FishBase. Available at: www.fishbase.org (accessed 14 May 2025).

Frohawk, F.W. (1924) *Natural History of British Butterflies*. Hutchinson, London, UK.

Frouz, J., Thébault, E., Pižl, V., Adl, S., Cajthaml, T. *et al.* (2013) Soil food web changes during spontaneous succession at post mining sites: A possible ecosystem engineering effect on food web organization? *PLOS One*, e79694.

Fry, E.L., Power, S.A. and Manning, P. (2014) Trait-based classification and manipulation of plant functional groups for biodiversity-ecosystem function experiments. *Journal of Vegetation Science*, 25, 248–261.

Fryxell, J.M., Falls, J.B., Falls, E.A. and Brooks, R.J. (1998) Long-term dynamics of small-mammal populations in Ontario. *Ecology*, 79, 213–225.

Fu, H., Zhong, J., Yuan, G., Ni, L., Xie, P. and Cao, T. (2014) Functional traits composition predict macrophytes community productivity along a water depth gradient in a freshwater lake. *Ecology and Evolution*, 4, 1516–1523.

Fujioka, E., Yoshimura, K., Ujino, T., Yoda, K., Fukui, D. and Hiryu, S. (2021) Foraging and commuting habitats of the greater horseshoe bat, revealed by high-resolution GPS tracking. Available at: https://assets-eu.researchsquare.com/files/rs-26350/v1/370ab978-9212-4772-

90d0-188369162ef5.pdf?c=1631833953 (accessed 14 May 2025).

Fujisaki, I., Pearlstine, E.V. and Mazzotti, F.J. (2010) The rapid spread of invasive Eurasian collared doves *Streptopelia decaocto* in the continental USA follows human-altered habitats. *Ibis*, 152, 622–632.

Fuller, M.R., Millspaugh, J.J., Church, K.E. and Kenward, R.E. (2005) Wildlife radiotelemetry. In: Braun, C.E. (ed.) *Techniques for Wildlife Investigations and Management*. The Wildlife Society, Bethesda, USA, pp. 377–417.

Fuller, R.J. (1982) *Bird Habitats in Britain*. T. and A.D. Poyser, Calton, UK.

Fuller, R.J. (ed.) (2012) *Birds and Habitat. Relationships in Changing Landscapes*. Cambridge University Press, Cambridge, UK.

Fuller, R.J., Hinsley, S.A. and Swetnam, R.D. (2004) The relevance of non-farmland habitats, uncropped areas and habitat diversity to the conservation of farmland birds. *Ibis*, 146, 22–31.

Fuller, R.J., Williamson, T., Barnes, G. and Dolman, P.M. (2017) Human activities and biodiversity opportunities in pre-industrial cultural landscapes: Relevance to conservation. *Journal of Applied Ecology*, 54, 459–469.

Fuller, R.M., Smith, G.M., Sanderson, J.M., Hill, R.A., and Thomson, A.G. (2002) The UK Land Cover Map 2000: Construction of a parcel-based vector map from satellite images. *Cartographic Journal*, 39, 15–25.

Funmilayo, O. (1977) Distribution and abundance of moles (*Talpa europaea* L.) in relation to physical habitat and food supply. *Oecologia*, 30, 277–283.

Furholt, M. (2018) Massive migrations? The impact of recent aDNA studies on our view of third millennium Europe. *European Journal of Archaeology*, 21, 159–191.

Fuss, J., Spassov, N., Begun, D.R. and Böhme, M. (2017) Potential hominin affinities of *Graecopithecus* from the Late Miocene of Europe. *PLOS One*, 12, e0177127.

Fuss, J., Spassov, N., Böhme, M., and Begun, D.R. (2018) Response to Benoit and Thackeray (2017): 'A cladistic analysis of *Graecopithecus*'. *South African Journal of Science*, 114, a0267.

Gabriel, R., Homem, N., Couto, A., Aranda, S.C. and Borges, P.A.V. (2011) Azorean bryophytes: A preliminary review of rarity patterns. In: Martins, A.M.F. and Carvalho, M.C. (eds) *Celebrating Darwin: Proceedings of the Symposium 'Darwin's Mistake and what we are doing to correct it'. Açoreana*, Suppl. 7, 149–206.

Galassi, D.M.P., Fiasca, B., Di Lorenzo, T., Porfirio, S., Montanari, S. and Fattorini, S. (2017) Groundwater biodiversity in a chemoautotrophic cave ecosystem: How geochemistry regulates microcrustacean community structure. *Aquatic Ecology*, 51, 75–90.

Galdikas-Brindamour, B. (1975) Orangutans, Indonesia's people of the forest. *National Geographic Magazine*, 148, 444–473.

Galiana, J. and Haseltine, W.A. (2019) *Aging Well: Solutions to the Most Pressing Global Challenges of Aging*. Palgrave Macmillan/Springer Nature Singapore, Singapore.

Gall, Y.M. and Konashev, M.B. (2001) The discovery of Gramicidin S: The intellectual transformation of G.F. Gause from biologist to researcher of antibiotics and on its meaning for the fate of Russian genetics. *History and Philosophy of the Life Sciences*, 23, 137–150.

Gallagher, R.V. and Leishman, M.R. (2012) A global analysis of trait variation and evolution in climbing plants. *Journal of Biogeography*, 39, 1757–1771.

Gallagher, R.V., Falster, D.S., Maitner, B.S., Salguero-Gómez, R., Vandvik, V. *et al.* (2020) Open Science principles for accelerating trait-based science across the Tree of Life. *Nature Ecology and Evolution*, 4, 294–303.

Gallagher, R.V., Butt, N., Carthey, A.J.R., Tulloch, A., Bland, L. *et al.* (2021) A guide to using species trait data in conservation. *One Earth*, 4, 927–936.

Gałuszka, A. and Migaszewski, Z.M. (2017) Chemical Signals of the Anthropocene. In: *Encyclopedia of the Anthropocene*. Elsevier, Amsterdam, The Netherlands.

Gandhi, M.K. (1960) *Trusteeship* (compiled by Kelekar, R.). Navajivan Publishing House, Ahemadabad, India.

Gao, Z., Karlsson, I., Geisen, S., Kowalchuk, G. and Jousset, A. (2019) Protists: Puppet masters of the rhizosphere microbiome. *Trends in Plant Science*, 24, 165–176.

Garbuzov, M. and Ratnieks, F.L.W. (2014) Ivy: An underappreciated key resource to flower-visiting insects in autumn. *Insect Conservation and Diversity*, 7, 91–102.

García-Llamas, P., Rangel, T.F., Calvo, L. and Suárez-Seoane, S. (2019) Linking species functional traits of terrestrial vertebrates and environmental filters: A case study in temperate mountain systems. *PLOS One*, 14, e0211760.

García-Martínez, M.Á., Martínez-Tlapa, D.L., Pérez-Toledo, G.R., Quiroz-Robledo, L.N., Castaño-Meneses, G. *et al.* (2015) Taxonomic, species and functional group diversity of ants in a tropical anthropogenic landscape. *Tropical Conservation Science*, 8, 1017–1032.

Gärdenfors, U., Hilton-Taylor, C., Mace, G.M. and Rodríguez, J.P. (2001) The application of IUCN Red List Criteria at regional levels. *Conservation Biology*, 15, 1206–1212.

Gardini, P., Fattorini, S., Audisio, P. and Sabatelli, S. (2025) Influence of different land-use types on soil arthropod communities in an urban area: A case study from Rome (Italy). *Land*, 14, 714.

Gardner, C.J., Radolalaina, P., Rajerison, M. and Greene, H.W. (2015) Cooperative rescue and predator fatality involving a group-living strepsirrhine, Coquerel's sifaka (*Propithecus coquereli*), and a Madagascar ground boa (*Acrantophis madagascariensis*). *Primates*, 56, 127–129.

Gare, A. (2000) Aleksandr Bogdanov's history, sociology and philosophy of science. *Studies in History and Philosophy of Science*, 31(2), 231–248.

Garey, J.R. and Schmidt-Rhaesa, A. (1998) The essential role of "minor" phyla in molecular studies of animal evolution. *American Zoologist*, 38, 907–917.

Garnier, E., Cortez, J., Billès, G., Navas, M.L., Roumet, C. *et al.* (2004) Plant functional markers capture ecosystem properties during secondary succession. *Ecology*, 85, 2630–2637.

Garrido-Benavent, I., Pérez-Ortega, S., Durán, J., Ascaso, C., Pointing, S.B. *et al.* (2020) Differential colonization and succession of microbial communities in rock and soil substrates on a maritime antarctic glacier forefield. *Frontiers in Microbiology*, 11, 126.

Gaston, K. (2010) Biodiversity. In: Sodhi, N.S. and Ehrlich, P.R. (eds) *Conservation Biology for All*. Oxford University Press, Oxford, UK.

Gaston, K.J. (1991) The magnitude of global insect species richness. *Conservation Biology*, 5,183–196.

Gaston, K.J. (1990) Patterns in the geographical ranges of species. *Biological Reviews*, 65, 105–129.

Gaston, K.J. (1994) *Rarity*. Chapman and Hall, London, UK.

Gaston, K.J. (2007) Latitudinal gradient in species richness. *Current Biology*, 17, R574.

Gaston, K.J. and Blackburn, T.M. (1996) Conservation implications of geographic range size-body size relationships. *Conservation Biology*, 10, 638–646.

Gaston, K.J. and Lawton, J.H. (1990) Effects of scale and habitat on the relationship between regional distribution and local abundance. *Oikos*, 58, 329–335.

Gaston, K.J., Chown, S.L. and Evans, K.L. (2008) Ecogeographical rules: Elements of a synthesis. *Journal of Biogeography*, 35, 483–500.

Gatesy, J. and Amato, G. (2008) The rapid accumulation of consistent molecular support for intergeneric crocodylian relationships. *Molecular Phylogenetics and Evolution*, 48, 1232–1237.

Gaudard, C.A., Robertson, M.P. and Bishop, T.R. (2019) Low levels of intraspecific trait variation in a keystone invertebrate group. *Oecologia*, 190, 725–735.

Gause, G.F. (1932) Experimental studies on the struggle for existence. *Journal of Experimental Biology*, 9, 389–402.

Gause, G.F. (1934) *The Struggle for Existence*. Williams and Wilkins, Baltimore, USA.

Gause, G.F. and Witt, A.A. (1935) Behavior of mixed populations and the problem of natural selection. *American Naturalist*, 69, 596–609.

Gauzere, P., Morin, X., Violle, C., Caspeta, I., Ray, C. and Blonder, B. (2020) Vacant yet invasible niches in forest community assembly. *Functional Ecology*, 34, 1945–1955.

Gavish, Y., Giladi, I. and Ziv, Y. (2019) Partitioning species and environmental diversity in fragmented landscapes: Do the alpha, beta and gamma components match? *Biodiversity and Conservation*, 28, 769–786.

Geeson, N.A., Brandt, J.B. and Thornes, J.B. (2002) *Mediterranean Desertification: A Mosaic of Processes and Responses*. John Wiley and Sons, Hoboken, USA.

Geiger, F., van der Lubbe, S.C.T.M., Brunsting, A.M.H. and de Snoo, G.R. (2010) Insect abundance in cow dung pats of different farming systems. *Entomologische Berichten*, 70, 106–110.

Geiselhardt, S.F., Peschke, K. and Nagel, P. (2007) A review of myrmecophily in ant nest beetles (Coleoptera: Carabidae: Paussinae): Linking early observations with recent findings. *Naturwissenschaften*, 94, 871–894.

Gentile, G., Bonelli, S. and Riva, F. (2021) Evaluating intraspecific variation in insect trait analysis. *Ecological Entomology*, 46, 11–18.

Gérikas Ribeiro, C., Marie, D., Lopes dos Santos, A., Pereira Brandini, F. and Vaulot, D. (2016) Estimating microbial populations by flow cytometry: Comparison between instruments. *Limnology Oceanography Methods*, 14, 750–758.

Gerlach, J., Borges, P.A.V., Fattorini, S., Porch, N. and Wilkins, V. (2024) Insect conservation on islands. In: Pryke, J.S., Samways, M.J., New, T.R., Cardoso, P. and Gaigher, R. (eds) *Routledge Handbook of Insect Conservation*, Routledge, Abingdon, UK, pp. 319–333.

Germaine, S.S. and Wakeling, B.F. (2001) Lizard species distributions and habitat occupation along an urban gradient in Tucson, Arizona, USA. *Biological Conservation*, 97, 229–237.

Gerwin, W., Raab, T., Birkhofer, K., Hinz, C., Letmathe, P. *et al.* (2023) Perspectives of lignite post-mining landscapes under changing environmental conditions: What can we learn from a comparison between the Rhenish and Lusatian region in Germany? *Environmental Sciences Europe*, 35, 36.

Getz, W.M. (2011) Biomass transformation webs provide a unified approach to consumer-resource modelling. *Ecology Letters*, 14, 113–124.

Geymonat, L. (1976) *Storia del Pensiero Filosofico e Scientifico*. Garzanti, Milano, Italy.

Giachello, S., Cantera, I., Carteron, A., Marta, S., Cipriano, C. *et al.* (2023) Toward a common set of functional traits for soil protists. *Soil Biology and Biochemistry*, 187, 109207.

Gibb, T.J. and Oseto, C. (2019) *Insect Collection and Identification: Techniques for the Field and Laboratory*, 2nd edn. Elsevier Science and Technology, San Diego, USA.

Gibson, C.C. and Watkinson, A.R. (1989) The host range and selectivity of a parasitic plant: *Rhinanthus minor* L. *Oecologia*, 78, 401–406.

Gibson, D.J. (1996) Textbook misconceptions: The climax concept of succession. *American Biology Teacher*, 58, 135–140.

Gibson, R. (2003) Go with the flow: Tidal migration in marine animals. *Hydrobiologia*, 503, 153–161.

Gibson-Reinemer, D. (2015) A vacant niche: How a central ecological concept emerged in the 19th century. *Bulletin of the Ecological Society of America*, 96, 324–335.

Gies, F. and Gies, J. (2016) *Life in a Medieval Village (Medieval Life)*. Harper and Row, New York, USA.

Giesecke, T., Brewer, S., Finsinger, W., Leydet, M. and Bradshaw, R.H.W. (2017) Patterns and dynamics of European vegetation change over the last 15,000 years. *Journal of Biogeography*, 44, 1441–1456.

Gilbert, F.S. (1985) Ecomorphological relationships in hoverflies (Diptera: Syrphidae). *Proceedings of the Royal Society B: Biological Sciences*, 224, 91–105.

Gilbertson, B., Kent, M. and Pyatt, F.B. (1985) *Practical Ecology for Geography and Biology Survey, Mapping and Data Analysis*. Chapman and Hall London, UK.

Gill, S.R., Pop, M., DeBoy, R.T., Eckburg, P.B., Turnbaugh, P.J. *et al.* (2006) Metagenomic analysis of the human distal gut microbiome. *Science*, 312, 1355–1359.

Giller, P.S. (1984) *Community Structure and The Niche*. Chapman and Hall, London, UK.

Gillespie, G.R., Brennan, K., Gentles, T., Hill, B., Low Choy, J. *et al.* (2015) *A Guide for the Use of Remote Cameras for Wildlife Survey in Northern Australia*. Charles Darwin University, Darwin, Australia.

Gillis, D.M. (2003) Ideal free distributions in fleet dynamics: A behavioral perspective on vessel movement in fisheries analysis. *Canadian Journal of Zoology*, 81, 177–187.

Gilpin, M.E. (1991) The genetic effective size of a metapopulation. In: Hanski, I.A. and Gilpin, M.E. (eds) *Metapopluation Biology*, Academic Press, London, UK, pp. 165–175.

Giménez Gómez, V.C., Verdú, J.R,. Gómez-Cifuentes, A., Vaz-de-Mello, F.Z. and Zurita, G.A. (2018) Influence of land use on the trophic niche overlap of dung beetles in the semideciduous Atlantic forest of Argentina. *Insect Conservation and Diversity*, 11, 554–564.

Girardello, M., Chapman, A., Dennis, R.L.H., Kaila, L., Borges, P.A.V., and Santangeli, A. (2019) Gaps in butterfly inventory data: A global analysis. *Biological Conservation*, 236, 289–295.

Giuggioli, L., Abramson, G., Kenkre, V.M., Parmenter, R.R. and Yates, T.L. (2006) Theory of home range estimation from displacement measurements of animal populations. *Journal of Theoretical Biology*, 240, 126–135.

Glass, B.P. (1982) Seasonal movements of Mexican free-tail bats *Tadarida brasiliensis mexicana* banded in the Great Plains. *Southwestern Naturalist*, 27, 127–133.

Glaudas, X., Rice, S.E., Clark, R.W. and Alexander, G.J. (2020) The intensity of sexual selection, body size and reproductive success in a mating system with male–male combat: Is bigger better? *Oikos*, 129, 998–1011.

Gleason, H.A. (1926) The individualistic concept of the plant association. *American Midland Naturalist*, 21, 92–110.

Godfrey-Smith, P. (2016) *Other Minds: The Octopus, the Sea, and the Deep Origins of Consciousness*. William Collins, Glasgow, UK.

Godfrey-Smith, P (2020) *Metazoa: Animal Life and the Birth of the Mind*. Farrar, Straus and Giroux, New York, USA.

Godwin, H. (1975) *The History of the British Flora*. Cambridge University Press, Cambridge, UK.

Goerck, J.M. (1997) Patterns of rarity in the birds of the Atlantic forest of Brazil. *Conservation Biology*, 11, 112–118.

Goldbogen, J.A. and Madsen, P.T. (2021) The largest of August Krogh animals: Physiology and biomechanics of the blue whale revisited. *Comparative Biochemistry and Physiology, Part A*, 254, 110894.

Goldbogen, J.A., Calambokidis, J., Croll, D.A., McKenna, M.F., Oleson, E. *et al.* (2011) Scaling of lunge-feeding performance in rorqual whales: Mass-specific energy expenditure increases with body size and progressively limits diving capacity. *Functional Ecology*, 26, 216–226.

Goldthwait, R.P., Loewe, F., Ugolini, F.C., Decker, F., DeLong, D.M. *et al.* (1966) Soil development and ecological succession in a deglaciated area of Muir Inlet, Southeast Alaska. In: Mirsky, A. (ed.) *Institute of Polar Studies Report No. 20*. Ohio State University, Columbus, USA.

Goldwasser, L. and Roughgarden, J. (1993) Construction of a large Caribbean food web. *Ecology*, 74, 1216–1233.

Goldyn, B., Hromade, M., Surmacki, A. and Tryjanowski, P. (2003) Habitat use and diet of the red fox *Vulpes vulpes* in an agricultural landscape in Poland. *Zeitschrift für Jagdwissenschaft*, 49, 191–200.

Gollin, D., Jedwab, R, and Vollrath, D. (2016) Urbanisation with and without industrialisation. *Journal of Economic Growth*, 21, 35–70.

Golubic, S., Radtke, G. and Le Campion-Alsumard, T. (2005) Endolithic fungi in marine ecosystems. *Trends in Microbiology*, 13, 229–235.

Gomez-Alvarez, V., King, G.M. and Nusslein, K. (2007) Comparative bacterial diversity in recent Hawaiian volcanic deposits of different ages. *FEMS Microbiology Ecology*, 60, 60e73.

Gómez-Godínez, L.J., Aguirre-Noyola, J.L., Martínez-Romero, E., Arteaga-Garibay, R.I., Ireta-Moreno, J. and Ruvalcaba-Gómez, J.M. (2023) A look at plant-growth-promoting bacteria. *Plants*, 12, 1668.

González-Bernal, E., Greenlees, M.J., Brown, G.P. and Shine, R. (2013) Interacting biocontrol programmes: Invasive cane toads reduce rates of breakdown of cow-pats by dung beetles. *Austral Ecology*, 38, 891–895.

Good, R. (1947) *The Geography of Flowering Plants*. Longmans, London, UK.

Goodacre, S. (2020) Gone with the wind. *The Biologist*, 67, 8–11.

Goodale, E., Beauchamp, G. and Ruxton, G.D. (2017) *Mixed Species Groups of Animals. Behaviour, Community Structure and Conservation*. Academic Press, London, UK.

Goodall, B. (1987) *The Penguin Dictionary of Human Geography*. Penguin, London, UK.

Goodall, J.M. (1969) *My Friends the Wild Chimpanzees*. National Geographic Society, Washington, DC, USA.

Goodall, J.M. (1971) *In the Shadow of Man*. Collins, London, UK.

Goodall, J.M. (1986) *The Chimpanzees of Gombe: Patterns of Behavior*. Belknap Press/Harvard University Press, Boston, USA.

Goodall, J.M. and van Lawick, H. (1971) *Innocent Killers*. Collins, London, UK.

Gooriah, L.D. and Chase, J.M. (2020) Sampling effects drive the species–area relationship in lake zooplankton. *Oikos*, 129, 124–132.

Gorta, S.B.Z., Callaghan, C.T., Samonte, F., Ooi, M.K.J., Mesaglio, T. *et al.* (2023) Multi-taxon biodiversity responses to the 2019–2020 Australian megafires. *Global Change Biology*, 29, 6727–6740.

Gorzelak, M.A., Asay, A.K., Pickles, B.J. and Simard, S.W. (2015) Inter-plant communication through mycorrhizal networks mediates complex adaptive behaviour in plant communities. *AoB PLANTS*, 7, plv050.

Gosz, J.R. (1993) Ecotone hierarchies. *Ecological Applications*, 3, 369–376.

Gotelli, N.J. (2000) Null model analysis of species co-occurrence patterns. *Ecology*, 81, 2606–2621.

Gotelli, N.J. (2008) *A Primer of Ecology*, 4th edn. Sinauer Associates, Sunderland, USA.

Gotelli, N.J. and Colwell, R.K. (2001) Quantifying biodiversity: Procedures and pitfalls in the measurement and comparison of species richness. *Ecology Letters*, 4, 379–391.

Gotelli, N.J. and Colwell, R. (2011) Estimating species richness. In: Magurran, A.E. and McGill, B.J. (eds) *Biological Diversity: Frontiers in Measurement and Assessment*. Oxford University Press, Oxford, UK, pp. 39–54.

Gotelli, N.J. and Graves, G.R. (1996) *Null Models in Ecology*. Smithsonian Institution Press, Washington, DC, USA.

Gotelli, N.J. and Kelly, W.G. (1993) A general model of metapopulation dynamics. *Oikos*, 68, 36–44.

Gottmann, J. (1961) *Megalopolis: The Urbanized Northeastern Seaboard of the United States*. MIT Press, Cambridge, USA.

Gottmann, J. (1976) Megalopolitan systems around the world. *Ekistics*, 41, 109–113.

Goudard, A. and Loreau, M. (2008) Nontrophic interactions, biodiversity, and ecosystem functioning: An interaction web model. *American Naturalist*, 171, 91–106.

Gould, L. and Gabriel, D.N. (2014) Wet and dry season diets of the endangered *Lemur catta* (ring-tailed lemur) in two mountainous rocky outcrop forest fragments in south-central Madagascar. *African Journal of Ecology*, 53, 320–330.

Gould, S.J. (1988) Kropotkin was no crackpot. *Natural History*, 97, 12–21.

Gould, S.J. (2002) *The Structure of Evolutionary Theory*. Harvard University Press, Cambridge, USA.

Gower, J.C. (1971) A general coefficient of similarity and some of its properties. *Biometrics*, 27, 857–871.

Graeter, G.J., Buhlmann, K.A., Wilkinson, L.R. and Gibbons J.W. (eds) (2013) *Inventory and Monitoring: Recommended Techniques for Reptiles and Amphibians*. PARC Technical Publication IM-1, Partners in Amphibian and Reptile Conservation, Birmingham, USA.

Graham, K.E. and Hobaiter, C. (2023) Towards a great ape dictionary: Inexperienced humans understand common nonhuman ape gestures. *PLOS Biology*, 21, e3001939.

Grant, A.J. and Fink, D.P.J. (1958) *Outlines of European History*. Longmans, London, UK.

Grant, P.R. and Grant, B.R. (2008) *How and Why Species Multiply: The Radiation of Darwin's Finches*. Princeton University Press, Princeton, USA.

Grassberger, M. and Frank, C. (2004) Initial study of arthropod succession on pig carrion in a central European urban habitat. *Journal of Medical Entomology*, 41, 511–523.

Gravel, D., Canham, C.D., Beaudet, M. and Messier, C. (2006) Reconciling niche and neutrality: The continuum hypothesis. *Ecological Letters*, 9, 399–409.

Gray, J.S., Bjørgesæter, A. and Ugland, K.I. (2006) On plotting species abundance distributions. *Journal of Animal Ecology*, 75, 752–756.

Gray, R.J. and Karlsson, C. (2022) 101 years of biofluorescent animal studies: Trends in literature, novel hypotheses, and best practices moving forward. *EcoEvoRxiv*. Available at: https://ecoevorxiv.org/repository/view/3834/ (accessed 16 May 2025).

Grayson, R.F. (1993) The distribution and conservation of the ponds of North-West England. *Lancashire Wildlife Journal* (1992–1993), 2 and 3, 23–51.

Greatorex-Davies, N., Sparks, T. and Woiwod, I. (2005) Changes in the Lepidoptera of Monks Wood NNR (1974–2003). In: Gardiner, C. and Sparks, T. (eds) *Ten Years of Change: Woodland Research at Monks Wood NNR, 1993–2003*. Proceedings of the 50th Anniversary Symposium, December 2003. English Nature Research Reports, 613. English Nature, Peterborough, UK.

Green, J.L. and Plotkin, J.B. (2007) A statistical theory for sampling species abundances. *Ecology Letters*, 10, 1037–1045.

Green, L.H. and Goldman, E. (eds) (2023) *Practical Handbook of Microbiology*, 4th edn. CRC Press, Boca Raton, USA.

Green, O.R. (2001) *A Manual of Practical Laboratory and Field Techniques in Palaeobiology*. Springer, Dordrecht, The Netherlands.

Greenberg, J. (2014) *A Feathered River Across the Sky: The Passenger Pigeon's Flight to Extinction*. Bloomsbury, New York, USA.

Gregory, R.D. and Marchant, J.H. (1996) Population trends of jays, magpies, jackdaws and carrion crows in the United Kingdom. *Bird Study*, 43, 28–37.

Greppi, C., Laursen, W. J., Budelli, G., Chang, E. C., Daniels, A. M. *et al.* (2020) Mosquito heat seeking is driven by an ancestral cooling receptor. *Science*, 367, 681–684.

Griffin, L.R. (1998) The distribution and abundance of the rook Corvus frugilegus L. as influenced by habitat suitability and competitive interactions. Durham theses, Durham University, Durham, UK. Available at: http://etheses.dur.ac.uk/4640/ (accessed 16 May 2025).

Griffin, L.R. (1999) Colonization patterns at rook *Corvus frugilegus* colonies: Implications for survey strategies. *Bird Study*, 46, 170–173.

Grigg, R.W. (1983) Community structure, succession and development of coral reefs in Hawaii. *Marine Ecology Progress Series*, 11, 1–14.

Grime, J.P. (1973) Competitive exclusion in herbaceous vegetation. *Nature*, 242, 344–347.

Grime, J.P. (1974) Vegetation classification by reference to strategies. *Nature*, 250, 26–31.

Grime, J.P. (1979) *Plant Strategies and Vegetation Processes*. Wiley, Chichester, UK.

Grime, J.P., Hodgson, J.G. and Hunt, R. (1988) *Comparative Plant Ecology. A Functional Approach to Common British Species*. Unwin Hyman, London, UK.

Grimm, A., Ramírez, A.M.P., Moulherat, S., Reynaud, J. and Henle, K. (2014) Life-history trait database of European reptile species. *Nature Conservation*, 9, 45–67.

Grinnell, J. (1917) The niche-relationships of the California thrasher. *The Auk*, 34, 427–433.

Grinnell, J. (1924) Geography and evolution. *Ecology*, 5, 225–229.

Grinnell, J. (1928) Presence and absence of animals. *University of California Chronicle*, 30, 429–450.

Grohmann, C., Oldeland, J., Stoyan, D. and Linsenmair, K.E. (2010) Multi-scale pattern analysis of a mound-building termite species. *Insectes Sociaux*, 57, 477–486.

Groucutt, H.S., White, T.S., Scerri, E.M.L., Andrieux, E., Clark-Wilson, R. *et al.* (2021) Multiple hominin dispersals into Southwest Asia over the past 400,000 years. *Nature*, 597, 376–380.

Grube, M., Cernava, T., Soh, J., Fuchs, S., Aschenbrenner, I. *et al.* (2015) Exploring functional contexts of symbiotic sustain within lichen-associated bacteria by comparative omics. *ISME Journal*, 9, 412–424.

Grunert, K., Holden, H., Jakobsen, E.R. and Stenseth, N.C. (2021) Evolutionarily stable strategies in stable and periodically fluctuating populations: The Rosenzweig–MacArthur predator–prey model, *Proceedings of the National Academy of Sciences of the United States of America*, 118, e2017463118.

Grüter, C., Jongepier, E. and Foitzik, S. (2018) Insect societies fight back: The evolution of defensive traits against social parasites. *Philosophical Transactions of the Royal Society B: Biological Sciences*, 373, 20170200.

Guerra, Á., González, Á., Pascual, S. and Dawe, E. (2011) The giant squid *Architeuthis*: An emblematic invertebrate that can represent concern for the conservation of marine biodiversity. *Biological Conservation*, 144, 1989–1998.

Guerrero, I., Morales, M.B., Oñate, J.J., Aavik, T., Bengtsson, J. *et al.* (2011) Taxonomic and functional diversity of farmland bird communities across Europe: Effects of biogeography and agricultural intensification. *Biodiversity and Conservation*, 20, 3663–3681.

Guiry, M.D. (2024) How many species of algae are there? A reprise. Four kingdoms, 14 phyla, 63 classes and still growing. *Journal of Phycology*, 60, 214–228.

Gunina, A., Ryzhova, I., Dorodnikov, M. and Kuzyakov, Y. (2015) Effect of plant communities on aggregate composition and organic matter stabilisation in young soils. *Plant and Soil*, 387, 265–275.

Günther, A. (1990) Distribution and bathymetric zonation of shell-boring endoliths in recent reef and shelf environments: Cozumel, Yucatan (Mexico). *Facies*, 22, 233–261.

Guo, B.P., Teneberg, S., Münch, R., Terunuma, D., Hatano, K. *et al.* (2009) Relapsing fever *Borrelia* binds to neolacto glycans and mediates rosetting of human erythrocytes. *Proceedings of the National Academy of Sciences of the United States of America*, 106, 19280–19285.

Guppy, C.S. (1986) The adaptive significance of alpine melanism in the butterfly *Parnassius phoebus* F. (Lepidoptera: Papilionidae). *Oecologia*, 70, 205–213.

Gupta, N., Keshari Das, S. and Siliwal, M. (2015) Natural history of the trapdoor spider *Idiops joida* Gupta *et al.* 2013 (Araneae: Idiopidae) from the Western Ghats in India. *Journal of Asia-Pacific Biodiversity*, 8, 38–42.

Guralnick, R. and Van Cleve, J. (2005) Strengths and weakness of museum and national survey data sets for predicting regional species richness: Comparative and combined approaches. *Diversity and Distributions*, 11, 349–359.

Gurnell, J., Wauters, L.A., Lurz, P.W.W. and Tosi, G. (2004) Alien species and interspecific competition: Effects of introduced eastern grey squirrels on red squirrel population dynamics. *Journal of Animal Ecology*, 73, 26–35.

Gurnell, J., Lurz, P. and Shuttleworth, C. (2016) Ecosystem impacts of an alien invader in Europe, the grey squirrel *Sciurus carolinensis*. In: Shuttleworth, C., Lurz, P. and Gurnell, J. (eds) *The Grey Squirrel: Ecology and Management of an Invasive Species in Europe*. European Squirrel Initiative, Bangor, UK, pp. 307–327.

Gusset, M. and Dick, G. (2011) The global reach of zoos and aquariums in visitor numbers and conservation expenditures. *Zoo Biology*, 30, 566–569.

Guthrie, C.G. and Moorhead, D.L. (2002) Density-dependent habitat selection: Evaluating isoleg theory with a Lotka-Volterra model. *Oikos*, 97, 184–194.

Gutiérrez, D., Thomas, C.D. and Léon-Cortés, J.L. (1999) Dispersal, distribution, patch-network and metapopulation dynamics of the dingy skipper butterfly (*Erynnis tages*). *Oecologia*, 121, 506–517.

Gwali, S., Okullo, P., Hafashimana, D. and Byabashaija, D.M. (2010) Taxonomic diversity, distinctness, and abundance of tree and shrub species in Kasagala Forest Reserve in Uganda: Implications for management and conservation policy decisions. *Tropical Conservation Science*, 3, 319–333.

Habel, J.C., Schmitt, T., Ulrich, W., Gros, P., Salcher, B. and Teucher, M. (2023) Landscape homogenisation and simplified butterfly community structure go on par across Northern Austria. *Landscape Ecology*, 38, 3237–3248.

Haberl, H., Winiwarter, V., Andersson, K., Ayres, R.U., Boone, C. *et al.* (2006). From LTER to LTSER: Conceptualizing the socioeconomic dimension of long-term socioecological research. *Ecology and Society*, 11, 13.

Haddad, C.R., de Jager, L.J.C. and Foord, S.H. (2019) Habitats and cardinal directions are key variables structuring spider leaf litter assemblages under *Searsia lancea*. *Pedobiologia*, 73, 10–19.

Haerter, J.O., Mitarai, N. and Sneppen, K. (2016) Food web assembly rules for generalized Lotka-Volterra equations. *PLOS Computational Biology*, 12, e1004727.

Hagmeier, E.M. (1966) A numerical analysis of the distributional patterns of North American Mammals. II. Re-evaluation of the provinces. *Systematic Biology*, 15, 279–299.

Hagmeier, E.M., Stults, C. and Dexter. A (1964) Numerical analysis of the distributional patterns of North American mammals. *Systematic Biology*, 13, 125–155.

Hairston, N.G. Jr. and Hairston, N.G. (1993) Cause-effect relationships in energy flow, trophic structure and interspecific interactions. *American Naturalist*, 142, 379–411.

Hakizimana, D., Hambuckers, A., Brotcorne, F. and Huynen, M.-C. (2015) Characterization of nest sites of chimpanzees (*Pan troglodytes schweinfurthii*) in Kibira National Park, Burundi. *African Primates*, 10, 1–12.

Hale, P.J. (2014) *Political Descent: Malthus, Mutualism, and the Politics of Evolution in Victorian England.* University of Chicago Press, Chicago, USA.

Hall, L.S., Krausman, P.R. and Morrison, M.L. (1997) The habitat concept and a plea for standard terminology. *Wildlife Society Bulletin*, 25, 173–182.

Hallmann, C.A., Sorg, M., Jongejans, E., Siepel, H., Hofland, N. *et al.* (2017) More than 75 percent decline over 27 years in total flying insect biomass in protected areas. *PLOS One*, 12, e0185809.

Hall-Spencer, J.M. and Moore, P.G. (2000) *Limaria hians* (Mollusca: Limacea): A neglected reef-forming keystone species. *Aquatic Conservation: Marine and Freshwater Ecosystems*, 10, 267–277.

Halme, P., Monkkonen, M., Kotiaho, J. S., Ylisirnio, A. L. and Markkanen, A. (2009) Quantifying the indicator power of an indicator species. *Conservation Biology*, 23, 1008–1016.

Halsey, L. (2018) A matter of life and... energy. *The Biologist*, 65, 18–21.

Hamilton, S. (2002) Between ritual and routine: Interpreting British prehistoric pottery production and distribution. In: Woodward, A. and Hill, J.D. (eds) *Prehistoric Britain: The Ceramic Basis*. Oxbow, Barnsley, UK, pp. 38–53.

Hamilton, W.D. (1971) Geometry of the selfish herd. *Journal of Theoretical Biology*, 31, 295–331.

Hancock, M.H. and Legg, C.J. (2012) Pitfall trapping bias and arthropod body mass. *Insect Conservation and Diversity*, 5, 312–318.

Hannus, J.-J. and Von Numers, M. (2008) Vascular plant species richness in relation to habitat diversity and island area in the Finnish archipelago. *Journal of Biogeography*, 35, 1077–1086.

Hansen, A.J., di Castri, F. and Naiman, R.J. (1988) Ecotones: What and why? *Biology International Special Issue*, 17, 9–46.

Hansen, T.F., Stenseth, N.C., Henttonen, H. and Tast, J. (1999) Interspecific and intraspecific competition as causes of direct and delayed density dependence in fluctuating vole population. *Proceedings of the National Academy of Sciences of the United States of America*, 96, 986–991.

Hanski, I. (1982) Dynamics of regional distribution: The core and satellite species hypothesis. *Oikos*, 38, 210–221.

Hanski, I. (1991) The dung insect community. In: Hanski, I. and Cambefort, Y. (eds) *Dung Beetle Ecology*. Princeton University Press, Princeton, USA, pp. 5–21.

Hanski, I.(1994a) A practical model of metapopulation dynamics. *Journal of Animal Ecology*, 63, 151–162.

Hanski, I. (1994b) Patch-occupancy dynamics in fragmented landscapes. *Trends in Ecology and Evolution*, 9, 131–135.

Hanski, I. (1997) Metapopulation dynamics. From concepts and observations to predictive models. In: Hanski, I. and Gilpin, M.E. (eds) *Metapopulation Biology. Ecology, Genetics, and Evolution*. Academic Press, London, UK, pp. 69–91.

Hanski, I. and Gilpin, M.E. (eds) (1997) *Metapopulation Biology. Ecology, Genetics, and Evolution*. Academic Press, London, UK.

Hanski, I. and Kuussaari, M. (1995) Butterfly metapopulation dynamics. In: Cappuccino, N. and Price, P.W. (eds) *Population Dynamics. New Approaches and Synthesis*. Academic Press, London, UK, pp. 149–171.

Hanski, I. and Simberloff, D. (1997) The metapopulation approach, its history, conceptual domain, and application to conservation. In: Hanski, I. and Gilpin, M.E. (eds) *Metapopulation Biology. Ecology, Genetics, and Evolution*. Academic Press, London, UK, pp. 5–26.

Hanski, I. and Thomas, C.D. (1994) Metapopulation dynamics and conservation: A spatially explicit model applied to butterflies. *Biological Conservation*, 68, 167–180.

Hanski, I., Kouki, J. and Halkka, A. (1993) Three explanations of the positive relationship between distribution and abundance of species. In: Ricklefs, E. and Schluter, D. (eds) *Species Diversity in Ecological Communities. Historical and Geographical Perspectives*. University of Chicago Press, Chicago, USA, pp. 108–116.

Haram, L.E., Carlton, J.T., Centurioni, L., Choong, H., Cornwell, B. *et al.* (2023) Extent and reproduction of coastal species on plastic debris in the North Pacific Subtropical Gyre. *Nature Ecology and Evolution*, 7, 687–697.

Hardesty-Moore, M., Deinet, S., Freeman, R., Titcomb, G.C., Dillon, E.M. *et al.* (2018) Migration in the Anthropocene: How collective navigation, environmental system and taxonomy shape the vulnerability of migratory species. *Philosophical Transactions of the Royal Society B: Biological Sciences*, 373, 20170017.

Hardin, G. (1957) The threat of clarity. *American Journal of Psychiatry*, 114, 392–396.

Hardy, P.B. (1998) *Butterflies of Greater Manchester*. PGL Enterprises, Sale, UK.

Hardy, P.B. and Dennis, R.L.H. (2010) A butterfly exploiting the matrix: *Pieris napi* (Linnaeus, 1758) (Lepidopetra: Pieridae) ovipositing amongst mown grass in a city park. *Entomologist Gazette*, 61, 155–158.

Hardy, P.B., Kinder, P.M., Sparks, T.H. and Dennis, R.L.H. (2010) Elevation and habitats: The potential of sites at different altitudes to provide refuges for phytophagous insects during climatic fluctuations. *Journal of Insect Conservation*, 14, 297–303.

Harel, R., Spiegel, O., Getz, W.M. and Nathan, R. (2017) Social foraging and individual consistency in following behaviour: Testing the information centre hypothesis in free-ranging vultures. *Proceedings of the Royal Society B: Biological Sciences*, 284, 20162654.

Harestad, A.S. and Bunnel, F.L. (1979) Home range and body weight – A reevaluation. *Ecology*, 60, 389–402.

Harmand, S., Lewis, J.E., Feibel, C.S., Lepre, C.J., Prat, S. *et al.* (2015) 3.3-million-year-old stone tools from Lomekwi 3, West Turkana, Kenya. *Nature*, 521, 310–315.

Harmer, R., Tucker, N. and Nickerson, R. (2004) Natural regeneration in storm damaged woods – 1987 storm sites revisited. *Quarterley Journal of Forestry*, 98, 183–190.

Harmon, M.E., Franklin, J.F., Swanson, F.J., Sollins, P., Gregory, S.V. *et al.* (1986) Ecology of coarse woody debris in temperate ecosystems. *Advances in Ecological Research*, 15, 133–302.

Harnik, P.G., Simpson, C., and Payne, J.L. (2012) Long-term differences in extinction risk among the seven forms of rarity. *Proceedings of the Royal Society B: Biological Sciences*, 279(1749), 4969–4976.

Harper, D., Abedon, S., Burrowes, B. and McConville, M. (2021) *Bacteriophages: Biology, Technology, Therapy*. Springer, Dordrecht, The Netherlands.

Harris, A.T., Woolnough, D.A. and Zanatta, D.T. (2011) Insular lake island biogeography: Using lake metrics to predict diversity in littoral zone mollusk communities. *Journal of the North American Benthological Society*, 30, 997–1008.

Harris, B., and Wilson, A.G. (1978) Equilibrium values and dynamics of attractiveness terms in production-constrained spatial interaction models. *Environment and Planning A: Economy and Space*, 10, 371–388.

Harris, C.D., and Ullman, E.L. (1945) The nature of cities. *Annals of the American Academy of Political and Social Science*, 242, 7–17.

Hart, A. (2018) Inside the killing jar. *The Biologist*, 65, 26–29.

Harte, J., Zillio, T., Conlisk, E. and Smith, A.B. (2008) Maximum entropy and the state-variable approach to macroecology. *Ecology*, 89, 2700–2711.

Hartemink, N., Vanwambeke, S.O., Purse, B.V., Gilbert, M. and Van Dyck, H. (2015) Towards a resource-based habitat approach for spatial modelling of vector-borne disease risks. *Biological Reviews*, 90, 1151–1162.

Hartnett, R. (2019) Variation in life-history traits among *Daphnia* and its relationship to species-level responses to phosphorus limitation. *Royal Society Open Science*, 6, 191024.

Harvey, D. (1969) *Explanation in Geography*. Edward Arnold, London, UK.

Harvey, J.A., Tougeron, K., Gols, R., Heinen, R., Abarca, M. *et al.* (2023) Scientists' warning on climate change and insects. *Ecological Monographs*, 93, e1553.

Harvey, P.H. and Clutton-Brock, T.H. (1981) Primate home-range size and metabolic needs. *Behavioral Ecology and Sociobiology*, 8, 151–155.

Haslett, J.R. (2007) *European Strategy for the Conservation of Invertebrates*. Nature and Environment No. 145. Council of Europe Publishing, Strasbourg, France.

Hatton, I.A., Galbraith, E.D., Merleau, N.S.C., Miettinen, T.P., Smith, B.M. and Shander, J.A. (2023) The human cell count and size distribution. *Proceedings of the National Academy of Sciences of the United States of America*, 120(39), e2303077120.

Hauer, F.R. and Lamberti, G.A. (eds) (2007) *Methods in Stream Ecology*, 2nd edn. Academic Press, London, UK.

Haurez, B., Petre, C-A., Vermeulen, C., Tagg, N. and Doucet, J-L. (2014) Western lowland gorilla density and nesting behavior in a Gabonese forest logged for 25 years: Implications for gorilla conservation. *Biodiversity Conservation*, 23, 2669–2687.

Häusler, H. (2018) Did anthropogeology anticipate the idea of the Anthropocene? *Anthropocene Review*, 5(1), 69–86.

Hawkins, H-J., Cargill, R.I.M., Van Nuland, M.E., Hagen, S.C., Field, K.J. *et al.* (2023) Mycorrhizal mycelium as a global carbon pool. *Current Biology*, 33, R560–R573.

Hayat, R., Ali, S., Amara, U., Khalid, R. and Ahmed, I. (2010) Soil beneficial bacteria and their role in plant growth promotion: A review. *Annals of Microbiology*, 60, 579–598.

Hayek, L.C. and Buzas, M.A. (2010) *Surveying Natural Populations. Quantitative Tools for Assessing Biodiversity*. Columbia University Press, New York, USA.

Hayes, M.P. (2015) The biology and ecology of the large blue butterfly *Phengaris* (*Maculinea*) *arion*: A review. *Journal of Insect Conservation*, 19, 1037–1051.

Häyhä, T. and Franzese, P.P. (2014) Ecosystem services assessment: A review under an ecological-economic and systems perspective. *Ecological Modelling*, 289, 124–132.

Hazen, R.M., Grew, E.S., Origlieri, M.J. and Downs, R.T. (2017) On the mineralogy of the "Anthropocene Epoch". *American Mineralogist*, 102, 595–611.

He, F. and Hu, X.-S. (2005) Hubbell's fundamental biodiversity parameter and the Simpson diversity index. *Ecology Letters*, 8, 386–390.

He, K., Liu, Q., Xu, D.-M., Qi, F.-Y., Bai, J. *et al.* (2021) Echolocation in soft-furred tree mice. *Science*, 372 (6548), eaay1513.

He, X., Jeffery, N.W., Stanley, R.R.E., Hamilton, L.C., Rubidge, E.M. and Abbott, C.L. (2023) eDNA metabarcoding enriches traditional trawl survey data for monitoring biodiversity in the marine environment. *ICES Journal of Marine Science: Journal Du Conseil*, 80(5), 1529–1538.

Heald, W.F. (1951) Sky islands of Arizona. *Natural History*, 60, 56–63, 95–96.

Hecht, L. (2023) Optimal population density: Trading off the quality and quantity of welfare. Wild Animal Initiative. Available at: www.wildanimalinitiative.org/blog/optimalpopulationdensity (accessed 16 May 2025).

Hedenström, H., Norevik, G., Warfvinge, K., Andersson, A., Backman, J. and Åkesson, S. (2016) Annual 10-month aerial life phase in the common swift *Apus apus*. *Current Biology*, 26, 3066–3070.

Hegarty, S. (2024) The real reason for the rise in male childlessness. Available at: www.bbc.com/news/articles/cp81ynn7r4mo (accessed 21 January 2025).

Hegde, N., Maddur, M.S., Kaveri, S.V. and Bayry, J. (2009) Reasons to include viruses in the tree of life. *Nature Reviews Microbiology*, 7, 615.

Heinloth, T., Uhlhorn, J. and Wernet, M.F. (2018) Insect responses to linearly polarized reflections: Orphan behaviors without neural circuits. *Frontiers in Cellular Neuroscience*, 12, 50.

Heino, J., Soininen, J., Lappalainen, J. and Virtanen, R. (2005) The relationship between species richness and taxonomic distinctness in freshwater organisms. *Limnology and Oceanography*, 50, 978–986.

Heino, J., Alahuhta, J. and Fattorini, S. (2015) Phylogenetic diversity of regional beetle faunas at high latitudes: Patterns, drivers and chance along ecological gradients. *Biodiversity and Conservation*, 24, 2751–2767.

Heiser, M. and Schmitt, T. (2010) Do different dispersal capacities influence the biogeography of the western Palearctic dragonflies (Odonata)? *Biological Journal of the Linnean Society*, 99, 177–195.

Hejcman, M., Hejcmanová, P., Pavlů, V. and Beneš, J. (2013) Origin and history of grasslands in Central Europe – A review. *Grass and Forage Science*, 68, 345–363.

Helfman, G.S., Collette, B.B., Facey, D.E. and Bowen, B.W. (2009) *The Diversity of Fishes. Biology, Evolution and Ecology*. Wiley-Blackwell, Oxford, UK.

Heller, J. (1990) Longevity in molluscs. *Malacologia*, 31, 259–295.

Helzer, C.J. and Jelinski, D.E. (1999) The relative importance of patch area and perimeter–area ratio to grassland breeding birds. *Ecological Applications*, 9, 1448–1458.

Henderson, P.A. (2003) *Practical Methods in Ecology*. Blackwell Publishing, Oxford, UK.

Henderson, P.A. (2021) *Southwood's Ecological Methods*, 5th edn. Oxford University Press, Oxford, UK.

Hendrickson, J.R. (1980) The ecological strategies of sea turtles. *American Zoologist*, 20, 597–608.

Hengl, T., Walsh, M.G., Sanderman, J., Wheeler, I., Harrison, S.P. and Prentice, I.C. (2018) Global mapping of potential natural vegetation: An assessment of machine learning algorithms for estimating land potential. *PeerJ*, 6, e5457.

Henkel, C.V., Burgerhout, E., de Wijze, D.L., Dirks, R.P., Minegishi, Y. *et al.* (2012). Primitive duplicate Hox clusters in the European eel's genome. *PLOS One*, 7, e32231.

Henn, J.J., Buzzard, V., Enquist, B.J., Halbritter, A.H., Klanderud, K. *et al.* (2018) Intraspecific trait variation and phenotypic plasticity mediate alpine plant species response to climate change. *Frontiers in Plant Science*, 9, 1548.

Henselowsky, F., Rölkens, J., Kelterbaum, D. and Bubenzer, O. (2021) Anthropogenic relief changes in a long-lasting lignite mining area ('Ville', Germany) derived from historic maps and digital elevation models. *Earth Surface Processes and Landforms*, 2021, 1–14.

Herberstein, M.E., McLean, D.J., Lowe, E., Wolff, J.O., Khan, M.K. *et al.* (2022) AnimalTraits – A curated animal trait database for body mass, metabolic rate and brain size. *Scientific Data*, 9, 265.

Herbold, B. and Moyle, P.B. (1986) Introduced species and vacant niches. *American Naturalist*, 128, 751–760.

Herlyn, H. (2016) The phylogenetic system of primates – Character evolution in the light of a consolidated tree. *Organisms Diversity and Evolution*, 16, 689–713.

Herngreen, G.F.W., Kedves, M., Rovnina, L.V. and Smirnova, S.B. (1996) Cretaceous palynofloral provinces: A review. In: Jansonius, J. and McGregor, D.C. (eds) *Palynology: Principles and Applications*, vol. 3. American Association of Stratigraphic Palynologists Foundation, Salt Lake City, USA, pp. 1157–1188.

Heyer, W.R., Donnelly, M.A., McDiarmid, R.W., Hayek, L.-A.C. and Foster M.S. (eds) (1994) *Measuring and Monitoring Biological Diversity: Standard Methods for Amphibians*. Smithsonian Books, Washington, DC, USA.

Hicks, N., Liu, X., Gregory, R., Kenny, J., Lucaci, A. *et al.* (2018) Temperature driven changes in benthic bacterial diversity influences biogeochemical cycling in coastal sediments. *Frontiers in Microbiology*, 9, 1730.

Hiddink, J. and Kaiser, M.J. (2005) Implications of Liebig's law of the minimum for the use of ecological indicators based on abundance. *Ecography*, 28, 264–271.

Hilborn, R. and Walters, C.J. (1992) *Quantitative Fisheries Stock Assessment: Choice, Dynamics and Uncertainty*. Springer Science+Business Media, Dordrecht, The Netherlands.

Hill, J.K. and Hamer, K.C. (2003) Using species abundance models as indicators of habitat disturbance in tropical forests. *Journal of Applied Ecology* 35, 458–460.

Hill, J.K., Thomas, C.D. and Lewis, O.T. (1996) Effects of habitat patch size and isolation on dispersal by *Hesperia comma* butterflies: Implications for metapopulation structure. *Journal of Animal Ecology*, 65, 725–735.

Hill, J.K., Collingham, Y.C., Thomas, C.D., Balkeley, D.S., Fox, R. *et al.* (2001) Impacts of landscape structure on butterfly range expansion. *Ecology Letters*, 4, 313–321.

Hill, J.K., Griffiths, H.M. and Thomas, C.D. (2008) Climate change and evolutionary adaptations at species' range margins. *Annual Review of Entomology*, 56, 143–159.

Hill, M.O. (1973) Diversity and evenness: A unifying notation and its consequences. *Ecology*, 54, 427–473.

Hill, M.O., Mountford, J.O., Roy, D.B. and Bunce, R.G.H. (1999) *Ellenberg's Indicator Values for British Plants. ECOFACT Volume 2 Technical Annex*. Centre for Ecology and Hydrology, Huntingdon, UK.

HilleRisLambers, J., Adler, P.B., Harpole, W.S., Levine, J.M. and Mayfield, M.M. (2012) Rethinking community assembly through the lens of coexistence theory. *Annual Review of Ecology, Evolution, and Systematics*, 43, 227–248.

Hinchliff, C.E., Smith, S.A., Allman, J.F., Burleigh, G., Chaudhary, R. *et al.* (2015) Synthesis of phylogeny and taxonomy into a comprehensive tree of life. *Proceedings of the National Academy of Sciences of the United States of America*, 112, 12764–12769.

Hixon, M.A., Pacala, S.W. and Sandin, S.A. (2002) Population regulation: Historical context and contemporary challenges of open vs. closed systems. *Ecology*, 83, 1490–1508.

Hjelmroos, M. and Franzén, L.G. (1994) Implications or recent long-distance pollen transport events for the interpretation of fossil pollen records in Fennoscandia. *Review of Palaeobotany and Palynology*, 82, 175–189.

Hochberg, M.E., Thomas, J.A. and Elmes, G.W. (1992) A modelling study of the population dynamics of a large blue butterfly, *Maculinea rebeli*, a parasite of red ant nests. *Journal of Animal Ecology*, 61, 397–409.

Hodder, I. (2007) Çatalhöyük in the context of the Middle Eastern Neolithic. *Annual Review of Anthropology*, 36, 105–120.

Hodder, K.H., Buckland, P.C., Kirby, K.J. and Bullock, J.M. (2009) Can the pre-Neolithic provide suitable models for re-wilding the landscape in Britain? *British Wildlife*, 20, 4–15.

Hodek, I. (2012) Diapause/dormancy. In: Hodek, I., van Emden, H.F. and Honěk, A. (eds) *Ecology and Behaviour of the Ladybird Beetles (Coccinellidae)*. Blackwell Publishing, Oxford, UK.

Hodgson, J.A., Bennie, J.J., Dale, G., Longley, N., Wilson, R.J. and Thomas, C.D. (2015) Predicting microscale shifts in the distribution of the butterfly *Plebejus argus* at the northern edge of its range. *Ecography*, 38, 998–1005.

Hodgson, J.G., Tallowin, J., Dennis, R.L.H., Thompson, K., Poschlod, P. *et al.* (2014) Changing leaf nitrogen and canopy height quantify processes leading to plant and butterfly diversity loss in agricultural landscapes. *Functional Ecology*, 28, 1284–1291.

Hodkinson, I.D. and Casson, D.A (1991) Lesser predilection for bugs: Hemiptera (Insecta) diversity in tropical forests. *Biological Journal of the Linnean Society*, 43, 101–119.

Hodkinson, T., Jones, M. and Waldren, S. (eds.) (2011). *Climate Change, Ecology and Systematics*. Cambridge University Press, Cambridge, UK.

Hoffmann, M., Belant, J.L., Chanson, J.S., Cox, N.A., Lamoreux, J. *et al.* (2011) The changing fates of the world's mammals. *Philosophical Transactions of the Royal Society B: Biological Sciences*, 366, 2598–2610.

Hofrichter, R. (ed.) (2000) *The Encyclopedia of Amphibians: The World of Frogs, Toads, Salamanders and Newts*. Firefly Books, Toronto, Canada.

Hohbein, R.R. and Conway, C.J. (2018) Pitfall traps: A review of methods for estimating arthropod abundance. *Wildlife Society Bulletin*, 42, 597–606.

Hohenlohe, P.A. (2002) Life history of *Littorina scutulata* and *L. plena*, sibling gastropod species with planktotrophic larvae. *Invertebrate Biology*, 121, 25–37.

Holderegger, R. and Spillmann, J. (2022) Do revisitation studies overestimate local extinction? *Conservation Science and Practice*, 4, e12797.

Holdridge, L. (1967) *Life Zone Ecology*, Tropical Science Center, San José, Costa Rica.

Holdridge, L.R. (1947) Determination of world plant formation from simple climate data. *Science*, 105, 367–368.

Hölldobler, B. and Wilson, E.O. (1990) *The Ants*. Harvard University Press, Cambridge, USA.

Holm, E. and Sholtz, C.H. (1980) Structure and pattern of Namib Desert dune ecosystem at Gobabe. *Madoqua*, 12, 3–39.

Holmes, G., Marriott, K., Briggs, C. and Wynne-Jones, S. (2020) What is rewilding, how should it be done, and why? A Q-method study of the views held by European rewilding advocates. *Conservation and Society*, 18, 77–88.

Holmgren, J. (2004) Roosting in tree foliage by common swifts *Apus apus*. *Ibis*, 146, 404–416.

Holsman, K.K., McDonald, P.S. and Armstrong, D.A. (2006) Intertidal migration and habitat use by subadult Dungeness crab *Cancer magister* in a NE Pacific estuary. *Marine Ecology Progress Series*, 308, 183–195.

Holt, B., Lessard, J.P., Borregaard, M.K., Fritz, S.A., Araujo, M.B. *et al.* (2013) An update of Wallace's zoogeographic regions of the world. *Science*, 339, 74–78.

Holt, R.D. (2009a) Prince Kropotkin meets the Hutchinsonian niche. *Israel Journal of Ecology and Evolution*, 55, 1–10.

Holt, R.D. (2009b) Bringing the Hutchinsonian niche into the 21st century: Ecological and evolutionary perspectives. *Proceedings of the National Academy of Sciences of the United States of America*, 106, 19659–19665.

Holt, R.D., Lawton, J.H., Gaston, K.J. and Blackburn, T.M. (1997) On the relationship between range size and local abundance: Back to basics. *Oikos*, 78, 183–190.

Homburg, K., Homburg, N., Schäfer, F., Schuldt, A. and Assmann, T. (2014) Carabids.org – a dynamic online database of ground beetle species traits (Coleoptera, Carabidae). *Insect Conservation and Diversity*, 7, 195–205.

Honegger, R., Edwards, D., Axe, L. and Strullu-Derrien, C. (2017) Fertile *Prototaxites taiti*: A basal ascomycete with inoperculate, polysporous asci lacking croziers. *Philosophical Transactions of the Royal Society B: Biological Sciences*, 373, 20170146.

Honkavaara. J., Koivula, M., Korpimaki, E., Siitari, H. and Viitala, J. (2002) Ultraviolet vision and foraging in terrestrial vertebrates. *Oikos*, 98, 505–511.

Hooper, D.U., Chapin, F.S. III, Ewel, J.J., Hector, A., Inchausti, P. *et al.* (2005) Effects of biodiversity on ecosystem functioning: A consensus of current knowledge. *Ecological Monographs*, 75, 3–35.

Hooper, D.U., Adair, E.C., Cardinale, B.J., Byrnes, J.E.K. and Hungate, B.A. (2012) A global synthesis reveals biodiversity loss as a major driver of ecosystem change. *Nature*, 486, 105–108.

Hoppert, M., Flies, C., Pohl, W., Gunzl, B. and Schneider, J. (2004) Colonization strategies of lithobiontic microorganisms on carbonate rocks. *Environmental Geology*, 46, 421e428.

Hortal, J., Borges, P.A.V. and Gaspar, C. (2006) Evaluating the performance of species richness estimators: Sensitivity to sample grain size. *Journal of Animal Ecology*, 75, 274–287.

Hortal, J., Triantis, K.A., Meiri, S., Thébault, E. and Sfenthourakis, S. (2009) Island species richness increases with habitat diversity. *American Naturalist*, 174, E205–217.

Hortal, J., Carrascal, L.M., Triantis, K.A., Thébault, E., Meiri, S. and Sfenthourakis, S. (2013) Species richness can decrease with altitude, but not with habitat diversity. *Proceedings of the National Academy of Sciences of the United States of America*, 110, E2149–E2150.

Hortal, J., de Bello, F., Diniz-Filho, J.A.F., Lewinsohn, T.M., Lobo, J.M. and Ladle, R.J. (2015) Seven shortfalls that beset large-scale knowledge of biodiversity. *Annual Review of Ecology, Evolution, and Systematics*, 46, 523–549.

Horton, K., van Doren, B.J., Stepanian, P.M., Farnsworth, A. and Kelly, J.F. (2016) Where in the air? Aerial habitat use of nocturnally migrating birds. *Biology Letters*, 12, 20160591.

Housharter, J. (2024) *Movement Distances of European Butterflies*. University of Vienna, Vienna, Austria. Available at: https://phaidra.univie.ac.at/detail/o:2044834 (accessed 16 May 2025).

Howard, M. (2009) *War in European History*. Updated edition. Oxford University Press, Oxford, UK.

Howarth, F.G. and Moldovan, O.T. (2018) The ecological classification of cave animals and their adaptations. In: Moldovan, O., Kováč, Ľ. and Halse, S. (eds) *Cave Ecology*. Springer, Cham, Switzerland.

Howe, H.F. (1977) Bird activity and seed dispersal of a tropical wet forest tree. *Ecology*, 58, 539–550.

Howell, K.L. (2010) A benthic classification system to aid in the implementation of marine protected area networks in the deep/high seas of the NE Atlantic. *Biological Conservation*, 143, 1041–1056.

Howse, P.E. (2013) Lepidopteran wing patterns and the evolution of satyric mimicry. *Biological Journal of the Linnean Society*, 109, 203–214.

Howse, P.E. and Allen, J.A. (1994) Satyric mimicry – The evolution of apparent imperfection. *Proceedings of the Royal Society B: Biological Sciences*, 257, 111–114.

Hoy, R.A. and Brereton, J.E. (2022) Does observer presence modify the behavior and enclosure use of captive Edwards' pheasants? *Journal of Zoological and Botanical Gardens*, 3, 147–157.

Hoyt, H. (1939) *The Structure and Growth of Residential Neighborhoods in American Cities*. Federal Housing Administration, Washington, DC, USA.

Hsieh, T.C., Ma, K.H. and Chao, A. (2016) iNEXT: An R package for interpolation and extrapolation of species diversity (Hill numbers). *Methods in Ecology and Evolution*, 7, 1451–1456.

Huang, R., Pimm, S.L. and Giri, C. (2020) Using metapopulation theory for practical conservation of mangrove endemic birds. *Conservation Biology*, 34, 266–275.

Hubálek, Z. (2000) Measures of species diversity in ecology: An evaluation. *Folia Zoologica*, 49, 241–260.

Hubbell, S.P. (1979) Tree dispersion abundance and diversity in a tropical tree forest. *Science*, 203, 1299–1309.

Hubbell, S.P. (1997) A unified theory of biogeography and relative species abundance and its application to tropical rain forests and coral reefs. *Coral Reefs*, 16(Suppl.), S9–S21.

Hubbell, S.P. (2001) *The Unified Neutral Theory of Biodiversity and Biogeography*. Princeton University Press, Princeton, USA.

Hubbell, S.P. and Foster, R.B. (1986) Biology, chance, and history and the structure of tropical rain forest tree communities. In: Diamond, J. and Case, T. (eds) *Community Ecology*. Harper and Row, New York, USA, pp. 314–329.

Hufkens, K., Scheunders, P. and Ceulemans, R. (2009) Ecotones in vegetation ecology: Methodologies and definitions revisited. *Ecological Research*, 24, 977–986.

Hug, L.A., Baker, B.J., Anantharaman, K., Brown, C.T., Probst, A.J. *et al.* (2016) A new view of the tree of life. *Nature Microbiology*, 1, 16048.

Hugenholtz, P., Chuvochina, M., Oren, A., Parks, D.H. and Soo, R. M. (2021) Prokaryotic taxonomy and nomenclature in the age of big sequence data. *ISME Journal*, 15(7), 1879–1892.

Hughes, A.C., Orr, M.C., Ma, K., Costello, M.J., Waller, J. *et al.* (2021) Sampling biases shape our view of the natural world. *Ecography*, 44, 1259–1269.

Hughes, J.B., Gretchen, C. and Ehrlich, P.R. (2000) Conservation of insect diversity: A habitat approach. *Conservation Biology*, 14, 1788–1797.

Hughes, J.D. (1985) Theophrastus as ecologist. *Environmental Review*, 9, 297–306.

Hugon, P., Dufour, J.-C., Colson, P., Fournier, P.-E., Sallah, K. and Raoult, D. (2015) A comprehensive repertoire of prokaryotic species identified in human beings. *Lancet Infectious Diseases*, 15, 1211–1219.

Hull, P.M. and Darroch, S.A.F. (2013) Mass extinctions and the structure and function of ecosystems. *Paleontological Society Papers*, 19, 115–156.

Humphrey, L. and Stringer, C. (2018) *Our Human Story*. Natural History Museum, London, UK.

Humphrey, S.R. (1975) Nursery roosts and community diversity of Nearctic bats. *Journal of Mammalogy*, 56, 321–346.

Hunter, P. (2023) The fungal grid: Fungal communication via electrical signals has inspired the hypothesis of a Wood Wide Web of plants and fungi. *EMBO reports*, 24, e57255.

Huntley, B. (1990) European post-glacial forests: Compositional changes in response to climatic change. *Journal of Vegetation Science*, 1, 507–518.

Huntley, B. (1991) How plants respond to climatic change: Migration rates, individualism and the consequences for plant communities. *Annals of Botany*, 67, 15–22.

Huntley, B. and Birks, H.J.B. (1983) *An Atlas of Past and Present Pollen Maps for Europe: 0 –13000 Years Ago*. Cambridge University Press, Cambridge, UK.

Hurd, L.E. and Fagan, W.F. (1992) Cursorial spiders and succession: Age or habitat structure? *Oecologia*, 92, 215–221.

Hurlbert, S.H. (1971) The nonconcept of species diversity: A critique and alternative parameters. *Ecology*, 52, 577–586.

Hurst, C.J., Crawford, R.L., Garland, J.L., Lipson, D.A., Mills A.L and Stetzenbach, L.D. (eds) (2007) *Manual of Environmental Microbiology*, 3rd edn. American Society for Microbiology, Washington, DC, USA.

Hurtado, P., Prieto, M., Aragón, G., de Bello, F. and Martínez, I. (2020) Intraspecific variability drives functional changes in lichen epiphytic communities across Europe. *Ecology*, 101, e03017.

Hussain, S.S., Markandya, A., Brander, L., McVittie, A., de Groot, R. *et al.* (2012) *The Challenge of Ecosystems and Biodiversity*. Copenhagen Consensus Paper 2012. Available at: https://copenhagenconsensus.com/sites/default/files/biodiversity.pdf (accessed 16 May 2025).

Hutcheson, K. (1970) A test for comparing diversity based on the Shannon formula. *Journal of Theoretical Biology*, 29, 151–154.

Hutchinson, G.E. (1944) Nitrogen in the biogeochemistry of the atmosphere. *American Scientist*, 32(3), 178–195.

Hutchinson, G.E. (1957) Concluding remarks. *Cold Spring Harbor Symposium on Quantitative Biology*, 22, 415–427.

Hutchinson, G.E. (1978) *An Introduction to Population Ecology*. Yale University Press, New Haven, USA.

Hutton, R. (1997) The Neolithic goddess: A study in modern tradition. *Antiquity*, 71, 91–99.

Huxley, C.R. (1978) The ant-plants *Myrmecodia* and *Hydnophytum* (Rubiaceae), and the relationships between their morphology, ant occupants, physiology and ecology. *New Phytologist*, 80, 231–268.

Huxley, C.R. and Wood, N.A. (1976) Aspects of the breeding of the moorhen in Britain. *Bird Study*, 23, 1–10.

Hyrenbach, K.D., Henry, M.F., Morgan, K.H., Welch, D.W. and Sydeman, W.J. (2007) Optimizing the width of strip transects for seabird surveys from vessels of opportunity. *Marine Ornithology*, 35, 29–37.

Ibanez, T., Keppel, G., Baider, C., Birkinshaw, C., Florens, F.B.V. *et al.* (2020) Tropical cyclones and island area shape species abundance distributions of local tree communities. *Oikos*, 129, 1856–1866.

IFAW (International Fund for Animal Welfare) (2016) *Killing for Trophies. An Analysis of Global Hunting Trade*. International Fund for Animal Welfare, Washington, DC, USA.

Ignacio-Espinoza, J.C., Solonenko, S.A. and Sullivan, M.B. (2013) The global virome: Not as big as we thought? *Current Opinion in Virology*, 3, 566–571.

Illies, J. (1967) *Limnofauna Europaea. Eine Zusammenstellung aller die europäischen Binnengewässer bewohnenden mehrzelligen Tierarten mit Angaben über ihre Verbreitung und Ökologie*. Fischer Verlag, Stuttgart, Germany.

Illies, J. (1978) *Limnofauna Europaea. Eine Zusammenstellung aller die europäischen Binnengewässer bewohnenden mehrzelligen Tiere mit Angaben über ihre Verbreitung und Ökologie*, 2nd edn. Fischer Verlag, Stuttgart, Germany.

Illies, J. and Botosaneanu, L. (1963) Problems and methods for the classification and ecological zoning of these currents, considering the faunistic point of view.

Mitteilungen – Internationale Vereinigung fuer Theoretische und Angewandte Limnologie, 12, 1–57.

Ims, R.A., Yoccoz, N.G. and Killengreen, S.T. (2011) Determinants of lemming outbreaks. *Proceedings of the National Academy of Sciences of the United States of America*, 108, 1970–1974.

Inglis, R., Wright, E. and Lill, J. (1994) The impact of hedges and farm woodlands on woodpigeon (*Columba palumbus*) nest densities. *Agriculture, Ecosystems and Environment*, 48, 257–262.

Inouye, D.W., Barr, B., Armitage, K.B. and Inouye, B.D. (2000) Climate change is affecting altitudinal migrants and hibernating species. *Proceedings of the National Academy of Sciences of the United States of America*, 97, 1630–1633.

Institute for Economics and Peace (2024) *Global Peace Index 2024: Measuring Peace in a Complex World*. Sydney, Australia. Available at: http://visionofhumanity.org/resources (accessed 16 May 2025).

International Committee on Taxonomy of Viruses (ICTV) (2024) Current ICTV Taxonomy Release. Available at: https://ictv.global/taxonomy (accessed 16 May 2025).

Introna, F., Suman, T.W. and Smialek, J.E. (1991) Sarcosaprophagous fly activity in Maryland (USA). *Journal of Forensic Sciences*, 36, 238–243.

IPCC (2022) *IPCC, 2019: Climate Change and Land. An IPCC Special Report on Climate Change, Desertification, Land Degradation, Sustainable Land Management, Food Security, and Greenhouse Gas Fluxes in Terrestrial Ecosystems*. Cambridge University Press, Cambridge, UK.

Irie, N., Satoh, N. and Kuratani, S. (2018) The phylum Vertebrata: A case for zoological recognition. *Zoological Letters*, 4, 32.

Irl, S.D.H., Harter, D.E.V., Steinbauer, M.J., Gallego Puyol, D., Fernandez-Palacios, J.M. *et al.* (2015) Climate vs. topography – Spatial patterns of plant species diversity and endemism on a high-elevation island. *Journal of Ecology*, 103, 1621–1633.

Isaac, J. (2009) Effects of climate change on life history: Implications for extinction risk in mammals. *Endangered Species Research*, 7, 115–123.

Isaac, J.L., Vanderwal, J., Johnson, C.N. and Williams, S.E. (2009) Resistance and resilience: Quantifying relative extinction risk in a diverse assemblage of Australian tropical rainforest vertebrates. *Diversity and Distributions*, 15, 280–288.

Isenberg, A.C. (2000) *The Destruction of the Bison: An Environmental History, 1750–1920*. Cambridge University Press, Cambridge, UK.

Israel, G. (1988) On the contribution of Volterra and Lotka to the development of modern biomathematics. *History and Philosophy of the Life Sciences*, 10(1), 37–49.

Israel, G. (1995) Volterra, D'Ancona e la biologia matematica, In: Volterra, V., D'Ancona, U. (1995) *Le associazioni biologiche studiate dal punto di vista matematico*, a cura di G. Israel, Teknos, Roma, Italy, pp. 7–47. [Italian translation of *Les associations biologiques au point de vue mathématique*, Paris, Hermann, 1935.]

IUCN (2002) *IUCN Technical Guidelines on the Management of Ex-Situ Populations for Conservation*. IUCN, Gland, Switzerland.

IUCN (2012) *Guidelines for Application of IUCN Red List Criteria at Regional and National Levels: Version 4.0*. IUCN, Gland, Switzerland.

IUCN (2024a) *Guidelines for Using the IUCN Red List Categories and Criteria*. Version 16. Available at: www.iucnredlist.org/documents/RedListGuidelines.pdf (accessed 16 May 2025).

IUCN (2024b) Summary Statistics. Available at: www.iucnredlist.org/resources/summary-statistics#Summary%20Tables (accessed 16 May 2025).

Ivanov, K.and Keiper, J. (2009) Effectiveness and biases of Winkler litter extraction and pitfall trapping for collecting ground-dwelling ants in northern temperate forests. *Environmental Entomology*, 38, 1724–1736.

Iversen, J. (1958) The bearing of glacial and interglacial epochs on the formation and extinction of plant taxa. *Uppsala Universiteit Arssk*, 6, 210–215.

Iversen, R. (2016) Arrowheads as indicators of interpersonal violence and group identity among the Neolithic Pitted Ware hunters of southwestern Scandinavia. *Journal of Anthropological Archaeology*, 44, 69–86.

Ives, A.R. and Carpenter, S.R. (2007) Stability and diversity of ecosystems. *Science*, 317, 58–62.

Ivošević, B., Han, Y-G., Cho, Y. and Kwon, O. (2015) The use of conservation drones in ecology and wildlife research. *Journal of Ecology and Environment*, 38, 113–118.

Iwamoto, T. and Dunbar, R.I.M. (1983) Thermoregulation, habitat quality and the behavioural ecology of gelada baboons. *Journal of Animal Ecology*, 52, 357–366.

Iwamoto, T., Mori, A., Kawai, M. and Bekele, A. (1996) Anti-predator behavior of gelada baboons. *Primates*, 37, 389–397.

Iyer, R., Caimano, M.J., Luthra, A., Axline, D. Jr., Corona, A. *et al.* (2015) Stage-specific global alterations in the transcriptomes of Lyme disease spirochetes during tick feeding and following mammalian host adaptation. *Molecular Microbiology*, 95, 509–538.

Jablonski, D. (2001) Lessons from the past: Evolutionary impacts of mass extinctions. *Proceedings of the National Academy of Sciences of the United States of America*, 98, 5393–5398.

Jablonski, D. (2005) Mass extinctions and macroevolution. *Paleobiology*, 31, 192–210.

Jackson, A. (2008) Common periwinkle (*Littorina littorea*). In: Tyler-Walters, H. and Hiscock, K. (eds) *Marine Life Information Network: Biology and Sensitivity Key Information Reviews*. Marine Biological Association, Plymouth, UK. Available at:

www.marlin.ac.uk/assets/pdf/species/marlin_species_1328_2019-03-21.pdf (accessed 16 May 2025).

Jackson, M.O. and Morelli, M. (2009) The reasons for wars – An updated survey. In: Coyne C. (ed.) *Handbook on the Political Economy of War*. Edward Elgar Publishing, Northampton, USA.

Jacobsen, L.B., Jensen, N.O., Willemoes, M., Hansen, L., Desholm, M. *et al.* (2017) Annual spatiotemporal migration schedules in three larger insectivorous birds: European nightjar, common swift and common cuckoo. *Animal Biotelemetry*, 5, 4.

Jaeger, J.A.G. (2000) Landscape division, splitting index, and effective mesh size: New measures of landscape fragmentation. *Landscape Ecology*, 15, 115–130.

Jagiello, Z., Dylewski, Ł. and Szulkin, M. (2024) The plastic homes of hermit crabs in the Anthropocene. *Science of the Total Environment*, 913, 168959.

Jain, M., Flynn, D.F.B., Prager, C.M., Hart, G.M., DeVan, C.M. *et al.* (2014) The importance of rare species: A trait-based assessment of rare species contributions to functional diversity and possible ecosystem function in tall-grass prairies. *Ecology and Evolution*, 4, 104–112.

Jambhekar, R. and Isvaran, K. (2022) Habitat specialisation and matrix resistance predict responses of butterfly populations to landscape features in tropical grassland–forest complexes. *Oecologia*, 199, 513–525.

James, S. (2010) *Lumbricus terrestris*. Available at: www.cabi.org/isc/datasheet/109385 (accessed 16 May 2025).

Janó, G. and Végvári, Z. (2016) Nest site selection of the great bustard (*Otis t. tarda*) in Körös-Maros National Park, Eastern Hungary. *Ornis Hungarica*, 24, 32–45.

Janowski, D. and Leski, T. (2022) Factors in the distribution of mycorrhizal and soil fungi. *Diversity*, 14, 1122.

Janz, N. and Nylin, S. (2008) The oscillation hypothesis of host plant range and speciation. In: Tilmon, K.J. (ed.) *Specialization, Speciation, and Radiation: The Evolutionary Biology of Herbivorous Insects*. University of California Press, Berkeley, USA, pp. 203–215.

Janzen, D. (1985) On ecological fitting. *Oikos*, 45, 308–310.

Jarić, I. and Cvijanović, G. (2012) The Tens Rule in invasion biology: Measure of a true impact or our lack of knowledge and understanding? *Environmental Management*, 50, 979–981.

Javan, G.T., Finley, S.J., Can, I., Wilkinson, J.E., Hanson, J.D. and Tarone, A.M. (2016) Human thanatomicrobiome succession and time since death. *Scientific Reports*, 6, 29598.

Jedrzejek, B., Drees, B., Daniëls, F.J., Hölzel, N. and Fraser, L. (2012) Vegetation discontinuities and altitudinal indicator species in mountains of West Greenland: Finding the best positions and traits to observe the impact of climate warming in the Arctic. *Applied Vegetation Science*, 15, 432–448.

Jeliazkov, A., Mijatovic, D., Chantepie, S., Andrew, N., Arlettaz, R. *et al.* (2020) A global database for metacommunity ecology, integrating species, traits, environment and space. *Scientific Data*, 7, 6.

Jenkins, D.G. (2014) Lakes and rivers as microcosms, Version 2.0. *Journal of Limnology*, 73(s1), 20–32.

Jensen, J.K., Ekroos, J., Watson, H., Salmón, P., Olsson, P. and Isaksson, C. (2023) Urban tree composition is associated with breeding success of a passerine bird, but effects vary within and between years. *Oecologia*, 201, 585–597.

Jensen, J.R. and Cowen, D.C. (1999) Remote sensing of urban/suburban infrastructure and socio-economic attributes. *Photogrammetry Engineering and Remote Sensing*, 65, 611–622.

Jeschke, J.M. and Pyšek, P. (2018) Tens rule. In: Jeschke, J.M. and Heger, T. (eds) *Invasion Biology: Hypotheses and Evidence*. CABI, Wallingford, UK.

Jessop, R., Collins, P. and Brown, M. (1998) The manufacture of leg flags in the light of experience. *Stilt*, 32, 50–52.

Jetz, W., Carbone, C., Fulford, J. and Brown, J.H. (2004) The scaling of animal space use. *Science*, 306, 266–268.

Jetz, W., Thomas, G.H., Joy, J.B., Hartmann, K. and Mooers, A.O. (2012) The global diversity of birds in space and time. *Nature*, 491, 444–448.

Jiménez-Valverde, A., Jiménez Mendoza, S., Martín Cano, J. and Munguira, M.L. (2006) Comparing relative model fit of several species accumulation functions to local Papilionoidea and Hesperoidea butterfly inventories of Mediterranean habitats. *Biodiversity and Conservation*, 15, 177–190.

Jin, L., Yu, J., Yuan, X. and Du, X. (2021) Fish classification using DNA barcode sequences through deep learning method. *Symmetry*, 13, 1599.

Jog, S.K. and Bried, J.T. (2021) Taxonomic distinctness poorly reflects floristic quality in a wetland study system. *Ecological Indicators*, 121, 107086.

John, E., Thomas, B., Başbay, O., Cebeci, Z. and Coutsis, J.G. (2018) The arrival of *Cacyreus marshalli* Butler, 1898 (Lepidoptera: Lycaenidae, Polyommatinae) in Crete, with additional notes on range expansion in Greece and along coastal areas of northwestern Turkey. *Entomologist's Gazette*, 69, 85–97.

Johns, D. (2019) *Conservation Politics: The Last Anti-colonial Battle*. Cambridge University Press, Cambridge, UK.

Johnson, C.G. (1969) *Migration and Dispersal of Insects by Flight*. Methuen, London, UK.

Johnson, D.D.P., Kays, R., Blackwell, P.G. and Macdonald, D.W. (2002) Does the resource dispersion hypothesis explain group living? *Trends in Ecology and Evolution*, 17, 563–570.

Johnson, E.A. and Miyanishi, K. (2008) Testing the assumptions of chronosequences in succession. *Ecology Letters*, 11, 419–431.

Johnson, J.A., Valero, K.A., Wang, S. and Tang, J. (2004) Thermal death kinetics of red flour beetle (Coleoptera: Tenebrionidae). *Journal of Economic Entomology*, 97, 1868–1873.

Johnson, M.P. (2005) Is there confusion over what is meant by 'open population'? *Hydrobiologia*, 544, 333–338.

Johnson, M.S. (2012) Epistasis, phenotypic disequilibrium and contrasting associations with climate in the land snail *Theba pisana*. *Heredity*, 108, 229–235.

Johnson, S., Domínguez-García, V. and Muñoz, M.A. (2013) Factors determining nestedness in complex networks. *PLOS One*, 8(9), e74025.

Johnson, S.K., Fitza, M.A., Lerner, D.A., Calhoun, D.M., Beldon, M.A. *et al.* (2018) Risky business: Linking *Toxoplasma gondii* infection and entrepreneurship behaviours across individuals and countries. *Proceedings of the Royal Society B: Biological Sciences*, 285(1883), 20180822.

Johnston, A., Hochachka, W.M., Strimas-Mackey, M.E., Gutierrez, V.R., Robinson, O.J. *et al.* (2021) Analytical guidelines to increase the value of community science data: An example using eBird data to estimate species distributions. *Diversity and Distributions*, 27, 1265–1277.

Johnston, D.W. and Odum, E.P. (1956) Breeding bird populations in relation to plant succession on the Piedmont in Georgia. *Ecology*, 37, 50–62.

Johnston, H. (1914) *Pioneers in South Africa*. Blackie and Son, London, UK.

Johst, K., Drechsler, M., Thomas, J. and Settele, J. (2006) Influence of mowing on the persistence of two endangered large blue butterfly species. *Journal of Applied Ecology*, 43, 333–342.

Jolly, A. (1966) *Lemur Behavior*. University of Chicago Press, Chicago, USA.

Jolly, A., Koyama, N., Rasamimanana, H., Crowley, H. and Williams, G. (2006) Berenty Reserve: A research site in southern Madagascar. In: Jolly, A., Sussman, R.W., Koyama, N. and Rasamimanana, H. (eds) *Ringtailed Lemur Biology:* Lemur catta *in Madagascar*. Springer, New York, USA, pp. 32–42.

Jones, C., Majerus, M and Timmins, R. (1993) Differential habitat selection in polymorphic Lepidoptera. *The Entomologist*, 112, 118–126.

Jones, C.G., Lawton, J.H. and Shachak, M. (1994) Organisms as ecosystem engineers. *Oikos*, 69, 373–386.

Jones, C.G., Lawton, J.H. and Shachak, M. (1997) Positive and negative effects of organisms as physical ecosystem engineers. *Ecology*, 78, 1946–1957.

Jones, D., Douglas, K. and Robson, D. (2018) Mysteries of our past. In: Brown, J. (ed.) *Becoming Human*, 2nd edn. New Scientist: The Collection, London, UK, pp. 17–23.

Jones, G. (2005) Echolocation. *Current Biology*, 15, R484–R488.

Jones, G. (2010) Molecular evolution: Gene convergence in echolocating mammals. *Current Biology*, 20, R62–R64.

Jones, G. and Teeling, E. (2006) The evolution of echolocation in bats. *Trends in Ecology and Evolution*, 21(3), 149–156.

Jones, J. (2001) Habitat selection studies in avian ecology: A critical review. *The Auk*, 118, 557–562.

Jones, L.P., Turvey, S.T. and Papworth, S.K. (2021) Is there evidence of shifting baseline syndrome in environmental managers? An assessment using perceptions of bird population targets in UK nature reserves. *Journal of Environmental Management*, 297, 113308.

Jones, M.C. and Cheung, W.W.L. (2015) Multi-model ensemble projections of climate change effects on global marine biodiversity. *ICES Journal of Marine Science: Journal du Conseil*, 72, 741–752.

Jonsson, H., Magnusdottir, E., Eggertsson, H.P., Stefansson, O.A., Arnadottir, G.A. *et al.* (2021) Differences between germline genomes of monozygotic twins. *Nature Genetics*, 53, 27–34.

Jönsson, K.I., Rabbow, E., Schill, R.O., Harms-Ringdahl, M. and Rettberg, P. (2008) Tardigrades survive exposure to space in low Earth orbit. *Current Biology*, 18(17), R729–R731.

Joppa, L.N., Roberts, D.L. and Pimm, S.L. (2011a) How many species of flowering plants are there? *Proceedings of the Royal Society B: Biological Sciences*, 278, 554–559.

Joppa, L.N., Roberts, D.L. and Pimm, S.L. (2011b) The population ecology and social behaviour of taxonomists. *Trends in Ecology and Evolution*, 26, 551–553.

Jordán, F. (2009) Keystone species and food webs. *Philosophical Transactions of the Royal Society B: Biological Sciences*, 364, 1733–1741.

Jorgensen, E.E. (2004) Small mammal use of microhabitat reviewed. *Journal of Mammalogy*, 85, 531–539.

Jørgensen, S.E. and Svirezhev, Y.M. (2004) *Towards a Thermodynamic Theory for Ecological Systems*. Elsevier, Amsterdam, The Netherlands.

Jørgensen, S.E., Halling-Sørensen, B. and Nielsen, S.N. (eds) (1996) *Handbook of Environmental and Ecological Modelling*. CRC Press, Boca Raton, USA.

Joseph, T.A., Shenhav, L., Xavier, J.B., Halperin, E. and Pe'er, I. (2020) Compositional Lotka-Volterra describes microbial dynamics in the simplex. *PLOS Computational Biology*, 16(5), e1007917.

Joshi, M.D. and Joshi, C. (2022) Areas of species diversity and endemicity of Nepal. *Ecosphere*, 13, e3969.

Jost, L. (2010) The relation between evenness and diversity. *Diversity*, 2, 207–232.

Jouquet, P., Harit, A.K., Cheik, S., Traoré, S. and Bottinelli, N. (2019) Termites: Soil engineers for ecological engineering. *Comptes Rendus Biologies*, 342, 258–259.

Jourdan, J., Riesch, R. and Cunze, S. (2021) Off to new shores: Climate niche expansion in invasive mosquitofish (*Gambusia* spp.). *Ecology and Evolution*, 11, 18369–18400.

Jouval, F., Bigot, L., Bureau, S., Quod, J-P., Penin, L. and Adjeroud, M. (2020) Diversity, structure and demography of coral assemblages on underwater lava flows of different ages at Reunion Island and implications for ecological succession hypotheses. *Scientific Reports*, 10, 20821.

Jung, V., Violle, C., Mondy, C., Hoffmann, L. and Muller, S. (2010) Intraspecific variability and trait-based community assembly. *Journal of Ecology*, 98, 1134–1140.

Jung, V., Albert, C.H., Violle, C., Kunstler, G., Loucougaray, G. and Spiegelberger, T. (2014) Intraspecific trait variability mediates the response of subalpine grassland communities to extreme drought events. *Journal of Ecology*, 102, 45–53.

Jurasinski, G. and Koch, M. (2011) Commentary: Do we have a consistent terminology for species diversity? We are on the way. *Oecologia*, 167, 893–902.

Jurskis, V. (2005) Decline of eucalypt forests as a consequence of unnatural fire regimes. *Australian Forestry*, 68, 257–262.

Juslén, A., Pykälä, J., Kuusela, S., Kaila, L., Kullberg, J. *et al.* (2016) Application of the Red List Index as an indicator of habitat change. *Biodiversity and Conservation*, 25, 569–585.

Kaiser, A., Merckx, T. and Van Dyck, H. (2018) Urbanisation and sex affect the consistency of butterfly personality across metamorphosis. *Behavioral Ecology and Sociobiology*, 72, 188.

Kaiser, A., Merckx, T. and Van Dyck, H. (2019) Personality traits influence contest outcome, and vice versa, in a territorial butterfly. *Scientific Reports*, 9, 2778.

Kalamulla, R., Karunarathna, S.C., Tibpromma, S., Gabappaththi, M.C.A., Suwannarach, N. *et al.* (2022) Arbuscular mycorrhizal fungi in sustainable agriculture. *Sustainability*, 14, 12250.

Kamboj, N., Kumar, A., Kamboj, V., Bisht, A., Pandey, N. and Bharti, M. (2021) Role of earthworm biodiversity in soil fertility and crop productivity improvement. In: Kumar, V., Kumar, S., Kamboj, N., Payum, T., Kumar, P. and Kumari, S. (eds) *Biological Diversity: Current Status and Conservation Policies*, vol. 1. Agro Environ Media, Agriculture and Environmental Science Academy, Haridwar, India, pp. 230–241.

Kamilar, J.M. and Muldoon, K.M. (2010) The climatic niche diversity of Malagasy primates: A phylogenetic perspective. *PLOS One*, 5, e11073.

Kamiya, T., O'Dwyer, K., Nakagawa, S. and Poulin, R. (2014) What determines species richness of parasitic organisms? A meta-analysis across animal, plant and fungal hosts. *Biological Reviews*, 89, 123–134.

Kamoun, S. and Hogenhout, S.A. (1996) Flightlessness and rapid terrestrial locomotion in tiger beetles of the *Cicindela* L. subgenus *Rivacindela* van Nidek from saline habitats of Australia (Coleoptera: Cicindelidae). *Coleopterist's Bulletin*, 50, 221–230.

Kang, C., Im, S., Lee, W.Y., Choi, Y., Stuart-Fox, D. and Huertas, B. (2021) Climate predicts both visible and near-infrared reflectance in butterflies. *Ecology Letters*, 24, 1869–1879.

Kark, S. (2013) Effects of ecotones on biodiversity. In: Levin S.A. (ed.) *Encyclopedia of Biodiversity*, 2nd edn, vol. 3. Academic Press, Waltham, USA, pp. 142–148.

Kark, S. and van Rensburg, B.J. (2006) Ecotones: Marginal or central areas of transition? *Israel Journal of Ecology and Evolution*, 52, 29–53.

Karlsson, M. (2022) *From Fossil to Fact: The Denisova Discovery as Science in Action*. Department of Culture and Society (IKOS), Linköping, Sweden.

Kasanin-Grubin, M. (2008) Badlands and environmental change. *Geographica Helvetica*, 63, 15–25.

Kasprzykowski, Z. (2003) Habitat preferences of foraging rooks *Corvus frugilegus* during the breeding period in the agricultural landscape of eastern Poland. *Acta Ornithologica*, 38, 27–31.

Kasprzykowski, Z. (2007) Reproduction of the rook, *Corvus frugilegus* in relation to the colony size and foraging habitats. *Folia Zoologica*, 56, 186–193.

Kattan, G. (1992) Rarity and vulnerability: The birds of the Cordillera Central of Colombia. *Conservation Biology*, 6, 64–70.

Kattge, J., Díaz, S., Lavorel, S., Prentice, I.C., Leadley, P. *et al.* (2011) TRY – A global database of plant traits. *Global Change Biology*, 17, 2905–2935.

Kattge, J., Bönisch, G., Díaz, S., Lavorel, S., Prentice, I.C. *et al.* (2020) TRY plant trait database – Enhanced coverage and open access. *Global Change Biology*, 26, 119–188.

Ke, H., Lewis, I.A., Morrisey, J.M., McLean, K.J., Ganesan, S.M. *et al.* (2015) Genetic investigation of tricarboxylic acid metabolism during the *Plasmodium falciparum* life cycle. *Cell Reports*, 11, 164–174.

Keane, R.M. and Crawley, M.J. (2002) Exotic plant invasions and the enemy release hypothesis. *Trends in Ecology and Evolution*, 17, 164–169.

Keddy, P. and Laughlin, D. (2022) *A Framework for Community Ecology: Species Pools, Filters and Traits*. Cambridge University Press, Cambridge, UK.

Keller, A., Ankenbrand, M.J., Bruelheide, H., Dekeyzer, S., Enquist, B.J. *et al.* (2023) Ten (mostly) simple rules to future-proof trait data in ecological and evolutionary sciences. *Methods in Ecology and Evolution*, 14, 444–458.

Keller, L. and Perrin, N. (1995) Quantifying the level of eusociality. *Proceedings: Biological Sciences*, 260(1359), 311–315.

Kellermann, V. and van Heerwaarden, B. (2019) Terrestrial insects and climate change: Adaptive responses in key traits. *Physiological Entomology*, 44, 99–115.

Kelly, C.L., Schwarzkopf, L., Gordon, I.J. and Hirsch, B. (2021) Population growth lags in introduced species. *Ecology and Evolution* 11, 4577–4587.

Kelt, D. and Van Vuren, D. (2001) The ecology and macroecology of mammalian home range area. *American Naturalist*, 157, 637–645.

Kemp, R., Hardy, P.B., Roy, D. and Dennis, R.L.H. (2008) The relative exploitation of annuals as larval host plants by phytophagous Lepidoptera. *Journal of Natural History*, 42, 1079–1093.

Kemp, W.P., Harvey, S.J. and O'Neill, K.M. (1990) Habitat and insect biology revisited: The search for patterns. *American Entomologist*, 36, 44–49.

Kempton, R.A. and Taylor, L.R. (1974) Log-series and log-normal parameters as diversity discriminants for the lepidoptera. *Journal of Animal Ecology*, 43, 381–399.

Kempton, R.A. and Taylor, L.R. (1976) Models and statistics for species diversity. *Nature*, 262, 818–820.

Kendall, D.G. (1948) On some modes of population growth leading to R. A. Fisher's logarithmic series distribution. *Biometrika*, 35, 6–15.

Kenis, M., Wegensteiner, R. and Griffin, C. (2004) Parasitoids, predators, nematodes and pathogens associated with bark weevil pests. In: Lieutier, F., Day, K.R., Grégoire, J.C. and Evans, H.F. (eds) *Bark and Wood Boring Insects in Living Trees in Europe, A Synthesis*. Kluwer Academic Publishers, The Netherlands, pp. 395–414.

Kenward, R.E. (2001) *A Manual of Wildlife Radio Tagging*. Academic Press, London, UK.

Keppel, G., Gillespie, T.W., Ormerod, P. and Fricker, G.A. (2016) Habitat diversity predicts orchid diversity in the tropical south-west Pacific. *Journal of Biogeography*, 43, 2332–2342.

Kerfeld, C.A., Sawaya, M.R., Tanaka, S., Nguyen, C.V., Phillips, M. *et al.* (2005) Protein structures forming the shell of primitive bacterial organelles. *Science*, 309, 936–938.

Kershaw, K.A. (1973) *Quantitative and Dynamic Plant Ecology*. Edward Arnold, London, UK.

Kéry, M. (2004) Extinction rate estimates for plant populations in revisitation studies: Importance of detectability. *Conservation Biology*, 18(2), 570–574.

Khalighifar, A., Jiménez, L., Nuñez-Penichet, C., Freeman, B., Ingenloff, K. *et al.* (2020) Inventory statistics meet big data: Complications for estimating numbers of species. *PeerJ*, 8, e8872.

Kichenin, E., Wardle, D.A., Peltzer, D.A., Morse, C.W. and Freschet, G.T. (2013) Contrasting effects of plant inter-and intraspecific variation on community-level trait measures along an environmental gradient. *Functional Ecology*, 27, 1254–1261.

Kifle, Z. and Bekele, A. (2021) Feeding ecology and diet of the southern geladas (*Theropithecus gelada obscurus*) in human-modified landscape, Wollo, Ethiopia. *Ecology and Evolution*, 11, 11373–11386.

Kifle, Z., Belay, G. and Bekele, A. (2013) Population size, group composition and behavioural ecology of geladas (*Theropithecus gelada*) and human–gelada conflict in Wonchit Valley, Ethiopia. *Pakistan Journal of Biological Sciences*, 16, 1248–1259.

Kilpatrick, A.M. and Ives, A.R. (2003) Species interactions can explain Taylor's power law for ecological time series. *Nature*, 422, 65–68.

Kim, D.-H., Cho, W.-S. and Chon, T.-S. (2013) Self-organizing map and species abundance distribution of stream benthic macroinvertebrates in revealing community patterns in different seasons. *Ecological Informatics*, 17, 14–29.

Kim, K.H., Ramadhar, T.R., Beemelmanns, C., Cao, S., Poulsen, M. *et al.* (2014) Natalamycin A, an ansamycin from a termite associated *Streptomyces* sp. *Chemical Science*, 5, 4333–4338.

Kimball, M.E. and Able, K.W. (2012) Tidal migrations of intertidal salt marsh creek nekton examined with underwater video. *Northeastern Naturalist*, 19, 475–486.

Kimura, M. (1983) *The Neutral Theory of Molecular Evolution*. Cambridge University Press, Cambridge, UK.

Kimura, M.T. (2021) Altitudinal migration of insects. *Entomological Science*, 24, 35–47.

Kinchin, I.M. (1995) *The Biology of the Tardigrades*. Portland Press, London, UK.

Kinchin, I.M. (2000) On the significance of a "minor" phylum (the Tardigrada) in the context of a constructivist view of knowledge. *Perspectives in Biology and Medicine*, 43, 243–251.

Kindvall, O. (1996) Habitat heterogeneity and survival in a bush cricket metapopulation. *Ecology*, 77, 207–214.

King, H., Ocko, S. and Mahadevan, L. (2015) Termite mounds harness diurnal temperature oscillations for ventilation. *Proceedings of the National Academy of Sciences of the United States of America*, 112, 11589–11593.

King, J.R. and Porter, S.D. (2005) Evaluation of sampling methods and species richness estimators for ants in upland ecosystems in Florida. *Environmental Entomology*, 34, 1566–1578.

Kingsland, S. (2015) Alfred J. Lotka and the origins of theoretical population ecology. *Proceedings of the National Academy of Sciences of the United States of America*, 112, 9493–9495.

Kirchhoff, T. (2020) The myth of Frederic Clements's mutualistic organicism, or: On the necessity to distinguish different concepts of organicism. *History and Philosophy of the Life Sciences*, 42, 24.

Kirchner, J.W. (1989) The Gaia hypothesis: Can it be tested? *Reviews of Geophysics*, 27, 223–235.

Kirk, D.A., Park, A.C., Smith, A.C., Howes, B.J., Prouse, B.K. *et al.* (2018) Our use, misuse, and abandonment of a concept: Whither habitat? *Ecology and Evolution*, 8(8), 4197–4208.

Kirkland, G.L. Jr. (1977) Responses of small mammals to the clearcutting of northern Appalachian forests. *Journal of Mammalogy*, 58, 600–609.

Kissel, M. and Kim, N.C. (2019) The emergence of human warfare: Current perspectives. *American Journal of Physical Anthropology*, 168, 141–163.

Kissling, D., Pattemore, D.E. and Hagen, M. (2014) Challenges and prospects in the telemetry of insects:

Insect telemetry. *Biological Reviews of the Cambridge Philosophical Society*, 89, 511–530.

Kistner, D.H. (1982) The social insects' bestiary. In: Herman, H.R. (ed.) *Social Insects*. Academic Press, New York, USA, pp. 2–244.

Kitzes, J. and Harte, J. (2015) Predicting extinction debt from community patterns. *Ecology*, 96, 2127–2136.

Kjær, K.H., Pedersen, M.W., De Sanctis, B., De Cahsan, B., Korneliussen, T.S. *et al.* (2022) A 2-million-year-old ecosystem in Greenland uncovered by environmental DNA. *Nature*, 612, 283–291.

Kleijn, D., Kohler, F., Báldi, A., Batáry, P., Concepción, E.D. *et al.* (2009) On the relationship between farmland biodiversity and land-use intensity in Europe. *Proceedings of the Royal Society B: Biological Sciences*, 276, 903–909.

Kleine, T., Touboul, M., Bourdon, B., Nimmo, F., Mezger, K. *et al.* (2009) Hf–W chronology of the accretion and early evolution of asteroids and terrestrial planets. *Geochimica et Cosmochimica Acta*, 73, 5150–5188.

Kleyer, M., Trinogga, J., Cebrián-Piqueras, M.A., Trenkamp, A., Fløjgaard, C. *et al.* (2019) Trait correlation network analysis identifies biomass allocation traits and stem specific length as hub traits in herbaceous perennial plants. *Journal of Ecology*, 107, 829–842.

Kleypas, J.A., Feely, R.A., Fabry, V.J., Langdon, C., Sabine, C.L. and Robbins, L.L. (2006) Impacts of ocean acidification on coral reefs and other marine calcifiers: A guide for future research. Report of a workshop held April 18–20, 2005, St Petersburg, FL, sponsored by NSF, NOAA, and the US Geological Survey. Available at: www.carboncyclescience.us/sites/default/files/documents/final-acidification.pdf (accessed 19 May 2025).

Kliman, R., Sheehy, B. and Schultz, J. (2008) Genetic drift and effective population size. *Nature Education*, 1(3), 3.

Klinger, L.F. (1996) The myth of the classic hydrosere model of bog succession. *Arctic and Alpine Research*, 28, 1–9.

Knell, R.J. (2009) Population density and the evolution of male aggression. *Journal of Zoology*, 278, 83–90.

Knighton, C.S. (ed.) (2004) *Pepys's Later Diaries*. Stroud, Sutton, UK.

Knoll, A.H. and Nowak, M.A. (2017) The timetable of evolution. *Science Advances*, 3, e1603076.

Knope, M.L., Bush, A.M., Frishkoff, L.O., Heim, N.A. and Payne, J.L. (2020) Ecologically diverse clades dominate the oceans via extinction resistance. *Science*, 367(6481), 1035–1038.

Kobel-Lamparski, von A. and Lamparski, F. (2004) *Lumbricus badensis* – ein Franzose, der in Baden groß wurde. *Mitteilungen des Badischen Landesvereins für Naturkunde und Naturschutz*, 18, 91–109.

Kočárek, P. (2003) Decomposition and Coleoptera succession on exposed carrion of small mammal in Opava, the Czech Republic. *European Journal of Soil Biology*, 39, 31–45.

Kocher, S. and Williams, E.H. (2000) The diversity and abundance of North American butterflies vary with habitat disturbance and geography. *Journal of Biogeography*, 27, 785–794.

Kocsis, Á.T., Reddin, C.J. and Kiessling, W. (2018) The stability of coastal benthic biogeography over the last 10 million years. *Global Ecology and Biogeography*, 27, 1106–1120.

Kohda, M., Yamanouchi, H., Hirata, T., Satoh, S. and Ota, K. (2017) A novel aspect of goby–shrimp symbiosis: Gobies provide droppings in their burrows as vital food for their partner shrimps. *Marine Biology*, 164, 22.

Köhler, F. (2000) *Totholzkäfer in Naturwaldzellen des nördlichen Rheinlands: vergleichende Studien zur Totholzkäferfauna Deutschlands und deutschen Naturwaldforschung*. Die Landesanstalt für Ökologie, Bodenordnung und Forsten NRW (LÖBF), Recklinghausen, Germany.

Kohls, S.J., Baker, D.D., van Kessel, C. and Dawson, J.O. (2003) An assessment of soil enrichment by actinosrizal N^2 fixation using $\delta_{15}N$ values in a chronosequence of deglaciation at Glacier Bay, Alaska. *Plant and Soil*, 254, 11–17.

Kohn, D.D. and Walsh, D.M. (1994) Plant species richness – the effect of island size and habitat diversity. *Journal of Ecology*, 82, 367–377.

Kolbert, E. (2014) *The Sixth Extinction: An Unnatural History*. Henry Holt, New York, USA.

Koleff, P., Gaston, K.J. and Lennon, J.J. (2003) Measuring beta diversity for presence-absence data. *Journal of Animal Ecology*, 72, 367–382.

Kolen, J. (1999) Hominids without homes. On the nature of Middle Palaeolithic settlement in Europe. In: Roebroeks, W. and Gamble, C. (eds) *The Middle Palaeolithic Occupation of Europe*. University of Leiden, Leiden, The Netherlands, pp. 139–175.

Koopman, B.O. (1980) *Search and Screening*. Pergamon, Oxford, UK.

Korányi, D., Gallé, R., Donkó, B., Chamberlain, D.E. and Batáry, P. (2021) Urbanization does not affect green space bird species richness in a mid-sized city. *Urban Ecosystems*, 24, 789–800.

Korňana, M. and Kropil, R. (2014) What are ecological guilds? Dilemma of guild concepts. *Russian Journal of Ecology*, 45, 445–447.

Kostitzin, V.A. (1935) *Evolution de l'atmosphère: Circulation Organique: Époques Glaciaires, Exposés de Biométrie et de Statistique Biologique*. Hermann, Paris, France.

Kotze, D.J., Lehvävirta, S., Koivula, M., O'Hara, R.B. and Spence, J.R. (2012) Effects of habitat edges and trampling on the distribution of ground beetles (Coleoptera, Carabidae) in urban forests. *Journal of Insect Conservation*, 16, 883–897.

Koutsidi, M., Moukas, C. and Tzanatos, E. (2020) Trait-based life strategies, ecological niches, and niche overlap in the nekton of the data-poor Mediterranean Sea. *Ecology and Evolution*, 10, 7129–7144.

Koyama, N., Soma, T., Ichino, S. and Takahata, Y. (2006) Home ranges of ringtailed lemur troops and the density of large trees at Berenty Reserve, Madagascar. In: Jolly, A., Sussman, R.W., Koyama, N. and Rasamimanana, H. (eds) *Ringtailed Lemur Biology. Developments in Primatology: Progress and Prospect.* Springer, Boston, USA, pp. 86–101.

Koziol, L. and Bever, J.D. (2019) Mycorrhizal feedbacks generate positive frequency dependence accelerating grassland succession. *Journal of Ecology*, 107, 622–632.

Kraft, N.J.B., Adler, P.B., Godoy, O., James, E.C., Fuller, S. and Levine, J.M. (2015a) Community assembly, coexistence and the environmental filtering metaphor. *Functional Ecology*, 29, 592–599.

Kraft, N.J.B., Godoy, O. and Levine, J.M. (2015b) Plant functional traits and the multidimensional nature of species coexistence. *Proceedings of the National Academy of Science of the United States of America*, 112, 797–802.

Kral-O'Brien, K.C., Karasch, B.M., Hovick, T.J., Moranz, R.A. and Harmon, J.P. (2020) Morphological traits determine detectability bias in North American grassland butterflies. *Ecosphere*, 11, e03304.

Kramer, A.M., Berec, L. and Drake, J.M. (2018) Editorial: Allee effects in ecology and evolution. *Journal of Animal Ecology*, 87, 7–10.

Krasnov, B.R., Shenbrot, G.I., Mouillot, D., Khokhlova, I.S. and Poulin, R. (2005) Spatial variation in species diversity and composition of flea assemblages in small mammalian hosts: Geographical distance or faunal similarity? *Journal of Biogeography*, 32, 633–644.

Krauss, J., Steffan-Dewenter, I., Muller, C.B. and Tscharntke, T. (2005) Relative importance of resource quantity, isolation and habitat quality for landscape distribution of a monophagous butterfly. *Ecography*, 28, 465–474.

Krebs, C.J. (1999) *Ecological Methodology*, 2nd edn. Benjamin/Cummings, Menlo Park, USA.

Krebs, C.J., Boonstra, R. and Boutin, S. (2018) Using experimentation to understand the 10-year snowshoe hare cycle in the boreal forest of North America. *Journal of Animal Ecology*, 87, 87–100.

Kreft, H. and Jetz, W. (2010) A framework for delineating biogeographical regions based on species distributions. *Journal of Biogeography*, 37, 2029–2053.

Kreft, H. and Jetz, W. (2013) Comment on An update of Wallace's zoogeographic regions of the world. *Science*, 341, 343.

Krishna, K., Grimaldi, D.A., Krishna, V. and Engel, M.S. (2013) Treatise on the Isoptera of the world: vol 1, Introduction. *Bulletin of the American Museum of Natural History*, 377, 1–200.

Kroon, H.D and Olff, H. (1995) On the use of the guild concept in plant ecology. *Folia Geobotanica and Phytotaxonomica*, 30, 519–528.

Kudrna, O. (2002) The distribution atlas of European butterflies. *Oedippus*, 20, 1–343.

Kunegel-Lion, M., Neilson, E.W., Mansuy, N. and Goodsman, D.W. (2022) Habitat quality does not predict animal population abundance on frequently disturbed landscapes. *Ecological Modelling*, 469, 109943.

Kuntal, B.K., Gadgil, C. and Mande, S.S. (2019) Web-gLV: A web based platform for Lotka-Volterra based modeling and simulation of microbial populations. *Frontiers in Microbiology*, 10, 288.

Kurek, P., Dobrowolska, D. and Wiatrowska, B. (2019) Dispersal distance and burial mode of acorns in Eurasian jays *Garrulus glandarius* in European temperate forests. *Acta Ornithologica*, 53, 155–162.

Kutschera, V. and Elliott, J.M. (2010) Charles Darwin's observation on the behaviour of earthworms and the evolutionary history of a giant endemic species from Germany, *Lumbricus badensis* (Oligochaeta: Lumbricidae). *Applied and Environmental Soil Science*, 2, 823047.

Labandeira, C.C., Johnson, K.R. and Wilf, P. (2002) Impact of the terminal Cretaceous event on plant–insect associations. *Proceedings of the National Academy of Sciences of the United States of America*, 99, 2061–2066.

Lack, D. (1943) *The Life of the Robin.* H.E. Witherby, London, UK. (Latest edition 2016, Pallas Athene, London.)

Lack, D. (1966a) *Swifts in a Tower.* Methuen, London, UK.

Lack, D. (1966b) *Population Studies of Birds.* Clarendon Press, Oxford, UK.

Ladd, P.G., Crosti, R. and Pignatti, S. (2005) Vegetative and seedling regeneration after fire in planted Sardinian pinewood compared with that in other areas of Mediterranean type climate. *Journal of Biogeography*, 32, 85–98.

Lafferty, K.D. and Suchanek, T.H. (2016) Revisiting Paine's 1966 sea star removal experiment, the most-cited empirical article in the American Naturalist. *American Naturalist*, 188, 365–378.

Lai, C-H., Lin, S-H., Tsai, C-Y. and Chen, S-H. (2018) Identifying farm pond habitat suitability for the common moorhen (*Gallinula chloropus*): A conservation-perspective approach. *Sustainability*, 10, 1352.

Laing, S.E., Buckland, S.T., Burn, R.W. and Lambie D. (2003) Dung and nest surveys: Estimating decay rates. *Journal of Applied Ecology*, 40, 1102–1111.

Lake, S., Liley, D., Still, R. and Swash, A. (2015) *Britain's Habitats. A Fieldguide to the Wildlife Habitats of Britain and Ireland.* WILDGuides Ltd, Old Basing, UK and Princeton University Press, Woodstock, UK.

Laliberté, E. and Legendre, P. (2010) A distance-based framework for measuring functional diversity from multiple traits. *Ecology*, 91, 299–305.

Lambert, M.R. and Donihue, C.M. (2020) Urban biodiversity management using evolutionary tools. *Nature Ecology and Evolution*, 4, 903–910.

Lamichhaney, S., Berglund, J., Almén, M.S., Maqbool, K., Grabherr, M. *et al.* (2015) Evolution of Darwin's finches and their beaks revealed by genome sequencing. *Nature*, 518, 371–375.

Lamparski, F. (1985) Der Einfluss der Regenwurmart *Lumbricus badensis* auf Waldböden im Sudschwarzwald. *Freiburger Bodenkundliche Abhandlungen*, 15, 1–206.

Lan, B., Zhou, X., Yang, N. and Sun, S. (2023) Spectral radius is a better metric than weighted NODF to detect network nestedness: Linking species coexistence to network structure using a plant–larval sawfly bipartite. *Food Webs*, 36, e00303.

Lande, R. (1996) Statistics and partitioning of species diversity, and similarity among multiple communities. *Oikos*, 76, 5–13.

Lande, R., Engen, S. and Saether, B.-E. (2003) *Stochastic Population Dynamics in Ecology and Conservation*. Oxford University Press, Oxord, UK.

Landi, P. and Piccardi, C. (2014) Community analysis in directed networks: In-, out-, and pseudocommunities. *Physical Review E*, 89, 012814.

Landi, P., Minoarivelo, H.O., Brännström, Å., Hui, C. and Dieckmann, U. (2018) Complexity and stability of ecological networks: A review of the theory. *Population Ecology*, 60, 319–345.

Landrigan, P.J., Stegeman, J.J., Fleming, L.E., Allemand, D., Anderson, D.M. *et al.* (2020) Human health and ocean pollution. *Annals of Global Health*, 86, 151.

Lanza, B. (2012) *Fauna d'Italia*. Vol. XLVII Mammalia V – Chiroptera. Calderini, Bologna, Italy.

Lapenis, A.G. (2002) Directed evolution of the biosphere: Biogeochemical selection or Gaia? *Professional Geographer*, 54, 379–391.

Lapola, D.M., Oyama, M.D., Nobre, C.A. and Sampaio, G. (2008) A new world natural vegetation map for global change studies. *Annals of the Brazilian Academy of Sciences*, 80, 397–408.

Lara, E., Dunmack, K., García-Martín, J.M., Kudryavtsev, A. and Kosakyan, A. (2020) Amorboid protist systematics: A report on the 'Systematics of amoeboid protists' symposium at the 8th ECOP/ISOP meeting, Rome, 2019. *European Journal of Protistology*, 76, 125727.

Larabee, F.J., Smith, A.A. and Suarez, A.V. (2018) Snap-jaw morphology is specialized for high-speed power amplification in the Dracula ant, *Mystrium camillae*. *Royal Society Open Science*, 5, 181447.

Larkin, J.L., Maehr, D.S., Krupa, J.J., Cox, J.J., Alexy, K. *et al.* (2008) Small mammal response to vegetation and spoil conditions on a reclaimed surface mine in eastern Kentucky. *Southeastern Naturalist*, 73, 401–412.

Larsen, B.B., Miller, E.C., Rhodes, M.K. and Wiens, J.J. (2017) Inordinate Fondness multiplied and redistributed: The number of species on Earth and the New Pie of Life. *Quarterly Review of Biology*, 92, 229–265.

Larsen, F.W., Bladt, J. and Rahbek, C. (2007) Improving the performance of indicator groups for the identification of important areas for species conservation. *Conservation Biology*, 21, 731–740.

Larson, J.E. and Funk, J.L. (2016) Seedling root responses to soil moisture and the identification of a below-ground trait spectrum across three growth forms. *New Phytologist*, 210, 827–838.

Laumer, I.B., Rahman, A., Rahmaeti, T. Azhari, U., Hermansyah *et al.* (2024) Active self-treatment of a facial wound with a biologically active plant by a male Sumatran orangutan. *Scientific Reports*, 14, 8932.

Laurance, W.F. (2008) Theory meets reality: How habitat fragmentation research has transcended island biogeographic theory. *Biological Conservation*, 141, 1731–1744.

Laurence, B.R. (1954) The larval inhabitants of cow pats. *Journal of Animal Ecology*, 23, 234–260.

Lavelle, P. (2012) Soil as a habitat. In: Wall, D.H. (ed.) *Soil Ecology and Ecosystem Services*. Oxford University Press, Oxford, UK, pp. 7–27.

Laver, P.N. and Kelly, M.J. (2008) A critical review of home range studies. *Journal of Wildlife Management*, 72, 290–298.

Lavorel, S., and Garnier, E. (2002) Predicting changes in community composition and ecosystem functioning from plant traits: revisiting the Holy Grail. *Functional Ecology*, 16, 545–556.

Lawler, J.J., White, D., Sifneos, J.C. and Master L.L. (2003) Rare species and the use of indicator groups for conservation planning. *Conservation Biology*, 17, 875–882.

Lawlor, L.R. (1980) Structure and stability in natural and randomly constructed competitive communities. *American Naturalist*, 116, 394–408.

Lawrence, D.H. (1997) [1921 first published] *Sea and Sardinia*. Edited by Mara Kalnins. Cambridge University Press, Cambridge, UK.

Lawson, C.R., Bennie, J., Thomas, C.D., Hodgson, J.A., Bernhard, T. *et al.* (2013) *The Status and Conservation of the Silver-spotted Skipper* Hesperia comma *in South-East England 2000–2009*. University of Exeter, Exeter, UK.

Lawton, J.H. (1999) Are there general laws in ecology? *Oikos*, 84, 177–192.

Lawton, J.H. and Gaston, K. (2001) Indicator species. In: Levin, S.A. (ed.) *Encyclopedia of Biodiversity*, vol. 3, Academic Press, San Diego, USA, pp. 437–450.

Lawton, J.H. and Warren, P.H. (1988) Static and dynamic explanations for patterns in food webs. *Trends in Ecology and Evolution*, 3, 242–245.

Layton, C., Shelamoff, V., Cameron, M.J., Tatsumi, M., Wright, J.T. and Johnson, C.R. (2019) Resilience and stability of kelp forests: The importance of patch dynamics and environment-engineer feedbacks. *PLOS One*, 14, e0210220.

Lazarina, M., Michailidou, D.-E., Tsianou, M. and Kallimanis, A.S. (2023) How biodiversity, climate and landscape drive functional redundancy of British butterflies. *Insects*, 14, 722.

Leather, S.R. (ed.). (2005) *Insect Sampling in Forest Ecosystems*. Blackwell, Oxford, UK.

Leather, S.R. (2017) "Ecological Armageddon" – More evidence for the drastic decline in insect numbers. *Annals of Applied Biology*, 172, 1–3.

Leather, S.R. (2021) Entomyopia and Entoalexia – Two potentially life-threatening conditions. *Antenna*, 45, 80–82.

Le Bagousse-Pinguet, Y., de Bello, F., Vandewalle, M., Lepš, J. and Sykes, M.T. (2014) Species richness of limestone grasslands increases with trait overlap: Evidence from within-and between-species functional diversity partitioning. *Journal of Ecology*, 102, 466–474.

Lebreton, L., Slat, B., Ferrari, F., Sainte-Rose, B., Aitken, J. *et al.* (2018) Evidence that the Great Pacific Garbage Patch is rapidly accumulating plastic. *Scientific Reports*, 8, 4666.

Lee, A. (2020) Sexual Eeling. The slippery subject of eel reproduction evaded human understanding for millennia. *History Today*, 70. Available at: www.historytoday.com/archive/natural-histories/sexual-eeling (accessed 19 May 2024).

Lee, C.M. and Wall R. (2006) Cow-dung colonization and decomposition following insect exclusion. *Bulletin of Entomological Research*, 96, 315–322.

Lees, A.C., Haskell, L., Allinson, T., Bezeng, S.B. Burfield, I.J. *et al.* (2022) State of the world's birds. *Annual Review of Environment and Resources*, 47, 231–260.

Lee-Yaw, J.A., McCune, J.L., Pironon, S. and Sheth, S.N. (2022) Species distribution models rarely predict the biology of real populations. *Ecography*, 2022, e05877.

Legendre, P. (2014) Replacement and richness difference components. *Global Ecology and Biogeography*, 23, 1324–1334.

Legendre, P. and Legendre, L. (2012) *Numerical Ecology*, 3rd edn. Elsevier, Amsterdam, The Netherlands.

Legendre, P., Borcard, D. and Peres-Neto, P.R. (2005) Analyzing beta diversity: Partitioning the spatial variation of community composition data. *Ecological Monographs*, 75,435–450.

Legendre, S., Schoener, T.W., Clobert, J. and Spiller, D.A. (2008) How is extinction risk related to population-size variability over time? A family of models for species with repeated extinction and immigration. *American Naturalist*, 172, 282–298.

Legras, G., Loiseau, N. and Gaertner, J.-C. (2018) Functional richness: Overview of indices and underlying concepts. *Acta Oecologica*, 87, 34–44.

Lehenberger, M., Foh, N., Göttlein, A., Six, D. and Biedermann, P.H.W. (2021) Nutrient-poor breeding substrates of ambrosia beetles are enriched with biologically important elements. *Frontiers in Microbiology*, 12, 664542.

Lehmann, J. and Boesch, C. (2004) To fission or to fusion: effects of community size on wild chimpanzee (*Pan troglodytes verus*) social organisation. *Behavioral Ecology and Sociobiology*, 56, 207–216.

Lehner, S., Schulz, S. and Dötterl, S. (2022) The mystery of the butterfly bush *Buddleja davidii*: How are the butterflies attracted? *Frontiers in Plant Science*, 13, 994851.

Lei, K., Pan, H. and Lin, C. (2016) A landscape approach towards ecological restoration and sustainable development of mining areas. *Ecological Engineering*, 90, 320–325.

Leibold, M.A. (1998) Similarity and local co-existence of species in regional biotas. *Evolutionary Ecology*, 12, 95–110.

Leigh, E.G. Jr (2007) Neutral theory: A historical perspective. *Journal of Evolutionary Biology*, 20, 2075–2091.

Leinster, T. and Cobbold, C. (2012) Measuring diversity: The importance of species similarity. *Ecology*, 93, 477–489.

Lemon, L.M., Smithm K.L. Jr. and Huffard, C.L. (2022) Abyssal epibenthic holothurians respond differently to food quantity and concentration fluctuations over a decade of daily observation (2007–2017). *Deep Sea Research Part I: Oceanographic Research Papers*, 188, 103853.

Lennon, J.J., Koleff, P., Greenwood, J.J.D. and Gaston, K.J. (2001) The geographical structure of British bird distributions: Diversity, spatial turnover and scale. *Journal of Animal Ecology*, 70, 966–979.

Leonard, M.L., Horn, A.G. and Eden, S.F. (1988) Parent–offspring aggression in moorhens. *Behavioral Ecology and Sociobiology*, 23, 265–270.

Leote, P., Cajaiba, R.L., Cabral, J.A., Brescovit, A.D. and Santos, M. (2020) Are data-mining techniques useful for selecting ecological indicators in biodiverse regions? Bridges between market basket analysis and indicator value analysis from a case study in the neotropics. *Ecological Indicators*, 109(105833), 105833.

Leprieur, F., Albouy, C., De Bortoli, J., Cowman, P.F., Bellwood, D.R. and Mouillot, D. (2012) Quantifying phylogenetic beta diversity: Distinguishing between 'true' turnover of lineages and phylogenetic diversity gradients. *PLOS One*, 7, e42760.

Lepš, J., de Bello, F., Šmilauer, P. and Doležal, J. (2011) Community trait response to environment: Disentangling species turnover vs intraspecific trait variability effects. *Ecography*, 34, 856–863.

Le Roy, E. (1927) *L'exigence idéaliste et le fait de l'évolution*. Boivin, Paris, France.

Leroy, A. and Leroy, J. (2005) Notes on the natural history of a trapdoor spider *Ancylotrypa* Simon (Araneae, Cyrtaucheniidae) that constructs a spherical burrow plug. *Journal of Arachnology*, 33(2), 558–561.

Leroy, B., Dias, M.S., Giraud, E., Hugueny, B., Jézéquel, C. *et al.* (2019) Global biogeographical regions of freshwater fish species. *Journal of Biogeography*, 46, 2407–2419.

Leveau, L.M., Ruggiero, A., Matthews, T.J. and Bellocq, M.I (2019) A global consistent positive effect of urban green area size on bird richness. *Avian Research*, 10, 30.

Levins, R. (1968) *Evolution in Changing Environments: Some Theoretical Explorations*. Princeton University Press, Princeton, USA.

Levins, R. (1969) Some demographic and genetic consequences of environmental heterogeneity for biological control. *Bulletin of the Entomological Society of America*, 15, 237–240.

Levins, R. (1970) Extinction. In: Gerstenhaber, M. (ed.) *Some Mathematical Problems in Biology*. American Mathematical Society, Providence, USA, pp. 75–107.

Levinson, S.C. and Enfield, N.J. (2006) *Roots of Human Sociality. Culture, Cognition and Interaction*. Routledge, New York, USA.

Levy, J.M. and Connor, E.F. (2004) Are gardens effective in butterfly conservation? A case study with the pipevine swallowtail, *Battus philenor*. *Journal of Insect Conservation*, 8, 323–330.

Lewis, B., Griffiths, R.A. and Barrios, Y. (2007) *Field Assessment of Great Crested Newt* Triturus cristatus *Mitigation Projects in England*. Natural England Research Report NERR001. University of Kent, Canterbury, UK.

Lewis, S. and Maslin, M. (2015) Defining the Anthropocene. *Nature*, 519, 171–180.

Leyhausen, P. (1979) *Cat Behavior: The Predatory and Social Behavior of Domestic and Wild Cats*. Garland STPM Press, New York, USA.

Leyte-Manrique, A., Buelna-Chontal, A.A., Torres-Diaz, M.A., Berriozabal-Islas, C. and Maciel-Mata, C.A. (2019) A comparison of amphibian and reptile diversity between disturbed and undisturbed environments of Salvatierra, Guanajuato, Mexico. *Tropical Conservation Science*, 12, 1–12.

Li, S., Cadotte, M.W., Meiners, S.J., Pu, Z., Fukami, T. and Jiang, L. (2016) Convergence and divergence in a long-term old-field succession: The importance of spatial scale and species abundance. *Ecology Letters*, 19, 1101–1109.

Li, W., Cheng, J.M., Yu, K.L., Epstein, H.E., Guo, L. *et al.* (2015) Plant functional diversity can be independent of species diversity: Observations based on the impact of 4-yrs of nitrogen and phosphorus additions in an alpine meadow. *PLOS One*, 10, e0136040.

Li, X. and Wiens, J.J. (2023) Estimating global biodiversity: The role of cryptic insect species. *Systematic Biology*, 72, 391–403.

Li, Y., Wilson, D., Grundel, R., Campbell, S., Knight, J. *et al.* (2023) Extinction risk modeling predicts range-wide differences of climate change impact on Karner blue butterfly (*Lycaeides melissa samuelis*). *PLOS One*, 18(11), e0262382.

Li, Y., Yang, F., Ou, Y., Zhang, D., Liu, J. *et al.* (2013) Changes in forest soil properties in different successional stages in lower tropical China. *PLOS One*, 8, e81359.

Liao, C., Xavier, J.B. and Zhu, Z. (2020) Enhanced inference of ecological networks by parameterizing ensembles of population dynamics models constrained with prior knowledge. *BMC Ecology and Evolution*, 20, 3.

Liddell, E., Sunnucks, P. and Cook, C.N. (2021) To mix or not to mix gene pools for threatened species management? Few studies use genetic data to examine the risks of both actions, but failing to do so leads disproportionately to recommendations for separate management. *Biological Conservation*, 256, 109072.

Lightfoot, D.C., Brantley, S.L. and Allen, C.D. (2008) Geographic patterns of ground-dwelling arthropods across an ecoregional transition in the North American Southwest. *Western North American Naturalist*, 68(1), 83–102.

Lin, B., Foxfoot, I.R., Miller, C.M, Venkatamaran, V.V., Kerby, J.T. *et al.* (2020) Leopard predation on gelada monkeys at Guassa, Ethiopia. *American Journal of Primatology*, 82, e23098.

Lin, L., Liu, Y., Lin, H. and Kang, B. (2024) Considering species functional and phylogenetic rarity in the conservation of fish biodiversity. *Diversity and Distributions*, 30, e13804.

Lincoln, R.J., Boxshall, G.A. and Clark, P.F. (1982) *A Dictionary of Ecology, Evolution and Systematics*. Cambridge University Press, Cambridge, UK.

Linder, H.P., de Klerk, H.M., Born, J., Burgess, N.D., Fjeldså, J. and Rahbek, C. (2012) The partitioning of Africa: Statistically defined biogeographical regions in sub-Saharan Africa. *Journal of Biogeography*, 39, 1189–1205.

Lindsey, P.A., Roulet, P.A. and Romañach, S.S. (2007) Economic and conservation significance of the trophy hunting industry in sub-Saharan Africa. *Biological Conservation*, 134, 455–469.

Linnaeus, C. (1735) *Systema Naturae, Sive Regna Tria Naturae Systematice Proposita Per Classes, Ordines, Genera, & Species*. T. Haak, Leiden, The Netherlands.

Linnakoski, R. and Forbes, K.M. (2019) Pathogens – the hidden face of forest invasions by wood-boring insect pests. *Frontiers in Plant Science*, 10, 90.

Lisicki, M., Velho Rodrigues, M.F., Goldstein, R.E. and Lauga, E. (2019) Swimming eukaryotic microorganisms exhibit a universal speed distribution. *eLife*, 8, e44907.

Lister, N.-M.E. (1998) A systems approach to biodiversity conservation planning. *Environmental Monitoring and Assessment*, 49, 123–155.

Liu, C., Li, Y., Yan, P. and He, N. (2021) How to improve the predictions of plant functional traits on ecosystem functioning? *Frontiers in Plant Science*, 12, 622260.

Liu, J., Coomes, D.A., Hu, G., Liu, J., Yu, J. *et al.* (2019) Larger fragments have more late-successional species of woody plants than smaller fragments after 50 years of secondary succession. *Journal of Ecology*, 107, 582–594.

Liu, Y.-N., Bairoliya, S., Zaiden, N. and Cao, B. (2024) Establishment of plastic-associated microbial community from superworm gut microbiome. *Environment International*, 183, 108349.

Livet, J. and Mousnier, R. (1980) *Histoire générale de l'Europe*. Presses Universitaires de France, Paris, France.

Lloyd-Smith, J.O. (2007) Maximum likelihood estimation of the negative binomial dispersion parameter for highly overdispersed data, with applications to infectious diseases. *PLOS One*, 2(2), e180.

Loarie, S.R., Duffy, P.B., Hamilton, H., Asner, G.P., Field, C.B. and Ackerly, D.D. (2009) The velocity of climate change. *Nature*, 462, 1052–1055.

Lobo, J.M. and Martín-Piera, F. (2002) Searching for a predictive model for species richness of Iberian dung beetle based on spatial and environmental variables. *Conservation Biology*, 16, 158–173.

Lobo, J.M., Gutiérrez-Cánovas, C., Fox, R., Dennis, R.L.H. and Sánchez-Fernández, D. (2021) Species traits influence the process of temporal trends in biodiversity inventorying: A case study using the British butterfly database. *Insect Conservation and Diversity*, 14, 748–755.

Locey, K.J. and Lennon, J.T. (2016) Scaling laws predict global microbial diversity. *Proceedings of the National Academy of Sciences of the United States of America*, 113, 5970–5975.

Lock, G. and Ralston, I.B.M. (2022) *Atlas of the Hillforts of Britain and Ireland*. Edinburgh University Press, Edinburgh, UK.

Lodé, T. (2006) Can low densities of carnivores result in genetic depletion? An investigation within French polecat populations. *Journal of Animal Breeding and Genetics*, 123, 152–158.

Logunov, D.M. and Dennis, R.L.H. (2016) The insects and other invertebrates of Alderley Edge (section C, Order Lepidoptera – butterflies). In: Prag, A.J.N.W. (ed.) *Living with the Edge: Alderley's Story*. Manchester University Press, Manchester, UK, pp. 243–253.

Loh, T.-L., Larson, E.R., David, S.R., de Souza, L.S. and Gericke, R. (2018) Quantifying the contribution of zoos and aquariums to peer-reviewed scientific research. *Facets*, 3, 287–299.

Lohbeck, M., Poorter, L., Paz, H., Pla, L., van Breugel, M. *et al.* (2012) Functional diversity changes during tropical forest succession. *Perspectives in Plant Ecology, Evolution and Systematics*, 14, 89–96.

Lohmann, K.J. (2010) Animal behaviour: Magnetic-field perception. *Nature*, 464, 1140–1142.

Lombardo, P. and Mjelde, M. (2014) Quantifying interspecific spatial overlap in aquatic macrophyte communities. *Hydrobiologia*, 737, 25–43.

Lomolino, M.V., Brown, J.H. and Davis, R. (1989) Island biogeography of montane forest mammals in the American Southwest. *Ecology*, 70, 180–194.

Lomolino, M.V., Riddle, B.R., Whittaker, R.J. and Brown, J.H. (2010) *Biogeography*, 4th edn. Sinauer Associates Inc., Sunderland, MA.

Londo, G. (1976) The decimal scale for relevés of permanent quadrats. *Vegetatio*, 33, 61–64.

Long, R.A., MacKay, P., Zielinski, W.J., and Ray, J.C. (eds) (2008) *Noninvasive Survey Methods for Carnivores*. Island Press, Washington, DC, USA.

Longland, W.S. and Price, M.V. (1991) Direct observations of owls and heteromyid rodents: Can predation risk explain microhabitat use? *Ecology*, 72, 2261–2273.

Lopes de Carvalho, L.M. and Linhares, A.X. (2001) Seasonality of insect succession and pig carcass decomposition in a natural forest area in southeastern Brazil. *Journal of Forensic Sciences*, 46, 604–608.

López-Barrera, F., Manson, R.H., González-Espinosa, M. and Newton, A.C. (2007) Effects of varying forest edge permeability on seed dispersal in a neotropical montane forest. *Landscape Ecology*, 22(2), 189–203.

Loreau, M. and Hector, A. (2001) Partitioning selection and complementarity in biodiversity experiments. *Nature*, 412, 72–76.

Loreau, M. and de Mazancourt, C. (2013) Biodiversity and ecosystem stability: A synthesis of underlying mechanisms. *Ecology Letters*, 16, 106–115.

Loron, C.C., Cooper, L.M., McMahon, S., Jordan, S.F., Gromov, A.V. *et al.* (2025) *Prototaxites* was an extinct lineage of multicellular terrestrial eukaryotes. *bioRxiv*. https://doi.org/10.1101/2025.03.14.643340

Losch, A. (1954) *The Economics of Location*. Yale University Press, New Haven, USA.

Lotka, A.J. (1910) Contribution to the theory of periodic reaction. *Journal of Physical Chemistry*, 14, 271–274.

Lotka, A.J. (1920) Analytical note on certain rhythmic relations in organic systems. *Proceedings of the National Academy of Sciences of the United States of America*, 6, 410–415.

Lotka, A.J. (1925) *Elements of Physical Biology*. Williams and Wilkins, Baltimore, USA.

Lotka, A.J. (1932) The growth of mixed populations: Two species competing for a common food supply. *Journal of the Washington Academy of Sciences*, 22, 461–469.

Lotman, Y.M. 1984. O semiosfere. *Trudy po znakovym sistemam*, 17, 5–23.

Lovelock, J.E. (1972) Gaia as seen through the atmosphere. *Atmospheric Environment*, 6, 579–580.

Lovelock, J.E. (1987) *Gaia: A New Look at Life on Earth*. Oxford University Press, Oxford, UK.

Lovelock, J.E. and Margulis, L. (1974) Atmospheric homeostasis by and for the biosphere: The Gaia hypothesis. *Tellus*, 26, 2–10.

Lowery, C.M., Bralower, T.J., Owens, J.D., Rodríguez-Tovar, F.J., Jones, H. *et al.* (2018) Rapid recovery of life at ground zero of the end-Cretaceous mass extinction. *Nature*, 558, 288–291.

Lowry, I.S. (1964) *A Model of Metropolis*. Memorandum RM-4035-RC. Rand Corporation, Santa Monica, CA. Available at: www.rand.org/content/dam/rand/pubs/research_memoranda/2006/RM4035.pdf (accessed 20 May 2025).

Lozupone, C. and Knight, R. (2005) UniFrac: A new phylogenetic method for comparing microbial communities. *Applied and Environmental Microbiology*, 71, 8228–8235.

Lozupone, C., Lladser, M.E., Knights, D., Stombaugh, J. and Knight, R. (2011) UniFrac: An effective distance metric for microbial community comparison. *ISME Journal*, 5, 169–172.

Lu, M., Winner, K. and Jetz, W. (2021) A unifying framework for quantifying and comparing n-dimensional hypervolumes. *Methods in Ecology and Evolution*, 12, 1953–1968.

Ludsin, S.A. and Wolfe, A.D. (2001) Biological invasion theory: Darwin's contributions from *The Origin of Species*. *BioScience*, 51, 780–789.

Ludwig, J.A. and Reynolds, J.F. (1988) *Statistical Ecology. A Primer on Methods and Computing*. John Wiley and Sons, New York, USA.

Luketa, S. (2012) New views on the megaclassification of life. *Protistology*, 7, 218–237.

Luley, C.J., Nowak, D. and Greenfield, E.J. (2009) Frequency and severity of trunk decay in street tree maples in four New York cities. *Arboriculture and Urban Forestry*, 35, 94–99.

Luncz, L.V., Tan, A., Haslam, M., Kulik, L., Proffitt, T. *et al.* (2017) Resource depletion through primate stone technology. *eLife*, 6, e23647

Lundquist, J.H., Anderton, D.L. and Yaukey, D. (2015) *Demography. The Study of Human Population*, 4th edn. Waveland Press, Long Grove, USA.

Luoto, M., Rekolainen, S., Salt, C.A. and Hansen, H.S. (2001) Managing radioactively contaminated land: Implications for habitat diversity. *Environmental Management*, 27, 595–608.

Lutz, P.L. and Musick, J.A. (1997) *The Biology of Sea Turtles*. Taylor and Francis, London, UK.

Lydekker, R. (1896) *The Royal Natural History*, 6. Frederick Warne, London, UK.

Lymbery, S.J., Webber, B.L. and Didham, R.K. (2023) Complex battlefields favor strong soldiers over large armies in social animal warfare. *Proceedings of the National Academy of Sciences of the United States of America*, 120, e2217973120.

Lynch Alfaro, J.W., Cortés-Ortiz, L., Di Fiore, A. and Boubli, J.P. (2015) Phylogeny and biogeography of Neotropical primates. *Molecular Phylogenetics and Evolution*, 82(Pt B), 345–530.

Lynch, J.M. (1990) *The Rhizosphere*. Wiley-Interscience, Chichester, UK.

Lyons, K.G., Brigham, C.A., Traut, B.H. and Schwartz, M.W. (2005) Rare species and ecosystem functioning. *Conservation Biology*, 19, 1019–1024.

Lyson, T.R., Miller, I.M., Bercovici, A.D., Weissenburger, K., Fuentes, A.J. *et al.* (2019) Exceptional continental record of biotic recovery after the Cretaceous–Paleogene mass extinction. *Science*, 366, 977–983.

MacArthur, R.H. (1957) On the relative abundance of bird species. *Proceedings of the National Academy of Sciences of the United States of America*, 43(3), 293–295.

MacArthur, R. and Levins, R. (1964) Competition, habitat selection, and character displacement in a patchy environment. *Proceedings of the National Academy of Sciences of the United States of America*, 51, 1207.

MacArthur, R. and Levins, R. (1967) The limiting similarity, convergence, and divergence of coexisting species. *American Naturalist*, 101, 377–385.

MacArthur, R.H. and Wilson, E.O. (1963) An equilibrium theory of insular zoogeography. *Evolution*, 17, 373–387.

MacArthur, R.H. and Wilson, E.O. (1967) *The Theory of Island Biogeography*. Princeton University Press, Princeton, USA.

Macdonald, D.W. (1983a) Predation on earthworms by terrestrial vertebrates. In: Satchell, J.E. (ed.) *Earthworm Ecology*. Chapman and Hall, London, UK, pp. 393–414.

Macdonald, D.W. (1983b) The ecology of carnivore social behavior. *Nature*, 301, 379–384.

Macdonald, D.W. and Feber, R.E. (2015) *Wildlife Conservation on Farmland*. Oxford University Press, Oxford, UK.

Macdonald, D.W. and Johnson, D.D.P. (2015) Patchwork planet: The resource dispersion hypothesis, society, and the ecology of life. *Journal of Ecology*, 295, 75–107.

Macdonald, D.W., Boitani, L., Dinerstein, E., Fritz, H. and Wrangham, R. (2013) Conserving large mammals. In: Macdonald, D.W. and Willis, K.J. (eds) *Key Topics in Conservation Biology 2*. Wiley-Blackwell, Chichester, UK.

Mace, G.M. and Baillie J.E.M. (2007) The 2010 biodiversity indicators: Challenges for science and policy. *Conservation Biology*, 21, 1406–1413.

Mace, G.M and Collar, N.J. (1995) Extinction risk assessment for birds through quantitative criteria. *Ibis*, 137, S240–S246.

Mace, G.M. and Kershaw, M. (1997) Extinction risk and rarity on an ecological timescale. In: Kunin, W.E. and Gaston, K.J. (eds) *The Biology of Rarity. Population and Community Biology Series*, vol 17. Springer, Dordrecht, The Netherlands.

Mace, G.M., Collar, N.J., Gaston, K.J., Hilton-Taylor, C., Akçakaya, H.R. *et al.* (2008) Quantification of extinction risk: IUCN's system for classifying threatened species. *Conservation Biology*, 22(6), 1424–1442.

Maciel, E.A. (2021) An index for assessing the rare species of a community. *Ecological Indicators*, 124, 107424.

Mackay-Hope, D. (2021) *The Secret Life of Fish: The Astonishing Truth about our Aquatic Cousins*. Ivy Press, Lewes, UK.

Mackey, R.L. and Currie, D.J. (2001) The diversity-disturbance relationship: Is it generally strong and peaked? *Ecology*, 82, 3479–3492.

MacNiven, I.S. (1998) *Lawrence Durrell, A Biography*. Faber and Faber, London, UK.

Macphail, R.I., Cruise, G.M., Jallen, M., Linderholm, J. and Reynolds, P. (2004) Archaeological soil and pollen analysis of experimental floor deposits; with special reference to Butser Ancient Farm, Hampshire, UK. *Journal of Archaeological Science*, 31, 175–191.

Madin, J.S., Anderson, K.D., Andreasen, M.H., Bridge, T.C.L., Cairns, S.D. *et al.* (2016) The Coral Trait Database, a curated database of trait information for coral species from the global oceans. *Scientific Data*, 3, 160017.

Maes, D. and Van Dyck, H. (2001) Butterfly diversity loss in Flanders (north Belgium): Europe's worst case scenario? *Biological Conservation*, 99, 263–276.

Maes, D., Bauwens, D., De Bruyn, L., Anselin, A., Vermeersch, G. *et al.* (2005) Species rich-ness coincidence: Conservation strategies based on predictive modelling. *Biodiversity and Conservation*, 14, 1345–1364.

Maes, D., Van Calster, H., Herremans, M. and Van Dyck, H. (2022) Challenges and bottlenecks for butterfly conservation in a highly anthropogenic region: Europe's worst case scenario revisited. *Biological Conservation*, 274, 109732.

Maggi, E., Bertocci, I., Vaselli, S. and Benedetti-Cecchi, L. (2011) Connell and Slatyer's models of succession in the biodiversity era. *Ecology*, 92, 1399–1406.

Magneville, C., Loiseau, N., Albouy, C., Casajus, N., Claverie, T. *et al.* (2022) mFD: An R package to compute and illustrate the multiple facets of functional diversity. *Ecography*, 2022, e05904.

Magrath, R.D., Haff, T.M., McLachlan, J.R. and Igic, B. (2015) Wild birds learn to eavesdrop on heterospecific alarm calls. *Current Biology*, 25, 2047–2050.

Magura, T. and Lövei, G.L. (2020) The permeability of natural versus anthropogenic forest edges modulates the abundance of ground beetles of different dispersal power and habitat affinity. *Diversity*, 12, 320.

Magurran, A.E. (1988) *Ecological Diversity and Its Measurement*. Princeton University Press: Princeton, USA.

Magurran, A.E. (2004) *Measuring Biological Diversity*. Blackwell Publishing, Oxford, UK.

Magurran, A.E. and Henderson, P.A. (2003) Explaining the excess of rare species in natural species abundance distributions. *Nature*, 422, 714–716.

Maher, C.R. and Lott, D.F. (1995) Definitions of territoriality used in the study of variation in vertebrate spacing systems. *Animal Behaviour*, 49, 1581–1597.

Maher, L.A. and Conkey, M. (2019) Homes for hunters? Exploring the concept of home at hunter-gatherer sites in Upper Paleolithic Europe and Epipaleolithic Southwest Asia. *Current Anthropology*, 60, 91–137.

Maheshwari, A. (2020) Extinction is inevitable in the globalized world. *Human Dimensions of Wildlife*, 26, 501–502.

Major, J. (1969) Historical development of the ecosystem concept. In: Van Dyne, G.M. (ed.) *The Ecosystem Concept in Natural Resource Management*. Academic Press, New York, USA, pp. 9–22.

Makarieva, A.M. and Gorshkov, V.G. (2007) Biotic pump of atmospheric moisture as driver of the hydrological cycle on land. *Hydrology and Earth System Sciences*, 11, 1013–1033.

Makoto, K. and Wilson, S.D. (2019) When and where does dispersal limitation matter in primary succession? *Journal of Ecology*, 107, 559–565.

Malaivijitnond, S., Lekprayoon, C., Tandavanittj, N., Panha, S., Cheewatham, C. and Hamada, Y. (2007) Stone-tool usage by Thai long-tailed macaques (*Macaca fascicularis*). *American Journal of Primatology*, 69, 227–233.

Malhi, Y., Franklin, J., Seddon, N., Solan, M., Turner, M.G. *et al.* (2020) Climate change and ecosystems: Threats, opportunities and solutions. *Philosophical Transactions of the Royal Society B: Biological Sciences*, 375, 20190104.

Malik, S., Kishore, S., Kumar, S.A. and Dhasmana, A. (2023) Role of bacteria in biological removal of environmental pollutants. In: Shah, M.P. and Vyas, B.R.M. (eds) *Emerging Technologies in Applied and Environmental Microbiology*. Academic Press, London, pp. 205–225.

Mallatt, J., Blatt, M.R., Draguhn, A., Robinson, D.G. and Taiz, L. (2021) Debunking a myth: Plant consciousness. *Protoplasma*, 258, 459–476.

Mallet, J. (2012) The struggle for existence. How the notion of carrying capacity, *K*, obscures the links between demography, Darwinian evolution and speciation. *Evolutionary Ecology Research*, 14, 627–665.

Malt, J. and Lank, D. (2007) Temporal dynamics of edge effects on nest predation risk for the marbled murrelet. *Biological Conservation*, 140(1–2), 160–173.

Malthus, T. (1798) *An Essay on the Principle of Population, as It Affects the Future Improvement of Society, with Remarks on the Speculations of Mr. Godwin, M. Condorcet, and Other Writers*. J. Johnson, London, UK.

Mammola, S., Carmona, C.P., Guillerme, T. and Cardoso, P. (2021) Concepts and applications in functional diversity. *Functional Ecology*, 35, 1869–1885.

Mammola, S., Pavlek, M., Huber, B.A., Isaia, M., Ballarin, F. *et al.* (2022) A trait database and updated checklist for European subterranean spiders. *Scientific Data*, 9, Article 236.

Mammola, S., Adamo, M., Antić, D., Calevo, J., Cancellario, T. *et al.* (2023) Drivers of species knowledge across the tree of life. *eLife*, 12, RP88251.

Mandal, F.B., Chatterjee, C. and Ghosh, A. (2011) Ecosystems and human well-being. *Journal of Environment and Sociobiology*, 8, 25–42.

Mani, M.S. (1968) *Ecology and Biogeography of High Altitude Insects*. Springer, Dordrecht, The Netherlands.

Manjunatha, B.S. and Soni, N.K. (2014) Estimation of age from development and eruption of teeth. *Journal of Forensic Dental Sciences*, 6, 73–76.

Manlick, P.J., Woodford, J.E., Gilbert, J.H., Eklund, D. and Pauli, J.N. (2017) Augmentation provides nominal genetic and demographic rescue for an endangered carnivore. *Conservation Letters*, 10, 178–185.

Mann, A. and Weiss, M. (1996) Hominoid phylogeny and taxonomy: A consideration of the molecular and fossil evidence in an historical perspective. *Molecular Phylogenetics and Evolution*, 5, 169–181.

Mann, R.W., Bass, W.M. and Meadows, L. (1990) Time since death and decomposition of the human body: Variables and observations in case and experimental field studies. *Journal of Forensic Science*, 35, 103–111.

Manne, L.L. and Pimm, S.L. (2001) Beyond eight forms of rarity: Which species are threatened and which will be next? *Animal Conservation*, 4, 221–229.

Manning, J.A. and Goldberg, C.S. (2010) Estimating population size using capture–recapture encounter histories created from point-coordinate locations of animals. *Methods in Ecology and Evolution*, 1, 389–397.

Mantoni, C., Di Musciano, M. and Fattorini, S. (2020) Use of microarthropods to evaluate the impact of fire on soil biological quality. *Journal of Environmental Management*, 266, 110624.

Mantoni, C., Pellegrini, M., Dapporto, L., Del Gallo, M.M., Pace, L. *et al.* (2021a) Comparison of soil biology quality in organically and conventionally managed agro-ecosystems using microarthropods. *Agriculture*, 2021, 11(10), 1022.

Mantoni, C., Tsafack, N., Palusci, E., Di Pietro, S. and Fattorini, S. (2021b) Diversity patterns of dung beetles along a Mediterranean elevational gradient. *Insects*, 12(9), 781.

Marcantonio, M., Voda, R., Da Re, D., Igot, Q., Dennis, R.L.H. *et al.* (2023) The effect of habitat on insect movements: Experimental evidence from wild-caught butterflies. *Insects*, 14, 737.

Marciniak, A., Barański, M.Z., Bayliss, A., Czerniak, L., Goslar, T. *et al.* (2015) Fragmenting times: Interpreting a Bayesian chronology for the Late Neolithic occupation of Çatalhöyük East, Turkey. *Antiquity*, 89, 154–176.

Marcon, E., Zhang, Z. and Hérault, B. (2014) *The Decomposition of Similarity-Based Diversity and Its Bias Correction*. Available at: https://hal.univ-antilles.fr/hal-00989454v2 (accessed 20 May 2025).

Maritan, M., Autin, L., Karr, J., Covert, M.W., Olson, A.J. and Goodsell, D.S. (2022) Building structural models of a whole mycoplasma cell. *Journal of Molecular Biology*, 434, 167351.

Markesteijn, L., Poorter, L., Paz, H., Sack, L. and Bongers, F. (2011) Ecological differentiation in xylem cavitation resistance is associated with stem and leaf structural traits. *Plant, Cell and Environment*, 34, 137–148.

Marks, P. (2023) Martian gamble. *New Scientist*, 3463, 21.

Marov, M.Y. (ed). (2015) The structure of the universe. In: *The Fundamentals of Modern Astrophysics*. Springer, New York, USA, pp. 279–294.

Marques, G.M., Augustine, S., Lika, K., Pecquerie, L., Domingos, T. and Kooijman, S.A.L.M. (2018) The AmP project: Comparing species on the basis of dynamic energy budget parameters. *PLOS Computational Biology*, 14(5), e1006100.

Marquet, P.A., Fernandez, M., Navarrete, S.A. and Valdovinos, C. (2004) Diversity emerging: Toward a deconstruction of biodiversity patterns. In: Lomolino, M.V. and Heaney, L.R. (eds) *Frontiers of Biogeography: New Directions in the Geography of Nature*. Sinauer Associates, Sunderland, USA, pp. 191–209.

Marsden, B.W., Engelhardt, K.A. and Neel, M.C. (2013) Genetic rescue versus outbreeding depression in Vallisneria americana: Implications for mixing seed sources for restoration. *Biological Conservation*, 167, 203–214.

Marshall, A. (1979) Axially-pitched longhouses from New Guinea and Neolithic Europe. *Archaeology and Physical Anthropology in Oceania*, 14, 99–106.

Marshall, D.C. and Cooley, J.R. (2000) Reproductive character displacement and speciation in periodical cicadas, with description of a new species, 13-year *Magicicada neotredecim*. *Evolution*, 54(4), 1313–1325.

Marshall, M. (2023) The elusive origins of life. *New Scientist*, 260, 36–39.

Martin, A.E. and Fahrig, L. (2018) Habitat specialist birds disperse farther and are more migratory than habitat generalist birds. *Ecology*, 99, 2058–2066.

Martín, M.J.R., De Pablo, C.L. and De Agar, P.M. (2006) Landscape changes over time: Comparison of land uses, boundaries and mosaics. *Landscape Ecology*, 21, 1075–1088.

Martin, R.D. (2012) Primates. *Current Biology*, 22, R785–R790.

Martin, S.J., Funch, R.R., Hanson, P.R. and Yoo, E-H. (2018) A vast 4,000-year-old spatial pattern of termite mounds. *Current Biology*, 28, R1292–R1293.

Martin, T.E. (1998) Are microhabitat preferences of coexisting species under selection and adaptive? *Ecology*, 79, 656–670.

Martin, Y.E. and Johnson, E.A. (2012) Biogeosciences survey: Studying interactions of the biosphere with the lithosphere, hydrosphere and atmosphere. *Progress in Physical Geography*, 36, 833–852.

Martinez, E., Duque, P. and Wolff, M. (2006) Succession pattern of carrion-feeding insects in Paramo, Colombia. *Forensic Science International*, 166, 182–189.

Martínez, M.L., Váazquez, G., Favila, M.E. and Álvarez-Molina, L. (2019) Variation in different measures of diversity during primary succession on a tropical coastal dune. *Plant Ecology and Diversity*, 12, 489–505.

Martinez, N.D. (1994) Scale-dependent constraints on food-web structure. *American Naturalist*, 144, 935–953.

Martínez-Roldán, H., Pérez-Crespo, M.J. and Lara, C. (2024) Unraveling habitat-driven shifts in alpha, beta, and gamma diversity of hummingbirds and their floral resource. *PeerJ*, 12, e17713.

Martínez-Sastre, R., Peña, R., González-Ibáñez, A., García, D. and Miñarro, M. (2021) Top-down and bottom-up regulation of codling moth populations in cider apple orchards. *Crop Protection*, 2021, 105545.

Martinko, E.A., Hagen, R.H. and Griffith, J.A. (2006) Successional change in the insect community of a fragmented landscape. *Landscape Ecology*, 21, 711–721.

Maser, C., Anderson, R.G., Cromack, K., Williams, J.T. and Martin, R.E. (1979) Dead and down woody material. In: Thomas, J.W. (ed.) *Wildlife Habitats in Managed Forests of the Blue Mountains of Oregon and Washington*. Agriculture Handbook No. 553. U.S. Department of Agriculture, Forest Service, Washington, DC, USA.

Maslow, A.H. (1943) A theory of human motivation. *Psychological Review*, 50, 370–396.

Maslow, A.H. (1987) *Motivation and Personality*, 3rd edn. Pearson Education, Delhi, India.

Mason, N.W.H. and Wilson, J.B. (2006) Mechanisms of species coexistence in a lawn community: Mutual corroboration between two independent assembly rules. *Community Ecology*, 7, 109–116.

Massenberg, J.R., Schiller, J., Schröter-Schlaack, C. and Schröter-Schlaack, C. (2022) Towards a holistic approach to rewilding in cultural landscapes. *People and Nature*, 5, 1–12.

Masson-Boivin, C. and Sachs, J.L. (2018) Symbiotic nitrogen fixation by rhizobia – The roots of a success story. *Current Opinion in Plant Biology*, 44, 7–15.

Mathakutha, R., Steyn, C., Le Roux, P., Greve, M., Chown, S.L. *et al.* (2019) Invasive species differ in key functional traits from native and non-invasive alien plant species. *Journal of Vegetation Science*, 30, 994–1006.

Mathewson, H.A. and Morrison, M.L. (2015) The misunderstanding of habitat. In: Morrison, M.L. and Mathewson, H.A. (eds) *Wildlife Habitat Conservation. Concepts, Challenges, and Solutions*. Johns Hopkins University Press, Baltimore, USA, pp. 3–8.

Mathieu, R., Freeman, C. and Aryal, J. (2007) Mapping private gardens in urban areas using object-oriented techniques and very high-resolution satellite imagery. *Landscape and Urban Planning*, 81, 179–192.

Matter, S.F. and Roland, J. (2002) An experimental examination of the effects of habitat quality on the dispersal and local abundance of the butterfly *Parnassius smintheus*. *Ecological Entomology*, 27, 308–316.

Matter, S.F., Roland, J., Keyghobadi, N. and Sabourin, K. (2003) The effects of isolation, habitat area and resources on the abundance, density and movement of the butterfly *Parnassius smintheus*. *American Midland Naturalist*, 150, 26–36.

Matthews, T.J. and Whittaker, R.J. (2014a) Fitting and comparing competing models of the species abundance distribution: Assessment and prospect. *Frontiers of Biogeography*, 6, 2.

Matthews, T.J. and Whittaker, R.J. (2014b) Neutral theory and the species abundance distribution: Recent developments and prospects for unifying niche and neutral perspectives. *Ecology and Evolution*, 4, 2263–2277.

Matthews, T.J. and Whittaker, R.J. (2015) On the species abundance distribution in applied ecology and biodiversity management. *Journal of Applied Ecology*, 52, 443–454.

Matthews, T.J., Borregaard, M.K., Ugland, K.I., Borges, P.A.V., Rigal, F. *et al.* (2014) The gambin model provides a superior fit to species abundance distributions with a single free parameter: Evidence, implementation and interpretation. *Ecography*, 37, 1002–1011.

Matthews, T.J., Cotte-Jones, H.E.W. and Whittaker, R.J. (2015) Quantifying and interpreting nestedness in habitat islands: A synthetic analysis of multiple datasets. *Diversity and Distributions*, 21, 392–404.

Matthews, T.J., Sadler, J.P., Kubota, Y., Woodall, C.W. and Pugh, T.A.M. (2019) Systematic variation in North American tree species abundance distributions along macroecological climatic gradients. *Global Ecology and Biogeography*, 28, 601–611.

Matthews, T.J., Triantis, K. and Whittaker, R. (eds) (2021) *The Species-Area Relationship: Theory and Application*. Cambridge University Press, Cambridge, UK.

Matthies, D., Bräuer, I., Maibom, W. and Tscharntke, T. (2004) Population size and the risk of local extinction: Empirical evidence from rare plants. *Oikos*, 105, 481–488.

Mattick, F. (1964) Übersicht über die Florenreiche und Florengebiete der Erde. In: Melchior, H. (ed.) *Englers Syllabus der Pflanzenfamilien. II*, 11th edn. Gebrüder Borntraeger Verlagsbuchhandlung, Berlin, Germany.

Matuszewski, S. and Szafałowicz, M. (2013) Temperature-dependent appearance of forensically useful beetles on carcasses. *Forensic Science International*, 229, 92–99.

Matuszewski, S., Hall, M.J.R., Moreau, G., Schoenly, K.G., Tarone, A.M. and Villet. M.H. (2020) Pigs vs people: The use of pigs as analogues for humans in forensic entomology and taphonomy research. *International Journal of Legal Medicine*, 134, 793–810.

Mauriac, M. (2011) Lascaux. The history of the discovery of an outstanding decorated cave. *Adoranten*, 2011, 5–25. Available at: www.rockartscandinavia.com/images/articles/lascaux_a11.pdf (accessed 20 May 2025).

Maus, V., Giljum, S., Gutschlhofer, J., da Silva, D.M., Probst, M. *et al.* (2020) A global-scale data set of mining areas. *Scientific Data*, 7, 289.

May, R. and McLean, A.R. (eds) (2007) *Theoretical Ecology: Principles and Applications*, 3rd edn. Oxford University Press, Oxford, UK.

May, R.M. (1988) How many species are there on earth? *Science*, 241, 1441–1449.

May, R.M. (1972) Will a large complex system be stable? *Nature*, 238, 413–414.

May, R.M. (1973a) On the relationships among various types of population models. *American Naturalist*, 107, 46–57.

May, R.M. (1973b) *Stability and Complexity in Model Ecosystems*. Princeton University Press, Princeton, USA.

May, R.M. (1974) On the theory of niche overlap. *Theoretical Population Biology*, 5, 297–332.

May, R.M. (1975) Patterns of species abundance and diversity. In: Cody, M.L. and Diamond, J.M. (eds) *Ecology and Evolution of Communities*. Harvard University Press, Cambridge, USA, pp. 81–120.

May, R.M. (1976a) Models for two interacting populations. In: May, R.M. (ed.) *Theoretical Ecology: Principles and Applications*. Blackwell Scientific Publications, Oxford, UK, pp. 49–70.

May, R.M. (1976b) Models for single populations. In: May, R.M. (ed.) *Theoretical Ecology: Principles and Applications*. Blackwell Scientific Publications, Oxford, UK, pp. 4–25.

May, R.M. (1978) Host-parasitoid systems in patchy environments: A phenomenological model. *Journal of Animal Ecology*, 47, 833–843.

Mayfield, M.M. and Levine, J.M. (2010) Opposing effects of competitive exclusion on the phylogenetic structure of communities. *Ecology Letters*, 13(9), 1085–1093.

Mayne, R., Roberts, N., Phillips, N., Weerasekera, R. and Adamatzky, A. (2023) Propagation of electrical signals by fungi. *BioSystems*, 229, 104933.

Mazalla, L. and Diekmann, M. (2022) Regression to the mean in vegetation science. *Journal of Vegetation Science*, 33, e13117.

Mazel, F., Wüest, R.O., Gueguen, M., Renaud, J., Ficetola, G.F. *et al.* (2017) The geography of ecological niche evolution in mammals. *Current Biology*, 27, 1369–1374.

Mazziotta, A., Granath, G., Rydin, H., Bengtsson, F. and Norberg, J. (2019) Scaling functional traits to ecosystem processes: Towards a mechanistic understanding in peat mosses. *Journal of Ecology*, 107, 843–859.

McCallum, M.L. (2015) Vertebrate biodiversity losses point to a sixth mass extinction. *Biodiversity and Conservation*, 24, 2497–2519.

McClain, C.R., Balk, M.A., Benfield, M.C., Branch, T.A., Chen, C. *et al.* (2015) Sizing ocean giants: Patterns of intraspecific size variation in marine megafauna. *PeerJ*, 3, e715.

McClintock, B.T. and Thomas, L. (2020) Estimating abundance or occupancy from unmarked populations. In: Murray, D.L. and Sandercock, B. (eds) *Population Ecology in Practice*. Wiley-Blackwell, London, UK, pp. 49–78.

McCormack, J.E., Huang, H. and Knowles, L.L. (2009) Sky islands. In: Gillespie, R. and Clague, D. (eds) *Encyclopedia of Islands*. University of California Press, Berkeley, USA, pp. 839–843.

McCracken, G.F. (1984) Communal nursing in Mexican free-tailed bat maternity colonies. *Science*, 223, 1090–1091.

McCullough, M.E., Kurzban, R. and Tabak, B.A. (2012) Cognitive systems for revenge and forgiveness. *Behavioural and Brain Sciences*, 36, 1–58.

McDiarmid, R.W., Foster, M.S., Guyer, C., Whitfield Gibbons, J. and Chernoff, N. (eds) (2012) *Reptile Biodiversity: Standard Methods for Inventory and Monitoring*. University of California Press, Berkeley, USA.

McGill, B.J. (2003) A test of the unified neutral theory of biodiversity. *Nature*, 422, 881–885.

McGill, B.J., Enquist, B.J., Weiher, E. and Westoby, M. (2006) Rebuilding community ecology from functional traits. *Trends in Ecology and Evolution*, 21, 178–185.

McGill, B.J., Etienne, R.S., Gray, J.S., Alonso, D., Anderson, M.J. *et al.* (2007) Species abundance distributions: Moving beyond single prediction theories to integration within an ecological framework. *Ecology Letters*, 10, 995–1015.

McGill, B.J., Chase, J., Hortal, J., Overcast, I., Rominger, A. *et al.* (2019) Unifying Macroecology and Macroevolution to answer fundamental questions about biodiversity. *Global Ecology and Biogeography*, 28, 1925–1936.

McGinnies, W.G. (1981) *Discovering the Desert. Legacy of the Carnegie Desert Botanical Laboratory*. University of Arizona Press, Tucson, USA.

McGlinn, D.J., Engel, T., Blowes, S.A., Gotelli, N.J., Knight, T.M. *et al.* (2021) A multiscale framework for disentangling the roles of evenness, density, and aggregation on diversity gradients. *Ecology*, 102(2), e03233.

McGowan, A., Sharp, S.P., Simeoni, M. and Hatchwell, B. (2006) Competing for position in the communal roosts of long-tailed tits. *Animal Behaviour*, 72, 1035–1043.

McIntire, E.J.B. and Fajardo, A. (2014) Facilitation as a ubiquitous driver of biodiversity. *New Phytologist*, 201, 403–416.

McIntosh, A.R., Greig, H.S. and Howard, S. (2022) Regulation of open populations of a stream insect through larval density dependence. *Journal of Animal Ecology*, 91, 1582–1595.

McKendrick, A.G. and Kesava Pai, M. (1911) XLV.—The rate of multiplication of micro-organisms: A mathematical study. *Proceedings of the Royal Society of Edinburgh*, 31, 649–653.

McKiernan, J. (2024) Bill to ban trophy hunting imports back in front of MPs. Available at: www.bbc.co.uk/news/uk-politics-68626052 (accessed 20 May 2025).

McLain, D.K. (1981) Resource partitioning by three species of hemipteran herbivores on the basis of host plant density. *Oecologia*, 48, 414–417.

McLester, E., Sweeney, K., Stewart, F.A. and Piel, A.K. (2018) Leopard (*Panthera pardus*) predation on a red-tailed monkey (*Cercopithecus ascanius*) in the Issa Valley, western Tanzania. *Primates*, 60, 15–19.

McMichael, A.J., Campbell-Lendrum, D.H., Corvalán, C.F., Ebi, K.L., Githeko, A.K. *et al.* (2003) *Climate Change and Human Health Risks and Responses.* World Health Organization, Geneva, Switzerland.

McNab, B.K. (1963) Bioenergetics and the determination of home range size. *American Naturalist*, 97, 133–140.

Meck, J.R., Jones, M.T., Willey, L.L. and Mays, J.D. (2020) Autecological study of Gulf Coast box turtles (*Terrapene carolina major*) in the Florida Panhandle, USA, reveals unique spatial and behavioral characteristics. *Herpetological Conservation and Biology*, 15, 293–305.

Medellin, R.A., Wiederholt, R. and Lopez-Hoffman, L. (2017) Conservation relevance of bat caves for biodiversity and ecosystem services. *Biological Conservation*, 211, 45–50.

Meehl, G.A., Stocker, T.F., Collins, W.D., Friedlingstein, P., Gaye, A.T. *et al.* (2007) Global climate projections. In: Solomon, S., Qin, D., Manning, M., Chen, Z., Marquis, M. *et al.* (eds) *Climate Change 2007: The Physical Science Basis. Contribution of Working Group I to the Fourth Assessment Report of the Intergovernmental Panel on Climate Change.* Cambridge University Press, Cambridge, UK.

Meek, W.R., Burman, P.J., Nowakowski, M., Sparks, T.H., Hill, R.A. *et al.* (2009) Habitat does not influence breeding performance in a long-term barn owl *Tyto alba* study. *Bird Study*, 56, 369–380.

Meeker, D.O. Jr and Merkel, D.L. (1984) Climax theories and a recommendation for vegetation classification: A viewpoint. *Journal of Range Management*, 37(5), 427–430.

Meier, R. and Dikow, T. (2004) Significance of specimen databases from taxonomic revisions for estimating and mapping the global species diversity of invertebrates and repatriating reliable specimen data. *Conservation Biology*, 18, 478–488.

Meineke, E.K. and Davies, T.J. (2019) Museum specimens provide novel insights into changing plant–herbivore interactions. *Philosophical Transactions of the Royal Society B: Biological Sciences*, 374(1763), 20170393.

Mellanby, K. (1967) Food and activity in the mole *Talpa Europaea*. *Nature*, 215, 1128–1130.

Melo, A., Pereira, R.A.S., Santos, A.J., Shepherd, G.J., Machado, G. *et al.* (2003) Comparing species richness among assemblages using sample units: Why not use extrapolation methods to standardize different sample size? *Oikos*, 101, 398–410.

Memmott, J., Waser, N.M. and Price, M.V. (2004) Tolerance of pollination networks to species extinctions. *Proceedings of the Royal Society B: Biological Sciences*, 271, 2605–2611.

Menéndez, R. (2007) How are insects responding to global warming? *Tijdschrift voor Entomologie*, 150, 355–365.

Menéndez, R., and Gutiérrez, D. (1999) Heterotrophic succession within dung-inhabiting beetle communities in northern Spain. *Acta Oecologica*, 20(5), 527–535.

Menger, J., Santorelli Jr, S., Emilio, T., Magnusson, W.E. and Anciães, M. (2024) Palms predict the distributions of birds in southwestern Amazonia and are potential surrogates for land-use planning by citizen scientists. *Biodiversity and Conservation*, 33(10), 2911–2924.

Merriam, C.H. and Stejneger, L.H. (1890) Results of a biological survey of the San Francisco Mountain region and the desert of the Little Colorado, Arizona. *North American Fauna*, 3, 1–136.

Mertl-Millhollen, A.S. (2006) Scent marking as resource defense by female *Lemur catta*. *American Journal of Primatology*, 68, 605–621.

Meskell, L., Nakamura, C., King, R. and Farid, S. (2008) Figured Lifeworlds and depositional practices at Çatalhöyük. *Cambridge Archaeological Journal*, 18, 139–161.

Mesquita, D.O., Costa, G.C., Figueredo, A.S., França, F.G.R., Garda, A.A. *et al.* (2015a) The autecology of *Anolis brasiliensis* (Squamata, Dactyloidae) in a Neotropical savanna. *Herpetological Journal*, 25, 233–244.

Mesquita, R. de C., Massoca, P.E.D., Jakovac, C.C., Bentos, T.V. and Williamson, G.B. (2015b) Amazon rain forest succession: Stochasticity or land-use legacy? *BioScience*, 65, 849–861.

Meyer, V.W., Braack, L.E.O., Biggs, H.C. and Ebersohn, C. (1999) Distribution and density of termite mounds in the northern Kruger National Park, with specific reference to those constructed by *Macrotermes* Holmgren (Isoptera: Termitidae). *African Entomology*, 7, 123–130.

Michel, J., Ebert, D. and Hall, M.D. (2016) The transgenerational impact of population density signals on host-parasite interactions. *BMC Evolutionary Biology*, 16, 254.

Middleton-Welling, J., Dapporto, L., García-Barros, E., Wiemers, M., Nowicki, P. *et al.* (2020) A new comprehensive trait database of European and Maghreb butterflies, Papilionoidea. *Scientific Data*, 7, 351.

Miele, V. and Matias, C. (2017) Revealing the hidden structure of dynamic ecological networks. *Royal Society Open Science*, 4, 170251.

Miles, C. (2015) The earwig collection (Dermaptera) of the Manchester Museum, UK, with a complete type catalogue. *European Journal of Taxonomy*, 141, 1–138.

Miles, J. (1987) Vegetation succession: Past and present perceptions. In: Gray, A.J., Crawley, M.J. and Edwards, P.J. (eds) *Colonization, Succession and Stability.* Blackwell Scientific Publications, Oxford, UK, pp. 1–29.

Millar, J.D., Mason, H. and Kidd, L. (2022) What is Q methodology? *Evidence Based Nursing*, 25, 77–78.

Millennium Ecosystem Assessment (2005) *Ecosystems and Human Well-being: Synthesis.* Island Press Washington, DC, USA.

Miller, E. and Coll, M. (2010) Spatial distribution and deviations from the IFD when animals forage over large resource patches. *Behavioral Ecology*, 21, 927–935.

Miller, J.E.D., Damschen, E.I. and Ives, A.R. (2019) Functional traits and community composition: A comparison among community-weighted means, weighted correlations, and multilevel models. *Methods in Ecology and Evolution*, 10, 415–425.

Miller, S.L. (1953) A production of amino acids under possible primitive Earth conditions. *Science*, 117, 528–529.

Miller, T.E.X. (2007) Demographic models reveal the shape of density dependence for a specialist insect herbivore on variable host plants. *Journal of Animal Ecology*, 76, 722–729.

Millhauser, J.K. and Earle, T. (2022) Biodiversity and the human past: Lessons for conservation biology. *Biological Conservation*, 272, 109599.

Mills, D. (1989) *Ecology and Management of Atlantic Salmon*. Chapman and Hall, London, UK.

Mills, L.S., Soulé, M.E. and Doak, D.F. (1993) The keystone species concept in ecology and conservation. *BioScience*, 43, 219–224.

Milner-Gulland, E.J., Fryxell, J.M. and Sinclair, A.R.E. (2011) *Animal Migration: A Synthesis*. Oxford University Press, New York, USA.

Milright, R.D.P. (2003) Migration routes, breeding areas and between-winter recurrence of nominate Redwings *Turdus iliacus iliacus* revealed by recoveries of winter ringed birds. *Ringing and Migration*, 21, 183–192.

Milton, K. and May, M.L. (1976) Body weight, diet and home range area in primates. *Nature*, 259, 459–462.

Mimura, N. (2013) Sea-level rise caused by climate change and its implications for society. *Proceedings of the Japan Academy Series B Physical and Biological Sciences*, 89, 281–301.

Minor, J. (2007) Ecotone. In: Robbins, P. (ed.) *Encyclopedia of Environment and Society*. SAGE Publications, Thousand Oaks, USA.

Miranda Cebrián, H., Font, X., Roquet, C., Pizarro Gavilán, M. and García, M.B. (2022) Phylogenetic patterns of rarity and vulnerability in the flora of a temperate mountain range. *Oikos*, 2022, e09365.

Mirtl, M., Borer, E.T., Djukic, I., Forsius, M., Haubold, H. *et al.* (2018). Genesis, goals and achievements of Long-Term Ecological Research at the global scale: A critical review of ILTER and future directions. *Science of the Total Environment*, 626, 1439–1462.

Mishra, R.K., Mohammad, N. and Roychoudhury, N. (2016) Soil pollution: Causes, effects and control. *Van Sangyan*, 3, 1–14.

Misirlioğlu, M., Reynolds, J.W., Stojanović, M., Trakić, T., Sekulić, J. *et al.* (2023) Earthworms (Clitellata, Megadrili) of the world: an updated checklist of valid species and families, with notes on their distribution. *Zootaxa*, 5255, 417–438.

Mistri, M., Ceccherelli, V.U. and Rossi, R. (2000) Taxonomic distinctness and diversity measures: Responses in lagoonal macrobenthic communities. *Italian Journal of Zoology*, 67(3), 297–301.

Mittermeier, R.A., Rylands, A.B. and Wilson, D.E. (2013) *The Mammals of the World. Pt 3: Primates*. Lynx Edicions, Barcelona, Spain.

Mizroch, S.A., Rice, D.W. and Breiwick, J.M. (1984) The blue whale, *Balaenoptera musculus*. *Marine Fisheries Review*, 46, 15–19.

Mizumoto, N., Bourguignon, T. and Kanao, T. (2022) Termite nest evolution fostered social parasitism by termitophilous rove beetles. *Evolution*, 76, 1064–1072.

Mlot, N.J., Tovey, C.A. and Hua, D.L. (2011) Fire ants self-assemble into waterproof rafts to survive floods. *Proceedings of the National Academy of Sciences of the United States of America*, 108, 7669–7673.

Mobley, K.B., Aykanat, T., Czorlich, Y., House, A., Kurko, J. *et al.* (2021) Maturation in Atlantic salmon (*Salmo salar*, Salmonidae): A synthesis of ecological, genetic, and molecular processes. *Review in Fish Biology and Fisheries*, 31, 523–571.

Moe, C.L. and Rheingans, R.D. (2006) Global challenges in water, sanitation and health. *Journal of Water and Health*, 04(Suppl), 41–57.

Moffett, M.W. (2011) Ants and the art of war. *Scientific American*, 305(6), 84–89.

Mohr, C.O. (1943) Cattle droppings as ecological units. *Ecological Monographs*, 13, 275–298.

Moilanen, A. and Nieninen, M. (2002) Simple connectivity measures in spatial ecology. *Ecology*, 83, 1131–1145.

Moleón, M., Sánchez-Zapata, J.A., Sebastián-González, E. and Owen-Smith, N. (2015) Carcass size shapes the structure and functioning of an African scavenging assemblage. *Oikos*, 124, 1391–1403.

Monastersky, R. (2015) First atomic blast proposed as start of Anthropocene. *Nature News*. Available at: https://doi.org/10.1038/nature.2015.16739 (accessed 21 May 2025).

Monfasani, J. (2015) *Renaissance Humanism, from the Middle Ages to Modern Times*. Routledge, New York, USA.

Mongkolsamrit, S., Kobmoo, N., Tasanathai, K., Khonsanit, A., Noisripoom W. *et al.* (2012) Life cycle, host range and temporal variation of *Ophiocordyceps unilateralis/Hirsutella formicarum* on Formicine ants. *Journal of Invertebrate Pathology*, 111, 217–224.

Montano, S., Fattorini, S., Parravicini, V., Berumen, M.L., Galli, P. *et al.* (2017) Corals hosting symbiotic hydrozoans are less susceptible to predation and disease. *Proceedings of the Royal Society B: Biological Sciences*, 284, 20172405.

Montgomery, G.A., Belitz, M.W., Guralnick, R.P. and Tingley, M.W. (2021) Standards and best practices for monitoring and benchmarking insects. *Frontiers in Ecology and Evolution*, 8, 579193.

Montgomery, G.A., Dunn, R.R., Fox, R., Jongejans, E., Leather, S.R. *et al.* (2020) Is the insect apocalypse upon us? How to find out. *Biological Conservation*, 241, 108327.

Moody, E.R.R., Álvarez-Carretero, S., Mahendrarajah, T.A., Clark, J.W., Betts, H.C. *et al.* (2024) The nature of the last universal common ancestor and its impact on the early Earth system. *Nature Ecology and Evolution*, 8, 1654–1666.

Moore, W., Meyer, W.M. 3rd, Eble, J.A., Franklin, K., Wiens, J.F. and Brusca, R.C. (2013) Introduction to the Arizona Sky Island Arthropod Project (ASAP): Systematics, biogeography, ecology, and population genetics of arthropods of the Madrean Sky Islands. *Proceedings of the Rocky Mountain Research Station*, 2013, 144–168.

Moorhead, S. (2013) *History of Roman Coinage in Britain*. Greenlight Publishing, Essex, UK.

Moorjani, P., Amorim, C.E.G., Arndt, P.F. and Przeworski, M. (2016) Variation in the molecular clock of primates. *Proceedings of the National Academy of Sciences of the United States of America*, 113, 10607–10612.

Moose, R.A., Schigel, D., Kirby, L.J. and Shumskaya, M. (2019) Dead wood fungi in North America: An insight into research and conservation potential. *Nature Conservation*, 32, 1–17.

Mora, C., Tittensor, D.P., Adl, S., Simpson, A.G.B. and Worm, B. (2011) How many species are there on earth and in the ocean? *PLOS Biology*, 9, e1001127.

Mora, M., Mahnert, A., Koskinen, K., Pausan, M.R., Oberauner-Wappis, L. *et al.* (2016) Microorganisms in confined habitats: Microbial monitoring and control of intensive care units, operating rooms, cleanrooms and the international space station. *Frontiers in Microbiology*, 7, 1573.

Moradi, H., Prado, P.I., Oldeland, J., Borges, P.A.V. and Fattorini, S. (2025) Use of cover data to model species abundance distributions through continuous probability functions. *Scientific Reports*, 15, 11361.

Moreno, C.E., Castillo-Campos, G. and Verdú, J.R. (2009) Taxonomic diversity as complementary information to assess plant species diversity in secondary vegetation and primary tropical deciduous forest. *Journal of Vegetation Science*, 20, 935–943.

Moreno de las Heras, M. and Gallart, F. (2018) The origin of badlands. In: Moreno de las Heras, M. and Gallart, F. (eds) *Badlands Dynamics in the Context of Global Change*. Elsevier, Amsterdam, The Netherlands, pp. 27–59.

Moretti, M., Dias, A.T.C., de Bello, F., Altermatt, F., Chown, S.L. *et al.* (2017) Handbook of protocols for standardized measurement of terrestrial invertebrate functional traits. *Functional Ecology*, 31, 558–567.

Morlon, H., White, E.P., Etienne, R.S., Green, J.L., Ostling, A., *et al.* (2009) Taking species abundance distributions beyond individuals. *Ecology Letters*, 12, 488–501.

Morris, D., Davidson, D.L. and Krebs, C.J. (2000) Measuring the ghost of competition: Insights from density-dependent habitat selection on the coexistence and dynamics of lemmings. *Evolutionary Ecology Research*, 2, 41–67.

Morris, D.W. (1987a) Spatial scale and the cost of density-dependent habitat selection. *Evolutionary Ecology*, 1, 379–388.

Morris, D.W. (1987b) Ecological scale and habitat use. *Ecology*, 68, 362–369.

Morris, D.W. (1988) Habitat-dependent population regulation and community structure. *Evolutionary Ecology*, 2, 253–269.

Morris, D.W. (2003) How can we apply theories of habitat selection to wildlife conservation and management? *Wildlife Research*, 30, 303–319.

Morris, E.K., Caruso, T., Buscot, F., Fischer, M., Hancock, C. *et al.* (2014) Choosing and using diversity indices: Insights for ecological applications from the German Biodiversity Exploratories. *Ecology and Evolution*, 4, 3514–3524.

Morrison, M.L., Marcot, B.G. and Mannan, R.W. (2006) *Wildlife-Habitat Relationships: Concepts and Applications*, 3rd edn. Island Press, Washington, DC, USA.

Morrison, R.E. (2021) Exact reduction of the generalized Lotka–Volterra equations via integral and algebraic substitutions. *Computation*, 9, 49.

Morrone, J.J. (2002) Biogeographical regions under track and cladistic scrutiny. A comment on C. Barry Cox (2001) The biogeographical regions reconsidered. *Journal of Biogeography*, 28, 511–523.

Morrone, J.J. (2009) *Evolutionary Biogeography: An Integrative Approach with Case Studies*. Columbia University Press, New York, USA.

Morrone, J.J. (2014) Biogeographical regionalisation of the Neotropical region. *Zootaxa*, 3782, 1–110.

Morrone, J.J. (2015a) Biogeographical regionalization of the world: A reappraisal. *Australian Systematic Botany*, 28, 81–81.

Morrone, J.J. (2015b) Biogeographical regionalisation of the Andean region. *Zootaxa*, 3936, 207–236.

Morrone, J.J. (2018) The spectre of biogeographical regionalization. *Journal of Biogeography*, 45, 282–288.

Morrone, J.J. and Ebach, M.C. (2022) Toward a terrestrial biogeographical regionalisation of the world: Historical notes, characterisation and area nomenclature. *Australian Systematic Botany*, 35, 89–126.

Morse, D.H. (2008) Structure and foraging patterns of flocks of tits and associated species in an English woodland during winter. *Ibis*, 120, 298–312.

Morton, A.C.G. (1985) The population biology of an insect with a restricted distribution: Cupido minimus *Fuessly (Lepidoptera: Lycaenidae)*. PhD thesis, University of Southampton, Southampton, UK.

Moss, S. (2018) *The Wren: A Biography*. Square Peg, London, UK.

Mouillot, D., Graham, N.A.J., Villéger, S., Mason, N.W.H. and Bellwood, D.R. (2013) A functional approach reveals community responses to disturbances. *Trends in Ecology and Evolution*, 28, 167–177.

Mouillot, D., Stubbs, W., Faure, M., Dumay, O., Tomasini, J.A. *et al.* (2005) Niche overlap estimates based on quantitative functional traits: A new family of non-parametric indices. *Oecologia*, 145, 345–353.

Mouillot, D., Villéger, S., Parravicini, V., Kulbicki, M., Arias-González, J.E. *et al.* (2014). Functional over-redundancy and high functional vulnerability in global fish faunas on tropical reefs. *Proceedings of the National Academy of Sciences of the United States of America*, 111, 13757–13762.

Mouquet, N., Gravel, D., Massol, F. and Calcagno, V. (2013) Extending the concept of keystone species to communities and ecosystems. *Ecology Letters*, 16, 1–8.

Mouritsen, H. (2013) The magnetic senses. In: Galizia, C.G. and Lledo, P.M. (eds) *Neurosciences – From Molecule to Behavior: A University Textbook*. Springer-Verlag, Berlin, Germany, pp. 427–443.

Movius, H.L. Jr. (1950) A wooden spear of third interglacial age from Lower Saxony. *Southwestern Journal of Anthropology*, 6, 139–142.

Mucina, L. (2019) Biome: Evolution of a crucial ecological and biogeographical concept. *New Phytologist*, 222, 97–114.

Mueller, G.M. and Schmit, J.P. (2007) Fungal biodiversity: What do we know? What can we predict? *Biodiversity and Conservation*, 16, 1–5.

Mugume, S., Isabirye-Basuta, G., Otali, E., Reyna-Hurtado, R. and Chapman, C.A. (2015) How do human activities influence the status and distribution of terrestrial mammals in forest reserves? *Journal of Mammalogy*, 96, 998–1004.

Müller, F., Baessler, C., Schubert, H. and Klotz, S. (eds) (2010) *Long-term Ecological Research: Between Theory and Application*. Springer Science + Business Media, Dordrecht, The Netherlands.

Müller, P. (1986) *Biogeography*. Harper and Row, New York, USA.

Mundra, S., Kjønaas, O.J., Morgado, L.N., Krabberød, A.K., Ransedokken, Y. and Kauserud, H. (2021) Soil depth matters: Shift in composition and inter-kingdom co-occurrence patterns of microorganisms in forest soils. *FEMS Microbiology Ecology*, 97(3), fiab022.

Munian, K., Azman, S.M., Ruzman, N.A., Fauzi, N.F.M. and Zakaria, A.N. (2020) Diversity and composition of volant and non-volant small mammals in northern Selangor State Park and adjacent forest of Peninsular Malaysia. *Biodiversity Data Journal*, 8, e50304.

Muñoz-Amezcua, C.E., Tapia-McClung, H. and Rao, D. (2023) When a glimpse is enough: Partial mimicry of jumping spiders by insects. *Behavioural Processes*, 213, 104956.

Muñoz-Galicia, D., Lara, C., Castillo-Guevara, C., Cuautle, M. and Rodríguez-Flores, C. (2023) Impacts of land use change on native plant-butterfly interaction networks from central Mexico. *PeerJ*, 11, e16205.

Murat, D., Byrne, M. and Komeili, A. (2010) Cell biology of prokaryotic organelles. *Cold Spring Harbor Perspectives in Biology*, 2, a000422.

Murcia, C. (1995) Edge effects in fragmented forests: Implications for conservation. *Trends in Ecology and Evolution*, 10(2), 58–62.

Murray, D.L. and Sandercock, B.K. (2020) *Population Ecology in Practice*. John Wiley and Sons, Chichester, UK.

Murray, K.A. and Baselga, A. (2015) Reply to Chen and Schmera: Partitioning beta diversity into replacement and nestedness-resultant components is not controversial. *Proceedings of the National Academy of Sciences of the United States of America*, 112(52), E7162.

Murray, S.N., Ambrose, R. and Dethier, M.N. (2006) *Monitoring Rocky Shores*. University of California Press, Berkeley, USA.

Murton, R.K. (1965) *The Wood Pigeon*. HarperCollins, London, UK.

Murton, R.K., Isaacson, A.J. and Westwood, N.J. (1971) The significance of gregarious feeding behaviour and adrenal stress in a population of Wood-pigeons *Columba palumbus*. *Journal of Zoology*, 165, 53–84.

Muscatine, L. and Porter, J.W. (1977) Reef corals: Mutualistic symbioses adapted to nutrient-poor environments. *Bioscience*, 27, 454–460.

Mutshinda, C.M., Mishra, A., Finkel, Z.V. and Irwin, A.J. (2023) Density regulation amplifies environmentally induced population fluctuations. *PeerJ*, 11, e14701.

Muvengwi, J. and Witkowski, E.T.F. (2020) Cascading effects of termite mounds in African savannas. *New Zealand Journal of Botany*, 58, 167–193.

Myers, J.H. (2018) Population cycles: Generalities, exceptions and remaining mysteries. *Proceedings of the Royal Society B: Biological Sciences*, 285, 20172841.

Myhrvold, N.P., Baldridge, E., Chan, B., Sivam, D., Freeman, D.L. and Ernest, S.K.M. (2015) An amniote life-history database to perform comparative analyses with birds, mammals, and reptiles. *Ecology*, 96, 3109.

Myler, R. and Cook, L.M. (2010) Butterfly diversity from a sequential survey. *Entomologist's Record and Journal of Variation*, 122, 23–27.

Naeem, S. (1998) Species redundancy and ecosystem reliability. *Conservation Biology*, 12, 39–45.

Naiman, R.J., Melillo, J.M. and Hobbie, J.E. (1986) Ecosystem aleration of boreal forest streams by beaver (*Castor canadensis*). *Ecology*, 67, 1254–1269.

Nalesso, R.C., Paresque, K., Piumbini, P.P., Tonini, J.F.R., Almeida, L.G. and Níckel, V.M. (2008) Oyster spat recruitment in Espírito Santo state, Brazil, using recycled materials. *Brazilian Journal of Oceanography*, 56, 281–288.

Nalmpanti, M., Chrysafi, A., Meeuwig, J.J. and Tsikliras, A.C. (2023) Monitoring marine fishes using underwater video techniques in the Mediterranean Sea. *Reviews in Fish Biology and Fisheries*, 33, 1291–1310.

Naranjo-Ortiz, M.A. and Gabaldón, T. (2019) Fungal evolution: Diversity, taxonomy and phylogeny of the Fungi. *Biological Reviews*, 94, 2101–2137.

Nash, D.R., Als, T.D., Maile, R., Jones, G.R., and Boomsma, J.J. (2008) A mosaic of chemical coevolution in a large blue butterfly. *Science*, 319, 88–90.

Naskrecki, P. (2013) Endangered terrestrial invertebrates. In: Levin, S.A. (ed.) *Encyclopedia of Biodiversity*, vol. 3, Academic Press, San Diego, USA, pp. 219–227.

Nayfach, S., Roux, S., Seshadri, R., Udwary, D., Varghese, N. *et al.* (2021) A genomic catalog of Earth's microbiomes. *Nature Biotechnology*, 39(4), 499–509.

Nei, M., Maruyama, T. and Chakraborty, R. (1975) The bottleneck effect and genetic variability in populations. *Evolution*, 29(1), 1–10.

Nel, A., Prokop, J., Pecharová, M., Engel, M.S. and Garrouste, R. (2018) Palaeozoic giant dragonfies were hawker predators. *Scientific Reports*, 8, 12141.

Nemergut, D.R., Anderson, S.P., Cleveland, C.C., Martin, A.P., Miller, A.E. *et al.* (2007) Microbial community succession in unvegetated, recently-deglaciated soils. *Microbiology Ecology*, 53, 110e122.

Nentwig, W. (2007) Human environmental impact in the Paleolithic and Neolithic. In: Henke, W. and Tattersall, I. (eds) *Handbook of Paleoanthropology*. Springer, Berlin, Germany, pp. 1881–1900.

Neruda, P. and Nerudová, Z. (2000) The Upper Palaeolithic Levallois industry from Hradsko (Melnik District, Czech Republic). *Anthropologie*, 38, 271–281.

Neves, K.G. (2024) Botanic gardens in biodiversity conservation and sustainability: History, contemporary engagements, decolonization challenges, and renewed potential. *Journal of Zoological and Botanical Gardens*, 5(2), 260–275.

Newell, G.E. (1958) An experimental analysis of the behaviour of *Littorina Littorea* (L.) under natural conditions and in the laboratory. *Journal of the Marine Biological Association*, 37, 241– 266.

Newman, K., Buckland, S.T., Morgan, B., King, R., Borchers, D.L. *et al.* (2014) *Modelling Population Dynamics: Model Formulation, Fitting and Assessment Using State-Space Methods*. Springer, London, UK.

Newman, M.E.J. (2006) Modularity and community structure in networks. *Network Science*, 103, 8577–8582.

Newman, M.E.J. (2010) *Networks: An Introduction*. Oxford University Press, Oxford, UK.

Newman, M.E.J. and Girvan, M. (2004) Finding and evaluating community structure in networks. *Physical Review E*, 69, 026113.

Newsome, T.M., Barton, B., Buck, J.C., DeBruyn, J., Spencer, E. *et al.* (2021) Monitoring the dead as an ecosystem indicator. *Ecology and Evolution*, 11, 5844–5856.

Newsome, T.M., Ballard, G.-A., Dickman, C.R., Fleming, P.J.S. and van de Ven, T. (2013) Home range, activity and sociality of a top predator, the dingo: A test of the Resource Dispersion Hypothesis. *Ecography*, 36, 914–925.

Newson, S.E., Woodburn, R.J.W., Noble, D.G., Baillie, S.R. and Gregory, R.D. (2005) Evaluating the Breeding Bird Survey for producing national population size and density estimates. *Bird Study*, 52, 42–54.

Newton, I. (1687) [published in English in 1728]. *Philosophiæ Naturalis Principia Mathematica*. Imprimatur. S. Pepys, Reg. Soc. Praeces, London, UK.

Newton, I. (2007) *The Migration Ecology of Birds*. Academic Press, London, UK.

Newton-Fisher, N.E. (2014) The hunting behavior and carnivory of wild chimpanzees. In: Henke, W. and Tattersall, I. (eds) *Handbook of Paleoanthropology*. Springer-Verlag, Berlin, Germany.

Nichols, D.J. and Johnson, K.R. (2008) *Plants and the K–T Boundary*. Cambridge University Press, Cambridge, UK.

Nicholson, A.J. (1933) The balance of animal populations. *Journal of Animal Ecology*, 2, 132–178.

Nicholson, A.J. (1954) An outline of the dynamics of animal populations. *Australian Journal of Zoology*, 2, 9–65.

Nicholson, A.J. (1957) The self-adjustment of populations to change. *Cold Spring Harbor Symposium on Quantitative Biology*, 22, 153–172.

Nicholson, A.J. and Bailey. V.A. (1935) The balance of animal populations, part I. *Proceedings of the Zoological Society of London*, 3, 551–598.

Nielsen, M., Langley, M.C., Shipton, C. and Kapitány, R. (2020) *Homo neanderthalensis* and the evolutionary origins of ritual in *Homo sapiens*. *Philosophical Transactions of the Royal Society B: Biological Sciences*, 375, 20190424.

Niemelä, J. (1993) Interspecific competition in ground-beetle assemblages (Carabidae): What have we learned? *Oikos*, 66, 325–335.

Nilon, C.H. (2011) Urban biodiversity and the importance of management and conservation. *Landscape and Ecological Engineering*, 7, 45–52.

Nimmo, D.G., MacNally, R., Cunningham, S.C., Haslem, A. and Bennett, A.F. (2015) Vive la résistance: reviving resistance for 21st century conservation. *Trends in Ecology and Evolution*, 30, 516–523.

Niu, Y., Ren, G., Lin, G., Di Biase, L. and Fattorini, S. (2019) Community structure of tenebrionid beetles in the Ulan Buh Desert (Inner Mongolia, China) (Coleoptera: Tenebrionidae). *Fragmenta Entomologica*, 51, 193–200.

Niu, Y., Ren, G., Lin, G., Di Biase, L. and Fattorini. S. (2020) Fine scale vegetation characteristics drives insect ensemble structures in a desert ecosystem: The tenebrionid beetles (Coleoptera: Tenebrionidae) inhabiting the Ulan Buh Desert (Inner Mongolia, China). *Insects*, 11, 410

Nogués-Bravo, D. and Martínez-Rica, J.P. (2004) Factors controlling the spatial species richness pattern of four groups of terrestrial vertebrates in an area between two different biogeographic regions in northern Spain. *Journal of Biogeography*, 31, 629–640.

Norden, N., Angarita, H.A., Bongers, F., Martínez-Ramos, M., Granzowde la Cerda, I. *et al.* (2015) Successional dynamics in Neotropical forests are as uncertain as they are predictable. *Proceedings of the National Academy of Sciences of the United States of America*, 112, 8013–8018.

Nores, M. (1995) Insular biogeography of birds on mountain-tops in north western Argentina. *Journal of Biogeography*, 22(1), 61–70.

Northrup, J.M., Vander Wal, E., Bonar, M., Fieberg, J., Laforge, M.P. *et al.* (2022) Conceptual and methodological advances in habitat-selection modeling: Guidelines for ecology and evolution. *Ecological Applications*, 32(1), e02470.

Norton-Griffiths, M. (1973) Counting the Serengeti migratory wildebeest using two-stage sampling. *African Journal of Ecology*, 11, 135–149.

Nowicki, P., Bonelli, S., Barbero, F. and Balletto, E. (2009) Relative importance of density-dependent regulation and environmental stochasticity for butterfly population dynamics. *Oecologia*, 161(2), 227–239.

Ntinou, M. and Badal, E. (2000) Local vegetation and charcoal analysis; an example from two late Neolithic sites in Northern Greece. In: Halstead, P. and Frederick, C. (eds) *Landscape and Land Use in Postglacial Greece*. Sheffield Academic Press, Sheffield, UK, pp. 38–51.

Numata, M. (1969) Progressive and retrogressive gradient of grassland vegetation measured by degree of succession—Ecological judgement of grassland condition and trend IV. *Vegetatio*, 19, 96–127.

Nunes, C.A., Beiroz, W., da Silva, P.G., Braga, R.F., Fernandes, G.W. and Neves, F.D.S. (2019) Fire? They don't give a dung! The resilience of dung beetles to fire in a tropical savanna. *Ecological Entomology*, 44, 315–323.

Nuutinen, V. and Butt, K.R. (1997) The mating behaviour of the earthworm *Lumbricus terrestris* (Oligochaeta: Lumbricidae). *Journal of Zoology*, 242, 783–798.

Nylin, S. (1988) Host plant specialization and seasonality in a polyphagous butterfly, *Polygonia c-album* (Nymphalidae). *Oikos*, 53, 381–386.

Nylin, S. (1992) Seasonal plasticity in life history traits: Growth and development in *Polygonia c-album*. *Biological Journal of the Linnean Society*, 47, 301–323.

Nylin, S., Janz, N. and Wedell, N. (1996) Oviposition plant preferences and offspring performance in the Comma butterfly: Correlations and conflicts. *Entomologia Experimentalis et Applicata*, 80, 141–144.

Oakley, K.P. (1972) *Man the Tool-maker*, 6th edn. Trustees of the British Museum (Natural History), London, UK.

Oates, M. (2020) *His Imperial Majesty*. Bloomsbury, London, UK.

O'Brien, K., Ashioya, L. and Faith, J.T. (2025) Non-analog behaviour of Eastern African herbivore communities during the Last Glacial Period. *Ecology Letters*, 28, e70041.

Ocampo, A., Vajda, V. and Buffetaut, E., (2006) Unravelling the Cretaceous–Paleogene (KT) turnover, evidence from flora, fauna and geology. In: Cockell, C., Gilmour, I. and Koeberl, C. (eds) *Biological Processes Associated with Impact Events*. Springer, Berlin, Germany, pp. 197–219.

Ocampo-Zuleta, K., Pausas, J.G. and Paula, S. (2024) FLAMITS: A global database of plant flammability traits. *Global Ecology and Biogeography*, 33, 412–425.

Ocko, S.A., Heyde, A. and Mahadevan, L. (2019) Morphogenesis of termite mounds. *Proceedings of the National Academy of Sciences of the United States of America*, 116, 3379–3384.

Odling-Smee, F.J., Laland, K.N. and Feldman, M.W. (2003) *Niche Construction: The Neglected Process in Evolution*. Princeton University Press, Princeton, USA.

Odling-Smee, F.J., Erwin, D., Palkovacs, E.P., Feldman, M.W. and Laland, K.N. (2013) Niche construction theory: A practical guide for ecologists. *Quarterly Review of Biology*, 88, 3–28.

O'Dowd, D.J. and Willson, M.F. (1991) Associations between mites and leaf domatia. *Trends in Ecology and Evolution*, 6, 179–182.

Odum, E.P. (1953) *Fundamentals of Ecology*. W.B. Saunders, Philadelphia, USA.

Odum, E.P. (1963) *Ecology*. Holt, Rinehart and Winston, New York, USA.

Odum, E.P. (1983) *Basic Ecology*. CBS College Publishing, New York, USA.

Odum, E.P. and Barrett, G.W. (2005) *Fundamentals of Ecology*. Cengage Learning, Nelson Education, Toronto, Canada.

Ofstad, E.G., Herfindal, I., Solberg, E.J. and Sæther, B-E. (2016) Home ranges, habitat and body mass: Simple correlates of home range size in ungulates. *Proceedings of the Royal Society B: Biological Sciences*, 283, 20161234.

Ohayon, S., Granot, I. and Belmaker, J. (2021) A meta-analysis reveals edge effects within marine protected areas. *Nature Ecology and Evolution*, 5, 1301–1308.

Ohayon, S., Homma, H., Malamud, S., Ostrovsky, I., Yahel, R., Mehner, T., Kanari, M. and Belmaker, J. (2023) Consistent edge effect patterns revealed using continuous surveys across an Eastern Mediterranean no-take marine protected area. *ICES Journal of Marine Science: Journal Du Conseil*, 80, 1594–1605.

Oikonomou, A. (2021) Freshwater biogeography in a nutshell. In: Guilbert, E. (ed.) *Biogeography. An Integrative Approach of the Evolution if Living*. Wiley, Hoboken, USA, pp. 219–243.

Oishi, T. (2013) Human-gorilla and gorilla-human: Dynamics of human-animal boundaries and interethnic relationships in the central African rainforest. *Revue de Primatologie*, 5, 63.

Oishi, Y. (2024) Bryophytes enhance nitrogen content in decaying wood via biological interactions. *Ecosphere*, 15, e4755.

Okuniewska-Nowaczyk, I., Rahmonov, O., Snytko, V.A. and Szczypek, T. (2018) Palynological record of the history of vegetation in the sandy areas of Southern Poland. *Geography and Natural Resources*, 39, 396–402.

Olalla-Kerstupp, A., Ruiz-Aymá, G., González-Rojas, J.I. and Guzmán-Velasco, A. (2015) High fidelity to wintering, stop-over and breeding sites shown by a long-billed curlew *Numenius americanus* tracked with satellite telemetry on migratory flights across North America. *Bird Study*, 62, 556–560.

Olea, P.P., Mateo-Tomás, P. and Sánchez-Zapata, J. A. (eds) (2019) *Carrion Ecology and Management*. Springer, Cham, Switzerland.

Oli, M.K. (2019) Population cycles in voles and lemmings: State of the science and future directions. *Mammal Review*, 49, 226–239.

Oliva, A. (2001) Insects of forensic significance in Argentina. *Forensic Science International*, 120, 145–154.

Oliva, M. (2018) The Little Ice Age, the climatic background of present-day warming in Europe. *Cuadernos de Investigación Geográfica*, 44, 7–13.

Oliveira, B.F., São-Pedro, V.A., Santos-Barrera, G., Penone, C. and Costa, G.C. (2017) AmphiBIO, a global database for amphibian ecological traits. *Scientific Data*, 4, 170123.

Oliveira, U., Soares-Filho, B.S., Santos, A.J., Paglia, A.P., Brescovit, A.D. *et al.* (2019) Modelling highly biodiverse areas in Brazil. *Scientific Reports*, 9, 6355.

Ollerton, J., Erenler, H., Edwards, M. and Crockett, R. (2014) Extinctions of aculeate pollinators in Britain and the role of large-scale agricultural changes. *Science*, 346, 1360–1362.

Olson, D.M., Dinerstein, E., Wikramanayake, E.D., Burgess, N.D., Powell, G.V.N. *et al.* (2001) Terrestrial ecoregions of the world: A new map of life on earth. *BioScience*, 51, 933–938.

Olson, J.S. (1958) Rates of succession and soil changes on Southen Lake Michigan sand dunes. *Botanical Gazette*, 119, 125–170.

Omelon, C.R. (2016) Endolithic microorganisms and their habitats. In: Hurst, C.J. (ed.) *Advances in Environmental Microbiology. Their World: A Diversity of Microbial Environments*. Springer, Cham, Switzerland.

O'Neil, S.T., Vucetich, J.A., Beyer Jr, D.E., Hoy, S.R. and Bump, J.K. (2020) Territoriality drives preemptive habitat selection in recovering wolves: Implications for carnivore conservation. *Journal of Animal Ecology*, 89, 1433–1447.

O'Neill, B.J. (2016) Community disassembly in ephemeral ecosystems. *Ecology*, 12, 3285–3292.

Opdam, P. (1991) Metapopulation theory and habitat fragmentation: A review of Holarctic breeding bird studies. *Landscape Ecology*, 5, 93–106.

Ordonez, A. and Williams, J.W. (2013) Climatic and biotic velocities for woody taxa distributions over the last 16000 years in eastern North America. *Ecology Letters*, 16, 773–781.

Orians, G.H. and Pearson, N.E. (1979) On the theory of central place foraging. In: Horn, D.J., Mitchell, R.D. and Stairs, G.R. (eds) *Analysis of Ecological Systems*. Ohio State University Press, Columbus, USA, pp. 154–177.

Oro, D. (2008) Living in a ghetto within a local population: An empirical example of an ideal despotic distribution. *Ecology*, 89, 838–846.

Orton, C. and Hughes, M. (2013) *Pottery in Archaeology*. Cambridge University Press, Cambridge, UK.

Oschmann, W., Grasshof, M. and Gudo, M. (2002) The early evolution of the planet Earth and the origin of life. *Senckenbergiana Lethaea*, 82, 285–294.

Outram, A.K., Stear, N.A., Bendrey, R., Olsen, S., Kasparov, A. *et al.* (2009) The earliest horse harnessing and milking. *Science*, 323, 1332–1335.

Ovaskainen, O. and Meerson, B. (2010) Stochastic models of population extinction. *Trends in Ecology and Evolution*, 25, 643–652.

Owens, B.E. and Carlton, C.E. (2015) "Berlese vs. Winkler": Comparison of two forest litter Coleoptera extraction methods and the ECOLI (Extraction of Coleoptera in Litter) protocol. *Coleopterists' Bulletin*, 69(4), 645–661.

Oyarte Galvez, L., Bisot, C., Bourrianne, P., Cargill, R., Klein, M. *et al.* (2025) A travelling-wave strategy for plant-fungal trade. *Nature*, 639, 172–180.

Ozaki, K., Isono, M., Kawahara, T., Iida, S., Kudo, T. and Fukuyama. K. (2006) A mechanistic approach to evaluation of umbrella species as conservation surrogates. *Conservation Biology*, 20, 1507–1515.

Pace, N.R. (1997) A molecular view of microbial diversity and the biosphere. *Science*, 276, 734–740.

Padgett, D.K. and Priyam, P. (2017) Pavement dwelling in Delhi, India: An ethnographic account of survival on the margins. *Human Organization*, 76, 73–81.

Padilla-Sanchez, V. (2021) Structural model of bacteriophage T4. *WikiJournal of Science*, 4(1), 5.

Padoa-Schioppa, E., Baietto, M., Massa, R. and Bottoni, L. (2006) Bird communities as bioindicators: The focal species concept in agricultural landscapes. *Ecological Indicators*, 6(1), 83–93.

Pagel, J.E., Anderson, C.M., Bell, D.A., Deal, E. and Kiff, L. (2018) Peregrine falcons: The neighbors upstairs. In: Boal, C.W. and Dykstra, C.R. (eds) *Urban Raptors*. Island Press, Washington, DC, USA.

Paine, R.T. (1966) Food web complexity and species diversity. *American Naturalist*, 100(910), 65–75.

Paine, R.T. (1969) The *Pisaster-Tegula* interaction: Prey patches, predator food preference, and intertidal community structure. *Ecology*, 50, 950–961.

Paine, R.T. (1996) Preface. In: Polis, G.A. and Winemiller, K.O. (eds) *Food Webs. Integration of Patterns and Dynamics*. Springer Science+Business Media, Dordrecht, The Netherlands, pp. ix–x.

Pallardy, R. (2021a) A dog eat dog world. *The Biologist*, 68, 23–25.

Pallardy, R. (2021b) Listening in. *The Biologist*, 68, 16–19.

Pallen, M.J., Telatin, A. and Oren, A. (2021) The next million names for Archaea and Bacteria. *Trends in Microbiology*, 29(4), 289–298.

Palmeirim, A.F., Farneda, F.Z., Vieira, M.V. and Peres, C.A. (2021) Forest area predicts all dimensions of small mammal and lizard diversity in Amazonian insular forest fragments. *Landscape Ecology*, 36, 3401–3418.

Palusci, E., Mantoni, C., Strona, G. and Fattorini, S. (2021) Wildfire does not affect the dung beetle diversity of high-altitude Mediterranean habitats. *International Journal of Wildland Fire*, 30, 636–642.

Pancerasa, M., Ambrosini, R., Saino, N. and Casagrandi, R. (2018) Barn swallows long-distance migration occurs between significantly temperature-correlated areas. *Scientific Reports*, 8, 12359.

Panitsa, M., Tzanoudakis, D., Triantis, K.A. and Sfenthourakis, S. (2006) Patterns of species richness on very small islands: The plants of the Aegean archipelago. *Journal of Biogeography*, 33, 1223–1234

Pansu, M. and Gautheyrou, J. (2006) *Handbook of Soil Analysis: Mineralogical, Organic and Inorganic Methods*. Springer, Berlin, Germany.

Pantel, J.H., Varengo, S. and Venturini, F. (2023) Peter Kropotkin and social ecology: Between biology and revolution. *Anarchist Studies*, 31(2), 73–95.

Papworth, S.K., Rist, J., Coad, L. and Milner-Gulland, E.J. (2008) Evidence for shifting baseline syndrome in conservation. *Conservation Letters*, 2, 93–100.

Parenti, L.R. and Ebach, M.C. (2009) *Comparative Biogeography: Discovering and Classifying Biogeographical Patterns of a Dynamic Earth*. University of California Press, Berkeley, USA.

Parisi, F., Pioli, S., Lombardi, F., Fravolini, G., Marchetti, M. and Tognetti, R. (2018) Linking deadwood traits with saproxylic invertebrates and fungi in European forests – A review. *iForest*, 11, 423–436.

Parker, G.A. (1974) Assessment strategy and the evolution of animal conflicts. *Journal of Theoretical Biology*, 47, 223–243.

Parker, J. (2016) Myrmecophily in beetles (Coleoptera): Evolutionary patterns and biological mechanisms. *Myrmecological News*, 22, 65–108.

Parr, C.S., Schulz, K.S., Hammock, J., Wilson, N., Leary, P. *et al.* (2016) TraitBank: Practical semantics for organism attribute data. *Semantic Web*, 7, 577–588.

Parsons, K.J., Rigg, A., Conith, A.J., Kitchener, A.C., Harris, S. and Zhu, H. (2020) Skull morphology diverges between urban and rural populations of red foxes mirroring patterns of domestication and macroevolution. *Proceedings of the Royal Society B: Biological Sciences*, 287, 20200763.

Parte, A.C., Sardà Carbasse, J., Meier-Kolthoff, J.P., Reimer, L.C. and Göker, M. (2020) List of Prokaryotic names with Standing in Nomenclature (LPSN) moves to the DSMZ. *International Journal of Systematic and Evolutionary Microbiology*, 70, 5607–5612.

Pascoal, F., Costa, R. and Magalhães, C. (2021) The microbial rare biosphere: Current concepts, methods and ecological principles. *FEMS Microbiology Ecology*, 97, fiaa227.

Pascual-Rico, R., Sánchez-Zapata, J.A., Navarro, J., Eguía, S., Anadón, J.D. and Botella, F. (2020) Ecological niche overlap between co-occurring native and exotic ungulates: Insights for a conservation conflict. *Biological Invasions*, 22, 2497–2508.

Pasolli, E., Asnicar, F., Manara, S., Zolfo, M., Karcher, N. *et al.* (2019) Extensive unexplored human microbiome diversity revealed by over 150,000 genomes from metagenomes spanning age, geography, and lifestyle. *Cell*, 176, 649–662.e20.

Passos, I., Figueiredo, A., Almeida, A.M. and Ribeiro, M.M. (2024) Uncertainties in plant species niche modeling under climate change scenarios. *Ecologies*, 5, 402–419.

Pastorok, R., Bartell, S.M., Ferson, S. and Ginzburg, L.R. (eds) (2001) *Ecological Modeling in Risk Assessment: Chemical Effects on Populations, Ecosystems and Landscapes*. CRC Press, Boca Raton, USA.

Pataro, L., Molina-Venegas, R., Calatayud, J., Moreno-Saiz, J.C. and Rodríguez, M.Á. (2021) An updated phylogenetic bioregionalization for the European fern flora. *Biodiversity and Conservation*, 30, 201–215.

Patrício, A.R., Hawkes, L.A., Monsinjon, J.R., Godley, B.J. and Fuentes, M.M.P.B. (2021) Climate change and marine turtles: Recent advances and future directions. *Endangered Species Research*, 44, 363–395.

Pätsch, R., Jašková, A., Chytrý, M., Kucherov, I.B., Schaminée, J.H.J. *et al.* (2019) Making them visible and usable—Vegetation-plot observations from Fennoscandia based on historical species-quantity scales. *Applied Vegetation Science*, 22, 465–473.

Patterson, C. (1956) Age of meteorites and the Earth. *Geochimica et Cosmochimica Acta*, 10, 230–237.

Pauly, D. (1995) Anecdotes and the shifting baseline syndrome of fisheries. *Trends in Ecology and Evolution*, 10, 430.

Pavoine, S. and Bonsall, M.B. (2011) Measuring biodiversity to explain community assembly: A unified approach. *Biological Reviews*, 86, 792–812.

Pawlowski, J., Audic, S., Adl, S., Bass, D., Belbahri, L. *et al.* (2012) CBOL Protist Working Group: Barcoding eukaryotic richness beyond the animal, plant, and fungal kingdoms. *PLOS Biology*, 10, e1001419.

Payne, J.A. (1965) A summer carrion study on the baby pig *Sus scrofa* L. *Ecology*, 46, 592–602.

Pazar, C.C. (2018) *Terraforming of Terrestrial Earth-sized Planetary Bodies*. B.A. Earth and Planetary Sciences, University of Colorado Boulder, Boulder, USA.

Peachey, C.A. (1980) The ecology of the butterfly community in Berwood Forest, UK. MPhil. thesis (CNAA), Oxford Polytechnic, Oxford, UK.

Peacock, E., Derocher, A.E., Lunn, N.J. and Obbar, M.E. (2010) Polar bear ecology and management in Hudson Bay in the face of climate change. In: Ferguson, S.H., Loseto, L. and Mallory, M.K. (eds) *A Little Less Arctic: Top Predators in the World's Largest Northern*

Inland Sea, Hudson Bay. Springer, Dordrecht, The Netherlands, pp. 93–115.

Pearce, F. (2020) Weather makers. Forests supply the world with rain. A controversial Russian theory claims they also make wind. *Science*, 368, 1302–1305.

Pearce, F., Carbone, C., Cowlishaw, G. and Isaac, N.J.B. (2013) Space-use scaling and home range overlap in primates. *Proceedings of the Royal Society B: Biological Sciences*, 280, 20122122.

Pearl, R. (1930) *Introduction to Medical Biometry and Statistics*, 2nd edn. W.B. Saunders, Philadelphia, USA.

Pearl, R. and Reed, L.J. (1920) On the rate of growth of the population of the United States since 1790 and its mathematical representation. *Proceedings of the National Academy of Sciences USA*, 6, 275–288.

Pearson, H.C., Savoca, M.S., Costa, D.P., Lomas, M.W., Molina, R. *et al.* (2023) Whales in the carbon cycle: Can recovery remove carbon dioxide? *Trends in Ecology and Evolution*, 38(3), 238–249.

Pecenka, J. (2017) The cattle dung arthropod community in eastern South Dakota: Their colonization, impact on degradation, and response to rangeland management. MSc thesis, South Dakota State University, Brookings, USA. Available at: http://openprairie.sdstate.edu/etd/1181 (accessed 21 May 2025).

Peck, H.L. (2013) Investigating ecological impacts of the non-native population of rose-ringed parakeets (*Psittacula krameri*) in the UK. PhD thesis, Imperial College, London, UK.

Pederson, R., Bocher, P., Garthe, S., Fort, J., Mercker, M. *et al.* (2022) Bird migration in space and time: Chain migration by Eurasian curlew *Numenius arquata arquata* along the East Atlantic Flyway. *Journal of Avian Biology*, 2022, e02924.

Peet, R.K. and Christensen, N.L. (1980) Succession: A population process. *Vegetatio*, 43, 131–140.

Pekár, S., Wolff, J.O., Černecká, Ľ., Birkhofer, K., Mammola, S. *et al.* (2021) The World Spider Trait database: A centralized global open repository for curated data on spider traits. *Database*, 2021, baab064.

Pełechaty, M., Pukacz, A., Apolinarska, K., Pełechata, A. and Siepak, M. (2013) The significance of Chara vegetation in the precipitation of lacustrine calcium carbonate. *Sedimentology*, 60(4), 1017–1035.

Pellens, R. and Grandcolas, P. (eds) (2016) *Biodiversity Conservation and Phylogenetic Systematics: Preserving our Evolutionary Heritage in an Extinction Crisis*. Springer, Cham, Switzerland.

Peltanová, A., Dvořák, L. and Juřičková, L. (2012) The spread of non-native *Cepaea nemoralis* and *Monacha cartusiana* (Gastropoda: Pulmonata) in the Czech Republic with comments on other land snail immigrants. *Biologia*, 67, 384–389.

Peltzer, D.A., Wardle, D.A., Allison, V.J., Baisden, W.T., Bardgett, R.D. *et al.* (2010) Understanding ecosystem retrogression. *Ecological Monographs*, 80, 509–529.

Pembury Smith, M.Q.R. and Ruxton, G.D. (2021) How fast is a snail's pace? The influences of size and substrate on gastropod speed of locomotion. *Journal of Zoology*, 314, 12–19.

Peng, W., Ma, N.L., Zhang, D., Zhou, Q., Yue, X. *et al.* (2020) A review of historical and recent locust outbreaks: Links to global warming, food security and mitigation strategies. *Environmental Research*, 191, 110046.

Pentecost, A. (2010) The marl lakes of the British Isles. *Freshwater Reviews*, 2, 167–197.

Perelman, P., Johnson, W.E., Roos, C., Seuánez, H.N., Horvath, J.E. *et al.* (2011) A molecular phylogeny of living primates. *PLOS Genetics*, 7, e1001342.

Perez, K.O., Carlson, R.L., Shulman, M.J. and Ellis, J.C. (2009) Why are intertidal snails rare in the subtidal? Predation, growth and the vertical distribution of *Littorina littorea* (L.) in the Gulf of Maine. *Journal of Experimental Marine Biology and Ecology*, 369, 79–86.

Pérez-Espona, S. (2021) *Eciton* army ants—Umbrella species for conservation in Neotropical forests. *Diversity*, 13, 136.

Perkins-McVey, M. (2022) A portrait of the neurophysiologist as a young man: Claus, Darwin, and Sigmund Freud's search for the testes of the eel (1875–1877). *History of Psychology*, 25(4), 367–384.

Perry, S. (2011) Social traditions and social learning in capuchin monkeys (*Cebus*). *Philosophical Transactions of the Royal Society B: Biological Sciences*, 366, 988–996.

Petchey, O.L. and Gaston, K.J. (2002) Functional diversity (FD), species richness and community composition. *Ecology Letters*, 5, 402–411.

Petri, G., Expert, P., Turkheimer, F., Carhart-Harris, R., Nutt, D. *et al.* (2014) Homological scaffolds of brain functional networks. *Journal of the Royal Society Interface*, 11, 20140873.

Pettersen, A.K., Hall, M.D., White, C.R. and Marshall, D.J. (2020) Metabolic rate, context-dependent selection, and the competition-colonization trade-off. *Evolution Letters*, 4, 333–344.

Pfadenhauer, J.S. and Klötzli, F. A. (2020) *Global Vegetation: Fundamentals, Ecology and Distribution* (trans. von Gadow, C.). Springer Nature, Cham, Switzerland.

Phillips, H.R.P., Guerra, C.A., Bartz, M.L.C., Briones, M.J.I., Brown, G. *et al.* (2019) Global distribution of earthworm diversity. *Science*, 366(6464), 480–485.

Phukhamsakda, C., Nilsson, R.H., Bhunjun, C.S., de Farias, A.R.G., Sun, Y.-R. *et al.* (2022) The numbers of fungi: Contributions from traditional taxonomic studies and challenges of metabarcoding. *Fungal Diversity*, 114, 327–386.

Pianka, E.R. (1970) On *r*- and *K*-selection. *American Naturalist*, 104, 592–597.

Pianka, E.R. (1973) The structure of lizard communities. *Annual Review of Ecology and Systematics*, 4, 53–74.

Pianka, E.R. (1974) Niche overlap and diffuse competition. *Proceedings of the National Academy of Sciences of the United States of America*, 71, 2141–2145.

Pianka, E.R. (1975) Niche relations of desert lizards. In: Cody, M. and Diamond, J. (eds) *Ecology and Evolution of Communities*. Harvard University Press, Cambridge, USA, pp. 292–314.

Pickering, T.R., Egeland, C.P., Dominguez-Rodrigo, M., Brain, C.K. and Schnell, A.G. (2008) Testing the "shift in the balance of power" hypothesis at Swartkrans, South Africa: Hominid cave use and subsistence behavior in the Early Pleistocene. *Journal of Anthropological Archaeology*, 27, 30–45.

Pickrell, J. (2019) The making of mammals. *Nature*, 574, 468–472.

Pidwirny, M. (2006) *Plant Succession. Fundamentals of Physical Geography*, 2nd edn. Available at: www.physicalgeography.net/fundamentals/9i.html (accessed 22 May 2025).

Pielou, E.C. (1979) *Biogeography*. John Wiley and Sons, New York, USA.

Pielou, E.C. (1992) *Biogeography*. Krieger Publishing Company, Malabar, USA.

Pike, C., Crook, V. and Gollock, M. (2020) European eel. *Anguilla anguilla*. IUCN Red List of Threatened Species 2020. Available at: https://dx.doi.org/10.2305/IUCN.UK.2020-2.RLTS.T60344A152845178.en (accessed 22 May 2025).

Pimm, S.L., Lawton, J.H. and Cohen, J.E. (1991) Food web patterns and their consequences. *Nature*, 350, 669–674.

Pimm, S.L., Russell, G.J., Gittleman, J.L. and Brooks, T.M. (1995) The future of biodiversity. *Science*, 269, 347–350.

Pimm, S.L., Jenkins, C., Abell, R., Brooks, T.M. Gittleman, J.L. *et al.* (2014) The biodiversity of species and their rates of extinction, distribution, and protection. *Science*, 344, 987–1007.

Pinker, S. (2011) *The Better Angels of Our Nature: Why Violence Has Declined*. Viking, New York, USA.

Pinto, A.V., Hansson, B., Patramanis, I., Morales, H.E. and van Oosterhout, C. (2024) The impact of habitat loss and population fragmentation on genomic erosion. *Conservation Genetics*, 25, 49–57.

Pisor, A.C. and Jones, J.H. (2021) Human adaptation to climate change: An introduction to the special issue. *American Journal of Human Biology*, 33, e23530.

Pitman, N.C.A., Terborgh, J., Silman, M.R. and Nunez, V. (1999) Tree species distributions in an upper Amazonian forest. *Ecology*, 80, 2651–2661.

Pitt, D. and Samson, P.R. (eds) (1998) *The Biosphere and Noösphere Reader: Global Environment, Society and Change*. Routledge, London, UK.

Pitts, M. (1996) The stone axe in Neolithic Britain. *Proceedings of the Prehistoric Society*, 62, 311–371.

Pitzalis, M., Fattorini, S., Trucchi, E. and Bologna, M.A. (2005) Comparative analysis of species diversity of Isopoda Oniscidea and Collembola communities in burnt and unburnt habitats in Central Italy. *Italian Journal of Zoology*, 72, 127–140.

Piza-Roca, C., van't Zelfde, M., La Haye, M.J.J., Jongejans, E. and Raes, N. (2018) European badger habitat requirements in the Netherlands – Combining ecological niche models with neighbourhood analysis. *Wildlife Biology*, 2018, wlb.00453.

Planck, M. (1900) Ueber irreversible Strahlungsvorgänge. *Annalen der Physik*, 306, 69–122.

Pocheville, A. (2015) The ecological niche: History and recent controversies. In: Heams, T., Huneman, P., Lecointre, G. and Silberstein, M. (eds) *Handbook of Evolutionary Thinking in the Sciences*. Springer, Dordrecht, The Netherlands, pp. 547–586.

Podani, J. (1999) Extending Gower's general coefficient of similarity to ordinal characters. *Taxon*, 48, 331–340.

Podani, J. and Schmera, D. (2011) A new conceptual and methodological framework for exploring and explaining pattern in presence-absence data. *Oikos*, 120, 1625–1638.

Podani, J., Ricotta, C. and Schmera, D. (2013) A general framework for analysing beta diversity, nestedness and related community-level phenomena based on abundance data. *Ecological Complexity*, 15, 52–61.

Podani, J., Ódor, P., Fattorini, S., Strona, G., Heino, J. and Schmera, D. (2018) Exploring multiple presence-absence data structures in ecology and biogeography. *Ecological Modelling*, 383, 41–51.

Poinar, H.N., Kuch, M., Sobolik, K.D., Barnes, I., Stankiewiczi, A.B. *et al.* (2001) A molecular analysis of dietary diversity for three archaic Native Americans. *Proceedings of the National Academy of Sciences of the United States of America*, 98, 4317–4322.

Polis, G.A. and McCormick, S.J. (1986) Scorpions, spiders and solpugids: predation and competition among distantly related taxa. *Oecologia*, 71(1), 111–116.

Polis, G.A. and Winemiller, K.O. (1996) *Food Webs: Integration of Pattern and Process*. Chapman and Hall, London, UK.

Pollan, M. (2006) *The Omnivore's Dilemma: The Search for a Perfect Meal in a Fast-Food World*. Penguin Press, New York, USA.

Pollard, E. and Yates, T.J. (1993a) *Monitoring Butterflies for Ecology and Conservation*. Chapman and Hall, London, UK.

Pollard, E. and Yates, T.J. (1993b) Population fluctuations of the holly blue butterfly *Celastrina argiolus* (L.) (Lepidoptera: Lycaenidae). *Entomologist's Gazette*, 44, 3–9.

Poloni, R., Iannella, M., Fusco, G. and Fattorini, S. (2022) Conservation biogeography of high-altitude longhorn beetles under climate change. *Insect Conservation and Diversity*, 15(4), 429–444.

Poole, K.G. and Heard, D.C. (2003) Seasonal habitat use and movements of mountain goats, *Oreamnos americanus*, in East-central British Columbia. *Canadian Field-Naturalist*, 117(4), 565.

Poorter, H., Niinemets, Ü., Poorter, L., Wright, I.J. and Villar, R. (2009) Causes and consequences of variation in leaf mass per area (LMA): A meta-analysis. *New Phytologist*, 182, 565–588.

Porter, E.E., Forschner, B.R. and Blair, R.B. (2001) Woody vegetation and canopy fragmentation along a forest-to-urban gradient. *Urban Ecosystems*, 5, 131–151.

Poschlod, P., Baumann, A. and Karlík, P. (2009) Origin and development of grasslands in Central Europe. In: Veen, P., Jefferson, R., de Smidt, J. and van der Straaten, J. (eds) *Grasslands in Europe of High Nature Value.* KNNV Publishing, Zeist, Netherlands, pp. 15–26.

Potter, C., Harwood, T., Knight, J.D. and Tomlinson, I. (2011) Learning from history, predicting the future: The UK Dutch elm disease outbreak in relation to contemporary tree disease threats. *Philosophical Transactions of the Royal Society B: Biological Sciences*, 366, 1966–1974.

Powell, K.E., Oliver, T.H., González-Suárez, M., Botham, M.S., Harrower, C.A. *et al.* (2024) Asynchrony in terrestrial insect abundance corresponds with species traits. *Ecology and Evolution*, 14, e10910.

Powell, L.A. and Gale, G.A. (2015) *Estimation of Parameters for Animal Populations. A Primer for the Rest of Us.* Caught Napping Publications, Lincoln, USA.

Powell, R.A. (2000) Animal home ranges and territories and home range estimators. In: Boitani, L. and Fuller, T. (eds) *Research Techniques in Animal Ecology: Controversies and Consequences*. Columbia University Press, New York, USA, pp. 65–110.

Powell, R.A. and Mitchell, M.S. (2012) What is a home range? *Journal of Mammalogy*, 93, 948–958.

Power, M.E., and Mills, L.S. (1995) The keystone cops meet in Hilo. *Trends in Ecology and Evolution*, 10, 182–184.

Power, M.E., Matthews, W.J. and Stewart, A.J. (1985) Grazing minnows, piscivorous bass, and stream algae: Dynamics of a strong interaction. *Ecology*, 66, 1448–1456.

Pozas-Schacre, C., Casey, J.M., Brandl, S.J., Kulbicki, M., Harmelin-Vivien, M. *et al.* (2021) Congruent trophic pathways underpin global coral reef food webs. *Proceedings of the National Academy of Sciences of the United States of America*, 118, e2100966118.

Prach, K. and Walker, L.R. (2019) Differences between primary and secondary plant succession among biomes of the world. *Journal of Ecology*, 107, 510–516.

Prado, P.I., Miranda, M.D. and Chalom, A. (2018) Package 'sads': Maximum likelihood models for species abundance distributions. Available at: https://cran.r-project.org/web/packages/sads/sads.pdf (accessed 22 May 2025).

Preininger, D., Schoas, B., Kramer, D. and Boeckle, M. (2019) Waste disposal sites as all-you-can eat buffets for carrion crows (*Corvus corone*). *Animals*, 9, 215.

Preston, F.W. (1948) The commonness, and rarity, of species. *Ecology*, 29, 254–283.

Prevedello, J.A., Dickman, C.R., Vieira, M.V. and Vieira, E.M. (2013) Population responses of small mammals to food supply and predators: A global meta-analysis. *Journal of Animal Ecology*, 82, 927–936.

Price, M.V. (1978) The role of microhabitat in structuring desert rodent communities. *Ecology*, 59, 910–921.

Price, P.W., Denno, R.F., Eubanks, M.D., Finke, D.L. and Kaplan, I. (2011) *Insect Ecology: Behavior, Populations and Communities*. Cambridge University Press, Cambridge, UK, p. 82.

Price, T.D., Wahl, J. and Bentley, R.A. (2006) Isotopic evidence for mobility and group organization among Neolithic farmers at Talheim, Germany, 5000 BC. *European Journal of Archaeology*, 9, 259–284.

Prichard, J.C. (1826) *Researches into the Physical History of Mankind*, 2nd edn. Houlfton and Stoneman, London, UK.

Primack, R.B. (2014) *Essentials of Conservation Biology*, 6th edn. Sinauer Associates, Sunderland, USA.

Priya, R.U., Geetha, D. and Darshan, S. (2016) Biology and cultivation of black ear mushroom – *Auricularia* spp. *Advances in Life Sciences*, 5, 10252–10254.

Prober, S.M., Doerr, V.A.J., Broadhurst, L.M., Williams, K.J. and Dickson, F. (2019) Shifting the conservation paradigm: A synthesis of options for renovating nature under climate change. *Ecological Monographs*, 89, e01333.

Procheş Ş. and Ramdhani, S. (2020) A global regionalisation based on the present-day distribution of broad plant lineages. *Phytotaxa*, 442, 20–26.

Proença, V. and Pereira, H.M. (2013) Species-area models to assess biodiversity change in multi-habitat landscapes: The importance of species habitat affinity. *Basic and Applied Ecology*,14, 102–114.

Prous, X., Ferreira, R.L. and Martins, R.P. (2004) Ecotone delimitation: Epigean–hypogean transition in cave ecosystems. *Austral Ecology*, 29, 374–382.

Pueyo, S., He, F. and Zillio, T. (2007) The maximum entropy formalism and the idiosyncratic theory of biodiversity. *Ecology Letters*, 10, 1017–1028.

Pujolar, J.M., De Leo, G.A., Ciccotti, E. and Zane, L. (2009) Genetic composition of Atlantic and Mediterranean recruits of European eel *Anguilla anguilla* based on EST-linked microsatellite loci. *Journal of Fish Biology*, 74 (9), 2034–2046.

Pulliam, H.R. and Danielson, B.J. (1991) Sources, sinks, and habitat selection: A landscape perspective on population dynamics. *American Naturalist*, 137, S50–S66.

Pullin, A.S. (2002) *Conservation Biology*. Cambridge University Press, Cambridge, UK.

Pulsford, S.A., Lindenmayer, D.B. and Driscoll, D.A. (2016) A succession of theories: Purging redundancy from disturbance theory. *Biological Reviews*, 91, 148–167.

Punttila, P., Haila, Y., Niemelä, N. and Pajunen, T. (1994) Ant communities in fragments of old-growth taiga and managed surroundings. *Annales Zoologici Fennici*, 31, 131–144.

Pusey, A.E., Pintea, L., Wilson M.L., Kamenya, S. and Goodall, J. (2007) The contribution of long-term research at Gombe National Park to chimpanzee conservation. *Conservation Biology*, 21, 623–634.

Putman, R., Langbein, J., Green, P. and Watson, P. (2011) Identifying threshold densities for wild deer in the UK above which negative impacts may occur. *Mammal Review*, 41, 175–196.

Putman, R.J. (1994) *Community Ecology*. Chapman and Hall, London, UK.

Pyke, G.H. (2019) Optimal foraging theory: An introduction. In: Choe, J.C. (ed.) *Encyclopedia of Animal Behavior*, 2nd edn. Elsevier, Amsterdam, The Netherlands, pp. 111–117.

Pyke, G.H. and Starr, C.K. (2021) Optimal foraging theory. In: Starr, C.K. (ed.) *Encyclopedia of Social Insects*. Springer, Cham, Switzerland, pp. 677–685.

Pyšek, P., Prach, K., Joyce, C.B., Mucina, L., Rapson, G.L. and Müllerová, J. (eds) (2001) The role of vegetation succession in ecosystem restoration. *Applied Vegetation Science*, 4, 1–114.

Qiu, W., Rutherford, S., Mao, A. and Chu, C. (2017) The pandemic and its impacts. *Health, Culture and Society*, 9, 1–11.

Rabinowitz, D., Cairns, S. and Dillon, T. (1986) Seven forms of rarity and their frequency in the flora of the British Isles. In: Soulé, M.E. (ed.) *Conservation Biology: The Science of Scarcity and Diversity*. Sinauer Associates, Sunderland, USA, pp. 182–204.

Rabinowitz, D.S. (1981) Seven forms of rarity. In: Synge, H. (ed.) *The Biological Aspects of Rare Plant Conservation*. Wiley, Chichester, UK, pp. 205–217.

Rabl, D., Alonso-Rodríguez, A.M., Brehm, G. and Fiedler, K. (2020) Trait variation in moths mirrors small-scaled ecological gradients in a tropical forest landscape. *Insects*, 11, 612.

Racicot, R.A., Boessenecker, R.W., Darroch, S.A.F. and Geisler, J.H. (2019) Evidence for convergent evolution of ultrasonic hearing in toothed whales. *Biology Letters*, 15, 20190083.

Rae, A. (2017) *A Land Cover Atlas of the United Kingdom*. University of Sheffield, Sheffield, UK. Available at: https://orda.shef.ac.uk/articles/journal_contribution/A_Land_Cover_Atlas_of_the_United_Kingdom_Document_/5266495?file=9231904 (accessed 22 May 2025).

Raerinne, J. (2023) Myths of past biases and progress in biology. *Theory in Biosciences*, 142, 383–399.

Rainey, F.A. and Oren, A. (2006) *Extremophiles*. Academic Press, Cambridge, USA.

Rajchard, J. (2009) Ultraviolet (UV) light perception by birds: A review. *Veterinární Medicína*, 54, 351–359.

Raman, T.R.S., Rawat, G.S. and Johnsingh, A.J.T. (1998) Recovery of tropical rainforest avifauna in relation to vegetation succession following shifting cultivation in Mizoram, north-east India. *Journal of Applied Ecology*, 35, 214–231.

Ramos-Fernandez, G., Smith Aguilar, S.E., Schaffner, C.M., Vick, L.G. and Aureli, F. (2013) Site fidelity in space use by spider monkeys (*Ateles geoffroyi*) in the Yucatan Peninsula, Mexico. *PLOS One*, 8(5), e62813.

Ramsey, K.M., Marcheva, B., Kohsaka, A. and Bass, J. (2007) The clockwork of metabolism. *Annual Review of Nutrition*, 27, 219–240.

Rao, C.R. (1982) Diversity and dissimilarity coefficients: A unified approach. *Theoretical Population Biology*, 21, 24–43.

Rao, Y., Cai, L., Chen, X., Zhou, X., Fu, S. and Huang, H. (2021) Responses of functional traits of macrobenthic communities to human activities in Daya Bay (A subtropical semi-enclosed bay), China. *Frontiers in Environmental Science*, 9, 766580.

Raoult, D., Audic, S., Robert, C., Abergel, C., Renesto, P. *et al.* (2004) The 1.2 megabase genome sequence of Mimivirus. *Science*, 306, 1344–1350.

Rapacciuolo, G., Young, A. and Johnson, R. (2021) Deriving indicators of biodiversity change from unstructured community-contributed data. *Oikos*, 130, 1225–1239.

Rasmont, P., Franzén, M., Lecocq, T., Harpke, A., Roberts, S.P.M. *et al.* (2015) Climatic risk and distribution atlas of European bumblebees. *Biorisk*, 10, 1–236.

Raunkiær, C. (1905) Types biologiques pour la géographie botanique. *Oversigt over det Kongelige Danske Videnskabernes Selskabs Forhandlinger*, 5, 347–438.

Ravenscroft, N.O.M. (1990) The ecology and conservation of the silver-studded blue butterfly *Plebejus argus* L. on the sandlings of East Anglia, England. *Biological Conservation*, 53, 21–36.

Ray, G.C. and Hayden, B.P. (1992) Coastal zone ecotones. In: Hansen, A.J. and DiCastri, F. (eds) *Landscape Boundaries. Consequences for Biotic Diversity and Ecological Flows*. Springer-Verlag, New York, USA.

Rayner, A. (1997) *Degrees of Freedom: Living in Dynamic Boundaries*. Imperial College Press, London, UK.

Rayner, A.D.M. and Boddy, L. (1988) Fungal communities in the decay of wood. In: Marshall, K.C. (ed.) *Advances in Microbial Ecology*. Plenum Press, New York, USA, pp. 115–166.

Razafindratsima, O.H., Brown, K.A., Carvalho, F., Johnson, S.E., Wright, P.C. and Dunham, A.E. (2018) Edge effects on components of diversity and above-ground biomass in a tropical rainforest. *Journal of Applied Ecology*, 55, 977–985.

Reeves, R.R., Smith, B.D., Crespo, E.A., and Notarbartolo di Sciara, G. (2003) *Dolphins, Whales and Porpoises. 2002–2010 Conservation Action Plan for the World's Cetaceans*. IUCN, Gland, Switzerland.

Reguero, M., Goin, F., Hospitaleche, C.A., Dutra, T. and Marenssi, S. (2013) *Late Cretaceous/Paleogene West Antarctica Terrestrial Biota and its Intercontinental Affinities*. Springer, Dordrecht, The Netherlands.

Rehan, S.M., Leys, R. and Schwarz, M.P. (2013) First evidence for a massive extinction event affecting bees close to the K-T boundary. *PLOS One*, 8, e76683.

Reich, D., Green, R.E., Kircher, M., Krause, J., Patterson, N. *et al.* (2010) Genetic history of an archaic hominin group from Denisova Cave in Siberia. *Nature*, 468, 1053–1060.

Reif, J. (2013) Long-term trends in bird populations: A review of patterns and potential drivers in North America and Europe. *Acta Ornithologica*, 48, 1–16.

Reilly, P.F., Tjahjadi, A., Miller, S.L., Akey, J.M. and Tucci, S. (2022) The contribution of Neanderthal introgression to modern human traits. *Current Biology*, 32(18), R970–R983.

Relton, J. (1972) Breeding biology of moorhens on Huntingdonshire farm ponds. *British Birds*, 65, 248–256.

Remien, C.H., Eckwright, M.J. and Ridenhour, B.J. (2021) Structural identifiability of the generalized Lotka–Volterra model for microbiome studies. *Royal Society Open Science*, 8(7), 201378.

Remy, W., Taylor, T.N., Hass, H. and Kerp, H. (1994) Four hundred-million-year-old vesicular arbuscular mycorrhizae. *Proceedings of the National Academy of Sciences of the United States of America*, 91, 11841–11843.

Renault, D., Leclerc, C., Colleu, M.-A., Boutet, A., Hotte, H. *et al.* (2022) The rising threat of climate change for arthropods from Earth's cold regions: Taxonomic rather than native status drives species sensitivity. *Global Change Biology*, 28, 5914–5927.

Renčo, M., Čerevková, A. and Homolova, Z. (2021) Nematode communities indicate the negative impact of *Reynoutria japonica* invasion on soil fauna in ruderal habitats of tatra national park in Slovakia. *Global Ecology and Conservation*, 26, e01470.

Rendoll-Cárcamo, J., Gañán, M., Madriz, R., Convey, P. and Contador, T. (2023) Wing reduction and body size variation along a steep elevation gradient: A case study with Magellanic sub-Antarctic mayflies and stoneflies. *Frontiers in Ecology and Evolution*, 11, 1188889.

Renfrew, C., Todd, I. and Tringham, R. (1974) Beyond a subsistence economy: The evolution of social organization in prehistoric Europe. In: Moore, C.B. (ed.) *Reconstructing Complex Societies: An Archaeological Colloquium* (supplement to the *Bulletin of the American Schools of Oriental Research*, 20), pp. 69–95.

Renkonen, O. (1938) Statisch-okologische Untersuchungen uber die terrestiche kaferwelt der finnischen bruchmoore. *Annales Botanici Societatis Zoologicae-Botanicae Fennicae Vanamo*, 6, 1–231.

Renn, D.F. (1968) *Norman Castles in Britain*. Humanities Press, New York, USA.

Renne, P.R., Deino, A.L., Hilgen, F.J., Kuiper, K.F., Mark, D.F. *et al.* (2013) Time scales of critical events around the Cretaceous-Paleogene boundary. *Science*, 339, 684–687.

Rényi, A. (1961) On measures of entropy and information. In: Neyman, J. (ed.) *Proceedings of the Fourth Berkeley Symposium on Mathematical Statistics and Probability*, vol. 1. University of California Press, Berkeley, USA, pp. 547–560.

Rey Benayas, J., Scheiner, S., García Sánchez-Colomer, M. and Levassor, C. (1999) Commonness and rarity: Theory and application of a new model to Mediterranean montane grasslands. *Conservation Ecology*, 3, 5.

Reyjol, Y., Hugueny, B., Pont, D., Bianco, P.G., Beier, U. *et al.* (2007) Patterns in species richness and endemism of European freshwater fish. *Global Ecology and Biogeography*, 16, 65–75.

Reynolds, C. (2019) The ecological framework: Water chemistry, hydrology and phytoplankton. In: Wall, T. and Wall, G. (eds) *Rostherne Mere: Aspects of a Wetland Nature Reserve*. Tom Wall, Shropshire, UK, pp. 155–172.

Reynolds, S. (2023) Insect declines and ecosystems. *Antenna*, 47, 59–65.

Rhodes, C.J. (2017) The whispering world of plants: 'The Wood Wide Web'. *Science Progress*, 100, 331–337.

Rhodes, O.E., Chesser, R.K. and Smith, M.H. (1996) *Population Dynamics in Ecological Space and Time*. University of Chicago Press, Chicago, USA.

Rice, C.G. (2008) Seasonal altitudinal movements of mountain goats. *Journal of Wildlife Management*, 72(8), 1706–1716.

Richards, M.P. and Trinkaus, E. (2009) Isotopic evidence for the diets of European Neanderthal and early modern humans. *Proceedings of the National Academy of Sciences of the United States of America*, 106, 16034–16039.

Richards, O.W. and Waloff, N. (1954) Studies on the biology and population dynamics of British grasshoppers. *Anti-locust Bulletin*, 17, 1–182.

Richardson, J., Lees, A.C. and Marsden, S. (2023) Landscape-scale predictors of persistence of an urban stock dove *Columba oenas* population. *Urban Ecosystems*, 26, 249–259.

Richner, H. and Heeb, P. (1996) Communal life: Honest signaling and the recruitment center hypothesis. *Behavioral Ecology*, 7, 115.

Ricklefs, R.E. and Lovette, I.J. (1999) The roles of island area per se and habitat diversity in the species–area relationships of four Lesser Antillean faunal groups. *Ecology*, 68, 1142–1160.

Ricotta, C. and Burrascano, S. (2009) Testing for differences in beta diversity with asymmetric dissimilarities. *Ecological Indicators*, 9, 719–724.

Ricotta, C., Carranza, M.L., Avena, G. and Blasi, C. (2002) Are potential natural vegetation maps a meaningful alternative to neutral landscape models? *Applied Vegetation Science*, 5, 271–275.

Ricotta, C., de Bello, F., Moretti, M., Caccianiga, M., Cerabolini, B.E.L. and Pavoine, S. (2016) Measuring the functional redundancy of biological communities: A quantitative guide. *Methods in Ecology and Evolution*, 7, 1386–1395.

Riddle, B.R., Ladle, R.J., Lourie, S.A. and Whittaker, R.J. (2011) Basic biogeography: estimating biodiversity and mapping nature. In: Ladle, R.J. and Whittaker, R.J. (eds) *Conservation Biogeography*. John Wiley and Sons, Chichester, UK, pp. 47–92.

Riechert, S.E. (1981) The consequences of being territorial: Spiders, a case study. *American Naturalist*, 117, 871–892.

Ries, L. and Sisk, T.D. (2004) A predictive model of edge effects. *Ecology*, 85(11), 2917–2926.

Ries, L., Fletcher, R.J. Jr, Battin, J. and Sisk, T.D. (2004) Ecological responses to habitat edges: Mechanisms, models, and variability explained. *Annual Review of Ecology, Evolution, and Systematics*, 35, 491–522.

Ries, L., Murphy, S.M., Wimp, G.M. and Fletcher, R.J. Jr. (2017) Closing persistent gaps in knowledge about edge ecology. *Current Landscape Ecology Reports*, 2, 30–41.

Rikardsen, A.H., Righton, D., Strøm, J.F., Thorstad, E.B., Gargan, P. *et al.* (2021) Redefining the oceanic distribution of Atlantic salmon. *Scientific Reports*, 11, 12266.

Ritchie, H. (2021) Wild mammals have declined by 85% since the rise of humans, but there is a possible future where they flourish. Available at: https://ourworldindata.org/wild-mammal-decline (accessed 22 May 2025).

Ritchie, H. (2022) There have been five mass extinctions in Earth's history. When did the "Big Five" mass extinctions happen, and what were their causes? Our World in Data. Available at: https://ourworldindata.org/mass-extinctions (accessed 22 May 2025).

Ritvo, S. (2022) Optimal foraging theory. In: Vonk, J. and Shackelford, T.K. (eds) *Encyclopedia of Animal Cognition and Behavior*. Springer, Cham, Switzerland, pp. 4833–4838.

Rizzolo, J.B. (2021) Wildlife tourism and consumption. *Journal of Sustainable Tourism*, 31, 1–14.

Robbins, M.M. (2008) Feeding competition and agonistic relationships among Bwindi *Gorilla beringei*. *International Journal of Primatology*, 29, 999–1018.

Robbins, M.M. and McNeilage, A. (2003) Home range and frugivory patterns of mountain gorillas in Bwindi Impenetrable National Park, Uganda. *International Journal of Primatology*, 24, 467–491.

Robbins, M.M., Bermejo, M., Cipolletta, C., Magliocca, F., Parnell, R.J. and Stokes, E. (2004) Social structure and life-history patterns in western gorillas (*Gorilla gorilla gorilla*). *American Journal of Primatology*, 64, 145–159.

Roberge, J.M. and Angelstam, P. (2004) Usefulness of the umbrella species concept as a conservation tool. *Conservation Biology*, 18, 76–85.

Roberts, E.M., Todd, C.N., Aanen, D.K., Nobre, T., Hilbert-Wolf, H.L. *et al.* (2016) Oligocene termite nests with in situ fungus gardens from the Rukwa rift basin, Tanzania, support a Paleogene African origin for insect agriculture. *PLOS One*, 11, e0156847.

Roberts, N. (2014) *The Holocene: An Environmental History*. Wiley-Blackwell, Chichester, UK.

Robinson, C., Steinberg, D.K., Anderson, T.R., Arístegui, J., Carlson, C.A. *et al.* (2010). Mesopelagic zone ecology and biogeochemistry – a synthesis. *Deep-Sea Research. Part II, Topical Studies in Oceanography*, 57(16), 1504–1518.

Robinson, D.G., Ammer, C., Polle, A., Bauhus. J., Aloni, R. *et al.* (2024) Mother trees, altruistic fungi, and the perils of plant personification. *Trends in Plant Science*, 29, 20–31.

Robinson, R.A., Freeman, S.N., Balmer, D.E. and Grantham, J.A. (2007) Cetti's warbler *Cettia cetti*: Analysis of an expanding population. *Bird Study*, 54, 230–235.

Robson, G.C. and Richards, O.W. (1936) *The Variation of Animals in Nature*. Longmans, Green and Co., London.

Rocchini, D., Tordoni, E., Marchetto, E., Marcantonio, M., Barbosa, A.M. *et al.* (2023) A quixotic view of spatial bias in modelling the distribution of species and their diversity. *NPJ Biodiversity*, 2, 10.

Rockström, J., Steffen, W., Noone, K., Persson, A., Chapin, F.S. 3rd *et al.* (2009) A safe operating space for humanity. *Nature*, 461(7263), 472–475.

Rockwood, L.L. (2015) *Introduction to Population Ecology*, 2nd edn. Wiley-Blackwell, Chichester, UK.

Rodriguez-Cabal, M.A., Barrios-Garcia, M.N., Amico, G.C., Aizen, M.A. and Sanders, N.J. (2013) Node-by-node disassembly of a mutualistic interaction web driven by species introductions. *Proceedings of the National Academy of Sciences of the United States of America*, 110, 16503–16507.

Rodwell, J.S. (2006) *National Vegetation Classification: Users' Handbook*. Joint Nature Conservation Committee, Peterborough, UK.

Roesch, L.F.W., Fulthorpe, R.R., Riva, A., Casella, G., Hadwin, A.K.M. *et al.* (2007) Pyrosequencing enumerates and contrasts soil microbial diversity. *ISME Journal*, 1, 283–290.

Roff, D.A. (1986) The evolution of wing dimorphism in insects. *Evolution*, 40(5), 1009–1020.

Roff, D.A. (1990) The evolution of flightlessness in insects. *Ecological Monographs*, 60(4), 389–421.

Rogers, A.R., Bohlender, R.J. and Huff, C.D. (2017) Early history of Neanderthals and Denisovans. *Proceedings of the National Academy of Sciences of the United States of America*, 114, 9859–9863.

Rohde, K. (2006) *Nonequilibrium Ecology*. Cambridge University Press, Cambridge, UK.

Rohwer, F. (2003) Global phage diversity. *Cell*, 113(2), 141.

Roll, U., Feldman, A., Novosolov, M., Allison, A., Bauer, A.M. *et.al.* (2017) The global distribution of tetrapods reveals a need for targeted reptile conservation. *Nature Ecology and Evolution*, 1, 1677–1682.

Roman, J., Estes, J.A., Morissette, L., Smith, C., Costa, D. *et al.* (2014) Whales as marine ecosystem engineers. *Frontiers in Ecology and the Environment*, 12, 377–385.

Romero, A. (2009) *Cave Biology: Life in Darkness*. Cambridge University Press, New York, USA.

Romero, J. (2022) What was on the prehistoric menu? *The Biologist*, 69, 16–19.

Rondinini, C., Di Marco, M., Visconti, P., Butchart, S.H.M. and Boitani, L. (2014) Update or outdate: Long-term viability of the IUCN Red List. *Conservation Letters*, 7, 126–130.

Root, R.B. (1967) The niche exploitation pattern of the blue-gray gnatcatcher. *Ecological Monographs*, 37, 317–350.

Root, R.B. (1973) Organisation of a plant-arthropod association in simple and diverse habitats: The fauna of collards (*Brassica oleracea*). *Ecological Monographs*, 43, 95–124.

Roper, C.F.E. (2016) Architeuthidae Pfeffer 1900, *Architeuthis* Steenstrup 1857, *Architeuthis dux* Steenstrup, 1857. Available at: http://tolweb.org/Architeuthis_dux/19408/2016.02.27 (accessed 22 May 2025).

Roper, C.F.E. and Shea, E.K. (2013) Unanswered questions about the giant squid *Architeuthis* (Architeuthidae) illustrate our incomplete knowledge of coleoid cephalopods. *American Malacological Bulletin*, 31(1), 109–122.

Roscher, C., Schumacher, J., Gubsch, M., Lipowsky, A., Weigelt, A. *et al.* (2012) Using plant functional traits to explain diversity-productivity relationships. *PLOS One*, 7, e36760.

Rosa, R., Lopes, V.M., Guerreiro, M., Bolstad, K. and Xavier, J.C. (2017) Biology and ecology of the world's largest invertebrate, the colossal squid (*Mesonychoteuthis hamiltoni*): a short review. *Polar Biology*, 40, 1871–1883.

Rose, R.K., Nadolny, R.M., Kiser, J., Rice, S.E., Salamone, H.G. *et al.* (2018) Compositional changes in two small mammal communities during succession in southeastern Virginia. *Virginia Journal of Science*, 69, 1–12.

Rosenberg, G. (2014) A new critical estimate of named species-level diversity of the recent Mollusca. *American Malacological Bulletin*, 32 (2), 308–322.

Rosenberger, A.L. and Hartwig, W.C. (2013) New World monkeys. In: *Encyclopedia of Life Sciences*. John Wiley and Sons, Chichester, UK, pp. 1–10.

Rosenfeld, J.S. (2002) Functional redundancy in ecology and conservation. *Oikos*, 98, 156–162.

Rosenzweig, M.L. (1979) Optimal habitat selection in two-species competitive systems. *Fortschritte der Zoologie*, 25, 283–293.

Rosenzweig, M.L. (1995) *Species Diversity in Space and Time*. Cambridge University Press, Cambridge, UK.

Rosenzweig, M.L. and MacArthur, R. (1963) Graphical representation and stability conditions of predator-prey interaction. *American Naturalist*, 97, 209–223.

Rosindell, J., Hubbell, S.P. and Etienne, R.S. (2011) The unified neutral theory of biodiversity and biogeography at age ten. *Trends in Ecology and Evolution*, 26, 340–348.

Roskam, J.C. (2019) *Plant Galls of Europe*. KNNV Uitgeverij, Zeist, The Netherlands.

Ross, D.J. (2001) Impact of the northern Pacific seastar *Asterias amurensis* on soft sediment assemblages, including commercial species, in southeast Tasmania. PhD thesis, University of Tasmania.

Ross, E.S. (2019) *Megalopolis*. In: Orum A.M. (ed.) *The Wiley Blackwell Encyclopedia of Urban and Regional Studies*. John Wiley and Sons, Chichester, UK, pp. 1–4.

Ross, K.G. and Matthews, R.W. (1991) *The Social Biology of Wasps*. Comstock, Ithaca, USA.

Rosvall, M. and Bergstrom, C.T. (2007) An information-theoretic framework for resolving community structure in complex networks. *Proceedings of the National Academy of Sciences of the United States of America*, 104(18), 7327–7331.

Rosvall, M. and Bergstrom, C.T. (2010) Mapping change in large networks. *PLOS One*, 5(1), e8694.

Roswell, M., Dushoff, J. and Winfree, R. (2021) A conceptual guide to measuring species diversity. *Oikos*, 130, 321–338.

Rowe, N. (1996) *The Pictorial Guide to the Living Primates*. Pogonias Press, Charlestown, USA.

Rowley, J.J.L., Dau, Q.V., Nguyen, T.T., Cao, T.T. and Nguyen, S.V. (2011) A new species of *Gracixalus* (Anura: Rhacophoridae) with a hyperextended vocal repertoire from Vietnam. *Zootaxa*, 3125, 22–38.

Roy, H., Bacon, J., Beckmann, B., Harrower, C.A. and Hill, M.O. (2012) *Non-Native Species in Great Britain: Establishment, Detection and Reporting to Inform Effective Decision Making*. Centre for Ecology and Hydrology, Wallingford, UK.

Royce, W.F. (1996) *Introduction to the Practice of Fishery Science*, 2nd edn. Academic Press, London, UK.

Royle, J.A. and Converse, S.J. (2020) Estimating abundance from capture-recapture data. In: Murray, D.L. and Sandercock, B.K. (eds) *Population Ecology in Practice*. John Wiley and Sons, Chichester, UK, pp. 103–122.

Rubenstein, M.A., Weiskopf, S.R., Bertrand, R., Carter, S.L., Comte, L. *et al.* (2023) Climate change and the global redistribution of biodiversity: Substantial variation in empirical support for expected range shifts. *Environmental Evidence*, 12,7.

Rubio, J.L., Safriel, U.N., Daussa, R., Blum, W.E.H. and Pedrazzini, F. (eds) (2009) *Water Scarcity, Land Degradation and Desertification in the Mediterranean Region. Environmental and Security Aspects*. NATO Science for Peace and Security Series C: Environmental Security. Springer, Dordrecht, The Netherlands.

Rubio-López, I., Pardos, F., Fontaneto, D., Martínez, A. and García-Gómez, G. (2023) Biases and distribution patterns in hard-bodied microscopic animals (Acari: Halacaridae): Size does not matter, but generalism and sampling effort do. *Diversity and Distributions*, 29, 821–833.

Rueda, M., Rodríguez, M.Á. and Hawkins, B.A. (2010) Towards a biogeographic regionalization of the

European biota. *Journal of Biogeography*, 37, 2067–2076.

Ruggiero, M.A., Gordon, D.P., Orrell, T.M., Bailly, N., Bourgoin, T. *et al.* (2015) A higher level classification of all living organisms. *PLOS One*, 10(4), e0119248.

Rull, V., Connor, S.E. and Elias, R.B. (2017) Potential natural vegetation and pre-anthropic pollen records on the Azores Islands in a Macaronesian context. *Journal of Biogeography*, 44, 2437–2440.

Runge, C.A., Withey, J.C., Naugle, D.E., Fargione, J.E., Helmstedt, K.J. *et al.* (2019) Single species conservation as an umbrella for management of landscape threats. *PLOS One*, 14, e0209619.

Runnels, C.N., Payne, C., Rifkind, N.V., White, C., Wolff, N.P. and LeBlanc, S.A. (2009) Warfare in Neolithic Thessaly: A case study. *Hesperia: Journal of the American School of Classical Studies at Athens*, 78, 165–194.

Runyan, C. and D'Odorico, P. (2016) *Global Deforestation*. Cambridge University Press, Cambridge, UK.

Ruprecht, E. (2005) Secondary succession in old-fields in the Transylvanian lowland (Romania). *Preslia*, 77, 145–157.

Russell, M.B., Fraver, S., Aakala, T., Gove, J.H., Woodall, C.W. *et al.* (2015) Quantifying carbon stores and decomposition in dead wood: A review. *Forest Ecology and Management*, 350, 107–128.

Russo, A. and Lax, G. (2022) Using Artificial Intelligence for space challenges: A survey. *Applied Sciences*, 12(10), 5106.

Rutherford, M.G. (2011) Viable land snail eggs in a bird pellet. *Living World, Journal of the Trinidad and Tobago Field Naturalists' Club*, 2011, 73–74.

Ruxton, G.D., Sherratt, T.N. and Speed, M.P. (2004) *Avoiding Attack: The Evolutionary Ecology of Crypsis, Warning Signals, and Mimicry*. Oxford University Press, Oxford, UK.

Ryan, T. (2019) *Marine Biology: An Ecological Approach*. ED-Tech Press, Waltham Abbey, UK.

Rydin, H. and Borgegård, S-O. (1991) Plant characteristics over a century of primary succession on islands: Lake Hjälmaren. *Ecology*, 72, 1089–1101.

Sabu, T.K., Vinod, K.V. and Vineesh, P.J. (2006) Guild structure, diversity and succession of dung beetles associated with Indian elephant dung in South Western Ghats forests. *Journal of Insect Science*, 6, 17.

Sabu, T.K., Shiju, R.T., Vinod, K. and Nithya, S. (2011) A comparison of the pitfall trap, Winkler extractor and Berlese funnel for sampling ground-dwelling arthropods in tropical montane cloud forests. *Journal of Insect Science*, 11, 28.

Saccheri, I., Kuussaari, M., Kankare, M., Vikman, P., Fortelius, W. and Hanski, I. (1998) Inbreeding and extinction in a butterfly metapopulation. *Nature*, 392, 491–494.

Saccheri, I.J., Wilson, I.J., Nichols, R.A., Bruford, M.W. and Brakefield, P.M. (1999) Inbreeding of bottle-necked butterfly populations: Estimation using the likelihood of changes in marker allele frequencies. *Genetics*, 151, 1053–1063.

Sadowski, R.F. (2017) The Neolithic revolution. In: Thompson, P.B. and Kaplan, D.M. (eds) *Encyclopedia of Food and Agricultural Ethics*, 2nd edn. Springer, Dordrecht, The Netherlands, pp. 1–8.

Sætersdal, M. and Birks, H.J.B. (1997) A comparative ecological study of Norwegian mountain plants in relation to possible future climatic change. *Journal of Biogeography*, 24, 127–152.

Sagan, D. (2010) Introduction: Umwelt after Uexküll. In: von Uexküll, J., von Uexküll, M. and O'Neil, J.D. (eds) *A Foray into the Worlds of Animals and Humans: With a Theory of Meaning*. University of Minnesota Press, Minneapolis, USA, pp. 1–34.

Saitoh, S., Suzuki, H., Hanzawa, N. and Tamate, HB. (2011) Species diversity and community structure of pelagic copepods in the marine lakes of Palau. *Hydrobiologia*, 666, 85–97.

Salditt, A., Karr, L., Salibi, E., Le Vay, K., Braun, D. and Mutschler, H. (2023) Ribozyme-mediated RNA synthesis and replication in a model Hadean microenvironment. *Nature Communications*, 14(1), 1495.

Sale, P.F. and Tolimieri, N. (2000) Density dependence at some time and place? *Oecologia*, 124, 166–171.

Sales, K., Vasudeva, R., Dickinson, M.E., Godwin, J.L., Lumley, A.J. *et al.* (2018) Experimental heatwaves compromise sperm function and cause transgenerational damage in a model insect. *Nature Communications*, 9, 4771.

Sales, L.P., Hayward, M.W. and Loyola, R. (2021) What do you mean by "niche"? Modern ecological theories are not coherent on rhetoric about the niche concept. *Acta Oecologica*, 110, 103701.

Salgueiro, P.A., Silva, C., Silva, A., Sá, C. and Mira, A. (2019) Can quarries provide novel conditions for a bird of rocky habitats? *Restoration Ecology*, 28, 988–994.

Salgueiro, P.A., Prach, K., Branquinho, C. and Mira, A. (2020) Enhancing biodiversity and ecosystem services in quarry restoration – Challenges, strategies and practice. *Restoration Ecology*, 28, 655–660.

Salo, J., Kalliola, R., Häkkinen, I., Mäkinen, Y., Niemelä, P. *et al.* (1986) River dynamics and the diversity of Amazon lowland forest. *Nature*, 322, 254–258.

Salvati, L., Manganaro, A., Fattorini, S. and Piattella, E. (1999) Population features of kestrels *Falco tinnunculus* in urban, suburban and rural areas in Central Italy. *Acta Ornithologica*, 34, 54–58.

Salvati, L., Manganaro, A. and Fattorini, S. (2000) Responsiveness of nesting Eurasian kestrels *Falco tinnunculus* to call playbacks. *Journal of Raptor Research*, 34, 319–321.

Salvati, L., Manganaro, A. and Fattorini, S. (2002) Breeding density, colony size, and colony spacing in relation to nest sites in an urban jackdaw *Corvus monedula* population. *Larus*, 48, 39–45.

Salvati, L., Zitti, M., and Perini, L. (2016) Fifty years on: Long-term patterns of land sensitivity to desertification in Italy. *Land Degradation and Development*, 27, 97–107.

Samonds, K.E., Godfrey, L.R., Ali, J.R., Goodman, S.M., Vences, M. *et al.* (2013) Imperfect isolation: Factors and filters shaping Madagascar's extant vertebrate fauna. *PLOS One*, 8, e62086.

Samraoui, F., Alfarhan, A.H. and Samraoui, B. (2013) Status and breeding ecology of the common moorhen *Gallinula chloropus* in Algeria. *Ostrich: Journal of Ornithology*, 84, 137–144.

Samuels, D.S. and Samuels, L.R.N. (2016) Gene regulation during the enzootic cycle of the Lyme disease spirochete. *Forum on Immunopathological Diseases and Therapeutics*, 7, 205–212.

Samuels, D.S., Lybecker, M.C., Yang, X.F., Ouyang, Z., Bourret, T.J. *et al.* (2021) Gene regulation and transcriptomics. In: Radolf, J.D. and Samuels, D.S. (eds) *Lyme Disease and Relapsing Fever Spirochetes, Genomics, Molecular Biology, Host Interactions, and Disease Pathogenesis.* Caister Academic Press, Norfolk, UK, pp. 87–129.

Samways, D. (2019) Book review. *Empty planet: The shock of global population decline* by Darell Bricker and John Ibbitson. *Journal of Population and Sustainability*, 3, 87–90.

Samways, M.J., New, T.R. and McGeoch, M.A. (2009) *Insect Conservation: A Handbook of Approaches and Methods.* Oxford University Press, Oxford, UK.

Sánchez-Fernández, D., Fox, R., Dennis, R.L.H. and Lobo, J.M. (2021) How complete are insect inventories? An assessment of the British butterfly database highlighting the influence of dynamic distribution shifts on sampling completeness. *Biodiversity and Conservation*, 30, 889–902.

Sandel, A.A. and Watts, D.P. (2021) Lethal coalitionary aggression associated with a community fission in chimpanzees (*Pan Troglodytes*) at Ngogo, Kibale National Park, Uganda. *International Journal of Primatology*, 42, 26–48.

Sander, E.L., Wootton, J.T. and Allesina, S. (2015) What can interaction webs tell us about species roles? *PLOS Computational Biology*, 11, e1004330.

Sanderson, J.G. and Pimm, S.L. (2015) *Patterns in Nature. The Analysis of Species Co-occurrences.* University of Chicago Press, Chicago, USA.

Sandom, C., Donlan, C.J., Svenning, J.-C. and Hansen, D. (2013) Rewilding. In: Macdonald, D.W. and Willis, K.J. (eds) *Key Topics in Conservation Biology 2.* Wiley-Blackwell, Chichester, UK, pp. 430–451.

Sandu, C., Brasoveanu, D., Silivestru, V., Filipescu, B. and Sandu, R.C. (2017) Technology for terraformation of Mars, other planets and natural satellites. Council for European Aerospace Societies, Aerospace Europe 6th CEAS Conference, CEAS 2017 paper no. 199, pp. 1–8.

Santos, J.C. and Fernandes, G.W. (eds) (2021) *Measuring Arthropod Biodiversity: A Handbook of Sampling Methods.* Springer, Cham, Switzerland.

Sanz, C.M., Strait, D., Ayina, C.E., Massamba, J.M., Ebombi, T.F. *et al.* (2022) Interspecific interactions between sympatric apes. *iScience*, 25, 105059.

Sargis, E.J., Millien, V., Woodman, N. and Olson, L.E. (2018) Rule reversal: Ecogeographical patterns of body size variation in the common treeshrew (Mammalia, Scandentia). *Ecology and Evolution*, 8, 1634–1645.

Säterberg, T., Sellman, S. and Ebenman, B. (2013) High frequency of functional extinctions in ecological networks. *Nature*, 499, 468–470.

Sauther, M.L., Cuozzo, F.P., Youssouf, J.I.A., Fish, K.D., LaFleur, M. *et al.* (2013) Limestone cliff-face and cave use by wild ring-tailed lemurs (*Lemur catta*) in southwestern Madagascar. *Madagascar Conservation and Development*, 8, 73–80.

Sauther, M.L., Gould, L., Cuozzo, F.P. and O'Mara, M.T. (2015) Ring-tailed lemurs: A species re-imagined. *Folia Primatologia*, 86, 5–13.

Savoca, M.S., Czapanskiy, M.F., Kahane-Rapport, S.R., Gough, W.T., Fahlbusch, J.A. *et al.* (2021) Baleen whale prey consumption based on high-resolution foraging measurements. *Nature*, 599, 85–90.

Savolainen, R. and Deslippe, R.J. (2001) Facultative and obligate slave making in *Formica* ants. *Naturwissenschaften*, 88, 347–350.

Sayol, F., Steinbauer, M.J., Blackburn, T.M., Antonelli, A. and Faurby, S. (2020) Anthropogenic extinctions conceal widespread evolution of flightlessness in birds. *Science Advances*, 6, eabb6095.

Scanes, C.G. (2018) Animal perception including differences with humans. In: Scanes, C.G. and Toukhsati, S.R. (eds) *Animals and Human Society.* Elsevier, Amsterdam, The Netherlands, pp. 1–11.

Schaller, G.B. (1963) *The Mountain Gorilla: Ecology and Behavior.* University of Chicago Press, Chicago, USA.

Schatzmann, E., Gerrard, R. and Barbour, A.D. (1986) Measures of niche overlap, I. *IMA Journal of Mathematics Applied in Medicine and Biology*, 3, 99–113.

Scheele, B.C. Heard, G.W., Cardillo. M., Duncan, R.P., Gillespie, G.R. *et al.* (2022) An invasive pathogen drives directional niche contractions in amphibians. *Nature Ecology and Evolution*, 7, 1–11.

Scheiner, S.M. and Willig, M.R. (2008) A general theory of ecology. *Theoretical Ecology*, 1, 21–28.

Schertler, A., Rabitsch, W., Moser, D., Wessely J. and Essl, F. (2020) The potential current distribution of the coypu (*Myocastor coypus*) in Europe and climate change induced shifts in the near future. *NeoBiota*, 58, 129–160.

Schieck, J. and Song, S.J. (2006) Changes in bird communities throughout succession following fire and harvest in boreal forests of western North America: Literature review and meta-analyses. *Canadian Journal of Forest Research*, 36, 1299–1318.

Schimitschek, E. (1952–1953) Forstentomologische Studien in Urwald Rotwald. *Zeitschrift für Angewandte Entomologie*, 34, 178–215, 513–542; 35, 1–54.

Schinner, F., Ohlinger, E.R., Kandeler, E. and Margesin, R. (2012) *Methods in Soil Biology*. Springer Science and Business Media, Berlin, Germany.

Schipper, J., Chanson, J.S., Chiozza, F., Cox, N.A., Hoffmann, M. *et al.* (2008) The status of the world's land and marine mammals: Diversity, threat, and knowledge. *Science*, 322, 225–230.

Schlegel, H.G. and Jannasch, H.W. (2006) Prokaryotes and their habitats. In: Dworkin, M., Falkow, S., Rosenberg, E., Schleifer, K.H. and Stackebrandt, E. (eds) *The Prokaryotes*. Springer, New York, USA, pp. 137–184.

Schleuter, D., Daufresne, M., Massol, F. and Argillier, C. (2010) A user's guide to functional diversity indices. *Ecological Monographs*, 80, 469–484.

Schlingmann, A., Graham, S., Benyei, P., Corbera, E., Martinez Sanesteban, I. *et al.* (2021) Global patterns of adaptation to climate change by indigenous peoples and local communities. A systematic review. *Current Opinion in Environmental Sustainability*, 51, 55–64.

Schmarda, K.L. (1853) *Die geographische Verbreitung der Thiere*. Carl Gerold and Son, Vienna, Austria.

Schmera, D., Ricotta, C. and Podani, J. (2023) Components of functional diversity revisited: A new classification and its theoretical and practical implications. *Ecology and Evolution*, 13, e10614.

Schmid, J.S., Taubert, F., Wiegand, T., Sun, I.-F. and Huth, A. (2020) Network science applied to forest megaplots: Tropical tree species coexist in small-world networks. *Scientific Reports*, 10(1), 1–10.

Schmidt, H.G. and Garbutt, D.J. (1985) Species abundance data from fouling communities conform to the gamma distribution. *Marine Ecology Progress Series*, 23, 287–290.

Schmidt, J. (1922) The breeding places of the eel. *Philosophical Transactions of the Royal Society B: Biological Sciences*, 211, 179–208.

Schmidt, N.M., Hardwick, B., Gilg, O., Høye, T.T., Krogh, P.H. *et al.* (2017) Interaction webs in arctic ecosystems: Determinants of arctic change? *Ambio*, 46(Suppl. 1), S12–S25.

Schmitt, T. and Varga, Z. (2012) Extra-Mediterranean refugia: The rule and not the exception? *Frontiers in Zoology*, 9, 22.

Schmitt, C.J., Cook, J.A., Zamudio, K.R. and Edwards, S.V. Museum (2018) specimens of terrestrial vertebrates are sensitive indicators of environmental change in the Anthropocene. *Philosphical Transactions of the Royal Society of London. Series B, Biological Sciences*, 374(1763), 20170387.

Schmitt, M.H., Stears, K., Wilmers, C.C. and Shrader, A.M. (2014) Determining the relative importance of dilution and detection for zebra foraging in mixed-species herds. *Animal Behaviour*, 96, 151–158.

Schmitt, T., Fritz, U., Delfino, M., Ulrich, W. and Habel, J.C. (2021) Biogeography of Italy revisited: Genetic lineages confirm major phylogeographic patterns and a pre-Pleistocene origin of its biota. *Frontiers in Zoology*, 18, 34.

Schneider, F.D., Fichtmueller, D., Gossner, M.M., Güntsch, A., Jochum, M. *et al.* (2019) Towards an ecological trait-data standard. *Methods in Ecology and Evolution*, 10, 2006–2019.

Schneider, H. and Sampaio, I. (2015) The systematics and evolution of New World primates – A review. *Molecular Phylogenetics and Evolution*, 82(Pt B), 348–357.

Schneider, S.C., García, A., Martín-Closas, C. and Chivas, A.R. (2015) The role of Charophytes (Charales) in past and present environments: An overview. *Aquatic Botany*, 120, 2–6.

Schnell, J.K., Harris, G.M., Pimm, S.L. and Russell, G.J. (2013) Estimating extinction risk with metapopulation models of large-scale fragmentation. *Conservation Biology*, 27, 520–530.

Schoener, T.W. (1974) Resource partitioning in ecological communities. *Science*, 185, 27–39.

Schoener, T.W. (1989a) The ecological niche. In: Cherrett, J.M. (ed.) *Ecological Concepts: The Contribution of Ecology to an Understanding of the Natural World*. Blackwell Scientific, Oxford, UK, pp. 79–114.

Schoener, T.W. (1989b) Food webs from the small to the large. *Ecology*, 70, 1559–1589.

Schoener, T.W., Spiller, D.A. and Losos, J.B. (2001) Natural restoration of the species–area relation for a lizard after a hurricane. *Science*, 294, 1525–1528.

Schönrogge, K., Wardlaw, J.C., Peters, A.J., Everett, S., Thomas, J.A. and Elmes, G.W. (2004) Changes in chemical signature and host specificity from larval retrieval to full social integration in the myrmecophilous butterfly. *Journal of Chemical Ecology*, 30, 91–107.

Schorger, A.W. (1955) *The Passenger Pigeon: Its Natural History and Extinction*. University of Oklahoma Press, Norman, USA.

Schowalter, T.D. (1995) Canopy arthropod communities in relation to forest age and alternative harvest practices in western Oregon. *Forest Ecology and Management*, 78, 115–125.

Schowalter, T.D. (2016) *Insect Ecology. An Ecosystem Approach*. Elsevier, New York, USA.

Schowalter, T.D. and Ganio, L.M. (2003) Diel, seasonal and disturbance-induced variation in invertebrate assemblages. In: Basset, Y., Novotny, V., Miller, S.E. and Kitching, R.L. (eds) *Arthropods of Tropical Forests*. Cambridge University Press, Cambridge, UK, pp. 315–328.

Schultz, C.B., Franco, A.M.A. and Crone, E.E. (2012) Response of butterflies to structural and resource boundaries. *Journal of Animal Ecology*, 81, 724–734.

Schultz, J. (2005) *The Ecozones of the World: The Ecological Divisions of the Geosphere*, 2nd edn. (trans. Ahnert, B.). Springer-Verlag, Berlin, Germany.

Schulz, S., Brankatschk, R., Dümig, A., Kögel-Knabner, I., Schloter, M. and Zeyer, J. (2013) The role of microorganisms at different stages of ecosystem development for soil formation. *Biogeosciences*, 10, 3983–3996.

Schütte, U.M.E., Abdo, Z., Bent, S.J., Williams, C.J., Schneider, G.M. *et al.* (2009) Bacterial succession in a glacier foreland of the high arctic. *ISME Journal*, 3, 1258e1268.

Schwartz-Marin, E., Gonzales-Romero, O., Murrieta-Fores, P., Hauskeller, C., and Favila-Vázquez, M. (2021) Psychedelic research and its biocolonial legacies. *PuLSe*, 50, 2–3.

Sclater, P.L. (1858) On the general geographical distribution of the members of the class Aves. *Journal of the Proceedings of the Linnean Society: Zoology*, 2, 130–145.

Scoble, M.J. (1992) *The Lepidoptera: Form, Function and Diversity*. Oxford University Press, Oxford, UK.

Scott, G.R. and Gibert, L. (2009) The oldest hand-axes in Europe. *Nature*, 461, 82–85.

Scott, M.P. (1998) The ecology and behavior of burying beetles. *Annual Review of Entomology*, 43, 595–618.

Scull, P., Palmer, M., Frey, F. and Kraly, E. (2012) A comparison of two home range modeling methods using Ugandan mountain gorilla data. *International Journal of Geographical Information Science*, 26, 2111–2121.

Seber, G.A.F. (1982) *The Estimation of Animal Abundance and Related Parameters*, 2nd edn. Macmillan, New York, USA.

Secor, D.H. (2015) *Migration Ecology of Marine Fishes*. Johns Hopkins University Press, Baltimore, USA.

Secretariat of the Convention on Biological Diversity (2003) *Handbook of the Convention on Biological Diversity*. Earthscan, London, UK.

Secretariat of the Convention on Biological Diversity (2011) *Convention on Biological Diversity. Text and Annexes*. United Nations Environment Programme, Montreal, Canada.

Segura, A. (2017) How does vegetation structure influence woodpeckers and secondary cavity nesting birds in African cork oak forest? *Acta Oecologica*, 83, 22–28.

Selås, V. (2016) Timing of population peaks of Norway lemming in relation to atmospheric pressure: A hypothesis to explain the spatial synchrony. *Scientific Reports*, 6, 27225.

Selosse, M.-A. (2002) Prototaxites: A 400 Myr-old giant fossil, a saprophytic holobasidionmycete, or lichen? *Mycological Research*, 106, 641–644.

Selva, N., Jędrzejewska, B., Jędrzejewski, W. and Wajrak, A. (2005) Factors affecting carcass use by a guild of scavengers in European temperate woodland. *Canadian Journal of Zoology*, 83, 1590–1601.

Sender, R., Fuchs, S. and Milo, R. (2016) Revised estimates for the number of human and bacteria cells in the body. *PLoS Biology*, 14, e1002533.

Seneviratne, S.I., Zhang, X., Adnan, M., Badi, W., Dereczynski, C. *et al.* (2021) Weather and climate extreme events in a changing climate. In: Masson-Delmotte, V., Zhai, P., Pirani, A., Connors, S.L., Péan, C. *et al.* (eds) *Climate Change 2021: The Physical Science Basis. Contribution of Working Group I to the Sixth Assessment Report of the Intergovernmental Panel on Climate Change*. Cambridge University Press, Cambridge, UK, pp. 1513–1766.

Seno, H. (2022) *A Primer on Population Dynamics Modeling: Basic Ideas for Mathematical Formulation*. SpringerNatureSingapore, Singapore.

Sepkoski, J.J. (1978) A kinetic model of Phanerozoic taxonomic diversity I. Analysis of marine orders. *Paleobiology*, 4, 223–251.

Sepkoski, J.J. (1984) A kinetic model of Phanerozoic taxonomic diversity. III. Post-Paleozoic families and mass extinctions. *Paleobiology*, 10, 246–267.

Sepkoski, J.J. (2002) A compendium of fossil marine animal genera. *Bulletins of American Paleontology*, 363, 1–560.

Sergio, F. and Hiraldo, F. (2008) Intraguild predation in raptor assemblages: A review. *Ibis*, 150, 132–145.

Serniak, L.T., Corbin, C.E., Pitt, A.L. and Rier, S.T. (2017) Effects of Japanese knotweed on avian diversity and function in riparian habitats. *Journal of Ornithology*, 158, 311–321.

Serong, M and Lill, A. (2012) Changes in bird assemblages during succession following disturbance in secondary wet forests in south-eastern Australia. *Emu*, 112, 117–128.

Serrat, A., Pons, P., Puig-Gironès, R. and Stefanescu, C. (2015) Environmental factors influencing butterfly abundance after a severe wildfire in Mediterranean vegetation. *Animal Biodiversity and Conservation*, 38, 207–220.

Settele, J., Kudrna, O., Harpke, A., Kühn, I., van Swaay, C. *et al.* (2008) Climate Risk Atlas of European Butterflies. *BioRisk*, 1, 1–712.

Seuront, L., Moisez, E., Hennion, C., Seuront-Scheffbuch, D. and Seuront, L.M.Y. (2018) *Littorina littorea* show small-scale persistent tidal height and habitat partitioning that is resilient to dislodgement through specific movement rates. *Journal of Experimental Marine Biology and Ecology*, 509, 24–35.

Sfenthourakis, S., Triantis, K.A., Proios, K. and Rigal, F. (2021) The role of ecological specialization in shaping patterns of insular communities. *Journal of Biogeography*, 48, 243–252.

Shafer, H.J. and Bryant, V.M. (1977) *Archeological and Botanical studies at Hinds Cave, Val Verde County, Texas*. Texas A&M University Anthropology Laboratory, College Station, USA.

Shannon, C.E. (1948) A Mathematical Theory of Communication. *Bell System Technical Journal*, 27, 379–423.

Shaw, R.A. (2021) *Global Lithium (Li) Mines, Deposits and Occurrences*. British Geological Survey, Keyworth, UK.

Shaykheeva, D., Panasyuk, M., Malganova, I. and Khairullin, I. (2016) World population estimates and projections: Data and methods. *Journal of Economics and Economic Education Research*, 17, 237–247.

Shea, J.J. (2013) *Stone Tools in the Paleolithic and Neolithic of the Near East: A Guide*. Cambridge University Press, Cambridge, UK.

Shea, J.J. (2014) Sink the Mousterian? Named stone tool industries (NASTIES) as obstacles to investigating hominin evolutionary relationships in the Later Middle Paleolithic Levant. *Quaternary International*, 350, 169–179.

Sheard, C., Neate-Clegg, M.H.C., Alioravainen, N., Jones, S E.I., Vincent, C. *et al.* (2020) Ecological drivers of global gradients in avian dispersal inferred from wing morphology. *Nature Communications*, 11, 2463.

Sheard, C., Street, S.E., Healy, S.D., Troisi, C.A., Clark, A.D. *et al.* (2024) Nest traits for the world's birds. *Global Ecology and Biogeography*, 33, 206–214.

Sheehy, E., Sutherland, C., O'Reilly, C. and Lambin, X. (2018) The enemy of my enemy is my friend: Native pine marten recovery reverses the decline of the red squirrel by suppressing grey squirrel populations. *Proceedings of the Royal Society B: Biological Sciences*, 285, 20172603.

Sheela, A.M., Manimekalai, B. and Dhinagaran, G. (2022) Review on the distribution of microplastics in the oceans and its impacts: Need for modeling-based approach to investigate the transport and risk of microplastic pollution. *Environmental Engineering Research*, 27, 210243.

Sheldrake, M. (2020) *Entangled Life*. Penguin/Random House, Dublin, Ireland.

Shelford, V.E. (1907) Preliminary note on the distribution of tiger beetles (*Cicindela*) and its relation to plant succession. *Biological Bulletin*, 14, 9–14.

Shelomi, M. (2012) Where are we now? Bergmann's rule sensu lato in insects. *American Naturalist*, 180, 511–519.

Sheth, S.N., Morueta-Holme, N. and Angert, A.L. (2020) Determinants of geographic range size in plants. *New Phytologist*, 226, 650–665.

Shi, Y.-W., Liu, X.-S., Wang, H.-Y. and Zhang, R.J. (2009) Seasonality of insect succession on exposed rabbit carrion in Guangzhou, China. *Insect Science*, 16, 425–439.

Shigesada, N. and Kawasaki, K. (1997) *Biological Invasions: Theory and Practice*. Oxford University Press, Oxford, UK.

Shimatani, K. (2001) On the measurement of species diversity incorporating species differences. *Oikos*, 93, 135–147.

Shipley, B. (2010) *From Plant Traits to Vegetation Structure: Chance and Selection in the Assembly of Ecological Communities*. Cambridge University Press, Cambridge, UK.

Shirey, V., Larsen, E., Doherty, A., Kim, C.A., Al-Sulaiman, F.T. *et al.* (2022) LepTraits 1.0 A globally comprehensive dataset of butterfly traits. *Scientific Data*, 9, 382.

Shorrocks, B., Rosewell, J., Edwards, K. and Atkinson, W. (1984) Interspecific competition is not a major organizing force in many insect communities. *Nature*, 310, 310–312.

Showler, A.T. (2008) Desert locust, *Schistocerca gregaria* Forskål (Orthoptera: Acrididae) plagues. In: Capinera, J.L. (ed.) *Encyclopedia of Entomology*. Springer, Dordrecht, The Netherlands, pp. 1181–1186.

Shreeve TG. (1986) Egg-laying by the speckled wood butterfly (*Pararge aegeria*) – the role of female behavior, host plant abundance and temperature. *Ecological Entomology*, 11, 229–236.

Shreeve, T.G. and Dennis, R.L.H. (2011) Landscape scale conservation: Resources, behaviour, the matrix and opportunities. *Journal of Insect Conservation*, 15, 179–188.

Shreeve, T.G., Dennis, R.L.H. and Pullin, A.S. (1996a) Marginality: Scale determined processes and the conservation of the British butterfly fauna. *Biodiversity and Conservation*, 5, 1131–1141.

Shreeve, T.G., Dennis, R.L.H. and Williams, W R. (1996b) Uniformity of wing spotting of *Maniola jurtina* (L.) (Lep. Satyrinae) in relation to environmental heterogeneity. *Nota Lepidopterologica*, 18, 77–92.

Shreeve, T.G., Dennis, R.L.H., Roy, D.B. and Moss, D. (2001) An ecological classification of British butterflies: Ecological attributes and biotope occupancy. *Journal of Insect Conservation*, 5, 145–161.

Shugart, H.H. and James, D. (1973) Ecological succession of breeding bird populations in northwestern Arkansas. *The Auk*, 90, 62–77.

Shupinski, A.B., Craffey, M., Smith, F.A. and Lyons, S. K. (2025) Different mammals, same structure: co-occurrence structure across the Plio-Pleistocene transition. *Paleobiology*, 1–10. doi:10.1017/pab.2024.53

Shurin, J.B., Arnott, S.E., Hillebrand, H., Longmuir, A., Pinel-Alloul, B. *et al.* (2007) Diversity–stability relationship varies with latitude in zooplankton. *Ecology Letters*, 10, 127–134.

Sielezniew, M., Dziekańska, I. and Stankiewicz-Fiedurek, A.M. (2010) Multiple host-ant use by the predatory social parasite *Phengaris (=Maculinea) arion* (Lepidoptera, Lycaenidae). *Journal of Insect Conservation*, 14, 141–149.

Sielezniew, M., Patricelli, D., Rutkowski, R., Witek, M., Bonelli, S. and Buś, M.M. (2015) Population genetics of the endangered obligatorily myrmecophilous butterfly *Phengaris (=Maculinea) arion* in two areas of its European range. *Insect Conservation and Diversity*, 8, 505–516.

Siemers, B.M., Schauermann, G., Turni, H. and von Merten, S. (2009) Why do shrews twitter? Communication or simple echo-based orientation. *Biology Letters*, 5, 593–596.

Siitonen, J. (2012) Microhabitats. In: Stokland, J.N., Siitonen, J. and Jonsson, B.G. (eds) *Biodiversity in Dead Wood*. Cambridge University Press, Cambridge, UK, pp. 150–182.

Siitonen, J., Stokland, J.N., Siitonen, J. and Jonsson, B.G. (2012) Threatened saproxylic species. In: Stokland, J.N., Siitonen, J. and Jonsson, B.G. (eds) *Biodiversity in Dead Wood*. Cambridge University Press, Cambridge, UK, pp. 356–379.

Sillero, N., Arenas-Castro, S., Enriquez-Urzelai, U., Vale C.G. and Sousa-Guedes, D. *et al.* (2022) Want to model a species niche? A step-by-step guideline on correlative ecological niche modelling. *Ecological Modelling*, 456, 109671.

Sillitoe, P. (1972) Warfare in New Guinea: A comparative study. Masters thesis, Durham University, Durham, UK. Available at: http://etheses.dur.ac.uk/9862/ (accessed 22 May 2025).

Silva, A.C.F., Boaventura, D.M., Thompson, R.C. and Hawkins, S.J. (2014) Spatial and temporal patterns of subtidal and intertidal crabs excursions. *Journal of Sea Research*, 85, 343–348.

Silva, I.A., Cianciaruso, M.V. and Batalha. M.A. (2010) Abundance distribution of common and rare plant species of Brazilian savannas along a seasonality gradient. *Acta Botanica Brasilica*, 24, 407–413.

Silva, M.S., Martins, R.P. and Ferreira, R.L. (2011) Cave lithology determining the structure of the invertebrate communities in the Brazilian Atlantic Rain Forest. *Biodiversity and Conservation*, 20, 1713–1729.

Silvertown, J. (2004) Plant coexistence and the niche. *Trends in Ecology and Evolution*, 19, 605–611.

Silvertown, J., Cook, L., Cameron, R., Dodd, M., McConway, K. *et al.* (2011) Citizen science reveals unexpected continental-scale evolutionary change in a model organism. *PLOS One*, 6, e18927.

Silvy, N.A., Lopez, R.R. and Peterson, M.J. (2005) Wildlife marking techniques. In: Braun, C.E. (ed.) *Techniques for Wildlife Investigations and Management*. The Wildlife Society, Bethesda, MD, pp. 339–376.

Simberloff, D. and Dayan, T. (1991) The guild concept and the structure of ecological communities. *Annual Review of Ecology and Systematics*, 22, 115–143.

Simberloff, D.S. and Wilson, E.O. (1969) Experimental zoogeography of islands: The colonisation of empty islands. *Ecology*, 50, 278–296.

Simcox, D.J., Randle, Z., Clarke, R.T. and Thomas, J.A. (2005) *Science and Socio-economically-based Management to Restore Species and Grassland Ecosystems of the Habitats Directive to Degraded Landscapes: The Case of* Maculinea arion *in Britain*. Pensoft, Sofia, Bulgaria.

Simms, E. (1978) *British Thrushes*. New Naturalist No. 63. Collins, London, UK.

Simms, E. (1990) [first edition, 1971] *Woodland Birds*. Godfrey Cave Associates Ltd, London, UK.

Simões, M.P., Dias, A., Belo, F., Pinto-Cruz, C. and Pinheiro, A. (2014) Natural vegetation management to conserve biodiversity and soil water in olive orchards. *Spanish Journal of Agricultural Research*, 12, 633–643.

Simon, B., Boziné Pullai, K., Selmeczi, D., Sebők, A., Tóthné Bogdányi, F. *et al.* (2022) Green corridors may sustain habitats for earthworms in a partially converted grassland. *Agronomy*, 12, 793.

Sinclair, A.R.E. and Arcese, P. (eds) (1995) *Serengeti II: Dynamics, Management and Conservation of an Ecosystem*. University of Chicago Press, Chicago, USA.

Sinclair, J.D. and Sampson, A.W. (1931) Establishment and succession of vegetation on different soil horizons. *Hilgardia*, 5, 155–174.

Singh, A. (2022) A comparative approach to stabilizing mechanisms between discrete- and continuous-time consumer-resource models. *PLOS One*, 17(4), e0265825.

Singh, J.S. (2012) Biodiversity: An overview. *Proceedings of the National Academy of Sciences, India, B Biological Sciences*, 82(Suppl. 2), 239–250.

Sisk, T.D., Haddad, N.M. and Ehrlich, P.R. (1997) Bird assemblages in patchy woodlands: Modeling the effects of edge and matrix habitat. *Ecological Applications* 7, 1170–1180.

Sisneros, R., Huang, J., Ostrouchov, G. and Hoffman, F. (2011) Visualizing life zone boundary sensitivities across climate. Models and temporal spans. *Procedia Computer Science*, 4, 1582–1591.

Sistiaga, A., Mallol, C., Galván, B. and Summons, R.E. (2014) The Neanderthal meal: A new perspective using faecal biomarkers. *PLOS One*, 9, e101045.

Sivakumar, K., Senthil Kumar, S., Kumar, R.S., Ramesh, C., Adhavan, D. *et al.* (2016) *Conservation Strategy and Action Plan for the Marine Turtles and their Habitats in Puducherry*. Wildlife Institute of India, Dehradun, India.

Six, D.L. and Wingfield, M.J. (2011) The role of phytopathogenicity in bark beetle-fungus symbioses: A challenge to the classic paradigm. *Annual Review of Entomology*, 56, 255–272.

Šizling, A.L., Storch, D., Šizlingová, E., Reif, J. and Gaston, K. J. (2009) Species abundance distribution results from a spatial analogy of central limit theorem. *Proceedings of the National Academy of Sciences of the United States of America*, 106, 6691–6695.

Skidmore, P. (1991) *Insects of the British Cow-dung Community*. Field Studies Council, AIDGAP, Shrewsbury, UK.

Skinner, B.J. and Murck, B.W. (2011) *The Blue Planet: An Introduction to Earth System Science*. Wiley, Hoboken, USA.

Skórka, P., Babiarz, T., Skórka, J. and Wójcik, J. (2006) Winter territoriality and fruit defence by the fieldfare (*Turdus pilaris*). *Journal of Ornithology*, 147, 371–375.

Sladecek, F.X.J., Hrcek, J., Klimes, P. and Konvicka, M. (2013) Interplay of succession and seasonality reflects resource utilization in an ephemeral habitat. *Acta Oecologica*, 46, 17–24.

Slik, J.W.F., Arroyo-Rodríguez, V., Aibac, S-I., Alvarez-Loayzad, P. and Alvese, L.F. (2015) An estimate of the number of tropical tree species. *Proceedings of the National Academy of Sciences of the United States of America*, 112, 7472–7477.

Slimak, L. and Giraud, Y. (2007) Circulations sur plusieurs centaines de kilomètres durant le Paléolithique moyen.

Contribution a la connaissance des sociétés néandertaliennes. *Comptes Rendus Palevol*, 6, 359–368.

Slobodchikoff, C.N. and Schulz, W.C. (1980) Measures of niche overlap. *Ecology*, 61, 1051–1055.

Smit, C. and Putman, R. (2011) Large herbivores as 'environmental engineers.' In: Putman, R., Apollonio, M. and Andersen, R. (eds) *Ungulate Management in Europe: Problems and Practices*. Cambridge University Press, Cambridge, UK, pp. 260–283.

Smith, B. and Wilson, J.B., (1996) A consumer's guide to evenness indices. *Oikos*, 76, 70–82.

Smith, B.D. and Zeder, M.A. (2013) The onset of the Anthropocene. *Anthropocene*, 4, 8–13.

Smith, C.Y., Warkentin, I.G. and Moroni, M.T. (2008) Snag availability for cavity nesters across a chronosequence of post-harvest landscapes in western Newfoundland. *Forest Ecology and Management*, 256, 641–647.

Smyth, J. (2006) The role of the house in early Neolithic Ireland. *European Journal of Archaeology*, 9, 229–257.

Sneath, P.H.A. and. Sokal, R.R. (1973) *Numerical Taxonomy. The Principles and Practice of Numerical Classification*. W.H. Freeman, San Francisco, USA.

Snow, B. and Snow, D. (1988) Birds and berries: A study of ecological interaction. *Ornithology*, 107, 227–228.

Soanes, K., Taylor, L., Ramalho, C.E., Maller, C., Paris, K. *et al.* 2023) Conserving urban biodiversity: Current practice, barriers, and enablers. *Conservation Letters*, 16, e12946.

Soberón, J. and Llorente, J. (1993) The use of species accumulation functions for the prediction of species richness. *Conservation Biology*, 7, 480–488.

Soberón, J. and Peterson, A.T. (2005) Interpretation of models of fundamental ecological niches and species' distributional areas. *Biodiversity Informatics*, 2, 1–10.

Soga, M. and Gaston, K.J. (2018) Shifting baseline syndrome: Causes, consequences, and implications. *Frontiers in Ecology and the Environment*, 16, 222–230.

Sokal, R.R. and Michener, C.D. (1958) A statistical method for evaluating systematic relationships. *University of Kansas Scientific Bulletin*, 38, 1409–1438.

Sokos, C.K., Birtsas, P.K., Connelly, J.W. and Papaspyropoulos, K.G. (2013) Hunting of migratory birds: Disturbance intolerant or harvest tolerant? *Wildlife Biology*, 19, 113–125.

Solé, R.V., Montoya, J.M. and Erwin, D.H. (2002) Recovery after mass extinction: Evolutionary assembly in large-scale biosphere dynamics. *Philosophical Transactions of the Royal Society B: Biological Sciences*, 357, 697–707.

Solem, A., Climo, F.M. and Roscoe, D.J. (1981) Sympatric species diversity of New Zealand land snails. *New Zealand Journal of Zoology*, 8, 453–485.

Somerfield, P.J. and Warwick, R.M. (2013) Meiofauna techniques. In: Eleftheriou, A. (ed.) *Methods for the Study of Marine Benthos*, 4th edn. John Wiley and Sons, Chichester, UK, pp. 253–284.

Sonnier, G., Shipley, B. and Navas, M.L. (2010) Quantifying relationships between traits and explicitly measured gradients of stress and disturbance in early successional plant communities. *Journal of Vegetation Science*, 21, 1014–1024.

Sørensen, T. (1936) Some ecosystematical characteristics determined by Raunkiær's circling method. In: *Nordiska (19. skandinaviska) naturforskarmötet i Helsingfors den 11-15 Agusti 1936*. [Proceedings of the 19th Scandinavian Naturalists' Meeting, Helsinki, Finland, August 11–15, 1936.] Nordiska naturforskarmötet, Helsinki, Finland, pp. 474–475.

Sousa, W.P. (1984) The role of disturbance in natural communities. *Annual Review of Ecology and Systematics*, 15, 353–391.

Southern, L.M., Deschner, T. and Pika, S. (2021) Lethal coalitionary attacks of chimpanzees (*Pan troglodytes troglodytes*) on gorillas (*Gorilla gorilla gorilla*) in the wild. *Scientifc Reports*, 11, 14673.

Southwood, T.R.E. (1976) Bionomic strategies and population parameters. In: May, R.M. (ed.) *Theoretical Ecology, Principles and Applications*. Blackwell Scientific Publications, Oxford, UK, pp. 26–48.

Southwood, T.R.E. and Henderson, P.A. (2000) *Ecological Methods*, 3rd edn. Blackwell Science, Oxford, UK.

Southwood, T.R.E., Brown, V.K., Reader, P.M. and Green, E.E. (1986) The use of different stages of a secondary succession by birds. *Bird Study*, 33, 159–163.

Spalding, A., Young, M.R. and Dennis, R.L.H. (2012) The importance of host plant-habitat substrate in the maintenance of a unique isolate of the sandhill rustic: Disturbance, shingle matrix and bare ground indicators. *Journal of Insect Conservation*, 16, 839–846.

Spalding, M.D., Fox, H.E., Allen, G.R., Davidson, N., Ferdaña, Z.A. *et al.* (2007) Marine ecoregions of the world: A bioregionalization of coastal and shelf areas. *Bioscience*, 57, 573–583.

Speight, M.C. (1989) *Saproxylic Invertebrates and their Conservation*. Council of Europe, Strasbourg, France.

Sperandii, M.G., Bazzichetto, M., Gatti, F. and Acosta, A.T.R. (2019) Back into the past: Resurveying random plots to track community changes in Italian coastal dunes. *Ecological Indicators*, 96, 572–578.

Spitzer, L., Beneš, J., Dandova, J., Jaskova, V. and Konvicka, M. (2009) The large blue butterfly, *Phengaris* [*Maculinea*] *arion*, as a conservation umbrella on a landscape scale: The case of the Czech Carpathians. *Ecological Indicators*, 9, 1056–1063.

Spotila, J.R., O'Connor, M.P. and Paladino, F.V. (1996) Thermal biology. In: Lutz, P.L.and Musick, J.A. (eds) *The Biology of Sea Turtles*. CRC Press, Boca Raton, USA.

Stachowicz, J.J. (2001) Mutualism, facilitation, and the structure of ecological communities: Positive interactions play a critical, but underappreciated, role in ecological communities by reducing physical or biotic stresses in existing habitats and by creating new habitats on which many species depend. *BioScience*, 51, 235–246.

Stahl, K., Lepczyk, C.A. and Christoffel, R.A. (2020) Evaluating conservation biology texts for bias in biodiversity representation. *PLOS One*, 15, e0234877.

Staley, J.T. (1999) Bacteria, their smallest representatives and subcellular structures, and the purported Precambrian fossil "Metallogenium". In: *Size Limits of Very Small Microorganisms. Proceedings of a Workshop*. National Academy Press, Washington, DC, USA, pp. 62–67.

Stampfli, G.M., Hochard, C., Vérard, C., Wilhem, C. and von Raumer, J. (2013) The formation of Pangea. *Tectonophysics*, 593, 1–19.

Staniczenko, P.P.A., Kopp, J.C. and Allesina, S. (2013) The ghost of nestedness in ecological networks. *Nature Communications*, 4, 1391.

Stankiewicz, A.M., Sielezniew, M. and Sawoniewicz, J. (2004) *Neotypus pusillus* Gregor, 1940 (Hymenoptera, Ichneumonidae) endoparasite of *Maculinea nausithous* (Bergsträsser, 1779) (Lepidoptera, Lycaenidae): New data on distribution in Poland with remarks on its biology. *Fragmenta Faunistica*, 47, 115–120.

Steensma, K.M.M., Lilley, L.P. and Zandberg, H.M. (2009) Life history and habitat requirements of the Oregon forestsnail, *Allogona townsendiana* (Mollusca, Gastropoda, Pulmonata, Polygyridae), in a British Columbia population. *Invertebrate Biology*, 128, 232–242.

Steenweg, R., Hebblewhite, M., Kays, R., Ahumada, J., Fisher, J.T. *et al.* (2017) Scaling-up camera traps: Monitoring the planet's biodiversity with networks of remote sensors. *Frontiers in Ecology and the Environment*, 15, 26–34.

Stefanescu, C., Herrando, S. and Páramo, F. (2004) Butterfly species-richness in the north-west Mediterranean Basin: The role of natural and human-induced factors. *Journal of Biogeography*, 31, 905–915.

Stefanescu, C., Alarcón, M. and Àvila, A. (2007) Migration of the painted lady butterfly, *Vanessa cardui*, to north-eastern Spain is aided by African wind currents. *Journal of Animal Ecology*, 76, 888–898.

Stefanescu, C., Alarcón, M., Izquierdo, R., Páramo, F. and Àvila, A. (2011) Moroccan source areas of the painted lady butterfly *Vanessa cardui* (Nymphlaidae: Nymphalinae) migrating into Europe in spring. *Journal of the Lepidopterists' Society*, 65, 15–26.

Stefanescu, C., Páramo, F., Åkesson, S., Alarcón, M., Ávila, A. *et al.* (2013) Multi-generational long-distance migration of insects: Studying the painted lady butterfly in the Western Palaearctic. *Ecography*, 36, 474–486.

Stefanescu, C., Soto, D.X., Talavera, G., Vila, R. and Hobson, K.A. (2016) Long-distance autumn migration across the Sahara by painted lady butterflies: Exploiting resource pulses in the tropical savannah. *Biology Letters*, 12, 20160561.

Steffen, W., Grinewald, J., Crutzen, P. and McNeill, J. (2011) The Anthropocene: Conceptual and historical perspectives. *Philosophical Transactions of the Royal Society A: Mathematical, Physical, and Engineering Sciences*, 369, 842–867.

Stein, A. and Kreft, H. (2014) Terminology and quantification of environmental heterogeneity in species-richness research. *Biological Reviews*, 90(3), 815–836.

Stein, A., Gerstner, K. and Kreft, H. (2014) Environmental heterogeneity as a universal driver of species richness across taxa, biomes and spatial scales. *Ecology Letters*, 17, 866–880.

Stein, W.E., Mannolini, F., Hernick, L.V., Landing, E. and Berry, C.M. (2007) Giant cladoxylopsid trees resolve the enigma of the Earth's earliest forest stumps at Gilboa. *Nature*, 446, 904–907.

Stenhouse, D.A. (2011) A history of *Agelastica alni* (L.) (Coleoptera: Chrysomelidae) in the British Isles. *Entomologist's Monthly Magazine*, 147, 197–212.

Stenseth, N.C., Falck, W., Bjornstad, O.N. and Krebs, C.J. (1997) Population regulation in snowshoe hare and Canadian lynx: Asymmetric food web configurations between hare and lynx. *Proceedings of the National Academy of Sciences of the United States of America*, 94, 5147–5152.

Stephens, D.W. and Krebs, J.R. (1987) *Foraging Theory*. Princeton University Press, Princeton, USA.

Sterck, E.H.M., Watts, D.P. and van Schaik, C.P. (1997) The evolution of female social relationships in nonhuman primates. *Behavioural Ecology and Sociobiology*, 41, 291–309.

Stevens, L., Brown, B. and Rowell, K. (2009) Foraging ecology of peregrine falcons (*Falco peregrinus*) along the Colorado River, Grand Canyon, Arizona. *Southwestern Naturalist*, 54, 284– 299.

Stevens, M. (2013) *Sensory Ecology, Behaviour and Evolution*. Oxford University Press, Oxford, UK.

Steves, I., Berliner, P. and Pinshow, B. (2021) Air temperature and humidity at the bottom of desert wolf spider burrows are not affected by surface conditions. *Insects*, 12, 943.

Stiner, M.C. (2001) Thirty years on the "Broad Spectrum Revolution" and Paleolithic demography. *Proceedings of the National Academy of Sciences of the United States of America*, 98, 6993–6996.

Stiner, M.C. and Kuhn, S.L. (2009) Paleolithic diet and the division of labor in Mediterranean Eurasia. In: Hublin, J.-J. and Richards, M.P. (eds) *The Evolution of Hominin Diets: Integrating Approaches to the Study of Palaeolithic Subsistence*, Springer Science + Business Media, Dordrecgt, The Netherlands, pp. 157–169.

Stoeck, T., Behnke, A., Christen, R., Amaral-Zettler, L., Rodriguez-Mora, M.J. *et al.* (2009) Massively parallel tag sequencing reveals the complexity of anaerobic marine protistan communities. *BMC Biology*, 7, 72.

Stone, L. and Roberts, A. (1990) The checkerboard score and species distributions. *Oecologia*, 85, 74–79.

Stoppani, A. (1873) *Corso di Geologia: Geologia Stratigrafica*. G. Bernardoni e G. Brigola, Milano, Italia.

Storch, D., Šímová, I., Smyčka, J., Bohdalková, E. Toszogyova, A. and Okie, J.G. (2022) Biodiversity dynamics in the Anthropocene: How human activities change equilibria of species richness. *Ecography*, e05778.

Stork, N.E., McBroom, J., Gely, C. and Hamilton, A.J. (2015) New approaches narrow global species estimates for beetles, insects, and terrestrial arthropods. *Proceedings of the National Academy of Sciences of the United States of America*, 112, 7519–7523.

Stough, J.M., Yutin, N., Chaban, Y.V., Moniruzzaaman, M., Gann, E.R. *et al.* (2019) Genome and environmental activity of a Chrysochromulina parva virus and its Virophages. *Frontiers in Microbiology*, 10, 703.

Strauch, A.M. (2013) The role of water quality in large mammal migratory behaviour in the Serengeti. *Ecohydrology*, 6, 343–354.

Strauss, S.Y., Webb, C.O. and Salamin, N. (2006) Exotic taxa less related to native species are more invasive. *Proceedings of the National Academy of Sciences of the United States of America*, 103, 5841–5845.

Strayer, D.L., Power, M.E., Fagan, W.F., Pickett, S.T.A. and Belnap, J. (2003) A classification of ecological boundaries. *Bioscience*, 53, 723–729.

Street, G.M., Erovenko, I.V. and Rowell, J.T. (2018) Dynamical facilitation of the ideal free distribution in nonideal populations. *Ecology and Evolution*, 8, 2471–2481.

Strona, G. (2015) Past, present and future of host–parasite co-extinctions. *International Journal for Parasitology: Parasites and Wildlife*, 4, 431–441.

Strona, G. (2022) *Hidden Pathways to Extinction*. Springer, Cham, Switzerland.

Strona, G. and Fattorini, S. (2014a) On the methods to assess significance in nestedness analyses. *Theory in Biosciences*, 133, 179–186.

Strona, G. and Fattorini, S. (2014b) A few good reasons why species-area relationships do not work for parasites. *BioMed Research International*, 2014, 271680.

Strona, G. and Fattorini, S. (2016) Are generalist parasites being lost from their hosts? *Journal of Animal Ecology*, 85, 621–623.

Strona, G. and Veech, J.A. (2015) A new measure of ecological network structure based on node overlap and segregation. *Methods in Ecology and Evolution*, 6(8), 907–915.

Strona, G., Galli, P. and Fattorini S. (2013a) Fish parasites unravel the paradox of missing coextinctions. *Nature Communications*, 4, 1718.

Strona, G., Stefani, F., Galli, P. and Fattorini, S. (2013b) A protocol to compare nestedness among submatrices. *Population Ecology*, 55, 227–239.

Strona, G., Nappo, D., Boccacci, F., Fattorini, S. and San-Miguel Ayanz, J. (2014a) A fast and unbiased procedure to randomize ecological binary matrices with fixed row and column totals. *Nature Communications*, 5, 4114.

Strona, G., Galli, P., Seveso, D., Montano, S. and Fattorini, S. (2014b) Nestedness for Dummies (NeD): A user friendly web interface for exploratory nestedness analysis. *Journal of Statistical Software*, 59, 3.

Strona, G., Mauri, A., Veech, J.A., Seufert, G., San-Miguel Ayanz, J. and Fattorini, S. (2016) Far from naturalness: How much does spatial ecological structure of European tree assemblages depart from potential natural vegetation? *PLOS One*, 11, e0165178.

Strona, G., Beck, P.S.A., Cabeza, M., Fattorini, S., Guilhaumon, F. *et al.* (2021) Ecological dependencies make remote reef fish communities most vulnerable to coral loss. *Nature Communications*, 4, 1718.

Strullu-Derrien, C., Goral, T., Spencer, A.R.T., Kenrick, P., Aime, C. *et al.* (2023) A fungal pathogen discovered in Devonian Rhynie chert. *Nature Communications*, 14, 7932.

Stubbs, W.J. and Wilson, J.B. (2004) Evidence for limiting similarity in a sand dune community. *Journal of Ecology*, 92, 557–567.

Stutz, A.J. (2018) Paleolithic. In: Trevathan, W. (ed.) *The International Encyclopedia of Biological Anthropology*. John Wiley and Sons, Hoboken, USA, pp. 1–109.

Sud, M. (2020) *Managing the Biodiversity Impacts of Fertiliser and Pesticide Use: Overview and Insights from Trends and Policies Across Selected OECD Countries*. Environment Working Paper 155. Organisation for Economic Co-operation and Development. Available at: www.oecd.org/content/dam/oecd/en/publications/reports/2020/03/managing-the-biodiversity-impacts-of-fertiliser-and-pesticide-use_e51ab39d/63942249-en.pdf (accessed 23 May 2025).

Suding, K., Lavorel, S., Chapin, F.S., Cornelissen, J.H.C., Díaz, S. *et al.* (2008) Scaling environmental change through the community-level: A trait-based response and effect framework for plants. *Global Change Biology*, 14, 1125–1140.

Sugihara, G. (1980) Minimal community structure: An explanation of species abundance patterns. *American Naturalist*, 116, 770–787.

Sugihara, G. and Ye, H. (2009) Cooperative network dynamics. *Nature*, 458, 979.

Sugihara, G., Bersier, L.-F., Southwood, T.R.E., Pimm, S.L. and May, R.M. (2003) Predicted correspondence between species abundances and dendrograms of niche similarities. *Proceedings of the National Academy of Sciences of the United States of America*, 100, 5246–5251.

Sukumar, R. (2006), A brief review of the status, distribution and biology of wild Asian elephants *Elephas maximus*. *International Zoo Yearbook*, 40, 1–8.

Sumudumali, R.G.I. and Jayawardana, J.M.C.K. (2021) A review of biological monitoring of aquatic ecosystems approaches: With special reference to macroinvertebrates and pesticide pollution. *Environmental Management*, 67, 263–276.

Sun, C., König, H.J., Uthes, S., Chen, C., Peng, L., and Hemminger, K. (2020) Protection effect of overwintering water bird habitat and defining the conservation priority area in Poyang Lake wetland, China. *Environmental Research Letters*, 15, 125013.

Sunday, J.M., Bates, A.E. and Dulvy, N.K. (2011) Global analysis of thermal tolerance and latitude in ectotherms. *Proceedings of the Royal Society B: Biological Sciences*, 278, 1823–1830.

Sunderland-Groves, J.L., Ekinde, A. and Mboh, H. (2009) Nesting behavior of *Gorilla gorilla diehli* at Kagwene Mountain, Cameroon: Implications for assessing group size and density. *International Journal of Primatology*, 30, 253–266.

Suntsova, M.V. and Buzdin, A.A. (2020) Differences between human and chimpanzee genomes and their implications in gene expression, protein functions and biochemical properties of the two species. *BMC Genomics*, 21, 535.

Suprunova, I.A. (2023) Historical and legal basis for the formation of the habitat concept of biodiversity conservation. *Uzhhorod National University Herald Series Law*, 79, 369–375. [in Ukranian]

Surovell, T.A. (2008) Extinctions of big game. *Encyclopedia of Archaeology*, 2, 1365–1374.

Sussman, R.W. and Ratsirarson, J. (2006) Beza Mahafaly Special Reserve: A research site in southwestern Madagascar. In: Jolly, A., Sussman, R.W., Koyama, N. and Rasamimanana, H. (eds) *Ringtailed Lemur Biology:* Lemur catta *in Madagascar*. Springer, New York, USA, pp. 43–51.

Sussman, R.W., Green, G.M., Porton, I., Andrianasolondraibe, O.L. and Ratsirarson, J. (2003) A survey of the habitat of *Lemur catta* in Southwestern and Southern Madagascar. *Primate Conservation*, 19, 32–57.

Sutcliffe, O.L., Thomas, C.D., Yates, T.J. and Greatorex-Davies, J.N. (1997) Correlated extinctions, colonizations and population fluctuations in a highly connected ringlet butterfly metapopulation. *Oecologia*, 109, 235–241.

Sutherland, W.J. (ed.) (2006) *Ecological Census Techniques*, 2nd edn. Cambridge University Press, Cambridge, UK.

Sutherland, W.J., Newton, I. and Green, R.E. (2004) *Bird Ecology and Conservation: A Handbook of Techniques*. Oxford University Press, Oxford, UK

Suthers, I., Rissik, D. and Richardson, A. (eds) (2019) *Plankton: Guide to Their Ecology and Monitoring for Water Quality*, 2nd edn. CRC Press, Boca Raton, USA.

Sutikna, T., Tocheri, M.W., Morwood, M.J., Saptomo, E.W., Jatmiko *et al.* (2016) Revised stratigraphy and chronology for *Homo floresiensis* at Liang Bua in Indonesia. *Nature*, 532, 366–369.

Sutton, A.L. and Beckley, L.E. (2017) Species richness, taxonomic distinctness and environmental influences on euphausiid zoogeography in the Indian Ocean. *Diversity*, 9, 23.

Suweis, S., Grilli, J., Banavar, J.R., Allesina, S. and Maritan, A. (2015) Effect of localization on the stability of mutualistic ecological networks. *Nature Communications*, 6, 10179.

Suzuki, Y., Webb, S.J., Kouduka, M., Kobayashi, H., Castillo, J. *et al.* (2024) Subsurface microbial colonization at mineral-filled veins in 2-billion-year-old mafic rock from the Bushveld Igneous Complex, South Africa. *Microbial Ecology*, 87, 116.

Swanson, H.K., Lysy, M., Power, M., Stasko, A.D., Johnson, J.D. and Reist, J.D. (2015) A New probabilistic method for quantifying n-dimensional ecological niches and niche overlap. *Ecology*, 96, 318–324.

Swedell, L. (2012) Primate sociality and social systems. *Nature Education Knowledge*, 3(10), 84.

Swengel, A.B. (2017) Habitat associations of sympatric violet-feeding fritillaries (*Euptoieta, Speyeria, Boloria*) (Lepidoptera: Nymphalidae) in tallgrass prairie. *Great Lakes Entomologist*, 30(1), 1.

Swenson, N.G. (2011) Phylogenetic beta diversity metrics, trait evolution and inferring the functional beta diversity of communities. *PLOS One*, 6, 21264.

Swenson, N.G., Mahler, D.L., Ferro, M. and Ritchie, A. (2007) The energetic determination, spatial dispersion and density dependence of *Myrmeleon* ant lion pits in Las Cruces, Costa Rica. *Biotropica*, 39, 774–777.

Swenson, N.G., Erickson, D.L., Mi, X., Bourg, N.A., Forero-Montaña, J. *et al.* (2012) Phylogenetic and functional alpha and beta diversity in temperate and tropical tree communities. *Ecology*, 93, S112–S125.

Swenson, N.G., Hulshof, C.M., Katabuchi, M. and Enquist, B.J. (2020) Long-term shifts in the functional composition and diversity of a tropical dry forest: A 30-yr study. *Ecological Monographs*, 90, e01408.

Syed, A.J. and Anderson, J.C. (2021) Applications of bioluminescence in biotechnology and beyond. *Chemical Society Review*, 50, 5668–5705.

Szabó, J., Dávid, L. and Lóczy, D. (2010) *Anthropogenic Geomorphology: A Guide to Man-made Landforms*. Springer Science+Business Media, Dordrecht, The Netherlands.

Szyszko-Podgórska, K., Dymitryszyn, I. and Kondras, M. (2023) Diversity in landscape management affects butterfly distribution. *Sustainability*, 15, 14775.

Taagepera, R. and Nemčok, M. (2023) World population growth over millennia: Ancient and present phases with a temporary halt in-between. *Anthropocene Review*, 11, 163–183.

Taberlet, P., Bonin, A., Zinger, L. and Coissac, É. (2018) *Environmental DNA: For Biodiversity Research and Monitoring*. Oxford Academic, Oxford, UK.

Tabor, G.M. and Aguirre, A.A. (2004) Ecosystem health and sentinel species: Adding an ecological element to the proverbial 'canary in the mineshaft'. *EcoHealth*, 1, 226–228.

Tabor, K.L., Fell, R.D. and Brewster, C.C. (2005) Insect fauna visiting carrion in Southwest Virginia. *Forensic Science International*, 150, 73–80.

Taheri, S., Naimi, B., Rahbek, R. and Araújo, M.B. (2021) Improvements in reports of species redistribution under climate change are required. *Science Advances*, 7(15), eabe1110.

Takahashi, J. (2008) A literature review of the spider monkey, *Ateles* sp., with special focus on risk for extinction. Thesis, Faculty of Veterinary Medicine and Animal Sciences, Swedish University of Agricultural

Sciences, Uppsala, Sweden. Available at: https://stud.epsilon.slu.se/10775/ (accessed 23 May 2025).

Takhtajan, A. (1986) *Floristic Regions of the World*. University of California Press, Berkeley, USA.

Tanabe, K. and Namba, T. (2005) Omnivory creates chaos in simple food web models. *Ecology*, 86, 3411–3414.

Tanner, J.T. (1966) Effects of population density on growth rates of animal populations. *Ecology*, 47, 733–745.

Tansley, A.G. (1935) The use and abuse of vegetational terms and concepts. *Ecology*, 16, 284–307.

Tansley, A.G. (1939) *The British Islands and Their Vegetation*. Cambridge University Press, Cambridge, UK.

Tansley, A.G. (1949) *Britain's Green Mantle*. George Allen and Unwin, London, UK.

Tantawi, T.I., El-Kady, E.M., Greenberg, B. and El-Ghaffar, H.A. (1996) Arthropod succession on exposed rabbit carrion in Alexandria, Egypt. *Journal of Medical Entomology*, 33, 566–580.

Tarkhnishvili, D., Gabelaia, M. and Adriaens, D. (2020) Phenotypic divergence, convergence and evolution of Caucasian rock lizards (*Darevskia*). *Biological Journal of the Linnean Society*, 130, 142–155.

Tarnita, C.E., Bonachela, J.A., Efrat Sheffer, E., Guyton, J.A., Coverdale, T.C. *et al.* (2017) A theoretical foundation for multi-scale regular vegetation patterns. *Nature*, 541, 398–402.

Tartally, A., Thomas, J.A., Anton, C., Balletto, E., Barbero, F. *et al.* (2019) Patterns of host use by brood parasitic *Maculinea* butterflies across Europe. *Philosophical Transactions of the Royal Society B: Biological Sciences*, 374, 20180202.

Tasse Taboue, G.C., Ntene Soh, B.C., Koumbo Tagagoum, U.B., Tchassem Fokoua, A.M., Ngueguim, J.R. *et al.* (2024) Local perceptions, hunting and export of the endangered Goliath frog *Conraua goliath* in Cameroon. *Oryx*, 58, 15–24.

Taulman, J.F. (2013) A comparison of fixed-width transects and fixed-radius point counts for breeding-bird surveys in a mixed hardwood forest. *Southeastern Naturalist*, 12(3), 457–477.

Tavşanoğlu, Ç. and Pausas, J.A (2018) Functional trait database for Mediterranean Basin plants. *Scientific Data*, 5, 180135.

Taylor, G.S., Braby, M.F., Moir, M.L., Harvey, M.S., Sands, D.P.A. *et al.* (2018) Strategic national approach for improving the conservation management of insects and allied invertebrates in Australia. *Austral Entomology*, 57, 124–149.

Taylor, L.R. (1961) Aggregation, variance and the mean. *Nature*, 139, 732–735.

Taylor, L.R. (1965) A natural law for the spatial disposition of insects. In: Freeman, P. (ed.) *Proceedings of the XII International Congress of Entomology*. Royal Entomological Society of London, London, pp. 396–397.

Taylor, L.R. (1971) Aggregation as a species characteristic. In: Patil, G.P., Pielou, E.C. and Waters, W.E. (eds) *Statistical Ecology*, vol. 1. Penn. State University Press, University Park, USA, pp. 356–372.

Taylor, L.R. (1978) Bates, Williams, Hutchinson – A variety of diversities. In: Mound, L.A. and Waloff, N. (eds) *Diversity of Insect Faunas*. Symposia of the Royal Entomological Society Number 9. Blackwell Scientific Publications, Oxford, UK, pp. 1–18.

Taylor, L.R. and Taylor, R.A.J. (1977) Aggregation, migration and population mechanics. *Nature*, 265, 415–421.

Taylor, L.R. and Taylor, R.A.J. (1979) A behavioural model for the evolution of spatial dynamics. In: Anderson, R.M., Turner, B.D. and Taylor, L.R. (eds) *Population Dynamics*. 20th Symposium of the British Ecological Society, London. Blackwell Scientific Publications, Oxford, UK, pp. 1–27.

Tedersoo, L., Mikryukov, V., Zizka, A., Bahram, M., Hagh-Doust, N. *et al.* (2022) Global patterns in endemicity and vulnerability of soil fungi. *Global Change Biology*, 28, 6696–6710.

Teichert, N., Lepage, M., Sagouis, A., Borja, A., Chust, G. *et al.* (2017) Functional redundancy and sensitivity of fish assemblages in European rivers, lakes and estuarine ecosystems. *Scientific Reports*, 7(1), 17611.

Teichroeb, J.A. and Sicotte, P. (2009) Test of the ecological-constraints model on ursine colobus monkeys (*Colobus vellerosus*) in Ghana. *American Journal of Primatology*, 71, 49–59.

Teittinen, A. and Soininen, J. (2015) Testing the theory of island biogeography for microorganisms-patterns for spring diatoms. *Aquatic Microbial Ecology*, 75(3), 239–250.

Telea, R.A., Topliceanu, T.-S., Bălăsoiu, D., Serbana, R. and Cogălniceanu, D. (2019) Biodiversity in quarries – A study case from Iglicioara quarry. *Academy of Romanian Scientists, Annals Series on Biological Sciences*, 8, 5–16.

Teletchea, F. (2019) *Animal Domestication*. IntechOpen. Available at: https://www.intechopen.com/books/6053 (accessed 1 June 2025).

Téllez-Hernández, E., Domínguez-Vega, H., Zuria, I., Marín-Togo, M.C. and Gómez-Ortiz, Y. (2023) Functional and ecological diversity of urban birds: Conservation and redesign of biocultural landscapes. *Journal for Nature Conservation*, 73(126395), 126395.

Tembe, D. and Mukaratirwa, S. (2021) Insect succession and decomposition pattern on pig carrion during warm and cold seasons in KwaZulu-Natal province of South Africa. *Journal of Medical Entomology*, 58(6), 2047–2057.

Terborgh, J. (1988) The big things that run the world — A sequel to E.O. Wilson. *Conservation Biology*, 2, 402–403.

Terborgh, J. and Winter, B. (1980) Some causes of extinction. In: Soulé, M.E. (ed.) *Conservation Biology: The Science of Scarcity and Diversity*. Sinauer Associates, Sunderland, USA, pp. 119–134.

Terborgh, J., Lopez, L., Nuñez, P., Rao, M., Shahabuddin, G. *et al.* (2001) Ecological meltdown in predator-free forest fragments. *Science*, 294, 1923–1926.

Terraube, J., Archaux, F., Deconchat, M., van Halder, I., Jactel, H. and Barbaro, L. (2016) Forest edges have high conservation value for bird communities in mosaic landscapes. *Ecology and Evolution*, 6, 5178–5189.

ter Steege, H., Pitman, N.C.A., Sabatier, D., Baraloto, C., Salomão, R.P. *et al.* (2013) Hyper-dominance in the Amazonian tree flora. *Science*, 342, 325–334.

Tesch, F.W. (2003) *The Eel*, 3rd edn. Blackwell Science, Oxford, UK.

Tews, J., Brose, U., Grimm, V., Tielbörger, K., Wichmann, M.C. *et al.* (2004) Animal species diversity driven by habitat heterogeneity/diversity: The importance of keystone structures. *Journal of Biogeography*, 31, 79–92.

Theissinger, K., Fernandes, C., Formenti, G., Bista, I., Berg, P.R. *et al.* (2023) How genomics can help biodiversity conservation. *Trends in Genetics*, 39, 545–559.

Thienemann, A. (1918) Lebensgemeinschaft und Lebensraum. *Naturwissenschaftliche Wochenschrift, N.F.*, 17, 282–290, 297–303.

Thierry, B. (2008) Primate socioecology, the lost dream of ecological determinism. *Evolutionary Anthropology*, 17, 93–96.

Thomas, B., Holland, J.D. and Minot, E.O, (2008) Elephant (*Loxodonta africana*) home ranges in Sabi Sand Reserve and Kruger National Park: A five-year satellite tracking study. *PLOS One*, 3, e3902.

Thomas, B.A. and Cleal, C.J. (2017) Distinguishing Pennsylvanian-age lowland, extra-basinal and upland vegetation. *Palaeobiodiversity and Palaeoenvironments*, 97, 273–293.

Thomas, C.D. (1984) Oviposition and egg load assessment by *Anthocharis cardamines* (L.) (Lepidoptera: Pieridae). *Entomologist's Gazette*, 35, 145–148.

Thomas, C.D. (1985) Specializations and polyphagy of *Plebejus argus* (Lepidoptera: Lycaenidae) in North Wales. *Ecological Entomology*, 10, 325–340.

Thomas, C.D. (2017) *Inheritors of the Earth: How Nature is Thriving in an Age of Extinction.* Allen Lane/Penguin Press, London, UK.

Thomas, C.D. and Hanski, I.A. (1997) Butterfly metapopulations. In: Hanski, I.A. and Gilpin, M.E. (eds) *Metapopulation Biology. Ecology, Genetics, and Evolution.* Academic Press, London, UK, pp. 359–386.

Thomas, C.D. and Harrison, S. (1992) Spatial dynamics of a patchily distributed butterfly species. *Journal of Animal Ecology*, 61, 437–446.

Thomas, C.D. and Jones, T.M. (1993) Partial recovery of skipper butterfly (*Hesperia comma*) from population refuges: Lessons for conservation in a fragmented landscape. *Journal of Animal Ecology*, 62, 472–481.

Thomas, C.D. and Kunin, W.E. (1999) The spatial structure of populations. *Journal of Animal Ecology*, 68, 647–657.

Thomas, C.D., Warren, M.S. and Thomas, J.A. (1992) Distributions of occupied and vacant butterfly habitats in fragmented landscapes. *Oecologia*, 92, 563–567.

Thomas, C.D., Glen, S.W.T., Lewis, O.T., Hill, J.K. and Blakeley, D.S. (1999) Population differentiation and conservation of endemic races: The butterfly, *Plebejus argus*. *Animal Conservation*, 2, 15–21.

Thomas, C.D., Bodsworth, E.J., Wilson, R.J., Simmons, A.D., Davies, Z.G. *et al.* (2001) Ecological and evolutionary processes at expanding range margins. *Nature*, 411, 577–581.

Thomas, C.D., Jones, T.H. and Hartley, S.E. (2019) "Insectageddon": A call for more robust data and rigorous analyses. *Global Change Biology*, 25, 1891–1892.

Thomas, J.A. (1980) Why did the large blue become extinct in Britain? *Oryx*, 15, 243–247.

Thomas, J.A. (1995) The ecology and conservation of *Maculinea arion* and other European species of butterfly. In: Pullin, A.S. (ed.) *The Ecology and Conservation of Butterflies*. Chapman and Hall, London, UK, pp. 180–197.

Thomas, J.A. (2002) Larval niche selection and evening exposure enhance adoption of a predacious social parasite, *Maculinea arion* (large blue butterfly), by *Myrmica* ants. *Oecologia*, 132, 531–537.

Thomas, J.A. and Elmes, G.W. (1993) Specialised searching and the hostile use of allomones by a parasitoid whose host, the butterfly *Maculinea rebeli*, inhabits ant nests. *Animal Behaviour*, 45, 593–602.

Thomas, J.A. and Elmes, G.W. (1998) Higher productivity at the cost of increased host-specificity when *Maculinea* butterfly larvae exploit ant colonies through trophallaxis rather than by predation. *Ecological Entomology*, 23, 457–464.

Thomas, J.A. and Lewington, R. (1991) *The Butterflies of Britain and Ireland.* Dorling Kindersley, London, UK.

Thomas, J.A. and Wardlaw, J.C. (1990) The effect of queen ants on the survival of *Maculinea arion* larvae in *Myrmica* ant nests. *Oecologia*, 85, 87–91.

Thomas, J.A. and Wardlaw, J.C. (1992) The capacity of a *Myrmica* ant nest to support a predacious species of *Maculinea* butterfly. *Oecologia*, 91, 101–109.

Thomas, J.A., Elmes, G.W., Wardlaw, J.C. and Woyciechowski, M. (1989) Host specificity among *Maculinea* butterflies in *Myrmica* ant nests. *Oecologia*, 79, 452–457.

Thomas, J.A., Clarke, R.T., Elmes, G.W. and Hochberg, M.E. (1998a) Population dynamics in the genus *Maculinea* (Lepidoptera: Lycaenidae). In: Dempster, J.P. and McLean, I.F. (eds) *Insect Populations: In Theory and Practice. 19th Symposium of the Royal Entomological Society 10–11 September 1997 at the University of Newcastle.* Springer Science+Business Media, Dordrecht, The Netherlands, pp. 261–290.

Thomas, J.A., Simcox, D.J., Wardlaw, J.C., Elmes, G.W., Hochberg, M.E. and Clarke, R.T. (1998b) Effects of latitude, altitude and climate on the habitat and conservation of the endangered butterfly *Maculinea arion* and its *Myrmica* ant hosts. *Journal of Insect Conservation*, 2, 39–46.

Thomas, J.A., Bourn, N.A.D., Clarke, R.T., Stewart, K.E., Simcox, D.J. *et al.* (2001) The quality and isolation of habitat patches both determine where butterflies persist in fragmented landscapes. *Proceedings of the Royal Society B: Biological Sciences*, 268, 1791–1796.

Thomas, J.A., Simcox, D.J. and Clarke, R.T. (2009). Successful conservation of a threatened *Maculinea* butterfly. *Science*, 325, 80–83.

Thomas, J.E., Carvalho, G.R., Haile, J., Rawlence, N.J., Martin, M.D. *et al.* (2019) Demographic reconstruction from ancient DNA supports rapid extinction of the great auk. *eLife*, 8, e47509.

Thomas, L., Buckland, S.T., Rexstad, E.A., Laake, J.L., Strindberg, S. *et al.* (2010) Distance software: Design and analysis of distance sampling surveys for estimating population size. *Journal of Applied Ecology*, 47, 5–14.

Thompson, D. (1995) *The Concise Oxford Dictionary*, 9th edn. Clarendon Press, Oxford, UK.

Thompson, G.G., Withers, P.C., Pianka, E.R. and Thompson, S.A. (2003) Assessing biodiversity with species accumulation curves; inventories of small reptiles by pit-trapping in Western Australia. *Austral Ecology*, 28, 361–383.

Thompson, M.S.A., Pontalier, H., Spence, M.A., Pinnegar, J.K., Greenstreet, S.P.R. *et al.* (2020) A feeding guild indicator to assess environmental change impacts on marine ecosystem structure and functioning. *Journal of Applied Ecology*, 57, 1769–1781.

Thomson, R.C., Spinks, P.Q. and Shaffer, H.B. (2021) A global phylogeny of turtles reveals a burst of climate-associated diversification on continental margins. *Proceedings of the National Academy of Sciences of the United States of America*, 118(7), e2012215118.

Thoreau, H.D. (1854) *Walden: or, Life in the Woods*. Ticknor and Fields, Boston, USA.

Thornton, I.W.B. (1997) *Krakatau: The Destruction and Reassembly of an Island Ecosystem*. Harvard University Press, Cambridge, USA.

Thornton, I.W.B., Zann, R.A., Rawlinson, P.A., Tidemann, C.R., Adikerana, A.S. and Widjoya, A.H.T. (1988) Colonization of the Krakatau islands by vertebrates: Equilibrium, succession, and possible delayed extinction. *Proceedings of the National Academy of Sciences of the United States of America*, 85, 515–518.

Thorogood, R., Mustonen, V., Aleixo, A., Aphalo, P.J., Asiegbu, F.O. *et al.* (2023) Understanding and applying biological resilience, from genes to ecosystems. *Npj Biodiversity*, 2, 16.

Threlfall, C.G., Mata, L., Mackie, J.A., Hahs, A.K., Stork, N.E. *et al.* (2017) Increasing biodiversity in urban green spaces through simple vegetation interventions. *Journal of Applied Ecology*, 54, 1874–1883.

Thuiller, W., Albert, C.H., Dubuis, A., Randin, C. and Guisan, A. (2010) Variation in habitat suitability does not always relate to variation in species' plant functional traits. *Biology Letters*, 6, 120–123.

Tihelka, E., Howard, R.J., Cai, C. and Lozano-Fernandez, J. (2022) Was there a Cambrian explosion on land? The case of arthropod terrestrialization. *Biology*, 11(10), 1516.

Tilley, R.J.D. (1995) Larval dispersion and food plant densities with respect to the Marsh Fritillary, *Eurodryas aurinia* (Rott.) (Lepidoptera: Nymphalidae). *Entomologist's Gazette*, 46, 163–167.

Tilley, R.J.D. (2016) Searching and finding: Prey and food plants. *Entomologist's Record and Journal of Variation*, 128, 129–133.

Tilley, R.J.D. and Dennis, R.L.H. (2017) Assessing probabilities of island location and occupation: A butterfly individual's perspective. *Entomologist's Gazette*, 68, 161–175.

Tilman, D. (1993) Community diversity and succession – The roles of competition, dispersal, and habitat modification. *Biodiversity and Ecosystem Function*, 99, 327–344.

Tilman, D., Reich, P., Knops, J., Wedin, D., Mielke, T. and Lehman, C. (2001) Diversity and productivity in a long-term grassland experiment. *Science*, 294, 843–845.

Tilman, D., Reich, P.B. and Knops, J.M.H. (2006) Biodiversity and ecosystem stability in a decade long grassland experiment. *Nature*, 441, 629–632.

Timofeev, S.F. (2001) Bergmann's principle and deep-water gigantism in marine crustaceans. *Biology Bulletin of the Russian Academy of Sciences*, 28(6), 646–650.

Tindale, N.B. (1974) *Aboriginal Tribes of Australia: Their Terrain, Environmental Controls, Distribution, Limits, and Proper Names*. Australian National University, Canberra, Australia.

Tintori, S.C., Sloat, S.A. and Rockman, M.V. (2022) Rapid isolation of wild nematodes by Baermann funnel. *Journal of Visualized Experiments*, 179, e63287.

Tiple, A.D., Khurad, A.M. and Dennis, R.L.H. (2007) Butterfly diversity in relation to a human impact gradient on an Indian university campus. *Nota Lepidopterologica*, 30, 179–188.

Tiple, A.D., Khurad, A.M. and Dennis, R.L.H. (2009) Adult butterfly feeding-nectar flower associations: Constraints of taxonomic affiliation, butterfly and nectar flower morphology. *Journal of Natural History*, 43, 855–884.

Tobias, J.A. (2022) A bird in the hand: Global-scale morphological trait datasets open new frontiers of ecology, evolution, and ecosystem science. *Ecology Letters*, 25, 573–580.

Tobler, W. (1970) A computer movie simulating urban growth in the Detroit region. *Economic Geography*, 46(Suppl), 234–240.

Tobler, W. (1999) Linear pycnophylactic reallocation – Comment on a paper by D. Martin. *International Journal of Geographical Information Science*, 13, 85–90.

Tobner, C.M., Paquette, A., Gravel, D., Reich, P.B., Williams, L.J. and Messier, C. (2016) Functional identity is the main driver of diversity effects in young tree communities. *Ecology Letters*, 19, 638–647.

Todd, P.W. (2006) Planetary biology and terraforming. *Gravitational and Space Biology*, 19, 79–84.

Todo, Y., Kitazato, H., Hashimoto, J. and Gooday, A.J. (2005) Simple foraminifera flourish at the ocean's deepest point. *Science*, 307, 689.

Toledo, L.F., Becker, C.G., Haddad, C.F.B. and Zamudio, K.R. (2014) Rarity as an indicator of endangerment in neotropical frogs. *Biological Conservation*, 179, 54–62.

Tolimieri, N. and Anderson, M.J. (2010) Taxonomic distinctness of demersal fishes of the California Current: Moving beyond simple measures of diversity for marine ecosystem-based management. *PLOS One*, 5, e10653.

Tong, W.H., Pavey, C., O'Handley, R. and Vyas, A. (2021) Behavioral biology of *Toxoplasma gondii* infection. *Parasites and Vectors*, 14, 1–6.

Tonkinson, R. (1991) *The Mardu Aborigines: Living the Dream in Australia's Desert.* Holt, Rinehart and Winston, New York, USA.

Toomey, D. (2016) Exploring how and why trees 'talk' to each other. *Yale Environment360*. Available at: https://e360.yale.edu/features/exploring_how_and_why_trees_talk_to_each_other (accessed 23 May 2025).

Torres, M. and Qiu, G. (2014) Automatic habitat classification using image analysis and random forest. *Ecological Informatics*, 23, 126–136.

Tórrez-Herrera, L.L., Davis, G.H. and Crofoot, M.C. (2020) Do monkeys avoid areas of home range overlap because they are dangerous? A test of the risk hypothesis in White-Faced Capuchin Monkeys (*Cebus capucinus*). *International Journal of Primatology*, 41, 246–264.

Tóth, Z., Szlavecz, K., Epp Schmidt, D.J., Hornunga, E., Setälä, H. *et al.* (2020) Earthworm assemblages in urban habitats across biogeographical regions. *Applied Soil Ecology*, 151, 103530.

Trap, J., Akpa-Vinceslas, M., Margerie, P., Boudsocq, S., Richard, F., Decaëns, T. and Aubert, M. (2017) Slow decomposition of leaf litter from mature *Fagus sylvatica* trees promotes offspring nitrogen acquisition by interacting with ectomycorrhizal fungi. *Journal of Ecology*, 105, 528–539.

Traveset, A., Kueffer, C. and Daehler, C.C. (2014) Global and regional nested patterns of non-native invasive floras on tropical islands. *Journal of Biogeography*, 41, 823–832.

Tree, I. (2018) *Wilding: The Return of Nature to a British Farm.* Pan Macmillan, London, UK.

Tree, I. and Burrell, C. (2023) *The Book of Wilding. A Practical Guide to Rewilding, Big and Small.* Bloomsbury, London, UK.

Trewavas, A., Baluška, F., Mancuso, S. and Calvo, P. (2020) Consciousness facilitates plant behavior. *Trends in Plant Science*, 25(3), 216–217.

Triantis, K.A., Mylonas, M., Weiser, M.D., Lika, K. and Vardinoyannis, K. (2005) Species richness, environmental heterogeneity and area: A case study based on land snails in Skyros archipelago (Aegean Sea, Greece). *Journal of Biogeography*, 32, 1727–1735.

Triantis, K.A., Vardinoyannis, K., Tsolaki, E.P., Botsaris, I., Lika, K. and Mylonas, M. (2006) Re-approaching the small island effect. *Journal of Biogeography*, 33, 914–923.

Trifonov, E.N. (2011) Vocabulary of definitions of life suggests a definition. *Journal of Biomolecular Structure and Dynamics*, 29(2), 259–266.

Trigg, A.B. (2001) Veblen, Bourdieu and conspicuous consumption. *Journal of Economic Issues*, 35, 99–115.

Trimble, M.J. and van Aarde, R.J. (2012) Geographical and taxonomic biases in research on biodiversity in human-modified landscapes. *Ecosphere*, 3(12),119.

Troudet, J., Grandcolas, P., Blin, A., Vignes-Lebbe, R. and Legendre, F. (2017) Taxonomic bias in biodiversity data and societal preferences. *Scientific Reports*, 7, 9132.

Troxell, B. and Yang, F. (2013) Metal-dependent gene regulation in the causative agent of Lyme disease. *Frontiers in Cellular and Infection Microbiology*, 3, 79.

Trudgill, S.T. (1977) *Soil and Vegetation Systems.* Clarendon Press, Oxford, UK.

Trussell, G.C., Johnson, A.S., Rudolph, S.G. and Gilfillan, E.S. (1993) Resistance to dislodgement: Habitat and size specific differences in morphology and tenacity in an intertidal snail. *Marine Ecology Progress Series*, 100, 135–144.

Tsafack, N., Borges, P.A.V., Xie, Y., Wang, X. and Fattorini, S. (2021a) Emergent rarity properties in carabid communities from Chinese steppes with different climatic conditions. *Frontiers in Ecology and Evolution*, 9, 603436.

Tsafack, N., Wang, X., Xie, Y. and Fattorini, S. (2021b) Niche overlap and species co-occurrence patterns in carabid communities of the northern Chinese steppes. In: Spence, J., Casale, A., Assmann, T., Liebherr, J.K. and Penev, L. (eds) *Systematic Zoology and Biodiversity Science: A tribute to Terry Erwin* (1940–2020). *ZooKeys*, 1044, 929–949.

Tsafack, N., Fang, W., Wang, X., Xie, Y., Wang, X. and Fattorini, S. (2022) Influence of grazing and solar panel installation on tenebrionid beetles (Coleoptera Tenebrionidae) of a central Asian steppe. *Journal of Environmental Management*, 320, 115791.

Tsafack, N., Pozsgai, G., Boieiro, M., Ros-Prieto, A., Nunes, R. *et al.* (2023) Edge effects constrain endemic but not introduced arthropod species in a pristine forest on Terceira (Azores, Portugal). *Forest Ecology and Management*, 528, 120646.

Tucker, C.M., Cadotte, M.W., Carvalho, S.B., Davies, T.J., Ferrier, S. *et al.* (2017) A guide to phylogenetic metrics for conservation, community ecology and macroecology. *Biological Reviews*, 92(2), 698–715.

Tudor, O., Dennis, R.L.H., Greatorex-Davies, J.N. and Sparks, T.H. (2004) Flower preferences of woodland butterflies in the UK: Nectaring specialists are species of conservation concern. *Biological Conservation*, 119, 397–403.

Tuomisto, H. (2010a) A diversity of beta diversities: Straightening up a concept gone awry. Part 1. Defining beta diversity as a function of alpha and gamma diversity. *Ecography*, 33, 2–22.

Tuomisto, H. (2010b) A diversity of beta diversities: Straightening up a concept gone awry. Part 2. Quantifying beta diversity and related phenomena. *Ecography*, 33, 23–45.

Tuomisto, H. (2012) An updated consumer's guide to evenness and related indices. *Oikos*, 121, 1203–1218.

Tuomisto, H. and Ruokolainen, K. (2006) Analyzing or explaining beta diversity? Understanding the targets of different methods of analysis. *Ecology*, 87(11), 2697–2708.

Turchin, P. (1995) Population regulation: Old arguments and a new synthesis. In: Cappuccino, N. and Price, P.W. (eds) *Population Dynamics: New Applications and Synthesis*. Academic Press, New York, pp. 19–39.

Turchin, P., Oksanen, L., Ekerholm, P., Oksanen, T. and Henttonen, H. (2000) Are lemmings prey or predators? *Nature*, 405, 562–565.

Turlure, C., Van Dyck, H., Schtickzelle, N. and Baguette, M. (2009) Resource based definition of the habitat, niche overlap and conservation of two glacial relict butterflies. *Oikos*, 118, 950–960.

Turlure, C., Schtickzelle, N., Dubois, Q., Baguette, M., Dennis, R.L.H. and Van Dyck, H. (2019) Suitability and between-site transferability of the resource-based habitat concept: A test with butterflies. *Frontiers in Ecology and Evolution*, 7, 127.

Turner, A. (2010) *The Barn Swallow*. Bloomsbury Publishing, London, UK.

Turner, B.L., Zemunik, G., Laliberté, E., Drake, J.J., Jones, F.A. and Saltonstall, K. (2019) Contrasting patterns of plant and microbial diversity during long-term ecosystem development. *Journal of Ecology*, 107, 606–621.

Turner, D., Kropinski, A.M. and Adriaenssens, E.M. (2021) A roadmap for genome-based phage taxonomy. *Viruses*, 13(3), 506.

Turner, J.S. (2001) On the mound of *Macrotermes michaelseni* as an organ of respiratory gas exchange. *Physiological and Biochemical Zoology*, 74, 798–822.

Turner, M.G. (1989) Landscape ecology: The effect of pattern on process. *Annual Review of Ecology, Evolution and Systemics*, 20, 171–197.

Turner, M.G. and Gardner, R.H. (2015) *Landscape Ecology in Theory and Practice: Pattern and Process*. Springer, New York, USA.

Turtle Taxonomy Working Group (2021) *Turtles of the World: Annotated Checklist and Atlas of Taxonomy, Synonymy, Distribution, and Conservation Status*, 9th edn. In: Rhodin, A.G.J., Iverson, J.B., van Dijk, P.P., Stanford, C.B., Goode, E.V., Buhlmann, K.A. and Mittermeier, R.A. (eds) *Conservation Biology of Freshwater Turtles and Tortoises: A Compilation Project of the IUCN/SSC Tortoise and Freshwater Turtle Specialist Group*. Chelonian Research Monographs, 8. Chelonian Research Foundation, Arlington VT, and Turtle Conservancy, Ojia, USA.

Turvey, S.T. and Cheke, A.S. (2008) Dead as a dodo: The fortuitous rise to fame of an extinction icon. *Historical Biology*, 20, 149–163.

Tuttle, R.H. (2014) *Apes and Human Evolution*. Harvard University, Press, Cambridge, USA.

Tuxen, R. (1933) Klimaxprobleme des nordwesteuropaischen Festlandes. *Nederlandsch Kruidkundig Archief*, 43, 293–309.

Tylianakis, J.M., Tscharntke, T. and Lewis, O.T. (2007) Habitat modification alters the structure of tropical host-parasitoid food webs. *Nature*, 445, 202–205.

Tzfira, T. and Citovsky, V. (eds) (2008) *Agrobacterium: From Biology to Biotechnology*. Springer, New York, USA.

Uchida, K., Blakey, R.V., Burger, J.R., Cooper, D.S., Niesner, C.A. and Blumstein, D.T. (2021) Urban biodiversity and the importance of scale. *Trends in Ecology and Evolution*, 36, 123–131.

Udvardy, M.D.F. (1969) *Dynamic Zoogeography, with Special Reference to Land Animals*. Van Nostrand Reinhold, New York, USA.

Udvardy, M.D.F. (1975) *A Classification of the Biogeographical Provinces of the World*. IUCN Occasional Paper no. 18. International Union for Conservation of Nature and Natural Resources, Morges, Switzerland.

Uexküll, J. von. (1957) A stroll through the worlds of animals and men: A picture book of invisible worlds. In: Schiller, C.H. (ed.) *Instinctive Behavior: The Development of a Modern Concept*. International Universities Press, New York, USA, pp. 5–80.

Ugland, K.I.D., Lambshead, P.J., McGill, B., Gray, J.S., O'Dea, N. *et al.* (2007) Modelling dimensionality in species abundance distributions: Description and evaluation of the Gambin model. *Evolutionary Ecology Research*, 9, 313–324.

Ulrich, W. and Gotelli, N.J. (2007) Null model analysis of species nestedness patterns. *Ecology*, 88, 1824–1831.

Ulrich, W., Almeida-Neto, M. and Gotelli, N.J. (2009) A consumer's guide to nestedness analysis. *Oikos*, 118, 3–17.

Ulrich, W., Ollik, M. and Ugland, K.I. (2010) A meta-analysis of species–abundance distributions. *Oikos*, 119, 1149–1155.

Ulrich, W., Santiago, S., Thomas, D.A., Dougill, J.A. and Maestre, T.F. (2016) Environmental correlates of species rank – Abundance distributions in global drylands. *Perspectives in Plant Ecology, Evolution and Systematics*, 20, 56–64.

Ulrich, W., Kubota, Y., Kusumoto, B., Baselga, A., Tuomisto, H. and Gotelli, N.J. (2018a) Species richness correlates of raw and standardized co-occurrence metrics. *Global Ecology and Biogeography*, 27, 395–399.

Ulrich, W., Nakadai, R., Matthews, T.J. and Kubota, Y. (2018b) The two-parameter Weibull distribution as a universal tool to model the variation in species relative abundances. *Ecological Complexity*, 36, 110–116.

Ulrich, W., Matthews, T.J. and Kubota, Y. (2019) Constraints on the distribution of species abundance indicate universal mechanisms of community assembly. *Ecological Research*, 35, 362–371.

Ulrich, W., Matthews, T.J., Biurrun, I., Campos, A.J., Czortek, P. *et al.* (2022) Environmental drivers and spatial scaling of species abundance distributions in Palaearctic grassland vegetation. *Ecology*, 103, e3725.

Ulrich, W., Borges, P.A.V., Fattorini, S., ter Steege, H., Gotelli, N.J. *et al.* (2025) Species and trait abundance distributions as tools for understanding disturbance effects and community assembly in applied ecology. *Journal of Applied Ecology*, 62, 2096–2110.

Umaña, M.N., Zhang, C., Cao, M., Lin, L. and Swenson, N.G. (2015) Commonness, rarity, and intraspecific variation in traits and performance in tropical tree seedlings. *Ecology Letters*, 18, 1329–1337.

Underwood, A.J. (1979) The ecology of intertidal gastropods. *Advances in Marine Biology*, 16, 111–210.

United Nations Department of Economic and Social Affairs, Population Division (2022) *World Population Prospects 2022: Summary of Results.* UN DESA/POP/2022/TR/NO.3. United Nations Department of Economic and Social Affairs, Population Division, New York, USA.

Unsworth, R.K.F., Peters, J.R., McCloskey, R.M. and Hinder, S.L. (2014) Optimising stereo baited underwater video for sampling fish and invertebrates in temperate coastal habitats. *Estuarine, Coastal and Shelf Science*, 150, 281–287.

Urban, D.L., O'Neill, R.V. and Shugart, H.H. (1987) Landscape ecology. *BioScience*, 37,119–127.

Usherwood, P.N.R. and Newth, D.R. (eds) (1975) *'Simple' Nervous Systems: An International Symposium.* Edward Arnold, London, UK.

Uuemaa, E., Antrop, M., Roosaare, J. Marja, A. and Mander, U. (2009) Landscape metrics and indices: An overview of their use in landscape research. *Living Reviews in Landscape Research*, 3, 1.

Vadas, P.A., Aarons, S.R., Butler, D.M. and Dougherty, W.J. (2011) A new model for dung decomposition and phosphorous transformations and loss in runoff. *Soil Research*, 49, 367–375.

Vajda, V. and Bercovici, A. (2014) The global vegetation pattern across the Cretaceous–Paleogene mass extinction interval: A template for other extinction events. *Global and Planetary Change*, 122, 29–49.

Vajda, V. and McLoughlin, S. (2007) Extinction and recovery patterns of the vegetation across the Cretaceous–Palaeogene boundary – A tool for unravelling the causes of the end-Permian mass-extinction. *Review of Palaeobotany and Palynology*, 144, 99–112.

Vajda, V., Cavalcante, L., Palmgren, K., Krüger, A. and Ivarsson, M. (2023) Prototaxites reinterpreted as mega-rhizomorphs, facilitating nutrient transport in early terrestrial ecosystems. *Canadian Journal of Microbiology*, 69(1), 17–31.

Vaksmaa, A., Guerrero-Cruz, S., Ghosh, P., Zeghal, E., Hernando-Morales, V. and Niemann, H. (2023) Role of fungi in biomediation of emerging pollutants. *Frontiers in Marine Science*, 10, 1070905.

Valeix, M., Loveridge, A.J. and Macdonald, D.W. (2012) Influence of prey dispersion on territory and group size of African lions: A test of the resource dispersion hypothesis. *Ecology*, 93, 2490–2496.

Valentine, P.C. (2019) *Sediment Classification and the Characterization, Identification, and Mapping of Geologic Substrates for the Glaciated Gulf of Maine Seabed and other Terrains, providing a Physical Framework for Ecological Research and Seabed Management.* US Geological Survey, Reston, USA.

Välimäki, P. and Itämies, J. (2003) Migration of the clouded Apollo butterfly *Parnassius mnemosyne* in a network of suitable habitats – Effects of patch characteristics. *Ecography*, 26, 679–691.

Vallauri, D., André, J., Dodelin, B., Eynard-Machet, R. and Rambaud, D. (2005) *Bois mort et à cavités – Une clé pour des forêts vivantes.* Éditions Techniques et Documentaires, Lavoisier, Paris, France.

Van Andel, J., Bakker, J.P. and Grootjans, A.P. (1993) Mechanisms of vegetation succession: A review of concepts and perspectives. *Acta Botanica Neerlandica*, 42, 413–433.

Van Andel, T. and Davies W. (eds) (2003) *Neanderthals and Modern Humans in the European Landscape during the Last Glaciation.* McDonald Institute for Archaeological Research, Cambridge, UK.

van Breugel, M., Craven, D., Lai, H.R., Baillon, M., Turner, B.L. and Hall, J.S. (2019) Soil nutrients and dispersal limitation shape compositional variation in secondary tropical forests across multiple scales. *Journal of Ecology*, 107, 566–581.

van Dam, A. (2019) Diversity and its decomposition into variety, balance and disparity. *Royal Society Open Science*, 6, 90452.

van der Maarel, E. (1979) Transformation of cover-abundance values in phytosociology and its effectson community similarity. *Vegetatio*, 39, 97–114.

van der Maarel, E. (1990) Ecotones and ecoclines are different. *Journal of Vegetation Science*, 1, 135–138.

Vandermeer, J.H. and Goldberg, D.E. (2013) *Population Ecology: First Principles*, 2nd edn. Princeton University Press, Princeton, USA.

van der Sluijs, J.P. (2020) Insect decline, an emerging global environmental risk. *Current Opinion in Environmental Sustainability*, 46, 39–42.

Vandewalle, M., de Bello, F., Berg, M.P., Bolger, T., Dolédec, S. *et al.* (2010) Functional traits as indicators of biodiversity response to land use changes across ecosystems and organisms. *Biodiversity and Conservation*, 19, 2921–2947.

Vandewoestijne, S., Baguette, M., Brakefield, P.M. and Saccheri, J. (2004) Phylogeography of *Aglais urticae* (Lepidoptera) based on DNA sequences of the mitochondrial COI gene and control region. *Molecular Phylogenetics and Evolution*, 31, 630–646.

van Dijk, K., Cibot, M. and McLennan, M.R. (2021) Chimpanzees (*Pan troglodytes*) adapt their nesting behavior after large-scale forest clearance and community decline. *American Journal of Primatology*, 83, e23323.

Vandiver, P.B. (2022) Upper Paleolithic ceramic figurines and similarities to some late Pleistocene pigment and pottery materials and technologies of Eurasia. *Quaternary International*, 608–609.

van Dobben, H.F., Wamelink, G.W.W., Slim, P.A., Kamiński, J. and Piórkowski, H. (2017) Species-rich grassland can persist under nitrogen-rich but phosphorus-limited conditions. *Plant and Soil*, 411, 451–466.

Van Dyck, H., Van Strien, A., Maes, D. and van Swaay, C.A.M. (2009) Declines in common, widespread butterflies in a landscape under intense human use. *Conservation Biology*, 23, 957–965.

Vane-Wright, R.I. (2014) What is life? And what might be said of the role of behaviour in its evolution? *Biological Journal of the Linnean Society*, 112, 219–241.

Vane-Wright, R.I. (2021) Book Review. Butterfly Biology Systems: Connections and interactions in life history and behaviour by Roger L. H. Dennis. *Entomologist's Gazette*, 72, 275–278.

van Ginneken, V., Durif, C., Balm, S.P., Boot, R., Verstegen, M. *et al.* (2007a) Silvering of European eel (*Anguilla anguilla L.*): Seasonal changes of morphological and metabolic parameters. *Animal Biology*, 57(1), 63–77.

van Ginneken, V., Durif, C., Dufour, S., Sbaihi, M., Boot, R. *et al.* (2007b) Endocrine profiles during silvering of the European eel (*Anguilla anguilla* L.) living in saltwater. *Animal Biology*, 57(4), 453–465.

van Holstein, L.A. and Foley, R.A. (2022) A process-based approach to hominin taxonomy provides new perspectives on hominin speciation. *Evolutionary Anthropology*, 31, 166–174.

van Leeuwen, K.L., Hill, R.A. and Korstjens, A.H. (2020) Classifying chimpanzee (*Pan troglodytes*) landscapes across large-scale environmental gradients in Africa. *International Journal of Primatology*, 41, 1152–1169.

Van Meerbeek, K., Jucker, T. and Svenning, J.-C. (2021) Unifying the concepts of stability and resilience in ecology. *Journal of Ecology*, 109, 3114–3132.

Vannote, R.L., Minshall, G. W., Cummins, K.W., Sedell, J.R. and Cushing, C.E. (1980) The river continuum concept. *Canadian Journal of Fisheries and Aquatic Sciences*, 37(1), 130–137.

van Nus, T. and Neto, J.M. (2017) Urban roost of wintering Barn Swallows *Hirundo rustica* in Aveiro, Portugal. *Ardea*, 105, 73–78.

van Schaik, C.P. (1999) The socioecology of fissionfusion sociality in orangutans. *Primates*, 40, 69–86.

van Tongeren, E., Sistri, G., Bonifacino, M., Menchetti, M., Pasquali, L. *et al.* (2023) Unstructured citizen science reduces the perception of butterfly local extinctions: The interplay between species traits and user effort. *Biodiversity and Conservation*, 32, 4701–4718.

Van Valkenburgh, B. (1988) Trophic diversity in past and present guilds of large predatory mammals. *Paleobiology*, 14, 155–173.

Van Valkenburgh, B. (1989) Carnivore dental adaptations and diet: A study of trophic diversity within guilds. In: Gittleman, J.L. (ed.) *Carnivore Behavior, Ecology, and Evolution*. Springer Science+Business Media, Dordrecht, The Netherlands.

Van Vuren, I.L.A. and van Vuren, D. (1996) Differential costs of locational and social dispersal and their consequences for female group-living primates. *Behaviour*, 133, 1–36.

Vanwambeke, S.O., Li, S. and Hartemink, N.A. (2016) A resource-based habitat concept for tick-borne diseases. *Ecology and Control of Vector-borne Diseases*, 4, 205–216.

van Winkle, W. (1975) Comparison of several probabilistic home-range models. *Journal of Wildlife Management*, 39, 118–123.

Varley, G.C., Gradwell, G.R. and Hassell, M.P. (1973) *Insect Population Ecology. An Analytic Approach*. Blackwell Science, Oxford, UK.

Vea Salvanes, A.G., Devine, J., Jensen, K.H., Hestetun, J.T., Sjotun, K. and Glenner, H. (eds). (2018) *Marine Ecological Field Methods: A Guide for Marine Biologists and Fisheries Scientists*. John Wiley and Sons, Chichester, UK.

Venables, A. (2018) Urbanisation in developing economies: Building cities that work. *Region*, 5, 91–100.

Verberk, W. (2011) Explaining general patterns in species abundance and distributions. *Nature Education Knowledge*, 3, 38.

Verberk, W.C.E.P., van der Velde, G. and Esselink, H. (2010) Explaining abundance-occupancy relationships in specialists and generalists: A case study on aquatic macroinvertebrates in standing waters. *Journal of Animal Ecology*, 79, 589–601.

Verhulst, P.-F. (1838) Notice sur la loi que la population suit dans son accroissement. *Correspondance Mathématique et Physique*, 10, 113–121.

Verhulst, P.-F. (1845) Recherches mathématiques sur la loi d'accroissement de la population. *Nouveaux Mémoires de l'Académie Royale des Sciences et Belles-Lettres de Bruxelles*, 18, 1–42.

Verhulst, P.-F. (1847) Deuxième mémoire sur la loi d'accroissement de la population. *Mémoires de l'Académie Royale des Sciences, des Lettres et des Beaux-Arts de Belgique*, 20, 1–32.

Vernadsky, V.I. (1945) The biosphere and the noönsphere. *American Scientist*, 33, 1–12.

Verner, T. (1977) On the adaptive significance of territoriality. *American Naturalist*, 111, 769–775.

Viana, D.S., Gangoso, L., Figuerola, J. and Bouten, W. (2016) Overseas seed dispersal by migratory birds.

Proceedings of the Royal Society B: Biological Sciences, 283(1822), 20152406.

Victorero, L., Samadi, S., O'Hara, T.D., Mouchet, M., Delavenne, J. *et al.* (2023) Global benthic biogeographical regions and macroecological drivers for ophiuroids. *Ecography*, 2023, e06627.

Vieira, M.C. and Almeida-Neto, M. (2015) A simple stochastic model for complex coextinctions in mutualistic networks: Robustness decreases with connectance. *Ecology Letters*, 18, 144–152.

Vieira, W.F., Kerry, C. and Hockings, K.J. (2019) A comparison of methods to determine chimpanzee home-range size in a forest–farm mosaic at Madina in Cantanhez National Park, Guinea-Bissau. *Primates*, 60, 355–365.

Villard, M.-A. (1998) On forest-interior species, edge avoidance, area sensitivity, and dogmas in avian conservation. *The Auk*, 115, 801–805.

Villéger, S., Mason, N. W. H. and Mouillot, D. (2008) New multidimensional functional diversity indices for a multifaceted framework in functional ecology. *Ecology*, 89, 2290–2301.

Villéger, S., Grenouillet, G. and Brosse, S. (2013) Decomposing functional β-diversity reveals that low functional β-diversity is driven by low functional turnover in European fish assemblages. *Global Ecology and Biogeography*, 22, 671–681.

Vincent, P. (1990) *The Biogeography of the British Isles. An Introduction*, Routledge London, UK.

Violle, C., Navas, M.L., Vile, D., Kazakou, E., Fortunel, C. *et al.* (2007) Let the concept of trait be functional! *Oikos*, 116, 882–892.

Violle, C., Enquist, B.J., McGill, B.J., Jiang, L., Albert, C.H. *et al.* (2012) The return of the variance: Intraspecific variability in community ecology. *Trends in Ecology and Evolution*, 27, 244–252.

Violle, C., Thuiller, W., Mouquet, N., Munoz, F., Kraft, N.J. *et al.* (2017) Functional rarity: The ecology of outliers. *Trends in Ecology and Evolution*, 32(5), 356–367.

Virani, R.S. (2020) Diversity of butterflies in Tipeshwar Wildlife Sanctuary of Maharashtra, India. *Bioscience Biotechnology Research Communications*, 13, 1390–1402.

Viviano, A. and Mori, E. (2021) Population counts and potential impact of two successful invaders in a town of Northern Italy: The case of ring-necked parakeet and Alexandrine parakeet in Reggio Emilia. *Natural History Sciences*, 8(1), 35–40.

Vogt, M.A.B. (2021) Agricultural wilding: Rewilding for agricultural landscapes through an increase in wild productive systems. *Journal of Environmental Management*, 284, 112050.

Volkov, I., Banavar, J.R., Hubbell, S.P. and Maritan, A. (2003) Neutral theory and relative species abundance in ecology. *Nature*, 424, 1035–1037.

Volland, J.-M., Gonzalez-Rizzo, S., Gros, O., Tyml, T., Ivanova, N. *et al.* (2022) A centimeter-long bacterium with DNA contained in metabolically active, membrane-bound organelles. *Science*, 376(6600), 1453–1458.

Volterra, V. (1926a). Fluctuations in the abundance of a species considered mathematically. *Nature*, 118, 558–560.

Volterra, V. (1926b) Variazioni e fluttuazioni del numero d'individui in specie animali conviventi. *Memorie della Accademia Nazionale dei Lincei*, 2, 31–113.

Volterra, V. (1931) Ricerche matematiche sulle associazioni biologiche. *Giornale dell'Istituto Italiano degli Attuari*, 2, 5–65.

Volterra, V. and D'Ancona, U. (1935) *Les Associations Biologiques au Point de Vue Mathématique*. Hermann, Paris, France.

von Beeren, C., Maruyama, M., Rosli, H. and Witte, V. (2011) Differential host defense of multiple parasites in ants. *Evolutionary Ecology*, 25, 259–276.

von Bertalanffy, L. (1950) An outline of general system theory. *British Journal for the Philosophy of Science*, 1, 134–165.

von Bertalanffy, L. (1968) *General System Theory: Foundations, Development, Applications*. Braziller, New York, USA.

von Clausewits, C. (1943) [1832] *On War*. (English translation of *Vom Kriege*; Matthijs Jolles, O.J. trans.). Random House, New York, USA.

von Hoermann, C., Benbow, M.E., Rottler-Hoermann, A.M., Lackner, T., Sommer, D. *et al.* (2023) Factors influencing carrion communities are only partially consistent with those of deadwood necromass. *Oecologia*, 201, 537–547.

von Humboldt, A. (1816) Sur les lois que l'on observe dans la distribution des formes végétales. *Annales de Chimie et de Physique*, 1, 225–239.

von Humboldt, A. (1817) *De Distributione Geographica Plantarum Secundum Coeli Temperiem et Altitudinem Montium: Prolegomena*. Libraria Graeco-Latino-Germanica, Paris, France.

von Humboldt, A. and Bonpland, A. (1805) *Essai sur la Géographie des Plantes*. Levrault, Schoell et Compagnie, Paris, France.

von Humboldt, A. and Bonpland, A. (1814-1829) *Personal Narrative of Travels to the Equinoctial Regions of the New Continent, During the Years 1799–1804* (trans. Williams, H.M.). Longman, Hurst, Rees, Orme, and Brown, London, UK.

von Liebig, J. (1840) *Die Organische Chemie in Ihrer Anwendung auf Agrikultur und Physiologie*. Friedrich Vieweg, Wiesbaden, Germany.

Vorzimmer, P. (1969) Darwin, Malthus, and the theory of natural selection. *Journal of the History of Ideas*, 30, 527–542.

Vranken, S., Robuchon, M., Dekeyzer, S., Bárbara, I., Bartsch, I. *et al.* (2023) AlgaeTraits: A trait database for (European) seaweeds, Earth Syst. *Scientific Data*, 15, 2711–2754.

Vrdoljak, S.M. and Samways, M.J. (2012) Optimising coloured pan traps to survey flower visiting insects. *Journal of Insect Conservation*, 16, 345–354.

Vuilleumier, F. (1973) Insular biogeography in continental regions. II. Cave faunas from Tessin, Southern Switzerland. *Systematic Zoology*, 22(1), 64–76.

Waddington, C. (2004) *The Joy of Flint. An Introduction to Stone Tools and Guide to the Museum of Antiquities Collection*. Museum of Antiquities, University of Newcastle upon Tyne, Newcastle upon Tyne, UK.

Wagner, D., Friedlander, A.M., Pyle, R.L., Brooks, C M., Gjerde, K.M. and 'Aulani Wilhelm, T. (2020) Coral reefs of the high seas: Hidden biodiversity hotspots in need of protection. *Frontiers in Marine Science*, 7, 567428.

Wagner, D.L. and Liebherr, J.K. (1992) Flightlessness in insects. *Trends in Ecology and Evolution*, 7, 216–220.

Wagner, D.L., Fox, R., Salcido, D.M. and Dyer, L.A. (2021) A window to the world of global insect declines: Moth biodiversity trends are complex and heterogeneous. *Proceedings of the National Academy of Sciences of the United States of America*, 118, e2002549117.

Waide, R.B. and Kingsland, S.E. (eds) (2021) *The Challenges of Long Term Ecological Research: A Historical Analysis*. Springer Nature, Cham, Switzerland.

Walker, B.H. (1992) Biodiversity and ecological redundancy. *Conservation Biology*, 6, 18–23.

Walker, J.J. and Pace, N.R. (2007) Endolithic microbial ecosystems. *Annual Review of Microbiology*, 61, 331–347.

Walker, L.R. and del Moral, R. (2003) *Primary Succession and Ecosystem Rehabilitation*. Cambridge University Press, Cambridge, UK.

Walker, L.R., Walker, J. and Hobbs, R.J. (eds) (2007) *Linking Restoration and Ecological Succession*. Springer-Verlag, New York, USA.

Walker, L.R., Wardle, D.A., Bardgett, R.D. and Clarkson, B.D. (2010) The use of chronosequences in studies of ecological succession and soil development. *Journal of Ecology*, 98, 725–736.

Walker, M.J.C., Bauer, A.M., Edgeworth, M., Ellis, E.C., Finney, S.C. *et al.* (2024) The Anthropocene is best understood as an ongoing, intensifying, diachronous event. *Boreas*, 53, 1–3.

Walker, P.G.T., Whittaker, C., Watson, O., Baguelin, M., Ainslie, K.E.C. *et al.* (2020) *Report 12: The Global Impact of COVID-19 and Strategies for Mitigation and Suppression*. Imperial College COVID-19 Response Team. Available at: https://spiral.imperial.ac.uk/handle/10044/1/77735 (accessed 25 May 2025).

Wallace, A.R. (1876) *The Geographical Distribution of Animals; With A Study of the Relations of Living and Extinct Faunas As Elucidating the Past Changes of the Earth's Surface*. Harper and Brothers, New York, USA.

Walter, G.H. and Hengeveld, R. (2014) *Autecology. Organisms, Interactions and Environmental Dynamics*. CRC Press, Boca Raton, USA.

Walter, H. (1954) Grundlagen der Pflanzenverbreitung. Part II. Arealkunde. Vol. III. In: Walter, H. (ed.) *Einführung in die Phytologie*. Ulmer, Stuttgart, Germany.

Walter, H. and Box, E. (1976) Global classification of natural terrestrial ecosystems. *Vegetatio*, 32(2), 75–81.

Walter, H. and Breckle, S.-W. (1985) *Ecological Systems of the Geobiosphere*. Springer-Verlag, Berlin, Germany.

Walther, J.V. (2014) *Earth's Natural Resources*. Jones and Bartlett Learning, Burlington, USA.

Waltner-Toews, D. (2012) *Origin of Feces. What Excrement Tells Us About Evolution, Ecology, and a Sustainable Society*. ECW Press, Toronto, Canada.

Wanders, N. and Wada, Y. (2015) Human and climate impacts on the 21st century hydrological drought. *Journal of Hydrology*, 526, 208–220.

Wang, B. and Qiu, Y.L. (2006) Phylogenetic distribution and evolution of mycorrhizas in land plants. *Mycorrhiza*, 16(5), 299–363.

Wang, J., Kan, J., Qian, G., Chen, J., Xia, Z. *et al.* (2019) Denitrification and anammox: Understanding nitrogen loss from Yangtze Estuary to the East China Sea (ECS). *Environmental Pollution*, 252,1659–1670.

Wang, R., Kass, J.M., Chaudhary, C., Economo, E.P. and Guénard, B. (2024) Global biogeographic regions for ants have complex relationships with those for plants and tetrapods. *Nature Communications*, 15, 5641.

Wang, R.H., Yang, S., Liu, Z., Zhang, Y., Wang, X. *et al.* (2023) PhageScope: A well-annotated bacteriophage database with automatic analyses and visualizations. *Nucleic Acids Research*, 52, D1, D756–D761.

Wang, S., Fei, P., Wu, Z., Luo, Z., Wu, Y. *et al.* (2023) Adult tiger beetles *Cicindela gemmata* modify their foraging strategy in different hunting contexts. *Insect Science*, 30, 1749–1758.

Wang, S.C. and Dodson, P. (2006) Estimating the diversity of dinosaurs. *Proceedings of the National Academy of Sciences of the United States of America*, 103(37), 13601–13605.

Wang, Y., Huang, B., Yan, G., Liu, G. Xing, Y. and Wang, Q. (2022) Effects of long-term nitrogen addition and seasonal variation on soil faunal community structure in a temperate natural secondary forest. *European Journal of Forest Research*, 141, 547–560.

Wang, Z. and Zeng, F. (2021) Impact of vegetation succession on leaf-litter-soil C:N:P stoichiometry and their intrinsic relationship in the Ziwuling area of China's loess plateau. *Journal of Forest Research*, 32, 697–711.

Warakai, D., Okena, D.S., Igag, P., Opiang, M. and Mack, A.L. (2013) Tree cavity-using wildlife and the potential of artificial nest boxes for wildlife management in New Guinea. *Tropical Conservation Science*, 6, 711–733.

Ward, P. and Zahavi, A. (1973) The importance of certain assemblages of birds as "information-centres" for food-finding. *Ibis*, 115, 517–534.

Wardle, D.A., Walker, L.R. and Bardgett, R.D. (2004) Ecosystem properties and forest decline in contrasting long-term chronosequences. *Science*, 305, 509–513.

Ward-Perkins, B. (2006) *The Fall of Rome and the End of Civilization*. Oxford University Press, Oxford, UK.

Ware, I.M., Van Nuland, M.E., Yang, Z.K., Schadt, C.W., Schweitzer, J.A. and Bailey, J.K. (2021) Climate-driven divergence in plant-microbiome interactions generates range-wide variation in bud break phenology. *Communications Biology*, 4, 748.

Warming, E. (1895) *Plantesamfund–Grundtræk af den økologiske Plantegeografi*. Philipsen, Copenhagen, Denmark.

Warming, E. and Vahl, M. (1909) *Oecology of Plants. An Introduction to the Study of Plant-Communities* [Extended and translated edition of Warming 1895]. Clarendon Press, Oxford, UK.

Warren, D.L., Glor, R.E. and Turelli, M. (2008) Environmental niche equivalency versus conservatism: Quantitative approaches to niche evolution. *Evolution*, 62, 2868–2883.

Warren, M.S. (1992) Butterfly populations. In: Dennis, R.L.H. (ed.) *Ecology of Butterflies in Britain*. Oxford University Press, Oxford, UK, pp. 73–92.

Warwick, R.M. and Clarke, K.R. (1995) New 'biodiversity' measures reveal a decrease in taxonomic distinctness with increasing stress. *Marine Ecology Progress Series*, 129, 301–305.

Waters, C.N., Syvitski, J.P.M., Gałuszka, A., Hancock, G.J., Zalasiewicz, J. *et al.* (2015) Can nuclear weapons fallout mark the beginning of the Anthropocene Epoch? *Bulletin of the Atomic Scientists*, 71, 46–57.

Waters, N. (2017) Tobler's first law of geography. In: Richardson, D., Castree, N., Goodchild, M.F., Kobayashi, A., Liu, W. and Marston R.A. (eds) *The International Encyclopaedia of Geography*. John Wiley and Sons, Hoboken, USA.

Watson, E. and Carlton, C. (2005) Succession of forensically significant carrion beetle larvae on large carcasses (Coleoptera: Silphidae). *Southeastern Naturalist*, 4(2), 335–346.

Watson, G.F. and Davis, P.H. (2019) Systematic review and meta-analysis of variation in *Toxoplasma gondii* cyst burden in the murine model. *Experimental Parasitology*, 196, 55–62.

Watson, J. and Crick, F. (1953) Molecular structure of nucleic acids: A structure for deoxyribose nucleic acid. *Nature*, 171, 737–738.

Watson, J.C., Payne, R.C., Chamberlain, A.T., Jones, R.K. and Sellers, W.I. (2008) The energetic costs of load-carrying and the evolution of bipedalism. *Journal of Human Evolution*, 54, 675–683.

Watt, A.S. (1947) Pattern and process in the plant community. *Journal of Ecology*, 35, 1–22.

Waugh, D. (2000) *Geography. An Integrated Approach*, 3rd edn. Nelson Thornes, Cheltenham, UK.

Webb, C.O. (2000) Exploring the phylogenetic structure of ecological communities: An example for rain forest trees. *American Naturalist*, 156, 145–155.

Webb, C.O., Ackerly, D.D., McPeek, M.A. and Donoghue, M.J. (2002) Phylogenies and community ecology. *Annual Review of Ecology, Evolution, and Systematics*, 33, 475–505.

Webster, M. (2007) A Cambrian peak in morphological variation within trilobite species. *Science*, 317(5837), 499–502.

Weddell, B.J. (2002) *Conserving Living Natural Resources*. Cambridge University Press, Cambridge, UK.

Weeks, A.R., Sgro, C.M., Young, A.G., Frankham, R., Mitchell, N.J. *et al.* (2011) Assessing the benefits and risks of translocations in changing environments: A genetic perspective. *Evolutionary Applications*, 4, 709–725.

Wegener, A. (1912) Die Entstehung der Kontinente. *Petermanns Geographische Mitteilungen*, 58, 185–195, 253–256, 305–308.

Wegner, C.-E., Stahl, R., Velsko, I., Hübner, A., Fagernäs, Z. *et al.* (2023) A glimpse of the paleome in endolithic microbial communities. *Microbiome*, 11, 210.

Weigelt, P., Jetz, W. and Kreft, H. (2013) Bioclimatic and physical characterization of the world's islands. *Proceedings of the National Academy of Sciences of the United States of America*, 110, 15307–15312.

Weigt, L.A., Driskell, A.C., Baldwin, C.C. and Ormos, A. (2012) DNA barcoding fishes. In: Kress J. and Erickson, D.L. (eds) *DNA Barcodes: Methods and Protocols*. Springer Science+Business Media, Dordrecht, The Netherlands, pp. 109–126.

Weiher, E. and Keddy, P.A. (1995) The assembly of experimental wetland plant communities. *Oikos*, 73, 323–335.

Weiher, E., Clarke, G.D.P. and Keddy, P.A. (1998) Community assembly rules, morphological dispersion, and the coexistence of plant species. *Oikos*, 81, 309–322.

Weir, J.C. (2021) Search image-driven apostatic selection and the evolution of phenotypic polymorphism. In: *Encyclopedia of Life Sciences*. John Wiley and Sons, Chichester, UK, pp. 1–10.

Weldeslassie, T., Naz, H., Singh, B. and Oves, M. (2018) Chemical contaminants for soil, air and aquatic ecosystem. In: Oves, M., Khan, M.Z. and Ismail, I.M.I. (eds) *Modern Age Environmental Problems and their Remediation*. Springer, Dordrecht, The Netherlands.

Wellington, W. (1945) Conditions governing the distribution of insects in the free atmosphere. *Canadian Entomologist*, 77(1), 7–15.

Wells, K.D. (2007) *The Ecology and Behavior of Amphibians*. University of Chicago Press, Chicago, USA.

West, N.M., Gibson, D.J. and Minchin, P.R. (2009) Characterizing the microhabitats of exotic species in Illinois shale barrens. *Plant Ecology*, 200, 255–265.

Westall, F., Brack, A., Fairén, A. G. and Schulte, M.D. (2023) Setting the geological scene for the origin of life and continuing open questions about its emergence. *Frontiers in Astronomy and Space Sciences*, 9, 1095701.

Westerband, A.C., Funk, J.L. and Barton, K.E. (2021) Intraspecific trait variation in plants: A renewed focus on its role in ecological processes. *Annals of Botany*, 127, 397–410.

Westoby, M. (1984) The self-thinning rule. *Advances in Ecological Research*, 14, 167–225.

Weygoldt, P. (1969) *The Biology of Pseudoscorpions*. Harvard University Press, Cambridge, USA.

WFO (2025) World Flora Online. Available at: https://about.worldfloraonline.org/ (accessed 25 May 2025).

Wheater, C.P., Bell, J.R. and Cook, P.A. (2011) *Practical Field Ecology. A Project Guide*. Wiley-Blackwell, Oxford, UK.

Whinkin, E., Opalka, M., Watters, C., Jaffe, A. and Aggarwal, S. (2023) Psilocybin in palliative care: An update. *Current Geriaticrics Reports*, 12, 50–59.

White, G. 1981 [1789 first published]. *The Natural History of Selborne*. Thames and Hudson, London, UK.

White, J. (1980) Demographic factors in populations of plants. In: Solbrig, O.T. (ed.) *Demography and Evolution in Plant Populations*. Blackwell, Oxford, UK.

White, T.D. (2018) The origin of our species. In: Brown, J. (ed.) *Becoming Human*, 2nd ed. New Scientist: The Collection, London, UK, pp. 6–9.

Whitehouse, M.E. and Jaffe, K. (1996) Ant wars: Combat strategies, territory and nest defence in the leaf-cutting ant Atta laevigata. *Animal Behaviour*, 51, 1207–1217.

Whittaker, R.H. (1951) A criticism of the plant association and climatic climax concepts. *Northwest Science*, 25, 17–31.

Whittaker, R.H. (1953) A consideration of climax theory: The climax as a population and pattern. *Ecological Monographs*, 23, 41–78.

Whittaker, R.H. (1965) Dominance and diversity in land plant communities: Numerical relations of species express the importance of competition in community function and evolution. *Science*, 147, 250–260.

Whittaker, R.H. (1967) Gradient analysis of vegetation. *Biological Reviews*, 42, 207–264.

Whittaker, R.H. (1969) New concepts of kingdoms or organisms. Evolutionary relations are better represented by new classifications than by the traditional two kingdoms. *Science*, 163, 150–160.

Whittaker, R.H. (1972) Evolution and measurement of species diversity. *Taxon*, 21, 213–251.

Whittaker, R.H. (1975) *Communities and Ecosystems*, 2nd edn. Macmillan, New York, USA.

Whittaker, R.J., Riddle, B.R., Hawkins, B.A. and Ladle, R.J. (2013) The geographical distribution of life and the problem of regionalization: 100 years after Alfred Russel Wallace. *Journal of Biogeography*, 40, 2209–2214.

Whittaker, R.J., Fernández- Palacios, J.M. and Matthews, T.J. (2023) *Island Biogeography. Geo-environmental Dynamics, Ecology, Evolution, Human Impact, and Conservation*. Oxford University Press, Oxford, UK.

Whittaker, R.H., Levin, S.A. and Root, R.B. (1973) Niche, habitat and ecotope. *American Naturalist*, 107, 321–338.

Whittow, J (1984) *Dictionary of Physical Geography*. Penguin, London, UK.

Wich, S.A. and Marshall, A.J. (eds.) (2016) *An introduction to primate conservation*. Oxford University Press Oxford, UK.

Wiens, J.A. (1989) Spatial scaling in ecology. *Functional Ecology*, 3, 385–397.

Wiklund, C. (1981) Generalist vs. specialist oviposition behaviour in *Papilio machaon* (Lepidoptera) and functional aspects on the hierarchy of oviposition preferences. *Oikos*, 36, 163–170.

Wiklund, C. and Ahrberg, C. (1978) Hostplants, nectar source plants and habitat selection of males and females of *Anthocharis cardamines* (Lepidoptera). *Oikos*, 31, 169–183.

Wilde, P., Llobrera, J. and Ver Ploeg, M. (2014) Population density, poverty, and food retail access in the United States: An empirical approach. *International Food and Agribusiness Management Review*, 17 (Special Issue A), 171–186.

Wiley, E.O. (1981) *Phylogenetics. The Theory and Practice of Phylogenetic Systematics*. John Wiley and Sons, Hoboken, USA.

Wilkinson, D.M. (1999) The disturbing history of intermediate disturbance. *Oikos*, 84(1), 145–147.

Wilkinson, D.M. (2021) *Ecology and Natural History*. Harper Collins Publishers, London, UK.

Wilkinson, M.G. (ed.) (2015) *Flow Cytometry in Microbiology: Technology and Applications*. Caister Academic Press, Norfolk. UK.

Willdenow, K.L. (1792) *Grundriss der Kräuterkunde zu Vorlesungen*. Haude und Spener, Berlin, Germany.

Williams, C.B. (1943) Area and the number of species. *Nature*, 152, 264–267.

Williams, C.B. (1958) *Insect Migration*. Collins, London, UK.

Williams, C.B. (1964) *Patterns in the Balance of Nature*. Academic Press, London, UK.

Williams, J.W. and Jackson, S.T. (2007) Novel climates, no-analog communities, and ecological surprises. *Frontiers in Ecology and the Environment*, 5, 475–482.

Williams, P.H. (1996) Mapping variations in the strength and breadth of biogeographic transition zones using species turnover. *Proceedings of the Royal Society B: Biological Sciences*, 263, 579–588.

Williams, R.J. and Martinez, N.D. (2000) Simple rules yield complex food web. *Nature*, 404, 180–183.

Williams, S.A. and Shattuck, M.R. (2015) Ecology, longevity and naked mole-rats: confounding effects of sociality? *Proceedings of the Royal Society B: Biological Sciences*, 282(1802), 20141664.

Williams, S.E. and Hero, J.-M. (1998) Rainforest frogs of the Australian wet tropics: Guild classification and the ecological similarity of declining species. *Proceedings of the Royal Society B: Biological Sciences*, 265, 597–602.

Williamson, J., Rowe, E., Rendell, T., Healey, J., Jones, D. and Nason, M. (2003) *Restoring Habitats of High Conservation Value after Quarrying: Best Practice Manual*. Institute of Environmental Science, University of Wales, Bangor, UK.

Williamson, M. (2011) Alien plants in Britain. Biological invasions. In: Pimental, D. (ed.) *Biological Invasions:*

Economic and Environmental Costs of Alien Plant, Animal, and Microbe Species, 2nd edn. CRC Press, Boca Raton, USA, pp. 107–128.

Williamson, M. and Fitter, A. (1996a) The varying success of invaders. *Ecology*, 77, 1661–1666.

Williamson, M. and Fitter, A. (1996b) The characters of successful invaders. *Biological Conservation*, 78, 163–170.

Willig, M.R. and Walker, L. R. (eds) (2016) *Long-term Ecological Research: Changing the Nature of Scientists*. Oxford University Press, Oxford, UK.

Willis, A.J. (1997) The ecosystem: An evolving concept viewed historically. *Functional Ecology*, 11, 268–271.

Willis, S. (2011) Samian ware and society in Roman Britain and beyond. *Britannia*, 42, 167–242.

Wilson, D.E., Cole, F.R., Nichols, J.D., Rudran, R. and Foster, M.S. (eds) (1996). *Measuring and Monitoring Biological Diversity: Standard Methods for Mammals*. Smithsonian Books, Washington, DC, USA.

Wilson, E.O. (1971) *The Insect Societies*. Belknap Press, Cambridge, USA.

Wilson, E.O. (1984) *Biophilia*. Harvard University Press, Cambridge, USA.

Wilson, E.O. (2000) *Sociobiology: The New Synthesis*. Twenty-fifth Anniversary Edition. Harvard University Press, Cambridge, USA.

Wilson, E.O. (2013) *The Social Conquest of Earth*. Liveright, New York, USA.

Wilson, J. (1973) Wader populations of Morecambe Bay, Lancashire. *Bird Study*, 20, 9–23.

Wilson, J.B. (1993) Would we recognise a broken-stick community if we found one? *Oikos*, 67, 181–183.

Wilson, J.B. (1999) Guilds, functional types and ecological groups. *Oikos*, 86, 507–522.

Wilson, M.L. and Wrangham, R.W. (2003) Intergroup relations in chimpanzees. *Annual Review of Anthropology*, 32, 363–392.

Winkler, D. (2005) Ecological succession of breeding bird communities in deciduous and coniferous forests in the Sopron Mountains, Hungary. *Acta Silvatica et Lignaria Hungarica*, 1, 49–58.

Winkler, D.W. (2006) Roosts and migrations of swallows. *Hornero*, 21(2), 85–97.

Winter, M., Johnson, D.H. and Faaborg, J. (2000) Evidence for edge effects on multiple levels in tallgrass prairie. *The Condor*, 102, 256–266.

Winter, S. and Möller, G.C. (2008) Microhabitats in lowland beech forests as monitoring tool for nature conservation. *Forest Ecology and Management*, 255, 1251–1261.

Winter, S., Bauer, T., Strauss, P., Kratschmer, S., Paredes, D. *et al.* (2018) Effects of vegetation management intensity on biodiversity and ecosystem services in vineyards: A meta-analysis. *Journal of Applied Ecology*, 55, 2484–2495.

Wintle, B.A., Kujala, H., Whitehead, A., Cameron, A., Veloz, S. *et al.* (2019) Global synthesis of conservation studies reveals the importance of small habitat patches for biodiversity. *Proceedings of the National Academy of Sciences of the United States of America*, 116(3), 909–914.

Wise, D.H. (1981) A removal experiment with darkling beetles: Lack of evidence for interspecific competition. *Ecology*, 62, 727–738.

With, K.A. (2019) *Essentials of Landscape Ecology*. Oxford University Press, Oxford, UK.

Witte, V., Janssen, R., Eppenstein, A. and Maschwitz, U. (2002) *Allopeas myrmekophilos* (Gastropoda, Pulmonata), the first myrmecophilous mollusc living in colonies of the ponerine army ant *Leptogenys distinguenda* (Formicidae, Ponerinae). *Insectes Sociaux*, 49, 301–305.

Woese, C.R. and Fox, G.E. (1977) Phylogenetic structure of the prokaryotic domain: The primary kingdoms. *Proceedings of the National Academy of Sciences of the United States of America*, 74, 5088–5090.

Woese, C.R., Balch, W.E., Magrum, L.J., Fox, G.E. and Wolfe, R.S. (1977) An ancient divergence among the bacteria. *Journal of Molecular Evolution*, 9, 305–311.

Woese, C.R., Kandler, O. and Wheelis, M.L. (1990) Towards a natural system of organisms: Proposal for the domains Archaea, Bacteria, and Eucarya. *Proceedings of the National Academy of Sciences of the United States of America*, 87, 4576–4579.

Wolff, M., Uribe, A., Ortiz, A. and Duque, P. (2001) A preliminary study of forensic entomology in Medellín, Colombia. *Forensic Science International*, 120, 53–59.

Wong, M.K.L., Guénard, B. and Lewis, O.T. (2019) Trait-based ecology of terrestrial arthropods. *Biological Review of the Cambridge Philosophical Society*, 94, 999–1022.

Wong, W.T. and Woo, W.C. (2010) Sea-level change – observations, causes and impacts. Hong Kong Institute of Engineers Conference on Climate Change. Available at: www.hko.gov.hk/hko/publica/reprint/r925.pdf (accessed 25 May 2025).

Wood, B. and Boyle E.K. (2016) Hominin taxic diversity: Fact or fantasy? *American Journal of Physical Anthropology*, 159, 37–78.

Woodson, C.B., Schramski, J.R. and Joye, S.B. (2018) A unifying theory for top-heavy ecosystem structure in the ocean. *Nature Communications*, 9, 23.

Woodward, F.I. (1987) *Climate and Plant Distribution*. Cambridge University Press, Cambridge, UK.

Woolley, C., Thomas, G., Blackshaw, R.P. and Goodacre, S.L. (2016) Aerial dispersal activity of spiders sampled from farmland in southern England. *Journal of Arachnology*, 44, 347–358.

Worboys, G.L., Francis, W.L. and Lockwood, M. (2010) (eds) *Connectivity Conservation Management: A Global Guide*. Earthscan Ltd, London, UK.

Worth, A.R., Thompson, R.A. and Lymbery, A. J. (2014) Reevaluating the evidence for *Toxoplasma gondii* induced behavioural changes in rodents. *Advances in Parasitology*, 85, 109–142.

Worton, B.J. (1987) A review of models of home range for animal movement. *Ecological Modelling*, 38, 277–298.

Worton, B.J. (1995) Using Monte Carlo simulation to evaluate kernel-based home range estimators. *Journal of Wildlife Management*, 59, 794–800.

Wrangham, R., Lundy, R., Crofoot, M. and Gilby, I. (2007) Use of overlap zones among group-living primates: A test of the risk hypothesis. *Behaviour*, 144(12), 1599–1619.

Wrangham, R.W. (1979) On the evolution of ape social systems. *Social Science Information*, 18, 335–368.

Wrangham, R.W. (1980) An ecological model of female-bonded primate groups. *Behaviour*, 75, 262–300.

Wright, S. (1931) Evolution in Mendelian populations. *Genetics*, 16, 97.

Wulf, A. (2015) *The Invention of Nature*. John Murray, London, UK.

WWF (2024) *Living Planet Report 2024 – A System in Peril*. WWF, Gland, Switzerland. Available at: https://livingplanet.panda.org/en-GB/ (accessed 25 May 2025).

Wyndham, F.S. (2000) The sphere of the mind: Reviving the noösphere concept for ecological anthropology. *Journal of Ecological Anthropology*, 4, 87–91.

Wynn, K. and Bloom, P. (2014) The moral baby. In: Killen, M. and Smetana, J.G. (eds) *Handbook of Moral Development*. Psychology Press, London, UK, pp. 435–453.

Xavier, J. and Kauffman, S. (2022) Small-molecule autocatalytic networks are universal metabolic fossils. *Philosophical Transactions of the Royal Society A: Mathematical, Physical, and Engineering Sciences*, 380, 20210244.

Xiao, H., Dee, L.E., Chadès, I., Peyrard, N., Sabbadin, R. *et al.* (2018) Win-wins for biodiversity and ecosystem service conservation depend on the trophic levels of the species providing services. *Journal of Applied Ecology*, 55, 2160–2170.

Xu, L., Kong, D., Wang, L., Gu, H., Kenney, T. and Xu, X. (2023) Proportional stochastic generalized Lotka–Volterra model with an application to learning microbial community structures. *Applied Mathematics and Computation*, 448, 127932.

Yachi, S. and Loreau, M. (1999) Biodiversity and ecosystem functioning in a fluctuating environment: The insurance hypothesis. *Proceedings of the National Academy of Sciences of the United States of America*, 96, 1463–1468.

Yamaura, Y., Kawahara, T., Lida, S. and Ozaki, K. (2008) Relative importance of the area and shape of patches to the diversity of multiple taxa. *Conservation Biology*, 22, 1513–1522.

Yang, P.J., Pham, J., Choo, J. and Hu, D.L. (2014) Duration of urination does not change with body size. *Proceedings of the National Academy of Sciences of the United States of America*, 111(33), 11932–11937.

Yang, W., Liu, D., You, Q., Chen, B., Jian, M. *et al.* (2021) Taxonomic bias in occurrence information of angiosperm species in China. *Science China Life Sciences*, 64, 584–592.

Yang, Y. and Hui, C. (2020) How competitive intransitivity and niche overlap affect spatial coexistence. *Oikos*, 130, 260–273.

Yang, Z. (2018) *Molecular Phylogenetics*. In: Pfennig, K. (ed.) *Evolutionary Biology*. Oxford University Press, Oxford, UK.

Yao, M., Zhang, S., Lu, Q., Chen, X., Zhang, S.-Y. *et al.* (2022) Fishing for fish environmental DNA: Ecological applications, methodological considerations, surveying designs, and ways forward. *Molecular Ecology*, 31, 5132–5164.

Yates, C.F., Trexler, R.V., Bonet, I., King, W.L., Hockett, K.L. and Bell, T.H. (2022) Rapid niche shifts in bacteria following conditioning in novel soil environments. *Functional Ecology*, 36, 3085–3095.

Yeom, D.J. and Kim, J.H. (2011) Comparative evaluation of species diversity indices in the natural deciduous forest of Mt. Jeombong. *Forest Science and Technology*, 7, 68–74.

Yihune, M., Bekele, A. and Tefera, Z. (2009) Human-gelada baboon conflict in and around the Simien Mountains National Park, Ethiopia. *African Journal of Ecology*, 47, 276–282.

Yong, E. (2016) *I Contain Multitudes. The Microbes Within Us and a Grander View of Life*. Ecco Press, New York, USA.

Yoshida, S., Hiraga, K., Takehana, T., Taniguchi, I., Yamaji, H. *et al.* (2016) A bacterium that degrades and assimilates poly(ethylene terephthalate). *Science*, 351 (6278), 1196–1199.

Yoshitake, S., Soutome, H. and Koizumi, H. (2014) Deposition and decomposition of cattle dung and its impact on soil properties and plant growth in a cool-temperate pasture. *Ecological Research*, 29, 673–684.

Young, A., Boyle, T. and Brown, T. (1996) The population genetic consequences of habitat fragmentation for plants. *Trends in Ecology and Evolution*, 11, 413–418.

Yu, H., Qi, W., Liu, C., Yang, L., Wang, L. *et al.* (2019) Different stages of aquatic vegetation succession driven by environmental disturbance in the last 38 years. *Water*, 11, 1412.

Zaidner, Y., Prévost, M., Shahack-Gross, R., Weissbrod, L., Yeshurun, R. *et al.* (2025) Evidence from Tinshemet Cave in Israel suggests behavioural uniformity across *Homo* groups in the Levantine mid-Middle Palaeolithic circa 130,000–80,000 years ago. *Nature Human Behaviour*, 9, 886–901.

Zalasiewicz, J., Williams, M., Steffen, W. and Crutzen, P. (2010) The new world of the Anthropocene. *Environmental Science and Technology*, 44(7), 2228–2231.

Zalasiewicz, J., Waters, C.N., Williams, M., Barnosky, A.D., Cearreta, A. *et al.* (2015) When did the Anthropocene begin? A mid-twentieth century boundary level is stratigraphically optimal. *Quaternary International*, 383, 196–203.

Zalewski, M., Dudek, D., Tiunov, A.V., Godeau, J.-F., Okuzaki, Y. *et al.* (2014) High niche overlap in the stable isotope space of ground beetles. *Annales Zoologici Fennici*, 51, 301–312.

Zamudio, K.R., Bell, R.C. and Mason, N.A. (2016) Phenotypes in phylogeography: Species' traits, environmental variation, and vertebrate diversification. *Proceedings of the National Academy of Sciences of the United States of America*, 13, 8041–8048.

Zanno, L.E., Drymala, S., Nesbitt, S.J. and Schneider, V.P. (2015) Early crocodylomorph increases top tier predator diversity during rise of dinosaurs. *Scientific Reports*, 5, 9276.

Zaragueta-Bagils, R., Bourdon, E., Ung, V., Vignes-Lebbe, R. and Malécot, V. (2009) On the International Code of Area Nomenclature (ICAN). *Journal of Biogeography*, 36, 1618–1619.

Zelený, D., Helsen, K. and Lee, Y.N. (2024) Extending the CWM approach to intraspecific trait variation: how to deal with overly optimistic standard tests? *Oecologia*, 205(2), 257–269.

Zenimoto, K., Sasai, Y., Sasaki, H. and Kimura, S. (2011) Estimation of larval duration in *Anguilla* spp., based on cohort analysis, otolith microstructure, and Lagrangian simulations. *Marine Ecology Progress Series*, 438, 219–228.

Zeuss, D., Bald, L., Gottwald, J., Becker, M., Bellafkir, H. *et al.* (2023) Nature 4.0: A networked sensor system for integrated biodiversity monitoring. *Global Change Biology*, 30, e17056.

Zhang, M. (2011) Solar magnetic field: The key to understanding the sun. *Chinese Science Bulletin*, 56(35), 3888–3889.

Zhang, M., Wang, J., Li, L., Feng, D. and Cao, E. (2020) Dynamic changes in landscape pattern in a large-scale opencast coal mine area from 1986 to 2015: A complex network approach. *Catena*, 194, 104738.

Zhang, P. and Watanabe, K. (2007) The habit of hot-spring bath in a free ranging group of Japanese macaque in the Jigokudani, Nagano Prefecture. *American Journal of Primatology*, 69, 1425–1430.

Zhang, Q., Buyantuev, A., Li, F.Y., Jiang, L., Niu, J. *et al.* (2017) Functional dominance rather than taxonomic diversity and functional diversity mainly affects community aboveground biomass in the Inner Mongolia grassland. *Ecology and Evolution*, 7, 1605–1615.

Zhang, Z.-Q. (ed.) (2013) Animal biodiversity: An outline of higher-level classification and survey of taxonomic richness (Addenda 2013). *Zootaxa*, 3703, 1–82.

Zhao, L., Rad, B.B. and Wang, X. (2019) Long-lasting vocal plasticity in adult marmoset monkeys. *Proceedings of the Royal Society B: Biological Sciences*, 286, 20190817.

Zhao, Q., Wang, Y., Ayele, G., Xu, Z. and Yu, Z. (2021) Only mass migration of fungi runs through the biotopes of soil, phyllosphere and faeces. *Journal of Soils and Sediments*, 21, 1151–1164.

Zhao, S., Fang, J., Peng, C., Tang, Z. and Piao, S. (2006) Patterns of fish species richness in China's lakes. *Global Ecology and Biogeography*, 15(4), 386–394.

Zheng, S., Didham, R.K., Yu, M. and Webber, B.L. (2022) Species-level CWM values mask contrasting intra- versus interspecific trait shifts at subtropical forest edges. *Ecography*, 2022, e05837.

Zheng, S., Hu, J., Ma, Z., Lindenmayer, D. and Liu, J. (2023) Increases in intraspecific body size variation are common among North American mammals and birds between 1880 and 2020. *Nature Ecology and Evolution*, 7, 347–354.

Zhilin, M. (2016) Mesolithic bone arrowheads from Ivanovskoye 7 (central Russia): Technology of the manufacture and use-wear traces. *Quaternary International*, 427, 230–244.

Zhou, R., Ci, X., Hu, J., Zhang, X., Cao, G. *et al.* (2023) Transitional areas of vegetation as biodiversity hotspots evidenced by multifaceted biodiversity analysis of a dominant group in Chinese evergreen broad-leaved forests. *Ecological Indicators*, 147, 110001.

Zografou, K., Swartz, M.T., Adamidis, G.C., Tilden, V.P., McKinney, E.N. and Sewall, B.J. (2021) Species traits affect phenological responses to climate change in a butterfly community. *Scientifc Reports*, 11, 3283.

Zonneveld, I.S. (1995) Vicinism and mass effect. *Journal of Vegetation Science*, 6, 441–444.

Zottola, A. and de Majo, C. (2022) The Anthropocene: The genesis of a term and popularization in the press. *Text and Talk*, 42, 453–473.

Index

Note: Page numbers in italics denote tables and figures

α-diversity 97, 195
  indices, incorporating species abundance 360–363
β-diversity 97, 123, 126, 128
  dissimilarity measures and 363–365
γ-diversity 97, 126, 195, 363, 364

abiotic cues 166
absolute density 340
Abundance-based Coverage Estimator (ACE) 356–356
abyssal zone 18
Acadoparadoxides briareus (Cambrian trilobites) *311*, *312*
accommodation model 78
accumulation curve 115, 355, 357
Acrocanthosaurus atokensis *311*, *312*
Acrocephalus familiaris (Laysan millerbird) 333
Acrocephalus familiaris kingi (Nihoa millerbird) 333
age/sex-specific pyramids 38
ageing criteria study 38
Agelastica alni (alder beetle) 223
age-sex pyramids *38*
Aglais urticae (tortoiseshell butterfly) 56, *142*, 143, 157, 178, 179, 180, 327
agricultural wilding 283
Agrobacterium genus 219
Ailuropoda melanoleuca (giant panda) *60*
airspace 18
Alderley Edge 141
Allee responses 187, 318, *321*
allelopathy 107
Allen's rule *8*
Alliaria petiolata (hedge garlic) 138, *139*, 43
Alliaria petiolate (garlic mustard) 116
alligators 249
allochthonous birds 229
Allogona townsendiana (Oregon forestsnail) 230, *231*
  research on 231
Allomerus ants 62
allopatric speciation 210
allopatry 185
Alnus glutinosa (alder trees) 239
Alpheus bellulus (shrimp) 50
Amblyeleotris japonica (goby) 50
ambrosia beetles 226, *227*, *228*
anadromous species 241, 242
Anguilla (European eel) 243–244, *243*
Anthocharis cardamines (orange tip butterfly) 44, 138, *139*, 141, 143, 157, 167, 180, 327
anthosphere 4
Anthropocene 2–3, 265, 274, 277, 278
  climate change and habitat loss and 285–289
  definition of 286
  ecosystem issues and 290–292
  genetic losses and 292
  human impact measurement on habitat and 300–304
  invasives and 292–296
  red list audit and 289–290
  species conservation and 299–300
  winners and losers in 296–299

anthroposphere 2
apes 151, 263–265
Aphantopus hyperantus (ringlet butterfly) 141, 210
aphotic zone 17, 18
aposematism 166
apostatic selection *321*
Apus apus (common swift) 246
aquatic habitats 17
Archaeopteris species 310, *311*, *312*
Archean 372
Architeuthis dux (giant squid) 244–246, *245*
area of occupancy (AOO) 103
area samplers 344
Argynnis paphia (silver-washed fritillary butterfly) 141, *203*
Aristotelia chilensis (understorey shrub) 333, *334*
arthropods 19, 56, 77, 86, 88, 90, *342*
  population surveying of 343–344
assemblage 79
Association of Zoos and Aquariums (AZA) 335
asymptotic models 357
atmosphere 1, 2, *3*, 18, 134, 241, 277, 285, 288, 302
augmentation 294, 331, 333
Austropotamobius pallipes (white-clawed crayfish) 295
autecology 28, 121, 122, 134, 156, *158*, 208, 324
autotrophic successions 76
autotrophs 33, 61, 66, 221
azonal biomes 11
Azorean Islands *71*

bacteriophages 215
Baermann funnel 346
baited traps and cages 345
Balaenoptera physalus (fin whales) 241
baleen whales 241
Baroni Urbani and Buser coefficient 128
barriers 22, 46, *49*
Baselga's framework 363, 364
bathyal zone 18
bathypelagic zone 1, 17
Batrachochytrium dendrobatidis (chytrid fungus) 336
bats 249
bauplans 54, 177
beating technique *342*, 344
Berger–Parker index *120*, 361
Bergmann's rule *8*
Berlese funnels *342*, 343
Bernwood Forest, Buckinghamshire, UK 23, *24*, 27
between-habitat studies 99
biodiversity conservation 307–312
  traditional habitat designation and 306–307
biodiversity loss, human activities causing *283*
biogeochemical cycles 3
biogeographical provinces 11
biogeographical regions 4–9, 72, 97, 229, 247
biogeographical studies 90
bioindicators 60, 232
biomes 4, 9–17, 21, 22, 32, 66, 72, 97, 168, *218*, 219, 229, 262, 264, 267, 292, 305
biome stacking 11
biosphere 2, 12, 32, 133, 217, 288, 302
biotic cues 166
biotic provinces 4, 11
biotopes 12, 80, 81, 117, 133, 137, 146, 163, 181, 200, 209, 210, 212, 219, 229, 235, 238–241, 258, 259, 260, 267, 288, 297, *301*, 305
bipartite networks 29, 32
  hypothetical *30*
bipedalism 255–256, 258
bird hides *342*, 346
Bison bison (North American bison) 105, 249
blue whales 241
Boorman–Levitt metapopulation model 188, *190*, 204
Bootstrap estimator 356
Borrelia burgdorferi 219, *220*
botanical gardens 335
boundaries
  change, over time *21*
  functions of *21*
  habitat, structural features of *16*
  origin and maintenance of *21*
  spatial structure of *21*
Brassica oleracea (cliff cabbage) 116
Brereton Heath *140*, 141
Brillouin index *120*
British Trust for Ornithology 92
broadcast indirect (inferential) methods 38
broken stick model *111*, 359
brotion 25
Brualdi and Sanderson Discrepancy (BR) 129
bubble net feeding 241, *245*
Burnham–Overton method 341
burrow habitats *20*
Butterfly Monitoring Scheme BMS UK sites 116
butterfly species 23, 26, 27, *57*, 207, 277, 317
  affinity among *24*
  biological and geographical traits with C-S-R strategy scores *58–59*
  diversity *127*
  outdoor cages to determine behaviour in *325*
  oviposition bias, on host plant patches *181*
  recommended actions and information types prior to and after release of *309–310*
  resource-based habitat defined for *136*
  resource-based habitat paradigm elements as case study of 138–143
  resource-based habitat principles relating to *149*
  resource dynamics for *152*
  variation in resource use in *174*
  see also individual species

caimans 249
Calamophyton sp. *311, 312*
Calluna vulgaris (ling/heather) 26, *140*, 239
camera trapping 346
Campostoma anomalum (minnows) 61
Canis lupus (grey wolf) 268
Capra caucasica (West Caucasian tur) 269
captive breeding 335
Carboniferous forest *311, 312*
Cardamine pratensis (cuckoo flower) 138, *139*, 143, 327
carnivores *35*
Carnufex carolinensis 310, *311, 312*
carposphere 4
carrion
  breakdown in 85
  recycling of 84
carrying capacity 40, 41, 350
cascades 31
cascading coextinctions 32, 333
cascading effects and introductions 331–334
Castanea sativa (chestnut) 65
castes and slaves 233–235
Cataclysta lemnata (China mark moth) 188
catadromous species 241, 243–244
Çatalhöyük settlement (Turkey) 270, *271*
Catalysed Reporter Deposition Fluorescence *In Situ Hybridisation* (CARD-FISH) 347
caulosphere 4
Cebus capucinus (white-faced capuchins) 260, 261, *262*
Cenozoic 372

central business district (CBD) 316
central-place-foraging hypothesis 235
Central Place Theory *275*
Centre for Ecology and Hydrology 13
Centrocercus spp. (sage grouse) 327
Cepaea nemoralis (mollusc) 95, 231
Cercopithecus Ascanius (redtail monkeys) 260
Cervus elaphus (red deer) 160
Chalcolithic culture 273
Chao1 estimator 356
Chao2 estimator 356
Chao-like index 362
chemical knockdown 344
chemical pollution *283*, 288
chemosynthetic organisms 33
chi-square (χ2) statistics 364
Chlamyphorus truncatus (pink fairy armadillo) 246
Chorthippus brunneus (field grasshopper) *39*
Christaller's model *275*
cities 315, 316
city landscape 301
clades 214, 221, 255, 259, 260, 328
Clausilia bidentata 230
Clemmys guttata (spotted turtle) 248
Clethrionomys (voles) 40
climate change *283*
 and shifts, in resource-based habitats 336–339
climatic climax *70*, 71
climax 21, 73, 88, 102
 concepts of 71
 vegetation 72
climax pattern theory 71
closed communities *79*, 80
closed habitats 307
clumped dispersion 107
clumping, of resources *171*, 172
coastal habitats *13*
coefficients of niche breadth and overlap 119–121
Coenonympha tullia (heath butterfly) 26, 88, *190*, 200, *201*, *202*, 209
Colobus guereza 259
colonisation ability 172, 183, *184*
colonisation structures, in industrial and urban environments *314*
colour traps *342*, 344
Columba livia domestica (feral rock dove) 335
Columba oenas (stock dove) 335
Columba palumbus (wood pigeon) 26, 238–239, 335
column samplers 344
Commonwealth Scientific and Industrial Research Organisation (CSIRO) 86
communal nesting *237*
communities 4, 7, 9, 15, 23, 42, 71, 72, 79, 92, 97, 114, 283, 305, 327, 340
 disassembly of *334*
 principles and rules influencing *101*
 super-organism concept of 78
community-weighted mean (CWM) 153, 155
compaction, of resource 146, 149–150, 161–162
competition 62, 102, 121, 178, 259, 267, 351
competitive, stress tolerant and ruderal strategies (C-S-R model) *56*, *57*, *58–59*, 76, 183
competitive exclusion 44, 54, 366
complementary resources *137*, *147*, *173*
composition, of resources 144–145, *145*, *149*, *152*
concentric zone model 316
confined habitats 307
connectance 128, *338*
connectivity, of resources 22, 46, 144, *145*, 146, *149*, *152*
Connochaetes gnou (African wildebeest) 249
Connochaetes taurinus (African wildebeest) 249
Conraua goliath (Goliath frog) *326*
conservation introductions 307, 331, 333
conservationism 303
construction opportunities for habitats and biodiversity 313–315
consumable resource 25, 66, 136
contingency coefficient 364
continuous probability *functions* 112
conurbations 316
Convention on Biological Diversity 290, 334
Convention on International Trade in Endangered Species (CITES) 282
Coordination of Information on the Environment (CORINE) 12
 land-cover classes 12, *14*, 15
coprophagous beetles 88
coral reefs 18
core samplers 344
corridors 22, 46, *49*, 205
Corvus corone (carrion crows) 237, 238
Corvus frugilegus (rooks) 95, 235, 237
Corvus monedula (jackdaws) 238
COSEWIC 231
crenon zone 18
Cretaceous–Palaeogene event 310
crocodiles 249
crocodylomorphs 310, *311*, *312*
crypsis 125, 166
C-score 123
Cuculus canorus (parasitic cuckoo) 241, 327
cues 166
Cyanistes caeruleus (blue tit) 313

Danaus Plexippus (monarch butterfly) 26, *60*
DAPI staining 347
Daubentonia madagascariensis (aye-ayes) 260
deadwood 81, 82, *83*
 recycling of 84
decomposition
 pedogenesis and 33
 of pig carcass *85*
 during regeneration cycle in woodlands *83*

deforestation *283*
demographic transition model (DTM) *38*, 285
Dendrocopos major (great spotted and pied woodpeckers) 238, 326
density 113
 measures of 340
 plant 341–343
 population 93, 102, 187, 259, 340
 of resources *147*
density-dependent factors 40–41
density-independent factors 40, 41
Devonian-Carboniferous tree *311, 312*
diatoms 194
Dice–Sørensen coefficient of dissimilarity 363
Dice–Sørensen similarity *124*
dippers 344
disassembly 333
disclimaxes 21
disparity 99, 116, 362
dispersion of resource and dispersion pattern of organism 168
disturbance 41, 72, 313
 anthropogenic 310
 levels, and organisms' responses 73–74
disturbance heterogeneity model 73
diurnal cycles 94–96
diversity 97, 126, 195, 363, 364
 components of 99
 functional 66, 131–132, 314, 365–367
 see also individual entries
diversity measures
 α-diversity indices and 360–363
 β-diversity and dissimilarity measures and 363–365
 asymptotic models and 357
 non-parametric estimators and 356–357
 phylogenetic diversity and 367
 SAD models and 357–359
 species richness and 355–356
 taxonomic diversity and distinctness and 367–368
diversity profile 118, *119*
domatia 74, 77
dominance measure *120*
downgrading 61
Dromiciops gliroides (marsupial) 333, *334*
Dryas octopetala × *integrifolia* (taxon) *37*
dulosis 234
dung 86, 88
 cow 86, *87*
 recycling of 86
Dung Beetle Ecosystem Engineers project 86
dung beetles 86, *87*, 88
dysphotic zone 17

earthworms 246–247
 lifecycle of *247*
Echinolittorina lineolata 232
ecoclines 20–21
ecogeographical rules 8
ecological constraints model 259
ecological diversity 305
ecological engineers 50, 52, 59, 62, 234
ecological niche 23, 358
 as hypervolume *42*
ecological successions 56, 66, 71, 72, 73, 115, 202, 204
ecoregions 11, 22
ecosystems 4, 9, 13, 28, 32, 59, 133, 181, 241, 263, 289, 305, 306, 322
 biodiverse, humans impacting *303–304*
 focal species in 59–62
 functioning, disturbance effects on *308*
 integration units and 28–37
 interconnections in 62–64
 resilience 31, 74, 132, 307, 308
 resistance 307, 308
 rewilding principles to restore 283
 trophic cascade features in *33*
ecosystem services 302–303, 314
 potential 315
ecotones 7, 15, *16*, 20–21, *46*, 95
 definition of 21
ecotopes 32, 33
ecozone 11–12
Ectopistes migratorius (passenger pigeon) 105, 298
edaphic life 17
edge effect 22, 141, *142*
edges 20–23, 46
 ecological responses to 22
eight-cell model 92
Elaenia albiceps 334
electrofishing technique 345
Ellenberg's indicator values 153
Ellenberg values 229
ellipses *148*
elm bark beetles 295
elm trees 295
emergence traps 344–345
emergent properties 319, 322
endoparasitoids 328
Enhydra lutris (sea otters) 61
entoalexia 310
entomyopia 310
environmental (living) conditions 136
environmental DNA (eDNA) 345
environmental heterogeneity 118
enzootic cycle 219
Eonycteris spelaea (Thai nectivore dawn bat) 95
ephemeral metapopulations 200
epikarst 17
epipelagic zone 17, 95
equilibrium theory of island biogeography (ETIB) 194
equiprobable-equiprobable (EE) algorithm 130
Equus ferus przewalskii (wild horse) 335
Erica sp. 239

Erica tetralix (cross-leaved heath) 26
Erinaceus europaeus (hedgehog) 26
Eriophorum hummocks 88
Eriophorum sp. 239
Eriophorum vaginatum (hare's-tail cotton grass) 26
Erithacus rubecula (robins) 235, *237*
Escherichia coli 219
estimators 115
eukaryotes 212–213, 369, 372
  habitats of 221–223
  heterotrophic 214
Euphausia superba (Antarctic krill) 241
euphotic zone 17
Euphydryas aurinia (marsh and fritillary butterfly) 126, 167
European Biogeographical Regions (EBRs) 7
European Nature Information System (EUNIS) 12
eusociality 224, 229, 233
evenness 99, 360
evenness index *120*, 360
evenness measure *120*
exponential growth model 39, 40, 348–349
ex situ conservation 335
extant insect orders *224*
extensionism 5
extent of occurrence (EOO) 103
extinction 31, 33
extinction debt 191
extrapolation 355, 356
extremophiles 213, 219

Fabriciana adippe (high brown fritillary butterfly) 169
facilitation 43, 62, 66
facts 100
Faith's phylogenetic diversity 367
Falco tinnunculus (kestrels) 238
fasciolosis 336
Fertile Crescent *271*
Fisher's α *120*, 360
fission–fusion social organisation 259
fixed-fixed (FF) algorithm 130
flagship species 59, 60, 289, 322, 327
flightlessness, evolution of 176
flow cytometry 347
focal species, in ecosystems *59–62*, 300
food chains 25, 32, 33
food webs 32, 33–34, 114, 334, 346
  network representation, hypothetical *29*
  Serengeti plant–mammal *51*, *53*
fossil markers, in human evolution *257*
Foster's rule *8*
foundation stone species *59*, 60–61, 289
fragmentation 191, 326
Fraxinus spp. (ash trees) 295
freshwater environment *5*
fresh waters *13*
Fulica atra (coots) 239
functional β-diversity 366–367
functional dendrograms *365*
functional dispersion 366
functional divergence 366
functional diversity 66, 131–132, 314, 365–367
functional evenness 366
functional groups 31
  guilds and 52–54
functional identity 366
functional mean nearest neighbour distance (FNND) 366
functional mean pairwise distance (FMPD) 366
functional originality 366
functional redundancy 131–132, 307, 321, 366
functional richness 366
functional specialisation 366
functional traits 103, 153–155, 229
fundamental niche 43, *45*
fungi 2, 4, 63, 66
  plants and 226–228
  sphere of *227–228*

Gaia hypothesis 3–4
Gallinula chloropus (moorhens) 239, 315
gambin model 359
Gamma distribution 359
gastropods
  land and sea *231*
  shore 232
Gause's formulation 354
Gause's principle of exclusion 25
Gee minnow traps 344
generalism 32, 146, 210, 260
generalists 30, 32, 90, 102, 105, 113, 161, 183, 290, 297, 300, 321, 331
General Systems Theory 319
generic diversity 123
genetic depletion 292
genetic diversity 305
genetic drift 99, 294
genetics and conservation 294
geological cycles 88–90
geological events, in species succession 372–374
geometric series 108, *111*, 359
geosphere 2, 18
giant dragonfly *311*, *312*
Gini–Simpson index 361
glacial–interglacial cycles *89*
Global Biodiversity Information Facility (GBIF) 216
Gloger's rule *8*
Gower's coefficient 128
Gower distance 131
grab sampler *342*
Gram-negative bacteria 219
grasslands *13*

Great Orme *330*
green walls *314*
Grinnell's axiom 25
ground-dwelling arthropods 343
guilds 50, 359
functional groups and 52–54
Gymnogyps californianus (California condor) 335

habitat 50, 66, 305
in biodiversity 210–211
boundaries, structural features of *16*
climate and land use changes and 207–208
compaction, and resource integration 161–162
definitions of *2*
diurnal cycles in and out of 94–96
genetic changes and 208–210
human impact measurement on 300–304
indices of species diversity and evenness for 116–119
limits to assessing species richness in 114–116
and niche compared *45*
population status and changes within 37–41
principles affecting resource quality and use in *180*
relationship measurement across 121, 123–124
resource compaction and intersection in 146, 149–150
same species distinctiveness across different 124–126
seasonal patterns and processes in 91–94
species richness patterns in 106–114
species' view of 23–27
stability of 176
suitability, changes in 150
suitability elements, and key resources 143–146
trajectories of 66–80
turnover between, with time 126
types, of UK 13–15
units, diversity and gradients within and beyond 126–131
unravelling relationships between species' traits and niches and 150–156
without vegetation 17–18
see also individual entries
habitat diversity 116, 117–118, 194
habitat islands 188, 191–197
examples of *190*
habitat loss *283*, 300, *301*
habitat preference 25, 165
habitat selection 25
cues and order in 165–166
habitat use 25, 165
Hadean 372
hadopelagic zone 17
hamlets 273, 315
hand sampling *342*
Harmonia axyridis (harlequin ladybird beetle) 295
haustorium 228
heathlands *13*
hedges 22
Helogale parvula (dwarf mongooses) 233
hemiparasite 228
herbivore herds *250*
herbivores *35*, *37*
herbivory 33
Hesperia comma (silver-spotted skipper butterfly) 202, 327
heterogeneity
environmental 118
of resources *148*
heterotrophic eukaryotes 214
heterotrophic successions 76–80, 81, *84*, 87, 91
heterotrophs 33, 66, 221
hierarchy of hypotheses (HoH) approach 295
Hill numbers 118, 361–362
phylogenetic diversity based on 367
Hirtella physophora (myrmecophyte) 62
histogram 108, *123*, 358
holistic concept 79
holobionts 66
Holocene 12, 74
home ranges 76, 105, 138, 143, 156, 163, *164*, 170, 171, 182, 211, 229, 230, *236*, 238, 241, 252, 259, 263, 273
distinguishing territorial defence from 159–160
group size affecting 172
link between lifetime track, resource-based habitat and 157–159
overlap of 262
hominid evolution models 255–258
hominins 233
homologies 9
homoplasy 9, 255
host–parasite networks 29
host–parasite relationships 197
human evolution
Industrial Revolution and 274–276
Metal Age, Medieval advances and modern industry beginnings and 273–274
need, greed, and war and 276–281
Neolithic humans and first settlements and 270–273
Palaeolithic and Mesolithic human home ranges and habitats and 265–270
human population status *287*
Hutchinson model 43
Hyacinthoides non-scripta (bluebell) 138
hydrosere model *75*
hydrosphere 2, 3, 18, 134, 241, 248, 277, 285, 302
Hymenoscyphus fraxineus (ascomycete) 295
hyperspace 43, 44, 50
niche 52
hyphae 226
hyporheic zone 17

ideal despotic distribution (IDD) model 178, 179
ideal free distribution (IFD) model 121, 175, 179, 237
ideal pre-emptive distribution (IPD) 175–176
ideal pre-emptive selection 177
impact tens rule 295
Impatiens glandulifera (Himalayan balsam) 295
inbreeding 99, 294, 296
inbreeding depression 294
Incidence-based Coverage Estimator (ICE) 356, 357
index of aggregation 106
indicator species 59, 60
individualistic concept 78, 79
Industrial Revolution 274–276
information-centre hypothesis 235
inhibition model 71
insurance hypothesis 321
integration, of resources 161–162, *173*
intensive farming *283*
interaction matrices *36*
interaction theory 44, 50
interbout arousals 338
intercontinental migration 172
Interim Biogeographic Regionalisation of Australia 7
intermediate disturbance hypothesis 73
International Union for Conservation of Nature (IUCN) 92, 244
    Red List 186, 206, 212, 289–290, *326*
interspecific competition 43, 353
interspecific variation 155
intertidal zone 17
intraguild predation 34
intraspecific competition 39
intraspecific variation 155
intrinsic rate of increase 39
introduction and conservation 331, 333
invasion tens rule 295
invasives *283*, 292–296
    affecting native flora and fauna *293*
island biogeography 188, 191
isodar theory 175
isoleg theory 121

Jaccard coefficient 363
Jackknife 1 estimator 356
Jackknife 2 estimator 356
Jolly–Seber method 341
Jordan's rule *8*

kernel analysis *164*
key factor analysis 178, 182
keystone
    communities 62
    ecosystems 62
    habitats 62
    species 59, 61–62, 221, 289, 298, 322, 327, 331
    structures 62
kingdoms of life 212–216
    basic elements for measurement in case studies and 228–230
    bias in biodiversity studies and 216
    eukaryote habitats and 221–223
    fidelity to and movement within locations and zones and 249
    group size and 235–238
    habitat structures in invertebrates and 223–224
    habitat structures in plants and fungi and 226–228
    habitat structures in vertebrates and 224–226
    occupation of biomes and biotopes and 238–241
    occupation of different spheres and activity zones and 241–248
    occupation of variety of spheres and zones and 248–249
    phagy (consumer) groups and 249–250
    prokaryote habitats and 216–221
    slow-moving and fast-moving organisms and 230–233
    social organisation complexity and 233–235
K-selected species 71
K-strategists *54–56*, *55*, 73, 76

Labidura herculeana (giant earwig) *284*
laimosphere 4
land habitats 1
    schemes, features of 12, 15
landscape ecology 46–49, 50, 65, 100, 118, 133, 169, 172, 188, 200, 210, 238, 259, 283, 288, 297, 302, 305, 306, 313, 327, 329–331, 366
    variability 329
landscape taming versus landscape wilding 312–313
Lasiommata megera (wall brown butterfly) 138, *140*, 141, 143, 157, 178, 315
last eukaryotic common ancestor (LECA) 221
latitude 308
laurel forests *71*
Laurus azorica 71
law, scientific 100
Leinster–Cobbold estimators 362–363
Lemmus lemmus (lemmings) 40
Lemur catta (ring-tailed lemur) 260
lemurs 260
lentic system 18, 192, 193
Lepidoptera *199*, 229, 232
leptocephali 244
Lepus americanus (snowshoe hares) 40
Levins index 122
Levins metapopulation 202
    classical *190*
Levins-type metapopulation models 188
life, hierarchical organisation of *320*

life conservation 305–306
biodiversity conservation and 307–312
biodiversity conservation and traditional habitat designation and 306–307
climate changes and shifts in resource-based habitats and 336–339
construction opportunities for habitats and biodiversity and 313–315
intimate community of species and 328–329
introductions and cascading effects and 331–334
landscape taming versus landscape wilding and 312–313
resource-based habitat and species conservation and 322–325
resource-based planning and 325–328
special landscape of species and 329–331
species distributions, limits and variations in resource-based habitats and 335–336
systems approach and 318–322
urban sites as fit for species and 315–318
zoos, botanical gardens and wildlife parks and 334–335
life forms 15, 52
life-history strategies *57*
traits and 54–59, 66, 76, 90, 205
lifetime track (LTT) 143, 163, 165, 166, 168, 182, 263
group size affecting 172
link between resource-based habitat, home range and 157–159
life zones 4, 9, *10*
lightsheeting technique *342*, 344
light traps 344, 345
Limenitis Camilla (white admiral) *179*
limnetic zone 18
limnofauna 7
Lincoln index *see* Petersen index
linear resources *147*
line-transect distance sampling 346
Lithic reduction 270
lithosphere 2, *3*, 134, 241, 248, 249, 277, 285
litter arthropods 343
littoral zone 17, 18
Littorina littorea (common periwinkle) 232
living walls *see* green walls
local colonisation 33
logistic equation 54
logistic growth model 40
logistic model 349–351
lognormal distribution 108, *110*, *111*, 318, 358
log-series 108, 357–358
long-term ecological research (LTER) 306
lotic system 18, 192, 193–194
Lotka–Volterra predator–prey equations 351–353
Loxodonta africana (African elephants) 249
Lumbricus badensis (giant worm) 248
Lumbricus friend 248
Lumbricus terrestris (lob worm) 247
Lycosa hyraculus *20*
Lynx canadensis (lynxes) 40
Lysandra bellargus (Adonis blue butterfly) *56*

Macaca fuscata (Japanese macaques) 251
Macaca sylvanus (Barbary macaque) 251
Macaca tonkeana (Tonkean macaques) 263
macroecological viewpoint 62
macrohabitat 19
Macrotermes natalensis (termites) 234
Magicicada spp. (periodical cicadas) 246
mainland– island model *see* Boorman–Levitt metapopulation model
Malaise traps *342*, 344
Maniola jurtina (meadow brown butterfly) 1, 95, 137, 138, 139, 141, 143, 157, 210, 211, 315
mapping 228
Margalef richness index *120*
marine environment 5
marine lakes 193
marine regionalisation 5
mark-recapture (MR) technique 95, 138, 141, 157, 158, 186, 340, 346
Martes foina (stone martin) 26
Maslow's theory of hierarchy of needs 276–277
mass extinction event 7, 90, 282, *284*, 288, 305, 310, 338, 372
life forms between *311*
materials pollution *283*
matrix 33, 44, 46, 102, 112, 188, 191, 237, 298, 317
interaction *36*
significance of 137–143
vacant 200
matrix temperature 129
McIntosh diversity index *120*
mean pairwise dissimilarity (MPD) 366
Medieval British agricultural village system model *275*
Megaptera novae-angliae (humpback whales) 241
bubble net feeding in *245*
meiofauna 341
Melaleuca 77
membrane filtration 347
Menhinick richness index *120*
Mesodinium rubrum (protist ciliate) 232–233
Mesonychoteuthis hamiltoni (colossal squid) 244
mesopelagic zone 17, 95
Mesozoic 372
MetaCommunity Ecology (CESTES) 323–324
metagenome-assembled genomes (MAGs) *218*
metagenomics *218*
metagenomic sequencing 347
Metal Age, Medieval advances and modern industry beginnings 273–274
metapopulation 43, 46, 50, 97, 102, 146, 263, 303, 318, 329, 332
habitat quality and 199–200
isolation and habitat unit area and 188–199

matrix quality principles for *205*
model principles for day-flying Lepidoptera *199*
models of 272, 273
shifting habitats and 200–202
sporadic *203*
microbial communities 219
microbiomes 17
microhabitats 1, 15, 18–20, 179, 219, 273
dynamics of 80–88
microhabitat sampling *342*
micro-organisms 1, 17, 77, 338, 346, *370*
biomass estimation of *347*
communities of 4
Microtus (voles) 40
mimicry 166
minimum amount of suitable habitat (MASH) 198
minimum viable metapopulation (MVM) 198
minimum viable population 104
mist nets *342*, 346
mobility, of organisms 7, 15, 25, 99, 146, 172, 188
modularity 31, 128, *338*
modular organisms 167
molecular phylogeographical techniques 7–8
monocentric city model 316
monoclimax 71
monophagous species 146
monophyly 213
Most Probable Number (MPN) 347
motility 219, 221
mobility and 172–173
mountains *13*
mountaintops 195, 197
movement capacity 172, 183, *184*
multi-nuclei model 316
multipartite population 188
multivariate analyses 365
multi-way contingency tables 364
mushroom morphology and life cycle *227*
Mustela nigripes (blackfooted ferret) 335
mutualism 43, *62*
mutualistic networks 29, 32
mutualistic symbiosis 50
mycelium 62, 226
mycorrhiza 63
mycorrhizal (mycelial) networks 62
mycorrhizal mycelium 226
Myrmecodia beccarii 77
myrmecodomatia 77
myrmecotrophy 62
Myrmica sabuleti (ant) 50, 328
Mytilus californianus (mussel) 61

National Vegetation Classification 15
Natufian culture 270, *271*
natural capital 303
natural populations, surveying 340–341
in amphibians and reptiles 345–346
in birds 346
in fish 345
in freshwater invertebrates 344–345
in mammals 346
in marine invertebrates 345
in microscopic organisms 346–347
in plants 341–343
in terrestrial arthropods 343–344
nearest taxon index 367
neighbourhood effects *147*
nemo 278
Neolithic humans and first settlements 270–273
neritic zone 17–18
nestedness 128, 129–130, 194, 363
definition of 32
Nestedness measure based on overlap and decreasing fill (NODF) 129, *338*
nest site limitation hypothesis 235
net relatedness index 367
nets and plankton samplers 345
networked sensor system 339
networks 62 336–338
architecture of 31
compartmentalisation and nestedness of 31
complexity of 30
stability and 29–32
neutral theory 43, 44, 49–52, 358
New World monkeys 261
niche 37, 74, 150–156, 168, 259, 296, 305, 313
breadth and overlap, coefficients of 119–123
construction, theory of 50
ecological 23, *42*, 358
and habitat compared *45*
hyperspace 52
hypervolumes 359
organism interactions and 41–44
overlap 44, 119–123, 166, 261
separation *45*
theory of 44
unique 102
niche pre-emption model *111*, 359
Nicholson–Bailey model 352
nodes 20, 29, 46
non-equilibrium model *190*
non-parametric estimators 356–357
noönsphere 286
noosphere 2, *3*
Numenius americanus (long-billed curlew 239
Numenius Arquata (curlews) 239, *240*

Ochlodes sylvanus (large skipper butterfly) 141, *142*, 157, 178, 180
Oldowan culture 268, 269
Old World monkeys 261–263
oligophagous species 146

open communities *79*, 80
open habitats 307
Ophiostoma novo-ulmi 295
opines 219
optimal foraging theory 353
Oreamnos americanus (mountain goats) 206
organismal 1
organismal diversity 305
orobiome 11
orthograde primate postures *256*
orthoseres 66
oscillation hypothesis 185, 260
outbreeding depression 294, 332
oyster beds 18

Pacifastacus leniusculus (American signal crayfish) 295
painted lady (*Vanessa cardui*) 159
Palaeolithic and Mesolithic human home ranges and habitats 265–270
Palaeozoic event 310, 372
palynofloral signature 90
palynology 207
pan traps *342*
Pan troglodytes (chimpanzees) 260
Paramecium species (red and grey squirrels) 44
parapatry 210
Pararge aegeria (speckled wood butterfly) 141, 143, 150, 208
parasitism 62
parr 242
partial mimicry 323
Parus major (great tit) 313
passive traps 346
past extinctions *291–292*
patches 46
patchy population model *190*
path analysis *117*
peacock butterfly (*Aglais io*) 56, 143, 157, 180
Pearson's point correlation coefficient 364
Pediculus humanus capitis 19
pedobiome 11
persistence 31
Petersen index 340–341
phagy (consumer) groups 249–250
phantom bottom 95
Phengaris (*Maculinea*) *arion* 249
Phengaris arion (large blue butterfly) 50, 327, 328, 332
phenology, change in 338
photic zone 17, 18
photoplots 343
photosynthetic organisms 33
phylloplane 4
Phylloscopus sibilatrix (wood warblers) 238
phyllosphere 4, 82
phylogenetic diversity 131, 210, 367
phylogenetic β-diversity 367
PhyloSor index 367
Physeter macrocephalus (sperm whale) 246
physiognomy, of resources 144, 146, *149*, *152*
phytogeography 5
Pianka index 122
Picus viridis (green woodpecker) 26
Pielou evenness *see* evenness index
pierid butterflies 166, 205
Pieris brassicae (large white butterfly) 116, 205
Pieris napi (green-veined butterflies) 44, 157, 161, 205
Pisaster ochraceus (starfish) 61
pitfall traps *342*, 343–344
plagioseres 66
Plantae kingdom 226–227
plant-butterfly interaction networks 336, *337*
    land use change impacts on *338*
plant functional traits 153–155, *154*
Plasmodium falciparum 221
Plasmodium life cycle, causing malaria in humans *222*
playback response techniques 346
Plebejus argus (silver-studded blue butterfly) 161, 202, 329, *330*, 331
Pleistocene 268
plots 108, 155, 173, 341
    permanent 343
Podani–Schmera– Carvalho decomposition 363–364, 367
Pogonocichla stellata (African montane bird) 197
point counts 346
point-intercept method 341, 343
Polioptila caerulea (blue-gray gnatcatcher) 52
polyclimatic climax 71
Polygonia c-album (comma butterfly) 336
Polyommatus icarus (common blue butterfly) 143, 157
polyphagous species 146, 290
polyphyletic grouping 221
polypore fungi *227*, *228*
pongid–hominid evolutionary tree 251–255
population bottlenecks 294
population dynamics 150, 348
    exponential growth and 348–349
    logistic model and 349–351
    Lotka–Volterra predator–prey equations and 351–353
    Volterra equations for competition and 353–354
population regulation 41
potamon zone 18
potential natural vegetation (PNV) 12, 71, 72, 74
potworms *247*
predation 33, 34, 62, 351
predation hypothesis 235
predator–prey system 351
presocial 223
Preston's veil line 109, *110*
primary consumers 33
primary succession 72

primates
  apes and 263–265
  classification of extant *253–254*
  hominid evolution models and 255–258
  living, home ranges and habitats in 259–260
  New World monkeys and 261
  Old World monkeys and 261–263
  pongid–hominid evolutionary tree and 251–255
  postures of *256*
  prosimians and 260–261
  socioecology of 265
  tree of *252*
principles 100
Priodontes maximus (giant armadillo) 246
process-response model *203*
profundal zone 18
progression succession 72
prokaryotes 2, 4, 66, 212–213, 214, 299, 346, 369
  habitats of 216–221
pronograde primate posture *256*
proportional-proportional (CE) algorithm 130
proportional species turnover 364
prosimians 260–261
Proterozoic 372
protists 2, 4, 299, 369
protist species 214
Prototaxites fungus 310
Prototaxites milwaukeensis (Devonian fungus) *311–312*
pseudosinks 198
Psittacula eupatria (Alexandrine parakeet) 229
Psittacula krameria (rose-ringed parakeet) 229
Pthirus pubis 19
Pyrgus malvae (grizzled skipper) 200
Pyronia tithonus (gatekeeper butterfly) 141, 143

Q-methodology 313
quadrats 15, 100, 106, 115, 341, *342*, 345
quantitative polymerase chain reactions 347

radiotelemetry 346
random dispersion 107
randomisation algorithms 122
random search model 167
rank-abundance plots 108, 359
  idealised shape of *111*
Rao's quadratic diversity index 362
Rao's quadratic entropy 365
Raphanus raphanistrum (sea radish) 77
Rapoport's rule *8*
rarefaction *115*, 116
rarefaction curve 355
rare species 25, 96, 100, 104, 109, *110*, 116, 296, 328
  anthropogenic reasons for *297*
  see also species rarity
realised niche 43
recovery 308
recruitment-centre hypothesis 235
red data book 1, 216
red lists 186, 206, *326*
  audit 289–290
Regaliceratops peterhewsi 311, 312
regeneration complex 76
regression to the mean 307
reinforcement 307
reintroduction 307, 331, 332, 333
relative density 340
relative species turnover 364
relaxation 191, 195
residential and retail location model 316
resilience 31, 74, 132, 307, 308
resistance 307, 308
resource-based habitat (RBH) 18, 20, 22, 25, 26, 43, 62, 77, 93, 94, 100, 112, 118, 128, 133, 252, 259, 263, 267, 270, 273, 294, 298, 305, 317, 318, 328, 369
  climate changes and shifts in 336–339
  components of 136–137
  cues and order in habitat selection and 165–166
  as defined for butterflies *135*, *136*
  dimensions, factors in 169–172
  distinguishing territorial defence from home range and 159–160
  early case studies of 156–157
  of focal species, guide to record data on *332*
  focus on species' habitat 133–136
  generalism and specialism continuum 182–183
  habitat and space-time scales and 168–169
  habitat factors in species evolution and extinction and 183–185
  habitat suitability changes and 150
  habitat suitability elements and 143–146
  habitat variability computing components and 166–168
  human dilemma and 276–281
  Industrial Revolution and Anthropocene and profusion in 274–276
  interactions in 173–179
  link between lifetime track, home range and 157–159
  marginality and 206–207
  matrix significance and 137–143
  minutiae selection of essential needs and 179–181
  motility and mobility in 172–173
  potential 169
  resource compaction and intersection in habitats and 146, 149–150
  resource integration and habitat compaction and 161–162
  semiosphere and 163–165
  species conservation and 322–323

resource-based habitat (RBH) (*continued*)
  species distributions, limits and variations in 335–336
  species' traits and identification of 323–325
  unravelling relationships between species' traits and niches and habitats 150–156
  variation in 160–161
  as working system 181–182
  see also kingdoms of life; unstable landscapes
resource-based planning 325–328
resource dispersion hypothesis (RDH) 171–172
resource dynamics 150, *152*
resource-holding potential 238
resource partitioning 54
retrogressive succession 72
rewilding 283, 312, 313
Reynoutria japonica (Japanese knotweed) 295
Rhinanthus minor (yellow rattle) 228
Rhinella marina (cane toads) 86, 333
Rhinolophus ferrumequinum (horseshoe bat) 95
rhithron zone 18
rhizosphere 4
rhodoliths 18
richness 99, 360
  see also species richness
risk hypothesis 260, 261
rivet popper model 322
robustness 128
rocky habitats *13*
roglobiont species richness *195*
r-selected species 71
r-strategists 54–56, *55*, 73, 76
ruggedness *148*
rules 8, 100, *101*

Salmo salar (Atlantic salmon) 242–243, *242*
sample coverage 116
sampling methods *342*
Santa Catalina Mountains *196*
saproxylic species 19, 82, *83*, *84*, 145, 297
satyric mimicry 323
Schnabel method 341
Schoenoplectus lacustris (lakeshore bulrush) 239
scrub *13*
Sea, Land and Air Malaise (SLAM) trap *342*, 344
sea grass 18
search theory 167
seasonal fidelity 94
sea turtles 248
  life cycle of *248*
secondary consumers 33
secondary succession 72, *78*
sector model 316
self-thinning rule 41
semiosphere 163–165
senses 166
sentinel species 59, 60
Sephanoides sephaniodes (southernmost hummingbird species) 333, *334*
seral stages 66
sere 71, 115
Serengeti plant–mammal food web *51*, *53*
sessile organisms 233
settlement patterns *275*
Shannon diversity 116, 117, *119*, *120*, 122, 194, 361
Shannon entropy 361
Shannon index 360
Sherman trap 346
shifting baseline syndrome 65, 289
similarity-sensitive indices 362–363
Simpson diversity 358
Simpson evenness *120*, 361
Simpson index 116, 118, *119*, *120*, 122, 123, 360–361, 365, 368
simultaneous hermaphrodites *247*
sinks 198
Sitta europaea (nuthatches) 326
sky islands 195
slave ants 234
slow-moving and fast-moving organisms 230–233
social insects 233
sociality 216
social parasitism 234
social structures 225, 259
soil 4, 17, 71, 76, 88, 246, 288
soil arthropods 343
soil emergence traps 344
Sokal and Michener simple matching coefficient 128
Solenopsis invicta (fire ant) 224
sources 198
space-time scaling, of Earth natural events *67*
spatial distribution patterns 107–108
specialism 146, 210, 260
specialism–generalism continuum 161, 172, *173*, 182–183, *184*, 185
specialists 30, 32, 90, 102, 105, 113, 161, 183, 297, 300, 321, 331
species abundance distribution (SAD) 37, 62, 100, 106, 108–109, 111–114, 116, 306, 318, 357–359
species-area relationship (SAR) 100, 117, 191–192, *192*, 197
species distribution modelling (SDM) 208
species diversity and evenness indices 116–119
species hunting and overexploitation *283*
species rarity 103
  stability and 318–322
  see also rare species
species richness 8, 132, 194, 355–356, 363, 367
  bird 197
  estimates of *214*
  and evenness, indices of 116–119
  limits to assessing 114–116

patterns of 106–114
Troglobiont *195*
species segregation 123
species translocations 307, 315, 324
types of 331
species vulnerability 104–105
spectral radius 129
stability, significance of 308
statistical significance 364–365
Sterna paradisaea (Arctic tern) 26
St Helena *284*
stone tools 265–267, 269–270
Palaeolithic *266*
strip-transects 346
Sturnus vulgaris (common starlings) 95
stygobites 194
substrates 1, 4, 18, 20, 23, 66, 128, 133, 298, 306, 313, 315, 325–326, 327
subterranean habitats 17
subzonal biomes 11
subzonobiomes 11
suction traps 344
sufficiency, of resources *147*
super-organism 78, 79, 234
supersaturation 191
supplementary resources *147*, *173*
Suricata suricatta (meerkats) 233
swallow (*Hirundo rustica*) 159, 241
sweep nets *342*, 344
sympatric speciation 210
sympatry 185
Syncerus caffer (South African buffalo) 249
synecological studies 28
systems approach 318–322
systems theory 319–320

Tadarida brasiliensis Mexicana (Mexican free-tailed bats) 249
Talpa europaea (mole) 246
Tamarix gallica (tamarisk) 239
taxa 90
taxocenes 52, 80, 112
taxocenose 80, 112
taxonomic diversity and distinctiveness 367–368
Taylor's power law 119
temporal aspects *145*
tens rule scheme 295
termites
eusociality and 233
nests of 233–234
termite bounds and *225*, 234
Termitomyces genus 234
territories 76, 159–160, 178, 182, 226, *236–237*, 237, 238, 260, 276
territory mapping 346
tertiary consumers 33
Tetrix undulata (pygmy grasshopper) 125
Theba pisana (mollusc) 125, 230
theory, scientific 100
Theropithecus gelada (geladas) 262–263
Thorson's rule *8*
three-model framework 71
Thymelicus sylvestris (small skipper buttefly) 141
Thymus serpyllum (wild thyme) 328
tiger beetles, of saline habitats 176
Tobler's first law of geography 296–297
tolerance, as component of stability 308
tolerance model 70
towns 315, 316
Toxoplasma gondii 222–223
traditional habitat model 1, 133
biogeographical regions and 4–9
biome view of 9–17
edges, ecotones, and ecoclines and 20–23
generalisations for 100–106
microhabitats and 18–20
species' habitat and 23–27
spheres of existence and 1–4
without vegetation 17–18
traits 9, 44, 74, 91, 132, 168, 177, 260, 296, 307, 313–314, 335
data life cycle *324*
functional 103, 153–155, 229, 314
life-history strategies and 54–59, 66, 76, 90, 205
resource-based habitats identification and 323–325
transects 100, 106, 115, 249, 341, *342*, 345, 346
transformations, in vegetations *70*
transition zones 7
translocation 315, 324
intentional 333
tactics 307
types of 331
trapdoor spiders 19, *20*
tree, of life *370*
tree-killing bark beetles 226, *227, 228*
tree microhabitats *20*
Triassic event 310, 338
Tristerix corymbosus (mistletoe) 333, *334*
Triturus cristatus (great-crested newts) 315
trophallaxis 328
trophic levels 32, 33, 114
trophic networks 29
trophic pyramids 35
Turdus merula (Blackbird) *236*
Turdus philomelos (song thrushes) 235
Typha angustifolia (bulrush) 239
Tyto alba (barn owl) 26

umbrella species 59, 60, 312, 322, 327, 331
Umwelt 165
underwater video techniques 345
uniform dispersion 107

unipartite network 29
unique fraction (UniFrac) metric 367
unstable landscapes 186–187
climate and land use changes and 207–208
factors driving local population to extinction and 202–205
habitat place in biodiversity and 210–211
habitats and genetic changes and 208–210
marginality and resource-based habitat status and 206–207
matrix as empty and 205–206
metapopulation and 188–202
population structure and dynamics and 187–188
urban areas 206, 276
urbanisation 315–318
urban sites, as fit for species 315–318
Urtica dioica (stinging nettle) 230
utility resource 25, 66, 136

vacant niche hypothesis 296
vagrants 80, 92, 93, 100, 112, *113*, 114, 116, 126, 131, 298
Vanessa Atalanta (red admiral) 166
Veblen goods 274, *275*
vegetation *317*
vegetational belts 4, 11
vegetation succession, terrestrial 66–76, *110*, 114, 126, 150, 188, 283, 310
features of *68–69*
hydrosere model *75*
plant community and ecosystem characteristics during *76*
veil line 109, *110*
Venn diagram *124*, 134
vertical gardens *see* green walls
Vespula germanica (wasp) *334*
vicariance 7
vicinism 25, 292
villages 272, 274, 276, 315
visitors 93
Volterra equations, for competition 353–354
von Humboldt's approach 4, 5
vulnerability 104–105, *338*
Vulpes vulpes (fox) 25, 26

war 278–280, *279*
in non-human animals 280
water habitats 1
water sampler *342*
water use *283*
Wattieza species 310
Weibull distribution 359
wetlands *13*
Whittaker index 364
Wilks's χ^2 statistics 364
Winkler extractor 343
within-habitat studies 99, 100
woodland birds, seasonality in *93*
woodlands *13*
wood wide web 62–64, *63*

Yule coefficient 128

zero-sum multinomial (ZSM) distribution 358
Zipf distribution 359
zonombiomes 11
zoogeographical regions 5
zoos, botanical gardens and wildlife parks 334–335
Zophobas atratus (tenebrionid beetle) 221

Printed and bound by CPI Group (UK) Ltd, Croydon, CR0 4YY
22/10/2025
01982205-0002